SIGNALS
AND
SYSTEMS

TARUN KUMAR RAWAT

Department of ECE
Netaji Subhas Institute of Technology
New Delhi

OXFORD
UNIVERSITY PRESS

OXFORD
UNIVERSITY PRESS

Oxford University Press is a department of the University of Oxford.
It furthers the University's objective of excellence in research, scholarship,
and education by publishing worldwide. Oxford is a registered trademark of
Oxford University Press in the UK and in certain other countries.

Published in India by
Oxford University Press
Ground Floor, 2/11, Ansari Road, Daryaganj, New Delhi, 110 002, India

First Edition published in 2010
14th impression 2019

ISBN-13: 978-0-19-806679-8
ISBN-10: 0-19-806679-1

Typeset in Times New Roman
by Dash Publishing Solutions, New Delhi
Printed in India by Repro India Ltd., Surat

To
My father
(Late) Sh. Devender Kumar

Preface

Signals are physical quantities that vary with time, space, or any other independent variable. They are represented mathematically as functions of one or more independent variables. On the other hand, a system is an entity that processes a set of input signals to yield another set of output signals. This book deals with the study of both signals and systems.

The concepts and theories of signals and systems are needed in almost all engineering disciplines. They form the foundation for further studies in areas such as analog and digital communication, analog and digital signal processing, and control systems. It is therefore essential for engineering students to grasp the basics of this subject. Keeping this in mind, this book is designed to cover the engineering curricula of major Indian universities as well as the curricula of AMIE, AMIETE, and other competitive examinations such as GATE and IES.

ABOUT THE BOOK

This book has been specially designed for undergraduate students of electronics and communication, electrical, computer science, and instrumentation and control engineering for a course on signals and systems. The book presents the concepts of the subject in a lucid and easy to understand format by providing the reader with a broad range of problems having varying levels of difficulty. The language has been kept simple to ensure that students can easily grasp the fundamentals of the subject. All the chapters in the book are designed to provide adequate practice through numerous solved examples, multiple-choice questions, and numerical problems.

The book deals with the subject comprehensively by providing separate extensive chapters on continuous-time and discrete-time Fourier series, continuous-time and discrete-time Fourier transforms, Laplace transform, sampling, z-transform, and state-space analysis. Each chapter introduces the objectives in the beginning and then sequentially proceeds from the basic to advanced topics. A detailed summary at the end of each chapter provides a point-wise recapitulation of all the important points in the chapter. Every key concept is illustrated using solved examples.

The salient features of the text include the following:

- A separate chapter on Hilbert transform including the concept of pre-envelope and complex envelope
- Appendices on mathematical relations, complex numbers, and partial fraction expansion
- A set of model question papers at the end of the book
- Numerous MATLAB programs

CONTENTS AND COVERAGE

This book has been divided into 12 chapters and 3 appendices.

Chapter 1 introduces the mathematical description and representation of both continuous-time and discrete-time signals and systems and their properties.

Chapter 2 develops the fundamental input-output relationship for linear time-invariant (LTI) systems, and explains the unit impulse response and step response of the system and convolution operation. Properties of convolution integral and convolution sum are also discussed. Additionally, the concept of correlation of energy and power signals is also presented along with their properties.

Chapter 3 explains continuous-time Fourier series (CTFS), its properties, and its applications to LTI systems. The concept of frequency-domain representations and how to decompose periodic signals into their frequency components are also discussed in this chapter.

Chapter 4 deals with discrete-time Fourier series (DTFS), its properties, and its applications to LTI systems.

Chapter 5 presents continuous-time Fourier transform (CTFT), its properties, and its applications to LTI systems. The concepts of energy spectral density and power spectral density are also discussed. Distortionless transmission of signals, and ideal and non-ideal filters of continuous-time type are covered in this chapter.

Chapter 6 explains the discrete-time Fourier transform (DTFT), its properties, and its applications to LTI systems. Distortionless transmission of signals, and ideal and non-ideal filters of discrete-time type are also discussed in this chapter.

Chapter 7 is devoted to the Hilbert transform (HT) of continuous-time and discrete-time signals, and their properties. Pre-envelope and complex envelope concepts are also discussed.

Chapter 8 is devoted to the sampling of continuous-time and discrete-time lowpass and bandpass signals.

Chapter 9 deals with the Laplace transform (LT), its properties, and its applications to continuous-time LTI systems analysis. Block diagram representation of a system and methods of realization of a given system function $H(s)$ through different realization structures are also discussed in this chapter.

Chapter 10 deals with the z-transform (ZT), its properties, and its applications to discrete-time LTI systems analysis. Block diagram representation of a system, and methods of realization of a given system function $H(z)$ through different realization structures are also discussed in this chapter.

Chapter 11 presents the state-space or state variable concept and analysis for both continuous-time and discrete-time systems.

Chapter 12 is devoted to MATLAB programs.

There are three appendices included at the end of the book. Appendix A provides a recapitulation of mathematical relations such as trigonometric identities, power series expansion, exponential series, derivatives, definite and indefinite integrals, exponential and logarithmic functions, and Taylor series. Appendix B contains complex numbers including their representations and operations, and Appendix C reviews partial fraction expansion. Two model question papers are also provided for practice.

ACKNOWLEDGEMENTS

It is a pleasure to acknowledge the assistance received from several individuals during the preparation of this book. I am greatly indebted to all the faculty members of ECE division of Netaji Subhas Institute of Technology (NSIT): Prof. Harish Parthasarathy, Dr Maneesha Gupta, Ms Sujata Sengar, Sh. D.V. Gadre, Dr S.P. Singh, Dr Parul Garg, Sh. D. Upadhyay, and Ms Jyotsna Singh. I am also thankful to Prof. Raj Senani, HOD ECE division and director NSIT, for providing all the required facilities as well as congenial atmosphere for the preparation of this manuscript. I express my sincere gratitude and thanks to the entire team of Oxford University Press for publishing this book.

I would also like to thank my mother Smt. Rajrani and my parents-in-law for their valuable support. In particular, I am extremely thankful to my father-in-law, Dr Bhupendra Singh for always being there during this project. My deepest gratitude goes to my wife Deepti and my son Titiksh who showed infinite patience and understanding in writing this book.

Finally, I thank all who have helped me directly or indirectly during this project. Any suggestions for improving this book are welcome.

Tarun Kumar Rawat

Contents

Introduction to Signals and Systems

1.1 INTRODUCTION

In this chapter, we begin our development of the analytical framework for signals and systems by introducing their mathematical descriptions and representations. In the chapters that follow, we build on this foundation in order to develop and describe additional concepts and methods that add considerably both to our understanding of signals and systems and to our ability to analyse and solve problems involving signals and systems.

A *signal* is defined as any physical quantity that varies with time, space, or any other independent variable or variables. Signals are represented mathematically as functions of one or more independent variables. For example, a speech signal can be represented mathematically by acoustic pressure as a functions of time, and a picture can be represented by brightness as a function of two spatial variables. In this book, we deal almost exclusively with signals that are functions of time.

Signals may be processed further by *systems,* which may modify them or extract additional information from them. A system is an entity that processes a set of input signals to yield another set of output signals. A system may be made up of physical components, as in electrical, mechanical, or hydraulic systems (hardware realization), or it may be an algorithm that computes an output from an input signal (software realization).

1.2 CLASSIFICATION OF SIGNALS

Depending on the independent variables and the value of the function defining the signal, various types of signals can be defined.

1.2.1 Continuous-time and Discrete-time Signals

A continuous-time or analog signal is defined for a continuum of values of the independent variable time t. A discrete-time signal is defined only at discrete-time n, and consequently for discrete-time signals, the independent variable takes on only at a discrete set of values. The amplitude of the discrete-time signal between two time instants is just not defined. A speech signal as a function of time and atmospheric pressure as a function of altitude are examples of continuous-time signals, whereas the quarterly gross domestic product (GDP), monthly sales of a corporation, and stock market daily averages are examples of discrete-time signals. Illustrations of a continuous-time signal $x(t)$ and a discrete-time signal $x(n)$ are shown in Fig. 1.1. It is important to note that the discrete-time signal $x(n)$ is defined *only* for integer values of the independent variable.

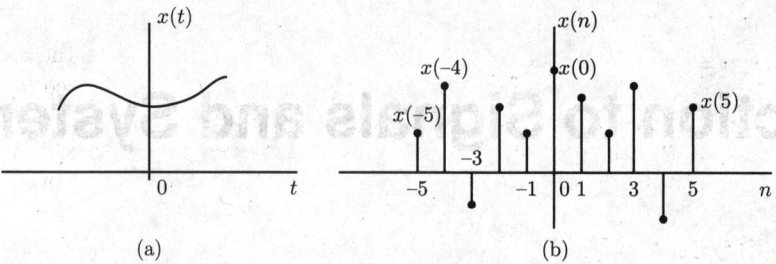

(a)   (b)

Fig. 1.1   Graphical representation of (a) continuous-time and (b) discrete-time signals

1.2.2   Continuous-valued and Discrete-valued Signals

The value of a continuous-time or discrete-time signal can be continuous or discrete. If a signal takes on all possible values on a finite or an infinite range, it is said to be a continuous-valued signal. Alternatively, if the signal takes on values from a finite set of possible values, it is said to be a discrete-time signal. Usually, these values are equidistant and hence can be expressed as an integer multiple of the distance between two successive values. A discrete-time signal having a set of discrete values is called a *digital signal*. Figure 1.2 shows a digital signal that takes on one of four possible values.

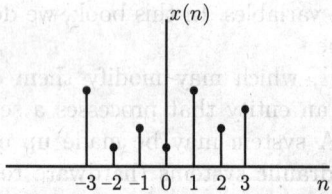

Fig. 1.2   Digital signal with four different amplitude levels

1.2.3   Multichannel and Multidimensional Signals

In some applications, signals are generated by multiple sources or multiple sensors. Such signals, in turn, can be represented in vector form. If $x_k(t)$, $k = 1, 2, 3$, denotes the electrical signal from the kth sensor as a function of time, the set of $k = 3$ signals can be represented by a vector $\mathbf{x}(t)$, where

$$\mathbf{x}(t) = \begin{pmatrix} x_1(t) \\ x_2(t) \\ x_3(t) \end{pmatrix}$$

We refer to such a vector of signals as a *multichannel signal*. In electrocardiography, e.g., 3-lead and 12-lead electrocardiograms are often used in practice, which result in 3-channel and 12-channel signals.

If a signal is a function of a single-independent variable, it is called a *one-dimensional* signal. On the other hand, a signal is called *M-dimensional* if its value is a function of M independent variables.

A picture is an example of a two-dimensional signal since the intensity or brightness $I(x, y)$ at each point is a function of two independent variables. On the other hand, a black-and-white television picture may be represented as $I(x, y, t)$ since the brightness is a function of time. Hence, the TV picture may be treated as a three-dimensional signal. In contrast, a colour TV picture may be described by three intensity functions of the form $I_r(x, y, t)$, $I_g(x, y, t)$, and $I_b(x, y, t)$, corresponding to the brightness of the three principal colours (red, green, and blue) as functions of time. Hence, the colour TV picture is a three-channel, three-dimensional signal, which can be represented by the vector

$$\mathbf{I}(x, y, t) = \begin{pmatrix} I_r(x, y, t) \\ I_g(x, y, t) \\ I_b(x, y, t) \end{pmatrix}$$

In this book we deal with single-channel, one-dimensional real- or complex-valued signals and we refer to them simply as signals.

1.2.4 Deterministic and Random Signals

A signal whose complete physical description is known, either in a mathematical form or in a graphical form, is a *deterministic signal*. The nature and amplitude of such a signal at any time can be predicted. The pattern of such a signal is regular. A signal whose values cannot be predicted precisely but are known only in terms of probabilistic description, such as mean value or mean-square value, is a *random signal*. A typical example of a random signal is thermal noise in an electrical circuit.

1.3 TRANSFORMATIONS OF THE INDEPENDENT VARIABLE (TIME)

Three types of transformations are possible on the independent variable: (i) time shifting, (ii) time scaling, and (iii) time reversal or folding.

1.3.1 Time Shifting

A signal $x(t)$ may be shifted in time by replacing the independent variable t by $t - t_0$, where t_0 is known as the shifting factor. If $t_0 > 0$, the signal is shifted to the right and the time shift results in a delay of the signal by t_0 units of time. If $t_0 < 0$, the signal is shifted to the left and the time shift results in an advance of the signal by $|t_0|$ units in time. A time shift in continuous time is illustrated in Fig. 1.3, in which we have three signals $x(t)$, $x(t - t_0)$, and $x(t + t_0)$ that are identical in shape but are displaced or shifted relative to each other. We will also encounter time shifts in discrete time, as illustrated in Fig. 1.4.

1.3.2 Time Scaling

A second modification of the independent variable involves replacing t by at, where a is known as scaling factor. If $a > 1$, the scaling results in compression (speeded up), and if

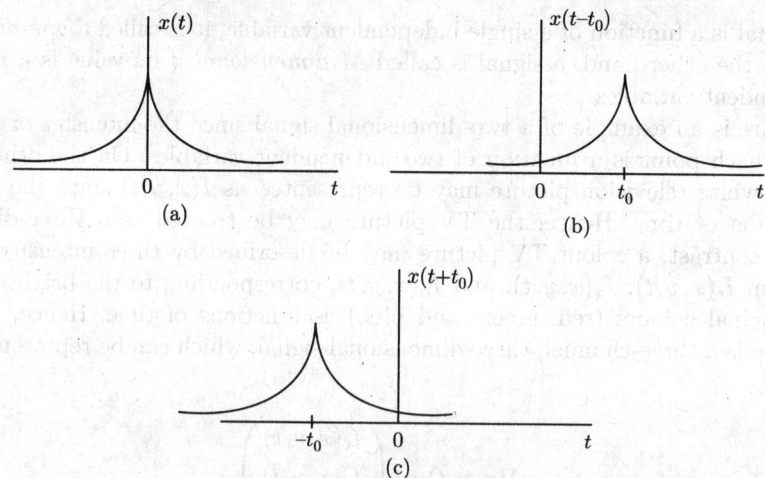

Fig. 1.3   Continuous-time signals related by time shift

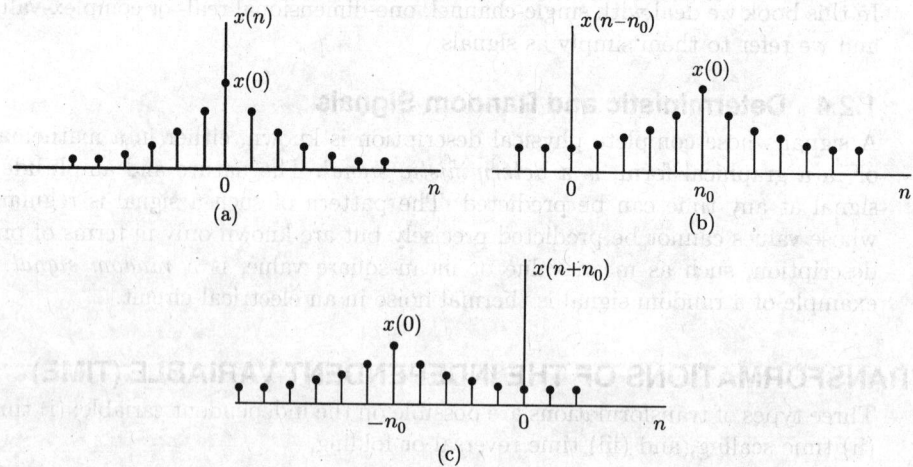

Fig. 1.4   Discrete-time signals related by time shift

$0 < a < 1$, the scaling results in expansion (slowed down). In Fig. 1.5 we have illustrated three signals, $x(t)$, $x(2t)$, and $x(t/2)$, that are related by linear scale changes in the independent variable.

1.3.3   Time Reversal

A third modification of the independent variable involves replacing t by $-t$. The results of this operation is *folding* or *reflection* or *time reversal* of the signal about the time origin $t = 0$. For example, as illustrated in Fig. 1.6, the signal $x(-t)$ is obtained from the signal by a re-election about $t = 0$ (i.e., by reversing the signal). Similarly, as depicted in Fig. 1.7, the signal $x(-n)$ is obtained from the signal $x(n)$ by a re-election about $n = 0$.

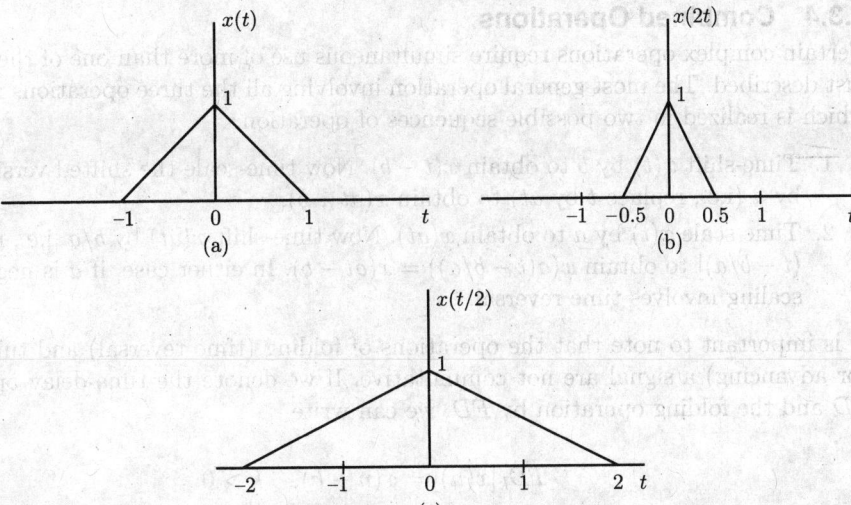

Fig. 1.5 Time-scaling operation

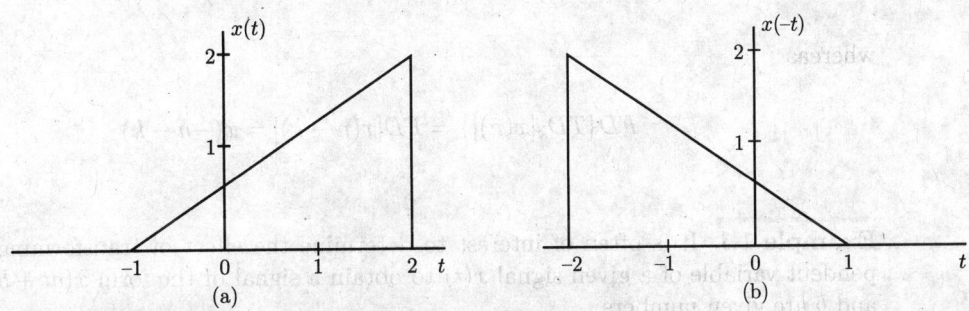

Fig. 1.6 Continuous-time signals related by time-reversal operation

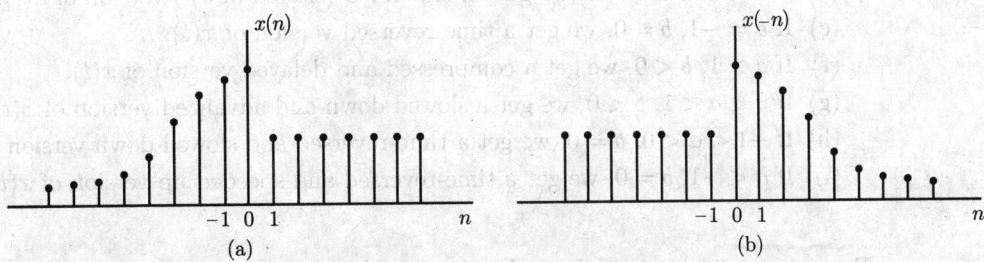

Fig. 1.7 Discrete-time signals related by time-reversal operation

1.3.4 Combined Operations

Certain complex operations require simultaneous use of more than one of the operations just described. The most general operation involving all the three operations is $x(at - b)$, which is realized in two possible sequences of operation:

1. Time-shift $x(t)$ by b to obtain $x(t - b)$. Now time-scale the shifted version $x(t - b)$ by a (i.e., replace t by at) to obtain $x(at - b)$.

2. Time-scale $x(t)$ by a to obtain $x(at)$. Now time-shift $x(at)$ by b/a [i.e., replace t by $(t - b/a)$] to obtain $x(a(t - b/a)) = x(at - b)$. In either case, if a is negative, time scaling involves time reversal.

It is important to note that the operations of folding (time reversal) and time delaying (or advancing) a signal are not commutative. If we denote the time-delay operation by TD and the folding operation by FD, we can write

$$TD_k[x(n)] = x(n - k), \quad k > 0 \tag{1.1}$$

$$FD[x(n)] = x(-n) \tag{1.2}$$

Now

$$TD_k\{FD[x(n)]\} = TD_k[x(-n)] = x(-(n - k)) = x(-n + k) \tag{1.3}$$

whereas

$$FD\{TD_k[x(n)]\} = FD[x(n - k)] = x(-n - k) \tag{1.4}$$

Example 1.1 It is often of interest to determine the effect of transforming the independent variable of a given signal $x(t)$ to obtain a signal of the form $x(at + b)$, where a and b are given numbers.

(a) If $a = 1$, $b < 0$, we get a time-delayed version of $x(t)$.

(b) If $a = 1$, $b > 0$, we get a time-advanced version of $x(t)$.

(c) If $a > 1$, $b = 0$, we get a compressed (speeded up) version of $x(t)$.

(d) If $0 < a < 1$, $b = 0$, we get an expanded (slowed down) version of $x(t)$.

(e) If $a = -1$, $b = 0$, we get a time-reversed version of $x(t)$.

(f) If $a > 1$, $b < 0$, we get a compressed and delayed version of $x(t)$.

(g) If $0 < a < 1$, $b < 0$, we get a slowed down and advanced version of $x(t)$.

(h) If $-1 < a < 0$, $b = 0$, we get a time-reversed and slowed down version of $x(t)$.

(i) If $a < -1$, $b = 0$, we get a time-reversed and speeded up version of $x(t)$.

Example 1.2 A triangular pulse signal $x(t)$ is depicted in Fig. 1.8. Sketch each of the following signals derived from $x(t)$: (a) $x(3t)$, (b) $x(3t + 2)$, and (c) $x(-2t - 1)$.

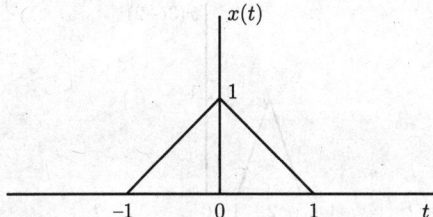

Fig. 1.8 Triangular pulse

Solution

The mathematical form of $x(t)$ is

$$x(t) = \begin{cases} 1+t, & -1 \le t \le 0 \\ 1-t, & 0 \le t \le 1 \end{cases}$$

(a) To obtain the mathematical form of $x(3t)$ from $x(t)$, replace t by $3t$ in $x(t)$ [see Fig. 1.9(a)]:

$$x(3t) = \begin{cases} 1+3t, & -1 \le 3t \le 0 \longrightarrow -\dfrac{1}{3} \le t \le 0 \\ 1-3t, & 0 \le 3t \le 1 \longrightarrow 0 \le t \le \dfrac{1}{3} \end{cases}$$

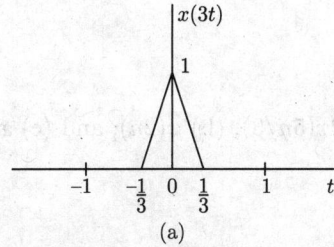

(a)

Fig. 1.9(a) Signal $x(3t)$

(b) The mathematical form of $x(3t+2)$ is

$$x(3t+2) = \begin{cases} 1+(3t+2) = 3+3t, & -1 \le (3t+2) \le 0 \longrightarrow -1 \le t \le -\dfrac{2}{3} \\ 1-(3t+2) = -1-3t, & 0 \le (3t+2) \le 1 \longrightarrow -\dfrac{2}{3} \le t \le -\dfrac{1}{3} \end{cases}$$

To obtain $x(3t+2)$ from $x(t)$, the time-shifting and time-scaling operations must be performed in the correct order. Time-shift $x(t)$ by -2 to obtain $x(t-(-2)) = x(t+2)$. Now, time-scale the shifted version $x(t+2)$ by 3 (i.e., replace t by $3t$) to obtain $x(3t+2)$, i.e., $x(t) \rightarrow x(t+2) \rightarrow x(3t+2)$ [see Fig. 1.9(b)].

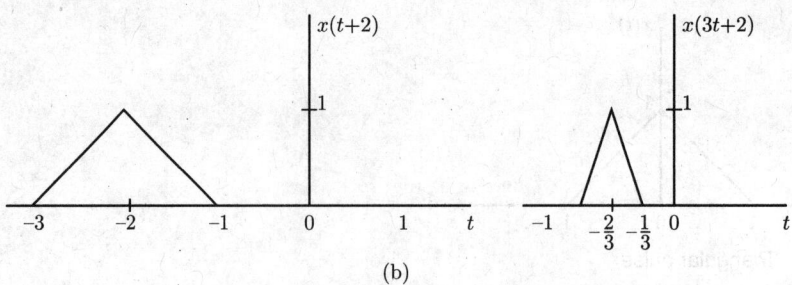

(b)

Fig. 1.9(b) Signal $x(t+2)$ and $x(3t+2)$

(c) To obtain $x(-2t - 1)$ from $x(t)$, the time-shifting operation is performed first on $x(t)$ resulting in an intermediate signal $x(t - 1)$. Now, time-scale the shifted version $x(t - 1)$ by 2 (i.e., replace t by $2t$) to obtain $x(2t - 1)$ and after that apply the time-reversal operation [i.e., replace t by $-t$ to obtain $x(-2t - 1)$], i.e., $x(t) \rightarrow x(t - 1) \rightarrow x(2t - 1) \rightarrow x(-2t - 1)$ [see Fig. 1.9(c)].

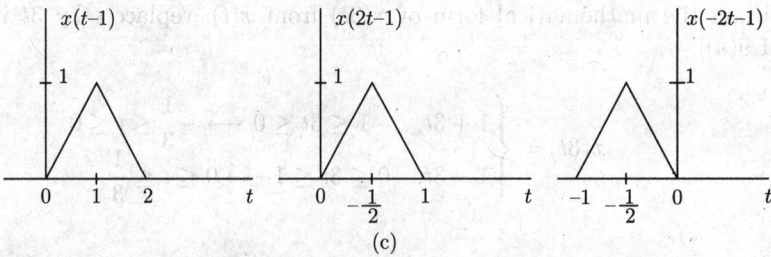

(c)

Fig. 1.9(c) Signal $x(t - 1)$, $x(2t - 1)$, and $x(-2t - 1)$

Example 1.3 Let $x(n) = e^{-n/2}u(n)$. Find (a) $2x(5n/3)$, (b) $x(2n)$, and (c) $x(n^2)$.

Solution
(a) With $w(n) = 2x(5n/3)$, we have

$$w(0) = 2x(0) = 2$$
$$w(1) = 2x(5/3) = 0$$
$$w(2) = 2x(10/3) = 0$$
$$w(3) = 2x(5) = 2e^{-5/2}$$
$$w(4) = 2x(20/3) = 0, \text{etc.}$$

Here we have assumed that $x(n)$ is zero if n is not an integer. It is clear that the general expression for $w(n)$ is

$$w(n) = 2x\left(\frac{5n}{3}\right) = \begin{cases} 2e^{-5n/6}, & n = 0, 3, 6, \text{etc.} \\ 0, & \text{otherwise} \end{cases}$$

(b) With $y(n) = x(2n)$, we have

$$y(0) = x(0) = 1$$

$$y(1) = x(2) = e^{-1}$$

$$y(2) = x(4) = e^{-2}$$

$$y(3) = x(6) = e^{-3}, \text{etc.}$$

The general expression for $y(n)$ is therefore

$$y(n) = x(2n) = \begin{cases} e^{-n}, & n \geq 0 \\ 0, & n < 0 \end{cases}$$

(c) With $z(n) = x(n^2)$, we have

$$z(0) = x(0) = 1$$

$$z(1) = x(1) = e^{-1/2}$$

$$z(2) = x(4) = e^{-2}$$

$$z(3) = x(9) = e^{-9/2}, \text{etc.}$$

The general expression for $z(n)$ is therefore

$$z(n) = x(n^2) = \begin{cases} e^{-n^2/2}, & n \geq 0 \\ 0, & n < 0 \end{cases}$$

The preceding example shows that for discrete-time signals, time scaling does not yield just a stretched or compressed version of the original signal but may give a totally different waveform.

1.4 SINGULARITY FUNCTIONS: UNIT STEP, UNIT IMPULSE, AND UNIT RAMP FUNCTIONS

In the area of signals and systems, step, impulse, and ramp functions play very important roles. They serve as a basis for representing other signals. All these functions are known as singularity functions. Singularity is a point at which a function does not possess a derivative. Each singularity function has a singular point at the origin and is zero elsewhere. Singularity functions are also defined as generalized functions. A *generalized function* is defined by its effect on other functions instead of by its value at every instant of time.

1.4.1 Unit Step Function

- Continuous-time unit step function is defined as

$$u(t) = \begin{cases} 1, & t > 0 \\ \dfrac{1}{2}, & t = 0 \\ 0, & t < 0 \end{cases} \tag{1.5}$$

and is shown in Fig. 1.10(a)

This signal is important for analytic studies, and it also has many practical applications. Note that the unit step function is continuous for all t, except for $t = 0$ where there is a discontinuity.

- Discrete-time unit step function is defined as

$$u(n) = \begin{cases} 1, & n \geq 0 \\ 0, & n < 0 \end{cases} \tag{1.6}$$

and is shown in Fig. 1.10(b).

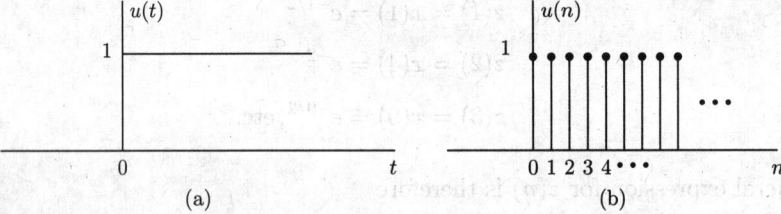

Fig. 1.10   (a) Continuous-time and (b) discrete-time unit step functions

1.4.2 Unit Impulse Function

Continuous-time Unit Impulse Function

Continuous-time unit impulse function, also known as *Dirac delta function* $\delta(t)$, was first defined by P.A.M. Dirac as

$$\delta(t) = 0, \qquad t \neq 0$$

$$\int_{-\infty}^{\infty} \delta(t)dt = 1, \qquad \text{area under unit impulse function is one} \tag{1.7}$$

and is shown in Fig. 1.11(a). The unit impulse can be regarded as a rectangular pulse with a width that has become infinitesimally small, a height that has become infinitely large, and an overall area that has been maintained at unity. The important feature of the unit impulse function is not its shape but the fact that its effective duration (pulse width) approaches zero while its area remains at unity. Pulses such as rectangular, exponential, triangular, and Gaussian types may be used in impulse approximation.

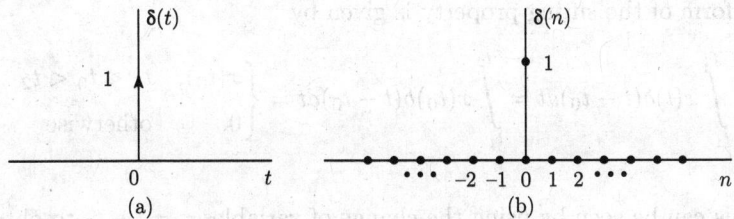

Fig. 1.11 (a) Continuous-time and (b) discrete-time unit impulse functions

Note that the definition of $\delta(t)$ in Eq. (1.7) does not make sense if $\delta(t)$ is an ordinary function. It is meaningful only if $\delta(t)$ is interpreted as a functional, i.e., as a process of assigning the value $x(0)$ to the signal $x(t)$. An ordinary function is specified by its value for all time t. The impulse function is zero everywhere except at $t = 0$, and at this, the only interesting part of its range, it is undefined. These difficulties are resolved by defining the impulse as a generalized function rather than an ordinary function. All the functions derived from the unit impulse function (successive derivatives and integrals) are called *singularity functions*. From Eq. (1.7), it follows that the function $K\delta(t) = 0$ for all $t \neq 0$, and its area is K. Thus, $K\delta(t)$ is an impulse function whose area is K.

Derivative of singularity functions Let $x(t)$ be any test function continuous at $t = 0$. If $g(t)$ is a generalized function, its derivative $dg(t)/dt = g'(t)$ is defined by the following relation:

$$\int_{-\infty}^{\infty} g'(t)x(t)dt = -\int_{-\infty}^{\infty} g(t)x'(t)dt \qquad (1.8)$$

where $x'(t)$ is the derivative of $x(t)$.

Properties of continuous-time unit impulse function

1. **Sifting property**

$$\int_{-\infty}^{\infty} x(t)\delta(t)\, dt = x(t)|_{t=0} = x(0) \qquad (1.9)$$

Proof The multiplication of a continuous-time function $x(t)$ with an impulse located at $t = 0$ results in an impulse, which is located at $t = 0$ and has strength $x(0)$ [the value of $x(t)$ at the location of the impulse].

$$\int_{-\infty}^{\infty} x(t)\delta(t)dt = \int_{-\infty}^{\infty} x(0)\delta(t)dt$$

$$= x(0)\int_{-\infty}^{\infty} \delta(t)dt = x(0)$$

The other form of the sifting property is given by

$$\int_{t_1}^{t_2} x(t)\delta(t-t_0)dt = \int_{t_1}^{t_2} x(t_0)\delta(t-t_0)dt = \begin{cases} x(t_0), & t_1 < t_0 < t_2 \\ 0, & \text{otherwise} \end{cases} \quad (1.10)$$

■

Proof This can be seen by using the change of variables $\tau = t - t_0$ to obtain

$$\int_{t_1}^{t_2} x(t)\delta(t-t_0)dt = \int_{t_1-t_0}^{t_2-t_0} x(\tau+t_0)\delta(\tau)d\tau$$

$$= x(t_0), \quad t_1 < t_0 < t_2$$

Note: This function is discontinuous at $t_0 = t_1$ and $t_0 = t_2$. The value of the function at t_1 or t_2 should be given by

$$\int_{t_1}^{t_2} x(t)\delta(t-t_0)dt = \frac{1}{2}x(t_0), \quad t_0 = t_1 \text{ or } t_0 = t_2$$

In general, the property can be written as

$$x(t) = \int_{-\infty}^{\infty} x(\tau)\delta(t-\tau)\,d\tau$$

which implies that the signal $x(t)$ can be expressed as a continuous sum of weighted impulses.

■

2. Scaling property

$$\delta(at) = \frac{1}{|a|}\delta(t) \quad (1.11)$$

Proof With a change of variable, $at = \tau$, hence $t = \tau/a$, $dt = (1/a)d\tau$, we obtain the following equations:

- If $a > 0$,

$$\int_{-\infty}^{\infty} \delta(at)x(t)\,dt = \frac{1}{a}\int_{-\infty}^{\infty} \delta(\tau)x\left(\frac{\tau}{a}\right)d\tau$$

$$= \frac{1}{a}x\left(\frac{\tau}{a}\right)\Big|_{\tau=0} = \frac{1}{|a|}x(0)$$

- If $a < 0$,

$$\int\limits_{-\infty}^{\infty} \delta(at)x(t)dt = -\frac{1}{a}\int\limits_{\infty}^{-\infty} \delta(\tau)x\left(-\frac{\tau}{a}\right)d\tau = \frac{1}{a}\int\limits_{-\infty}^{\infty} \delta(\tau)x\left(-\frac{\tau}{a}\right)d\tau$$

$$= \frac{1}{a}x\left(-\frac{\tau}{a}\right)\bigg|_{\tau=0} = \frac{1}{|a|}x(0)$$

Thus for any a,

$$\int\limits_{-\infty}^{\infty} \delta(at)x(t)dt = \frac{1}{|a|}x(0) = \frac{1}{|a|}\int\limits_{-\infty}^{\infty} \delta(t)x(t)dt = \int\limits_{-\infty}^{\infty} \frac{1}{|a|}\delta(t)x(t)dt$$

for any $x(t)$. From the above equation, we get

$$\delta(at) = \frac{1}{|a|}\delta(t)$$

∎

3. Impulse function is an even function

$$\delta(-t) = \delta(t) \tag{1.12}$$

Proof Substituting $a = -1$ into Eq. (1.11), we obtain

$$\delta(-t) = \frac{1}{|-1|}\delta(t) = \delta(t)$$

which shows that $\delta(t)$ is an even function.

∎

4. Sampling property or multiplication property
Multiplying the signal $x(t)$ by a unit impulse samples the value of the signal at the point at which the impulse is located, i.e.,

$$x(t)\delta(t) = x(t)|_{t=0}\delta(t) = x(0)\delta(t) \tag{1.13}$$

Proof If $x(t)$ is continuous at $t = 0$, then

$$\int\limits_{-\infty}^{\infty} [x(t)\delta(t)]\phi(t)dt = \int\limits_{-\infty}^{\infty} \delta(t)[x(t)\phi(t)]dt$$

$$= x(0)\phi(0) = x(0)\int\limits_{-\infty}^{\infty} \delta(t)\phi(t)dt$$

Therefore,

$$\int\limits_{-\infty}^{\infty} [x(t)\delta(t)]\phi(t)dt = \int\limits_{-\infty}^{\infty} [x(0)\delta(t)]\phi(t)dt$$

for all $\phi(t)$. From the above equation, we get

$$x(t)\delta(t) = x(0)\delta(t)$$

∎

The other form of the multiplication property is given by

$$x(t)\delta(t - t_0) = x(t_0)\delta(t - t_0) \tag{1.14}$$

Proof If $x(t)$ is continuous at $t = t_0$, then

$$\int_{-\infty}^{\infty} [x(t)\delta(t - t_0)]\phi(t)dt = \int_{-\infty}^{\infty} \delta(t - t_0)[x(t)\phi(t)]dt$$

$$= x(t_0)\phi(t_0) = x(t_0) \int_{-\infty}^{\infty} \delta(t - t_0)\phi(t)dt$$

Therefore,

$$\int_{-\infty}^{\infty} [x(t)\delta(t - t_0)]\phi(t)dt = \int_{-\infty}^{\infty} [x(t_0)\delta(t - t_0)]\phi(t)dt$$

for all $\phi(t)$. From the above equation, we get

$$x(t)\delta(t - t_0) = x(t_0)\delta(t - t_0)$$

∎

5. **Differentiation property**

$$\int_{-\infty}^{\infty} x(t)\delta'(t)dt = -x'(0) \tag{1.15}$$

Proof The derivative of the singularity function is defined as

$$\int_{-\infty}^{\infty} \delta'(t)x(t)dt = - \int_{-\infty}^{\infty} \delta(t)x'(t)dt = -x'(0)$$

∎

6. **Amplitude reversal**

$$t\delta'(t) = -\delta(t) \tag{1.16}$$

Proof Consider

$$\int_{-\infty}^{\infty} [t\delta'(t)]\phi(t)dt = \int_{-\infty}^{\infty} \delta'(t)[t\phi(t)]dt = - \int_{-\infty}^{\infty} \delta(t)\frac{d}{dt}[t\phi(t)]dt$$

$$= - \int_{-\infty}^{\infty} \delta(t)[\phi(t) + t\phi'(t)]dt = -[\phi(t) + t\phi'(t)]|_{t=0}$$

$$= -\phi(0) = - \int_{-\infty}^{\infty} \delta(t)\phi(t)dt$$

Therefore,

$$\int\limits_{-\infty}^{\infty} [t\delta'(t)]\phi(t)dt = \int\limits_{-\infty}^{\infty} [-\delta(t)]\phi(t)dt$$

From the above equation, we get

$$t\delta'(t) = -\delta(t) \qquad\blacksquare$$

Derivative of impulse functions (doublet function) Derivatives of all orders of the impulse function are also singularity functions. The first derivative $d\delta(t)/dt = \delta'(t)$ is referred to as a *doublet function*. A doublet function is defined as

$$\frac{d\delta(t)}{dt} = \delta'(t) = 0, \quad t \neq 0$$

$$\int\limits_{-\infty}^{\infty} \delta'(t)dt = 0, \quad \text{area under unit doublet function is zero} \tag{1.17}$$

$$\int\limits_{-\infty}^{\infty} x(t)\delta'(t)dt = -\int\limits_{-\infty}^{\infty} x'(t)\delta(t)dt = -x'(t)\Big|_{t=0} = -x'(0)$$

where $x(t)$ is any continuous function having a continuous first derivative at $t = 0$.

Example 1.4 Prove the following properties of a doublet function:

(a) $\displaystyle\int\limits_{t_1}^{t_2} x(t)\delta'(t - t_0)dt = -x'(t_0), \quad t_1 < t_0 < t_2$ \hfill (1.18)

(b) $x(t)\delta'(t - t_0) = x(t_0)\delta'(t - t_0) - x'(t_0)\delta(t - t_0)$ \hfill (1.19)

(c) $\displaystyle\int\limits_{-\infty}^{t} \delta'(\tau - t_0)d\tau = \delta(t - t_0)$ \hfill (1.20)

(d) $\delta'(at + b) = \dfrac{1}{a|a|}\delta'\left(t + \dfrac{b}{a}\right)$ \hfill (1.21)

Solution

(a) The given results can be demonstrated using integration by parts as follows:

$$\int\limits_{t_1}^{t_2} x(t)\delta'(t - t_0)dt = \int\limits_{t_1}^{t_2} x(t)d[\delta'(t - t_0)]$$

$$= x(t)\delta(t - t_0)|_{t_1}^{t_2} - \int\limits_{t_1}^{t_2} x'(t)\delta(t - t_0)dt$$

$$= 0 - 0 - x(t)\Big|_{t=t_0} = -x'(t_0)$$

(b) Let $\psi(t)$ be a test signal. Therefore,

$$\int_{-\infty}^{\infty} [x(t)\delta'(t-t_0)]\psi(t)dt = -\int_{-\infty}^{\infty} [\psi(t)x(t)]'\delta(t-t_0)dt$$

$$= -[\psi'(t)x(t) + \psi(t)x'(t)]\Big|_{t=t_0}$$

$$= -\psi'(t_0)x(t_0) - \psi(t_0)x'(t_0)$$

$$= x(t_0)\int_{-\infty}^{\infty} \delta'(t-t_0)\psi(t)dt - x'(t_0)\int_{-\infty}^{\infty} \delta(t-t_0)\psi(t)dt$$

$$\int_{-\infty}^{\infty} [x(t)\delta'(t-t_0)]\psi(t)dt = \int_{-\infty}^{\infty} [x(t_0)\delta'(t-t_0) - x'(t_0)\delta(t-t_0)]\psi(t)dt$$

Comparing both sides of the above equation, we get

$$x(t)\delta'(t-t_0) = x(t_0)\delta'(t-t_0) - x'(t_0)\delta(t-t_0)$$

(c) Consider the integral

$$\int_{-\infty}^{t} \delta'(\tau-t_0)d\tau = \delta(\tau-t_0)\Big|_{-\infty}^{t} = \delta(t-t_0)$$

(d) We know that

$$\delta(at+b) = \delta\left(a\left(t+\frac{b}{a}\right)\right)$$

$$= \frac{1}{|a|}\delta\left(t+\frac{b}{a}\right)$$

Differentiating both sides with respect to t, we get

$$a\delta'(at+b) = \frac{1}{|a|}\delta'\left(t+\frac{b}{a}\right)$$

$$\delta'(at+b) = \frac{1}{a|a|}\delta'\left(t+\frac{b}{a}\right)$$

Higher-order derivatives of the impulse function give

$$\int_{-\infty}^{\infty} x(t)\frac{d^k\delta(t)}{dt^k}\,dt = (-1)^k \frac{d^k x(t)}{dt^k}\Big|_{t=0} \tag{1.22}$$

Doublets and higher-order derivatives of the impulse function are not often used as signal models in system analysis.

Discrete-time Unit Impulse Function

Discrete-time unit impulse function (or unit sample function) is defined as

$$\delta(n) = \begin{cases} 1, & n = 0 \\ 0, & n \neq 0 \end{cases} \tag{1.23}$$

and is shown in Fig. 1.11(b).

Properties of discrete-time unit impulse function

1.
$$\delta(kn) = \delta(n) \tag{1.24}$$

where k is an integer.

Proof By definition,

$$\delta(n) = \begin{cases} 1, & n = 0 \\ 0, & n \neq 0 \end{cases}$$

$$\delta(kn) = \begin{cases} 1, & kn = 0 \longrightarrow n = 0 \\ 0, & kn \neq 0 \longrightarrow n \neq 0 \end{cases}$$

$$= \delta(n)$$

2. Unit impulse function $\delta(n)$ and unit step function $u(n)$ are related by

$$\delta(n) = u(n) - u(n-1) \tag{1.25}$$

Proof From the definition of the unit step function,

$$u(n) = \begin{cases} 1, & n \geq 0 \\ 0, & n < 0 \end{cases}$$

$$= \delta(n) + \delta(n-1) + \delta(n-2) + \cdots$$

$$= \sum_{k=0}^{\infty} \delta(n-k)$$

$$= \delta(n) + \sum_{k=1}^{\infty} \delta(n-k)$$

and

$$u(n-1) = \begin{cases} 1, & n-1 \geq 0 \longrightarrow n \geq 1 \\ 0, & n-1 < 0 \longrightarrow n < 1 \end{cases}$$

$$= \delta(n-1) + \delta(n-2) + \delta(n-3) + \cdots$$

$$= \sum_{k=1}^{\infty} \delta(n-k)$$

From the definition of $u(n)$ and $u(n-1)$,

$$u(n) - u(n-1) = \delta(n) + \sum_{k=1}^{\infty} \delta(n-k) - \sum_{k=1}^{\infty} \delta(n-k)$$

$$= \delta(n)$$

∎

3. For any arbitrary sequence $x(n)$,

$$x(n)\delta(n-k) = x(k)\delta(n-k) \tag{1.26}$$

Proof From the definition of discrete impulse function,

$$\delta(n-k) = \begin{cases} 1, & (n-k) = 0 \longrightarrow n = k \\ 0, & (n-k) \neq 0 \longrightarrow n \neq k \end{cases}$$

Therefore,

$$x(n)\delta(n-k) = x(n)|_{n=k}\delta(n-k)$$

$$= x(k)\delta(n-k)$$

∎

4.

$$\sum_{k=-\infty}^{n} \delta(k) = \sum_{k=0}^{\infty} \delta(n-k) = u(n) = \begin{cases} 1, & n \geq 0 \\ 0, & n < 0 \end{cases} \tag{1.27}$$

Proof Consider the sum of δ function from $-\infty$ to n

$$= \sum_{k=-\infty}^{n} \delta(k)$$

By replacing k by $(n-m)$, we can write the above equation as

$$= \sum_{m=0}^{\infty} \delta(n-m)$$

$$= \delta(n) + \delta(n-1) + \delta(n-2) + \cdots$$

$$= u(n)$$

∎

1.4.3 Unit Ramp Function

The ramp function shown in Fig. 1.12(a) is defined by

$$r(t) = tu(t) = \begin{cases} t, & t \geq 0 \\ 0, & t < 0 \end{cases} \tag{1.28}$$

The discrete-time version of the ramp function shown in Fig. 1.12(b) is defined by

$$r(n) = nu(n) = \begin{cases} n, & n \geq 0 \\ 0, & n < 0 \end{cases} \tag{1.29}$$

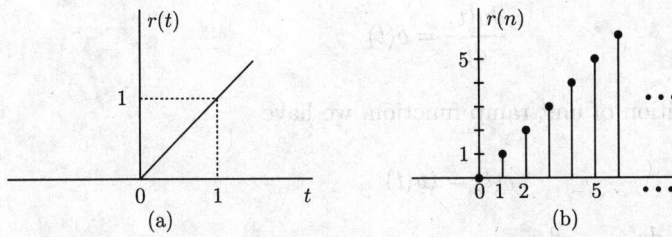

Fig. 1.12   (a) Continuous-time and (b) discrete-time unit ramp functions

---

Example 1.5   Prove the following relationship between unit impulse function $\delta(t)$, unit step function $u(t)$, and unit ramp function $r(t)$.

(a) $\dfrac{du(t)}{dt} = \delta(t)$ or $u(t) = \displaystyle\int_{-\infty}^{t} \delta(\tau)\, d\tau$

(b) $\dfrac{dr(t)}{dt} = u(t)$ or $r(t) = \displaystyle\int_{-\infty}^{t} u(\tau)\, d\tau$

Solution

(a) Let $x(t)$ be any test function continuous at $t = 0$. Now consider the integral

$$\int_{-\infty}^{\infty} \frac{du(t)}{dt} x(t)\, dt = \int_{-\infty}^{\infty} u'(t)x(t)\, dt$$

$$= u(t)x(t)\Big|_{-\infty}^{\infty} - \int_{-\infty}^{\infty} x'(t)u(t)\, dt$$

$$= x(\infty) - 0 - \int_0^\infty x'(t)\, dt$$

$$= x(\infty) - x(t)\big|_0^\infty$$

$$= x(\infty) - [x(\infty) - x(0)]$$

$$= x(0)$$

$$\int_{-\infty}^{\infty} \frac{du(t)}{dt} x(t)\, dt = \int_{-\infty}^{\infty} \delta(t) x(t)\, dt$$

Therefore,
$$\frac{du(t)}{dt} = \delta(t)$$

(b) From the definition of unit ramp function, we have

$$r(t) = tu(t)$$

$$\frac{dr(t)}{dt} = \frac{d}{dt}[tu(t)] = t\delta(t) + u(t) = 0 + u(t) = u(t)$$

Example 1.6 Draw the waveforms of the following signals:

(a) $x_1(t) = u(t+2)$, (b) $x_2(t) = u(t-2)$, (c) $x_3(t) = u(-t)$, (d) $x_4(t) = u(-2t+1)$,
(e) $x_5(t) = u(-2t-1)$, (f) $x_6(t) = x_1(t) - x_2(t)$, (g) $x_7(t) = u(t) - 2u(t-1) + u(t-2)$,
(h) $x_8(t) = dx_6(t)/dt$, and (i) $x_9(t) = dx_7(t)/dt$, where $u(t)$ is the unit step signal.

Solution
(a) $u(t+2)$ is a unit step signal shifted towards left by 2 units [Fig. 1.13(a)].

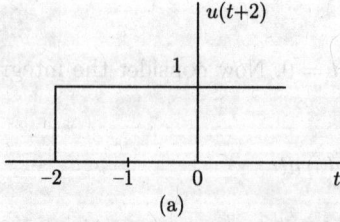

(a)

Fig. 1.13(a)

(b) $u(t-2)$ is a unit step signal shifted towards right by 2 units [Fig. 1.13(b)].

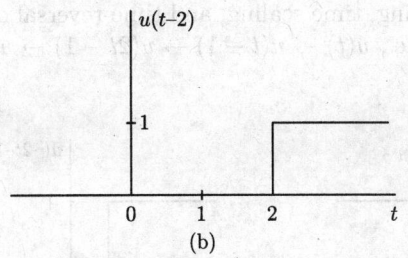

Fig. 1.13(b)

(c) $u(-t)$ is a folded version of $u(t)$ [Fig. 1.13(c)].

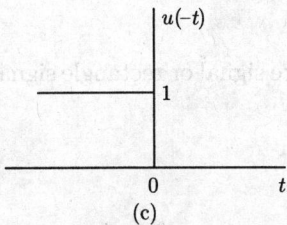

Fig. 1.13(c)

(d) The mathematical form of $u(-2t + 1)$ is

$$u(-2t + 1) = \begin{cases} 1, & (-2t + 1) > 0 \longrightarrow t < \dfrac{1}{2} \\[2mm] 0, & (-2t + 1) < 0 \longrightarrow t > \dfrac{1}{2} \end{cases}$$

To obtain $u(-2t + 1)$ from $u(t)$, time-shifting, time-scaling, and time-reversal operations must be performed in the correct order, i.e., $u(t) \to u(t + 1) \to u(2t + 1) \to u(-2t + 1)$ [Fig. 1.13(d)].

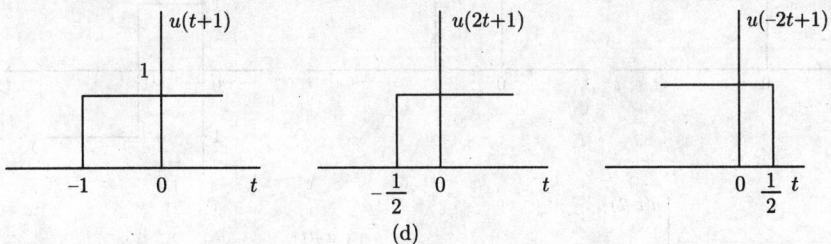

Fig. 1.13(d)

(e) The mathematical form of $u(-2t - 1)$ is

$$u(-2t - 1) = \begin{cases} 1, & (-2t - 1) > 0 \longrightarrow t < -\dfrac{1}{2} \\[2mm] 0, & (-2t - 1) < 0 \longrightarrow t > -\dfrac{1}{2} \end{cases}$$

To obtain $u(-2t - 1)$ from $u(t)$, time-shifting, time scaling, and time-reversal operations must be performed in the correct order, i.e., $u(t) \to u(t-1) \to u(2t-1) \to u(-2t-1)$ [Fig. 1.13(e)].

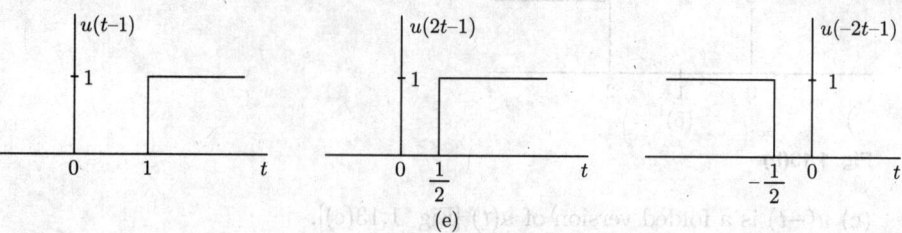

Fig. 1.13(e)

(f) $x_6(t) = x_1(t) - x_2(t) = u(t+2) - u(t-2)$ is a gate signal or rectangle signal as shown in Fig. 1.13(f).

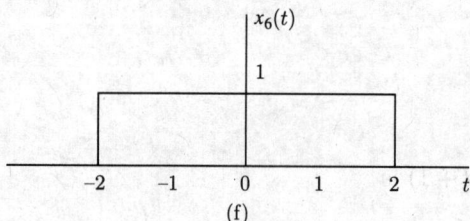

Fig. 1.13(f)

(g) $x_7(t) = u(t) - 2u(t-1) + u(t-2)$ is shown in Fig. 1.13(g).

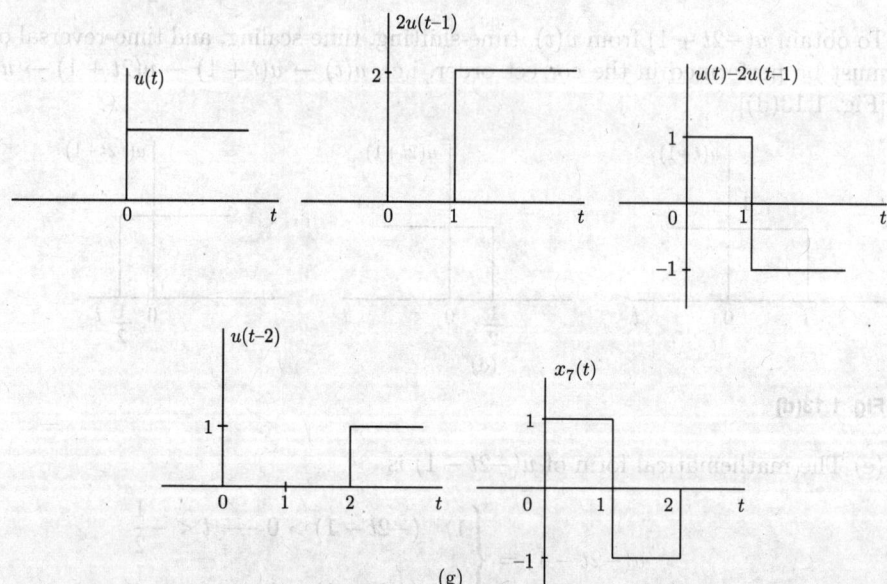

Fig. 1.13(g)

(h) The derivative of $x_6(t)$ is determined as

$$x_8(t) = \frac{dx_6(t)}{dt} = \frac{d}{dt}\big[x_1(t) - x_2(t)\big]$$

substituting the values of $x_1(t)$ and $x_2(t)$ from parts (a) and (b), we get [Fig. 1.13(h)]

$$x_8(t) = \frac{d}{dt}u(t+2) - \frac{d}{dt}u(t-2) = \delta(t+2) - \delta(t-2)$$

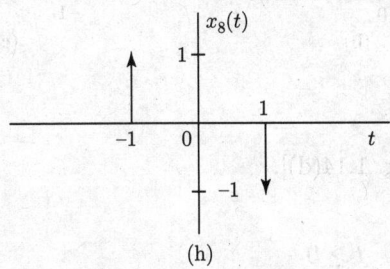

(h)

Fig. 1.13(h)

(i) The derivative of $x_7(t)$ is determined as [Fig. 1.13(i)]

$$x_9(t) = \frac{dx_7(t)}{dt} = \frac{d}{dt}\big[u(t) - 2u(t-1) + u(t-2)\big] = \delta(t) - 2\delta(t-1) + \delta(t-2)$$

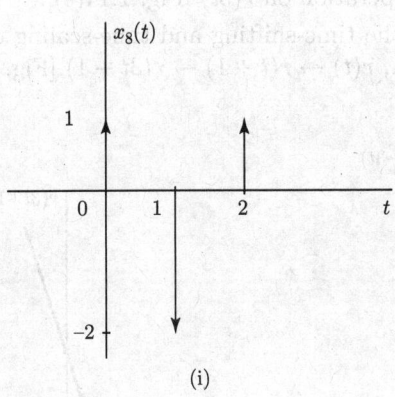

(i)

Fig. 1.13(i)

Example 1.7 Draw the waveforms of the following signals:
(a) $x_1(t) = r(t-1)$, (b) $x_2(t) = r(t+1)$, (c) $x_3(t) = r(-t)$, (d) $x_4(t) = r(3t)$,
(e) $x_5(t) = r(-3t)$, (f) $x_6(t) = r(3t+1)$, (g) $x_7(t) = r(-3t-1)$, (h) $x_8(t) = r(t) - r(t-1)$,
(i) $x_9(t) = x_8(t) - r(t-1) + r(t-2)$, (j) $x_{10}(t) = dx_8(t)/dt$, and (k) $x_{11}(t) = dx_9(t)/dt$,
where $r(t)$ is the unit ramp signal.

Solution
(a) $r(t-1)$ is a ramp signal shifted towards right by 1 unit [Fig. 1.14(a)].
(b) $r(t+1)$ is a ramp signal shifted towards left by 1 unit [Fig. 1.14(b)].

(c) $r(-t)$ is a folded version of $r(t)$ [Fig. 1.14(c)].

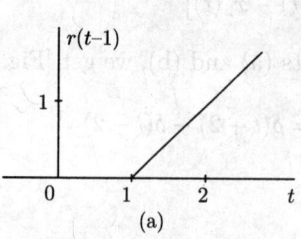
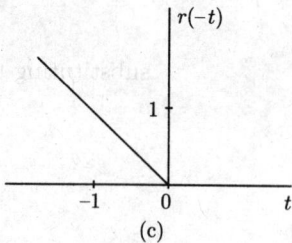

Fig. 1.14(a), (b), (c)

(d) $r(3t)$ is a time-scaled version of $r(t)$ [Fig. 1.14(d)].

$$r(t) = \begin{cases} t, & t \geq 0 \\ 0, & t < 0 \end{cases}$$

$$x_4(t) = r(3t) = \begin{cases} 3t, & 3t \geq 0 \longrightarrow t \geq 0 \\ 0, & 3t < 0 \longrightarrow t < 0 \end{cases}$$

(e) To obtain $r(-3t)$, apply time-reversal operation on $r(3t)$ [Fig. 1.14(e)].

(f) To obtain $x_6(t) = r(3t + 1)$ from $r(t)$, the time-shifting and time-scaling operations must be performed in the correct order, i.e., $r(t) \to r(t+1) \to r(3t+1)$ [Fig. 1.14(f)].

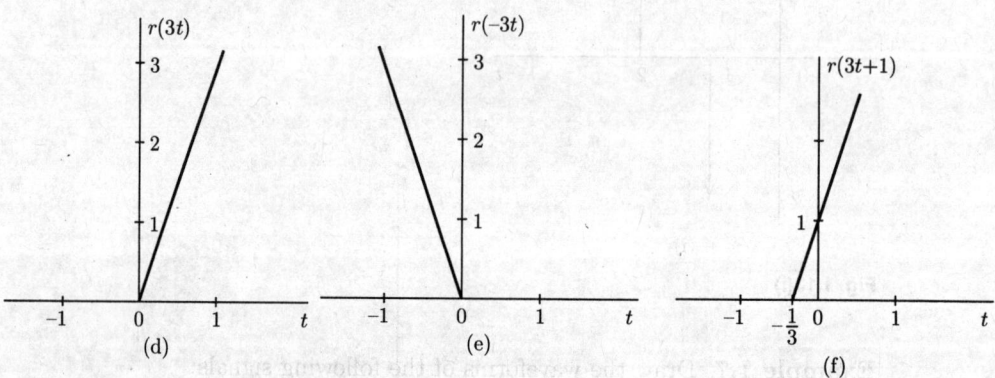

Fig. 1.14(d), (e), (f)

(g) To obtain $x_7(t) = r(-3t - 1)$ from $r(t)$, time-shifting, time-scaling, and time-reversal operations must be performed in the correct order, i.e., $r(t) \to r(t-1) \to r(3t-1) \to r(-3t-1)$ [Fig. 1.14(g)].

(h) $x_8(t) = r(t) - r(t-1)$ is shown in Fig. 1.14(h).

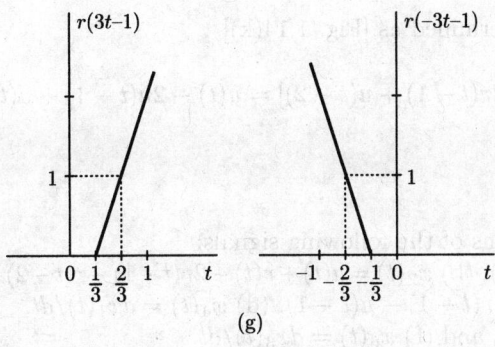 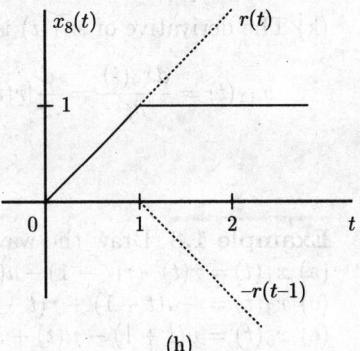

Fig. 1.14(g), (h)

(i) $x_9(t)$ is shown in Fig. 1.14(i).

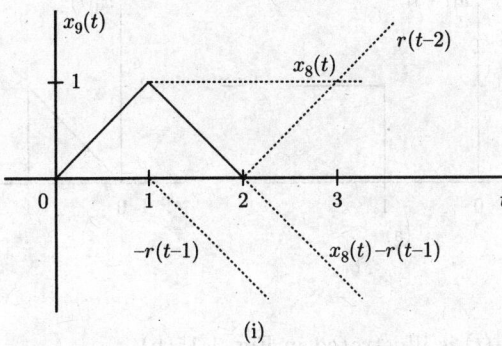

Fig. 1.14(i)

(j) The derivative of $x_8(t)$ is determined as [Fig. 1.14(j)]

$$x_{10}(t) = \frac{dx_8(t)}{dt} = \frac{d}{dt}[r(t) - r(t-1)] = u(t) - u(t-1)$$

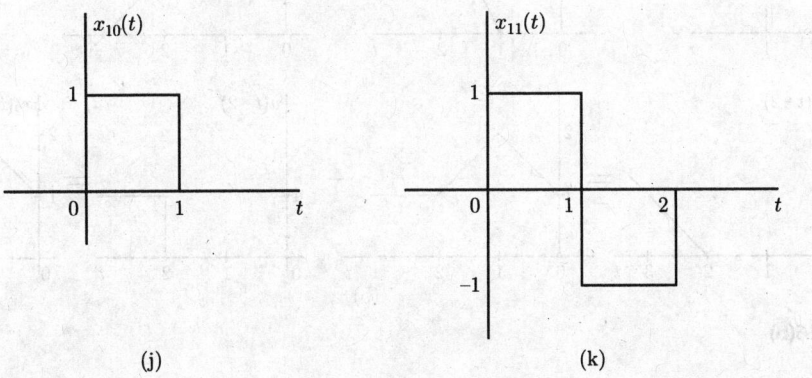

Fig. 1.14(j), (k)

(k) The derivative of $x_{11}(t)$ is determined as [Fig. 1.14(k)]

$$x_{11}(t) = \frac{dx_9(t)}{dt} = \frac{d}{dt}[r(t) - 2r(t-1) + u(t-2)] = u(t) - 2u(t-1) + u(t-2)$$

Example 1.8 Draw the waveforms of the following signals:
(a) $x_1(t) = r(t) - r(t-1) - u(t-1)$, (b) $x_2(t) = u(t) + r(t) - 2r(t-1) + r(t-2) - u(t-2)$,
(c) $x_3(t) = -u(t+1) + r(t+1) - r(t-1) - u(t-1)$, (d) $x_4(t) = dx_1(t)/dt$,
(e) $x_5(t) = u(t+1) - u(t) + x_1(t)$, and (f) $x_6(t) = dx_3(t)/dt$.

Solution
(a) The procedure of obtaining $x_1(t)$ is illustrated in Fig. 1.15(a).

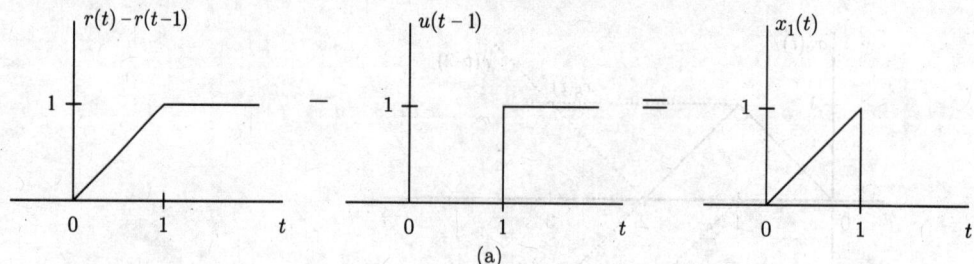

(a)

Fig. 1.15(a)

(b) The procedure of obtaining $x_2(t)$ is illustrated in Fig. 1.15(b).

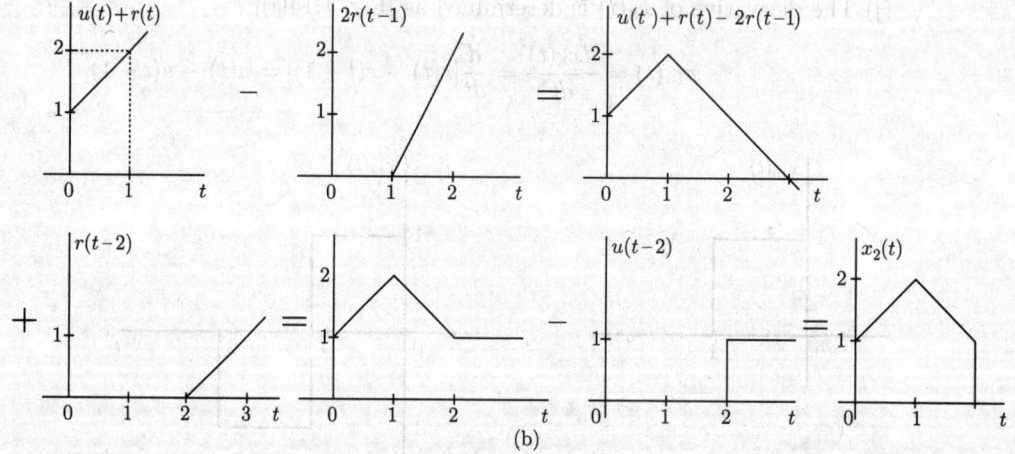

(b)

Fig. 1.15(b)

(c) The procedure of obtaining $x_3(t)$ is illustrated in Fig. 1.15(c).

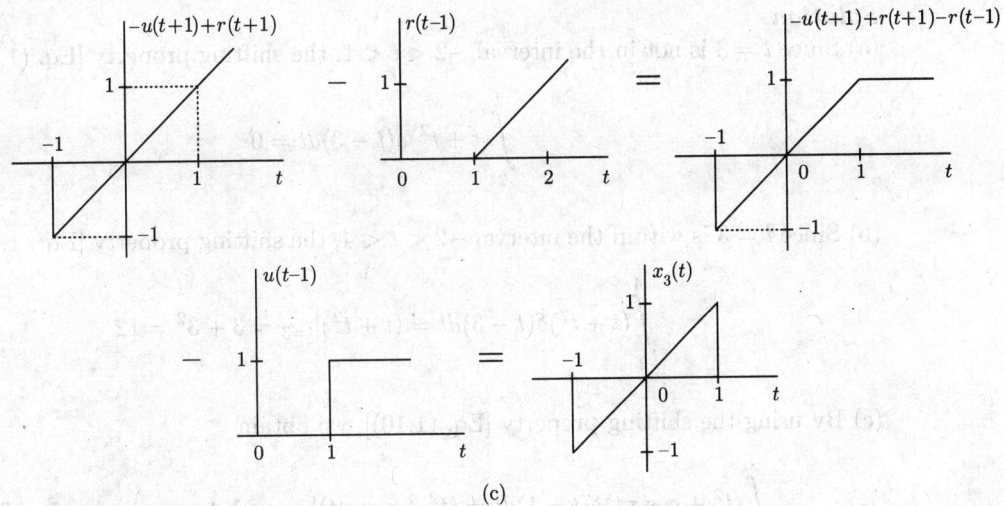

Fig. 1.15(c)

(d) The derivative of $x_1(t)$ is determined as [Fig. 1.15(d)]

$$x_4(t) = \frac{dx_1(t)}{dt} = \frac{d}{dt}[r(t) - r(t-1) - u(t-1)] = u(t) - u(t-1) - \delta(t-1)$$

(e) $x_5(t)$ is shown in Fig. 1.15(e).
(f) The derivative of $x_3(t)$ is determined as [Fig. 1.15(f)]

$$x_6(t) = \frac{dx_3(t)}{dt} = \frac{d}{dt}[-u(t+1) + r(t+1) - r(t-1) - u(t-1)]$$

$$= -\delta(t+1) + u(t+1) - u(t-1) - \delta(t-1)$$

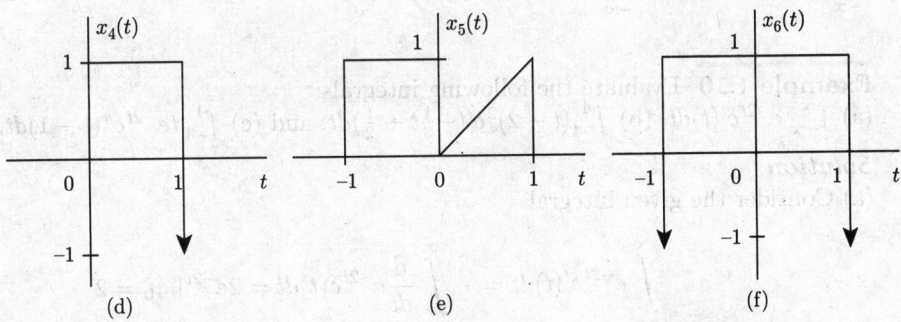

Fig. 1.15(d), (e), (f)

Example 1.9 Evaluate the following integrals:
(a) $\int_{-2}^{1}(t+t^2)\delta(t-3)dt$, (b) $\int_{-2}^{4}(t+t^2)\delta(t-3)dt$, (c) $\int_{-\infty}^{\infty}(t^2+\cos \pi t)\delta(t-1)dt$,
(d) $\int_{0}^{3}e^{t-2}\delta(2t-4)dt$, and (e) $\int_{-\infty}^{\infty}e^{-t}\delta(2t-2)dt$.

Solution

(a) Since $t = 3$ is not in the interval $-2 < t < 1$, the shifting property [Eq. (1.10)] yields

$$\int_{-2}^{1} (t + t^2)\delta(t - 3)dt = 0$$

(b) Since $t = 3$ is within the interval $-2 < t < 4$, the shifting property [Eq. (1.10)] yields

$$\int_{-2}^{4} (t + t^2)\delta(t - 3)dt = (t + t^2)|_{t=3} = 3 + 3^2 = 12$$

(c) By using the shifting property [Eq. (1.10)], we obtain

$$\int_{-\infty}^{\infty} (t^2 + \cos \pi t)\delta(t - 1)dt = (t^2 + \cos \pi t)|_{t=1} = 1 + \cos \pi = 1 - 1 = 0$$

(d) By using the scaling property [Eq. (1.11)] and then the shifting property [Eq. (1.10)], we get

$$\int_{0}^{3} e^{t-2}\delta(2t - 4)dt = \int_{0}^{3} e^{t-2}\delta[2(t - 2)] = \int_{0}^{3} e^{t-2}\frac{1}{2}\delta(t - 2)dt = \frac{1}{2}e^{t-2}|_{t=2} = \frac{1}{2}e^{0} = \frac{1}{2}$$

(e) By using the scaling property [Eq. (1.11)] and then the shifting property [Eq. (1.10)], we get

$$\int_{-\infty}^{\infty} e^{-t}\delta(2t - 2)dt = \int_{-\infty}^{\infty} e^{-t}\delta(2(t - 1))dt = \int_{-\infty}^{\infty} e^{-t}\frac{1}{2}\delta(t - 1)dt = \frac{1}{2}e^{-t}|_{t=1} = \frac{1}{2}e^{-1} = \frac{1}{2e}$$

Example 1.10 Evaluate the following integrals:
(a) $\int_{-\infty}^{\infty} e^{-2t}\delta'(t)dt$, (b) $\int_{-4}^{4}(t - 2)^2\delta'(-\frac{1}{3}t + \frac{1}{2})dt$, and (c) $\int_{-4}^{1} te^{-2t}\delta''(t - 1)dt$.

Solution

(a) Consider the given integral

$$\int_{-\infty}^{\infty} e^{-2t}\delta'(t)dt = -\int_{-\infty}^{\infty} \frac{d}{dt}e^{-2t}\delta(t)dt = 2e^{-2t}|_{t=0} = 2$$

(b) Consider the given integral

$$\int_{-4}^{4} (t - 2)^2\delta'\left(-\frac{1}{3}t + \frac{1}{2}\right)dt = \int_{-4}^{4} (t - 2)^2\delta'\left(-\frac{1}{3}\left(t - \frac{3}{2}\right)\right)dt$$

$$= \int_{-4}^{4} 3(t-2)^2 \delta'\left(t - \frac{3}{2}\right) dt$$

$$= -3 \int_{-4}^{4} \frac{d}{dt}(t-2)^2 \delta\left(t - \frac{3}{2}\right) dt$$

$$= -6(t-2)|_{t=3/2} = 3$$

(c) Consider the given integral

$$\int_{-4}^{1} te^{-2t} \delta''(t-1) dt = (-1)^2 \frac{d^2}{dt^2}(te^{-2t})|_{t=1} = (4t-4)e^{-2t}|_{t=1} = 0$$

1.5 PERIODIC AND APERIODIC SIGNALS

A periodic signal has a waveform that repeats over and over, with the time between the repeats defined as the period of the signal. The continuous-time signal $x(t)$ is periodic if and only if

$$x(t+T) = x(t), \quad \text{for all } t \tag{1.30}$$

where T is positive and is the period of the signal. In other words, a periodic signal $x(t)$ remains unchanged when time-shifted by one period. For this reason a periodic signal must start at $t = -\infty$; if it started at some finite instant say $t = 0$, the time-shifted signal $x(t+T)$ would start at $t = -T$ and $x(t+T) \neq x(t)$. Therefore a periodic signal, by definition, must start at $t = -\infty$ and continued forever. It should be noted that no physical signals are actually periodic since they all begin at some time and/or cease to exist at some later time. The fundamental period T_0 of $x(t)$ is the smallest positive value of T for which Eq. (1.30) holds. This definition of fundamental period works, except if $x(t)$ is a constant. In this case the fundamental period is undefined since $x(t)$ is periodic for any choice of T. A signal $x(t)$ that is not periodic will be referred to as an *aperiodic* signal.

The discrete-time signal $x(n)$ is periodic if and only if

$$x(n+N) = x(n), \quad \text{for all } n \tag{1.31}$$

where N is a positive integer and is the period of the signal measured in terms of number of sample spacings (samples/cycle). The smallest value of N for which the definition hold is called the fundamental period of the periodic signal.

Example 1.11 Show that

(a) the complex exponential signal $x(t) = e^{j\omega_0 t}$ is periodic and its fundamental period is $2\pi/\omega_0$.

(b) the complex exponential signal $x(n) = e^{j\omega_0 n}$ is periodic only if $2\pi/\omega_0$ is a ratio number.

Solution

(a) By Eq. (1.30), $x(t)$ will be periodic if

$$x(t) = x(t + T)$$

$$e^{j\omega_0 t} = e^{j\omega_0 (t+T)}$$

$$= e^{j\omega_0 t} \underbrace{e^{j\omega_0 T}}_{=1}$$

Thus, we must have

$$e^{j\omega_0 T} = 1 = e^{j2m\pi}, \quad m = \text{positive integer}$$

Comparing the powers of the exponent, we get

$$\omega_0 T = 2m\pi$$

$$T = \frac{2m\pi}{\omega_0}$$

Thus, the fundamental period (smallest positive value) $T = 2\pi/\omega_0$.

(b) By Eq. (1.31), $x(n)$ will be periodic if

$$x(n) = x(n + N)$$

$$e^{j\omega_0 n} = e^{j\omega_0 (n+N)}$$

$$= e^{j\omega_0 n} \underbrace{e^{j\omega_0 N}}_{=1}$$

Thus, we must have

$$e^{j\omega_0 N} = 1 = e^{j2m\pi}, \quad m = \text{a positive integer}$$

Comparing the powers of the exponent, we get

$$\omega_0 N = 2m\pi$$

$$\frac{\omega_0}{2\pi} = \frac{m}{N} = \text{a rational number}$$

Thus, $x(n)$ is periodic only if $\omega_0/2\pi$ is a rational number.

1.5.1 Properties of Periodic Signals

Following are the properties of a periodic signal:

1. A periodic signal is an everlasting signal.
2. A periodic signal $x(t)$ can be generated by periodic extension of any segment of $x(t)$ of duration T (the period). As a result we can generate $x(t)$ from any segment of $x(t)$ having a duration of one period by replacing this segment and the reproduction thereof end to end ad infinitum on either side.

3. Area under $x(t)$ over any interval of duration T is the same, i.e., for any real numbers a and b

$$\int\limits_{a}^{a+T} x(t)dt = \int\limits_{b}^{b+T} x(t)dt \qquad (1.32)$$

This results from the fact that a periodic signal takes the same values at the intervals of T.

4. The sum of M periodic continuous-time signals is not necessarily periodic. It is periodic with period T if and only if the condition

$$\frac{T}{T_i} = n_i, \quad 1 \le i \le M \qquad (1.33)$$

is satisfied, where T_i is the period of the ith signal in the sum and n_i is an integer. Simultaneous solution for the first signal in the sum paired with each additional signal in the sum gives

$$\frac{T_1}{T_i} = \frac{n_i}{n_1}, \quad 2 \le i \le M \qquad (1.34)$$

Thus, an equivalent necessary and sufficient condition for a sum of periodic continuous-time signals to be periodic is that all the ratios of the periodic of the first signal in the sum to the period of another signal in the sum be *rational*, i.e., ratios of two integers.

The following steps can be taken to determine the period of the sum signal, if it is periodic. First convert each ratio from Eq. (1.34) to a ratio of integers. If this conversion is not possible, the sum signal is not periodic. Then factor the greatest common divisor (GCD) from the numerator and denominator of each individual ratio. The least common multiple (LCM) of the denominators of the resulting ratios is the value of n_1. The period of the sum signal is computed as

$$T = T_1 n_1 \qquad (1.35)$$

The preceding development is also applicable to the sum of discrete-time periodic signals if T is replaced by N and each T_i is replaced by N_i.

5. A sum of discrete-time periodic signals is always periodic because the period ratios N_1/N_i are always rational.

Example 1.12 Assume $x_1(t)$ and $x_2(t)$ are periodic signals with periods T_1 and T_2, respectively. Under what conditions is the sum

$$x(t) = x_1(t) + x_2(t)$$

periodic and what will be the period of $x(t)$ if it is periodic?

Solution

Given that $x_1(t)$ and $x_2(t)$ are periodic with periods T_1 and T_2, respectively. Thus, $x_1(t)$ and $x_2(t)$ may be written as

$$x_1(t) = x_1(t + T_1) = x_1(t + mT_1)$$

and

$$x_2(t) = x_2(t + T_2) = x_2(t + nT_2)$$

where m and n are integers. Now if T_1 and T_2 are such that

$$T = mT_1 = nT_2$$

Then

$$x(t + T) = x_1(t + T) + x_2(t + T)$$
$$= x_1(t + mT_1) + x_2(t + nT_2)$$
$$= x_1(t) + x_2(t)$$
$$x(t + T) = x(t)$$

i.e., $x(t)$ is periodic in this case. Therefore, the condition for $x(t)$ to be periodic is

$$\frac{T_1}{T_2} = \frac{n}{m} = \text{a rational number}$$

From the above analysis, it is clear that the smallest common period is the LCM of T_1 and T_2 and is given by the equation

$$mT_1 = nT_2 = T$$

if the integers m and n are relatively prime. On the other hand, if the ratio T_1/T_2 is an irrational number, then the signals $x_1(t)$ and $x_2(t)$ will not have a common period and so $x(t)$ will not be periodic.

Example 1.13   Determine whether the continuous-time signals $x(t)$ and $y(t)$ and the discrete-time signal $z(n)$ are periodic. Compute the period for any that are periodic.

(a) For $x(t)$,

$$x(t) = x_1(t) + x_2(t) + x_3(t)$$

where $x_1(t)$, $x_2(t)$, and $x_3(t)$ have periods of 4, 1.25, and $\sqrt{2}$ s, respectively.
(b) For $y(t)$,

$$y(t) = y_1(t) + y_2(t) + y_3(t)$$

where $y_1(t)$, $y_2(t)$, and $y_3(t)$ have periods of 1.08, 3.6, and 2.025 s, respectively.
(c) For $z(n)$,

$$z(n) = z_1(n) + z_2(n)$$

where $z_1(n)$ and $z_2(n)$ have periods of 90 and 54 s, respectively.

Solution
(a) For $x(t)$,

$$\frac{T_1}{T_2} = \frac{4}{1.25} = \frac{400}{125} = \frac{16}{5} = \text{rational number}$$

$$\frac{T_1}{T_3} = \frac{4}{\sqrt{2}} = \text{irrational number}$$

The signal $x(t)$ is not periodic since T_1/T_3 is not a rational number.
(b) For $y(t)$,

$$\frac{T_1}{T_2} = \frac{1.08}{3.6} = \frac{108}{360} = \frac{3}{10} = \frac{3}{(2)(5)} = \text{rational number}$$

$$\frac{T_1}{T_3} = \frac{1.08}{2.025} = \frac{1080}{2025} = \frac{8}{15} = \frac{8}{(3)(5)} = \text{rational number}$$

The signal $y(t)$ is periodic since the ratios are rational.
$n_1 = $ LCM of denominators $= (2)(3)(5) = 30$

$$T = T_1 n_1 = (1.08)(30) = 32.4 \text{ s}$$

(c) The discrete-time signal $z(n)$ is periodic since $z_1(n)$ and $z_2(n)$ are periodic.

$$\frac{N_1}{N_2} = \frac{90}{54} = \frac{5}{3} = \text{rational number}$$

$$n_1 = \text{LCM of denominators} = 3$$

$$N = N_1 n_1 = (90)(3) = 270$$

Example 1.14 Determine whether or not each of the following signals is periodic. If a signal is periodic, specify its fundamental period. (a) $x_1(t) = je^{j10t}$, (b) $x_2(t) = e^{(-1+j)t}$, (c) $x_3(n) = e^{j7\pi n}$, (d) $x_4(n) = 3e^{j3\pi(n+1/2)/5}$, and (e) $x_5(n) = 3e^{j3/5(n+1/2)}$.

Solution
(a) $x_1(t)$ is a periodic complex exponential signal:

$$x_1(t) = je^{j10t} = e^{j(10t+(\pi/2))} = e^{j(\omega_0 t + \phi)}$$

which gives $\omega_0 = 10$. Therefore, the fundamental period of $x_1(t)$ is

$$T_0 = \frac{2\pi}{\omega_0} = \frac{2\pi}{10} = \frac{\pi}{5}$$

(b) $x_2(t)$ is a complex exponential multiplied by a decaying exponential. Therefore, $x_2(t)$ is not a periodic signal.

$$x_2(t) = e^{(-1+j)t} = e^{j(1+j)t} = e^{-t}e^{jt} = e^{j\omega_0 t}$$

which gives $\omega_0 = 1 + j$, a complex number. The frequency of a signal can never be complex; it must have a real value. So $x_2(t)$ is not periodic.

(c) $x_3(n)$ is a periodic signal:

$$x_3(n) = e^{j7\pi n} = e^{j\omega_0 n}$$

which gives

$$\omega_0 = 7\pi$$

$$2 \times \omega_0 = 7 \times 2\pi$$

$$\frac{\omega_0}{2\pi} = \frac{m}{N} = \frac{7}{2} = \text{rational number}$$

Thus, $x_3(n)$ is periodic with fundamental period $N = 2$.

(d) $x_4(n)$ is periodic:

$$x_4(n) = 3e^{j3\pi(n+1/2)/5} = 3e^{j(3\pi/10)}e^{j(3\pi/5)n} = Ke^{j\omega_0 n}$$

which gives

$$\omega_0 = \frac{3\pi}{5}$$

$$2 \times \omega_0 = \frac{2\pi \times 3}{5}$$

$$\frac{\omega_0}{2\pi} = \frac{m}{N} = \frac{3}{10} = \text{rational number}$$

Thus, $x_4(n)$ is periodic with fundamental period $N = 10$.

(e) $x_5(n)$ is not periodic:

$$x_5(n) = 3e^{j3/5(n+1/2)} = 3e^{j(3/10)}e^{j(3/5)n} = Ke^{j\omega_0 n}$$

which gives

$$\omega_0 = \frac{3}{5}$$

$$\frac{\omega_0}{2\pi} = \frac{3}{10\pi} = \text{irrational number}$$

Since $\omega_0/2\pi$ is an irrational number, $x_5(n)$ is not periodic.

Example 1.15 Determine whether or not each of the following signals is periodic:

(a) $x_1(t) = 2e^{j(t+\pi/4)}u(t)$,

(b) $x_2(n) = u(n) + u(-n)$, and

(c) $x_3(n) = \sum_{k=-\infty}^{\infty} \left[\delta(n - 4k) - \delta(n - 1 - 4k)\right]$.

Solution

(a) $x_1(t)$ is not a periodic signal because it is zero for $t < 0$, i.e., $x_1(t)$ is not an everlasting signal.

(b) $x_2(n)$, as shown in Fig. 1.16, is not a periodic signal:

$$x_2(n) = u(n) + u(-n)$$

$$x_2(n + N) = u(n + N) + u(-(n + N)), \quad N = \text{an integer}$$

$$= u(n + N) + u(-n - N) \neq x_2(n)$$

$x_2(n)$ is not periodic because $x_2(n) \neq x_2(n + N)$.

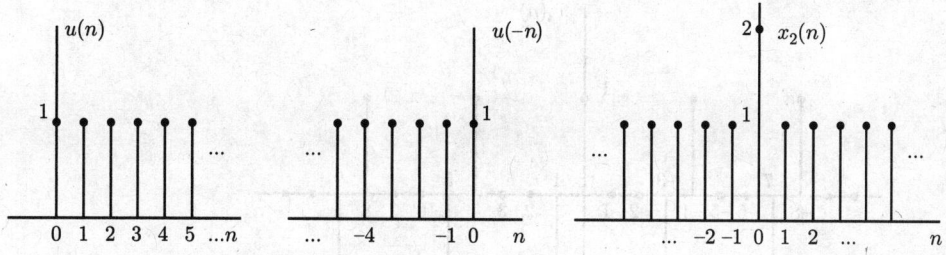

Fig. 1.16 Signal $u(n)$, $u(-n)$, and $x_2(n) = u(n) + u(-n)$

(c) **Method 1:** Given that

$$x_3(n) = \sum_{k=-\infty}^{\infty} \left[\delta(n - 4k) - \delta(n - 1 - 4k)\right]$$

$$x_3(n + N) = \sum_{k=-\infty}^{\infty} \left[\delta(n + N - 4k) - \delta(n + N - 1 - 4k)\right]$$

$$= \sum_{k=-\infty}^{\infty} \left[\delta\left(n - 4\left(k - \frac{N}{4}\right)\right) - \delta\left(n - 1 - 4\left(k - \frac{N}{4}\right)\right)\right]$$

A change of variable is performed by letting $m = k - (N/4)$. Therefore,

$$x_3(n + N) = \sum_{m=-\infty}^{\infty} \left[\delta(n - 4m) - \delta(n - 1 - 4m)\right]$$

$$= x_3(n)$$

Hence $x_3(n)$ is periodic if $m = k - (N/4)$ is an integer. For m to be an integer, the minimum value of N should be 4.

Method 2: $x_3(n)$ is periodic with period $N = 4$.

$$x_3(n) = \sum_{k=-\infty}^{\infty} \left[\delta(n - 4k) - \delta(n - 1 - 4k)\right]$$

$$x_3(n) = \cdots + \delta(n + 8) - \delta(n + 7) + \delta(n + 4) - \delta(n + 3) + \delta(n) - \delta(n - 1) + \delta(n - 4)$$
$$- \delta(n - 5) + \cdots$$
$$x_3(n + 4) = \cdots + \delta(n + 8) - \delta(n + 7) + \delta(n + 4) - \delta(n + 3) + \delta(n) - \delta(n - 1) + \delta(n - 4)$$
$$- \delta(n - 5) + \cdots$$
$$= x_3(n)$$

$x_3(n)$ is periodic with period $N = 4$ (see Fig. 1.17).

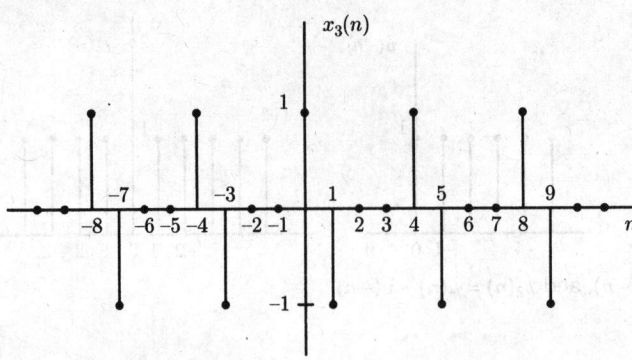

Fig. 1.17 Signal $x_3(n)$

1.6 SOME ELEMENTARY SIGNALS

Several elementary signals that occur frequently in applications also serve as a basis for representing other signals. Many of these signals have features that make them particularly useful in solving engineering problems and, therefore, of importance in our subsequent studies.

1.6.1 Real Exponential Signals

- **Continuous-time exponential signal** A real exponential signal, in its most general form, is written as

$$x(t) = Ce^{at} \tag{1.36}$$

where both C and a are real parameters. The parameter C is the amplitude of the exponential signal measured at time $t = 0$. Depending on whether the other parameter a is positive or negative, we may identify two special cases:

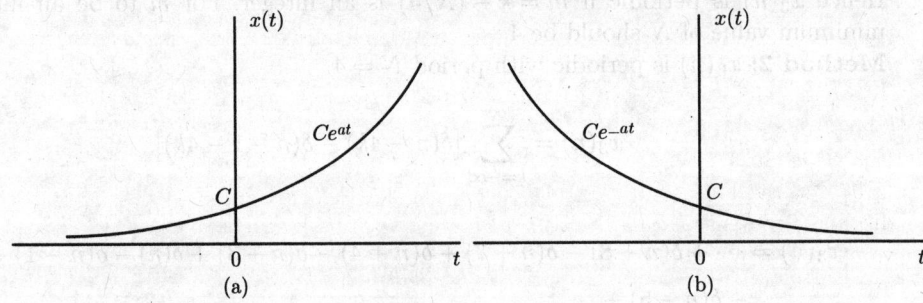

Fig. 1.18 (a) Growing exponential and (b) decaying exponential

Growing exponential If a is positive $(a > 0)$, then as t increases $x(t)$ is a growing exponential—a form that is used in describing many different physical processes, including chain reactions in atomic explosions and complex chemical reactions.

Decaying exponential If a is negative $(a < 0)$, then $x(t)$ is a decaying exponential—a signal that is also used to describe a wide variety of phenomenon, including the process of radioactive decay and the responses of *RC* circuits and damped mechanical systems.

The two forms of an exponential signal are illustrated in Fig. 1.18.

• **Discrete-time exponential signal** In discrete time, the complex exponential signal or sequence is defined as

$$x(n) = Cr^n \tag{1.37}$$

If C and r are real, then $x(n)$ is known as real exponential signal.

This could alternately be expressed as

$$x(n) = Ce^{\alpha n} \tag{1.38}$$

where

$$r = e^{\alpha}$$

As illustrated in Fig. 1.19, if $|r| > 1$, the magnitude of the signal grows exponentially with n, while if $|r| < 1$, we have a decaying exponential. Furthermore, if r is positive, all the values of Cr^n are of the same sign, but if r is negative, then the sign of $x(n)$ alternates.

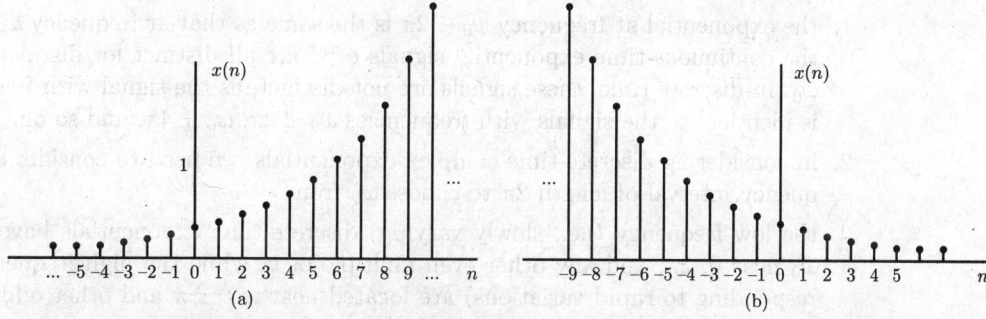

Fig. 1.19 (a) Growing exponential and (b) decaying exponential

1.6.2 Complex Exponential Signals

The mathematical forms of complex exponential signals are the same as those shown in Eqs (1.36) and (1.37), with the following differences: In the continuous-time case, in Eq. (1.36), parameter C or parameter a or both assume complex values. Similarly, in the discrete-time case, in Eq. (1.37), parameter C or parameter r or both assume complex values. Two commonly encountered examples of complex exponential signals are $e^{j\omega t}$ and $e^{j\omega n}$.

Consider

$$x(t) = e^{j\omega_0 t}$$

An important property of this signal is that it is periodic. To verify this, we recall from Eq. (1.30) that $x(t)$ will be periodic with period T if

$$x(t) = x(t + T)$$

$$e^{j\omega_0 t} = e^{j\omega_0(t+T)}$$

$$= e^{j\omega_0 t} \underbrace{e^{j\omega_0 T}}_{=1}$$

Thus, it follows that for periodicity, we must have

$$e^{j\omega_0 T} = 1 = e^{j2\pi m} \tag{1.39}$$

where $m = \pm 1, \pm 2, \ldots$ is an integer. If $\omega_0 = 0$, then $x(t) = 1$, which is periodic for any value of T. If $\omega_0 \neq 0$, then the fundamental period T_0 of $x(t)$, i.e., the smallest positive value of T for which Eq. (1.39) holds, is

$$T_0 = \frac{2\pi}{|\omega_0|} \tag{1.40}$$

Thus, the signals $e^{j\omega_0 t}$ and $e^{-j\omega_0 t}$ have the same fundamental period. Note that (1) the larger the magnitude of ω_0, the higher is the rate of oscillation in the signal and (2) $e^{j\omega_0 t}$ is periodic for any value of ω_0.

Consider the discrete-time exponential signals $e^{j\omega_0 n}$ and $e^{j(\omega_0 + 2\pi)n}$:

$$e^{j(\omega_0 + 2\pi)n} = e^{j2\pi n}e^{j\omega_0 n} = e^{j\omega_0 n} \tag{1.41}$$

From Eq. (1.41), we see that

1. the exponential at frequency $\omega_0 + 2\pi$ is the same as that at frequency ω_0, whereas the continuous-time exponential signals $e^{j\omega_0 t}$ are all distinct for distinct values of ω_0. In discrete-time, these signals are not distinct, as the signal with frequency ω_0 is identical to the signals with frequencies $\omega_0 \pm 2\pi$, $\omega_0 \pm 4\pi$, and so on.

2. in considering discrete-time complex exponentials, we need to consider only a frequency interval of length 2π to choose ω_0 from.

3. the low frequency (i.e., slowly varying) discrete-time exponentials have values of ω_0 near 0, 2π, and any other even multiple of π, while the high frequencies (corresponding to rapid variations) are located near $\omega = \pm \pi$ and other odd multiples of π.

Note in particular that for $\omega_0 = \pi$ or any other odd multiple of π,

$$e^{j\pi n} = (e^{j\pi})^n = (-1)^n \tag{1.42}$$

so that this signal oscillates rapidly, changing sign at each point in time.

Consider the periodicity of discrete-time complex exponential signal $e^{j\omega_0 n}$. In order for the signal $e^{j\omega_0 n}$ to be periodic with period $N > 0$, we must have

$$e^{j\omega_0 n} = e^{j\omega_0(n+N)}$$

$$= e^{j\omega_0 n} \underbrace{e^{j\omega_0 N}}_{=1}$$

Thus, it follows that for periodicity, we must have

$$e^{j\omega_0 N} = 1 = e^{j2\pi m}, \quad m = \text{a positive integer}$$

Equating the powers of the exponential signal, we have

$$\omega_0 N = 2\pi m$$

or equivalently,

$$\frac{\omega_0}{2\pi} = \frac{m}{N} = \text{a rational number} \tag{1.43}$$

According to Eq. (1.43), the signal $e^{j\omega_0 n}$ is periodic if $\omega_0/2\pi$ is a rational number and is not periodic otherwise. Or, we can say that for $e^{j\omega_0 n}$ to be periodic, both m and N must be integers.

1.6.3 Signum Function

The signum function (written *sgn*) shown in Fig. 1.20 is defined by

$$\text{sgn}(t) = \begin{cases} 1, & t > 0 \\ 0, & t = 0 \\ -1, & t < 0 \end{cases} \tag{1.44}$$

The signum function can be expressed in terms of the unit step function as

$$\text{sgn}(t) = -1 + 2u(t) \tag{1.45}$$

The signum function is one of the most often used signals in communication and control theory.

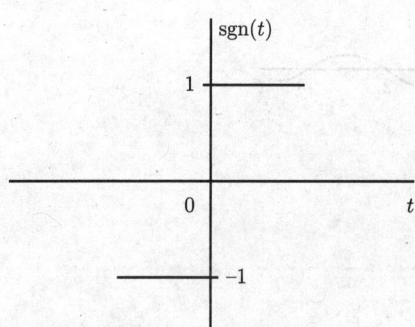

Fig. 1.20 Signum function

1.6.4 Sampling Function

A sampling function $\text{Sa}(t)$ is defined as

$$\text{Sa}(t) = \frac{\sin t}{t} \tag{1.46}$$

Since the denominator is an increasing function of t and the numerator is bounded, i.e., $|\sin t| \le 1$, $\text{Sa}(t)$ is simply a damped sine wave. Figure 1.21(a) shows that $\text{Sa}(t)$ is an even function of t having its peak at $t = 0$ and zero-crossing at $t = \pm n\pi$. The value of the function at $t = 0$ is established by using L'Hôspital's rule. A closely related function is sinc t, which is defined by

$$\text{sinc } t = \frac{\sin \pi t}{\pi t} = \text{Sa}(\pi t) \tag{1.47}$$

and is shown in Fig. 1.21(b). Note that sinc t is a compressed version of Sa(t); the compression factor is π.

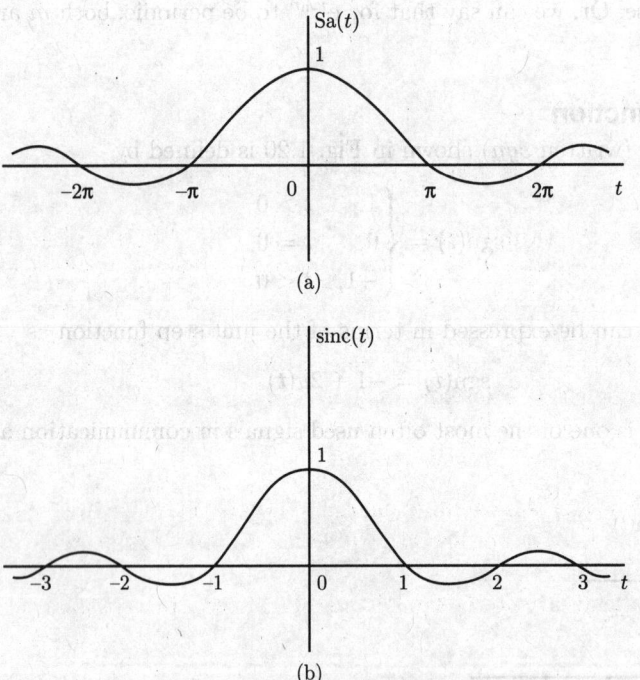

Fig. 1.21 The sampling function

1.7 ENERGY AND POWER SIGNALS

The energy of a continuous-time and a discrete-time signal is defined as follows:

Continuous-time signal

$$E_x = \int\limits_{-\infty}^{+\infty} |x(t)|^2 dt \tag{1.48}$$

Discrete-time signal

$$E_x = \sum_{n=-\infty}^{+\infty} |x(n)|^2 \tag{1.49}$$

The power of a continuous-time and a discrete-time signal is defined as follows:

Continuous-time signal

$$P_x = \lim_{T \to \infty} \frac{1}{T} \int_{-T/2}^{T/2} |x(t)|^2 dt \tag{1.50}$$

or, it may be defined as

$$P_x = \lim_{T \to \infty} \frac{1}{2T} \int_{-T}^{T} |x(t)|^2 dt \tag{1.51}$$

Discrete-time signal

$$P_x = \lim_{N \to \infty} \frac{1}{2N+1} \sum_{n=-N}^{+N} |x(n)|^2 \tag{1.52}$$

If $0 < E_x < \infty$, i.e., if E_x is finite and $P_x = 0$, then $x(t)$ [or $x(n)$] is referred to as an *energy* signal. If $E_x = \infty$, but $0 < P_x < \infty$, i.e., P_x is finite, then $x(t)$ [or $x(n)$] is referred to as a *power* signal.

Observe that power is the time average of energy. Since the averaging is over an infinitely large interval, a signal with finite energy has zero power, and a signal with finite power has infinite energy. Therefore, a signal cannot both be an energy signal and a power signal. On the other hand, there are signals that are neither energy nor power signals.

Note: (i) All practical signals have finite energies and are therefore energy signals. It is impossible to generate a true power signal in practice because such a signal has infinite duration and infinite energy.

(ii) All finite periodic signals are power signals; however, not all power signals are periodic, e.g., unit step signal.

Example 1.16 For an energy signal $x(t)$ with energy E_x, show that

(a) the energy of the signals $x_1(t) = -x(t)$, $x_2(t) = x(-t)$, and $x_3(t) = x(t-T)$ is E_x.

(b) the energy of $x_1(t) = x(at)$ as well as $x_2(t) = x(at-b)$ is E_x/a.

(c) the energy of $x_1(t) = ax(t)$ is $a^2 E_x$.

Solution

By definition

$$E_x = \int_{-\infty}^{\infty} |x(t)|^2 dt$$

(a) (i) The energy contained in the signal $x_1(t) = -x(t)$ is given by

$$E_{x_1} = \int_{-\infty}^{\infty} |x_1(t)|^2 dt = \int_{-\infty}^{\infty} |-x(t)|^2 dt = \int_{-\infty}^{\infty} |x(t)|^2 dt = E_x$$

(ii) The energy contained in the signal $x_2(t) = x(-t)$ is given by

$$E_{x_2} = \int\limits_{-\infty}^{\infty} |x_2(t)|^2 dt = \int\limits_{-\infty}^{\infty} |x(-t)|^2 dt$$

A change of variables is performed by letting $-t = \tau$, which also yields $dt = -d\tau$, $\tau = \infty$ as $t = -\infty$, and $\tau = -\infty$ as $t = \infty$. Therefore,

$$E_{x_2} = -\int\limits_{\infty}^{-\infty} |x(\tau)|^2 d\tau = \int\limits_{-\infty}^{\infty} |x(\tau)|^2 d\tau = E_x$$

(iii) The energy contained in the signal $x_3(t) = x(t - T)$ is given by

$$E_{x_3} = \int\limits_{-\infty}^{\infty} |x_3(t)|^2 dt = \int\limits_{-\infty}^{\infty} |x(t - T)|^2 dt$$

A change of variables is performed by letting $(t - T) = \tau$, which also yields $dt = d\tau$, $\tau = -\infty$ as $t = -\infty$, and $\tau = \infty$ as $t = \infty$. Therefore,

$$E_{x_3} = \int\limits_{-\infty}^{\infty} |x(\tau)|^2 d\tau = E_x$$

(b) (i) The energy contained in the signal $x_1(t) = x(at)$ is given by

$$E_{x_1} = \int\limits_{-\infty}^{\infty} |x_1(t)|^2 dt = \int\limits_{-\infty}^{\infty} |x(at)|^2 dt$$

A change of variables is performed by letting $at = \tau$, which also yields $dt = (1/a)\, d\tau$, $\tau = -\infty$ as $t = -\infty$, and $\tau = \infty$ as $t = \infty$. Therefore,

$$E_{x_1} == \frac{1}{a} \int\limits_{-\infty}^{\infty} |x(\tau)|^2 d\tau = \frac{E_x}{a}$$

(ii) The energy contained in the signal $x_2(t) = x(at - b)$ is given by

$$E_{x_2} = \int\limits_{-\infty}^{\infty} |x_2(t)|^2 dt = \int\limits_{-\infty}^{\infty} |x(at - b)|^2 dt$$

A change of variables is performed by letting $(at - b) = \tau$, which also yields $dt = (1/a)d\tau$, $\tau = -\infty$ as $t = -\infty$, and $\tau = \infty$ as $t = \infty$. Therefore,

$$E_{x_2} = \frac{1}{a} \int\limits_{-\infty}^{\infty} |x(\tau)|^2 d\tau = \frac{E_x}{a}$$

(c) The energy contained in the signal $x_1(t) = ax(t)$ is given by

$$E_{x_1} = \int\limits_{-\infty}^{\infty} |x_1(t)|^2 dt = \int\limits_{-\infty}^{\infty} |ax(t)|^2 dt = a^2 \int\limits_{-\infty}^{\infty} |x(t)|^2 dt = a^2 E_x$$

Note: The time inversion and time shifting does not affect signal energy. On the other hand, time compression of a signal $(a > 1)$ reduces the energy and time expansion of a signal $(a < 1)$ increases the energy.

Example 1.17 Show that all finite periodic signals are power signals. Or, show that if $x(t)$ is periodic with fundamental period T, then the power P_x defined by Eq. (1.50) is the same as the average power $x(t)$ over any interval of length T, i.e.,

$$P_x = \frac{1}{T} \int\limits_{0}^{T} |x(t)|^2 dt$$

Solution
Method 1: Power contained in a signal $x(t)$ is defined as

$$P_x = \lim_{n \to \infty} \frac{1}{2nT} \int\limits_{-nT}^{nT} |x(t)|^2 dt$$

If $x(t)$ is periodic with period T, then the integral in the above equation is the same over any interval of length T. Therefore,

$$P_x = \lim_{n \to \infty} \frac{1}{2nT} \left[\int\limits_{-nT}^{(-n+1)T} |x(t)|^2 dt + \cdots + \int\limits_{-T}^{0} |x(t)|^2 dt + \int\limits_{0}^{T} |x(t)|^2 dt + \cdots \right.$$

$$\left. + \int\limits_{(n-1)t}^{nT} |x(t)|^2 dt \right]$$

$$= \lim_{n \to \infty} \frac{1}{2nT} \left[\sum_{k=-n}^{n-1} \int\limits_{kT}^{(k+1)T} |x(t)|^2 dt \right]$$

$$P_x = \lim_{n \to \infty} \frac{1}{2nT} \left[\sum_{k=-n}^{n-1} \int\limits_{0}^{T} |x(t+kT)|^2 dt \right]$$

Since $x(t)$ is periodic, i.e., $x(t) = x(t + kT)$, we get

$$P_x = \lim_{n \to \infty} \frac{1}{2nT} \left[\sum_{k=-n}^{n-1} \int_0^T |x(t)|^2 dt \right]$$

$$= \lim_{n \to \infty} \frac{1}{2nT} \left[2n \int_0^T |x(t)|^2 dt \right]$$

$$P_x = \frac{1}{T} \int_0^T |x(t)|^2 dt < \infty$$

Since the power contained in a periodic signal is finite, all finite periodic signals are power signals.

Method 2: The signal energy in one period of a periodic signal with period T is

$$E_{x1} = \int_0^T |x(t)|^2 dt$$

Since all periods are alike, the energy in n periods is

$$E_{xn} = n \int_0^T |x(t)|^2 dt$$

The average signal power for all time (i.e., over all periods) is the signal power

$$P_x = \lim_{n \to \infty} \frac{E_{xn}}{nT} = \frac{1}{T} \int_0^T |x(t)|^2 dt$$

We see that the signal power of a periodic signal is equal to the average signal power in any period of the signal. If the signal energy in one period of a periodic signal is finite, then the total energy of the periodic signal is infinite, because it contains an infinite number of periods, and the signal power of the periodic signal is finite. Therefore, the signal is a power signal.

Example 1.18 Determine the values of P_x and E_x for each of the following signals:
(a) $x_1(t) = e^{-2t} u(t)$, (b) $x_2(t) = e^{j(2t + \pi/4)}$, (c) $x_3(t) = A \cos(t)$, (d) $x_1(n) = (1/2)^n u(n)$,
(e) $x_2(n) = e^{j((\pi/2)n + \pi/8)}$, and (f) $x_3(n) = \cos((\pi/4)n)$.

Solution

(a) Given that

$$x_1(t) = e^{-2t}u(t)$$

$$|x_1(t)|^2 = e^{-4t}u(t)$$

Therefore,

$$E_x = \int_{-\infty}^{\infty} |x_1(t)|^2 dt = \int_{0}^{\infty} e^{-4t} dt = \frac{1}{4}$$

$$P_x = \lim_{T \to \infty} \frac{1}{T} \int_{-T/2}^{T/2} |x_1(t)|^2 dt = \lim_{T \to \infty} \frac{1}{T} \int_{-T/2}^{T/2} e^{-4t}u(t) dt = \lim_{T \to \infty} \frac{1}{T} \int_{0}^{T/2} e^{-4t} dt$$

$$= \lim_{T \to \infty} \frac{1}{T} \frac{e^{-4t}}{-4} \bigg|_{0}^{T/2} = \lim_{T \to \infty} \frac{1}{T}[e^{-4T} - 1] = 0$$

Since $P_x = 0$ and $0 < E_x < \infty$, $x_1(t)$ is an energy signal.

(b) Given that

$$x_2(t) = e^{j(2t+\pi/4)}$$

$$|x_2(t)|^2 = 1$$

Therefore,

$$E_x = \int_{-\infty}^{\infty} (1) dt = \infty$$

$$P_x = \lim_{T \to \infty} \frac{1}{T} \int_{-T/2}^{T/2} |x_2(t)|^2 dt = \lim_{T \to \infty} \frac{1}{T} \int_{-T/2}^{T/2} dt = \lim_{T \to \infty} \frac{1}{T} T = 1$$

Since $E_x = \infty$ and $0 < P_x < \infty$, $x_2(t)$ is a power signal.

(c) Given that

$$x_3(t) = A \cos(t)$$

$$|x_3(t)|^2 = A^2 \cos^2(t)$$

Therefore,

$$E_x = \int\limits_{-\infty}^{\infty} A^2 \cos^2(t)dt = \int\limits_{-\infty}^{\infty} \frac{A^2}{2}\left[1 + \cos(2t)\right]dt = \infty$$

$$P_x = \lim_{T\to\infty} \frac{1}{T} \int\limits_{-T/2}^{T/2} A^2 \cos^2(t)dt = \lim_{T\to\infty} \frac{1}{T} \int\limits_{-T/2}^{T/2} \frac{A^2}{2}\left[1 + \cos(2t)\right]dt = \frac{A^2}{2} + 0 = \frac{A^2}{2}$$

Since $E_x = \infty$ and $0 < P_x < \infty$, $x_3(t)$ is a power signal.

(d) Given that

$$x_1(n) = \left(\frac{1}{2}\right)^n u(n)$$

$$|x_1(n)|^2 = \left(\frac{1}{4}\right)^n u(n)$$

Therefore,

$$E_x = \sum_{n=-\infty}^{\infty} |x_1(n)|^2 = \sum_{n=-\infty}^{\infty} \left(\frac{1}{4}\right)^n u(n) = \sum_{n=0}^{\infty} \left(\frac{1}{4}\right)^n = \frac{1}{1-(1/4)} = \frac{4}{3}$$

$$P_x = \lim_{N\to\infty} \frac{1}{2N+1} \sum_{n=-N}^{N} |x_1(n)|^2 = \lim_{N\to\infty} \frac{1}{2N+1} \sum_{n=-N}^{N} \left(\frac{1}{4}\right)^n u(n) = \lim_{N\to\infty} \frac{1}{2N+1} \sum_{n=0}^{N} \left(\frac{1}{4}\right)^n$$

$$= \lim_{N\to\infty} \frac{1}{2N+1} \left(\frac{1-(1/4)^{N+1}}{1-(1/4)}\right) = 0$$

Since $0 < E_x < \infty$ and $P_x = 0$, $x_1(n)$ is an energy signal.

(e) Given that

$$x_2(n) = e^{j((\pi/2)n + \pi/8)}$$

$$|x_2(n)|^2 = 1$$

Therefore,

$$E_x = \sum_{n=-\infty}^{\infty} |x_2(n)|^2 = \sum_{n=-\infty}^{\infty} 1 = \infty$$

$$P_x = \lim_{N\to\infty} \frac{1}{2N+1} \sum_{n=-N}^{N} |x_2(n)|^2 = \lim_{N\to\infty} \frac{1}{2N+1} \sum_{n=-N}^{N} 1 = \lim_{N\to\infty} \frac{1}{2N+1} 2N + 1 = 1$$

Since $E_x = \infty$ and $0 < P_x < \infty$, $x_2(n)$ is a power signal.

(f) Given that

$$x_3(n) = \cos\left(\frac{\pi}{4}n\right)$$

$$|x_3(n)|^2 = \cos^2\left(\frac{\pi}{4}n\right) = \frac{1 + \cos((\pi/2)n)}{2}$$

Therefore,

$$E_x = \sum_{n=-\infty}^{\infty} |x_3(n)|^2 = \sum_{n=-\infty}^{\infty} \frac{1 + \cos((\pi/2)n)}{2} = \sum_{n=-\infty}^{\infty} \frac{1}{2} + \sum_{n=-\infty}^{\infty} \frac{\cos((\pi/2)n)}{2} = \infty$$

$$P_x = \lim_{N\to\infty} \frac{1}{2N+1} \sum_{n=-N}^{N} |x_3(n)|^2 = \lim_{N\to\infty} \frac{1}{2N+1} \sum_{n=-N}^{N} \frac{1 + \cos(\pi/2)n}{2}$$

$$= \left(\lim_{N\to\infty} \frac{1}{2N+1} \sum_{n=-N}^{N} \frac{1}{2}\right) + \lim_{N\to\infty} \frac{1}{2N+1} \sum_{n=-N}^{N} \frac{1}{2}\cos(\pi/2n)$$

$$= \left(\lim_{N\to\infty} \frac{1}{2} \frac{1}{(2N+1)}(2N+1)\right) + 0 = \frac{1}{2}$$

Since $E_x = \infty$ and $0 < P_x < \infty$, $x_3(n)$ is a power signal.

Example 1.19 Determine whether the following signals are power or energy signals or neither.

(a) $x(t) = A \sin t, \quad -\infty < t < \infty$
(b) $x(t) = A[u(t + a) - u(t - a)], \quad a > 0$
(c) $x(t) = e^{-a|t|}, \quad a > 0$
(d) $x(t) = u(t)$
(e) $x(t) = r(t) = tu(t)$

Solution
(a) The given signal $x(t) = A \sin t$ is a periodic signal with period $T = 2\pi$ and hence it is a power signal:

$$P_x = \lim_{T\to\infty} \frac{1}{T} \int_{-T/2}^{T/2} |x(t)|^2 dt = \frac{1}{2\pi} \int_{-\pi}^{\pi} A^2 \sin^2 t\, dt = \frac{A^2}{2\pi} \int_{-\pi}^{\pi} \frac{1}{2}(1 - \cos 2t) dt = \frac{A^2}{2} < \infty$$

(b) The given signal $x(t) = A[u(t + a) - u(t - a)]$, where $a > 0$, is a finite duration signal and hence it is an energy signal:

$$E_x = \int_{-\infty}^{+\infty} |x(t)|^2 dt = \int_{-a}^{a} A^2 dt = 2aA^2$$

(c) Consider the given signal

$$x(t) = e^{-a|t|} = \begin{cases} e^{-at}, & t > 0 \\ e^{at}, & t < 0 \end{cases}$$

The energy contained in the signal $x(t)$ is given by

$$E_x = \int_{-\infty}^{+\infty} |x(t)|^2 dt = \int_{-\infty}^{\infty} e^{-2a|t|} dt = \int_{-\infty}^{0} e^{2at} dt + \int_{0}^{\infty} e^{-2at} dt = \frac{1}{a} < \infty$$

Thus, $x(t)$ is an energy signal.

(d) Consider the given signal

$$x(t) = u(t) = \begin{cases} 1, & t > 0 \\ 0, & t < 0 \end{cases}$$

The energy contained in the signal $x(t)$ is given by

$$E_x = \int_{-\infty}^{\infty} |x(t)|^2 dt = \int_{-\infty}^{\infty} |u(t)|^2 dt = \int_{-\infty}^{\infty} 1^2 dt = \infty$$

The power contained in the signal $x(t)$ is given by

$$P_x = \lim_{T \to \infty} \frac{1}{T} \int_{-T/2}^{T/2} |x(t)|^2 dt = \lim_{T \to \infty} \frac{1}{T} \int_{-T/2}^{T/2} |u(t)|^2 dt = \lim_{T \to \infty} \frac{1}{T} \int_{0}^{T/2} 1^2 dt = \lim_{T \to \infty} \frac{1}{T} \frac{T}{2} = \frac{1}{2}$$

Thus, the unit step function $u(t)$ is a power signal.

(e) Consider the given signal

$$x(t) = r(t) = tu(t) = \begin{cases} t, & t \geq 0 \\ 0, & t < 0 \end{cases}$$

The energy contained in the signal $x(t)$ is given by

$$E_x = \int_{-\infty}^{\infty} |x(t)|^2 dt = \int_{0}^{\infty} t^2 dt = \infty$$

The power contained in the signal $x(t) = r(t)$ is given by

$$P_x = \lim_{T \to \infty} \frac{1}{T} \int_{-T/2}^{T/2} |x(t)|^2 dt = \lim_{T \to \infty} \frac{1}{T} \int_{0}^{T/2} t^2 dt = \lim_{T \to \infty} \frac{1}{T} \frac{(T/2)^3}{3} = \lim_{T \to \infty} \frac{T^2}{24} = \infty$$

Thus, $x(t) = r(t)$, i.e., ramp signal is neither an energy signal nor a power signal.

Example 1.20 Consider the continuous-time signal

$$x(t) = \delta(t+2) - \delta(t-2)$$

Calculate the value of E_y for the following signal:

$$y(t) = \int_{-\infty}^{t} x(\tau)d\tau$$

Solution

Given that

$$y(t) = \int_{-\infty}^{t} x(\tau)d\tau$$

$$= \int_{-\infty}^{t} \Big(\delta(\tau+2) - \delta(\tau-2)\Big)d\tau$$

$$= \int_{-\infty}^{t} \delta(\tau+2)d\tau - \int_{-\infty}^{t} \delta(\tau-2)d\tau$$

$$= u(t+2) - u(t-2) = \begin{cases} 1, & -2 < t < 2 \\ 0, & \text{otherwise} \end{cases}$$

Therefore,

$$E_y = \int_{-\infty}^{\infty} |y(t)|^2 dt = \int_{-2}^{2} 1\,dt = 4$$

1.8 EVEN AND ODD SIGNALS

A signal $x(t)$ or $x(n)$ is referred to as an *even* signal if it is identical with its reflection about the origin. In continuous time a signal is even if

$$x(t) = x(-t) \tag{1.53}$$

while a discrete-time signal is even if

$$x(n) = x(-n) \tag{1.54}$$

An even signal has the same value at instants t (or n) and $-t$ (or $-n$) for all values of t (or n). Clearly, an even signal is symmetrical about the vertical axis.

A signal is referred to as odd if

$$x(t) = -x(-t) \tag{1.55}$$

or

$$x(n) = -x(-n) \tag{1.56}$$

The value of an odd signal at the instant t (or n) is the negative of its value at the instant $-t$ (or $-n$). Therefore, an odd signal is antisymmetric about the vertical axis. An odd signal must necessarily be 0 at $t = 0$ or $n = 0$ since Eqs (1.55) and (1.56) require that $x(0) = -x(0)$.

1.8.1   Even and Odd Components of a Signal

Every signal can be expressed as a sum of even and odd components.

Continuous-time signal

$$x(t) = \mathcal{E}\{x(t)\} + \mathcal{O}\{x(t)\} = x_e(t) + x_o(t) \tag{1.57}$$

where $\mathcal{E}\{x(t)\}$ and $\mathcal{O}\{x(t)\}$ are an even and odd components of $x(t)$, respectively.

Similarly, for **discrete-time signal**

$$x(n) = \mathcal{E}\{x(n)\} + \mathcal{O}\{x(n)\} = x_e(n) + x_o(n) \tag{1.58}$$

The even part of $x(t)$ is given by

$$\mathcal{E}\{x(t)\} = x_e(t) = \frac{x(t) + x(-t)}{2} \tag{1.59}$$

The even part of $x(n)$ is given by

$$\mathcal{E}\{x(n)\} = x_e(n) = \frac{x(n) + x(-n)}{2} \tag{1.60}$$

The odd part of $x(t)$ is given by

$$\mathcal{O}\{x(t)\} = x_o(t) = \frac{x(t) - x(-t)}{2} \tag{1.61}$$

The odd part of $x(n)$ is given by

$$\mathcal{O}\{x(n)\} = x_o(n) = \frac{x(n) - x(-n)}{2} \tag{1.62}$$

1.8.2   Properties of Continuous-time Even and Odd Signals

Following are the properties of continuous-time even and odd signals:

1. If $x_1(t)$ is an odd signal $\left[x_1(t) = -x_1(-t)\right]$ and $x_2(t)$ is an even signal $\left[x_2(t) = x_2(-t)\right]$, then

(a) $x_3(t) = x_1(t)x_2(t)$ is an odd signal, i.e.,

$$\text{even signal} \times \text{odd signal} = \text{odd signal}$$

Proof Consider the signal $x_3(t)$:

$$x_3(t) = x_1(t)x_2(t)$$
$$x_3(-t) = x_1(-t)x_2(-t) = -x_1(t)x_2(t) = -x_3(t)$$

∎

(b) $x_4(t) = x_1(t)x_1(t)$ is an even signal, i.e.,

$$\text{odd signal} \times \text{odd signal} = \text{even signal}$$

Proof Consider the signal $x_4(t)$:

$$x_4(t) = x_1(t)x_1(t)$$
$$x_4(-t) = x_1(-t)x_1(-t) = -x_1(t) \times -x_1(t) = x_1(t)x_1(t) = x_4(t)$$

∎

(c) $x_5(t) = x_2(t)x_2(t)$ is an even signal, i.e.,

$$\text{even signal} \times \text{even signal} = \text{even signal}$$

Proof Consider the signal $x_5(t)$:

$$x_5(t) = x_2(t)x_2(t)$$
$$x_5(-t) = x_2(-t)x_2(-t) = x_2(t)x_2(t) = x_5(t)$$

∎

2. If $x(t)$ is an even signal, then

$$\int_{-\infty}^{\infty} x(t)dt = 2\int_{0}^{\infty} x(t)dt \tag{1.63}$$

Proof Consider the LHS of Eq. (1.63):

$$\int_{-\infty}^{\infty} x(t)dt = \int_{-\infty}^{0} x(t)dt + \int_{0}^{\infty} x(t)dt$$

$$= \int_{0}^{\infty} x(-t)dt + \int_{0}^{\infty} x(t)dt$$

Since $x(t)$ is an even signal, i.e., $[x(t) = x(-t)]$, we have

$$\int_{-\infty}^{\infty} x(t)dt = \int_{0}^{\infty} x(t)dt + \int_{0}^{\infty} x(t)dt = 2\int_{0}^{\infty} x(t)dt$$

∎

3. If $x(t)$ is an odd signal, then

$$\int_{-\infty}^{\infty} x(t)dt = 0 \qquad (1.64)$$

Proof Consider the LHS of Eq. (1.64):

$$\int_{-\infty}^{\infty} x(t)dt = \int_{-\infty}^{0} x(t)dt + \int_{0}^{\infty} x(t)dt$$

$$= \int_{0}^{\infty} x(-t)dt + \int_{0}^{\infty} x(t)dt$$

Since $x(t)$ is an odd signal, i.e., $[x(t) = -x(-t)]$, we have

$$\int_{-\infty}^{\infty} x(t)dt = -\int_{0}^{\infty} x(t)dt + \int_{0}^{\infty} x(t)dt = 0$$

∎

4. If $x(t)$ is an even signal, then its derivative is an odd signal.

Proof Since $x(t)$ is an even signal, we have

$$x(t) = x(-t)$$

$$\frac{dx(t)}{dt} = -\frac{dx(-t)}{dt}$$

∎

5. If $x(t)$ is an odd signal, then its derivative is an even signal.

Proof Since $x(t)$ is an odd signal, we have

$$x(t) = -x(-t)$$

$$\frac{dx(t)}{dt} = \frac{dx(-t)}{dt}$$

∎

6. If $x(t)$ is an arbitrary signal with its even and odd parts denoted by $x_e(t) = \mathcal{E}\{x(t)\}$ and $x_o(t) = \mathcal{O}\{x(t)\}$, respectively, then

$$\int_{-\infty}^{\infty} x^2(t)dt = \int_{-\infty}^{\infty} x_e^2(t)dt + \int_{-\infty}^{\infty} x_o^2(t)dt \qquad (1.65)$$

Proof Consider the LHS of Eq. (1.65),

$$\int_{-\infty}^{\infty} x^2(t)dt = \int_{-\infty}^{\infty} \left[x_e(t) + x_o(t)\right]^2 dt$$

$$= \int_{-\infty}^{\infty} x_e^2(t)dt + \int_{-\infty}^{\infty} x_o^2(t)dt + 2\int_{-\infty}^{\infty} x_e(t)x_o(t)dt$$

$$= \int_{-\infty}^{\infty} x_e^2(t)dt + \int_{-\infty}^{\infty} x_o^2(t)dt + 2\left[\int_{-\infty}^{0} x_e(t)x_o(t)dt + \int_{0}^{\infty} x_e(t)x_o(t)dt\right]$$

$$= \int_{-\infty}^{\infty} x_e^2(t)dt + \int_{-\infty}^{\infty} x_o^2(t)dt + 2\left[\int_{0}^{\infty} x_e(-t)x_o(-t)dt + \int_{0}^{\infty} x_e(t)x_o(t)dt\right]$$

$$= \int_{-\infty}^{\infty} x_e^2(t)dt + \int_{-\infty}^{\infty} x_o^2(t)dt + 2\left[-\int_{0}^{\infty} x_e(t)x_o(t)dt + \int_{0}^{\infty} x_e(t)x_o(t)dt\right]$$

$$\int_{-\infty}^{\infty} x^2(t)dt = \int_{-\infty}^{\infty} x_e^2(t)dt + \int_{-\infty}^{\infty} x_o^2(t)dt + 0 = \int_{-\infty}^{\infty} x_e^2(t)dt + \int_{-\infty}^{\infty} x_o^2(t)dt \qquad ■$$

1.8.3 Properties of Discrete-time Even and Odd Signals

Following are the properties of discrete-time even and odd signals
1. Although part (a) in the previous section has been stated in terms of continuous-time signal, the analogous property is also valid in discrete-time.
2. If $x(n)$ is an even signal, then

$$\sum_{n=-\infty}^{\infty} x(n) = x(0) + 2\sum_{n=1}^{\infty} x(n) \qquad (1.66)$$

Proof Consider the LHS of Eq. (1.66)

$$\sum_{n=-\infty}^{\infty} x(n) = \sum_{n=-\infty}^{-1} x(n) + \sum_{n=0}^{\infty} x(n)$$

$$= \sum_{n=1}^{\infty} x(-n) + \sum_{n=0}^{\infty} x(n)$$

Since $x(n)$ is an even signal, i.e., $\left[x(n) = x(-n)\right]$, we have

$$\sum_{n=-\infty}^{\infty} x(n) = \sum_{n=1}^{\infty} x(n) + x(0) + \sum_{n=1}^{\infty} x(n) = x(0) + 2\sum_{n=1}^{\infty} x(n) \qquad ■$$

3. If $x(n)$ is an odd signal, then

$$\sum_{n=-\infty}^{\infty} x(n) = 0 \qquad (1.67)$$

Proof Consider the LHS of Eq. (1.67)

$$\sum_{n=-\infty}^{\infty} x(n) = \sum_{n=-\infty}^{-1} x(n) + \sum_{n=0}^{\infty} x(n)$$

$$= \sum_{n=1}^{\infty} x(-n) + \sum_{n=0}^{\infty} x(n)$$

Since $x(n)$ is an odd signal, i.e., $[(x(n) = -x(-n)]$, $x(0) = 0$, we have

$$\sum_{n=-\infty}^{\infty} x(n) = -\sum_{n=1}^{\infty} x(n) + x(0) + \sum_{n=1}^{\infty} x(n) = 0 \qquad \blacksquare$$

4. If $x(n)$ is an arbitrary signal with its even and odd parts denoted by $x_e(n) = \mathcal{E}\{x(n)\}$ and $x_o(n) = \mathcal{O}\{x(n)\}$, respectively, then

$$\sum_{n=-\infty}^{\infty} x^2(n) = \sum_{n=-\infty}^{\infty} x_e^2(n) + \sum_{n=-\infty}^{\infty} x_o^2(n) \qquad (1.68)$$

Proof Consider the LHS of Eq. (1.68), and the fact that $x(n) = x_e(n) + x_o(n)$, we have

$$\sum_{n=-\infty}^{\infty} x^2(n) = \sum_{n=-\infty}^{\infty} \left[x_e(n) + x_o(n)\right]^2$$

$$= \sum_{n=-\infty}^{\infty} x_e^2(n) + \sum_{n=-\infty}^{\infty} x_o^2(n) + 2\sum_{n=-\infty}^{\infty} x_e(n)x_o(n)$$

$$= \sum_{n=-\infty}^{\infty} x_e^2(n) + \sum_{n=-\infty}^{\infty} x_o^2(n) + 2\left[\sum_{n=-\infty}^{-1} x_e(n)x_o(n) + \sum_{n=0}^{\infty} x_e(n)x_o(n)\right]$$

$$= \sum_{n=-\infty}^{\infty} x_e^2(n) + \sum_{n=-\infty}^{\infty} x_o^2(n) + 2\left[\sum_{n=1}^{\infty} x_e(-n)x_o(-n) + \sum_{n=0}^{\infty} x_e(n)x_o(n)\right]$$

Since $x_e(n) = x_e(-n)$, $x_0(n) = -x_0(-n)$, and $x_o(0) = 0$, we have

$$\sum_{n=-\infty}^{\infty} x^2(n) = \sum_{n=-\infty}^{\infty} x_e^2(n) + \sum_{n=-\infty}^{\infty} x_o^2(n) + 2\left[-\sum_{n=1}^{\infty} x_e(n)x_o(n) + x_e(0)x_o(0) + \sum_{n=1}^{\infty} x_e(n)x_o(n)\right]$$

$$= \sum_{n=-\infty}^{\infty} x_e^2(n) + \sum_{n=-\infty}^{\infty} x_o^2(n) + 0 = \sum_{n=-\infty}^{\infty} x_e^2(n) + \sum_{n=-\infty}^{\infty} x_o^2(n) \qquad \blacksquare$$

Example 1.21   Verify Eqs (1.59) and (1.61).

Solution

From Eq. (1.57), we have

$$x(t) = x_e(t) + x_o(t)$$

$$x(-t) = x_e(-t) + x_o(-t) = x_e(t) - x_o(-t)$$

Thus, the addition of the above two equations gives Eq. (1.59), i.e.,

$$x(t) + x(-t) = 2x_e(t)$$

$$x_e(t) = \frac{x(t) + x(-t)}{2}$$

and the subtraction gives Eq. (1.61), i.e.,

$$x(t) - x(-t) = 2x_o(t)$$

$$x_o(t) = \frac{x(t) - x(-t)}{2}$$

Example 1.22 Determine the even and odd parts of the unit step signal $u(n)$.

Solution

The unit step signal is defined as

$$u(n) = \begin{cases} 1, & n \geq 0 \\ 0, & n < 0 \end{cases}$$

The even and odd parts of $u(n)$ are

$$\mathcal{E}\{u(n)\} = \frac{u(n) + u(-n)}{2} \quad \text{and} \quad \mathcal{O}\{u(n)\} = \frac{u(n) - u(-n)}{2}$$

The even and odd parts of $u(n)$ as shown in Fig. 1.22 can be determined as follows:

n	$\cdots$	-3	-2	-1	0	1	2	3	$\cdots$
$u(n)$		0	0	0	0	1	1	1	1
$u(n)$		1	1	1	1	1	0	0	0
$\mathcal{E}\{u(n)\}$		$\frac{1}{2}$	$\frac{1}{2}$	$\frac{1}{2}$	$\frac{1}{2}$	1	$\frac{1}{2}$	$\frac{1}{2}$	$\frac{1}{2}$	$\frac{1}{2}$
$\mathcal{O}\{u(n)\}$		$-\frac{1}{2}$	$-\frac{1}{2}$	$-\frac{1}{2}$	$-\frac{1}{2}$	0	$\frac{1}{2}$	$\frac{1}{2}$	$\frac{1}{2}$	$\frac{1}{2}$

Clearly, the even part is

$$\mathcal{E}\{u(n)\} = u_e(n) = \begin{cases} \dfrac{1}{2}, & n > 0 \\ 1, & n = 0 \\ \dfrac{1}{2}, & n < 0 \end{cases}$$

and the odd part is

$$\mathcal{O}\{u(n)\} = u_o(n) = \begin{cases} \dfrac{1}{2}, & n > 0 \\ 0, & n = 0 \\ -\dfrac{1}{2}, & n < 0 \end{cases}$$

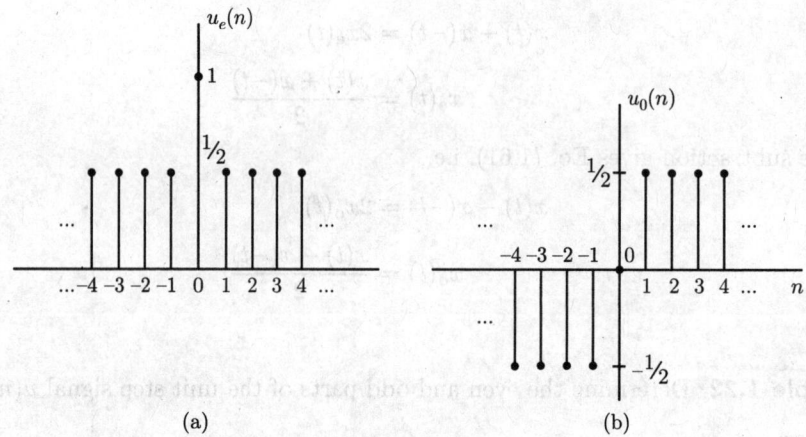

Fig. 1.22   (a) Even and (b) odd parts of $u(n)$

Example 1.23   Find the even and odd components of the signal $x(t) = e^{-2t} \cos(t)$.

Solution
Replacing t with $-t$ in the expression for $x(t)$ yields

$$x(-t) = e^{2t} \cos(-t) = e^{2t} \cos(t)$$

Hence, by applying Eqs (1.59) and (1.61) to the problem at hand, we get

$$x_e(t) = \frac{1}{2}[x(t) + x(-t)] = \frac{1}{2}[e^{-2t} \cos t + e^{2t} \cos t] = \cos t \left[\frac{e^{-2t} + e^{2t}}{2}\right] = \cos t \cosh(2t)$$

and

$$x_o(t) = \frac{1}{2}[x(t) - x(-t)] = \frac{1}{2}[e^{-2t} \cos t - e^{2t} \cos t] = \cos t \left[\frac{e^{-2t} - e^{2t}}{2}\right] = -\cos t \sinh(2t)$$

where $\cosh(2t)$ and $\sinh(2t)$, respectively, denote the hyperbolic cosine and sine of time t.

Example 1.24   Find the even and odd components of the following signals:

(a) $x(t) = u(t) = \begin{cases} 1, & t > 0 \\ 0, & t < 0 \end{cases}$ and   (b) $x(t) = \begin{cases} Ae^{-\alpha t}, & t > 0 \\ 0, & t < 0 \end{cases}$

Solution

(a) Replacing t with $-t$ in the expression for $x(t)$ yields

$$x(-t) = \begin{cases} 1, & -t > 0 \longrightarrow t < 0 \\ 0, & -t < 0 \longrightarrow t > 0 \end{cases}$$

Hence, by applying Eqs (1.59) and (1.61) to the problem at hand, we get

$$x_e(t) = \frac{1}{2}[x(t) + x(-t)] = \frac{1}{2}, \qquad \text{all } t \text{ except } t = 0$$

and

$$x_o(t) = \frac{1}{2}[x(t) - x(-t)] = \begin{cases} \dfrac{1}{2}, & t > 0 \\ -\dfrac{1}{2}, & t < 0 \end{cases}$$

The only problem here is the value of these functions at $t = 0$. If we define $x(0) = 1/2$ (the definition here is consistent with our definition of the signal at a point of discontinuity), then

$$x_e(0) = \frac{1}{2} \quad \text{and} \quad x_o(0) = 0$$

Signals $x(t) = u(t)$, for $x_e(t)$ and $x_o(t)$, are plotted in Fig. 1.23.

(b) Replacing t with $-t$ in the expression for $x(t)$ yields

$$x(-t) = \begin{cases} Ae^{\alpha t}, & -t > 0 \longrightarrow t < 0 \\ 0, & -t < 0 \longrightarrow t > 0 \end{cases}$$

Hence, by applying Eqs (1.59) and (1.61) to the problem at hand, we get

$$x_e(t) = \frac{1}{2}[x(t) + x(-t)] = \begin{cases} \dfrac{1}{2}Ae^{-\alpha t}, & t > 0 \\ \dfrac{1}{2}Ae^{\alpha t}, & t < 0 \end{cases}$$

$$= \frac{1}{2}Ae^{-\alpha|t|}$$

and

$$x_o(t) = \frac{1}{2}[x(t) - x(-t)] = \begin{cases} \dfrac{1}{2}Ae^{-\alpha t}, & t > 0 \\ -\dfrac{1}{2}Ae^{\alpha t}, & t < 0 \end{cases}$$

Signals $x(t)$, for $x_e(t)$ and $x_o(t)$, are plotted in Fig. 1.24.

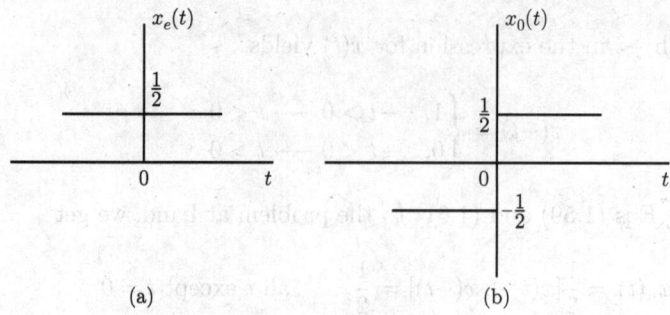

Fig. 1.23 (a) Even and (b) odd parts of $u(t)$

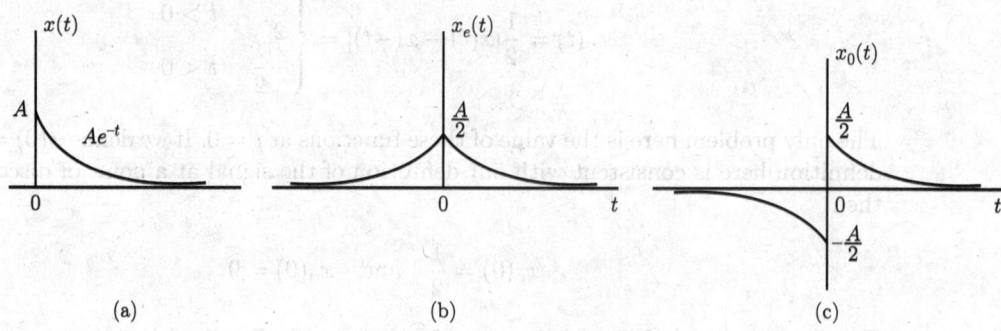

Fig. 1.24 (a) The signal $Ae^{-\alpha t}$, (b) its even part, and (c) its odd part

1.9 CAUSAL, ANTICAUSAL, AND NONCAUSAL SIGNALS

A signal that does not start before $t = 0$ is a causal signal. In other words, $x(t)$ is a causal signal if

$$x(t) = 0, \quad t < 0 \tag{1.69}$$

A signal that starts before $t = 0$ is a *noncausal* signal. The signal is *anticausal* if

$$x(t) = 0, \quad t \geq 0 \tag{1.70}$$

Any signal that does not contain any singularities (a delta function or its derivatives) at $t = 0$ can be written as the sum of a causal part $x^+(t)$ and anticausal part $x^-(t)$, i.e.,

$$x(t) = x^+(t) + x^-(t) \tag{1.71}$$

For example, the exponential $x(t) = e^{-at}$ can be written as

$$x(t) = e^{-at}u(t) + e^{-at}u(-t)$$

where the first term represents causal part of $x(t)$ and the second term represents the anticausal part of $x(t)$. Note that multiplying the signal by the unit step ensures that the resulting signal is causal.

1.10 CONTINUOUS-TIME AND DISCRETE-TIME SYSTEMS

Every physical system is broadly characterized by its ability to accept an input such as voltage, current, force, pressure, displacement, etc., and to produce an output in response to this input. For example, a radar receiver is an electronic system whose input is the reflection of an electromagnetic signal from the target and whose output is a video signal displayed on the radar screen. Similarly, a robot is a system whose input is an electric control signal and whose output is a motion or action on the part of the robot. A third example is a filter whose input is a signal corrupted by noise and interference and whose output is the desired signal. In brief, a system can be viewed as a process that results in transforming input signals into output signals.

We are interested in both continuous-time and discrete-time systems. A continuous-time system is one in which continuous-time input signals are transformed into continuous-time output signals. Such a system is represented pictorially as shown in Fig. 1.25(a), where $x(t)$ is the input and $y(t)$ is the output. Alternatively, we will often represent the input–output relation of a continuous-time system by the notation

$$x(t) \rightarrow y(t) \tag{1.72}$$

Similarly, a discrete-time system, i.e., a system that transforms discrete-time inputs into discrete-time outputs, is depicted in Fig. 1.25(b) and will sometimes be represented symbolically as

$$x(n) \rightarrow y(n) \tag{1.73}$$

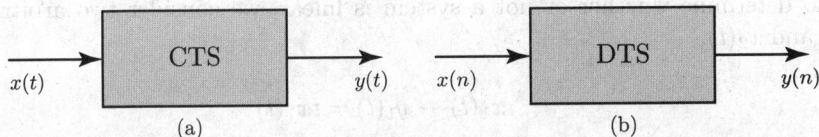

Fig. 1.25 (a) Continuous-time system (CTS) and (b) discrete-time system (DTS)

1.11 BASIC SYSTEM PROPERTIES

Our intent in this section is to lend additional substance to the concept of systems by discussing their classification according to the way the system interacts with the input signal. This interaction, which defines the model for the system, can be linear or nonlinear, time invariant or time varying, memoryless or with memory, static or dynamic, causal or noncausal, and stable or unstable.

1.11.1 Linear and Nonlinear Systems

When the system is linear, the superposition principle can be applied. Superposition simply implies that the response resulting from several input signals can be computed as the sum of the responses resulting from each input signal acting alone. Mathematically, the superposition principle can be stated as follows: Let $y_1(t)$ be the response of a continuous-time system to an input $x_1(t)$ and let $y_2(t)$ be the response corresponding to an input $x_2(t)$. Then the system is linear (follows the principle of superposition) if

- the response to $x_1(t) + x_2(t)$ is $y_1(t) + y_2(t)$ and
- the response to $\alpha x_1(t)$ is $\alpha y_1(t)$, where α is any arbitrary constant.

The first property is referred to as the *additivity* property; the second is referred to as the *scaling* or *homogeneity* property. These two properties defining a linear system can be combined into a single statement as

- continuous time: $\alpha x_1(t) + \beta x_2(t) \rightarrow \alpha y_1(t) + \beta y_2(t)$
- discrete time: $\alpha x_1(n) + \beta x_2(n) \rightarrow \alpha y_1(n) + \beta y_2(n)$

Here α and β are any complex constants. When a system violates either the principle of superposition or the property of homogeneity, the system is said to be *nonlinear*.

A direct consequence of the superposition property is that, for linear systems, an input that is zero for all time results in an output that is zero for all time. For example, if $x(n) \rightarrow y(n)$, then the homogeneity property tells us that

$$0 = 0.x(n) \rightarrow 0.y(n) = 0 \tag{1.74}$$

Example 1.25 For each of the following input–output relationships, determine whether the corresponding system is linear:
(a) $y(t) = tx(t)$, (b) $y(t) = x^2(t)$, (c) $y(n) = \mathcal{Re}\{x(n)\}$, and (d) $y(n) = 2x(n) - 3$.

Solution
(a) To determine whether or not a system is linear, we consider two arbitrary inputs $x_1(t)$ and $x_2(t)$.

$$x_1(t) \rightarrow y_1(t) = tx_1(t)$$
$$x_2(t) \rightarrow y_2(t) = tx_2(t)$$
$$x_3(t) \rightarrow y_3(t) = tx_3(t)$$

Let $x_3(t)$ be a linear combination of $x_1(t)$ and $x_2(t)$, i.e.,

$$x_3(t) = ax_1(t) + bx_2(t)$$

where a and b are arbitrary scalars. If the system is linear, then

$$y_3(t) = ay_1(t) + by_2(t)$$

Consider the LHS of the above equation

$$y_3(t) = tx_3(t) = t\big[ax_1(t) + bx_2(t)\big] = atx_1(t) + btx_2(t) = ay_1(t) + by_2(t) = \text{RHS}$$

Since LHS = RHS, we conclude that the system is linear.

(b) To determine whether or not a system is linear, we consider two arbitrary inputs $x_1(t)$ and $x_2(t)$.

$$x_1(t) \rightarrow y_1(t) = x_1^2(t)$$

$$x_2(t) \rightarrow y_2(t) = x_2^2(t)$$

$$x_3(t) \rightarrow y_3(t) = x_3^2(t)$$

Let $x_3(t)$ be a linear combination of $x_1(t)$ and $x_2(t)$, i.e.,

$$x_3(t) = ax_1(t) + bx_2(t)$$

where a and b are arbitrary scalars. If the system is linear, then

$$y_3(t) = ay_1(t) + by_2(t)$$

Consider the LHS of the above equation

$$y_3(t) = x_3^2(t) = \left[ax_1(t) + bx_2(t) \right]^2 = a^2 x_1^2(t) + b^2 x_2^2(t) + 2abx_1(t)x_2(t)$$

$$= a^2 y_1(t) + b^2 y_2(t) + 2abx_1(t)x_2(t) \neq \text{RHS}$$

Since LHS $\neq$ RHS, we conclude that the system is nonlinear.

(c) When checking the linearity of a system, it is important to remember that the system must satisfy both the additivity and homogeneity properties and that the signals as well as any scaling constants are allowed to be complex. Consider the system specified by

$$y(n) = \mathcal{R}e\{x(n)\}$$

Now consider two arbitrary inputs $x_1(n)$ and $x_2(n)$

$$x_1(n) \rightarrow y_1(n) = \mathcal{R}e\{x_1(n)\}$$

$$x_2(n) \rightarrow y_2(n) = \mathcal{R}e\{x_2(n)\}$$

$$x_3(n) \rightarrow y_3(n) = \mathcal{R}e\{x_3(n)\}$$

Let $x_3(n)$ be a linear combination of $x_1(n)$ and $x_2(n)$, i.e.,

$$x_3(n) = x_1(n) + x_2(n)$$

and so $y_3(n) = \mathcal{R}e\{x_1(n) + x_2(n)\} = \mathcal{R}e\{x_1(n)\} + \mathcal{R}e\{x_2(n)\} = y_1(n) + y_2(n)$

This system is additive; however, it does not satisfy the homogeneity property, as we now demonstrate.

Let

$$x_1(n) = r(n) + js(n)$$

be an arbitrary complex input with real and imaginary parts $r(n)$ and $s(n)$, respectively, so that the corresponding output is

$$y_1(n) = r(n)$$

Now consider scaling $x_1(n)$ by a complex number, e.g., $a = j$, i.e., consider the input

$$x_2(n) = jx_1(n) = j[r(n) + js(n)] = -s(n) + jr(n)$$

The output corresponding to $x_2(n)$ is

$$y_2(n) = \mathcal{R}e\{x_2(n)\} = -s(n)$$

which is not equal to the scaled version of $y_1(n)$

$$ay_1(n) = jr(n)$$

We conclude that the system violates the homogeneity property and hence is not linear.
(d) To determine whether or not a system is linear, we consider two arbitrary inputs $x_1(n)$ and $x_2(n)$.

$$x_1(n) \rightarrow y_1(n) = 2x_1(n) - 3$$
$$x_2(n) \rightarrow y_2(n) = 2x_2(n) - 3$$
$$x_3(n) \rightarrow y_3(n) = 2x_3(n) - 3$$

Let $x_3(n)$ be a linear combination of $x_1(n)$ and $x_2(n)$, i.e.,

$$x_3(n) = ax_1(n) + bx_2(n)$$

where a and b are arbitrary scalars. If the system is linear, then

$$y_3(n) = ay_1(n) + by_2(n)$$

Consider the LHS of the above equation

$$y_3(n) = 2x_3(n) - 3 = 2[ax_1(n) + bx_2(n)] - 3 = a2x_1(n) + b2x_2(n) - 3 \neq \text{RHS}$$

Since LHS $\neq$ RHS, we conclude that the system is nonlinear. Alternatively, since $y(n) = 3$ if $x(n) = 0$, we see that the system violates the 'zero-in/zero-out' property of linear systems given in Eq. (1.74).

Example 1.26 For the system described by the following equations, with the input $x(t)$ and output $y(t)$, determine which of the systems are linear and which are nonlinear:

(a) $\dfrac{dy(t)}{dt} + 3y(t) = x(t)$ (1.75)

(b) $\dfrac{dy(t)}{dt} + 2y(t) = x^2(t)$ (1.76)

(c) $\dfrac{d^2y(t)}{dt^2} + 2y(t) = x(t)$ (1.77)

(d) $\dfrac{dy(t)}{dt} + 3y(t) + 4 = x(t)$ (1.78)

Solution

(a) Let the system response to the inputs $x_1(t)$ and $x_2(t)$ be $y_1(t)$ and $y_2(t)$, respectively. Then

$$\frac{dy_1(t)}{dt} + 3y_1(t) = x_1(t) \quad \text{and} \quad \frac{dy_2(t)}{dt} + 3y_2(t) = x_2(t)$$

By multiplying the first equation by a, the second with b, and then adding both, we get

$$\frac{d}{dt}\big[ay_1(t) + by_2(t)\big] + 3\big[ay_1(t) + by_2(t)\big] = ax_1(t) + bx_2(t)$$

But this equation, is the system equation, i.e., Eq. (1.75) with $x(t) = ax_1(t) + bx_2(t)$ and $y(t) = ay_1(t) + by_2(t)$. Therefore, when the input is $ax_1(t) + bx_2(t)$, the system response is $ay_1(t) + by_2(t)$. Consequently, the system is linear.

(b) Let the system response to the inputs $x_1(t)$ and $x_2(t)$ be $y_1(t)$ and $y_2(t)$, respectively. Then

$$\frac{dy_1(t)}{dt} + 2y_1(t) = x_1^2(t) \quad \text{and} \quad \frac{dy_2(t)}{dt} + 2y_2(t) = x_2^2(t)$$

By multiplying the first equation by a, the second with b, and then adding both, we get

$$\frac{d}{dt}\big[ay_1(t) + by_2(t)\big] + 2\big[ay_1(t) + by_2(t)\big] = ax_1^2(t) + bx_2^2(t) \tag{i}$$

But this equation is not the system equation, i.e., Eq. (1.76) with $x(t) = ax_1(t) + bx_2(t)$ and $y(t) = ay_1(t) + by_2(t)$ that is,

$$\frac{d}{dt}\big[ay_1(t) + by_2(t)\big] + 2\big[ay_1(t) + by_2(t)\big] = \big[ax_1(t) + bx_2(t)\big]^2$$

$$= a^2 x_1^2(t) + b^2 x_2^2(t) + 2ab x_1(t) x_2(t) \tag{ii}$$

Since Eq. (i) is not equal to Eq. (ii), the system is nonlinear.

(c) Let the system response to the inputs $x_1(t)$ and $x_2(t)$ be $y_1(t)$ and $y_2(t)$, respectively. Then

$$\frac{d^2 y_1(t)}{dt^2} + 2y_1(t) = x_1(t) \quad \text{and} \quad \frac{d^2 y_2(t)}{dt^2} + 2y_2(t) = x_2(t)$$

By multiplying the first equation by a, the second with b, and then adding both, we get

$$\frac{d^2}{dt^2}\big[ay_1(t) + by_2(t)\big] + 2\big[ay_1(t) + by_2(t)\big] = ax_1(t) + bx_2(t)$$

But this equation is the system equation, i.e., Eq. (1.77) with $x(t) = ax_1(t) + bx_2(t)$ and $y(t) = ay_1(t) + by_2(t)$. Therefore, when the input $x(t) = ax_1(t) + bx_2(t)$, the system response is $y(t) = ay_1(t) + by_2(t)$. Consequently, the system is linear.

(d) Let the system response to the inputs $x_1(t)$ and $x_2(t)$ be $y_1(t)$ and $y_2(t)$, respectively. Then

$$\frac{dy_1(t)}{dt} + 3y_1(t) + 4 = x_1(t) \quad \text{and} \quad \frac{dy_2(t)}{dt} + 3y_2(t) + 4 = x_2(t)$$

By multiplying the first equation by a, the second with b, and then adding both, we get

$$\frac{d}{dt}\big[ay_1(t) + by_2(t)\big] + 3\big[ay_1(t) + by_2(t)\big] + 4(a + b) = ax_1(t) + bx_2(t) \qquad \text{(iii)}$$

But this equation is not the system equation, i.e., Eq. (1.78) with $x(t) = ax_1(t) + bx_2(t)$ and $y(t) = ay_1(t) + by_2(t)$ that is,

$$\frac{d}{dt}\big[ay_1(t) + by_2(t)\big] + 3\big[ay_1(t) + by_2(t)\big] + 4 = ax_1(t) + bx_2(t) \qquad \text{(iv)}$$

Since Eq. (iii) is not equal to Eq. (iv), the system is nonlinear.

1.11.2   Time-varying and Time-invariant Systems

Conceptually, a system is time invariant if the behaviour and characteristics of the system are fixed over time. In other words, a system is said to be time invariant if a time shift in the input signal causes an identical time shift in the output signal. For example, if $y(t)$ is the output of a continuous-time system corresponding to the input $x(t)$, a time-invariant system will have $y(t - t_0)$ as the output when $x(t - t_0)$ is the input, i.e.,

$$\text{if } x(t) \longrightarrow y(t), \quad \text{then } x(t - t_0) \longrightarrow y(t - t_0)$$

Similarly, in discrete time with $y(n)$ the output corresponding to the input $x(n)$, a time-invariant system will have $y(n - n_0)$ as the output when $x(n - n_0)$ is the input, i.e.,

$$\text{if } x(n) \longrightarrow y(n), \quad \text{then } x(n - n_0) \longrightarrow y(n - n_0)$$

The procedure for testing whether a system is time invariant is summarized in the following steps:

1. Let $y_1(t)$ be the output corresponding to $x_1(t)$.
2. Consider a second input $x_2(t)$ obtained by shifting $x_1(t)$

$$x_2(t) = x_1(t - t_0)$$

and find the output $y_2(t)$ corresponding to the input $x_2(t)$.
3. From Step 1, find $y_1(t - t_0)$ and compare with $y_2(t)$.
4. If $y_2(t) = y_1(t - t_0)$, then the system is time invariant; otherwise it is a time-varying system.

Example 1.27   For each of the following input–output relationships, determine whether the corresponding system is time-invariant:
(a) $y(n) = nx(n)$, (b) $y(n) = x(-n)$, (c) $y(t) = \sin\big[x(t)\big]$, and (d) $y(t) = x(2t)$.

Solution

(a) To determine whether a system is time-invariant or not, we consider an arbitrary input $x_1(n)$:

$$x_1(n) \longrightarrow y_1(n) = nx_1(n)$$
$$x_2(n) \longrightarrow y_2(n) = nx_2(n)$$

Let $x_2(n)$ be obtained by shifting $x_1(n)$ in time, i.e.,

$$x_2(n) = x_1(n - n_0)$$

If the system is time-invariant, then

$$y_2(n) = y_1(n - n_0)$$

Considering the LHS of the above equation, we obtain

$$y_2(n) = nx_2(n) = nx_1(n - n_0)$$

Now, consider

$$y_1(n) = nx_1(n)$$
$$y_1(n - n_0) = (n - n_0)x_1(n - n_0)$$

Clearly, $y_2(n) \neq y_1(n - n_0)$, and hence, this system is not time-invariant.

(b) To determine whether a system is time-invariant or not, we consider an arbitrary input $x_1(n)$:

$$x_1(n) \longrightarrow y_1(n) = x_1(-n)$$
$$x_2(n) \longrightarrow y_2(n) = x_2(-n)$$

Let $x_2(n)$ be obtained by shifting $x_1(n)$ in time, i.e.,

$$x_2(n) = x_1(n - n_0)$$

If the system is time-invariant, then

$$y_2(n) = y_1(n - n_0)$$

Considering the LHS of the above equation, we obtain

$$y_2(n) = x_2(-n) = nx_1(-n - n_0)$$

Now, consider

$$y_1(n) = x_1(-n)$$
$$y_1(n - n_0) = x_1(-(n - n_0)) = x_1(-n + n_0)$$

Clearly, $y_2(n) \neq y_1(n - n_0)$, and hence, this system is not time-invariant.

(c) To determine whether a system is time-invariant or not, we consider an arbitrary input $x_1(t)$:

$$x_1(t) \longrightarrow y_1(t) = \sin[x_1(t)]$$
$$x_2(t) \longrightarrow y_2(t) = \sin[x_2(t)]$$

Let $x_2(t)$ be obtained by shifting $x_1(t)$ in time, i.e.,

$$x_2(t) = x_1(t - t_0)$$

If the system is time-invariant, then

$$y_2(t) = y_1(t - t_0)$$

Considering the LHS of the above equation, we obtain

$$y_2(t) = \sin[x_2(t)] = \sin[x_1(t - t_0)]$$

Now, consider

$$y_1(t) = \sin[x_1(t)]$$

$$y_1(t - t_0) = \sin[x_1(t - t_0)]$$

Clearly, $y_2(t) = y_1(t - t_0)$, and hence, this system is time-invariant.

(d) To determine whether a system is time-invariant or not, we consider an arbitrary input $x_1(t)$.

$$x_1(t) \longrightarrow y_1(t) = x_1(2t)$$

$$x_2(t) \longrightarrow y_2(t) = x_2(2t)$$

Let $x_2(t)$ be obtained by shifting $x_1(t)$ in time, that is

$$x_2(t) = x_1(t - t_0)$$

If the system is time-invariant, then

$$y_2(t) = y_1(t - t_0)$$

Considering the LHS of the above equation, we obtain

$$y_2(t) = x_2(2t) = x_1(2t - t_0)$$

Now, consider

$$y_1(t) = x_1(2t)$$

$$y_1(t - t_0) = x_1(2(t - t_0)) = x_1(2t - 2t_0)$$

Clearly, $y_2(t) \neq y_1(t - t_0)$, and hence, this system is not time-invariant.

1.11.3 Causal Systems

A system is causal, or nonanticipatory (also known as physically realizable), if the output at any time t_0 or n_0 depends only on the values of the input at the present time and in the past, i.e., for $t \leq t_0$ or $n \leq n_0$. Equivalently, if two inputs to a causal system are identical up to some time t_0 or n_0, the corresponding outputs must also be equal up to this time since a causal system cannot predict if the two inputs will be different after t_0 or n_0 (in the future). Mathematically, if

$$x_1(n) = x_2(n), \quad n \leq n_0$$

and the system is causal, then

$$y_1(n) = y_2(n), \quad n \le n_0$$

All memoryless systems are causal systems since the output responds only to the current value of the input.

Example 1.28 For each of the following input–output relationships, determine whether the corresponding system is causal: (a) $y(n) = nx(n)$, (b) $y(n) = x(-n)$, (c) $y(t) = \sin(x(t))$, (d) $y(t) = x(2t)$, and (e) $y(t) = x(t)\cos(t+1)$.

Solution
(a) Consider the output $y(n)$ at a positive time n_0:

$$y(n)\big|_{n=n_0} = y(n_0) = nx(n)\big|_{n=n_0} = n_0 x(n_0)$$

Now, consider the output $y(n)$ at a negative time $-n_0$:

$$y(n)\big|_{n=-n_0} = y(-n_0) = nx(n)\big|_{n=-n_0} = -n_0 x(-n_0)$$

In both the above cases the present output depends upon the present input, and hence, the system is causal.

(b) Consider the output $y(n)$ at a positive time n_0:

$$y(n)\big|_{n=n_0} = y(n_0) = x(-n)\big|_{n=n_0} = x(-n_0)$$

Now, consider the output $y(n)$ at a negative time $-n_0$:

$$y(n)\big|_{n=-n_0} = y(-n_0) = x(-n)\big|_{n=-n_0} = x(n_0)$$

In the first case, the present output depends upon the past input but in the second case the present output depends upon the future inputs, and hence, the system is noncausal.

(c) In this system the current value of the input $x(t)$ influences the current value of the output $y(t)$, and so the system is causal.

(d) Consider the output $y(t)$ at a positive time t_0:

$$y(t)\big|_{t=t_0} = y(t_0) = x(2t)\big|_{t=t_0} = x(2t_0)$$

Now, consider the output $y(t)$ at a negative time $-t_0$:

$$y(t)\big|_{t=-t_0} = y(-t_0) = x(2t)\big|_{t=-t_0} = x(-2t_0)$$

In first case, the present output depends upon the future input but in second case the present output depends upon the past input, and so the system is noncausal.

(e) In this system, the output at any time t_0 equals the input at that same time multiplied by a number that varies with time. Thus, only the current value of the input $x(t)$ influences the current value of the output $y(t)$, and we conclude that this system is causal.

1.11.4 Stable Systems

Stability is another important system property. Informally, a stable system is one in which small inputs lead to responses that do not diverge. A system is said to be *bounded-input, bounded-output* (BIBO) stable if and only if every bounded input results in a bounded output. The output of such a system does not diverge if the input does not diverge.

Signal $x(t)$ is said to be bounded if its magnitude does not grow without bound, i.e.,

$$|x(t)| \leq B_x < \infty, \quad \text{for all } t$$

A system is BIBO stable if, for any bounded input $x(t)$, the response $y(t)$ is also bounded. That is,

$$|x(t)| \leq B_x < \infty \text{ implies } |y(t)| \leq B_y < \infty \tag{1.79}$$

Example 1.29 For each of the following input–output relationships, determine whether the corresponding system is stable: (a) $y(t) = tx(t)$, (b) $y(t) = e^{x(t)}$, (c) $y(n) = 1/3[x(n) + x(n-1) + x(n-2)]$, (d) $y(n) = r^n x(n)$, where $r > 1$, and (e) $y(n) = \sum_{k=0}^{\infty} \rho^k x(n-k)$.

Solution
(a) Assume that

$$|x(t)| \leq B_x < \infty, \quad \text{for all } t$$

Using the given input–output relation, we have

$$y(t) = tx(t)$$

and so we may write

$$|y(t)| = |tx(t)| = |t||x(t)| = |t|B_x$$

As $t \to \infty$ the output $|y(t)| \to \infty$, which is unbounded, since no matter what finite constant we pick, $|y(t)|$ will exceed that constant for some time t. The condition that the input signal is bounded is not sufficient to guarantee a bounded output signal, and so the system is unstable. To prove stability, we need to establish that all bounded inputs produce a bounded output.

(b) Assume that

$$|x(t)| \leq B_x < \infty, \quad \text{for all } t$$

or

$$-B_x \leq x(t) \leq B_x, \quad \text{for all } t$$

Using the given input–output relation, we have

$$y(t) = e^{x(t)}$$

and so we may write

$$e^{-B_x} \leq y(t) \leq e^{B_x}$$

We conclude that in this system if any input is bounded by an arbitrary positive number B_x, the corresponding output is guaranteed to be bounded by e^{B_x}. Thus, the given system is stable.

(c) Assume that

$$|x(n)| \leq B_x < \infty, \quad \text{for all } n$$

Using the input–output relation, we have

$$y(n) = \frac{1}{3}\big[x(n) + x(n-1) + x(n-2)\big]$$

$$|y(n)| = \left|\frac{1}{3}\big[x(n) + x(n-1) + x(n-2)\big]\right|$$

$$\leq \frac{1}{3}\big[|x(n)| + |x(n-1)| + |x(n-2)|\big]$$

$$\leq \frac{1}{3}[B_x + B_x + B_x] = B_x$$

Hence, the absolute value of the output signal $y(n)$ is always less than the maximum absolute value of the input signal $x(n)$ for all n, which shows that the moving-average system is stable.

(d) Assume that the input signal $x(n)$ satisfies the condition

$$|x(n)| \leq B_x < \infty, \quad \text{for all } n$$

We then find that

$$|y(n)| = |r^n x(n)| = |r^n||x(n)|$$

With $r > 1$, the multiplying factor r^n diverges for increasing n. Accordingly, the condition that the input signal is bounded is not sufficient to guarantee a bounded output signal, and so the system is unstable.

(e) Assume that the input signal $x(n)$ satisfies the condition

$$|x(n)| \leq B_x < \infty, \quad \text{for all } n$$

We then find that

$$|y(n)| = \left|\sum_{k=0}^{\infty} \rho^k x(n-k)\right|$$

$$\leq \sum_{k=0}^{\infty} |\rho^k||x(n-k)|$$

$$\leq \sum_{k=0}^{\infty} |\rho^k| B_x$$

Case I When $|\rho| \geq 1$, the output will be unbounded $[y(n) \to \infty]$ and the system is unstable.

Case II when $|\rho| < 1$,

$$|y(n)| = B_x \frac{1}{1 - \rho}$$

with $|\rho| < 1$, the output will be bounded and the system is stable.

1.11.5   Systems With and Without Memory

A system is said to possess *memory* (with memory system or dynamic system) if its output signal depends on the past or future values of the input signal. The temporal extent of the past or future values, on which the output depends, defines how far the memory of the system extends into the past or future. In contrast, a system is said to be *memoryless* (or static or instantaneous) if its output signal depends only on the present value of the input signal. For example, the system specified by the relationship

$$y(n) = 2x(n) + 3x^2(n) \tag{1.80}$$

is memoryless, as the value of $y(n)$ at any particular time n_0 depends only on the value of $x(n)$ at that time. Similarly, a resistor is a memoryless system: with the input $x(t)$ taken as the current and with the voltage taken as the output $y(t)$, the input–output relationship of a resistor is

$$y(t) = Rx(t) \tag{1.81}$$

where R is the resistance.

An example of a discrete-time system with memory is an *accumulator* or *summer*

$$y(n) = \sum_{k=-\infty}^{n} x(k) \tag{1.82}$$

and a second example is a delay

$$y(n) = x(n - 1) \tag{1.83}$$

A capacitor is an example of a continuous-time system with memory. With input taken as the current and output as the voltage, the input–output relationship in the case of the capacitor is

$$y(t) = \frac{1}{C} \int_{-\infty}^{t} x(\tau) d\tau \tag{1.84}$$

where C is the capacitance. It is obvious that the output at any time t depends on the entire history of the input.

1.11.6   Invertibility and Inverse Systems

A system is invertible if by observing the output, we can determine its input. That is, we can construct an inverse system that when cascaded with the given system, as illustrated in Fig. 1.26, yields an output equal to the original input of the given system. In other words, the inverse system 'undoes' what the given system does to input. So the effect of the given system can be eliminated by cascading it with its inverse system. When several different inputs result in the same output, it is impossible to obtain the input from the output, and the system is *noninvertible*. Therefore, for an invertible system, it is essential that every input has a unique output so that there is a one-to-one mapping between an input and the corresponding output.

A rectifier, specified by an equation $y(t) = |x(t)|$, is noninvertible because the rectification operation cannot be undone.

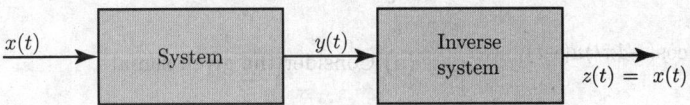

Fig. 1.26 Invertibility and inverse system

SOLVED EXAMPLES

Example 1.30 Determine the fundamental period of the following signals:
(a) $x(t) = 2\cos(10t + 1) - \sin(4t - 1)$
(b) $x(n) = 1 + e^{j4\pi n/7} - e^{j2\pi n/5}$

Solution
(a)
$$x(t) = 2\cos(10t + 1) - \sin(4t - 1)$$

Comparing with

$$x(t) = A_1 \cos(\omega_1 t + \phi_1) - A_2 \cos(\omega_2 t + \phi_2)$$

we get

$$\omega_1 = 10, \quad \omega_2 = 4$$

$$T_1 = \frac{2\pi}{10}, \quad T_2 = \frac{2\pi}{4}$$

$$\frac{T_1}{T_2} = \frac{2}{5} = \text{a rational number}$$

Therefore, $x(t)$ is periodic with fundamental period $T = 5T_1 = 2T_2 = \pi$.

(b)
$$x(n) = 1 + e^{j4\pi n/7} - e^{j2\pi n/5}$$

The first term on the RHS is a constant. The second and third terms on the RHS has frequencies $\omega_1 = 4\pi/7$ and $\omega_2 = 2\pi/5$, respectively.

$$\omega_1 = \frac{4\pi}{7}, \quad \omega_2 = \frac{2\pi}{5}$$

$$7 = \frac{2\pi}{\omega_1} 2, \quad 5 = \frac{2\pi}{\omega_2}$$

$$N_1 = \frac{2\pi}{\omega_1} m_1, \quad N_2 = \frac{2\pi}{\omega_2} m_2$$

$$\frac{N_1}{N_2} = \frac{7}{5} = \text{a rational number}$$

Therefore, $x(n)$ is periodic with fundamental period $N = 5N_1 = 7N_2 = 35$.

Example 1.31 Determine whether or not each of the following signals is periodic. If the signal is periodic, determine its fundamental period.
(a) $x(t) = \left[\cos\left(2t - \frac{\pi}{3}\right)\right]^2$
(b) $x(t) = \mathcal{E}\{\cos(4\pi t)\, u(t)\}$
(c) $x(t) = \mathcal{E}\{\sin(4\pi t)\, u(t)\}$
(d) $x(t) = e^{-2t} u(t)$
(e) $x(n) = \cos\left(\frac{\pi}{8} n^2\right)$
(f) $x(n) = \cos\left(\frac{\pi}{2} n\right) \cos\left(\frac{\pi}{4} n\right)$

Solution
(a) Consider the given signal
$$x(t) = \left[\cos\left(2t - \frac{\pi}{3}\right)\right]^2$$

$$= \frac{1}{2}\left[1 + \cos\left(4t - \frac{2\pi}{3}\right)\right]$$

The first term on the RHS is a constant and the frequency of the second term is $\omega = 4$. Therefore, $x(t)$ is a periodic signal with time period

$$T = \frac{2\pi}{4} = \frac{\pi}{2}$$

(b) Consider the given signal

$$x(t) = \mathcal{E}\{\cos(4\pi t)\, u(t)\}$$

$$= \frac{1}{2}\left[\cos(4\pi t)\, u(t) + \cos(-4\pi t)u(-t)\right]$$

$$= \frac{1}{2}\left[\cos(4\pi t)\, u(t) + \cos(4\pi t)u(-t)\right]$$

$$= \frac{1}{2}\cos(4\pi t) = \frac{1}{2}\cos(\omega t)$$

Therefore,

$$\omega = 4\pi$$

$$T = \frac{2\pi}{4\pi} = \frac{1}{2}$$

Thus, $x(t)$ is a periodic signal with period $T = 1/2$ (Fig. 1.27).

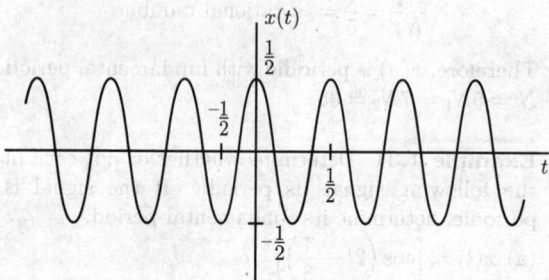

Fig. 1.27

(c) Consider the given signal

$$x(t) = \mathcal{E}\{\sin(4\pi t)\, u(t)\}$$

$$= \frac{1}{2}\left[\sin(4\pi t)u(t) + \sin(-4\pi t)u(-t)\right]$$

$$= \frac{1}{2}\left[\sin(4\pi t)u(t) - \sin(4\pi t)u(-t)\right]$$

$$= \frac{1}{2}\sin(4\pi t)[u(t) - u(-t)]$$

$x(t)$ is not a periodic signal (as shown in Fig. 1.28).

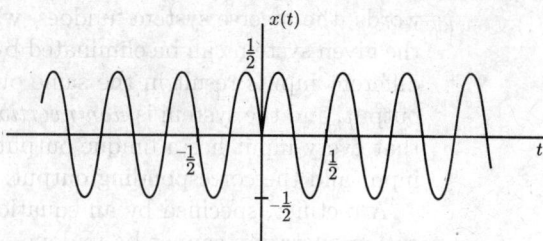

Fig. 1.28

(d) Consider the given signal

$$x(t) = e^{-2t}u(t)$$

$x(t)$ is not a periodic signal because it is zero for $t < 0$ (Fig. 1.29).

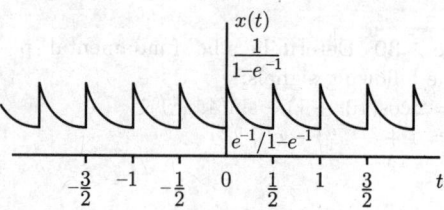

Fig. 1.29

(e) Consider the given signal

$$x(n) = \cos\left(\frac{\pi}{8}n^2\right)$$

$$x(n+N) = \cos\left(\frac{\pi}{8}(n+N)^2\right)$$

for $x(n)$ to be periodic

$$x(n+N) = x(n)$$

$$\cos\left(\frac{\pi}{8}(n+N)^2\right) = \cos\left(\frac{\pi}{8}n^2\right)$$

$$\frac{\pi}{8}(n+N)^2 = \frac{\pi}{8}n^2 + 2k\pi$$

$$\frac{\pi}{8}(N^2 + 2nN) = 2k\pi$$

$$N^2 + 2Nn = 16k$$

where n and k are integers. By inspection, $N = 8$, $16, 32, \ldots$ will satisfy the above equation. Consequently, $x(n)$ is periodic with period $N = 8$.

(f) Consider the given signal

$$x(n) = \cos\left(\frac{\pi}{2}n\right)\cos\left(\frac{\pi}{4}n\right)$$

$$= \frac{1}{2}\left(\cos\left(\frac{3\pi}{4}n\right) + \cos\left(\frac{\pi}{4}n\right)\right)$$

$$\omega_1 = \frac{3\pi}{4}, \quad \omega_2 = \frac{\pi}{4}$$

$$8 = \frac{2\pi}{\omega_1}3, \quad 8 = \frac{2\pi}{\omega_2}$$

$$N_1 = \frac{2\pi}{\omega_1}m_1, \quad N_2 = \frac{2\pi}{\omega_2}m_2$$

Both $N_1 = 8$ and $N_2 = 8$ are equal integers. $x(n)$ is periodic with period $N = 8$.

Example 1.32 Consider the discrete-time signal

$$x(n) = 1 - \sum_{k=3}^{\infty} \delta(n - 1 - k)$$

Determine the values of the integers M and n_0 so that $x(n)$ may be expressed as

$$x(n) = u(Mn - n_0)$$

Solution
Consider the given signal

$$x(n) = 1 - \sum_{k=3}^{\infty} \delta(n - 1 - k)$$

$$= \sum_{k=-\infty}^{\infty} \delta(n - k) - \sum_{k=4}^{\infty} \delta(n - k)$$

$$= \sum_{k=-\infty}^{3} \delta(n - k) + \sum_{k=4}^{\infty} \delta(n - k) - \sum_{k=4}^{\infty} \delta(n - k)$$

$$= \sum_{k=-\infty}^{3} \delta(n - k)$$

$$= u(-n + 3) = u(Mn - n_0)$$

This implies that $M = -1$ and $n_0 = -3$. The signal $x(n)$ is shown in Fig. 1.30. $x(n)$ can be obtained by shifting $u(n)$ by 3 towards the left and then flipping (time-reversal) the shifted signal.

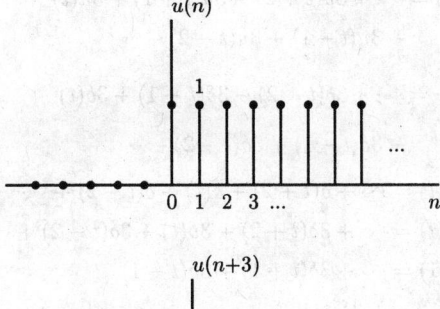

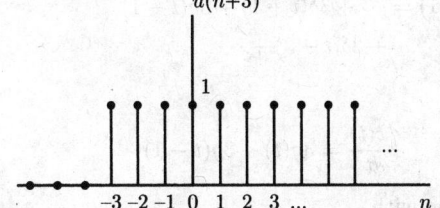

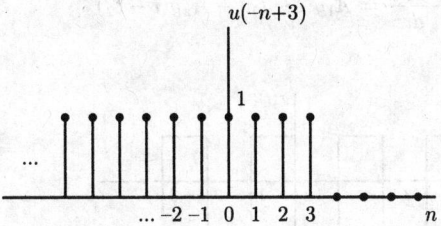

Fig. 1.30

Example 1.33 Consider a periodic signal

$$x(t) = \begin{cases} 1, & 0 \le t \le 1 \\ -2, & 1 < t < 2 \end{cases}$$

with period $T = 2$. The derivative of this signal is related to the 'impulse train'

$$g(t) = \sum_{k=-\infty}^{\infty} \delta(t - 2k)$$

with period $T = 2$. It can be shown that

$$\frac{dx(t)}{dt} = A_1 g(t - t_1) + A_2 g(t - t_2)$$

Determine the values of A_1, t_1, A_2, and t_2.

Solution
Consider the given signal

$$x(t) = \begin{cases} 1, & 0 \le t \le 1 \\ -2, & 1 < t < 2 \end{cases}$$

$$x(t) = \cdots + 3u(t+2) - 3u(t+1) + 3u(t)$$
$$- 3u(t-1) + 3u(t-2) - \cdots$$

$$\frac{dx(t)}{dt} = \cdots + 3\delta(t+2) - 3\delta(t+1) + 3\delta(t)$$
$$- 3\delta(t-1) + 3\delta(t-2) - \cdots$$

and $g(t) = \cdots + \delta(t+2) + \delta(t) + \delta(t-2) + \cdots$
$$3g(t) = \cdots + 3\delta(t+2) + 3\delta(t) + 3\delta(t-2) + \cdots$$
$$3g(t-1) = \cdots + 3\delta(t+1) + 3\delta(t-1)$$
$$+ 3\delta(t-3) + \cdots$$

Thus,

$$\frac{dx(t)}{dt} = 3g(t) + 3g(t-1)$$

Comparing with

$$\frac{dx(t)}{dt} = A_1 g(t-t_1) + A_2 g(t-t_2)$$

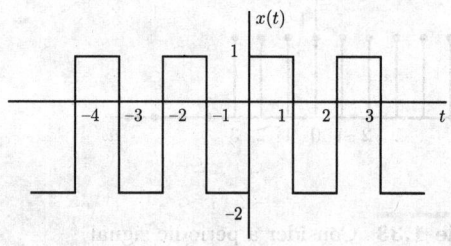

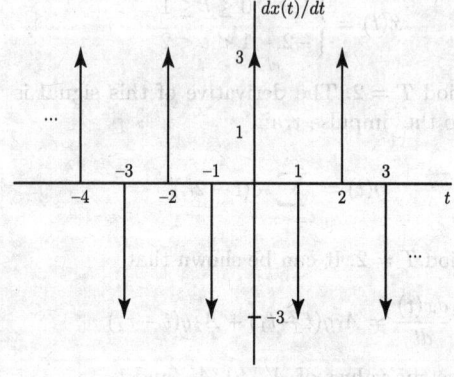

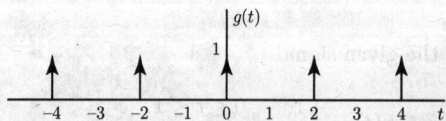

Fig. 1.31

we get $A_1 = 3$, $t_1 = 0$, $A_2 = 3$, and $t_2 = 1$. The signal $x(t)$, its derivative $dx(t)/dt$, and $g(t)$ are shown in Fig. 1.31.

Example 1.34 Consider a discrete-time system with input $x(n)$ and output $y(n)$ related by

$$y(n) = \sum_{k=n-n_0}^{n+n_0} x(k)$$

where n_0 is a finite positive integer.
(a) Is this system linear?
(b) Is this system time-invariant?
(c) If $x(n)$ is known to be bounded by a finite integer B_x [i.e., $|x(n)| < B_x$ for all n], it can be shown that $y(n)$ is bounded by a finite number C. We conclude that the given system is stable. Express C in terms of B_x and n_0.

Solution
(a) To determine whether or not a system is linear, we consider two arbitrary inputs $x_1(n)$ and $x_2(n)$.

$$x_1(n) \rightarrow y_1(n) = \sum_{k=n-n_0}^{n+n_0} x_1(k)$$

$$x_2(n) \rightarrow y_2(n) = \sum_{k=n-n_0}^{n+n_0} x_2(k)$$

$$x_3(n) \rightarrow y_3(n) = \sum_{k=n-n_0}^{n+n_0} x_3(k)$$

Let $x_3(n)$ be a linear combination of $x_1(n)$ and $x_2(n)$. That is,

$$x_3(n) = ax_1(n) + bx_2(n)$$

where a and b are arbitrary scalars. If the system is linear, then

$$y_3(n) = ay_1(n) + by_2(n)$$

Consider the LHS of the above equation

$$y_3(n) = \sum_{k=n-n_0}^{n+n_0} x_3(k)$$

$$= \sum_{k=n-n_0}^{n+n_0} \big(ax_1(k) + bx_2(k)\big)$$

$$= a \sum_{k=n-n_0}^{n+n_0} x_1(k) + b \sum_{k=n-n_0}^{n+n_0} x_2(k)$$

$$= ay_1(n) + by_2(n) = \text{RHS}$$

Since LHS = RHS, we conclude that the system is linear.

(b) Let $x_1(n)$ be an arbitrary input to the system.

$$x_1(n) \longrightarrow y_1(n) = \sum_{k=n-n_0}^{n+n_0} x_1(k)$$

$$x_2(n) \longrightarrow y_2(n) = \sum_{k=n-n_0}^{n+n_0} x_2(k)$$

Let $x_2(n)$ be obtained by shifting $x_1(n)$ in time, i.e.,

$$x_2(n) = x_1(n - m)$$

If the system is time-invariant, then

$$y_2(n) = y_1(n - m)$$

Consider the LHS of the above equation,

$$y_2(n) = \sum_{k=n-n_0}^{n+n_0} x_2(k)$$

$$= \sum_{k=n-n_0}^{n+n_0} x_1(k - m)$$

Now, consider

$$y_1(n) = \sum_{k=n-n_0}^{n+n_0} x_1(k)$$

$$y_1(n - m) = \sum_{k=n-m-n_0}^{n-m+n_0} x_1(k)$$

$$= \sum_{k=n-n_0}^{n+n_0} x_1(k - m)$$

Clearly, $y_2(n) = y_1(n - m)$; thus, this system is time-invariant.

(c) $x(n)$ is bounded by a finite integer B_x, i.e.,

$$|x(n)| < B_x \quad \text{for all } n$$

We then find that

$$|y(n)| = \left| \sum_{k=n-n_0}^{n+n_0} x(k) \right|$$

$$\leq \sum_{k=n-n_0}^{n+n_0} |x(k)|$$

$$\leq \sum_{k=n-n_0}^{n+n_0} B_x$$

$$\leq B_x\big[(n + n_0) - (n - n_0) + 1\big]$$

$$\leq B_x(2n_0 + 1) = C$$

Example 1.35 Consider a continuous-time system with input $x(t)$ and output $y(t)$ related by

$$y(t) = x\big(\sin(t)\big)$$

(a) Is this system causal?
(b) Is this system linear?

Solution
(a) Case I: Consider the output $y(t)$ at time $t = -\pi$

$$y(t)\Big|_{t=-\pi} = x\big(\sin(-\pi)\big)$$

$$y(-3.14) = x(0)$$

In this case, the present output depends on the future input.
Case II: Consider the output at time $t = 0$

$$y(t)\Big|_{t=0} = x\big(\sin(0)\big)$$

$$y(0) = x(0)$$

In this case, the present output depends on the present input.
Case III: Consider the output at time $t = \pi$

$$y(t)\Big|_{t=\pi} = x\big(\sin(\pi)\big)$$

$$y(3.14) = x(0)$$

In this case, the present output depends on the past input.

The present output of this system depends on the present, past, and future inputs, and so this system is noncausal.

(b) To determine whether or not a system is linear, we consider two arbitrary inputs $x_1(t)$ and $x_2(t)$.

$$x_1(t) \rightarrow y_1(t) = x_1(\sin(t))$$

$$x_2(t) \rightarrow y_2(t) = x_2(\sin(t))$$

$$x_3(t) \rightarrow y_3(t) = x_3(\sin(t))$$

Let $x_3(t)$ be a linear combination of $x_1(t)$ and $x_2(t)$. That is,

$$x_3(t) = ax_1(t) + bx_2(t)$$

where a and b are arbitrary scalars. If the system is linear, then

$$y_3(t) = ay_1(t) + by_2(t)$$

Consider the LHS of the above equation

$$y_3(t) = x_3(\sin(t))$$
$$= ax_1(\sin(t)) + bx_2(\sin(t))$$
$$= ay_1(t) + by_2(t) = \text{RHS}$$

Since LHS = RHS, we conclude that the system is linear.

Example 1.36   The system that follow have input $x(t)$ or $x(n)$ and output $y(t)$ or $y(n)$. For each system, determine whether it is (i) memoryless, (ii) stable, (iii) causal, (iv) linear, or (v) time-invariant.

(a) $y(t) = \int\limits_{-\infty}^{2t} x(\tau)d\tau$

(b) $y(t) = \dfrac{dx(t)}{dt}$

(c) $y(t) = \begin{cases} 0, & t < 0 \\ x(t) + x(t-2), & t \geq 0 \end{cases}$

(d) $y(t) = \begin{cases} 0, & x(t) < 0 \\ x(t) + x(t-2), & x(t) \geq 0 \end{cases}$

(e) $y(n) = \mathcal{E}\{x(n-1)\}$

(f) $y(n) = x(n^2)$

Solution

(a) (i) This system integrates the input from $-\infty$ to $2t$. The present output depends on the past, present, and future inputs. We may conclude that this system is a *with-memory* system.

(ii) Assume that the input signal $x(t)$ satisfies the condition

$$|x(t)| \leq B_x < \infty \quad \text{for all } t.$$

We then find that

$$|y(t)| = \left| \int\limits_{-\infty}^{2t} x(\tau)d\tau \right|$$

$$= \int\limits_{-\infty}^{2t} |x(\tau)|d\tau$$

$$= \int\limits_{-\infty}^{2t} B_x d\tau$$

$$= \infty$$

We conclude that in this system if any input is bounded by an arbitrary positive number B_x, the corresponding output is unbounded. Thus, the given system is *unstable*.

(iii) In this system, the present output anticipates with the future inputs (the upper limit of the integration is $2t$). Consequently, the given system is *noncausal*.

(iv) To determine whether or not a system is linear, we consider two arbitrary inputs $x_1(t)$ and $x_2(t)$.

$$x_1(t) \rightarrow y_1(t) = \int\limits_{-\infty}^{2t} x_1(\tau)d\tau$$

$$x_2(t) \rightarrow y_2(t) = \int\limits_{-\infty}^{2t} x_2(\tau)d\tau$$

$$x_3(t) \rightarrow y_3(t) = \int\limits_{-\infty}^{2t} x_3(\tau)d\tau$$

Let $x_3(t)$ be a linear combination of $x_1(t)$ and $x_2(t)$. That is,

$$x_3(t) = ax_1(t) + bx_2(t)$$

where a and b are arbitrary scalars. If the system is linear, then

$$y_3(t) = ay_1(t) + by_2(t)$$

Consider the LHS of the above equation

$$y_3(t) = \int\limits_{-\infty}^{2t} x_3(\tau)d\tau$$

$$= \int\limits_{-\infty}^{2t} \big(ax_1(\tau) + bx_2(\tau)\big)d\tau$$

$$= a \int\limits_{-\infty}^{2t} x_1(\tau)d\tau + b \int\limits_{-\infty}^{2t} x_2(\tau)d\tau$$

$$= ay_1(t) + by_2(t) = \text{RHS}$$

Since LHS = RHS, we conclude that the system is *linear*.

(v) Let $x_1(t)$ be an arbitrary input to the system, and let

$$y_1(t) = \int\limits_{-\infty}^{2t} x_1(\tau)d\tau$$

be the corresponding output. Then consider a second input obtained by shifting $x_1(t)$ in time:

$$x_2(t) = x_1(t - t_0)$$

The output corresponding to this input is

$$y_2(t) = \int\limits_{-\infty}^{2t} x_2(\tau)d\tau$$

$$= \int\limits_{-\infty}^{2t} x_1(\tau - t_0)d\tau$$

$$= \int\limits_{-\infty}^{2t-t_0} x_1(\tau)d\tau$$

Now, consider

$$y_1(t) = \int\limits_{-\infty}^{2t} x_1(\tau)d\tau$$

$$y_1(t - t_0) = \int\limits_{-\infty}^{2(t-t_0)} x_1(\tau)d\tau$$

We see that $y_2(t) \neq y_1(t - t_0)$, and therefore this system is not time-invariant.

(b) (i)

$$y(t) = \frac{dx(t)}{dt}$$

$$y(t) = \lim_{\Delta t \to \infty} \frac{x(t) - x(t - \Delta t)}{\Delta t}$$

The present output depends on the present and past inputs. Consequently, this system is a *with memory* system.

(ii) Consider the bounded input

$$x(t) = u(t) \quad \text{unit step function}$$

then, the corresponding output is

$$y(t) = \frac{du(t)}{dt}$$

$$= \delta(t)$$

The bounded input produces an unbounded output. We may conclude that the given system is *unstable*.

(iii) The present output depends on the present and past values of the input. Consequently, this system is *causal*.

(iv) To determine whether or not a system is linear, we consider two arbitrary inputs $x_1(t)$ and $x_2(t)$.

$$x_1(t) \to y_1(t) = \frac{dx_1(t)}{dt}$$

$$x_2(t) \to y_2(t) = \frac{dx_2(t)}{dt}$$

$$x_3(t) \to y_3(t) = \frac{dx_3(t)}{dt}$$

Let $x_3(t)$ be a linear combination of $x_1(t)$ and $x_2(t)$. That is,

$$x_3(t) = ax_1(t) + bx_2(t)$$

where a and b are arbitrary scalars. If the system is linear, then

$$y_3(t) = ay_1(t) + by_2(t)$$

Consider the LHS of the above equation

$$y_3(t) = \frac{dx_3(t)}{dt}$$

$$= \frac{d}{dt}\big(ax_1(t) + bx_2(t)\big)$$

$$= a\frac{dx_1(t)}{dt} + b\frac{dx_2(t)}{dt}$$

$$= ay_1(t) + by_2(t) = \text{RHS}$$

Since LHS = RHS, we conclude that the system is *linear*.

(v) Let $x_1(t)$ be an arbitrary input to the system, and let

$$y_1(t) = \frac{dx_1(t)}{dt}$$

be the corresponding output. Then consider a second input obtained by shifting $x_1(t)$ in time:

$$x_2(t) = x_1(t - t_0)$$

The output corresponding to this input is

$$y_2(t) = \frac{dx_2(t)}{dt}$$

$$= \frac{dx_1(t - t_0)}{dt}$$

Now, consider

$$y_1(t) = \frac{dx_1(t)}{dt}$$

$$y_1(t - t_0) = \frac{dx_1(t - t_0)}{dt}$$

We see that $y_2(t) = y_1(t - t_0)$, and therefore this system is time-invariant.

(c) (i) This system has memory, since the value of the output signal $y(t)$ at time t depends on the present and one past input signal $x(t)$.

(ii) Assume that the input signal $x(t)$ satisfies the condition

$$|x(t)| \leq B_x < \infty, \quad \text{for all } t$$

We then find that

$$|y(t)| = \begin{cases} 0, & t < 0 \\ |x(t) + x(t - 2)|, & t \geq 0 \end{cases}$$

$$\leq \begin{cases} 0, & t < 0 \\ |x(t)| + |x(t - 2)|, & t \geq 0 \end{cases}$$

$$\leq \begin{cases} 0, & t < 0 \\ 2B_x, & t \geq 0 \end{cases}$$

We conclude that in this system if any input is bounded by an arbitrary positive number B_x, the corresponding output is guaranteed to be bounded. Thus, the given system is stable.

(iii) This system is causal, since the present output $y(t)$ at time t depends on the present and past values of the input signal $x(t)$.

(iv) To determine whether or not a system is linear, we consider two arbitrary inputs $x_1(t)$ and $x_2(t)$.

$$x_1(t) \to y_1(t) = \begin{cases} 0, & t < 0 \\ x_1(t) + x_1(t - 2), & t \geq 0 \end{cases}$$

$$x_2(t) \to y_2(t) = \begin{cases} 0, & t < 0 \\ x_2(t) + x_2(t - 2), & t \geq 0 \end{cases}$$

$$x_3(t) \to y_3(t) = \begin{cases} 0, & t < 0 \\ x_3(t) + x_3(t - 2), & t \geq 0 \end{cases}$$

Let $x_3(t)$ be a linear combination of $x_1(t)$ and $x_2(t)$. That is,

$$x_3(t) = ax_1(t) + bx_2(t)$$

where a and b are arbitrary scalars. If the system is linear, then

$$y_3(t) = ay_1(t) + by_2(t)$$

Consider the LHS of the above equation

$$y_3(t) = \begin{cases} 0, & t < 0 \\ x_3(t) + x_3(t - 2), & t \geq 0 \end{cases}$$

$$= \begin{cases} 0, & t < 0 \\ [ax_1(t) + bx_2(t)], & \\ \quad + [ax_1(t - 2) + bx_2(t - 2)], & t \geq 0 \end{cases}$$

$$= \begin{cases} 0, & t < 0 \\ a[x_1(t) + x_1(t - 2)], & \\ \quad + b[x_2(t) + x_2(t - 2)], & t \geq 0 \end{cases}$$

$$= ay_1(t) + by_2(t) = \text{RHS}$$

Since LHS = RHS, we conclude that the system is *linear*.

(v) Let $x_1(t)$ be an arbitrary input to the system, and let

$$y_1(t) = \begin{cases} 0, & t < 0 \\ x_1(t) + x_1(t - 2), & t \geq 0 \end{cases}$$

be the corresponding output. Then consider a second input obtained by shifting $x_1(t)$ in time:

$$x_2(t) = x_1(t - t_0)$$

The output corresponding to this input is

$$y_2(t) = \begin{cases} 0, & t < 0 \\ x_2(t) + x_2(t-2), & t \geq 0 \end{cases}$$

$$= \begin{cases} 0, & t < 0 \\ x_1(t - t_0) + x_1(t - t_0 - 2), & t \geq 0 \end{cases}$$

Now, consider

$$y_1(t) = \begin{cases} 0, & t < 0 \\ x_1(t) + x_1(t-2), & t \geq 0 \end{cases}$$

$$y_1(t - t_0) = \begin{cases} 0, & t - t_0 < 0 \\ x_1(t - t_0) & \\ +x_1(t - t_0 - 2), & t - t_0 \geq 0 \end{cases}$$

$$= \begin{cases} 0, & t < t_0 \\ x_1(t - t_0) & \\ +x_1(t - t_0 - 2), & t \geq t_0 \end{cases}$$

We see that $y_2(t) \neq y_1(t - t_0)$, and therefore this system is not time-invariant.

(d) (i) This system has memory, since the value of the output signal $y(t)$ at time t depends on the present and one past input signal $x(t)$.

(ii) Assume that the input signal $x(t)$ satisfies the condition

$$|x(t)| \leq B_x < \infty, \quad \text{for all } t$$

We then find that

$$|y(t)| = \begin{cases} 0, & x(t) < 0 \\ |x(t) + x(t-2)|, & x(t) \geq 0 \end{cases}$$

$$\leq \begin{cases} 0, & x(t) < 0 \\ |x(t)| + |x(t-2)|, & x(t) \geq 0 \end{cases}$$

$$\leq \begin{cases} 0, & x(t) < 0 \\ 2B_x, & x(t) \geq 0 \end{cases}$$

We conclude that in this system if any input is bounded by an arbitrary positive number B_x, the corresponding output is guaranteed to be bounded. Thus, the given system is stable.

(iii) This system is causal, since the present output $y(t)$ at time t depends on the present and past values of the input signal $x(t)$.

(iv) To determine whether or not a system is linear, we consider two arbitrary inputs $x_1(t)$ and $x_2(t)$.

$$x_1(t) \rightarrow y_1(t) = \begin{cases} 0, & x_1(t) < 0 \\ x_1(t) + x_1(t-2), & x_1(t) \geq 0 \end{cases}$$

$$x_2(t) \rightarrow y_2(t) = \begin{cases} 0, & x_2(t) < 0 \\ x_2(t) + x_2(t-2), & x_2(t) \geq 0 \end{cases}$$

$$x_3(t) \rightarrow y_3(t) = \begin{cases} 0, & x_3(t) < 0 \\ x_3(t) + x_3(t-2), & x_3(t) \geq 0 \end{cases}$$

Let $x_3(t)$ be a linear combination of $x_1(t)$ and $x_2(t)$. That is,

$$x_3(t) = ax_1(t) + bx_2(t)$$

where a and b are arbitrary scalars. If the system is linear, then

$$y_3(t) = ay_1(t) + by_2(t)$$

Consider the LHS of the above equation

$$y_3(t) = \begin{cases} 0, & x_3(t) < 0 \\ x_3(t) + x_3(t-2), & x_3(t) \geq 0 \end{cases}$$

$$= \begin{cases} 0, & (ax_1(t) + bx_2(t)) < 0 \\ [ax_1(t) + bx_2(t)] & \\ +[ax_1(t-2) & \\ +bx_2(t-2)], & (ax_1(t) + bx_2(t)) \geq 0 \end{cases}$$

$$= \begin{cases} 0, & (ax_1(t) + bx_2(t)) < 0 \\ a[x_1(t) + x_1(t-2)] & \\ +b[x_2(t) & \\ +x_2(t-2)], & (ax_1(t) + bx_2(t)) \geq 0 \end{cases}$$

$$\neq \text{RHS}$$

Since LHS $\neq$ RHS, we conclude that the system is *nonlinear*.

(v) Let $x_1(t)$ be an arbitrary input to the system, and let

$$y_1(t) = \begin{cases} 0, & x_1(t) < 0 \\ x_1(t) + x_1(t-2), & x_1(t) \geq 0 \end{cases}$$

be the corresponding output. Then consider a second input obtained by shifting $x_1(t)$ in time:

$$x_2(t) = x_1(t - t_0)$$

The output corresponding to this input is

$$y_2(t) = \begin{cases} 0, & x_2(t) < 0 \\ x_2(t) + x_2(t - 2), & x_2(t) \geq 0 \end{cases}$$

$$= \begin{cases} 0, & x_1(t - t_0) < 0 \\ x_1(t - t_0) \\ \quad + x_1(t - t_0 - 2), & x_1(t - t_0) \geq 0 \end{cases}$$

Now, consider

$$y_1(t) = \begin{cases} 0, & x_1(t) < 0 \\ x_1(t) + x_1(t - 2), & x_1(t) \geq 0 \end{cases}$$

$$y_1(t - t_0) = \begin{cases} 0, & x_1(t - t_0) < 0 \\ x_1(t - t_0) \\ \quad + x_1(t - t_0 - 2), & x_1(t - t_0) \geq 0 \end{cases}$$

We see that $y_2(t) = y_1(t - t_0)$, and therefore this system is time-invariant.

(e)

$$y(n) = \mathcal{E}\{x(n - 1)\}$$

or, equivalently

$$y(n) = \frac{x(n - 1) + x(-n - 1)}{2}$$

(i) This system has memory, since the value of the output signal $y(n)$ at time n depends on the present and one past input signal $x(n)$.

(ii) Assume that the input signal $x(n)$ satisfies the condition

$$|x(n)| \leq B_x < \infty, \quad \text{for all } n$$

We then find that

$$|y(n)| = \left| \frac{x(n - 1) + x(-n - 1)}{2} \right|$$

$$\leq \left| \frac{x(n - 1)}{2} \right| + \left| \frac{x(-n - 1)}{2} \right|$$

$$\leq B_x$$

We conclude that in this system if any input is bounded by an arbitrary positive number B_x, the corresponding output is guaranteed to be bounded. Thus, the given system is stable.

(iii) This system is noncausal, since the present output $y(n)$ at time n depends on the future values of the input signal $x(n)$.

(iv) To determine whether or not a system is linear, we consider two arbitrary inputs $x_1(n)$ and $x_2(n)$.

$$x_1(n) \rightarrow y_1(n) = \frac{x_1(n - 1) + x_1(-n - 1)}{2}$$

$$x_2(n) \rightarrow y_2(n) = \frac{x_2(n - 1) + x_2(-n - 1)}{2}$$

$$x_3(n) \rightarrow y_3(n) = \frac{x_3(n - 1) + x_3(-n - 1)}{2}$$

Let $x_3(n)$ be a linear combination of $x_1(n)$ and $x_2(n)$. That is,

$$x_3(n) = ax_1(n) + bx_2(n)$$

where a and b are arbitrary scalars. If the system is linear, then

$$y_3(n) = ay_1(n) + by_2(n)$$

Consider the LHS of the above equation

$$y_3(n) = \frac{x_3(n - 1) + x_3(-n - 1)}{2}$$

$$= \frac{1}{2}[ax_1(n - 1) + bx_2(n - 1)]$$

$$\quad + [ax_1(-n - 1) + bx_2(-n - 1)]$$

$$= a[x_1(n - 1) + x_1(-n - 1)]$$

$$\quad + b[x_2(n - 1) + x_2(-n - 1)]$$

$$= ay_1(n) + by_2(n) = \text{RHS}$$

Since LHS = RHS, we conclude that the system is *linear*.

(v) Let $x_1(n)$ be an arbitrary input to the system, and let

$$y_1(n) = \frac{x_1(n - 1) + x_1(-n - 1)}{2}$$

be the corresponding output. Then consider a second input obtained by shifting $x_1(n)$ in time:

$$x_2(n) = x_1(n - n_0)$$

The output corresponding to this input is

$$y_2(n) = \frac{x_2(n-1) + x_2(-n-1)}{2}$$

$$= \frac{x_1(n-n_0-1) + x_1(-n-n_0-1)}{2}$$

Now, consider

$$y_1(n) = \frac{x_1(n-1) + x_1(-n-1)}{2}$$

$$y_1(n-n_0) = \frac{x_1(n-n_0-1) + x_1(-n+n_0-1)}{2}$$

We see that $y_2(n) \neq y_1(n-n_0)$, and therefore this system is *not time-invariant*.

(f) (i) This system has memory, since the value of the output signal $y(n)$ at time n depends on the future input.

$$y(n)|_{n=n_0} = y(n_0) = x(n_0^2)$$

and $\qquad y(n)|_{n=-n_0} = y(-n_0) = x(n_0^2)$

(ii) Assume that the input signal $x(n)$ satisfies the condition

$$|x(n)| \leq B_x < \infty, \quad \text{for all } n$$

We then find that

$$|y(n)| = |x(n^2)|$$

$$= B_x$$

We conclude that in this system if any input is bounded by an arbitrary positive number B_x, the corresponding output is guaranteed to be bounded. Thus, the given system is stable.

(iii) This system is noncausal, since the present output $y(n)$ at time n depends on the future values of the input signal $x(n)$.

(iv) To determine whether or not a system is linear, we consider two arbitrary inputs $x_1(n)$ and $x_2(n)$.

$$x_1(n) \to y_1(n) = x_1(n^2)$$

$$x_2(n) \to y_2(n) = x_2(n^2)$$

$$x_3(n) \to y_3(n) = x_3(n^2)$$

Let $x_3(n)$ be a linear combination of $x_1(n)$ and $x_2(n)$. That is,

$$x_3(n) = ax_1(n) + bx_2(n)$$

where a and b are arbitrary scalars. If the system is linear, then

$$y_3(n) = ay_1(n) + by_2(n)$$

Consider the LHS of the above equation

$$y_3(n) = x_3(n^2)$$

$$= ax_1(n^2) + bx_2(n^2)$$

$$= ay_1(n) + by_2(n) = \text{RHS}$$

Since LHS = RHS, we conclude that the system is *linear*.

(v) Let $x_1(n)$ be an arbitrary input to the system, and let

$$y_1(n) = x_1(n^2)$$

be the corresponding output. Then consider a second input obtained by shifting $x_1(n)$ in time:

$$x_2(n) = x_1(n-n_0)$$

The output corresponding to this input is

$$y_2(n) = x_2(n^2)$$

$$= x_1(n^2 - n_0)$$

Now, consider

$$y_1(n) = x_1(n^2)$$

$$y_1(n-n_0) = x_1((n-n_0)^2)$$

We see that $y_2(n) \neq y_1(n-n_0)$, and therefore this system is *not time-invariant*.

Example 1.37 (a) An arbitrary real-valued continuous-time signal $x(t)$ occupies the entire interval $-\infty < x(t) < \infty$. Show that the energy of the signal $x(t)$ is equal to the sum of the energy of the even component $x_e(t)$ and the energy of the odd component $x_o(t)$.

(b) Show that an arbitrary real-valued discrete-time signal $x(n)$ satisfies a relationship similar to that satisfied by the continuous signal in part (a).

Solution

(a) See Property 6 of continuous-time even and odd signals.

(b) See Property 4 of discrete-time even and odd signals.

Example 1.38 Let $x(t)$ be a continuous-time signal, and let

$$y_1(t) = x(2t) \quad \text{and} \quad y_2(t) = x(t/2)$$

The signal $y_1(t)$ represents a speeded up version of $x(t)$ in the sense that the duration of the signal is cut in half. Similarly, $y_2(t)$ represents a slowed down version of $x(t)$ in the sense that the duration of the signal is doubled. Consider the following statements:

(1) If $x(t)$ is periodic, then $y_1(t)$ is periodic.

(2) If $y_1(t)$ is periodic, then $x(t)$ is periodic.

(3) If $x(t)$ is periodic, then $y_2(t)$ is periodic.

(4) If $y_2(t)$ is periodic, then $x(t)$ is periodic.

For each of these statements, determine whether it is true and, if so, determine the relationship between the fundamental periods of the two signals considered in the statement.

Solution

(1) Let $x(t)$ be periodic with period T, i.e.,

$$x(t) = x(t + T)$$
$$x(2t) = x(2t + T)$$

and $\qquad y_1(t) = x(2t)$

$$y_1(t + T_0) = x(2(t + T_0))$$
$$= x(2t + 2T_0)$$
$$= x(2t + T) \quad \text{for } T = 2T_0$$
$$= x(2t) = y_1(t)$$

$y_1(t)$ is periodic with period $T_0 = T/2$.

(2) Let $y_1(t)$ be periodic with period T_0.

$$y_1(t) = y_1(t + T_0)$$

Now, consider

$$x(2t) = y_1(t)$$
$$x(2(t + T/2)) = y_1(t + T/2)$$
$$x(2t + T) = y_1(t + T/2)$$
$$= y_1(t + T_0) \quad \text{for } T_0 = T/2$$
$$= y_1(t) = x(2t)$$

$x(t)$ is periodic with period $T = 2T_0$.

(3) Let $x(t)$ be periodic with period T, i.e.,

$$x(t) = x(t + T)$$
$$x(t/2) = x(t/2 + T)$$

and $\qquad y_2(t) = x(t/2)$

$$y_2(t + T_0) = x((t + T_0)/2)$$
$$= x(t/2 + T_0/2)$$
$$= x(t/2 + T) \quad \text{for } T = T_0/2$$
$$= x(t/2) = y_2(t)$$

$y_2(t)$ is periodic with period $T_0 = 2T$.

(4) Let $y_2(t)$ be periodic with period T_0.

$$y_2(t) = y_2(t + T_0)$$

Now, consider

$$x(t/2) = y_2(t)$$
$$x((t + 2T)/2) = y_2(t + 2T)$$
$$x(t/2 + T) = y_2(t + 2T)$$
$$= y_2(t + T_0) \quad \text{for } T_0 = 2T$$

$x(t)$ is periodic with period $T = T_0/2$.

Example 1.39 Let $x(n)$ be a discrete-time signal, and let

$$y_1(n) = x(2n)$$

and $\qquad y_2(n) = \begin{cases} x(n/2), & n = \text{an even number} \\ 0, & n = \text{an odd number} \end{cases}$

The signals $y_1(n)$ and $y_2(n)$ respectively represent in some sense the speeded up and slowed down versions of $x(n)$. Consider the following statements:

(1) If $x(n)$ is periodic, then $y_1(n)$ is periodic.

(2) If $y_1(n)$ is periodic, then $x(n)$ is periodic.

(3) If $x(n)$ is periodic, then $y_2(n)$ is periodic.

(4) If $y_2(n)$ is periodic, then $x(n)$ is periodic.

For each of these statements, determine whether it is true and, if so, determine the relationship between the fundamental periods of the two signals considered in the statement.

Solution

(1) True. Let $x(n)$ be periodic with period N.

$$x(n) = x(n + mN), \quad m = \text{an integer}$$

$$x(2n) = x(2n + mN)$$

Now, consider

$$y_1(n) = x(2n)$$
$$y_1(n + N_0) = x(2(n + N_0))$$
$$= x(2n + 2N_0)$$
$$= x(2n + mN), \quad \text{for } mN = 2N_0$$
$$= x(2n) = y_1(n)$$

$y_1(n)$ is periodic with period

$$N_0 = \frac{mN}{2}$$

$$= \begin{cases} \frac{N}{2}, & \text{if } N \text{ is even } (m = 1) \\ N, & \text{if } N \text{ is odd } (m = 2) \end{cases}$$

(2) False. $y_1(n)$ being periodic does not imply $x(n)$ is periodic, i.e., let $x(n) = g(n) + h(n)$, where

$$g(n) = \begin{cases} 1, & n = \text{an even number} \\ 0, & n = \text{an odd number} \end{cases}$$

and

$$h(n) = \begin{cases} 0, & n = \text{an even number} \\ (1/2)^n, & n = \text{an odd number} \end{cases}$$

Then $y_1(n) = x(2n)$ is periodic but $x(n)$ is clearly not periodic (Fig. 1.32).

(3) True. Let $x(n)$ be periodic with period N.

$$x(n) = x(n + mN) \quad m = \text{an integer}$$

$$x(n/2) = x(n/2 + mN)$$

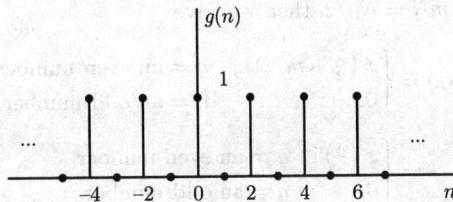

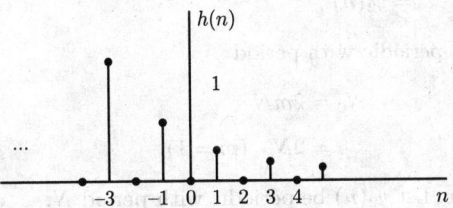

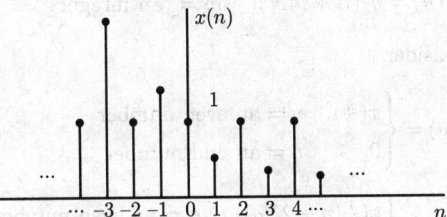

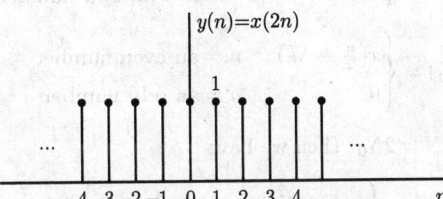

Fig. 1.32

Now, consider

$$y_2(n) = \begin{cases} x(n/2), & n = \text{an even number} \\ 0, & n = \text{an odd number} \end{cases}$$

$$y_2(n + N_0) = \begin{cases} x\left(\frac{1}{2}(n + N_0)\right), & n = \text{an even number} \\ 0, & n = \text{an odd number} \end{cases}$$

$$= \begin{cases} x\left(\frac{n}{2} + \frac{N_0}{2}\right), & n = \text{an even number} \\ 0, & n = \text{an odd number} \end{cases}$$

Assume $mN = N_0/2$, then we have

$$y_2(n + N_0) = \begin{cases} x\left(\frac{n}{2} + mN)\right), & n = \text{an even number} \\ 0, & n = \text{an odd number} \end{cases}$$

$$= \begin{cases} x\left(\frac{n}{2}\right), & n = \text{an even number} \\ 0, & n = \text{an odd number} \end{cases}$$

$$= y_2(n)$$

$y_2(n)$ is periodic with period

$$N_0 = 2mN$$
$$= 2N \quad (m = 1)$$

(4) True. Let $y_2(n)$ be periodic with period N.

$$y_2(n) = y_2(n + mN), \quad m = \text{an integer}$$

Now, consider

$$y_2(n) = \begin{cases} x\left(\frac{n}{2}\right), & n = \text{an even number} \\ 0, & n = \text{an odd number} \end{cases}$$

$$y_2(n + 2N_0) = \begin{cases} x\left(\frac{1}{2}(n + 2N_0)\right), & n = \text{an even number} \\ 0, & n = \text{an odd number} \end{cases}$$

$$y_2(n + 2N_0) = \begin{cases} x\left(\frac{n}{2} + N_0\right), & n = \text{an even number} \\ 0, & n = \text{an odd number} \end{cases}$$

Assume $N = 2N_0$, then we have

$$y_2(n + N) = \begin{cases} x\left(\frac{n}{2} + N_0\right), & n = \text{an even number} \\ 0, & n = \text{an odd number} \end{cases}$$

$$y_2(n) = \begin{cases} x\left(\frac{n}{2} + N_0\right), & n = \text{an even number} \\ 0, & n = \text{an odd number} \end{cases}$$

$x(n)$ is periodic with period $N_0 = N/2$.

Example 1.40 A discrete-time system is both linear and time-invariant. Suppose the output $y(n)$ due to an input $x(n) = \delta(n)$ is as shown in Fig. 1.33(a).
(a) Find the output $y_1(n)$ due to an input $x_1(n) = \delta(n-1)$.
(b) Find the output $y_2(n)$ due to an input $x_2(n) = 2\delta(n) - \delta(n-2)$.
(c) Find the output $y_3(n)$ due to the input depicted in Fig. 1.33(b).

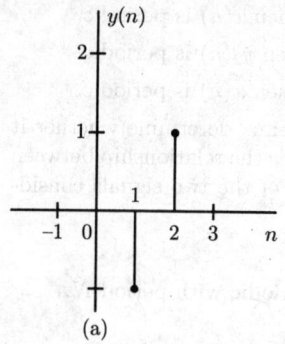

(a)

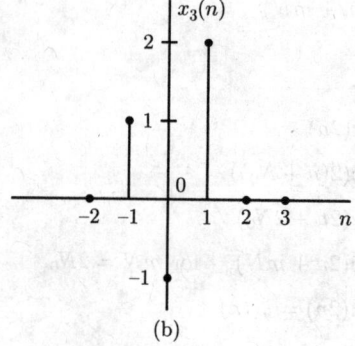

(b)

Fig. 1.33

Solution
(a) The given system is time-invariant. Therefore, a time shift in the input signal causes an identical time shift in the output signal [Fig. 1.34(a)]. The output corresponding to the input $x_1(n) = x(n-1)$ is $y_1(n) = y(n-1)$.

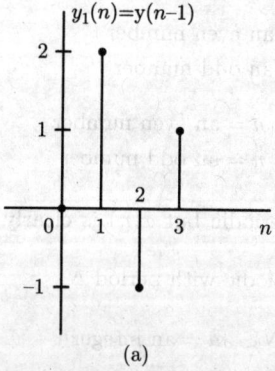

(a)

Fig. 1.34(a)

(b) The given system is both linear and time-invariant. The output [shown in Fig. 1.34(b)] corresponding to the input (i) $2\delta(n)$ is $2y(n)$, (ii) $\delta(n-2)$ is $y(n-2)$, and (iii) $2\delta(n) - \delta(n-2)$ is $y_2(n) = 2y(n) - y(n-2)$.

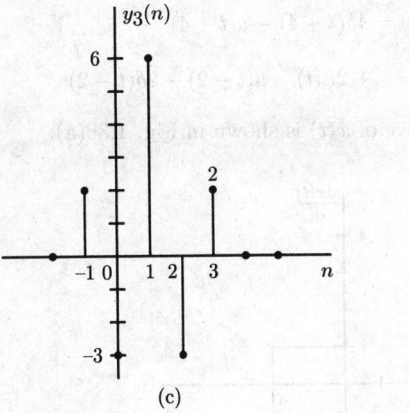

(c)

Fig. 1.34(c)

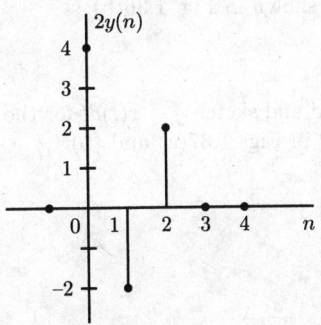

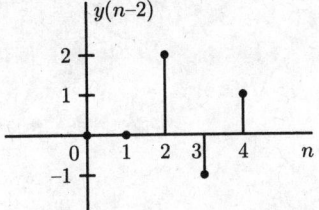

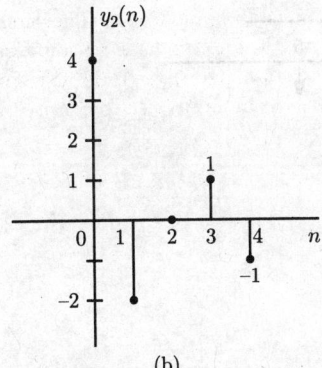

(b)

Fig. 1.34(b)

(c) From Fig. 1.33(b), the output [shown in Fig. 1.34(c)] corresponding to the input $x_3(n) = \delta(n+1) - \delta(n) + 2\delta(n-1)$ is $y_3(n) = y(n+1) - y(n) + 2y(n-1)$.

Example 1.41 (a) Find and sketch dx/dt for the signal $x(t)$ shown in Fig. 1.35(a).
(b) Find and sketch d^2x/dt^2 for the signal $x(t)$ depicted in Fig. 1.35(b).

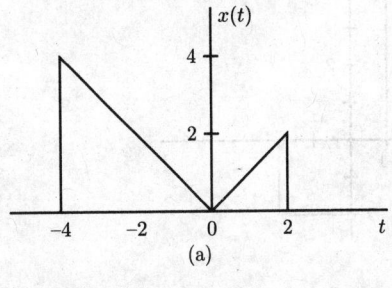

(a)

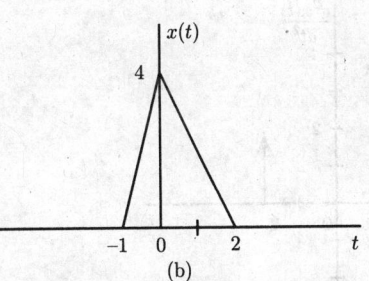

(b)

Fig. 1.35

Solution
(a) From Fig. 1.35(a)

$$x(t) = 4u(t+4) - r(t+4)$$
$$+ 2r(t) - r(t-2) - 2u(t-2)$$

$$\frac{d}{dt}x(t) = 4\delta(t+4) - u(t+4)$$

$$+ 2u(t) - u(t-2) - 2\delta(t-2)$$

The derivative of $x(t)$ is shown in Fig. 1.36(a).

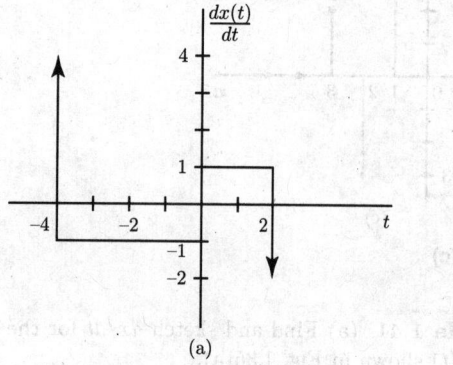

(a)

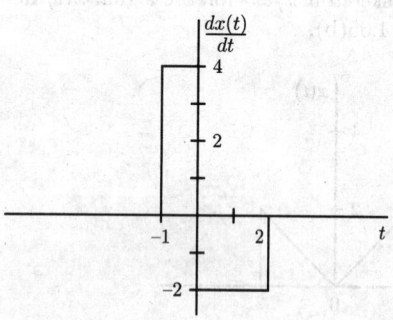

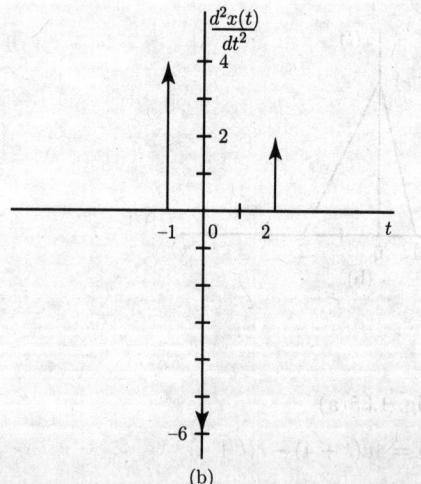

Fig. 1.36

(b) From Fig. 1.35(b)

$$x(t) = 4r(t+1) - 4r(t) - 2r(t) + 2r(t-2)$$

$$\frac{d}{dt}x(t) = 4u(t+1) - 4u(t) - 2u(t) + 2u(t-2)$$

$$\frac{d^2}{dt^2}x(t) = 4\delta(t+1) - 4\delta(t) - 2\delta(t) + 2\delta(t-2)$$

dx/dt and d^2x/dt^2 is shown in Fig. 1.36(b).

Example 1.42 Find and sketch $\int_{-\infty}^{t} x(t)dt$ for the signal $x(t)$ illustrated in Figs 1.37(a) and (b).

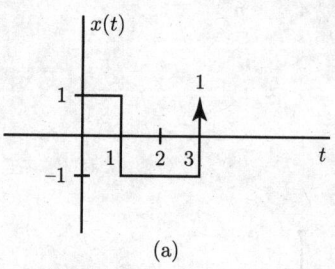

(a)

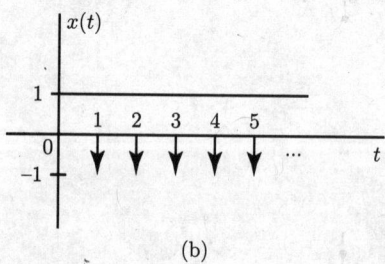

(b)

Fig. 1.37

Solution
(a) From Fig. 1.37(a)

$$x(t) = u(t) - 2u(t-1) + u(t-3) + \delta(t-3)$$

$$\int_{-\infty}^{t} x(t)dt = r(t) - 2r(t-1) + r(t-3) + u(t-3)$$

The integrated version of Fig. 1.37(a) is shown in Fig. 1.38(a).

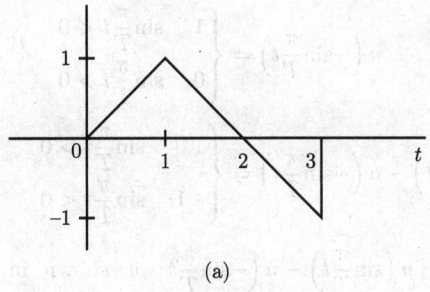

(a)

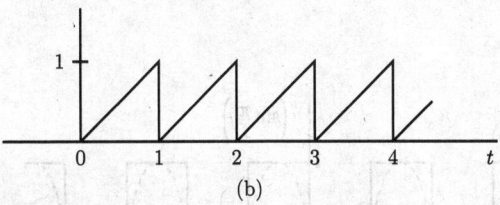

(b)

Fig. 1.38

From Fig. 1.37(b)

$$x(t) = u(t) - \delta(t-1) - \delta(t-2) - \delta(t-3)$$

$$\int_{-\infty}^{t} x(t)dt = r(t) - u(t-1) - u(t-2) - u(t-3)$$

The integrated version of Fig. 1.37(b) is shown in Fig. 1.38(b).

Example 1.43 Sketch the following signals:

(a) $x_1(t) = \delta(\cos t)$

(b) $x_2(t) = \text{sgn}\left(\sin\frac{\pi}{T}t\right)$

(c) $x_3(t) = t\,\text{sgn}(\cos t)\quad 0 \le t \le 2\pi$

(d) $x_4(t) = u\left(\sin\frac{\pi}{T}t\right) - u\left(-\sin\frac{\pi}{T}t\right)$

Solution

(a) Since $\qquad\qquad \delta(t) = 0,\quad t \ne 0$

we get $\qquad x_1(t) = \delta(\cos t) = 0,\quad \cos t \ne 0$

$x_1(t)$ is shown in Fig. 1.39(a).

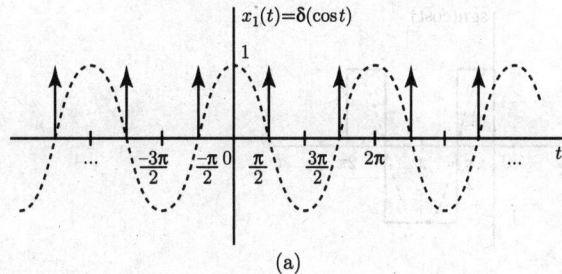

(a)

Fig. 1.39(a)

(b) Since $\qquad \text{sgn}(t) = \begin{cases} 1, & t > 0 \\ -1, & t < 0 \end{cases}$

we get

$$x_2(t) = \text{sgn}\left(\sin\frac{\pi}{T}t\right) = \begin{cases} 1, & \sin\frac{\pi}{T}t > 0 \\ -1, & \sin\frac{\pi}{T}t < 0 \end{cases}$$

$x_2(t)$ is shown in Fig. 1.39(b)

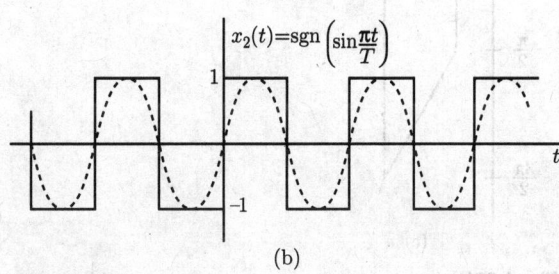

(b)

Fig. 1.39(b)

(c) Since $\qquad \text{sgn}(t) = \begin{cases} 1, & t > 0 \\ -1, & t < 0 \end{cases}$

we get

$$t\,\text{sgn}(\cos t) = \begin{cases} t, & \cos t > 0 \\ -t, & \cos t < 0 \end{cases}$$

The signal $t\,\text{sgn}(\cos t)$ is shown in Fig. 1.39(c).

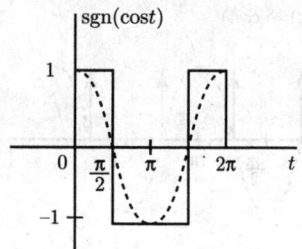

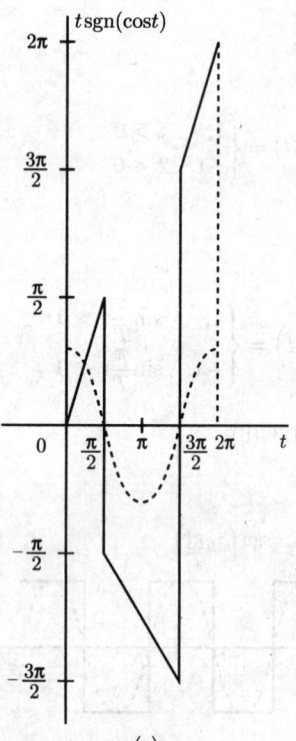

(c)

Fig. 1.39(c)

(d) Since

$$u(t) = \begin{cases} 1, & t > 0 \\ 0, & t < 0 \end{cases}$$

we get

$$u\left(\sin\frac{\pi}{T}t\right) = \begin{cases} 1, & \sin\frac{\pi}{T}t > 0 \\ 0, & \sin\frac{\pi}{T}t < 0 \end{cases}$$

$$u\left(-\sin\frac{\pi}{T}t\right) = \begin{cases} 1, & \sin\frac{\pi}{T}t < 0 \\ 0, & \sin\frac{\pi}{T}t > 0 \end{cases}$$

$$u\left(\sin\frac{\pi}{T}t\right) - u\left(-\sin\frac{\pi}{T}t\right) = \begin{cases} 1, & \sin\frac{\pi}{T}t > 0 \\ -1, & \sin\frac{\pi}{T}t < 0 \end{cases}$$

The signal $u\left(\sin\frac{\pi}{T}t\right) - u\left(-\sin\frac{\pi}{T}t\right)$ is shown in Fig. 1.39(d).

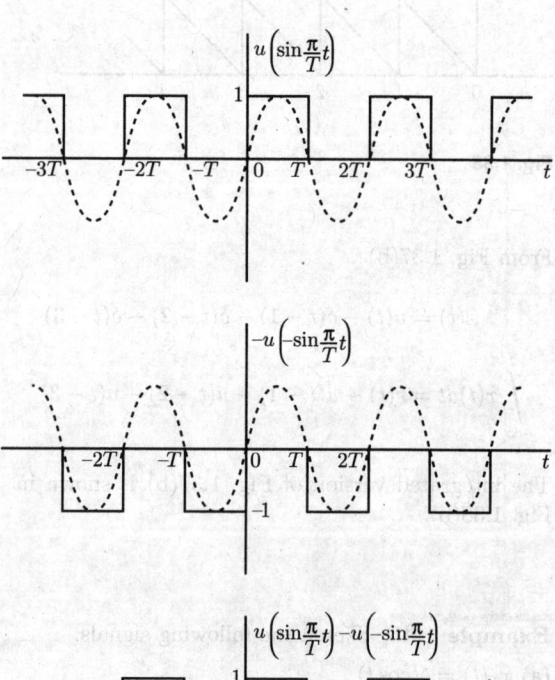

(d)

Fig. 1.39(d)

Example 1.44 Write the equation for the waveforms in Fig. 1.40 using shifted step functions.

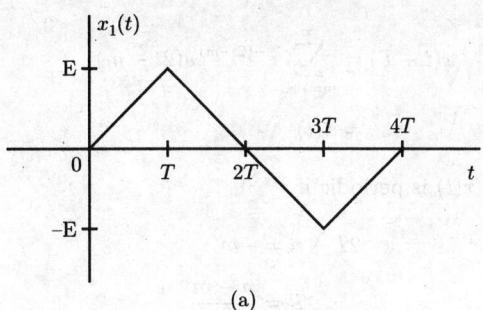

(a)

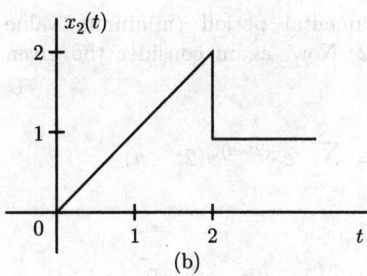

(b)

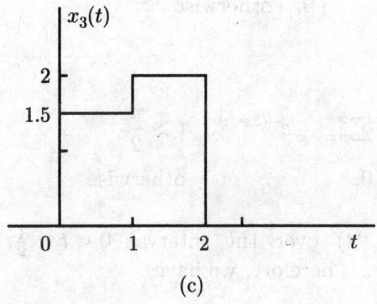

(c)

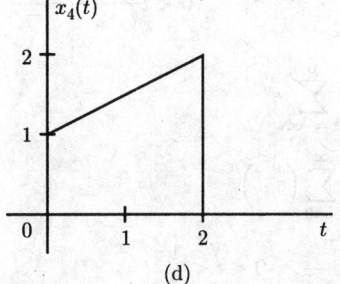

(d)

Fig. 1.40

Solution

(a)

$$x_1(t) = \frac{E}{T}r(t) - \frac{2E}{T}r(t-T) + \frac{2E}{T}r(t-3T)$$

$$- \frac{E}{T}r(t-4T)$$

$$= \frac{E}{T}tu(t) - \frac{2E}{T}(t-T)u(t-T)$$

$$+ \frac{2E}{T}(t-3T)u(t-3T)$$

$$- \frac{E}{T}(t-4T)u(t-4T)$$

(b)

$$x_2(t) = r(t) - r(t-2) - u(t-2)$$

$$= tu(t) - (t-2)u(t-2) - u(t-2)$$

$$= tu(t) - (t-1)u(t-2)$$

(c)

$$x_3(t) = 1.5u(t) + 0.5u(t-1) - 2u(t-2)$$

(d)

$$x_4(t) = u(t) + \frac{1}{2}r(t) - \frac{1}{2}r(t-2) - 2u(t-2)$$

$$= u(t) + \frac{1}{2}tu(t) - \frac{1}{2}(t-2)u(t-2)$$

$$- 2u(t-2)$$

$$= \left(\frac{1}{2}t + 1\right)u(t) - \left(\frac{1}{2}t + 1\right)u(t-2)$$

Example 1.45 The waveform $x(t)$ in Fig. 1.41 is defined as

$$x(t) = \begin{cases} \frac{3}{\epsilon^3}(t-\epsilon)^2, & 0 \le t \le \epsilon \\ 0, & \text{elsewhere} \end{cases}$$

Show that as $\epsilon \to 0$, $x(t)$ becomes a unit impulse.

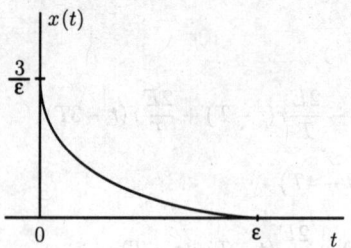

Fig. 1.41

Solution

Area under the unit impulse function is unity.

$$\int_{-\infty}^{\infty} x(t)dt = \int_{0}^{\epsilon} \frac{3}{\epsilon^3}(t - \epsilon^2)dt$$

$$= \frac{3}{\epsilon^3}\left[\frac{t^3}{3} + \epsilon^2 t - \epsilon t^2\right]_{0}^{\epsilon}$$

$$= 1$$

As $\epsilon \to 0$, the width of $x(t) \to 0$ and height approaches ∞ and its area remains at unity. Therefore, as $\epsilon \to 0$, $x(t)$ becomes a unit impulse function.

Example 1.46 Sketch and check whether the given signal

$$x(t) = \sum_{n=-\infty}^{\infty} e^{-(2t-n)}u(2t - n)$$

is periodic? If yes, compute its average power.

Solution

Consider the given signal

$$x(t) = \sum_{n=-\infty}^{\infty} e^{-(2t-n)}u(2t - n)$$

For $x(t)$ to be periodic, $x(t + T) = x(t)$; therefore,

$$x(t + T) = \sum_{n=-\infty}^{\infty} e^{-\left(2(t+T)-n\right)}u\left(2(t + T) - n\right)$$

$$= \sum_{n=-\infty}^{\infty} e^{-(2t+2T-n)}u(2t + 2T - n)$$

A change of variables is performed by letting $2T - n = -m$ (m-an integer), which also yields $m = -\infty$ as $n = -\infty$, and $m = \infty$ as $n = \infty$. Therefore,

$$x(t + T) = \sum_{m=-\infty}^{\infty} e^{-(2t-m)}u(2t - m)$$

$$= x(t)$$

Thus, $x(t)$ is periodic if

$$2T - n = -m$$

$$T = \frac{n - m}{2}$$

Thus, the fundamental period (minimum value of T) is $T = 1/2$. Now, again consider the given signal

$$x(t) = \sum_{n=-\infty}^{\infty} e^{-(2t-n)}u(2t - n)$$

Since $\quad u(2t - n) = \begin{cases} 1, & (2t - n) > 0 \longrightarrow t > \dfrac{n}{2} \\ 0, & \text{otherwise} \end{cases}$

Therefore,

$$x(t) = \begin{cases} \sum_{n=-\infty}^{\infty} e^{-(2t-n)}, & t > \dfrac{n}{2} \\ 0, & \text{otherwise} \end{cases}$$

To determine $x(t)$ over the interval $0 < t < \frac{1}{2}$, $n = 0, -1, -2, \ldots$. Therefore, we have

$$x(t) = \sum_{n=-\infty}^{0} e^{-(2t-n)}, \qquad 0 < t < \frac{1}{2}$$

$$= e^{-2t}\sum_{n=-\infty}^{0} e^{n}$$

$$= e^{-2t}\sum_{n=0}^{\infty}\left(e^{-1}\right)^{n}$$

$$x(t) = \frac{e^{-2t}}{1 - e^{-1}}, \qquad 0 < t < \frac{1}{2}$$

21. A continuous-time system is a transformation that operates on a continuous-time input signal to produce a continuous-time output signal.

22. A system is linear if it follows the principle of superposition.

23. A system is time-invariant if a time shift in the input signal causes an identical time shift in the output.

24. The systems satisfying both linearity and time-invariant properties are known as linear time-invariant or LTI systems.

25. A system is memoryless if the present value of the output $y(t)$ depends only on the present value of the input $x(t)$.

26. A system is causal if the output $y(t_0)$ depends only on values of the input $x(t)$ for $t \leq t_0$.

27. A system is invertible if the distinct inputs lead to the distinct outputs.

28. A system is BIBO stable if bounded inputs result in bounded output.

29. Concepts such as linearity, memory, time invariance, and causality in discrete-time systems are similar to those in continuous-time systems.

MULTIPLE-CHOICE QUESTIONS

1. Let $u(n)$ be a unit step sequence. The sequence $u(N - n)$ can be described as

   (a) $x(n) = \begin{cases} 1, & n < N \\ 0, & \text{otherwise} \end{cases}$

   (b) $x(n) = \begin{cases} 1, & n \leq N \\ 0, & \text{otherwise} \end{cases}$

   (c) $x(n) = \begin{cases} 1, & n > N \\ 0, & \text{otherwise} \end{cases}$

   (d) $x(n) = \begin{cases} 1, & n \geq N \\ 0, & \text{otherwise} \end{cases}$

2. Given a unit step function $u(t)$. Its time derivative is
   (a) a unit impulse
   (b) another step function
   (c) a unit ramp function
   (d) a sine function

3. The function $\sin(\pi u)/\pi u$ is denoted by
   (a) $\text{sinc}(\pi u)$
   (b) $\text{sinc}(u)$
   (c) signum
   (d) none of these

4. A periodic signal $x(n)$ of period N_1 is added to another periodic signal of period N_2. Then the period of the resulting signal is always
   (a) $N_1 + N_2$
   (b) $N_1 N_2$
   (c) LCM of N_1 and N_2
   (d) GCD of N_1 and N_2

5. The system having input $x(n)$ related to output $y(n)$ as $y(n) = \log_{10}|x(n)|$ is
   (a) nonlinear, causal, stable

   (b) linear, noncausal, stable
   (c) nonlinear, causal, not stable
   (d) linear, noncausal, not stable

6. To obtain $x(4 - 2n)$ from the given signal $x(n)$, the following precedence (or priority) rule is used for operations on the independent variable n:
   (a) Time scaling→Time shifting→Reflection
   (b) Reflection→Time scaling→Time shifting
   (c) Time scaling→Reflection→Time shifting
   (d) Time shifting→Time scaling→Reflection

7. The system described by $y(n) = nx(n)$ is
   (a) linear, time-varying, and stable
   (b) nonlinear, time-invariant, and unstable
   (c) nonlinear, time-varying, and stable
   (d) linear, time-varying, and unstable

8. The period of the signal $x(t) = 10\sin(12\pi t) + 4\cos(18\pi t)$ is
   (a) $\pi/4$
   (b) $1/6$
   (c) $1/9$
   (d) $1/3$

9. The area under the curve $\int_{-\infty}^{\infty} \delta(t)dt$ is
   (a) ∞
   (b) unity
   (c) 0
   (d) undefined

10. Given $x(n) = a^{|n|}, |a| < 1$ is
    (a) an energy signal
    (b) a power signal
    (c) neither an energy nor a power signal
    (d) an energy as well as a power signal

11. The system characterized by the equation $y(t) = ax(t) + b$ is

(a) linear for any value of b
(b) linear if $b > 0$
(c) linear if $b < 0$
(d) nonlinear

12. The discrete-time signal $x(n) = (-1)^n$ is periodic with fundamental period
    (a) 6
    (b) 4
    (c) 2
    (d) 0

13. A useful property of the unit impulse $\delta(t)$ is that
    (a) $\delta(at) = a\delta(t)$
    (b) $\delta(at) = \delta(t)$
    (c) $\delta(at) = \dfrac{1}{|a|}\delta(t)$
    (d) $\delta(at) = [\delta(t)]^a$

14. Any signal $x(t)$ can be represented as
    (a) $x_e(t) + x_o(t)$
    (b) $x_e(t) - x_o(t)$
    (c) $\dfrac{x_e(t)}{x_o(t)}$
    (d) $x_e(t) \times x_o(t)$

15. Periodic signals are
    (a) $x(t + T) = x(t)$
    (b) $x(t - T) = x(t)$
    (c) $x(n = mN) = x(n)$
    (d) all the above

16. Invertible systems are systems in which
    (a) output can be uniquely obtained from the knowledge of input
    (b) input signal can be uniquely determined by observing the output signal
    (c) both (a) and (b)
    (d) system output is always constant

17. Choose the false statement:
    (a) $t\delta(t) = 0$
    (b) $\cos(t)\delta(t - \pi) = -\delta(t - \pi)$
    (c) $\delta(t) = \int u(t)dt$
    (d) $t\delta'(t) = -\delta(t)$

PROBLEMS

1.1 Prove each of the following statements. In each case, let energy signal $x_1(t)$ have energy $E[x_1(t)]$, let energy signal $x_2(t)$ have energy $E[x_2(t)]$, and let T be a nonzero, finite, real-valued constant.
(i) Prove $E[Tx_1(t)] = T^2 E[x_1(t)]$. That is, amplitude scaling a signal by constant T scales the signal energy by T^2.
(ii) Prove $E[x_1(t)] = E[x_1(t - T)]$. That is, shifting a signal does not affect its energy.
(iii) If $(x_1(t) \neq 0) \Rightarrow (x_2(t) = 0)$ and $(x_2(t) \neq 0) \Rightarrow (x_1(t) = 0)$, then prove $E[x_1(t) + x_2(t)] = E[x_1(t)] + E[x_2(t)]$. That is, the energy of the sum of two nonoverlapping signals is the sum of the two individual energies.
(iv) Prove $E[x_1(Tt)] = 1/|T|E[x_1(t)]$. That is, time-scaling a signal by T reciprocally scales the signal energy by $1/|T|$.

1.2 Show that the power of a signal

$$x(t) = \sum_{k=m}^{n} C_k e^{j\omega_k t} \text{ is } P_x = \sum_{k=m}^{n} |C_k|^2$$

assuming all frequencies to be distinct, that is, $\omega_i \neq \omega_k$ for all $i \neq k$.

1.3 For the signal $x(t)$ illustrated in Fig. 1.35(a), sketch
(i) $x(t - 4)$
(ii) $x(t/1.5)$
(iii) $x(-t)$
(iv) $x(2t - 4)$
(v) $x(2 - t)$

1.4 In Fig. 1.42, express signals $x_1(t), x_2(t)$, and $x_3(t)$ in terms of signal $x(t)$ and its time-shifted, time-scaled, or time-reversed versions.

1.5 Simplify the following expressions:
(i) $\left(\dfrac{\sin t}{t^2 + 2}\right)\delta(t)$

(ii) $\left(\dfrac{\omega + 2}{\omega^2 + 9}\right)\delta(\omega)$

(iii) $[e^{-t}\cos(3t - 60°)]\delta(t)$

(iv) $\left(\dfrac{1}{j\omega + 2}\right)\delta(\omega + 3)$

(v) $\left(\dfrac{\sin k\omega}{\omega}\right)\delta(\omega)$

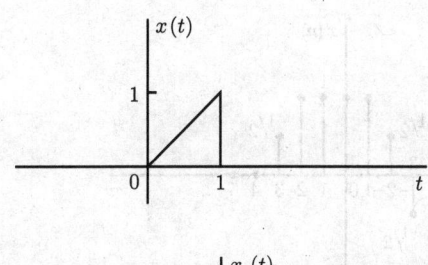

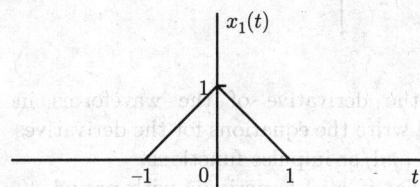

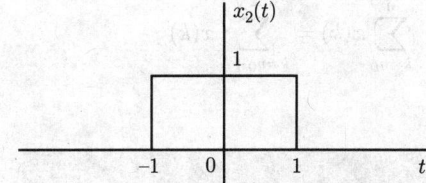

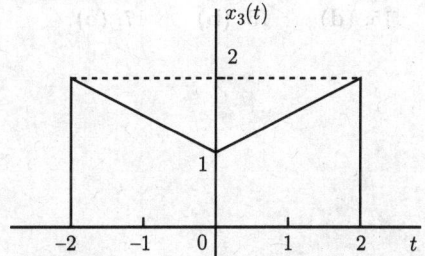

Fig. 1.42

1.6 Evaluate the following integrals:

(i) $\displaystyle\int_{-\infty}^{\infty} \left(\frac{2}{3}t - \frac{3}{2}\right)\delta(t-1)dt$

(ii) $\displaystyle\int_{-\infty}^{\infty} (t-1)\delta\left(\frac{2}{3}t - \frac{3}{2}\right)dt$

(iii) $\displaystyle\int_{-3}^{-2} \left[e^{-t+1} + \sin\left(\frac{2\pi}{3}t\right)\right]\delta\left(t - \frac{3}{2}\right)dt$

(iv) $\displaystyle\int_{-3}^{2} \left[e^{-t+1} + \sin\left(\frac{2\pi}{3}t\right)\right]\delta\left(t - \frac{3}{2}\right)dt$

(v) $\displaystyle\int_{-\infty}^{\infty} e^{-5t+1}\delta'(t-5)dt$

1.7 (a) If $x_e(t)$ and $x_0(t)$ are even and odd components of a real signal $x(t)$, then show that

$$\int_{-\infty}^{\infty} x_e(t)x_0(t)dt = 0$$

(b) Show that

$$\int_{-\infty}^{\infty} x(t)dt = \int_{-\infty}^{\infty} x_e(t)dt$$

1.8 Find and sketch the even and odd components of the following:
(i) $u(t)$
(ii) $r(t) = tu(t)$
(iii) $\sin(\omega_0 t)u(t)$
(iv) $\cos(\omega_0 t)u(t)$

1.9 An aperiodic signal is defined as $x(t) = \sin(\pi t)u(t)$, where $u(t)$ is the continuous-time step function. Is the odd portion of this signal, $x_0(t)$, periodic? Justify your answer.

1.10 The output of the system $y(t)$ is related to the input $x(t)$ as

$$y(t) = \cos\left(2\pi f_c t + k\int_{-\infty}^{t} x(\tau)d\tau\right)$$

where k is a constant parameter.
(a) Show that the system is nonlinear.
(b) Show that the system is time-variant.

1.11 The systems that follow have input $x(n)$ and output $y(n)$. For each system, determine whether it is (i) memoryless, (ii) stable, (iii) causal, (iv) linear, or (v) time invariant.
(a) $y(n) = \log_e[x(n)]$
(b) $y(n) = \log_{10}[|x(n)|]$
(c) $y(n) = \text{median}\{x(n-1), x(n), x(n+1)\}$
(d) $y(n) = Trun[x(n)]$, where $Trun[x(n)]$ denotes the integer part of $x(n)$ obtained by truncation.
(e) $y(n) = \text{Round}[x(n)]$, where $\text{Round}[x(n)]$ denotes the integer part of $x(n)$ obtained by rounding.
(f) $y(n) = |x(n)|$
(g) $y(n) = \text{sgn}[x(n)]$

1.12 Sketch the following signals:
(i) $x_1(t) = u(t) + 5u(t-1) - 2u(t-2)$
(ii) $x_2(t) = r(t) - r(t-1) - u(t-2)$
(iii) $x_3(t) = 2u(t) + \delta(t-1)$
(iv) $x_4(t) = u(t)u(t-2)$
(v) $x_1(t)x_2(t + \frac{1}{2})$

1.13 A discrete-time signal is shown in Fig. 1.43. Sketch and label each of the following signals:
(i) $x(n-4)$
(ii) $x(3-n)$
(iii) $x(3n)$
(iv) $x(3n+1)$
(v) $x(n)u(3-n)$
(vi) $x(n-2)\delta(n-2)$
(vii) $\frac{1}{2}x(n) + \frac{1}{2}(-1)^n x(n)$
(viii) $x(n-1)^2$

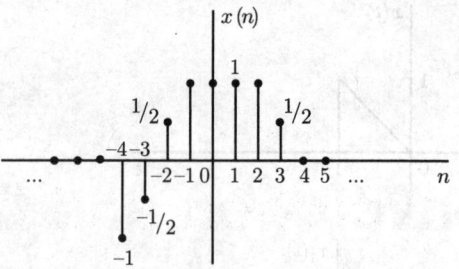

Fig. 1.43

1.14 Find the derivative of the waveforms in Fig. 1.40 and write the equations for the derivatives using shifted and/or impulse functions.

1.15 Show that if $x(n)$ is periodic with period N, then

$$\sum_{k=n_0}^{n} x(k) = \sum_{k=n_0+N}^{n+N} x(k)$$

ANSWERS TO MULTIPLE-CHOICE QUESTIONS

1. **(b)**   2. **(a)**   3. **(b)**   4. **(c)**   5. **(c)**   6. **(d)**   7. **(d)**   8. **(d)**   9. **(b)**
10. **(a)**   11. **(d)**   12. **(c)**   13. **(c)**   14. **(a)**   15. **(d)**   16. **(b)**   17. **(c)**

Convolution and Correlation

2.1 INTRODUCTION

In the previous chapter we discussed a number of basic system properties. Of these, linearity and time invariance play a fundamental role in signal and system analysis because of the many physical phenomena that can be modelled by linear time-invariant (LTI) systems and because a mathematical analysis of the behaviour of such systems can be carried out in a fairly straightforward manner. In this chapter, we begin the development by deriving and examining a fundamental and extremely useful representation for LTI systems and by introducing an important class of these systems.

One of the primary reasons LTI systems are amenable to analysis is that any such system possesses the superposition property described in Section 1.11.1. As a consequence, if we can represent the input to an LTI system in terms of a linear combination of a set of basic signals, we can then use superposition to compute the output of the system in terms of its response to these basic signals.

A fundamental problem in system analysis is determining the response to some specified input. Analytically, this can be answered in many different ways. One obvious way is to solve the differential equation describing the system, subject to the specified input and initial conditions. In the following section, we introduce a second method that exploits the linearity and time invariance of the system. We begin by characterizing an LTI system in terms of its *impulse response*, defined as the output of an LTI system due to a unit impulse signal input applied at time $t = 0$ or $n = 0$. The impulse response completely characterizes the behaviour of any LTI system. Given the impulse response, we determine the output due to an arbitrary input signal by expressing the input as a weighted superposition of time-shifted impulses. For linearity and time invariance, the output signal must be a weighted superposition of time-shifted impulse responses. This weighted superposition is termed the *convolution sum* for discrete-time systems and the *convolution integral* for continuous-time systems.

2.2 CONTINUOUS-TIME LTI SYSTEMS: THE CONVOLUTION INTEGRAL

The goal of this section is to obtain a complete characterization of a continuous-time LTI system in terms of its unit impulse response.

2.2.1 Unit Impulse Response

The impulse response is defined as the output of an LTI system due to a unit impulse signal input applied at time $t = 0$ or $n = 0$ (Fig. 2.1).

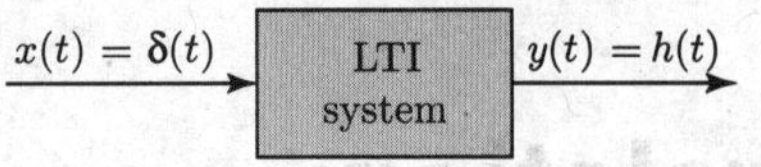

$$x(t) = \delta(t) \rightarrow \boxed{\begin{array}{c} \text{LTI} \\ \text{system} \end{array}} \rightarrow y(t) = h(t)$$

Fig. 2.1   Continuous-time LTI system

$$x(t) \rightarrow y(t)$$
$$\delta(t) \rightarrow h(t)$$

where $\delta(t)$ is the unit impulse function and $h(t)$ is the unit impulse response of a continuous-time LTI system (CTS). Similarly for a discrete-time system (DTS), $\delta(n)$ is the unit impulse or unit sample input and $h(n)$ is the unit impulse response or unit sample response (Fig. 2.2).

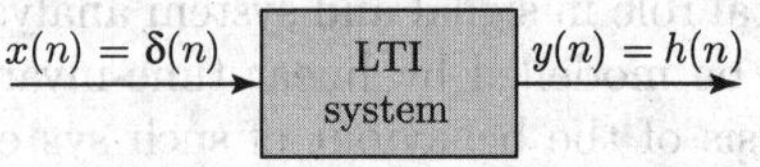

$$x(n) = \delta(n) \rightarrow \boxed{\begin{array}{c} \text{LTI} \\ \text{system} \end{array}} \rightarrow y(n) = h(n)$$

Fig. 2.2   Discrete-time LTI system

$$x(n) \rightarrow y(n)$$
$$\delta(n) \rightarrow h(n)$$

2.2.2 Convolution Integral

Linear systems are governed by the superposition principle. Let the responses of the system to two inputs $x_1(t)$ and $x_2(t)$ be $y_1(t)$ and $y_2(t)$, respectively. The system is linear if the response to the input $x(t) = a_1 x_1(t) + a_2 x_2(t)$ is equal to $y(t) = a_1 y_1(t) + a_2 y_2(t)$. More generally, if the input $x(t)$ is the weighted sum of any set signals $x_i(t)$, if the response to $x_i(t)$ is $y_i(t)$, and if the system is linear, the output $y(t)$ will be the weighted sum of the response $y_i(t)$. That is, if

$$x(t) = a_1 x_1(t) + a_2 x_2(t) + \cdots + a_N x_N(t) = \sum_{i=1}^{N} a_i x_i(t)$$

we have

$$y(t) = a_1 y_1(t) + a_2 y_2(t) + \cdots + a_N y_N(t) = \sum_{i=1}^{N} a_i y_i(t)$$

In Section 1.4.2 [Property 1, i.e., Eq. (1.10)], we demonstrated that the unit impulse function can be used as basic building blocks to represent arbitrary signals. In fact, the shifting property of the δ function,

$$x(t) = \int_{-\infty}^{\infty} x(\tau)\, \delta(t - \tau)\, d\tau$$

shows that any signal $x(t)$ can be expressed as a continuum of weighted impulses.

Now consider a continuous-time LTI system with input $x(t)$. Using the superposition and time-invariant property of LTI systems, we can express output $y(t)$ as a linear combination of the responses of the system to shifted impulse signals.

$$x(t) \longrightarrow y(t)$$

$$\delta(t) \longrightarrow h(t)$$

$$\delta(t - \tau) \longrightarrow h(t - \tau)$$

$$x(\tau)\delta(t - \tau) \longrightarrow x(\tau)h(t - \tau)$$

$$x(t) = \int_{-\infty}^{\infty} x(\tau)\delta(t - \tau)\, d\tau \longrightarrow y(t) = \int_{-\infty}^{\infty} x(\tau)h(t - \tau)\, d\tau$$

Thus, we obtain the system output $y(t)$ to an arbitrary input $x(t)$ in terms of the unit impulse response $h(t)$,

$$y(t) = \int_{-\infty}^{\infty} x(\tau)h(t - \tau)\, d\tau \tag{2.1}$$

The integral relationship expressed in Eq. (2.1) is called the *convolution integral* of signals $x(t)$ and $h(t)$. This operation is represented symbolically as

$$y(t) = x(t) * h(t) \tag{2.2}$$

Continuous-convolution (Convolution Integral) Evaluation

It is useful to present examples of the evaluation of continuous convolution before indicating its properties. Our first example considers two functions, each defined by a single equation for all values of the independent variable. The convolution integral is conceptually very easy to evaluate because it consists of a single direct evaluation of the integral and produces a single equation for all values of the independent variable.

Example 2.1 Find the convolution of the two continuous-time signals

$$f(t) = e^{-t^2} \quad \text{and} \quad g(t) = 3t^2, \quad \text{for all } t$$

Solution
By definition

$$y(t) = f(t) * g(t) = \int_{-\infty}^{\infty} f(\tau)g(t - \tau)\, d\tau$$

$$= \int_{-\infty}^{\infty} e^{-\tau^2}[3(t - \tau)^2]\, d\tau$$

$$= 3t^2 \int_{-\infty}^{\infty} e^{-\tau^2} \, d\tau - 6t \int_{-\infty}^{\infty} \tau e^{-\tau^2} \, d\tau + 3 \int_{-\infty}^{\infty} \tau^2 e^{-\tau^2} \, d\tau$$

$$= 3t^2 \sqrt{\pi} - 0 + 3 \frac{\sqrt{\pi}}{2}$$

$$= 5.31t^2 + 2.659, \quad \text{for all } t$$

Next we will consider the convolution of a piecewise-defined function and a function defined by a single equation for all values of the independent variable. In this case the integral must be evaluated in two or more parts. However, the resulting convolution is a function defined by a single equation; hence, the integral needs to be evaluated only once to yield a result that is valid for all values of the independent variable.

Example 2.2 Find the convolution of the two continuous-time signals

$$f(t) = 3\cos 2t, \quad \text{for all } t \quad \text{and} \quad g(t) = e^{-|t|} = \begin{cases} e^t, & t < 0 \\ e^{-t}, & t \ge 0 \end{cases}$$

Solution
By definition

$$y(t) = f(t) * g(t) = \int_{-\infty}^{\infty} f(\tau)g(t - \tau) \, d\tau$$

Now

$$f(\tau) = 3\cos 2\tau, \quad \text{for all } \tau \quad \text{and} \quad g(t - \tau) = \begin{cases} e^{(t-\tau)}, & (t - \tau) < 0 \to \tau > t \\ e^{-(t-\tau)}, & (t - \tau) \ge 0 \to \tau \le t \end{cases}$$

Therefore,

$$y(t) = \int_{-\infty}^{t} (3\cos 2\tau)(e^{(\tau-t)}) \, d\tau + \int_{t}^{\infty} (3\cos 2\tau)(e^{(t-\tau)}) \, d\tau$$

$$= 3e^{-t} \int_{-\infty}^{t} e^{\tau} \cos 2\tau d\tau + 3e^{t} \int_{t}^{\infty} e^{-\tau} \cos 2\tau \, d\tau$$

$$= 3e^{-t} \left[\frac{e^{\tau}(\cos 2\tau + 2\sin 2\tau)}{5} \right]_{-\infty}^{t} + 3e^{t} \left[\frac{e^{-\tau}(-\cos 2\tau + 2\sin 2\tau)}{5} \right]_{t}^{\infty}$$

$$= \frac{6}{5}\cos 2t, \quad \text{for all } t$$

The convolution of two functions that are piecewise defined produces a convolution integral that must be evaluated in two or more parts. Since the resulting convolution is also a piecewise-defined function, the convolution integral must be evaluated as many times as there are pieces in the resulting function to find the mathematical expressions for those pieces. For the evaluation of the convolution of two piecewise-defined functions, it is helpful to express carefully the functions being multiplied to form the integrand and to sketch these functions in positions corresponding to each of the integral evaluations that must be made. We list in what follows the steps in evaluating the convolution defined by Eq. (2.1):

Step 1. Write equations and draw sketches for $x(t)$ and $h(t)$.
Step 2. Write equations and draw sketches for $x(\tau)$ and $h(t - \tau)$.
Step 3. Perform integrations to find $y(t)$ for all t.

The following examples illustrate the use of the evaluation steps.

Example 2.3 Find the convolution of the two continuous-time signals

$$x(t) = e^{-|t|}, \text{ for all } t \quad \text{and} \quad h(t) = \begin{cases} e^{-2t}, & t \geq 1 \\ 0, & t < 1 \end{cases}$$

Solution
By definition

$$y(t) = x(t) * h(t) = \int_{-\infty}^{\infty} x(\tau)h(t - \tau) \, d\tau$$

Step 1: Write the equations and draw sketches for $x(t)$ and $h(t)$ as shown in Fig. 2.3.

$$x(t) = \begin{cases} e^{-t}, & t \geq 0 \\ e^{t}, & t < 0 \end{cases} \qquad h(t) = \begin{cases} e^{-2t}, & t \geq 1 \\ 0, & t < 1 \end{cases}$$

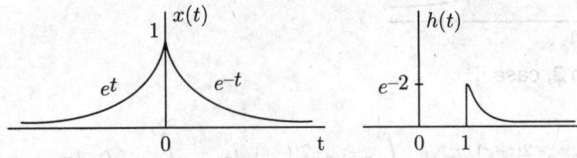

Fig. 2.3 Step 1 for Example 2.3

Step 2: Write the equations and draw sketches for $x(\tau)$ and $h(t - \tau)$ [apply precedence rule: $h(t) \to h(\tau) \to h(\tau + t) \to h(-\tau + t) = h(t - \tau)$] as shown in Fig. 2.4.

$$x(\tau) = \begin{cases} e^{-\tau}, & \tau \geq 0 \\ e^{\tau}, & \tau < 0 \end{cases} \qquad h(t - \tau) = \begin{cases} e^{-2(t-\tau)}, & (t - \tau) \geq 1 \to \tau \leq (t - 1) \\ 0, & (t - \tau) < 1 \to \tau > (t - 1) \end{cases}$$

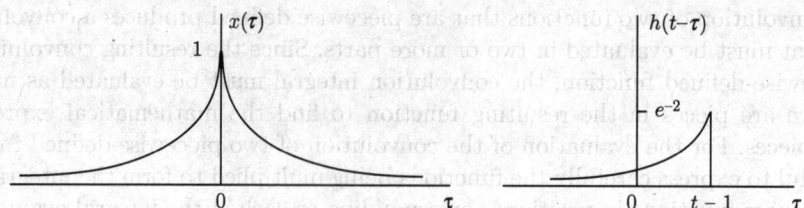

Fig. 2.4   Step 2 for Example 2.3

Step 3: Perform integration to find $y(t)$ for all values of t.

Case I: $(t-1) < 0 \longrightarrow t < 1$. Figure 2.5 shows $x(\tau)$ and $h(t-\tau)$. The two functions overlap only over the range $-\infty \leq \tau \leq t-1$ (shaded interval) such that

$$y(t) = \int_{-\infty}^{t-1} e^\tau e^{-2(t-\tau)}\, d\tau + \int_{t-1}^{\infty} x(\tau) \times 0\, d\tau = e^{-2t} \int_{-\infty}^{t-1} e^{3\tau}\, d\tau + 0 = \frac{1}{3} e^{(t-3)} \quad t < 1$$

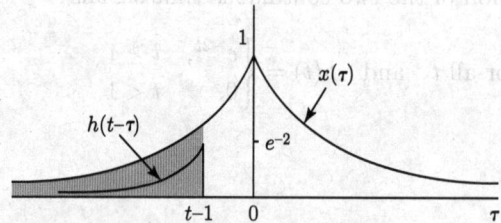

Fig. 2.5   Signal sketch for step 3, case I

Case II: $(t-1) \geq 0 \longrightarrow t \geq 1$. Figure 2.6 shows $x(\tau)$ and $h(t-\tau)$. The two functions overlap only over the range $-\infty \leq \tau \leq t-1$ (shaded interval) such that

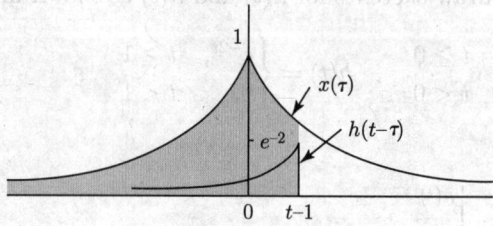

Fig. 2.6   Signal sketch for step 3, case II

$$y(t) = \int_{-\infty}^{0} e^\tau e^{-2(t-\tau)}\, d\tau + \int_{0}^{t-1} e^{-\tau} e^{-2(t-\tau)}\, d\tau + \int_{t-1}^{\infty} e^{-\tau} 0\, d\tau$$

$$= e^{-2t} \left(\int_{-\infty}^{0} e^{3\tau}\, d\tau + \int_{0}^{t-1} e^\tau\, d\tau \right)$$

$$= e^{-(t+1)} - \frac{2}{3} e^{-2t}, \quad t \geq 1$$

In summary,

$$y(t) = \begin{cases} e^{-(t+1)} - \dfrac{2}{3}e^{-2t}, & t \geq 1 \\ \dfrac{1}{3}e^{(t-3)}, & t < 1 \end{cases}$$

The solution is illustrated in Fig. 2.7.

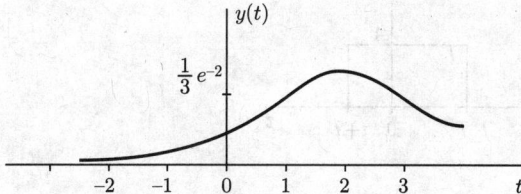

Fig. 2.7 Convolution result

Example 2.4 Find the convolution of a rectangle signal (or gate function) $x(t)$ with itself.

$$x(t) = A \operatorname{rect}\left(\frac{t}{2T}\right) = A\,\Pi\left(\frac{t}{2T}\right) = \begin{cases} A, & -T < t < T \\ 0, & \text{otherwise} \end{cases}$$

Or, evaluate the convolution integral for a system with input $x(t)$ and impulse response $h(t)$, respectively, given by

$$x(t) = h(t) = A\left[u(t+T) - u(t-T)\right]$$

Solution
By definition

$$y(t) = x(t) * h(t) = \int\limits_{-\infty}^{\infty} x(\tau)h(t-\tau)\,d\tau$$

Step 1: Write the equations and draw sketches for $x(t)$ and $h(t)$ as shown in Fig. 2.8.

$$x(t) = \begin{cases} A, & -T < t < T \\ 0, & \text{otherwise} \end{cases} \qquad h(t) = \begin{cases} A, & -T < t < T \\ 0, & \text{otherwise} \end{cases}$$

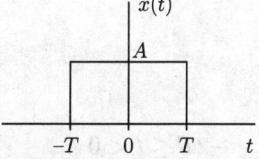 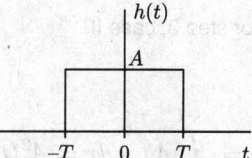

Fig. 2.8 Step 1 for Example 2.4

Step 2: Write the equations and draw sketches for $x(\tau)$ and $h(t - \tau)$ [apply precedence rule: $h(t) \rightarrow h(\tau) \rightarrow h(\tau + t) \rightarrow h(-\tau + t) = h(t - \tau)$] as shown in Fig. 2.9.

$$x(\tau) = \begin{cases} A, & -T < \tau < T \\ 0, & \text{otherwise} \end{cases} \qquad h(t - \tau) = \begin{cases} A, & -T < (t - \tau) < T \longrightarrow (t - T) < \tau < (t + T) \\ 0, & \text{otherwise} \end{cases}$$

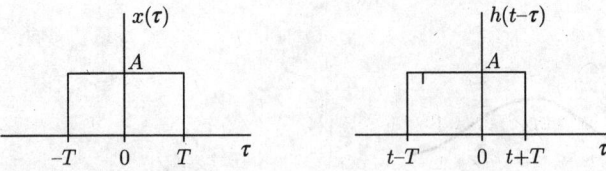

Fig. 2.9 Step 2 for Example 2.4

Step 3: Perform integration to find $y(t)$ for all values of t.

Case I: $t + T < -T \longrightarrow t < -2T$. Figure 2.10 shows $x(\tau)$ and $h(t - \tau)$. We see that the signals do not overlap (Fig. 2.10); hence, the product of the integrand is zero. Thus,

$$y(t) = \int\limits_{-\infty}^{\infty} x(\tau)h(t - \tau)\,d\tau = 0$$

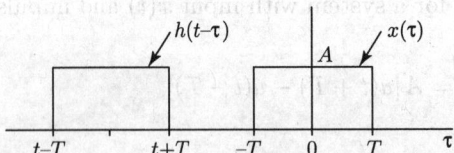

Fig. 2.10 Signal sketch for step 3, case I

Case II: $-T \le (t + T) < T \longrightarrow -2T \le t < 0$. Figure 2.11 shows $x(\tau)$ and $h(t - \tau)$. The two functions overlap only over the interval $-T \le \tau \le (t + T)$ (shaded interval) such that

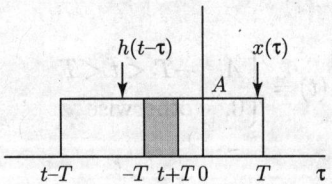

Fig. 2.11 Signal sketch for step 3, case II

$$y(t) = \int\limits_{-T}^{t+T} A \cdot A\,d\tau = A^2(t + 2T), \qquad -2T \le t < 0$$

Case III: $-T \leq t - T \leq T \longrightarrow 0 \leq t \leq 2T$. Figure 2.12 shows $x(\tau)$ and $h(t - \tau)$. The two functions overlap only over the interval $(t - T) \leq \tau \leq T$ (shaded interval) such that

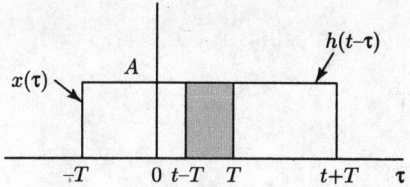

Fig. 2.12 Signal sketch for step 3, case III

$$y(t) = \int_{t-T}^{T} A \cdot A\, d\tau = A^2(2T - t), \quad 0 \leq t \leq 2T$$

Case IV: $t - T > T \longrightarrow t > 2T$. Figure 2.13 shows $x(\tau)$ and $h(t - \tau)$. The two functions do not overlap (Fig. 2.13); hence, the product of the integrand is zero.

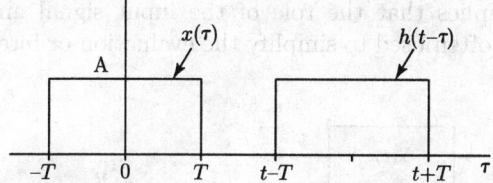

Fig. 2.13 Signal sketch for step 3, case IV

$$y(t) = \int_{-\infty}^{\infty} x(\tau)h(t - \tau)\, d\tau = 0, \quad t > 2T$$

In summary

$$y(t) = 2TA^2 \triangle \left(\frac{t}{2T} \right) = \begin{cases} 0, & t < -2T \\ A^2(t + 2T), & -2T \leq t < 0 \\ A^2(2T - t), & 0 \leq t \leq 2T \\ 0, & t > 2T \end{cases}$$

or, in more compact form

$$y(t) = \begin{cases} A^2(2T - |t|), & |t| \leq 2T \\ 0, & |t| \geq 2T \end{cases}$$

This signal (see Fig. 2.14) is called the triangular signal. We use the notation $\triangle (t/2T)$ to denote the triangular signal that is of unit height, centred around $t = 0$, and has base of length $4T$.

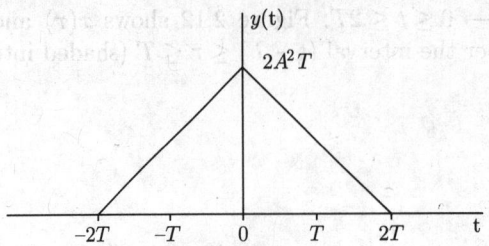

Fig. 2.14 Convolution result

2.3 PROPERTIES OF CONVOLUTION INTEGRAL

Continuous-time convolution satisfies the following properties:

2.3.1 Commutative Property

The convolution operation possesses the commutative property, or

$$x(t) * h(t) = h(t) * x(t) \tag{2.3}$$

This property (see Fig. 2.15), implies that the role of the input signal and impulse response are interchangeable. It is often used to simplify the evaluation or interpretation of the convolution integral.

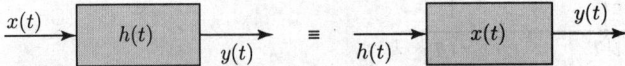

Fig. 2.15

Proof By definition

$$x(t) * h(t) = \int\limits_{-\infty}^{\infty} x(\tau)h(t-\tau)\,d\tau$$

If we perform a change of variables by letting $(t - \tau) = \alpha$, then $\tau = (t - \alpha)$, $d\alpha = -d\tau$, $\alpha \to \infty$ as $\tau \to -\infty$, and $\alpha \to -\infty$ as $\tau \to \infty$. Therefore,

$$x(t) * h(t) = -\int\limits_{\infty}^{-\infty} x(t-\alpha)h(\alpha)\,d\alpha = \int\limits_{-\infty}^{\infty} h(\alpha)x(t-\alpha)\,d\alpha = h(t) * x(t) \qquad \blacksquare$$

2.3.2 Associative Property

The continuous convolution is associative, i.e.,

$$x(t) * [h_1(t) * h_2(t)] = [x(t) * h_1(t)] * h_2(t) \tag{2.4}$$

Associativity, as shown in Fig. 2.16, implies that a cascade combination of LTI systems can be replaced by a single system whose impulse response is the convolution of the individual impulse responses.

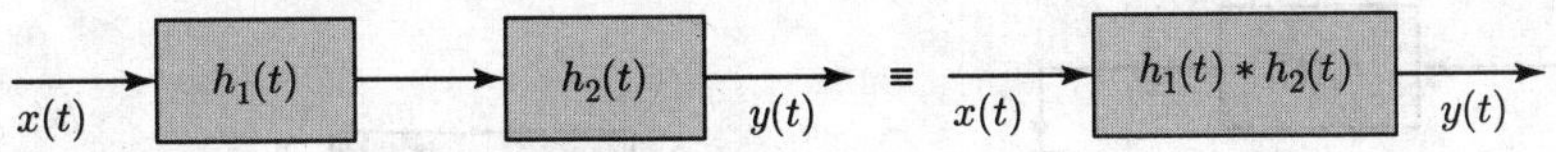

Fig. 2.16

Proof By definition

$$x(t) * [h_1(t) * h_2(t)] = x(t) * \left[\int\limits_{-\infty}^{\infty} h_1(\tau) h_2(t - \tau) \, d\tau \right]$$

$$= \int\limits_{-\infty}^{\infty} x(\alpha) \left[\int\limits_{-\infty}^{\infty} h_1(\tau) h_2(t - \alpha - \tau) \, d\tau \right] d\alpha$$

A change of variables is performed by letting $\tau = \beta - \alpha$, which also yields $d\tau = d\beta$, $\beta \to \infty$ as $\tau \to \infty$, and $\beta \to -\infty$ as $\tau \to -\infty$. Therefore,

$$x(t) * [h_1(t) * h_2(t)] = \int\limits_{-\infty}^{\infty} \int\limits_{-\infty}^{\infty} x(\alpha) h_1(\beta - \alpha) h_2(t - \beta) \, d\beta \, d\alpha$$

Interchanging the order of integration gives the desired result

$$x(t) * [h_1(t) * h_2(t)] = \int\limits_{-\infty}^{\infty} \left[\int\limits_{-\infty}^{\infty} x(\alpha) h_1(\beta - \alpha) \, d\alpha \right] h_2(t - \beta) \, d\beta$$

$$= \int\limits_{-\infty}^{\infty} [x(\beta) * h_1(\beta)] \, h_2(t - \beta) d\beta = [x(t) * h_1(t)] * h_2(t)$$

Since continuous convolution is associative, there is no need to indicate which convolution is to be performed first; hence, the square brackets can be removed from Eq. (2.4). Since continuous convolution is also commutative, then

$$x(t) * h_1(t) * h_2(t) = x(t) * h_2(t) * h_1(t) = \cdots = h_2(t) * h_1(t) * x(t) \qquad \blacksquare$$

2.3.3 Distributive Property

The convolution operation possesses the distributive property, or

$$x(t) * [h_1(t) + h_2(t)] = x(t) * h_1(t) + x(t) * h_2(t) \qquad (2.5)$$

Distributivity, as shown in Fig. 2.17, states that a parallel combination of LTI systems is equivalent to a single system whose impulse response is the sum of the individual impulse responses in the parallel configuration.

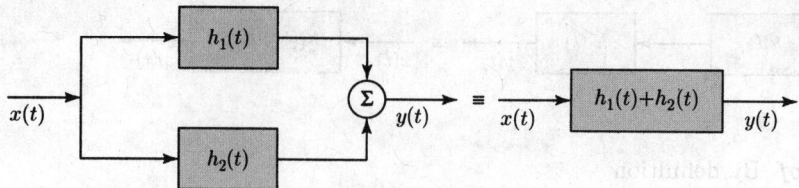

Fig. 2.17

Proof By definition

$$x(t) * [h_1(t) + h_2(t)] = \int_{-\infty}^{\infty} x(\tau)[h_1(t - \tau) + h_2(t - \tau)] \, d\tau$$

$$= \int_{-\infty}^{\infty} x(\tau)h_1(t - \tau) \, d\tau + \int_{-\infty}^{\infty} x(\tau)h_2(t - \tau) \, d\tau$$

$$= [x(t) * h_1(t)] + [x(t) * h_2(t)]$$

2.3.4 Shift Property
According to the shift property,

if $x(t) * h(t) = y(t)$, then $x(t) * h(t - t_0) = x(t - t_0) * h(t) = y(t - t_0)$ (2.6)

and

$$x(t - t_1) * h(t - t_2) = y(t - t_1 - t_2) \qquad (2.7)$$

Proof By definition

$$x(t) * h(t) = \int_{-\infty}^{\infty} x(\tau)h(t - \tau) \, d\tau = y(t)$$

$$x(t) * h(t - t_0) = \int_{-\infty}^{\infty} x(\tau)h(t - t_0 - \tau) \, d\tau = y(t - t_0)$$

Equation (2.7) follows from Eq. (2.6).

2.3.5 Convolution with an Impulse
The convolution of a signal $x(t)$ with a unit impulse function results in the signal $x(t)$ itself

$$x(t) * \delta(t) = x(t) \qquad (2.8)$$

Proof By definition

$$x(t) * \delta(t) = \int_{-\infty}^{\infty} x(\tau)\delta(t - \tau) \, d\tau$$

From the sampling property of the impulse function, we have

$$x(\tau)\delta(t-\tau) = x(t)\delta(t-\tau)$$

and therefore

$$x(t) * \delta(t) = \int\limits_{-\infty}^{\infty} x(t)\delta(t-\tau)\,d\tau = x(t)\int\limits_{-\infty}^{\infty}\delta(t-\tau)\,d\tau = x(t)$$

■

2.3.6 Width Property

The width of the nonzero extent (the interval of time between the first and last nonzero values) of the continuous convolution of two functions equals the sum of the widths of the nonzero extents of the two functions. In other words, if the durations (widths) of $x(t)$ and $h(t)$ are finite, given by W_x and W_h, respectively, then the duration (width) of $x(t) * h(t)$ is $W_x + W_h$.

Example 2.5 Two continuous-time signals $x(t)$ and $h(t)$ are shown in Fig. 2.18. Where will the right and left edges of the convolution of $x(t)$ and $h(t)$ be located? What will be the width of the nonzero extent of the convolution?

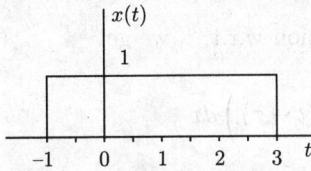

Fig. 2.18 Signal $x(t)$ and $h(t)$

Solution

Right edge of $x(t) = \text{RE}_x = 3$

Right edge of $h(t) = \text{RE}_h = 4$

Left edge of $x(t) = \text{LE}_x = -1$

Left edge of $h(t) = \text{LE}_h = 0$

Duration (width) of $x(t) = W_x = \text{RE}_x - \text{LE}_x = 3 - (-1) = 4$

Duration (width) of $h(t) = W_h = \text{RE}_h - \text{LE}_h = 4 - 0 = 4$

$y(t) = x(t) * h(t)$

Right edge of $y(t) = \text{RE}_y = \text{RE}_x + \text{RE}_h = 3 + 4 = 7$

Left edge of $y(t) = \text{LE}_y = \text{LE}_x + \text{LE}_h = (-1) + 0 = -1$

Duration (width) of $y(t) = W_x + W_h = 4 + 4 = 8$.

The result of the convolution of $x(t)$ and $h(t)$ is shown in Fig. 2.19. One can verify that it is correct.

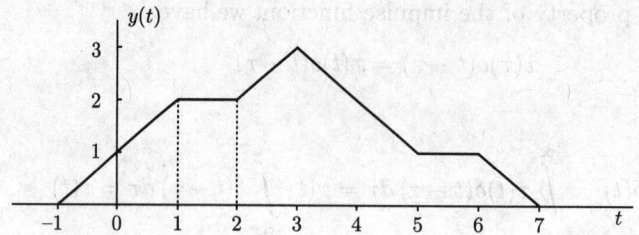

Fig. 2.19 Convolution result

2.3.7 Differentiation Property

According to the differentiation property,

$$\text{if}\quad x(t) * h(t) = y(t), \quad \text{then}\quad \left(\frac{d}{dt}x(t)\right) * h(t) = x(t) * \left(\frac{d}{dt}h(t)\right) = \frac{d}{dt}y(t) \qquad (2.9)$$

Proof By definition

$$y(t) = x(t) * h(t) = \int\limits_{-\infty}^{\infty} x(\tau)h(t-\tau)\,d\tau$$

By differentiating both sides of the above equation w.r.t. t, we get

$$\frac{d}{dt}y(t) = \int\limits_{-\infty}^{\infty} x(\tau)\left(\frac{d}{dt}h(t-\tau)\right)d\tau$$

$$= x(t) * \left(\frac{d}{dt}h(t)\right)$$

∎

2.3.8 Time-scaling Property

According to the time-scaling property,

$$\text{if}\quad x(t) * h(t) = y(t), \quad \text{then}\quad x(at) * h(at) = \frac{1}{|a|}y(at) \qquad (2.10)$$

This property states that if both $x(t)$ and $h(t)$ are time-scaled by a, their convolution is also time-scaled by a and multiplied by $1/|a|$.

Proof **Case I:** $a > 0$. By definition

$$x(t) * h(t) = \int\limits_{-\infty}^{\infty} x(\tau)h(t-\tau)\,d\tau = y(t)$$

$$x(at) * h(at) = \int\limits_{-\infty}^{\infty} x(a\tau)h(a(t-\tau))\,d\tau = \int\limits_{-\infty}^{\infty} x(a\tau)h(at-a\tau)\,d\tau$$

A change of variables is performed by letting $a\tau = \alpha$, which also yields $d\tau = (1/a)d\alpha$, $\alpha \to \infty$ as $\tau \to \infty$, and $\alpha \to -\infty$ as $\tau \to -\infty$. Therefore,

$$x(at) * h(at) = \frac{1}{a} \int_{-\infty}^{\infty} x(\alpha)h(at - \alpha)\, d\alpha = \frac{1}{a}y(at)$$

Case II: $a < 0$. By definition

$$x(t) * h(t) = \int_{-\infty}^{\infty} x(\tau)h(t - \tau)\, d\tau = y(t)$$

$$x(-at) * h(-at) = \int_{-\infty}^{\infty} x(-a\tau)h(-a(t - \tau))\, d\tau = \int_{-\infty}^{\infty} x(-a\tau)h(-at + a\tau)\, d\tau$$

A change of variables is performed by letting $-a\tau = \alpha$, which also yields $d\tau = -(1/a)d\alpha$, $\alpha \to \infty$ as $\tau \to -\infty$, and $\alpha \to -\infty$ as $\tau \to \infty$. Therefore,

$$x(at) * h(at) = \frac{1}{a} \int_{-\infty}^{\infty} x(\alpha)h(at - \alpha)\, d\alpha = \frac{1}{a}y(at)$$

From the above two cases it is evident that

$$x(at) * h(at) = \frac{1}{|a|}y(at)$$ ■

Example 2.6 Show that
(a) the convolution of an odd and an even function is an odd function.
(b) the convolution of two odd functions is an even function.
(c) the convolution of two even functions is an even function.

Solution
(a) Let $x(t)$ and $h(t)$ be an odd and even function, respectively, i.e., $x(t) = -x(-t)$ and $h(t) = h(-t)$. Using time-scaling property, we have

$$x(t) * h(t) = y(t)$$

$$x(at) * h(at) = \frac{1}{|a|}y(at)$$

Substituting $a = -1$ into the above equation, we get

$$x(-t) * h(-t) = y(-t)$$

$$-x(t) * h(t) = y(-t)$$

$$x(t) * h(t) = -y(-t) = y(t) \quad \text{an odd function}$$

(b) Let both $x(t)$ and $h(t)$ be odd functions, i.e., $x(t) = -x(-t)$ and $h(t) = -h(-t)$. Using time-scaling property, we have

$$x(t) * h(t) = y(t)$$

$$x(at) * h(at) = \frac{1}{|a|} y(at)$$

Substituting $a = -1$ into the above equation, we get

$$x(-t) * h(-t) = y(-t)$$

$$-x(t) * -h(t) = y(-t)$$

$$x(t) * h(t) = y(-t) = y(t) \quad \text{an even function}$$

(c) Let both $x(t)$ and $h(t)$ be even functions, i.e., $x(t) = x(-t)$ and $h(t) = h(-t)$. Using time-scaling property, we have

$$x(t) * h(t) = y(t)$$

$$x(at) * h(at) = \frac{1}{|a|} y(at)$$

Substituting $a = -1$ into the above equation, we get

$$x(-t) * h(-t) = y(-t)$$

$$x(t) * h(t) = y(-t) = y(t) \quad \text{an even function}$$

Example 2.7 The convolution has the property that the area of the convolution integral is equal to the product of the two signals entering into the convolution. We define the area under a continuous-time signal $y(t)$ as

$$A_y = \int_{-\infty}^{\infty} y(t)\, dt$$

Show that if $y(t) = x(t) * h(t)$, then $A_y = A_x A_h$

Solution
By definition

$$y(t) = x(t) * h(t)$$

$$y(t) = \int_{-\infty}^{\infty} x(\tau) h(t - \tau)\, d\tau$$

$$\int_{-\infty}^{\infty} y(t)\, dt = \int_{-\infty}^{\infty} \left[\int_{-\infty}^{\infty} x(\tau) h(t - \tau)\, d\tau \right] dt$$

Interchanging the orders of the integration results in

$$\int_{-\infty}^{\infty} y(t)\,dt = \int_{-\infty}^{\infty} x(\tau) \underbrace{\left[\int_{-\infty}^{\infty} h(t-\tau)dt\right]}_{A_h} d\tau$$

$$A_y = \int_{-\infty}^{\infty} x(\tau) A_h\,d\tau = \int_{-\infty}^{\infty} x(\tau)d\tau\; A_h = A_x A_h$$

Example 2.8 Prove the following convolution integrals:

(a) $x(t) * \delta(t - t_0) = x(t - t_0)$

(b) $x(t) * \delta(t + t_0) = x(t + t_0)$

(c) $x(t) * u(t) = \displaystyle\int_{-\infty}^{t} x(\tau)d\tau$

(d) $x(t) * u_1(t) = x(t) * \dfrac{d\delta(t)}{dt} = \dfrac{dx(t)}{dt}$

(e) $u(t) * u(t) = tu(t) = r(t)$

Solution

(a) By definition

$$x(t) * \delta(t - t_0) = \int_{-\infty}^{\infty} x(\tau)\delta(t - \tau - t_0)\,d\tau$$

From the sampling property of the impulse function, we have

$$x(\tau)\delta(t - \tau - t_0) = x(t - t_0)\delta(t - \tau - t_0)$$

and therefore

$$x(t) * \delta(t - t_0) = \int_{-\infty}^{\infty} x(t - t_0)\delta(t - \tau - t_0)\,d\tau = x(t - t_0) \underbrace{\int_{-\infty}^{\infty} \delta(t - \tau - t_0)\,d\tau}_{=1} = x(t - t_0)$$

(b) By definition

$$x(t) * \delta(t + t_0) = \int_{-\infty}^{\infty} x(\tau)\delta(t - \tau + t_0)\,d\tau$$

From the sampling property of the impulse function, we have

$$x(\tau)\delta(t - \tau + t_0) = x(t + t_0)\delta(t - \tau + t_0)$$

and therefore

$$x(t) * \delta(t + t_0) = \int_{-\infty}^{\infty} x(t + t_0)\delta(t - \tau + t_0)\,d\tau = x(t + t_0) \underbrace{\int_{-\infty}^{\infty} \delta(t - \tau + t_0)\,d\tau}_{=1} = x(t + t_0)$$

(c) By definition

$$x(t) * u(t) = \int_{-\infty}^{\infty} x(\tau) u(t - \tau) \, d\tau$$

Since

$$u(t - \tau) = \begin{cases} 1, & t - \tau > 0 \longrightarrow \tau < t \\ 0, & \text{otherwise} \end{cases}$$

Therefore, we have

$$x(t) * u(t) = \int_{-\infty}^{t} x(\tau) \, d\tau$$

(d) By definition

$$x(t) * u_1(t) = x(t) * \frac{d\delta(t)}{dt} = x(t) * \delta'(t) = \int_{-\infty}^{\infty} x(\tau) \delta'(t - \tau) \, d\tau$$

$$= -\int_{-\infty}^{\infty} x'(\tau)[-\delta(t - \tau)] \, d\tau$$

$$= \int_{-\infty}^{\infty} x'(\tau) \delta(t - \tau) \, d\tau$$

$$= x'(\tau)\Big|_{\tau=t} = x'(t)$$

(e) By definition

$$u(t) * u(t) = \int_{-\infty}^{\infty} u(\tau) u(t - \tau) \, d\tau$$

Since

$$u(t - \tau) = \begin{cases} 1, & t - \tau > 0 \longrightarrow \tau < t \\ 0, & \text{otherwise} \end{cases}$$

Therefore, we have

$$u(t) * u(t) = \int_{-\infty}^{t} u(\tau) \, d\tau = \int_{0}^{t} (1) \, d\tau = \begin{cases} t, & t \geq 0 \\ 0, & t < 0 \end{cases} = r(t)$$

Example 2.9 Determine and sketch the convolution of the following two signals:

$$x(t) = \begin{cases} 1, & -1 < t < 1 \\ 0, & \text{elsewhere} \end{cases} \quad \text{and} \quad h(t) = \delta(t+1) + 2\delta(t+2)$$

Solution
Using the result of part (b) of Example 2.8, we have

$$y(t) = x(t) * h(t) = x(t) * [\delta(t+1) + 2\delta(t+2)]$$
$$= [x(t) * \delta(t+1)] + [x(t) * 2\delta(t+2)]$$
$$= x(t+1) + 2x(t+2)$$

The signals $x(t)$, $h(t)$, and $y(t)$ are shown in Fig. 2.20.

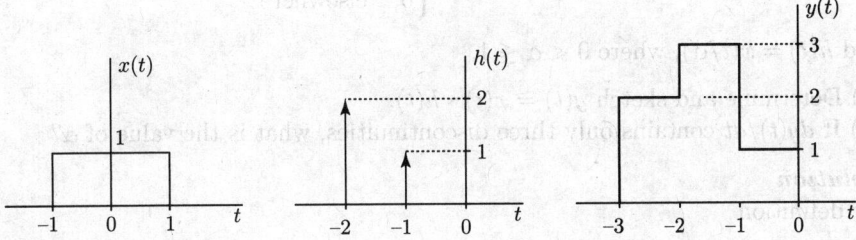

Fig. 2.20

Example 2.10 Determine and sketch the convolution of the following two signals:

$$x(t) = \begin{cases} t+1, & 0 \le t \le 1 \\ 2-t, & 1 < t \le 2 \\ 0, & \text{elsewhere} \end{cases} \quad \text{and} \quad h(t) = \delta(t+2) + 2\delta(t+1)$$

Solution
Using the result of part (b) of Example 2.8, we have

$$y(t) = x(t) * h(t) = x(t) * [\delta(t+2) + 2\delta(t+1)]$$
$$= x(t) * \delta(t+2) + x(t) * 2\delta(t+1)$$
$$y(t) = x(t+2) + 2x(t+1)$$

The signals $x(t)$, $h(t)$, and $y(t)$ are shown in Fig. 2.21. Using these plots, we can easily show that

$$y(t) = \begin{cases} t+3, & -2 < t \le 1 \\ t+4, & -1 < t \le 0 \\ 2-2t, & 0 < t \le 1 \\ 0, & \text{otherwise} \end{cases}$$

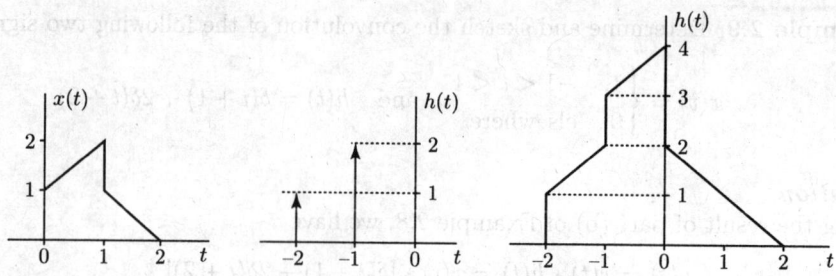

Fig. 2.21

Example 2.11 Suppose that

$$x(t) = \begin{cases} 1, & 0 \le t \le 1 \\ 0, & \text{elsewhere} \end{cases}$$

and $h(t) = x(t/\alpha)$, where $0 < \alpha \le 1$.

(a) Determine and sketch $y(t) = x(t) * h(t)$.
(b) If $dy(t)/dt$ contains only three discontinuities, what is the value of α?

Solution
By definition

$$y(t) = x(t) * h(t) = \int_{-\infty}^{\infty} x(\tau)h(t - \tau)\, d\tau$$

Step 1: Write the equations and draw sketches for $x(t)$ and $h(t)$ as shown in Fig. 2.22.

$$x(t) = \begin{cases} 1, & 0 \le t \le 1 \\ 0, & \text{elsewhere} \end{cases} \qquad h(t) = x\left(\frac{t}{\alpha}\right) = \begin{cases} 1, & 0 \le t/\alpha \le 1 \to 0 \le t \le \alpha \\ 0, & \text{elsewhere} \end{cases}$$

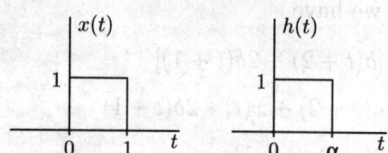

Fig. 2.22   Step 1 for Example 2.11

Step 2: Write the equations and draw sketches for $x(\tau)$ and $h(t - \tau)$ [apply precedence rule: $h(t) \to h(\tau) \to h(\tau + t) \to h(-\tau + t) = h(t - \tau)$] as shown in Fig. 2.23.

$$x(\tau) = \begin{cases} 1, & 0 \le \tau \le 1 \\ 0, & \text{elsewhere} \end{cases} \qquad h(t - \tau) = \begin{cases} 1, & 0 \le (t - \tau) \le \alpha \to (t - \alpha) \le \tau \le t \\ 0, & \text{elsewhere} \end{cases}$$

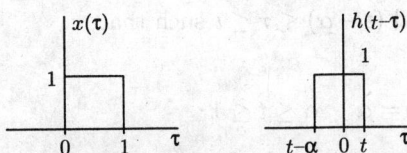

Fig. 2.23   Step 2 for Example 2.11

Step 3:   Perform integration to find $y(t)$ for all values of t.
Case I: $t < 0$. Figure 2.24 shows $x(\tau)$ and $h(t - \tau)$.

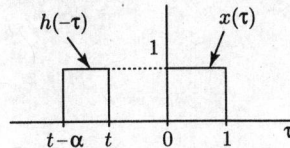

Fig. 2.24   Signal sketch for step 3, case I

We see that the signals do not overlap (Fig. 2.24); hence, the product of the integrand is zero. Thus,

$$y(t) = \int_{-\infty}^{\infty} x(\tau)h(t - \tau)\,d\tau = 0$$

Case II: $t \geq 0$ and $(t - \alpha) < 0 \longrightarrow 0 \leq t < \alpha$. Figure 2.25 shows $x(\tau)$ and $h(t - \tau)$.

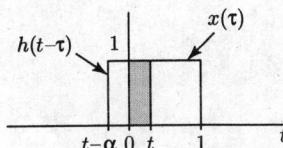

Fig. 2.25   Signal sketch for step 3, case II

The two signals overlap over the interval $0 \leq \tau \leq t$ (shaded interval) such that

$$y(t) = \int_{0}^{t} 1\,d\tau = t, \qquad 0 \leq t < \alpha$$

Case III: $(t - \alpha) \geq 0$ and $t \leq 1 \longrightarrow \alpha \leq t \leq 1$. Figure 2.26 shows $x(\tau)$ and $h(t - \tau)$.

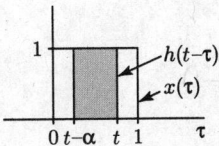

Fig. 2.26   Signal sketch for step 3, case III

The two signals overlap over the interval $(t - \alpha) \leq \tau \leq t$ such that

$$y(t) = \int_{t-\alpha}^{t} 1 \, d\tau = \alpha, \quad \alpha \leq t \leq 1$$

Case IV: $t - \alpha \leq 1$ and $t > 1 \longrightarrow 1 < t \leq (\alpha + 1)$. Figure 2.27 shows $x(\tau)$ and $h(t - \tau)$.

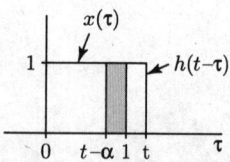

Fig. 2.27 Signal sketch for step 3, case IV

The two signals overlap over the interval $(t - \alpha) \leq \tau \leq 1$ such that

$$y(t) = \int_{t-\alpha}^{1} 1 \, d\tau = 1 - t + \alpha, \quad 1 < t \leq (\alpha + 1)$$

Case V: $(t - \alpha) > 1 \longrightarrow t > (\alpha + 1)$. Figure 2.28 shows $x(\tau)$ and $h(t - \tau)$.

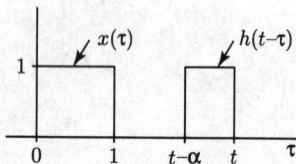

Fig. 2.28 Signal sketch for step 3, case V

We see that the signals do not overlap (Fig. 2.28); hence, the product of the integrand is zero. Thus,

$$y(t) = 0, \quad t > (\alpha + 1)$$

In summary

$$y(t) = \begin{cases} 0, & t < 0 \\ t, & 0 \leq t < \alpha \\ \alpha, & \alpha \leq t < 1 \\ 1 - t + \alpha, & 1 \leq t \leq (\alpha + 1) \\ 0, & t > (\alpha + 1) \end{cases}$$

The signal $y(t)$ is shown in Fig. 2.29.

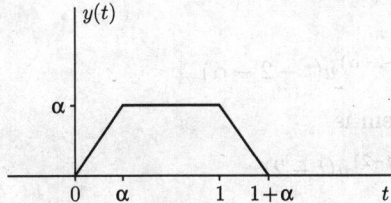

Fig. 2.29 Convolution result

(b) The derivative $(dy(t)/dt)$ of $y(t)$ is shown in Fig. 2.30. From the plot of $(dy(t)/dt)$ it is clear that it has discontinuities at 0, α, 1, and $(1+\alpha)$. If we want only three discontinuities, then we need to ensure that $\alpha = 1$.

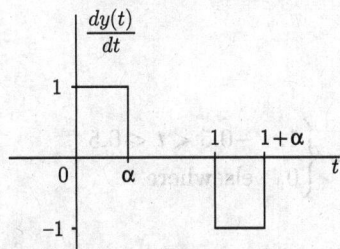

Fig. 2.30 Derivative of $y(t)$

Example 2.12 Consider an LTI system with input and output related through the following equation:

$$y(t) = \int\limits_{-\infty}^{t} e^{-(t-\tau)} x(\tau - 2)\, d\tau$$

What is the impulse response $h(t)$ for this system?

Solution
Given that

$$y(t) = \int\limits_{-\infty}^{t} e^{-(t-\tau)} x(\tau - 2)\, d\tau$$

A change of variables is performed by letting $\alpha = (\tau - 2)$, which also yields $d\alpha = d\tau$, $\alpha \to -\infty$ as $\tau \to -\infty$, and $\alpha \to (t-2)$ as $\tau \to t$. Therefore,

$$y(t) = \int\limits_{-\infty}^{t-2} e^{-(t-2-\alpha)} x(\alpha)\, d\alpha = \int\limits_{-\infty}^{\infty} e^{-(t-2-\alpha)} u(t-2-\alpha) x(\alpha)\, d\alpha$$

Comparing with

$$y(t) = \int\limits_{-\infty}^{\infty} h(t-\alpha) x(\alpha)\, d\alpha$$

we get

$$h(t - \alpha) = e^{-(t-2-\alpha)} u(t - 2 - \alpha)$$

Therefore, the impulse response of the system is

$$h(t) = e^{-(t-2)} u(t - 2)$$

Example 2.13 Suppose that the signal $x(t) = u(t + 0.5) - u(t - 0.5)$ and the signal $h(t) = e^{j\omega_0 t}$.

(a) Determine a value of ω_0 which ensures that $y(0) = 0$, where $y(t) = x(t) * h(t)$.
(b) Is your answer to the previous part unique?

Solution
Given that

$$x(t) = u(t + 0.5) - u(t - 0.5) = \begin{cases} 1, & -0.5 < t < 0.5 \\ 0, & \text{elsewhere} \end{cases}$$

By definition

$$y(t) = x(t) * h(t) = \int_{-\infty}^{\infty} x(\tau) h(t - \tau) \, d\tau = \int_{-0.5}^{0.5} e^{j\omega_0(t-\tau)} \, d\tau$$

Therefore,

$$y(t)\Big|_{t=0} = y(0) = \int_{-0.5}^{0.5} e^{-j\omega_0 \tau} \, d\tau = \frac{2}{\omega_0} \sin\left(\frac{\omega_0}{2}\right)$$

(a) If $\omega_0 = 2\pi$, then $y(0) = 0$.
(b) Our answer to part (a) is not unique. For any $\omega_0 = 2n\pi$, $n = \pm 1, \pm 2, \ldots$, $y(0) = 0$.

Example 2.14 Consider an LTI system S and a signal $x(t) = 2e^{-3t} u(t - 1)$. If

$$x(t) \longrightarrow y(t) \quad \text{and} \quad \frac{dx(t)}{dt} \longrightarrow -3y(t) + e^{-2t} u(t)$$

then determine the impulse response $h(t)$ of S.

Solution
Given that

$$x(t) = 2e^{-3t} u(t - 1)$$

By differentiating both sides of the above equation w.r.t. t, we get

$$\frac{dx(t)}{dt} = -6e^{-3t} u(t - 1) + 2e^{-3t} \delta(t - 1) = -3x(t) + 2e^{-3} \delta(t - 1)$$

The given system S is an LTI system. Therefore,

$$x(t) \longrightarrow y(t)$$

$$\delta(t) \longrightarrow h(t)$$

$$2\delta(t-1) \longrightarrow 2h(t-1)$$

$$\frac{dx(t)}{dt} = -3x(t) + 2e^{-3}\delta(t-1) \longrightarrow -3y(t) + 2e^{-3}h(t-1)$$

From the given information, we have

$$-3y(t) + 2e^{-3}h(t-1) = -3y(t) + e^{-2t}u(t)$$

$$2e^{-3}h(t-1) = e^{-2t}u(t)$$

$$h(t-1) = \frac{1}{2}e^3 e^{-2t}u(t)$$

$$h(t) = \frac{1}{2}e^3 e^{-2(t+1)}u(t+1)$$

$$= \frac{1}{2}e\, e^{-2t}u(t+1)$$

$$h(t) = \frac{1}{2}e^{-2t+1}u(t+1)$$

2.4 DISCRETE-TIME LTI SYSTEMS: THE CONVOLUTION SUM

2.4.1 Convolution Sum

In this section we follow a procedure parallel to that used in the continuous-time case by expressing an arbitrary input $x(n)$ as a sum of impulse components. In Eq. (1.26) we demonstrated the product of $x(n)$ and a time-shifted impulse sequence, i.e., $x(n)\delta(n-k) = x(k)\delta(n-k)$. The component of $x(n)$ at $n=k$ is $x(k)\delta(n-k)$. We see that the multiplication of a signal by a time-shifted impulse results in a time-shifted impulse with amplitude given by the value of the signal at that time the impulse occurs. This property allows us to express $x(n)$ as the following weighted sum of time-shifted impulses:

$$x(n) = \cdots + x(-2)\delta(n+2) + x(-1)\delta(n+1) + x(0)\delta(n) + x(1)\delta(n-1)$$

$$+ x(2)\delta(n-2) + \cdots$$

$$x(n) = \sum_{k=-\infty}^{\infty} x(k)\delta(n-k) \tag{2.11}$$

A graphical representation of Eq. (2.11) is given in Fig. 2.31.

For a linear system, knowing the system response to impulse $\delta(n)$, the system response to any arbitrary input could be obtained by summing the system response to various

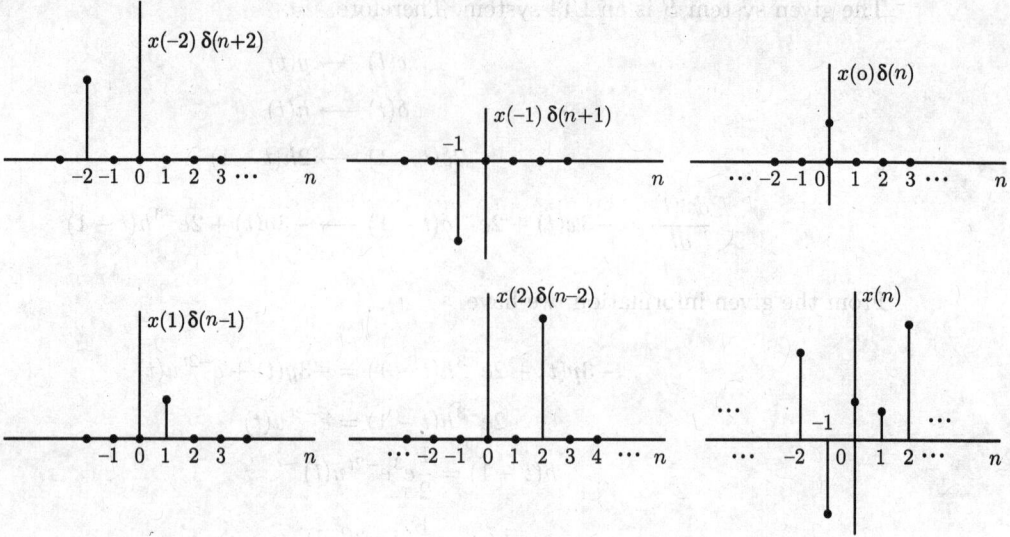

Fig. 2.31 Decomposition of a discrete-time signal into a weighted sum of shifted impulses

impulse components. Let $h(n)$ be the system response to impulse input $\delta(n)$. We shall use the notation

$$x(n) \rightarrow y(n)$$

to indicate the input and the corresponding response of the system. Using the superposition and time-invariant property of LTI systems, we can express output $y(t)$ as a linear combination of the responses of the system to shifted impulse signals.

$$\delta(n) \longrightarrow h(n)$$
$$\delta(n-k) \longrightarrow h(n-k)$$
$$x(k)\delta(n-k) \longrightarrow x(k)h(n-k)$$
$$x(n) = \sum_{k=-\infty}^{\infty} x(k)\delta(n-k) \longrightarrow y(n) = \sum_{k=-\infty}^{\infty} x(k)h(n-k)$$

Thus, we have obtained the system output $y(n)$ to an arbitrary input $x(n)$ in terms of the unit impulse response $h(n)$:

$$y(n) = \sum_{k=-\infty}^{\infty} x(k)h(n-k) \tag{2.12}$$

The relationship expressed in Eq. (2.12) is called the convolution sum of sequences $x(n)$ and $h(n)$. This operation is represented symbolically as

$$y(n) = x(n) * h(n) \tag{2.13}$$

Discrete-convolution (convolution sum) Evaluation

The following examples serve to illustrate techniques that can be used to evaluate discrete convolution. The first example uses two sequences, each defined by a single equation for all values of n.

Example 2.15 Find the convolution of the two sequences

$$x(n) = e^{-n^2}, \text{ for all } n \text{ and } h(n) = 3n^2, \text{ for all } n$$

Solution
By definition

$$y(n) = x(n) * h(n) = \sum_{k=-\infty}^{\infty} x(k)h(n-k)$$

$$= \sum_{k=-\infty}^{\infty} e^{-k^2} \left[3(n-k)^2\right]$$

$$= \sum_{k=-\infty}^{\infty} e^{-k^2} \left[3(n^2 - 2nk + k^2)\right]$$

$$y(n) = 3n^2 \sum_{k=-\infty}^{\infty} e^{-k^2} - 6n \sum_{k=-\infty}^{\infty} k\, e^{-k^2} + 3 \sum_{k=-\infty}^{\infty} k^2\, e^{-k^2}$$

$$= 3n^2 \sum_{k=-\infty}^{\infty} e^{-k^2} - 0 + 3 \sum_{k=-\infty}^{\infty} k^2\, e^{-k^2}$$

$$= 5.318n^2 + 2.654, \quad \text{for all } n$$

Example 2.16 Find the convolution of the two sequences

$$x(n) = e^{-n^2}, \text{ for all } n \text{ and } h(n) = \begin{cases} e^{n}, & n < 0 \\ e^{-n}, & n \geq 0 \end{cases}$$

Solution
By definition

$$y(n) = x(n) * h(n) = \sum_{k=-\infty}^{\infty} x(k)h(n-k)$$

From the given $x(n)$ and $h(n)$, we get

$$x(k) = e^{-k^2}, \text{ for all } k \text{ and } h(n-k) = \begin{cases} e^{n-k}, & (n-k) < 0 \longrightarrow k > n \\ e^{-(n-k)}, & (n-k) \geq 0 \longrightarrow k \leq n \end{cases}$$

Therefore,

$$y(n) = \sum_{k=-\infty}^{n} e^{-k^2} e^{-(n-k)} + \sum_{k=n+1}^{\infty} e^{-k^2} e^{(n-k)} = e^{-n} \sum_{k=-\infty}^{n} e^{(k-k^2)} + e^{n} \sum_{k=n+1}^{\infty} e^{-(k+k^2)}$$

In the previous two examples the need to compute infinite sums creates a difficulty in obtaining solutions for discrete convolutions. The infinite sums do not occur if sequences have zero values prior to some starting time. We show that if

$$x(n) = 0, \ n < k_1 \quad \text{and} \quad h(n) = 0, \ n < k_2$$

then

$$x(k) = 0, \ k < k_1 \quad \text{and} \quad h(n-k) = 0, \ (n-k) < k_2 \longrightarrow k > (n-k_2)$$

Therefore,

$$y(n) = x(n) * h(n) = \begin{cases} \sum_{k=k_1}^{n-k_2} x(k)h(n-k), & n \geq k_1 + k_2 \\ 0, & n < k_1 + k_2 \end{cases}$$

which is a finite sum.

Example 2.17 Find the convolution of the two sequences

$$x(n) = \begin{cases} 0, & n < -5 \\ \left(\dfrac{1}{2}\right)^n, & n \geq -5 \end{cases} \quad \text{and} \quad h(n) = \begin{cases} 0, & n < 3 \\ \left(\dfrac{1}{3}\right)^n, & n \geq 3 \end{cases}$$

Solution
By definition

$$y(n) = x(n) * h(n) = \sum_{k=k_1}^{n-k_2} x(k)h(n-k)$$

where $k_1 = -5$ and $k_2 = 3$. Therefore,

$$y(n) = 0, \quad n < (k_1 + k_2) \longrightarrow n < (-5+3) = -2$$

and

$$y(n) = \sum_{k=k_1}^{n-k_2} x(k)h(n-k), \qquad n \geq k_1 + k_2$$

$$= \sum_{k=-5}^{n-3} \left(\frac{1}{2}\right)^k \left(\frac{1}{3}\right)^{(n-k)}, \qquad n \geq (-5+3) = -2$$

A change of variables is performed by letting $m = k + 5$, which also yields $m \to 0$ as $k \to -5$ and $m \to (n+2)$ as $k \to (n-3)$. Therefore,

$$y(n) = \sum_{m=0}^{n+2} \left(\frac{1}{2}\right)^{(m-5)} \left(\frac{1}{3}\right)^{(n-m+5)} = \left(\frac{1}{3}\right)^n \left(\frac{2}{3}\right)^5 \sum_{m=0}^{n+2} \left(\frac{3}{2}\right)^m = \left(\frac{1}{3}\right)^n \left(\frac{2}{3}\right)^5 \frac{1 - (3/2)^{n+3}}{1 - 3/2}$$

$$= -2 \left(\frac{2}{3}\right)^5 \left(\frac{1}{3}\right)^n \left[1 - \left(\frac{3}{2}\right)^{n+3}\right] = -2 \left(\frac{2}{3}\right)^5 \left[\left(\frac{1}{3}\right)^n - \left(\frac{1}{2}\right)^n \left(\frac{3}{2}\right)^3\right]$$

$$= -64 \left(\frac{1}{3}\right)^{n+5} + 2 \left(\frac{2}{3}\right)^2 \left(\frac{1}{2}\right)^n$$

Graphical Method for the Convolution Sum

The convolution sum of two signals $x(n)$ and $h(n)$ is given by

$$y(n) = \sum_{k=-\infty}^{\infty} x(k) h(n-k)$$

The convolution operation is performed as follows:

1. We first plot $x(k)$ and $h(k)$ as functions of k (not n) because summation is over k.

2. *Shifting* Shift $h(k)$ by n units to obtain $h(k+n)$.

3. *Folding* Invert $h(k+n)$ about the vertical axis $(k=0)$ to obtain $h(-k+n) = h(n-k)$.

4. *Multiplication and addition* Next we multiply $x(k)$ and $h(n-k)$ and add all the products to obtain $y(n)$. The procedure is repeated for each value of n over the range $-\infty$ to ∞.

We shall demonstrate by an example the graphical procedure for finding the convolution sum.

Example 2.18 Find the convolution sum of a rectangle signal (or gate function) with itself.

$$x(n) = \text{rect}\left(\frac{n}{2N}\right) = \Pi\left(\frac{n}{2N}\right) = \begin{cases} 1, & -N \le n \le N \\ 0, & \text{otherwise} \end{cases}$$

Or, evaluate the convolution sum for a system with input $x(n)$ and impulse response $h(n)$, respectively, given by

$$x(n) = h(n) = [u(n+N) - u(n-N-1)]$$

Solution
By definition

$$y(n) = x(n) * h(n) = \sum_{k=-\infty}^{\infty} x(k) h(n-k)$$

Step 1: Write the equations and draw sketches for $x(n)$ and $h(n)$ as shown in Fig. 2.32.

$$x(n) = \begin{cases} 1, & -N \leq n \leq N \\ 0, & \text{otherwise} \end{cases} \qquad h(n) = \begin{cases} 1, & -N \leq n \leq N \\ 0, & \text{otherwise} \end{cases}$$

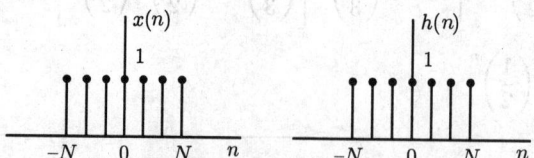

Fig. 2.32   Step 1 for Example 2.18

Step 2: Write the equations and draw sketches for $x(k)$ and $h(n-k)$ [apply precedence rule: $h(n) \rightarrow h(k) \rightarrow h(k+n) \rightarrow h(-k+n) = h(n-k)$] as shown in Fig. 2.33.

$$x(k) = \begin{cases} 1, & -N \leq k \leq N \\ 0, & \text{otherwise} \end{cases} \qquad h(n-k) = \begin{cases} 1, & -N \leq (n-k) \leq N \longrightarrow (n-N) \leq k \leq (n+N) \\ 0, & \text{otherwise} \end{cases}$$

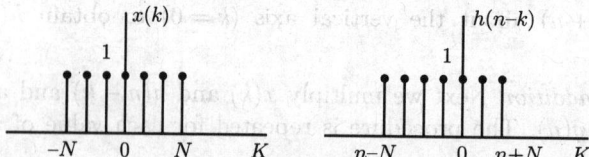

Fig. 2.33   Step 2 for Example 2.18

Step 3: Perform summation to find $y(n)$ for all values of n.

Case I: $(n+N) < -N \longrightarrow n < -2N$. Figure 2.34 shows $x(k)$ and $h(n-k)$. We see that the signals do not overlap (Fig. 2.34); hence, the product of the summand is zero.

$$y(n) = \sum_{k=-\infty}^{\infty} x(k)h(n-k) = 0$$

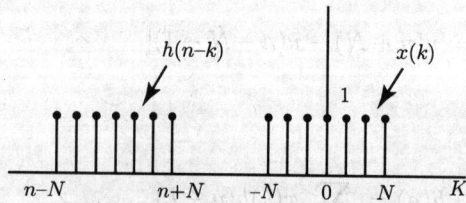

Fig. 2.34   Signal sketch for step 3, case-I

Case II: $-N \le (n+N) < N \longrightarrow -2N \le n < 0$. Figure 2.35 shows $x(k)$ and $h(n-k)$. The two signals overlap over the interval $-N \le k \le (n+N)$ such that

$$y(n) = \sum_{k=-N}^{n+N} 1 = (n+N) - (-N) + 1 = n + 2N + 1, \qquad -2N \le n < 0$$

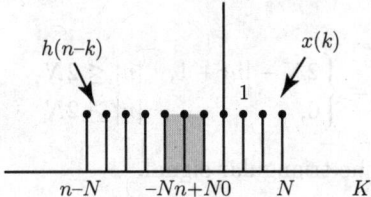

Fig. 2.35 Signal sketch for step 3, case II

Case III: $-N \le (n-N) \le N \longrightarrow 0 \le n \le 2N$. Figure 2.36 shows $x(k)$ and $h(n-k)$. The two signals overlap over the interval $(n-N) \le k \le N$, such that

$$y(n) = \sum_{k=n-N}^{N} 1 = (N) - (n-N) + 1 = 2N - n + 1, \qquad 0 \le n \le 2N$$

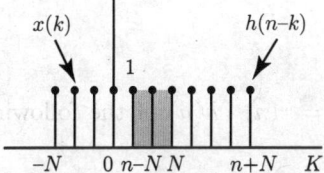

Fig. 2.36 Signal sketch for step 3, case III

Case IV: $(n-N) > N \longrightarrow n > 2N$. Figure 2.37 shows $x(k)$ and $h(n-k)$.

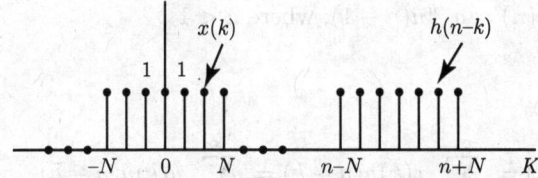

Fig. 2.37 Signal sketch for step 3, case IV

We see that the signals do not overlap (Fig. 2.37); hence, the product of the summand is zero.

$$y(n) = \sum_{k=-\infty}^{\infty} x(k)h(n-k) = 0, \qquad n > 2N$$

In summary

$$y(n) = \begin{cases} 0, & n < -2N \\ 2N + n + 1, & -2N \le n < 0 \\ 2N - n + 1, & 0 \le n \le 2N \\ 0, & n > 2N \end{cases}$$

or, in more compact form

$$y(n) = (2N + 1)\mathrm{tri}\left(\frac{n}{2N + 1}\right) = \begin{cases} 2N - |n| + 1, & |n| \le 2N \\ 0, & |n| \ge 2N \end{cases}$$

This signal (as shown in Fig. 2.38) is called the triangular signal.

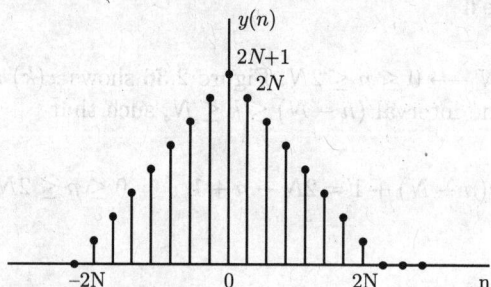

Fig. 2.38   Convolution result

Example 2.19 Compute the convolution $y(n) = x(n) * h(n)$ of the following pairs of signals:

(a) $x(n) = h(n) = u(n)$
(b) $x(n) = (0.8)^n u(n)$ and $h(n) = (0.4)^n u(n)$
(c) $x(n) = h(n) = a^n u(n)$
(d) $x(n) = u(n - 1)$ and $h(n) = \alpha^n u(n - 1)$
(e) $x(n) = r(n) = nu(n)$ and $h(n) = a^{-n} u(n - 1)$, where $a < 1$

Solution
By definition
(a)

$$y(n) = x(n) * h(n) = \sum_{k=-\infty}^{\infty} x(k)h(n - k) = \sum_{k=-\infty}^{\infty} u(k)u(n - k)$$

The lower limit on the convolution sum simplifies to $k = 0$ [because $u(k) = 0$, $k < 0$], the upper limit to $k = n$ [because $u(n - k) = 0$, $k > n$], and we get

$$y(n) = \sum_{k=0}^{n} 1 = (n + 1)u(n) = r(n + 1)$$

(b) By definition

$$y(n) = x(n) * h(n) = \sum_{k=-\infty}^{\infty} x(k)h(n-k) = \sum_{k=-\infty}^{\infty} (0.8)^k u(k)(0.4)^{n-k} u(n-k)$$

The lower limit on the convolution sum simplifies to $k = 0$ [because $u(k) = 0$, $k < 0$], the upper limit to $k = n$ [because $u(n-k) = 0$, $k > n$], and we get

$$y(n) = \sum_{k=0}^{n} (0.8)^k (0.4)^{n-k} = (0.4)^n \sum_{k=0}^{n} 2^k = (0.4)^n \frac{1 - 2^{n+1}}{1 - 2} = (0.4)^n (2^{n+1} - 1)$$

$$= [2(0.8)^n - (0.4)^n] u(n)$$

(c) By definition

$$y(n) = x(n) * h(n) = \sum_{k=-\infty}^{\infty} x(k)h(n-k) = \sum_{k=-\infty}^{\infty} a^k u(k) a^{n-k} u(n-k)$$

The lower limit on the convolution sum simplifies to $k = 0$ [because $u(k) = 0$, $k < 0$], the upper limit to $k = n$ [because $u(n-k) = 0$, $k > n$], and we get

$$y(n) = \sum_{k=0}^{n} a^k a^{n-k} = a^n \sum_{k=0}^{n} 1 = (n+1) a^n u(n)$$

(d) Given that

$$x(n) = u(n-1) = \begin{cases} 1, & n \geq 1 \\ 0, & n < 1 \end{cases} \quad \text{and} \quad h(n) = \alpha^n u(n-1) = \begin{cases} \alpha^n, & n \geq 1 \\ 0, & n < 1 \end{cases}$$

By definition

$$y(n) = x(n) * h(n) = \sum_{k=k_1}^{n-k_2} x(k)h(n-k)$$

where $k_1 = 1$ and $k_2 = 1$.

$$y(n) = 0 \quad n < (k_1 + k_2) \to n < 2$$

and

$$y(n) = \sum_{k=k_1}^{n-k_2} x(k)h(n-k) \quad n \geq (k_1 + k_2)$$

$$= \sum_{k=1}^{n-1} \alpha^{n-k} \quad n \geq 1 + 1 = 2$$

$$= \alpha^n \sum_{k=1}^{n-1} \alpha^{-k} = \alpha^n \sum_{m=0}^{n-2} \alpha^{-(m+1)} = \alpha^n \alpha^{-1} \sum_{m=0}^{n-2} \alpha^{-(m)} = \frac{\alpha - \alpha^n}{1 - \alpha}$$

Therefore,

$$y(n) = \begin{cases} \dfrac{\alpha - \alpha^n}{1 - \alpha}, & n \geq 2 \\ 0, & n < 0 \end{cases} \quad \text{or, more compactly} \quad y(n) = \dfrac{\alpha - \alpha^n}{1 - \alpha} u(n - 2)$$

(e) Given that

$$x(n) = nu(n) = \begin{cases} n, & n \geq 0 \\ 0, & n < 0 \end{cases} \quad \text{and} \quad h(n) = a^{-n}u(n - 1) = \begin{cases} a^{-n}, & n \geq 1 \\ 0, & n < 1 \end{cases}$$

By definition

$$y(n) = x(n) * h(n) = \sum_{k=k_1}^{n-k_2} x(k)h(n - k)$$

where $k_1 = 0$ and $k_2 = 1$.

$$y(n) = 0 \quad n < (k_1 + k_2) \rightarrow n < 1$$

and

$$y(n) = \sum_{k=k_1}^{n-k_2} x(k)h(n - k) \quad n \geq (k_1 + k_2)$$

$$= \sum_{k=0}^{n-1} ka^{-(n-k)} \quad n \geq (0 + 1) = 1$$

$$= a^{-n} \sum_{k=0}^{n-1} ka^k = \frac{a^{-n+1}}{(1 - a)^2}[1 - na^{n-1} + (n - 1)a^n]$$

Therefore,

$$y(n) = \begin{cases} \dfrac{a^{-n+1}}{(1 - a)^2}[1 - na^{n-1} + (n - 1)a^n], & n \geq 1 \\ 0, & n < 1 \end{cases}$$

or, more compactly

$$y(n) = \frac{a^{-n+1}}{(1 - a)^2}[1 - na^{n-1} + (n - 1)a^n]u(n - 1)$$

Example 2.20 Suppose that the unit impulse response of an LTI system is a unit ramp,

$$h(n) = r(n) = nu(n)$$

Compute the response of this system to a unit step input $x(n) = u(n)$.
Or, determine the step response $s(n)$ of the LTI system characterized by the impulse response $h(n) = nu(n)$.

Solution

Given that

$$x(n) = u(n) = \begin{cases} 1, & n \geq 0 \\ 0, & n < 0 \end{cases} \quad \text{and} \quad h(n) = nu(n) = \begin{cases} n, & n \geq 0 \\ 0, & n < 0 \end{cases}$$

By definition

$$y(n) = x(n) * h(n) = \sum_{k=k_1}^{n-k_2} x(k)h(n-k)$$

where $k_1 = 0$ and $k_2 = 0$.

$$y(n) = 0 \quad n < (k_1 + k_2) \rightarrow n < 0$$

and

$$y(n) = \sum_{k=k_1}^{n-k_2} x(k)h(n-k), \quad n \geq k_1 + k_2$$

$$= \sum_{k=0}^{n} (n-k), \quad n \geq 0$$

$$y(n) = n + (n-1) + (n-2) + \cdots + 1 + 0 = \frac{n(n+1)}{2}$$

Therefore,

$$y(n) = \begin{cases} \dfrac{n(n+1)}{2}, & n \geq 0 \\ 0, & n < 0 \end{cases} \quad \text{or, more compactly} \quad y(n) = \frac{n(n+1)}{2}u(n)$$

2.5   PROPERTIES OF THE CONVOLUTION SUM

The structure of the convolution sum is similar to that of the convolution integral. Moreover, the properties of the convolution sum are similar to those of the convolution integral. We shall enumerate these properties here without proofs. The proofs are similar to those for the convolution integral.

2.5.1   Commutative Property

The convolution sum possesses the commutative property:

$$x(n) * h(n) = h(n) * x(n) \tag{2.14}$$

2.5.2   Associative Property

The convolution sum possesses the associative property:

$$x(n) * [h_1(n) * h_2(n)] = [x(n) * h_1(n)] * h_2(n) \tag{2.15}$$

2.5.3 Distributive Property

The convolution sum possesses the distributive property:

$$x(n) * [h_1(n) + h_2(n)] = [x(n) * h_1(n)] + [x(n) * h_2(n)] \tag{2.16}$$

2.5.4 Shifting Property

According to the shift property:

$$\text{if } x(n) * h(n) = y(n), \quad \text{then } x(n) * h(n - n_0) = x(n - n_0) * h(n) = y(n - n_0) \tag{2.17}$$

and

$$x(n - n_1) * h(n - n_2) = y(n - n_1 - n_2) \tag{2.18}$$

2.5.5 Convolution with an Impulse

Convolution of a signal $x(n)$ with a unit impulse function results in the signal $x(n)$ itself:

$$x(n) * \delta(n) = x(n) \tag{2.19}$$

2.5.6 Width Property

The discrete convolution of two sequences has the same type of edge properties as the continuous convolution of two continuous-time signals. That is, the location of the first nonzero sequence value (left edge) of the discrete convolution of two sequences equals the sum of the locations of the first nonzero sequence values (left edges) of the two sequences (i.e., $\text{LE}_y = \text{LE}_x + \text{LE}_h$). The same is true for the last nonzero sequence values (right edges) (i.e., $\text{RE}_y = \text{RE}_x + \text{RE}_h$).

We earlier found that the width of the nonzero extent of the continuous convolution of two signals is the sum of the widths of the nonzero extents of the two signals. A similar property that holds for discrete convolution is that the number of samples in the nonzero extent of a discrete convolution of two sequences is one less than the sum of the number of samples in the nonzero extents of the two sequences. Let N_x and N_h be the number of samples in the nonzero extents of the sequences $x(n)$ and $h(n)$, respectively. The number of samples in the nonzero extents of the sequence $y(n) = x(n) * h(n)$ is

$$N_y = N_x + N_h - 1 \tag{2.20}$$

Proof The number of samples in the nonzero extents of the sequence $x(n)$ is

$$N_x = \text{RE}_x - \text{LE}_x + 1$$

The number of samples in the nonzero extents of the sequence $h(n)$ is

$$N_h = \text{RE}_h - \text{LE}_h + 1$$

Since

$$\text{RE}_y = \text{RE}_x + \text{RE}_h \quad \text{and} \quad \text{LE}_y = \text{LE}_x + \text{LE}_h$$

The number of samples in the nonzero extents of the sequence $y(n) = x(n) * h(n)$ is

$$
\begin{aligned}
N_y &= \text{RE}_y - \text{LE}_y + 1 \\
&= \text{RE}_x + \text{RE}_h - \text{LE}_x - \text{LE}_h + 1 \\
&= (\text{RE}_x - \text{LE}_x + 1) + (\text{RE}_h - \text{LE}_h + 1) - 1 \\
&= N_x + N_h - 1
\end{aligned}
$$

∎

2.5.7 Sum Property

The sum of the impulses in a convolution sum of two discrete-time sequences is the product of the sums of the impulses in the two individual sequences. Let the sum of all the impulses in the signals $y(n)$, $x(n)$, and $h(n)$ be S_y, S_x, and S_h, respectively. Then

$$S_y = S_x S_h \qquad (2.21)$$

Proof By definition

$$y(n) = x(n) * h(n) = \sum_{k=-\infty}^{\infty} x(k)h(n-k)$$

$$S_y = \sum_{n=-\infty}^{\infty} y(n) = \sum_{n=-\infty}^{\infty} \sum_{k=-\infty}^{\infty} x(k)h(n-k)$$

Interchanging the order of the summation,

$$S_y = \underbrace{\sum_{k=-\infty}^{\infty} x(k)}_{S_x} \underbrace{\sum_{n=-\infty}^{\infty} h(n-k)}_{S_h} = S_x S_h \qquad \blacksquare$$

2.6 RELATIONSHIP BETWEEN LTI SYSTEM PROPERTIES AND THE IMPULSE RESPONSE

The impulse response completely characterizes the input–output behaviour of an LTI system. Hence, the properties of the system, such as memory, causality, and stability, are related to the system's impulse response.

2.6.1 LTI Systems With and Without Memory

The present output of a memoryless LTI system depends only on the present input. Exploiting the commutative property of convolution, we may express the output of a discrete-time LTI system as

$$y(n) = h(n) * x(n)$$

$$= \sum_{k=-\infty}^{\infty} h(k)x(n-k)$$

$$= \cdots + h(-2)x(n+2) + h(-1)x(n+1) + h(0)x(n) + h(1)x(n-1) + h(2)x(n-2) + \cdots$$

For this system to be memoryless, $y(n)$ must depend only on $x(n)$ and therefore cannot depend on $x(n-k)$ for $k \neq 0$. Hence, every term in the above equation must be zero, except $h(0)x(n)$. This condition implies that $h(k) = 0$ for $k \neq 0$; thus a discrete-time LTI system is memoryless if and only if

$$h(n) = K\delta(n) \qquad (2.22)$$

where $K = h(0)$ is a constant and the convolution sum reduces to the relation

$$y(n) = Kx(n)$$

If a discrete-time LTI system has an impulse response $h(n)$ that is not identically zero for $n \neq 0$, then the system has memory.

From Eq. (2.1), we can deduce similar properties for continuous-time LTI systems with and without memory. In particular, a continuous-time LTI system is memoryless if $h(t) = 0$ for $t \neq 0$, and such a memoryless LTI system has the form

$$y(t) = Kx(t)$$

for some constant K and has the impulse response

$$h(t) = K\delta(t) \tag{2.23}$$

2.6.2 Causality for LTI Systems

The output of a causal system depends only on the present and past values of the input to the system. Again, we write the convolution sum as

$$y(n) = \sum_{k=-\infty}^{\infty} h(k)x(n-k) = \left[\sum_{k=-\infty}^{-1} h(k)x(n-k)\right] + \left[\sum_{k=0}^{\infty} h(k)x(n-k)\right]$$

$$y(n) = \left[\cdots + h(-2)x(n+2) + h(-1)x(n+1)\right] + \left[h(0)x(n) + h(1)x(n-1)\right.$$
$$\left. + h(2)x(n-2) + \cdots\right]$$

We see that past and present values of the input $x(n), x(n-1), x(n-2), \ldots$ are associated with indices $k \geq 0$ in the impulse response $h(k)$, while future values of the input, $x(n+1), x(n+2), \ldots$ are associated with indices $k < 0$. In order, then, for $y(n)$ to depend only on past and present values of the input, we require that $h(k) = 0$ for $k < 0$. Hence, for a discrete-time causal LTI system,

$$h(n) = 0, \quad \text{for } n < 0 \tag{2.24}$$

and the discrete convolution takes the new form

$$y(n) = \sum_{k=0}^{\infty} h(k)x(n-k)$$

The alternative equivalent form is

$$y(n) = \sum_{k=-\infty}^{n} x(k)h(n-k)$$

Similarly, a continuous-time LTI system is causal if

$$h(t) = 0, \quad \text{for } t < 0 \tag{2.25}$$

and in this case the continuous convolution is given by

$$y(t) = \int_{0}^{\infty} h(\tau)x(t - \tau)\, d\tau = \int_{-\infty}^{t} x(\tau)h(t - \tau)\, d\tau$$

2.6.3 Stability for LTI Systems

A system is stable if every bounded input produces a bounded output. We shall now derive conditions on $h(n)$ that guarantee stability of the system by bounding the convolution sum. The magnitude of the output is given by

$$|y(n)| = |h(n) * x(n)|$$

$$= \left| \sum_{k=-\infty}^{\infty} h(k)x(n - k) \right|$$

$$\leq \sum_{k=-\infty}^{\infty} |h(k)||x(n - k)|$$

If we assume that the input is bounded, or $|x(n)| \leq B_x < \infty$, then $|x(n - k)| \leq B_x$, and it follows that

$$|y(n)| \leq B_x \sum_{k=-\infty}^{\infty} |h(k)|$$

From the above equation, we can conclude that if the impulse response is *absolutely summable*, i.e., if

$$\sum_{k=-\infty}^{\infty} |h(k)| < \infty \tag{2.26}$$

then $y(n)$ is bounded in magnitude, and hence, the system is stable.

In continuous time, we obtain an analogous characterization of stability in terms of the impulse response of an LTI system. Specifically, if $|x(t)| \leq B_x$ for all t, then

$$|y(t)| = \left| \int_{-\infty}^{\infty} h(\tau)x(t - \tau)\, d\tau \right|$$

$$\leq \int_{-\infty}^{\infty} |h(\tau)||x(t - \tau)|\, d\tau$$

$$\leq B_x \int_{-\infty}^{\infty} |h(\tau)|\, d\tau$$

Therefore, the system is stable if the impulse response is *absolutely integrable*, i.e., if

$$\int_{-\infty}^{\infty} |h(\tau)|d\tau < \infty \tag{2.27}$$

2.6.4 Invertibility for LTI Systems

Consider a continuous-time LTI system with impulse response $h(t)$. This system is invertible only if an inverse system exists that, when connected in series with the original system, produces an output equal to the input to the first system. Furthermore, if an LTI system is invertible, then it has an LTI inverse.

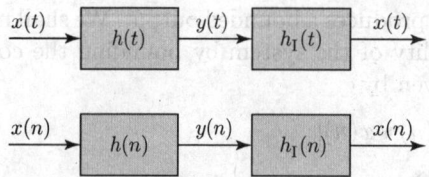

Fig. 2.39 Concept of a general invertible system

The relationship between the impulse response, $h(t)$, and that of the corresponding inverse system, $h_I(t)$, is easily derived. The impulse response of the cascade connection in Fig. 2.39 is the convolution of $h(t)$ and $h_I(t)$. We require the output of the cascade to equal the input, or

$$x(t) * [h(t) * h_I(t)] = x(t)$$

This requirement implies that

$$h(t) * h_I(t) = \delta(t) \qquad (2.28)$$

similarly, the impulse response of a discrete-time LTI inverse system, $h_I(n)$, must satisfy

$$h(n) * h_I(n) = \delta(n) \qquad (2.29)$$

Table 2.1 summarizes the relationship between LTI system properties and impulse response characteristics.

Table 2.1 Properties of the Impulse Response Representation for LTI System

Property	Continuous-time System	Discrete-time System
Memoryless	$h(t) = K\delta(t)$	$h(n) = K\delta(n)$
Causality	$h(t) = 0$, for $t < 0$	$h(n) = 0$, for $n < 0$
Stability	$\int\limits_{-\infty}^{\infty} |h(t)|dt < \infty$	$\sum_{n=-\infty}^{\infty} |h(n)| < \infty$
Invertibility	$h(t) * h_I(t) = \delta(t)$	$h(n) * h_I(n) = \delta(n)$

2.6.5 Unit Step Response of an LTI System

Step input signals are often used to characterize the response of an LTI system to sudden changes in the input. The step response $s(n)$ or $s(t)$ is defined as the output due to a unit step input signal $x(n) = u(n)$ or $x(t) = u(t)$. From the convolution sum representation, the step response of a discrete-time LTI system is the convolution of the unit step with the impulse response, i.e.,

$$s(n) = h(n) * u(n) = \sum_{k=-\infty}^{\infty} h(k)u(n-k)$$

Now, since

$$u(n-k) = \begin{cases} 1, & k \le n \\ 0, & k > n \end{cases}$$

we have

$$s(n) = \sum_{k=-\infty}^{n} h(k) \tag{2.30}$$

From this equation, it is clear that $h(n)$ can be recovered from $s(n)$:

$$s(n) = h(n) + \sum_{k=-\infty}^{n-1} h(k)$$

$$s(n) = h(n) + s(n-1)$$

$$h(n) = s(n) - s(n-1) \tag{2.31}$$

The step response of a discrete-time LTI system is the running sum of its impulse response. Conversely, the impulse response of a discrete-time LTI system is the first difference of its step response. Similarly, in continuous time, the step response of an LTI system with impulse response $h(t)$ is given by

$$s(t) = \int_{-\infty}^{\infty} h(\tau)u(t-\tau)\,d\tau$$

$$s(t) = \int_{-\infty}^{t} h(\tau)\,d\tau \tag{2.32}$$

i.e., the unit step response of a continuous-time LTI system is the running integral of its impulse response, and from Eq. (2.32), the unit impulse response is the first derivative of the unit step response or

$$h(t) = \frac{ds(t)}{dt} \tag{2.33}$$

Example 2.21 (a) Demonstrate the validity of the following relationships:

$$y(t) = \left(\int_{-\infty}^{t} x(\tau)\,d\tau \right) * h'(t) = \int_{-\infty}^{t} [x'(\tau) * h(\tau)]\,d\tau = x'(t) * \left(\int_{-\infty}^{t} h(\tau)\,d\tau \right)$$

(b) Let $s(t)$ be the unit step response of a continuous-time LTI system. Show that the response $y(t)$ to the input $x(t)$ is

$$y(t) = x'(t) * s(t) = \int\limits_{-\infty}^{\infty} x'(\tau)s(t-\tau)\,d\tau$$

Solution

(a) We know that

$$y(t) = x(t) * h(t)$$

$$y(t) * u(t) = [x(t) * u(t)] * h(t) \tag{2.34}$$

$$\int\limits_{-\infty}^{t} y(\tau)\,d\tau = \left[\int\limits_{-\infty}^{t} x(\tau)\,d\tau\right] * h(t)$$

Using the differentiation property of the convolution

$$\frac{d}{dt}\left(\int\limits_{-\infty}^{t} y(\tau)\,d\tau\right) = \left[\int\limits_{-\infty}^{t} x(\tau)d\tau\right] * \frac{d}{dt}h(t)$$

$$y(t) = \left[\int\limits_{-\infty}^{t} x(\tau)d\tau\right] * h'(t)$$

Now, from Eq. (2.34), we have

$$y(t) * u(t) = x(t) * [h(t) * u(t)]$$

$$\int\limits_{-\infty}^{t} y(\tau)\,d\tau = x(t) * \left[\int\limits_{-\infty}^{t} h(\tau)\,d\tau\right]$$

Using the differentiation property of the convolution

$$\frac{d}{dt}\left(\int\limits_{-\infty}^{t} y(\tau)\,d\tau\right) = \frac{d}{dt}x(t) * \left[\int\limits_{-\infty}^{t} h(\tau)\,d\tau\right]$$

$$y(t) = x'(t) * \left[\int\limits_{-\infty}^{t} h(\tau)\,d\tau\right] \tag{2.35}$$

Now, by definition,

$$y(t) = \int\limits_{-\infty}^{\infty} x'(\alpha) \left[\int\limits_{-\infty}^{t-\alpha} h(\tau)\, d\tau \right] d\alpha = \int\limits_{-\infty}^{\infty} x'(\alpha) \left[\int\limits_{-\infty}^{t} h(\tau - \alpha)\, d\tau \right] d\alpha$$

$$= \int\limits_{-\infty}^{t} \left[\int\limits_{-\infty}^{\infty} x'(\alpha) h(\tau - \alpha)\, d\alpha \right] d\tau = \int\limits_{-\infty}^{t} [x'(\tau) * h(\tau)]\, d\tau$$

(b) Using Eqs (2.32) and (2.35), we have

$$y(t) = x'(t) * s(t) = \int\limits_{-\infty}^{\infty} x'(\tau) s(t - \tau)\, d\tau$$

2.7 CORRELATION OF SIGNALS

A mathematical operation that closely resembles convolution is correlation. Just as in the case of convolution two signals are involved in correlation as well. In contrast to convolution, however, our objective in computing the correlation between the two signals is to measure the degree to which the two signals are similar.

2.7.1 Crosscorrelation Function of Energy Signals

Suppose that we have two signals $x(t)$ and $y(t)$ each of which has finite energy. The crosscorrelation of $x(t)$ and $y(t)$ is a signal $R_{xy}(\tau)$, which is defined as

$$R_{xy}(\tau) = \int\limits_{-\infty}^{\infty} x(t) y^*(t - \tau)\, dt = \int\limits_{-\infty}^{\infty} x(t + \tau) y^*(t)\, dt \qquad (2.36)$$

or, if both signals $x(t)$ and $y(t)$ are real, then

$$R_{xy}(\tau) = \int\limits_{-\infty}^{\infty} x(t) y(t - \tau)\, dt = \int\limits_{-\infty}^{\infty} x(t + \tau) y(t)\, dt \qquad (2.37)$$

The index τ is the (time) shift (or lag) parameter and the subscripts xy on the crosscorrelation function $R_{xy}(\tau)$ indicate the signals being correlated. The order of the subscripts, with x preceding y, indicates the direction in which one signal is shifted relative to the other.

If we reverse the roles of $x(t)$ and $y(t)$ in Eq. (2.37) and therefore reverse the order of the indices xy, we obtain the crosscorrelation function

$$R_{yx}(\tau) = \int\limits_{-\infty}^{\infty} y(t) x(t - \tau)\, dt = \int\limits_{-\infty}^{\infty} y(t + \tau) x(t)\, dt \qquad (2.38)$$

By comparing Eq. (2.37) with Eq. (2.38), we conclude that

$$R_{xy}(\tau) = R_{yx}(-\tau) \qquad (2.39)$$

Therefore, $R_{xy}(\tau)$ is simply the folded version of $R_{yx}(\tau)$, where the folding is done w.r.t. $\tau = 0$. Hence, $R_{xy}(\tau)$ provides the same information as $R_{yx}(\tau)$ w.r.t. the similarity of $x(t)$ to $y(t)$.

The similarities between the computation of the crosscorrelation of two signals and the convolution integral of two signals is apparent. For energy signals, there is a simple mathematical relationship between correlation and convolution integral:

$$R_{xy(\tau)} = x(\tau) * y(-\tau) \qquad (2.40)$$

Proof

$$R_{xy(\tau)} = x(\tau) * y(-\tau) = \int_{-\infty}^{\infty} x(t)y\big(-(\tau - t)\big)\, dt = \int_{-\infty}^{\infty} x(t)y(t - \tau)\, dt \qquad \blacksquare$$

2.7.2 Crosscorrelation Function of Power Signals

The crosscorrelation function between two power signals $x(t)$ and $y(t)$ is mathematically defined by

$$R_{xy}(\tau) = \lim_{T \to \infty} \frac{1}{T} \int_{-T/2}^{T/2} x(t)y^*(t - \tau)\, dt = \lim_{T \to \infty} \frac{1}{T} \int_{-T/2}^{T/2} x(t + \tau)y^*(t)\, dt \qquad (2.41)$$

If $x(t)$ and $y(t)$ are both real, then

$$R_{xy}(\tau) = \lim_{T \to \infty} \frac{1}{T} \int_{-T/2}^{T/2} x(t)y(t - \tau)\, dt = \lim_{T \to \infty} \frac{1}{T} \int_{-T/2}^{T/2} x(t + \tau)y(t)\, dt \qquad (2.42)$$

An important special case of correlation of power signals is the correlation between two periodic signals whose fundamental periods are such that the product of the two signals is also periodic. This will happen any time the fundamental periods of the two periodic signals have an LCM.

For two signals whose product has a period T, the general form of the correlation function (for real power signals) is given by

$$R_{xy}(\tau) = \lim_{T \to \infty} \frac{1}{T} \int_{-T/2}^{T/2} x(t)y(t - \tau)\, dt$$

can be replaced by

$$R_{xy}(\tau) = \frac{1}{T} \int_{-T/2}^{T/2} x(t)y(t - \tau)\, dt \qquad (2.43)$$

because the integral over one period of the product, divided by the period (which is the average of the integrand over one period) is the same as the average over any integer number of periods, including infinitely many periods.

2.7.3 Autocorrelation Function of Continuous-time Signals

A very important special case of the correlation function is the correlation of a function with itself. This type of correlation function is called the *autocorrelation* function. If $x(t)$ is an energy signal, its autocorrelation is

$$R_{xx}(\tau) = \int_{-\infty}^{\infty} x(t)x(t-\tau)\,dt = \int_{-\infty}^{\infty} x(t+\tau)x(t)\,dt \qquad (2.44)$$

If $x(t)$ is a power signal, its autocorrelation is

$$R_{xx}(\tau) = \lim_{T\to\infty} \frac{1}{T} \int_{-T/2}^{T/2} x(t)x(t-\tau)\,dt = \lim_{T\to\infty} \frac{1}{T} \int_{-T/2}^{T/2} x(t+\tau)x(t)\,dt \qquad (2.45)$$

2.7.4 Properties of Crosscorrelation and Autocorrelation Functions

Following are the properties of crosscorrelation and autocorrelation functions:

1. We have already demonstrated that the crosscorrelation function satisfies the property

$$R_{xy}(\tau) = R_{yx}(-\tau)$$

With $y(t) = x(t)$, this relation results in the following important property for the autocorrelation function:

$$R_{xx}(\tau) = R_{xx}(-\tau) \qquad (2.46)$$

Hence, the autocorrelation function is an even function.

Proof

$$R_{xx}(\tau) = \int_{-\infty}^{\infty} x(t)x(t-\tau)\,dt$$

A change of variables is performed by letting $(t-\tau) = \alpha$, which also yields $dt = d\alpha$, $\alpha \to \infty$ as $t \to \infty$, and $\alpha \to -\infty$ as $t \to -\infty$. Therefore,

$$R_{xx}(\tau) = \int_{-\infty}^{\infty} x(\alpha+\tau)x(\alpha)\,d\alpha = \int_{-\infty}^{\infty} x(\alpha)x(\alpha+\tau)\,d\alpha = \int_{-\infty}^{\infty} x(\alpha)x(\alpha-(-\tau))\,d\alpha$$

$$= R_{xx}(-\tau)$$

2. Relation to signal energy and signal power:
   Case 1 If $x(t)$ is an energy signal, then

$$R_{xx}(0) = E_x = \int_{-\infty}^{\infty} x^2(t)\, dt \qquad (2.47)$$

Proof

$$R_{xx}(\tau) = \int_{-\infty}^{\infty} x(t)x(t-\tau)\, dt$$

$$R_{xx}(\tau)|_{\tau=0} = \int_{-\infty}^{\infty} x(t)x(t)\, dt$$

$$R_{xx}(0) = \int_{-\infty}^{\infty} x^2(t)\, dt = E_x$$

∎

Case 2 If $x(t)$ is a power signal, then

$$R_{xx}(0) = P_x = \lim_{T \to \infty} \frac{1}{T} \int_{-T/2}^{T/2} x^2(t)\, dt \qquad (2.48)$$

Proof

$$R_{xx}(\tau) = \lim_{T \to \infty} \frac{1}{T} \int_{-T/2}^{T/2} x(t)x(t-\tau)\, dt$$

$$R_{xx}(\tau)|_{\tau=0} = \lim_{T \to \infty} \frac{1}{T} \int_{-T/2}^{T/2} x(t)x(t)\, dt$$

$$R_{xx}(0) = \lim_{T \to \infty} \frac{1}{T} \int_{-T/2}^{T/2} x^2(t)\, dt = P_x$$

∎

3. The crosscorrelation satisfies the following condition:

$$|R_{xy}(\tau)| \leq \sqrt{R_{xx}(0)R_{yy}(0)} = \sqrt{E_x E_y} \qquad (2.49)$$

The autocorrelation satisfies the following condition:

$$|R_{xx}(\tau)| \leq R_{xx}(0) = E_x \qquad (2.50)$$

Proof Let $x(t)$ and $y(t)$ be the two finite energy signals from which we form the linear combination,

$$ax(t) + by(t - \tau)$$

where a and b are two constants and τ is some time shift. The energy in this signal is

$$\int\limits_{-\infty}^{\infty} [ax(t) + by(t - \tau)]^2 \, dt = a^2 \int\limits_{-\infty}^{\infty} x^2(t) \, dt + b^2 \int\limits_{-\infty}^{\infty} y^2(t - \tau) \, dt + 2ab \int\limits_{-\infty}^{\infty} x(t)y(t - \tau) \, dt$$

$$= a^2 R_{xx}(0) + b^2 R_{yy}(0) + 2ab R_{xy}(\tau)$$

First, we note that $R_{xx}(0) = E_x$ and $R_{yy}(0) = E_y$, which are the energies of $x(t)$ and $y(t)$, respectively. It is obvious that

$$a^2 R_{xx}(0) + b^2 R_{yy}(0) + 2ab R_{xy}(\tau) \geq 0$$

Now, assuming that $b \neq 0$, we can divide the above equation by b^2 to obtain

$$R_{xx}(0) \left(\frac{a}{b}\right)^2 + 2R_{xy}(\tau) \left(\frac{a}{b}\right) + R_{yy}(0) \geq 0$$

We view this equation as a quadratic equation with coefficients $R_{xx}(0)$, $2R_{xy}(\tau)$, and $R_{yy}(0)$. Since the quadratic is nonnegative, it follows that its discriminant must be nonpositive (the discriminant of quadratic polynomial $ax^2 + bx + c$ is $b^2 - 4ac$), i.e.,

$$4[R_{xy}^2(\tau) - R_{xx}(0)R_{yy}(0)] \leq 0$$

$$|R_{xy}(\tau)| \leq \sqrt{R_{xx}(0)R_{yy}(0)} = \sqrt{E_x E_y}$$

For autocorrelation function where $y(t) = x(t)$, the above equation reduces to

$$|R_{xx}(\tau)| \leq R_{xx}(0) = E_x$$

This means that the autocorrelation function of a signal attains its maximum value at zero lag. ■

4. The autocorrelation of a periodic signal with period T is periodic with the same period:

$$\text{if } x(t) = x(t \pm T), \quad \text{then } R_{xx}(\tau) = R_{xx}(\tau \pm T) \tag{2.51}$$

Proof Let $x(t)$ be a periodic signal with period T. Then

$$x(t) = x(t + T)$$

$$x(t - \tau) = x(t - \tau + T)$$

Now,

$$R_{xx}(\tau) = \frac{1}{T} \int_{-T/2}^{T/2} x(t)x(t-\tau)\, dt$$

$$R_{xx}(\tau - T) = \frac{1}{T} \int_{-T/2}^{T/2} x(t)x(t-(\tau-T))\, dt$$

$$= \frac{1}{T} \int_{-T/2}^{T/2} x(t)x(t-\tau+T)\, dt = \frac{1}{T} \int_{-T/2}^{T/2} x(t)x(t-\tau)\, dt = R_{xx}(\tau) \quad \blacksquare$$

5. If $R_{xy}(0) = 0$, then $x(t)$ and $y(t)$ are orthogonal.

Proof By definition

$$R_{xy}(\tau) = \int_{-\infty}^{\infty} x(t)y(t-\tau)\, dt$$

$$R_{xy}(\tau)|_{\tau=0} = R_{xy}(0) = \int_{-\infty}^{\infty} x(t)y(t)\, dt = 0 \quad \blacksquare$$

2.7.5   Crosscorrelation Sequence of Discrete-time Energy Signals

Suppose that we have two signals $x(n)$ and $y(n)$ each of which has finite energy. The crosscorrelation of $x(n)$ and $y(n)$ is a signal $R_{xy}(m)$, which is defined as

$$R_{xy}(m) = \sum_{n=-\infty}^{\infty} x(n)y^*(n-m) = \sum_{n=-\infty}^{\infty} x(n+m)y^*(n) \tag{2.52}$$

or, if both signals $x(n)$ and $y(n)$ are real, then

$$R_{xy}(m) = \sum_{n=-\infty}^{\infty} x(n)y(n-m) = \sum_{n=-\infty}^{\infty} x(n+m)y(n) \tag{2.53}$$

The index m is the (time) shift (or lag) parameter and the subscripts xy on the crosscorrelation function $R_{xy}(m)$ indicate the signals being correlated. The order of the subscripts, with x preceding y, indicates the direction in which one signal is shifted relative to the other.

If we reverse the roles of $x(n)$ and $y(n)$ in Eq. (2.53) and therefore reverse the order of the indices xy, we obtain the crosscorrelation function

$$R_{yx}(m) = \sum_{n=-\infty}^{\infty} y(n)x(n-m) = \sum_{n=-\infty}^{\infty} y(n+m)x(n) \tag{2.54}$$

By comparing Eq. (2.53) with Eq. (2.54), we conclude that

$$R_{xy}(m) = R_{yx}(-m) \tag{2.55}$$

Therefore, $R_{xy}(m)$ is simply the folded version of $R_{yx}(m)$, where the folding is done w.r.t. $m = 0$. Hence, $R_{xy}(m)$ provides the same information as $R_{yx}(m)$ w.r.t. the similarity of $x(n)$ to $y(n)$.

The similarities between the computation of the crosscorrelation of two signals and the convolution sum of two sequences is apparent. For energy signals, there is a simple mathematical relationship between correlation and convolution sum:

$$R_{xy(m)} = x(m) * y(-m) \tag{2.56}$$

Proof

$$R_{xy(m)} = x(m) * y(-m) = \sum_{n=-\infty}^{\infty} x(n)y\big(-(m-n)\big) = \sum_{n=-\infty}^{\infty} x(n)y(n-m) \qquad \blacksquare$$

2.7.6 Crosscorrelation Sequence of Power Signals

The crosscorrelation function between two power signals $x(n)$ and $y(n)$ is mathematically defined by

$$R_{xy}(m) = \lim_{N \to \infty} \frac{1}{2N+1} \sum_{n=-N}^{N} x(n)y^*(n-m) = \lim_{N \to \infty} \frac{1}{2N+1} \sum_{n=-N}^{N} x(n+m)y^*(n) \tag{2.57}$$

If $x(n)$ and $y(n)$ are both real,

$$R_{xy}(m) = \lim_{N \to \infty} \frac{1}{2N+1} \sum_{n=-N}^{N} x(n)y(n-m) = \lim_{N \to \infty} \frac{1}{2N+1} \sum_{n=-N}^{N} x(n+m)y(n) \tag{2.58}$$

An important special case of correlation of power signals is the correlation between two periodic signals whose fundamental periods are such that the product of the two signals is also periodic. This will happen any time the fundamental periods of the two periodic signals have an LCM.

For two signals whose product has a period N, the general form of the correlation function (for real power signals)

$$R_{xy}(m) = \lim_{N \to \infty} \frac{1}{2N+1} \sum_{n=-N}^{N} x(n)y(n-m)$$

can be replaced by

$$R_{xy}(m) = \frac{1}{2N+1} \sum_{n=-N}^{N} x(n)y(n-m) \tag{2.59}$$

2.7.7 Autocorrelation Sequence of Discrete-time Signals

In the special case where $y(n) = x(n)$, we have the autocorrelation sequence of $x(n)$. If $x(n)$ is an energy signal, its autocorrelation is

$$R_{xx}(m) = \sum_{n=-\infty}^{\infty} x(n)x(n-m) = \sum_{n=-\infty}^{\infty} x(n+m)x(n) \qquad (2.60)$$

If $x(n)$ is a power signal, its autocorrelation is

$$R_{xx}(m) = \lim_{N\to\infty} \frac{1}{2N+1} \sum_{n=-N}^{N} x(n)x(n-m) = \lim_{N\to\infty} \frac{1}{2N+1} \sum_{n=-N}^{N} x(n+m)x(n) \quad (2.61)$$

2.7.8 Properties of Crosscorrelation and Autocorrelation Sequences

Following are the properties of crosscorrelation and autocorrelation sequences:

1. We have already demonstrated that the crosscorrelation sequence satisfies the property

$$R_{xy}(m) = R_{yx}(-m)$$

With $y(n) = x(n)$, this relation results in the following important property for the autocorrelation sequence:

$$R_{xx}(m) = R_{xx}(-m) \qquad (2.62)$$

Hence, The autocorrelation sequence is an even function.

Proof By definition

$$R_{xx}(m) = \sum_{n=-\infty}^{\infty} x(n)x(n-m)$$

A change of variables is performed by letting $n - m = k$, which also yields $k \to \infty$ as $n \to \infty$ and $k \to -\infty$ as $n \to -\infty$. Therefore,

$$R_{xx}(m) = \sum_{n=-\infty}^{\infty} x(k+m)x(k) = \sum_{n=-\infty}^{\infty} x(k)x(k+m) = \sum_{n=-\infty}^{\infty} x(k)x\big(k-(-m)\big)$$

$$= R_{xx}(-m) \qquad\qquad \blacksquare$$

2. Relation to signal energy and signal power:
   Case 1: If $x(n)$ is an energy signal, then

$$R_{xx}(0) = E_x = \sum_{n=-\infty}^{\infty} x^2(n) \qquad (2.63)$$

Proof By definition

$$R_{xx}(m) = \sum_{n=-\infty}^{\infty} x(n)x(n-m)$$

$$R_{xx}(m)\Big|_{m=0} = R_{xx}(0) = \sum_{n=-\infty}^{\infty} x(n)x(n) = \sum_{n=-\infty}^{\infty} x^2(n) = E_x \qquad \blacksquare$$

Case 2: If $x(n)$ is a power signal, then

$$R_{xx}(0) = P_x = \lim_{N\to\infty} \frac{1}{2N+1} \sum_{n=-N}^{N} x^2(n) \qquad (2.64)$$

Proof By definition

$$R_{xx}(m) = \lim_{N\to\infty} \frac{1}{2N+1} \sum_{n=-N}^{N} x(n)x(n-m)$$

$$R_{xx}(m)\Big|_{m=0} = R_{xx}(0) = \lim_{N\to\infty} \frac{1}{2N+1} \sum_{n=-N}^{N} x(n)x(n) = \lim_{N\to\infty} \frac{1}{2N+1} \sum_{n=-N}^{N} x^2(n) = P_x$$

$$\blacksquare$$

3. The crosscorrelation sequence satisfies the following condition:

$$|R_{xy}(m)| \le \sqrt{R_{xx}(0)R_{yy}(0)} = \sqrt{E_x E_y} \qquad (2.65)$$

The autocorrelation sequence satisfies the following condition:

$$|R_{xx}(m)| \le R_{xx}(0) = E_x \qquad (2.66)$$

Proof Let $x(n)$ and $y(n)$ be the two finite energy discrete-time signals from which we form the linear combination

$$ax(n) + by(n-m)$$

where a and b are two constants and m is some time shift. The energy in this signal is

$$\sum_{n=-\infty}^{\infty} [ax(n) + by(n-m)]^2 = a^2 \sum_{n=-\infty}^{\infty} x^2(n)dt + b^2 \sum_{n=-\infty}^{\infty} y^2(n-m)$$

$$+ 2ab \sum_{n=-\infty}^{\infty} x(n)y(n-m)$$

$$= a^2 R_{xx}(0) + b^2 R_{yy}(0) + 2abR_{xy}(m)$$

First, we note that $R_{xx}(0) = E_x$ and $R_{yy}(0) = E_y$, which are the energies of $x(n)$ and $y(n)$ respectively. It is obvious that

$$a^2 R_{xx}(0) + b^2 R_{yy}(0) + 2ab R_{xy}(m) \geq 0$$

Now, assuming that $b \neq 0$, we can divide the above equation by b^2 to obtain

$$R_{xx}(0) \left(\frac{a}{b}\right)^2 + 2R_{xy}(m) \left(\frac{a}{b}\right) + R_{yy}(0) \geq 0$$

We view this equation as a quadratic equation with coefficients $R_{xx}(0)$, $2R_{xy}(m)$, and $R_{yy}(0)$. Since the quadratic is nonnegative, it follows that its discriminant must be nonpositive, i.e.,

$$4[R_{xy}^2(m) - R_{xx}(0)R_{yy}(0)] \leq 0$$

$$|R_{xy}(m)| \leq \sqrt{R_{xx}(0)R_{yy}(0)} = \sqrt{E_x E_y}$$

For the autocorrelation function where $y(n) = x(n)$, the above equation reduces to

$$|R_{xx}(m)| \leq R_{xx}(0) = E_x$$

SOLVED EXAMPLES

Example 2.22 Let

$$x(t) = u(t - 3) - u(t - 5) \text{ and } h(t) = e^{-3t}u(t)$$

(a) Compute $y(t) = x(t) * h(t)$.

(b) Compute $g(t) = \left(\dfrac{d}{dt}x(t)\right) * h(t)$.

(c) How is $g(t)$ related to $y(t)$?

Solution

$$y(t) = x(t) * h(t) = \int_{-\infty}^{\infty} x(\tau)h(t - \tau)\,d\tau$$

Step 1:   Write the equations and draw sketches for $x(t)$ and $h(t)$ as shown in Fig. 2.40.

$$x(t) = u(t - 3) - u(t - 5) = \begin{cases} 1, & 3 < t < 5 \\ 0, & \text{otherwise} \end{cases}$$

$$h(t) = \begin{cases} e^{-3t}, & t \geq 0 \\ 0, & t < 0 \end{cases}$$

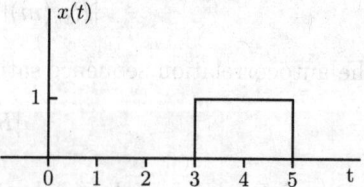

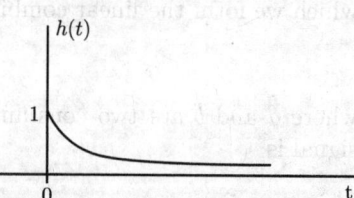

Fig. 2.40   Step 1 for Example 2.22

Step 2:   Write the equations and draw sketches for $x(\tau)$ and $h(t - \tau)$ [apply precedence rule: $h(t) \to h(\tau) \to h(\tau + t) \to h(-\tau + t) = h(t - \tau)$] as shown in Fig. 2.41.

$$x(\tau) = \begin{cases} 1, & 3 < \tau < 5 \\ 0, & \text{otherwise} \end{cases}$$

$$h(t-\tau) = \begin{cases} e^{-3(t-\tau)}, & (t-\tau) \geq 0 \to \tau \leq t \\ 0, & (t-\tau) < 0 \to \tau > t \end{cases}$$

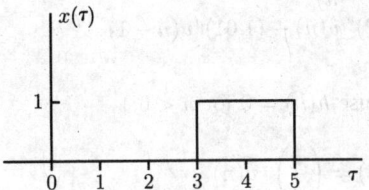

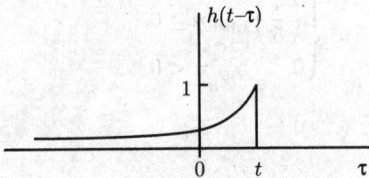

Fig. 2.41 Step 2 for Example 2.22

Step 3: Perform integration to find $y(t)$ for all values of t.

Case I: $t < 3$. Figure 2.42 shows $x(\tau)$ and $h(t - \tau)$. We see that the signals do not overlap (Fig. 2.42); hence, the product of the integrand is zero.

$$y(t) = \int_{-\infty}^{\infty} x(\tau)h(t-\tau)\,d\tau = 0$$

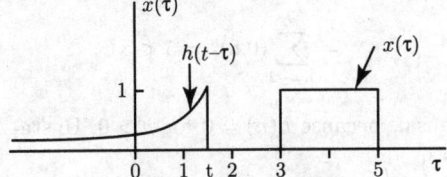

Fig. 2.42 Signal sketch for step 3, case I

Case II: $3 \leq t < 5$. Figure 2.43 shows $x(\tau)$ and $h(t - \tau)$. The two signals overlap over the interval $3 \leq \tau \leq t$ such that

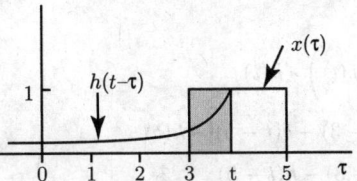

Fig. 2.43 Signal sketch for step 3, case II

$$y(t) = \int_{3}^{t} e^{-3(t-\tau)}\,d\tau = e^{-3t}\int_{3}^{t} e^{3\tau}\,d\tau$$

$$= \frac{1 - e^{-3(t-3)}}{3}$$

Case III: $t \geq 5$. Figure 2.44 shows $x(\tau)$ and $h(t - \tau)$.

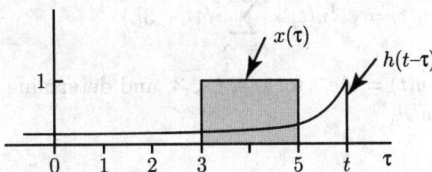

Fig. 2.44 Signal sketch for step 3, case III

The two signals overlap over the interval $3 \leq \tau \leq 5$ such that

$$y(t) = \int_{3}^{5} e^{-3(t-\tau)}\,d\tau = e^{-3t}\int_{3}^{5} e^{3\tau}\,d\tau$$

$$= \frac{\left(1 - e^{-6}\right)e^{-3(t-5)}}{3}$$

In summary

$$y(t) = \begin{cases} 0, & t < 3 \\ \dfrac{1 - e^{-3(t-3)}}{3}, & 3 \leq t < 5 \\ \dfrac{(1 - e^{-6})e^{-3(t-5)}}{3}, & t \geq 5 \end{cases}$$

(b) Differentiating $x(t)$ w.r.t. t, we get

$$\frac{dx(t)}{dt} = \delta(t-3) - \delta(t-5)$$

Therefore,

$$g(t) = \left(\frac{d}{dt}x(t)\right) * h(t)$$

$$= [\delta(t-3) - \delta(t-5)] * h(t)$$

$$= h(t-3) - h(t-5)$$

$$= e^{-3(t-3)}u(t-3) - e^{-3(t-5)}u(t-5)$$

(c) From the differentiation property, i.e., Eq. (2.9) of the convolution

$$\frac{dy(t)}{dt} = \frac{dx(t)}{dt} * h(t)$$

Hence,

$$g(t) = \frac{dy(t)}{dt}$$

Example 2.23 Let

$$y(t) = e^{-t}u(t) * \sum_{k=-\infty}^{\infty} \delta(t-3k)$$

Show that $y(t) = Ae^{-t}$ for $0 \le t < 3$, and determine the value of A.

Solution

$$y(t) = e^{-t}u(t) * \sum_{k=-\infty}^{\infty} \delta(t-3k)$$

$$= \sum_{k=-\infty}^{\infty} e^{-(t-3k)}u(t-3k)$$

$$= \begin{cases} \sum_{k=-\infty}^{\infty} e^{-t+3k}, & (t-3k) > 0 \longrightarrow t > 3k \\ 0, & \text{otherwise} \end{cases}$$

To determine $y(t)$ over the interval $0 \le t < 3$, $k = 0, -1, -2, \ldots$. Therefore, we have

$$y(t) = \sum_{k=-\infty}^{0} e^{-(t-3k)}, \qquad 0 \le t < 3$$

$$= e^{-t} \sum_{k=-\infty}^{0} e^{3k}$$

$$= e^{-t} \sum_{k=0}^{\infty} \left(e^{-3}\right)^{k}$$

$$y(t) = \frac{e^{-t}}{1-e^{-3}}, \qquad 0 \le t < 3$$

Therefore, $A = 1/(1-e^{-3})$.

Example 2.24 The following are the impulse responses of discrete-time LTI systems. Determine whether each system is causal and/or stable. Justify your answers.
(a) $h(n) = (1/5)^{n} u(n)$
(b) $h(n) = (0.8)^{n}u(n+2)$
(c) $h(n) = (1/2)^{n} u(-n)$
(d) $h(n) = 5^{n}u(3-n)$
(e) $h(n) = (-1/2)^{n} u(n) + (1.01)^{n}u(n-1)$

Solution
(a) Causal, because $h(n) = 0$ for $n < 0$.

$$h(n) = \left(\frac{1}{5}\right)^{n} u(n)$$

$$= \begin{cases} \left(\frac{1}{5}\right)^{n}, & n \ge 0 \\ 0, & n < 0 \end{cases}$$

Stable, because

$$\sum_{n=-\infty}^{\infty} h(n) = \sum_{n=-\infty}^{\infty} \left(\frac{1}{5}\right)^{n} u(n)$$

$$= \sum_{n=0}^{\infty} \left(\frac{1}{5}\right)^{n} = \frac{5}{4} < \infty$$

(b) Noncausal, because $h(n) \ne 0$ for $n < 0$.

$$h(n) = \begin{cases} (0.8)^{n}, & n \ge -2 \\ 0, & n < -2 \end{cases}$$

Stable, because

$$\sum_{n=-\infty}^{\infty} h(n) = \sum_{n=-\infty}^{\infty} (0.8)^{n}u(n+2)$$

$$= \sum_{n=-2}^{\infty} (0.8)^{n} = 5 < \infty$$

(c) Anticausal, because $h(n) = 0$ for $n > 0$. Unstable, because

$$\sum_{n=-\infty}^{\infty} h(n) = \sum_{n=-\infty}^{\infty} \left(\frac{1}{2}\right)^{n} u(-n)$$

$$= \sum_{n=-\infty}^{0} \left(\frac{1}{2}\right)^{n} = \infty$$

(d) Noncausal, because $h(n) \neq 0$ for $n < 0$.

$$h(n) = \begin{cases} 5^n, & n \leq 3 \\ 0, & n > 3 \end{cases}$$

Stable, because

$$\sum_{n=-\infty}^{\infty} h(n) = \sum_{n=-\infty}^{\infty} 5^n u(3-n)$$

$$= \sum_{n=-\infty}^{3} 5^n = \frac{625}{4} < \infty$$

(e) Causal, because $h(n) = 0$ for $n < 0$.

$$h(n) = \begin{cases} 1, & n = 0 \\ \left(-\frac{1}{2}\right)^n + (1.01)^n, & n \geq 1 \end{cases}$$

Unstable, because

$$\sum_{n=-\infty}^{\infty} h(n)$$

$$= \sum_{n=-\infty}^{\infty} \left[\left(-\frac{1}{2}\right)^n u(n) + (1.01)^n u(n-1) \right]$$

$$= \sum_{n=-\infty}^{\infty} \left(-\frac{1}{2}\right)^n u(n) + \sum_{n=-\infty}^{\infty} (1.01)^n u(n-1)$$

$$= \sum_{n=0}^{\infty} \left(-\frac{1}{2}\right)^n + \sum_{n=1}^{\infty} (1.01)^n = \infty$$

Example 2.25 The following are the impulse responses of continuous-time LTI systems. Determine whether each system is causal and/or stable. Justify your answers.
(a) $h(t) = e^{-4t} u(t-2)$
(b) $h(t) = e^{-6t} u(3-t)$
(c) $h(t) = e^{-6|t|}$
(d) $h(t) = te^{-t} u(t)$

Solution
(a) Causal, because $h(t) = 0$ for $t < 0$. Stable, because

$$\int_{-\infty}^{\infty} |h(t)| dt = \int_{-\infty}^{\infty} e^{-4t} u(t-2) dt$$

$$= \int_{2}^{\infty} e^{-4t} dt = \frac{e^{-8}}{4} < \infty$$

(b) Noncausal, because $h(t) \neq 0$ for $t < 0$. Unstable, because $\int_{-\infty}^{\infty} |h(t)| dt = \infty$
(c) Noncausal, because $h(t) \neq 0$ for $t < 0$.

$$h(t) = e^{-6|t|}$$

$$= \begin{cases} e^{-6t}, & t > 0 \\ e^{6t}, & t < 0 \end{cases}$$

$$= e^{-6t} u(t) + e^{6t} u(-t)$$

Stable, because

$$\int_{-\infty}^{\infty} |h(t)| dt = \int_{-\infty}^{\infty} e^{-6|t|} dt$$

$$= \int_{-\infty}^{0} e^{6t} dt + \int_{0}^{\infty} e^{-6t} dt = \frac{1}{3} < \infty$$

(d) Causal, because $h(t) = 0$ for $t < 0$. Stable, because

$$\int_{-\infty}^{\infty} |h(t)| dt = \int_{-\infty}^{\infty} te^{-t} u(t) dt$$

$$= \int_{0}^{\infty} te^{-t} dt = 1 < \infty$$

Example 2.26 Evaluate the step response for the LTI systems represented by the following impulse responses:
(a) $h(n) = (-1/2)^n u(n)$
(b) $h(n) = \delta(n) - \delta(n-2)$
(c) $h(n) = (-1)^n [u(n+2) - u(n-3)]$
(d) $h(t) = e^{-|t|}$
(e) $h(t) = 1/4 (u(t) - u(t-4))$

Solution
(a) From Eq. (2.30), we have

$$s(n) = \sum_{k=-\infty}^{n} h(k) = \sum_{k=-\infty}^{n} \left(-\frac{1}{2}\right)^k u(k)$$

$$= \sum_{k=0}^{n} \left(-\frac{1}{2}\right)^k = \frac{1 - (-1/2)^{n+1}}{1 - (-1/2)}$$

$$= \frac{2}{3} + \frac{1}{3} \left(-\frac{1}{2}\right)^n$$

(b) From Eq. (2.30), we have

$$s(n) = \sum_{k=-\infty}^{n} h(k)$$

$$= \sum_{k=-\infty}^{n} [\delta(k) - \delta(k-2)]$$

$$= \sum_{k=-\infty}^{n} \delta(k) - \sum_{k=-\infty}^{n} \delta(k-2)$$

$$= u(n) - u(n-2)$$

(c) Again from Eq. (2.30), we have

$$s(n) = \sum_{k=-\infty}^{n} h(k)$$

$$= \sum_{k=-\infty}^{n} (-1)^k [u(k+2) - u(k-3)]$$

$$= \sum_{k=-\infty}^{n} (-1)^k u(k+2) - \sum_{k=-\infty}^{n} (-1)^k u(k-3)$$

$$= \sum_{k=-2}^{n} (-1)^k - \sum_{k=3}^{n} (-1)^k$$

$$= \sum_{m=0}^{n+2} (-1)^{m-2} - \sum_{m=0}^{n-3} (-1)^{m+3}$$

$$= \frac{1}{2}[1 + (-1)^n] + \frac{1}{2}[1 - (-1)^n] = 1$$

(d) From Eq. (2.32), we have

$$s(t) = \int_{-\infty}^{t} h(\tau) d\tau = \int_{-\infty}^{t} e^{-|\tau|} d\tau$$

$$= \int_{-\infty}^{0} e^{\tau} d\tau + \int_{0}^{t} e^{-\tau} d\tau$$

$$= 2 - e^{-t}$$

(e) From Eq. (2.32), we have

$$s(t) = \int_{-\infty}^{t} h(\tau) d\tau$$

$$= \int_{-\infty}^{t} \frac{1}{4}(u(\tau) - u(\tau - 4)) d\tau$$

$$= \frac{1}{4}\left[\int_{-\infty}^{t} u(\tau) d\tau - \int_{-\infty}^{t} u(\tau - 4) d\tau \right]$$

$$= \frac{1}{4}[r(t) - r(t-4)]$$

Example 2.27 Consider the signal

$$x(n) = \alpha^n u(n)$$

(a) Determine the signal $g(n) = x(n) - \alpha x(n-1)$.
(b) Use the result of part (a) in conjunction with properties of convolution in order to determine a sequence $h(n)$ such that

$$x(n) * h(n) = \left(\frac{1}{2}\right)^n [u(n+2) - u(n-2)]$$

Solution
(a) We may write

$$g(n) = x(n) - \alpha x(n-1)$$

$$= \alpha^n u(n) - \alpha \alpha^{(n-1)} u(n-1)$$

$$= \alpha^n u(n) - \alpha^n u(n-1)$$

$$= \alpha^n [u(n) - u(n-1)]$$

$$= \alpha^n \delta(n) = \delta(n)$$

(b) Given that

$$x(n) * h(n)$$

$$= \left(\frac{1}{2}\right)^n [u(n+2) - u(n-2)]$$

$$= \begin{cases} \left(\frac{1}{2}\right)^n, & -2 \le n \le 1 \\ 0, & \text{otherwise} \end{cases}$$

$$= \left(\frac{1}{2}\right)^n [\delta(n+2) + \delta(n+1)$$

$$+ \delta(n) + \delta(n-1)]$$

$$= \left(\frac{1}{2}\right)^n \delta(n+2) + \left(\frac{1}{2}\right)^n \delta(n+1)$$

$$+ \left(\frac{1}{2}\right)^n \delta(n) + \left(\frac{1}{2}\right)^n \delta(n-1)]$$

$$= 4\delta(n+2) + 2\delta(n+1) + \delta(n) + \frac{1}{2}\delta(n-1)$$

Note that $g(n) = x(n) * [\delta(n) - \alpha\delta(n-1)]$. Therefore from part (a), we know that

$$x(n) * [\delta(n) - \alpha\delta(n-1)] = \delta(n)$$

Using this and the shift property of the convolution, we may write

$$x(n) * [\delta(n-1) - \alpha\delta(n-2)] = \delta(n-1)$$

$$x(n) * [\delta(n+1) - \alpha\delta(n)] = \delta(n+1)$$

$$x(n) * [\delta(n+2) - \alpha\delta(n+1)] = \delta(n+2)$$

Therefore,

$$x(n) * h(n)$$

$$= 4x(n) * [\delta(n+2) - \alpha\delta(n+1)]$$

$$+ 2x(n) * [\delta(n+1) - \alpha\delta(n)]$$

$$+ x(n) * [\delta(n) - \alpha\delta(n-1)]$$

$$+ \frac{1}{2}x(n) * [\delta(n) - \alpha\delta(n-1)]$$

$$= x(n) * \Big[4\delta(n+2) - 4\alpha\delta(n+1)$$

$$+ 2\delta(n+1) - 2\alpha\delta(n) + \delta(n)$$

$$- \alpha\delta(n-1) + \frac{1}{2}\delta(n-1) - \frac{1}{2}\alpha\delta(n-2)\Big]$$

$$= x(n) * \Big[4\delta(n+2) + (2-4\alpha)\delta(n+1)$$

$$+ (1-2\alpha)\delta(n) + (1/2 - \alpha)\delta(n-1)$$

$$+ (1/2)\delta(n-2)\Big]$$

Hence,

$$h(n) = 4\delta(n+2) + (2-4\alpha)\delta(n+1)$$

$$+ (1-2\alpha)\delta(n) + (1/2 - \alpha)\delta(n-1)$$

$$+ (1/2)\delta(n-2)$$

Example 2.28 Consider the cascade of the following two systems S_1 and S_2 as depicted in Fig. 2.45:

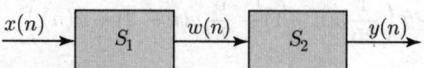

Fig. 2.45

S_1: causal LTI

$$w(n) = \frac{1}{2}w(n-1) + x(n)$$

S_2: causal LTI

$$y(n) = \alpha y(n-1) + \beta w(n)$$

The difference equation relating $x(n)$ and $y(n)$ is

$$y(n) = -\frac{1}{8}y(n-2) + \frac{3}{4}y(n-1) + x(n)$$

(a) Determine α and β.
(b) Show the impulse response of the cascade connection of S_1 and S_2.

Solution
Consider the difference equation relating $y(n)$ and $w(n)$ for S_2:

$$y(n) = \alpha y(n-1) + \beta w(n)$$

From this equation we may write

$$w(n) = \frac{1}{\beta}y(n) - \frac{\alpha}{\beta}y(n-1)$$

and

$$w(n-1) = \frac{1}{\beta}y(n-1) - \frac{\alpha}{\beta}y(n-2)$$

Substituting this in the difference equation relating $w(n)$ and $x(n)$ for S_1:

$$w(n) = \frac{1}{2}\left(\frac{1}{\beta}y(n-1) - \frac{\alpha}{\beta}y(n-2)\right) + x(n)$$

$$= \frac{1}{2\beta}y(n-1) - \frac{\alpha}{2\beta}y(n-2) + x(n)$$

Substituting this in the difference equation relating $y(n)$ and $w(n)$ for S_2:

$$y(n) = \alpha y(n-1)$$

$$+ \beta\left(\frac{1}{2\beta}y(n-1) - \frac{\alpha}{2\beta}y(n-2) + x(n)\right)$$

$$y(n) = \left(\alpha + \frac{1}{2}\right)y(n-1) - \frac{\alpha}{2}y(n-2) + \beta x(n)$$

Comparing with the given equation relating $y(n)$ and $x(n)$, we obtain

$$\alpha = \frac{1}{4} \quad \text{and} \quad \beta = 1$$

Example 2.29 Determine whether each of the following statements concerning LTI systems is true or false. Justify your answers.
(a) If $h(t)$ is the impulse response of an LTI system and $h(t)$ is periodic and nonzero, the system is unstable.
(b) The inverse of a causal LTI system is always causal.
(c) If $|h(n)| \leq K$ for each n, where K is a given number, then the LTI system with $h(n)$ as its impulse response is stable.
(d) If a discrete-time LTI system has an impulse response $h(n)$ of finite duration, the system is stable.
(e) If an LTI system is causal, it is stable.
(f) The cascade of a noncausal LTI system with a causal one is necessarily noncausal.
(g) A continuous-time LTI system is stable if and only if its step response $s(t)$ is absolutely integrable, i.e., if and only if

$$\int_{-\infty}^{\infty} |s(t)|\, dt < \infty$$

(h) A discrete-time LTI system is causal if and only if its step response $s(n)$ is zero for $n < 0$.

Solution
(a) True. If $h(t)$ is periodic and nonzero, then

$$\int_{-\infty}^{\infty} |h(t)|\, dt = \infty$$

Therefore, $h(t)$ is unstable.
(b) False. For example, inverse of $h(n) = \delta(n-k)$ is $g(n) = \delta(n+k)$, which is noncausal.
(c) False. For example, $h(n) = u(n)$ implies that

$$\sum_{n=-\infty}^{\infty} |h(n)| = \sum_{n=0}^{\infty} u(n) = \sum_{n=0}^{\infty} 1 = \infty$$

This is an unstable system.
(d) True. Assuming that $h(n)$ is bounded and nonzero in the range $n_1 \leq n \leq n_2$,

$$\sum_{n=n_1}^{n_2} |h(n)| < \infty$$

This implies that the system is stable.
(e) False. For example, $h(t) = e^t u(t)$ is causal but unstable.
(f) False. For example, the cascade of a causal system with impulse response $h_1(n) = \delta(n-1)$ and a noncausal system with impulse response $h_2(n) = \delta(n+1)$ leads to a system with overall impulse response $h(n) = h_1(n) * h_2(n) = \delta(n)$.
(g) False. For example, if $h(t) = e^{-t}u(t)$, then $s(t) = \int_{-\infty}^{t} h(\tau)\, d\tau = (1 - e^{-t})u(t)$, and

$$\int_{-\infty}^{\infty} |s(t)|dt = \int_{-\infty}^{\infty} |(1 - e^{-t})u(t)|\, dt$$

$$= \int_{0}^{\infty} (1 - e^{-t})\, dt$$

$$= [t + e^{-t}]_0^{\infty} = \infty$$

Although the system is stable, the step response is not absolutely integrable.
(h) True. We may write $u(n) = \sum_{k=0}^{\infty} \delta(n-k)$. Therefore,

$$s(n) = \sum_{k=0}^{\infty} h(n-k)$$

If $s(n) = 0$ for $n < 0$, then $h(n) = 0$ for $n < 0$ and the system is causal.

Example 2.30 Find and sketch the autocorrelation function $R_{xx}(\tau)$ for

$$x(t) = e^{-at}u(t), \quad a > 0$$

Solution
By definition

$$R_{xx}(\tau) = \int_{-\infty}^{\infty} x(t)x(t-\tau)\,dt$$

$$= \int_{-\infty}^{\infty} e^{-at}u(t)e^{-a(t-\tau)}u(t-\tau)\,dt$$

$$= e^{a\tau}\int_{-\infty}^{\infty} e^{-2at}u(t)u(t-\tau)\,dt$$

For $\tau > 0$,

$$u(t)u(t-\tau) = \begin{cases} 1, & t > \tau \\ 0; & t < \tau \end{cases}$$

$$R_{xx}(\tau) = e^{a\tau}\int_{\tau}^{\infty} e^{-2at}\,dt$$

$$= \frac{1}{2a}e^{-a\tau}$$

Since $R_{xx}(\tau)$ is an even function of τ, we conclude that

$$R_{xx}(\tau) = \frac{1}{2a}e^{-a|\tau|}, \quad a > 0$$

which is sketched in Fig. 2.46.

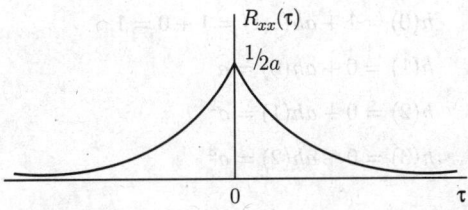

Fig. 2.46

Example 2.31 Find the autocorrelation function $R_{xx}(\tau)$ of sine wave signal

$$x(t) = A\sin(\omega_0 t + \phi), \quad \omega_0 = \frac{2\pi}{T}$$

Solution
By definition

$$R_{xx}(\tau)$$

$$= \lim_{T\to\infty}\frac{1}{T}\int_{-T/2}^{T/2} x(t)x(t-\tau)\,dt$$

$$= \lim_{T\to\infty}\frac{A^2}{T}\int_{-T/2}^{T/2}\sin(\omega_0 t + \phi)$$

$$\sin(\omega_0(t-\tau)+\phi)\,dt$$

$$= \lim_{T\to\infty}\frac{A^2}{2T}\int_{-T/2}^{T/2}\big[\cos(\omega_0\tau)$$

$$- \cos(2\omega_0 t + 2\phi - \omega_0 t)\big]\,dt$$

$$= \lim_{T\to\infty}\frac{A^2}{2T}\int_{-T/2}^{T/2}\cos(\omega_0\tau)\,dt$$

$$- \lim_{T\to\infty}\frac{A^2}{2T}\int_{-T/2}^{T/2}\cos(2\omega_0 t + 2\phi - \omega_0 t)\,dt$$

$$= \frac{A^2}{2T}\cos(\omega_0\tau)\int_{-T/2}^{T/2} dt - 0$$

$$= \frac{A^2}{2}\cos(\omega_0\tau)$$

Now at $\tau = 0$,

$$R_{xx}(\tau)\Big|_{\tau=0} = R_{xx}(0) = P_x = \frac{A^2}{2}$$

Example 2.32 Determine the overall impulse response of the system shown in Fig. 2.47. Given that

$$h_1(n) = \delta(n) - a\delta(n-1)$$

$$h_2(n) = \left(\frac{1}{2}\right)^n u(n)$$

$$h_3(n) = a^n u(n)$$

$$h_4(n) = (n-1)u(n)$$

$$h_5(n) = \delta(n) + nu(n-1) + \delta(n-2)$$

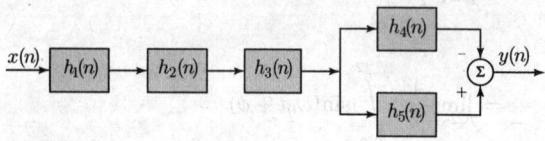

Fig. 2.47

Solution

It is clear from the figure that

$$h(n) = h_1(n) * h_2(n) * h_3(n) * [h_5(n) - h_4(n)]$$

To evaluate $h(n)$, we first perform the convolution $h_1(n) * h_3(n)$ as

$$h_1(n) * h_3(n) = [\delta(n) - a\delta(n-1)] * a^n u(n)$$

$$= a^n u(n) - a^n u(n-1)$$

$$= a^n [u(n) - u(n-1)]$$

$$= a^n \delta(n) = \delta(n)$$

Also,

$$h_5(n) - h_4(n) = \delta(n) + nu(n-1) + \delta(n-2)$$

$$- (n-1)u(n)$$

$$= \delta(n) + \delta(n-2) + u(n)$$

so that

$$h(n) = \delta(n) * h_2(n) * [\delta(n) + \delta(n-2) + u(n)]$$

$$= h_2(n) + h_2(n-2) + s_2(n)$$

where $s_2(n)$ represents the step response corresponding to $h_2(n)$, i.e.,

$$s_2(n) = \left(\frac{1}{2}\right)^n u(n) * u(n)$$

$$= \sum_{k=0}^{n} \left(\frac{1}{2}\right)^k, \quad n \geq 0$$

$$= \frac{1 - (1/2)^{n+1}}{1 - (1/2)}$$

$$= 2\left[1 - \left(\frac{1}{2}\right)^{n+1}\right], \quad n \geq 0$$

$$s_2(n) = 2u(n) - \left(\frac{1}{2}\right)^n u(n)$$

Thus,

$$h(n) = \left(\frac{1}{2}\right)^n u(n) + \left(\frac{1}{2}\right)^{n-2} u(n-2)$$

$$+ 2u(n) - \left(\frac{1}{2}\right)^n u(n)$$

$$= \left(\frac{1}{2}\right)^{n-2} u(n-2) + 2u(n)$$

Example 2.33 A system is described by the difference equation $y(n) = x(n) + ay(n-1)$. Find the impulse response of the inverse of this system. From the impulse response, find the difference equation of the inverse system. Assume the system is initially relaxed.

Solution

First we find the impulse response of this system by setting $x(n) = \delta(n)$ such that $y(n)$ corresponds to the impulse response $h(n)$. Thus, we may write

$$h(n) = \delta(n) + ah(n-1)$$

$$h(0) = 1 + ah(-1) = 1 + 0 = 1$$

$$h(1) = 0 + ah(0) = a$$

$$h(2) = 0 + ah(1) = a^2$$

$$h(3) = 0 + ah(2) = a^3$$

$$\vdots$$

$$h(n) = a^n$$

Therefore,

$$h(n) = a^n u(n)$$

The inverse system must satisfy

$$h(n) * h_I(n) = \delta(n)$$

Note that

$$h(n) = \delta(n) + ah(n-1)$$

$$h(n) - ah(n-1) = \delta(n)$$

$$h(n) * [\delta(n) - a\delta(n-1)] = \delta(n)$$

comparing with

$$h(n) * h_I(n) = \delta(n)$$

we get

$$h_I(n) = \delta(n) - a\delta(n-1)$$

The input to the inverse system is $y(n)$ and output is $x(n)$.

$$x(n) = h_I(n) * y(n)$$
$$= [\delta(n) - a\delta(n-1)] * y(n)$$
$$= y(n) - ay(n-1)$$

Example 2.34 A signal $x(n)$ with known autocorrelation $R_{xx}(m)$ is applied to an LTI system with impulse response $h(n)$, producing the output signal

$$y(n) = x(n) * h(n) = \sum_{k=-\infty}^{\infty} x(k)h(n-k)$$

(a) Find the expression for $R_{yx}(m)$ in terms of $R_{xx}(m)$.

(b) Find the expression for $R_{xy}(m)$ in terms of $R_{xx}(m)$.

(c) Find the expression for $R_{yy}(m)$ in terms of $R_{xx}(m)$.

Solution

(a) The crosscorrelation between output and the input signal is

$$R_{yx}(m) = y(m) * x(-m)$$
$$= [x(m) * h(m)] * x(-m)$$
$$= [x(m) * x(-m)] * h(m)$$
$$= R_{xx}(m) * h(m)$$

Hence the crosscorrelation between the input and the output of the system is the convolution of the impulse response with the autocorrelation of the input sequence. Alternatively, $R_{yx}(m)$ may be viewed as the output of the LTI system when the input sequence is $R_{xx}(m)$.

(b) The crosscorrelation between the input and the output signal is

$$R_{xy}(m) = x(m) * y(-m)$$
$$= x(m) * [h(-m) * x(-m)]$$
$$= [x(m) * x(-m)] * h(-m)$$
$$= R_{xx}(m) * h(-m)$$

(c) The autocorrelation of the output signal is

$$R_{yy}(m) = y(m) * y(-m)$$
$$= [x(m) * h(m)] * [x(-m) * h(-m)]$$
$$= [x(m) * x(-m)] * [h(m) * h(-m)]$$
$$= R_{xx}(m) * R_{hh}(-m)$$

The autocorrelation $R_{hh}(m)$ of the impulse response $h(n)$ exists if the system is stable.

SUMMARY

1. A linear time-invariant (LTI) system is completely characterized by its impulse response $h(t)$.

2. The output $y(t)$ of an LTI system is the convolution of the input $x(t)$ with the impulse response of the system

$$y(t) = x(t) * h(t) = \int_{\infty}^{\infty} x(\tau)h(_t-\tau)\, d\tau$$

3. The convolution operation gives only the zero-state response (i.e., forced response) of the system.

4. The convolution operator is commutative, distributive, and associative.

5. The step response of a continuous-time linear system with impulse response $h(t)$ is

$$s(t) = \int_{-\infty}^{t} h(\tau) \, d\tau$$

6. The step response of a discrete-time linear system with impulse response $h(n)$ is

$$s(n) = \sum_{k=-\infty}^{n} h(k)$$

7. A continuous-time LTI system is causal if $h(t) = 0$ for $t < 0$.

8. A discrete-time LTI system is causal if $h(n) = 0$ for $n < 0$.

9. A continuous-time LTI system is stable if and only if

$$\int_{-\infty}^{\infty} |h(\tau)| \, d\tau < \infty$$

10. A discrete-time LTI system is stable if and only if

$$\sum_{n=-\infty}^{\infty} h(n) < \infty$$

11. For energy signals, there is a simple mathematical relationship between correlation and convolution sum:

$$R_{xy(m)} = x(m) * y(-m)$$

12. Correlation is used to measure the similarity between the two signals.

MULTIPLE-CHOICE QUESTIONS

1. A continuous-time periodic signal $x(t)$, having a period T, is convolved with itself. The resulting signal is
   (a) not periodic
   (b) periodic having a period T
   (c) periodic having a period $2T$
   (d) periodic having a period $T/2$

2. The impulse response of a system described by the differential equation
   $d^2 y(t)/(dt) + y(t) = x(t)$ will be
   (a) a constant
   (b) an impulse function
   (c) a sinusoid
   (d) an exponentially decaying function

3. The condition $\int_{-\infty}^{\infty} |h(t)| dt < \infty$ must be satisfied by a system that is
   (a) memoryless
   (b) BIBO stable
   (c) causal
   (d) invertible.

4. For a system with input $x(n) = \delta(n-1)$ and impulse response $h(n) = \delta(n+1)$, the output is
   (a) $\delta(n-1)$
   (b) $\delta(n-2)$
   (c) $\delta(n)$
   (d) $\delta(n+2)$

5. The unit step response of an LTI system with impulse response $h(n) = \delta(n) - \delta(n-1)$ is
   (a) $\delta(n-1)$
   (b) $\delta(n)$
   (c) $u(n-1)$
   (d) $u(n)$

6. The convolution of a finite sequence with an infinite sequence
   (a) may be a finite or infinite sequence
   (b) is always a finite sequence
   (c) is always an infinite sequence
   (d) cannot be found

7. Convolution is used to find
   (a) the impulse response of an LTI system
   (b) frequency response of a system
   (c) the time response of an LTI system
   (d) the phase response of an LTI system

8. The step response of an LTI system when the impulse response $h(n)$ is unit step $u(n)$ is
   (a) $(n+1)u(n)$
   (b) $nu(n)$
   (c) $(n-1)u(n)$
   (d) $n^2 u(n)$

9. The convolution of a rectangle pulse with itself is

(a) another rectangle pulse
(b) square pulse
(c) triangular pulse
(d) sinc pulse

10. The autocorrelation of a rectangular pulse is
(a) another rectangle pulse
(b) square pulse
(c) triangular pulse
(d) sinc pulse

11. The impulse response of a system is $h(n) = \alpha^n u(n)$. The condition for the system to be BIBO stable is
(a) α is real and positive
(b) α is real and negative
(c) $|\alpha| > 1$
(d) $|\alpha| < 1$

12. The unit impulse response of an LTI system is the unit step function $u(t)$. For $t > 0$, the response of the system to an excitation $e^{-at}u(t)$, $a > 0$, will be
(a) ae^{-at}
(b) $\dfrac{1 - e^{-at}}{a}$

(c) $a(1 - e^{-at})$
(d) $1 - e^{-at}$

13. When the impulse response $h(t)$ of an LTI system is integrated over the whole time domain, it gives
(a) steady-state response of the system
(b) dc response of the system
(c) transient response of the system
(d) steady-state error of the system

14. The convolution of $u(n)$ with $u(n-4)$ at $n = 5$ is
(a) 5
(b) 2
(c) 1
(d) 0

15. A first-order all-pass filter impulse response is given by $h(t) = -\delta(t) + 2e^{-t}u(t)$. The zero state response of this filter for the input $e^t u(-t)$ is
(a) $e^{-t}u(t)$
(b) $e^{-t}u(t) + 2e^{-t}u(-t)$
(c) $e^{-2t}u(t)$
(d) $e^{-t}u(-t)$

PROBLEMS

2.1 Find the convolution $y(n) = x(n) * h(n)$ of the following signals:

(a)

$$x(n) = \begin{cases} -1, & -5 \leq n \leq -1 \\ 1, & 0 \leq n \leq 4 \end{cases}$$

$$h(n) = 2u(n)$$

(b)

$$x(n) = \left(\frac{1}{2}\right)^n u(n)$$

$$h(n) = \delta(n) + \delta(n-1) + \left(\frac{1}{3}\right)^n u(n)$$

(c)

$$x(n) = u(n)$$

$$h(n) = 1, \quad 0 \leq n \leq 9$$

(d)

$$x(n) = \left(\frac{1}{3}\right)^n u(n)$$

$$h(n) = \delta(n) + \left(\frac{1}{3}\right)^n u(n)$$

2.2 Determine the impulse response for the cascade of two LTI systems having impulse responses

$$h_1(n) = \left(\frac{1}{2}\right)^n u(n)$$

and

$$h_2(n) = \left(\frac{1}{4}\right)^n u(n)$$

2.3 Determine the range of values of the parameter α for which the LTI system with impulse response

$$h(n) = \alpha^n u(n)$$

is stable.

2.4 Determine the range of values of α and β for which the LTI system with impulse response

$$h(n) = \begin{cases} \alpha^n, & n \geq 0 \\ \beta^n, & n < 0 \end{cases}$$

is stable.

2.5 Compute the autocorrelation of the signal

$$x(n) = a^n u(n), \quad 0 < a < 1$$

2.6 Determine and sketch the convolution $y(n)$ of the signals

$$x(n) = \begin{cases} \dfrac{1}{3}n, & 0 \leq n \leq 6 \\ 0, & \text{elsewhere} \end{cases}$$

$$h(n) = \begin{cases} 1, & -2 \leq n \leq 2 \\ 0, & \text{elsewhere} \end{cases}$$

(a) Graphically

(b) Analytically

2.7 The discrete-time system

$$y(n) = ny(n-1) + x(n)$$

is at rest [i.e., $y(-1) = 0$]. Check if the system is LTI and BIBO stable.

2.8 Find the convolution $y(n) = x(n) * h(n)$ for each of the following pairs of finite sequences:

(a) $x(n) = \left\{ 1, -\dfrac{1}{2}, \dfrac{1}{4}, -\dfrac{1}{8}, \dfrac{1}{16} \right\}$

$h(n) = \{1, -1, 1, -1\}$
$\phantom{h(n) = \{1,}\uparrow$

(b) $x(n) = \{1, 2, 3, 0, -1\}$
$\phantom{x(n) = \{}\uparrow$

$h(n) = \{2, -1, 3, 1, -2\}$
$\phantom{h(n) = \{}\uparrow$

(c) $x(n) = \left\{ 3, \dfrac{1}{2}, -\dfrac{1}{4}, 1, 4 \right\}$
$\phantom{x(n) = \{ 3, }\uparrow$

$h(n) = \left\{ 2, -1, \dfrac{1}{2}, -\dfrac{1}{2} \right\}$
$\uparrow$

where the arrow indicates the value for $n = 0$. In this notation, it is assumed that not all values listed are zero.

2.9 Let

$$x(n) = \delta(n) + 2\delta(n-1) - \delta(n-3)$$

and

$$h(n) = 2\delta(n+1) + 2\delta(n-1)$$

Compute and plot each of the following convolutions:

(a) $y_1(n) = x(n) * h(n)$

(b) $y_2(n) = x(n+2) * h(n)$

(c) $y_3(n) = x(n) * h(n+2)$

2.10 Consider an input $x(n)$ and a unit impulse response $h(n)$ given by

$$x(n) = \left(\frac{1}{2}\right)^{n-2} u(n-2)$$

$$h(n) = u(n+2)$$

Determine and plot the output $y(n) = x(n) * h(n)$.

2.11 (a) Find the impulse response of the system shown in Fig. 2.48. Assume that

$$h_1(n) = h_2(n) = \left(\frac{1}{3}\right)^n u(n)$$

$$h_3(n) = u(n)$$

$$h_4(n) = \left(\frac{1}{2}\right)^n u(n)$$

(b) Find the response of the system to the unit step input.

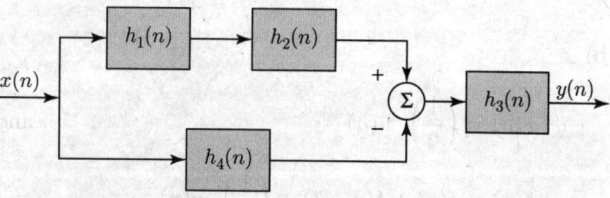

Fig. 2.48

2.12 Determine the step response of an LTI system whose impulse response is given by

$$h(n) = (-1)^n [u(n+2) - u(n-3)]$$

2.13 Let $y(t) = x(t) * h(t)$ and $g(t) = x(t/a) * h(t/a)$. Show that $g(t) = Ay(Bt)$. Determine the values of A and B.

2.14 Find the autocorrelation of the following signals:
(a) $x(t) = e^{-at}u(t)$
(b) $x(t) = A\Pi\left(\frac{t}{T}\right)$

2.15 Find the expression for the impulse response relating the input $x(t)$ to the output $y(t)$ in terms of the impulse response of each subsystem for the system shown in Fig. 2.49.

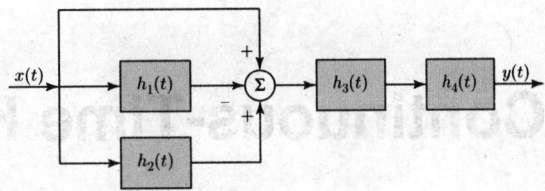

Fig. 2.49

ANSWERS TO MULTIPLE-CHOICE QUESTIONS

1. **(b)**   2. **(c)**   3. **(b)**   4. **(c)**   5. **(b)**   6. **(c)**   7. **(c)**   8. **(a)**   9. **(c)**   10. **(c)**
11. **(d)**   12. **(b)**   13. **(a)**   14. **(b)**   15. **(d)**

Continuous-Time Fourier Series

3.1 INTRODUCTION

As we have seen in the previous chapter, we can obtain the response of a linear system to an arbitrary input by representing it in terms of basic signals. The specific signals used were the shifted δ-functions. Often, it is convenient to choose a set of orthogonal waveforms as the basic signals. There are several reasons for doing this. First, it is mathematically convenient to represent an arbitrary signal as a weighted sum of orthogonal waveforms, since many of the calculations involving signals are simplified by using such a representation. Second, it is possible to visualize the signal as a vector in an orthogonal coordinate system, with the orthogonal waveforms being coordinates. Finally, representations in terms of orthogonal basis functions provides a convenient means of solving for the response of linear systems to arbitrary inputs.

For periodic signals, a convenient choice for an orthogonal basis is the set of harmonically related complex exponentials. The choice of these waveforms is appropriate since such complex exponentials are periodic, are relatively easy to manipulate mathematically, and yield results that have a meaningful physical interpretation. *The representation of a non-sinusoidal periodic signal in terms of complex exponentials, or equivalently, in terms of sine and cosine waveforms, leads to the Fourier series.* Or, in other words, *the Fourier series is a mathematical tool that allows the representation of any periodic signal as the sum of harmonically related sinusoids.* The Fourier series is named after the French physicist Joseph Fourier, who was the first to suggest that periodic signals could be represented by a sum of sinusoids.

So far, we have considered only time domain descriptions of continuous-time signals and systems. In this chapter, we introduce the concept of frequency domain representations and we learn how to decompose periodic signals into their frequency components.

3.1.1 A Vector View of Signals: Orthogonal Representations

Orthogonal representation of signals are important in solving many engineering problems. Two of the reasons this is so are that it is mathematically convenient to represent arbitrary signals as a weighted sum of orthogonal waveforms, since many of the calculations involving signals are simplified by using such a representation, and that it is possible to visualize the signal as a vector in an orthogonal coordinate system, with the orthogonal waveforms being the unit coordinates.

We define an N-dimensional *orthogonal space* as a space characterized by a set of N linearly independent functions $\{\phi_k(t)\}$ called *basis functions*. Any arbitrary function in the space can be generated by a linear combinations of these functions. The basis functions must satisfy the conditions

$$\int_0^T \phi_k(t)\phi_m^*(t)\,dt = C_k\delta(k-m), \qquad 0 \le t \le T \quad k,m = 1,2,\ldots,N \qquad (3.1)$$

where $\phi_k^*(t)$ stands for the complex conjugate of the signal and $\delta(k-m)$, called the Kronecker delta function, is defined as

$$\delta(k-m) = \begin{cases} 1, & k=m \\ 0, & k \ne m \end{cases}$$

When the C_k constants are nonzero, the signal space is called orthogonal. When the basis functions are normalized so that each $C_k = 1$, the space is called an orthonormal space. The principal requirement for orthogonality can be stated as follows: Each $\phi_k(t)$ function of the set of basis functions must be independent of the other members of the set. From a geometric point of view, each $\phi_k(t)$ is mutually perpendicular to each of the other $\phi_k(t)$ for $k \ne m$. An example of such a space with $N = 3$ is shown in Fig. 3.1, where the mutually perpendicular axes are designated $\phi_1(t)$, $\phi_2(t)$, and $\phi_3(t)$. If $\phi_k(t)$ corresponds to a real-valued voltage or current waveform component, associated with a $1\,\Omega$ resistive load, then the energy dissipated in the load in T seconds, due to $\phi_k(t)$, is

$$E_k = \int_0^T \phi_k^2(t)\,dt = C_k \qquad (3.2)$$

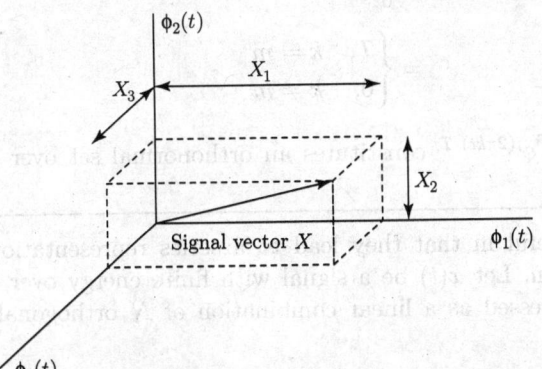

Fig. 3.1 Vectorial representation of the signal $x(t)$

Example 3.1 Show that the signals $\phi_m(t) = \sin mt$, $m = 1, 2, 3, \ldots$, form an orthogonal set on the interval $-\pi < t < \pi$.

Solution

The signals $\phi_m(t) = \sin mt$, $i = 1, 2, 3, \ldots$, form an orthogonal set on the interval $-\pi < t < \pi$ because

$$\int_{-\pi}^{\pi} \phi_m(t)\phi_n(t)\, dt = \int_{-\pi}^{\pi} (\sin mt)(\sin nt)\, dt$$

$$= \frac{1}{2}\int_{-\pi}^{\pi} \cos(m-n)t\, dt - \frac{1}{2}\int_{-\pi}^{\pi} \cos(m+n)t\, dt$$

$$= \begin{cases} \pi, & m = n \\ 0, & m \neq n \end{cases}$$

Since the energy in each signal equals π, the following set of signals constitutes an orthonormal set over the interval $-\pi < t < \pi$:

$$\frac{\sin t}{\sqrt{\pi}}, \frac{\sin 2t}{\sqrt{\pi}}, \frac{\sin 3t}{\sqrt{\pi}}, \ldots$$

Example 3.2 Show that the signals $\phi_k(t) = e^{j(2\pi kt)/T}$, $k = 0, \pm 1, \pm 2, \ldots$, form an orthogonal set on the interval $0 < t < T$.

Solution:

The signals $\phi_k(t) = e^{j(2\pi kt)/T}$, $k = 0, \pm 1, \pm 2, \ldots$, form an orthogonal set on the interval $0 < t < T$ because

$$\int_{-\pi}^{\pi} \phi_k(t)\phi_m(t)\, dt = \int_{0}^{T} e^{j(2\pi kt)/T} e^{-j(2\pi mt)/T}\, dt$$

$$= \begin{cases} T, & k = m \\ 0, & k \neq m \end{cases}$$

and hence, the signal $1/\sqrt{T}\, e^{j(2\pi kt)/T}$ constitutes an orthonormal set over the interval $0 < t < T$.

Orthonormal sets are useful in that they lead to a series representation of signals in a relatively simple fashion. Let $x(t)$ be a signal with finite energy over the interval $0 < t < T$. $x(t)$ can be expressed as a linear combination of N orthogonal waveforms $\phi_1(t), \phi_2(t), \ldots, \phi_N(t)$ as

$$x(t) = X_1\phi_1(t) + X_2\phi_2(t) + \cdots + X_N\phi_N(t)$$

This relationship is expressed in more compact notation as

$$x(t) = \sum_{k=1}^{N} X_k \phi_k(t) \tag{3.3}$$

where

$$X_k = \frac{1}{C_k} \int_0^T x(t)\phi_k(t)\,dt, \quad k = 1,\ldots,N \tag{3.4}$$

The coefficient X_k is the value of the $\phi_k(t)$ component of signal $x(t)$. The signal $x(t)$ can be viewed as a set of vectors, $\mathbf{X} = \{X_1, X_2, \ldots, X_N\}$. If, for example, $N = 3$, we may plot the vector $\mathbf{X}$ corresponding to the waveform

$$x(t) = X_1\phi_1(t) + X_2\phi_2(t) + X_3\phi_3(t)$$

as a point in a three-dimensional Euclidean space with coordinates (X_1, X_2, X_3) as shown in Fig. 3.1. In general, once a set of N orthogonal functions has been adopted, $x(t)$ is completely determined by the vector of its coefficients:

$$\mathbf{X} = (X_1, X_2, \ldots, X_N) \tag{3.5}$$

The normalized energy E_x of $x(t)$ over the interval $0 < t < T$ is

$$E_x = \int_0^T |x(t)|^2\,dt = \int_0^T \left| \sum_{k=1}^{N} X_k\phi_k(t) \right|^2 dt$$

$$= \sum_{j=1}^{N} \sum_{k=1}^{N} X_j X_k^* \int_0^T \phi_j(t)\phi_k^*(t)\,dt$$

$$= \sum_{j=1}^{N} \sum_{k=1}^{N} X_j X_k^* C_k \delta(j - k)$$

$$= \sum_{k=1}^{N} |X_k|^2 C_k$$

This result relates the energy of the signal $x(t)$ to the sum of the squares of the orthogonal series coefficients. If orthonormal functions are used (i.e., $C_k = 1$), the normalized energy is given by

$$E_x = \sum_{k=1}^{N} |X_k|^2$$

In terms of the coefficient vector $\mathbf{X}$, the normalized energy can be written as

$$E_x = (\mathbf{X}^*)^T \mathbf{X}$$

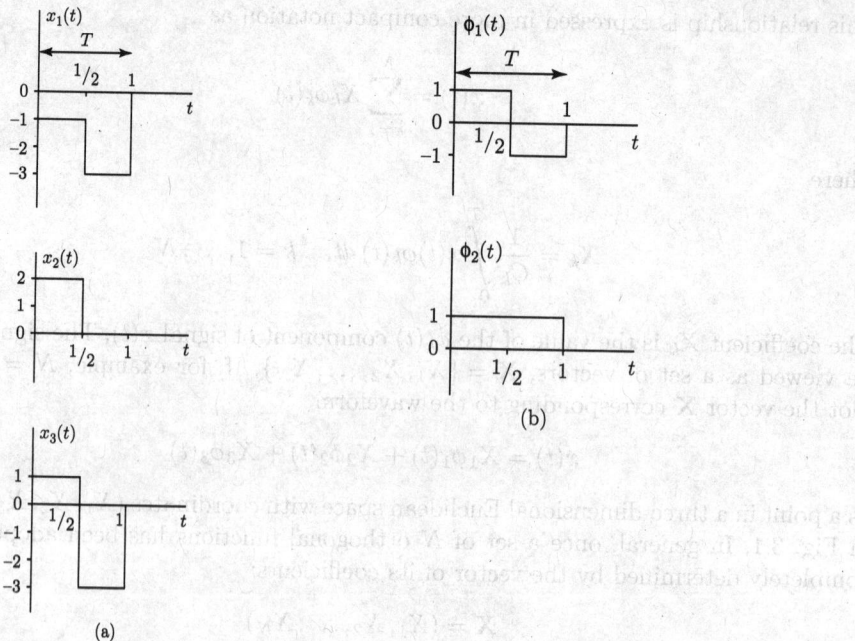

Fig. 3.2 (a) Arbitrary signal set and (b) a set of orthogonal basis functions

Example 3.3

(a) Figure 3.2(a) shows a set of three waveforms, $x_1(t)$, $x_2(t)$, and $x_3(t)$. Demonstrate that these waveforms do not form an orthogonal set.

(b) Figure 3.2(b) shows a set of two waveforms, $\phi_1(t)$ and $\phi_2(t)$. Verify that these waveforms form an orthogonal set.

(c) Show how the nonorthogonal waveforms set in part (a) can be expressed as a linear combination of the orthogonal set in part (b).

Solution

(a) $x_1(t)$, $x_2(t)$, and $x_3(t)$ are clearly not orthogonal, since they do not meet the requirements of Eq. (3.1); that is, the time integrated value (over a symbol duration) of the cross product of any two of the three waveforms is not zero. Let us verify this for $x_1(t)$ and $x_2(t)$:

$$\int_0^T x_1(t)x_2(t)\,dt = \int_0^{T/2} x_1(t)x_2(t)\,dt + \int_{T/2}^T x_1(t)x_2(t)\,dt$$

$$= \int_0^{T/2} (-1)(2)\,dt + \int_{T/2}^T (-3)(0)\,dt$$

$$= -T = -1$$

Similarly, for $x_2(t)$ and $x_3(t)$:

$$\int_0^T x_2(t)x_3(t)\,dt = \int_0^{T/2} x_2(t)x_3(t)\,dt + \int_{T/2}^T x_2(t)x_3(t)\,dt$$

$$= \int_0^{T/2} (2)(1)\,dt + \int_{T/2}^T (0)(-3)\,dt$$

$$= T = 1$$

Similarly, for $x_1(t)$ and $x_3(t)$:

$$\int_0^T x_1(t)x_3(t)\,dt = \int_0^{T/2} x_1(t)x_3(t)\,dt + \int_{T/2}^T x_1(t)x_3(t)\,dt$$

$$= \int_0^{T/2} (-1)(1)\,dt + \int_{T/2}^T (-3)(-3)\,dt = 4T = 4$$

The integral over the interval T of each of the cross products $x_1(t)x_2(t)$, $x_2(t)x_3(t)$, and $x_1(t)x_3(t)$ results in nonzero values. Hence, the waveform set $\{x_i(t)\}(i = 1, 2, 3)$ in Fig. 3.2(a) is not an orthogonal set.

(b) Using Eq. (3.1), we verify that $\phi_1(t)$ and $\phi_2(t)$ form an orthogonal set as follows:

$$\int_0^T \phi_1(t)\phi_2(t)\,dt = \int_0^{T/2} (1)(1)\,dt + \int_{T/2}^T (-1)(1)\,dt = 0$$

The integral over the interval T of the cross product $\phi_1(t)\phi_2(t)$ results in zero value. Hence, the waveforms $\phi_1(t)\phi_2(t)$ in Fig. 3.2(b) form an orthogonal set.

(c) Using Eq. (3.2),

$$C_1 = \int_0^T \phi_1^2(t)\,dt = \int_0^{T/2} (1)^2\,dt + \int_{T/2}^T (-1)^2\,dt = T$$

and

$$C_2 = \int_0^T \phi_2^2(t)\,dt = \int_0^T (1)^2\,dt = T$$

now using Eq. (3.4) with $C_1 = C_2 = T = 1$, we can express the nonorthogonal set $\{x_i(t)\}(i = 1, 2, 3)$ as a linear combination of the orthogonal basis waveforms $\phi_1(t)$

and $\phi_2(t)$:

$$x_1(t) = X_{11}\phi_1(t) + X_{12}\phi_2(t)$$
$$x_2(t) = X_{21}\phi_1(t) + X_{22}\phi_2(t)$$
$$x_3(t) = X_{31}\phi_1(t) + X_{32}\phi_2(t)$$

where, from Eq. (3.4), X_{1k} is

$$X_{1k} = \frac{1}{C_k} \int_0^T x_1(t)\phi_k(t)\, dt$$

Therefore,

$$X_{11} = \frac{1}{C_1} \int_0^T x_1(t)\phi_1(t)\, dt = \frac{1}{T} \int_0^{T/2} (-1)(1)\, dt + \int_{T/2}^T (-3)(-1)\, dt = 1$$

$$X_{12} = \frac{1}{C_2} \int_0^T x_1(t)\phi_2(t)\, dt = \frac{1}{T} \int_0^{T/2} (-1)(1)\, dt + \int_{T/2}^T (-3)(1)\, dt = -2$$

Now, from Eq. (3.4), X_{2k} is

$$X_{2k} = \frac{1}{C_k} \int_0^T x_2(t)\phi_k(t)\, dt$$

Therefore,

$$X_{21} = \frac{1}{C_1} \int_0^T x_2(t)\phi_1(t)\, dt = \frac{1}{T} \int_0^{T/2} (2)(1)\, dt + \int_{T/2}^T (0)(-1)\, dt = 1$$

$$X_{22} = \frac{1}{C_2} \int_0^T x_2(t)\phi_2(t)\, dt = \frac{1}{T} \int_0^{T/2} (2)(1)\, dt + \int_{T/2}^T (0)(1)\, dt = 1$$

Again, from Eq. (3.4), X_{3k} is

$$X_{3k} = \frac{1}{C_k} \int_0^T x_3(t)\phi_k(t)\, dt$$

Therefore,

$$X_{31} = \frac{1}{C_1} \int_0^T x_3(t)\phi_1(t)\,dt = \frac{1}{T} \left[\int_0^{T/2} (1)(1)\,dt + \int_{T/2}^T (-3)(-1)\,dt \right] = 2$$

$$X_{32} = \frac{1}{C_2} \int_0^T x_3(t)\phi_2(t)\,dt = \frac{1}{T} \left[\int_0^{T/2} (1)(1)\,dt + \int_{T/2}^T (-3)(1)\,dt \right] = -1$$

The nonorthogonal set $\{x_i(t)\}(i = 1, 2, 3)$ is a linear combination of the orthogonal basis waveforms $\phi_1(t)$ and $\phi_2(t)$ as

$$x_1(t) = \phi_1(t) - 2\phi_2(t)$$
$$x_2(t) = \phi_1(t) + \phi_2(t)$$
$$x_3(t) = 2\phi_1(t) - \phi_2(t)$$

3.2 FOURIER SERIES

The Fourier series is a mathematical tool that allows the representation of any periodic signal as the sum of harmonically related sinusoids. Any periodic signal, i.e., one for which $x(t) = x(t + T)$, can be expressed by a Fourier series provided that

1. if it is discontinuous, there are a finite number of discontinuities in the period T;
2. it has a finite average value over the period T;
3. it has a finite number of positive and negative maxima in the period T.

When these *Dirichlet conditions* are satisfied, the Fourier series exist. The Fourier series is of two types: (i) trigonometric Fourier series and (ii) exponential Fourier series.

3.2.1 Trigonometric Fourier Series

The trigonometric Fourier series is expressed as

$$x(t) = a_0 + \sum_{n=1}^{\infty} \left[a_n \cos(n\omega_0 t) + b_n \sin(n\omega_0 t) \right] \tag{3.6}$$

where a_0, a_n, and b_n are the trigonometric Fourier series coefficients. The coefficients may be obtained from $x(t)$ using

$$a_0 = \frac{1}{T} \int_0^T x(t)\,dt \tag{3.7}$$

$$a_n = \frac{2}{T} \int_0^T x(t) \cos(n\omega_0 t)\, dt \tag{3.8}$$

$$b_n = \frac{2}{T} \int_0^T x(t) \sin(n\omega_0 t)\, dt \tag{3.9}$$

The integrations can be carried out from $-T/2$ to $T/2$ or over any other full period that might simplify the calculation. The series with coefficients obtained from the above evaluation integrals converges uniformly to the function at all points of continuity and converges to the mean value at points of discontinuity. The series of Eq. (3.6) may be written in a number of apparently different although equivalent forms (polar form).

3.2.2 Polar Form Representation of the Fourier series

There are two ways to represent the Fourier series in a polar form:

Case I Assume $a_n = c_n \cos(\theta_n)$ and $b_n = -c_n \sin(\theta_n)$, where c_n and θ_n are related to a_n and b_n as

$$c_0 = a_0 \quad \text{and} \quad c_n = \sqrt{a_n^2 + b_n^2}, \quad \text{for } n \geq 1 \tag{3.10}$$

$$\theta_n = \tan^{-1} \frac{-b_n}{a_n} \tag{3.11}$$

Substituting $a_n = c_n \cos(\theta_n)$ and $b_n = -c_n \sin(\theta_n)$ into Eq. (3.6) gives

$$x(t) = a_0 + \sum_{n=1}^{\infty} \left[c_n \cos(\theta_n) \cos(n\omega_0 t) - c_n \sin(\theta_n) \sin(n\omega_0 t) \right]$$

$$= a_0 + \sum_{n=1}^{\infty} c_n \cos(n\omega_0 t + \theta_n)$$

$$x(t) = c_0 + \sum_{n=1}^{\infty} c_n \cos(n\omega_0 t + \theta_n) \tag{3.12}$$

with c_n and θ_n defined by Eqs (3.10) and (3.11). The coefficient c_n is the amplitude and θ_n the phase of the nth harmonic. Observe that if we know that a Fourier series is to be constructed in the form of Eq. (3.12), then the set of numbers c_n and θ_n contains all the needed information. The plot of c_n as a function of n is known as the amplitude spectrum; the plot of θ_n as a function of n is the phase spectrum.

Case II Assume $a_n = c_n \sin(\phi_n)$ and $b_n = c_n \cos(\phi_n)$, where c_n and ϕ_n are related to a_n and b_n as

$$c_0 = a_0 \quad \text{and} \quad c_n = \sqrt{a_n^2 + b_n^2}, \text{ for } n \geq 1 \tag{3.13}$$

$$\phi_n = \tan^{-1} \frac{a_n}{b_n} \tag{3.14}$$

Substituting $a_n = c_n \sin(\phi_n)$ and $b_n = c_n \cos(\phi_n)$ in Eq. (3.6) gives

$$x(t) = a_0 + \sum_{n=1}^{\infty} \left[c_n \sin(\phi_n) \cos(n\omega_0 t) + c_n \cos(\phi_n) \sin(n\omega_0 t) \right]$$

$$= a_0 + \sum_{n=1}^{\infty} c_n \sin(n\omega_0 t + \phi_n)$$

$$x(t) = c_0 + \sum_{n=1}^{\infty} c_n \sin(n\omega_0 t + \phi_n) \tag{3.15}$$

with c_n and ϕ_n defined by Eqs (3.13) and (3.14).

3.2.3   Evaluation of Fourier Series Coefficients

In this section we will derive Eqs (3.7), (3.8), and (3.9). The evaluation of a_0, a_n, and b_n coefficients in Eq. (3.6) is accomplished by using simple integral equations, which may be derived by using the orthogonality property of the set of functions involved, namely $\cos(n\omega_0 t)$ and $\sin(n\omega_0 t)$ with integer values for n and m. These functions are orthogonal over the interval from t_0 to $t_0 + T$ for any t_0. We will often use the value $t_0 = 0$ or $t_0 = -T/2$, but with the understanding that any period may be used. First observe that

$$\int_{0}^{T} \sin(m\omega_0 t)\, dt = 0, \quad \text{for all } m \tag{3.16}$$

and

$$\int_{0}^{T} \cos(n\omega_0 t)\, dt = 0, \quad \text{for all } n \neq 0 \tag{3.17}$$

since the average value of a sinusoid over m and n complete cycles in the period T is zero. The following three cross product terms are also zero for the stated relationships of m and n:

$$\int_{0}^{T} \sin(m\omega_0 t) \cos(n\omega_0 t)\, dt = 0, \quad \text{for all } m, n \tag{3.18}$$

$$\int_{0}^{T} \sin(m\omega_0 t) \sin(n\omega_0 t)\, dt = \begin{cases} 0, & m \neq n \\ \dfrac{T}{2}, & m = n \end{cases} \tag{3.19}$$

and

$$\int_{0}^{T} \cos(m\omega_0 t) \cos(n\omega_0 t)\, dt = \begin{cases} 0, & m \neq n \\ \dfrac{T}{2}, & m = n \end{cases} \tag{3.20}$$

Case I Proof of Eq. (3.7), i.e., $a_0 = \dfrac{1}{T} \displaystyle\int_0^T x(t)\,dt$

From Eq. (3.6), we have

$$x(t) = a_0 + [a_1 \cos(\omega_0 t) + b_1 \sin(\omega_0 t)] + [a_2 \cos(2\omega_0 t) + b_2 \sin(2\omega_0 t)] + \cdots$$
$$+ [a_n \cos(n\omega_0 t) + b_n \sin(n\omega_0 t)] + \cdots$$

Integrating the above equation over the interval of time from 0 to T gives

$$\int_0^T x(t)\,dt = \int_0^T a_0\,dt + \int_0^T a_1 \cos(\omega_0 t)\,dt + \int_0^T b_1 \sin(\omega_0 t)\,dt + \int_0^T a_2 \cos(2\omega_0 t)\,dt$$
$$+ \int_0^T b_2 \sin(2\omega_0 t)\,dt + \cdots + \int_0^T a_n \cos(n\omega_0 t)\,dt + \int_0^T b_n \sin(n\omega_0 t)\,dt + \cdots$$

Using the results of Eqs (3.16) and (3.17), all the terms on the RHS of the above equation are found to have a zero value, except the first term. Therefore,

$$\int_0^T x(t)\,dt = \int_0^T a_0\,dt + 0 + 0 + \cdots$$

$$\int_0^T x(t)\,dt = a_0 T$$

$$a_0 = \frac{1}{T} \int_0^T x(t)\,dt$$

The coefficient a_0 is simply the average value of $x(t)$ over a period, sometimes also known as the dc value of the signal.

Case II Proof of Eq. (3.8), i.e., $a_n = \dfrac{2}{T} \displaystyle\int_0^T x(t) \cos(n\omega_0 t)\,dt$

From Eq. (3.6), we have

$$x(t) = a_0 + \big[a_1 \cos(\omega_0 t) + b_1 \sin(\omega_0 t)\big] + \big[a_2 \cos(2\omega_0 t) + b_2 \sin(2\omega_0 t)\big] + \cdots$$
$$+ \big[a_n \cos(n\omega_0 t) + b_n \sin(n\omega_0 t)\big] + \cdots$$

By multiplying both sides of the above equation by $\cos(n\omega_0 t)$ and integrating over the interval of time from 0 to T, we get

$$\int_0^T x(t)\cos(n\omega_0 t)\,dt = \int_0^T a_0 \cos(n\omega_0 t)\,dt + \int_0^T a_1 \cos(\omega_0 t)\cos(n\omega_0 t)\,dt$$

$$+ \int_0^T b_1 \sin(\omega_0 t)\cos(n\omega_0 t)\,dt + \int_0^T a_2 \cos(2\omega_0 t)\cos(n\omega_0 t)\,dt$$

$$+ \int_0^T b_2 \sin(2\omega_0 t)\cos(n\omega_0 t)\,dt + \cdots$$

$$+ \int_0^T a_n \cos^2(n\omega_0 t)\,dt + \int_0^T b_n \sin(n\omega_0 t)\cos(n\omega_0 t)\,dt + \cdots$$

Using the results of Eqs (3.17), (3.18), and (3.20), all the terms on the RHS of the above equation are found to have a zero value, except the one involving the integration of $\cos^2(n\omega_0 t)$, which has the value $T/2$. Therefore,

$$\int_0^T x(t)\cos(n\omega_0 t)\,dt = a_n \frac{T}{2}$$

$$a_n = \frac{2}{T}\int_0^T x(t)\cos(n\omega_0 t)\,dt$$

Case III Proof of Eq. (3.9), i.e., $b_n = \dfrac{2}{T}\displaystyle\int_0^T x(t)\sin(n\omega_0 t)\,dt$

From Eq.(3.6), we have

$$x(t) = a_0 + \big[a_1 \cos(\omega_0 t) + b_1 \sin(\omega_0 t)\big] + \big[a_2 \cos(2\omega_0 t) + b_2 \sin(2\omega_0 t)\big] + \cdots$$

$$+ \big[a_n \cos(n\omega_0 t) + b_n \sin(n\omega_0 t)\big] + \cdots$$

By multiplying both sides of the above equation by $\sin(n\omega_0 t)$ and integrating over the interval of time from 0 to T, we get

$$\int_0^T x(t)\sin(n\omega_0 t)\,dt = \int_0^T a_0 \sin(n\omega_0 t)\,dt + \int_0^T a_1 \sin(\omega_0 t)\sin(n\omega_0 t)\,dt$$

$$+ \int_0^T b_1 \sin(\omega_0 t)\sin(n\omega_0 t)\,dt$$

$$+ \int_0^T a_2 \cos(2\omega_0 t) \sin(n\omega_0 t)\, dt + \int_0^T b_2 \sin(2\omega_0 t) \sin(n\omega_0 t)\, dt + \cdots$$

$$+ \int_0^T a_n \cos(n\omega_0 t) \sin(n\omega_0 t)\, dt + \int_0^T b_n \sin^2(n\omega_0 t)\, dt + \cdots$$

Using the results of Eqs (3.16), (3.18), and (3.19), all the terms on the RHS of the above equation are found to have a zero value, except the one involving the integration of $\sin^2(n\omega_0 t)$, which has the value $T/2$. Therefore,

$$\int_0^T x(t) \sin(n\omega_0 t)\, dt = b_n \frac{T}{2}$$

$$b_n = \frac{2}{T} \int_0^T x(t) \sin(n\omega_0 t)\, dt$$

Example 3.4 Find the trigonometric Fourier series for the waveform shown in Fig. 3.3.

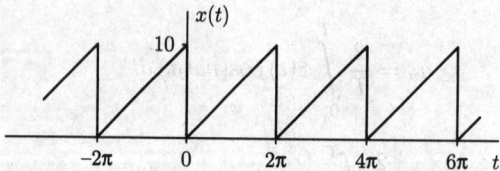

Fig. 3.3 A periodic signal

Solution

The waveform is periodic with period $T = 2\pi$ and fundamental frequency $\omega_0 = 2\pi/T = 1$. The given waveform for one period may be written as

$$x(t) = \frac{10}{2\pi} t, \quad 0 < t < 2\pi$$

From Eq. (3.6), the trigonometric Fourier series of $x(t)$ is

$$x(t) = a_0 + \sum_{n=1}^{\infty} \left[a_n \cos(n\omega_0 t) + b_n \sin(n\omega_0 t) \right]$$

where

$$a_0 = \frac{1}{T} \int\limits_0^T x(t)\, dt = \frac{1}{2\pi} \int\limits_0^{2\pi} \frac{10}{2\pi} t\, dt = \frac{10}{(2\pi)^2} \left[\frac{t^2}{2}\right]_0^{2\pi} = 5$$

$$a_n = \frac{2}{T} \int\limits_0^T x(t) \cos(n\omega_0 t)\, dt = \frac{2}{2\pi} \int\limits_0^{2\pi} \frac{10}{2\pi} t \cos(nt)\, dt$$

$$= \frac{10}{2\pi^2} \left[\frac{t}{n} \sin(nt) + \frac{1}{n^2} \cos(nt)\right]_0^{2\pi} = \frac{10}{2n^2\pi^2}[\cos(2n\pi) - \cos(0)] = 0$$

and

$$b_n = \frac{2}{T} \int\limits_0^T x(t) \sin(n\omega_0 t)\, dt = \frac{2}{2\pi} \int\limits_0^{2\pi} \frac{10}{2\pi} t \sin(nt)\, dt$$

$$= \frac{10}{2\pi^2} \left[-\frac{t}{n} \cos(nt) + \frac{1}{n^2} \sin(nt)\right]_0^{2\pi} = -\frac{10}{n\pi}$$

Substituting these coefficients (a_0, a_n, and b_n) into Eq. (3.6), we get

$$x(t) = 5 - \sum_{n=1}^{\infty} \left[0 + \frac{10}{n\pi} \sin(nt)\right]$$

$$= 5 - \frac{10}{\pi} \sin(t) - \frac{10}{2\pi} \sin(2t) - \frac{10}{3\pi} \sin(3t) - \cdots$$

3.2.4 Symmetry Conditions

The series obtained in Example 3.4 contained only sine terms in addition to a constant term. Other waveforms will have only cosine terms; and sometimes only odd harmonics are present in the series, whether the series contains sine, cosine, or both types of terms. This is the result of certain types of symmetry exhibited by the waveform. Knowledge of such symmetry results in reduced calculations in determining the Fourier series.

1. **$x(t)$ is an even signal, i.e., $x(t) = x(-t)$**

   The signal $x(t) = 2 + t^2 + t^4$ is an example of even signal since the functional values for t and $-t$ are equal. Even signals of special interest to us are the functions $\cos(n\omega_0 t)$. The sum or product of two or more even signals is an even signal, and with the addition of a constant, the even nature of the signal is still preserved. In Fig. 3.4, the waveforms shown represent even functions of t. They are symmetrical w.r.t. the vertical axis.

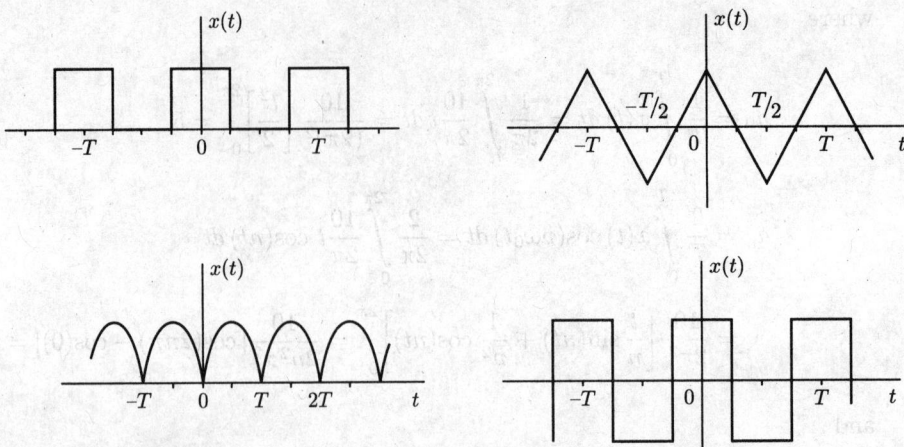

Fig. 3.4 Waveforms with even symmetry

If $x(t)$ has an even symmetry, then the Fourier series coefficients a_0, a_n, and b_n are defined as (we choose our interval for integration from $-T/2$ to $T/2$)

$$a_0 = \frac{2}{T} \int_0^{T/2} x(t)\,dt$$

$$a_n = \frac{4}{T} \int_0^{T/2} x(t)\cos(n\omega_0 t)\,dt \qquad (3.21)$$

$$b_n = 0$$

The Fourier series expansion of an even periodic signal contains only cosine terms and a constant.

Proof

(a) From Eq. (3.7), we have

$$a_0 = \frac{1}{T} \int_{-T/2}^{T/2} x(t)\,dt$$

$$= \frac{1}{T} \left(\int_{-T/2}^{0} x(t)\,dt + \int_{0}^{T/2} x(t)\,dt \right)$$

$$= \frac{1}{T} \left(\int_{0}^{T/2} x(-t)\,dt + \int_{0}^{T/2} x(t)\,dt \right)$$

Since, $x(t)$ is an even signal, i.e., $x(t) = x(-t)$, we have

$$a_0 = \frac{1}{T} \left(\int\limits_0^{T/2} x(t)\, dt + \int\limits_0^{T/2} x(t)\, dt \right)$$

$$a_0 = \frac{2}{T} \int\limits_0^{T/2} x(t)\, dt$$

(b) Now consider Eq. (3.8)

$$a_n = \frac{2}{T} \int\limits_{-T/2}^{T/2} x(t) \cos(n\omega_0 t)\, dt$$

$$= \frac{2}{T} \left(\int\limits_{-T/2}^{0} x(t) \cos(n\omega_0 t)\, dt + \int\limits_0^{T/2} x(t) \cos(n\omega_0 t)\, dt \right)$$

$$= \frac{2}{T} \left(\int\limits_0^{T/2} x(-t) \cos(-n\omega_0 t)\, dt + \int\limits_0^{T/2} x(t) \cos(n\omega_0 t)\, dt \right)$$

Both $x(t)$ and $\cos(n\omega_0 t)$ are even signals. The product of two even signals results in an even signal, and therefore

$$a_n = \frac{2}{T} \left(\int\limits_0^{T/2} x(t) \cos(n\omega_0 t)\, dt + \int\limits_0^{T/2} x(t) \cos(n\omega_0 t)\, dt \right)$$

$$a_n = \frac{4}{T} \int\limits_0^{T/2} x(t) \cos(n\omega_0 t)\, dt$$

(c) From Eq. (3.9), we have

$$b_n = \frac{2}{T} \int\limits_{-T/2}^{T/2} x(t) \sin(n\omega_0 t)\, dt$$

$$= \frac{2}{T} \left(\int\limits_{-T/2}^{0} x(t) \sin(n\omega_0 t)\, dt + \int\limits_0^{T/2} x(t) \sin(n\omega_0 t)\, dt \right)$$

$$= \frac{2}{T} \left(\int\limits_0^{T/2} x(-t) \sin(-n\omega_0 t)\, dt + \int\limits_0^{T/2} x(t) \sin(n\omega_0 t)\, dt \right)$$

$x(t)$ is an even signal and $\sin(n\omega_0 t)$ is an odd signal. The product of an odd and an even signal results in an odd signal, and therefore

$$b_n = \frac{2}{T}\left(-\int_0^{T/2} x(t)\sin(n\omega_0 t)\,dt + \int_0^{T/2} x(t)\sin(n\omega_0 t)\,dt\right)$$

$$b_n = 0 \qquad\qquad\blacksquare$$

2. **$x(t)$ is an odd signal, i.e., $x(t) = -x(-t)$**

The signal $x(t) = t + t^3 + t^5$ is an example of an odd signal since the values of the signal for t and $-t$ are of opposite sign. Odd signals of special interest to us are the functions $\sin(n\omega_0 t)$. The sum of two or more odd signals is an odd signal, but the addition of a constant removes the odd nature of the function. The product of two odd signals is an even signal. The waveforms shown in Fig. 3.5 represent odd functions of t.

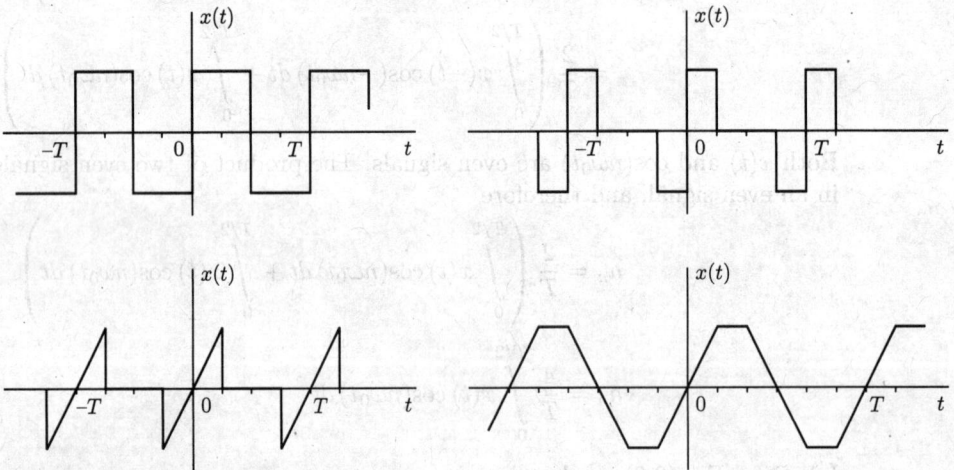

Fig. 3.5 Waveforms with odd symmetry

If $x(t)$ has an odd symmetry, then the Fourier series coefficients a_0, a_n, and b_n are defined as

$$a_0 = 0$$

$$a_n = 0 \qquad\qquad (3.22)$$

$$b_n = \frac{4}{T}\int_0^{T/2} x(t)\sin(n\omega_0 t)\,dt$$

The Fourier series expansion of an odd periodic signal contains only sine terms.

Proof

(a) From Eq. (3.7), we have

$$a_0 = \frac{1}{T} \int_{-T/2}^{T/2} x(t)\, dt$$

$$= \frac{1}{T} \left(\int_{-T/2}^{0} x(t)\, dt + \int_{0}^{T/2} x(t)\, dt \right)$$

$$= \frac{1}{T} \left(\int_{0}^{T/2} x(-t)\, dt + \int_{0}^{T/2} x(t)\, dt \right)$$

Since, $x(t)$ is an odd signal, i.e., $x(t) = -x(-t)$, we have

$$a_0 = \frac{1}{T} \left(-\int_{0}^{T/2} x(t)\, dt + \int_{0}^{T/2} x(t)\, dt \right) = 0$$

(b) From Eq. (3.8), we have

$$a_n = \frac{2}{T} \int_{-T/2}^{T/2} x(t) \cos(n\omega_0 t)\, dt$$

$$= \frac{2}{T} \left(\int_{-T/2}^{0} x(t) \cos(n\omega_0 t)\, dt + \int_{0}^{T/2} x(t) \cos(n\omega_0 t)\, dt \right)$$

$$= \frac{2}{T} \left(\int_{0}^{T/2} x(-t) \cos(-n\omega_0 t)\, dt + \int_{0}^{T/2} x(t) \cos(n\omega_0 t)\, dt \right)$$

$x(t)$ is an odd signal and $\cos(n\omega_0 t)$ is an even signal. The product of an odd and an even signal results in an odd signal, and therefore

$$a_n = \frac{2}{T} \left(-\int_{0}^{T/2} x(t) \cos(n\omega_0 t)\, dt + \int_{0}^{T/2} x(t) \cos(n\omega_0 t)\, dt \right) = 0$$

(c) From Eq. (3.9), we have

$$b_n = \frac{2}{T} \int_{-T/2}^{T/2} x(t) \sin(n\omega_0 t) \, dt$$

$$= \frac{2}{T} \left(\int_{-T/2}^{0} x(t) \sin(n\omega_0 t) \, dt + \int_{0}^{T/2} x(t) \sin(n\omega_0 t) \, dt \right)$$

$$= \frac{2}{T} \left(\int_{0}^{T/2} x(-t) \sin(-n\omega_0 t) \, dt + \int_{0}^{T/2} x(t) \sin(n\omega_0 t) \, dt \right)$$

Both $x(t)$ and $\sin(n\omega_0 t)$ are odd signals. The product of two odd signals results in an even signal, and therefore

$$b_n = \frac{2}{T} \left(\int_{0}^{T/2} x(t) \sin(n\omega_0 t) \, dt + \int_{0}^{T/2} x(t) \sin(n\omega_0 t) \, dt \right)$$

$$b_n = \frac{4}{T} \int_{0}^{T/2} x(t) \sin(n\omega_0 t) \, dt$$

3. **Half-wave symmetry, i.e., $x(t) = -x\left(t \pm \dfrac{T}{2}\right)$**

This symmetry is described by

$$x(t) = -x\left(t \pm \frac{T}{2}\right)$$

and is illustrated in Fig. 3.6, showing that the waveform with t increasing from $-T/2$ to 0 is the negative of the waveform with t increasing from 0 to $T/2$. Clearly, this waveform is neither even nor odd, so it must be both.

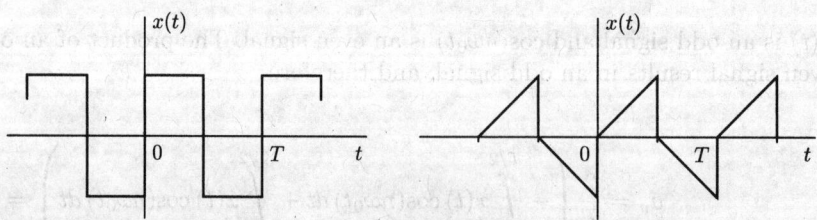

Fig. 3.6 Waveforms with half-wave symmetry

If $x(t)$ has a half-wave symmetry, then the Fourier series coefficients a_0, a_n, and b_n are defined as

$$a_0 = 0, \qquad a_n = b_n = 0 \quad n \text{ even}$$

$$a_n = \frac{4}{T} \int_0^{T/2} x(t) \cos(n\omega_0 t)\, dt \quad n \text{ odd} \tag{3.23}$$

$$b_n = \frac{4}{T} \int_0^{T/2} x(t) \sin(n\omega_0 t)\, dt \quad n \text{ odd}$$

The Fourier series expansion of a periodic signal with half-wave symmetry contains only odd harmonics.

Proof

(a) From Eq. (3.7), we have

$$a_0 = \frac{1}{T} \int_{-T/2}^{T/2} x(t)\, dt$$

$$= \frac{1}{T} \left(\int_{-T/2}^{0} x(t)\, dt + \int_{0}^{T/2} x(t)\, dt \right)$$

$$= \frac{1}{T} \left(\int_{0}^{T/2} x(t - T/2)\, dt + \int_{0}^{T/2} x(t)\, dt \right)$$

$x(t)$ has a half-wave symmetry, $-x(t) = x(t - T/2)$.

$$a_0 = \frac{1}{T} \left(- \int_{0}^{T/2} x(t)\, dt + \int_{0}^{T/2} x(t)\, dt \right) = 0$$

(b) From Eq. (3.8)

$$a_n = \frac{2}{T} \int_{-T/2}^{T/2} x(t) \cos(n\omega_0 t)\, dt$$

$$= \frac{2}{T} \left(\int_{-T/2}^{0} x(t) \cos(n\omega_0 t)\, dt + \int_{0}^{T/2} x(t) \cos(n\omega_0 t)\, dt \right)$$

$$= \frac{2}{T} \left(\int_0^{T/2} x(t - T/2) \cos{(n\omega_0(t - T/2))}\, dt + \int_0^{T/2} x(t) \cos(n\omega_0 t)\, dt \right)$$

$$= \frac{2}{T} \left(-\int_0^{T/2} x(t) \cos{(n\omega_0(t - T/2))}\, dt + \int_0^{T/2} x(t) \cos(n\omega_0 t)\, dt \right)$$

Since, $x(t)$ has a half-wave symmetry, i.e., $-x(t) = x(t - T/2)$ and

$$\cos(n\omega_0(t - T/2)) = \cos\left(n\omega_0 t - n\omega_0 \frac{T}{2} \right)$$

$$= \cos\left(n\omega_0 t - n\frac{2\pi}{T}\frac{T}{2} \right)$$

$$= \cos{(n\omega_0 t - n\pi)}$$

$$= \begin{cases} \cos(n\omega_0 t), & n \text{ even} \\ -\cos(n\omega_0 t), & n \text{ odd} \end{cases}$$

therefore, we have

$$a_n = \begin{cases} \dfrac{2}{T} \left(-\displaystyle\int_0^{T/2} x(t) \cos(n\omega_0 t)\, dt + \int_0^{T/2} x(t) \cos(n\omega_0 t)\, dt \right), & n \text{ even} \\[4mm] \dfrac{2}{T} \left(\displaystyle\int_0^{T/2} x(t) \cos(n\omega_0 t)\, dt + \int_0^{T/2} x(t) \cos(n\omega_0 t)\, dt \right), & n \text{ odd} \end{cases}$$

$$a_n = \begin{cases} 0, & n \text{ even} \\[2mm] \dfrac{4}{T} \displaystyle\int_0^{T/2} x(t) \cos(n\omega_0 t)\, dt, & n \text{ odd} \end{cases}$$

(c) From Eq. (3.9)

$$b_n = \frac{2}{T} \int_{-T/2}^{T/2} x(t) \sin(n\omega_0 t)\, dt$$

$$= \frac{2}{T} \left(\int_{-T/2}^{0} x(t) \sin(n\omega_0 t)\, dt + \int_0^{T/2} x(t) \sin(n\omega_0 t)\, dt \right)$$

$$= \frac{2}{T} \left(\int_0^{T/2} x(t - T/2) \sin\left(n\omega_0(t - T/2)\right) \, dt + \int_0^{T/2} x(t) \sin(n\omega_0 t) \, dt \right)$$

$$= \frac{2}{T} \left(-\int_0^{T/2} x(t) \sin\left(n\omega_0(t - T/2)\right) \, dt + \int_0^{T/2} x(t) \sin(n\omega_0 t) \, dt \right)$$

Since, $x(t)$ has a half-wave symmetry, i.e., $-x(t) = x(t - T/2)$ and

$$\sin\left(n\omega_0(t - T/2)\right) = \sin\left(n\omega_0 t - n\omega_0 \frac{T}{2}\right)$$

$$= \sin\left(n\omega_0 t - n\frac{2\pi}{T}\frac{T}{2}\right)$$

$$= \sin(n\omega_0 t - n\pi)$$

$$= \begin{cases} \sin(n\omega_0 t), & n \text{ even} \\ -\sin(n\omega_0 t), & n \text{ odd} \end{cases}$$

therefore, we have

$$b_n = \begin{cases} \dfrac{2}{T} \left(-\displaystyle\int_0^{T/2} x(t) \sin(n\omega_0 t) \, dt + \int_0^{T/2} x(t) \sin(n\omega_0 t) \, dt \right), & n \text{ even} \\ \dfrac{2}{T} \left(\displaystyle\int_0^{T/2} x(t) \sin(n\omega_0 t) \, dt + \int_0^{T/2} x(t) \sin(n\omega_0 t) \, dt \right), & n \text{ odd} \end{cases}$$

$$b_n = \begin{cases} 0, & n \text{ even} \\ \dfrac{4}{T} \displaystyle\int_0^{T/2} x(t) \sin(n\omega_0 t) \, dt, & n \text{ odd} \end{cases}$$

4. **Quarter-wave symmetry**

A periodic function possesses a quarter-wave symmetry if

 (i) it has either odd or even symmetry and
(ii) it has half-wave symmetry.

A waveform that satifies the above conditions and if it is shifted towards right or left by $T/4$ period gives a new waveform with a half-wave symmetry, but the even (odd) symmetry of the original unshifted waveform has changed into odd (even) symmetry, is said to be having a quarter-wave symmetry.

(a) If a periodic function is even, and has quarter-wave symmetry, the Fourier coefficients are

$$a_0 = 0, \quad b_n = 0, \quad \text{for all } n$$

$$a_n = 0, \quad \text{for even } n \tag{3.24}$$

$$a_n = \frac{8}{T} \int_0^{T/4} x(t) \cos(n\omega_0 t)\, dt, \quad \text{for odd } n$$

(b) If a periodic function is odd, and has quarter-wave symmetry, the Fourier coefficients are

$$a_0 = 0, \quad a_n = 0, \quad \text{for all } n$$

$$b_n = 0, \quad \text{for even } n \tag{3.25}$$

$$b_n = \frac{8}{T} \int_0^{T/4} x(t) \sin(n\omega_0 t)\, dt, \quad \text{for odd } n$$

Table 3.1 summarizes the symmetry conditions.

Table 3.1 Fourier Series Representation

Name of symmetry	Condition	Property	a_0	a_n ($n\neq0$)	b_n
Even	$x(t) = x(-t)$	cosine terms only	$\frac{4}{T}\int_0^{T/2} x(t)\,dt$	$\frac{4}{T}\int_0^{T/2} x(t)\cos(n\omega_0 t)\,dt$	0
Odd	$x(t) = -x(-t)$	sine terms only	0	0	$\frac{4}{T}\int_0^{T/2} x(t)\sin(n\omega_0 t)\,dt$
Half-wave	$x(t) = -x(t \pm T/2)$	odd n only	0	$\frac{4}{T}\int_0^{T/2} x(t)\cos(n\omega_0 t)\,dt$	$\frac{4}{T}\int_0^{T/2} x(t)\sin(n\omega_0 t)\,dt$

Example 3.5 Find the trigonometric Fourier series for the square wave shown in Fig. 3.7 and plot the line spectrum.

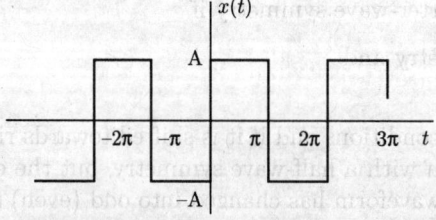

Fig. 3.7 Square wave

Solution

The waveform is periodic with period $T = 2\pi$ and fundamental frequency $\omega_0 = 2\pi/T = 1$. The given waveform for one period may be written as

$$x(t) = \begin{cases} A, & 0 < t < \pi \\ -A, & \pi < t < 2\pi \end{cases}$$

The wave is an odd function, so the symmetry condition requires that [Eq. (3.22)]

$$a_0 = 0, \qquad a_n = 0$$

$$b_n = \frac{4}{T} \int\limits_0^{T/2} x(t) \sin(n\omega_0 t)\, dt = \frac{4}{2\pi} \int\limits_0^{\pi} A \sin(nt)\, dt = \frac{2A}{\pi} \left[-\frac{1}{n} \cos(nt) \right]_0^{\pi}$$

$$= \frac{2A}{n\pi}(1 - \cos(n\pi))$$

$$b_n = \begin{cases} \dfrac{4A}{n\pi}, & n = 1, 3, 5, \ldots \\ 0, & n = 2, 4, 6, \ldots \end{cases}$$

The trigonometric Fourier series for the square wave is

$$x(t) = \sum_{n=1}^{\infty} b_n \sin(nt)$$

$$= \frac{4A}{\pi} \sin(t) + \frac{4A}{3\pi} \sin(3t) + \frac{4A}{5\pi} \sin(5t) + \cdots$$

The series contains only odd-harmonic sine terms. Since the wave in Fig. 3.7 is odd, its series contains only sine terms, and since it also has a half-wave symmetry, only odd harmonics are present.

The line spectrum for this series is shown in Fig. 3.8. From Eq.(3.13), we have

$$c_n = \sqrt{a_n^2 + b_n^2}$$

$$c_n = \sqrt{0 + b_n^2} = |b_n|$$

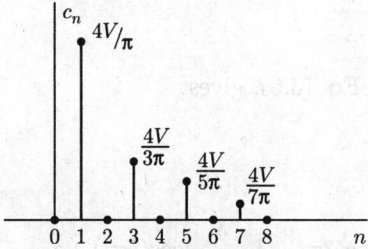

Fig. 3.8 Line spectrum

Example 3.6 Find the trigonometric Fourier series for the triangular wave shown in Fig. 3.9 and plot the line spectrum.

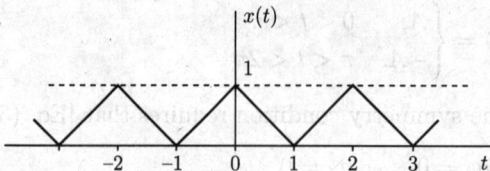

Fig. 3.9 Triangular wave

Solution

The waveform is periodic with period $T = 2$ and fundamental frequency $\omega_0 = 2\pi/T = \pi$. The given waveform for one period may be written as

$$x(t) = \begin{cases} 1 - t, & 0 < t < 1 \\ t - 1, & 1 < t < 2 \end{cases}$$

The wave is an even function, so the symmetry condition requires that [Eq. (3.21)] $b_n = 0$,

$$a_0 = \frac{2}{T} \int\limits_0^{T/2} x(t)\, dt = \frac{2}{2} \int\limits_0^1 (1 - t)\, dt = \left[t - \frac{t^2}{2} \right] = \frac{1}{2}$$

and

$$a_n = \frac{4}{T} \int\limits_0^{T/2} x(t) \cos(n\omega_0 t)\, dt = \frac{4}{2} \int\limits_0^1 (1 - t) \cos(n\pi t)\, dt$$

$$= 2 \left(\left[(1 - t) \frac{\sin(n\pi t)}{n\pi} \right]_0^1 - \left[\frac{\cos(n\pi t)}{n^2 \pi^2} \right]_0^1 \right)$$

$$= \frac{2}{n^2 \pi^2} (1 - \cos(n\pi))$$

$$a_n = \begin{cases} \dfrac{4}{n^2 \pi^2}, & n = 1, 3, 5, \ldots \\ 0, & n = 2, 4, 6, \ldots \end{cases}$$

Substituting the values of a_0, a_n, and b_n into Eq. (3.6), gives

$$x(t) = a_0 + \sum_{n=1}^{\infty} a_n \cos(n\omega_0 t)$$

$$= \frac{1}{2} + \frac{4}{\pi^2} \cos(\pi t) + \frac{4}{(3\pi)^2} \cos(3\pi t) + \frac{4}{(5\pi)^2} \cos(5\pi t) + \cdots$$

The series contains only odd-harmonic cosine terms. The coefficients decreases as $1/n^2$, and thus the series converges rapidly. This fact is evident from the line spectrum shown in Fig. 3.10. From Eq. (3.10), we have

$$c_n = \sqrt{a_n^2 + b_n^2}$$

$$c_n = \sqrt{a_n^2 + 0} = |a_n|$$

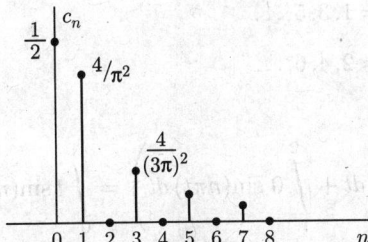

Fig. 3.10   Line spectrum

Example 3.7   Find the trigonometric Fourier series for the waveform shown in Fig. 3.11 and sketch the line spectrum.

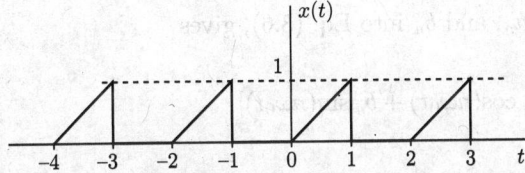

Fig. 3.11

Solution

The waveform is periodic with period $T = 2$ and fundamental frequency $\omega_0 = 2\pi/T = \pi$. The given waveform for one period may be written as

$$x(t) = \begin{cases} t, & 0 < t < 1 \\ 0, & 1 < t < 2 \end{cases}$$

Since the wave is neither even nor odd, the series will contain both sine and cosine terms.

$$a_0 = \frac{1}{T} \int_0^T x(t)\, dt = \frac{1}{2} \left(\int_0^1 t\, dt + \int_1^2 0\, dt \right) = \frac{1}{4}$$

Now we determine a_n and b_n:

$$a_n = \frac{2}{T} \int_0^T x(t) \cos(n\omega_0 t)\, dt = \frac{2}{2} \left(\int_0^1 t \cos(n\pi t)\, dt + \int_1^2 0 \cos(n\pi t)\, dt \right) = \int_0^1 t \cos(n\pi t)\, dt + 0$$

$$= \left[t \frac{\sin(n\pi t)}{n\pi} \right]_0^1 + \left[\frac{\cos(n\pi t)}{(n\pi)^2} \right]_0^1 = \frac{1}{(n\pi)^2} (\cos(n\pi) - 1)$$

$$a_n = \begin{cases} -\dfrac{2}{(n\pi)^2}, & n = 1, 3, 5, \ldots \\ 0, & n = 2, 4, 6, \ldots \end{cases}$$

$$b_n = \frac{2}{T} \int_0^T x(t) \sin(n\omega_0 t)\, dt = \frac{2}{2} \left(\int_0^1 t \sin(n\pi t)\, dt + \int_1^2 0 \sin(n\pi t)\, dt \right) = \int_0^1 t \sin(n\pi t)\, dt + 0$$

$$= \left[-t \frac{\cos(n\pi t)}{n\pi} \right]_0^1 + \left[\frac{\sin(n\pi t)}{(n\pi)^2} \right]_0^1 = -\frac{1}{n\pi} \cos(n\pi)$$

$$b_n = \begin{cases} \dfrac{1}{n\pi}, & n = 1, 3, 5, \ldots \\ -\dfrac{1}{n\pi}, & n = 2, 4, 6, \ldots \end{cases}$$

Substituting the values of a_0, a_n, and b_n into Eq. (3.6), gives

$$x(t) = a_0 + \sum_{n=1}^{\infty} [a_n \cos(n\omega_0 t) + b_n \sin(n\omega_0 t)]$$

$$= \frac{1}{4} - \frac{2}{\pi^2} \cos(\pi t) - \frac{2}{(3\pi)^2} \cos(3\pi t) - \frac{2}{(5\pi)^2} \cos(5\pi t) - \ldots$$

$$+ \frac{1}{\pi} \sin(\pi t) - \frac{1}{2\pi} \sin(2\pi t) + \frac{1}{3\pi} \sin(3\pi t) - \ldots$$

The line spectrum is shown in Fig. 3.12. The even harmonic amplitudes are given directly by $|b_n|$ since there are no even-harmonic cosine terms. However, the odd-harmonic amplitudes must be computed using $c_n = \sqrt{a_n^2 + b_n^2}, n \geq 1$ and $c_0 = a_0$. Thus,

$$c_1 = \sqrt{\left(\frac{2}{\pi} \right)^2 + \left(\frac{1}{\pi} \right)^2} = 0.377$$

$$c_3 = 0.109$$

$$c_5 = 0.064$$

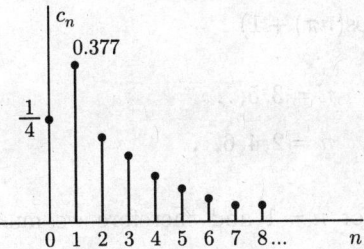

Fig. 3.12 Line spectrum

Example 3.8 Find the trigonometric Fourier series for the half-wave rectified sine wave shown in Fig 3.13 and sketch the line spectrum.

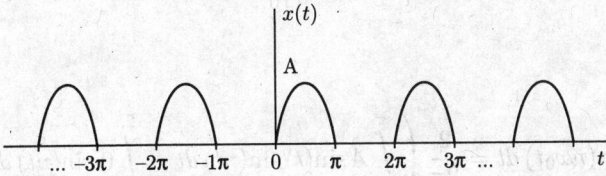

Fig. 3.13 Half-wave rectified sine wave

Solution

The waveform is periodic with period $T = 2\pi$ and fundamental frequency $\omega_0 = 2\pi/T = 1$. The given waveform for one period may be written as

$$x(t) = \begin{cases} A\sin(t), & 0 < t < \pi \\ 0, & \pi < t < 2\pi \end{cases}$$

The wave shows no symmetry and we therefore expect the series to contain both sine and cosine terms.

$$a_0 = \frac{1}{T}\int\limits_0^T x(t)\, dt = \frac{1}{2\pi}\left(\int\limits_0^\pi A\sin(t)\, dt + \int\limits_\pi^{2\pi} 0\, dt\right) = \frac{A}{2\pi}[-\cos(t)]_0^\pi = \frac{A}{\pi}$$

Now we determine a_n:

$$a_n = \frac{2}{T}\int\limits_0^T x(t)\cos(n\omega_0 t)\, dt = \frac{2}{2\pi}\left(\int\limits_0^\pi A\sin(t)\cos(nt)\, dt + \int\limits_\pi^{2\pi} 0 \times \cos(nt)\, dt\right)$$

$$= \frac{A}{\pi}\int\limits_0^\pi \sin(t)\cos(nt)\, dt + 0 = \frac{A}{\pi}\left[\frac{-n\sin(t)\sin(nt) - \cos(nt)\cos(t)}{1 - n^2}\right]_0^\pi$$

$$a_n = \frac{A}{\pi(1 - n^2)}(\cos(n\pi) + 1)$$

$$a_n = \begin{cases} 0, & n = 3, 5, \ldots \\ \dfrac{2A}{\pi(1 - n^2)}, & n = 2, 4, 6, \ldots \end{cases}$$

However, this expression is indeterminate for $n = 1$ and therefore we must integrate separately for a_1:

$$a_1 = \frac{2}{T}\int_0^T x(t)\cos(\omega_0 t)\, dt = \frac{2}{2\pi}\left(\int_0^\pi A\sin(t)\cos(t)\, dt + \int_\pi^{2\pi} 0\cos(t)\, dt\right)$$

$$= \frac{A}{2\pi}\int_0^\pi \sin(2t)\, dt + 0 = \frac{A}{2\pi}\left[-\frac{\cos(2t)}{2}\right]_0^\pi = 0$$

Now we evaluate b_n:

$$b_n = \frac{2}{T}\int_0^T x(t)\sin(n\omega_0 t)\, dt = \frac{2}{2\pi}\left(\int_0^\pi A\sin(t)\sin(nt)\, dt + \int_\pi^{2\pi} 0\sin(nt)\, dt\right)$$

$$= \frac{A}{\pi}\int_0^\pi \sin(t)\sin(nt)\, dt + 0 = \frac{A}{\pi}\left[\frac{n\sin(t)\cos(nt) - \sin(nt)\cos(t)}{1 - n^2}\right]_0^\pi = 0$$

This expression is also indeterminate for $n = 1$, and b_1 is evaluated separately:

$$b_1 = \frac{2}{T}\int_0^T x(t)\sin(\omega_0 t)\, dt = \frac{2}{2\pi}\left(\int_0^\pi A\sin^2(t)\, dt + \int_\pi^{2\pi} 0\sin(t)\, dt\right)$$

$$= \frac{A}{\pi}\int_0^\pi \sin^2(t)\, dt = \frac{A}{2\pi}\left[t - \frac{\sin(2t)}{2}\right]_0^\pi = \frac{A}{2}$$

Substituting the values of a_0, a_n, and b_n into Eq. (3.6) gives

$$x(t) = \frac{A}{\pi}\left(1 + \frac{\pi}{2}\sin(t) - \frac{2}{3}\cos(2t) - \frac{2}{15}\cos(4t) - \cdots\right)$$

The line spectrum, shown in Fig. 3.14, is determined using $c_n = \sqrt{a_n^2 + b_n^2}$ for $n \geq 1$ and $c_0 = a_0$.

$$c_0 = \frac{A}{\pi}, \qquad c_1 = \sqrt{a_1^2 + b_1^2} = \frac{A}{2}, \qquad c_n = |a_n|\ n \geq 2$$

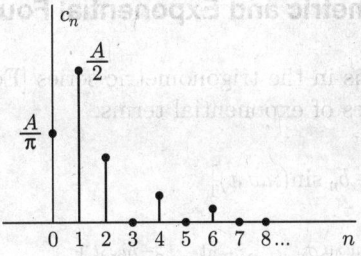

Fig. 3.14 Line spectrum

3.2.5 Gibbs Phenomenon

The Gibbs phenomenon is present only when there is a jump discontinuity in $x(t)$. When a continuous function $x(t)$ is synthesized by using the first N terms of the Fourier series, the synthesized function approaches $x(t)$ for all t as $N \to \infty$. No Gibbs phenomenon appears.

The Gibbs phenomenon occurs near points of discontinuity of a periodic signal $x(t)$ that is approximated by a Fourier series in which only a finite number of terms are kept. When we use only the first N terms in the Fourier series to synthesize a signal, we are abruptly terminating the series, giving a unit weight to the first N harmonics and zero weight to all the remaining harmonics beyond N. This abrupt termination of the series causes the Gibbs phenomenon in synthesis of discontinuous functions. Near a point of discontinuity, the Fourier series approximation oscillates about the numerical value it should achieve according to the Fourier convergence theorem, which is valid in the infinite series limit. Further, the overshoot near the discontinuity does not vanish as more and more modes are retained. Instead, the overshoot is finite no matter what finite number of modes N are retained, even though the region of overshoot gets progressively smaller as $N \to \infty$.

3.2.6 Exponential Fourier Series

The exponential Fourier series is expressed as

$$x(t) = \sum_{n=-\infty}^{\infty} X_n \, e^{jn\omega_0 t} \tag{3.26}$$

where

$$X_n = \frac{1}{T} \int_0^T x(t) \, e^{-jn\omega_0 t} \, dt \tag{3.27}$$

are the exponential Fourier series coefficients of the signal $x(t)$. The integrations can be carried out from $-T/2$ to $T/2$, or over any other full period that might simplify the calculation. Equation (3.26) is referred to as the *synthesis* equation and Eq. (3.27) as the *analysis* equation.

The Fourier series coefficients are known as a *frequency domain representation* of $x(t)$ because each Fourier series coefficient is associated with a complex sinusoid of a different frequency.

3.2.7 Relationship Between Trigonometric and Exponential Fourier Series

If we express each of the sine and cosine terms in the trigonometric series [Eq. (3.6)] by its exponential equivalent, the result is a series of exponential terms:

$$x(t) = a_0 + \sum_{n=1}^{\infty} \left[a_n \cos(n\omega_0 t) + b_n \sin(n\omega_0 t) \right]$$

$$= a_0 + \sum_{n=1}^{\infty} \left[a_n \frac{e^{jn\omega_0 t} + e^{-jn\omega_0 t}}{2} + b_n \frac{e^{jn\omega_0 t} - e^{-jn\omega_0 t}}{2j} \right]$$

$$= a_0 + \sum_{n=1}^{\infty} \left[a_n \frac{e^{jn\omega_0 t} + e^{-jn\omega_0 t}}{2} - jb_n \frac{e^{jn\omega_0 t} - e^{-jn\omega_0 t}}{2} \right]$$

$$x(t) = a_0 + \sum_{n=1}^{\infty} \left[\left(\frac{a_n - jb_n}{2} \right) e^{jn\omega_0 t} + \left(\frac{a_n + jb_n}{2} \right) e^{-jn\omega_0 t} \right] \tag{3.28}$$

Assume

$$X_n = \frac{a_n - jb_n}{2} \tag{3.29}$$

Substituting the values of a_n and b_n from Eqs (3.8) and (3.9) into the above equation gives

$$X_n = \frac{1}{2} \left(\frac{2}{T} \int_0^T x(t) \cos(n\omega_0 t)\, dt - j\frac{2}{T} \int_0^T x(t) \sin(n\omega_0 t)\, dt \right)$$

$$= \frac{1}{2} \frac{2}{T} \int_0^T x(t) \left(\cos(n\omega_0 t) - j \sin(n\omega_0 t) \right) dt$$

$$X_n = \frac{1}{T} \int_0^T x(t) e^{-jn\omega_0 t}\, dt$$

Substituting $n = 0$ in the above equation gives

$$X_0 = \frac{1}{T} \int_0^T x(t)\, dt = a_0$$

$$X_0 = a_0 \tag{3.30}$$

Again, consider

$$X_n = \frac{a_n - jb_n}{2}$$

$$X_{-n} = \frac{a_{-n} - jb_{-n}}{2}$$

we know that

$$a_n = \frac{2}{T} \int_0^T x(t) \cos(n\omega_0 t) \, dt$$

$$a_{-n} = \frac{2}{T} \int_0^T x(t) \cos(-n\omega_0 t) \, dt$$

$$a_{-n} = \frac{2}{T} \int_0^T x(t) \cos(n\omega_0 t) \, dt = a_n$$

$$a_{-n} = a_n \quad \text{an even Fourier series coefficient} \tag{3.31}$$

and

$$b_n = \frac{2}{T} \int_0^T x(t) \sin(n\omega_0 t) \, dt$$

$$b_{-n} = \frac{2}{T} \int_0^T x(t) \sin(-n\omega_0 t) \, dt$$

$$= -\frac{2}{T} \int_0^T x(t) \sin(n\omega_0 t) \, dt = -b_n$$

$$b_{-n} = -b_n \quad \text{an odd Fourier series coefficient} \tag{3.32}$$

Substituting $a_{-n} = a_n$ and $b_{-n} = -b_n$ in the expression of X_{-n} gives

$$X_{-n} = \frac{a_n + jb_n}{2} \tag{3.33}$$

from Eqs (3.28), (3.29), (3.30), and Eq. (3.33), we have

$$x(t) = X_0 + \sum_{n=1}^{\infty} \left[X_n e^{jn\omega_0 t} + X_{-n} e^{-jn\omega_0 t} \right]$$

$$x(t) = \sum_{n=-\infty}^{\infty} X_n e^{jn\omega_0 t}$$

The exponential Fourier series is just another way of representing trigonometric Fourier series (or vice versa). The two form carry identical information: no more, no less. The reasons for preferring the exponential form are that it is more compact, and the expression for deriving the exponential coefficients is also more compact, than those in the trigonometric series. Furthermore, the LTI system response to exponential signals is also simpler than the system response to sinusoids. In addition, the exponential form proves to be much easier than the trigonometric form to manipulate mathematically.

A minor disadvantage of the exponential form is that it cannot be visualized as easily as sinusoids. For intuitive and qualitative understanding, the sinusoids have an edge over exponentials. Fortunately, this difficulty can be overcome readily because of the close connection between exponential and Fourier spectra. Using Eqs (3.29), (3.30), and (3.33), we may convert the exponential Fourier series into the trigonometric Fourier series. Note that

$$a_0 = X_0 \tag{3.34}$$

$$a_n = 2\operatorname{Re}\{X_n\} \quad \text{or} \quad a_n = X_n + X_{-n} \tag{3.35}$$

$$b_n = -2\operatorname{Im}\{X_n\} \quad \text{or} \quad b_n = j(X_n - X_{-n}) \tag{3.36}$$

3.2.8   Line Spectrum

A plot showing each of the harmonic amplitudes in the wave is called the *line spectrum*. The lines decrease rapidly for waves with rapidly convergent series. Waves with discontinuities, such as the sawtooth and square wave, have spectra with slowly decreasing amplitudes since their series have strong high harmonics. The harmonic content and the line spectrum of a wave are part of the very nature of that wave and never change, regardless of the method of analysis. Shifting the origin gives the trigonometric series a completely different appearance, and the exponential series coefficients also change greatly. However, the same harmonic always appear in the series and their amplitudes remains the same.

From Eq. (3.29), the *magnitude spectrum* is

$$|X_n| = |X_{-n}| = \frac{\sqrt{a_n^2 + b_n^2}}{2} = \frac{c_n}{2} \tag{3.37}$$

The *magnitude spectrum* is an even function.

$$c_n = \begin{cases} 2|X_n|, & n \geq 1 \\ X_0, & n = 0 \end{cases} \tag{3.38}$$

and, the *phase spectrum* is

$$\angle X_n = \tan^{-1}\left(\frac{-b_n}{a_n}\right) = \theta_n \tag{3.39}$$

$$\angle X_{-n} = \tan^{-1}\left(\frac{-b_{-n}}{a_{-n}}\right)$$

$$= \tan^{-1}\left(\frac{b_n}{a_n}\right) = -\tan^{-1}\left(\frac{-b_n}{a_n}\right) = -\theta_n = -\angle X_n$$

$$= -\angle X_n \tag{3.40}$$

The phase spectrum is an odd function.

3.2.9 Concept of Negative Frequency

The existence of the spectrum at negative frequencies is somewhat disturbing because, by definition, the frequency is a positive quantity. How do we interpret a negative frequency?

Negative frequencies are present in the exponential Fourier series because the mathematical model of a signal requires the use of negative frequencies. The exponential spectra are a graphical representation of coefficients X_n as a function of ω. Existence of the spectrum at $\omega = -n\omega_0$ is merely an indication that an exponential component $e^{-jn\omega_0 t}$ exists in the series. The function $e^{-jn\omega_0 t}$ of frequency $-n\omega_0$ (negative frequency) represents the counterpart of the signal $e^{jn\omega_0}$ of frequency $n\omega_0$ (positive frequency), and when these two exponentials are combined, they provide a real function (sinusoid) of frequency $n\omega_0$.

Example 3.9 Find the exponential Fourier series for the waveform shown in Fig. 3.3. Using the coefficients of this exponential series, obtain a_n and b_n of the trigonometric series.

Solution
The waveform is periodic with period $T = 2\pi$ and fundamental frequency $\omega_0 = 2\pi/T = 1$. The given waveform for one period may be written as

$$x(t) = \frac{10}{2\pi}t, \quad 0 < t < 2\pi$$

From Eq. (3.27), the exponential Fourier series coefficient is

$$X_n = \frac{1}{T}\int_0^T x(t)\,e^{-jn\omega_0 t}\,dt = \frac{1}{2\pi}\int_0^{2\pi}\frac{10}{2\pi}t\,e^{-jnt}\,dt$$

$$= \frac{10}{(2\pi)^2}\left[-\frac{1}{jn}t\,e^{-jnt} - \frac{1}{(jn)^2}e^{-jnt}\right]_0^{2\pi} = j\frac{10}{2n\pi}$$

The average value is

$$X_0 = a_0 = \frac{1}{T}\int_0^T x(t)\,dt = \frac{1}{2\pi}\int_0^{2\pi}\frac{10}{2\pi}t\,dt = \frac{10}{(2\pi)^2}\left[\frac{t^2}{2}\right]_0^{2\pi} = 5$$

The exponential Fourier series for the given waveform is

$$x(t) = \sum_{n=-\infty}^{\infty}X_n e^{jn\omega_0 t} = \sum_{n=-\infty}^{\infty}X_n e^{jnt}$$

$$= \cdots - j\frac{10}{4\pi}e^{-j2t} - j\frac{10}{2\pi}e^{-jt} + 5 + j\frac{10}{2\pi}e^{jt} + j\frac{10}{4\pi}e^{j2t} + \cdots$$

The trigonometric Fourier series coefficients are [from Eqs (3.34), (3.35), and (3.36)]

$$a_0 = X_0 = 5, \qquad a_n = 0, \qquad b_n = -\frac{10}{n\pi}$$

and so

$$x(t) = 5 - \frac{10}{\pi}\sin(t) - \frac{10}{2\pi}\sin(2t) - \frac{10}{3\pi}\sin(3t) - \cdots$$

Example 3.10 Find the exponential Fourier series for the half-wave rectified sine wave shown in Fig. 3.13 and sketch the line spectrum.

Solution

The waveform is periodic with period $T = 2\pi$ and fundamental frequency $\omega_0 = 2\pi/T = 1$. The given waveform for one period may be written as

$$x(t) = \begin{cases} A\sin(t), & 0 < t < \pi \\ 0, & \pi < t < 2\pi \end{cases}$$

From Eq. (3.27), the exponential Fourier series coefficient is

$$X_n = \frac{1}{T}\int_0^T x(t)e^{-jn\omega_0 t}\,dt = \frac{1}{2\pi}\left(\int_0^\pi A\sin(t)e^{-jnt}\,dt + \int_\pi^{2\pi} 0\,e^{-jnt}\,dt\right)$$

$$= \frac{A}{2\pi}\int_0^\pi \sin(t)e^{-jnt}\,dt = \frac{A}{2\pi}I$$

where

$$I = \int_0^\pi e^{-jnt}\sin(t)\,dt$$

$$I = \left[-e^{-jnt}\cos(t)\right]_0^\pi - jn\int_0^\pi e^{-jnt}\cos(t)\,dt$$

$$I = \left[-e^{-jnt}\cos(t)\right]_0^\pi + \left[-jn\,e^{-jnt}\sin(t)\right]_0^\pi + n^2\int_0^\pi e^{-jnt}\sin(t)\,dt$$

$$I = \left[-e^{-jnt}\cos(t)\right]_0^\pi + \left[-jn\,e^{-jnt}\sin(t)\right]_0^\pi + n^2 I$$

$$I(1-n^2) = \left[e^{-jnt}(-\cos(t) - jn\sin(t))\right]_0^\pi$$

$$I = \frac{(e^{-jn\pi} + 1)}{(1 - n^2)}$$

Therefore,

$$X_n = \frac{A}{2\pi}I = \frac{A(e^{-jn\pi} + 1)}{2\pi(1 - n^2)} = \begin{cases} \dfrac{A(e^{-jn\pi} + 1)}{\pi(1 - n^2)}, & n = \pm 2, \pm 4, \pm 6, \ldots \\ 0, & n = \pm 3, \pm 5, \pm 7, \ldots \end{cases}$$

For $n = \pm 1$, the expression for X_n becomes indeterminate. L'Hospital's rule may be applied, i.e., the numerator and denominator are separately differentiated w.r.t. n, after which n is allowed to approach ± 1.

$$X_1 = \lim_{n \to 1} \frac{d/dn A(e^{-jn\pi} + 1)}{d/dn 2\pi(1 - n^2)} = -j\frac{A}{4}$$

Similarly,

$$X_{-1} = \lim_{n \to -1} \frac{d/dn A(e^{-jn\pi} + 1)}{d/dn 2\pi(1 - n^2)} = j\frac{A}{4}$$

The average value is

$$X_0 = a_0 = \frac{1}{T} \int_0^T x(t) \, dt = \frac{1}{2\pi} \left(\int_0^\pi A \sin(t) \, dt + 0 \right) = \frac{A}{2\pi} [-\cos(t)]_0^\pi = \frac{A}{\pi}$$

The exponential Fourier series for the given waveform is

$$x(t) = \sum_{n=-\infty}^{\infty} X_n e^{jn\omega_0 t} = \sum_{n=-\infty}^{\infty} X_n e^{jnt}$$

$$= \cdots - \frac{A}{15\pi} e^{-j4t} - \frac{A}{3\pi} e^{-j2t} + j\frac{A}{4} e^{-jt} + \frac{A}{\pi} - j\frac{A}{4} e^{jt} - \frac{A}{3\pi} e^{j2t} - \frac{A}{15\pi} e^{j4t} - \cdots$$

The harmonic amplitudes are [Eq. (3.38)]

$$c_n = \begin{cases} \dfrac{A}{\pi}, & n = 0 \\[2mm] \dfrac{A}{2}, & n = 1 \\[2mm] \dfrac{2A}{\pi(n^2 - 1)}, & n = 2, 4, 6, \ldots \\[2mm] 0, & n = 3, 5, 7, \ldots \end{cases}$$

The line spectrum is exactly as plotted in Fig. 3.14.

Example 3.11 Find the exponential Fourier series and sketch the corresponding spectra for the impulse train $x(t) = \delta_{T_0}(t)$ depicted in Fig. 3.15. From this result sketch the line spectrum and write the trigonometric Fourier series for $\delta_{T_0}(t)$.

Solution
The waveform is periodic with period T_0 and fundamental frequency $\omega_0 = 2\pi/T_0$.

$$x(t) = \sum_{n=-\infty}^{\infty} \delta(t - nT_0) \tag{3.41}$$

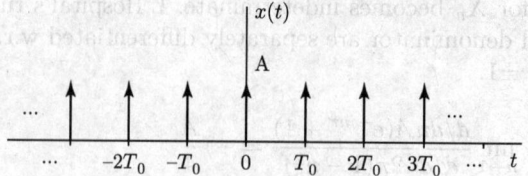

Fig. 3.15 Impulse train

By choosing the interval of integration $(-T_0/2, T_0/2)$, the given waveform for one period may be written as

$$x(t) = \delta(t), \quad -\frac{T_0}{2} < t < \frac{T_0}{2}$$

From Eq. (3.27), the exponential Fourier series coefficient is

$$X_n = \frac{1}{T_0} \int_{-T_0/2}^{T_0/2} x(t) \, e^{-jn\omega_0 t} \, dt$$

$$= \frac{1}{T_0} \int_{-T_0/2}^{T_0/2} \delta(t) \, e^{-jn\omega_0 t} \, dt$$

From the sampling property (Eq. 1.13), the integral on the RHS is the value of $e^{-jn\omega_0 t}$ at $t = 0$ (where the impulse is located). Therefore,

$$X_n = \frac{1}{T_0}$$

Substitution of this value in Eq. (3.26) yields the desired exponential Fourier series:

$$x(t) = \frac{1}{T_0} \sum_{n=-\infty}^{\infty} e^{jn\omega_0 t} \tag{3.42}$$

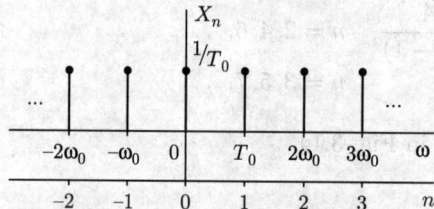

Fig. 3.16 Exponential Fourier spectra

Figure 3.16 shows the exponential spectrum. To sketch the line spectrum, we use Eq. (3.38) to obtain

$$c_0 = X_0 = \frac{1}{T_0}$$

$$c_n = 2|X_n| = \frac{2}{T_0}, \quad n \geq 1$$

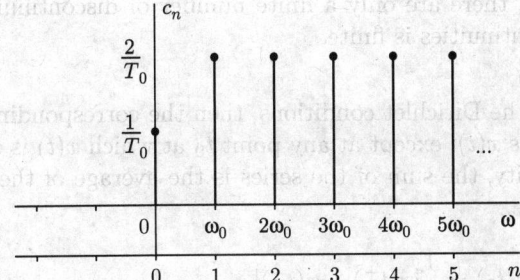

Fig. 3.17 Line spectrum

Figure 3.17 shows the line spectrum. Using Eqs (3.34), (3.35), and (3.36), we may convert an exponential Fourier series into a trigonometric Fourier series:

$$a_0 = X_0 = \frac{1}{T_0}$$

$$a_n = 2Re\{X_n\} = \frac{2}{T_0}$$

$$b_n = 0$$

3.3 DIRICHLET CONDITIONS

Dirichlet showed that if $x(t)$ satisfies certain conditions (*Dirichlet conditions*), its Fourier series is guaranteed to converge pointwise at all points where $x(t)$ is continuous. These conditions are as follows:

1. Over any period, $x(t)$ must be *absolutely integrable*, i.e.,

$$\int_0^T |x(t)|\, dt < \infty \tag{3.43}$$

This guarantees that each coefficient X_n will be finite since

$$|X_n| = \frac{1}{T} \left| \int_0^T x(t)\, e^{-jn\omega_0 t}\, dt \right|$$

$$\leq \frac{1}{T} \int_0^T \left| x(t)\, e^{-jn\omega_0 t} \right|\, dt = \frac{1}{T} \int_0^T |x(t)|\, dt$$

So if

$$\int_0^T |x(t)|\, dt < \infty$$

then

$$|X_n| < \infty$$

2. In any finite interval of time, $x(t)$ is of bounded variation, i.e., there are no more than a finite number of maxima and minima during any single period of the signal.

3. In any finite interval of time, there are only a finite number of discontinuities. Furthermore, each of these discontinuities is finite.

Thus if a signal $x(t)$ satisfies the Dirichlet conditions, then the corresponding Fourier series is convergent and its sum is $x(t)$, except at any point t_0 at which $x(t)$ is discontinuous. At the point of discontinuity, the sum of the series is the average of the left- and right-hand limits of $x(t)$ at t_0, i.e.,

$$x(t_0) = \frac{1}{2}[x(t_0^+) + x(t_0^-)]$$

3.4 PROPERTIES OF CONTINUOUS-TIME FOURIER SERIES

Fourier series representations possesses a number of important properties that are useful for developing conceptual insights into such representations, and they can also help to reduce the complexity of the evaluation of the Fourier series of many signals.

We will use a shorthand notation to indicate the relationship between a periodic signal and its Fourier series coefficients. Specifically, suppose that $x(t)$ is periodic signal with period T and fundamental frequency $\omega_0 = 2\pi/T$. Then if the Fourier series coefficients of $x(t)$ are denoted by X_n, we will use the notation

$$x(t) \longleftrightarrow X_n$$

to signify the pairing of a periodic signal with its Fourier series coefficients.

3.4.1 Linearity

If $x(t)$ and $y(t)$ denote two periodic signals with period T and

$$x(t) \longleftrightarrow X_n$$
$$y(t) \longleftrightarrow Y_n$$

then

$$z(t) = ax(t) + by(t) \longleftrightarrow Z_n = aX_n + bY_n \qquad (3.44)$$

Proof The exponential Fourier series coefficient of $z(t)$ is

$$Z_n = \frac{1}{T}\int_0^T z(t)\, e^{-jn\omega_0 t}\, dt$$

$$= \frac{1}{T}\int_0^T [ax(t) + by(t)]\, e^{-jn\omega_0 t}\, dt$$

$$= a\frac{1}{T}\int_0^T x(t)\, e^{-jn\omega_0 t}\, dt + b\frac{1}{T}\int_0^T y(t)\, e^{-jn\omega_0 t}\, dt$$

$$= aX_n + bY_n \qquad \blacksquare$$

3.4.2 Time Shifting

When a time shift is applied to a periodic signal $x(t)$, the period T of the signal is preserved. If

$$x(t) \longleftrightarrow X_n$$

then

$$y(t) = x(t - t_0) \longleftrightarrow Y_n = X_n e^{-jn\omega_0 t_0} \qquad (3.45)$$

One consequence of this property is that when a signal is shifted in time the *magnitudes* of its Fourier series coefficients remain unaltered, i.e., $|Y_n| = |X_n|$.

Proof By definition

$$Y_n = \frac{1}{T} \int_0^T y(t)\, e^{-jn\omega_0 t}\, dt = \frac{1}{T} \int_0^T x(t - t_0)\, e^{-jn\omega_0 t}\, dt$$

A change of variables is performed by letting $\tau = t - t_0$, which also yields $d\tau = dt$, $\tau \to -t_0$ as $t \to 0$, and $\tau \to T - t_0$ as $t \to T$. Therefore,

$$Y_n = \frac{1}{T} \int_{-t_0}^{T-t_0} x(\tau)\, e^{-jn\omega_0(\tau + t_0)} d\tau$$

$$Y_n = \underbrace{\frac{1}{T} \int_{-t_0}^{T-t_0} x(\tau)\, e^{-jn\omega_0 \tau} d\tau}_{X_n}\ e^{-jn\omega_0 t_0} = X_n e^{-jn\omega_0 t_0}$$

$$|Y_n| = |X_n|$$

∎

3.4.3 Frequency Shifting

If

$$x(t) \longleftrightarrow X_n$$

then

$$y(t) = e^{jm\omega_0 t} x(t) \longleftrightarrow Y_n = X_{n-m} \qquad (3.46)$$

Proof By definition

$$Y_n = \frac{1}{T} \int_0^T y(t)\, e^{-jn\omega_0 t}\, dt = \frac{1}{T} \int_0^T e^{jm\omega_0 t} x(t)\, e^{-jn\omega_0 t}\, dt = \frac{1}{T} \int_0^T x(t)\, e^{-j(n-m)\omega_0 t}\, dt = X_{n-m}$$

Hence, a frequency shift corresponds to multiplication in time domain by a complex sinusoid whose frequency is equal to the time shift.

∎

3.4.4 Time Reversal

If

$$x(t) \longleftrightarrow X_n$$

then

$$y(t) = x(-t) \longleftrightarrow Y_n = X_{-n} \tag{3.47}$$

In other words time reversal applied to a continuous-time signal results in a time reversal of the corresponding sequence of Fourier series coefficients.

Proof By definition

$$Y_n = \frac{1}{T} \int_0^T y(t)\, e^{-jn\omega_0 t}\, dt = \frac{1}{T} \int_0^T x(-t)\, e^{-jn\omega_0 t}\, dt = \frac{1}{T} \int_0^T x(t)\, e^{-j(-n)\omega_0 t}\, dt = X_{-n}$$

An interesting consequence of the time-reversal property is that if $x(t)$ is even then its Fourier series coefficients are also even, i.e.,

$$\text{if}\quad x(-t) = x(t), \quad \text{then}\quad X_{-n} = X_n \tag{3.48}$$

Similarly, if $x(t)$ is odd, then so are its Fourier series coefficients, i.e.,

$$\text{if}\quad x(-t) = -x(t), \quad \text{then}\quad X_{-n} = -X_n \tag{3.49}$$

∎

3.4.5 Time Scaling

If

$$x(t) \longleftrightarrow X_n$$

then

$$y(t) = x(at) \longleftrightarrow Y_n = X_n \tag{3.50}$$

Proof Time scaling is an operation that in general changes the period of the underlying signal. Specifically, if $x(t)$ is periodic with period T and fundamental frequency $\omega_0 = 2\pi/T$, then $y(t) = x(at)$, where a is a positive real number, is periodic with period T/a and fundamental frequency $a\omega_0$.

$$Y_n = \frac{1}{T/a} \int_0^{T/a} y(t)\, e^{-jna\omega_0 t}\, dt$$

$$= \frac{1}{T/a} \int_0^{T/a} x(at)\, e^{-jna\omega_0 t}\, dt$$

A change of variables is performed by letting $\tau = at$, which also yields $d\tau = a\, dt$, $\tau \to 0$ as $t \to 0$, and $\tau \to T$ as $t \to T/a$. Therefore,

$$Y_n = \frac{1}{T} \int_0^T x(\tau)\, e^{-jn\omega_0 \tau}\, d\tau = X_n$$

We emphasize that while the Fourier series coefficients have not changed, the Fourier series representation has changed because of the change in the fundamental frequency.

3.4.6   Periodic Convolution

If

$$x(t) \longleftrightarrow X_n$$

$$y(t) \longleftrightarrow Y_n$$

then

$$z(t) = x(t) \circledast y(t) = \frac{1}{T} \int_0^T x(\tau)y(t-\tau)d\tau \longleftrightarrow Z_n = X_n Y_n \qquad (3.51)$$

Proof For periodic signals with the same period, a special form of convolution, known as periodic convolution, is defined by the integral:

$$z(t) = \frac{1}{T} \int_0^T x(\tau)y(t-\tau)d\tau$$

It is easy to show that $z(t)$ is periodic with period T and the periodic convolution is commutative and associative. Thus, we can write $z(t)$ in a Fourier series representation with coefficients:

$$Z_n = \frac{1}{T} \int_0^T z(t)\, e^{-jn\omega_0 t}\, dt$$

$$= \frac{1}{T} \int_0^T \left(\frac{1}{T} \int_0^T x(\tau)y(t-\tau)d\tau \right) e^{-jn\omega_0 t}\, dt$$

$$= \frac{1}{T} \int_0^T x(\tau) \left(\frac{1}{T} \int_0^T y(t-\tau)\, e^{-jn\omega_0 t}\, dt \right) d\tau$$

From the time-shifting property, i.e., if $y(t) \longleftrightarrow Y_n$, then $y(t-\tau) \longleftrightarrow Y_n\, e^{-jn\omega_0\tau}$, we have

$$Z_n = \underbrace{\frac{1}{T} \int_0^T x(\tau)\, e^{-jn\omega_0\tau} d\tau}_{X_n}\, Y_n = X_n Y_n$$

The convolution in time transforms to multiplication of the frequency domain representations.

3.4.7   Multiplication

If $x(t)$ and $y(t)$ are periodic signal with the same period T and

$$x(t) \longleftrightarrow X_n$$

$$y(t) \longleftrightarrow Y_n$$

then

$$z(t) = x(t)y(t) \longleftrightarrow Z_n = \sum_{k=-\infty}^{\infty} X_k Y_{n-k} \qquad (3.52)$$

Z_n may be interpreted as the discrete-time convolution of the X_n and Y_n:

Proof Consider the signal $z(t)$:

$$z(t) = x(t)y(t) = \sum_{k=-\infty}^{\infty} X_k e^{jk\omega_0 t} \sum_{m=-\infty}^{\infty} Y_m e^{jm\omega_0 t} = \sum_{m=-\infty}^{\infty} \sum_{k=-\infty}^{\infty} X_k Y_m e^{j(k+m)\omega_0 t}$$

A change of variables is performed by letting $n = k + m$, which also yields $m = n - k$, $n \to -\infty$ as $m \to -\infty$, and $n \to \infty$ as $m \to \infty$. Therefore,

$$z(t) = \sum_{n=-\infty}^{\infty} \left(\sum_{k=-\infty}^{\infty} X_k Y_{n-k} \right) e^{jn\omega_0 t} = \sum_{n=-\infty}^{\infty} Z_n e^{jn\omega_0 t}$$

Thus,

$$Z_n = \sum_{k=-\infty}^{\infty} X_k Y_{n-k}$$

3.4.8 Differentiation

If

$$x(t) \longleftrightarrow X_n$$

then

$$y(t) = \frac{dx(t)}{dt} \longleftrightarrow Y_n = jn\omega_0 X_n \qquad (3.53)$$

Proof If $x(t)$ is a periodic signal, then we have the Fourier series representation:

$$x(t) = \sum_{n=-\infty}^{\infty} X_n e^{jn\omega_0 t}$$

Differentiating both sides of this equation gives

$$y(t) = \frac{dx(t)}{dt} = \sum_{n=-\infty}^{\infty} [X_n jn\omega_0] e^{jn\omega_0 t} = \sum_{n=-\infty}^{\infty} Y_n e^{jn\omega_0 t}$$

and thus we conclude that

$$y(t) = \frac{dx(t)}{dt} \longleftrightarrow Y_n = jn\omega_0 X_n$$

Differentiation forces the time-averaged value of the differentiated signal to be zero; hence, the Fourier series coefficient for $n = 0$ is zero.

3.4.9   Integration

If

$$x(t) \longleftrightarrow X_n$$

and $X_0 = 0$ [so that $\int_{-\infty}^{t} x(t)\, dt$ is finite valued and periodic], then

$$y(t) = \int_{-\infty}^{t} x(t)\, dt \longleftrightarrow Y_n = \frac{1}{jn\omega_0} X_n \qquad (3.54)$$

Proof   If $x(t)$ is a periodic signal, then we have the Fourier series representation:

$$x(t) = \sum_{n=-\infty}^{\infty} X_n\, e^{jn\omega_0 t}$$

Integrating both sides of this equation gives

$$y(t) = \int_{-\infty}^{t} x(t)\, dt = \sum_{n=-\infty}^{\infty} \left(X_n \frac{1}{jn\omega_0} \right) e^{jn\omega_0 t} = \sum_{n=-\infty}^{\infty} Y_n\, e^{jn\omega_0 t}$$

and thus we conclude that

$$y(t) = \int_{-\infty}^{t} x(t)\, dt \longleftrightarrow Y_n = \frac{1}{jn\omega_0} X_n$$

The integration attenuates (de-emphasizes) the magnitude of the high-frequency components of the signal. High-frequency components of the signal are the main contributors to its sharp details, such as those occurring at the points of discontinuity. Hence, integration smoothes the signal, and this is one of the reasons it is sometimes called a *smoothing operation*.

3.4.10   Conjugation and Conjugate Symmetry

If

$$x(t) \longleftrightarrow X_n$$

then

$$y(t) = x^*(t) \longleftrightarrow Y_n = X_{-n}^* \qquad (3.55)$$

Taking the complex conjugate of a periodic signal $x(t)$ has the effect of complex conjugation and time reversal on the corresponding Fourier series coefficients.

Proof By definition

$$Y_n = \frac{1}{T} \int_0^T y(t) \, e^{-jn\omega_0 t} \, dt = \frac{1}{T} \int_0^T x^*(t) \, e^{-jn\omega_0 t} \, dt$$

$$= \left(\frac{1}{T} \int_0^T x(t) \, e^{jn\omega_0 t} \, dt \right)^* = \left(\underbrace{\frac{1}{T} \int_0^T x(t) \, e^{-j(-n)\omega_0 t} \, dt}_{X_{-n}} \right)^*$$

$$= (X_{-n})^* = X_{-n}^*$$

Case I If $x(t)$ is real, i.e.,
if

$$x^*(t) = x(t)$$

then

$$X_{-n}^* = X_n$$

$$X_{-n} = X_n^* \tag{3.56}$$

the Fourier series coefficients will be *conjugate symmetric*. Also if $x(t)$ is real and even, then from Eqs (3.48) and (3.56), we have

$$X_{-n} = X_n^* = X_n \tag{3.57}$$

That is, *if $x(t)$ is real and even, then so are its Fourier series coefficients.*

Case II *If $x(t)$ is real and odd, then its Fourier series coefficients are purely imaginary and odd.* Using Eqs (3.49) and (3.56), we get

$$X_{-n} = X_n^* = -X_n \tag{3.58}$$

Case III Even and odd decomposition of real signals: If $x(t)$ is real and

$$x(t) \longleftrightarrow X_n$$

then

1.

$$x_e(t) = \mathcal{E}\{x(t)\} \longleftrightarrow \mathrm{Re}\{X_n\} \tag{3.59}$$

The even part of a signal $x(t)$ is defined as

$$x_e(t) = \frac{1}{2}[x(t) + x(-t)]$$

Using the linearity property and Eq. (3.54), we get

$$x_e(t) \longleftrightarrow \frac{1}{2}[X_n + X_{-n}]$$

$$x_e(t) \longleftrightarrow \frac{1}{2}[X_n + X_n^*]$$

$$x_e(t) \longleftrightarrow \frac{1}{2}2Re\{X_n\}$$

$$x_e(t) \longleftrightarrow Re\{X_n\}$$

and

2.

$$x_o(t) = \mathcal{O}\{x(t)\} \longleftrightarrow j\,\text{Im}\{X_n\} \tag{3.60}$$

The odd part of a signal $x(t)$ is defined as

$$x_o(t) = \frac{1}{2}[x(t) - x(-t)]$$

Using the linearity property and Eq. (3.54), we get

$$x_o(t) \longleftrightarrow \frac{1}{2}[X_n - X_{-n}]$$

$$x_0(t) \longleftrightarrow \frac{1}{2}[X_n - X_n^*]$$

$$x_0(t) \longleftrightarrow \frac{1}{2}2j\,\text{Im}\{X_n\}$$

$$x_0(t) \longleftrightarrow j\,\text{Im}\{X_n\}$$

3.4.11 Parseval's Theorem for Power Signals

If

$$x(t) \longleftrightarrow X_n$$

then

$$\frac{1}{T}\int_0^T |x(t)|^2\,dt = \sum_{n=-\infty}^{\infty} |X_n|^2 = X_0^2 + \sum_{n=1}^{\infty} 2|X_n|^2 \tag{3.61}$$

Parseval's relation states that the total average power in a periodic signal equals the sum of the average powers in all of its harmonic components.

Proof Consider the LHS of Eq. (3.61):

$$\frac{1}{T}\int_0^T |x(t)|^2\, dt = \frac{1}{T}\int_0^T x(t)x^*(t)\, dt$$

$$= \frac{1}{T}\int_0^T x(t)\left(\sum_{n=-\infty}^{\infty} X_n\, e^{jn\omega_0 t}\right)^*\, dt$$

$$= \frac{1}{T}\int_0^T x(t)\left(\sum_{n=-\infty}^{\infty} X_n^*\, e^{-jn\omega_0 t}\right)\, dt$$

$$= \sum_{n=-\infty}^{\infty} X_n^*\left(\frac{1}{T}\int_0^T x(t)\, e^{-jn\omega_0 t}\, dt\right)$$

$$= \sum_{n=-\infty}^{\infty} X_n^* X_n$$

$$\frac{1}{T}\int_0^T |x(t)|^2\, dt = \sum_{n=-\infty}^{\infty} |X_n|^2 = X_0^2 + \sum_{n=1}^{\infty} 2|X_n|^2$$

The result indicates that the total average power of $x(t)$ is the sum of the average power in each harmonic component. ∎

3.5 SYSTEMS WITH PERIODIC INPUTS

The response of an LTI system to a sinusoidal input leads to a characterization of system behaviour that is termed the *frequency response* of the system. This characterization is obtained in terms of the impulse response by using convolution and a complex sinusoidal input signal. Let the impulse response of a system be $h(t)$ and the input be $x(t) = e^{j\omega_0 t}$. Then, the convolution integral gives the output as

$$y(t) = h(t) * x(t)$$

$$= \int_{-\infty}^{\infty} h(\tau)x(t-\tau)d\tau$$

$$= \int_{-\infty}^{\infty} h(\tau)\, e^{j\omega_0(t-\tau)}d\tau$$

$$= \int_{-\infty}^{\infty} h(\tau)\, e^{-j\omega_0 \tau}d\tau\, e^{j\omega_0 t}$$

$$y(t) = H(\omega_0)\, e^{j\omega_0 t}$$

where we define

$$H(\omega_0) = \int_{-\infty}^{\infty} h(\tau)\, e^{-j\omega_0 \tau}d\tau$$

The output of the system is thus a complex sinusoid of the same frequency as the input multiplied by the complex number $H(\omega_0)$ [Fig. 3.18(a)]. An intuitive interpretation of

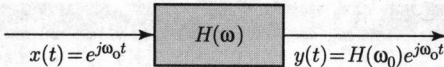

$$x(t) = e^{j\omega_0 t} \qquad H(\omega) \qquad y(t) = H(\omega_0) e^{j\omega_0 t}$$

Fig. 3.18(a) Response of an LTI system to complex sinusoid

the sinusoidal steady-state response is obtained by writing the complex-valued frequency response $H(\omega_0)$ in polar form. We have

$$H(\omega_0) = |H(\omega_0)| \, e^{j\angle H(\omega_0)} \tag{3.62}$$

where $|H(\omega_0)|$ is now termed the *magnitude response* and $\angle H(\omega_0)$ is termed the *phase response* of the system. By substituting this polar form into Eq. (3.62), we may express the output as

$$y(t) = |H(\omega_0)| \, e^{j(\omega_0 t + \angle H(\omega_0))} \tag{3.63}$$

The system thus modifies the amplitude of the input by $|H(\omega_0)|$ and the phase by $\angle H(\omega_0)$. Knowing $H(\omega_0)$, we can determine whether the system amplifies or attenuates a given sinusoidal component of the input and how much of a phase shift the system adds to that particular component.

We say that the complex sinusoid $x(t) = e^{j\omega_0 t}$ is an *eigenfunction* of the system associated with the *eigenvalue* $H(\omega_0)$. Signals that are eigenfunctions of systems play an important role in LTI system theory. By representing arbitrary signals as weighted superpositions of eigenfunctions, we transform the operation of convolution to multiplication. Thus, if the input to an LTI system is given by

$$x(t) = \sum_{n=-\infty}^{\infty} X_n \, e^{jn\omega_0 t} \tag{3.64}$$

then, the output of the system is given by [Fig. 3.18(b)]

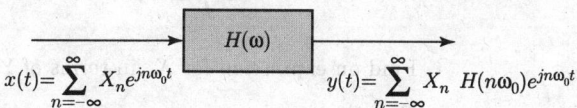

$$x(t) = \sum_{n=-\infty}^{\infty} X_n e^{jn\omega_0 t} \qquad H(\omega) \qquad y(t) = \sum_{n=-\infty}^{\infty} X_n \, H(n\omega_0) e^{jn\omega_0 t}$$

Fig. 3.18(b) Response of an LTI system to a periodic signal

$$y(t) = \sum_{n=-\infty}^{\infty} X_n H(n\omega_0) \, e^{jn\omega_0 t} = \sum_{n=-\infty}^{\infty} Y_n \, e^{jn\omega_0 t} \tag{3.65}$$

Therefore, the Fourier series coefficients of the output of an LTI system is given by

$$Y_n = X_n H(n\omega_0) \tag{3.66}$$

SOLVED EXAMPLES

Example 3.12 A continuous-time periodic signal $x(t)$ is real-valued and has a fundamental period $T = 8$. The nonzero Fourier series coefficients for $x(t)$ are

$$X_1 = X_{-1} = 2, \quad X_3 = X_{-3}^* = 4j$$

Express $x(t)$ in the form

$$x(t) = \sum_{n=0}^{\infty} A_n \cos(\omega_n t + \phi_n)$$

Solution
The signal $x(t)$ is periodic with fundamental frequency $\omega_0 = 2\pi/8 = \pi/4$. Using the Fourier synthesis Eq. (3.26), we have

$$x(t) = \sum_{n=-\infty}^{\infty} X_n e^{jn\omega_0 t}$$

$$= X_1 e^{j\omega_0 t} + X_{-1} e^{-j\omega_0 t}$$
$$+ X_3 e^{j3\omega_0 t} + X_{-3} e^{-3j\omega_0 t}$$

$$= 2 e^{j\frac{\pi}{4}t} + 2 e^{-j\frac{\pi}{4}t} + 4j e^{j3\frac{\pi}{4}t} - 4j e^{-j3\frac{\pi}{4}t}$$

$$= 4\left(\frac{e^{j\frac{\pi}{4}t} + e^{-j\frac{\pi}{4}t}}{2} \right) - 8\left(\frac{e^{j3\frac{\pi}{4}t} - e^{-j3\frac{\pi}{4}t}}{2j} \right)$$

$$= 4\cos\left(\frac{\pi}{4}t\right) - 8\sin\left(\frac{3\pi}{4}t\right)$$

$$= 4\cos\left(\frac{\pi}{4}t\right) + 8\cos\left(\frac{3\pi}{4}t + \frac{\pi}{2}\right)$$

Example 3.13 For the continuous-time periodic signal

$$x(t) = 2 + \cos\left(\frac{2\pi}{3}t\right) + 4\sin\left(\frac{5\pi}{3}t\right),$$

determine the fundamental frequency ω_0 and the Fourier series coefficients X_n such that

$$x(t) = \sum_{n=-\infty}^{\infty} X_n e^{jn\omega_0 t}$$

Solution
Consider the given signal

$$x(t) = 2 + \cos\left(\frac{2\pi}{3}t\right) + 4\sin\left(\frac{5\pi}{3}t\right)$$

$$= 2 + \left(\frac{e^{j\frac{2\pi}{3}t} + e^{-j\frac{2\pi}{3}t}}{2} \right) + 4\left(\frac{e^{j\frac{5\pi}{3}t} - e^{-j\frac{5\pi}{3}t}}{2j} \right)$$

$$= 2 + \frac{1}{2} e^{j2\frac{\pi}{3}t} + \frac{1}{2} e^{-j2\frac{\pi}{3}t} + \frac{2}{j} e^{j5\frac{\pi}{3}t} - \frac{2}{j} e^{-j5\frac{\pi}{3}t}$$

$$x(t) = 2 + \frac{1}{2} e^{j2\frac{\pi}{3}t} + \frac{1}{2} e^{-j2\frac{\pi}{3}t} - 2j e^{j5\frac{\pi}{3}t} + 2j e^{-j5\frac{\pi}{3}t}$$

Comparing with

$$x(t) = X_0 + X_2 e^{j2\omega_0 t} + X_{-2} e^{-j2\omega_0 t}$$
$$+ X_5 e^{j5\omega_0 t} + X_{-5} e^{-j5\omega_0 t}$$

The nonzero Fourier series coefficients are

$$X_0 = 2, \qquad X_2 = X_{-2} = \frac{1}{2}, \qquad X_5 = X_{-5}^* = -2j$$

Example 3.14 Suppose the periodic signal $x(t)$ has fundamental period T and Fourier coefficients X_n. In a variety of situations, it is easier to calculate the Fourier series coefficients Y_n for $y(t) = dx(t)/dt$ as apposed to calculating X_n directly. Given that

$$\int_{T}^{2T} x(t)\, dt = 2$$

Find an expression for X_n in terms of Y_n and T.

Solution
Given that

$$x(t) \longleftrightarrow X_n$$

Using the differentiation property, Eq. (3.53), we have

$$y(t) = \frac{dx(t)}{dt} \longleftrightarrow Y_n = jn\omega_0 X_n$$

Therefore,

$$X_n = \frac{Y_n}{jn\omega_0}, \quad n \neq 0$$

From Eq. (3.27), we have

$$X_n = \frac{1}{T} \int_T^{2T} x(t)\, e^{-jn\omega_0 t}\, dt$$

for $n = 0$, we have

$$X_n = \frac{1}{T} \int_T^{2T} x(t)\, dt = \frac{2}{T}$$

Therefore,

$$X_n = \begin{cases} \dfrac{2}{T}, & n = 0 \\ \dfrac{Y_n}{jn\omega_0}, & n \neq 0 \end{cases}$$

Example 3.15 Find the compact trigonometric Fourier series for the periodic signal $x(t)$ shown in Fig. 3.19. Sketch the amplitude and phase spectra for $x(t)$.

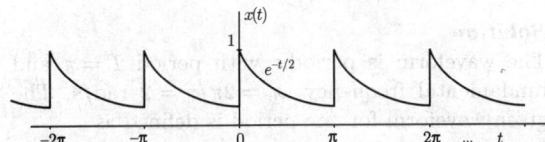

Fig. 3.19 Periodic signal

Solution

The given signal $x(t)$ is periodic with period $T = \pi$ and fundamental frequency $\omega_0 = 2\pi/\pi = 2$ rad/s. The given waveform for one period is defined as

$$x(t) = e^{-\frac{t}{2}}, \quad 0 < t < \pi$$

Since the wave is neither even nor odd, the series will contain both sine and cosine terms.

$$a_0 = \frac{1}{T} \int_0^T x(t)\, dt = \frac{1}{\pi} \int_0^\pi e^{-\frac{t}{2}}\, dt$$

$$= \frac{1}{\pi} \left[\frac{e^{-t/2}}{-1/2} \right]_0^\pi = \frac{2}{\pi}[1 - e^{-\pi/2}] = 0.504$$

Now we determine a_n:

$$a_n = \frac{2}{T} \int_0^T x(t) \cos(n\omega_0 t)\, dt$$

$$= \frac{2}{\pi} \int_0^\pi e^{-t/2} \cos(2nt)\, dt$$

$$= \frac{2}{\pi} I$$

where

$$I = \int_0^\pi e^{-t/2} \cos(2nt)\, dt$$

$$= \left[\frac{e^{-t/2} \sin(2nt)}{2n} \right]_0^\pi$$

$$\quad + \frac{1}{4n} \int_0^\pi e^{-t/2} \sin(2nt)\, dt$$

$$= 0 + \frac{1}{4n} \left[-\frac{e^{-t/2} \cos(2nt)}{2n} \right]_0^\pi$$

$$\quad - \frac{1}{16n^2} \int_0^\pi e^{-t/2} \cos(2nt)\, dt$$

$$I = \frac{1}{4n} \left[-\frac{e^{-t/2} \cos(2nt)}{2n} \right]_0^\pi - \frac{1}{16n^2} I$$

$$I\left(1 + \frac{1}{16n^2}\right) = \frac{1}{8n^2}[1 - e^{-\pi/2}]$$

$$I\left(\frac{16n^2 + 1}{16n^2}\right) = \frac{1}{8n^2}[1 - e^{-\pi/2}]$$

$$I = 2\frac{[1 - e^{-\pi/2}]}{16n^2 + 1}$$

Substituting I in the expression of a_n gives

$$a_n = 2\frac{\frac{2}{\pi}[1 - e^{-\pi/2}]}{16n^2 + 1}$$

$$= 0.504\left(\frac{2}{16n^2 + 1}\right)$$

Now we determine b_n:

$$b_n = \frac{2}{T} \int_0^T x(t) \sin(n\omega_0 t)\, dt$$

$$= \frac{2}{\pi} \int_0^\pi e^{-t/2} \sin(n\pi t)\, dt$$

$$= 0.504\left(\frac{8n}{16n^2 + 1}\right)$$

Substituting the values of a_0, a_n, and b_n into Eq (3.6), gives

$$x(t) = a_0 + \sum_{n=1}^{\infty} [a_n \cos(n\omega_0 t) + b_n \sin(n\omega_0 t)]$$

$$x(t) = 0.504 \left[1 + \sum_{n=1}^{\infty} \frac{2}{1 + 16n^2} \Big(\cos(2nt) + 4n \sin(2nt) \Big) \right]$$

Also from Eqs (3.10) and (3.11)

$$c_0 = a_0 = 0.504$$

$$c_n = \sqrt{a_n^2 + b_n^2}$$

$$= 0.504 \sqrt{\frac{4}{(1 + 16n^2)^2} + \frac{64n^2}{(1 + 16n^2)^2}}$$

$$= 0.504 \left(\frac{2}{\sqrt{1 + 16n^2}} \right)$$

$$\theta_n = \tan^{-1} \left(\frac{-b_n}{a_n} \right) = \tan^{-1}(4n) = -\tan^{-1}(4n)$$

Amplitude and phases of the dc and the first six harmonics are computed from the above equations and displayed in Table 3.2. The amplitude and phase spectra are shown in Figs 3.20(a) and (b), respectively.

Table 3.2

n	c_n	θ_n
0	0.504	0
1	0.244	−75.96
2	0.125	−82.87
3	0.084	−85.24
4	0.063	−86.42
5	0.0504	−87.14
6	0.042	−87.61

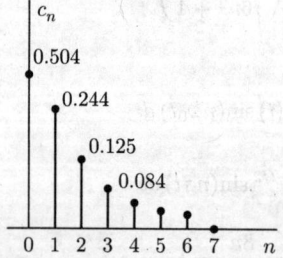

Fig. 3.20(a)   Amplitude spectra

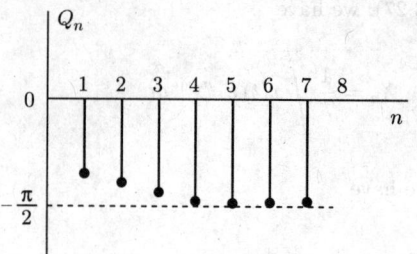

Fig. 3.20(b)   Phase spectra

The compact or polar form representation of Fourier series is

$$x(t) = 0.504$$

$$+ 0.504 \sum_{n=1}^{\infty} \frac{2}{\sqrt{1 + 16n^2}} \cos(2nt - \tan^{-1}(4n))$$

Example 3.16 Find the exponential Fourier series for the signal in Fig. 3.19.

Solution
The waveform is periodic with period $T = \pi$ and fundamental frequency $\omega_0 = 2\pi/\pi = 2$ rad/s. The given waveform for one period is defined as

$$x(t) = e^{-t/2}, \quad 0 < t < \pi$$

From Eq. (3.27), the exponential Fourier series coefficient is

$$X_n = \frac{1}{T} \int_0^T x(t) e^{-jn\omega_0 t}\, dt = \frac{1}{\pi} \int_0^{\pi} e^{-\frac{t}{2}} e^{-j2nt}\, dt$$

$$X_n = \frac{1}{\pi} \int_0^{\pi} e^{-\left(\frac{1}{2} + j2n \right)t}\, dt$$

$$= -\frac{1}{\pi(\frac{1}{2} + j2n)} \left[e^{-\left(\frac{1}{2} + j2n \right)t} \right]_0^{\pi}$$

$$X_n = \frac{0.504}{1 + j4n}$$

Substituting $n = 0, \pm 1, \pm 2, \ldots$ gives

$$X_0 = 0.504$$

$$X_1 = \frac{0.504}{1 + j4} = 0.122 \, e^{-j75.96}$$

$$|X_1| = 0.122, \angle X_1 = -75.96$$

$$X_{-1} = \frac{0.504}{1 - j4} = 0.122 \, e^{j75.96}$$

$$|X_{-1}| = 0.122, \angle X_{-1} = 75.96$$

$$X_2 = \frac{0.504}{1 + j8} = 0.0625 \, e^{-j82.87}$$

$$|X_2| = 0.0625, \angle X_2 = -82.87$$

$$X_{-2} = \frac{0.504}{1 - j8} = 0.0625 \, e^{j82.87}$$

$$|X_{-2}| = 0.0625, \angle X_{-2} = 82.87$$

Note that X_n and X_{-n} are conjugates, as expected [Eq. (3.55)]. Figure 3.21(a) shows the magnitude spectrum and Fig.3.21(b) shows the phase spectrum.

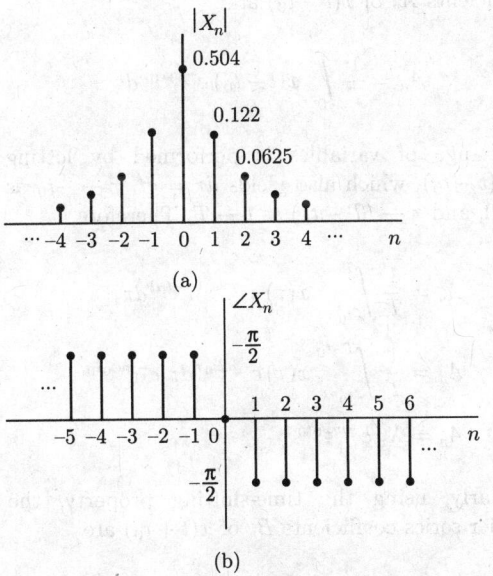

(a)

(b)

Fig. 3.21   (a) Magnitude spectrum, (b) Phase spectrum

The exponential Fourier series for the given waveform is

$$x(t) = \sum_{n=-\infty}^{\infty} X_n \, e^{jn\omega_0 t} = 0.504 \sum_{n=-\infty}^{\infty} \frac{1}{1 + j4n} \, e^{j2nt}$$

Example 3.17 Suppose we are given the following information about a signal $x(t)$:

1. $x(t)$ is real and odd.
2. $x(t)$ is periodic with period $T = 2$ and has Fourier coefficients X_n.
3. $X_n = 0$ for $|n| > 1$.
4. $\dfrac{1}{2} \displaystyle\int_0^2 |x(t)|^2 \, dt = 1$.

Specify two different signals that satisfy these conditions.

Solution
Since $x(t)$ is real and odd (clue 1), its Fourier series coefficients are purely imaginary and odd [Eq. (3.58)]. Therefore,

$$X_n = -X_{-n} \quad \text{and} \quad X_0 = 0$$

Also, since it is given that $X_n = 0$ for $|n| > 1$, the only unknown Fourier series coefficients are X_1 and X_{-1}. Using Parseval's relation,

$$\frac{1}{T} \int_0^T |x(t)|^2 \, dt = \sum_{n=-\infty}^{\infty} |X_n|^2$$

for the given signal we have

$$\frac{1}{2} \int_0^2 |x(t)|^2 \, dt = \sum_{n=-1}^{1} |X_n|^2 = 1$$

$$|x_1|^2 + |x_{-1}|^2 = 1$$

$$2|X_1| = 1$$

$$|X_1|^2 = \frac{1}{2}$$

Therefore,

$$X_1 = -X_{-1} = j\frac{1}{\sqrt{2}} \quad \text{or} \quad X_1 = -X_{-1} = -j\frac{1}{\sqrt{2}}$$

The two possible signals which satisfy the given information are

$$x_1(t) = j\frac{1}{\sqrt{2}}e^{j\pi t} - j\frac{1}{\sqrt{2}}e^{-j\pi t}$$

$$= j\frac{1}{\sqrt{2}}2j\left(\frac{e^{j\pi t} - e^{-j\pi t}}{2j}\right)$$

$$= -\sqrt{2}\sin(\pi t)$$

and

$$x_2(t) = -j\frac{1}{\sqrt{2}}e^{j\pi t} + j\frac{1}{\sqrt{2}}e^{-j\pi t}$$

$$= -j\frac{1}{\sqrt{2}}2j\left(\frac{e^{j\pi t} - e^{-j\pi t}}{2j}\right)$$

$$= \sqrt{2}\sin(\pi t)$$

Example 3.18 Let $x(t)$ be a periodic signal whose Fourier series coefficients are

$$X_n = \begin{cases} 2, & n = 0 \\ j\left(\dfrac{1}{2}\right)^{|n|}, & \text{otherwise} \end{cases}$$

Use Fourier series properties to answer the following questions:
(a) Is $x(t)$ real?
(b) Is $x(t)$ even?
(c) Is $dx(t)/dt$ even?

Solution
(a) If $x(t)$ is real, then $X_n = X_{-n}^*$. Since $X_n \neq X_{-n}^*$, $x(t)$ is not real.
(b) If $x(t)$ is even, then $X_n = X_{-n}$.

$$X_n = X_{-n} = \begin{cases} 2, & n = 0 \\ j\left(\dfrac{1}{2}\right)^{|k|}, & \text{otherwise} \end{cases}$$

Since $X_n = X_{-n}$, $x(t)$ is even.
(c) We have

$$y(t) = \frac{dx(t)}{dt} \longleftrightarrow Y_n = jn\omega_0 X_n$$

Therefore,

$$Y_n = \begin{cases} 0, & n = 0 \\ -n\left(\dfrac{1}{2}\right)^{|n|}\omega_0, & \text{otherwise} \end{cases}$$

and

$$Y_{-n} = \begin{cases} 0, & n = 0 \\ n\left(\dfrac{1}{2}\right)^{|n|}\omega_0, & \text{otherwise} \end{cases}$$

Since $Y_n \neq Y_{-n}$, $y(t) = dx(t)/dt$ is not even.

Example 3.19 Let $x(t)$ be a periodic signal with fundamental period T and Fourier series coefficients X_n. Derive the Fourier series coefficients of each of the following signals in terms of X_n:
(a) $x(t - t_0) + x(t + t_0)$
(b) $\mathcal{E}\{x(t)\} = x_e(t)$
(c) $\text{Re}\{x(t)\}$
(d) $\dfrac{d^2x(t)}{dt^2}$
(e) $x(3t - 1)$

Solution
The Fourier series coefficients of $x(t)$ is X_n, i.e.,

$$x(t) \longleftrightarrow X_n$$

(a) $x(t - t_0)$ is also periodic with period T and fundamental frequency $\omega_0 = 2\pi/T$. The Fourier series coefficients A_n of $x(t - t_0)$ are

$$A_n = \frac{1}{T}\int_0^T x(t - t_0)e^{-jn\omega_0 t}\,dt$$

A change of variables is performed by letting $\tau = (t - t_0)$, which also yields $d\tau = dt$, $\tau \to -t_0$ as $t \to 0$, and $\tau \to (T - t_0)$ as $t \to T$. Therefore,

$$A_n = \frac{1}{T}\int_{-t_0}^{T-t_0} x(\tau)e^{-jn\omega_0(\tau + t_0)}\,d\tau$$

$$A_n = \frac{1}{T}\int_{-t_0}^{T-t_0} x(\tau)e^{-jn\omega_0 \tau}\,d\tau\, e^{-jn\omega_0 t_0}$$

$$A_n = X_n e^{-jn\omega_0 t_0}$$

Similarly, using the time-shifting property, the Fourier series coefficients B_n of $x(t + t_0)$ are

$$B_n = X_n e^{jn\omega_0 t_0}$$

Finally, the Fourier series coefficients Y_n of $x(t - t_0) + x(t + t_0)$ are

$$Y_n = A_n + B_n = X_n e^{-jn\omega_0 t_0} + X_n e^{jn\omega_0 t_0}$$

$$= 2X_n \left(\frac{e^{-jn\omega_0 t_0} + e^{jn\omega_0 t_0}}{2} \right)$$

$$= 2\cos(n\omega_0 t_0) X_n$$

(b) Note that

$$\mathcal{E}\{x(t)\} = x_e(t) = \frac{x(t) + x(-t)}{2}$$

The Fourier series coefficients of $x(-t)$ are

$$A_n = \frac{1}{T} \int_0^T x(-t) e^{-jn\omega_0 t} dt$$

$$= \frac{1}{T} \int_0^T x(t) e^{-j(-n)\omega_0 t} dt$$

$$= X_{-n}$$

Therefore, the Fourier series coefficients Y_n of $x_e(t)$ are

$$Y_n = \frac{X_n + X_{-n}}{2}$$

(c) Note that

$$\text{Re}\{x(t)\} = \frac{x(t) + x^*(t)}{2}$$

The Fourier series coefficients A_n of $x^*(t)$ are

$$A_n = \frac{1}{T} \int_0^T x^*(t) e^{-jn\omega_0 t} dt$$

$$= \left(\frac{1}{T} \int_0^T x(t) e^{jn\omega_0 t} dt \right)^*$$

$$= \left(\frac{1}{T} \int_0^T x(t) e^{-j(-n)\omega_0 t} dt \right)^*$$

$$A_n = (X_{-n})^* = X_{-n}^*$$

Therefore, the Fourier series coefficients of $\text{Re}\{x(t)\}$ are

$$Y_n = \frac{X_n + A_n}{2}$$

$$= \frac{X_n + X_{-n}^*}{2}$$

(d) The Fourier series synthesis equation gives

$$x(t) = \sum_{n=-\infty}^{\infty} X_n e^{jn\omega_0 t}$$

Differentiating both sides w.r.t. t twice gives

$$\frac{d^2 x(t)}{dt^2} = \sum_{n=-\infty}^{\infty} (-n^2 \omega_0^2 X_n) e^{jn\omega_0 t}$$

By inspection, we know that the Fourier series coefficients of $d^2 x(t)/dt^2$ are $-n^2 \omega_0^2 X_n$.

(e) The period of $x(3t - 1)$ is $T/3$ and fundamental frequency $3\omega_0$. The Fourier series coefficients Y_n of $y(t) = x(3t - 1)$ are

$$Y_n = \frac{1}{T/3} \int_0^{T/3} y(t) e^{-jn3\omega_0 t} dt$$

$$= \frac{1}{T/3} \int_0^{T/3} x(3t - 1) e^{-jn3\omega_0 t} dt$$

A change of variables is performed by letting $\tau = (3t - 1)$, which also yields $dt = 1/3 d\tau$, $\tau \to -1$ as $t \to 0$, and $\tau \to T - 1$ as $t \to T/3$. Therefore,

$$Y_n = \frac{1}{T} \int_{-1}^{T-1} x(\tau) e^{-jn3\omega_0 \frac{(\tau+1)}{3}} d\tau$$

$$= \frac{1}{T} \int_{-1}^{T-1} x(\tau) e^{-jn\omega_0 \tau} d\tau \, e^{-jn\omega_0}$$

$$= X_n e^{-jn\omega_0}$$

Example 3.20 Suppose we are given the following information about a continuous-time periodic signal with period 3 and Fourier coefficients X_n:

1. $X_n = X_{n+2}$.
2. $X_n = X_{-n}$.
3. $\int_{-0.5}^{0.5} x(t) dt = 1$.
4. $\int_{0.5}^{1.5} x(t) dt = 2$.

Determine $x(t)$.

Solution
The continuous-time signal $x(t)$ is periodic with period $T = 3$ and fundamental frequency $\omega_0 = 2\pi/T = 2\pi/3$. Since $X_n = X_{-n}$, we require that $x(t) = x(-t)$,

i.e., $x(t)$ must be an even signal. Also, note that

$$X_n = X_{n+2}$$

$$\frac{1}{T} \int_0^T x(t) e^{-jn\omega_0 t} dt = \frac{1}{T} \int_0^T x(t) e^{-j(n+2)\omega_0 t} dt$$

$$\frac{1}{T} \int_0^T x(t) e^{-jn\omega_0 t} dt = \frac{1}{T} \int_0^T [x(t) e^{-j2\omega_0 t}] e^{-jn\omega_0 t} dt$$

we require that

$$x(t) = x(t) e^{-j2\omega_0 t}$$

or

$$e^{-j2\omega_0 t} = 1$$

$$e^{-j2\omega_0 t} = 1 = e^{j2m\pi}, \quad m = 0, \pm 1, \pm 2, \cdots$$

The above equation indicates that

$$e^{-j2\omega_0 t} = \begin{cases} 1, & t = 0, \pm \frac{3}{2}, 3, \pm \frac{9}{2}, \cdots \\ 0, & \text{otherwise} \end{cases}$$

Since $\int_{-0.5}^{0.5} x(t) dt = 1$, we may conclude that $x(t) = \delta(t)$ for $-0.5 \le t \le 0.5$. Also since $\int_{0.5}^{1.5} x(t) dt = 2$, we may conclude that $x(t) = 2\delta(t - 3/2)$ in the range $0.5 \le t \le 1.5$. Therefore, $x(t)$ may be written as

$$x(t) = \sum_{k=-\infty}^{\infty} \delta(t - 3k) + 2 \sum_{k=-\infty}^{\infty} \delta(t - 3k - 3/2)$$

Example 3.21 Show that the Fourier series coefficients of the signal

$$z(t) = x(t)y(t) = \sum_{n=-\infty}^{\infty} Z_n e^{jn\omega_0 t}$$

are given by the discrete convolution

$$Z_n = \sum_{k=-\infty}^{\infty} X_k Y_{n-k}$$

Solution

$$z(t) = x(t)y(t)$$

$$= \sum_{k=-\infty}^{\infty} X_k e^{jk\omega_0 t} \sum_{n=-\infty}^{\infty} Y_n e^{jn\omega_0 t}$$

$$= \sum_{n=-\infty}^{\infty} \sum_{k=-\infty}^{\infty} X_k Y_n e^{j(k+n)\omega_0 t}$$

A change of variables is performed by letting $m = (k + n)$, which also yields $n = (m - k)$, $m \to -\infty$ as $n \to -\infty$, and $m \to \infty$ as $n \to \infty$. Therefore,

$$z(t) = \sum_{m=-\infty}^{\infty} \left(\sum_{k=-\infty}^{\infty} X_k Y_{m-k} \right) e^{jm\omega_0 t}$$

$$= \sum_{m=-\infty}^{\infty} Z_m e^{jm\omega_0 t}$$

Thus,

$$Z_n = \sum_{k=-\infty}^{\infty} X_k Y_{n-k}$$

Example 3.22 Let $x(t)$ be a real-valued signal with fundamental period T and Fourier series coefficient X_n.
(a) Show that $X_n = X_{-n}^*$ and X_0 must be real.
(b) Show that if $x(t)$ is even, then its Fourier coefficients must be real and even.
(c) Show that if $x(t)$ is odd, then its Fourier series coefficients are imaginary and odd and $X_0 = 0$.
(d) Show that the Fourier series coefficients of the even part of $x(t)$ are equal to

$$\text{Re}\{x(t)\}$$

(e) Show that the Fourier series coefficients of the odd part of $x(t)$ are equal to

$$j\text{Im}\{x(t)\}$$

Solution
(a) From conjugation and conjugate symmetry property, i.e., Eq. (3.55), we know that Fourier series coefficients of $x^*(t)$ are X_{-n}^*, and we also know that if $x(t)$ is real, then $x(t) = x^*(t)$. Therefore $X_n = X_{-n}^*$.

This implies $X_0 = X_0^*$. Therefore, X_0 must be real.
(b) Let $x(t)$ be a real and even signal:

$$x(t) = x(-t)$$

From Eq. (3.47), the Fourier series coefficients of $x(-t)$ are X_{-n}. Therefore,

$$X_n = X_n^* = X_{-n}$$

(c) Let $x(t)$ be a real and odd signal:

$$x(t) = -x(-t)$$

From Eq. (3.47), the Fourier series coefficients of $-(-t)$ are $-X_{-n}$. Therefore,

$$X_n = X_n^* = -X_{-n}$$

Note that for $n = 0$, the above equation requires that $X_0 = -X_0$, we may conclude that $X_0 = 0$.
(d) See the proof of Eq. (3.59).
(e) See the proof of Eq. (3.60).

Example 3.23 Suppose we are given the following information about a signal $x(t)$:

1. $x(t)$ is real.
2. $x(t)$ is periodic with period $T = 6$ and has Fourier coefficients X_n.
3. $X_n = 0$ for $n = 0$ and $n > 2$.
4. $x(t) = -x(t - 3)$.
5. $\dfrac{1}{6} \displaystyle\int_{-3}^{3} |x(t)|^2 \, dt = \dfrac{1}{2}$.
6. X_1 is a positive real number.

Show that $x(t) = A\cos(Bt + C)$, and determine the values of the constants A, B, and C.

Solution
The given signal $x(t)$ is periodic with period $T = 6$ and fundamental frequency $\omega_0 = 2\pi/6 = \pi/3$. Since $x(t)$ is real, $X_n = X_{-n}^*$. Since $X_n = 0$ for $n = 0$ and $n > 2$, the only unknown Fourier series coefficients are X_1, X_{-1}, X_2, and X_{-2}. Now $x(t)$ is

$$x(t) = X_1 e^{j\omega_0 t} + X_{-1} e^{-j\omega_0 t}$$
$$+ X_2 e^{j2\omega_0 t} + X_{-2} e^{-j2\omega_0 t}$$

Since X_1 is a positive real number, $X_1 = X_{-1}$.

$$x(t) = 2X_1 \left(\frac{e^{j\omega_0 t} + e^{-j\omega_0 t}}{2} \right)$$
$$+ X_2 e^{j2\omega_0 t} + X_{-2} e^{-j2\omega_0 t}$$

$$x(t) = 2X_1 \cos(\omega_0 t) + X_2 e^{j2\omega_0 t} + X_{-2} e^{-j2\omega_0 t}$$

$$x(t) = 2X_1 \cos(\pi t/3) + X_2 e^{j2\frac{\pi}{3}t} + X_{-2} e^{-j2\frac{\pi}{3}t}$$

From this, we get

$$x(t - 3) = 2X_1 \cos\left(\frac{\pi}{3}(t - 3) \right)$$
$$+ X_2 e^{j2\frac{\pi}{3}(t-3)} + X_{-2} e^{-j2\frac{\pi}{3}(t-3)}$$

$$x(t - 3) = 2X_1 \cos\left(\frac{\pi}{3}t - \pi \right)$$
$$+ X_2 e^{j2\frac{\pi}{3}t} e^{-j2\pi} + X_{-2} e^{-j2\frac{\pi}{3}t} e^{j2\pi}$$

$$x(t - 3) = -2X_1 \cos(\pi t/3)$$
$$+ X_2 e^{j2\frac{\pi}{3}t} + X_{-2} e^{-j2\frac{\pi}{3}t}$$

Now, if we need $x(t) = -x(t-3)$, then $X_2 = X_{-2} = 0$, and

$$x(t) = 2X_1 \cos(\pi t/3)$$

Now using Parseval's relation on clue 5, we get

$$\frac{1}{T} \int_{-T/2}^{T/2} |x(t)|^2 \, dt = \sum_{n=-\infty}^{\infty} |X_n|^2$$

$$\frac{1}{6} \int_{-3}^{3} |x(t)|^2 \, dt = |X_1|^2 + |X_{-1}|^2 = \frac{1}{2}$$

$$2|X_1|^2 = \frac{1}{2}$$

$$|X_1| = \frac{1}{2}$$

Since X_1 is positive, we have $X_1 = X_{-1} = 1/2$. Therefore,

$$x(t) = \cos(\pi t/3)$$

Example 3.24 Consider a continuous-time system whose frequency response is

$$H(j\omega) = \begin{cases} 1, & |\omega| \le 100 \\ 0, & |\omega| > 100 \end{cases}$$

When the input to this system is a signal $x(t)$ with the fundamental period $T = \pi/6$ and Fourier series coefficient X_n, it is found that
(a) $y(t) = x(t)$. Comment on the Fourier series coefficients of $x(t)$.
(b) $y(t) = 0$. Comment on the Fourier series coefficients of $x(t)$.

Solution
The given signal $x(t)$ is periodic with period $T = \pi/6$, fundamental frequency $\omega_0 = 12$, and $x(t)$ is represented as

$$x(t) = \sum_{n=-\infty}^{\infty} X_n \, e^{jn\omega_0 t}$$

From Eqs (3.65) and (3.66), the output of the system is

$$y(t) = \sum_{n=-\infty}^{\infty} X_n H(jn\omega_0) \, e^{jn\omega_0 t}$$

(a) Since $H(j\omega) = 0$ for $\omega > 100$, the largest value of $|n|$ for which $y(t) = x(t)$ or for which $X_n \neq 0$ should be such that

$$|n|\omega_0 \leq 100$$

$$12|n| \leq 100$$

$$|n| \leq 8$$

Therefore, the Fourier series coefficients $X_n \neq 0$ for $n \leq 8$.
(b) For $y(t) = 0$, the Fourier series coefficients $X_n = 0$ for $n \leq 8$.

Example 3.25 The trigonometric Fourier spectra (line spectrum) of a certain periodic signal $x(t)$ are shown in Fig. 3.22. By inspecting these spectra, sketch the corresponding exponential Fourier spectra and verify your results analytically.

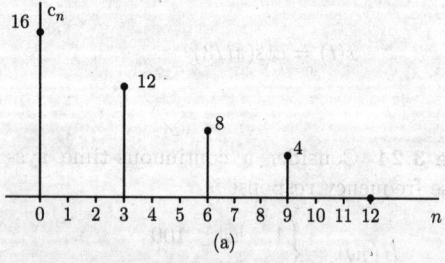

Fig. 3.22(a)   Amplitude spectrum

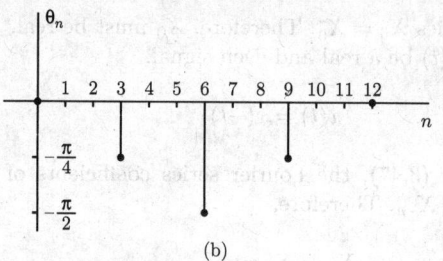

(b)

Fig. 3.22(b)   Phase spectrum

Solution
The trigonometric spectral components exist at frequencies 0, 3, 6, and 9. The exponential spectral components exist at 0, 3, 6, 9 and −3, −6, −9.
First consider the amplitude spectrum. From Eq. (3.38), we have

$$X_0 = c_0 = 16.$$

Now $|X_n|$ is an even function of ω, and from Eq. (3.37), we have

$$|X_n| = |X_{-n}| = c_n/2$$

Thus, all the remaining spectrum $|X_n|$ for positive n is half the trigonometric amplitude spectrum c_n, and the spectrum $|X_n|$ for negative n is a reflection about the vertical axis of the spectrum for positive n, as shown in Fig. 3.23(a).

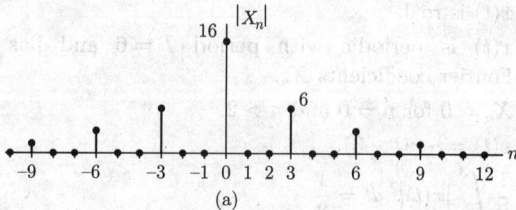

Fig. 3.23(a)   Magnitude spectrum

The phase spectrum is an odd function. Therefore, the phase spectrum $\angle X_n = \theta_n$ for positive n and is $-\theta_n$ for negative n as depicted in Fig. 3.23(b).

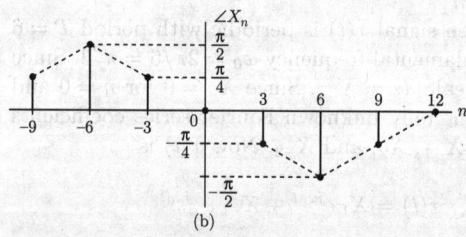

Fig. 3.23(b)   Phase spectrum

We shall now verify that both sets of spectra represent the same signal. The amplitude and the phase of the component of frequency 3 are 12 and $-\pi/4$, respectively. Therefore, this component can be expressed as $12\cos(3t - \pi/4)$.

The amplitude and the phase of the component of frequency 6 are 8 and $-\pi/2$, respectively. Therefore, this component can be expressed as $8\cos(6t - \pi/2)$.

The amplitude and the phase of the component of frequency 9 are 4 and $-\pi/4$, respectively. Therefore, this component can be expressed as $4\cos(9t - \pi/4)$. We can write the Fourier series for $x(t)$ as

$$x(t) = 16 + 12\cos(3t - \pi/4) + 8\cos(6t - \pi/2)$$
$$+ 4\cos(9t - \pi/4)$$

Now consider the exponential spectra shown in Fig. 3.23(a) and (b). They contain components of frequencies 0, ± 3, ± 6, and ± 9. The dc component is $X_0 = 16$. The component e^{j3t} (frequency 3) has magnitude 6 and angle $-\pi/4$. Therefore, this component strength is $6e^{j-\pi/4}$, and it can be expressed as $(6e^{-j\pi/4})e^{j3t}$. Similarly, the component of frequency -3 is $(6e^{j\pi/4})e^{-j3t}$. Proceeding in this manner, the signal corresponding to the spectra shown in Fig. 3.23, is

$$x(t) = 16 + [6\,e^{-j\pi/4}\,e^{j3t} + 6\,e^{j\pi/4}\,e^{-j3t}]$$
$$+ [4\,e^{-j\pi/2}\,e^{j6t} + 4\,e^{j\pi/2}\,e^{-j6t}]$$
$$+ [2\,e^{-j\pi/4}\,e^{j9t} + 2\,e^{j\pi/4}\,e^{-j9t}]$$
$$= 16 + 6[e^{j(3t-\pi/4)} + e^{-j(3t-\pi/4)}]$$
$$+ 4[e^{j(6t-\pi/2)} + e^{-j(6t-\pi/2)}]$$
$$+ 2[e^{j(9t-\pi/4)} + e^{-j(9t-\pi/4)}]$$
$$= 16 + 12\cos(3t - \pi/4) + 8\cos(6t - \pi/2)$$
$$+ 4\cos(9t - \pi/4)$$

Example 3.26 Find the exponential Fourier series and sketch the corresponding Fourier spectrum X_n versus ω for the full-wave rectified sine wave depicted in Fig. 3.24.

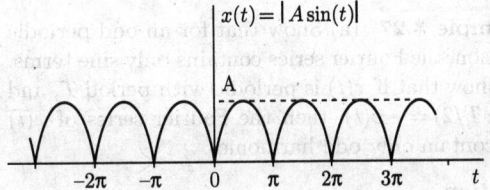

Fig. 3.24 Full-wave rectified sine wave

Solution

The waveform is periodic with period $T = \pi$ and fundamental frequency $\omega_0 = 2\pi/\pi = 2$ rad/s. The given waveform for one period is defined as

$$x(t) = A\sin(t), \quad 0 < t < \pi$$

From Eq. (3.27), the exponential Fourier series coefficient is

$$X_n = \frac{1}{T}\int_0^T x(t)\,e^{-jn\omega_0 t}\,dt$$

$$= \frac{1}{\pi}\int_0^\pi A\sin(t)\,e^{-j2nt}\,dt$$

$$= \frac{A}{\pi}\int_0^\pi e^{-j2nt}\sin(t)\,dt$$

$$X_n = \frac{2A}{\pi(1 - 4n^2)}$$

Figure 3.25 shows the magnitude spectrum.

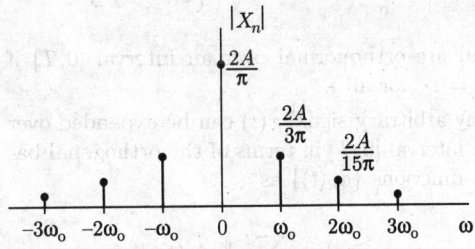

Fig. 3.25 Magnitude spectrum

The exponential Fourier series for the given waveform is

$$x(t) = \sum_{n=-\infty}^{\infty} X_n\,e^{jn\omega_0 t} = \frac{2A}{\pi}\sum_{n=-\infty}^{\infty}\frac{1}{1 - 4n^2}\,e^{j2nt}$$

Example 3.27 (a) Show that for an odd periodic function, the Fourier series contains only sine terms.
(b) Show that if $x(t)$ is periodic with period T, and $x(t \pm T/2) = -x(t)$, then the Fourier series of $x(t)$ will contain only odd harmonics.

Solution

(a) See the proof of Eq. (3.22) in Section 3.2.4.

(b) See the proof of Eq. (3.23) in Section 3.2.4.

Example 3.28 Let $z(t)$ be a real signal and $z(t) = x(t) + y(t)$. Find a condition so that

$$\int_{-\infty}^{\infty} |z(t)|^2\, dt = \int_{-\infty}^{\infty} |x(t)|^2\, dt + \int_{-\infty}^{\infty} |y(t)|^2\, dt$$

or

$$E_z = E_x + E_y$$

Solution

$$E_z = \int_{-\infty}^{\infty} |z(t)|^2\, dt$$

$$= \int_{-\infty}^{\infty} |x(t) + y(t)|^2\, dt$$

$$= \int_{-\infty}^{\infty} |x(t)|^2\, dt + \int_{-\infty}^{\infty} |y(t)|^2\, dt$$

$$+ \int_{-\infty}^{\infty} x(t)y^*(t)\, dt + \int_{-\infty}^{\infty} x^*(t)y(t)\, dt$$

If $x(t)$ and $y(t)$ are orthogonal, then the two integrals of the products $x(t)y^*(t)$ and $x^*(t)y(t)$ are zero, and

$$\int_{-\infty}^{\infty} |z(t)|^2\, dt = \int_{-\infty}^{\infty} |x(t)|^2\, dt + \int_{-\infty}^{\infty} |y(t)|^2\, dt$$

$$E_z = E_x + E_y$$

The energy of the sum of two orthogonal signals is equal to the sum of the energies of the two signals.

SUMMARY

1. Two functions $\phi_i(t)$ and $\phi_j(t)$ are orthogonal over an interval $[0, T]$ if

$$\int_0^T \phi_i(t)\phi_j(t)\, dt = \begin{cases} E_i, & i = j \\ 0, & i \neq j \end{cases}$$

and are orthonormal over an interval $[0, T]$ if $E_i = 1$ for all i.

2. Any arbitrary signal $x(t)$ can be expanded over an interval $[0, T]$ in terms of the orthogonal basis functions $\{\phi_n(t)\}$ as

$$x(t) = \sum_{n=-\infty}^{\infty} X_n \phi_n(t)\, dt$$

where

$$X_n = \frac{1}{E_n} \int_0^T x(t)\phi_n(t)\, dt$$

3. The complex exponentials

$$\phi_n(t) = e^{jn\omega_0 t}, \quad \omega_0 = \frac{2\pi}{T}$$

are orthogonal over the interval $[0, T]$.

4. A periodic signal $x(t)$, of period T, can be expanded in an exponential Fourier series as

$$x(t) = \sum_{n=-\infty}^{\infty} X_n e^{jn\omega_0 t}$$

where X_n are called Fourier series coefficients and are given by

$$X_n = \frac{1}{T} \int_0^T x(t)\, e^{-jn\omega_0 t}\, dt$$

5. The plot of $|X_n|$ versus $n\omega_0$ is called the magnitude spectrum.

6. The plot of $\angle X_n$ versus $n\omega_0$ is called the phase spectrum.

Table 3.3 Fourier Series Representation

Series form	Coefficient computation	Conversion formulae
Trigonometric $$x(t) = a_0 + \sum_{n=1}^{\infty} \left[a_n \cos(n\omega_0 t) + b_n \sin(n\omega_0 t) \right]$$	$$a_0 = \frac{1}{T} \int_0^T x(t)\,dt$$ $$a_n = \frac{2}{T} \int_0^T x(t) \cos(n\omega_0 t)\,dt$$ $$b_n = \frac{2}{T} \int_0^T x(t) \sin(n\omega_0 t)\,dt$$	$$a_0 = c_0 = X_0$$ $$a_n - jb_n = c_n\, e^{j\theta_n} = 2X_n$$ $$a_n + jb_n = c_n\, e^{-j\theta_n} = 2X_{-n}$$
Compact trigonometric (polar form) $$x(t) = c_0 + \sum_{n=1}^{\infty} c_n \cos(n\omega_0 t + \theta_n)$$	$$c_0 = a_0$$ $$c_n = \sqrt{a_n^2 + b_n^2}$$ $$\theta_n = \tan^{-1}\left(\frac{-b_n}{a_n}\right)$$	$$c_0 = X_0$$ $$c_n = 2|X_n|, \quad n \geq 1$$ $$\theta_n = \angle X_n$$
Exponential $$x(t) = \sum_{n=-\infty}^{\infty} X_n\, e^{jn\omega_0 t}$$	$$X_n = \frac{1}{T} \int_0^T x(t)\, e^{-jn\omega_0 t}\,dt$$	

7. For periodic signals, both the magnitude and phase spectra are line spectra. For real-valued signals, the magnitude spectrum has even symmetry and the phase has odd symmetry.

8. If signal $x(t)$ is real-valued signal, then it can be expanded in trigonometric series of the form

$$x(t) = a_0 + \sum_{n=1}^{\infty} \left[a_n \cos(n\omega_0 t) + b_n \sin(n\omega_0 t) \right]$$

where a_0, a_n, and b_n are the Fourier series coefficients.

9. The relation between the trigonometric-series coefficients and the exponential series coefficients is given by

$$a_0 = X_0$$

$$a_n = 2\text{Re}\{X_n\} \quad \text{or} \quad a_n = X_n + X_{-n}$$

$$b_n = -2\text{Im}\{X_n\} \quad \text{or} \quad b_n = j(X_n - X_{-n})$$

$$X_n = \frac{a_n - jb_n}{2}$$

10. An alternative form (the polar form) of the Fourier series is

$$x(t) = c_0 + \sum_{n=1}^{\infty} c_n \cos(n\omega_0 t + \theta_n)$$

where

$$c_0 = a_0$$

$$c_n = \sqrt{a_n^2 + b_n^2} \quad \text{for } n \geq 1$$

$$\theta_n = \tan^{-1}\frac{-b_n}{a_n}$$

or

$$c_n = \begin{cases} 2|X_n|, & n \geq 1 \\ X_0, & n = 0 \end{cases}$$

$$\angle X_n = \theta_n = \tan^{-1}\frac{-b_n}{a_n}$$

11. For the Fourier series to converge, the signal $x(t)$ must be absolutely integrable, have only a

finite number of maxima and minima, and have a finite number of discontinuities over any period. This set of conditions is known as Dirichlet conditions.

12. The response $y(t)$ of an LTI system to the periodic input $x(t)$ is

$$y(t) = \sum_{n=-\infty}^{\infty} X_n H(n\omega_0) e^{jn\omega_0 t}$$

where $H(jn\omega_0)$ is the frequency response of the system at $\omega = n\omega_0$ and X_n are the Fourier series coefficients of the input $x(t)$.

13. Representing $x(t)$ by a finite series results in an overshoot behaviour at the points of discontinuity. This phenomenon is known as Gibbs's phenomenon.

14. Table 3.3 shows the Fourier series representation of a periodic signal of period T and fundamental frequency $\omega_0 = 2\pi/T$.

MULTIPLE-CHOICE QUESTIONS

1. If the Fourier series coefficients of a signal are periodic, then the signal must be
   (a) continuous time, periodic
   (b) discrete time, periodic
   (c) continuous time, nonperiodic
   (d) discrete time, nonperiodic

2. If a periodic function $f(t)$ of period T satisfies $f(t) = -f\left(t + \frac{T}{2}\right)$, then in its Fourier series expansion,
   (a) the constant term will be zero
   (b) there will be no cosine terms
   (c) there will be no sine terms
   (d) there will be no even harmonics

3. A periodic signal which can be expanded in Fourier series
   (a) is a power signal
   (b) is an energy signal
   (c) is neither a power signal or an energy signal
   (d) can be either a power signal or an energy signal depending on the nature

4. The Fourier series of a real, even periodic signal will contain only
   (a) cosine terms
   (b) sine terms
   (c) even terms
   (d) odd harmonics

5. Fourier series of any periodic signal $x(t)$ can only be obtained if
   (a) $\int_0^T |x(t)| \, dt < \infty$
   (b) finite number of discontinuities within finite time interval T
   (c) both (a) and (b)
   (d) infinite number of discontinuities

6. Even signal satisfies
   (a) $x(t) = x(-t)$
   (b) $x(t) = -x(-t)$
   (c) $x(t) = -x\,(t + T/2)$
   (d) $x(t) = x\,(t + T/2)$

7. Odd signal satisfies
   (a) $x(t) = x(-t)$
   (b) $x(t) = -x(-t)$
   (c) $x(t) = -x\,(t + T/2)$
   (d) $x(t) = x\,(t + T/2)$

8. For half-wave (odd) symmetry, with T_0 is the period of $x(t)$, which one of the following is correct?
   (a) $x\,(t \pm T_0/2) = -x(t)$
   (b) $x\,(t \pm T_0/2) = x(t)$
   (c) $x(t \pm T_0) = -x(t)$
   (d) $x(t \pm T_0) = x(t)$

9. The Fourier series representations are based on using
   (a) constant coefficients
   (b) only cosine functions
   (c) only sine functions
   (d) orthogonal functions

10. What is the magnitude of the exponential Fourier series coefficient of the fundamental term of the periodic signal

$$x(t) = 3 + 2\cos(3t) + \sin(3t)$$

   (a) $3/2$
   (b) $\sqrt{5}/2$
   (c) $1/2$
   (d) 1

PROBLEMS

3.1 Consider the functions

$$\phi_1(t) = e^{-|t|} \quad \text{and} \quad \phi_2(t) = 1 - A e^{-2|t|}$$

Determine the constant A such that $\phi_1(t)$ and $\phi_2(t)$ are orthogonal over the interval $(-\infty, \infty)$.

3.2 Compute the exponential Fourier series coefficient of

$$x(t) = \sum_{m=-\infty}^{\infty} \left[\delta\left(t - \frac{m}{2}\right) + \delta\left(t - 3\frac{m}{2}\right) \right]$$

3.3 Find the trigonometric Fourier series for periodic signals shown in Figs 3.26(a) and (b). Sketch their amplitude and phase spectra.

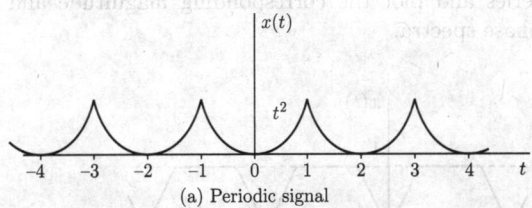

(a) Periodic signal

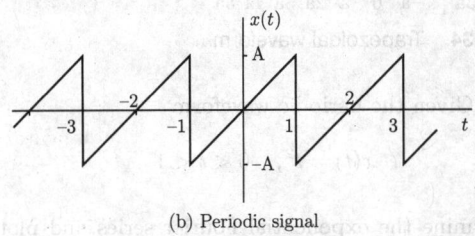

(b) Periodic signal

Fig. 3.26

3.4 Find the trigonometric Fourier series for periodic signals shown in Figs 3.27(a) and (b).

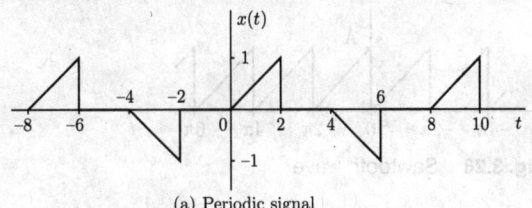

(a) Periodic signal

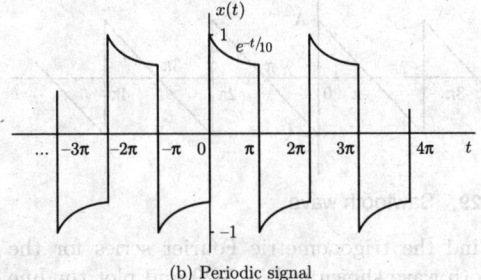

(b) Periodic signal

Fig. 3.27

3.5 Consider a continuous-time LTI system with impulse response

$$h(t) = e^{-4|t|}$$

Find the Fourier series representation of the output $y(t)$ for each of the following inputs:
(a) $x(t) = \sum_{n=-\infty}^{\infty} \delta(t - n)$
(b) $x(t) = \sum_{n=-\infty}^{\infty} (-1)^n \delta(t - n)$

3.6 Let

$$x(t) = \begin{cases} t, & 0 \le t \le 1 \\ 2 - t, & 1 \le t \le 2 \end{cases}$$

be a periodic signal with fundamental period $T = 2$ and Fourier series coefficients X_n.
(a) Determine the value of X_0.
(b) Determine the Fourier series representation of $dx(t)/dt$.
(c) Use the result of part (b) and the differentiation property of the Fourier series to determine the Fourier series coefficients of $x(t)$.

3.7 Find the trigonometric Fourier series for the sawtooth wave shown in Fig. 3.28 and plot the line spectrum.

224 *Signals and Systems*

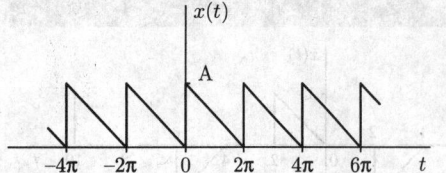

Fig. 3.28   Sawtooth wave

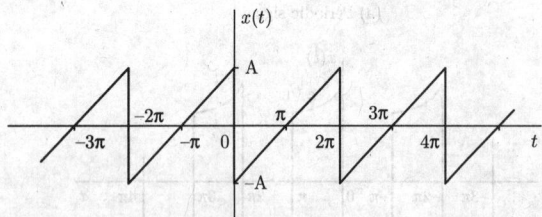

Fig. 3.29   Sawtooth wave

3.8 Find the trigonometric Fourier series for the sawtooth wave shown in Fig. 3.29 and plot the line spectrum.

3.9 Find the trigonometric Fourier series for the waveform shown in Fig. 3.30 and plot the line spectrum.

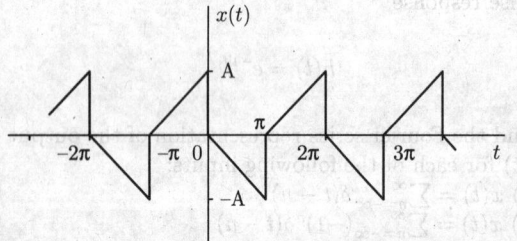

Fig. 3.30   Periodic signal

3.10 Find the trigonometric Fourier series for the waveforms shown in Fig. 3.31.

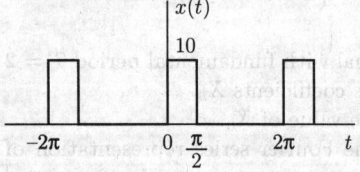

Fig. 3.31   Periodic signal

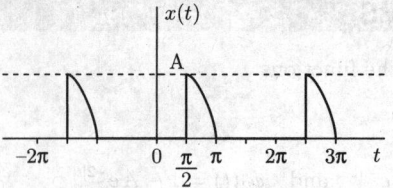

Fig. 3.32   Periodic signal

3.11 Find the trigonometric Fourier series for the waveform shown in Fig. 3.32.

3.12 Find the exponential Fourier series for the waveform shown in Fig. 3.33.

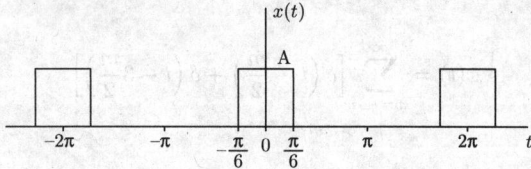

Fig. 3.33   Periodic signal

3.13 The waveform shown in Fig. 3.34 is a trapezoidal in form and of period $T = 4a$. For this periodic function, determine the exponential Fourier series and plot the corresponding magnitude and phase spectra.

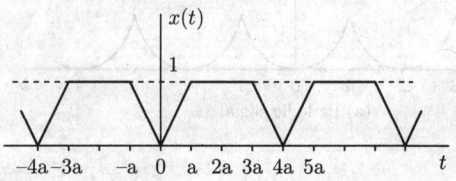

Fig. 3.34   Trapezoidal waveform

3.14 Given the periodic waveform

$$x(t) = t^2, \quad 0 < t < 1$$

Determine the exponential Fourier series and plot the magnitude and phase spectra.

Discrete-Time Fourier Series

4.1 INTRODUCTION

In the previous chapter, we studied the ways of representing a continuous-time periodic signal as a sum of sinusoids or exponentials. In this chapter, we will focus on discrete-time signals. While the discussion closely resembles the description used for continuous-time signals, there are some important differences. In particular, the Fourier series representation of a discrete-time periodic signal is a finite series, as opposed to the infinite series representation required for continuous-time periodic signals. As a consequence, there are no mathematical issues of convergence.

4.2 DISCRETE-TIME FOURIER SERIES (DTFS)

A discrete-time signal $x(n)$ is periodic with period N if

$$x(n) = x(n+N) \tag{4.1}$$

for some positive integer N. The smallest value of N for which this equation holds is the *fundamental period*. The fundamental frequency is $\omega_0 = 2\pi/N$ rad/sample. A periodic signal $x(n)$ can be represented by a discrete-time Fourier series made up of sinusoids (or exponentials) of fundamental frequency $\omega_0 = 2\pi/N$ and its harmonics. Because of its compactness and ease of mathematical manipulations, the exponential form is preferable to the trigonometric. The complex exponential $e^{j(2\pi/N)n}$ is periodic with period N. The exponential Fourier series consists of the exponentials e^{j0n}, $e^{\pm j\omega_0 n}$, $e^{\pm j2\omega_0 n}$, ..., $e^{jk\omega_0 n}$ ... and so on. The set of all discrete-time complex exponential signals that are periodic with period N is given by

$$\phi_k(n) = e^{jk\omega_0 n} = e^{jk(2\pi/N)n}, \quad k = 0, \pm 1, \pm 2, \ldots \tag{4.2}$$

All these signals have fundamental frequencies that are multiples of $2\pi/N$ and thus are harmonically related. The discrete-time periodic signal $x(n)$ can be represented by a summation of complex exponentials $\phi_k(n)$ of the form

$$x(n) = \sum_k X_k \phi_k(n) = \sum_k X_k e^{jk\omega_0 n} = \sum_k X_k e^{jk(2\pi/N)n} \tag{4.3}$$

As mentioned in Section 1.6.2 [Eq. (1.41)], the discrete-time exponentials whose frequencies are separated by 2π (or integer multiples of 2π) are identical.

Specifically,

$$\phi_0(n) = e^{j0(2\pi/N)n} = \phi_N(n) = e^{jN(2\pi/N)n}$$

$$\phi_1(n) = e^{j(2\pi/N)n} = \phi_{N+1}(n) = e^{j(N+1)(2\pi/N)n}$$

and, in general

$$\phi_k(n) = \phi_{k+N}(n) \tag{4.4}$$

Thus, the first harmonic is identical to the $(N+1)$st harmonic, the second harmonic is identical to the $(N+2)$nd harmonic, and so on. In other words, there are only N independent harmonics, and their frequencies range over an interval 2π (because the harmonics are separated by $\omega_0 = 2\pi/N$). This means, that the discrete-time Fourier series has only a finite number N of terms, unlike their continuous-time counterpart. This means, we may choose N independent harmonics $e^{jk\omega_0 n}$ over $0 \le k \le (N-1)$, or over $-1 \le k \le (N-2)$, or over $1 \le k \le N$, or over any other suitable choice for that matter. Everyone of these sets will have the same harmonics, although in different order. Therefore, the Fourier series of a periodic signal $x(n)$ consists of only N harmonics and can be expressed as

$$x(n) = \sum_{k=k_0}^{k_0+N-1} X_k e^{jk\omega_0 n} \tag{4.5}$$

where k_0 is arbitrary. Since k_0 is arbitrary, we can use the notation

$$x(n) = \sum_{k=\langle N \rangle} X_k e^{jk\omega_0 n} \tag{4.6}$$

where $\omega_0 = 2\pi/N$, and $\displaystyle\sum_{k=\langle N \rangle}$ means summation over any range of consecutive k's exactly N in length. Equation (4.6) is referred to as the *discrete-time Fourier series* and the coefficients X_k as the *Fourier series coefficients*.

4.2.1 Evaluation of DTFS Coefficients

To determine the Fourier series coefficients X_k, we replace the summation variable k by m on the right side of Eq. (4.6) and multiply both sides by $e^{-jk\omega_0 n}$ to get

$$x(n)e^{-j\omega_0 kn} = \sum_{m=\langle N \rangle} X_m e^{j\omega_0(m-k)n}$$

Then, we sum over values of n in $[0, N-1]$ to get

$$\sum_{n=0}^{N-1} x(n)e^{-j\omega_0 kn} = \sum_{n=0}^{N-1} \sum_{m=\langle N \rangle} X_m e^{j\omega_0(m-k)n} \tag{4.7}$$

By interchanging the order of summation in Eq. (4.7), we can write

$$\sum_{n=0}^{N-1} x(n)e^{-j\omega_0 kn} = \sum_{m=\langle N \rangle} X_m \sum_{n=0}^{N-1} e^{j\omega_0(m-k)n} \tag{4.8}$$

We know that
$$\sum_{n=0}^{N-1} \alpha^n = \frac{1 - \alpha^N}{1 - \alpha}, \quad \alpha \neq 1 \tag{4.9}$$

For $\alpha = 1$, we have
$$\sum_{n=0}^{N-1} \alpha^n = N \tag{4.10}$$

If $m - k$ is not an integer multiple of N (i.e., $(m - k) \neq rN$ for $r = 0, \pm 1, \pm 2, \cdots$), we can let $\alpha = e^{j\omega_0(m-k)n}$ in Eq. (4.9) to get

$$\sum_{n=0}^{N-1} e^{j\omega_0(m-k)n} = \frac{1 - e^{j\omega_0(m-k)N}}{1 - e^{j\omega_0(m-k)}} = \frac{1 - e^{j2\pi/N(m-k)N}}{1 - e^{j2\pi/N(m-k)}} = 0 \tag{4.11}$$

If $(m - k)$ is an integer multiple of N, we can use Eq. (4.10), so that we have

$$\sum_{n=0}^{N-1} e^{j\omega_0(m-k)n} = N \tag{4.12}$$

Combining Eqs (4.11) and (4.12), we write

$$\sum_{n=0}^{N-1} e^{j\omega_0(m-k)n} = N\delta(m - k - rN) \tag{4.13}$$

where $\delta(m - k - rN)$ is the unit sample occurring at $m = k + rN$. Substitution into Eq. (4.8) then yields

$$\sum_{n=0}^{N-1} x(n)e^{-j\omega_0 kn} = \sum_{m=\langle N \rangle} X_m N\delta(m - k - rN) \tag{4.14}$$

Since the summation on the right side is carried out over N consecutive values of m for a fixed value of k, it is clear that the only value that r can take in the given range of summation is $r = 0$. Thus, the only nonzero value in the sum corresponds to $m = k$, and the right-hand side of Eq. (4.14) evaluates to NX_k, so that

$$X_k = \frac{1}{N} \sum_{n=0}^{N-1} x(n)e^{-j\omega_0 kn} \tag{4.15}$$

Because each of the terms in the summation in Eq. (4.15) is periodic with period N, the summation can be taken over any N successive values of n. We thus have the pair of equations

$$x(n) = \sum_{k=\langle N \rangle} X_k e^{jk\omega_0 n} \tag{4.16}$$

and

$$X_k = \frac{1}{N} \sum_{n=\langle N \rangle} x(n)e^{-jk\omega_0 n} \tag{4.17}$$

which together form the discrete-time Fourier series pair. Equation (4.16) is referred to as the *synthesis* equation and Eq. (4.17) as the *analysis* equation. As in continuous-time periodic signal, the discrete-time Fourier series coefficients X_k are often referred to as the *spectral coefficients* of $x(n)$. These coefficients specify a decomposition of $x(n)$ into a sum of N harmonically related complex exponentials.

Because the Fourier series for discrete-time periodic signals is a finite sum defined entirely by the values of the signal over one period, the series always converges. The Fourier series provides an exact alternative representation of the time signal, and issues such as convergence or the Gibbs phenomenon do not arise.

4.2.2 Magnitude and Phase Spectrum of Discrete-Time Periodic Signals (Fourier Spectra)

The Fourier series consists of N components

$$X_0, X_1 e^{j\omega_0 n}, X_2 e^{j2\omega_0 n}, \ldots, X_{N-1} e^{j(N-1)\omega_0 n}$$

The frequencies of these components are $0, \omega_0, 2\omega_0, \ldots, (N-1)\omega_0$ where $\omega_0 = 2\pi/N$. The amount of the kth harmonic is X_k (the Fourier coefficient). We can plot this amount X_k as a function of index k or ω. Such a plot is called the *Fourier spectrum* of $x(n)$.

In general, the Fourier coefficients X_k, are complex, and they can be represented in the polar form as

$$X_k = |X_k| e^{j\angle X_k} \tag{4.18}$$

The plot of $|X_k|$ versus ω is called the magnitude spectrum and that of $\angle X_k$ versus ω is called the angle (or phase) spectrum. These two plots together are the frequency spectra of $x(n)$. Knowing these spectra, we can reconstruct or synthesize $x(n)$ according to Eq. (4.16).

The results are very similar to the representation of a continuous-time periodic signal by an exponential Fourier series except that, the bandwidth of the continuous-time periodic signal is infinite and consists of an infinite number of exponential components (harmonics). The spectrum of the discrete-time periodic signal, in contrast, is bandlimited and has at most N components.

The DTFS coefficients X_k are periodic with period N, i.e.,

$$X_{k+N} = X_k \tag{4.19}$$

Proof From Eq. (4.17), we have

$$X_k = \frac{1}{N} \sum_{n=\langle N \rangle} x(n) e^{-jk\omega_0 n}$$

$$X_{k+N} = \frac{1}{N} \sum_{n=\langle N \rangle} x(n) e^{-j(k+N)\omega_0 n}$$

$$= \frac{1}{N} \sum_{n=\langle N \rangle} x(n) e^{-jk\omega_0 n} e^{j(2\pi/N)Nn}$$

$$= \frac{1}{N} \sum_{n=\langle N \rangle} x(n) e^{-jk\omega_0 n} e^{j2\pi n}$$

$$= \frac{1}{N} \sum_{n=\langle N \rangle} x(n) e^{-jk\omega_0 n}$$

$$X_{k+N} = X_k$$

Along the k scale, X_k repeats at intervals of N, and along the ω scale, X_k repeats every 2π intervals.

Example 4.1 Determine the Fourier series coefficients of the signal $x(n)$ and plot its magnitude and phase spectrum.

$$x(n) = 1 + \sin\left(\frac{2\pi}{N}n\right) + 3\cos\left(\frac{2\pi}{N}n\right) + \cos\left(\frac{4\pi}{N}n + \frac{\pi}{2}\right)$$

Solution

The given signal is periodic with period N and frequency $\omega_0 = 2\pi/N$ and it can be written as

$$x(n) = 1 + \sin(\omega_0 n) + 3\cos(\omega_0 n) + \cos\left(2\omega_0 n + \frac{\pi}{2}\right)$$

$$x(n) = 1 + \left(\frac{e^{j\omega_0 n} - e^{-j\omega_0 n}}{2j}\right) + 3\left(\frac{e^{j\omega_0 n} + e^{-j\omega_0 n}}{2}\right) + \left(\frac{e^{j2\omega_0 n + \pi/2} + e^{-j2\omega_0 n + \pi/2}}{2}\right)$$

$$= 1 + \left(\frac{3}{2} + \frac{1}{2j}\right) e^{j\omega_0 n} + \left(\frac{3}{2} - \frac{1}{2j}\right) e^{-j\omega_0 n} + \left(\frac{1}{2}e^{j\pi/2}\right) e^{j2\omega_0 n} + \left(\frac{1}{2}e^{-j\pi/2}\right) e^{-j2\omega_0 n}$$

Comparing the above equation with

$$x(n) = X_0 + X_1 e^{j\omega_0 n} + X_{-1} e^{-j\omega_0 n} + X_2 e^{j2\omega_0 n} + X_{-2} e^{-j2\omega_0 n}$$

we get

$$X_0 = 1$$

$$X_1 = \frac{3}{2} + \frac{1}{2j} = \frac{3}{2} - \frac{1}{2}j$$

$$X_{-1} = \frac{3}{2} - \frac{1}{2j} = \frac{3}{2} + \frac{1}{2}j$$

$$X_2 = \frac{1}{2}j$$

$$X_{-2} = -\frac{1}{2}j$$

All other Fourier series coefficients in the interval of summation in the synthesis equation are zero. The magnitude of the Fourier coefficients are

$$|X_0| = 1, \quad |X_1| = |X_{-1}| = \frac{\sqrt{10}}{2}, \quad |X_2| = |X_{-2}| = \frac{1}{2}$$

The phase of the Fourier coefficients are

$$\angle X_0 = 0, \quad \angle X_1 = -\tan^{-1}(1/3), \quad \angle X_{-1} = \tan^{-1}(1/3)$$

$$\angle X_2 = \frac{\pi}{2}, \quad \angle X_{-2} = -\frac{\pi}{2}$$

The magnitude spectrum and the phase spectrum are depicted in Figs 4.1(a) and (b), respectively.

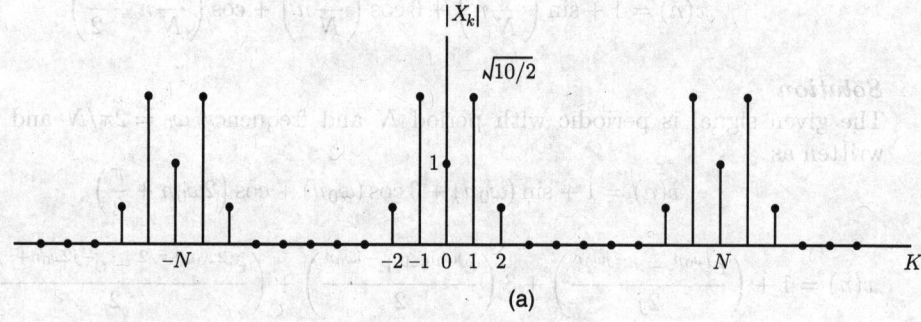

(a)

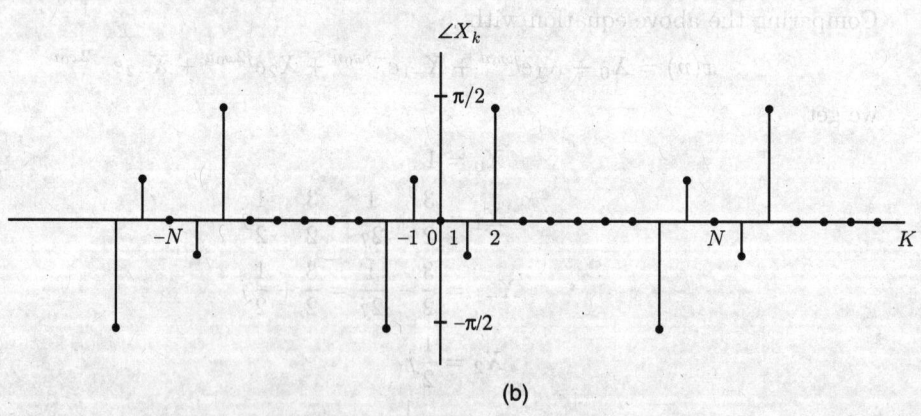

(b)

Fig. 4.1   (a) Magnitude spectrum, (b) Phase spectrum

Example 4.2 Evaluate the Fourier series for the discrete-time periodic square wave $x(n)$ shown in Fig. 4.2.

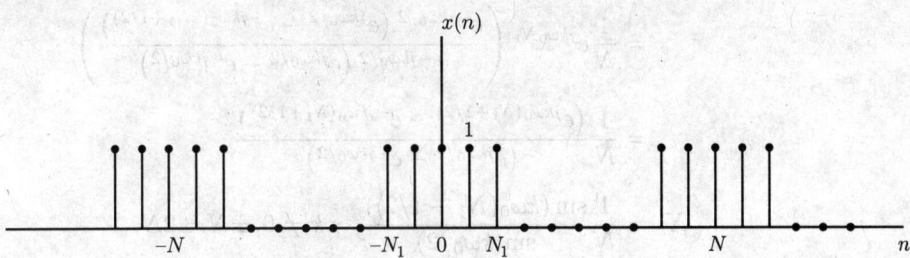

Fig. 4.2 Discrete-time periodic square wave

Solution
The given signal is periodic with period N and frequency $\omega_0 = 2\pi/N$ and it can be written as

$$x(n) = 1, \quad \text{for } -N_1 \leq n \leq N_1$$

It is convenient to choose the length-N interval of summation in Eq. (4.17) so that it includes the range $-N_1 \leq n \leq N_1$. The Fourier coefficients are

$$X_k = \frac{1}{N} \sum_{n=\langle N \rangle} x(n) e^{-jk\omega_0 n} = \frac{1}{N} \sum_{n=-N_1}^{N_1} x(n) e^{-jk\omega_0 n} = \frac{1}{N} \sum_{n=-N_1}^{N_1} e^{-jk\omega_0 n}$$

A change of variables is performed by letting $m = (n + N_1)$, which also yields $n = (m - N_1)$, $m = 0$ as $n = -N_1$, and $m = 2N_1$ as $n = N_1$. Therefore,

$$X_k = \frac{1}{N} \sum_{m=0}^{2N_1} e^{-jk\omega_0 (m-N_1)}$$

$$X_k = \frac{1}{N} e^{jk\omega_0 N_1} \sum_{m=0}^{2N_1} e^{-jk\omega_0 m} \tag{4.20}$$

Now, for $k = 0, \pm N, \pm 2N, \cdots$, we have $e^{jk\omega_0} = e^{-jk\omega_0} = 1$, and Eq. (4.20) becomes

$$X_k = \frac{1}{N} \sum_{m=0}^{2N_1} 1 = \frac{2N_1 + 1}{N}, \quad k = 0, \pm N, \pm 2N, \ldots$$

For $k \neq 0, \pm N, \pm 2N, \cdots$, we may sum the geometric series in Eq. (4.20) to obtain

$$X_k = \frac{1}{N} e^{jk\omega_0 N_1} \left(\frac{1 - e^{-jk\omega_0(2N_1+1)}}{1 - e^{-jk\omega_0}} \right)$$

$$= \frac{1}{N} e^{jk\omega_0 N_1} \left(\frac{e^{-jk\omega_0/2} \left(e^{jk\omega_0/2} - e^{-jk\omega_0(2N_1+1/2)} \right)}{e^{-jk\omega_0/2} \left(e^{jk\omega_0/2} - e^{-jk\omega_0/2} \right)} \right)$$

$$= \frac{1}{N} \frac{\left(e^{jk\omega_0(N_1+1/2)} - e^{-jk\omega_0(N_1+1/2)} \right)}{\left(e^{jk\omega_0/2} - e^{-jk\omega_0/2} \right)}$$

$$X_k = \frac{1}{N} \frac{\sin\left(k\omega_0(N_1 + 1/2) \right)}{\sin(k\omega_0/2)}, \quad k \neq 0, \pm N, \pm 2N, \ldots$$

An alternative expression for X_k is obtained by substituting $\omega_0 = 2\pi/N$, yielding

$$X_k = \begin{cases} \dfrac{1}{N} \dfrac{\sin(k\pi(2N_1 + 1)/N)}{\sin(k\pi/N)}, & k \neq 0, \pm N, \pm 2N, \ldots \\[3mm] \dfrac{2N_1 + 1}{N}, & k = 0, \pm N, \pm 2N, \ldots \end{cases}$$

Example 4.3 Find the DTFS coefficients of the N-periodic impulse train

$$x(n) = \sum_{m=-\infty}^{\infty} \delta(n - mN)$$

as shown in Fig. 4.3.

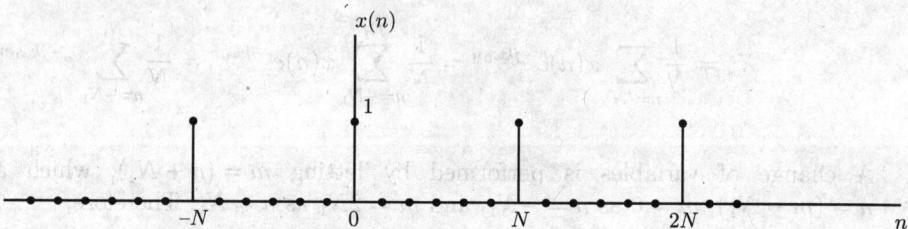

Fig. 4.3 A discrete-time impulse train with time period N

Solution

The signal is periodic with period N and frequency $\omega_0 = 2\pi/N$. The given signal for one period may be written as

$$x(n) = \begin{cases} 1, & n = 0 \\ 0, & 1 \leq n \leq N-1 \end{cases}$$

$$= \delta(n)$$

Therefore, the Fourier series coefficients are

$$X_k = \frac{1}{N} \sum_{n=0}^{N-1} x(n) e^{-jk\omega_0 n} = \frac{1}{N} \sum_{n=0}^{N-1} \delta(n) e^{-jk 2\pi/N n} = \frac{1}{N} \left[e^{-jk(2\pi/N)n} \right]_{n=0} = \frac{1}{N}$$

Example 4.4 Find the frequency domain representation of the signal depicted in Fig. 4.4.

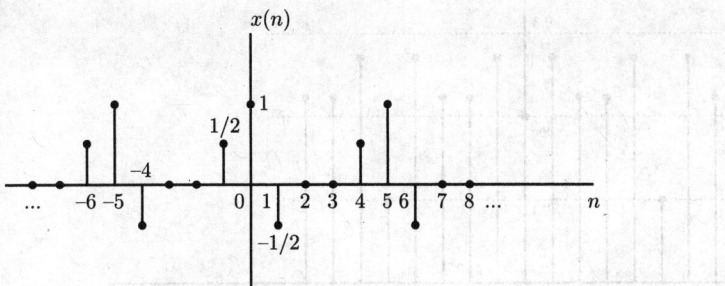

Fig. 4.4 Periodic signal

Solution

The signal has period $N = 5$, so $\omega_0 = \dfrac{2\pi}{5}$. Also, the signal has odd symmetry, so we sum over $n = -2$ to $n = 2$ in Eq. (4.17) to obtain X_k.

$$X_k = \frac{1}{N} \sum_{n=\langle N \rangle} x(n) e^{-jk\omega_0 n}$$

$$= \frac{1}{5} \sum_{n=-2}^{2} x(n) e^{-jk2\pi/5n}$$

$$= \frac{1}{5} \left[x(-2)e^{jk4\pi/5} + x(-1)e^{jk2\pi/5} + x(0) + x(1)e^{-jk2\pi/5} + x(2)e^{-jk4\pi/5} \right]$$

Using the values of $x(n)$, we get

$$X_k = \frac{1}{5} \left[1 + \frac{1}{2} e^{jk2\pi/5} - \frac{1}{2} e^{-jk2\pi/5} \right] = \frac{1}{5} \left[1 + j\sin\left(k\frac{2\pi}{5} \right) \right]$$

The one period of the DTFS coefficients X_k, $k = -2$ to $k = 2$ are

$$X_{-2} = \frac{1}{5} - j\frac{\sin(4\pi/5)}{5} = 0.232 e^{-j0.531}$$

$$X_{-1} = \frac{1}{5} - j\frac{\sin(2\pi/5)}{5} = 0.276 e^{-j0.760}$$

$$X_0 = \frac{1}{5} = 0.2 e^{j0}$$

$$X_1 = \frac{1}{5} + j\frac{\sin(2\pi/5)}{5} = 0.276 e^{j0.760}$$

$$X_2 = \frac{1}{5} + j\frac{\sin(4\pi/5)}{5} = 0.232 e^{j0.531}$$

Figure 4.5 depicts the magnitude and phase of X_k as functions of frequency index k.

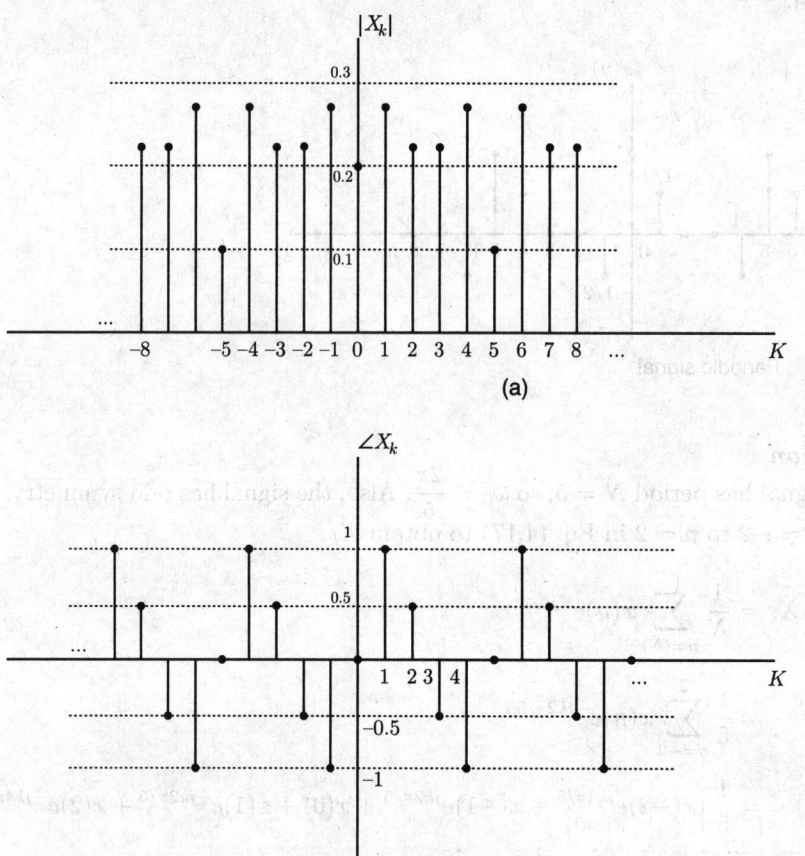

Fig. 4.5   (a) Magnitude spectrum, (b) Phase spectrum

Now suppose we calculate X_k using $n = 0$ to $n = 4$ for the limits in Eq. (4.17)

$$X_k = \frac{1}{N} \sum_{n=\langle N \rangle} x(n)e^{-jk\omega_0 n}$$

$$X_k = \frac{1}{5} \sum_{n=0}^{4} x(n)e^{-jk(2\pi/5)n}$$

$$= \frac{1}{5} \left[x(0) + x(1)e^{-jk(2\pi/5)} + x(2)e^{-jk(4\pi/5)} + x(3)e^{-jk(6\pi/5)} + x(4)e^{-jk(8\pi/5)} \right]$$

Using the values of $x(n)$, we get

$$X_k = \frac{1}{5}\left[1 - \frac{1}{2}e^{-jk(2\pi/5)} + \frac{1}{2}e^{-jk(8\pi/5)}\right]$$

Note that

$$e^{-jk(8\pi/5)} = e^{-jk2\pi}e^{jk(2\pi/5)} = e^{jk(2\pi/5)}$$

Therefore,

$$X_k = \frac{1}{5}\left[1 + \frac{1}{2}e^{jk(2\pi/5)} - \frac{1}{2}e^{-jk(2\pi/5)}\right]$$

$$= \frac{1}{5}\left[1 + j\sin\left(k\frac{2\pi}{5}\right)\right]$$

we see that both intervals, $-2 \leq n \leq 2$ and $0 \leq n \leq 4$ yield equivalent expressions for the DTFS coefficients.

Example 4.5 Determine the time-domain signal $x(n)$ from the DTFS coefficients depicted in Fig. 4.6:

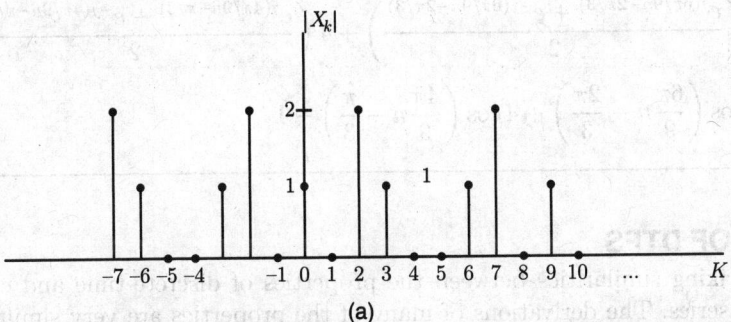

(a)

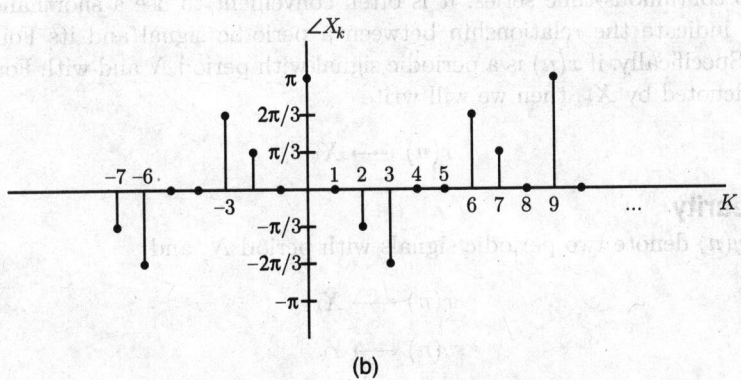

(b)

Fig. 4.6 (a) Magnitude spectrum, (b) Phase spectrum

Solution

The DTFS coefficients have period $N = 9$, so $\omega_0 = 2\pi/9$. From Fig. 4.6, the DTFS coefficients over the interval $k = -4$ to $k = 4$ are

$$X_k = |X_k|e^{j\angle X_k}, \quad -4 \le k \le 4$$

$$X_{-4} = 0, \qquad X_{-3} = 1e^{j2\pi/3}, \quad X_{-2} = 2e^{j\pi/3}, \quad X_{-1} = 0,$$

$$X_0 = 1e^{j\pi}, \qquad X_1 = 0, \qquad X_2 = 2e^{-j\pi/3}, \quad X_3 = 1e^{-j2\pi/3}, \quad X_4 = 0$$

It is convenient to evaluate Eq. (4.16) over the interval $k = -4$ to $k = 4$ to obtain $x(n)$.

$$x(n) = \sum_{k=\langle N \rangle} X_k e^{jk\omega_0 n}$$

$$= \sum_{k=-4}^{4} X_k e^{jk2\pi/9n}$$

$$= X_{-4}e^{-j8\pi/9n} + X_{-3}e^{-j6\pi/9n} + X_{-2}e^{-j4\pi/9n} + X_{-1}e^{-j2\pi/9n}$$

$$\quad + X_0 + X_1 e^{j2\pi/9n} + X_2 e^{j4\pi/9n} + X_3 e^{j6\pi/9n} + X_4 e^{j8\pi/9n}$$

$$= e^{j2\pi/3}e^{-j6\pi/9n} + 2e^{j\pi/3}e^{-j4\pi/9n} + e^{j\pi} + 2e^{-j\pi/3}e^{j4\pi/9n} + e^{-j2\pi/3}e^{j6\pi/9n}$$

$$= 2\left(\frac{e^{j(6\pi/9n - 2\pi/3)} + e^{-j(6\pi/9n - 2\pi/3)}}{2}\right) + 4\left(\frac{e^{j(4\pi/9n - \pi/3)} + e^{-j(4\pi/9n - \pi/3)}}{2}\right) - 1$$

$$x(n) = 2\cos\left(\frac{6\pi}{9}n - \frac{2\pi}{3}\right) + 4\cos\left(\frac{4\pi}{9}n - \frac{\pi}{3}\right) - 1$$

4.3 PROPERTIES OF DTFS

There are striking similarities between the properties of discrete-time and continuous-time Fourier series. The derivations of many of the properties are very similar to those of the corresponding properties of continuous-time Fourier series.

Similar to continuous-time series, it is often convenient to use a shorthand notation in DTFS to indicate the relationship between a periodic signal and its Fourier series coefficients. Specifically, if $x(n)$ is a periodic signal with period N and with Fourier series coefficients denoted by X_k, then we will write

$$x(n) \longleftrightarrow X_k$$

4.3.1 Linearity

If $x(n)$ and $y(n)$ denote two periodic signals with period N, and

$$x(n) \longleftrightarrow X_k$$

$$y(n) \longleftrightarrow Y_k$$

then

$$z(n) = ax(n) + by(n) \longleftrightarrow Z_k = aX_k + bY_k \qquad (4.21)$$

Proof The Fourier series coefficients of $z(n)$ is given by

$$Z_k = \frac{1}{N} \sum_{n=\langle N \rangle} x(n)e^{-jk\omega_0 n} = \frac{1}{N} \sum_{n=\langle N \rangle} [ax(n) + by(n)]e^{-jk\omega_0 n}$$

$$= a \underbrace{\frac{1}{N} \sum_{n=\langle N \rangle} x(n)e^{-jk\omega_0 n}}_{X_k} + b \underbrace{\frac{1}{N} \sum_{n=\langle N \rangle} y(n)e^{-jk\omega_0 n}}_{Y_k}$$

$$= aX_k + bY_k$$

4.3.2 Time Shifting

When a time shift is applied to a periodic signal $x(n)$, the period N of the signal is preserved. If

$$x(n) \longleftrightarrow X_k$$

then
$$y(n) = x(n - n_0) \longleftrightarrow Y_k = X_k e^{-jk\omega_0 n_0} \tag{4.22}$$

One consequence of this property is that, when a signal is shifted in time, the *magnitudes* of its Fourier series coefficients remain unaltered. That is, $|Y_k| = |X_k|$.

Proof By definition,

$$Y_k = \frac{1}{N} \sum_{n=\langle N \rangle} y(n)e^{-jk\omega_0 n} = \frac{1}{N} \sum_{n=0}^{N-1} y(n)e^{-jk\omega_0 n} = \frac{1}{N} \sum_{n=0}^{N-1} x(n - n_0)e^{-jk\omega_0 n}$$

A change of variables is performed by letting $m = (n - n_0)$, which also yields $(m \to -n_0)$ as $(n \to 0)$, and $(m \to (N - 1 - n_0))$ as $(n \to (N - 1))$. Therefore,

$$Y_k = \frac{1}{N} \sum_{m=-n_0}^{N-1-n_0} x(m)e^{-jk\omega_0(m+n_0)}$$

$$Y_k = \frac{1}{N} \sum_{m=-n_0}^{N-1-n_0} x(m)e^{-jk\omega_0 m} \, e^{-jk\omega_0 n_0}$$

$$Y_k = X_k e^{-jk\omega_0 n_0}$$

$$|Y_k| = |X_k|$$

4.3.3 Frequency Shifting

If
$$x(n) \longleftrightarrow X_k$$

then
$$y(n) = e^{jM\omega_0 n}x(n) \longleftrightarrow Y_k = X_{k-M} \tag{4.23}$$

Proof By definition,

$$Y_k = \frac{1}{N} \sum_{n=\langle N \rangle} y(n) e^{-jk\omega_0 n} = \frac{1}{N} \sum_{n=0}^{N-1} y(n) e^{-jk\omega_0 n}$$

$$= \frac{1}{N} \sum_{n=0}^{N-1} e^{jM\omega_0 n} x(n) e^{-jk\omega_0 n}$$

$$= \frac{1}{N} \sum_{n=0}^{N-1} x(n) e^{-j(k-M)\omega_0 n} = X_{k-M}$$

Hence, a frequency shift corresponds to multiplication in time domain by a complex sinusoid whose frequency is equal to the time shift. ∎

4.3.4 Time Reversal

If
$$x(n) \longleftrightarrow X_k$$

then
$$y(n) = x(-n) \longleftrightarrow Y_k = X_{-k} \tag{4.24}$$

In other words, the time reversal applied to a discrete-time signal results in a time reversal of the corresponding sequence of Fourier series coefficients.

Proof

$$Y_k = \frac{1}{N} \sum_{n=\langle N \rangle} y(n) e^{-jk\omega_0 n} = \frac{1}{N} \sum_{n=0}^{N-1} y(n) e^{-jk\omega_0 n}$$

$$= \frac{1}{N} \sum_{n=0}^{N-1} x(-n) e^{-jk\omega_0 n} = \frac{1}{N} \sum_{m=-(N-1)}^{0} x(m) e^{-j(-k)\omega_0 m} = X_{-k}$$

An interesting consequence of the time-reversal property is that if $x(n)$ is even then its Fourier series coefficients are also even, i.e., if

$$x(-n) = x(n)$$

then
$$X_{-k} = X_k \tag{4.25}$$

Similarly, if $x(n)$ is odd, then so are its Fourier series coefficients, i.e., if

$$x(-n) = -x(n)$$

then
$$X_{-k} = -X_k \tag{4.26}$$

4.3.5 Time Scaling

Because of the discrete-time nature of the time index for discrete-time signals, the relation between time scaling and Fourier series coefficients in discrete-time takes on a different form from its continuous-time counterpart. If we try to define the signal $x(an)$, we run into difficulties if a is not an integer. Therefore, we cannot slow down the signal by choosing $a < 1$. On the other hand, if we let a be an integer other than ± 1, we do not merely speed up the original signal. For example, since n can take on only integer values, the signal $x(2n)$ consists of the even samples of $x(n)$ alone.

Let m be a positive integer, and the signal is defined as

$$x_{(m)}(n) = \begin{cases} x(n/m), & \text{if } n \text{ is a multiple of } m \\ 0, & \text{if } n \text{ is not a multiple of } m \end{cases} \qquad (4.27)$$

$x_m(n)$ can be obtained from $x(n)$ by placing $(m-1)$ zeros between successive values of the original signal. Intuitively, we can think of $x_{(m)}(n)$ as a slowed-down version of $x(n)$. If $x(n)$ is periodic with period N, then $y(n) = x_{(m)}(n)$ is also periodic with period mN. Now, if

$$x(n) \longleftrightarrow X_k$$

then
$$y(n) = x_{(m)}(n) \longleftrightarrow Y_k = \frac{1}{m} X_k \qquad (4.28)$$

The Fourier series coefficients $Y_k = (1/m) X_k$ are also periodic with period mN.

Proof The Fourier series coefficients of $y(n) = x_{(m)}(n)$ [$y(n)$ is periodic with period mN and fundamental frequency ω_0/m] are given by

$$Y_k = \frac{1}{mN} \sum_{n=0}^{m(N-1)} y(n) e^{-jk(\omega_0/m)n}$$

$$= \frac{1}{mN} \sum_{n=0}^{m(N-1)} x_{(m)}(n) e^{-jk(\omega_0/m)n}$$

$$= \frac{1}{mN} \sum_{n=0}^{m(N-1)} x\left(\frac{n}{m}\right) e^{-jk(\omega_0/m)n}$$

A change of variables is performed by letting $r = n/m$, which also yields $r = 0$ as $n = 0$, and $r = N - 1$ as $n = m(N-1)$. Therefore,

$$Y_k = \frac{1}{m} \underbrace{\frac{1}{N} \sum_{r=0}^{N-1} x(r) e^{-jk\omega_0 r}}_{X_k} = \frac{1}{m} X_k$$

4.3.6 Periodic Convolution

If
$$x(n) \longleftrightarrow X_k,$$
$$y(n) \longleftrightarrow Y_k$$

then
$$z(n) = x(n) \circledast y(n) = \sum_{r=\langle N \rangle} x(r) y(n-r) \longleftrightarrow Z_k = N X_k Y_k \qquad (4.29)$$

where $*$ denotes the periodic convolution.

Proof For periodic signals with the same period, a special form of convolution, known as periodic convolution, is defined as

$$z(n) = \sum_{r=\langle N \rangle} x(r)y(n-r)$$

It is easy to show that $z(n)$ is periodic with period N and the periodic convolution is commutative and associative. Thus, we can write $z(n)$ in a Fourier series representation with coefficients

$$Z_k = \frac{1}{N} \sum_{n=\langle N \rangle} z(n)e^{-jk\omega_0 n} = \frac{1}{N} \sum_{n=\langle N \rangle} \left(\sum_{r=\langle N \rangle} x(r)y(n-r) \right) e^{-jk\omega_0 n}$$

$$Z_k = \sum_{r=\langle N \rangle} x(r) \left(\frac{1}{N} \sum_{n=\langle N \rangle} y(n-r)e^{-jk\omega_0 n} \right)$$

From the time shifting property, i.e., if

$$y(n) \longleftrightarrow Y_k,$$

then

$$y(n-r) \longleftrightarrow Y_k e^{-jr\omega_0 n}$$

we have

$$Z_k = N \underbrace{\frac{1}{N} \sum_{r=\langle N \rangle} x(r)e^{-jr\omega_0 n}}_{X_k} Y_k = N X_k Y_k$$

The convolution in time transforms to multiplication of the frequency domain representations. ∎

4.3.7 Multiplication

If $x(n)$ and $y(n)$ are periodic signals with the same period N, and

$$x(n) \longleftrightarrow X_k,$$
$$y(n) \longleftrightarrow Y_k,$$

then

$$z(n) = x(n)y(n) \longleftrightarrow Z_k = \sum_{r=\langle N \rangle} X_r Y_{k-r} \tag{4.30}$$

Z_k may be interpreted as the periodic convolution between the two periodic sequences of Fourier coefficients X_k and Y_k.

Proof Consider the signal $z(n)$.

$$z(n) = x(n)y(n) = \sum_{r=\langle N \rangle} X_r e^{jr\omega_0 n} \sum_{m=\langle N \rangle} Y_m e^{jm\omega_0 n}$$

$$= \sum_{r=\langle N \rangle} X_r \sum_{m=\langle N \rangle} Y_m e^{j(m+r)\omega_0 n}$$

A change of variables is performed by letting $k = m + r$, which also yields $m = (k - r)$, $(k \rightarrow r)$ as $(m \rightarrow 0)$, and $[k \rightarrow (r + N - 1)]$ as $[m \rightarrow (N - 1)]$. Therefore,

$$z(n) = \sum_{r=\langle N \rangle} X_r \sum_{k=r}^{r+N-1} Y_{k-r} e^{jk\omega_0 n}$$

$$= \sum_{r=\langle N \rangle} X_r \sum_{k=\langle N \rangle} Y_{k-r} e^{jk\omega_0 n}$$

$$z(n) = \sum_{k=\langle N \rangle} \left(\sum_{r=\langle N \rangle} X_r Y_{k-r} \right) e^{jk\omega_0 n} = \sum_{k=\langle N \rangle} Z_k e^{jk\omega_0 n}$$

Thus,
$$Z_k = \sum_{r=\langle N \rangle} X_r Y_{k-r}$$

4.3.8 First Difference

The discrete-time equivalent to the differentiation property of the continuous-time Fourier series involves the use of the first-difference operation, which is defined as $y(n) = x(n) - x(n-1)$. If $x(n)$ is periodic with period N, then so is $y(n)$, since shifting $x(n)$ or linearly combining $x(n)$ with another periodic signal whose period is N always results in a periodic signal with period N. Also, if

$$x(n) \longleftrightarrow X_k,$$

then
$$y(n) = x(n) - x(n-1) \longleftrightarrow Y_k = (1 - e^{-jk\omega_0})X_k \tag{4.31}$$

A common use of this property is in situations where the evaluation of Fourier series coefficients is easier for the first difference than for the original sequence.

Proof

Given that
$$x(n) \longleftrightarrow X_k$$

Using the time-shifting property, we get

$$x(n-1) \longleftrightarrow X_k e^{-jk\omega_0}$$

Now, using the linearity property, we get

$$x(n) - x(n-1) \longleftrightarrow X_k - X_k e^{-jk\omega_0}$$

$$x(n) - x(n-1) \longleftrightarrow (1 - e^{-jk\omega_0})X_k$$

4.3.9 Running Sum or Accumulation

If
$$x(n) \longleftrightarrow X_k$$
$$y(n) \longleftrightarrow Y_k$$

then
$$y(n) = \sum_{k=-\infty}^{n} x(k) \longleftrightarrow Y_k = \left(\frac{1}{1 - e^{-jk\omega_0}}\right) X_k, \quad k \neq 0 \qquad (4.32)$$

The discrete-time Fourier series coefficient Y_k of the running sum $y(n) = \sum_{k=-\infty}^{n} x(k)$ is finite-valued and periodic only if $X_0 = 0$.

Proof Consider the running sum

$$y(n) = \sum_{k=-\infty}^{n} x(k)$$

$$y(n) = x(n) + \sum_{k=-\infty}^{n-1} x(k)$$

$$y(n) = x(n) + y(n-1)$$

$$y(n) - y(n-1) = x(n)$$

Using the time-shifting and linearity properties, we get

$$Y_k - Y_k e^{-jk\omega_0} = X_k$$

$$Y_k = \left(\frac{1}{1 - e^{-jk\omega_0}}\right) X_k$$

4.3.10 Conjugation and Conjugate Symmetry

If
$$x(n) \longleftrightarrow X_k$$

then
$$y(n) = x^*(n) \longleftrightarrow Y_k = X_{-k}^* \qquad (4.33)$$

Taking the complex conjugate of a periodic signal $x(n)$ has the effect of complex conjugation and time reversal on the corresponding Fourier series coefficients.

Proof

$$Y_k = \frac{1}{N} \sum_{n=\langle N \rangle} y(n) e^{-jk\omega_0 n} = \frac{1}{N} \sum_{n=\langle N \rangle} x^*(n) e^{-jk\omega_0 n}$$

$$= \left(\frac{1}{N} \sum_{n=\langle N \rangle} x(n) e^{jk\omega_0 n} \right)^* = \left(\underbrace{\frac{1}{N} \sum_{n=\langle N \rangle} x(n) e^{-j(-k)\omega_0 n}}_{X_{-k}} \right)^*$$

$$= (X_{-k})^* = X_{-k}^*$$

Case I If $x(n)$ is real, i.e., if

$$x^*(n) = x(n)$$

then

$$X_{-k}^* = X_k$$

Therefore,

$$X_{-k} = X_k^* \tag{4.34}$$

In other words, the Fourier series coefficients will be *conjugate symmetric*. Also if $x(n)$ is real and even, then from Eqs (4.25) and (4.34), we have

$$X_{-k} = X_k^* = X_k \tag{4.35}$$

That is, if $x(t)$ is *real and even*, then so are its Fourier series coefficients.

Case II If $x(n)$ is *real and odd*, then its Fourier series coefficients are purely imaginary and odd. Using Eqs (4.26) and (4.34), we get

$$X_{-k} = X_k^* = -X_k \tag{4.36}$$

Case III Even and odd decomposition of real signals: If $x(n)$ is real and

$$x(n) \longleftrightarrow X_k$$

then

1.
$$x_e(n) = \mathcal{E}\{x(n)\} \longleftrightarrow Re\{X_k\} \tag{4.37}$$

The even part of a signal $x(n)$ is defined as

$$x_e(n) = \frac{1}{2}[x(n) + x(-n)]$$

Proof

Using the linearity property, we get

$$x_e(n) \longleftrightarrow \frac{1}{2} 2Re\{X_k\}$$

$$x_e(n) \longleftrightarrow Re\{X_k\}$$

$$x_e(n) \longleftrightarrow \frac{1}{2}[X_k + X_{-k}]$$

$$x_e(n) \longleftrightarrow \frac{1}{2}[X_k + X_k^*]$$

and

2.
$$x_o(n) = \mathcal{O}\{x(n)\} \longleftrightarrow jIm\{X_k\} \qquad (4.38)$$

The odd part of a signal $x(n)$ is defined as

$$x_o(n) = \frac{1}{2}[x(n) - x(-n)]$$

Using the linearity property, we get

$$x_o(n) \longleftrightarrow \frac{1}{2}[X_k - X_{-k}]$$

$$x_o(n) \longleftrightarrow \frac{1}{2}[X_k - X_k^*]$$

$$x_o(n) \longleftrightarrow \frac{1}{2} 2jIm\{X_k\}$$

$$x_o(n) \longleftrightarrow jIm\{X_k\}$$

4.3.11 Parseval's Relation

If
$$x(n) \longleftrightarrow X_k$$

then
$$\frac{1}{N} \sum_{n=\langle N \rangle} |x(n)|^2 = \sum_{k=\langle N \rangle} |X_k|^2 \qquad (4.39)$$

As in the continuous-time case, the left-hand side of Parseval's relation is the average power in one period for the periodic signal $x(n)$. Similarly, $|X_k|^2$ is the average power in the kth harmonic component of $x(n)$. Thus, Parseval's relation states that the average power in a periodic signal equals the sum of the average powers in all of its harmonic components. In discrete-time, of course, there are only N distinct harmonic components, and since the X_k are periodic with period N, the sum on the right-hand side of Eq. (4.39) can be taken over any N consecutive values of k.

Proof Consider the LHS of Eq. (4.39), we have

$$\frac{1}{N}\sum_{n=\langle N\rangle}|x(n)|^2 = \frac{1}{N}\sum_{n=\langle N\rangle}x(n)\,x^*(n)$$

$$= \frac{1}{N}\sum_{n=\langle N\rangle}x(n)\left(\sum_{k=\langle N\rangle}X_k e^{jk\omega_0 n}\right)^*$$

$$= \frac{1}{N}\sum_{n=\langle N\rangle}x(n)\left(\sum_{k=\langle N\rangle}X_k^* e^{-jk\omega_0 n}\right)$$

$$= \sum_{k=\langle N\rangle}X_k^*\left(\frac{1}{N}\sum_{n=\langle N\rangle}x(n)^{-jk\omega_0 n}\right)$$

$$= \sum_{k=\langle N\rangle}X_k^* X_k$$

$$= \sum_{k=\langle N\rangle}|X_k|^2$$

4.4 SYSTEMS WITH PERIODIC INPUTS

The response of an LTI system to a sinusoidal input leads to a characterization of system behaviour that is termed the *frequency response* of the system. This characterization is obtained in terms of the impulse response by using convolution and a complex sinusoidal input signal. Let the impulse response of a system be $h(n)$ and the input be $x(n) = e^{j\omega_0 n}$. Then, the convolution integral gives the output as

$$y(n) = h(n) * x(n) = \sum_{m=-\infty}^{\infty}h(m)x(n-m) = \sum_{m=-\infty}^{\infty}h(m)e^{j\omega_0(n-m)}$$

$$= \underbrace{\sum_{m=-\infty}^{\infty}h(m)e^{-j\omega_0 m}}_{H(e^{j\omega_0})}\, e^{j\omega_0 n}$$

$$y(n) = H(e^{j\omega_0})e^{j\omega_0 n} \tag{4.40}$$

where we define

$$H(e^{j\omega_0}) = \sum_{m=-\infty}^{\infty}h(m)e^{-j\omega_0 m}$$

The output of the system is thus a complex sinusoid of the same frequency as the input multiplied by the complex number $H(e^{j\omega_0})$ [Fig. 4.7(a)]. An intuitive interpretation of the sinusoidal steady-state response is obtained by writing the complex-valued frequency response $H(e^{j\omega_0})$ in polar form. We have

$$H(e^{j\omega_0}) = |H(e^{j\omega_0})|e^{j\angle H(e^{j\omega_0})} \tag{4.41}$$

where $|H(e^{j\omega_0})|$ is now termed the *magnitude response* and $\angle H(e^{j\omega_0})$ is termed the *phase response* of the system. Substituting this polar form into Eq. (4.39), we may express the output as

$$y(n) = |H(e^{j\omega_0})|e^{j(\omega_0 n + \angle H(e^{j\omega_0}))} \tag{4.42}$$

The system thus modifies the amplitude of the input by $|H(e^{j\omega_0})|$ and the phase by $\angle H(e^{j\omega_0})$. Knowing $H(e^{j\omega_0})$, we can determine whether the system amplifies or attenuates a given sinusoidal component of the input and how much of a phase shift the system adds to that particular component.

We say that the complex sinusoid $x(n) = e^{j\omega_0 n}$ is an *eigenfunction* of the system associated with the *eigenvalue* $H(e^{j\omega_0})$.

Signals that are eigenfunctions of systems play an important role in LTI system theory. By representing arbitrary signals as weighted superpositions of eigenfunctions, we transform the operation of convolution to multiplication. Thus, if the input to an LTI system is given by

$$x(n) = \sum_{k=\langle N \rangle} X_k e^{jk\omega_0 n} \tag{4.43}$$

then, the output of the system is given by [Fig. 4.7(b)]

$$y(n) = \sum_{k=\langle N \rangle} X_k H(e^{jk\omega_0})e^{jk\omega_0 n} = \sum_{k=\langle N \rangle} Y_k e^{jk\omega_0 n} \tag{4.44}$$

where $\omega_0 = \dfrac{2\pi}{N}$ and $Y_k = X_k H(e^{jk\omega_0})$.

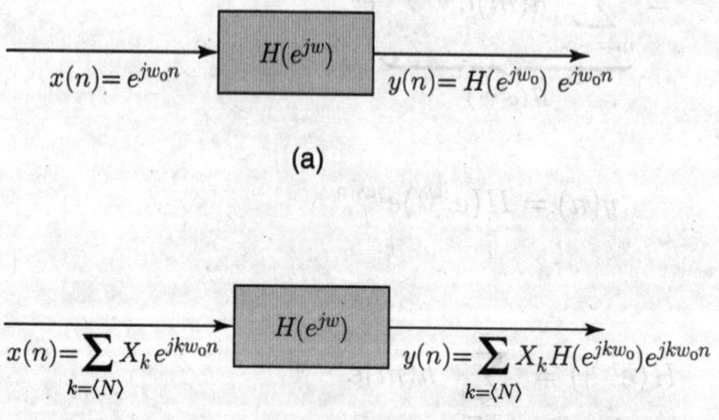

(a)

(b)

Fig. 4.7 Response of an LTI system to complex sinusoid

SOLVED EXAMPLES

Example 4.6 A discrete-time periodic signal $x(n)$ is real valued and has a fundamental period $N = 5$. The nonzero Fourier series coefficients for $x(n)$ are

$$X_0 = 1, \quad X_2 = X_{-2}^* e^{j\pi/4}, \quad X_4 = X_{-4}^* = 2e^{j\pi/3}$$

Express $x(n)$ in the form

$$x(n) = A_0 + \sum_{k=1}^{\infty} A_k \sin(\omega_k n + \phi_k)$$

Solution

The signal $x(n)$ is periodic with period $N = 5$ and fundamental frequency $\omega_0 = 2\pi/N = 2\pi/5$. Using the Fourier synthesis Eq. (4.16), we have

$$x(n) = \sum_{k=\langle N \rangle} X_k e^{jk\omega_0 n}$$

$$= X_0 + X_2 e^{j2\omega_0 n} + X_{-2} e^{-j2\omega_0 n}$$
$$\quad + X_4 e^{j4\omega_0 n} + X_{-4} e^{-j4\omega_0 n}$$

$$= X_0 + X_2 e^{j22\pi/Nn} + X_{-2} e^{-j22\pi/Nn}$$
$$\quad + X_4 e^{j42\pi/Nn} + X_{-4} e^{-j42\pi/Nn}$$

$$= 1 + e^{j\pi/4} e^{j22\pi/5n} + e^{-j\pi/4} e^{-j22\pi/5n}$$
$$\quad + 2e^{j\pi/3} e^{j42\pi5/n} + 2e^{-j\pi/3} e^{-j42\pi/5n}$$

$$= 1 + 2\left(\frac{e^{j(4\pi/5n+\pi/4)} + e^{-j(4\pi/5n+\pi/4)}}{2} \right)$$
$$\quad + 4\left(\frac{e^{j(8\pi/5n+\pi/3)} + e^{-j(8\pi/5n+\pi/3)}}{2} \right)$$

$$= 1 + 2\cos\left(\frac{4\pi}{5}n + \frac{\pi}{4} \right) + 4\cos\left(\frac{8\pi}{5}n + \frac{\pi}{3} \right)$$

$$= 1 + 2\sin\left(\frac{4\pi}{5}n + \frac{3\pi}{4} \right) + 4\sin\left(\frac{8\pi}{5}n + \frac{5\pi}{6} \right)$$

$$= A_0 + \sum_{k=1}^{2} A_k \sin(\omega_k n + \phi_k)$$

Example 4.7 Evaluate the numerical values of one period of the Fourier series coefficients of the periodic signal

$$x(n) = \sum_{m=-\infty}^{\infty} [4\delta(n - 4m) + 8\delta(n - 1 - 4m)]$$

as shown in Fig. 4.8.

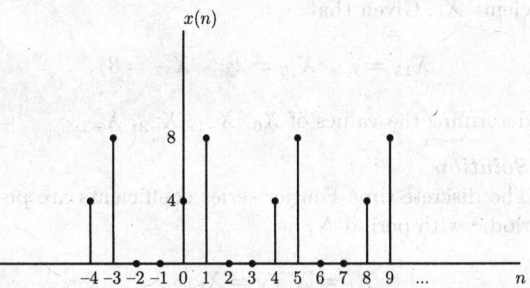

Fig. 4.8 Periodic signal

Solution

The signal is periodic with period $N = 4$ and frequency $\omega_0 = 2\pi/N = 2\pi/4$. The given signal for one period may be written as

$$x(n) = \begin{cases} 4, & n = 0 \\ 8, & n = 1 \\ 0, & n = 2, 3 \end{cases}$$

$$= 4\delta(n) + 8\delta(n-1)$$

Therefore, the Fourier series coefficients are

$$X_k = \frac{1}{N} \sum_{n=0}^{N-1} x(n) e^{-jk\omega_0 n}$$

$$= \frac{1}{4} \sum_{n=0}^{3} x(n) e^{-jk2\pi/4n}$$

$$= \frac{1}{4}[4 + 8e^{-jk\pi/2}]$$

$$X_k = 1 + 2e^{-jk\pi/2}$$

The Fourier series coefficients X_k are also periodic $(X_{k+N} = X_k)$ with period $N = 4$. The numerical values of one period of X_k, $0 \le k \le 3$ are

$$X_0 = 1 + 2 = 3$$

$$X_1 = 1 + 2e^{-j\pi/2}$$
$$= 1 + 2(\cos(\pi/2) - j\sin(\pi/2)) = 1 - 2j$$

$$X_2 = 1 + 2e^{-j\pi}$$
$$= 1 + 2(\cos(\pi) - j\sin(\pi)) = 1 - 2 = -1$$

$$X_3 = 1 + 2e^{-j3\pi/2}$$
$$= 1 + 2(\cos(3\pi/2) - j\sin(3\pi/2)) = 1 + 2j$$

Example 4.8 Let $x(n)$ be a real and odd periodic signal with period $N = 7$ and Fourier series coefficients X_k. Given that

$$X_{15} = j, \quad X_{16} = 2j, \quad X_{17} = 3j$$

determine the values of $X_0, X_{-1}, X_{-2}, X_{-3}$.

Solution
The discrete-time Fourier series coefficients are periodic with period N, i.e.,

$$X_k = X_{k+N} = X_{k+2N}$$

Therefore, for $N = 7$, we have

$$X_1 = X_8 = X_{15} = j$$
$$X_2 = X_9 = X_{16} = 2j$$
$$X_3 = X_{10} = X_{17} = 3j$$

Since the given signal $x(n)$ is real and odd, the Fourier series coefficients X_k will be purely imaginary and odd ($X_k = -X_{-k}$). Therefore,

$$X_0 = 0$$
$$X_1 = -X_{-1}$$
$$X_2 = -X_{-2}$$
$$X_3 = -X_{-3}$$

Finally, we have

$$X_{-1} = -X_1 = -j$$
$$X_{-2} = -X_2 = -2j$$
$$X_{-3} = -X_3 = -3j$$

Example 4.9 Suppose we are given the following information about a signal $x(n)$:

1. $x(n)$ is a real and even signal.
2. $x(n)$ has a period $N = 10$ and Fourier coefficients X_k.
3. $X_{11} = 5$.
4. $\frac{1}{10} \sum_{n=0}^{9} |x(n)|^2 = 50$.

Show that $x(n) = A\cos(Bn + C)$, and specify numerical values for the constants A, B, and C.

Solution
The Fourier series coefficients are periodic with period $N = 10$, we have

$$X_k = X_{k+N}$$
$$X_k = X_{k+10}$$
$$X_1 = X_{11} = 5$$
$$X_{-1} = X_9$$

Since, the given signal $x(n)$ is real and even, The Fourier coefficients X_k are also real and even. Therefore,

$$X_k = X_{-k}$$
$$X_1 = X_{-1} = X_9 = 5$$

Using Parseval's relation,

$$\frac{1}{N} \sum_{n=0}^{N-1} |x(n)|^2 = \sum_{k=0}^{N-1} |X_k|^2$$

$$\frac{1}{10} \sum_{n=0}^{9} |x(n)|^2 = \sum_{k=0}^{9} |X_k|^2 = 50$$

$$|x_0|^2 + |X_1|^2 + |X_9|^2 + \sum_{k=2}^{8} |X_k|^2 = 50$$

$$|x_0|^2 + 25 + 25 + \sum_{k=2}^{8} |X_k|^2 = 50$$

$$|X_0|^2 + \sum_{k=2}^{8} |X_k|^2 = 0$$

Therefore, for one period the Fourier coefficients are

$$X_k = \begin{cases} 5, & k = 1, 9 \\ 0, & k = 0 \text{ and } 2 \leq k \leq 8 \end{cases}$$

Now using the synthesis Eq. (4.16), we have

$$x(n) = \sum_{k=\langle N \rangle} X_k e^{jk\omega_0 n}$$

$$= \sum_{k=0}^{N-1} X_k e^{jk2\pi/Nn} = \sum_{k=0}^{9} X_k e^{jk2\pi/10n}$$

$$= X_1 e^{j2\pi/10n} + X_9 e^{j92\pi/10n}$$

Substituting $X_9 = X_{-1}$, we get

$$x(n) = X_1 e^{j2\pi/10n} + X_{-1} e^{-j2\pi/10n}$$

$$= 5e^{j\pi/5n} + 5e^{-j\pi/5n}$$

$$= 10 \left(\frac{e^{j\pi/5n} + e^{-j\pi/5n}}{2} \right)$$

$$= 10 \cos \left(\frac{n\pi}{5} \right)$$

Example 4.10 Each of the two sequences $x(n)$ and $y(n)$ has a period $N = 4$, and the corresponding Fourier series coefficients are specified as

$$x(n) \longleftrightarrow X_k, \quad y(n) \longleftrightarrow Y_k$$

where

$$X_0 = X_3 = \frac{1}{2} X_1 = \frac{1}{2} X_2 = 1$$

$$Y_0 = Y_1 = Y_2 = Y_3 = 1$$

Using the multiplication property, determine the Fourier series coefficients Z_k for the signal $z(n) = x(n)y(n)$.

Solution
The given signal $z(n) = x(n)y(n)$ is periodic with period $N = 4$. Using the multiplication property [Eq. (4.30)], we have

$$z(n) = x(n)y(n)$$

$$Z_k = \sum_{r=\langle N \rangle} X_r Y_{k-r}$$

$$Z_k = \sum_{r=0}^{N-1} X_r Y_{k-r}$$

$$Z_k = \sum_{r=0}^{3} X_r Y_{k-r}$$

$$Z_k = X_0 Y_k + X_1 Y_{k-1} + X_2 Y_{k-2} + X_3 Y_{k-3}$$

$$Z_k = Y_k + 2Y_{k-1} + 2Y_{k-2} + Y_{k-3}$$

Since $Y_k = Y_{k+N} = 1$ for all values of k, therefore

$$Z_k = 1 + 2 + 2 + 1 = 6 \quad \text{for all values of } k$$

Example 4.11 When the impulse train

$$x(n) = \sum_{k=-\infty}^{\infty} \delta(n - 4k)$$

is the input to a particular LTI system with frequency response $H(e^{j\omega})$, the output of the system is found to be

$$y(n) = \cos \left(\frac{5\pi}{2} n + \frac{\pi}{4} \right)$$

Determine the values of $H(e^{jk\pi/2})$ for $k = 0, 1, 2$ and 3.

Solution
The given input signal $x(n)$ is periodic with period $N = 4$ and frequency $\omega_0 = 2\pi/N = 2\pi/4$. The given signal for one period may be written as

$$x(n) = \begin{cases} 1, & n = 0 \\ 0, & n = 1, 2, 3 \end{cases}$$

$$= \delta(n)$$

Therefore, the Fourier series coefficients are

$$X_k = \frac{1}{N} \sum_{n=0}^{N-1} x(n) e^{-jk\omega_0 n}$$

$$= \frac{1}{4} \sum_{n=0}^{3} x(n) e^{-jk2\pi/4n}$$

$$= \frac{1}{4} \sum_{n=0}^{3} \delta(n) e^{-jk2\pi/4n}$$

$$X_k = \frac{1}{4}, \quad \text{for all } k$$

From Eq. (4.44), we know that the output $y(n)$ of an LTI system to a periodic input is given by,

$$y(n) = \sum_{k=\langle N \rangle} X_k H(e^{jk\omega_0}) e^{jk\omega_0 n}$$

$$= \sum_{k=0}^{N-1} X_k H(e^{jk2\pi/N}) e^{jk2\pi/Nn}$$

$$= \sum_{k=0}^{3} X_k H(e^{jk2\pi/4}) e^{jk2\pi/4n}$$

$$y(n) = X_0 H(e^{j0}) e^{j0n} + X_1 H(e^{j\pi/2}) e^{j\pi/2n}$$

$$+ X_2 H(e^{j\pi n}) e^{j\pi n} + X_3 H(e^{j3\pi/2}) e^{j3\pi/2n}$$

$$y(n) = \frac{1}{4}H(e^{j0})e^{j0n} + \frac{1}{4}H(e^{j\pi/2})e^{j\pi/2n}$$

$$+ \frac{1}{4}H(e^{j\pi})e^{j\pi n} + \frac{1}{4}H(e^{j3\pi/2})e^{j3\pi/2n}$$

$$\text{(4.45)}$$

Given that

$$y(n) = \cos\left(\frac{5\pi}{2}n + \frac{\pi}{4}\right)$$

$$= \cos\left(2\pi n + \frac{\pi}{2}n + \frac{\pi}{4}\right)$$

$$= \cos\left(\frac{\pi}{2}n + \frac{\pi}{4}\right)$$

$$y(n) = \frac{1}{2}e^{j(\pi/2n+\pi/4)} + \frac{1}{2}e^{-j(\pi/2n+\pi/4)}$$

$$y(n) = \frac{1}{2}e^{j(\pi/2n+\pi/4)} + \frac{1}{2}e^{j(3\pi/2n-\pi/4)}$$

$$y(n) = \frac{1}{2}e^{j\pi/4}e^{j(\pi/2n)} + \frac{1}{2}e^{-\pi/4}e^{j(3\pi/2n)}$$

Comparing the above equation with Eq. (4.45), we get

$$H(e^{j0}) = H(e^{j\pi}) = 0,$$

$$H(e^{j\pi/2}) = 2e^{j\pi/4}, \quad H(e^{j3\pi/2}) = 2e^{-j\pi/4}$$

Example 4.12 Determine the output of the filter shown in Fig. 4.9 for the following periodic inputs:

(a) $x_1(n) = (-1)^n$

(b) $x_2(n) = 1 + \sin\left(\frac{3\pi}{8}n + \frac{\pi}{4}\right)$

(c) $x_3(n) = \sum_{k=-\infty}^{\infty}\left(\frac{1}{2}\right)^{n-4k}u(n-4k)$

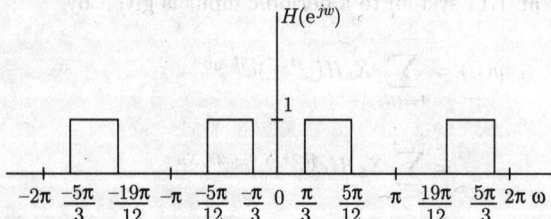

Fig. 4.9

Solution

(a) Consider the given signal $x_1(n)$.

$$x_1(n) = (-1)^n = e^{j\pi n} = e^{j2\pi/2n}$$

Therefore, $x_1(n)$ is periodic with period $N = 2$, and frequency $\omega_0 = \dfrac{2\pi}{N} = \dfrac{2\pi}{2} = \pi$. From Eq. (4.16), we have

$$x_1(n) = \sum_{k=\langle N\rangle} X_k e^{jk\omega_0 n}$$

$$= \sum_{k=0}^{N-1} X_k e^{jk\omega_0 n} = \sum_{k=0}^{1} X_k e^{jk2\pi/2n}$$

$$x_1(n) = X_0 + X_1 e^{j2\pi/2n}$$

Comparing with

$$x_1(n) = e^{j2\pi/2n}$$

we get

$$X_0 = 0, \quad X_1 = 1$$

From Eq. (4.44), the output $y_1(n)$ of the system is given by

$$y_1(n) = \sum_{k=\langle N\rangle} X_k H(e^{jk\omega_0})e^{jk\omega_0 n}$$

$$= \sum_{k=0}^{N-1} X_k H(e^{jk2\pi/2})e^{jk2\pi/2n}$$

$$= \sum_{k=0}^{1} X_k H(e^{jk2\pi/2})e^{jk2\pi/2n}$$

$$= 0 + X_1 H(e^{j\pi})e^{j\pi n}$$

$$= 0 \quad \text{[from Fig. 4.9, } H(e^{j\pi}) = 0]$$

(b) Consider the given signal $x_2(n)$

$$x_2(n) = 1 + \sin\left(\frac{3\pi}{8}n + \frac{\pi}{4}\right)$$

$$= 1 + \sin(\omega n + \phi)$$

The frequency ω of the given signal is

$$\omega = \frac{3\pi}{8}$$

$$\frac{2\pi}{\omega} = \frac{16}{3} = \frac{N}{m} \quad \text{a rational number}$$

Therefore, $x_2(n)$ is periodic with period $N = 16$ and frequency $\omega_0 = 2\pi/N = 2\pi/16$. The signal $x_2(n)$

may be written as

$$x_2(n) = 1 + \frac{e^{j(3\pi/8n+\pi/4)} - e^{-j(3\pi/8n+\pi/4)}}{2j}$$

$$= e^{j02\pi/16n} - \frac{j}{2}e^{j\pi/4}e^{j32\pi/16n}$$

$$+ \frac{j}{2}e^{-j\pi/4}e^{-j32\pi/16n}$$

$$= X_0 e^{j02\pi/16n} + X_3 e^{j32\pi/16n}$$

$$+ X_{-3} e^{-j32\pi/16n}$$

Noting that the DTFS coefficients are periodic with period $N = 16$, we have

$$X_k = X_{k+N}$$

$$X_{-3} = X_{13}$$

Therefore,

$$x_2(n) = X_0 e^{j02\pi/16n} + X_3 e^{j32\pi/16n} + X_{13} e^{j132\pi/16n}$$

The nonzero Fourier coefficients of $x_2(n)$ in the range $0 \le k \le 15$ are

$$X_0 = 1, \quad X_3 = -\frac{j}{2}e^{j\pi/4}, \quad X_{13} = \frac{j}{2}e^{-j\pi/4}$$

From Eq. (4.44), the output $y_2(n)$ of the system is given by

$$y_2(n) = \sum_{k=\langle N \rangle} X_k H(e^{jk\omega_0}) e^{jk\omega_0 n}$$

$$= \sum_{k=0}^{N-1} X_k H(e^{jk2\pi/16}) e^{jk2\pi/16n}$$

$$= \sum_{k=0}^{15} X_k H(e^{jk2\pi/16}) e^{jk2\pi/16n}$$

$$= X_0 H(e^{j02\pi/16}) e^{j02\pi/16n}$$

$$+ X_3 H(e^{j32\pi/16}) e^{j32\pi/16n}$$

$$+ X_{13} H(e^{j132\pi/16}) e^{j132\pi/16n}$$

$$= X_0 H(e^{j0\pi/8}) e^{j02\pi/16n} + X_3 H(e^{j3\pi/8}) e^{j32\pi/16n}$$

$$+ X_{13} H(e^{j13\pi/8}) e^{j132\pi/16n}$$

from Fig. 4.9, we have

$$H(e^{j0\pi/8}) = 0 \quad \text{and} \quad H(e^{j3\pi/8}) = H(e^{j13\pi/8}) = 1$$

Therefore,

$$y_2(n) = 0 - \frac{j}{2}e^{j\pi/4}e^{j32\pi/16n} + \frac{j}{2}e^{-j\pi/4}e^{-j32\pi/16n}$$

$$= \frac{e^{j(3\pi/8n+\pi/4)} - e^{-j(3\pi/8n+\pi/4)}}{2j}$$

$$y_2(n) = \sin\left(3\pi/8n + \pi/4\right)$$

(c) The given signal $x_3(n)$ may be written as

$$x_3(n) = \sum_{k=-\infty}^{\infty} \left(\frac{1}{2}\right)^{n-4k} u(n-4k)$$

$$x_3(n) = \left(\frac{1}{2}\right)^n u(n) * \left(\sum_{k=-\infty}^{\infty} \delta(n-4k)\right)$$

$$x_3(n) = g_1(n) * g_2(n)$$

where

$$g_1(n) = \left(\frac{1}{2}\right)^n u(n)$$

$$g_2(n) = \left(\sum_{k=-\infty}^{\infty} \delta(n-4k)\right)$$

Let $h(n)$ be the impulse response of the given filter. The output $y_3(n)$ of the filter is

$$y_3(n) = x_3(n) * h(n)$$

$$= [g_1(n) * g_2(n)] * h(n)$$

$$y_3(n) = g_1(n) * [g_2(n) * h(n)] = g_1(n) * z(n)$$

Therefore, $y_3(n)$ may be obtained by passing the signal $g_2(n)$ through the filter with impulse response $h(n)$ (frequency response $H(e^{j\omega})$), and then convolving the result with $g_1(n)$.

The signal $g_2(n)$ is periodic with period $N = 4$ and frequency $\omega_0 = 2\pi/N = 2\pi/4$. The given signal for one period may be written as

$$g_2(n) = \begin{cases} 1, & n = 0 \\ 0, & n = 1, 2, 3 \end{cases}$$

$$= \delta(n)$$

Therefore, the Fourier series coefficients are

$$X_k = \frac{1}{N} \sum_{n=0}^{N-1} g_2(n) e^{-jk\omega_0 n}$$

$$= \frac{1}{4} \sum_{n=0}^{3} g_2(n) e^{-jk2\pi/4n}$$

$$= \frac{1}{4} \sum_{n=0}^{3} \delta(n) e^{-jk2\pi/4n}$$

$$X_k = \frac{1}{4} \quad \text{for all } k$$

The output $z(n)$ obtained by passing the periodic signal $g_2(n)$ through the filter with frequency response $H(e^{j\omega})$ is given by Eq. (4.44).

$$z(n) = \sum_{k=\langle N \rangle} X_k H(e^{jk\omega_0}) e^{jk\omega_0 n}$$

$$= \sum_{k=0}^{N-1} X_k H(e^{jk2\pi/N}) e^{jnk2\pi/N}$$

$$= \sum_{k=0}^{3} X_k H(e^{jk2\pi/4}) e^{jnk2\pi/4}$$

$$z(n) = X_0 H(e^{j0}) e^{j0} + X_1 H(e^{j\pi/2}) e^{jn\pi/2}$$
$$+ X_2 H(e^{j\pi}) e^{j\pi n} + X_3 H(e^{j3\pi/2}) e^{j3n\pi/2}$$

$$= \frac{1}{4} \Big[H(e^{j0}) e^{j0} + H(e^{j\pi/2}) e^{jn\pi/2}$$
$$+ H(e^{j\pi}) e^{j\pi n} + H(e^{j3\pi/2}) e^{j3n\pi/2} \Big]$$

From Fig. 4.9, we have

$$H(e^{j0}) = H(e^{j\frac{\pi}{2}}) = H(e^{j\pi}) = H(e^{j\frac{3\pi}{2}}) = 0$$

Therefore,

$$z(n) = 0$$

Now, the final output is

$$y_3(n) = g_1(n) * z(n)$$
$$= g_1(n) * 0 = 0$$

Example 4.13 In each of the following, we specify the DTFS coefficients of a signal that is periodic with period $N = 8$. Determine the signal $x(n)$ in each case.

(a) $X_k = \cos\left(\dfrac{k\pi}{4}\right) + \sin\left(\dfrac{3k\pi}{4}\right)$

(b) $X_k = \begin{cases} \sin\left(\dfrac{k\pi}{3}\right), & 0 \le k \le 6 \\ 0, & k = 7 \end{cases}$

Solution

(a) The signal $x(n)$ is periodic with period $N = 8$ and frequency $\omega_0 = 2\pi/N = 2\pi/8$. The DTFS coefficients for one period $0 \le k \le 7$ are

$$X_k = \cos\left(\frac{k\pi}{4}\right) + \sin\left(\frac{3k\pi}{4}\right)$$

$$X_0 = 1, \quad X_1 = \sqrt{2}, \quad X_2 = -1, \quad X_3 = 0$$
$$X_4 = -1, \quad X_5 = -\sqrt{2}, \quad X_6 = 1, \quad X_7 = 0$$

From the synthesis Eq. (4.16), over one period $0 \le n \le 7$, we have

$$x(n) = \sum_{k=\langle N \rangle} X_k e^{jk\omega_0 n}$$

$$= \sum_{k=0}^{7} X_k e^{jk\omega_0 n}$$

$$= X_0 + X_1 e^{j\omega_0 n} + X_2 e^{j2\omega_0 n} + X_3 e^{j3\omega_0 n}$$
$$+ X_4 e^{j4\omega_0 n} + X_5 e^{j5\omega_0 n} + X_6 e^{j6\omega_0 n}$$
$$+ X_7 e^{j7\omega_0 n}$$

$$= 1 + \sqrt{2} e^{j\omega_0 n} - e^{j2\omega_0 n} - e^{j4\omega_0 n}$$
$$- \sqrt{2} e^{j5\omega_0 n} + e^{j6\omega_0 n}$$

$$x(n) = 1 + \sqrt{2} e^{j2\pi/8 n} - e^{j22\pi/8 n} - e^{j42\pi/8 n}$$
$$- \sqrt{2} e^{j52\pi/8 n} + e^{j62\pi/8 n}$$

$$x(n) = \begin{cases} 4, & n = 1, 7 \\ 4j, & n = 3 \\ -4j, & n = 5 \\ 0, & n = 0, 2, 4, 6 \end{cases}$$

Over one period $0 \le n \le 7$, $x(n)$ may be written as

$$x(n) = 4\delta(n-1) + 4j\delta(n-3) - 4j\delta(n-5)$$
$$+ 4\delta(n-7)$$

(b) The signal $x(n)$ is periodic with period $N = 8$ and frequency $\omega_0 = 2\pi/N = 2\pi/8$. The DTFS coefficients for one period $0 \le k \le 7$ are

$$X_k = \begin{cases} \sin\left(\dfrac{k\pi}{3}\right), & 0 \le k \le 6 \\ 0, & k = 7 \end{cases}$$

From the synthesis Eq. (4.16), over one period $0 \le n \le 7$, we have

$$x(n) = \sum_{k=\langle N \rangle} X_k e^{jk\omega_0 n}$$

$$= \sum_{k=0}^{7} X_k e^{jk\omega_0 n}$$

$$= \sum_{k=0}^{6} \sin\left(\frac{k\pi}{3}\right) e^{jk\omega_0 n}$$

$$x(n) = \sum_{k=0}^{6} \frac{1}{2j}\left(e^{jk\pi/3} - e^{-jk\pi/3}\right) e^{jk2\pi/8n}$$

$$= \frac{1}{2j}\sum_{k=0}^{6} e^{jk\pi/3}e^{jk\pi/4n} - \frac{1}{2j}\sum_{k=0}^{6} e^{-jk\pi/3}e^{jk\pi/4n}$$

$$= \frac{1}{2j}\sum_{k=0}^{6} e^{j(\pi/4n+\pi/3)k} - \frac{1}{2j}\sum_{k=0}^{6} e^{j(\pi/4n-\pi/3)k}$$

$$= \frac{1}{2j}\frac{1 - e^{j7(\pi/4n+\pi/3)}}{1 - e^{j(\pi/4n+\pi/3)}} - \frac{1}{2j}\frac{1 - e^{j7(\pi/4n-\pi/3)}}{1 - e^{j(\pi/4n-\pi/3)}}$$

$$x(n) = \frac{1}{2j}\left[\frac{-e^{j3n\pi/4}\sin\frac{7}{2}\left(\pi/4n+\pi/3\right)}{\sin\frac{1}{2}\left(\pi/4n+\pi/3\right)}\right.$$

$$\left. + \frac{e^{j3n\pi/4}\sin\frac{7}{2}\left(\pi/4n-\pi/3\right)}{\sin\frac{1}{2}\left(\pi/4n-\pi/3\right)}\right]$$

$$x(n) = \frac{e^{j3n\pi/4}}{2j}\left[\frac{\sin\frac{7}{2}\left(\pi/4n-\pi3\right)}{\sin\frac{1}{2}\left(\pi/4n-\pi/3\right)}\right.$$

$$\left. - \frac{\sin\frac{7}{2}\left(\pi/4n+\pi/3\right)}{\sin\frac{1}{2}\left(\pi/4n+\pi/3\right)}\right]$$

Example 4.14 Consider the following three discrete-time signals with a fundamental period of $N = 6$:

$$x(n) = 1 + \cos\left(\frac{2\pi}{6}n\right),$$

$$y(n) = \sin\left(\frac{2\pi}{6}n + \frac{\pi}{4}\right),$$

$$z(n) = x(n)y(n)$$

(a) Determine the Fourier series coefficients of $x(n)$.

(b) Determine the Fourier series coefficients of $y(n)$.

(c) Use the results of parts (a) and (b), along with the multiplication property of the discrete-time Fourier series, to determine the Fourier coefficients of $z(n) = x(n)y(n)$.

Solution

(a) The given signal $x(n)$ is periodic with period $N = 6$ and frequency $\omega_0 = 2\pi/6$. Consider the given signal $x(n)$,

$$x(n) = 1 + \cos\left(2\pi/6n\right) = 1 + \frac{1}{2}e^{j2\pi/6n}$$

$$+ \frac{1}{2}e^{-j2\pi/6n}$$

$$x(n) = X_0 + X_1 e^{j2\pi/6n} + X_{-1}e^{-j2\pi/6n}$$

Noting that the DTFS coefficients are periodic with period $N = 6$, we have

$$X_k = X_{k+N}$$

$$X_{-1} = X_5$$

Therefore, over the interval $0 \le k \le 5$ the nonzero DTFS coefficients of $x(n)$ are

$$X_0 = 1, \quad X_1 = \frac{1}{2}, \quad X_5 = \frac{1}{2}$$

(b) The given signal $y(n)$ is periodic with period $N = 6$ and frequency $\omega_0 = 2\pi/6$. Consider the given signal $y(n)$,

$$y(n) = \sin\left(\frac{2\pi}{6}n + \frac{\pi}{4}\right)$$

$$= \frac{1}{2j}e^{j(2\pi/6n+\pi/4)} - \frac{1}{2j}e^{-j(2\pi/6n+\pi/4)}$$

$$= \frac{1}{2j}e^{j\pi/4}e^{j2\pi/6n} - \frac{1}{2j}e^{-j\pi/4}e^{-j2\pi/6n}$$

$$= Y_1 e^{j2\pi/6n} + Y_{-1}e^{-j2\pi/6n}$$

254 *Signals and Systems*

Noting that the DTFS coefficients are periodic with period $N = 6$, we have

$$Y_k = Y_{k+N}$$
$$Y_{-1} = Y_5$$

Therefore, over the interval $0 \le k \le 5$ the nonzero DTFS coefficients of $y(n)$ are

$$Y_1 = \frac{1}{2j}e^{j\pi/4}, \quad Y_5 = -\frac{1}{2j}e^{-j\pi/4}$$

(c) The given signal $z(n) = x(n)y(n)$ is periodic with period $N = 6$. Using the multiplication property Eq. (4.30), we have

$$z(n) = x(n)y(n)$$

$$Z_k = \sum_{r=\langle N \rangle} X_r Y_{k-r}$$

$$Z_k = \sum_{r=0}^{N-1} X_r Y_{k-r}$$

$$Z_k = \sum_{r=0}^{5} X_r Y_{k-r}$$

$$Z_k = X_0 Y_k + X_1 Y_{k-1} + X_2 Y_{k-2}$$
$$+ X_3 Y_{k-3} + X_4 Y_{k-4} + X_5 Y_{k-5}$$

$$Z_k = X_0 Y_k + X_1 Y_{k-1} + X_5 Y_{k-5}$$

Making the use of $Y_k = Y_{k+N}$, we get the DTFS coefficients of $z(n)$ over the interval $0 \le k \le 5$.

$$Z_k = \begin{cases} \frac{1}{2}\sin\left(\frac{\pi}{4}\right), & k = 0 \\ \frac{1}{2j}e^{j\frac{\pi}{4}}, & k = 1 \\ \frac{1}{4j}e^{j\frac{\pi}{4}}, & k = 2 \\ 0, & k = 3 \\ -\frac{1}{4j}e^{-j\frac{\pi}{4}}, & k = 4 \\ -\frac{1}{2j}e^{-j\frac{\pi}{4}}, & k = 5 \end{cases}$$

Example 4.15 Suppose we are given the following information about a periodic signal $x(n)$ with period $N = 8$ and Fourier series coefficients X_k:
1. $X_k = -X_{k-4}$
2. $x(2n+1) = (-1)^n$
Sketch one period of $x(n)$.

Solution
We know that if

$$x(n) \longleftrightarrow X_k$$

then, from the frequency shifting property, i.e., Eq. (4.23), we have

$$(-1)^n x(n) = e^{j\pi n}x(n) = e^{j2\pi/NN/2n}x(n) \longleftrightarrow X_{k-N/2}$$

In this case $N = 8$, therefore

$$(-1)^n x(n) \longleftrightarrow X_{k-4}$$

Since, it is given that $X_k = -X_{k-4}$, we have

$$(-1)^n x(n) \longleftrightarrow -X_k$$
$$(-1)^n x(n) = -x(n)$$
$$x(n)[1 + (-1)^n] = 0$$

This implies that

$$x(n) = 0, \text{ for } n = 0, \pm2, \pm4, \pm6, \cdots$$

From the given information $x(2n+1) = (-1)^n$, we get

$$x(1) = 1, \quad x(3) = -1, \quad x(5) = 1, \quad x(7) = -1$$

Therefore, over one period $0 \le n \le 7$, $x(n)$ is defined as

$$x(n) = \begin{cases} 0, & n = 0, 2, 4, 6 \\ 1, & n = 1, 5 \\ -1, & n = 3, 7 \end{cases}$$

One period of $x(n)$ is shown in Fig. 4.10.

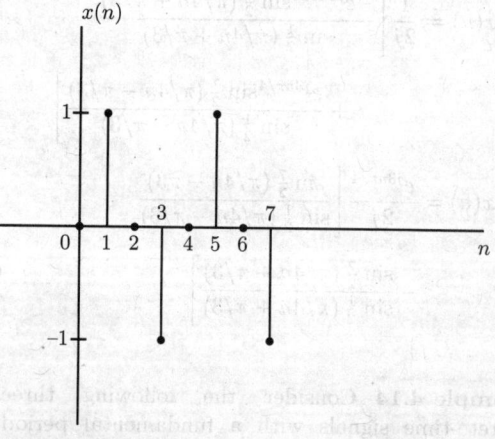

Fig. 4.10

Example 4.16 Let $x(n)$ be a periodic signal with period $N = 8$ and Fourier series coefficients $X_k = -X_{k-4}$. A signal

$$y(n) = \left(\frac{1 + (-1)^n}{2}\right) x(n-1)$$

with period $N = 8$ is generated. Denoting the Fourier series coefficients of $y(n)$ by Y_k, find a function such that

$$Y_k = f(k)X_k$$

Solution
We know that if

$$x(n) \longleftrightarrow X_k$$

then, from the frequency shifting property, i.e., Eq. (4.23), we have

$$(-1)^n x(n) = e^{j\pi n}x(n) = e^{j2\pi/N \, N/2n} \longleftrightarrow X_{k-\frac{N}{2}}$$

In this case $N = 8$, therefore

$$(-1)^n x(n) \longleftrightarrow X_{k-4}$$

Since, it is given that $X_k = -X_{k-4}$, we have

$$(-1)^n x(n) \longleftrightarrow -X_k$$
$$(-1)^n x(n) = -x(n)$$
$$-(-1)^n x(n) = x(n)$$
$$(-1)^{n+1} x(n) = x(n)$$
$$(-1)^n x(n-1) = x(n-1)$$

Now consider

$$y(n) = \left(\frac{1 + (-1)^n}{2}\right) x(n-1)$$

$$= \frac{1}{2}x(n-1) + \frac{1}{2}(-1)^n x(n-1)$$

We have already proved that $(-1)^n x(n-1) = x(n-1)$, therefore

$$y(n) = \frac{1}{2}x(n-1) + \frac{1}{2}x(n-1)$$

$$y(n) = x(n-1)$$

Using the time-shifting property, we get

$$Y_k = X_k e^{-jk2\pi/8} = f(k)X_k$$

Thus,

$$f(k) = e^{-jk2\pi/8}$$

SUMMARY

1. A periodic discrete-time signal $x(n)$ with period N can be represented by the discrete-time Fourier series (DTFS)

$$x(n) = \sum_{k=\langle N \rangle} X_k e^{jk\omega_0 n}$$

where $\omega_0 = 2\pi/N$ and $\sum_{k=\langle N \rangle}$ means summation over any range of consecutive k's exactly N in length.

2. The DTFS coefficients X_k are given by

$$X_k = \frac{1}{N} \sum_{k=\langle N \rangle} x(n) e^{-jk\omega_0 n}$$

3. The coefficients X_k are periodic with period N, so that

$$X_k = X_{k+N}$$

4. The DTFS is a finite sum over only N terms. It provides an exact alternative representation of the time signal, and issues such as convergence or the Gibbs phenomenon do not arise.

5. If X_k are the DTFS coefficients of the signal $x(n)$, then the coefficients of $x(n-m)$ are equal to $X_k e^{-j\omega_0 km}$.

6. If the periodic sequence $x(n)$ with DTFS coefficients X_k is input into an LTI system with impulse response $h(n)$, the DTFS coefficients Y_k of the output $y(n)$ are given by

$$Y_k = X_k H(e^{j\omega_0 k})$$

where

$$H(e^{j\omega}) = \sum_{n=-\infty}^{\infty} h(n) e^{-j\omega n}$$

7. If an LTI system is excited by a periodic signal, the response is also a periodic signal with the same fundamental period.

MULTIPLE-CHOICE QUESTIONS

1. If the Fourier series coefficients of a signal are periodic then the signal must be
   (a) continuous-time, periodic
   (b) discrete-time, periodic
   (c) continuous-time, non-periodic
   (d) discrete-time, non-periodic

2. The DTFS coefficients of a real and even periodic signal are
   (a) real and odd   (b) imaginary and even
   (c) real and even   (d) imaginary and odd

3. The DTFS coefficients of a real and odd periodic signal are
   (a) real and odd   (b) imaginary and even
   (c) real and even   (d) imaginary and odd

4. Let X_k represent the DTFS coefficients of the periodic sequence $x(n)$ with period N. The DTFS coefficients of the signal $(-1)^n x(n)$ in terms of X_k are
   (a) X_k (b) X_{-k} (c) $X_{k+\frac{N}{2}}$ (d) $X_{k-\frac{N}{2}}$

5. Let X_k represent the DTFS coefficients of the periodic sequence $x(n)$ with period N. The DTFS coefficients of the signal $x(n-1)$ in terms of X_k are
   (a) $X_k e^{-jk2\pi/N}$ (b) $X_k e^{jk2\pi/N}$
   (c) X_{k+1} (d) X_{k-1}

6. Let X_k represent the DTFS coefficients of the periodic sequence $x(n)$ with period N. The DTFS coefficients of the signal $x(n+1)$ in terms of X_k are
   (a) $X_k e^{-jk2\pi/N}$ (b) $X_k e^{jk2\pi/N}$
   (c) X_{k+1} (d) X_{k-1}

7. Let X_k represent the DTFS coefficients of the periodic sequence $x(n)$ with period N. The DTFS coefficients of the signal $x\left(\dfrac{n}{2}\right)$ in terms of X_k are
   (a) X_{2k} (b) $X_{k/2}$ (c) $2X_k$ (d) $\dfrac{X_k}{2}$

PROBLEMS

4.1 One period of the DTFS coefficients of a signal is given by

$$X_k = \left(\frac{1}{2}\right)^k, \quad 0 \le k \le 9$$

Find the time-domain signal $x(n)$ assuming $N = 10$.

4.2 Find the time-domain signal corresponding to the DTFS coefficients

$$X_k = \cos\left(\frac{k4\pi}{11}\right) + 2j\sin\left(\frac{k6\pi}{11}\right)$$

4.3 Determine the DTFS coefficients for the following signals:
(a) $x(n) = 1 + \sin(n\pi/12 + 3\pi/8)$
(b) $x(n) = \cos(n\pi/30) + 2\sin(n\pi/90)$

4.4 Find the DTFS coefficients of the signal depicted in Fig. 4.11.

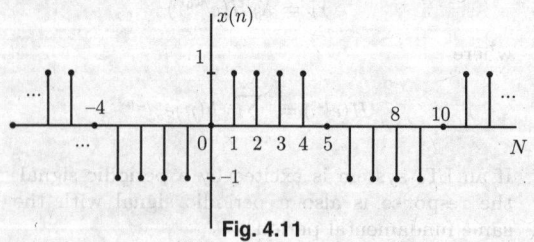

Fig. 4.11

4.5 Consider a discrete-time LTI system with impulse response

$$h(n) = \left(\frac{1}{2}\right)^{|n|}$$

Find the Fourier series representation of the output $y(n)$ for each of the following inputs:
(a) $x(n) = \sum_{k=-\infty}^{\infty} \delta(n - 4k)$
(b) $x(n)$ is periodic with period 6 and

$$x(n) = \begin{cases} 1, & n = 0, \pm1 \\ 0, & n = \pm2, \pm3 \end{cases}$$

4.6 Consider a discrete-time LTI system with impulse response

$$h(n) = \begin{cases} 1, & 0 \le n \le 2 \\ -1, & -2 \le n \le -1 \\ 0, & \text{otherwise} \end{cases}$$

Given that the input to this system is

$$x(n) = \sum_{k=-\infty}^{\infty} \delta(n - 4k)$$

determine the Fourier series coefficients of the output $y(n)$.

4.7 Consider a discrete-time LTI system whose frequency response is

$$H(e^{j\omega}) = \begin{cases} 1, & |\omega| \leq \pi/8 \\ 0, & \pi/8 < |\omega| < \pi \end{cases}$$

Show that if the input $x(n)$ to this system has a period $N = 3$, the output $y(n)$ has only one nonzero Fourier series coefficient per period.

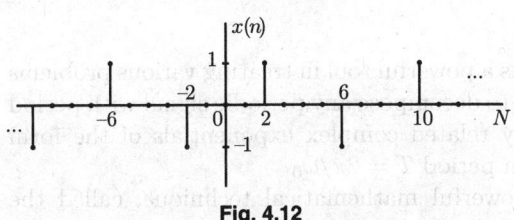

Fig. 4.12

4.8 Evaluate the DTFS coefficients of the following signals:

(a) $x(n) = \cos\left(\dfrac{6\pi}{17}n + \dfrac{\pi}{3}\right)$

(b) $x(n) = 2\sin\left(\dfrac{14\pi}{19}n\right) + \cos\left(\dfrac{10\pi}{19}n\right) + 1$

(c) $x(n) = \sum_{m=-\infty}^{\infty}(-1)^m\left[\delta(n-2m) + \delta(n+3m)\right]$

(d) $x(n)$ as depicted in Fig. 4.12.

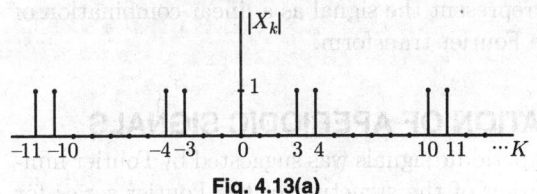

Fig. 4.13(a)

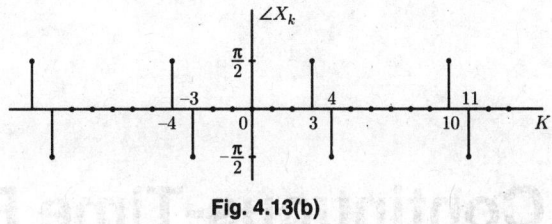

Fig. 4.13(b)

4.9 Evaluate the time-domain signals represented by the following DTFS coefficients:

(a) $X_k = \cos\left(\dfrac{8\pi}{21}k\right)$

(b) $X_k = \cos\left(\dfrac{10\pi}{19}k\right) + j2\sin\left(\dfrac{4\pi}{19}k\right)$

(c) $X_k = \displaystyle\sum_{m=-\infty}^{\infty}(-1)^m[\delta(k-2m)$
$\quad -2\delta(k+3m)]$

(d) X_k as depicted in Fig. 4.13.

4.10 One period of the DTFS coefficients of a signal is given by

$$X_k = \left(\frac{1}{2}\right)^k, \quad 0 \leq k \leq 9$$

Find the time-domain signal $x(n)$ assuming $N = 10$.

4.11 Let X_k represent the DTFS coefficients of the periodic sequence $x(n)$ with period N. Find the DTFS coefficients of each of the following signals in terms of X_k:

(a) $x(n - n_0)$

(b) $x(-n)$

(c) $(-1)^n x(n)$

(d) $y(n) = \left(\dfrac{1+(-1)^n}{2}\right)x(n) = \begin{cases} x(n), & n \text{ even} \\ 0, & n \text{ odd} \end{cases}$

(e) $y(n) = \begin{cases} x(n), & n \text{ odd} \\ 0, & n \text{ even} \end{cases}$

4.12 Show that for a real periodic sequence $x(n)$, $X_k = X_{N-k}^*$.

ANSWERS TO MULTIPLE-CHOICE QUESTIONS

1. **(b)**   2. **(c)**   3. **(d)**   4. **(d)**   5. **(a)**   6. **(b)**   7. **(d)**

Continuous-Time Fourier Transform

5.1 INTRODUCTION

We saw in Chapter 3 that the Fourier series is a powerful tool in treating various problems involving periodic signals. We were also able to decompose any periodic signal with period T in terms of infinitely many harmonically related complex exponentials of the form $e^{jn\omega_0 t}$. All such harmonics have the common period $T = 2\pi/\omega_0$.

In this chapter, we consider another powerful mathematical technique, called the Fourier transform, for describing both periodic and nonperiodic signals for which no Fourier series exists. Like the Fourier series coefficients, the Fourier transform specifies the spectral content of a signal, thus providing a frequency domain description of the signal. Besides being useful in analytically representing aperiodic signals, the Fourier transform is a valuable tool in the analysis of LTI systems.

An aperiodic signal can be viewed as a periodic signal with an infinite period. More precisely, in the Fourier series representation of a periodic signal, as the period increases the fundamental frequency decreases and the harmonically related components become closer in frequency. As the period becomes infinite, the frequency components form a continuum and the Fourier series sum becomes an integral. The resulting spectrum of coefficients in this representation is called the Fourier transform, and the synthesis integral itself, which uses these coefficients to represent the signal as a linear combination of complex exponentials, is called the inverse Fourier transform.

5.2 FOURIER TRANSFORM REPRESENTATION OF APERIODIC SIGNALS

The generalization of the Fourier series to aperiodic signals was suggested by Fourier himself and can be deduced from an examination of the structure of the Fourier series for periodic signals as the period $T \to \infty$. After taking the limit, we will find that the magnitude spectrum of an aperiodic signal is not a line spectrum (as with a periodic signal), but instead occupies a continuum of frequencies. The same is true of the corresponding phase spectrum.

To clarify how the change from discrete to continuous spectra takes place, consider the periodic signal $\tilde{x}(t)$ shown in Fig. 5.1. Now think of keeping the waveform of one period of $\tilde{x}(t)$ unchanged, but carefully and intentionally increase T. In the limit as $T \to \infty$, only a single pulse remains because the nearest neighbours have been moved to infinity. We

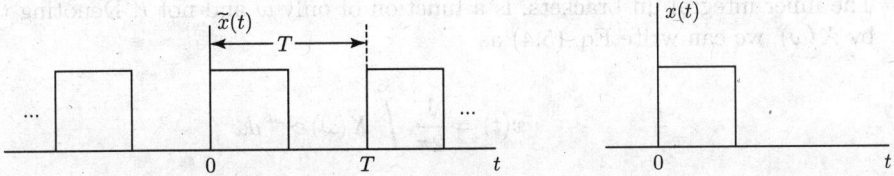

Fig. 5.1 Allowing the period T of $\widetilde{x}(t)$ to increase to obtain the aperiodic signal $x(t)$.

saw in Chapter 3 (Section 3.4.5, time-scaling property) that as the period T increases the fundamental frequency $\omega_0 = 2\pi/T$ decreases and the harmonically related components become closer in frequency. As $T \to \infty$, the spacing between lines approaches zero. This means that the spectral lines move closer, eventually becoming a continuum. The overall shapes of the magnitude and phase spectra are determined by the shape of the single pulse that remains in the new signal $x(t)$, which is aperiodic.

To investigate what happens mathematically, we use the exponential form of the Fourier series representation for $\widetilde{x}(t)$, i.e.,

$$\widetilde{x}(t) = \sum_{n=-\infty}^{\infty} X_n e^{jn\omega_0 t} \tag{5.1}$$

where

$$X_n = \frac{1}{T} \int_{-T/2}^{T/2} \widetilde{x}(t)\, e^{-jn\omega_0 t}\, dt \tag{5.2}$$

In the limit as $T \to \infty$, we see that $\omega_0 = 2\pi/T$ becomes an infinitesimally small quantity, $d\omega$, so that

$$\frac{1}{T} \to \frac{d\omega}{2\pi}$$

We argue that in the limit, $n\omega_0$ should be a continuous variable. Then, from Eq. (5.2), the Fourier coefficients per unit frequency interval are

$$\frac{X_n}{d\omega} = \frac{1}{2\pi} \int_{-\infty}^{\infty} \widetilde{x}(t)\, e^{-j\omega t}\, dt \tag{5.3}$$

Substituting Eq. (5.3) into Eq. (5.1), and recognizing that in the limit the sum becomes an integral and $\widetilde{x}(t)$ approaches $x(t)$, we obtain

$$x(t) = \int_{-\infty}^{\infty} \left(\int_{-\infty}^{\infty} x(t)\, e^{-j\omega t}\, dt \right) e^{j\omega t} \frac{d\omega}{2\pi} \tag{5.4}$$

The inner integral, in brackets, is a function of only ω and not t. Denoting the integral by $X(\omega)$, we can write Eq. (5.4) as

$$x(t) = \frac{1}{2\pi} \int_{-\infty}^{\infty} X(\omega) \, e^{j\omega t} \, d\omega \tag{5.5}$$

where

$$X(\omega) = \int_{-\infty}^{\infty} x(t) \, e^{-j\omega t} \, dt \tag{5.6}$$

Equations (5.5) and (5.6) are referred to as the *Fourier transform pair*, with the function $X(\omega)$ referred to as the *Fourier transform* or Fourier integral of $x(t)$ and Eq. (5.5) as the *inverse Fourier transform* equation. $X(\omega)$ plays the same role for aperiodic signals that X_n plays for periodic signals. Thus, $X(\omega)$ is the spectrum of $x(t)$ and is a continuous function defined for all values of ω, whereas X_n is defined only for discrete frequencies. Therefore, an aperiodic signal has a continuous spectrum rather than a line spectrum.

We call $X(\omega)$ the Fourier transform of $x(t)$, and $x(t)$ the inverse Fourier transform of $X(\omega)$. The same information is conveyed by the statement that $x(t)$ and $X(\omega)$ are a Fourier transform pair. Symbolically, this statement is expressed as

$$X(\omega) = \mathcal{F}[x(t)] \quad \text{and} \quad x(t) = \mathcal{F}^{-1}[X(\omega)]$$

or

$$x(t) \longleftrightarrow X(\omega)$$

$X(\omega)$, in general, is a complex function of the variable ω. Thus, it can be written as

$$X(\omega) = X_R(\omega) + jX_I(\omega)$$

$$X(\omega) = |X(\omega)| \, e^{j\angle X(\omega)} \tag{5.7}$$

where $|X(\omega)| = \sqrt{X_R^2(\omega) + X_I^2(\omega)}$ is the magnitude and $\angle X(\omega) = \tan^{-1}\left[\dfrac{X_I(\omega)}{X_R(\omega)}\right]$ is the angle (or phase) of $X(\omega)$.

The $|X(\omega)|$ plotted against ω is called the *magnitude spectrum* of $x(t)$, and the $\angle X(\omega)$ plotted against ω is called the *phase spectrum*. For real $x(t)$, the magnitude spectrum $|X(\omega)|$ is an even function, and the phase spectrum $\angle X(\omega)$ is an odd function of ω, i.e.,

$$|X(-\omega)| = |X(\omega)| \tag{5.8}$$

and

$$\angle X(-\omega) = -\angle X(\omega) \tag{5.9}$$

These results were derived earlier for the line spectrum of periodic signal (Section 3.2.8).

It is important at this point to comment on conventional notation. In communication system literature, Fourier optics, and image processing, the functional form $X(f)$ is usually used for the transform of $x(t)$. In control system literature, one can find the use of both $X(\omega)$ and $X(j\omega)$. The Fourier transform in terms of cyclic frequency f is written as

$$X(f) = \mathcal{F}[x(t)] = \int\limits_{-\infty}^{\infty} x(t)\, e^{-j2\pi ft}\, dt \qquad (5.10)$$

The inverse Fourier transform in terms of cyclic frequency f is written as

$$x(t) = \mathcal{F}^{-1}[X(f)] = \int\limits_{-\infty}^{\infty} X(f)\, e^{j2\pi ft}\, dt \qquad (5.11)$$

The first definition [Eqs (5.5) and (5.6)] is written in terms of the radian frequency variable ω instead of cyclic frequency f. Radian frequency has a somewhat more direct relationship to the time constants and resonant frequencies of real systems, and, as a result, the transform of some system functions are somewhat simpler using this form. Either definition can be converted to the other by using the relationship $\omega = 2\pi f$. It is natural to wonder at this point what the physical significance of $X(f)$ is. One way to understand it is to find the units of $X(f)$, which depend on what the units of $x(t)$ are. To make the idea concrete, suppose for the moment that the units of $x(t)$ are volts (V). The transformation process begins by multiplying $x(t)$ by the complex exponential $e^{-j2\pi ft}$. The exponent of $x(t)$ consists of three dimensionless numbers, $-j$, 2, and π along with f and t, which are frequency and time, respectively. Frequency has units of hertz or 1/seconds, and time has units of seconds. Therefore, the exponent of e is dimensionless and so is $e^{-j2\pi ft}$. Then we multiply by dt which has units of seconds. Thus the integration process accumulates the area under the product of $x(t)$ and $e^{-j2\pi ft}$. This area has units of volt seconds. Therefore, $X(f)$ *has units of volt seconds or volts/hertz*. Similarly, $X(\omega)$ *would have units of volts/radian/seconds*.

The function $X(\omega)$ or $X(f)$ is sometimes called the *amplitude spectral density* or just the *spectrum* of $x(t)$. It expresses the variation of complex sinusoids with frequency which, when added, form $x(t)$. The word *spectral* refers to the variation with respect to frequency. The word *density* comes from the units, volts/hertz.

5.3 CONVERGENCE OF FOURIER TRANSFORM

Dirichlet showed that if $x(t)$ satisfies certain conditions (*Dirichlet conditions*), its Fourier transform is guaranteed to converge pointwise at all points where $x(t)$ is continuous. These conditions are as follows:

1. $x(t)$ be *absolutely integrable*, i.e.,

$$\int\limits_{-\infty}^{\infty} |x(t)|\, dt < \infty \qquad (5.12)$$

This guarantees that $X(\omega)$ will be finite since

$$|X(\omega)| = \left| \int_{-\infty}^{\infty} x(t)\, e^{-j\omega t}\, dt \right|$$

$$|X(\omega)| \leq \int_{-\infty}^{\infty} \left| x(t)\, e^{-j\omega t} \right| dt = \int_{-\infty}^{\infty} |x(t)|\, dt$$

So

$$\text{if } \int_{-\infty}^{\infty} |x(t)|\, dt < \infty, \text{ then } |X(\omega)| < \infty$$

2. In any finite interval of time, $x(t)$ is of bounded variation, i.e., $x(t)$ have a finite number of maxima and minima.

3. In any finite interval of time, there are only a finite number of discontinuities. Furthermore, each of these discontinuities is finite.

Therefore, absolutely integrable signals that are continuous or that have a finite number of discontinuities have Fourier transform. At a point of discontinuity, t_0, the inverse Fourier transform in Eq. (5.5) converges to $1/2[x(t_0^+) + x(t_0^-)]$.

The conditions just given for the existence of the Fourier transform of $x(t)$ are sufficient conditions. This means that there are signals that violate either one or both conditions and yet possess a Fourier transform. Examples are power signals (unit step signal, periodic signals, etc.) that are neither absolutely integrable nor square integrable over an infinite interval, but still have Fourier transforms.

Example 5.1 Find the Fourier transform (FT) of the continuous-time signal

$$x(t) = e^{-at} u(t), \quad a > 0$$

shown in Fig. 5.2. Plot the magnitude and phase spectrum of $x(t)$.

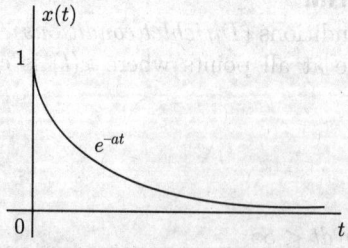

Fig. 5.2 Real-time domain exponential signal

Solution

By definition [Eq. (5.6)],

$$X(\omega) = \int_{-\infty}^{\infty} x(t) e^{-j\omega t}\, dt = \int_{-\infty}^{\infty} e^{-at} u(t)\, e^{-j\omega t}\, dt$$

$$= \int_{0}^{\infty} e^{-(a+j\omega)t}\, dt$$

$$= -\frac{1}{a+j\omega} e^{-(a+j\omega)t}\Big|_{0}^{\infty}$$

$$X(\omega) = -\frac{1}{a+j\omega}(0-1) = \frac{1}{a+j\omega}$$

$$e^{-at} u(t) \longleftrightarrow \frac{1}{a+j\omega} \tag{5.13}$$

Since this Fourier transform is complex valued, to plot it as a function of ω, we express $X(\omega)$ in terms of its magnitude and phase:

$$|X(\omega)| = \frac{1}{\sqrt{a^2 + \omega^2}}$$

$$\angle X(\omega) = -\tan^{-1}\left(\frac{\omega}{a}\right)$$

The magnitude spectrum $|X(\omega)|$ and phase spectrum $\angle X(\omega)$ are depicted in Fig. 5.3. Observe that $|X(\omega)|$ is an even function of ω, and $\angle X(\omega)$ is an odd function of ω, as expected.

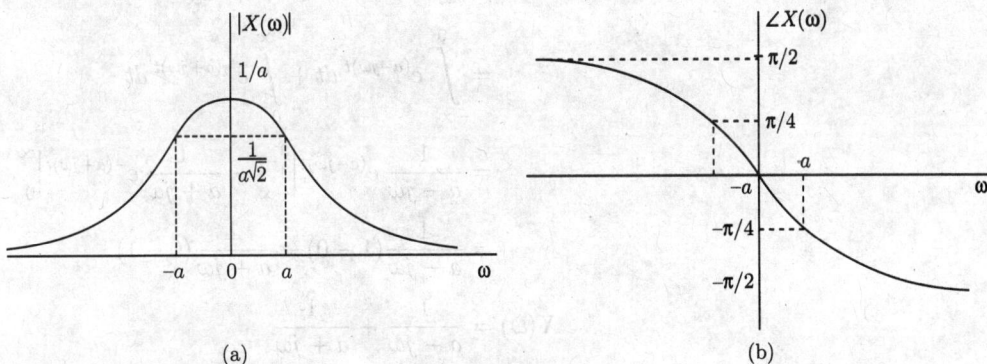

(a)   (b)

Fig. 5.3   (a) Magnitude spectrum and (b) phase spectrum

Example 5.2   Define the continuous-time signal [as shown in Fig. 5.4(a)]

$$x(t) = e^{-a|t|}, \quad a > 0$$

Find its Fourier transform.

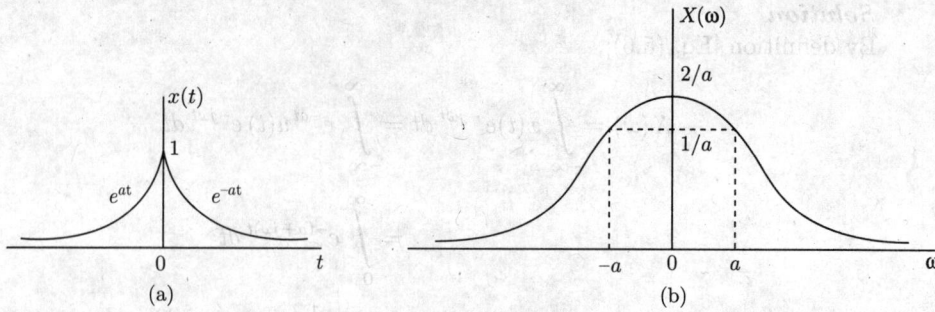

Fig. 5.4   (a) Time-domain signal and (b) its spectrum

Solution
The signal $x(t)$ is defined as

$$x(t) = e^{-a|t|} = \begin{cases} e^{at}, & t < 0 \\ e^{-at}, & t \geq 0 \end{cases}$$

$$= e^{at}u(-t) + e^{-at}u(t)$$

By definition, we have

$$X(\omega) = \int_{-\infty}^{\infty} x(t)\, e^{-j\omega t}\, dt = \int_{-\infty}^{\infty} [e^{at}u(-t) + e^{-at}u(t)]\, e^{-j\omega t}\, dt$$

$$= \int_{-\infty}^{\infty} e^{at}u(-t)\, e^{-j\omega t}\, dt + \int_{-\infty}^{\infty} e^{-at}u(t)\, e^{-j\omega t}\, dt$$

$$= \int_{-\infty}^{0} e^{(a-j\omega)t}\, dt + \int_{0}^{\infty} e^{-(a+j\omega)t}\, dt$$

$$= \frac{1}{a - j\omega} e^{(a-j\omega)t}\Big|_{-\infty}^{0} - \frac{1}{a + j\omega} e^{-(a+j\omega)t}\Big|_{0}^{\infty}$$

$$= \frac{1}{a - j\omega}(1 - 0) - \frac{1}{a + j\omega}(0 - 1)$$

$$X(\omega) = \frac{1}{a - j\omega} + \frac{1}{a + j\omega}$$

$$X(\omega) = \frac{2a}{a^2 + \omega^2}$$

$$e^{-a|t|} \longleftrightarrow \frac{2a}{a^2 + \omega^2} \tag{5.14}$$

$X(\omega)$ is real and it is illustrated in Fig. 5.4(b).

Example 5.3 Determine the Fourier transform of the unit impulse

$$x(t) = \delta(t)$$

Solution
By definition [Eq. (5.6)],

$$X(\omega) = \int_{-\infty}^{\infty} x(t)\, e^{-j\omega t}\, dt = \int_{-\infty}^{\infty} \delta(t)\, e^{-j\omega t}\, dt = e^{-j\omega t}\Big|_{t=0} = 1$$

$$\delta(t) \longleftrightarrow 1 \tag{5.15}$$

That is, the unit impulse has a Fourier transform consisting of equal contributions at all frequencies. The unit impulse function and its Fourier transform are depicted in Fig. 5.5.

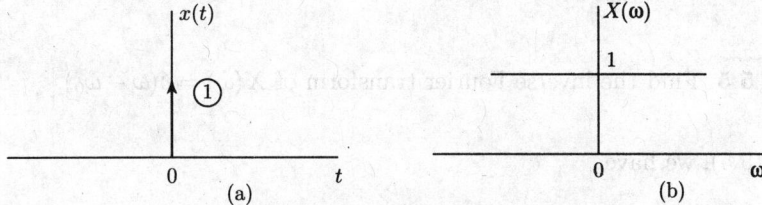

Fig. 5.5 (a) Unit impulse function and (b) its Fourier transform

Example 5.4 Find the inverse Fourier transform of $X(\omega) = \delta(\omega)$.

Solution
From Eq. (5.5), we have

$$x(t) = \mathcal{F}^{-1}[X(\omega)] = \frac{1}{2\pi} \int_{-\infty}^{\infty} X(\omega)\, e^{j\omega t} d\omega$$

$$\mathcal{F}^{-1}[\delta(\omega)] = \frac{1}{2\pi} \int_{-\infty}^{\infty} \delta(\omega)\, e^{j\omega t} d\omega$$

From the sampling property of the impulse function, we have

$$\mathcal{F}^{-1}[\delta(\omega)] = \frac{1}{2\pi}\, e^{j\omega t}\Big|_{\omega=0} = \frac{1}{2\pi}$$

or

$$\delta(\omega) = \mathcal{F}\left[\frac{1}{2\pi}\right]$$

Therefore,

$$\frac{1}{2\pi} \longleftrightarrow \delta(\omega) \tag{5.16}$$

or

$$1 \longleftrightarrow 2\pi\delta(\omega) \tag{5.17}$$

This result shows that the spectrum of a constant signal $x(t) = 1$ is an impulse $2\pi\delta(\omega)$, as illustrated in Fig. 5.6.

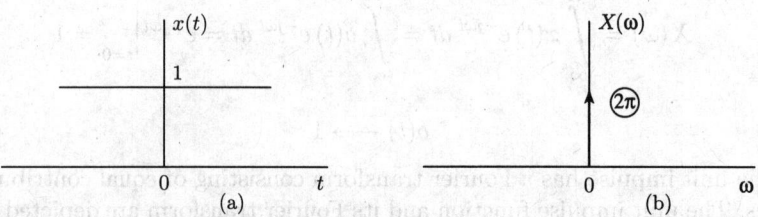

Fig. 5.6 (a) Constant signal and (b) its Fourier transform

Example 5.5 Find the inverse Fourier transform of $X(\omega) = \delta(\omega - \omega_0)$.

Solution
From Eq. (5.5), we have

$$x(t) = \mathcal{F}^{-1}[X(\omega)] = \frac{1}{2\pi} \int\limits_{-\infty}^{\infty} X(\omega)\, e^{j\omega t} d\omega$$

$$\mathcal{F}^{-1}[\delta(\omega - \omega_0)] = \frac{1}{2\pi} \int\limits_{-\infty}^{\infty} \delta(\omega - \omega_0)\, e^{j\omega t} d\omega$$

From the sampling property of the impulse function, we have

$$\mathcal{F}^{-1}[\delta(\omega - \omega_0)] = \frac{1}{2\pi} e^{j\omega t}\Big|_{\omega=\omega_0} = \frac{1}{2\pi} e^{j\omega_0 t}$$

or

$$\delta(\omega - \omega_0) = \mathcal{F}\left[\frac{1}{2\pi} e^{j\omega_0 t}\right]$$

Therefore,

$$\frac{1}{2\pi} e^{j\omega_0 t} \longleftrightarrow \delta(\omega - \omega_0) \tag{5.18}$$

or

$$e^{j\omega_0 t} \longleftrightarrow 2\pi\delta(\omega - \omega_0) \tag{5.19}$$

Similarly, we have

$$e^{-j\omega_0 t} \longleftrightarrow 2\pi\delta(\omega + \omega_0) \tag{5.20}$$

Example 5.6 Consider the rectangular pulse (gate pulse) signal depicted in Fig. 5.7. and defined as

$$x(t) = A \text{ rect}\left(\frac{t}{2T_0}\right) = A \prod\left(\frac{t}{2T_0}\right) = \begin{cases} A, & |t| < T_0 \\ 0, & |t| > T_0 \end{cases}$$

Find the Fourier transform of $x(t)$.

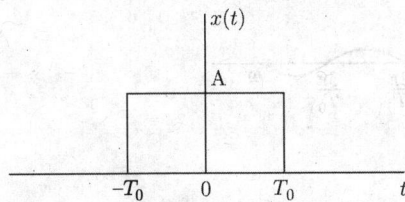

Fig. 5.7 Rectangular pulse or gate pulse

Solution
By definition [Eq. (5.6)],

$$X(\omega) = \int_{-\infty}^{\infty} x(t)\,e^{-j\omega t}\,dt = \int_{-T_0}^{T_0} A\,e^{-j\omega t}\,dt$$

$$= A\frac{1}{-j\omega}\,e^{-j\omega t}\Big|_{-T_0}^{T_0}$$

$$= \frac{A}{-j\omega}(e^{-j\omega T_0} - e^{j\omega T_0})$$

$$= \frac{2A}{\omega}\left(\frac{e^{j\omega T_0} - e^{-j\omega T_0}}{2j}\right)$$

$$= \frac{2A}{\omega}\sin(\omega T_0)$$

$$X(\omega) = 2AT_0\frac{\sin(\pi\omega T_0/\pi)}{(\pi\omega T_0/\pi)}$$

$$X(\omega) = 2AT_0 \text{ sinc}\left(\frac{\omega T_0}{\pi}\right)$$

$$A \text{ rect}\left(\frac{t}{2T_0}\right) \longleftrightarrow 2AT_0 \text{ sinc}\left(\frac{\omega T_0}{\pi}\right) \tag{5.21}$$

$\text{sinc}(\omega T_0/\pi) = 0$ when $\sin(\omega T_0) = 0$ except at $\omega T_0 = 0$, where it appears to be indeterminate. This means that $\text{sinc}(\omega T_0/\pi) = 0$ for $\omega T_0 = \pm n\pi$, $n = 1,2,3,\ldots$. The Fourier transform $X(\omega)$ shown in Fig. 5.8(a) exhibits positive and negative values.

A negative amplitude can be considered to be a positive amplitude with a phase of $-\pi$ or π. We use this observation to plot the magnitude spectrum $|X(\omega)| = |2AT_0 \operatorname{sinc}(\omega T_0/\pi)|$ [Fig. 5.8(b)] and the phase spectrum $\angle X(\omega)$ [Fig. 5.8(c)].

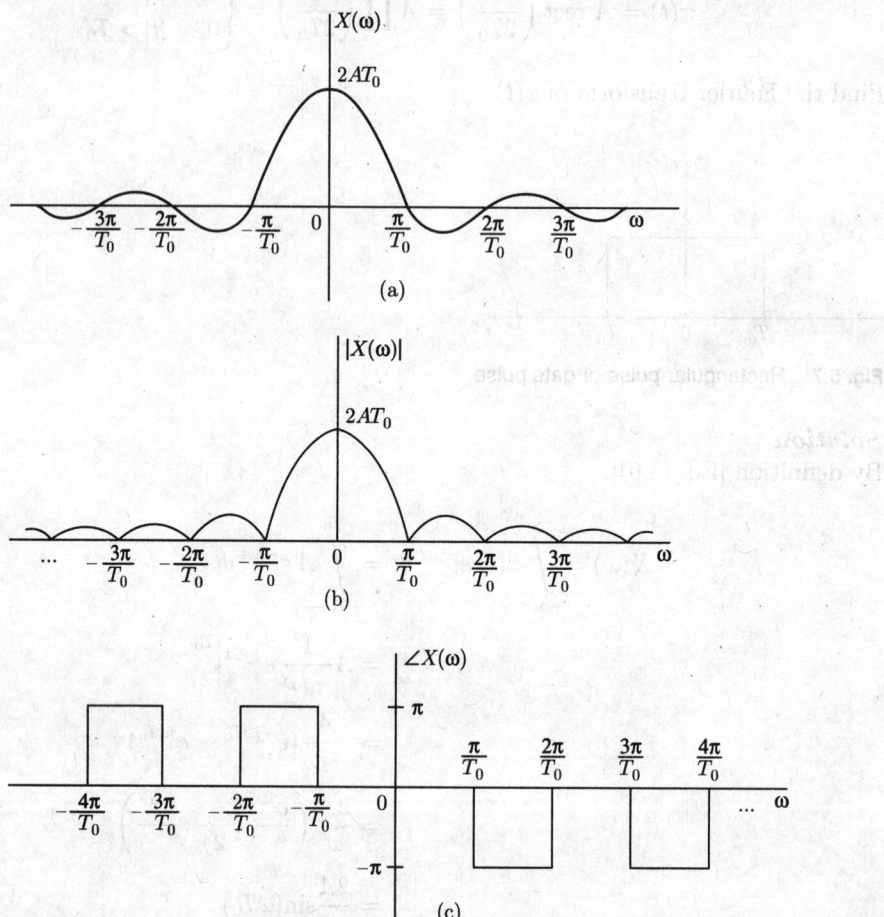

(a)

(b)

(c)

Fig. 5.8 (a) Fourier transform of a gate pulse, (b) its magnitude spectrum, and (c) its phase spectrum

---

Example 5.7 Consider the triangular pulse signal depicted in Fig. 5.9(a) and defined as

$$x(t) = A \triangle \left(\frac{t}{T}\right) = \begin{cases} A\left(1 - \frac{|t|}{T}\right), & |t| \leq T \\ 0, & |t| \geq T \end{cases} = \begin{cases} A\left(1 + \frac{t}{T}\right), & -T \leq t < 0 \\ A\left(1 - \frac{t}{T}\right), & 0 \leq t \leq T \\ 0, & |t| \geq T \end{cases}$$

Find the Fourier transform of $x(t)$.

Solution
By definition [Eq. (5.6)],

$$X(\omega) = \int\limits_{-\infty}^{\infty} x(t)\, e^{-j\omega t}\, dt = \int\limits_{-T}^{0} A\left(1 + \frac{t}{T}\right) e^{-j\omega t}\, dt + \int\limits_{0}^{T} A\left(1 - \frac{t}{T}\right) e^{-j\omega t}\, dt$$

$$= \int\limits_{0}^{T} A\left(1 - \frac{t}{T}\right) e^{j\omega t}\, dt + \int\limits_{0}^{T} A\left(1 - \frac{t}{T}\right) e^{-j\omega t}\, dt$$

$$= 2A \int\limits_{0}^{T}\left(1 - \frac{t}{T}\right) \frac{e^{j\omega t} + e^{-j\omega t}}{2}\, dt$$

$$= 2A \int\limits_{0}^{T}\left(1 - \frac{t}{T}\right) \cos(\omega t)\, dt$$

$$= \left[2A\left(1 - \frac{t}{T}\right)\frac{\sin(\omega t)}{\omega}\right]_{0}^{T} + \frac{2A}{\omega T}\int\limits_{0}^{T} \sin(\omega t)\, dt$$

$$= 0 - \frac{2A}{\omega^2 T}\cos(\omega t)\Big|_{0}^{T}$$

$$= \frac{2A}{\omega^2 T}(1 - \cos(\omega T))$$

$$X(\omega) = \frac{4A}{\omega^2 T}\sin^2\left(\frac{\omega T}{2}\right)$$

$$= AT\left(\frac{\sin(\omega T/2)}{(\omega T/2)}\right)^2$$

$$= AT\left(\frac{\sin(\pi \omega T/2\pi)}{(\pi \omega T/2\pi)}\right)^2$$

$$X(\omega) = AT\,\mathrm{sinc}^2\left(\frac{\omega T}{2\pi}\right)$$

$$A\triangle\left(\frac{t}{T}\right) \longleftrightarrow AT\,\mathrm{sinc}^2\left(\frac{\omega T}{2\pi}\right) \tag{5.22}$$

$\mathrm{sinc}^2(\omega T/2\pi) = 0$ when $\sin(\omega T/2) = 0$ except at $\omega T/2 = 0$, where it appears to be indeterminate. This means that $\mathrm{sinc}^2(\omega T/2\pi) = 0$ for $\omega T/2 = \pm n\pi$, $n = 1, 2, 3, \ldots$. The Fourier transform $X(\omega)$ is depicted in Fig. 5.9(b).

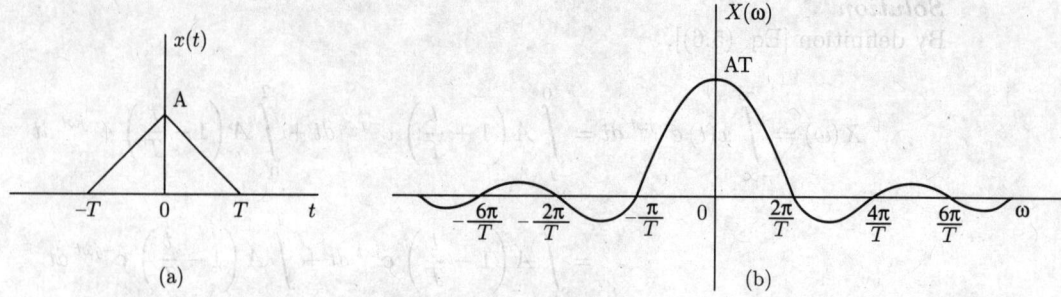

Fig. 5.9 (a) Triangular pulse and (b) its Fourier transform

Example 5.8 Consider the Gaussian pulse signal depicted in Fig. 5.10(a) and defined as

$$x(t) = e^{-\pi t^2}$$

Find the Fourier transform of $x(t)$.

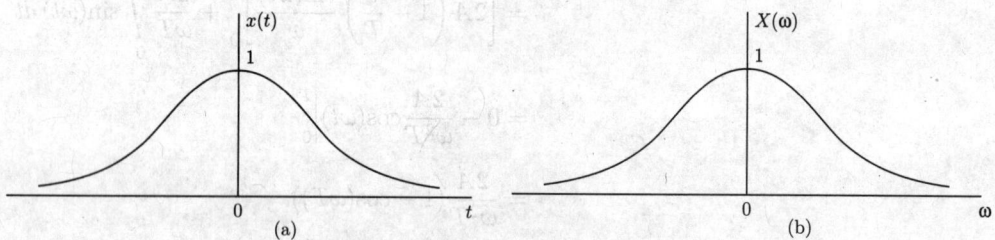

Fig. 5.10 (a) Gaussian pulse and (b) its Fourier transform

Solution
By definition [Eq. (5.6)],

$$X(\omega) = \int_{-\infty}^{\infty} x(t)\, e^{-j\omega t}\, dt = \int_{-\infty}^{\infty} e^{-\pi t^2}\, e^{-j\omega t}\, dt = \int_{-\infty}^{\infty} e^{-(\pi t^2 + j\omega t)}\, dt$$

Substituting $\pi t^2 + j\omega t = \left(\sqrt{\pi}\, t + j\omega/2\sqrt{\pi}\right)^2 + \omega^2/4\pi$ gives

$$X(\omega) = \int_{-\infty}^{\infty} e^{-\left(\sqrt{\pi}t + j\omega/2\sqrt{\pi}\right)^2} e^{-\omega^2/4\pi}\, dt$$

$$= e^{-\omega^2/4\pi} \int_{-\infty}^{\infty} e^{-\left(\sqrt{\pi}t + j\omega/2\sqrt{\pi}\right)^2}\, dt$$

A change of variables is performed by letting $u = \sqrt{\pi}t + j\omega/2\sqrt{\pi}$, which also yields $dt = du/\sqrt{\pi}$, $u \to -\infty$ as $t \to -\infty$, and $u \to \infty$ as $t \to \infty$. Therefore,

$$X(\omega) = e^{-\omega^2/4\pi} \int_{-\infty}^{\infty} e^{-u^2} \frac{du}{\sqrt{\pi}}$$

$$= e^{-\omega^2/4\pi} \frac{2}{\sqrt{\pi}} \int_{0}^{\infty} e^{-u^2} du$$

Since $\int_0^\infty e^{-u^2} du = \sqrt{\pi}/2$, we have

$$X(w) = e^{-\omega^2/4\pi} \frac{2}{\sqrt{\pi}} \frac{\sqrt{\pi}}{2}$$

$$X(\omega) = e^{-\omega^2/4\pi}$$

$$e^{-\pi t^2} \longleftrightarrow e^{-\omega^2/4\pi} \tag{5.23}$$

or

$$e^{-\pi t^2} \longleftrightarrow e^{-\pi f^2} \tag{5.24}$$

The Fourier transform $X(\omega)$ of a Gaussian pulse is a Gaussian pulse [as shown in Fig. 5.10(b)].

Example 5.9 Find the Fourier transform of $x(t) = \cos(\omega_0 t)$.

Solution
We know that (Euler's formula)

$$x(t) = \cos(\omega_0 t) = \frac{e^{j\omega_0 t} + e^{-j\omega_0 t}}{2}$$

Taking the Fourier transform, we have

$$\mathcal{F}[x(t)] = X(\omega) = \mathcal{F}\left[\frac{e^{j\omega_0 t} + e^{-j\omega_0 t}}{2}\right] = \frac{1}{2}\left(\mathcal{F}[e^{j\omega_0 t}] + \mathcal{F}[e^{-j\omega_0 t}]\right)$$

Using Eqs (5.19) and (5.20), we have

$$X(\omega) = \frac{1}{2}[2\pi\delta(\omega - \omega_0) + 2\pi\delta(\omega + \omega_0)]$$

$$X(\omega) = \pi[\delta(\omega - \omega_0) + \delta(\omega + \omega_0)]$$

$$\cos(\omega_0 t) \longleftrightarrow \pi[\delta(\omega - \omega_0) + \delta(\omega + \omega_0)] \tag{5.25}$$

The Fourier transform of $\cos(\omega_0 t)$ consists of two impulses at ω_0 and $-\omega_0$, as shown in Fig. 5.11.

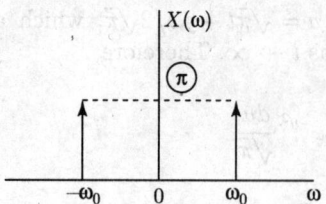

Fig. 5.11

Example 5.10 Find the Fourier transform of $x(t) = \sin(\omega_0 t)$.

Solution

We know that (Euler formula)

$$x(t) = \sin(\omega_0 t) = \frac{e^{j\omega_0 t} - e^{-j\omega_0 t}}{2j}$$

By taking the Fourier transform, we have

$$\mathcal{F}[x(t)] = X(\omega) = \mathcal{F}\left[\frac{e^{j\omega_0 t} - e^{-j\omega_0 t}}{2j}\right] = \frac{1}{2j}\left(\mathcal{F}[e^{j\omega_0 t}] - \mathcal{F}[e^{-j\omega_0 t}]\right)$$

Using Eqs (5.19) and (5.20), we have

$$X(\omega) = \frac{1}{2j}[2\pi\delta(\omega - \omega_0) - 2\pi\delta(\omega + \omega_0)]$$

$$X(\omega) = \frac{\pi}{j}[\delta(\omega - \omega_0) - \delta(\omega + \omega_0)]$$

$$\sin(\omega_0 t) \longleftrightarrow \frac{\pi}{j}[\delta(\omega - \omega_0) - \delta(\omega + \omega_0)] \tag{5.26}$$

Example 5.11 Find the Fourier transform of the signum function $x(t) = \text{sgn}(t)$.

$$\text{sgn}(t) = \begin{cases} 1, & t > 0 \\ 0, & t = 0 \\ -1, & t < 0 \end{cases}$$

Solution

The given signal $x(t) = \text{sgn}(t)$ can be rewritten as

$$\text{sgn}(t) = u(t) - u(-t)$$

This signal is not absolutely integrable. So we approach this problem by considering $\text{sgn}(t)$ to be a sum of exponentials $e^{-at}u(t) - e^{at}u(-t)$ in the limit as $a \to 0$ (Fig. 5.12).

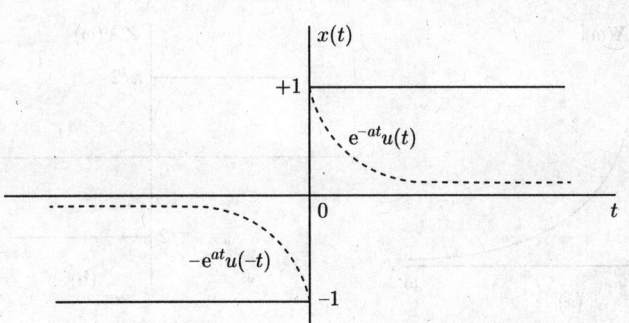

Fig. 5.12 Signum function

Thus,

$$\text{sgn}(t) = \lim_{a \to 0}[e^{-at}u(t) - e^{at}u(-t)]$$

and

$$X(\omega) = \mathcal{F}[\text{sgn}(t)] = \lim_{a \to 0} \mathcal{F}[e^{-at}u(t) - e^{at}u(-t)]$$

$$= \lim_{a \to 0} \left(\mathcal{F}[e^{-at}u(t)] - \mathcal{F}[e^{at}u(-t)] \right)$$

$$= \lim_{a \to 0} \left(\frac{1}{a + j\omega} - \frac{1}{a - j\omega} \right)$$

$$= \lim_{a \to 0} \left(\frac{-2j\omega}{a^2 + \omega^2} \right)$$

$$= \frac{-2j\omega}{\omega^2}$$

$$X(\omega) = \mathcal{F}[\text{sgn}(t)] = \frac{2}{j\omega}$$

$$\text{sgn}(t) \longleftrightarrow \frac{2}{j\omega} \qquad (5.27)$$

The magnitude spectrum is

$$|X(\omega)| = \left| \frac{2}{\omega} \right|$$

and the phase spectrum is

$$\angle X(\omega) = -\tan^{-1}\left(\frac{\omega}{0} \right) = \begin{cases} -\dfrac{\pi}{2}, & \omega > 0 \\ \dfrac{\pi}{2}, & \omega < 0 \end{cases}$$

Figure 5.13 shows the magnitude and phase spectrum.

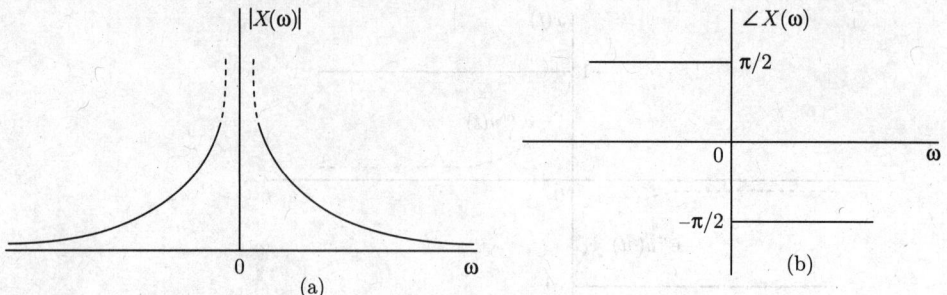

Fig. 5.13 (a) Magnitude spectrum and (b) phase spectrum of the signum function

Example 5.12 Find the Fourier transform of the unit step function $u(t)$.

Solution

Method 1: The unit step function can be expressed as

$$2u(t) = 1 + \text{sgn}(t)$$

$$u(t) = \frac{1}{2} + \frac{1}{2}\text{sgn}(t)$$

$$\mathcal{F}[u(t)] = U(\omega) = \mathcal{F}\left[\frac{1}{2} + \frac{1}{2}\text{sgn}(t)\right]$$

$$U(\omega) = \frac{1}{2}\mathcal{F}[1] + \frac{1}{2}\mathcal{F}[\text{sgn}(t)]$$

$$U(\omega) = \frac{1}{2}2\pi\delta(\omega) + \frac{1}{2}\frac{2}{j\omega}$$

$$U(\omega) = \pi\delta(\omega) + \frac{1}{j\omega}$$

$$u(t) \longleftrightarrow \pi\delta(\omega) + \frac{1}{j\omega} \tag{5.28}$$

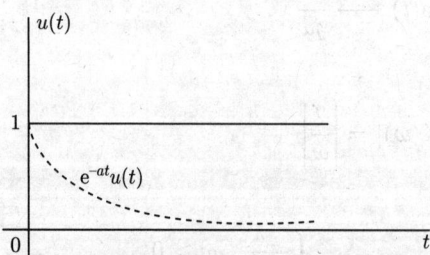

Fig. 5.14 Unit step function

Method 2: The unit step signal is not absolutely integrable. So we approach this problem by considering $u(t)$ to be a decaying exponential $e^{-at}u(t)$ in the limit as $a \to 0$

(Fig. 5.14). Thus,

$$u(t) = \lim_{a \to 0} e^{-at} u(t)$$

$$\mathcal{F}[u(t)] = U(\omega) = \mathcal{F}[\lim_{a \to 0} e^{-at} u(t)] = \lim_{a \to 0} \mathcal{F}[e^{-at} u(t)] = \lim_{a \to 0} \frac{1}{a + j\omega}$$

Expressing the RHS in terms of its real and imaginary parts yields

$$U(\omega) = \lim_{a \to 0} \left[\frac{a}{a^2 + \omega^2} - j \frac{\omega}{a^2 + \omega^2} \right] = \lim_{a \to 0} \left[\frac{a}{a^2 + \omega^2} \right] + \frac{1}{j\omega}$$

The function $a/a^2 + \omega^2$ has interesting properties. First, the area under this function is π regardless of the value of a:

$$\int_{-\infty}^{\infty} \frac{a}{a^2 + \omega^2} d\omega = \left[\tan^{-1} \frac{\omega}{a} \right]_{-\infty}^{\infty} = \frac{\pi}{2} + \frac{\pi}{2} = \pi$$

Second, when $a \to 0$, this function approaches zero for all $\omega \neq 0$, and all its area (π) is concentrated at a single point $\omega = 0$. Clearly, this function approaches as impulse of strength π, i.e.,

$$\lim_{a \to 0} \left[\frac{a}{a^2 + \omega^2} \right] = \pi \delta(\omega)$$

Thus,

$$U(\omega) = \pi \delta(\omega) + \frac{1}{j\omega}$$

5.4 PROPERTIES OF FOURIER TRANSFORM

Fourier transform possesses a number of important properties that are useful for developing conceptual insights into the transform and into the relationship between the time-domain and frequency-domain descriptions of a signal. In addition, they can also help to reduce the complexity of the evaluation of the Fourier transform of many signals.

We will use a shorthand notation

$$x(t) \longleftrightarrow X(\omega)$$

to indicate the relationship between a time-domain signal $x(t)$ and its Fourier transform $X(\omega)$.

5.4.1 Linearity

If

$$x_1(t) \longleftrightarrow X_1(\omega) \quad \text{and} \quad x_2(t) \longleftrightarrow X_2(\omega)$$

then

$$ax_1(t) + bx_2(t) \longleftrightarrow aX_1(\omega) + bX_2(\omega) \tag{5.29}$$

Proof The Fourier transform of $ax_1(t) + bx_2(t)$ is given by

$$\mathcal{F}[ax_1(t) + bx_2(t)] = \int_{-\infty}^{\infty} [ax_1(t) + bx_2(t)] e^{-j\omega t} \, dt$$

$$\mathcal{F}[ax_1(t) + bx_2(t)] = a \underbrace{\int_{-\infty}^{\infty} x_1(t) e^{-j\omega t} \, dt}_{X_1(\omega)} + b \underbrace{\int_{-\infty}^{\infty} x_2(t) e^{-j\omega t} \, dt}_{X_2(\omega)} = aX_1(\omega) + bX_2(\omega)$$

5.4.2 Time Shifting

If

$$x(t) \longleftrightarrow X(\omega)$$

then

$$x(t - t_0) \longleftrightarrow X(\omega) e^{-j\omega t_0} \tag{5.30}$$

Proof The Fourier transform of $x(t - t_0)$ is given by

$$\mathcal{F}[x(t - t_0)] = \int_{-\infty}^{\infty} x(t - t_0) e^{-j\omega t} \, dt$$

A change of variables is performed by letting $\tau = (t - t_0)$, which also yields $d\tau = dt$, $\tau \to -\infty$ as $t \to -\infty$, and $\tau \to \infty$ as $t \to \infty$. Therefore,

$$\mathcal{F}[x(t - t_0)] = \int_{-\infty}^{\infty} x(\tau) e^{-j\omega(\tau + t_0)} d\tau$$

$$= e^{-j\omega t_0} \int_{-\infty}^{\infty} x(\tau) e^{-j\omega \tau} d\tau$$

$$\mathcal{F}[x(t - t_0)] = X(\omega) e^{-j\omega t_0}$$

One consequence of this property is that when a signal is shifted in time, the *magnitudes* of its Fourier transform remain unaltered. That is, if we express $X(\omega)$ in polar form as

$$\mathcal{F}[x(t)] = X(\omega) = |X(\omega)| e^{j\angle X(\omega)}$$

then

$$\mathcal{F}[x(t - t_0)] = X(\omega) e^{-j\omega t_0} = |X(\omega)| e^{j(\angle X(\omega) - \omega t_0)}$$

Thus, the effect of a time shift on a signal is to introduce into its Fourier transform a phase shift, ωt_0, which is a linear function of ω.

5.4.3 Frequency Shifting

If

$$x(t) \longleftrightarrow X(\omega)$$

then

$$x(t)\, e^{j\omega_0 t} \longleftrightarrow X(\omega - \omega_0) \qquad (5.31)$$

Proof The Fourier transform of $x(t)\, e^{j\omega_0 t}$ is given by

$$\mathcal{F}[x(t)\, e^{j\omega_0 t}] = \int\limits_{-\infty}^{\infty} [x(t)\, e^{j\omega_0 t}]\, e^{-j\omega t}\, dt$$

$$= \int\limits_{-\infty}^{\infty} x(t)\, e^{-j(\omega - \omega_0)t}\, dt$$

$$\mathcal{F}[x(t)\, e^{j\omega_0 t}] = X(\omega - \omega_0)$$

Hence, a frequency shift corresponds to multiplication in time domain by a complex sinusoid whose frequency is equal to the time shift. ∎

5.4.4 Time and Frequency Scaling

If

$$x(t) \longleftrightarrow X(\omega)$$

then

$$x(at) \longleftrightarrow \frac{1}{|a|} X\left(\frac{\omega}{a}\right) \qquad (5.32)$$

Proof The Fourier transform of $x(at)$ is given by

$$\mathcal{F}[x(at)] = \int\limits_{-\infty}^{\infty} x(at)\, e^{-j\omega t}\, dt$$

Case I For a positive real constant a,

$$\mathcal{F}[x(at)] = \int\limits_{-\infty}^{\infty} x(at)\, e^{-j\omega t}\, dt$$

A change of variables is performed by letting $\tau = at$, which also yields $d\tau = a\, dt$, $\tau \to -\infty$ as $t \to -\infty$ and $\tau \to \infty$ as $t \to \infty$. Therefore,

$$\mathcal{F}[x(at)] = \frac{1}{a} \int\limits_{-\infty}^{\infty} x(\tau)\, e^{-j(\omega/a)\tau}\, d\tau = \frac{1}{a} X\left(\frac{\omega}{a}\right)$$

Case II For a negative real constant $-a$,

$$\mathcal{F}[x(-at)] = \int\limits_{-\infty}^{\infty} x(-at)\, e^{-j\omega t}\, dt$$

A change of variables is performed by letting $\tau = -at$, which also yields $d\tau = -a\,dt$, $\tau \to \infty$ as $t \to -\infty$ and $\tau \to -\infty$ as $t \to \infty$. Therefore,

$$\mathcal{F}[x(-at)] = -\frac{1}{a} \int_{\infty}^{-\infty} x(\tau)\, e^{j(\omega/a)\tau}\, d\tau$$

$$= \frac{1}{a} \int_{-\infty}^{\infty} x(\tau)\, e^{-j(-\omega/a)\tau}\, d\tau$$

$$= \frac{1}{a} X\left(-\frac{\omega}{a}\right)$$

Combining the two cases, we have

$$\mathcal{F}[x(at)] = \frac{1}{|a|} X\left(\frac{\omega}{a}\right)$$

The scaling property states that time compression of a signal results in its spectral expansion, and time expansion of the signal results in its spectral compression. The relation between compression in one domain and expansion in the other is the basis for an idea called the *uncertainty principle* of Fourier analysis.

5.4.5 Time Reversal

If

$$x(t) \longleftrightarrow X(\omega)$$

then

$$x(-t) \longleftrightarrow X(-\omega) \tag{5.33}$$

Proof Substituting $a = -1$ in Eq. (5.32), we have

$$\mathcal{F}[x(-t)] = \frac{1}{|-1|} X\left(\frac{\omega}{-1}\right)$$

$$\mathcal{F}[x(-t)] = X(-\omega)$$

The time-reversal property states that reversing a signal in time also reverses its Fourier transform. An interesting consequence of the time-reversal property is that if $x(t)$ is even, then its Fourier transform is also even, i.e.,

$$\text{if } x(-t) = x(t), \quad \text{then } X(-\omega) = X(\omega) \tag{5.34}$$

Similarly, if $x(t)$ is odd, then so is its Fourier transform, i.e.,

$$\text{if } x(-t) = -x(t), \quad \text{then } X(-\omega) = -X(\omega) \tag{5.35}$$

5.4.6 Differentiation in Time Domain

If

$$x(t) \longleftrightarrow X(\omega)$$

then

$$\frac{dx(t)}{dt} \longleftrightarrow j\omega X(\omega) \qquad (5.36)$$

Proof From Eq. (5.5), we have

$$x(t) = \mathcal{F}^{-1}[X(\omega)] = \frac{1}{2\pi} \int\limits_{-\infty}^{\infty} X(\omega)\, e^{j\omega t} d\omega$$

Differentiating both sides gives

$$\frac{dx(t)}{dt} = \frac{1}{2\pi} \int\limits_{-\infty}^{\infty} [j\omega X(\omega)]\, e^{j\omega t} d\omega$$

$$\frac{dx(t)}{dt} = \mathcal{F}^{-1}[j\omega X(\omega)]$$

Therefore,

$$\frac{dx(t)}{dt} \longleftrightarrow j\omega X(\omega)$$

Similarly, the differentiation property can be extended to yield

$$\frac{d^2 x(t)}{dt^2} \longleftrightarrow (j\omega)^2 X(\omega) \qquad (5.37)$$

∎

Example 5.13 Consider the triangular pulse signal depicted in Fig. 5.15(a) and defined as

$$x(t) = A \triangle \left(\frac{t}{T}\right) = \begin{cases} A\left(1 - \dfrac{|t|}{T}\right), & |t| \le T \\ 0, & |t| \ge T \end{cases} = \begin{cases} A\left(1 + \dfrac{t}{T}\right), & -T \le t < 0 \\ A\left(1 - \dfrac{t}{T}\right), & 0 \le t \le T \\ 0, & |t| \ge T \end{cases}$$

Find the Fourier transform of $x(t)$ using the differentiation property.

Solution

Using the ramp function $r(t)$, the given triangular pulse can be written as

$$x(t) = \frac{A}{T} r(t+T) - \frac{2A}{T} r(t) + \frac{A}{T} r(t-T)$$

The first derivative of the triangular pulse is [Fig. 5.15(b)]

$$\frac{dx(t)}{dt} = \frac{A}{T} u(t+T) - \frac{2A}{T} u(t) + \frac{A}{T} u(t-T)$$

and the second derivative is [Fig. 5.15(c)]

$$\frac{d^2 x(t)}{dt^2} = \frac{A}{T} \delta(t+T) - \frac{2A}{T} \delta(t) + \frac{A}{T} \delta(t-T)$$

280 *Signals and Systems*

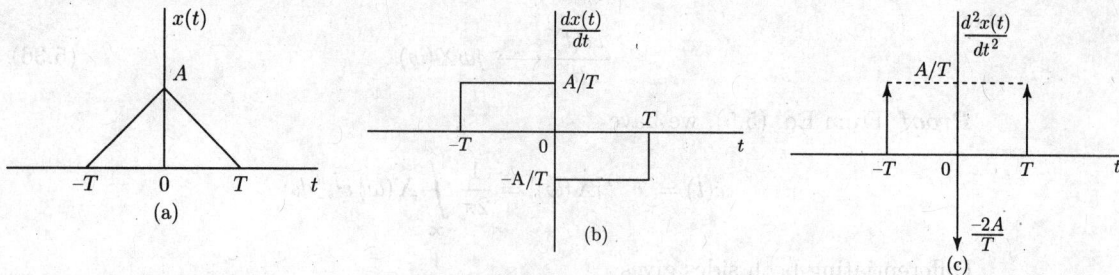

Fig. 5.15 (a) A triangular pulse, (b) its first derivative, and (c) its second derivative

The Fourier transform of the second derivative is

$$\mathcal{F}\left[\frac{d^2 x(t)}{dt^2}\right] = \mathcal{F}\left[\frac{A}{T}\delta(t+T) - \frac{2A}{T}\delta(t) + \frac{A}{T}\delta(t-T)\right]$$

$$= \frac{A}{T}\mathcal{F}[\delta(t+T)] - \frac{2A}{T}\mathcal{F}[\delta(t)] + \frac{A}{T}\mathcal{F}[\delta(t-T)]$$

Now, using the differentiation and the time-shifting property, we get

$$(j\omega)^2 X(\omega) = \frac{A}{T}e^{j\omega T} - \frac{2A}{T} + \frac{A}{T}e^{-j\omega T}$$

$$= \frac{2A}{T}\left(\frac{e^{j\omega T} + e^{-j\omega T}}{2}\right) - \frac{2A}{T}$$

$$(j\omega)^2 X(\omega) = \frac{2A}{T}[\cos(\omega T) - 1]$$

$$= \frac{2A}{(j\omega)^2 T}[\cos(\omega T) - 1] = \frac{2A}{\omega^2 T}[1 - \cos(\omega T)]$$

$$X(\omega) = \frac{4A}{\omega^2 T}\sin^2\left(\frac{\omega T}{2}\right) = AT\left(\frac{\sin(\omega T/2)}{(\omega T/2)}\right)^2$$

$$X(\omega) = AT\left(\frac{\sin(\pi\omega T/2\pi)}{(\pi\omega T/2\pi)}\right)^2 = AT\,\text{sinc}^2\left(\frac{\omega T}{2\pi}\right)$$

5.4.7 Convolution Property

If

$$x_1(t) \longleftrightarrow X_1(\omega) \quad\text{and}\quad x_2(t) \longleftrightarrow X_2(\omega)$$

then

$$x_1(t) * x_2(t) \longleftrightarrow X_1(\omega)X_2(\omega) \tag{5.38}$$

The Fourier transform maps the convolution of two signals into the product of their Fourier transforms.

Proof The Fourier transform of $x_1(t) * x_2(t)$ is given by

$$\mathcal{F}[x_1(t) * x_2(t)] = \int\limits_{-\infty}^{\infty} [x_1(t) * x_2(t)] \, e^{-j\omega t} \, dt$$

$$= \int\limits_{-\infty}^{\infty} \left(\int\limits_{-\infty}^{\infty} x_1(\tau) x_2(t-\tau) d\tau \right) e^{-j\omega t} \, dt$$

Interchanging the order of integration and noting that $x_1(\tau)$ does not depend on t gives

$$\mathcal{F}[x_1(t) * x_2(t)] = \int\limits_{-\infty}^{\infty} x_1(\tau) \left(\int\limits_{-\infty}^{\infty} x_2(t-\tau) \, e^{-j\omega t} \, dt \right) d\tau$$

By the time-shifting property, Eq. (5.30), the bracketed term is $X_2(\omega) \, e^{-j\omega\tau}$. Substituting this into the above equation yields

$$\mathcal{F}[x_1(t) * x_2(t)] = \int\limits_{-\infty}^{\infty} x_1(\tau) \left(X_2(\omega) \, e^{-j\omega\tau} \right) d\tau$$

$$= X_2(\omega) \int\limits_{-\infty}^{\infty} x_1(\tau) \, e^{-j\omega\tau} d\tau$$

$$\mathcal{F}[x_1(t) * x_2(t)] = X_2(\omega) X_1(\omega) = X_1(\omega) X_2(\omega)$$

Therefore,

$$x_1(t) * x_2(t) \longleftrightarrow X_1(\omega) X_2(\omega)$$

The convolution property states that convolution in the time domain corresponds to multiplication in the frequency domain. ∎

5.4.8 Multiplication (or Modulation) Property

If

$$x_1(t) \longleftrightarrow X_1(\omega) \quad \text{and} \quad x_2(t) \longleftrightarrow X_2(\omega)$$

then

$$x_1(t) x_2(t) \longleftrightarrow \frac{1}{2\pi} [X_1(\omega) * X_2(\omega)] \tag{5.39}$$

The Fourier transform maps the multiplication of two signals into the convolution of their Fourier transforms.

Proof The Fourier transform of $x_1(t) x_2(t)$ is

$$\mathcal{F}[x_1(t) x_2(t)] = \int\limits_{-\infty}^{\infty} [x_1(t) x_2(t)] \, e^{-j\omega t} \, dt$$

$$= \int\limits_{-\infty}^{\infty} \left(\frac{1}{2\pi} \int\limits_{-\infty}^{\infty} X_1(\theta) \, e^{j\theta t} \, d\theta \right) x_2(t) \, e^{-j\omega t} \, dt$$

Interchanging the order of integration and noting that $X_1(\theta)$ does not depend on t yields

$$\mathcal{F}[x_1(t)x_2(t)] = \frac{1}{2\pi} \int_{-\infty}^{\infty} X_1(\theta) \left(\int_{-\infty}^{\infty} [x_2(t)\, e^{j\theta t}]\, e^{-j\omega t}\, dt \right) d\theta$$

By the frequency-shifting property, Eq. (5.31), the bracketed term is $X_2(\omega - \theta)$. Substituting this into the above equation yields

$$\mathcal{F}[x_1(t)x_2(t)] = \frac{1}{2\pi} \int_{-\infty}^{\infty} X_1(\theta)X_2(\omega - \theta)\, d\theta$$

$$= \frac{1}{2\pi} [X_1(\omega) * X_2(\omega)]$$

Therefore,

$$x_1(t)x_2(t) \longleftrightarrow \frac{1}{2\pi} [X_1(\omega) * X_2(\omega)]$$

■

Example 5.14   Consider the pulse signal $g(t)$ shown in Fig. 5.16(a), which consists of a sinusoidal wave of amplitude A and frequency ω_c extending in duration from $t = -T/2$ to $t = T/2$. This signal is sometimes referred to as an *RF pulse* when the frequency ω_c falls in the radiofrequency band. Find the Fourier transform of the RF pulse $g(t)$.

Solution
The signal $g(t)$ of Fig. 5.16(a) may be expressed mathematically as the multiplication of two time-domain signals: rectangle signal of duration T and a sinusoidal signal of frequency ω_c:

$$g(t) = A \operatorname{rect}\left(\frac{t}{T}\right) \cos(\omega_c t)$$

$$= x(t) \cos(\omega_c t)$$

where $x(t) = A \operatorname{rect}(t/T)$. The Fourier transform of the rectangle signal $x(t)$ is given by [Eq. (5.21)]:

$$X(\omega) = AT \frac{\sin(\omega T/2)}{(\omega T/2)} = AT \operatorname{sinc}\left(\frac{\omega T}{2\pi}\right)$$

Now, consider the RF pulse $g(t)$:

$$g(t) = x(t) \cos(\omega_c t)$$

Applying the multiplication property gives

$$\mathcal{F}[g(t)] = \frac{1}{2\pi}\Big(\mathcal{F}[x(t)] * \mathcal{F}[\cos(\omega_c t)]\Big)$$

$$G(\omega) = \frac{1}{2\pi}\Big(X(\omega) * \pi[\delta(\omega - \omega_c) + \delta(\omega + \omega_c)]\Big)$$

$$G(\omega) = \frac{1}{2}[X(\omega - \omega_c) + X(\omega + \omega_c)]$$

Substituting $X(\omega) = AT \operatorname{sinc}(\omega T/2\pi)$ into the above equation yields

$$G(\omega) = \frac{1}{2}\Big[AT \operatorname{sinc}\Big(\frac{(\omega - \omega_c)T}{2\pi}\Big) + AT \operatorname{sinc}\Big(\frac{(\omega + \omega_c)T}{2\pi}\Big)\Big]$$

$$G(\omega) = \frac{AT}{2}\Big[\operatorname{sinc}\Big(\frac{(\omega - \omega_c)T}{2\pi}\Big) + \operatorname{sinc}\Big(\frac{(\omega + \omega_c)T}{2\pi}\Big)\Big]$$

The magnitude spectrum of the RF pulse is shown in Fig. 5.16(b).

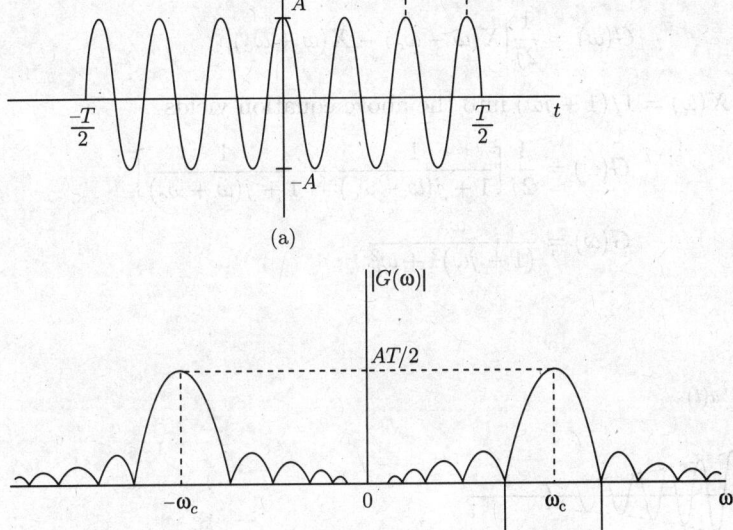

Fig. 5.16   (a) RF pulse and (b) its magnitude spectrum

Example 5.15 Consider an exponentially damped sinusoidal wave defined by (Fig. 5.17)

$$g(t) = e^{-t}\sin(\omega_c t)u(t)$$

Find the Fourier transform of $g(t)$.

Solution

The given exponentially damped sinusoidal wave is a multiplication of two time-domain signals: an exponentially decaying signal and a sinusoidal signal of frequency ω_c.

$$g(t) = e^{-t}\sin(\omega_c t)u(t) = e^{-t}u(t)\sin(\omega_c t) = x(t)\sin(\omega_c t)$$

where $x(t) = e^{-t}u(t)$. The Fourier transform of $x(t)$ is given by [substitute $a = 1$ into Eq. (5.13)]

$$X(\omega) = \frac{1}{1 + j\omega}$$

Now, consider the exponentially damped sinusoidal wave $g(t)$:

$$g(t) = x(t)\,\sin(\omega_c t)$$

Applying the multiplication property gives

$$\mathcal{F}[g(t)] = \frac{1}{2\pi}\Big(\mathcal{F}[x(t)] * \mathcal{F}[\sin(\omega_c t)]\Big)$$

$$G(\omega) = \frac{1}{2\pi}\Big(X(\omega) * \frac{\pi}{j}[\delta(\omega - \omega_c) - \delta(\omega + \omega_c)]\Big)$$

$$G(\omega) = \frac{1}{2j}[X(\omega - \omega_c) - X(\omega + \omega_c)]$$

Substituting $X(\omega) = 1/(1 + j\omega)$ into the above equation yields

$$G(\omega) = \frac{1}{2j}\Big[\frac{1}{1 + j(\omega - \omega_c)} - \frac{1}{1 + j(\omega + \omega_c)}\Big]$$

$$G(\omega) = \frac{\omega_c}{(1 + j\omega)^2 + \omega_c^2}$$

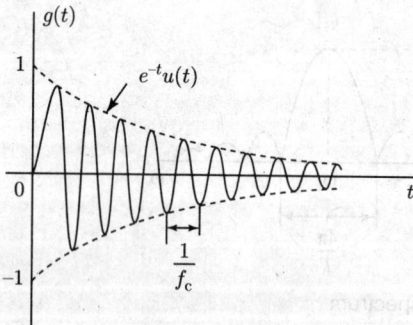

Fig. 5.17 Damped sinusoidal wave

Example 5.16 Consider an exponentially damped sinusoidal wave defined by

$$g(t) = e^{-at}\cos(\omega_c t)u(t)$$

Find the Fourier transform of $g(t)$.

Solution
The given exponentially damped sinusoidal wave is a multiplication of two time-domain signals: an exponentially decaying signal and a sinusoidal signal of frequency ω_c.

$$g(t) = e^{-at}\cos(\omega_c t)u(t) = e^{-at}u(t)\cos(\omega_c t) = x(t)\cos(\omega_c t)$$

where $x(t) = e^{-at}u(t)$. The Fourier transform of $x(t)$ is given by [Eq. (5.13)]

$$X(\omega) = \frac{1}{a + j\omega}$$

Now, consider the exponentially damped sinusoidal wave $g(t)$:

$$g(t) = x(t)\cos(\omega_c t)$$

Applying the multiplication property gives

$$\mathcal{F}[g(t)] = \frac{1}{2\pi}\Big(\mathcal{F}[x(t)] * \mathcal{F}[\cos(\omega_c t)]\Big)$$

$$G(\omega) = \frac{1}{2\pi}\Big(X(\omega) * \pi[\delta(\omega - \omega_c) + \delta(\omega + \omega_c)]\Big)$$

$$G(\omega) = \frac{1}{2}[X(\omega - \omega_c) + X(\omega + \omega_c)]$$

Substituting $X(\omega) = 1/(a + j\omega)$ into the above equation yields

$$G(\omega) = \frac{1}{2}\Big[\frac{1}{a + j(\omega - \omega_c)} + \frac{1}{a + j(\omega + \omega_c)}\Big]$$

$$G(\omega) = \frac{a + j\omega}{(a + j\omega)^2 + \omega_c^2}$$

Example 5.17 Find the Fourier transform of $x(t) = \cos(\omega_c t)u(t)$.

Solution
Consider the given signal

$$x(t) = \cos(\omega_c t)u(t)$$

Applying the multiplication property gives

$$\mathcal{F}[x(t)] = \frac{1}{2\pi}\Big(\mathcal{F}[\cos(\omega_c t)] * \mathcal{F}[u(t)]\Big)$$

$$X(\omega) = \frac{1}{2\pi}\Big(\pi[\delta(\omega - \omega_c) + \delta(\omega + \omega_c)] * \Big[\pi\delta(\omega) + \frac{1}{j\omega}\Big]\Big)$$

$$X(\omega) = \frac{1}{2\pi}\pi^2[\delta(\omega - \omega_c) + \delta(\omega + \omega_c)] + \frac{1}{2\pi}\pi\Big[\frac{1}{j(\omega - \omega_c)} + \frac{1}{j(\omega + \omega_c)}\Big]$$

$$X(\omega) = \frac{\pi}{2}[\delta(\omega - \omega_c) + \delta(\omega + \omega_c)] + \Big[\frac{j\omega}{\omega_c^2 - \omega^2}\Big]$$

5.4.9   Differentiation in Frequency Domain

If

$$x(t) \longleftrightarrow X(\omega)$$

then

$$-jtx(t) \longleftrightarrow \frac{dX(\omega)}{d\omega} \tag{5.40}$$

or

$$tx(t) \longleftrightarrow j\frac{dX(\omega)}{d\omega} \tag{5.41}$$

Proof From Eq. (5.6), we have

$$X(\omega) = \int\limits_{-\infty}^{\infty} x(t)\, e^{-j\omega t}\, dt$$

Differentiating both sides w.r.t. ω gives

$$\frac{dX(\omega)}{d\omega} = \int\limits_{-\infty}^{\infty} [-jtx(t)]\, e^{-j\omega t}\, dt$$

$$\frac{dX(\omega)}{d\omega} = \mathcal{F}[-jtx(t)]$$

Therefore,

$$-jtx(t) \longleftrightarrow \frac{dX(\omega)}{d\omega}$$

or

$$tx(t) \longleftrightarrow j\frac{dX(\omega)}{d\omega} \tag{5.42}$$

∎

Example 5.18 Find the Fourier transform of the signal $g(t)$ defined by

$$g(t) = t\,e^{-at}u(t)$$

Solution
Consider the given signal

$$g(t) = t\,e^{-at}u(t)$$

$$g(t) = t\,x(t)$$

From Eq. (5.42), we have

$$\mathcal{F}[g(t)] = \mathcal{F}[t\,x(t)] = j\frac{dX(\omega)}{d\omega}$$

Here, $x(t) = e^{-at}u(t)$ and its Fourier transform $X(\omega) = 1/(a + j\omega)$.

$$G(\omega) = \mathcal{F}[t\,e^{-at}u(t)] = j\frac{d}{d\omega}\left(\frac{1}{a + j\omega}\right)$$

$$G(\omega) = \frac{1}{(a + j\omega)^2}$$

Therefore,

$$t\,e^{-at}u(t) \longleftrightarrow \frac{1}{(a + j\omega)^2} \tag{5.43}$$

5.4.10 Integration

If

$$x(t) \longleftrightarrow X(\omega)$$

then

$$\int\limits_{-\infty}^{t} x(\tau)\,d\tau \longleftrightarrow \frac{1}{j\omega}X(\omega) + \pi X(0)\delta(\omega) \tag{5.44}$$

Proof The convolution of a signal $x(t)$ with a unit step function $u(t)$ is given by

$$x(t) * u(t) = \int\limits_{-\infty}^{\infty} x(\tau)u(t - \tau)\,d\tau$$

Since

$$u(t - \tau) = \begin{cases} 1, & (t - \tau) > 0 \longrightarrow \tau < t \\ 0, & (t - \tau) < 0 \longrightarrow \tau > t \end{cases}$$

we get

$$x(t) * u(t) = \int\limits_{-\infty}^{t} x(\tau)\,d\tau$$

The convolution of a signal with a unit step is the same as the cumulative integral of the signal. Now we can prove the integration property of the Fourier transform.

$$\int\limits_{-\infty}^{t} x(\tau)\,d\tau = x(t) * u(t)$$

$$\mathcal{F}\left[\int\limits_{-\infty}^{t} x(\tau)\,d\tau\right] = \mathcal{F}[x(t) * u(t)]$$

Using the convolution property [Eq. (5.38)] and Eq. (5.28), we get

$$\mathcal{F}\left[\int_{-\infty}^{t} x(\tau)\,d\tau\right] = X(\omega)U(\omega) = X(\omega)\left(\frac{1}{j\omega} + \pi\delta(\omega)\right)$$

$$= \frac{1}{j\omega}X(\omega) + \pi X(\omega)\delta(\omega)$$

$$= \frac{1}{j\omega}X(\omega) + \pi X(0)\delta(\omega)$$

Therefore,

$$\int_{-\infty}^{t} x(\tau)d\tau \longleftrightarrow \frac{1}{j\omega}X(\omega) + \pi X(0)\delta(\omega)$$

5.4.11   Duality

If

$$x(t) \longleftrightarrow X(\omega)$$

then

$$X(t) \longleftrightarrow 2\pi x(-\omega) \tag{5.45}$$

Proof By definition

$$x(t) = \frac{1}{2\pi}\int_{-\infty}^{\infty} X(\omega)\,e^{j\omega t}\,d\omega$$

$$2\pi x(t) = \int_{-\infty}^{\infty} X(\omega)\,e^{j\omega t}\,d\omega$$

replacing t with $-t$ in the above equation gives

$$2\pi x(-t) = \int_{-\infty}^{\infty} X(\omega)\,e^{-j\omega t}\,d\omega$$

Now interchanging the variables t and ω yields

$$2\pi x(-\omega) = \int_{-\infty}^{\infty} X(t)\,e^{-j\omega t}\,dt$$

$$2\pi x(-\omega) = \mathcal{F}[X(t)]$$

Therefore,

$$X(t) \longleftrightarrow 2\pi x(-\omega)$$

Example 5.19 Find the Fourier transform $G(\omega)$ of the signal

$$g(t) = \frac{1}{1 + jt}$$

Solution

Define $X(\omega) = 1/(1 + j\omega)$ by replacing t with ω in the expression of $g(t)$. From Eq. (5.13), we have

$$e^{-at}u(t) \longleftrightarrow \frac{1}{a + j\omega}$$

By substituting $a = 1$, we get

$$e^{-t}u(t) \longleftrightarrow \frac{1}{1 + j\omega}$$

$$x(t) \longleftrightarrow X(\omega)$$

$$x(t) = e^{-t}u(t) \quad \text{and} \quad X(\omega) = \frac{1}{1 + j\omega}$$

$$x(-\omega) = e^{\omega}u(-\omega) \quad \text{and} \quad X(t) = \frac{1}{1 + jt}$$

The duality property of the Fourier transform states that if

$$x(t) \longleftrightarrow X(\omega)$$

then

$$X(t) \longleftrightarrow 2\pi x(-\omega)$$

$$\frac{1}{1 + jt} \longleftrightarrow 2\pi e^{\omega}u(-\omega)$$

Therefore,

$$\mathcal{F}\left[\frac{1}{1 + jt}\right] = 2\pi\, e^{\omega}u(-\omega) \qquad (5.46)$$

Example 5.20 Find the Fourier transform $G(\omega)$ of the signal [Fig. 5.18(a)]

$$g(t) = \frac{1}{1 + t^2}$$

Solution

Define $X(\omega) = 1/(1 + \omega^2)$ by replacing t with ω in the expression of $g(t)$. From Eq. (5.14), we have

$$e^{-a|t|} \longleftrightarrow \frac{2a}{a^2 + \omega^2}$$

By substituting $a = 1$, we get

$$e^{-|t|} \longleftrightarrow \frac{2}{1 + \omega^2}$$

$$\frac{1}{2} e^{-|t|} \longleftrightarrow \frac{1}{1 + \omega^2}$$

$$x(t) \longleftrightarrow X(\omega)$$

$$x(t) = \frac{1}{2} e^{-|t|} \quad \text{and} \quad X(\omega) = \frac{1}{1 + \omega^2}$$

$$x(-\omega) = \frac{1}{2} e^{-|-\omega|} = \frac{1}{2} e^{-|\omega|} \quad \text{and} \quad X(t) = \frac{1}{1 + t^2}$$

The duality property of the Fourier transform states that if

$$x(t) \longleftrightarrow X(\omega)$$

then

$$X(t) \longleftrightarrow 2\pi x(-\omega)$$

$$\frac{1}{1 + t^2} \longleftrightarrow 2\pi \frac{1}{2} e^{-|\omega|}$$

$$g(t) = \frac{1}{1 + t^2} \longleftrightarrow \pi e^{-|\omega|} = G(\omega)$$

Therefore,

$$\mathcal{F}\left[\frac{1}{1 + t^2}\right] = \pi e^{-|\omega|} \tag{5.47}$$

The Fourier transform $G(\omega)$ is shown in Fig. 5.18(b).

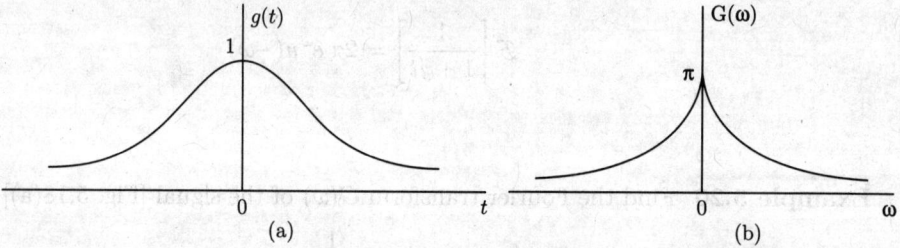

Fig. 5.18

Example 5.21 Find the Fourier transform $G(\omega)$ of the signal [Fig. 5.19(a)]

$$g(t) = \frac{1}{\pi t}$$

Solution

Define $X(\omega) = (1/\pi\omega)$ by replacing t with ω in the expression of $g(t)$. From Eq. (5.27), we have

$$\text{sgn}(t) \longleftrightarrow \frac{2}{j\omega}$$

$$\frac{j}{2\pi}\text{sgn}(t) \longleftrightarrow \frac{1}{\pi\omega}$$

$$x(t) \longleftrightarrow X(\omega)$$

$$x(t) = \frac{j}{2\pi}\text{sgn}(t) \quad \text{and} \quad X(\omega) = \frac{1}{\pi\omega}$$

$$x(-\omega) = \frac{j}{2\pi}\text{sgn}(-\omega) = -\frac{j}{2\pi}\text{sgn}(\omega) \quad \text{and} \quad X(t) = \frac{1}{\pi t}$$

The duality property of the Fourier transform states that if

$$x(t) \longleftrightarrow X(\omega)$$

then

$$X(t) \longleftrightarrow 2\pi x(-\omega)$$

$$\frac{1}{\pi t} \longleftrightarrow -2\pi \frac{j}{2\pi}\text{sgn}(\omega)$$

$$\frac{1}{\pi t} \longleftrightarrow -j\,\text{sgn}(\omega)$$

Therefore,

$$\mathcal{F}\left[\frac{1}{\pi t}\right] = -j\,\text{sgn}(\omega) = \begin{cases} -j, & \omega > 0 \\ j, & \omega < 0 \end{cases} \tag{5.48}$$

The Fourier transform $G(\omega)$ is shown in Fig. 5.19(b).

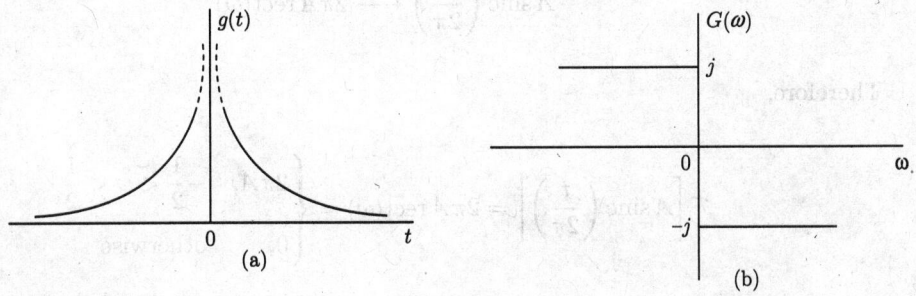

Fig. 5.19

Example 5.22 Find the Fourier transform $G(\omega)$ of the signal [Fig. 5.20(a)]

$$g(t) = A \operatorname{sinc}\left(\frac{t}{2\pi}\right)$$

Solution
Define $X(\omega) = A \operatorname{sinc}(\omega/2\pi)$ by replacing t with ω in the expression of $g(t)$. From Eq. (5.21), we have

$$A \operatorname{rect}\left(\frac{t}{2T_0}\right) \longleftrightarrow 2AT_0 \operatorname{sinc}\left(\frac{\omega T_0}{\pi}\right)$$

Substituting $2T_0 = 1$ yields

$$A \operatorname{rect}(t) \longleftrightarrow A \operatorname{sinc}\left(\frac{\omega}{2\pi}\right)$$

$$x(t) \longleftrightarrow X(\omega)$$

$$x(t) = A \operatorname{rect}(t) \quad \text{and} \quad X(\omega) = A \operatorname{sinc}\left(\frac{\omega}{2\pi}\right)$$

$$x(-\omega) = A \operatorname{rect}(-\omega) = A \operatorname{rect}(\omega) \quad \text{and} \quad X(t) = A \operatorname{sinc}\left(\frac{t}{2\pi}\right)$$

The duality property of the Fourier transform states that if

$$x(t) \longleftrightarrow X(\omega)$$

then

$$X(t) \longleftrightarrow 2\pi x(-\omega)$$

$$A \operatorname{sinc}\left(\frac{t}{2\pi}\right) \longleftrightarrow 2\pi A \operatorname{rect}(\omega)$$

Therefore,

$$\mathcal{F}\left[A \operatorname{sinc}\left(\frac{t}{2\pi}\right)\right] = 2\pi A \operatorname{rect}(\omega) = \begin{cases} 2\pi A, & -\dfrac{1}{2} < \omega < \dfrac{1}{2} \\ 0, & \text{otherwise} \end{cases} \tag{5.49}$$

The Fourier transform $G(\omega)$ is shown in Fig. 5.20(b).

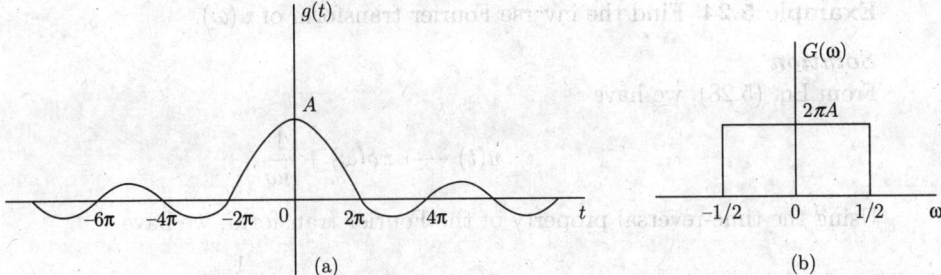

Fig. 5.20

Example 5.23 Find the Fourier transform $G(\omega)$ of the signal [Fig. 5.21(a)]

$$g(t) = 1, \quad \text{for all } t$$

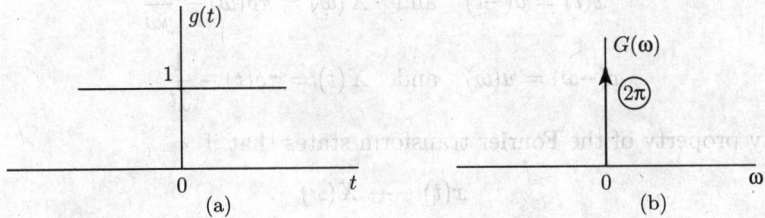

Fig. 5.21

Solution

Define $X(\omega) = 1$ for all ω, by replacing t with ω in the expression of $g(t)$. From Eq. (5.15), we have

$$\delta(t) \longleftrightarrow 1$$

$$x(t) \longleftrightarrow X(\omega)$$

$$x(t) = \delta(t) \quad \text{and} \quad X(\omega) = 1$$

$$x(-\omega) = \delta(-\omega) = \delta(\omega) \quad \text{and} \quad X(t) = 1$$

The duality property of the Fourier transform states that if

$$x(t) \longleftrightarrow X(\omega)$$

then

$$X(t) \longleftrightarrow 2\pi x(-\omega)$$

$$1 \longleftrightarrow 2\pi\delta(\omega)$$

Therefore,

$$\mathcal{F}[1] = 2\pi\delta(\omega) \tag{5.50}$$

The Fourier transform $G(\omega)$ is shown in Fig. 5.21(b).

Example 5.24 Find the inverse Fourier transform of $u(\omega)$.

Solution

From Eq. (5.28), we have

$$u(t) \longleftrightarrow \pi\delta(\omega) + \frac{1}{j\omega}$$

Using the time-reversal property of the Fourier transform, we have

$$u(-t) \longleftrightarrow \pi\delta(-\omega) - \frac{1}{j\omega}$$

$$u(-t) \longleftrightarrow \pi\delta(\omega) - \frac{1}{j\omega}$$

$$x(t) \longleftrightarrow X(\omega)$$

$$x(t) = u(-t) \quad \text{and} \quad X(\omega) = \pi\delta(\omega) - \frac{1}{j\omega}$$

$$x(-\omega) = u(\omega) \quad \text{and} \quad X(t) = \pi\delta(t) - \frac{1}{jt}$$

The duality property of the Fourier transform states that if

$$x(t) \longleftrightarrow X(\omega)$$

then

$$X(t) \longleftrightarrow 2\pi x(-\omega)$$

$$\pi\delta(t) - \frac{1}{jt} \longleftrightarrow 2\pi u(\omega)$$

$$\frac{1}{2}\delta(t) - \frac{1}{j2\pi t} \longleftrightarrow u(\omega)$$

Therefore,

$$\mathcal{F}\left[\frac{1}{2}\delta(t) - \frac{1}{j2\pi t}\right] = u(\omega)$$

or

$$\mathcal{F}^{-1}[u(\omega)] = \frac{1}{2}\delta(t) - \frac{1}{j2\pi t} \tag{5.51}$$

5.4.12 Conjugation and Conjugate Symmetry

If

$$x(t) \longleftrightarrow X(\omega)$$

then

$$x^*(t) \longleftrightarrow X^*(-\omega) \tag{5.52}$$

Proof The Fourier transform of $x^*(t)$ is given by

$$\mathcal{F}[x^*(t)] = \int_{-\infty}^{\infty} x^*(t)\, e^{-j\omega t}\, dt = \left[\int_{-\infty}^{\infty} x(t)\, e^{j\omega t}\, dt \right]^*$$

$$= \left[\int_{-\infty}^{\infty} x(t)\, e^{-j(-\omega)t}\, dt \right]^*$$

$$= [X(-\omega)]^*$$

$$\mathcal{F}[x^*(t)] = X^*(-\omega)$$

Case I If $x(t)$ is real, i.e., if

$$x^*(t) = x(t)$$

then

$$\mathcal{F}[x^*(t)] = \mathcal{F}[x(t)]$$

$$X^*(-\omega) = X(\omega)$$

$$X(-\omega) = X^*(\omega) \tag{5.53}$$

the Fourier transform will be *conjugate symmetric*. Also if $x(t)$ is real and even, then from Eqs (5.34) and (5.53), we have

$$X(-\omega) = X^*(\omega) = X(\omega) \tag{5.54}$$

That is, *if $x(t)$ is real and even, then so is its Fourier transform.*
Case II *If $x(t)$ is real and odd, then its Fourier transform is purely imaginary and odd.* Using Eqs (5.35) and (5.53), we get

$$X(-\omega) = X^*(\omega) = -X(\omega) \tag{5.55}$$

Case III *Even and odd decomposition of real signals*: If $x(t)$ is real and

$$x(t) \longleftrightarrow X(\omega)$$

then

1.

$$x_e(t) = \mathcal{E}\{x(t)\} \longleftrightarrow \mathrm{Re}\{X(\omega)\} \tag{5.56}$$

The even part of a signal $x(t)$ is defined as

$$x_e(t) = \frac{1}{2}[x(t) + x(-t)]$$

Taking the Fourier transform, we get

$$\mathcal{F}[x_e(t)] = \mathcal{F}\left[\frac{1}{2}[x(t) + x(-t)]\right] = \frac{1}{2}\Big(\mathcal{F}[x(t)] + \mathcal{F}[x(-t)]\Big) = \frac{1}{2}\Big(X(\omega) + X^*(\omega)\Big)$$

$$= \frac{1}{2}2\mathrm{Re}\{X(\omega)\}$$

$$\mathcal{F}[x_e(t)] = \mathrm{Re}\{X(\omega)\}$$

Therefore,

$$x_e(t) = \mathcal{E}\{x(t)\} \longleftrightarrow \mathrm{Re}\{X(\omega)\}$$

and

2.

$$x_o(t) = \mathcal{O}\{x(t)\} \longleftrightarrow j\mathrm{Im}\{X(\omega)\} \qquad (5.57)$$

The odd part of a signal $x(t)$ is defined as

$$x_o(t) = \frac{1}{2}[x(t) - x(-t)]$$

Taking the Fourier transform, we get

$$\mathcal{F}[x_o(t)] = \mathcal{F}\left[\frac{1}{2}[x(t) - x(-t)]\right] = \frac{1}{2}\Big(\mathcal{F}[x(t)] - \mathcal{F}[x(-t)]\Big)$$

$$= \frac{1}{2}\Big(X(\omega) - X^*(\omega)\Big)$$

$$= \frac{1}{2}2j\mathrm{Im}\{X(\omega)\}$$

$$\mathcal{F}[x_o(t)] = j\mathrm{Im}\{X(\omega)\}$$

Therefore,

$$x_0(t) = \mathcal{O}\{x(t)\} \longleftrightarrow j\mathrm{Im}\{X(\omega)\}$$

5.4.13   Area Under $x(t)$

If

$$x(t) \longleftrightarrow X(\omega)$$

then

$$\int_{-\infty}^{\infty} x(t)\, dt = X(0) \qquad (5.58)$$

That is, the area of a signal $x(t)$ is equal to the value of its Fourier transform $X(\omega)$ at $\omega = 0$.

Proof By definition

$$X(\omega) = \int\limits_{-\infty}^{\infty} x(t)\, e^{-j\omega t}\, dt$$

$$X(\omega)\Big|_{\omega=0} = X(0) = \int\limits_{-\infty}^{\infty} x(t)\, dt$$

■

5.4.14 Area Under $X(\omega)$

If

$$x(t) \longleftrightarrow X(\omega)$$

then

$$x(0) = \frac{1}{2\pi} \int\limits_{-\infty}^{\infty} X(\omega)\, d\omega \tag{5.59}$$

That is, the value of the function $x(t)$ at $t = 0$ is equal to the area under its Fourier transform $X(\omega)/2\pi$.

Proof From Eq. (5.5), we have

$$x(t) = \frac{1}{2\pi} \int\limits_{-\infty}^{\infty} X(\omega)\, e^{j\omega t}\, d\omega$$

$$x(t)\Big|_{t=0} = x(0) = \frac{1}{2\pi} \int\limits_{-\infty}^{\infty} X(\omega)\, d\omega$$

■

Example 5.25 Find the total area under the function

$$x(t) = 10\, \mathrm{sinc}\left(\frac{t+4}{7}\right)$$

Solution
From Eq. (5.49), we have ·

$$\mathcal{F}\left[A\, \mathrm{sinc}\left(\frac{t}{2\pi}\right)\right] = 2\pi A\, \mathrm{rect}(\omega)$$

$$A\, \mathrm{sinc}\left(\frac{t}{2\pi}\right) \longleftrightarrow 2\pi A\, \mathrm{rect}(\omega)$$

Using the time-scaling property (choose time-scaling constant $a = 2\pi$) Eq. (5.32), we get

$$A\,\text{sinc}(t) \longleftrightarrow A\,\text{rect}\left(\frac{\omega}{2\pi}\right)$$

Substitute $A = 10$ in the above equation:

$$10\,\text{sinc}(t) \longleftrightarrow 10\,\text{rect}\left(\frac{\omega}{2\pi}\right)$$

Using the time-shifting property, we get

$$10\,\text{sinc}\left(t + \frac{4}{7}\right) \longleftrightarrow 10\,\text{rect}\left(\frac{\omega}{2\pi}\right) e^{j\frac{4}{7}\omega}$$

Again using the time-scaling property $\left(a = \dfrac{1}{7}\right)$, we get

$$10\,\text{sinc}\left(\frac{1}{7}t + \frac{4}{7}\right) \longleftrightarrow 70\,\text{rect}\left(\frac{7\omega}{2\pi}\right) e^{j4\omega}$$

$$10\,\text{sinc}\left(\frac{t + 4}{7}\right) \longleftrightarrow 70\,\text{rect}\left(\frac{7\omega}{2\pi}\right) e^{j4\omega}$$

From Eq. (5.58), we have

$$\int\limits_{-\infty}^{\infty} 10\,\text{sinc}\left(\frac{t + 4}{7}\right) dt = \left[70\,\text{rect}\left(\frac{7\omega}{2\pi}\right) e^{j4\omega}\right]\Big|_{\omega=0} = 70$$

5.4.15 Parseval's Relation

Let $x(t)$ be an energy signal and if

$$x(t) \longleftrightarrow X(\omega)$$

then

$$E_x = \int\limits_{-\infty}^{\infty} |x(t)|^2\, dt = \frac{1}{2\pi} \int\limits_{-\infty}^{\infty} |X(\omega)|^2 d\omega \qquad (5.60)$$

Parseval's theorem states that the signal energies of an energy signal and its Fourier transform are equal.

Proof Consider the LHS of Eq. (5.60).

$$E_x = \int\limits_{-\infty}^{\infty} |x(t)|^2\, dt = \int\limits_{-\infty}^{\infty} x(t)x^*(t)\, dt$$

$$= \int\limits_{-\infty}^{\infty} x(t) \left(\frac{1}{2\pi} \int\limits_{-\infty}^{\infty} X(\omega)\, e^{j\omega t} d\omega \right)^* dt$$

$$= \int\limits_{-\infty}^{\infty} x(t) \left(\frac{1}{2\pi} \int\limits_{-\infty}^{\infty} X^*(\omega)\, e^{-j\omega t} d\omega \right) dt$$

$$= \frac{1}{2\pi} \int\limits_{-\infty}^{\infty} X^*(\omega) \left(\int\limits_{-\infty}^{\infty} x(t)\, e^{-j\omega t}\, dt \right) d\omega$$

$$= \frac{1}{2\pi} \int\limits_{-\infty}^{\infty} X^*(\omega)X(\omega)d\omega$$

$$\int\limits_{-\infty}^{\infty} |x(t)|^2\, dt = \frac{1}{2\pi} \int\limits_{-\infty}^{\infty} |X(\omega)|^2 d\omega$$

5.5 FOURIER TRANSFORM FOR PERIODIC SIGNALS

We can construct the Fourier transform of a periodic signal directly from its Fourier series representation. The resulting transform consists of a train of impulses in the frequency domain, with the areas of the impulses proportional to the Fourier series coefficients.

Consider the periodic signal $x(t)$ with period T and fundamental frequency $\omega_0 = (2\pi/T)$. The signal $x(t)$ has the Fourier series representation

$$x(t) = \sum_{n=-\infty}^{\infty} X_n\, e^{jn\omega_0 t}$$

Taking the Fourier transform on both sides yields

$$\mathcal{F}[x(t)] = \mathcal{F}\left[\sum_{n=-\infty}^{\infty} X_n\, e^{jn\omega_0 t} \right]$$

$$X(\omega) = \sum_{n=-\infty}^{\infty} X_n \mathcal{F}[e^{jn\omega_0 t}]$$

Using Eq. (5.19), we have $\mathcal{F}[e^{jn\omega_0 t}] = 2\pi\delta(\omega - n\omega_0)$, and therefore

$$X(\omega) = \sum_{n=-\infty}^{\infty} X_n[2\pi\delta(\omega - n\omega_0)]$$

$$= 2\pi \sum_{n=-\infty}^{\infty} X_n\delta(\omega - n\omega_0)$$

$$X(\omega) = 2\pi \sum_{n=-\infty}^{\infty} X_n\delta(\omega - n\omega_0) \tag{5.61}$$

Thus, the Fourier transform of a periodic signal is simply an impulse train with impulses located at $\omega = n\omega_0$, each of which has a strength $2\pi X_n$, and all impulses are separated from each other by ω_0. Note that because the signal $x(t)$ is periodic, the magnitude spectrum $|X(\omega)|$ is a train of impulses of strength $2\pi|X_n|$, whereas the spectrum obtained through the use of the Fourier series is a line spectrum with lines of finite magnitude $|X_n|$.

Example 5.26 Find and sketch the Fourier transform of the impulse train $x(t) = \delta_{T_0}(t)$ depicted in Fig. 5.22(a).

Solution
The impulse train $x(t)$ is periodic with period $T = T_0$ and fundamental frequency $\omega_0 = (2\pi/T_0)$.

$$x(t) = \sum_{n=-\infty}^{\infty} \delta(t - nT_0) \tag{5.62}$$

The impulse train for one period may be written as

$$x(t) = \delta(t), \quad -T_0/2 < t < T_0/2$$

The exponential Fourier series coefficients are

$$X_n = \frac{1}{T} \int_{-T/2}^{T/2} x(t)\, e^{-jn\omega_0 t}\, dt$$

$$X_n = \frac{1}{T_0} \int_{-T_0/2}^{T_0/2} x(t)\, e^{-jn\omega_0 t}\, dt$$

$$= \frac{1}{T_0} \int\limits_{-T_0/2}^{T_0/2} \delta(t)\, e^{-jn\omega_0 t}\, dt$$

$$X_n = \frac{1}{T_0}$$

From Eq. (5.61), the Fourier transform of a periodic signal $x(t)$ is

$$X(\omega) = 2\pi \sum_{n=-\infty}^{\infty} X_n \delta(\omega - n\omega_0)$$

Substituting $X_n = (1/T_0)$ into the above equation yields

$$X(\omega) = 2\pi \sum_{n=-\infty}^{\infty} \frac{1}{T_0} \delta(\omega - n\omega_0)$$

$$= \frac{2\pi}{T_0} \sum_{n=-\infty}^{\infty} \delta(\omega - n\omega_0)$$

$$X(\omega) = \omega_0 \sum_{n=-\infty}^{\infty} \delta(\omega - n\omega_0)$$

Thus, the Fourier transform of a periodic impulse in the time domain with period T_0 is a periodic impulse train in the frequency domain with period $\omega_0 = (2\pi/T)$, as sketched in Fig. 5.22(b).

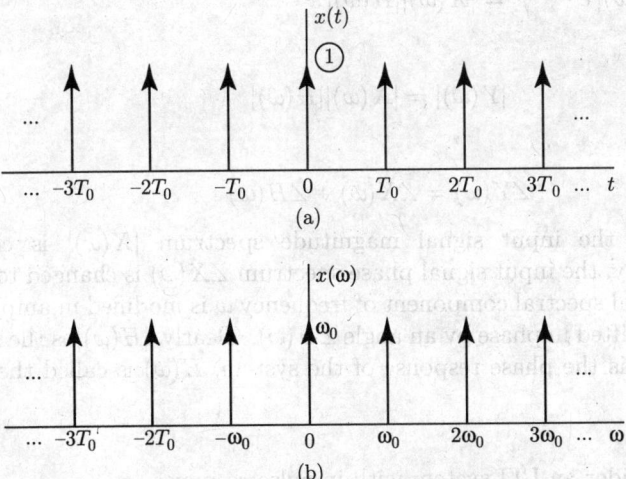

Fig. 5.22 (a) An impulse train and (b) its Fourier transform

5.6 SIGNAL TRANSMISSION THROUGH LTI SYSTEMS

If $x(t)$ and $y(t)$ are the input and output of an LTI system with impulse response $h(t)$, then

$$y(t) = x(t) * h(t)$$

Application of the time convolution property to the above equation yields [assuming that both $x(t)$ and $h(t)$ are Fourier transformable]

$$\mathcal{F}[y(t)] = \mathcal{F}[x(t)] \ \mathcal{F}[h(t)]$$

$$Y(\omega) = X(\omega)H(\omega) \tag{5.63}$$

or

$$H(\omega) = \frac{Y(\omega)}{X(\omega)} \tag{5.64}$$

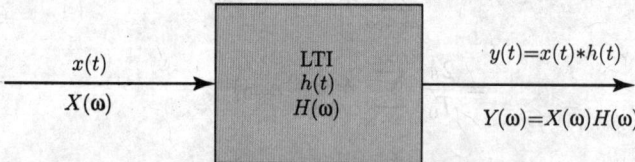

Fig. 5.23 Convolution property of LTI system response

The use of the convolution property for LTI systems is demonstrated in Fig. 5.23. Equation (5.63) can be expressed in polar form as

$$|Y(\omega)| \, e^{j\angle Y(\omega)} = \left(|X(\omega)| \, e^{j\angle X(\omega)} \right) \left(|H(\omega)| \, e^{j\angle H(\omega)} \right)$$

$$|Y(\omega)| \, e^{j\angle Y(\omega)} = |X(\omega)||H(\omega)| \, e^{j[\angle X(\omega) + \angle H(\omega)]}$$

Therefore,

$$|Y(\omega)| = |X(\omega)||H(\omega)| \tag{5.65}$$

and

$$\angle Y(\omega) = \angle X(\omega) + \angle H(\omega) \tag{5.66}$$

During transmission, the input signal magnitude spectrum $|X(\omega)|$ is changed to $|X(\omega)||H(\omega)|$. Similarly, the input signal phase spectrum $\angle X(\omega)$ is changed to $\angle X(\omega) + \angle H(\omega)$. An input signal spectral component of frequency ω is modified in amplitude by a factor $|H(\omega)|$ and is shifted in phase by an angle $\angle H(\omega)$. Clearly, $|H(\omega)|$ is the magnitude response, and $\angle H(\omega)$ is the phase response of the system. $H(\omega)$ is called the *frequency response* of the system.

Example 5.27 Consider an LTI system with impulse response

$$h(t) = e^{-at}u(t)$$

Find the response $y(t)$ of this system when the input is the unit step function, i.e., $x(t) = u(t)$.

Solution

From Eq. (5.63), we have

$$Y(\omega) = X(\omega)H(\omega) = \left[\pi\delta(\omega) + \frac{1}{j\omega}\right]\left(\frac{1}{a+j\omega}\right)$$

$$= \frac{\pi}{a}\delta(\omega) + \frac{1}{j\omega(a+j\omega)}$$

$$= \frac{\pi}{a}\delta(\omega) + \frac{1}{aj\omega} - \frac{1}{a(a+j\omega)}$$

$$Y(\omega) = \frac{1}{a}\left[\pi\delta(\omega) + \frac{1}{j\omega}\right] - \frac{1}{a}\frac{1}{a+j\omega}$$

Taking the inverse Fourier transform of both sides results in

$$y(t) = \frac{1}{a}u(t) - \frac{1}{a}e^{-at}u(t) = \frac{1}{a}[1 - e^{-at}]u(t)$$

Example 5.28 The output of an LTI system in response to an input $x(t) = e^{-2t}u(t)$ is $y(t) = e^{-t}u(t)$. Find the frequency response and impulse response of this system.

Solution

We take the Fourier transform of $x(t)$ and $y(t)$, obtaining

$$X(\omega) = \frac{1}{2+j\omega} \quad \text{and} \quad Y(\omega) = \frac{1}{1+j\omega}$$

Now, we use Eq. (5.64) to obtain the frequency response $H(\omega)$:

$$H(\omega) = \frac{Y(\omega)}{X(\omega)} = \frac{2+j\omega}{1+j\omega} = \frac{1+1+j\omega}{1+j\omega}$$

$$= \frac{1+j\omega}{1+j\omega} + \frac{1}{1+j\omega}$$

$$H(\omega) = 1 + \frac{1}{1+j\omega}$$

The inverse Fourier transform of the above equation yields the impulse response $h(t)$ of the system:

$$h(t) = \delta(t) + e^{-t}u(t)$$

Example 5.29 Consider a stable LTI system characterized by the differential equation

$$\frac{dy(t)}{dt} + ay(t) = x(t), \quad a > 0$$

Find the frequency response $H(\omega)$ and the impulse response $h(t)$ of the system.

Solution

Given that

$$\frac{dy(t)}{dt} + ay(t) = x(t), \quad a > 0$$

Taking the Fourier transform of both sides of the above equation yields

$$j\omega Y(\omega) + aY(\omega) = X(\omega)$$

$$Y(\omega)(a + j\omega) = X(\omega)$$

$$H(\omega) = \frac{Y(\omega)}{X(\omega)} = \frac{1}{a + j\omega}$$

Now take the inverse Fourier transform to obtain the impulse response of the system:

$$h(t) = e^{-at}u(t)$$

Example 5.30 Consider a stable LTI system characterized by the differential equation

$$\frac{d^2y(t)}{dt^2} + 4\frac{dy(t)}{dt} + 3y(t) = \frac{dx(t)}{dt} + 2x(t)$$

(a) Find the frequency response $H(\omega)$ and the impulse response $h(t)$ of the system.
(b) What is the response of this system if the input $x(t) = e^{-t}u(t)$?

Solution

(a) Given that

$$\frac{d^2y(t)}{dt^2} + 4\frac{dy(t)}{dt} + 3y(t) = \frac{dx(t)}{dt} + 2x(t)$$

Taking the Fourier transform on both sides yields

$$(j\omega)^2 Y(\omega) + 4j\omega Y(\omega) + 3Y(\omega) = j\omega X(\omega) + 2X(\omega)$$

$$Y(\omega)\left((j\omega)^2 + 4j\omega + 3\right) = X(\omega)(2 + j\omega)$$

$$H(\omega) = \frac{Y(\omega)}{X(\omega)} = \frac{2 + j\omega}{(j\omega)^2 + 4j\omega + 3} = \frac{2 + j\omega}{(1 + j\omega)(3 + j\omega)}$$

Using partial fraction expansion, we get

$$H(\omega) = \frac{1}{2}\frac{1}{1 + j\omega} + \frac{1}{2}\frac{1}{3 + j\omega}$$

The inverse Fourier transform of the above equation yields the impulse response $h(t)$ of the system:

$$\mathcal{F}^{-1}[H(\omega)] = h(t) = \frac{1}{2}e^{-t}u(t) + \frac{1}{2}e^{-3t}u(t)$$

(b) From Eq. (5.63), we have

$$Y(\omega) = X(\omega)H(\omega) = \left[\frac{1}{1+j\omega}\right]\left[\frac{2+j\omega}{(1+j\omega)(3+j\omega)}\right] = \frac{2+j\omega}{(1+j\omega)^2(3+j\omega)}$$

Using partial fraction expansion, we get

$$Y(\omega) = \frac{1}{4}\frac{1}{1+j\omega} + \frac{1}{2}\frac{1}{(1+j\omega)^2} - \frac{1}{4}\frac{1}{3+j\omega}$$

The inverse Fourier transform for each term in the above equation can be obtained by inspection. The first and third terms are of the same type that we have encountered in the previous part. To obtain the inverse Fourier transform of the second term, put $a = 1$ into the Eq. (5.43). Therefore, the response of the system is given by

$$\mathcal{F}^{-1}[Y(\omega)] = y(t) = \frac{1}{4}\mathcal{F}^{-1}\left[\frac{1}{1+j\omega}\right] + \frac{1}{2}\mathcal{F}^{-1}\left[\frac{1}{(1+j\omega)^2}\right] - \frac{1}{4}\mathcal{F}^{-1}\left[\frac{1}{3+j\omega}\right]$$

$$y(t) = \left[\frac{1}{4}e^{-t} + \frac{1}{2}te^{-t} - \frac{1}{4}e^{-3t}\right]u(t)$$

Example 5.31 Consider a causal LTI system with frequency response

$$H(\omega) = \frac{1}{3+j\omega}$$

For a particular input $x(t)$, this system is observed to produce the output

$$y(t) = e^{-3t}u(t) - e^{-4t}u(t)$$

Determine $x(t)$.

Solution

Consider the given output

$$y(t) = e^{-3t}u(t) - e^{-4t}u(t)$$

Taking the Fourier transform gives

$$Y(\omega) = \frac{1}{3+j\omega} - \frac{1}{4+j\omega} = \frac{1}{(3+j\omega)(4+j\omega)}$$

We know that the frequency response of an LTI system is given by

$$H(\omega) = \frac{Y(\omega)}{X(\omega)}$$

$$X(\omega) = \frac{Y(\omega)}{H(\omega)} = \frac{(3+j\omega)}{(3+j\omega)(4+j\omega)} = \frac{1}{4+j\omega}$$

Taking the inverse Fourier transform yields

$$x(t) = e^{-4t}u(t)$$

Example 5.32 A causal and stable LTI system S has the frequency response

$$H(\omega) = \frac{4 + j\omega}{6 - \omega^2 + 5j\omega}$$

(a) Determine a differential equation relating the input $x(t)$ and output $y(t)$ of S.
(b) Determine the impulse response $h(t)$ of S.
(c) What is the output of S when the input is

$$x(t) = e^{-4t}u(t) - t e^{-4t}u(t)$$

Solution
(a) Consider the given frequency response

$$H(\omega) = \frac{4 + j\omega}{6 - \omega^2 + 5j\omega}$$

$$H(\omega) = \frac{Y(\omega)}{X(\omega)} = \frac{4 + j\omega}{6 + (j\omega)^2 + 5j\omega}$$

$$(j\omega)^2 Y(\omega) + 5j\omega Y(\omega) + 6Y(\omega) = j\omega X(\omega) + 4X(\omega)$$

Taking the inverse Fourier transform to obtain the differential equation relating the input $x(t)$ and output $y(t)$ of S:

$$\frac{d^2y(t)}{dt^2} + 5\frac{dy(t)}{dt} + 6y(t) = \frac{dx(t)}{dt} + 4x(t)$$

(b) Given that

$$H(\omega) = \frac{4 + j\omega}{6 + (j\omega)^2 + 5j\omega} = \frac{4 + j\omega}{(2 + j\omega)(3 + j\omega)}$$

Using partial fraction expansion, we get

$$H(\omega) = \frac{2}{2 + j\omega} - \frac{1}{3 + j\omega}$$

The inverse Fourier transform yields the impulse response $h(t)$ of the system:

$$h(t) = 2 e^{-2t}u(t) - e^{-3t}u(t)$$

(c) Consider the given input

$$x(t) = e^{-4t}u(t) - t e^{-4t}u(t)$$

Taking the Fourier transform yields

$$X(\omega) = \frac{1}{4 + j\omega} - \frac{1}{(4 + j\omega)^2} = \frac{3 + j\omega}{(4 + j\omega)^2}$$

From Eq. (5.63), we have

$$Y(\omega) = X(\omega)H(\omega) = \frac{1}{(2+j\omega)(4+j\omega)} = \frac{1}{2}\frac{1}{2+j\omega} - \frac{1}{2}\frac{1}{4+j\omega}$$

Taking the inverse Fourier transform yields

$$y(t) = \frac{1}{2}\,e^{-2t}u(t) - \frac{1}{2}\,e^{-4t}u(t)$$

$$= \frac{1}{2}(e^{-2t} - e^{-4t})u(t)$$

5.6.1 Linear and Nonlinear Phase

Consider a continuous-time LTI system with impulse response $h(t)$ and frequency response $H(\omega)$. Let a signal $x(t)$ with Fourier transform $X(\omega)$ be applied to the input of the system. Let a signal $y(t)$ with Fourier transform $Y(\omega)$ denote the output of the system. In several applications, such as signal amplification or message signal transmission over a communication channel, we require that the output waveform be a replica of the input waveform, i.e., distortionless transmission. Transmission is said to be distortionless if the input and the output have identical waveshapes within a multiplicative constant and a constant time delay. Thus, in distortionless transmission, the input $x(t)$ and the output $y(t)$ satisfy the condition (Fig. 5.24):

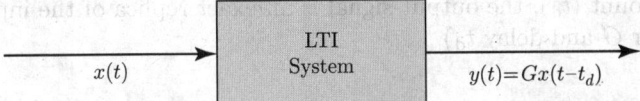

Fig. 5.24 Time-domain condition for distortionless transmission of a signal through an LTI system

$$y(t) = Gx(t - t_\mathrm{d}) \tag{5.67}$$

where the constant G accounts for a change in amplitude and the constant t_d accounts for a delay in transmission. Taking the Fourier transform of the above equation yields

$$Y(\omega) = GX(\omega)\,e^{-j\omega t_\mathrm{d}}$$

$$\frac{Y(\omega)}{X(\omega)} = H(\omega) = G\,e^{-j\omega t_\mathrm{d}}$$

This is the frequency response required of a system for distortionless transmission. From this equation it follows that

1. the magnitude response $|H(\omega)|$ must be a constant, i.e., we must have

$$|H(\omega)| = G \tag{5.68}$$

for some constant G.

2. the phase response $\angle H(\omega)$ must be a linear function of ω with slope $-t_\mathrm{d}$ and intercept zero, i.e., we must have

$$\angle H(\omega) = -\omega t_\mathrm{d} \tag{5.69}$$

These two conditions are illustrated in Figs 5.25(a) and (b), respectively.

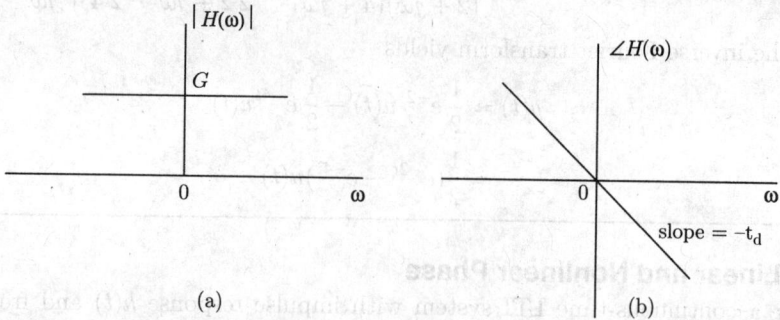

Fig. 5.25 Frequency response for distortionless transmission through an LTI system: (a) magnitude response and (b) phase response

The gain $|H(\omega)| = G$ means that every spectral component is multiplied by a constant G. A linear phase $\angle H(\omega) = -\omega t_d$ means that every spectral component is delayed by t_d seconds. For distortionless transmission, we require a linear phase characteristic. The phase is not only a linear function of ω, but it should also pass through the origin $\omega = 0$. Because each spectral component is attenuated by the same factor (G) and delayed by exactly the same amount (t_d), the output signal is an exact replica of the input (except for attenuating factor G and delay t_d).

5.6.2 Phase Delay and Group Delay

The two important parameters that characterize the form of the response $y(t)$ of an LTI system excited by an input signal $x(t)$ composed of a weighted sum of sinusoidal signals are phase delay and group delay. These two parameters are associated with the frequency response $H(\omega)$ of the system. For a linear phase system (whose phase varies linearly with frequency), both the phase delay and the group delay are constant.

Phase Delay

The time delay experienced by a single-frequency signal (i.e., a sinusoidal signal) when the signal passes through a system is referred to as the system phase delay. Now, let us assume that the input signal is the single-frequency signal

$$x(t) = A \cos(\omega_0 t + \theta)$$

$$x(t) = \frac{A}{2} e^{j(\omega_0 t + \theta)} + \frac{A}{2} e^{-j(\omega_0 t + \theta)}$$

and the system frequency response is

$$H(\omega) = |H(\omega)| e^{j \angle H(\omega)}$$

Then, the system output signal is [Eq. (3.64)]

$$y(t) = \frac{A}{2}|H(\omega_0)| e^{j(\omega_0 t + \theta + \angle H(\omega_0))} + \frac{A}{2}|H(-\omega_0)| e^{-j(\omega_0 t + \theta - \angle H(-\omega_0))}$$

Since $|H(\omega_0)| = |H(-\omega_0)|$ and $\angle H(\omega_0) = -\angle H(-\omega_0)$, we have

$$y(t) = \frac{A}{2}|H(\omega_0)| e^{j(\omega_0 t + \theta + \angle H(\omega_0))} + \frac{A}{2}|H(\omega_0)| e^{-j(\omega_0 t + \theta + \angle H(\omega_0))}$$

$$y(t) = A|H(\omega_0)| \cos(\omega_0 t + \angle H(\omega_0) + \theta)$$

$$= A|H(\omega_0)| \cos\left(\omega_0\left(t + \frac{\angle H(\omega_0)}{\omega_0}\right) + \theta\right)$$

$$y(t) = A|H(\omega_0)| \cos\left(\omega_0(t - \tau_\mathrm{p}(\omega_0)) + \theta\right)$$

where

$$\tau_\mathrm{p}(\omega_0) = -\frac{\angle H(\omega_0)}{\omega_0} \tag{5.70}$$

is the time delay experienced by the single-frequency signal with frequency ω_0 when it passes through the system. Thus, $\tau_\mathrm{p}(\omega_0)$ is the system phase delay for a signal with frequency ω_0. A negative system phase response at positive frequencies indicates that a signal is delayed in time when it passes through the system, whereas a positive system phase response at positive frequencies indicates that a signal is advanced in time when it passes through the system.

Group Delay

When the input signal contains many sinusoidal components with different frequencies that are not harmonically related, each component will go through different phase delays when processed by a frequency-selective LTI system, and the signal delay is determined using a different parameter called the group delay:

$$\tau_\mathrm{g}(\omega) = -\frac{d\angle H(\omega)}{d\omega} \tag{5.71}$$

We now derive Eq. (5.71) that defines the group delay by using a single-frequency modulating and carrier signals with zero phase for simplicity.

The assumed input signal (double-side band-suppressed carrier, i.e., DSB-SC modulated signal) is

$$s(t) = A\cos(\omega_\mathrm{m}t)\cos(\omega_\mathrm{c}t) \tag{5.72}$$

where ω_c is the carrier frequency and ω_m is the frequency of the modulating signal. This input signal is illustrated in Fig. 5.26.

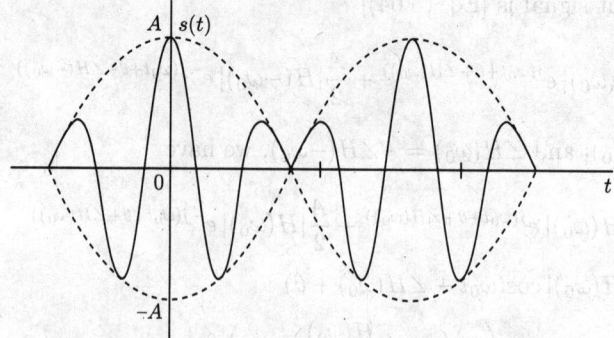

Fig. 5.26 Input signal

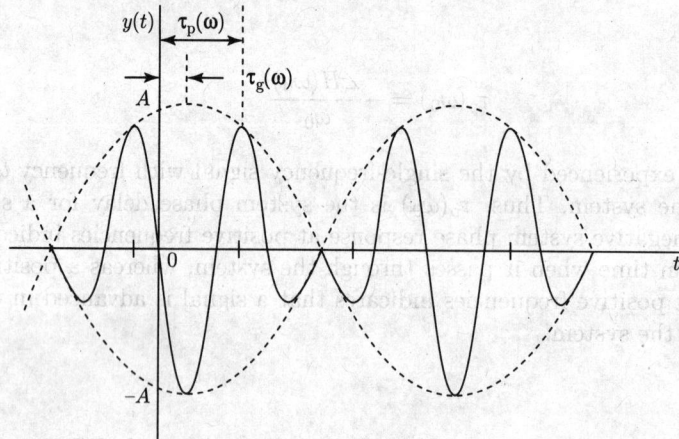

Fig. 5.27 Output signal

We use the cosine-product trigonometric identity to rewrite the input signal as

$$s(t) = \frac{A}{2}\cos[(\omega_c + \omega_m)t] + \frac{A}{2}\cos[(\omega_c - \omega_m)t]$$

$$s(t) = \frac{A}{2}\cos(\omega_1 t) + \frac{A}{2}\cos(\omega_2 t)$$

where

$$\omega_1 = \omega_c + \omega_m \tag{5.73}$$

and

$$\omega_2 = \omega_c - \omega_m \tag{5.74}$$

Now let the signal $s(t)$ passed through the system with frequency response $H(\omega)$. We assume that the system magnitude response and phase response are $|H(\omega)| = 1$ and $\angle H(\omega) = \phi(\omega)$, respectively. The system output signal is

$$y(t) = \frac{A}{2}|H(\omega_1)| \cos[\omega_1 t + \angle H(\omega_1)] + \frac{A}{2}|H(\omega_2)| \cos[\omega_2 t + \angle H(\omega_2)]$$

$$y(t) = \frac{A}{2} \cos[\omega_1 t + \phi(\omega_1)] + \frac{A}{2} \cos[\omega_2 t + \phi(\omega_2)]$$

where $\phi(\omega_1)$ and $\phi(\omega_2)$ are the phase shifts produced by the system at frequencies ω_1 and ω_2, respectively. Equivalently, we may express $y(t)$ as

$$y(t) = \frac{A}{2}\cos\left(\omega_c t + \omega_m t + \frac{\phi(\omega_1) + \phi(\omega_2)}{2} + \frac{\phi(\omega_1) - \phi(\omega_2)}{2}\right)$$

$$+ \frac{A}{2}\cos\left(\omega_c t - \omega_m t + \frac{\phi(\omega_1) + \phi(\omega_2)}{2} - \frac{\phi(\omega_1) - \phi(\omega_2)}{2}\right)$$

$$= \frac{A}{2}\cos\left(\left[\omega_c t + \frac{\phi(\omega_1) + \phi(\omega_2)}{2}\right] + \left[\omega_m t + \frac{\phi(\omega_1) - \phi(\omega_2)}{2}\right]\right)$$

$$+ \frac{A}{2}\cos\left(\left[\omega_c t + \frac{\phi(\omega_1) + \phi(\omega_2)}{2}\right] - \left[\omega_m t + \frac{\phi(\omega_1) - \phi(\omega_2)}{2}\right]\right)$$

$$y(t) = A\cos\left(\omega_c t + \frac{\phi(\omega_1) + \phi(\omega_2)}{2}\right)\cos\left(\omega_m t + \frac{\phi(\omega_1) - \phi(\omega_2)}{2}\right)$$

$$y(t) = A\cos\left(\omega_c\left[t + \frac{\phi(\omega_1) + \phi(\omega_2)}{2\omega_c}\right]\right)\cos\left(\omega_m\left[t + \frac{\phi(\omega_1) - \phi(\omega_2)}{2\omega_m}\right]\right)$$

$$y(t) = A\cos\left(\omega_c\left[t - \tau_p(\omega_c)\right]\right)\cos\left(\omega_m\left[t - \tau_g(\omega_c)\right]\right) \tag{5.75}$$

Comparing the output signal $y(t)$ in Eq. (5.75) with the input signal $s(t)$ in Eq. (5.72), we make the following two observations:

1. The carrier component at frequency ω_c in $y(t)$ lags its counterpart in $s(t)$ by $(\phi(\omega_1) + \phi(\omega_2)/2)$, which represents a time delay

$$\tau_p(\omega_c) = -\frac{\phi(\omega_1) + \phi(\omega_2)}{2\omega_c} = -\frac{\phi(\omega_1) + \phi(\omega_2)}{\omega_1 + \omega_2} \tag{5.76}$$

2. The modulating signal component at frequency ω_m in $y(t)$ lags its counterpart in $s(t)$ by $(\phi(\omega_1) - \phi(\omega_2)/2)$, which represents a time delay

$$\tau_g(\omega_c) = -\frac{\phi(\omega_1) - \phi(\omega_2)}{2\omega_m} = -\frac{\phi(\omega_1) - \phi(\omega_2)}{\omega_1 - \omega_2} \tag{5.77}$$

Suppose that the modulating signal frequency ω_m is small compared with the carrier frequency ω_c, which implies that the frequencies ω_1 and ω_2 are close together, with ω_c

between them. Such a modulated signal is known as narrowband signal. Then we may approximate the phase response $\phi(\omega)$ in the vicinity of $\omega = \omega_c$ by the two-term Taylor expansion

$$\phi(\omega) = \phi(\omega_c) + \frac{d\phi(\omega)}{d\omega}\bigg|_{\omega=\omega_c} (\omega - \omega_c) \tag{5.78}$$

Evaluating $\phi(\omega_1)$ and $\phi(\omega_2)$ using Eq. (5.78), we get

$$\phi(\omega_1) = \phi(\omega_c) + \frac{d\phi(\omega)}{d\omega}\bigg|_{\omega=\omega_c} (\omega_1 - \omega_c) \tag{5.79}$$

and

$$\phi(\omega_2) = \phi(\omega_c) + \frac{d\phi(\omega)}{d\omega}\bigg|_{\omega=\omega_c} (\omega_2 - \omega_c) \tag{5.80}$$

Now, from Eqs (5.73), (5.74), (5.76), (5.79), and (5.80), we get

$$\tau_p(\omega_c) = -\frac{\phi(\omega_c)}{\omega_c} \tag{5.81}$$

Similarly using the Eqs (5.73), (5.74), (5.77), (5.79), and (5.80), we get

$$\tau_g(\omega_c) = -\frac{d\phi(\omega)}{d\omega}\bigg|_{\omega=\omega_c} = -\frac{d\angle H(\omega)}{d\omega}\bigg|_{\omega=\omega_c} \tag{5.82}$$

The time delay $\tau_g(\omega)$ is called the group delay or envelope delay. Thus, the group delay at each frequency equals the negative of the slope of the phase at that frequency. If $\tau_g(\omega)$ is constant, all the components are delayed by the same interval. The group delay and phase delay are illustrated by the system output signal shown in Fig. 5.27.

Example 5.33   Find the impulse response of the system whose frequency response is given by

$$|H(\omega)| = \begin{cases} 1, & -\omega_c < \omega < \omega_c \\ 0, & \text{otherwise} \end{cases} \quad \text{and} \quad \angle H(\omega) = \begin{cases} \dfrac{\pi}{2}, & \omega > 0 \\ -\dfrac{\pi}{2}, & \omega < 0 \end{cases}$$

Solution
Using the magnitude response $|H(\omega)|$ and phase response $\angle H(\omega)$, the frequency response of the system is given by

$$H(\omega) = |H(\omega)| e^{j\angle H(\omega)}$$

Substituting the values of $|H(\omega)|$ and $\angle H(\omega)$ into the above equation yields

$$H(\omega) = \begin{cases} e^{j(\pi/2)}, & 0 < \omega < \omega_c \\ e^{-j(\pi/2)}, & -\omega_c < \omega < 0 \end{cases}$$

$$H(\omega) = \begin{cases} j, & 0 < \omega < \omega_c \\ -j, & -\omega_c < \omega < 0 \end{cases}$$

$$H(\omega) = j\left[\text{rect}\left(\frac{\omega}{\omega_c} - \frac{1}{2}\right) - \text{rect}\left(\frac{\omega}{\omega_c} + \frac{1}{2}\right)\right] \qquad (5.83)$$

We know that [Eq. (5.49)]

$$A\,\text{sinc}\left(\frac{t}{2\pi}\right) \longleftrightarrow 2\pi A\,\text{rect}(\omega)$$

$$\frac{\sin(t/2)}{(t/2)} \longleftrightarrow 2\pi\,\text{rect}(\omega)$$

$$\frac{\sin(t/2)}{\pi t} \longleftrightarrow \text{rect}(\omega)$$

Using frequency-shifting property, we have

$$\frac{\sin(t/2)}{\pi t}\,e^{j(1/2)t} \longleftrightarrow \text{rect}\left(\omega - \frac{1}{2}\right)$$

Now, using the time-scaling property, we have

$$\frac{\sin(\omega_c t/2)}{\pi\omega_c t}\,e^{j(\omega_c/2)t} \longleftrightarrow \frac{1}{\omega_c}\text{rect}\left(\frac{\omega}{\omega_c} - \frac{1}{2}\right)$$

$$\frac{\sin(\omega_c t/2)}{\pi t}\,e^{j(\omega_c/2)t} \longleftrightarrow \text{rect}\left(\frac{\omega}{\omega_c} - \frac{1}{2}\right)$$

Similarly, we have

$$\frac{\sin(\omega_c t/2)}{\pi t}\,e^{-j(\omega_c/2)t} \longleftrightarrow \text{rect}\left(\frac{\omega}{\omega_c} + \frac{1}{2}\right)$$

Now, taking the inverse Fourier transform of Eq. (5.83) and using the above results gives

$$h(t) = j\left[\frac{\sin(\omega_c t/2)}{\pi t}\,e^{j(\omega_c/2)t} - \frac{\sin(\omega_c t/2)}{\pi t}\,e^{-j(\omega_c/2)t}\right]$$

$$= 2j^2\frac{\sin(\omega_c t/2)}{\pi t}\left[\frac{e^{j(\omega_c/2)t} - e^{-j\omega_c/2t}}{2j}\right]$$

$$h(t) = -2\frac{\sin(\omega_c t/2)}{\pi t}\sin\left(\frac{\omega_c t}{2}\right) = -2\frac{\sin^2(\omega_c t/2)}{\pi t}$$

Example 5.34 Consider the following frequency response for a causal and stable LTI system:

$$H(\omega) = \frac{1 - j\omega}{1 + j\omega}$$

(a) Show that $|H(\omega)| = A$, and determine the value of A.

(b) Determine which of the following statements is true about $\tau_g(\omega)$, the group delay.

    (i) $\tau_g(\omega) = 0$, for $\omega > 0$

    (ii) $\tau_g(\omega) > 0$, for $\omega > 0$

    (iii) $\tau_g(\omega) < 0$, for $\omega > 0$

Solution

(a) Given that

$$H(\omega) = \frac{1 - j\omega}{1 + j\omega}$$

$$|H(\omega)| = \frac{\sqrt{1 + \omega^2}}{\sqrt{1 + \omega^2}} = 1$$

Therefore $A = 1$.

(b) The phase response of the system is

$$\angle H(\omega) = \tan^{-1}(-\omega) - \tan^{-1}(\omega)$$

$$= -2 \tan^{-1}(\omega)$$

Therefore, the group delay is

$$\tau_g(\omega) = -\frac{d\angle(H(\omega))}{d\omega} = \frac{d}{d\omega} 2 \tan^{-1}(\omega) = \frac{2}{1 + \omega^2}$$

Since $\tau_g(\omega) > 0$ for $\omega > 0$, statement (ii) is true.

Example 5.35 A causal LTI system has the frequency response $H(\omega)$ shown in Fig. 5.28. For each of the input signals given below, determine the filtered output signal $y(t)$.

(a) $x(t) = e^{jt}$

(b) $x(t) = \sin(\omega_0 t)u(t)$

(c) $X(\omega) = \dfrac{1}{j\omega(6 + j\omega)}$

(d) $X(\omega) = \dfrac{1}{2 + j\omega}$

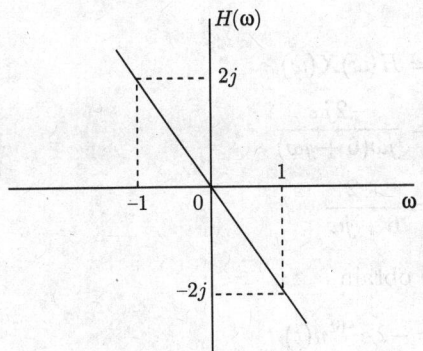

Fig. 5.28   Frequency response

Solution

From Fig. 5.28, the frequency response of the filter can be expressed as

$$H(\omega) = -2j\omega$$

We know that the filtered output signal is given by [Eq. (5.63)]

$$Y(\omega) = H(\omega)X(\omega)$$

$$Y(\omega) = -2j\omega X(\omega)$$

Taking the inverse Fourier transform gives

$$y(t) = -2\frac{dx(t)}{dt}$$

(a) Given that

$$x(t) = e^{jt}$$

Therefore,

$$y(t) = -2\frac{dx(t)}{dt} = -2\frac{d}{dt}e^{jt} = -2j\,e^{jt}$$

(b) Given that

$$x(t) = \sin(\omega_0 t)u(t)$$

Therefore,

$$y(t) = -2\frac{dx(t)}{dt} = -2\frac{d}{dt}\sin(\omega_0 t)u(t)$$

$$= -2[\omega_0 \cos(\omega_0 t)u(t) + \sin(\omega_0 t)\delta(t)]$$

$$= -2\omega_0 \cos(\omega_0 t)u(t)$$

(c) We know that

$$Y(\omega) = H(\omega)X(\omega)$$

$$= \frac{-2j\omega}{j\omega(6 + j\omega)}$$

$$Y(\omega) = \frac{-2}{6 + j\omega}$$

Taking the inverse Fourier transform, we obtain

$$y(t) = -2\,e^{-6t}u(t)$$

(d) Given that

$$X(\omega) = \frac{1}{2 + j\omega}$$

Taking the inverse Fourier transform, we obtain

$$x(t) = e^{-2t}u(t)$$

The filtered output signal is given by

$$y(t) = -2\frac{dx(t)}{dt} = -2\frac{d}{dt}\,e^{-2t}u(t)$$

$$= 4\,e^{-2t}u(t) - 2\,e^{-2t}\delta(t)$$

$$= 4\,e^{-2t}u(t) - 2\delta(t)$$

Example 5.36 Consider a continuous-time low-pass filter whose impulse response $h(t)$ is known to be real and whose frequency response magnitude is given as

$$|H(\omega)| = \begin{cases} 1, & |\omega| \leq 200\pi \\ 0, & \text{otherwise} \end{cases}$$

(a) Determine and sketch the real-valued impulse response $h(t)$ for this filter when the corresponding group delay function is specified as
(i) $\tau_g(\omega) = 5$ (ii) $\tau_g(\omega) = 5/2$ (iii) $\tau_g(\omega) = -5/2$
(b) If the impulse response $h(t)$ would had not been specified to be real, would knowledge of $|H(\omega)|$ and $\tau_g(\omega)$ be sufficient to determine $h(t)$ uniquely? Justify your answer.

Solution
We know that

$$\tau_g(\omega) = -\frac{d\angle H(\omega)}{d\omega}$$

If $\tau_g(\omega) = \alpha$, where α is a constant, then

$$\alpha = -\frac{d\angle H(\omega)}{d\omega}$$

$$\angle H(\omega) = -\alpha\omega + \beta$$

where β is another constant.

(a) We know that if $h(t)$ is real, then $\angle H(\omega)$ has to be an odd function. Therefore, the value of $\beta = 0$. Let us define $H_0(\omega) = |H(\omega)|$. Then

$$|H(\omega)| = H_0(\omega) = \begin{cases} 1, & |\omega| \le 200\pi \\ 0, & \text{otherwise} \end{cases}$$

Taking the inverse Fourier transform, we have

$$h_0(t) = \frac{1}{2\pi} \int_{-\infty}^{\infty} H_0(\omega)\, e^{j\omega t}\, d\omega = \frac{1}{2\pi} \int_{-200\pi}^{200\pi} e^{j\omega t}\, d\omega = \frac{\sin(200\pi t)}{\pi t}$$

(i) Given that $\tau_g(\omega) = \alpha = 5$. Hence, $\angle H(\omega) = -5\omega$. Then

$$H(\omega) = |H(\omega)|\, e^{j\angle H(\omega)}$$
$$= H_0(\omega)\, e^{-5j\omega}$$

Taking the inverse Fourier transform, we get

$$h(t) = h_0(t - 5) = \frac{\sin(200\pi(t - 5))}{\pi(t - 5)}$$

(ii) Given that $\tau_g(\omega) = \alpha = 5/2$. Hence, $\angle H(\omega) = -(5/2)\omega$. Then

$$H(\omega) = |H(\omega)|\, e^{j\angle H(\omega)}$$
$$= H_0(\omega)\, e^{-j(5/2)\omega}$$

Taking the inverse Fourier transform, we get

$$h(t) = h_0\left(t - \frac{5}{2}\right) = \frac{\sin(200\pi(t - 5/2))}{\pi(t - 5/2)}$$

(iii) Given that $\tau_g(\omega) = \alpha = 5/2$. Hence, $\angle H(\omega) = (5/2)\omega$. Then

$$H(\omega) = |H(\omega)|\, e^{j\angle H(\omega)}$$
$$= H_0(\omega)\, e^{j(5/2)\omega}$$

Taking the inverse Fourier transform, we get

$$h(t) = h_0\left(t + \frac{5}{2}\right) = \frac{\sin(200\pi(t + 5/2))}{\pi(t + 5/2)}$$

(b) If the impulse response $h(t)$ is not specified to be real, then $\angle H(\omega)$ does not have to be odd function. Therefore, $\beta \neq 0$. Knowledge of $|H(\omega)|$ and $\tau_g(\omega)$ would not be sufficient to determine β uniquely. Therefore, $h(t)$ cannot be determined uniquely.

5.7   IDEAL AND PRACTICAL FILTERS

Filtering is a process by which the essential and useful part of a signal is separated from the extraneous and undesirable components. The idea of filtering using LTI system is based on the convolution property of the Fourier transform, which states that for LTI systems, the Fourier transform of the output $Y(\omega)$ is the product of the Fourier transform of the input $X(\omega)$ and the frequency response $H(\omega)$, i.e., $Y(\omega) = X(\omega)H(\omega)$. In a sense, $H(\omega)$ acts as a weighting function or spectral-shaping function to the different frequency components in the input signal. Ideal filters allow distortionless transmission of a certain band of frequencies and completely suppress the remaining frequencies. The passband of a filter is the band of frequencies that are passed by the system, while the stopband refers to the range of frequencies that are attenuated by the system. In the ideal case, $|H(\omega)| = 1$ in a passband, while $|H(\omega)| = 0$ in a stopband. Filters are usually classified according to their frequency-domain characteristics as low-pass, high-pass, bandpass, bandstop or band elimination, and all-pass filters. The following lists definitions for each of these:

1. low-pass filters are those characterized by a passband that extends from $\omega = 0$ to $\omega = \omega_c$, where ω_c is called the cutoff frequency of the filter [Fig. 5.29(a)]. A low-pass filter attenuates high-frequency components of the input and passes the lower frequency components.
2. Highpass filters are characterized by a stopband that extends from $\omega = 0$ to $\omega = \omega_c$ and a passband that extends from $\omega = \omega_c$ to infinity [Fig. 5.29(b)]. A high-pass filter attenuates low frequencies and passes the high frequencies.

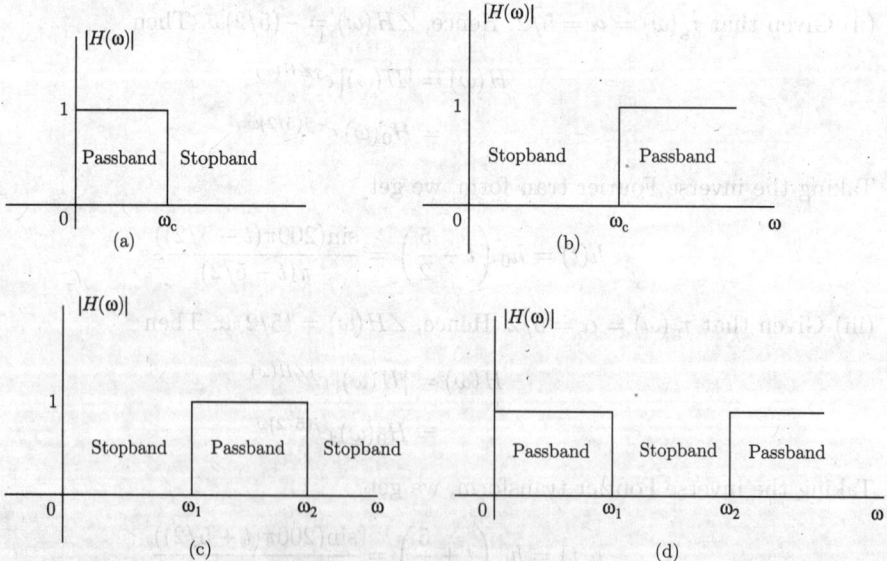

Fig. 5.29   Most common classes of filters: (a) LPF, (b) HPF, (c) BPF, and (d) BSF

3. Bandpass filters are characterized by a passband that extends from $\omega = \omega_1$ to $\omega = \omega_2$ and all other frequencies are stopped [Fig. 5.29(c)]. A bandpass filter passes signals within a certain frequency band and attenuates signals outside that band.

4. Bandstop filters stop frequencies extending from $\omega = \omega_1$ to $\omega = \omega_2$ and pass all other frequencies [Fig. 5.29(d)]. A bandstop filter attenuates signals within a certain frequency band and passes signals outside that band.

5. An allpass filter is characterized by a magnitude response that is constant for all frequencies. The characteristics of an allpass filter are completely determined by its phase-shift characteristics. An allpass filter has nonconstant group delay and different frequencies in the input are delayed by different amounts.

The filters described so far are referred to as ideal filters because they pass one set of frequencies without any change and completely stop others. Since it is impossible to realize filters with characteristic like those shown in Fig. 5.29, with abrupt changes from passband to stopband and vice versa, most of the filters we deal with in practice have some transition band. Ideal filters are noncausal filters and therefore physically unrealizable. For example, consider an ideal low-pass filter whose magnitude response and phase response are (as shown in Fig. 5.30) given by

$$|H(\omega)| = \text{rect}\,(\omega/2\omega_c) = \begin{cases} 1, & -\omega_c < \omega < \omega_c \\ 0, & \text{otherwise} \end{cases} \quad \text{and} \quad \angle H(\omega) = -\omega t_d$$

Therefore,

$$H(\omega) = |H(\omega)|\,e^{j\angle H(\omega)} = \text{rect}\left(\frac{\omega}{2\omega_c}\right)e^{-j\omega t_d}$$

Let us define $H_0(\omega) = |H(\omega)|$. Then

$$|H(\omega)| = H_0(\omega) = \begin{cases} 1, & -\omega_c < \omega < \omega_c \\ 0, & \text{otherwise} \end{cases}$$

Taking the inverse Fourier transform, yields

$$h_0(t) = \frac{1}{2\pi}\int\limits_{-\infty}^{\infty} H_0(\omega)\,e^{j\omega t}d\omega = \frac{1}{2\pi}\int\limits_{-\omega_c}^{\omega_c} e^{j\omega t}d\omega = \frac{1}{2\pi}\frac{e^{j\omega_c t}}{jt}\Bigg|_{-\omega_c}^{\omega_c} = \frac{\sin(\omega_c t)}{\pi t}$$

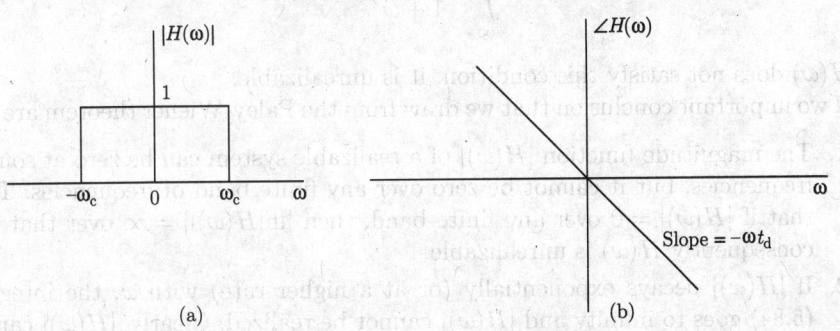

(a)   (b)

Fig. 5.30   (a) Magnitude response and (b) phase response

Now, consider

$$H(\omega) = |H(\omega)| \, e^{j\angle H(\omega)} = H_0(\omega) \, e^{-j\omega t_{\mathrm{d}}}$$

Taking the inverse Fourier transform gives

$$h(t) = h_0(t - t_{\mathrm{d}})$$

$$h(t) = \frac{\sin(\omega_{\mathrm{c}}(t - t_{\mathrm{d}}))}{\pi(t - t_{\mathrm{d}})}$$

Recall that $h(t)$ is the system response to impulse input $\delta(t)$, which is applied at $t = 0$. Figure 5.31 depicts a curious fact: the response $h(t)$ begins even before the input is applied (at $t = 0$). Clearly, the filter is noncausal and therefore physically unrealizable. Similarly, one can show that other ideal filters are also physically unrealizable.

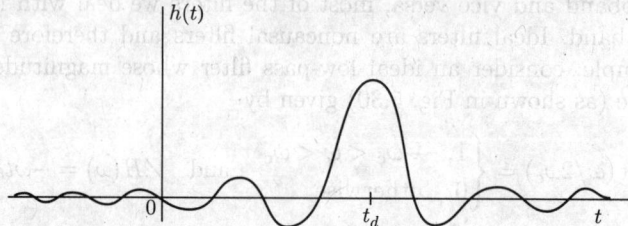

Fig. 5.31   Impulse response of an ideal LPF

5.7.1 Paley-Wiener Criterion

For a physically realizable system, the impulse response $h(t)$ must be causal, i.e.,

$$h(t) = 0, \quad t < 0$$

In the frequency domain, this condition is equivalent to the well-known Paley-Wiener criterion, which states that the necessary and sufficient condition for the magnitude response $|H(\omega)|$ to be realizable is

$$\int\limits_{-\infty}^{\infty} \frac{|\ln|H(\omega)||}{1 + \omega^2} \, d\omega < \infty \tag{5.84}$$

If $H(\omega)$ does not satisfy this condition, it is unrealizable.

Two important conclusion that we draw from the Paley-Wiener theorem are as follows:

1. The magnitude function $|H(\omega)|$ of a realizable system can be zero at some discrete frequencies, but it cannot be zero over any finite band of frequencies. This means that if $|H(\omega)| = 0$ over any finite band, then $|\ln|H(\omega)|| = \infty$ over that band, and consequently $H(\omega)$ is unrealizable.

2. If $|H(\omega)|$ decays exponentially (or at a higher rate) with ω, the integral in Eq. (5.84) goes to infinity and $|H(\omega)|$ cannot be realized. Clearly, $|H(\omega)|$ cannot decay too fast with ω.

Apparently, causality imposes some tight constraints on a LTI system. In addition to the Paley-Wiener condition, causality also implies a strong relationship between $H_R(\omega) = \text{Re}\{H(\omega)\}$ and $H_I(\omega) = \text{Im}\{H(\omega)\}$, the real and imaginary components of the frequency response $H(\omega)$. To illustrate this dependence, we decompose $h(t)$ into its even and odd parts, i.e.,

$$h(t) = h_e(t) + h_o(t)$$

where

$$h_e(t) = \frac{1}{2}[h(t) + h(-t)]$$

and

$$h_o(t) = \frac{1}{2}[h(t) - h(-t)]$$

Now, if $h(t)$ is causal, it is possible to recover $h(t)$ from its even part $h_e(t)$ or from its odd part $h_o(t)$. Indeed, it can be easily seen that

$$h(t) = 2h_e(t)u(t) = 2h_o(t)u(t) \tag{5.85}$$

$$h_e(t) = h_o(t)\text{sgn}(t) \tag{5.86}$$

and

$$h_o(t) = h_e(t)\text{sgn}(t) \tag{5.87}$$

Since $h_o(t) = h_e(t)$ for $t > 0$, there is a strong relationship between $h_o(t)$ and $h_e(t)$.

If $h(t)$ is absolutely integrable (i.e., BIBO stable), the frequency response $H(\omega)$ exists, and

$$H(\omega) = H_R(\omega) + jH_I(\omega)$$

In addition if $h(t)$ is real valued and causal, the conjugate property of the Fourier transform implies that

$$h_e(t) \longleftrightarrow H_R(\omega) \quad \text{and} \quad h_o(t) \longleftrightarrow jH_I(\omega)$$

Since $h(t)$ is completely specified by $h_e(t)$, it follows that $H(\omega)$ is completely determined if we know $H_R(\omega)$. Alternatively, $H(\omega)$ is completely determined from $H_I(\omega)$. In short, $H_R(\omega)$ and $H_I(\omega)$ are interdependent and cannot be specified independently if the system is causal. Equivalently, the magnitude and phase responses of a causal filter are interdependent and hence cannot be specified independently.

Taking the Fourier transform of Eq. (5.86) gives

$$\mathcal{F}[h_e(t)] = \frac{1}{2\pi}\Big[\mathcal{F}[h_o(t)] * \mathcal{F}[\text{sgn}(t)]\Big]$$

$$H_R(\omega) = \frac{1}{2\pi}\Big[jH_I(\omega) * \frac{2}{j\omega}\Big]$$

$$H_R(\omega) = \frac{1}{2\pi} \int_{-\infty}^{\infty} jH_I(\theta) \frac{2}{j(\omega - \theta)} d\theta$$

$$H_R(\omega) = \frac{1}{\pi} \int_{-\infty}^{\infty} \frac{H_I(\theta)}{\omega - \theta} d\theta \tag{5.88}$$

Similarly, taking the Fourier transform of Eq. (5.87) gives

$$\mathcal{F}[h_o(t)] = \frac{1}{2\pi} \left[\mathcal{F}[h_e(t)] * \mathcal{F}[\text{sgn}(t)] \right]$$

$$jH_I(\omega) = \frac{1}{2\pi} \left[H_R(\omega) * \frac{2}{j\omega} \right]$$

$$jH_I(\omega) = \frac{1}{2\pi} \int_{-\infty}^{\infty} H_R(\theta) \frac{2}{j(\omega - \theta)} d\theta$$

$$H_I(\omega) = -\frac{1}{\pi} \int_{-\infty}^{\infty} \frac{H_R(\theta)}{\omega - \theta} d\theta \tag{5.89}$$

Equations (5.88) and (5.89) are known as Hilbert transform pair.

To summarize, causality has very important implications in the design of frequency-selective filters. These are as follows:

1. The frequency response $H(\omega)$ cannot be zero except at a finite set of points in frequency.

2. The magnitude $|H(\omega)|$ cannot be constant in any finite range of frequencies and the transition from passband to stopband cannot be infinitely sharp.

3. The real and imaginary parts of $H(\omega)$ are interdependent and are related by the Hilbert transform. As a consequence, the magnitude $|H(\omega)|$ and phase $\angle H(\omega)$ cannot be chosen arbitrarily.

Example 5.37 Using Paley-Wiener criterion, prove that $|H(\omega)| = e^{-\omega^2}$ is not a suitable magnitude response for a causal LTI system.

Solution
According to the Paley-Wiener criterion, the condition for the magnitude response $|H(\omega)|$ to be realizable is

$$\int_{-\infty}^{\infty} \frac{|\ln |H(\omega)||}{1 + \omega^2} d\omega < \infty$$

where

$$|H(\omega)| = e^{-\omega^2}$$

Therefore, we have

$$
\int\limits_{-\infty}^{\infty} \frac{|\ln|e^{-\omega^2}||}{1+\omega^2}\,d\omega = \int\limits_{-\infty}^{\infty} \frac{\omega^2}{1+\omega^2}\,d\omega
$$

$$
= 2\int\limits_{0}^{\infty} \frac{\omega^2}{1+\omega^2}\,d\omega
$$

$$
= 2\int\limits_{0}^{\infty} \left(1 - \frac{1}{1+\omega^2}\right)d\omega
$$

$$
= 2[\omega - \tan^{-1}\omega]\Big|_{0}^{\infty}
$$

$$
= \infty
$$

Since the Paley-Wiener criterion is not satisfied, the given magnitude function is not realizable.

5.8 ENERGY SPECTRAL DENSITY (ESD)

Parseval's theorem relates the total signal energy in a signal $x(t)$ to its Fourier transform through

$$
E_x = \int\limits_{-\infty}^{\infty} |x(t)|^2\,dt = \frac{1}{2\pi}\int\limits_{-\infty}^{\infty} |X(\omega)|^2\,d\omega = \int\limits_{-\infty}^{\infty} |X(\omega)|^2 df
$$

Parseval's theorem states that the total energy E_x may be determined either by computing the energy per unit time ($|x(t)|^2$) and integrating over all time or by computing the energy per unit frequency $|X(\omega)|^2$ and integrating over all frequencies. For this reason, $|X(\omega)|^2$ represents energy per unit bandwidth and is often referred to as the *energy spectral density* (per unit bandwidth in hertz) or *energy density spectrum* of the signal $x(t)$ and is denoted by $\Psi_x(\omega)$. Hence,

$$
\Psi_x(\omega) = |X(\omega)|^2 \tag{5.90}
$$

The units of ESD depend on the units of the underlying signal $x(t)$. For example, if the signal unit is volts (V), its Fourier transform has units of V/Hz and its ESD has units of $(V/Hz)^2$ or $(V\,s)^2$.

5.8.1 Relationship Between Input and Output Energy Spectral Densities of an LTI System

Consider an LTI system with frequency response $H(\omega)$, input $x(t)$, and output $y(t)$. If $x(t)$ and $y(t)$ are energy signals, then their energy spectral densities are $\Psi_x(\omega) = |X(\omega)|^2$ and $\Psi_y(\omega) = |Y(\omega)|^2$, respectively. Since, we know that

$$
Y(\omega) = H(\omega)X(\omega)
$$

it follows that

$$|Y(\omega)|^2 = |H(\omega)X(\omega)|^2$$

$$|Y(\omega)|^2 = |H(\omega)|^2|X(\omega)|^2$$

$$\Psi_y(\omega) = |H(\omega)|^2\Psi_x(\omega) \tag{5.91}$$

5.8.2 Relation of ESD to Autocorrelation

The autocorrelation function $R_{xx}(\tau)$ of an energy signal is defined as [Eq. (2.44)]:

$$R_{xx}(\tau) = \int\limits_{-\infty}^{\infty} x(t)x(t-\tau)\,dt$$

$$R_x x(\tau) = x(\tau) * x(-\tau)$$

Taking the Fourier transform of the above equation gives

$$\mathcal{F}[R_{xx}(\tau)] = X(\omega)X(-\omega)$$

$$= X(\omega)X^*(\omega)$$

$$\mathcal{F}[R_{xx}(\tau)] = |X(\omega)|^2$$

$$\mathcal{F}[R_{xx}(\tau)] = \Psi_x(\omega)$$

$$R_{xx}(\tau) \longleftrightarrow \Psi_x(\omega) \tag{5.92}$$

Thus, the autocorrelation function $R_{xx}(\tau)$ and ESD makes a Fourier transform pair.

5.9 POWER SPECTRAL DENSITY (PSD)

PSD has the same relation to power signals as ESD has to energy signals. The expression of PSD may be obtained by assuming the power signal as a limiting case of an energy signal.

Consider a power signal of infinite duration. Since the total energy of a power signal cannot be found, let us first find the ESD of a truncated version $x_T(t)$ of the infinite duration signal $x(t)$. The signal $x_T(t)$ may be expressed as

$$x_T(t) = \begin{cases} x(t), & -\dfrac{T}{2} < t < \dfrac{T}{2} \\ 0, & \text{otherwise} \end{cases} \tag{5.93}$$

$$x_T(t) = \text{rect}\left(\dfrac{t}{T}\right)x(t)$$

The truncated signal $x_T(t)$ is of finite duration therefore it is an energy signal. Now, if $x_T(t) \longleftrightarrow X_T(\omega)$, then using the Parseval's theorem, we have

$$E_{x_T} = \int\limits_{-\infty}^{\infty} |x_T(t)|^2 \, dt = \frac{1}{2\pi} \int\limits_{-\infty}^{\infty} |X_T(\omega)|^2 \, d\omega$$

$$\int\limits_{-\infty}^{\infty} |x_T(t)|^2 \, dt = \int\limits_{-\infty}^{\infty} |X_T(\omega)|^2 \, df$$

Substituting the value of $x_T(t)$ from Eq. (5.93) into the above equation yields

$$\int\limits_{-T/2}^{T/2} |x(t)|^2 \, dt = \int\limits_{-\infty}^{\infty} |X_T(\omega)|^2 \, df$$

$$\lim_{T \to \infty} \frac{1}{T} \int\limits_{-T/2}^{T/2} |x(t)|^2 \, dt = \lim_{T \to \infty} \frac{1}{T} \int\limits_{-\infty}^{\infty} |X_T(\omega)|^2 \, df$$

The LHS of the above equation represents the average power P_x of the signal $x(t)$. Therefore,

$$P_x = \lim_{T \to \infty} \frac{1}{T} \int\limits_{-\infty}^{\infty} |X_T(\omega)|^2 \, df$$

$$= \int\limits_{-\infty}^{\infty} \lim_{T \to \infty} \frac{|X_T(\omega)|^2}{T} \, df$$

$$P_x = \int\limits_{-\infty}^{\infty} G_x(\omega) \, df \tag{5.94}$$

where

$$G_x(\omega) = \lim_{T \to \infty} \frac{|X_T(\omega)|^2}{T} \tag{5.95}$$

is the PSD (average power per unit bandwidth). The units of PSD depend on the units of the underlying signal $x(t)$. If the signal unit is amperes (A), the units of PSD are A^2/Hz. If the signal unit is volts (V), the units of PSD are V^2/Hz.

5.9.1 Relationship Between Input and Output Power Spectral Densities of an LTI System

Consider an LTI system with frequency response $H(\omega)$, input $x_T(t)$, and output $y_T(t)$. If $x(t)$ and $y(t)$ are power signals, then their power spectral densities are $G_x(\omega) = \lim_{T\to\infty} \dfrac{|X_T(\omega)|^2}{T}$ and $G_y(\omega) = \lim_{T\to\infty} \dfrac{|Y_T(\omega)|^2}{T}$, respectively. Since, we know that

$$Y_T(\omega) = H(\omega)X_T(\omega)$$

it follows that

$$|Y_T(\omega)|^2 = |H(\omega)X_T(\omega)|^2$$

$$|Y_T(\omega)|^2 = |H(\omega)|^2 |X_T(\omega)|^2$$

$$\lim_{T\to\infty} \frac{|Y_T(\omega)|^2}{T} = |H(\omega)|^2 \lim_{T\to\infty} \frac{|X_T(\omega)|^2}{T}$$

$$G_y(\omega) = |H(\omega)|^2 G_x(\omega) \tag{5.96}$$

5.9.2 Relation of PSD to Autocorrelation

The autocorrelation function $R_{xx}(\tau)$ of a power signal is defined as [Eq. (2.45)]:

$$R_{xx}(\tau) = \lim_{T\to\infty} \frac{1}{T} \int_{-T/2}^{T/2} x(t)x(t-\tau)\, dt$$

$$= \lim_{T\to\infty} \frac{1}{T} \int_{-\infty}^{\infty} x_T(t)x_T(t-\tau)\, dt$$

$$R_{xx}(\tau) = \lim_{T\to\infty} \frac{1}{T}[x_T(\tau) * x_T(-\tau)]$$

Taking the Fourier transform of the above equation, we get

$$\mathcal{F}[R_{xx}(\tau)] = \lim_{T\to\infty} \frac{1}{T} X_T(\omega)X_T(-\omega)$$

$$= \lim_{T\to\infty} \frac{1}{T} X_T(\omega)X_T^*(\omega)$$

$$\mathcal{F}[R_{xx}(\tau)] = \lim_{T\to\infty} \frac{1}{T}|X_T(\omega)|^2$$

$$\mathcal{F}[R_{xx}(\tau)] = G_x(\omega)$$

$$R_{xx}(\tau) \longleftrightarrow G_x(\omega) \tag{5.97}$$

Thus, the autocorrelation function $R_{xx}(\tau)$ and PSD $G_x(\omega)$ makes a Fourier transform pair.

5.10 PSD OF PERIODIC SIGNALS

The PSD of the periodic signal $x(t)$ with period T is

$$G_x(\omega) = 2\pi \sum_{n=-\infty}^{\infty} |X_n|^2 \delta(\omega - n\omega_0) \tag{5.98}$$

where X_n are the Fourier series coefficients of $x(t)$ and $\omega_0 = 2\pi/T$.

Proof We begin by defining the truncated signals $x_T(t)$ as the product $\text{rect}\left(\dfrac{t}{T}\right) x(t)$.

$$x_T(t) = \text{rect}\left(\frac{t}{T}\right) x(t)$$

Using the multiplication property, we have

$$X_T(\omega) = \frac{1}{2\pi} \left[\mathcal{F}\left[\text{rect}\left(\frac{t}{T}\right)\right] * X(\omega) \right]$$

$$= \frac{1}{2\pi} \left[T \, \text{sinc}\left(\frac{\omega T}{2\pi}\right) * 2\pi \sum_{n=-\infty}^{\infty} X_n \delta(\omega - n\omega_0) \right]$$

$$X_T(\omega) = \sum_{n=-\infty}^{\infty} X_n T \, \text{sinc}\left(\frac{T}{2\pi}(\omega - n\omega_0)\right) \tag{5.99}$$

From Eq. (5.99), we have

$$X_T^*(\omega) = \sum_{m=-\infty}^{\infty} X_m^* T \, \text{sinc}\left(\frac{T}{2\pi}(\omega - m\omega_0)\right) \tag{5.100}$$

From Eqs (5.99) and (5.100), we get

$$X_T(\omega) X_T^*(\omega) = \sum_{n=-\infty}^{\infty} \sum_{m=-\infty}^{\infty} X_n X_m^* T \, \text{sinc}\left(\frac{T}{2\pi}(\omega - n\omega_0)\right) T \, \text{sinc}\left(\frac{T}{2\pi}(\omega - m\omega_0)\right)$$

$$\frac{|X_T(\omega)|^2}{T} = \sum_{n=-\infty}^{\infty} \sum_{m=-\infty}^{\infty} X_n X_m^* \left[T \, \text{sinc}\left(\frac{T}{2\pi}(\omega - n\omega_0)\right) \text{sinc}\left(\frac{T}{2\pi}(\omega - m\omega_0)\right) \right]$$

Taking the limit of the above equation as $T \to \infty$, we have

$$\lim_{T\to\infty} \frac{|X_T(\omega)|^2}{T} = \sum_{n=-\infty}^{\infty} \sum_{m=-\infty}^{\infty} X_n X_m^* \lim_{T\to\infty} \left[T \, \text{sinc}\left(\frac{T}{2\pi}(\omega - n\omega_0)\right) \text{sinc}\left(\frac{T}{2\pi}(\omega - m\omega_0)\right) \right] \tag{5.101}$$

It has been observed earlier that as $T \to \infty$, the transform of the rectangular signal approaches to an impulse function and therefore

$$\lim_{T\to\infty} \left[T \, \text{sinc}\left(\frac{T}{2\pi}(\omega - n\omega_0)\right) \text{sinc}\left(\frac{T}{2\pi}(\omega - m\omega_0)\right) \right] = \begin{cases} 2\pi\delta(\omega - n\omega_0), & n = m \\ 0, & \text{otherwise} \end{cases}$$

Substituting the above value in Eq. (5.101) gives

$$\lim_{T \to \infty} \frac{|X_T(\omega)|^2}{T} = 2\pi \sum_{n=-\infty}^{\infty} |X_n|^2 \delta(\omega - n\omega_0)$$

$$G_x(\omega) = 2\pi \sum_{n=-\infty}^{\infty} |X_n|^2 \delta(\omega - n\omega_0)$$

SOLVED EXAMPLES

Example 5.38 Given that $x(t)$ has the Fourier transform $X(\omega)$, express the Fourier transforms of the signal listed below in terms of $X(\omega)$.

(a) $x_1(t) = x(1 - t) + x(-1 - t)$

(b) $x_2(t) = x(3t - 6)$

(c) $x_3(t) = \dfrac{d^2}{dt^2} x(t - 1)$

Solution

Given that

$$x(t) \longleftrightarrow X(\omega)$$

(a) Using the time-shifting property, we have

$$x(t + 1) \longleftrightarrow X(\omega) e^{j\omega}$$

and

$$x(t - 1) \longleftrightarrow X(\omega) e^{-j\omega}$$

Now, using the time-reversal property, we have

$$x(-t + 1) \longleftrightarrow X(-\omega) e^{-j\omega}$$

and

$$x(-t - 1) \longleftrightarrow X(-\omega) e^{j\omega}$$

Therefore,

$$x_1(t) \longleftrightarrow X_1(\omega)$$

$$x(1 - t) + x(-1 - t) \longleftrightarrow X(-\omega) e^{-j\omega} + X(-\omega) e^{j\omega}$$

$$x(1 - t) + x(-1 - t) \longleftrightarrow 2X(-\omega) \cos\omega$$

$$\mathcal{F}[x_1(t)] = X_1(\omega) = 2X(-\omega) \cos\omega$$

(b) Using the time-shifting property, we have

$$x(t - 6) \longleftrightarrow X(\omega) e^{-j6\omega}$$

Now, using the time-scaling property, we get

$$x_2(t) = x(3t - 6) \longleftrightarrow X_2(\omega) = \frac{1}{3} X\left(\frac{\omega}{3}\right) e^{-j6\frac{\omega}{3}}$$

$$X_2(\omega) = \frac{1}{3} X\left(\frac{\omega}{3}\right) e^{-j2\omega}$$

(c) Using the differentiation in time-domain property, we get

$$\frac{dx(t)}{dt} \longleftrightarrow j\omega X(\omega)$$

Applying this property again, we get

$$\frac{d^2 x(t)}{dt^2} \longleftrightarrow (j\omega)^2 X(\omega)$$

Now applying the time-shifting property, we get

$$x_3(t) = \frac{d^2 x(t - 1)}{dt^2} \longleftrightarrow X_3(\omega) = (j\omega)^2 X(\omega) e^{-j\omega}$$

$$X_3(\omega) = -\omega^2 X(\omega) e^{-j\omega}$$

Example 5.39 Find the Fourier transform of the trapezoidal pulse shown in Fig 5.32(a).

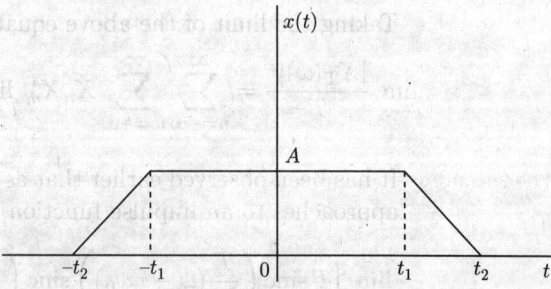

Fig. 5.32(a)   Trapezoidal pulse

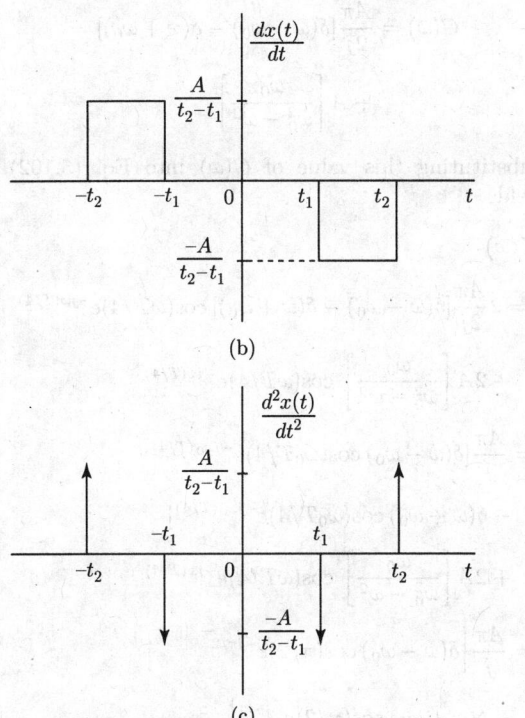

(b)

(c)

Fig. 5.32(b), (c) (b) First derivative and (c) second derivative

Solution

Using the ramp function $r(t)$, the given trapezoidal pulse can be written as

$$x(t) = \frac{A}{t_2 - t_1} r(t + t_2) - \frac{A}{t_2 - t_1} r(t + t_1)$$

$$- \frac{A}{t_2 - t_1} r(t - t_1) + \frac{A}{t_2 - t_1} r(t - t_2)$$

The first derivative of the given trapezoidal pulse is given by [Fig. 5.32(b)]

$$\frac{dx(t)}{dt} = \frac{A}{t_2 - t_1} \left[u(t + t_2) - u(t + t_1) \right.$$

$$\left. - u(t - t_1) + u(t - t_2) \right]$$

The second derivative is given by [Fig. 5.32(c)]

$$\frac{d^2 x(t)}{dt^2} = \frac{A}{t_2 - t_1} \left[\delta(t + t_2) - \delta(t + t_1) \right.$$

$$\left. - \delta(t - t_1) + \delta(t - t_2) \right]$$

The Fourier transform of the second derivative is

$$\mathcal{F} \left[\frac{d^2 x(t)}{dt^2} \right] = \frac{A}{t_2 - t_1} \mathcal{F}[\delta(t + t_2) - \delta(t + t_1)$$

$$- \delta(t - t_1) + \delta(t - t_2)]$$

$$\mathcal{F} \left[\frac{d^2 x(t)}{dt^2} \right] = \frac{A}{t_2 - t_1} \Big(\mathcal{F}[\delta(t + t_2)] - \mathcal{F}[\delta(t + t_1)]$$

$$- \mathcal{F}[\delta(t - t_1)] + \mathcal{F}[\delta(t - t_2)] \Big)$$

Now, taking the Fourier transform of the above equation and using the differentiation in time domain and the time-shifting property, we get

$$(j\omega)^2 X(\omega)$$

$$= \frac{A}{t_2 - t_1} \left[e^{j\omega t_2} - e^{j\omega t_1} - e^{-j\omega t_1} + e^{-j\omega t_2} \right]$$

$$= \frac{2A}{t_2 - t_1} \left[\frac{e^{j\omega t_2} + e^{-j\omega t_2}}{2} - \frac{e^{j\omega t_1} + e^{-j\omega t_1}}{2} \right]$$

$$(j\omega)^2 X(\omega) = \frac{2A}{t_2 - t_1} [\cos(\omega t_2) - \cos(\omega t_1)]$$

$$X(\omega) = \frac{2A}{\omega^2 (t_2 - t_1)} [\cos(\omega t_1) - \cos(\omega t_2)]$$

Example 5.40 Find the Fourier transform of the sinusoidal pulse shown in Fig 5.33(a).

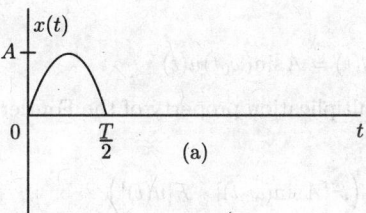

Fig. 5.33(a) Sinusoidal pulse

Solution

Defining a signal $g(t)$ of frequency $\omega_0 = 2\pi/T$ [as shown in Fig. 5.33(b)]

$$g(t) = A \sin(\omega_0 t) u(t)$$

$$g\left(t - \frac{T}{2}\right) = A \sin\left(\omega_0(t - T/2)\right) u(t - T/2)$$

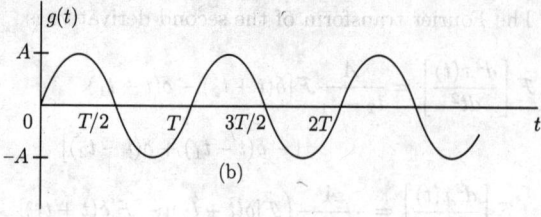

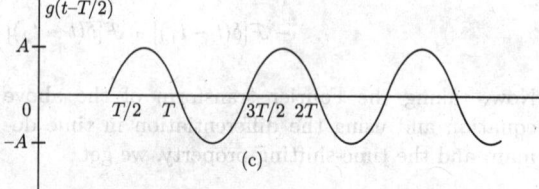

Fig. 5.33(b), (c) (b) Signal $g(t)$ and (c) time-shifted signal

The time-shifted signal $g(t - T/2)$ is shown in Fig. 5.33(c). The given sinusoidal pulse can be expressed as

$$x(t) = g(t) + g(t - T/2)$$

$$X(\omega) = G(\omega) + G(\omega)e^{-j\omega(T/2)}$$

$$= G(\omega)(1 + e^{-j\omega(T/2)})$$

$$= 2G(\omega)\left(\frac{e^{j\omega(T/4)} - e^{-j\omega(T/4)}}{2}\right)e^{-j\omega(T/4)}$$

$$X(\omega) = 2G(\omega)\cos(\omega T/4)e^{-j\omega(T/4)} \quad (5.102)$$

Now, consider

$$g(t) = A\sin(\omega_0 t)u(t)$$

Applying the multiplication property of the Fourier transform gives

$$\mathcal{F}[g(t)] = \frac{1}{2\pi}\Big(\mathcal{F}[A\sin(\omega_0 t)] * \mathcal{F}[u(t)]\Big)$$

$$G(\omega) = \frac{1}{2\pi}\left(\frac{A\pi}{j}[\delta(\omega - \omega_0) - \delta(\omega + \omega_0)]\right.$$

$$\left. * \left[\pi\delta(\omega) + \frac{1}{j\omega}\right]\right)$$

$$G(\omega) = \frac{1}{2\pi}\frac{A\pi^2}{j}[\delta(\omega - \omega_0) - \delta(\omega + \omega_0)]$$

$$+ \frac{1}{2\pi}\frac{A\pi}{j}\left[\frac{1}{j(\omega - \omega_0)} - \frac{1}{j(\omega + \omega_0)}\right]$$

$$G(\omega) = \frac{A\pi}{2j}[\delta(\omega - \omega_0) - \delta(\omega + \omega_0)]$$

$$+ A\left[\frac{\omega_0}{\omega_0^2 - \omega^2}\right]$$

Substituting this value of $G(\omega)$ into Eq. (5.102) yields

$$X(\omega)$$

$$= 2\frac{A\pi}{2j}[\delta(\omega - \omega_0) - \delta(\omega + \omega_0)]\cos(\omega T/4)e^{-j\omega(T/4)}$$

$$+ 2A\left[\frac{\omega_0}{\omega_0^2 - \omega^2}\right]\cos(\omega T/4)e^{-j\omega(T/4)}$$

$$= \frac{A\pi}{j}[\delta(\omega - \omega_0)\cos(\omega_0 T/4)e^{-j\omega_0(T/4)}$$

$$- \delta(\omega + \omega_0)\cos(\omega_0 T/4)e^{-j\omega_0(T/4)}]$$

$$+ 2A\left[\frac{\omega_0}{\omega_0^2 - \omega^2}\right]\cos(\omega T/4)e^{-j\omega(T/4)}$$

$$= \frac{A\pi}{j}[\delta(\omega - \omega_0)\cos(\pi/2)e^{-j\pi/2}$$

$$- \delta(\omega + \omega_0)\cos(\pi/2)e^{-j\pi/2}]$$

$$+ 2A\left[\frac{\omega_0}{\omega_0^2 - \omega^2}\right]\cos(\omega T/4)e^{-j\omega(T/4)}$$

$$= 0 + 2A\left[\frac{\omega_0}{\omega_0^2 - \omega^2}\right]\cos(\omega T/4)e^{-j\omega(T/4)}$$

Therefore,

$$X(\omega) = 2A\left[\frac{\omega_0}{\omega_0^2 - \omega^2}\right]\cos(\omega T/4)e^{-j\omega(T/4)}$$

Example 5.41 Find the magnitude and phase spectrum of the pulse shown in Fig 5.34(a).

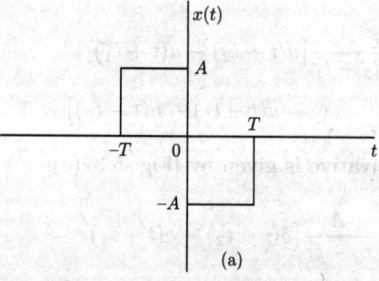

Fig. 5.34(a)

Solution

Analytically, the given pulse can be written as

$$x(t) = \begin{cases} A, & -T < t < 0 \\ -A, & 0 < t < T \\ 0, & \text{otherwise} \end{cases}$$

The Fourier transform $X(\omega)$ of $x(t)$ is given by

$$X(\omega) = \int_{-\infty}^{\infty} x(t)\, e^{-j\omega t}\, dt$$

$$= \int_{-T}^{0} A\, e^{-j\omega t} dt - \int_{0}^{T} A\, e^{-j\omega t}\, dt$$

$$= A \left[\frac{e^{-j\omega t}}{-j\omega}\right]_{-T}^{0} - A \left[\frac{e^{-j\omega t}}{-j\omega}\right]_{0}^{T}$$

$$= \frac{A}{-j\omega}[1 - e^{j\omega T}] + \frac{jA}{\omega}[e^{-j\omega T} - 1]$$

$$= \frac{jA}{\omega}[2 - e^{j\omega T} - e^{-j\omega T}]$$

$$= \frac{jA}{\omega}\left[2 - 2\frac{e^{j\omega T} + e^{-j\omega T}}{2}\right]$$

$$= \frac{2jA}{\omega}[1 - \cos(\omega T)]$$

Therefore, the magnitude spectrum $|X(\omega)|$ of $x(t)$ is given by

$$|X(\omega)| = \left|\frac{2A}{\omega}[1 - \cos(\omega T)]\right|$$

Thus, the magnitude spectrum is zero when $1 - \cos(\omega T) = 0$, i.e.,

$$1 - \cos(\omega T) = 0$$

$$\cos(\omega T) = 1 = \cos(2n\pi)$$

or

$$\omega T = 2n\pi$$

$$\omega = \frac{2n\pi}{T}$$

The phase spectrum of $x(t)$ is

$$\angle X(\omega) = \tan^{-1}\left[\frac{X_I(\omega)}{X_R(\omega)}\right]$$

$$\angle X(\omega) = \tan^{-1}\left[\frac{2A/\omega[1 - \cos(\omega T)]}{0}\right]$$

$$\angle X(\omega) = \begin{cases} \dfrac{\pi}{2}, & \omega > 0 \\[2mm] -\dfrac{\pi}{2}, & \omega < 0 \end{cases}$$

The magnitude and phase spectrum are shown in Figs 5.34(b) and (c).

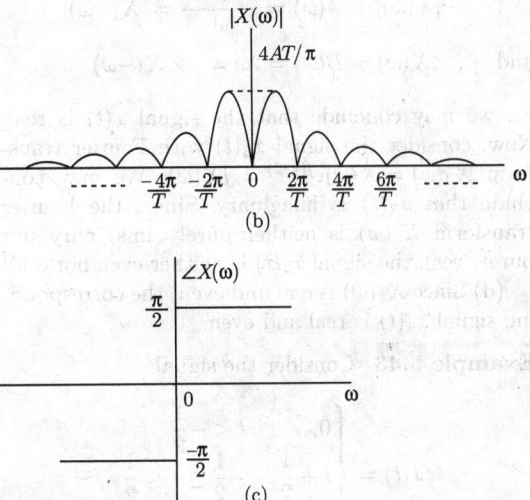

Fig. 5.34(b) and (c)

Example 5.42 For each of the following Fourier transform, use Fourier transform properties to determine whether the corresponding time-domain signal is (i) real, imaginary, or neither and (ii) even, odd, or neither. Do this without evaluating the inverse of any of the given transforms.

(a) $X_1(\omega) = u(\omega) - u(\omega - 2)$

(b) $X_2(\omega) = \cos(2\omega)\sin(\omega/2)$

(c) $X_3(\omega) = A(\omega)\, e^{jB(\omega)}$, where $A(\omega) = \sin(2\omega)/\omega$ and $B(\omega) = 2\omega + \pi/2$

(d) $X_4(\omega) = \sum_{k=-\infty}^{\infty} (1/2)^{|k|}\, \delta\left(\omega - k\pi/4\right)$

Solution

(a) We know that the Fourier transform of a real, signal is conjugate symmetric. Since $X_1(\omega)$ is not conjugate symmetric, the corresponding signal $x_1(t)$ is not real. Since $X_1(\omega)$ is neither even nor odd, the corresponding signal $x_1(t)$ is neither even nor odd.

(b) We know that the Fourier transform of a real and odd signal is purely imaginary and odd. Therefore, we may conclude that the Fourier transform of a purely imaginary and odd signal is real and odd. The given transform $X_2(\omega)$ is real and odd, we may therefore conclude that the corresponding signal $x_2(t)$ is purely imaginary and odd.

(c) We know that if $x(t)$ is real, then its $|X(\omega)| = |X(-\omega)|$ and $\angle X(\omega) = -\angle X(-\omega)$. Here

$$|X(\omega)| = A(\omega) = \frac{\sin(2\omega)}{\omega} = |X(-\omega)|$$

and $\quad \angle X(\omega) = B(\omega) = 2\omega = -\angle X(-\omega)$

so, we may conclude that the signal $x(t)$ is real. Now, consider the signal $x_3(t)$ with Fourier transform $X_3(\omega) = X(\omega)e^{j(\pi/2)} = jX(\omega)$. We may conclude that $x_3(t)$ is imaginary. Since, the Fourier transform $X_3(\omega)$ is neither purely imaginary nor purely real, the signal $x_3(t)$ is neither even nor odd.

(d) Since $X_4(\omega)$ is real and even, the corresponding signal $x_4(t)$ is real and even.

Example 5.43 Consider the signal

$$x(t) = \begin{cases} 0, & t < -\dfrac{1}{2} \\ t + \dfrac{1}{2}, & -\dfrac{1}{2} \le t \le \dfrac{1}{2} \\ 1, & t > \dfrac{1}{2} \end{cases}$$

(a) Use the differentiation and integration properties and the Fourier transform pairs to find a closed form expression for $X(\omega)$.

(b) What is the Fourier transform of $g(t) = x(t) - 1/2$?

Solution
The given signal $x(t)$ is as shown in the Fig 5.35(a). We may express $x(t)$ using ramp function as

$$x(t) = r\left(t + \frac{1}{2}\right) - r\left(t - \frac{1}{2}\right)$$

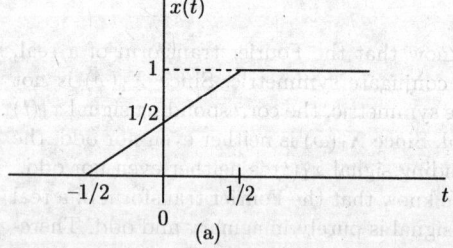

Fig. 5.35(a)

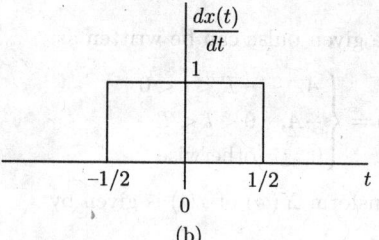

Fig. 5.35(b)

Taking the derivative of $x(t)$ [as shown in Fig. 5.35(b)], we get

$$\frac{dx(t)}{dt} = u\left(t + \frac{1}{2}\right) - u\left(t - \frac{1}{2}\right)$$

Taking the Fourier transform of the above equation and using the time-shifting property, we have

$$j\omega X(\omega) = U(\omega)e^{j(\omega/2)} - U(\omega)e^{-j(\omega/2)}$$

$$= 2jU(\omega)\frac{e^{j(\omega/2)} - e^{-j(\omega/2)}}{2j}$$

$$j\omega X(\omega) = 2jU(\omega)\sin\left(\frac{\omega}{2}\right)$$

$$X(\omega) = \frac{2j}{j\omega}\sin\left(\frac{\omega}{2}\right)\left[\pi\delta(\omega) + \frac{1}{j\omega}\right]$$

$$X(\omega) = \frac{2}{\omega}\sin\left(\frac{\omega}{2}\right)\left[\pi\delta(\omega) + \frac{1}{j\omega}\right]$$

$$= \frac{\sin(\omega/2)}{\omega/2}\pi\delta(\omega) + 2\frac{\sin(\omega/2)}{j\omega^2}$$

$$X(\omega) = \pi\delta(\omega) + 2\frac{\sin(\omega/2)}{j\omega^2}$$

(b) Given that

$$g(t) = x(t) - \frac{1}{2}$$

Taking the Fourier transform of the above equation, we get

$$G(\omega) = X(\omega) - \frac{1}{2}2\pi\delta(\omega)$$

$$= \pi\delta(\omega) + 2\frac{\sin(\omega/2)}{j\omega^2} - \pi\delta(\omega)$$

$$G(\omega) = 2\frac{\sin(\omega/2)}{j\omega^2}$$

Example 5.44 (a) Use the Fourier transform properties to determine the Fourier transform of the following signal:

$$x(t) = t\left(\frac{\sin(t)}{\pi t}\right)^2 = t\, g^2(t)$$

where $g(t) = \sin(t)/\pi t$.

(b) Use Parseval's relation and the result of the previous part to determine the numerical value of

$$A = \int_{-\infty}^{\infty} t^2 \left(\frac{\sin(t)}{\pi t}\right)^4 dt$$

Solution

(a) We know that [Eq. (5.49)]

$$A\,\mathrm{sinc}\left(\frac{t}{2\pi}\right) \longleftrightarrow 2\pi A\,\mathrm{rect}(\omega)$$

$$\frac{\sin(t/2)}{t/2} \longleftrightarrow 2\pi\,\mathrm{rect}(\omega)$$

Using the time-scaling property, we have

$$\frac{\sin(t)}{t} \longleftrightarrow \pi\,\mathrm{rect}\left(\frac{\omega}{2}\right)$$

$$g(t) = \frac{\sin(t)}{\pi t} \longleftrightarrow G(\omega) = \mathrm{rect}\left(\frac{\omega}{2}\right)$$

$G(\omega)$ is shown in Fig. 5.36(a), which is a rectangle function.

Now, using the multiplication property, we have

$$g^2(t) = g(t)g(t) \longleftrightarrow \frac{1}{2\pi}[G(\omega) * G(\omega)]$$

$$\underbrace{\left(\frac{\sin(t)}{\pi t}\right)^2}_{g_1(t)} \longleftrightarrow \underbrace{\frac{1}{2\pi}\left[\mathrm{rect}\left(\frac{\omega}{2}\right) * \mathrm{rect}\left(\frac{\omega}{2}\right)\right]}_{G_1(\omega)}$$

We know that the convolution of a rectangle function with itself is a triangle function [$G_1(\omega)$ is shown in Fig. 5.36(b)]. Therefore, we have

$$\left(\frac{\sin(t)}{\pi t}\right)^2 \longleftrightarrow \frac{1}{\pi}\triangle\left(\frac{\omega}{2}\right)$$

where

$$G_1(\omega) = \frac{1}{\pi}\triangle\left(\frac{\omega}{2}\right) = \begin{cases} \frac{1}{\pi}\left(1+\frac{\omega}{2}\right), & -2 \le \omega < 0 \\ \frac{1}{\pi}\left(1-\frac{\omega}{2}\right), & 0 \le \omega \le 2 \\ 0, & \text{otherwise} \end{cases}$$

Using the differentiation in frequency domain property, we have

$$t\, g_1(t) \longleftrightarrow j\frac{dG_1(\omega)}{d\omega}$$

$$\underbrace{t\left(\frac{\sin(t)}{\pi t}\right)^2}_{x(t)} \longleftrightarrow X(\omega) = \begin{cases} \dfrac{j}{2\pi}, & -2 \le \omega < 0 \\ \dfrac{-j}{2\pi}, & 0 \le \omega \le 2 \\ 0, & \text{otherwise} \end{cases}$$

$X(\omega)$ is shown in Fig. 5.36(c).

(b) Using Parseval's theorem, we have

$$\int_{-\infty}^{\infty} |x(t)|^2\, dt = \frac{1}{2\pi}\int_{-\infty}^{\infty} |X(\omega)|^2\, d\omega$$

$$\int_{-\infty}^{\infty} t\left(\frac{\sin(t)}{\pi t}\right)^2 dt = \frac{1}{2\pi}\left[\int_{-2}^{0}\frac{1}{4\pi^2}\,d\omega + \int_{0}^{2}\frac{1}{4\pi^2}\,d\omega\right]$$

$$\int_{-\infty}^{\infty} t\left(\frac{\sin(t)}{\pi t}\right)^2 dt = \frac{1}{2\pi^3}$$

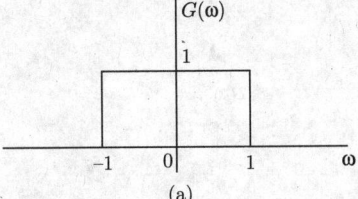

(a)

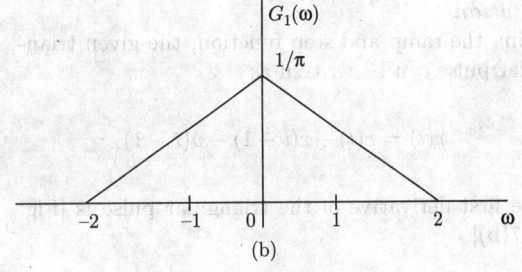

(b)

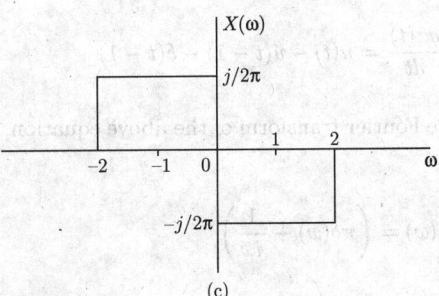

(c)

Fig. 5.36

Example 5.45 Find the Fourier transform of the triangular pulse $x(t)$ shown in Fig. 5.37(a).

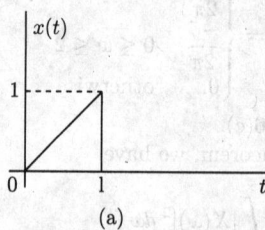

(a)

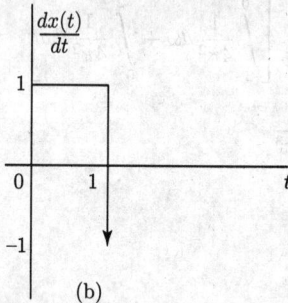

(b)

Fig. 5.37

Solution

Using the ramp and step function, the given triangular pulse can be written as

$$x(t) = r(t) - r(t-1) - u(t-1)$$

The first derivative of the triangular pulse is [Fig. 5.37(b)]

$$\frac{dx(t)}{dt} = u(t) - u(t-1) - \delta(t-1)$$

Taking the Fourier transform of the above equation, we have

$$j\omega X(\omega) = \left(\pi\delta(\omega) + \frac{1}{j\omega} \right)$$

$$- \left(\pi\delta(\omega) + \frac{1}{j\omega} \right) e^{-j\omega} - e^{-j\omega}$$

$$j\omega X(\omega) = \pi\delta(\omega)(1 - e^{-j\omega}) + \frac{1}{j\omega}$$

$$- \frac{1}{j\omega}e^{-j\omega} - e^{-j\omega}$$

$$X(\omega) = 0 - \frac{1}{\omega^2} + \frac{1}{\omega^2}e^{-j\omega} - \frac{1}{j\omega}e^{-j\omega}$$

$$X(\omega) = \frac{1}{\omega^2}(e^{-j\omega} + j\omega e^{-j\omega} - 1)$$

Example 5.46 The Fourier transform of the triangular pulse $x(t)$ in Fig. 5.38(a) is expressed as

$$X(\omega) = \frac{1}{\omega^2} \left(e^{-j\omega} + j\omega e^{-j\omega} - 1 \right)$$

Use this information and the time-shifting and time-scaling properties to find the Fourier transforms of the signals (a) $x_1(t)$ and (b) $x_2(t)$ shown in Figs 5.38(b) and (c), respectively.

Solution

(a) The relationship between $x_1(t)$ and $x(t)$ can be described as

$$x_1(t) = x(t+1) + x(-t+1)$$

Taking the Fourier transform of the above equation, we have

$$X_1(\omega) = X(\omega)e^{j\omega} + X(-\omega)e^{-j\omega}$$

$$= \frac{1}{\omega^2}(e^{-j\omega} + j\omega e^{-j\omega} - 1)e^{j\omega}$$

$$+ \frac{1}{\omega^2}(e^{j\omega} - j\omega e^{j\omega} - 1)e^{-j\omega}$$

$$= \frac{1}{\omega^2}(1 + j\omega - e^{-j\omega}) + \frac{1}{\omega^2}(1 - j\omega - e^{j\omega})$$

$$= \frac{1}{\omega^2}(2 - 2\cos(\omega))$$

$$= \frac{2}{\omega^2}(1 - \cos(\omega))$$

$$X_1(\omega) = \frac{2}{\omega^2}2\sin^2(\omega/2) = \frac{\sin^2(\omega/2)}{(\omega/2)^2}$$

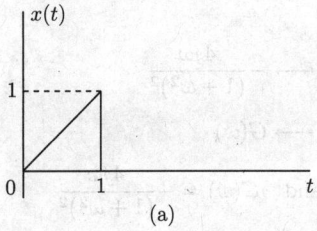

(a)

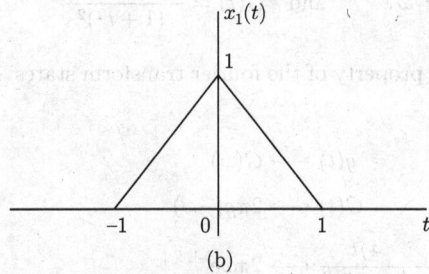

(b)

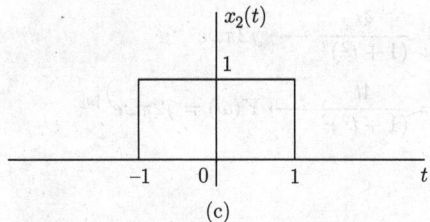

(c)

Fig. 5.38

(b) First, we express $x_2(t)$ in terms of $x(t)$. The derivative of $x(t)$ and its folded version is shown in Figs 5.39(a) and (b), respectively. Therefore, the relationship between $x_2(t)$ and $x(t)$ can be expressed as

$$x_2(t) = \frac{dx(t)}{dt} + \frac{dx(-t)}{dt} + \delta(t+1) + \delta(t-1)$$

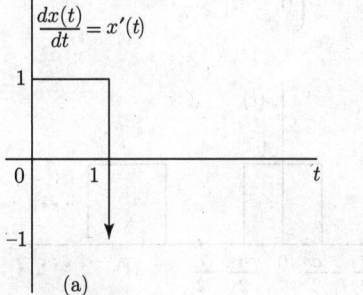

Fig. 5.39(a)

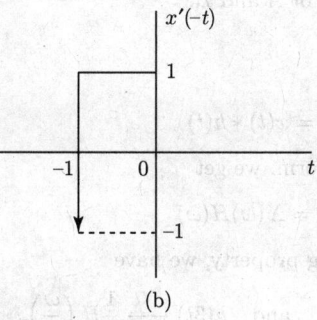

(b)

Fig. 5.39(b)

Taking the Fourier transform of the above equation, we have

$$X_2(\omega) = j\omega X(\omega) - j\omega X(-\omega) + e^{j\omega} + e^{-j\omega}$$

$$= j\omega[X(\omega) - X(-\omega)] + 2\frac{e^{j\omega} + e^{-j\omega}}{2}$$

$$= \frac{j\omega}{\omega^2}\left[e^{-j\omega} + j\omega e^{-j\omega} - 1 - e^{j\omega}\right.$$

$$\left. + j\omega e^{j\omega} + 1\right] + 2\cos(\omega)$$

$$= \frac{j}{\omega}\left[-2j\frac{e^{j\omega} - e^{-j\omega}}{2j} + j2\omega\frac{e^{j\omega} + e^{-j\omega}}{2}\right]$$

$$+ 2\cos(\omega)$$

$$= \frac{j}{\omega}\left[-2j\sin(\omega) + j2\omega\cos(\omega)\right] + 2\cos(\omega)$$

$$= 2\frac{\sin(\omega)}{\omega} - 2\cos(\omega) + 2\cos(\omega)$$

$$X_2(\omega) = 2\frac{\sin(\omega)}{\omega}$$

Example 5.47 Given the relationships

$$y(t) = x(t) * h(t)$$

and $$g(t) = x(3t) * h(3t)$$

and given that $x(t)$ has Fourier transform $X(\omega)$ and $h(t)$ has Fourier transform $H(\omega)$, use Fourier transform properties to show that $g(t)$ has the form

$$g(t) = Ay(Bt)$$

Determine the values of A and B.

Solution
Given that

$$y(t) = x(t) * h(t)$$

Taking Fourier transform, we get

$$Y(\omega) = X(\omega)H(\omega)$$

Using the time-scaling property, we have

$$x(3t) \longleftrightarrow \frac{1}{3}X\left(\frac{\omega}{3}\right) \quad \text{and} \quad h(3t) \longleftrightarrow \frac{1}{3}H\left(\frac{\omega}{3}\right)$$

Taking the Fourier transform of $g(t) = x(3t) * h(3t)$, we get

$$G(\omega) = \mathcal{F}[x(3t)]\mathcal{F}[h(3t)]$$

$$= \frac{1}{9}X\left(\frac{\omega}{3}\right) H\left(\frac{\omega}{3}\right)$$

$$G(\omega) = \frac{1}{9}Y\left(\frac{\omega}{3}\right)$$

Taking the inverse Fourier transform, we obtain

$$g(t) = \frac{1}{3}y(3t) = Ay(Bt)$$

Therefore, $A = 1/3$ and $B = 3$.

Example 5.48 (a) Use the appropriate Fourier transform properties to find the Fourier transform of $g(t) = te^{-|t|}$.
(b) Use the result of part (a), along with the duality property, to determine the Fourier transform of

$$y(t) = \frac{4t}{(1+t^2)^2}$$

Solution
(a) We know that

$$e^{-a|t|} \longleftrightarrow \frac{2a}{a^2 + \omega^2}$$

and substituting $a = 1$ gives

$$e^{-|t|} \longleftrightarrow \frac{2}{1 + \omega^2}$$

Using the differentiation in frequency domain property, we have

$$te^{-|t|} \longleftrightarrow j\frac{d}{d\omega}\left(\frac{2}{1+\omega^2}\right) = -\frac{4j\omega}{(1+\omega^2)^2}$$

(b) We have

$$te^{-|t|} \longleftrightarrow -\frac{4j\omega}{(1+\omega^2)^2}$$

$$g(t) \longleftrightarrow G(\omega)$$

$$g(t) = te^{-|t|} \quad \text{and} \quad G(\omega) = -\frac{4j\omega}{(1+\omega^2)^2}$$

$$g(-\omega) = \omega e^{-|\omega|} \quad \text{and} \quad G(t) = -\frac{4jt}{(1+t^2)^2}$$

The duality property of the fourier transform states that if

$$g(t) \longleftrightarrow G(\omega)$$

then

$$G(t) \longleftrightarrow 2\pi g(-\omega)$$

$$-\frac{4jt}{(1+t^2)^2} \longleftrightarrow 2\pi\omega e^{-|\omega|}$$

$$\frac{4t}{(1+t^2)^2} \longleftrightarrow j2\pi\omega e^{-|\omega|}$$

$$y(t) = \frac{4t}{(1+t^2)^2} \longleftrightarrow Y(\omega) = j2\pi\omega e^{-|\omega|}$$

Example 5.49 Find the Fourier transform of the periodic signal $x(t)$ as shown in Fig. 5.40.

Solution
The waveform is periodic with period T and fundamental frequency $\omega_0 = 2\pi/T$. The given waveform for one period may be written as

$$x(t) = \begin{cases} A, & |t| < \dfrac{a}{2} \\ 0, & |t| > \dfrac{a}{2} \end{cases}$$

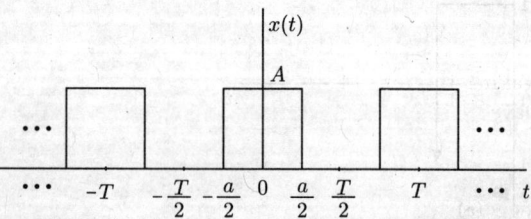

Fig. 5.40

The exponential Fourier series coefficient is

$$X_n = \frac{1}{T} \int_{-T/2}^{T/2} x(t) \, e^{-jn\omega_0 t} \, dt$$

$$= \frac{1}{T} \int_{-a/2}^{a/2} A \, e^{-jn\omega_0 t} \, dt$$

$$= \frac{A}{T} \left[\frac{e^{-jn\omega_0 t}}{-jn\omega_0} \right]_{-a/2}^{a/2}$$

$$X_n = \frac{Aa}{T} \frac{\sin(n\omega_0 a/2)}{n\omega_0 a/2}$$

From Eq. (5.61), the Fourier transform of a periodic signal $x(t)$ is given by

$$X(\omega) = 2\pi \sum_{n=-\infty}^{\infty} X_n \delta(\omega - n\omega_0)$$

$$X(\omega) = 2\pi \sum_{n=-\infty}^{\infty} \frac{Aa}{T} \frac{\sin(n\omega_0 a/2)}{n\omega_0 a/2} \delta(\omega - n\omega_0)$$

Example 5.50 Let $x(t)$ be a signal whose Fourier transform is

$$X(\omega) = \delta(\omega) + \delta(\omega - \pi) + \delta(\omega - 5)$$

and let

$$h(t) = u(t) - u(t - 2)$$

(a) Is $x(t)$ periodic?

(b) Is $y(t) = x(t) * h(t)$ periodic?

(c) Can the convolution of two aperiodic signals be periodic?

Solution
Given that

$$X(\omega) = \delta(\omega) + \delta(\omega - \pi) + \delta(\omega - 5)$$

Taking the inverse Fourier transform of $X(\omega)$, we have

$$x(t) = \frac{1}{2\pi} + \frac{1}{2\pi} e^{j\pi t} + \frac{1}{2\pi} e^{j5t}$$

The signal $x(t)$ is therefore a constant summed with two complex exponentials whose fundamental fre-

quencies are

$$\omega_1 = \pi \text{ rad/s} \quad \text{and} \quad \omega_2 = 5 \text{ rad/s}$$

$$T_1 = \frac{2\pi}{\omega_1} = \frac{2\pi}{\pi} = 2 \quad \text{and} \quad T_2 = \frac{2\pi}{\omega_2} = \frac{2\pi}{5}$$

The ratio $T_1/T_2 = 5/\pi$ is not a rational number and so $x(t)$ is not periodic.

(b) Given that

$$h(t) = u(t) - u(t - 2)$$

Taking the Fourier transform, we have

$$H(\omega) = U(\omega) - U(\omega) \, e^{-j2\omega}$$

$$= 2jU(\omega) \, e^{-j\omega} \frac{e^{j\omega} - e^{-j\omega}}{2j}$$

$$= 2j \, e^{-j\omega} \sin(\omega) U(\omega)$$

$$= 2j \, e^{-j\omega} \sin(\omega) \left[\pi\delta(\omega) + \frac{1}{j\omega} \right]$$

$$H(\omega) = \frac{2\sin(\omega)}{\omega} \, e^{-j\omega}$$

Consider the signal $y(t) = x(t) * h(t)$. From the convolution property, we have

$$Y(\omega) = X(\omega) H(\omega)$$

$$= [\delta(\omega) + \delta(\omega - \pi) + \delta(\omega - 5)] H(\omega)$$

$$Y(\omega) = H(0)\delta(\omega) + H(\pi)\delta(\omega - \pi) + H(5)\delta(\omega - 5)$$

Since $H(\pi) = 0$ and $H(0) = 1$, we get

$$Y(\omega) = \delta(\omega) + H(5)\delta(\omega - 5)$$

Taking the inverse Fourier transform, we get

$$y(t) = \frac{1}{2\pi} + \frac{1}{2\pi} H(5) \, e^{j5t}$$

Therefore, $y(t)$ is a constant summed with a complex exponential whose fundamental frequency is $\omega = 5$ rad/s and fundamental period $T = 2\pi/5$. We know that adding a constant to a complex exponential does not affect its periodicity. Therefore, $y(t) = x(t) * h(t)$ is periodic.

(c) Yes, it is obvious from parts (a) and (b) that the convolution of two aperiodic signals can be periodic.

Example 5.51 Consider a signal $x(t)$ with Fourier transform $X(\omega)$. Suppose we are given the following facts:

1. $x(t)$ is real and nonnegative.
2. $\mathcal{F}^{-1}[(1+j\omega)X(\omega)] = A\,e^{-2t}u(t)$, where A is independent of t.
3. $\int_{-\infty}^{\infty} |X(\omega)|^2 d\omega = 2\pi$.

Determine the closed-form expression for $x(t)$.

Solution
From fact no. 2, we have

$$\mathcal{F}^{-1}[(1+j\omega)X(\omega)] = A\,e^{-2t}u(t)$$

Taking the Fourier transform of the above equation, we obtain

$$(1+j\omega)X(\omega) = \frac{A}{2+j\omega}$$

$$X(\omega) = \frac{A}{(1+j\omega)(2+j\omega)}$$

Using partial fraction expansion, we have

$$X(\omega) = A\left(\frac{1}{1+j\omega} - \frac{1}{2+j\omega}\right)$$

Taking the inverse Fourier transform of the above equation, we get

$$x(t) = A\,e^{-t}u(t) - A\,e^{-2t}u(t)$$

Using Parseval's theorem, we have

$$\int_{-\infty}^{\infty} |x(t)|^2 dt = \frac{1}{2\pi} \int_{-\infty}^{\infty} |X(\omega)|^2 d\omega$$

$$\int_{-\infty}^{\infty} |x(t)|^2 dt = \frac{1}{2\pi} 2\pi$$

$$\int_{-\infty}^{\infty} |x(t)|^2 dt = 1$$

Substituting the expression of $x(t)$ in the above equation gives

$$\int_{-\infty}^{\infty} |A\,e^{-t}u(t) - A\,e^{-2t}u(t)|^2 dt = 1$$

$$\int_{-\infty}^{\infty} [A^2\,e^{-2t} - 2A^2\,e^{-3t} + A^2\,e^{-4t}]u(t)\,dt = 1$$

$$\int_{0}^{\infty} [A^2\,e^{-2t} - 2A^2\,e^{-3t} + A^2\,e^{-4t}]u(t)\,dt = 1$$

$$\frac{A^2}{12} = 1$$

$$A = \pm\sqrt{12}$$

Since $x(t)$ is nonnegative, we choose $A = \sqrt{12}$. Therefore,

$$x(t) = \sqrt{12}[e^{-t} - e^{-2t}]u(t)$$

Example 5.52 Let $x(t)$ be a signal with Fourier transform $X(\omega)$. Suppose we are given the following facts:

1. $x(t)$ is real.
2. $x(t) = 0$ for $t \le 0$.
3. $\frac{1}{2\pi} \int_{-\infty}^{\infty} \Re\{X(\omega)\}\,e^{j\omega t}d\omega = |t|\,e^{-|t|}$.

Determine the closed-form expression for $x(t)$.

Solution
From fact no. 3, we have

$$\frac{1}{2\pi} \int_{-\infty}^{\infty} \Re\{X(\omega)\}\,e^{j\omega t}d\omega = |t|\,e^{-|t|}$$

$$\mathcal{F}^{-1}[\Re\{X(\omega)\}] = |t|\,e^{-|t|}$$

We know that $\mathcal{E}\{x(t)\} = x_e(t) \longleftrightarrow \Re\{X(\omega)\}$, therefore, we have

$$x_e(t) = |t|\,e^{-|t|}$$

$$\frac{x(t) + x(-t)}{2} = |t|\,e^{-|t|}$$

Given that $x(t) = 0$ for $t < 0$. This implies that $x(-t) = 0$ for $t > 0$. So, we may conclude that

$$x(t) = 2|t|\, e^{-|t|}, \quad t \geq 0$$

Therefore,

$$x(t) = 2t\, e^{-t} u(t)$$

Example 5.53 Consider the signal

$$x(t) = \sum_{k=-\infty}^{\infty} \frac{\sin(k\pi/4)}{(k\pi/4)} \delta\left(t - k\frac{\pi}{4}\right)$$

(a) Determine $g(t)$ such that

$$x(t) = \left(\frac{\sin(t)}{\pi t}\right) g(t)$$

(b) Use the multiplication property of the Fourier transform to argue that $X(\omega)$ is periodic. Specify $X(\omega)$ over one period.

Solution
(a) Consider the given signal

$$x(t) = \sum_{k=-\infty}^{\infty} \frac{\sin(k\pi/4)}{(k\pi/4)} \delta\left(t - k\frac{\pi}{4}\right)$$

$$x(t) = \frac{\sin(t)}{\pi t} \sum_{k=-\infty}^{\infty} \pi\delta\left(t - k\frac{\pi}{4}\right)$$

$$x(t) = \left(\frac{\sin(t)}{\pi t}\right) g(t)$$

Therefore,

$$g(t) = \sum_{k=-\infty}^{\infty} \pi\delta\left(t - k\frac{\pi}{4}\right)$$

(b) From part (a), we have

$$x(t) = \left(\frac{\sin(t)}{\pi t}\right) g(t)$$

Taking the Fourier transform, we get

$$X(\omega) = \frac{1}{2\pi}\left[\mathcal{F}\left(\frac{\sin(t)}{\pi t}\right) * G(\omega)\right]$$

We know that

$$\frac{\sin(t)}{\pi t} \longleftrightarrow \operatorname{rect}\left(\frac{\omega}{2}\right)$$

Since $g(t)$ is an impulse train, its Fourier transform $G(\omega)$ is also an impulse train:

$$G(\omega) = \pi\frac{2\pi}{T_0} \sum_{k=-\infty}^{\infty} \delta\left(\omega - k\frac{2\pi}{T_0}\right)$$

Here, $T_0 = \pi/4$. Therefore,

$$G(\omega) = \pi\frac{2\pi}{\pi/4} \sum_{k=-\infty}^{\infty} \delta\left(\omega - k\frac{\pi}{4}\right)$$

$$G(\omega) = 8\pi \sum_{k=-\infty}^{\infty} \delta(\omega - 8k)$$

We see that $G(\omega)$ is periodic with period 8. Now, consider

$$X(\omega) = \frac{1}{2\pi}\left[\mathcal{F}\left(\frac{\sin(t)}{\pi t}\right) * G(\omega)\right]$$

$$= \frac{1}{2\pi}\left[\operatorname{rect}\left(\frac{\omega}{2}\right) * 8\pi \sum_{k=-\infty}^{\infty} \delta(\omega - 8k)\right]$$

$$X(\omega) = 4 \sum_{k=-\infty}^{\infty} \operatorname{rect}\left(\frac{\omega - 8k}{2}\right)$$

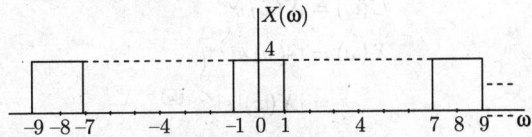

Fig. 5.41

$X(\omega)$ is shown in Fig. 5.41. Therefore, we may specify $X(\omega)$ over one period as

$$X(\omega) = \begin{cases} 1 & |\omega| \leq 1 \\ 0 & 1 < |\omega| \leq 4 \end{cases}$$

Example 5.54 Let $X(\omega)$ denote the Fourier transform of the signal $x(t)$ depicted in Fig. 5.42.
(a) Find $\angle X(\omega)$.
(b) Find $X(0)$.
(c) Find $\displaystyle\int_{-\infty}^{\infty} X(\omega)d\omega$.
(d) Evaluate $\displaystyle\int_{-\infty}^{\infty} X(\omega)\frac{2\sin(\omega)}{\omega} e^{j2\omega}d\omega$.
(e) Evaluate $\displaystyle\int_{-\infty}^{\infty} |X(\omega)|^2 d\omega$.
(f) Sketch the inverse Fourier transform of $\Re\{X(\omega)\}$.

Perform all these calculations without explicitly evaluating $X(\omega)$.

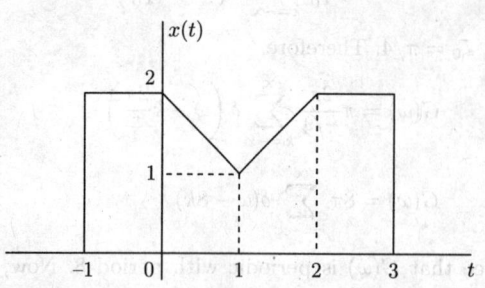

Fig. 5.42

Solution

(a) Note that $g(t) = x(t+1)$ is a real and even signal. Therefore, its Fourier transform $G(\omega)$ is also real and even. This implies that $\angle G(\omega) = 0$. Taking the Fourier transform of $g(t)$, we have

$$G(\omega) = X(\omega)\, e^{j\omega}$$
$$X(\omega) = G(\omega)\, e^{-j\omega}$$
$$= |X(\omega)|\, e^{j\angle X(\omega)}$$

Therefore,

$$\angle X(\omega) = -\omega$$

(b) From Eq. (5.58) [area under $x(t)$ property], we have

$$X(0) = \int_{-\infty}^{\infty} x(t)\, dt$$

$$= \left(\int_{-1}^{0} 2\, dt + \int_{0}^{1} (2-t)\, dt + \int_{1}^{2} t\, dt + \int_{2}^{3} 2\, dt \right)$$

$$= 2t \Big|_{-1}^{0} + \left[2t - \frac{t^2}{2} \right]_{0}^{1} + \frac{t^2}{2} \Big|_{1}^{2} + 2t \Big|_{2}^{3}$$

$$X(0) = 7$$

(c) From Eq. (5.59) [area under $X(\omega)$ property], we have

$$\int_{-\infty}^{\infty} X(\omega)\, d\omega = 2\pi x(0)$$

$$= 4\pi$$

(d) Assume $Y(\omega) = \dfrac{2\sin(\omega)}{\omega}\, e^{j2\omega}$. We know that

$$\text{rect}\left(\frac{t}{2} \right) \longleftrightarrow 2\frac{\sin(\omega)}{\omega}$$

$$\text{rect}\left(\frac{t+2}{2} \right) \longleftrightarrow 2\frac{\sin(\omega)}{\omega}\, e^{j2\omega}$$

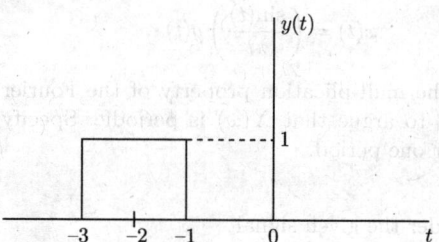

Fig. 5.43

The signal $y(t)$ is as shown in Fig. 5.43.

$$y(t) = \text{rect}\left(\frac{t+2}{2} \right) = \begin{cases} 1, & -3 < t < -1 \\ 0, & \text{otherwise} \end{cases}$$

Consider the given integral

$$\int_{-\infty}^{\infty} X(\omega)\frac{2\sin(\omega)}{\omega}\, e^{j2\omega}\, d\omega$$

$$= \int_{-\infty}^{\infty} X(\omega)Y(\omega)\, d\omega$$

$$= 2\pi[x(t) * y(t)]_{t=0}$$

$$= 2\pi \left[\int_{-\infty}^{\infty} x(\tau)y(t-\tau)\, d\tau \right]\Bigg|_{t=0}$$

$$= 2\pi \left[\int_{-\infty}^{\infty} x(\tau)y(-\tau)\, d\tau \right]$$

Since the product $x(\tau)y(-\tau)$ is nonzero only for the range $1 < t < 3$. Therefore,

$$\int_{-\infty}^{\infty} X(\omega) \frac{2\sin(\omega)}{\omega} e^{j2\omega}\, d\omega$$

$$= 2\pi \left[\int_{1}^{3} x(\tau)y(-\tau)\, d\tau \right]$$

$$= 2\pi \left[\int_{1}^{2} \tau\, d\tau + \int_{2}^{3} 2\, d\tau \right]$$

$$= 7\pi$$

(e) Using Parseval's theorem, we have

$$\int_{-\infty}^{\infty} |X(\omega)|^2\, d\omega = 2\pi \int_{-\infty}^{\infty} |x(t)|^2\, dt$$

$$= \left(\int_{-1}^{0} 4\, dt + \int_{0}^{1} (2-t)^2\, dt \right.$$

$$\left. + \int_{1}^{2} t^2\, dt + \int_{2}^{3} 4\, dt \right)$$

$$= 26\pi$$

(f) We know that

$$x_e(t) \longleftrightarrow \Re\{X(\omega)\}$$

Therefore,

$$\mathcal{F}^{-1}[\Re\{X(\omega)\}] = x_e(t) = \frac{x(t) + x(-t)}{2}$$

Using step and ramp functions, $x(t)$ can be written as

$$x(t) = 2u(t+1) - r(t) + 2r(t-1)$$
$$- r(t-2) - 2u(t-3)$$

The time-reversal signal is shown in Fig. 5.44(a) and can be written as

$$x(-t) = 2u(t+3) - r(t+2) + 2r(t+1)$$
$$- r(t) - 2u(t-1)$$

Therefore,

$$x_e(t) = \frac{x(t) + x(-t)}{2}$$

$$= u(t+3) - \frac{1}{2}r(t+2) + r(t+1)$$

$$+ u(t+1) - r(t) + r(t-1) - u(t-1)$$

$$- \frac{1}{2}r(t-2) - u(t-3)$$

This is as shown in Fig. 5.44(b).

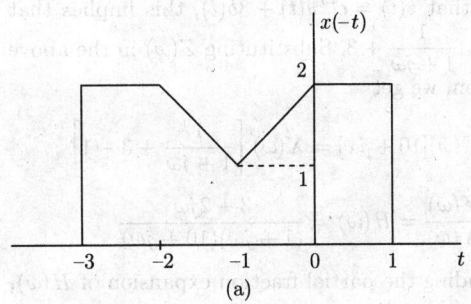

(a)

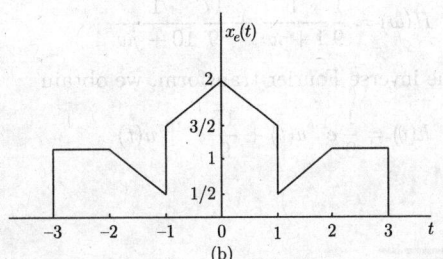

(b)

Fig. 5.44

Example 5.55 The output $y(t)$ of a causal LTI system is related to the input $x(t)$ by the equation

$$\frac{dy(t)}{dt} + 10y(t) = \int_{-\infty}^{\infty} x(\tau)z(t-\tau)\, d\tau - x(t)$$

where $z(t) = e^{-t}u(t) + 3\delta(t)$.
(a) Find the frequency response $H(\omega)$ of this system.
(b) Determine the impulse response of the system.

Solution

(a) Consider the equation

$$\frac{dy(t)}{dt} + 10y(t) = \int\limits_{-\infty}^{\infty} x(\tau)z(t-\tau)\,d\tau - x(t)$$

$$\frac{dy(t)}{dt} + 10y(t) = [x(t) * z(t)] - x(t)$$

Taking the Fourier transform of the above equation, we have

$$j\omega Y(\omega) + 10Y(\omega) = X(\omega)Z(\omega) - X(\omega)$$

$$Y(\omega)[10 + j\omega] = X(\omega)[Z(\omega) - 1]$$

Given that $z(t) = e^{-t}u(t) + 3\delta(t)$, this implies that $Z(\omega) = \dfrac{1}{1 + j\omega} + 3$. Substituting $Z(\omega)$ in the above equation, we get

$$Y(\omega)[10 + j\omega] = X(\omega)\left[\frac{1}{1 + j\omega} + 3 - 1\right]$$

$$\frac{Y(\omega)}{X(\omega)} = H(\omega) = \frac{3 + 2j\omega}{(1 + j\omega)(10 + j\omega)}$$

(b) Finding the partial fraction expansion of $H(\omega)$, we obtain

$$H(\omega) = \frac{1}{9}\frac{1}{1 + j\omega} + \frac{17}{9}\frac{1}{10 + j\omega}$$

Taking the inverse Fourier transform, we obtain

$$h(t) = \frac{1}{9}\,e^{-t}u(t) + \frac{17}{9}\,e^{-10t}u(t)$$

Example 5.56 For the LTI system described by the impulse response $h(t) = \delta(t) - 2\,e^{-2t}u(t)$. Determine and sketch the frequency response. Name the type of filter the system represents.

Solution

Given that

$$h(t) = \delta(t) - 2\,e^{-2t}u(t)$$

Taking the Fourier transform of $h(t)$, we obtain

$$H(\omega) = 1 - \frac{2}{2 + j\omega}$$

$$H(\omega) = \frac{j\omega}{2 + j\omega}$$

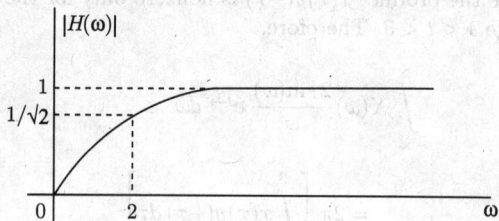

Fig. 5.45 Magnitude response of HPF

Taking the magnitude of $H(\omega)$, we obtain

$$|H(\omega)| = \frac{\omega}{\sqrt{4 + \omega^2}}$$

$$|H(\omega)| = \frac{1}{\sqrt{1 + \dfrac{4}{\omega^2}}}$$

The magnitude response is plotted in Fig. 5.45. At lower frequencies, the magnitude response is also low, and at higher frequencies the magnitude response is also high and so this filter blocks low frequencies and passes high frequencies. This is a high-pass filter.

Example 5.57 Determine and sketch the magnitude and phase response of the LTI causal system described by the differential equation

$$\frac{dy(t)}{dt} + y(t) = \frac{dx(t)}{dt} - x(t)$$

Solution

Consider the differential equation

$$\frac{dy(t)}{dt} + y(t) = \frac{dx(t)}{dt} - x(t)$$

Taking the Fourier transform of the above equation, we obtain

$$j\omega Y(\omega) + Y(\omega) = j\omega X(\omega) - X(\omega)$$

$$Y(\omega)[j\omega + 1] = X(\omega)[j\omega - 1]$$

$$\frac{Y(\omega)}{X(\omega)} = H(\omega) = \frac{j\omega + 1}{j\omega - 1}$$

The magnitude response is given by

$$|H(\omega)| = \frac{\sqrt{\omega^2 + 1}}{\sqrt{\omega^2 + 1}} = 1$$

and the phase response is given by

$$\angle H(\omega) = \tan^{-1}(\omega) - \tan^{-1}\left(\frac{\omega}{-1}\right)$$

$$= \tan^{-1}(\omega) + \tan^{-1}(\omega)$$

$$\angle H(\omega) = 2\tan^{-1}(\omega)$$

Both magnitude and phase response are shown in Fig. 5.46.

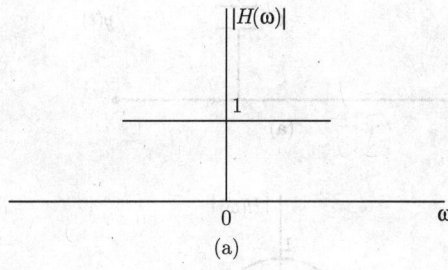

(a)

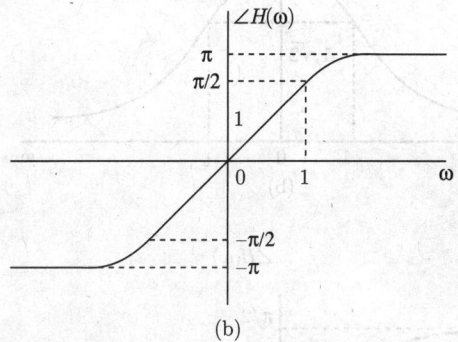

(b)

Fig. 5.46 (a) Magnitude response and (b) phase response

Example 5.58 Let

$$g(t) = x(t)\cos^2(t) * \frac{\sin(t)}{\pi t}$$

Assuming that $x(t)$ is real and $X(\omega) = 0$ for $|\omega| \geq 1$, show that there exists an LTI system S such that

$$x(t) \longrightarrow g(t)$$

Solution
Assume

$$y_1(t) = \cos^2(t) = \frac{1 + \cos(2t)}{2}$$

Taking its Fourier transform, we obtain

$$Y_1(\omega) = \frac{1}{2}2\pi\delta(\omega) + \frac{1}{2}\pi[\delta(\omega - 2) + \delta(\omega + 2)]$$

$$Y_1(\omega) = \pi\delta(\omega) + \frac{\pi}{2}\delta(\omega - 2) + \frac{\pi}{2}\delta(\omega + 2)$$

Now, let

$$y_2(t) = x(t)y_1(t) = x(t)\cos^2(t)$$

Taking the Fourier transform, we obtain

$$Y_2(\omega) = \frac{1}{2\pi}[X(\omega) * Y_1(\omega)]$$

$$= \frac{1}{2\pi}\left[X(\omega) * \left(\pi\delta(\omega) + \frac{\pi}{2}\delta(\omega - 2)\right.\right.$$

$$\left.\left. + \frac{\pi}{2}\delta(\omega + 2)\right)\right]$$

$$Y_2(\omega) = \frac{1}{2}X(\omega) + \frac{1}{4}X(\omega - 2) + \frac{1}{4}X(\omega + 2)$$

Now consider

$$\underbrace{\frac{\sin(t)}{\pi t}}_{y_3(t)} \longleftrightarrow \underbrace{\text{rect}\left(\frac{\omega}{2}\right)}_{Y_3(\omega)} = \begin{cases} 1, & |\omega| < 1 \\ 0, & \text{otherwise} \end{cases}$$

$X(\omega)$, $Y_2(\omega)$, and $Y_3(\omega)$ are as shown in Figs 5.47(a), (b), and (c) respectively.
Now, consider

$$g(t) = y_2(t) * y_3(t)$$

Taking the Fourier transform, we obtain

$$G(\omega) = Y_2(\omega)Y_3(\omega)$$

From Figs 5.47(b) and (c), it is evident that

$$G(\omega) = \frac{1}{2}X(\omega)$$

$$\frac{G(\omega)}{X(\omega)} = H(\omega) = \frac{1}{2}$$

Taking its inverse Fourier transform, we have

$$h(t) = \frac{1}{2}\delta(t)$$

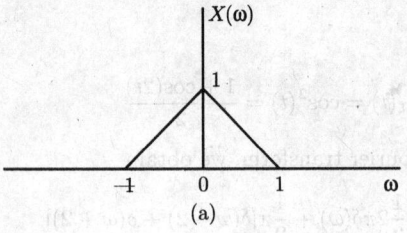

(a)

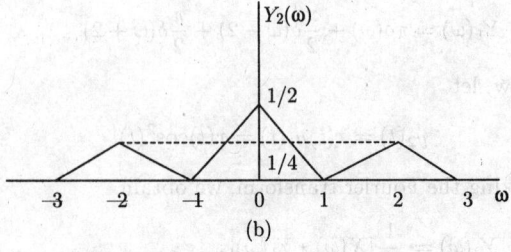

(b)

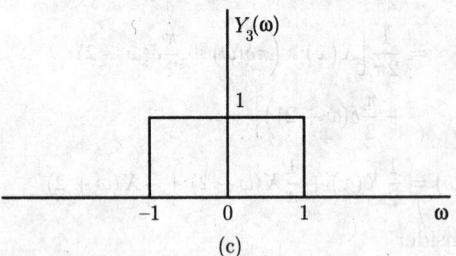

(c)

Fig. 5.47

Example 5.59 A signal $x(t) = 2e^{-t}u(t)$ is passed through an RC LPF [shown in Fig. 5.48(a)] having a cutoff frequency $\omega_c = 1$ rad/s. Find

(a) the transfer function of the LPF.

(b) the spectrum of the input signal $x(t)$.

(c) ESD at the input of the LPF.

(d) spectrum of the output signal $y(t)$.

(e) ESD at the output of the LPF.

(f) energy contained in the input and output signals. Are these two energies equal?

Solution

(a) The frequency response of the RC LPF is given by

$$H(\omega) = \frac{Y(\omega)}{X(\omega)} = \frac{\frac{1}{j\omega C}}{R + \frac{1}{j\omega C}}$$

$$= \frac{1}{1 + j\omega RC}$$

$$H(\omega) = \frac{1}{1 + j\frac{\omega}{\omega_c}}$$

where $\omega_c = 1/RC = 1$ rad/s. The magnitude and phase response are [as shown in Fig. 5.48(b) and (c)] given by

$$|H(\omega)| = \frac{1}{\sqrt{1 + \omega^2}}$$

$$\angle H(\omega) = -\tan^{-1}(\omega)$$

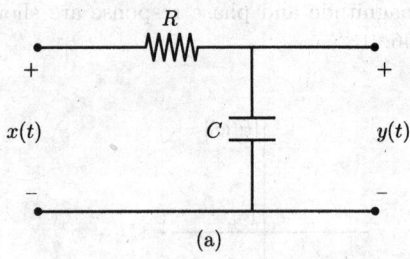

(a)

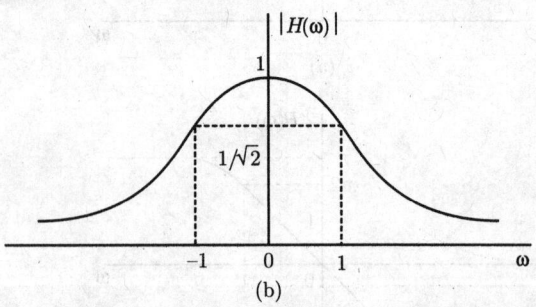

(b)

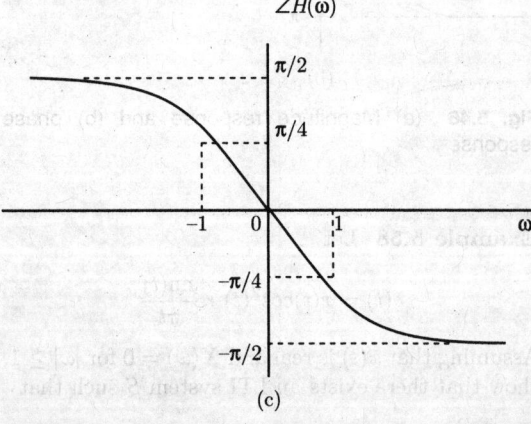

(c)

Fig. 5.48

(b) Consider the given signal

$$x(t) = 2e^{-t}u(t)$$

$$X(\omega) = \frac{2}{1+j\omega}$$

The magnitude and phase spectrum are [as shown in Figs 5.49(a) and (b)] given by

$$|X(\omega)| = \frac{2}{\sqrt{1+\omega^2}}$$

$$\angle X(\omega) = -\tan^{-1}(\omega)$$

(c) The ESD at the input of the LPF is given

$$\Psi_x(\omega) = |X(\omega)|^2$$

$$= \left| \frac{2}{1+j\omega} \right|^2$$

$$\Psi_x(\omega) = \frac{4}{1+\omega^2}$$

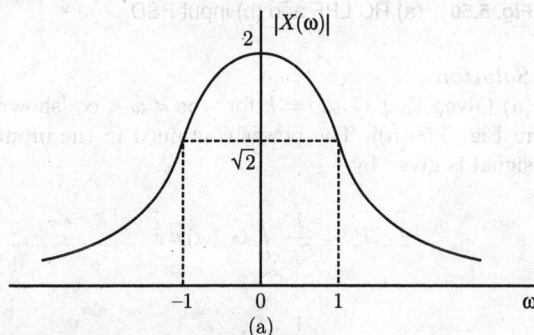

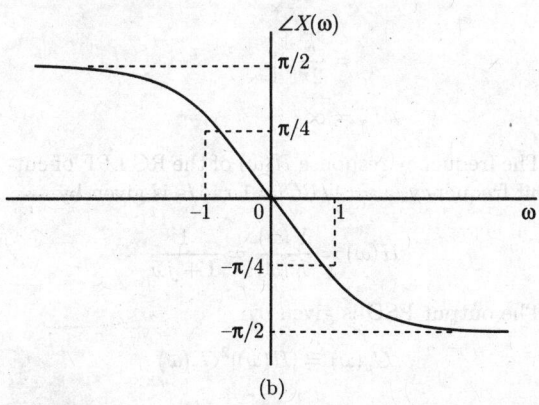

Fig. 5.49

(d) We know that

$$Y(\omega) = X(\omega)H(\omega)$$

$$= \frac{2}{1+j\omega} \frac{1}{1+j\omega}$$

$$= \frac{2}{(1+j\omega)^2}$$

(e) The ESD at the output of the LPF is given by

$$\Psi_y(\omega) = |Y(\omega)|^2$$

$$= \left| \frac{2}{(1+j\omega)^2} \right|^2$$

$$\Psi_y(\omega) = \frac{4}{(1+\omega^2)^2}$$

The energy contained in the input signal is given by

$$E_x = \frac{1}{2\pi} \int_{-\infty}^{\infty} \Psi_x(\omega)\, d\omega$$

$$= \frac{1}{2\pi} \int_{-\infty}^{\infty} \frac{4}{1+\omega^2}\, d\omega$$

$$= \frac{2}{2\pi} \int_{0}^{\infty} \frac{4}{1+\omega^2}\, d\omega$$

$$= \frac{4}{\pi} \tan^{-1}(\omega) \Big|_{0}^{\infty}$$

$$E_x = \frac{4}{\pi} \left[\frac{\pi}{2} - 0 \right]$$

$$E_x = 2$$

The energy contained in the output signal is given by

$$E_y = \frac{1}{2\pi} \int_{-\infty}^{\infty} \Psi_y(\omega)\, d\omega$$

$$= \frac{1}{2\pi} \int_{-\infty}^{\infty} \frac{4}{(1+\omega^2)^2}\, d\omega$$

A change of variables is performed by letting $\omega = \tan(\alpha)$, which also yields $d\omega = \sec^2(\alpha)\,d\alpha$, $\alpha \to -\pi/2$ as $\omega \to -\infty$, and $\alpha \to \pi/2$ as $\omega \to \infty$. Therefore,

$$E_y = \frac{4}{2\pi} \int\limits_{-\pi/2}^{\pi/2} \frac{\sec^2(\alpha)}{\sec^4(\alpha)}\,d\alpha$$

$$= \frac{2}{\pi} \int\limits_{-\pi/2}^{\pi/2} \cos^2(\alpha)\,d\alpha$$

$$= \frac{2}{\pi} \int\limits_{-\pi/2}^{\pi/2} \frac{1}{2}[1 + \cos 2(\alpha)]d\alpha$$

$$= \frac{1}{\pi} \left[\alpha + \frac{\sin(2\alpha)}{2} \right]_{-\pi/2}^{\pi/2}$$

$$E_y = \frac{1}{\pi} \left[\frac{\pi}{2} + \frac{\pi}{2} + 0 \right]$$

$$E_y = 1$$

Since the cutoff frequency of the LPF is finite, therefore all the frequency components greater than this cutoff frequency will not be allowed to pass through the LPF and the energy associated with such frequency components will be attenuated. So, the output energy will always be less than the input energy.

Example 5.60 Find the power contained at the output of an RC LPF [shown in Fig. 5.50(a)] if the input voltage has the PSD:

(a) $G_x(\omega) = k$ for $-\infty < \omega < \infty$

(b) $G_x(\omega)$ is a gate function of height 1 and width 2 placed symmetrically about the vertical axis.

(c) $G_x(\omega) = \pi[\delta(\omega + 1) + \delta(\omega - 1)]$.

In each case also find the power contained in the input signal and find whether power at the input and output are equal or not?

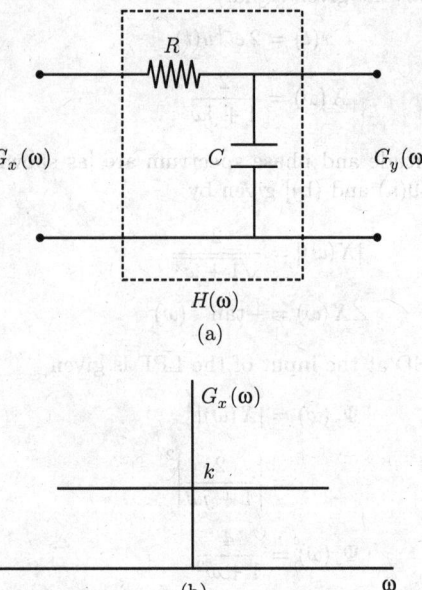

Fig. 5.50   (a) RC LPF and (b) input PSD

Solution
(a) Given that $G_x(\omega) = k$ for $-\infty < \omega < \infty$ [shown in Fig. 5.50(b)]. The power contained in the input signal is given by

$$P_x = \frac{1}{2\pi} \int\limits_{-\infty}^{\infty} G_x(\omega)\,d\omega$$

$$= \frac{1}{2\pi} \int\limits_{-\infty}^{\infty} k\,d\omega$$

$$= \frac{k}{2\pi} \omega \Big|_{-\infty}^{\infty}$$

$$P_x = \infty$$

The frequency response $H(\omega)$ of the RC LPF of cutoff frequency $\omega_c = 1/RC = 1$ rad/s is given by

$$H(\omega) = \frac{Y(\omega)}{X(\omega)} = \frac{1}{1 + j\omega}$$

The output PSD is given by

$$G_y(\omega) = |H(\omega)|^2 G_x(\omega)$$

$$= \frac{k}{1 + \omega^2}$$

The output power is given by

$$P_y = \frac{1}{2\pi} \int_{-\infty}^{\infty} G_y(\omega)\, d\omega$$

$$= \frac{1}{2\pi} \int_{-\infty}^{\infty} \frac{k}{1+\omega^2}\, d\omega$$

$$= \frac{k}{2\pi} \tan^{-1}(\omega)\Big|_{-\infty}^{\infty}$$

$$= \frac{k}{2\pi} \left[\frac{\pi}{2} + \frac{\pi}{2}\right]$$

$$P_y = \frac{k}{2}$$

The output power is less than the input power.
(b) The input PSD is defined as

$$G_x(\omega) = \begin{cases} 1, & -1 < \omega < 1 \\ 0, & \text{otherwise} \end{cases}$$

The power contained in the input signal is given by

$$P_x = \frac{1}{2\pi} \int_{-\infty}^{\infty} G_x(\omega)\, d\omega$$

$$= \frac{1}{2\pi} \int_{-1}^{1} 1\, d\omega$$

$$= \frac{1}{2\pi} \omega\Big|_{-1}^{1}$$

$$P_x = \frac{1}{\pi}$$

The output PSD is given by

$$G_y(\omega) = |H(\omega)|^2 G_x(\omega)$$

$$G_y(\omega) = \frac{1}{1+\omega^2}, \qquad -1 < \omega < 1$$

The output power is given by

$$P_y = \frac{1}{2\pi} \int_{-\infty}^{\infty} G_y(\omega)\, d\omega$$

$$= \frac{1}{2\pi} \int_{-1}^{1} \frac{1}{1+\omega^2}\, d\omega$$

$$P_y = \frac{1}{2\pi} \tan^{-1}(\omega)\Big|_{-1}^{1}$$

$$= \frac{1}{2\pi} \left[\frac{\pi}{4} + \frac{\pi}{4}\right]$$

$$P_y = \frac{1}{4}$$

The output power is less than the input power.
(c) The input PSD is given by

$$G_x(\omega) = \pi[\delta(\omega+1) + \delta(\omega-1)]$$

The power contained in the input signal is given by

$$P_x = \frac{1}{2\pi} \int_{-\infty}^{\infty} G_x(\omega)\, d\omega$$

$$= \frac{1}{2\pi} \int_{-\infty}^{\infty} \pi[\delta(\omega+1) + \delta(\omega-1)]\, d\omega$$

$$= \frac{\pi}{2\pi} \left[\int_{-\infty}^{\infty} \delta(\omega+1)\, d\omega + \int_{-\infty}^{\infty} \delta(\omega-1)\, d\omega\right]$$

$$P_x = \frac{1}{2}[1+1]$$

$$P_x = 1$$

The output PSD is given by

$$G_y(\omega) = |H(\omega)|^2 G_x(\omega)$$

$$G_y(\omega) = \frac{\pi}{1+\omega^2}[\delta(\omega+1) + \delta(\omega-1)]$$

The output power is given by

$$P_y = \frac{1}{2\pi} \int_{-\infty}^{\infty} G_y(\omega)\, d\omega$$

$$= \frac{1}{2\pi} \int_{-\infty}^{\infty} \frac{\pi}{1+\omega^2}[\delta(\omega+1) + \delta(\omega-1)]\, d\omega$$

$$= \frac{1}{2} \left[\int_{-\infty}^{\infty} \frac{1}{1+\omega^2}\delta(\omega+1)\, d\omega \right.$$

$$\left. + \int_{-\infty}^{\infty} \frac{1}{1+\omega^2}\delta(\omega-1)\, d\omega\right]$$

$$= \frac{1}{2}\left[\left.\frac{1}{1+\omega^2}\right|_{\omega=-1} + \left.\frac{1}{1+\omega^2}\right|_{\omega=1}\right]$$

$$= \frac{1}{2}\left[\frac{1}{2} + \frac{1}{2}\right]$$

$$P_y = \frac{1}{2}$$

The output power is less than the input power.

SUMMARY

1. The Fourier transform of $x(t)$ is defined by

$$X(\omega) = \int_{-\infty}^{\infty} x(t)\, e^{-j\omega t}\, dt$$

2. The inverse Fourier transform of $X(\omega)$ is defined by

$$x(t) = \frac{1}{2\pi}\int_{-\infty}^{\infty} X(\omega)\, e^{j\omega t}\, d\omega$$

3. The magnitude of $X(\omega)$ plotted against ω is called the magnitude spectrum of $x(t)$ and $|X(\omega)|^2$ is called the energy spectrum.

4. The angle of $X(\omega)$ plotted versus ω is called the phase spectrum.

5. Parseval's theorem states that

$$E_x = \int_{-\infty}^{\infty} |x(t)|^2\, dt = \frac{1}{2\pi}\int_{-\infty}^{\infty} |X(\omega)|^2\, d\omega$$

6. The PSD of $x(t)$ is defined by

$$G_x(\omega) = \lim_{T\to\infty} \frac{|X_T(\omega)|^2}{T}$$

where

$$x_T(t) \longleftrightarrow X_T(\omega)$$

and

$$x_T(t) = \begin{cases} x(t), & -T/2 < t < T/2 \\ 0, & \text{otherwise} \end{cases}$$

7. The more a signal is localized in one domain (time or frequency), the less it is localized in the other domain.

8. Convolution and multiplication of functions are dual in the time and frequency domains.

9. The Fourier transform of a periodic signal consists only of impulses.

10. Signal energy is conserved in the Fourier transformation process.

11. The correlation function indicates how correlated two signals are as a function of how much one of them is shifted in time.

12. Correlation and convolution are closely related mathematical processes.

13. The correlation of a signal with a shifted version of itself is called autocorrelation.

14. Autocorrelation is closely related to signal energy or power and contains important information about how rapidly a signal varies in time.

15. ESD and PSD are the frequency-domain counterparts of autocorrelation, related through the Fourier transform.

16. ESD and PSD indicated how the energy or power of a signal varies with frequency.

MULTIPLE-CHOICE QUESTIONS

1. The Fourier transform of a signal $x(t) = e^{2t}u(-t)$ is given by

   (a) $\dfrac{1}{2 - j\omega}$

   (b) $\dfrac{2}{1 - j\omega}$

   (c) $\dfrac{1}{j2 - \omega}$

   (d) $\dfrac{2}{j2 - \omega}$

2. The Fourier transform of a unit step function

   (a) does not exist
   (b) is another unit step
   (c) contains impulse functions
   (d) is $\dfrac{1}{j\omega}$

3. If the notation $*$ is used to denote the convolution, and $x(t) \longleftrightarrow X(\omega)$, $y(t) \longleftrightarrow Y(\omega)$ makes the Fourier transform pair, then $x(t)y(t) \longleftrightarrow F(\omega)$ is given by

(a) $X(\omega) * Y(\omega)$

(b) $X(\omega)Y(\omega)$

(c) $\dfrac{1}{2\pi} X(\omega)Y(\omega)$

(d) $\dfrac{1}{2\pi}[X(\omega) * Y(\omega)]$

4. The Fourier transform of a rectangular pulse is

(a) another rectangular pulse

(b) triangular pulse

(c) sinc function

(d) impulse.

5. The property of Fourier transform that states that the compression in time domain is equivalent to expansion in the frequency domain is

(a) duality

(b) time shifting

(c) time scaling

(d) frequency shifting

6. The spectrum of $x(t)$ extends from $-\omega_0$ to $+\omega_0$, while that of $h(t)$ extends from $-2\omega_0$ to $+2\omega_0$. The spectrum of

$$y(t) = \int_{-\infty}^{\infty} h(\tau)x(t-\tau)\, d\tau \text{ extends from}$$

(a) $-4\omega_0$ to $+4\omega_0$

(b) $3\omega_0$ to $+3\omega_0$

(c) $-2\omega_0$ to $+2\omega_0$

(d) $-\omega_0$ to $+\omega_0$

7. Inverse Fourier transform of $U(\omega)$ is

(a) $\dfrac{1}{2}\delta(t) + \dfrac{j}{2\pi t}$

(b) $\dfrac{1}{2}\delta(t)$

(c) $2\delta(t) + \dfrac{j}{\pi t}$

(d) $\delta(t) + \text{sgn}(t)$

8. The Fourier transform of a function $x(t)$ is $X(f)$. The Fourier transform of $\dfrac{dx(t)}{dt}$ will be

(a) $\dfrac{dX(f)}{df}$

(b) $j2\pi f X(f)$

(c) $jfX(f)$

(d) $\dfrac{X(f)}{jf}$

9. The frequency of a continuous-time signal $x(t)$ changes on transformation from $x(t)$ to $x(\alpha t)$, $\alpha > 0$, by a factor

(a) α

(b) $\dfrac{1}{\alpha}$

(c) α^2

(d) $\sqrt{\alpha}$

10. The Fourier transform of the exponential signal $e^{j\omega_0 t}$ is

(a) a constant

(b) a rectangular pulse

(c) an impulse

(d) a series of impulses

11. The Fourier transform of $u(t)$ is

(a) $\dfrac{1}{j2\pi f}$

(b) $j2\pi f$

(c) $\dfrac{1}{1 + j2\pi f}$

(d) none of these

12. The Fourier transform of a real and even signal $x(t)$ is

(a) real and even function of frequency.

(b) real and odd function of frequency.

(c) complex, in general.

(d) purely imaginary.

13. The Fourier transform of a real and odd signal $x(t)$ is

(a) imaginary and odd function of frequency.

(b) real and odd function of frequency.

(c) complex, in general.

(d) purely imaginary.

14. Fourier transform of a d.c. signal with unity strength is

(a) zero

(b) 1

(c) $2\pi\delta(\omega)$

(d) $2\delta(\omega)$

15. For distortionless transmission through an LTI system, amplitude of $H(\omega)$ is

(a) constant

(b) one

(c) zero

(d) linearly dependent on ω

16. For distortionless transmission through an LTI system, phase of $H(\omega)$ is

(a) constant

(b) one

(c) zero

(d) linearly dependent on ω

17. The Fourier transform of sgn(t) is

(a) $\dfrac{2}{j\omega}$

(b) $-\dfrac{2}{j\omega}$

(c) 1

(d) zero

PROBLEMS

5.1 If the unit impulse response of a causal LTI system contains no impulses at the origin, then show that with

$$H(\omega) = H_R(\omega) + jH_I(\omega)$$

$H_R(\omega)$ and $H_I(\omega)$ satisfy the following conditions:

$$H_R(\omega) = \frac{1}{\pi} \int_{-\infty}^{\infty} \frac{H_I(\theta)}{\omega - \theta} \, d\theta$$

$$H_I(\omega) = -\frac{1}{\pi} \int_{-\infty}^{\infty} \frac{H_R(\theta)}{\omega - \theta} \, d\theta$$

5.2 The input and the output of a causal LTI system are related by the differential equation

$$\frac{d^2 y(t)}{dt^2} + 6\frac{dy(t)}{dt} + 8y(t) = 2x(t)$$

(a) Find the impulse response of this system.

(b) What is the response of this system if $x(t) = t\,e^{-2t}u(t)$?

(c) Repeat part (a) for the causal LTI system described by the equation

$$\frac{d^2 y(t)}{dt^2} + \sqrt{2}\frac{dy(t)}{dt} + y(t) = 2\frac{d^2 x(t)}{dt^2} - 2x(t)$$

5.3 Consider an LTI system whose response to the input

$$x(t) = [e^{-t} + e^{-3t}]u(t)$$

is

$$y(t) = [2e^{-t} - 2\,e^{-4t}]u(t)$$

(a) Find the frequency response of this system.

(b) Determine the systems' impulse response.

(c) Find the differential equation relating the input and output of this system.

5.4 Consider the signal

$$x(t) = \begin{cases} 0, & |t| > 1 \\ \dfrac{t+1}{2}, & -1 \le t \le 1 \end{cases}$$

(a) Determine the closed form expression for $X(\omega)$.

(b) Take the real part of your answer to part (a), and verify that it is the Fourier transform of the even part of $x(t)$.

(c) What is the Fourier transform of the odd part of $x(t)$.

5.5 Find the impulse response of a system with the frequency response

$$H(\omega) = \frac{\sin^2(3\omega)\cos(\omega)}{\omega^2}$$

5.6 Determine the inverse Fourier transform of $X(\omega) = (4/\omega^2)\sin^2(\omega)$, using the convolution property of the Fourier transform.

5.7 Let $X(\omega) = \text{rect}\,(\omega - 1/2)$. Using the Fourier transform properties, find the Fourier transform of the following functions:

(a) $x(-t)$

(b) $tx(t)$

(c) $x(t+1)$

(d) $x(-2t+4)$

(e) $(t-1)x(t+1)$

(f) $\dfrac{dx(t)}{dt}$

(g) $t\frac{dx(t)}{dt}$

(h) $x(2t-1)\,e^{-j2t}$

(i) $x(t)\,e^{-j2t}$

5.8 Let $x(t) = e^{-2t}u(t)$ and let $y(t) = x(t+1) + x(t-1)$. Find $Y(\omega)$.

5.9 Find the energy of the following signals, using Parseval's theorem.

(a) $x(t) = e^{-2t}u(t)$

(b) $x(t) = u(t) - u(t-5)$

(c) $x(t) = \dfrac{\sin(\pi t)}{\pi t}$

5.10 A Gaussian-shaped signal

$$x(t) = A\,e^{-at^2}$$

is applied to a system whose input/output relationship is

$$y(t) = x^2(t)$$

Find the Fourier transform of the output $y(t)$.

5.11 The signal $x(t) = e^{-at}u(t)$ is input into a system with impulse response $h(t) = \dfrac{\sin(2t)}{\pi t}$.

(a) Find the Fourier transform $Y(\omega)$ of the output.

(b) For what value of a does the energy in the output equal one-half the input signal energy.

5.12 A signal $x(t)$ can be expressed as the sum of even and odd components:

$$x(t) = x_e(t) + x_o(t)$$

(a) If $x(t) \longleftrightarrow X(\omega)$, show that for real $x(t)$,

$$x_e(t) \longleftrightarrow \Re\{X(\omega)\}$$

and

$$x_o(t) \longleftrightarrow j\Im\{X(\omega)\}$$

(b) Verify these results by finding the Fourier transform of the even and odd components of the following signals: (i) $u(t)$ and (ii) $e^{-at}u(t)$.

5.13 Show that the energy of the Gaussian pulse

$$x(t) = \frac{1}{\sigma\sqrt{2\pi}}\, e^{-\frac{t^2}{2\sigma^2}}$$

is $\dfrac{1}{2\sigma\sqrt{\pi}}$. Verify this result by using Parseval's theorem to derive the energy E_x from $X(\omega)$.

5.14 If $x(t) = 25\ \mathrm{rect}\left(\dfrac{t-4}{10}\right)$, find the Fourier transform of $x(t)$.

5.15 What is the total area under the function,

$$x(t) = 100\,\mathrm{sinc}\left(\frac{t-8}{30}\right).$$

5.16 Find the Fourier transform of the signal $x(t) = e^{-a|t|}\,\mathrm{sgn}(t)$.

ANSWERS TO MULTIPLE-CHOICE QUESTIONS

1. (a)   2. (c)   3. (d)   4. (c)   5. (b)   6. (d)   7. (a)   8. (b)   9. (b)
10. (c)   11. (d)   12. (a)   13. (a)   14. (c)   15.(a)   16. (d)   17. (a)

Discrete-Time Fourier Transform

6.1 INTRODUCTION

In the previous chapter, we introduced the continuous-time Fourier transform and developed the many characteristics of that transform which makes the methods of Fourier analysis of such great value in analysing and understanding the properties of continuous-time signals and systems. In this chapter, we complete our development of the basic tools of Fourier analysis by introducing and examining the discrete-time Fourier transform.

6.2 FOURIER TRANSFORM REPRESENTATION OF APERIODIC DISCRETE-TIME SIGNALS

This derivation proceeds along a path analogous to the derivation of the CTFT from the CTFS. The DTFS representation of a periodic signal $x(n)$ (with period N and frequency $\omega_0 = 2\pi/N$) can be written as

$$x(n) = \sum_{k=\langle N \rangle} X_k \, e^{jk\omega_0 n}$$

Substituting $X_k = 1/N \sum_{n=\langle N \rangle} x(n) \, e^{-jk\omega_0 n}$ in the DTFS definition yields

$$x(n) = \sum_{k=\langle N \rangle} \left(\frac{1}{N} \sum_{n=\langle N \rangle} x(n) \, e^{-jk\omega_0 n} \right) e^{jk\omega_0 n}$$

or

$$x(n) = \sum_{k=\langle N \rangle} \left(\sum_{n=\langle N \rangle} x(n) \, e^{-jk\omega_0 n} \right) e^{jk\omega_0 n} \frac{\omega_0}{2\pi}$$

Since the inner summation is over any arbitrary range of n of width N, let the range be $-(N/2) \leq n \leq (N/2) - 1$ for N even or $-(N - 1/2) \leq n \leq (N - 1/2)$ for N odd. The outer summation is over any arbitrary range of k of width N, so let its range be

$k_0 \leq k \leq (k_0 + N - 1)$. Then

$$x(n) = \sum_{k=k_0}^{k_0+N-1} \left(\sum_{n=-\frac{N}{2}}^{\frac{N}{2}-1} x(n)\, e^{-jk\omega_0 n} \right) e^{jk\omega_0 n} \frac{\omega_0}{2\pi}, \quad \text{for } N \text{ even} \tag{6.1}$$

or

$$x(n) = \sum_{k=k_0}^{k_0+N-1} \left(\sum_{n=-\frac{N-1}{2}}^{\frac{N-1}{2}} x(n)\, e^{-jk\omega_0 n} \right) e^{jk\omega_0 n} \frac{\omega_0}{2\pi}, \quad \text{for } N \text{ odd} \tag{6.2}$$

Now let the fundamental period $N \to \infty$. In that limit $\omega_0/2\pi \to d\omega/2\pi$ and $k\omega_0 \to \omega$. The inner summation covers an infinite range. The outer summation approaches an integral in $\omega = k\omega_0$ covering a range of

$$k_0 \leq k \leq (k_0 + N - 1)$$
$$k_0 \leq k < (k_0 + N)$$
$$k_0 \leq \frac{\omega}{\omega_0} < (k_0 + N)$$
$$k_0\omega_0 < \omega < (k_0\omega_0 + N\omega_0)$$
$$k_0\omega_0 < \omega < (k_0\omega_0 + 2\pi)$$

Therefore, Eqs (6.1) and (6.2) both become

$$x(n) = \frac{1}{2\pi} \int_{2\pi} \left(\sum_{n=-\infty}^{\infty} x(n)\, e^{-j\omega n} \right) e^{j\omega n}\, d\omega \tag{6.3}$$

where $\int_{2\pi}$ indicates integration over any continuous interval of 2π. The inner summation of the above equation is a function of only ω and not n. Denoting the summation by $X(e^{j\omega})$, we can write Eq. (6.3) as

$$x(n) = \frac{1}{2\pi} \int_{2\pi} X(e^{j\omega})\, e^{j\omega n}\, d\omega \tag{6.4}$$

where

$$X(e^{j\omega}) = \sum_{n=-\infty}^{\infty} x(n)\, e^{-j\omega n} \tag{6.5}$$

Equation (6.4) is usually referred to as the *synthesis equation* because it synthesizes an arbitrary signal from its complex exponential components. On the other hand, Eq. (6.5) is referred to as the *analysis equation* because it analyses how much of each complex exponential signal is present in the original signal. We call $X(e^{j\omega})$ the discrete-time Fourier transform (DTFT) of $x(n)$, and $x(n)$ the inverse discrete-time Fourier transform (IDTFT) of $X(e^{j\omega})$. This nomenclature can be represented as

$$X(e^{j\omega}) = \text{DTFT}[x(n)] = \mathcal{F}[x(n)] \quad \text{and} \quad x(n) = \text{IDTFT}[X(e^{j\omega})] = \mathcal{F}^{-1}[X(e^{j\omega})]$$

The same information is conveyed by the statement that $x(n)$ and $X(e^{j\omega})$ are a (discrete-time) Fourier transform pair. Symbolically, this is expressed as

$$x(n) \longleftrightarrow X(e^{j\omega})$$

In general the Fourier transform $X(e^{j\omega})$ is a complex function of the real variable ω and can be written in the rectangular form as

$$X(e^{j\omega}) = X_R(e^{j\omega}) + jX_I(e^{j\omega}) \tag{6.6}$$

where $X_R(e^{j\omega})$ and $X_I(e^{j\omega})$ are, respectively, the real and imaginary parts of $X(e^{j\omega})$. From Eq. (6.6), it follows that

$$X_R(e^{j\omega}) = \frac{1}{2}[X(e^{j\omega}) + X^*(e^{j\omega})] \tag{6.7}$$

$$X_I(e^{j\omega}) = \frac{1}{2j}[X(e^{j\omega}) - X^*(e^{j\omega})] \tag{6.8}$$

where $X^*(e^{j\omega})$ denotes the complex conjugate of $X(e^{j\omega})$.

The Fourier transform $X(e^{j\omega})$ can alternatively be expressed in the polar form as

$$X(e^{j\omega}) = |X(e^{j\omega})| \, e^{j\theta(\omega)} \tag{6.9}$$

where

$$\theta(\omega) = \angle X(e^{j\omega}) \tag{6.10}$$

The quantity $|X(e^{j\omega})|$ is called the *magnitude spectrum* and the quantity $\theta(\omega) = \angle X(e^{j\omega})$ is called the *phase spectrum*.

The relations between the rectangular and polar forms of $X(e^{j\omega})$ follow from Eqs (6.6) and (6.9) and are given by

$$X_R(e^{j\omega}) = |X(e^{j\omega})| \cos(\theta(\omega))$$

$$X_I(e^{j\omega}) = |X(e^{j\omega})| \sin(\theta(\omega))$$

$$|X(e^{j\omega})| = \sqrt{X_R^2(e^{j\omega}) + X_I^2(e^{j\omega})} \tag{6.11}$$

$$\theta(\omega) = \angle X(e^{j\omega}) = \tan^{-1}\left[\frac{X_I(e^{j\omega})}{X_R(e^{j\omega})}\right] \tag{6.12}$$

Thus, for a real signal, it follows from the Eq. (6.11) that

$$|X(e^{-j\omega})| = \sqrt{X_R^2(e^{-j\omega}) + X_I^2(e^{-j\omega})}$$

$$|X(e^{-j\omega})| = \sqrt{X_R^2(e^{j\omega}) + X_I^2(e^{j\omega})}$$

$$|X(e^{-j\omega})| = |X(e^{j\omega})| \tag{6.13}$$

The magnitude spectrum $|X(e^{j\omega})|$ is an even function of ω. Likewise, for a real signal, we note from Eq. (6.12) that

$$\angle X(e^{-j\omega}) = \tan^{-1}\left[\frac{X_I(e^{-j\omega})}{X_R(e^{-j\omega})}\right] = \tan^{-1}\left[-\frac{X_I(e^{j\omega})}{X_R(e^{j\omega})}\right] = -\tan^{-1}\left[\frac{X_I(e^{j\omega})}{X_R(e^{j\omega})}\right] = -\angle X(e^{j\omega})$$

$$\angle X(e^{-j\omega}) = -\angle X(e^{j\omega}) \tag{6.14}$$

The above equation implies that the phase spectrum $\angle X(e^{j\omega})$ is an odd function of ω.

6.3 PERIODICITY OF THE DTFT

The DTFT is a periodic function in ω with a period 2π, i.e.,

$$X(e^{j(\omega+2\pi)}) = X(e^{j\omega}) \tag{6.15}$$

Proof We know that

$$X(e^{j\omega}) = \sum_{n=-\infty}^{\infty} x(n)\, e^{-j\omega n}$$

For any integer k, we have

$$X(e^{j(\omega+2\pi k)}) = \sum_{n=-\infty}^{\infty} x(n)\, e^{-j(\omega+2\pi k)n}$$

$$= \sum_{n=-\infty}^{\infty} x(n)\, e^{-j\omega n}\, e^{-j2\pi k n}$$

$$X(e^{j(\omega+2\pi k)}) = \sum_{n=-\infty}^{\infty} x(n)\, e^{-j\omega n}$$

$$X(e^{j(\omega+2\pi k)}) = X(e^{j\omega})$$

where we have used the fact that $e^{-j2\pi k n} = 1$. Hence $X(e^{j\omega})$ is periodic with period 2π. But this property is just a consequence of the fact that frequency range for any discrete-time signal is unique over the frequency interval of $(-\pi, \pi)$ or, equivalently, $(0, 2\pi)$, and any frequency outside this interval is equivalent to a frequency within this interval.

6.4 CONVERGENCE OF DTFT

An infinite series of Eq. (6.5) may or may not converge. The Fourier transform $X(e^{j\omega})$ of $x(n)$ is said to exist if the series in Eq. (6.5) converges in some sense. Let

$$X_K(e^{j\omega}) = \sum_{n=-K}^{K} x(n)\, e^{-j\omega n} \tag{6.16}$$

denote the partial sum of the weighted complex exponentials in Eq. (6.5). Then for uniform convergence of $X(e^{j\omega})$,

$$\lim_{K \to \infty} [X(e^{j\omega}) - X_K(e^{j\omega})] = 0$$

$$\lim_{K \to \infty} X_K(e^{j\omega}) = X(e^{j\omega})$$

Uniform convergence is guaranteed if $x(n)$ is absolutely summable. Indeed, if

$$\sum_{n=-\infty}^{\infty} |x(n)| < \infty \tag{6.17}$$

then

$$\left| X(e^{j\omega}) \right| = \left| \sum_{n=-\infty}^{\infty} x(n) e^{-j\omega n} \right|$$

$$\leq \sum_{n=-\infty}^{\infty} |x(n)| \left| e^{-j\omega n} \right|$$

$$\left| X(e^{j\omega}) \right| \leq \sum_{n=-\infty}^{\infty} |x(n)| < \infty$$

for all values of ω guaranteeing the existence of $X(e^{j\omega})$. Thus, Eq. (6.17) is a sufficient condition for the existence of the Fourier transform $X(e^{j\omega})$ of the sequence $x(n)$. We note that this is the discrete-time counterpart of the first Dirichlet condition for the Fourier transform of continuous-time signals. The last two conditions do not apply because of the discrete-time nature of $x(n)$.

A large class of sequences encountered in practice are of finite length with finite sample values. These sequences are absolutely summable, and hence, their Fourier transform converge uniformly. Some sequences are not absolutely summable, but they are square summable. Since

$$\sum_{n=-\infty}^{\infty} |x(n)|^2 \leq \left(\sum_{n=-\infty}^{\infty} |x(n)| \right)^2$$

an absolutely summable sequence always has a finite energy. However, a finite-energy sequence is not necessarily absolutely summable. We would like to define the Fourier transform of finite-energy sequences, but we may relax the condition of uniform convergence. For such sequences we can impose a mean-square convergence condition:

$$\lim_{K \to \infty} \int_{-\pi}^{\pi} \left| X(e^{j\omega}) - X_K(e^{j\omega}) \right|^2 d\omega = 0 \tag{6.18}$$

Thus, the energy in the error $X(e^{j\omega}) - X_K(e^{j\omega})$ tends towards zero, but the error $|X(e^{j\omega}) - X_K(e^{j\omega})|$ does not necessarily tend to zero. In this way we can include finite-energy signals in the class of signals for which the Fourier transform exists. These can also be defined for a certain class of sequences that are neither absolutely summable nor square-summable. Example of such sequences are the unit step sequence, the sinusoidal sequence, and the complex exponential sequence, which are neither absolutely summable nor square-summable.

6.4.1 Gibbs Phenomenon

Let us consider an example from the class of finite-energy signals. Suppose that

$$X(e^{j\omega}) = \begin{cases} 1, & |\omega| \leq \omega_c \\ 0, & \omega_c < |\omega| \leq \pi \end{cases}$$

The inverse DTFT of $X(e^{j\omega})$ is given by

$$x(n) = \frac{1}{2\pi} \int_{-\pi}^{\pi} X(e^{j\omega}) e^{j\omega n} d\omega$$

$$= \frac{1}{2\pi} \int_{-\omega_c}^{\omega_c} e^{j\omega n} d\omega$$

$$x(n) = \frac{1}{2\pi} \frac{e^{j\omega n}}{jn} \Big|_{-\omega_c}^{\omega_c}$$

$$x(n) = \frac{\sin(\omega_c n)}{n\pi}, \quad n \neq 0$$

For $n = 0$, the inverse Fourier transform expression reduces to

$$x(0) = \frac{1}{2\pi} \int_{-\pi}^{\pi} X(e^{j\omega}) d\omega = \frac{1}{2\pi} \int_{-\omega_c}^{\omega_c} d\omega = \frac{\omega_c}{\pi}$$

Hence

$$x(n) = \begin{cases} \dfrac{\omega_c}{\pi}, & n = 0 \\ \dfrac{\sin(\omega_c n)}{n\pi}, & n \neq 0 \end{cases} \tag{6.19}$$

It should be noted that often the above sequence is expressed in a compact form as

$$x(n) = \frac{\sin(\omega_c n)}{n\pi}, \quad -\infty < n < \infty \tag{6.20}$$

with the understanding that at $n = 0$, $x(n) = \omega_c/\pi$. Now let us consider the determination of the Fourier transform of the sequence given by Eq. (6.20). The sequence $x(n)$ is not absolutely summable. Hence the infinite series

$$\sum_{n=-\infty}^{\infty} x(n) e^{-j\omega n} = \sum_{n=-\infty}^{\infty} \frac{\sin(\omega_c n)}{n\pi} e^{-j\omega n} \tag{6.21}$$

does not converge uniformly for all ω. We shall show later that the energy of the above sequence is given by ω_c/π, and therefore, $x(n)$ is a finite-energy sequence. Hence, the sum in the Eq. (6.21) is guaranteed to converge to the $X(e^{j\omega})$ in the mean-square sense.

To elaborate on this point, let us consider the finite sum

$$X_K(e^{j\omega}) = \sum_{n=-K}^{K} \frac{\sin(\omega_c n)}{n\pi} e^{-j\omega n} \qquad (6.22)$$

Figure 6.1 shows the function $X_K(e^{j\omega})$ for several values of K. We note that there is a significant oscillatory overshoot at $\omega = \omega_c$, independent of the value of K. As K increases, the oscillations become more rapid, but the size of the ripple remains the same. One can show that as $K \to \infty$, the oscillations converges to the point of the discontinuity at $\omega = \omega_c$, but their amplitude does not go to zero. However, Eq. (6.18) is satisfied, and therefore, $X_K(e^{j\omega})$ converges to $X(e^{j\omega})$ in the mean-square sense.

The oscillatory behaviour of the approximation $X_K(e^{j\omega})$ to the function $X(e^{j\omega})$ at a point of discontinuity of $X(e^{j\omega})$ is called the *Gibbs phenomenon*.

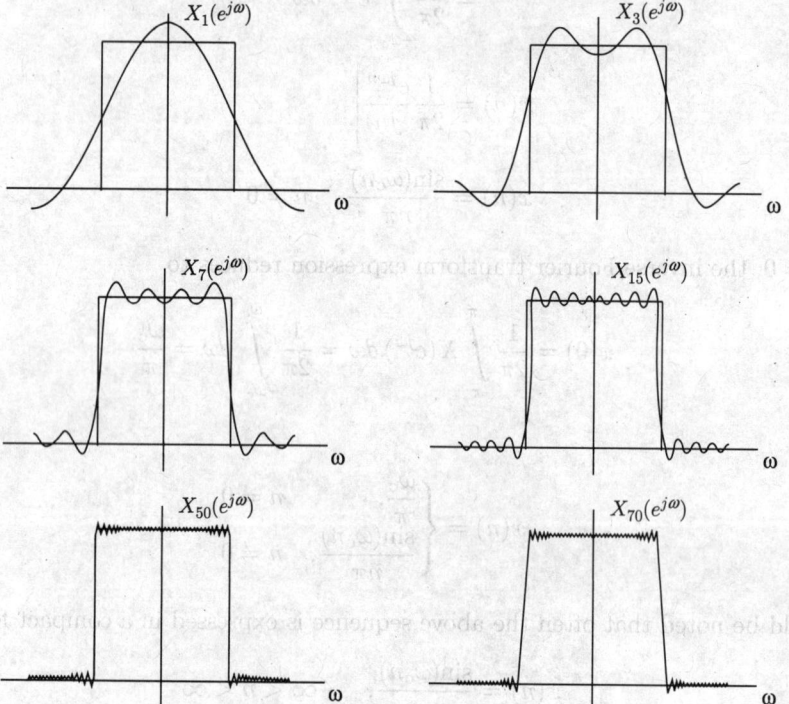

Fig. 6.1 Illustration of Gibbs' phenomenon at the point of discontinuity

Example 6.1 Find the Fourier transform of the unit sample sequence $x(n) = \delta(n)$.

Solution

From Eq. (6.5), the Fourier transform $X(e^{j\omega})$ of the sequence $x(n)$ is given by

$$X(e^{j\omega}) = \sum_{n=-\infty}^{\infty} x(n) e^{-j\omega n} = \sum_{n=-\infty}^{\infty} \delta(n) e^{-j\omega n} = e^{-j\omega n}\Big|_{n=0} = 1$$

$$\delta(n) \longleftrightarrow 1 \qquad (6.23)$$

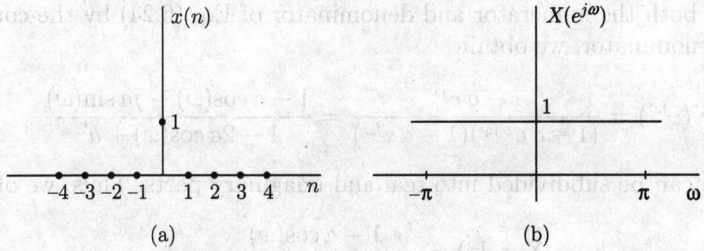

Fig. 6.2 (a) Unit impulse sequence and (b) its Fourier transform

The unit sample sequence has a Fourier transform representation consisting of equal contributions at all frequencies. The unit sample sequence and its Fourier transform are shown in Figs 6.2(a) and (b), respectively.

Example 6.2 Find the Fourier transform of the causal sequence [shown in Fig. 6.3(a)]

$$x(n) = a^n u(n), \quad |a| < 1.$$

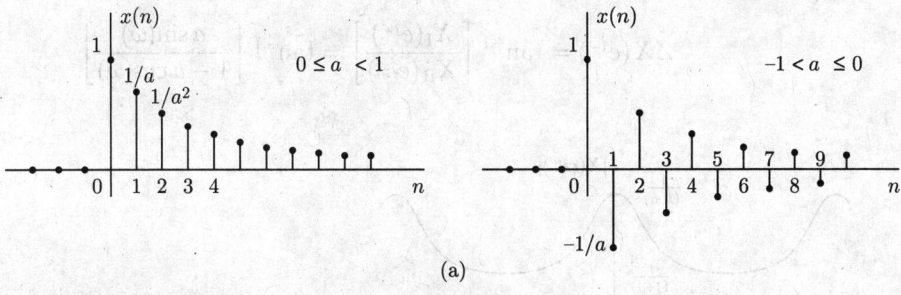

(a)

Fig. 6.3(a) Signal $x(n)$

Solution

From Eq.(6.5), the Fourier transform $X(e^{j\omega})$ of the sequence $x(n)$ is given by

$$X(e^{j\omega}) = \sum_{n=-\infty}^{\infty} x(n)\, e^{-j\omega n} = \sum_{n=-\infty}^{\infty} a^n u(n)\, e^{-j\omega n}$$

$$= \sum_{n=0}^{\infty} a^n\, e^{-j\omega n}$$

$$X(e^{j\omega}) = \sum_{n=0}^{\infty} (a\, e^{-j\omega})^n$$

$$X(e^{j\omega}) = \frac{1}{1 - a\, e^{-j\omega}}$$

$$a^n u(n) \longleftrightarrow \frac{1}{1 - a\, e^{-j\omega}}, \quad |a| < 1 \qquad (6.24)$$

By multiplying both the numerator and denominator of Eq. (6.24) by the complex conjugate of the denominator, we obtain

$$X(e^{j\omega}) = \frac{1 - a\,e^{j\omega}}{(1 - a\,e^{-j\omega})(1 - a\,e^{j\omega})} = \frac{1 - a\cos(\omega) - ja\sin(\omega)}{1 - 2a\cos(\omega) + a^2}$$

This expression can be subdivided into real and imaginary parts. Thus, we obtain

$$X_R(e^{j\omega}) = \frac{1 - a\cos(\omega)}{1 - 2a\cos(\omega) + a^2}$$

$$X_I(e^{j\omega}) = -\frac{a\sin(\omega)}{1 - 2a\cos(\omega) + a^2}$$

Using $X_R(e^{j\omega})$ and $X_I(e^{j\omega})$, the magnitude spectrum is given by

$$|X(e^{j\omega})| = \sqrt{X_R^2(e^{j\omega}) + X_I^2(e^{j\omega})} = \frac{1}{\sqrt{1 - 2a\cos(\omega) + a^2}}$$

and the phase spectrum is given by

$$\angle X(e^{j\omega}) = \tan^{-1}\left[\frac{X_I(e^{j\omega})}{X_R(e^{j\omega})}\right] = \tan^{-1}\left[\frac{a\sin(\omega)}{1 - a\cos(\omega)}\right]$$

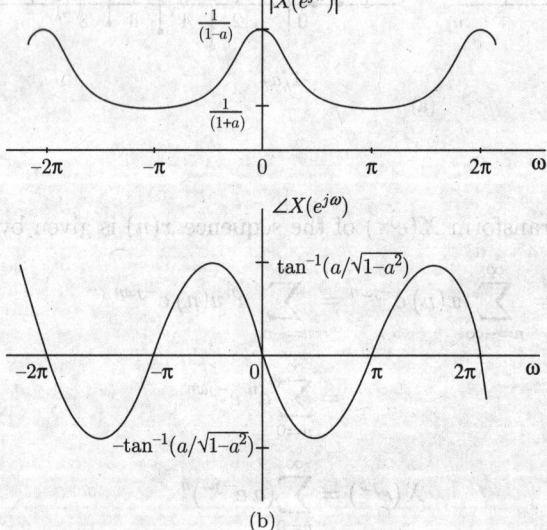

(b)

Fig. 6.3(b)   Magnitude and phase of the Fourier transform of Example 6.2 for *0<a<1*

The magnitude and phase spectrum are shown in Fig. 6.3(b) for $0 < a < 1$ and in Fig. 6.3(c) for $-1 < a < 0$. Note that the magnitude and phase spectrum are periodic in ω with period 2π.

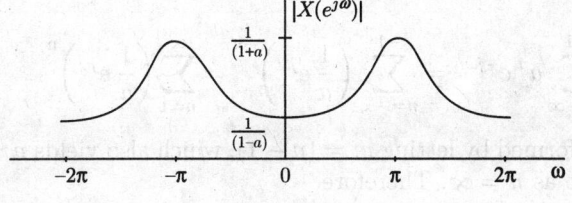

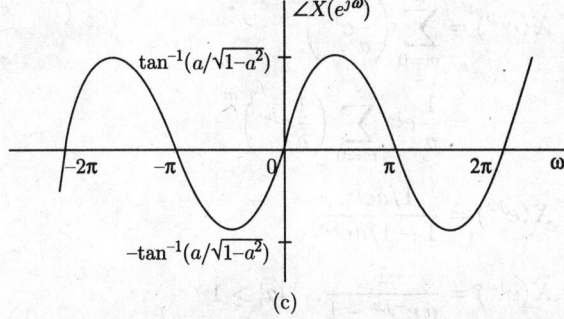

(c)

Fig. 6.3(c) Magnitude and phase of the Fourier transform for $-1 < a < 0$

Example 6.3 Find the Fourier transform of the sequence [shown in Fig. 6.4]

$$x(n) = a^n u(-n-1), \quad |a| > 1.$$

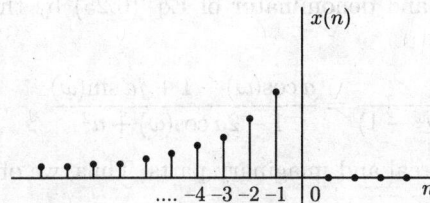

Fig. 6.4

Solution

From Eq. (6.5), the Fourier transform $X(e^{j\omega})$ of the sequence $x(n)$ is given by

$$X(e^{j\omega}) = \sum_{n=-\infty}^{\infty} x(n)\, e^{-j\omega n} = \sum_{n=-\infty}^{\infty} a^n u(-n-1)\, e^{-j\omega n}$$

Since

$$u(-n-1) = \begin{cases} 1, & (-n-1) \geq 0 \longrightarrow n \leq -1 \\ 0, & (-n-1) < 0 \longrightarrow n > -1 \end{cases}$$

We have

$$X(e^{j\omega}) = \sum_{n=-\infty}^{-1} a^n e^{-j\omega n} = \sum_{n=-\infty}^{-1} \left(\frac{1}{a} e^{j\omega}\right)^{-n} = \sum_{n=1}^{\infty} \left(\frac{1}{a} e^{j\omega}\right)^n$$

A change of variables is performed by letting $m = (n-1)$, which also yields $n = (m+1)$, $m = 0$ as $n = 1$ and $m = \infty$ as $n = \infty$. Therefore,

$$X(e^{j\omega}) = \sum_{m=0}^{\infty} \left(\frac{1}{a} e^{j\omega}\right)^{m+1}$$

$$= \frac{1}{a} e^{j\omega} \sum_{m=0}^{\infty} \left(\frac{1}{a} e^{j\omega}\right)^m$$

$$X(e^{j\omega}) = \frac{1/a e^{j\omega}}{1 - 1/a e^{j\omega}}$$

$$X(e^{j\omega}) = \frac{1}{a e^{-j\omega} - 1}, \quad |a| > 1$$

$$a^n u(-n-1) \longleftrightarrow \frac{1}{a e^{-j\omega} - 1}, \quad |a| > 1 \qquad (6.25)$$

Except for the change of sign, this Fourier transform is identical to that of $x(n) = a^n u(n)$, $|a| < 1$. Yet there is no ambiguity in determining the IDTFT of $X(e^{j\omega}) = \dfrac{1}{a e^{-j\omega} - 1}$ because of the restrictions on the value of a in each case. If $|a| < 1$, then its IDTFT is $x(n) = -a^n u(n)$. If $|a| > 1$, its IDTFT is $x(n) = a^n u(-n-1)$.

By multiplying both the numerator and denominator of Eq. (6.25) by the complex conjugate of the denominator, we obtain

$$X(e^{j\omega}) = \frac{a e^{j\omega} - 1}{(a e^{-j\omega} - 1)(a e^{j\omega} - 1)} = \frac{a \cos(\omega) - 1 + j a \sin(\omega)}{1 - 2a \cos(\omega) + a^2}$$

This expression can be subdivided into real and imaginary parts. Thus, we obtain

$$X_R(e^{j\omega}) = \frac{a \cos(\omega) - 1}{1 - 2a \cos(\omega) + a^2}$$

$$X_I(e^{j\omega}) = \frac{a \sin(\omega)}{1 - 2a \cos(\omega) + a^2}$$

Using $X_R(e^{j\omega})$ and $X_I(e^{j\omega})$, the magnitude spectrum is given by

$$|X(e^{j\omega})| = \sqrt{X_R^2(e^{j\omega}) + X_I^2(e^{j\omega})} = \frac{1}{\sqrt{1 - 2a \cos(\omega) + a^2}}$$

and the phase spectrum is given by

$$\angle X(e^{j\omega}) = \tan^{-1}\left[\frac{X_I(e^{j\omega})}{X_R(e^{j\omega})}\right] = \tan^{-1}\left[\frac{a \sin(\omega)}{a \cos(\omega) - 1}\right]$$

Example 6.4 Find the Fourier transform of the anticausal sequence

$$x(n) = a^{-n}u(-n-1), \qquad |a| < 1$$

Solution
From Eq. (6.5), the Fourier transform $X(e^{j\omega})$ of the sequence $x(n)$ is given by

$$X(e^{j\omega}) = \sum_{n=-\infty}^{\infty} x(n)\, e^{-j\omega n} = \sum_{n=-\infty}^{\infty} a^{-n}u(-n-1)\, e^{-j\omega n}$$

Since

$$u(-n-1) = \begin{cases} 1, & (-n-1) \ge 0 \longrightarrow n \le -1 \\ 0, & (-n-1) < 0 \longrightarrow n > -1 \end{cases}$$

We have

$$X(e^{j\omega}) = \sum_{n=-\infty}^{-1} a^{-n}\, e^{-j\omega n} = \sum_{n=-\infty}^{-1} (a\, e^{j\omega})^{-n} = \sum_{n=1}^{\infty} (a\, e^{j\omega})^{n}$$

A change of variables is performed by letting $m = (n-1)$, which also yields $n = (m+1)$, $m = 0$ as $n = 1$ and $m = \infty$ as $n = \infty$. Therefore,

$$X(e^{j\omega}) = \sum_{m=0}^{\infty} (a\, e^{j\omega})^{m+1} = a\, e^{j\omega} \sum_{m=0}^{\infty} (a\, e^{j\omega})^{m} = \frac{a\, e^{j\omega}}{1 - a\, e^{j\omega}}$$

$$a^{-n}u(-n-1) \longleftrightarrow \frac{a\, e^{j\omega}}{1 - a\, e^{j\omega}}, \qquad |a| < 1 \qquad (6.26)$$

Example 6.5 Find the Fourier transform of the noncausal sequence

$$x(n) = a^{|n|}, \qquad |a| < 1$$

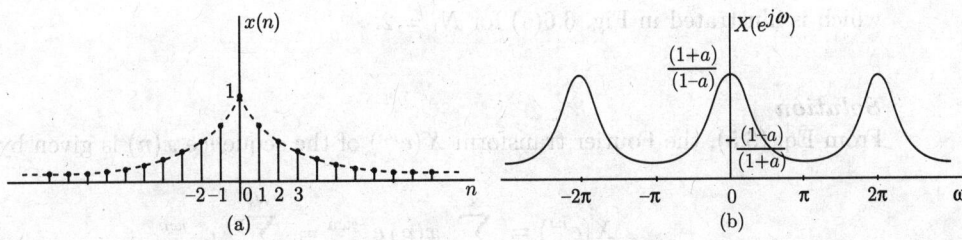

Fig. 6.5 (a) Signal $x(n) = a^{|n|}$ and (b) its Fourier transform

Solution

Consider the given sequence

$$x(n) = a^{|n|}, \quad |a| < 1$$

It can be written as

$$x(n) = \begin{cases} a^n, & n \geq 0 \\ a^{-n}, & n \leq -1 \end{cases}$$

$$x(n) = a^n u(n) + a^{-n} u(-n-1)$$

The signal $x(n)$ is plotted for $0 < a < 1$ in Fig. 6.5(a). Taking the Fourier transform of the above equation, we have

$$\mathcal{F}[x(n)] = \mathcal{F}[a^n u(n)] + \mathcal{F}[a^{-n} u(-n-1)]$$

Using Eqs (6.24) and (6.26), we have

$$X(e^{j\omega}) = \frac{1}{1 - a e^{-j\omega}} + \frac{a e^{j\omega}}{1 - a e^{j\omega}}$$

$$X(e^{j\omega}) = \frac{1 - a^2}{1 - 2a \cos(\omega) + a^2}$$

$$a^{|n|} \longleftrightarrow \frac{1 - a^2}{1 - 2a \cos(\omega) + a^2}, \quad |a| < 1 \tag{6.27}$$

In this case $X(e^{j\omega})$ is real and is illustrated in Fig. 6.5(b).

Example 6.6 Find the Fourier transform of the rectangular pulse

$$x(n) = u(n + N_1) - u(n - N_1 - 1) = \begin{cases} 1, & |n| \leq N_1 \\ 0, & |n| > N_1 \end{cases}$$

which is illustrated in Fig. 6.6(a) for $N_1 = 2$.

Solution

From Eq. (6.5), the Fourier transform $X(e^{j\omega})$ of the sequence $x(n)$ is given by

$$X(e^{j\omega}) = \sum_{n=-\infty}^{\infty} x(n) e^{-j\omega n} = \sum_{n=-N_1}^{N_1} e^{-j\omega n}$$

A change of variables is performed by letting $m = (n + N_1)$, which also yields $n = (m - N_1)$, $m = 0$ as $n = -N_1$ and $m = 2N_1$ as $n = N_1$. Therefore,

$$X(e^{j\omega}) = \sum_{m=0}^{2N_1} e^{-j\omega(m-N_1)}$$

$$X(e^{j\omega}) = e^{j\omega N_1} \sum_{m=0}^{2N_1} e^{-j\omega m}$$

$$= e^{j\omega N_1} \left(\frac{1 - e^{-j\omega(2N_1+1)}}{1 - e^{-j\omega}} \right)$$

$$= e^{j\omega N_1} \left(\frac{e^{-j\omega/2} \left(e^{j\omega/2} - e^{-j\omega(2N_1+1/2)} \right)}{e^{-j\omega/2} \left(e^{j\omega/2} - e^{-j\omega/2} \right)} \right)$$

$$= \frac{\left(e^{j\omega(N_1+1/2)} - e^{-j\omega(N_1+1/2)} \right)}{\left(e^{j\omega/2} - e^{-j\omega/2} \right)}$$

$$X(e^{j\omega}) = \frac{\sin\left(\omega(N_1 + 1/2) \right)}{\sin(\omega/2)}$$

$$x(n) = \begin{cases} 1, & |n| \le N_1 \\ 0, & |n| > N_1 \end{cases} \longleftrightarrow X(e^{j\omega}) = \frac{\sin\left(\omega(N_1 + 1/2) \right)}{\sin(\omega/2)} \qquad (6.28)$$

The Fourier transform is sketched in Fig. 6.6(b) for $N_1 = 2$. Since $X(e^{j\omega})$ is real, the magnitude and phase spectrum are given by

$$|X(e^{j\omega})| = \left| \frac{\sin\left(\omega(N_1 + 1/2) \right)}{\sin(\omega/2)} \right|$$

and

$$\angle X(e^{j\omega}) = \begin{cases} 0, & X(e^{j\omega}) > 0 \\ \pi, & X(e^{j\omega}) < 0 \end{cases}$$

The magnitude and phase spectrum are plotted, respectively, in Figs 6.6(c) and (d) for $N_1 = 2$.

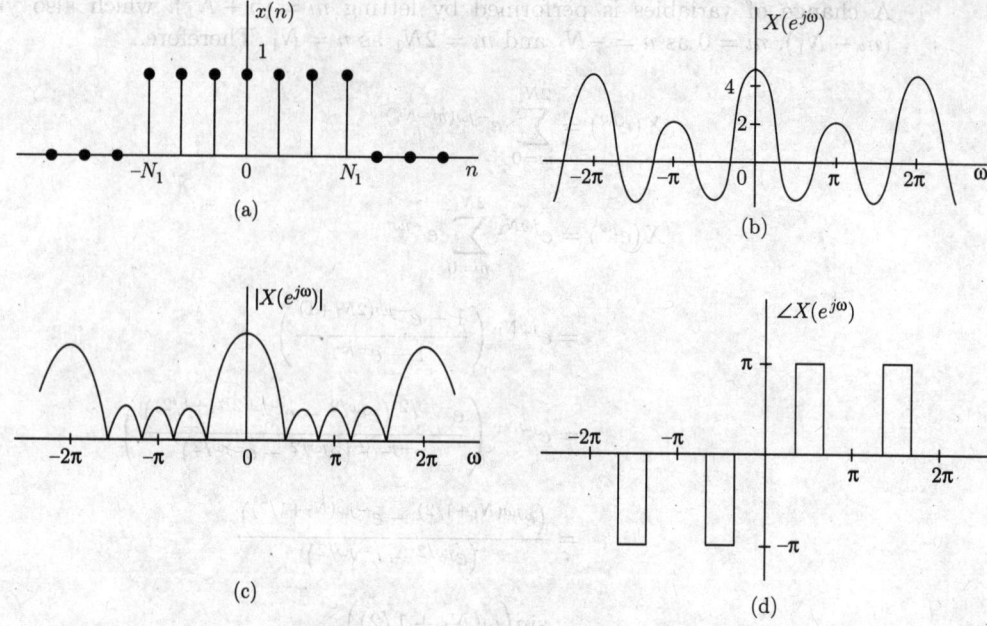

Fig. 6.6 (a) Rectangular pulse signal, (b) its Fourier transform, (c) magnitude spectrum and (d) its phase spectrum

Example 6.7 Find the inverse Fourier transform of

$$X(e^{j\omega}) = \sum_{m=-\infty}^{\infty} 2\pi\delta(\omega - \omega_0 - 2\pi m)$$

which is illustrated in Fig. 6.7.

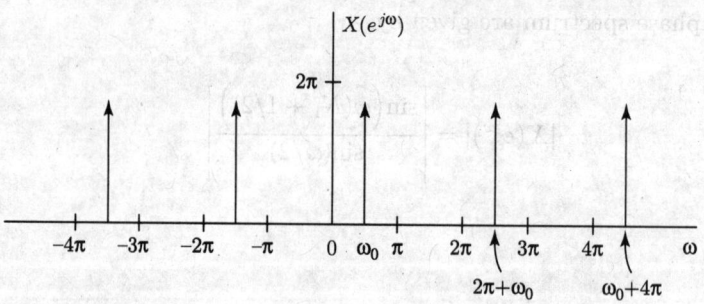

Fig. 6.7

Solution
From Fig. 6.7, we can observe that

$$X(e^{j\omega}) = 2\pi\delta(\omega - \omega_0), \quad 0 \le \omega < 2\pi$$

Now, from Eq. (6.4), we have

$$x(n) = \frac{1}{2\pi} \int_{2\pi} X(e^{j\omega}) \, e^{j\omega n} \, d\omega$$

where $\int_{2\pi}$ indicates integration over any continuous interval of 2π. Therefore,

$$x(n) = \frac{1}{2\pi} \int_0^{2\pi} X(e^{j\omega}) \, e^{j\omega n} \, d\omega$$

$$= \frac{1}{2\pi} \int_0^{2\pi} 2\pi\delta(\omega - \omega_0) \, e^{j\omega n} \, d\omega$$

$$x(n) = e^{j\omega n}\Big|_{\omega=\omega_0} = e^{j\omega_0 n}$$

$$e^{j\omega_0 n} \longleftrightarrow \sum_{m=-\infty}^{\infty} 2\pi\delta(\omega - \omega_0 - 2\pi m) \tag{6.29}$$

Example 6.8 Find the DTFT of the unit step function $u(n)$.

Solution
From Eq. (6.5), the Fourier transform $X(e^{j\omega})$ of $x(n)$ is defined by

$$X(e^{j\omega}) = \sum_{n=-\infty}^{\infty} x(n) \, e^{-j\omega n} = \sum_{n=0}^{\infty} e^{-j\omega n}$$

$$= \lim_{N\to\infty} \sum_{n=0}^{N-1} e^{-j\omega n}$$

$$X(e^{j\omega}) = \lim_{N\to\infty} \frac{1 - e^{-j\omega N}}{1 - e^{-j\omega}}$$

$$X(e^{j\omega}) = \frac{1}{1 - e^{-j\omega}} - \lim_{N\to\infty} \frac{e^{-j\omega N}}{1 - e^{-j\omega}} \tag{6.30}$$

Consider the second term $\lim_{N\to\infty} \dfrac{e^{-j\omega N}}{1 - e^{-j\omega}}$ on the RHS of the above equation. For any test function $f(\omega)$, we have

$$\int_{-\infty}^{\infty} f(\omega) \left(\lim_{N\to\infty} \frac{e^{-j\omega N}}{1 - e^{-j\omega}} \right) d\omega = \lim_{N\to\infty} \int_{-\infty}^{\infty} f(\omega) \frac{e^{-j\omega N}}{1 - e^{-j\omega}} \, d\omega$$

A change of variables is performed by letting $\omega N = y$, which also yields $\omega = y/N$, $y = -\infty$ as $\omega = -\infty$ and $y = \infty$ as $\omega = \infty$. Therefore,

$$\lim_{N\to\infty} \int_{-\infty}^{\infty} f(\omega)\frac{e^{-j\omega N}}{1 - e^{-j\omega}}\, d\omega = \lim_{N\to\infty} \int_{-\infty}^{\infty} f(y/N)\frac{e^{-jy}}{1 - e^{-jy/N}}\frac{dy}{N}$$

$$= f(0) \int_{-\infty}^{\infty} \left(\lim_{N\to\infty} \frac{e^{-jy}}{1 - e^{-jy/N}}\frac{dy}{N} \right)$$

$$= f(0) \int_{-\infty}^{\infty} \left(\lim_{N\to\infty} \frac{e^{-jy}}{N(1 - e^{-jy/N})} \right) dy$$

$$= f(0) \int_{-\infty}^{\infty} \left(\lim_{N\to\infty} \frac{e^{-jy}}{N(1-(1-jy/N + (-jy/N)^2)/2! - \cdots)} \right) dy$$

$$= f(0) \int_{-\infty}^{\infty} \frac{e^{-jy}}{jy}\, dy$$

$$\lim_{N\to\infty} \int_{-\infty}^{\infty} f(\omega)\frac{e^{-j\omega N}}{1 - e^{-j\omega}}\, d\omega = f(0) \int_{-\infty}^{\infty} \frac{\cos(y) - j\sin(y)}{jy}\, dy$$

$$= f(0)\left[\int_{-\infty}^{\infty} \frac{\cos(y)}{jy}\, dy - \int_{-\infty}^{\infty} \frac{j\sin(y)}{jy}\, dy \right]$$

$$= f(0)\left[\int_{-\infty}^{\infty} \frac{\cos(y)}{jy}\, dy - \int_{-\infty}^{\infty} \frac{\sin(y)}{y}\, dy \right]$$

The first term on the RHS is odd and so its integration is zero:

$$\lim_{N\to\infty} \int_{-\infty}^{\infty} f(\omega)\frac{e^{-j\omega N}}{1 - e^{-j\omega}}\, d\omega = f(0)\left[0 - \int_{-\infty}^{\infty} \frac{\sin(y)}{y}\, dy \right] = -f(0) \int_{-\infty}^{\infty} \frac{\sin(y)}{y}\, dy = -f(0)\pi$$

$$\int_{-\infty}^{\infty} \left[\lim_{N\to\infty} \frac{e^{-j\omega N}}{1 - e^{-j\omega}} \right] f(\omega)\, d\omega = -f(0)\pi = \int_{-\infty}^{\infty} [-\pi\delta(\omega)]f(\omega)\, d\omega$$

Therefore,

$$\lim_{N \to \infty} \frac{e^{-j\omega N}}{1 - e^{-j\omega}} = -\pi\delta(\omega)$$

Substituting $\lim_{N \to \infty}(e^{-j\omega N})/(1 - e^{-j\omega}) = -\pi\delta(\omega)$ into Eq. (6.30), we obtain

$$X(e^{j\omega}) = \frac{1}{1 - e^{-j\omega}} + \pi\delta(\omega)$$

However, the DTFT must be periodic in ω with period 2π. Therefore,

$$X(e^{j\omega}) = U(e^{j\omega}) = \frac{1}{1 - e^{-j\omega}} + \pi \sum_{m=-\infty}^{\infty} \delta(\omega - 2\pi m)$$

$$u(n) \longleftrightarrow \frac{1}{1 - e^{-j\omega}} + \pi \sum_{m=-\infty}^{\infty} \delta(\omega - 2\pi m) \tag{6.31}$$

Example 6.9 Find the inverse DTFT of the rectangular pulse spectrum defined only for $-\pi \le \omega \le \pi$

$$X(e^{j\omega}) = rect\left(\frac{\omega}{2\omega_c}\right) = \begin{cases} 1, & |\omega| < \omega_c \\ 0, & \omega_c < |\omega| \le \pi \end{cases}$$

which is depicted in Fig. 6.8(a). Alternatively, we can define $X(e^{j\omega})$ over all ω by writing it as an infinite sum of rectangle functions shifted by integer multiple of 2π

$$X(e^{j\omega}) = \sum_{m=-\infty}^{\infty} rect\left(\frac{\omega - 2\pi m}{2\omega_c}\right)$$

Solution
The inverse DTFT of $X(e^{j\omega})$ is given by

$$x(n) = \frac{1}{2\pi} \int_{-\pi}^{\pi} X(e^{j\omega}) e^{j\omega n} \, d\omega = \frac{1}{2\pi} \int_{-\omega_c}^{\omega_c} e^{j\omega n} \, d\omega$$

$$x(n) = \frac{1}{2\pi} \frac{e^{j\omega n}}{jn} \Bigg|_{-\omega_c}^{\omega_c}$$

$$x(n) = \frac{\sin(\omega_c n)}{n\pi}, \quad n \ne 0$$

For $n = 0$, the inverse Fourier transform expression reduces to

$$x(0) = \frac{1}{2\pi} \int\limits_{-\pi}^{\pi} X(e^{j\omega}) \, d\omega = \frac{1}{2\pi} \int\limits_{-\omega_c}^{\omega_c} d\omega = \frac{\omega_c}{\pi}$$

Hence

$$x(n) = \begin{cases} \dfrac{\omega_c}{\pi}, & n = 0 \\ \dfrac{\sin(\omega_c n)}{n\pi}, & n \neq 0 \end{cases} \tag{6.32}$$

It should be noted that often the above sequence is expressed in a compact form as

$$x(n) = \frac{\sin(\omega_c n)}{n\pi}, \quad -\infty < n < \infty \tag{6.33}$$

with the understanding that at $n = 0$, $x(n) = \omega_c/\pi$. Using the sinc function, we may also write

$$x(n) = \frac{\sin(\omega_c n)}{n\pi}$$

$$x(n) = \frac{\omega_c}{\pi} \text{sinc}\left(\frac{\omega_c n}{\pi}\right)$$

Therefore,

$$\frac{\omega_c}{\pi} \text{sinc}\left(\frac{\omega_c n}{\pi}\right) \longleftrightarrow \sum_{m=-\infty}^{\infty} \text{rect}\left(\frac{\omega - 2\pi m}{2\omega_c}\right) \tag{6.34}$$

The signal $x(n)$ is depicted in Fig. 6.8(b).

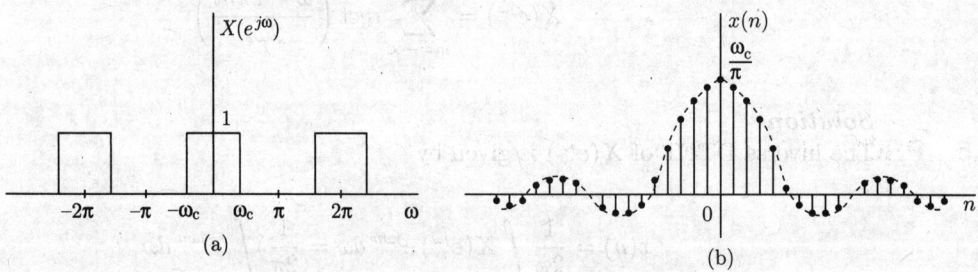

Fig. 6.8 (a) Rectangular pulse spectrum and (b) its inverse Fourier transform

Example 6.10 Find the inverse DTFT of

$$X(e^{j\omega}) = \delta(\omega), \quad -\pi < \omega \leq \pi$$

which is depicted in Fig. 6.9. Alternatively, we can define $X(e^{j\omega})$ over all ω by writing it as an infinite sum of delta functions shifted by integer multiple of 2π.

$$X(e^{j\omega}) = \sum_{m=-\infty}^{\infty} \delta(\omega - 2\pi m)$$

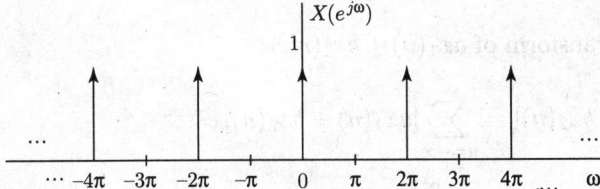

Fig. 6.9

Solution

The inverse DTFT of $X(e^{j\omega})$ is given by

$$x(n) = \frac{1}{2\pi} \int_{-\pi}^{\pi} X(e^{j\omega}) \, e^{j\omega n} \, d\omega = \frac{1}{2\pi} \int_{-\pi}^{\pi} \delta(\omega) \, e^{j\omega n} \, d\omega = \frac{1}{2\pi} e^{j\omega n} \Big|_{\omega=0} = \frac{1}{2\pi}$$

Therefore, we may write

$$\frac{1}{2\pi} \longleftrightarrow \sum_{m=-\infty}^{\infty} \delta(\omega - 2\pi m) \tag{6.35}$$

or

$$1 \longleftrightarrow 2\pi \sum_{m=-\infty}^{\infty} \delta(\omega - 2\pi m) \tag{6.36}$$

6.5 PROPERTIES OF DTFT

Fourier transform possesses a number of important properties that are useful for developing conceptual insights into the transform and into the relationship between the time-domain and frequency-domain descriptions of a signal. In addition, they can also help to reduce the complexity of the evaluation of the Fourier transform of many sequences. We will use a shorthand notation

$$x(n) \longleftrightarrow X(e^{j\omega})$$

to indicate the relationship between a discrete-time sequence $x(n)$ and its Fourier transform $X(e^{j\omega})$.

6.5.1 Linearity

If

$$x_1(n) \longleftrightarrow X_1(e^{j\omega}) \quad \text{and} \quad x_2(n) \longleftrightarrow X_2(e^{j\omega})$$

then

$$ax_1(n) + bx_2(n) \longleftrightarrow aX_1(e^{j\omega}) + bX_2(e^{j\omega}) \tag{6.37}$$

Proof The Fourier transform of $ax_1(n) + bx_2(n)$ is

$$\mathcal{F}[ax_1(n) + bx_2(n)] = \sum_{n=-\infty}^{\infty} [ax_1(n) + bx_2(n)] \, e^{-j\omega n}$$

$$= a \underbrace{\sum_{n=-\infty}^{\infty} x_1(n) \, e^{-j\omega n}}_{X_1(e^{j\omega})} + b \underbrace{\sum_{n=-\infty}^{\infty} x_2(n) \, e^{-j\omega n}}_{X_2(e^{j\omega})}$$

$$= aX_1(e^{j\omega}) + bX_2(e^{j\omega})$$

6.5.2 Time Shifting

If

$$x(n) \longleftrightarrow X(e^{j\omega})$$

then

$$y(n) = x(n - n_0) \longleftrightarrow Y(e^{j\omega}) = X(e^{j\omega})e^{-j\omega n_0} \tag{6.38}$$

Proof The Fourier transform of $x(n - n_0)$ is

$$\mathcal{F}[x(n - n_0)] = \sum_{n=-\infty}^{\infty} x(n - n_0) \, e^{-j\omega n}$$

A change of variables is performed by letting $m = (n - n_0)$, which also yields $n = (m + n_0)$, $m = -\infty$ as $n = -\infty$ and $m = \infty$ as $n = \infty$. Therefore,

$$\mathcal{F}[x(n - n_0)] = \sum_{m=-\infty}^{\infty} x(m) \, e^{-j\omega(m+n_0)}$$

$$= e^{-j\omega n_0} \sum_{m=-\infty}^{\infty} x(m) \, e^{-j\omega m}$$

$$\mathcal{F}[x(n - n_0)] = X(e^{j\omega}) \, e^{-j\omega n_0}$$

One consequence of this property is that, when a signal is shifted in time, the *magnitudes* of its Fourier transform remain unaltered. That is, if we express $X(e^{j\omega})$

in polar form as

$$\mathcal{F}[x(n)] = X(e^{j\omega}) = |X(e^{j\omega})|\, e^{j\angle X(e^{j\omega})}$$

then

$$\mathcal{F}[x(n - n_0)] = X(e^{j\omega})\, e^{-j\omega n_0} = |X(e^{j\omega})|\, e^{j(\angle X(e^{j\omega}) - \omega n_0)}$$

Thus, the effect of a time shift on a signal is to introduce into its Fourier transform a phase shift, ωn_0, which is a linear function of ω. ∎

Example 6.11 Find the DTFT of the sequence

$$x(n) = \frac{1}{4}\mathrm{sinc}\left(\frac{1}{4}(n - 2)\right)$$

shown in Fig. 6.10 (a).

Solution
From Eq. (6.34), we have

$$\frac{\omega_c}{\pi}\mathrm{sinc}\left(\frac{\omega_c n}{\pi}\right) \longleftrightarrow \sum_{m=-\infty}^{\infty} \mathrm{rect}\left(\frac{\omega - 2\pi m}{2\omega_c}\right)$$

Substituting $\omega_c = \pi/4$ into the above equation gives

$$\frac{1}{4}\mathrm{sinc}\left(\frac{1}{4}n\right) \longleftrightarrow \sum_{m=-\infty}^{\infty} \mathrm{rect}\left(\frac{\omega - 2\pi m}{\pi/2}\right)$$

The time-shifting property [Eq. (6.38)] yields

$$\frac{1}{4}\mathrm{sinc}\left(\frac{1}{4}(n - 2)\right) \longleftrightarrow \sum_{m=-\infty}^{\infty} \mathrm{rect}\left(\frac{\omega - 2\pi m}{\pi/2}\right) e^{-j2\omega}$$

The magnitude spectrum of the shifted signal is shown in Fig. 6.10(b).

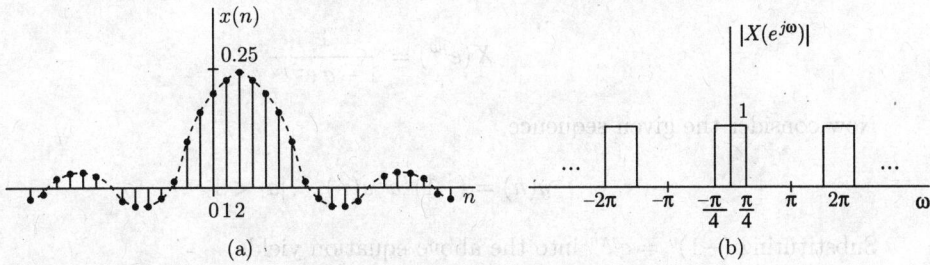

Fig. 6.10 Signal $x(n)$ of Example 6.11 and (b) its DTFT

6.5.3 Frequency Shifting

If

$$x(n) \longleftrightarrow X(e^{j\omega})$$

then

$$y(n) = x(n)\, e^{j\omega_0 n} \longleftrightarrow Y(e^{j\omega}) = X(e^{j(\omega-\omega_0)}) \tag{6.39}$$

Proof The Fourier transform of $x(n)\, e^{j\omega_0 n}$ is given by

$$\mathcal{F}[x(n)\, e^{j\omega_0 n}] = \sum_{n=-\infty}^{\infty} [x(n)\, e^{j\omega_0 n}]\, e^{-j\omega n}$$

$$= \sum_{n=-\infty}^{\infty} x(n)\, e^{-j(\omega-\omega_0)n}$$

$$\mathcal{F}[x(n)\, e^{j\omega_0 n}] = X(e^{j(\omega-\omega_0)})$$

Hence, a frequency shift corresponds to multiplication in time domain by a complex sinusoid whose frequency is equal to the frequency shift. ∎

Example 6.12 Find the Fourier transform of an exponential sequence with alternating sign

$$y(n) = (-1)^n a^n u(n), \quad |a| < 1$$

Solution
Define a sequence

$$x(n) = a^n u(n), \quad |a| < 1$$

From Eq. (6.24), its Fourier transform is given by

$$X(e^{j\omega}) = \frac{1}{1 - a\, e^{-j\omega}}$$

Now consider the given sequence

$$y(n) = (-1)^n a^n u(n), \quad |a| < 1$$

Substituting $(-1)^n = e^{j\pi n}$ into the above equation yields

$$y(n) = e^{j\pi n} a^n u(n) = e^{j\pi n} x(n)$$

Now apply the frequency-shifting property [Eq. (6.39)] to get the Fourier transform of $y(n)$:

$$Y(e^{j\omega}) = X(e^{j(\omega-\pi)})$$

$$= \frac{1}{1 - a\,e^{-j(\omega-\pi)}}$$

$$Y(e^{j\omega}) = \frac{1}{1 - a\,e^{j\pi}\,e^{-j\omega}}$$

$$Y(e^{j\omega}) = \frac{1}{1 + a\,e^{-j\omega}}$$

$$(-1)^n a^n u(n) \longleftrightarrow \frac{1}{1 + a\,e^{-j\omega}}, \qquad |a| < 1 \tag{6.40}$$

6.5.4 Time Reversal

If

$$x(n) \longleftrightarrow X(e^{j\omega})$$

then

$$y(n) = x(-n) \longleftrightarrow Y(e^{j\omega}) = X(e^{-j\omega}) \tag{6.41}$$

Proof The Fourier transform of $y(n)$ is given by

$$Y(e^{j\omega}) = \sum_{n=-\infty}^{\infty} y(n)\,e^{-j\omega n}$$

$$= \sum_{n=-\infty}^{\infty} x(-n)\,e^{-j\omega n}$$

Substituting $m = -n$ into the above equation gives

$$Y(e^{j\omega}) = \sum_{n=-\infty}^{\infty} x(m)\,e^{j\omega m}$$

$$= \sum_{n=-\infty}^{\infty} x(m)\,e^{-j(-\omega)m}$$

$$Y(e^{j\omega}) = X(e^{-j\omega})$$

An interesting consequence of the time-reversal property is that if $x(n)$ is even, then its Fourier transform is also even, i.e.,

$$\text{if } x(-n) = x(n), \text{ then } X(e^{-j\omega}) = X(e^{j\omega}) \tag{6.42}$$

Similarly, if $x(n)$ is odd, then so is its Fourier transform, i.e.,

$$\text{if } x(-n) = -x(n), \text{ then } X(e^{-j\omega}) = -X(e^{j\omega}) \tag{6.43}$$

6.5.5 Time Expansion

Because of the discrete-time nature of the time index for discrete-time signals, the relation between time and frequency scaling in discrete time takes on a somewhat different form from its continuous-time counterpart. If we try to define the signal $x(an)$, we run into difficulties if a is not an integer. Therefore, we cannot slow down the signal by choosing $a < 1$. On the other hand, if we let a be an integer other than ± 1, we do not merely speed up the original signal. For example, since n can take on only integer values, the signal $x(2n)$ consists of the even samples of $x(n)$ alone.

Let m be a positive integer, and define the signal

$$x_{(m)}(n) = \begin{cases} x\left(\dfrac{n}{m}\right), & \text{if } n \text{ is a multiple of } m \\ 0, & \text{if } n \text{ is not a multiple of } m. \end{cases} \tag{6.44}$$

$x_m(n)$ can be obtained from $x(n)$ by placing $(m-1)$ zeros between successive values of the original signal. Intuitively, we can think of $x_{(m)}(n)$ as a slowed-down version of $x(n)$. Now, if

$$x(n) \longleftrightarrow X(e^{j\omega})$$

then

$$y(n) = x_{(m)}(n) \longleftrightarrow Y(e^{j\omega}) = X(e^{jm\omega}) \tag{6.45}$$

Proof The Fourier transform of $y(n) = x_{(m)}(n)$ is given by

$$Y(e^{j\omega}) = \sum_{n=-\infty}^{\infty} y(n)\, e^{-j\omega n} = \sum_{n=-\infty}^{\infty} x_{(m)}(n)\, e^{-j\omega n} = \sum_{n=-\infty}^{\infty} x\left(\frac{n}{m}\right) e^{-j\omega n}$$

A change of variables is performed by letting $r = n/m$, which also yields $r = -\infty$ as $n = -\infty$ and $r = \infty$ as $n = \infty$. Therefore,

$$Y(e^{j\omega}) = \sum_{r=-\infty}^{\infty} x(r)\, e^{-j\omega m r}$$

$$Y(e^{j\omega}) = X(e^{jm\omega})$$

Note that as the signal spread out and slowed down in time by taking $m > 1$, its Fourier transform is compressed.

6.5.6 Differencing in Time Domain

The discrete-time parallel to the differentiation property of the continuous-time Fourier transform involves the use of the first-difference operation, which is defined as

$y(n) = x(n) - x(n - 1)$. Now, if

$$x(n) \longleftrightarrow X(e^{j\omega})$$

then

$$y(n) = x(n) - x(n - 1) \longleftrightarrow Y(e^{j\omega}) = (1 - e^{-j\omega})X(e^{j\omega}) \tag{6.46}$$

A common use of this property is in situations where evaluation of the Fourier transform is easier for the first difference than for the original sequence.

Proof Given that

$$x(n) \longleftrightarrow X(e^{j\omega})$$

Using the time-shifting property, we get

$$x(n - 1) \longleftrightarrow X(e^{j\omega}) e^{-j\omega}$$

Now, using the linearity property, we get

$$x(n) - x(n - 1) \longleftrightarrow X(e^{j\omega}) - X(e^{j\omega}) e^{-j\omega}$$
$$x(n) - x(n - 1) \longleftrightarrow (1 - e^{-j\omega})X(e^{j\omega})$$

6.5.7 Differentiation in Frequency Domain

If

$$x(n) \longleftrightarrow X(e^{j\omega})$$

then

$$-jnx(n) \longleftrightarrow \frac{dX(e^{j\omega})}{d\omega} \tag{6.47}$$

or

$$nx(n) \longleftrightarrow j\frac{dX(e^{j\omega})}{d\omega} \tag{6.48}$$

Proof From Eq. (6.5), we have

$$X(e^{j\omega}) = \sum_{n=-\infty}^{\infty} x(n) e^{-j\omega n}$$

By differentiating both sides w.r.t. ω, we obtain

$$\frac{dX(e^{j\omega})}{d\omega} = \sum_{n=-\infty}^{\infty} [-jnx(n)] e^{-j\omega n}$$

$$\frac{dX(e^{j\omega})}{d\omega} = \mathcal{F}[-jnx(n)]$$

Therefore,

$$-jnx(n) \longleftrightarrow \frac{dX(e^{j\omega})}{d\omega}$$

Multiplying both sides by j yields

$$nx(n) \longleftrightarrow j\frac{dX(e^{j\omega})}{d\omega}$$

Example 6.13 Determine the Fourier transform of the sequence

$$y(n) = na^n u(n), \quad |a| < 1$$

Solution
Define a sequence

$$x(n) = a^n u(n), \quad |a| < 1$$

From Eq. (6.24), its Fourier transform is given by

$$X(e^{j\omega}) = \frac{1}{1 - a\,e^{-j\omega}}$$

Now consider the given sequence

$$y(n) = na^n u(n) = nx(n)$$

Now apply the differentiation in frequency-domain property [Eq. (6.48)] to get the Fourier transform of $y(n)$:

$$Y(e^{j\omega}) = j\frac{dX(e^{j\omega})}{d\omega} = j\frac{d}{d\omega}\left(\frac{1}{1 - a\,e^{-j\omega}}\right) = \frac{a\,e^{-j\omega}}{(1 - a\,e^{-j\omega})^2}$$

$$na^n u(n) \longleftrightarrow \frac{a\,e^{-j\omega}}{(1 - a\,e^{-j\omega})^2}, \quad |a| < 1 \tag{6.49}$$

Example 6.14 Determine the inverse DTFT of

$$X(e^{j\omega}) = \frac{1}{(1 - a\,e^{-j\omega})^2}, \quad |a| < 1$$

Solution
From Eq. (6.49), we have

$$na^n u(n) \longleftrightarrow \frac{a\,e^{-j\omega}}{(1 - a\,e^{-j\omega})^2}, \quad |a| < 1$$

Using the time-shifting property, we obtain

$$(n+1)a^{n+1}u(n+1) \longleftrightarrow \left[\frac{a\,e^{-j\omega}}{(1-a\,e^{-j\omega})^2}\right]e^{j\omega}$$

$$(n+1)a^{n+1}u(n+1) \longleftrightarrow \frac{a}{(1-a\,e^{-j\omega})^2}$$

$$(n+1)a^n u(n+1) \longleftrightarrow \frac{1}{(1-a\,e^{-j\omega})^2}$$

Therefore,

$$x(n) = (n+1)a^n u(n+1)$$

It is worth noting that although the RHS is multiplied by a step function that begins at $n = -1$, the sequence $(n+1)a^n u(n+1)$ is still zero prior to $n = 0$, since the factor $n+1$ is zero at $n = -1$. Thus, we can alternatively express $x(n)$ as

$$x(n) = (n+1)a^n u(n)$$

and hence,

$$(n+1)a^n u(n) \longleftrightarrow \frac{1}{(1-a\,e^{-j\omega})^2} \qquad (6.50)$$

6.5.8 Convolution Property

If

$$x_1(n) \longleftrightarrow X_1(e^{j\omega}) \quad \text{and} \quad x_2(n) \longleftrightarrow X_2(e^{j\omega})$$

then

$$x_1(n) * x_2(n) \longleftrightarrow X_1(e^{j\omega})X_2(e^{j\omega}) \qquad (6.51)$$

The Fourier transform maps the convolution of two signals into the product of their Fourier transforms.

Proof The Fourier transform of $x_1(n) * x_2(n)$ is

$$\mathcal{F}[x_1(n) * x_2(n)] = \sum_{n=-\infty}^{\infty} [x_1(n) * x_2(n)]\,e^{-j\omega n}$$

$$= \sum_{n=-\infty}^{\infty} \left(\sum_{m=-\infty}^{\infty} x_1(m)x_2(n-m)\right) e^{-j\omega n}$$

By interchanging the order of summation, we have

$$\mathcal{F}[x_1(n) * x_2(n)] = \sum_{m=-\infty}^{\infty} x_1(m) \left(\sum_{n=-\infty}^{\infty} x_2(n-m)\,e^{-j\omega n}\right)$$

By applying the time-shifting property [Eq. (6.38)], the bracketed term is $X_2(e^{j\omega})e^{-j\omega m}$. Substituting this into the above equation yields

$$\mathcal{F}[x_1(n) * x_2(n)] = \sum_{m=-\infty}^{\infty} x_1(m)\left(X_2(e^{j\omega})\,e^{-j\omega m}\right)$$

$$= X_2(e^{j\omega}) \sum_{m=-\infty}^{\infty} x_1(m)\,e^{-j\omega m}$$

$$\mathcal{F}[x_1(n) * x_2(n)] = X_2(e^{j\omega})X_1(e^{j\omega}) = X_1(e^{j\omega})X_2(e^{j\omega})$$

Therefore,

$$x_1(n) * x_2(n) \longleftrightarrow X_1(e^{j\omega})X_2(e^{j\omega})$$

The convolution property states that convolution in the time domain corresponds to multiplication in the frequency domain.

Example 6.15 Using the convolution property determine the convolution $x(n) = x_1(n) * x_2(n)$ of the sequences

$$x_1(n) = x_2(n) = \{1, \underset{\uparrow}{1}, 1\} = \delta(n+1) + \delta(n) + \delta(n-1) = \begin{cases} 1, & -1 \le n \le 1 \\ 0, & \text{otherwise} \end{cases}$$

Solution
The Fourier transform of $x_1(n) = x_2(n)$ is given by

$$X_1(e^{j\omega}) = X_2(e^{j\omega}) = \sum_{n=-\infty}^{\infty} x_1(n)\,e^{-j\omega n} = \sum_{n=-1}^{1} e^{-j\omega n}$$

$$= e^{j\omega n} + 1 + e^{-j\omega n}$$

$$= 1 + 2\left(\frac{e^{j\omega n} + e^{-j\omega n}}{2}\right)$$

$$= 1 + 2\cos(\omega)$$

Now using the convolution property, we have

$$\mathcal{F}[x_1(n) * x_2(n)] = X(e^{j\omega}) = X_1(e^{j\omega})X_2(e^{j\omega})$$

$$= [1 + 2\cos(\omega)]^2$$

$$X(e^{j\omega}) = 3 + 4\cos(\omega) + 2\cos(2\omega)$$

$$= 3 + 2(e^{j\omega} + e^{-j\omega}) + (e^{j2\omega} + e^{-j2\omega})$$

$$X(e^{j\omega}) = e^{j2\omega} + 2\,e^{j\omega} + 3 + 2\,e^{-j\omega} + e^{-j2\omega}$$

Taking the inverse Fourier transform of the above equation, we obtain

$$x(n) = \delta(n+2) + 2\delta(n+1) + 3\delta(n) + 2\delta(n-1) + \delta(n-2) = \{1, 2, \underset{\uparrow}{3}, 2, 1\}$$

6.5.9 Accumulation Property

If

$$x(n) \longleftrightarrow X(e^{j\omega})$$

then

$$\sum_{k=-\infty}^{n} x(k) \longleftrightarrow \frac{1}{1 - e^{-j\omega}} X(e^{j\omega}) + \pi X(e^{j0}) \sum_{m=-\infty}^{\infty} \delta(\omega - 2\pi m) \qquad (6.52)$$

Proof Convolving a signal $x(n)$ with a unit step function $u(n)$ yields

$$x(n) * u(n) = \sum_{k=-\infty}^{\infty} x(k) u(n-k)$$

Since

$$u(n-k) = \begin{cases} 1, & (n-k) \geq 0 \longrightarrow k \leq n \\ 0, & (n-k) < 0 \longrightarrow k > n \end{cases}$$

we have

$$x(n) * u(n) = \sum_{k=-\infty}^{n} x(k)$$

Now we can prove the accumulation property of the Fourier transform:

$$\sum_{k=-\infty}^{n} x(k) = x(n) * u(n)$$

$$\mathcal{F}\left[\sum_{k=-\infty}^{n} x(k)\right] = \mathcal{F}[x(n) * u(n)]$$

Using the convolution property [Eq. (6.51)] and Eq. (6.31), we obtain

$$\mathcal{F}\left[\sum_{k=-\infty}^{n} x(k)\right] = X(e^{j\omega})U(e^{j\omega})$$

$$= X(e^{j\omega})\left(\frac{1}{1-e^{-j\omega}} + \pi \sum_{m=-\infty}^{\infty} \delta(\omega - 2\pi m)\right)$$

$$= \frac{1}{1-e^{-j\omega}}X(e^{j\omega}) + \pi X(e^{j\omega}) \sum_{m=-\infty}^{\infty} \delta(\omega - 2\pi m)$$

$$= \frac{1}{1-e^{-j\omega}}X(e^{j\omega}) + \pi X(e^{j0}) \sum_{m=-\infty}^{\infty} \delta(\omega - 2\pi m)$$

therefore,

$$\sum_{k=-\infty}^{n} x(k) \longleftrightarrow \frac{1}{1-e^{-j\omega}}X(e^{j\omega}) + \pi X(e^{j0}) \sum_{m=-\infty}^{\infty} \delta(\omega - 2\pi m)$$

6.5.10 Multiplication (or Modulation or Windowing) Property

If

$$x_1(n) \longleftrightarrow X_1(e^{j\omega}) \quad \text{and} \quad x_2(n) \longleftrightarrow X_2(e^{j\omega})$$

then

$$x_1(n)x_2(n) \longleftrightarrow \frac{1}{2\pi}[X_1(e^{j\omega}) \circledast X_1(e^{j\omega})] = \frac{1}{2\pi}\int_{2\pi} X_1(e^{j\theta})X_2(e^{j(\omega-\theta)})d\theta \tag{6.53}$$

where $\circledast$ denotes the periodic convolution. The Fourier transform maps the multiplication of two signals into the convolution of their Fourier transforms.

Proof The Fourier transform of $x_1(n)x_2(n)$ is given by

$$\mathcal{F}[x_1(n)x_2(n)] = \sum_{n=-\infty}^{\infty} [x_1(n)x_2(n)]\, e^{-j\omega n}$$

$$= \sum_{n=-\infty}^{\infty} \left(\frac{1}{2\pi}\int_{2\pi} X_1(e^{j\theta})\, e^{j\theta n}d\theta\right) x_2(n)\, e^{-j\omega n}$$

Interchanging the order of integration and summation and noting that $X_1(e^{j\theta})$ does not depend on n, we obtain

$$\mathcal{F}[x_1(n)x_2(n)] = \frac{1}{2\pi}\int_{2\pi} X_1(e^{j\theta}) \left(\sum_{n=-\infty}^{\infty} [x_2(n)\, e^{j\theta n}]\, e^{-j\omega n}\right) d\theta$$

By applying the frequency-shifting property [Eq. (6.39)], the bracketed term is $X_2(e^{j(\omega - \theta)})$. Substituting this into the above equation yields

$$\mathcal{F}[x_1(n)x_2(n)] = \frac{1}{2\pi} \int\limits_{2\pi} X_1(e^{j\theta}) X_2(e^{j(\omega - \theta)}) d\theta$$

$$= \frac{1}{2\pi} [X_1(e^{j\omega}) \circledast X_2(e^{j\omega})]$$

Therefore,

$$x_1(n)x_2(n) \longleftrightarrow \frac{1}{2\pi} [X_1(e^{j\omega}) \circledast X_2(e^{j\omega})]$$

The convolution integral in Eq. (6.53) is known as the periodic convolution of $X_1(e^{j\omega})$ and $X_2(e^{j\omega})$ because it is the convolution of two periodic functions having the same period. We note that the limits of integration extend over a single period. Furthermore, we note that due to the periodicity of the Fourier transform for discrete-time signals, there is no 'perfect' duality between the time and frequency domains w.r.t. the convolution operation, as in the case of continuous-time signals. Indeed, convolution in the time domain (aperiodic summation) is equivalent to multiplication of continuous periodic Fourier transforms. However, multiplication of aperiodic sequences is equivalent to periodic convolution of their Fourier transforms.

6.5.11 Conjugation and Conjugate Symmetry
If

$$x(n) \longleftrightarrow X(e^{j\omega})$$

then

$$x^*(n) \longleftrightarrow X^*(e^{-j\omega}) \tag{6.54}$$

Proof From Eq. (6.5) the Fourier transform of $x^*(n)$ is

$$\mathcal{F}[x^*(n)] = \sum_{n=-\infty}^{\infty} x^*(n)e^{-j\omega n} = \left[\sum_{n=-\infty}^{\infty} x(n)e^{j\omega n} \right]^*$$

$$= \left[\sum_{n=-\infty}^{\infty} x(n)e^{-j(-\omega)n} \right]^*$$

$$= [X(e^{-j\omega})]^*$$

$$\mathcal{F}[x^*(n)] = X^*(e^{-j\omega})$$

Case I If $x(n)$ is real, i.e., if

$$x^*(n) = x(n)$$

then

$$\mathcal{F}[x^*(n)] = \mathcal{F}[x(n)]$$

$$X^*(e^{-j\omega}) = X(e^{j\omega})$$

$$X(e^{-j\omega}) = X^*(e^{j\omega}) \tag{6.55}$$

So, if $x(n)$ is real, the Fourier transform will be *conjugate symmetric*. Also if $x(n)$ is real and even, then from Eqs (6.42) and (6.55), we have

$$X(e^{-j\omega}) = X^*(e^{j\omega}) = X(e^{j\omega}) \tag{6.56}$$

That is, *if $x(n)$ is real and even, then so is its Fourier transform.*

Case II *If $x(n)$ is real and odd, then its Fourier transform is purely imaginary and odd.* Using Eqs (6.43) and (6.55), we get

$$X(e^{-j\omega}) = X^*(e^{j\omega}) = -X(e^{j\omega}) \tag{6.57}$$

Case III *Even and odd decomposition of real signals*: If $x(n)$ is real and

$$x(n) \longleftrightarrow X(e^{j\omega})$$

then

1.
$$x_e(n) = \mathcal{E}\{x(n)\} \longleftrightarrow \text{Re}\{X(e^{j\omega})\} = X_R(e^{j\omega}) \tag{6.58}$$

Proof The even part of a signal $x(n)$ is defined as

$$x_e(n) = \frac{1}{2}[x(n) + x(-n)]$$

Taking the Fourier transform, we obtain

$$\mathcal{F}[x_e(n)] = \mathcal{F}\left[\frac{1}{2}[x(n) + x(-n)]\right]$$

$$= \frac{1}{2}\Big(\mathcal{F}[x(n)] + \mathcal{F}[x(-n)]\Big)$$

$$= \frac{1}{2}\Big(X(e^{j\omega}) + X^*(e^{j\omega})\Big)$$

$$= \frac{1}{2}2X_R(e^{j\omega})$$

$$\mathcal{F}[x_e(n)] = X_R(e^{j\omega})$$

Therefore,

$$x_e(n) = \mathcal{E}\{x(n)\} \longleftrightarrow X_R(e^{j\omega})$$

and

2.
$$x_o(n) = \mathcal{O}\{x(n)\} \longleftrightarrow j\text{Im}\{X(e^{j\omega})\} = jX_I(e^{j\omega}) \tag{6.59}$$

Proof The odd part of a signal $x(n)$ is defined as

$$x_o(n) = \frac{1}{2}[x(n) - x(-n)]$$

Taking the Fourier transform, we obtain

$$\mathcal{F}[x_\text{o}(n)] = \mathcal{F}\left[\frac{1}{2}[x(n) - x(-n)]\right]$$

$$= \frac{1}{2}\left(\mathcal{F}[x(n)] - \mathcal{F}[x(-n)]\right)$$

$$= \frac{1}{2}\left(X(e^{j\omega}) - X^*(e^{j\omega})\right)$$

$$= \frac{1}{2}2jX_\text{I}(e^{j\omega})$$

$$\mathcal{F}[x_\text{o}(n)] = jX_\text{I}(e^{j\omega})$$

Therefore,

$$x_\text{o}(n) = \mathcal{O}\{x(n)\} \longleftrightarrow jX_\text{I}(e^{j\omega})$$

∎

6.5.12 Parseval's Relation

Let $x(n)$ be an energy signal, and if

$$x(n) \longleftrightarrow X(e^{j\omega})$$

then

$$E_x = \sum_{n=-\infty}^{\infty} |x(n)|^2 = \frac{1}{2\pi}\int_{2\pi} |X(e^{j\omega})|^2 d\omega \tag{6.60}$$

Proof Consider the LHS of Eq. (6.60), we have

$$E_x = \sum_{n=-\infty}^{\infty} |x(n)|^2 = \sum_{n=-\infty}^{\infty} x(n)x^*(n)$$

$$= \sum_{n=-\infty}^{\infty} x(n)\left(\frac{1}{2\pi}\int_{2\pi} X(e^{j\omega})\,e^{j\omega n}\,d\omega\right)^*$$

$$= \sum_{n=-\infty}^{\infty} x(n)\left(\frac{1}{2\pi}\int_{2\pi} X^*(e^{j\omega})\,e^{-j\omega n}\,d\omega\right)$$

$$= \frac{1}{2\pi}\int_{2\pi} X^*(e^{j\omega})\left(\sum_{n=-\infty}^{\infty} x(n)\,e^{-j\omega t}\right)d\omega$$

$$= \frac{1}{2\pi}\int_{2\pi} X^*(e^{j\omega})X(e^{j\omega})\,d\omega$$

$$\sum_{n=-\infty}^{\infty} |x(n)|^2 = \frac{1}{2\pi}\int_{2\pi} |X(e^{j\omega})|^2 d\omega$$

∎

Example 6.16 Find the energy of the sequence

$$x(n) = \text{sinc}\left(\frac{\omega_c n}{\pi}\right)$$

assuming $\omega_c < \pi$.

Solution
From Eq. (6.34), we have

$$\frac{\omega_c}{\pi}\text{sinc}\left(\frac{\omega_c n}{\pi}\right) \longleftrightarrow \sum_{m=-\infty}^{\infty} \text{rect}\left(\frac{\omega - 2\pi m}{2\omega_c}\right)$$

$$x(n) = \text{sinc}\left(\frac{\omega_c n}{\pi}\right) \longleftrightarrow X(e^{j\omega}) = \frac{\pi}{\omega_c}\sum_{m=-\infty}^{\infty} \text{rect}\left(\frac{\omega - 2\pi m}{2\omega_c}\right)$$

Hence, from the Parseval's relation in Eq. (6.60), we have

$$E_x = \sum_{n=-\infty}^{\infty} |x(n)|^2 = \frac{1}{2\pi}\int_{2\pi} |X(e^{j\omega})|^2 \, d\omega$$

$$E_x = \frac{1}{2\pi}\int_{-\pi}^{\pi} \frac{\pi^2}{\omega_c^2}\left|\text{rect}\left(\frac{\omega}{2\omega_c}\right)\right|^2 d\omega \tag{6.61}$$

We know that

$$\text{rect}\left(\frac{\omega}{2\omega_c}\right) = \begin{cases} 1, & |\omega| < \omega_c \\ 0, & \omega_c < |\omega| \le \pi \end{cases}$$

Substituting the value of $\text{rect}\,(\omega/2\omega_c)$ into Eq. (6.61) yields

$$E_x = \frac{1}{2\pi}\frac{\pi^2}{\omega_c^2}\int_{-\omega_c}^{\omega_c} 1 \, d\omega = \frac{\pi}{2\omega_c^2}2\omega_c = \frac{\pi}{\omega_c}$$

6.6 SOME IMPORTANT RESULTS

Let $X(e^{j\omega})$ denote the DTFT of $x(n)$.

1.

$$X(e^{j0}) = \sum_{n=-\infty}^{\infty} x(n) \tag{6.62}$$

Proof From Eq. (6.5), we have

$$X(e^{j\omega}) = \sum_{n=-\infty}^{\infty} x(n)\,e^{-j\omega n}$$

Substituting $\omega = 0$ into the above equation gives

$$X(e^{j\omega})\Big|_{\omega=0} = X(e^{j0}) = \sum_{n=-\infty}^{\infty} x(n)$$

2.

$$X(e^{j\pi}) = \sum_{n=-\infty}^{\infty} (-1)^n x(n) \qquad (6.63)$$

Proof By definition

$$X(e^{j\omega}) = \sum_{n=-\infty}^{\infty} x(n) e^{-j\omega n}$$

Substituting $\omega = \pi$ into the above equation yields

$$X(e^{j\omega})\Big|_{\omega=\pi} = X(e^{j\pi}) = \sum_{n=-\infty}^{\infty} x(n) e^{-j\pi n} = \sum_{n=-\infty}^{\infty} x(n)(e^{-j\pi})^n$$

$$= \sum_{n=-\infty}^{\infty} x(n)(\cos(\pi) - j\sin(\pi))^n$$

$$X(e^{j\pi}) = \sum_{n=-\infty}^{\infty} x(n)(-1 - j0)^n$$

$$X(e^{j\pi}) = \sum_{n=-\infty}^{\infty} (-1)^n x(n)$$

3.

$$\int_{-\pi}^{\pi} X(e^{j\omega}) \, d\omega = 2\pi x(0) \qquad (6.64)$$

Proof From Eq. (6.4), we have

$$x(n) = \frac{1}{2\pi} \int_{-\pi}^{\pi} X(e^{j\omega}) e^{j\omega n} \, d\omega$$

Substituting $n = 0$ into the above equation yields

$$x(0) = \frac{1}{2\pi} \int_{-\pi}^{\pi} X(e^{j\omega}) \, d\omega$$

$$2\pi x(0) = \int_{-\pi}^{\pi} X(e^{j\omega}) \, d\omega$$

4.

$$\int_{-\pi}^{\pi} |X(e^{j\omega})|^2 \, d\omega = 2\pi \sum_{n=-\infty}^{\infty} |x(n)|^2 \qquad (6.65)$$

Proof From the Parseval's relation in Eq. (6.60), we have

$$\sum_{n=-\infty}^{\infty} |x(n)|^2 = \frac{1}{2\pi} \int_{-\pi}^{\pi} |X(e^{j\omega})|^2 d\omega$$

$$2\pi \sum_{n=-\infty}^{\infty} |x(n)|^2 = \int_{-\pi}^{\pi} |X(e^{j\omega})|^2 d\omega$$

5.

$$\int_{-\pi}^{\pi} \left| \frac{dX(e^{j\omega})}{d\omega} \right|^2 d\omega = 2\pi \sum_{n=-\infty}^{\infty} |nx(n)|^2 \qquad (6.66)$$

Proof From the differentiation in frequency-domain property [Eq. (6.48)], we have

$$y(n) = nx(n) \longleftrightarrow Y(e^{j\omega}) = j\frac{dX(e^{j\omega})}{d\omega}$$

Now, from the Parseval's relation in Eq. (6.60), we have

$$\sum_{n=-\infty}^{\infty} |y(n)|^2 = \frac{1}{2\pi} \int_{-\pi}^{\pi} |Y(e^{j\omega})|^2 d\omega$$

$$\sum_{n=-\infty}^{\infty} |nx(n)|^2 = \frac{1}{2\pi} \int_{-\pi}^{\pi} \left| j\frac{dX(e^{j\omega})}{d\omega} \right|^2 d\omega$$

$$2\pi \sum_{n=-\infty}^{\infty} |nx(n)|^2 = \int_{-\pi}^{\pi} \left| \frac{dX(e^{j\omega})}{d\omega} \right|^2 d\omega$$

Example 6.17 Let $X(e^{j\omega})$ denote the Fourier transform of the signal $x(n)$ depicted in Fig. 6.11. Perform the following calculations without explicitly evaluating $X(e^{j\omega})$:

(a) Evaluate $X(e^{j0})$.

(b) Find $\angle X(e^{j\omega})$.

(c) Evaluate $\int_{-\pi}^{\pi} X(e^{j\omega}) \, d\omega$.

(d) Find $X(e^{j\pi})$.

(e) Determine and sketch the signal whose Fourier transform is $\Re\{X(e^{j\omega})\}$.

(f) Evaluate $\int_{-\pi}^{\pi} |X(e^{j\omega})|^2 d\omega$.

(g) Evaluate $\int_{-\pi}^{\pi} \left| \dfrac{dX(e^{j\omega})}{d\omega} \right|^2 d\omega$.

Solution
From Fig. 6.11, we can observe that

$$x(n) = \{-1 \quad 0 \quad 1 \quad \underset{\uparrow}{2} \quad 1 \quad 0 \quad 1 \quad 2 \quad 1 \quad 0 \quad -1\}, \quad -3 \le n \le 7$$

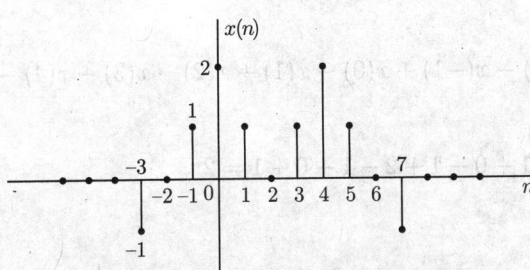

Fig. 6.11   Signal $x(n)$ for Example 6.17

(a) From Eq. (6.62), we have

$$X(e^{j0}) = \sum_{n=-\infty}^{\infty} x(n) = \sum_{n=-3}^{7} x(n) = 6$$

(b) We can observe that $y(n) = x(n+2)$ is real and even. Therefore (using the properties of the DTFT) the DTFT $Y(e^{j\omega}) = \mathcal{F}[y(n)]$ is real and even. Hence, $Y(e^{j\omega})$ has zero phase. We also know that

$$y(n) = x(n+2)$$
$$\mathcal{F}[y(n)] = \mathcal{F}[x(n+2)]$$
$$Y(e^{j\omega}) = X(e^{j\omega}) e^{j2\omega}$$

or, equivalently,

$$X(e^{j\omega}) = Y(e^{j\omega}) e^{-j2\omega}$$

Since $Y(e^{j\omega})$ has zero phase, we have

$$\angle X(e^{j\omega}) = -2\omega$$

(c) From Eq. (6.64), we have

$$\int_{-\pi}^{\pi} X(e^{j\omega}) \, d\omega = 2\pi x(0) = 4\pi$$

(d) From Eq. (6.63), we have

$$X(e^{j\pi}) = \sum_{n=-\infty}^{\infty} (-1)^n x(n)$$

$$= \sum_{n=-3}^{7} (-1)^n x(n)$$

$$= -x(-3) + x(-2) - x(-1) + x(0) - x(1) + x(2) - x(3) + x(4) - x(5)$$
$$+ x(6) - x(7)$$

$$= 1 + 0 - 1 + 2 - 1 + 0 - 1 + 2 - 1 + 0 + 1 = 2$$

(e) From Eq. (6.58), we have

$$x_e(n) \longleftrightarrow \Re\{X(e^{j\omega})\}$$

Therefore, the desired signal is

$$x_e(n) = \frac{x(n) + x(-n)}{2}$$

n	-7	-6	-5	-4	-3	-2	-1	0	1	2	3	4	5	6	7
$x(n)$	0	0	0	0	-1	0	1	2	1	0	1	2	1	0	-1
$x(-n)$	-1	0	1	2	1	0	1	2	1	0	-1	0	0	0	0
$\mathcal{E}\{x(n)\} = x_e(n)$	$-1/2$	0	$1/2$	1	0	0	1	2	1	0	0	1	$1/2$	0	$-1/2$

The signal is as shown in Fig. 6.12.

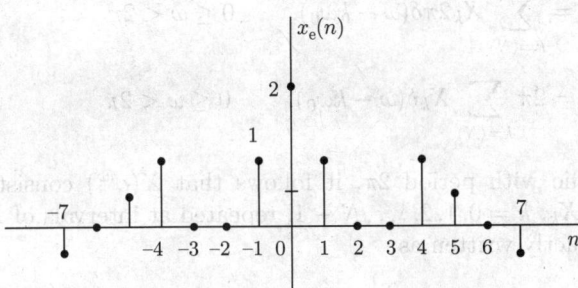

Fig. 6.12

(f) From Eq. (6.65), we have

$$\int_{-\pi}^{\pi} |X(e^{j\omega})|^2 \, d\omega = 2\pi \sum_{n=-\infty}^{\infty} |x(n)|^2 = 2\pi \sum_{n=-3}^{7} |x(n)|^2 = 28\pi$$

(g) From Eq. (6.66), we have

$$\int_{-\pi}^{\pi} \left| \frac{dX(e^{j\omega})}{d\omega} \right|^2 \, d\omega = 2\pi \sum_{n=-\infty}^{\infty} |nx(n)|^2 = 2\pi \sum_{n=-\infty}^{\infty} |n|^2 |x(n)|^2 = 2\pi \sum_{n=-3}^{7} |n|^2 |x(n)|^2 = 316\pi$$

6.7 FOURIER TRANSFORM OF PERIODIC SIGNALS

We can construct the Fourier transform of a periodic signal directly from its Fourier series representation. The resulting transform consists of a train of impulses in the frequency domain, with the areas of the impulses proportional to the Fourier series coefficients.

Consider the periodic signal $x(n)$ with period N and fundamental frequency $\omega_0 = 2\pi/N$. The signal $x(n)$ has the discrete-time Fourier series representation

$$x(n) = \sum_{k=\langle N \rangle} X_k \, e^{jk\omega_0 n}$$

Taking the Fourier transform on both sides yields

$$\mathcal{F}[x(n)] = \mathcal{F}\left[\sum_{k=\langle N \rangle} X_k \, e^{jk\omega_0 n} \right]$$

$$X(e^{j\omega}) = \sum_{k=\langle N \rangle} X_k \mathcal{F}[e^{jk\omega_0 n}]$$

Using Eq. (6.29), we have

$$\mathcal{F}[e^{jk\omega_0 n}] = 2\pi\delta(\omega - k\omega_0) \quad 0 \le \omega < 2\pi$$

so that

$$X(e^{j\omega}) = \sum_{k=\langle N \rangle} X_k 2\pi \delta(\omega - k\omega_0), \qquad 0 \le \omega < 2\pi$$

$$X(e^{j\omega}) = 2\pi \sum_{k=\langle N \rangle} X_k \delta(\omega - k\omega_0), \qquad 0 \le \omega < 2\pi$$

Since the DTFT is periodic with period 2π, it follows that $X(e^{j\omega})$ consists a set of N impulses of strength $2\pi X_k$, $k = 0, 1, 2, \ldots, N - 1$, repeated at intervals of $N\omega_0 = 2\pi$. Thus, $X(e^{j\omega})$ can be compactly written as

$$X(e^{j\omega}) = 2\pi \sum_{k=-\infty}^{\infty} X_k \delta(\omega - k\omega_0) \qquad (6.67)$$

Thus, the Fourier transform of a periodic signal is simply an impulse train with impulses located at $\omega = k\omega_0$, each of which has a strength $2\pi X_k$, and all impulses are separated from each other by ω_0.

Example 6.18 Find and sketch the Fourier transform of the discrete-time impulse train

$$x(n) = \sum_{m=-\infty}^{\infty} \delta(n - mN)$$

depicted in Fig. 6.13.

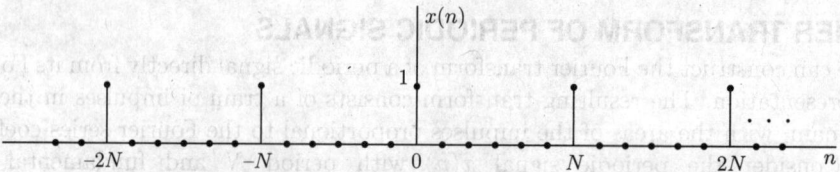

Fig. 6.13 Discrete-time impulse train

Solution

The signal is periodic with period N and frequency $\omega_0 = 2\pi/N$. The given signal for one period may be written as

$$x(n) = \begin{cases} 1, & n = 0 \\ 0, & 1 \le n \le (N - 1) \end{cases}$$

$$= \delta(n)$$

Therefore, the Fourier series coefficients are

$$X_k = \frac{1}{N} \sum_{n=0}^{N-1} x(n) \, e^{-jk\omega_0 n} = \frac{1}{N} \sum_{n=0}^{N-1} \delta(n) \, e^{-jk(2\pi/N)n} = \frac{1}{N} e^{-jk(2\pi/N)n}|_{n=0} = \frac{1}{N}$$

The DTFT of a periodic signal is given by [Eq. (6.67)]

$$X(e^{j\omega}) = 2\pi \sum_{k=-\infty}^{\infty} X_k \delta(\omega - k\omega_0)$$

Substituting $X_k = 1/N$ into the above equation gives

$$X(e^{j\omega}) = 2\pi \sum_{k=-\infty}^{\infty} \frac{1}{N} \delta(\omega - k\omega_0)$$

$$= \frac{2\pi}{N} \sum_{k=-\infty}^{\infty} \delta\left(\omega - \frac{2\pi k}{N}\right)$$

which is illustrated in Fig. 6.14.

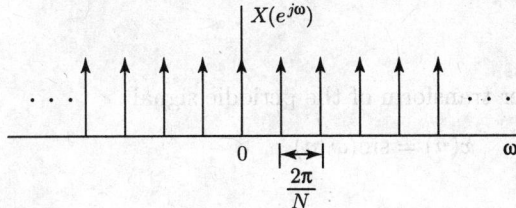

Fig. 6.14   DTFT of a discrete-time impulse train

Example 6.19 Find the Fourier transform of the periodic signal

$$x(n) = \cos(\omega_0 n)$$

with $\omega_0 = 2\pi/5$.

Solution
Consider the given signal

$$x(n) = \cos(\omega_0 n) = \frac{1}{2} e^{j\omega_0 n} + \frac{1}{2} e^{-j\omega_0 n}$$

We know that [Eq. (6.29)]

$$e^{j\omega_0 n} \longleftrightarrow \sum_{m=-\infty}^{\infty} 2\pi \delta(\omega - \omega_0 - 2\pi m)$$

Therefore,

$$X(e^{j\omega}) = \frac{1}{2}\mathcal{F}[e^{j\omega_0 n}] + \frac{1}{2}\mathcal{F}[e^{-j\omega_0 n}]$$

$$X(e^{j\omega}) = \frac{1}{2}\sum_{m=-\infty}^{\infty} 2\pi\delta(\omega - \omega_0 - 2\pi m) + \frac{1}{2}\sum_{m=-\infty}^{\infty} 2\pi\delta(\omega + \omega_0 - 2\pi m)$$

$$X(e^{j\omega}) = \frac{1}{2}\sum_{m=-\infty}^{\infty} 2\pi\delta\left(\omega - \frac{2\pi}{5} - 2\pi m\right) + \frac{1}{2}\sum_{m=-\infty}^{\infty} 2\pi\delta\left(\omega + \frac{2\pi}{5} - 2\pi m\right)$$

The DTFT $X(e^{j\omega})$ is illustrated in Fig. 6.15.

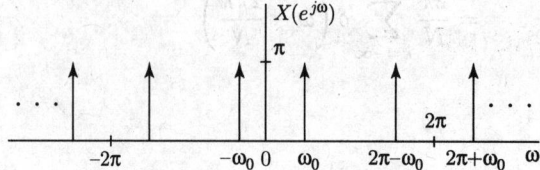

Fig. 6.15 DTFT of $x(n) = \cos(\omega_o n)$

Example 6.20 Find the Fourier transform of the periodic signal

$$x(n) = \sin(\omega_0 n)$$

with $\omega_0 = 2\pi/5$.

Solution
Consider the given signal

$$x(n) = \sin(\omega_0 n) = \frac{1}{2j}e^{j\omega_0 n} - \frac{1}{2j}e^{-j\omega_0 n}$$

We know that [Eq. (6.29)]

$$e^{j\omega_0 n} \longleftrightarrow \sum_{m=-\infty}^{\infty} 2\pi\delta(\omega - \omega_0 - 2\pi m)$$

Therefore,

$$X(e^{j\omega}) = \frac{1}{2j}\mathcal{F}[e^{j\omega_0 n}] - \frac{1}{2j}\mathcal{F}[e^{-j\omega_0 n}]$$

$$X(e^{j\omega}) = \frac{1}{2j}\sum_{m=-\infty}^{\infty} 2\pi\delta(\omega - \omega_0 - 2\pi m) - \frac{1}{2j}\sum_{m=-\infty}^{\infty} 2\pi\delta(\omega + \omega_0 - 2\pi m)$$

$$X(e^{j\omega}) = \frac{1}{2j}\sum_{m=-\infty}^{\infty} 2\pi\delta\left(\omega - \frac{2\pi}{5} - 2\pi m\right) \frac{1}{2j}\sum_{m=-\infty}^{\infty} 2\pi\delta\left(\omega + \frac{2\pi}{5} - 2\pi m\right)$$

The DTFT $X(e^{j\omega})$ is illustrated in Fig. 6.16.

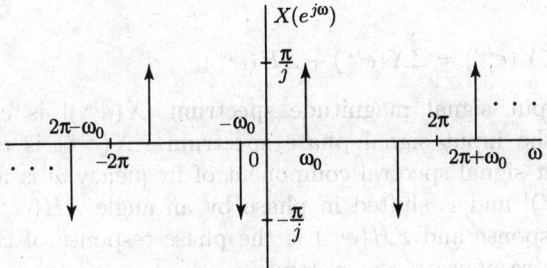

Fig. 6.16   DTFT of $x(n) = \sin(w_o n)$

6.8   SIGNAL TRANSMISSION THROUGH LTI SYSTEMS

If $x(n)$ and $y(n)$ are the input and output of an LTI system with impulse response $h(n)$, then

$$y(n) = x(n) * h(n)$$

Application of the time convolution property to the above equation yields [assuming that both $x(n)$ and $h(n)$ are Fourier transformable]

$$\mathcal{F}[y(n)] = \mathcal{F}[x(n)] * \mathcal{F}[h(n)]$$

$$Y(e^{j\omega}) = X(e^{j\omega})H(e^{j\omega}) \tag{6.68}$$

or

$$H(e^{j\omega}) = \frac{Y(e^{j\omega})}{X(e^{j\omega})} \tag{6.69}$$

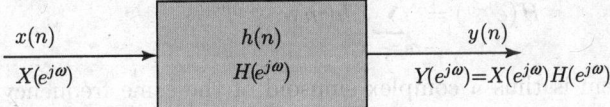

Fig. 6.17

The use of the convolution property for LTI systems is demonstrated in Fig. 6.17. Equation (6.68) can be expressed in the polar form as

$$|Y(e^{j\omega})| \, e^{j\angle Y(e^{j\omega})} = \left[|X(e^{j\omega})|e^{j\angle X(e^{j\omega})}\right] \left[|H(e^{j\omega})|e^{j\angle H(e^{j\omega})}\right]$$

$$|Y(e^{j\omega})| \, e^{j\angle Y(e^{j\omega})} = |X(e^{j\omega})||H(e^{j\omega})| \, e^{j[\angle X(e^{j\omega}) + \angle H(e^{j\omega})]}$$

Therefore,

$$|Y(e^{j\omega})| = |X(e^{j\omega})||H(e^{j\omega})| \tag{6.70}$$

and

$$\angle Y(e^{j\omega}) = \angle X(e^{j\omega}) + \angle H(e^{j\omega}) \tag{6.71}$$

During transmission, the input signal magnitude spectrum $|X(e^{j\omega})|$ is changed to $|X(e^{j\omega})||H(e^{j\omega})|$. Similarly, the input signal phase spectrum $\angle X(e^{j\omega})$ is changed to $\angle X(e^{j\omega}) + \angle H(e^{j\omega})$. An input signal spectral component of frequency ω is modified in amplitude by a factor $|H(e^{j\omega})|$ and is shifted in phase by an angle $\angle H(e^{j\omega})$. Clearly, $|H(e^{j\omega})|$ is the magnitude response and $\angle H(e^{j\omega})$ is the phase response of the system. $H(e^{j\omega})$ is called the frequency response of the system.

6.8.1 Response to Complex Exponentials

The response of an LTI system to an exponential input leads to a characterization of system behaviour that is termed the *frequency response* of the system. This characterization is obtained in terms of the impulse response by using convolution and an exponential input signal. Let the impulse response of a system be $h(n)$ and the input be $x(n) = e^{j\omega_0 n}$. Then, the convolution integral gives the output as

$$y(n) = h(n) * x(n) = \sum_{m=-\infty}^{\infty} h(m)x(n-m)$$

$$= \sum_{m=-\infty}^{\infty} h(m)\, e^{j\omega_0(n-m)}$$

$$= \sum_{m=-\infty}^{\infty} h(m)\, e^{-j\omega_0 m}\, e^{j\omega_0 n}$$

$$y(n) = H(e^{j\omega_0})\, e^{j\omega_0 n} \tag{6.72}$$

where we define

$$H(e^{j\omega_0}) = \sum_{m=-\infty}^{\infty} h(m)\, e^{-j\omega_0 m}$$

The output of the system is thus a complex sinusoid of the same frequency as the input multiplied by the complex number $H(e^{j\omega_0})$. An intuitive interpretation of the sinusoidal steady-state response is obtained by writing the complex-valued frequency response $H(e^{j\omega_0})$ in polar form. We have

$$H(e^{j\omega_0}) = |H(e^{j\omega_0})|\, e^{j\angle H(e^{j\omega_0})}$$

where $|H(e^{j\omega_0})|$ is now termed the *magnitude response* and $\angle H(e^{j\omega_0})$ is termed the *phase response* of the system. Substituting this polar form into Eq. (6.72), we may express the output as

$$y(n) = |H(e^{j\omega_0})|\, e^{j(\omega_0 n + \angle H(e^{j\omega_0}))}$$

The system thus modifies the amplitude of the input by $|H(e^{j\omega_0})|$ and the phase by $\angle H(e^{j\omega_0})$. Knowing $H(e^{j\omega_0})$, we can determine whether the system amplifies or attenuates a given sinusoidal component of the input and how much of a phase shift the system adds to that particular component.

We say that the complex sinusoid $x(n) = e^{j\omega_0 n}$ is an *eigenfunction* of the system associated with the *eigenvalue* $H(e^{j\omega_0})$. An eigenfunction of a system is an input signal that produces an output that differs from the input by a constant multiplicative factor.

6.8.2 Steady-State and Transient Responses

The output $y(n)$ of an LTI discrete-time system characterized by a constant coefficient difference equation is a sum of two parts: the complementary solution $y_c(n)$ (zero-input response) resulting with the input $x(n) = 0$ and a particular solution $y_p(n)$ (forced response) resulting from the specified input $x(n)$. Consider an LTI discrete-time system characterized by a linear constant coefficient difference equation of the form

$$\sum_{k=0}^{N} d_k y(n-k) = \sum_{k=0}^{M} x(n-k) \tag{6.73}$$

where $x(n)$ and $y(n)$ are, respectively, the input and output of the system, and d_k and p_k are constants. The *order* of the discrete-time system is given by $\max(N, M)$, which is the order of the difference equation characterizing the system. The complementary solution $y_c(n)$ is the solution of Eq. (6.73) with the input $x(n) = 0$, i.e., it is the solution of the homogeneous difference equation:

$$\sum_{k=0}^{N} d_k y(n-k) = 0 \tag{6.74}$$

We assume that the complementary solution $y_c(n)$ is of the form

$$y_c(n) = \lambda^n$$

Substituting the above in Eq. (6.74), we arrive at

$$\sum_{k=0}^{N} d_k y(n-k) = \sum_{k=0}^{N} d_k \lambda^{n-k} = 0$$

$$\lambda^{n-N}(d_0 \lambda^N + d_1 \lambda^{N-1} + \cdots + d_{N-1}\lambda + d_N) = 0$$

The polynomial $\sum_{k=0}^{N} d_k \lambda^{n-k}$ is called the characteristic polynomial of the discrete-time system of Eq. (6.73). Let $\lambda_1, \lambda_2, \ldots, \lambda_N$ denote its N roots. If these roots are all distinct, then the general form of the complementary solution is given by

$$y_c(n) = \alpha_1 \lambda_1^n + \alpha_2 \lambda_2^n + \cdots + \alpha_N \lambda_N^n \tag{6.75}$$

where $\alpha_1, \alpha_2, \ldots, \alpha_N$ are constants determined from the specified initial conditions of the discrete-time system and λ_i are the roots of the characteristic polynomial. Moreover, for a stable system, $|\lambda_i| < 1$, and as a result, the complementary solution $y_c(n)$ decays to zero for large values of n. The complementary solution part $y_c(n)$ of the output is also referred to as the *transient response*. We now examine the behaviour of the particular solution $y_p(n)$ for two specific types of input.

1. Consider first an input sequence with a constant amplitude starting at some time instant, say n_0, and then continuing forever afterwards. The output of a causal stable LTI discrete-time system will then be composed of a *steady-state response* (the particular solution), which is also a constant amplitude sequence, and a transient response with zero-valued samples after some time instant $n_1 > n_0$, resulting in a constant amplitude output after the time instant n_1.

2. Consider a constant amplitude sinusoidal input sequence. The output of a causal stable LTI system with a causal input, i.e., a constant amplitude sinusoidal sequence, will have after some time instant a steady-state output that is also a constant amplitude sinusoidal sequence of the same angular frequency as that of the input. In this case also, before the steady-state output is reached, the initial samples of the output will be composed of samples of the transient response and the steady-state response.

It is straightforward to develop the expression for the steady-state response of a LTI discrete-time system with a real impulse response $h(n)$ for a sinusoidal input in terms of its frequency response $H(e^{j\omega})$. Let the input to the LTI system be

$$x(n) = A\cos(\omega_0 n), \quad -\infty < n < \infty \qquad (6.76)$$

$$x(n) = \frac{A}{2}e^{j\omega_0 n} + \frac{A}{2}e^{-j\omega_0 n}$$

Because of the linearity, the response $y(n)$ to the input $x(n)$ is given by

$$\begin{aligned}
y(n) &= \frac{A}{2}H(e^{j\omega_0})\,e^{j\omega_0 n} + \frac{A}{2}H(e^{-j\omega_0})\,e^{-j\omega_0 n} \\
&= \frac{A}{2}|H(e^{j\omega_0})|\,e^{j\angle H(e^{j\omega_0})}\,e^{j\omega_0 n} + \frac{A}{2}|H(e^{-j\omega_0})|\,e^{j\angle H(e^{-j\omega_0})}\,e^{-j\omega_0 n} \\
&= \frac{A}{2}|H(e^{j\omega_0})|\,e^{j\angle H(e^{j\omega_0})}\,e^{j\omega_0 n} + \frac{A}{2}|H(e^{j\omega_0})|\,e^{-j\angle H(e^{j\omega_0})}\,e^{-j\omega_0 n}
\end{aligned}$$

where we have used the fact that the magnitude response of the LTI system with a real impulse response is an even function of ω, i.e., $|H(e^{j\omega_0})| = |H(e^{-j\omega_0})|$, and the phase response is an odd function of ω, i.e., $\angle H(e^{j\omega_0}) = -\angle H(e^{-j\omega_0})$. The output of the LTI system thus can be expressed as

$$y(n) = \frac{A}{2}|H(e^{j\omega_0})|\,e^{j(\omega_0 n + \angle H(e^{j\omega_0}))} + \frac{A}{2}|H(e^{j\omega_0})|\,e^{-j(\omega_0 n + \angle H(e^{j\omega_0}))}$$

$$y(n) = A\,|H(e^{j\omega_0})|\cos(\omega_0 n + \angle H(e^{j\omega_0})) \qquad (6.77)$$

Thus, the output signal $y(n)$ has the same sinusoidal waveform as the input signal $x(n)$ of Eq. (6.76), with two differences:

1. The amplitude is multiplied by $|H(e^{j\omega_0})|$, the value of the magnitude response of the discrete-time system at $\omega = \omega_0$.

2. The output signal has a phase lag relative to the input by an amount $\angle H(e^{j\omega_0})$, the value of the phase response of the discrete-time system at $\omega = \omega_0$. The expression

developed in Eq. (6.77) is the response of an LTI system to a sinusoidal input signal applied to the system at $n = -\infty$. We usually called such signal eternal sinusoids because they were applied at $n = -\infty$. In such a case, the response $y(n)$ is the steady-state response. There is no transient response in this case.

6.8.3 Response to a Causal Exponential Sequence

Consider a causal sequence applied at the input of a causal LTI discrete-time system at time $n = 0$. The output for such an input when observed at time $n = 0$ will consists of a transient part along with a steady-state component and thus will be different from the one shown in Eq. (6.77). To demonstrate this behaviour, we next develop the expression for the output $y(n)$ of a causal LTI discrete-time system when the input is a causal exponential sequence at $n = 0$, i.e.,

$$x(n) = e^{j\omega_0 n} u(n) \tag{6.78}$$

where $u(n)$ is the unit step sequence. Since $x(n) = 0$, for $n < 0$, we have $y(n) = 0$, for $n < 0$. For $n \geq 0$, the output $y(n)$ is given by

$$y(n) = h(n) * x(n)$$

$$= \sum_{m=0}^{\infty} h(m) x(n - m)$$

$$= \sum_{m=0}^{\infty} h(m) \, e^{j\omega_0(n-m)} u(n - m)$$

$$= \sum_{m=0}^{n} h(m) \, e^{-j\omega_0 m} \, e^{j\omega_0 n}$$

$$= \left(\sum_{m=0}^{n} h(m) \, e^{-j\omega_0 m} \right) e^{j\omega_0 n} + \left(\sum_{m=n+1}^{\infty} h(m) \, e^{-j\omega_0 m} \right) e^{j\omega_0 n}$$

$$- \left(\sum_{m=n+1}^{\infty} h(m) \, e^{-j\omega_0 m} \right) e^{j\omega_0 n}$$

$$y(n) = \left(\sum_{m=0}^{\infty} h(m) \, e^{-j\omega_0 m} \right) e^{j\omega_0 n} - \left(\sum_{m=n+1}^{\infty} h(m) \, e^{-j\omega_0 m} \right) e^{j\omega_0 n}$$

$$y(n) = H(e^{j\omega_0}) \, e^{j\omega_0 n} - \left(\sum_{m=n+1}^{\infty} h(m) \, e^{-j\omega_0 m} \right) e^{j\omega_0 n} \quad n \geq 0 \tag{6.79}$$

The first term in the above equation is the same as that given by Eq. (6.72) and is the *steady-state response:*

$$y_{\text{sr}}(n) = H(e^{j\omega_0}) \, e^{j\omega_0 n} \tag{6.80}$$

The second term in Eq. (6.79) is the *transient response*:

$$y_{tr}(n) = - \left(\sum_{m=n+1}^{\infty} h(m) e^{-j\omega_0 m} \right) e^{j\omega_0 n} \tag{6.81}$$

To determine the effect of the second term on the output response, we observe that

$$|y_{tr}(n)| = \left| \sum_{m=n+1}^{\infty} h(m) e^{-j\omega_0 (m-n)} \right| \leq \sum_{m=n+1}^{\infty} |h(m)| \leq \sum_{m=0}^{\infty} |h(m)| \tag{6.82}$$

For a causal and stable LTI discrete-time system, the impulse response is absolutely summable, and as a result the transient response $y_{tr}(n)$ is a bounded sequence. Moreover, as $n \to \infty$, $\sum_{m=n+1}^{\infty} |h(m)| \to 0$ and hence, the transient response decays to zero as n gets very large. In most practical cases, the transient response becomes negligibly small after some finite amount of time, and the system can be assumed to be in a steady state. It should be noted that transients will occur whenever an input signal is applied or changed.

Example 6.21 Consider a causal LTI system that is characterized by the difference equation

$$y(n) - \frac{3}{4}y(n-1) + \frac{1}{8}y(n-2) = 2x(n)$$

(a) Find the frequency response $H(e^{j\omega})$ and the impulse response $h(n)$ of the system.
(b) Find the response $y(n)$ if the input to this system is

$$x(n) = \left(\frac{1}{4} \right)^n u(n)$$

Solution
(a) Consider the given difference equation

$$y(n) - \frac{3}{4}y(n-1) + \frac{1}{8}y(n-2) = 2x(n)$$

Taking the DTFT of the above equation gives

$$Y(e^{j\omega}) - \frac{3}{4}Y(e^{j\omega}) e^{-j\omega} + \frac{1}{8}Y(e^{j\omega}) e^{-j2\omega} = 2X(e^{j\omega})$$

$$Y(e^{j\omega}) \left(1 - \frac{3}{4} e^{-j\omega} + \frac{1}{8} e^{-j2\omega} \right) = 2X(e^{j\omega})$$

From Eq. (6.69), the frequency response of this system is

$$H(e^{j\omega}) = \frac{Y(e^{j\omega})}{X(e^{j\omega})} = \frac{2}{1 - \frac{3}{4} e^{-j\omega} + \frac{1}{8} e^{-j2\omega}}$$

Now, by factorizing the denominator of the above equation, we obtain

$$H(e^{j\omega}) = \frac{2}{\left(1 - \frac{1}{2} e^{-j\omega}\right)\left(1 - \frac{1}{4} e^{-j\omega}\right)}$$

For discrete-time transforms, it is most convenient to substitute p for $e^{-j\omega}$. By making this substitution, we obtain the rational system function

$$H(e^{j\omega}) = \frac{2}{\left(1 - \frac{1}{2}p\right)\left(1 - \frac{1}{4}p\right)}$$

Using the partial fraction expansion, we obtain

$$H(p) = \frac{A_1}{1 - \frac{1}{2}p} + \frac{A_2}{1 - \frac{1}{4}p}$$

$$A_1 = \left[\left(1 - \frac{1}{2}p\right) H(p)\right]\bigg|_{p=2} = \frac{2}{1 - 1/2} = 4$$

$$A_2 = \left[\left(1 - \frac{1}{4}p\right) H(p)\right]\bigg|_{p=4} = \frac{2}{1 - 2} = -2$$

Thus,

$$H(p) = \frac{4}{1 - \left(\frac{1}{2}\right)p} - \frac{2}{1 - \left(\frac{1}{4}\right)p}$$

$$H(e^{j\omega}) = \frac{4}{1 - \left(\frac{1}{2}\right) e^{-j\omega}} - \frac{2}{1 - \left(\frac{1}{4}\right) e^{-j\omega}}$$

and taking the inverse DTFT of the above equation, we obtain the unit impulse response:

$$h(n) = 4 \left(\frac{1}{2}\right)^n u(n) - 2 \left(\frac{1}{4}\right)^n u(n)$$

(b) The DTFT of the given input $x(n) = (1/4)^n u(n)$ is given by

$$X(e^{j\omega}) = \frac{1}{1 - (1/4) e^{-j\omega}}$$

Using Eq. (6.68), we obtain

$$Y(e^{j\omega}) = X(e^{j\omega})H(e^{j\omega}) = \left[\frac{2}{\left(1 - \left(\frac{1}{2}\right) e^{-j\omega}\right)\left(1 - \left(\frac{1}{4}\right) e^{-j\omega}\right)}\right]\left[\frac{1}{1 - \left(\frac{1}{4}\right) e^{-j\omega}}\right]$$

$$= \frac{2}{\left(1 - \left(\frac{1}{2}\right) e^{-j\omega}\right)\left(1 - \left(\frac{1}{4}\right) e^{-j\omega}\right)^2}$$

Using the partial fraction expansion, we obtain

$$Y(e^{j\omega}) = -\frac{4}{1 - \left(\frac{1}{4}\right) e^{-j\omega}} - \frac{2}{\left(1 - \left(\frac{1}{4}\right) e^{-j\omega}\right)^2} + \frac{8}{1 - \left(\frac{1}{2}\right) e^{-j\omega}}$$

$$Y(e^{j\omega}) = Y_1(e^{j\omega}) + Y_2(e^{j\omega}) + Y_3(e^{j\omega})$$

Taking the inverse DTFT gives

$$y(n) = -y_1(n) - y_2(n) + y_3(n)$$

where $y_1(n) = 4(1/4)^n u(n)$ and $y_3(n) = 8(1/2)^n u(n)$, and to determine $y_2(n)$, substitute $a = 1/4$ in Eq. (6.50) to obtain

$$(n+1)\left(\frac{1}{4}\right)^n u(n) \longleftrightarrow \frac{1}{\left(1 - \frac{1}{4}e^{-j\omega}\right)^2}$$

$$2(n+1)\left(\frac{1}{4}\right)^n u(n) \longleftrightarrow \frac{2}{\left(1 - \frac{1}{4}e^{-j\omega}\right)^2}$$

Therefore,

$$y_2(n) = 2(n+1)\left(\frac{1}{4}\right)^n u(n)$$

and

$$y(n) = -4\left(\frac{1}{4}\right)^n u(n) - 2(n+1)\left(\frac{1}{4}\right)^n u(n) + 8\left(\frac{1}{2}\right)^n u(n)$$

$$y(n) = \left[-4\left(\frac{1}{4}\right)^n - 2(n+1)\left(\frac{1}{4}\right)^n + 8\left(\frac{1}{2}\right)^n\right]u(n)$$

Example 6.22 Suppose that we want to design a discrete-time LTI system with the property that if the input is

$$x(n) = \left(\frac{1}{2}\right)^n u(n) - \frac{1}{4}\left(\frac{1}{2}\right)^{n-1} u(n-1)$$

then the output is

$$y(n) = \left(\frac{1}{3}\right)^n u(n)$$

(a) Find the impulse response and frequency response of a discrete-time LTI system that has the foregoing property.

(b) Find a difference equation relating $x(n)$ and $y(n)$ that characterizes the system.

Solution

(a) Taking the DTFT of the given input $x(n)$, we obtain

$$X(e^{j\omega}) = \frac{1}{1 - \left(\frac{1}{2}\right)e^{-j\omega}} - \frac{1}{4}\frac{1}{1 - \left(\frac{1}{2}\right)e^{-j\omega}}e^{-j\omega} = \frac{1 - \left(\frac{1}{4}\right)e^{-j\omega}}{1 - \left(\frac{1}{2}\right)e^{-j\omega}}$$

Taking the DTFT of the output $y(n)$, we obtain

$$Y(e^{j\omega}) = \frac{1}{1 - \left(\frac{1}{3}\right)e^{-j\omega}}$$

The frequency response of the system is given by

$$H(e^{j\omega}) = \frac{Y(e^{j\omega})}{X(e^{j\omega})} = \frac{1 - \left(\frac{1}{2}\right)e^{-j\omega}}{\left(1 - \left(\frac{1}{3}\right)e^{-j\omega}\right)\left(1 - \left(\frac{1}{4}\right)e^{-j\omega}\right)}$$

(b) From part (a), we have

$$H(e^{j\omega}) = \frac{Y(e^{j\omega})}{X(e^{j\omega})} = \frac{1 - \left(\frac{1}{2}\right)e^{-j\omega}}{\left(1 - \left(\frac{1}{3}\right)e^{-j\omega}\right)\left(1 - \left(\frac{1}{4}\right)e^{-j\omega}\right)}$$

$$\frac{Y(e^{j\omega})}{X(e^{j\omega})} = \frac{1 - \frac{1}{2}e^{-j\omega}}{1 - \left(\frac{7}{12}\right)e^{-j\omega} + \left(\frac{1}{12}\right)e^{-j2\omega}}$$

$$Y(e^{j\omega})\left[1 - \frac{7}{12}e^{-j\omega} + \frac{1}{12}e^{-j2\omega}\right] = X(e^{j\omega})\left[1 - \frac{1}{2}e^{-j\omega}\right]$$

$$Y(e^{j\omega}) - \frac{7}{12}e^{-j\omega}Y(e^{j\omega}) + \frac{1}{12}e^{-j2\omega}Y(e^{j\omega}) = X(e^{j\omega}) - \frac{1}{2}e^{-j\omega}X(e^{j\omega})$$

Taking the inverse DTFT yields

$$y(n) - \frac{7}{12}y(n-1) + \frac{1}{12}y(n-2) = x(n) - \frac{1}{2}x(n-1)$$

Example 6.23 Suppose that a system has the response $(1/4)^n u(n)$ to the input $(n+2)$ $(1/2)^n u(n)$. If the output of this system is $\delta(n) - (-1/2)^n u(n)$, what is the input?

Solution
Given that

$$x(n) = (n+2)\left(\frac{1}{2}\right)^n u(n) = n\left(\frac{1}{2}\right)^n u(n) + 2\left(\frac{1}{2}\right)^n u(n)$$

Taking the DTFT of the above equation, we obtain

$$X(e^{j\omega}) = \frac{\left(\frac{1}{2}\right)e^{-j\omega}}{\left(1 - \left(\frac{1}{2}\right)e^{-j\omega}\right)^2} + 2\frac{1}{1 - \left(\frac{1}{2}\right)e^{-j\omega}} = \frac{2\left(1 - \left(\frac{1}{4}\right)e^{-j\omega}\right)}{\left(1 - \left(\frac{1}{2}\right)e^{-j\omega}\right)^2}$$

The response of the system is given by

$$y(n) = \left(\frac{1}{4}\right)^n u(n)$$

Taking its DTFT, we obtain

$$Y(e^{j\omega}) = \frac{1}{1 - \left(\frac{1}{4}\right)e^{-j\omega}}$$

The frequency response of the system is given by

$$H(e^{j\omega}) = \frac{Y(e^{j\omega})}{X(e^{j\omega})} = \frac{\left(1 - \left(\frac{1}{2}\right)e^{-j\omega}\right)^2}{2\left(1 - \left(\frac{1}{4}\right)e^{-j\omega}\right)^2}$$

We want to find $x(n)$ when

$$y(n) = \delta(n) - \left(-\frac{1}{2}\right)^n u(n)$$

$$Y(e^{j\omega}) = 1 - \frac{1}{1 + \left(\frac{1}{2}\right)e^{-j\omega}} = \frac{\left(\frac{1}{2}\right)e^{-j\omega}}{1 + \left(\frac{1}{2}\right)e^{-j\omega}}$$

From the expression of the frequency response $H(e^{j\omega})$, we have

$$X(e^{j\omega}) = \frac{2\left(1 - \left(\frac{1}{4}\right)e^{-j\omega}\right)^2}{\left(1 - \left(\frac{1}{2}\right)e^{-j\omega}\right)^2} Y(e^{j\omega}) = \frac{2\left(1 - \left(\frac{1}{4}\right)e^{-j\omega}\right)^2}{\left(1 - \left(\frac{1}{2}\right)e^{-j\omega}\right)^2} \frac{\left(\frac{1}{2}\right)e^{-j\omega}}{\left(1 + \left(\frac{1}{2}\right)e^{-j\omega}\right)}$$

$$X(e^{j\omega}) = \frac{e^{-j\omega}\left(1 - \left(\frac{1}{4}\right)e^{-j\omega}\right)^2}{\left(1 - \left(\frac{1}{2}\right)e^{-j\omega}\right)^2\left(1 + \left(\frac{1}{2}\right)e^{-j\omega}\right)}$$

Using the partial fraction expansion, we obtain

$$X(e^{j\omega}) = \frac{\left(\frac{3}{8}\right)e^{-j\omega}}{1 + \left(\frac{1}{2}\right)e^{-j\omega}} + \frac{\left(\frac{3}{8}\right)e^{-j\omega}}{1 - \left(\frac{1}{2}\right)e^{-j\omega}} + \frac{\left(\frac{1}{8}\right)e^{-j\omega}}{\left(1 - \left(\frac{1}{2}\right)e^{-j\omega}\right)^2}$$

Taking the inverse DTFT, we obtain

$$x(n) = \frac{3}{8}\left(-\frac{1}{2}\right)^{n-1} u(n-1) + \frac{3}{8}\left(\frac{1}{2}\right)^{n-1} u(n-1) + \frac{1}{4}n\left(\frac{1}{2}\right)^n u(n)$$

$$x(n) = \frac{3}{8}\left(-\frac{1}{2}\right)^{n-1} u(n-1) + \frac{3}{8}\left(\frac{1}{2}\right)^{n-1} u(n-1) + \frac{1}{8}n\left(\frac{1}{2}\right)^{n-1} u(n-1)$$

Example 6.24 Determine the response of the system with impulse response

$$h(n) = \left(\frac{1}{2}\right)^n u(n)$$

when the input is the complex exponential sequence

$$x(n) = A\,e^{j(\pi/2)n}, \quad -\infty < n < \infty$$

Solution
Given that

$$h(n) = \left(\frac{1}{2}\right)^n u(n)$$

Taking its DTFT, we obtain

$$H(e^{j\omega}) = \frac{1}{1 - \left(\frac{1}{2}\right)e^{-j\omega}}$$

At $\omega = \pi/2$, the above equation yields

$$H(e^{j\pi/2}) = \frac{1}{1 - (1/2)\,e^{-j\pi/2}} = \frac{1}{1 + j(1/2)} = \frac{2}{\sqrt{5}}\,e^{-j26.6°}$$

The response of the system to the complex exponential is given by [Eq. (6.72)]

$$y(n) = AH(e^{j\omega_0})\,e^{j\omega_0 n} = AH(e^{j\pi/2})\,e^{j\pi/2 n}$$

$$= A\left(\frac{2}{\sqrt{5}}\,e^{-j26.6°}\right)\,e^{j\pi/2 n}$$

$$y(n) = A\frac{2}{\sqrt{5}}\,e^{j(\pi n/2 - 26.6°)}$$

Example 6.25 Determine the response of the system with impulse response

$$h(n) = \left(\frac{1}{2}\right)^n u(n)$$

to the input signal

$$x(n) = 5 - 5\sin\left(\frac{\pi}{2}n\right) + 10\cos(\pi n), \quad -\infty < n < \infty$$

Solution
Given that

$$h(n) = \left(\frac{1}{2}\right)^n u(n)$$

Taking its DTFT, we obtain

$$H(e^{j\omega}) = \frac{1}{1 - (1/2)\,e^{-j\omega}}$$

The first term in the input signal $x(n)$ is a constant signal corresponding to $\omega = 0$, thus

$$H(e^{j0}) = \frac{1}{1 - 1/2} = 2$$

The second term in $x(n)$ has a frequency $\pi/2$. At this frequency, the frequency response of the system is

$$H(e^{j\pi/2}) = \frac{1}{1 - \frac{1}{2}e^{-j\pi/2}} = \frac{1}{1 + j1/2} = \frac{2}{\sqrt{5}}\,e^{-j26.6°}$$

Finally, the third term in $x(n)$ has a frequency $\omega = \pi$. At this frequency, the frequency response of the system is

$$H(e^{j\pi}) = \frac{1}{1 - (1/2)\,e^{-j\pi}} = \frac{1}{1 + (1/2)} = \frac{1}{3/2} = \frac{2}{3}$$

Hence the response of the system to $x(n)$ is

$$y(n) = 5H(e^{j0}) - H(e^{j\pi/2})5\sin\left(\frac{\pi}{2}n\right) + H(e^{j\pi})10\cos(\pi n)$$

$$y(n) = 10 - \frac{10}{\sqrt{5}}\sin\left(\frac{\pi}{2}n - 26.6°\right) + \frac{20}{3}\cos(\pi n)$$

6.8.4 Linear and Nonlinear Phase

Consider a discrete-time LTI system with impulse response $h(n)$ and frequency response $H(e^{j\omega})$. Let a signal $x(n)$ with Fourier transform $X(e^{j\omega})$ be applied to the input of the system. Let a signal $y(n)$ with Fourier transform $Y(e^{j\omega})$ denote the output of the system. In several applications, such as signal amplification or message signal transmission over a communication channel, we require that the output waveform be a replica of the input waveform, i.e., distortionless transmission. As in the continuous-time case, transmission is said to be distortionless if the input and the output have identical waveshapes within a multiplicative constant and a constant time delay. Thus, in distortionless transmission, the input $x(n)$ and the output $y(n)$ satisfy the condition (Fig. 6.18):

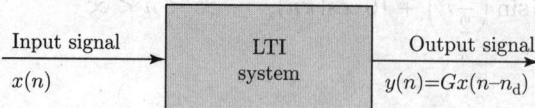

Fig. 6.18

$$y(n) = Gx(n - n_{\mathrm{d}}) \tag{6.83}$$

where the constant G accounts for a change in amplitude and the constant n_{d} accounts for a delay in transmission. Taking the Fourier transform of the above equation yields

$$Y(e^{j\omega}) = GX(e^{j\omega})e^{-j\omega n_{\mathrm{d}}}$$

$$\frac{Y(e^{j\omega})}{X(e^{j\omega})} = H(e^{j\omega}) = Ge^{-j\omega n_{\mathrm{d}}}$$

This is the frequency response required of a system for distortionless transmission. From this equation it follows that

1. the magnitude response $|H(e^{j\omega})|$ must be a constant, i.e., we must have

$$|H(e^{j\omega})| = G \tag{6.84}$$

for some constant G.

2. the phase response $\angle H(e^{j\omega})$ must be a linear function of ω with slope $-n_{\mathrm{d}}$ and intercept zero, i.e., we must have

$$\angle H(e^{j\omega}) = -\omega n_{\mathrm{d}} \tag{6.85}$$

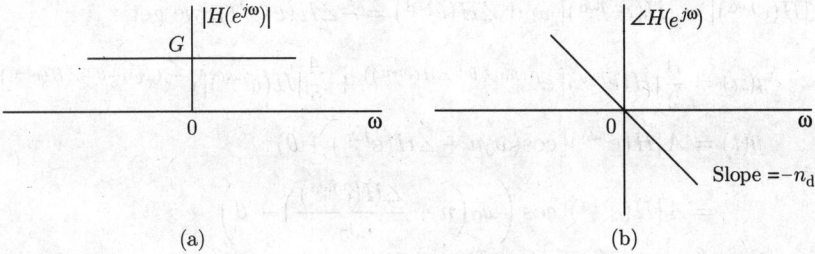

Fig. 6.19 (a) Magnitude response and (b) phase response of a system for distortionless transmission

These two conditions are illustrated in Figs 6.19(a) and (b), respectively. The gain $|H(e^{j\omega})| = G$ means that every spectral component is multiplied by a constant G. A linear phase $\angle H(e^{j\omega}) = -\omega n_d$ means that every spectral component is delayed by n_d seconds. For distortionless transmission, we require a linear-phase characteristic. The phase is not only a linear function of ω but it should also pass through the origin $\omega = 0$. Because each spectral component is attenuated by the same factor (G) and delayed by exactly the same amount (n_d), the output signal is an exact replica of the input (except for attenuating factor G and delay n_d).

6.8.5 Phase Delay and Group Delay

The two important parameters that characterize the form of the response $y(n)$ of an LTI system excited by an input signal $x(n)$ composed of a weighted sum of sinusoidal signals are phase delay and group delay. These two parameters are associated with the frequency response $H(e^{j\omega})$ of the system. For a linear phase system (whose phase varies linearly with frequency), both the phase delay and the group delay are constant.

Phase Delay

The time delay experienced by a single-frequency signal (i.e., a sinusoidal signal) when the signal passes through a system is referred to as the system phase delay. Now, let us assume that the input signal is the single-frequency signal

$$x(n) = A\cos(\omega_0 n + \theta)$$

$$x(n) = \frac{A}{2} e^{j(\omega_0 n + \theta)} + \frac{A}{2} e^{-j(\omega_0 n + \theta)}$$

and the system frequency response is

$$H(e^{j\omega}) = |H(e^{j\omega})| e^{j\angle H(e^{j\omega})}$$

Then, the system output signal is [Eq. (4.42)]

$$y(n) = \frac{A}{2}|H(e^{j\omega_0})| e^{j(\omega_0 n + \theta + \angle H(e^{j\omega_0}))} + \frac{A}{2}|H(e^{-j\omega_0})| e^{-j(\omega_0 n + \theta - \angle H(e^{-j\omega_0}))}$$

Since $|H(e^{j\omega_0})| = |H(e^{-j\omega_0})|$ and $\angle H(e^{j\omega_0}) = -\angle H(e^{-j\omega_0})$, we get

$$y(n) = \frac{A}{2}|H(e^{j\omega_0})|\, e^{j(\omega_0 n + \theta + \angle H(e^{j\omega_0}))} + \frac{A}{2}|H(e^{j\omega_0})|\, e^{-j(\omega_0 n + \theta + \angle H(e^{j\omega_0}))}$$

$$y(t) = A|H(e^{j\omega_0})| \cos(\omega_0 n + \angle H(e^{j\omega_0}) + \theta)$$

$$= A|H(e^{j\omega_0})| \cos\left(\omega_0\left(n + \frac{\angle H(e^{j\omega_0})}{\omega_0}\right) + \theta\right)$$

$$y(n) = A|H(e^{j\omega_0})| \cos\left(\omega_0(n - \tau_{\mathrm{p}}(\omega_0)) + \theta\right)$$

where

$$\tau_{\mathrm{p}}(\omega_0) = -\frac{\angle H(e^{j\omega_0})}{\omega_0} \tag{6.86}$$

is the time delay experienced by the single-frequency signal with frequency ω_0 when it passes through the system. Thus, $\tau_{\mathrm{p}}(\omega_0)$ is the system phase delay for a signal with frequency ω_0. A negative system phase response at positive frequencies indicates that a signal is delayed in time when it passes through the system, whereas a positive system phase response at positive frequencies indicates that a signal is advanced in time when it passes through the system.

Group Delay

When the input signal contains many sinusoidal components with different frequencies that are not harmonically related, each component will go through different phase delays when processed by a frequency selective LTI system, and the signal delay is determined using a different parameter called the group delay, as defined below:

$$\tau_{\mathrm{g}}(\omega) = -\frac{d\angle H(e^{j\omega})}{d\omega} \tag{6.87}$$

The time delay $\tau_{\mathrm{g}}(\omega)$ is called the group delay or envelope delay. Thus, the group delay at each frequency equals the negative of the slope of the phase at that frequency. If $\tau_{\mathrm{g}}(\omega)$ is constant, all the components are delayed by the same interval.

Example 6.26 Consider a discrete-time LTI system with frequency response $H(e^{j\omega})$ and real impulse response $h(n)$. The group-delay function for such a system is defined as

$$\tau_{\mathrm{g}}(\omega) = -\frac{d\angle H(e^{j\omega})}{d\omega}$$

where $\angle H(e^{j\omega})$ has no discontinuities. Suppose that, for this system,

$$|H(e^{j\pi/2})| = 2, \quad \angle H(e^{j0}) = 0, \quad \tau_{\mathrm{g}}\left(\frac{\pi}{2}\right) = 2$$

Determine the output of the system for each of the following inputs:
(a) $\cos(\pi/2n)$ (b) $\sin(7\pi/2n + \pi/4)$

Solution
(a) The input signal $x(n) = \cos(\pi/2n)$ has a frequency $\omega_0 = \pi/2$. Hence, the response of the system to $x(n)$ is

$$y(n) = |H(e^{j\omega_0})|\cos(\omega_0(n - \tau_g(\omega_0)))$$

$$= |H(e^{j\pi/2})|\cos\left(\frac{\pi}{2}(n - \tau_g(\pi/2))\right)$$

$$y(n) = 2\cos\left(\frac{\pi}{2}(n - 2)\right)$$

$$y(n) = 2\cos\left(\frac{\pi}{2}n - \pi\right)$$

(b) Consider the given input signal

$$x(n) = \sin\left(\frac{7\pi}{2}n + \frac{\pi}{4}\right)$$

$$x(n) = -\sin\left(\frac{\pi}{2}n - \frac{\pi}{4}\right)$$

The input signal $x(n)$ has a frequency $\omega_0 = \pi/2$. Hence, the response of the system to $x(n)$ is

$$y(n) = |H(e^{j\omega_0})|\sin(\omega_0(n - \tau_g(\omega_0)) + \theta)$$

$$= |H(e^{j\pi/2})|\sin\left(\frac{7\pi}{2}(n - \tau_g(\pi/2)) + \frac{\pi}{4}\right)$$

$$y(n) = 2\sin\left(\frac{7\pi}{2}(n - 2) + \frac{\pi}{4}\right)$$

$$y(n) = 2\sin\left(\frac{7\pi}{2}n - 7\pi + \frac{\pi}{4}\right)$$

$$= 2\sin\left(\frac{7\pi}{2}n - \pi + \frac{\pi}{4}\right)$$

$$y(n) = 2\sin\left(\frac{7\pi}{2}n - \frac{3\pi}{4}\right)$$

Example 6.27 Consider a nonrecursive filter with the impulse response $h(n)$ shown in Fig. 6.20. What is the group delay as a function of frequency for this filter?

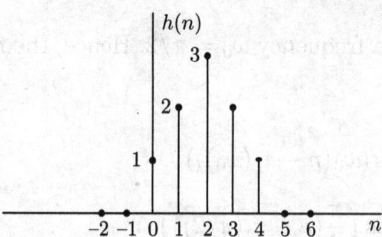

Fig. 6.20

Solution

We can observe that $h_1(n) = h(n+2)$ is real and even. Therefore, (using the properties of the DTFT) the DTFT $H_1(e^{j\omega}) = \mathcal{F}[h_1(n)]$ is real and even. Hence, $H_1(e^{j\omega})$ has zero phase. We also know that

$$h_1(n) = h(n+2)$$
$$\mathcal{F}[h_1(n)] = \mathcal{F}[h(n+2)]$$
$$H_1(e^{j\omega}) = H(e^{j\omega})\, e^{j2\omega}$$

or, equivalently,

$$H(e^{j\omega}) = H_1(e^{j\omega})\, e^{-j2\omega}$$

Since $H_1(e^{j\omega})$ has zero phase, we have

$$\angle H(e^{j\omega}) = -2\omega$$

Therefore, the group delay is

$$\tau_{\mathrm{g}}(\omega) = -\frac{d\angle H(e^{j\omega})}{d\omega} = 2$$

Example 6.28 Consider a discrete-time low-pass filter whose impulse response $h(n)$ is known to be real and whose frequency response magnitude in the region $-\pi \le \omega \le \pi$ is given as

$$|H(e^{j\omega})| = \begin{cases} 1, & |\omega| \le \pi/4 \\ 0, & \text{otherwise} \end{cases}$$

Determine the real-valued impulse response $h(n)$ for this filter when the corresponding group-delay function is specified as
(a) $\tau_{\mathrm{g}}(\omega) = 5$, (b) $\tau_{\mathrm{g}}(\omega) = 5/2$, (c) $\tau_{\mathrm{g}}(\omega) = -5/2$

Solution

Let $H_0(e^{j\omega}) = |H(e^{j\omega})|$. Then the inverse DTFT of $H_0(e^{j\omega})$ is given by

$$h_0(n) = \frac{1}{2\pi} \int_{-\pi}^{\pi} H_0(e^{j\omega}) e^{j\omega n} \, d\omega = \frac{1}{2\pi} \int_{-\pi/4}^{\pi/4} e^{j\omega n} \, d\omega$$

$$= \frac{1}{2\pi} \left. \frac{e^{j\omega n}}{jn} \right|_{-\pi/4}^{\pi/4}$$

$$= \frac{1}{n\pi} \frac{e^{j\pi/4n} - e^{-j\pi/4n}}{2j}$$

$$h_0(n) = \frac{\sin(\pi/4n)}{n\pi}$$

If $\tau_g(\omega) = -\dfrac{d\angle H(e^{j\omega})}{d\omega} = \alpha$ (where α is a constant), then $\angle H(e^{j\omega}) = -\alpha\omega + \beta$, where β is a constant. Given that $h(n)$ is real, therefore $\angle H(e^{j\omega})$ is an odd function, and hence for $\angle H(e^{j\omega})$ to be an odd function, the constant $\beta = 0$. Therefore,

$$H(e^{j\omega}) = |H(e^{j\omega})| e^{j\angle H(e^{j\omega})} = H_0(e^{j\omega}) e^{-j\alpha\omega}$$

Taking the inverse DTFT, we obtain

$$h(n) = h_0(n - \alpha)$$

$$h(n) = \frac{\sin[\pi/4(n - \alpha)]}{(n - \alpha)\pi}$$

(a) Given that $\tau_g(\omega) = 5 = \alpha$. Substituting $\alpha = 5$ in the expression of $h(n)$ yields

$$h(n) = \frac{\sin[\pi/4(n - 5)]}{(n - 5)\pi}$$

(b) Given that $\tau_g(\omega) = 5/2 = \alpha$. Substituting $\alpha = 5/2$ in the expression of $h(n)$ yields

$$h(n) = \frac{\sin[\pi/4(n - 5/2)]}{(n - 5/2)\pi}$$

(c) Given that $\tau_g(\omega) = -5/2 = \alpha$. Substituting $\alpha = -5/2$ in the expression of $h(n)$ yields

$$h(n) = \frac{\sin[\pi/4(n + 5/2)]}{(n + 5/2)\pi}$$

6.9 IDEAL AND PRACTICAL FILTERS

Filtering is a process by which the essential and useful part of a signal is separated from the extraneous and undesirable components. The idea of filtering using LTI system is based on the convolution property of the DTFT which states that for LTI systems, the DTFT of the output $Y(e^{j\omega})$ is the product of the DTFT of the input $X(e^{j\omega})$ and the frequency response $H(e^{j\omega})$, i.e., $Y(e^{j\omega}) = X(e^{j\omega})H(e^{j\omega})$. In a sense, $H(e^{j\omega})$ acts as a weighting function or spectral shaping function to the different frequency components in the input signal. Ideal filters allow distortionless transmission of a certain band of frequencies and completely suppress the remaining frequencies. The passband of a filter is the band of frequencies that are passed by the system, while the stopband refers to the range of frequencies that are attenuated by the system. In the ideal case, $|H(e^{j\omega})| = 1$ in a passband, while $|H(e^{\omega})| = 0$ in a stopband. Filters are usually classified according to their frequency-domain characteristics as low-pass, high-pass, bandpass, bandstop or band elimination, and all-pass filters. Following are the characteristics of these filters:

1. Low-pass filters are those characterized by a passband that extends from $\omega = 0$ to $\omega = \omega_c$, where ω_c is called the cutoff frequency of the filter [Fig. 6.21(a)]. A low-pass filter attenuates high-frequency components of the input and passes the low-frequency components.

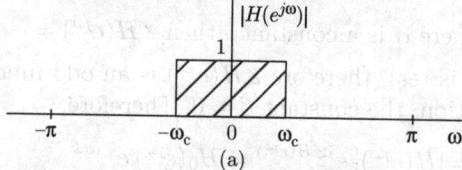

Fig. 6.21 (a) Low-pass filter

2. High-pass filters are characterized by a stopband that extends from $\omega = 0$ to $\omega = \omega_c$ and a passband that extends from $\omega = \omega_c$ to π [Fig. 6.21(b)]. A high-pass filter attenuates low frequencies and passes the high frequencies.

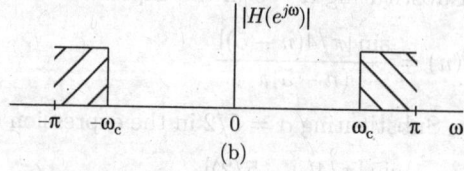

Fig. 6.21 (b) High-pass filter

3. Bandpass filters are characterized by a passband that extends from $\omega = \omega_1$ to $\omega = \omega_2$ and all other frequencies are stopped [Fig. 6.21(c)]. A bandpass filter passes signals within a certain frequency band and attenuates signals outside that band.

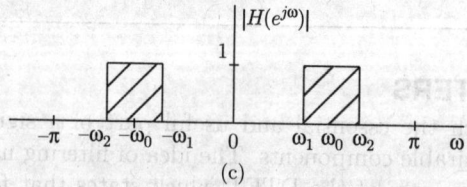

Fig. 6.21 (c) Bandpass filter

4. Bandstop filters stop frequencies extending from $\omega = \omega_1$ to $\omega = \omega_2$ and pass all other frequencies [Fig. 6.21(d)]. A bandstop filter attenuates signals within a certain frequency band and passes signals outside that band.

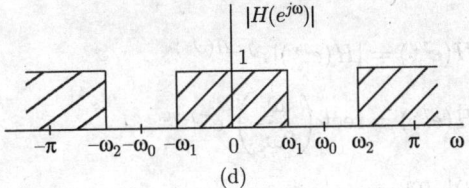

Fig. 6.21 (d) Bandstop filter

5. An all-pass filter is characterized by a magnitude response that is constant for all frequencies [Fig. 6.21(e)]. The characteristics of an all-pass filter are completely determined by its phase-shift characteristics. An all-pass filter has nonconstant group delay and different frequencies in the input are delayed by different amounts.

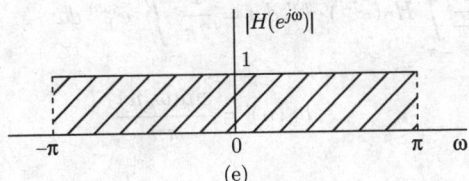

Fig. 6.21 (e) All-pass filter

The filters described so far are referred to as ideal filters because they pass one set of frequencies without any change and completely stop others. Since it is impossible to realize filters with characteristic like those shown in Fig. 6.21, with abrupt changes from passband to stopband and vice versa, most of the filters we deal with in practice have some transition band. Ideal filters are noncausal filters and therefore physically unrealizable. For example, consider an ideal low-pass filter whose magnitude response and phase response are (as shown in Fig. 6.22) given by

$$|H(e^{j\omega})| = \text{rect}\left(\frac{\omega}{2\omega_c}\right) = \begin{cases} 1, & -\omega_c < \omega < \omega_c \\ 0, & \text{otherwise} \end{cases}$$

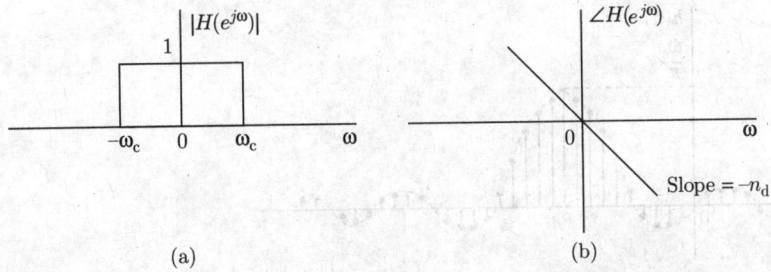

Fig. 6.22 (a) Magnitude response and (b) phase response of an ideal low-pass filter

and

$$\angle H(e^{j\omega}) = -\omega n_{\mathrm{d}}$$

Therefore,

$$H(e^{j\omega}) = |H(e^{j\omega})| \, e^{j\angle H(e^{j\omega})}$$

$$H(e^{j\omega}) = \mathrm{rect}\left(\frac{\omega}{2\omega_{\mathrm{c}}}\right) e^{-j\omega n_{\mathrm{d}}}$$

Let us define $H_0(e^{j\omega}) = |H(e^{j\omega})|$. Then,

$$|H(e^{j\omega})| = H_0(e^{j\omega}) = \begin{cases} 1, & -\omega_{\mathrm{c}} < \omega < \omega_{\mathrm{c}} \\ 0, & \text{otherwise} \end{cases}$$

Taking the inverse DTFT, we have

$$h_0(n) = \frac{1}{2\pi} \int_{-\pi}^{\pi} H_0(e^{j\omega}) \, e^{j\omega n} d\omega = \frac{1}{2\pi} \int_{-\omega_{\mathrm{c}}}^{\omega_{\mathrm{c}}} e^{j\omega n} \, d\omega$$

$$h_0(n) = \frac{\sin(\omega_{\mathrm{c}} n)}{\pi n}$$

Now, consider

$$H(e^{j\omega}) = |H(e^{j\omega})| \, e^{j\angle H(e^{j\omega})}$$

$$= H_0(e^{j\omega}) \, e^{-j\omega n_{\mathrm{d}}}$$

Taking the inverse DTFT, we get

$$h(n) = h_0(n - n_{\mathrm{d}}) = \frac{\sin(\omega_{\mathrm{c}}(n - n_{\mathrm{d}}))}{\pi(n - n_{\mathrm{d}})}$$

Recall that $h(n)$ is the system response to impulse input $\delta(n)$, which is applied at $n = 0$. Figure 6.23 depicts a curious fact: the response $h(n)$ begins even before the input is applied (at $n = 0$). Clearly, the filter is noncausal and therefore physically unrealizable. Similarly, one can show that other ideal filters are also physically unrealizable.

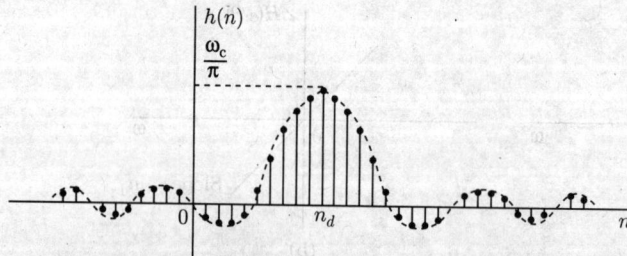

Fig. 6.23 Impulse response

6.9.1 Paley-Wiener Criterion

For a physically realizable system, the impulse response $h(n)$ must be causal and has finite energy, i.e.,

$$h(n) = 0, \quad n < 0$$

In the frequency domain, this condition is equivalent to the well-known Paley-Wiener criterion, which states that the necessary and sufficient condition for the magnitude response $|H(e^{j\omega})|$ to be realizable is

$$\int\limits_{-\pi}^{\pi} |\ln|H(e^{j\omega})|| \, d\omega < \infty \tag{6.88}$$

If $H(e^{j\omega})$ does not satisfy this condition, it is unrealizable.

Two important conclusion that we draw from the Paley-Wiener theorem are as follows:

1. The magnitude function $|H(e^{j\omega})|$ of a realizable system can be zero at some discrete frequencies, but it cannot be zero over any finite band of frequencies. This means that if $|H(e^{j\omega})| = 0$ over any finite band, then $|\ln|H(e^{j\omega})|| = \infty$ over that band and consequently $H(e^{j\omega})$ is unrealizable.

2. The magnitude function $|H(e^{j\omega})|$ cannot be constant in any finite range of frequencies and the transition from passband to stopband cannot be infinitely sharp [this is a consequence of the Gibbs phenomenon, which results from the truncation of $h(n)$ to achieve causality].

Apparently, causality imposes some tight constraints on a LTI system. In addition to the Paley-Wiener condition, causality also implies a strong relationship between $H_R(e^{j\omega}) = \Re\{H(e^{j\omega})\}$ and $H_I(e^{j\omega}) = \Im\{H(e^{j\omega})\}$, the real and imaginary components of the frequency response $H(e^{j\omega})$. To illustrate this dependence, we decompose $h(n)$ into its even and an odd part, i.e.,

$$h(n) = h_e(n) + h_o(n)$$

where

$$h_e(n) = \frac{1}{2}[h(n) + h(-n)]$$

and

$$h_o(n) = \frac{1}{2}[h(n) - h(-n)]$$

Now, if $h(n)$ is causal, it is possible to recover $h(n)$ from its even part $h_e(n)$ for $0 \leq n \leq \infty$ or from its odd part $h_o(n)$ for $1 \leq n \leq \infty$. Indeed, it can be easily seen that

$$h(n) = 2h_e(n)u(n) - h_e(0)\delta(n), \quad n \geq 0 \tag{6.89}$$

and

$$h(n) = 2h_o(n)u(n) + h(0)\delta(n), \quad n \geq 1 \tag{6.90}$$

Since $h_o(n) = 0$ for $n = 0$, we cannot recover $h(0)$ from $h_o(n)$ and hence we also must know $h(0)$. In any case, it is apparent that $h_o(n) = h_e(n)$ for $n \geq 1$ and so there is a strong relationship between $h_o(n)$ and $h_e(n)$.

If $h(n)$ is absolutely summable (i.e., BIBO stable), the frequency response $H(e^{j\omega})$ exists, and

$$H(e^{j\omega}) = H_R(e^{j\omega}) + jH_I(e^{j\omega})$$

In addition, if $h(n)$ is real valued and causal, the conjugate property of the Fourier transform imply that

$$h_e(n) \longleftrightarrow H_R(e^{j\omega}) \quad \text{and} \quad h_o(n) \longleftrightarrow jH_I(e^{j\omega})$$

Since $h(n)$ is completely specified by $h_e(n)$, it follows that $H(e^{j\omega})$ is completely determined if we know $H_R(e^{j\omega})$. Alternatively, $H(e^{j\omega})$ is completely determined from $H_I(e^{j\omega})$. In short, $H_R(e^{j\omega})$ and $H_I(e^{j\omega})$ are interdependent and cannot be specified independently if the system is causal. Equivalently, the magnitude and phase responses of a causal filter are interdependent and hence cannot be specified independently.

Taking the Fourier transform of Eq. (6.89), we have

$$\mathcal{F}[h(n)] = 2\mathcal{F}[h_e(n)u(n)] - h_e(0)\mathcal{F}[\delta(n)]$$

$$H(e^{j\omega}) = H_R(e^{j\omega}) + jH_I(e^{j\omega}) = 2[H_R(e^{j\omega}) * U(e^{j\omega})] - h_e(0)$$

$$H_R(e^{j\omega}) + jH_I(e^{j\omega}) = 2\frac{1}{2\pi} \int_{-\pi}^{\pi} H_R(e^{j\theta})U(e^{j(\omega-\theta)})d\theta - h_e(0)$$

$$H_R(e^{j\omega}) + jH_I(e^{j\omega}) = \frac{1}{\pi} \int_{-\pi}^{\pi} H_R(e^{j\theta})U(e^{j(\omega-\theta)})d\theta - h_e(0) \qquad (6.91)$$

where $U(e^{j\omega})$ is the DTFT of the unit step sequence $u(n)$. The DTFT of $u(n)$ is given by

$$U(e^{j\omega}) = \pi\delta(\omega) + \frac{1}{1 - e^{-j\omega}}, \quad -\pi < \omega < \pi$$

$$U(e^{j\omega}) = \pi\delta(\omega) + \frac{1}{2} - j\frac{1}{2}\cot\left(\frac{\omega}{2}\right) \qquad (6.92)$$

Substituting Eq. (6.92) into Eq. (6.91) gives

$$H_R(e^{j\omega}) + jH_I(e^{j\omega}) = \frac{1}{\pi} \int_{-\pi}^{\pi} H_R(e^{j\theta})\left[\pi\delta(\omega - \theta) + \frac{1}{2} - j\frac{1}{2}\cot\left(\frac{\omega - \theta}{2}\right)\right] d\theta - h_e(0)$$

$$= \frac{1}{\pi} \int_{-\pi}^{\pi} H_R(e^{j\theta})\pi\delta(\omega - \theta)d\theta + \frac{1}{\pi} \int_{-\pi}^{\pi} H_R(e^{j\theta})\frac{1}{2}d\theta - \frac{1}{\pi} \int_{-\pi}^{\pi} H_R(e^{j\theta})j\frac{1}{2}\cot\left(\frac{\omega - \theta}{2}\right) d\theta - h_e(0)$$

$$= \int\limits_{-\pi}^{\pi} H_{\mathrm{R}}(e^{j\theta})\delta(\omega - \theta)d\theta + \frac{1}{2\pi} \int\limits_{-\pi}^{\pi} H_{\mathrm{R}}(e^{j\theta})d\theta - j\frac{1}{2\pi} \int\limits_{-\pi}^{\pi} H_{\mathrm{R}}(e^{j\theta})\cot\left(\frac{\omega - \theta}{2}\right)d\theta - h_{\mathrm{e}}(0)$$

$$H_{\mathrm{R}}(e^{j\omega}) + jH_{\mathrm{I}}(e^{j\omega}) = H_{\mathrm{R}}(e^{j\theta})\Big|_{\theta=\omega} + h_{\mathrm{e}}(0) - j\frac{1}{2\pi} \int\limits_{-\pi}^{\pi} H_{\mathrm{R}}(e^{j\theta})\cot\left(\frac{\omega - \theta}{2}\right)d\theta - h_{\mathrm{e}}(0)$$

$$jH_{\mathrm{I}}(e^{j\omega}) = -j\frac{1}{2\pi} \int\limits_{-\pi}^{\pi} H_{\mathrm{R}}(e^{j\theta})\cot\left(\frac{\omega - \theta}{2}\right)d\theta$$

$$H_{\mathrm{I}}(e^{j\omega}) = -\frac{1}{2\pi} \int\limits_{-\pi}^{\pi} H_{\mathrm{R}}(e^{j\theta})\cot\left(\frac{\omega - \theta}{2}\right)d\theta \tag{6.93}$$

Similarly, taking the Fourier transform of Eq. (6.90) yields

$$\mathcal{F}[h(n)] = 2\mathcal{F}[h_{\mathrm{o}}(n)u(n)] + h(0)\mathcal{F}[\delta(n)]$$

$$H(e^{j\omega}) = H_{\mathrm{R}}(e^{j\omega}) + jH_{\mathrm{I}}(e^{j\omega}) = 2[jH_{\mathrm{I}}(e^{j\omega}) * U(e^{j\omega})] + h(0)$$

$$H_{\mathrm{R}}(e^{j\omega}) + jH_{\mathrm{I}}(e^{j\omega}) = 2j\frac{1}{2\pi} \int\limits_{-\pi}^{\pi} H_{\mathrm{I}}(e^{j\theta})U(e^{j(\omega-\theta)})\,d\theta + h(0)$$

$$H_{\mathrm{R}}(e^{j\omega}) + jH_{\mathrm{I}}(e^{j\omega}) = j\frac{1}{\pi} \int\limits_{-\pi}^{\pi} H_{\mathrm{I}}(e^{j\theta})U(e^{j(\omega-\theta)})\,d\theta + h(0) \tag{6.94}$$

where $U(e^{j\omega})$ is the DTFT of the unit step sequence $u(n)$. Substituting $U(e^{j\omega})$ from Eq. (6.92) into Eq. (6.94) gives

$$H_{\mathrm{R}}(e^{j\omega}) + jH_{\mathrm{I}}(e^{j\omega}) = j\frac{1}{\pi} \int\limits_{-\pi}^{\pi} H_{\mathrm{I}}(e^{j\theta})\left[\pi\delta(\omega - \theta) + \frac{1}{2} - j\frac{1}{2}\cot\left(\frac{\omega - \theta}{2}\right)\right]d\theta + h(0)$$

$$= j\frac{1}{\pi} \int\limits_{-\pi}^{\pi} H_{\mathrm{I}}(e^{j\theta})\pi\delta(\omega - \theta)\,d\theta + j\frac{1}{\pi} \int\limits_{-\pi}^{\pi} H_{\mathrm{I}}(e^{j\theta})\frac{1}{2}d\theta - j\frac{1}{\pi} \int\limits_{-\pi}^{\pi} H_{\mathrm{I}}(e^{j\theta})j\frac{1}{2}\cot\left(\frac{\omega - \theta}{2}\right)d\theta + h(0)$$

$$= j\int\limits_{-\pi}^{\pi} H_{\mathrm{I}}(e^{j\theta})\delta(\omega - \theta)d\theta + j\frac{1}{2\pi} \int\limits_{-\pi}^{\pi} H_{\mathrm{I}}(e^{j\theta})d\theta + \frac{1}{2\pi} \int\limits_{-\pi}^{\pi} H_{\mathrm{I}}(e^{j\theta})\cot\left(\frac{\omega - \theta}{2}\right)d\theta + h(0)$$

Since $H_I(e^{j\omega})$ is an odd function, we have $\int_{-\pi}^{\pi} H_I(e^{j\theta})d\theta = 0$, therefore

$$H_R(e^{j\omega}) + jH_I(e^{j\omega}) = jH_I(e^{j\theta})\Big|_{\theta=\omega} + 0 + \frac{1}{2\pi}\int_{-\pi}^{\pi} H_I(e^{j\theta})\cot\left(\frac{\omega-\theta}{2}\right) d\theta + h(0) .$$

$$H_R(e^{j\omega}) = \frac{1}{2\pi}\int_{-\pi}^{\pi} H_I(e^{j\theta})\cot\left(\frac{\omega-\theta}{2}\right) d\theta + h(0)$$

$$H_R(e^{j\omega}) = \frac{1}{2\pi}\int_{-\pi}^{\pi} H_I(e^{j\theta})\cot\left(\frac{\omega-\theta}{2}\right) d\theta + h(0) \tag{6.95}$$

Equations (6.93) and (6.95) are known as Hilbert transform pair.

To summarize, causality has very important implications in the design of frequency-selective filters. These implications are listed as follows:

1. The frequency response $H(e^{j\omega})$ cannot be zero except at a finite set of points in frequency.
2. The magnitude $|H(e^{j\omega})|$ cannot be constant in any finite range of frequencies and the transition from passband to stopband cannot be infinitely sharp.
3. The real and imaginary parts of $H(e^{j\omega})$ are interdependent and are related by the Hilbert transform. As a consequence, the magnitude $|H(e^{j\omega})|$ and phase $\angle H(e^{j\omega})$ cannot be chosen arbitrarily.

6.10 ENERGY SPECTRAL DENSITY (ESD)

Parseval's theorem relates the total signal energy in a signal $x(n)$ to its Fourier transform through

$$E_x = \sum_{n=-\infty}^{\infty} |x(n)|^2 = \frac{1}{2\pi}\int_{-\pi}^{\pi} |X(e^{j\omega})|^2 \, d\omega = \int_{-1/2}^{1/2} |X(e^{j\omega})|^2 \, df$$

Parseval's theorem states that the total energy E_x may be determined either by computing the energy per unit time $(|x(n)|^2)$ and summing over all time or by integrating the energy per unit frequency $|X(e^{j\omega})|^2/2\pi$ over a full 2π interval of distinct discrete-time frequencies. For this reason $|X(e^{j\omega})|^2$ represents energy per unit bandwidth and is often referred to as the *energy spectral density* or *energy density spectrum* of the signal $x(n)$ and is denoted by $\Psi_x(e^{j\omega})$. Hence,

$$\Psi_x(e^{j\omega}) = |X(e^{j\omega})|^2 \tag{6.96}$$

For discrete-time signals, the unit of ESD is simply the square of the signal unit (whatever that may be).

6.10.1 Relationship Between Input and Output ESDs of an LTI System

Consider an LTI system with frequency response $H(e^{j\omega})$, input $x(n)$, and output $y(n)$. If $x(n)$ and $y(n)$ are energy signals, then their ESDs are $\Psi_x(e^{j\omega}) = |X(e^{j\omega})|^2$ and $\Psi_y(e^{j\omega}) = |Y(e^{j\omega})|^2$, respectively. Since, we know that

$$Y(e^{j\omega}) = H(e^{j\omega})X(e^{j\omega})$$

it follows that

$$|Y(e^{j\omega})|^2 = |H(e^{j\omega})X(e^{j\omega})|^2$$

$$|Y(e^{j\omega})|^2 = |H(e^{j\omega})|^2|X(e^{j\omega})|^2$$

$$\Psi_y(e^{j\omega}) = |H(e^{j\omega})|^2\Psi_x(e^{j\omega}) \tag{6.97}$$

6.10.2 Relation of ESD to Autocorrelation

The autocorrelation function $R_{xx}(\tau)$ of a real energy signal is defined as [Eq. (2.60)]:

$$R_{xx}(m) = \sum_{n=-\infty}^{\infty} x(n)x(n-m)$$

$$R_{xx}(m) = x(m) * x(-m)$$

taking the Fourier transform of the above equation, we have

$$\mathcal{F}[R_{xx}(m)] = X(e^{j\omega})X(e^{-j\omega})$$

$$= X(e^{j\omega})X^*(e^{j\omega})$$

$$\mathcal{F}[R_{xx}(m)] = |X(e^{j\omega})|^2$$

$$\mathcal{F}[R_{xx}(m)] = \Psi_x(e^{j\omega})$$

$$R_{xx}(m) \longleftrightarrow \Psi_x(e^{j\omega}) \tag{6.98}$$

Thus, the autocorrelation function $R_{xx}(m)$ and ESD makes a Fourier transform pair.

6.11 POWER SPECTRAL DENSITY (PSD)

PSD has the same relation to power signals as ESD has to energy signals. The expression of PSD may be obtained by assuming the power signal as a limiting case of an energy signal.

Consider a power signal of infinite duration. Since the total energy of a power signal cannot be found, let us first find the ESD of a truncated version $x_N(n)$ of the infinite duration signal $x(n)$. The signal $x_N(n)$ may be expressed as

$$x_N(n) = \begin{cases} x(n), & -N \le n \le N \\ 0, & \text{otherwise} \end{cases} \tag{6.99}$$

The truncated signal $x_N(n)$ is of finite duration, therefore, it is an energy signal. Now, if $x_N(n) \longleftrightarrow X_N(e^{j\omega})$, then using the Parseval's theorem, we have

$$E_{x_N} = \sum_{n=-\infty}^{\infty} |x_N(n)|^2 = \frac{1}{2\pi} \int_{-\pi}^{\pi} |X_N(e^{j\omega})|^2 \, d\omega$$

Substituting the value of $x_N(n)$ from Eq. (6.99) into the above equation gives

$$\sum_{n=-N}^{N} |x(n)|^2 = \frac{1}{2\pi} \int_{-\pi}^{\pi} |X_N(e^{j\omega})|^2 \, d\omega$$

$$\lim_{N\to\infty} \frac{1}{2N+1} \sum_{n=-N}^{N} |x(n)|^2 = \lim_{N\to\infty} \frac{1}{2N+1} \frac{1}{2\pi} \int_{-\pi}^{\pi} |X_N(e^{j\omega})|^2 \, d\omega$$

$$\lim_{N\to\infty} \frac{1}{2N+1} \sum_{n=-N}^{N} |x(n)|^2 = \frac{1}{2\pi} \int_{-\pi}^{\pi} \lim_{N\to\infty} \frac{|X_N(e^{j\omega})|^2}{2N+1} \, d\omega$$

The LHS of the above equation represents the average power P_x of the signal $x(n)$. Therefore,

$$P_x = \lim_{N\to\infty} \frac{1}{2N+1} \sum_{n=-N}^{N} |x(n)|^2$$

$$= \frac{1}{2\pi} \int_{-\pi}^{\pi} \lim_{N\to\infty} \frac{|X_N(e^{j\omega})|^2}{2N+1} \, d\omega$$

$$P_x = \frac{1}{2\pi} \int_{-\pi}^{\pi} G_x(e^{j\omega}) \, d\omega \tag{6.100}$$

where

$$G_x(e^{j\omega}) = \lim_{N\to\infty} \frac{|X_N(e^{j\omega})|^2}{2N+1} \tag{6.101}$$

is the PSD. The units of PSD depend on the units of the underlying signal $x(n)$.

6.11.1 Relationship Between Input and Output PSDs of an LTI System

Consider an LTI system with frequency response $H(e^{j\omega})$, input $x_N(n)$, and output $y_N(n)$. If $x(n)$ and $y(n)$ are power signals, then their PSDs are

$$G_x(e^{j\omega}) = \lim_{N \to \infty} \frac{|X_N(e^{j\omega})|^2}{2N+1}$$

and

$$G_y(e^{j\omega}) = \lim_{N \to \infty} \frac{|Y_N(e^{j\omega})|^2}{2N+1},$$

respectively. Since, we know that

$$Y_N(e^{j\omega}) = H(e^{j\omega})X_N(e^{j\omega})$$

it follows that

$$|Y_N(e^{j\omega})|^2 = |H(e^{j\omega})X_N(e^{j\omega})|^2$$

$$|Y_N(e^{j\omega})|^2 = |H(e^{j\omega})|^2|X_N(e^{j\omega})|^2$$

$$\lim_{N \to \infty} \frac{|Y_N(e^{j\omega})|^2}{2N+1} = |H(e^{j\omega})|^2 \lim_{N \to \infty} \frac{|X_N(e^{j\omega})|^2}{2N+1}$$

$$G_y(e^{j\omega}) = |H(e^{j\omega})|^2 G_x(e^{j\omega}) \tag{6.102}$$

6.11.2 Relation of PSD to Autocorrelation

The autocorrelation function $R_{xx}(\tau)$ of a power signal is defined as [Eq. (2.61)]:

$$R_{xx}(m) = \lim_{N \to \infty} \frac{1}{2N+1} \sum_{n=-N}^{N} x(n)x(n-m)$$

$$= \lim_{N \to \infty} \frac{1}{2N+1} \sum_{n=-\infty}^{\infty} x_N(n)x_N(n-m)$$

$$R_{xx}(m) = \lim_{N \to \infty} \frac{1}{2N+1} [x_N(m) * x_N(-m)]$$

taking the Fourier transform of the above equation, we obtain

$$\mathcal{F}[R_{xx}(m)] = \lim_{N \to \infty} \frac{1}{2N+1} X_N(e^{j\omega})X_N(e^{-j\omega})$$

$$= \lim_{N \to \infty} \frac{1}{2N+1} X_N(e^{j\omega})X_N^*(e^{j\omega})$$

$$\mathcal{F}[R_{xx}(m)] = \lim_{N \to \infty} \frac{1}{2N+1} |X_N(e^{j\omega})|^2$$

$$\mathcal{F}[R_{xx}(m)] = G_x(e^{j\omega})$$

$$R_{xx}(m) \longleftrightarrow G_x(e^{j\omega}) \tag{6.103}$$

Thus, the autocorrelation function $R_{xx}(m)$ and PSD makes a Fourier transform pair.

SOLVED EXAMPLES

Example 6.29 Determine the Fourier transform for $-\pi \le \omega < \pi$ in the case of each of the following periodic signals:

(a) $x_1(n) = \sin(\pi/3 n + \pi/4)$

(b) $x_2(n) = 2 + \cos(\pi/6 n + \pi/8)$

Solution

We know that a periodic signal $x(n)$ with Fourier series representation

$$x(n) = \sum_{k=\langle N \rangle} X_k e^{jk2\pi/Nn}$$

has the DTFT $X(e^{j\omega})$ given by

$$X(e^{j\omega}) = 2\pi \sum_{k=-\infty}^{\infty} X_k \delta\left(\omega - k\frac{2\pi}{N}\right)$$

(a) Consider the given signal $x_1(n) = \sin(\pi/3 n + \pi/4)$. The fundamental period of $x_1(n)$ is $N = 6$.

$$x_1(n) = \sin\left(\frac{\pi}{3}n + \frac{\pi}{4}\right)$$

$$= \frac{1}{2j} e^{j(\pi/3n + \pi/4)} - \frac{1}{2j} e^{-j(\pi/3n + \pi/4)}$$

$$= \frac{1}{2j} e^{j\pi/4} e^{j\pi/3n} - \frac{1}{2j} e^{-j\pi/4} e^{-j\pi/3n}$$

$$x_1(n) = \frac{1}{2j} e^{j\pi/4} e^{j2\pi/6n} - \frac{1}{2j} e^{-j\pi/4} e^{-j2\pi/6n}$$

$$= X_1 e^{j2\pi/6n} + X_{-1} e^{-j2\pi/6n}$$

The nonzero Fourier series coefficients X_k of $x_1(n)$ in the range $-2 \le k \le 3$ are

$$X_1 = \frac{1}{2j} e^{j\pi/4}, \quad X_{-1} = -\frac{1}{2j} e^{-j\pi/4}$$

Substituting X_1 and X_{-1} in the expression of the DTFT of the periodic signals gives

$$X_1(e^{j\omega}) = 2\pi X_1 \delta\left(\omega - \frac{2\pi}{6}\right) + 2\pi X_{-1} \delta\left(\omega + \frac{2\pi}{6}\right)$$

$$= 2\pi \frac{1}{2j} e^{j\pi/4} \delta\left(\omega - \frac{2\pi}{6}\right)$$

$$- 2\pi \frac{1}{2j} e^{-j\pi/4} \delta\left(\omega + \frac{2\pi}{6}\right)$$

$$X_1(e^{j\omega}) = \frac{\pi}{j}\left[e^{j\pi/4}\delta\left(\omega - \frac{2\pi}{6}\right)\right.$$

$$\left. - e^{-j\pi/4}\delta\left(\omega + \frac{2\pi}{6}\right)\right], \quad -\pi \le \omega < \pi$$

(b) Consider the given signal $x_2(n) = 2 + \cos(\pi/6n + \pi/8)$. The fundamental period of $x_2(n)$ is $N = 12$.

$$x_2(n) = 2 + \cos\left(\frac{\pi}{6}n + \frac{\pi}{8}\right)$$

$$= 2 + \frac{1}{2} e^{j(\pi/6n + \pi/8)} + \frac{1}{2} e^{-j(\pi/6n + \pi/8)}$$

$$= 2 + \frac{1}{2} e^{j\pi/8} e^{j\pi/6n} + \frac{1}{2} e^{-j\pi/8} e^{-j\pi/6n}$$

$$x_2(n) = 2 + \frac{1}{2} e^{j\pi/8} e^{j2\pi/12n} + \frac{1}{2} e^{-j\pi/8} e^{-j2\pi/12n}$$

$$= X_0 + X_1 e^{j2\pi/12n} + X_{-1} e^{-j2\pi/12n}$$

The nonzero Fourier series coefficients X_k of $x_2(n)$ in the range $-5 \le k \le 6$ are

$$X_0 = 2, \quad X_1 = \frac{1}{2} e^{j\pi/8}, \quad X_{-1} = \frac{1}{2} e^{-j\pi/8}$$

Substituting X_0, X_1, and X_{-1} in the expression of the DTFT of the periodic signals gives

$$X_2(e^{j\omega}) = 2\pi X_0 \delta(\omega) + 2\pi X_1 \delta\left(\omega - \frac{2\pi}{12}\right)$$

$$+ 2\pi X_{-1} \delta\left(\omega + \frac{2\pi}{12}\right)$$

$$= 4\pi\delta(\omega) + 2\pi\frac{1}{2} e^{j\pi/8}\delta\left(\omega - \frac{2\pi}{12}\right)$$

$$+ 2\pi\frac{1}{2} e^{-j\pi/8}\delta\left(\omega + \frac{2\pi}{12}\right)$$

$$X_2(e^{j\omega}) = 4\pi\delta(\omega) + \pi\left[e^{j\pi/8}\delta\left(\omega - \frac{2\pi}{12}\right)\right.$$

$$\left. + e^{-j\pi/8}\delta\left(\omega + \frac{2\pi}{12}\right)\right],$$

$$-\pi \le \omega < \pi$$

Example 6.30 Determine the inverse Fourier transform of

(a)

$$X_1(e^{j\omega}) = \sum_{m=-\infty}^{\infty} \left[2\pi\delta(\omega - 2\pi m) \right.$$
$$+ \pi\delta\left(\omega - \frac{\pi}{2} - 2\pi m\right)$$
$$\left. + \pi\delta\left(\omega + \frac{\pi}{2} - 2\pi m\right) \right]$$

(b)

$$X_2(e^{j\omega}) = \begin{cases} 2j, & 0 < \omega \le \pi \\ -2j, & -\pi < \omega \le 0 \end{cases}$$

Solution
(a) The inverse DTFT $x_1(n)$ of $X_1(e^{j\omega})$ is given by

$$x_1(n) = \frac{1}{2\pi} \int_{-\pi}^{\pi} X_1(e^{j\omega}) e^{j\omega n} d\omega$$

$$= \frac{1}{2\pi} \int_{-\pi}^{\pi} \left[2\pi\delta(\omega) + \pi\delta\left(\omega - \frac{\pi}{2}\right) \right.$$
$$\left. + \pi\delta\left(\omega + \frac{\pi}{2}\right) \right] e^{j\omega n} d\omega$$

$$= \int_{-\pi}^{\pi} \delta(\omega) e^{j\omega n} d\omega$$
$$+ \frac{1}{2} \int_{-\pi}^{\pi} \delta\left(\omega - \frac{\pi}{2}\right) e^{j\omega n} d\omega$$
$$+ \frac{1}{2} \int_{-\pi}^{\pi} \delta\left(\omega + \frac{\pi}{2}\right) e^{j\omega n} d\omega$$

$$= e^{j\omega n}\Big|_{\omega=0} + \frac{1}{2} e^{j\omega n}\Big|_{\omega=\pi/2} + \frac{1}{2} e^{j\omega n}\Big|_{\omega=-\pi/2}$$

$$= e^{j0} + \frac{1}{2} e^{j\pi/2n} + \frac{1}{2} e^{-j\pi/2n}$$

$$x_1(n) = 1 + \cos\left(\frac{\pi}{2}n\right)$$

(b) The inverse DTFT $x_2(n)$ of $X_2(e^{j\omega})$ is given by

$$x_2(n) = \frac{1}{2\pi} \int_{-\pi}^{\pi} X_2(e^{j\omega}) e^{j\omega n} d\omega$$

$$= -\frac{1}{2\pi} \int_{-\pi}^{0} 2j\, e^{j\omega n} d\omega + \frac{1}{2\pi} \int_{0}^{\pi} 2j\, e^{j\omega n} d\omega$$

$$= -\frac{j}{\pi} \frac{e^{j\omega n}}{jn}\Big|_{-\pi}^{0} + \frac{j}{\pi} \frac{e^{j\omega n}}{jn}\Big|_{0}^{\pi}$$

$$= -\frac{1}{n\pi}[1 - e^{-jn\pi}] + \frac{1}{n\pi}[e^{jn\pi} - 1]$$

$$= \frac{1}{n\pi}[e^{jn\pi} - 1 - 1 + e^{-jn\pi}]$$

$$= \frac{1}{n\pi}[e^{jn\pi} + e^{-jn\pi} - 2]$$

$$= \frac{1}{n\pi}[2\cos(n\pi) - 2]$$

$$= \frac{2}{n\pi}[\cos(n\pi) - 1]$$

$$x_2(n) = \frac{4}{n\pi}\sin^2\left(\frac{n\pi}{2}\right)$$

Example 6.31 Find the DTFT of the following:

(a) $x_1(n) = \{1, -1, 2, 2\}$
(b) $x_2(n) = \delta(n-1) + \delta(n+1)$
(c) $x_3(n) = (0.5)^n u(n) + 2^n u(-n-1)$

Solution
(a) The DTFT $X_1(e^{j\omega})$ of $x_1(n)$ is given by

$$X_1(e^{j\omega}) = \sum_{n=-\infty}^{\infty} x_1(n) e^{-j\omega n}$$

$$= \sum_{n=0}^{3} x_1(n) e^{-j\omega n}$$

$$= x(0) + x(1) e^{-j\omega} + x(2) e^{-j2\omega}$$
$$+ x(3) e^{-j3\omega}$$

$$X_1(e^{j\omega}) = 1 - e^{-j\omega} + 2e^{-j2\omega} + 2e^{-j3\omega}$$

$$= 2j\,e^{-j\omega/2}\left(\frac{e^{j\omega/2} - e^{-j\omega/2}}{2j}\right)$$

$$+ 4e^{-j2.5\omega}\left(\frac{e^{j\omega/2} + e^{-j\omega/2}}{2}\right)$$

$$X_1(e^{j\omega}) = 2j\sin\left(\frac{\omega}{2}\right)e^{-j\omega/2} + 4\cos\left(\frac{\omega}{2}\right)e^{-j2.5\omega}$$

(b) Taking the Fourier transform of $x_2(n)$, we obtain

$$X_2(e^{j\omega}) = \mathcal{F}[\delta(n-1)] + \mathcal{F}[\delta(n+1)]$$

Using the time-shifting property of the DTFT, we obtain

$$X_2(e^{j\omega}) = e^{-j\omega} + e^{j\omega}$$

$$= 2\cos(\omega)$$

(c) Taking the Fourier transform of $x_3(n)$, we obtain

$$X_3(e^{j\omega}) = \mathcal{F}[(0.5)^n u(n)] + \mathcal{F}[2^n u(-n-1)]$$

$$= \frac{1}{1 - 0.5\,e^{-j\omega}} + \mathcal{F}\left[\left(\frac{1}{2}\right)^{-n} u(-n-1)\right]$$

$$= \frac{1}{1 - 0.5\,e^{-j\omega}} + \mathcal{F}\left[(0.5)^{-n} u(-n-1)\right]$$

$$= \frac{1}{1 - 0.5\,e^{-j\omega}} + \frac{0.5\,e^{j\omega}}{1 - 0.5\,e^{j\omega}}$$

$$= \frac{1 - (0.5)^2}{1 - \cos(\omega) + (0.5)^2}$$

$$X_3(e^{j\omega}) = \frac{0.75}{1 - \cos(\omega) + 0.25}$$

Example 6.32   Given that $x(n)$ has Fourier transform $X(e^{j\omega})$, express the Fourier transform of the following signals in terms of $X(e^{j\omega})$:

(a) $x_1(n) = x(1-n) + x(-1-n)$

(b) $x_2(n) = \dfrac{x^*(n) + x(n)}{2}$

(c) $x_3(n) = (n-1)^2 x(n)$

(d) $x_4(n) = x(2n+1)$

(e) $x_5(n) = e^{j\pi/2n} x(n+2)$

Solution
Given that

$$x(n) \longleftrightarrow X(e^{j\omega})$$

(a) Using the time-shifting property [Eq. (6.38)], we have

$$x(n+1) \longleftrightarrow X(e^{j\omega})\,e^{j\omega}$$

and $\qquad x(n-1) \longleftrightarrow X(e^{j\omega})\,e^{-j\omega}$

Using the time-reversal property [Eq. (6.41)] on this, we have

$$x(-n+1) = x(1-n) \longleftrightarrow X(e^{-j\omega})\,e^{-j\omega}$$

and $\quad x(-n-1) = x(-1-n) \longleftrightarrow X(e^{-j\omega})\,e^{j\omega}$

Therefore,

$$\mathcal{F}[x_1(n)] = \mathcal{F}[x(1-n) + x(-1-n)]$$

$$X(e^{-j\omega}) = X(e^{-j\omega})\,e^{-j\omega} + X(e^{-j\omega})\,e^{j\omega}$$

$$X_1(e^{j\omega}) = 2X(e^{-j\omega})\cos(\omega)$$

(b) Using the conjugation property [Eq. (6.54)], we have

$$x^*(n) \longleftrightarrow X^*(e^{-j\omega})$$

Using the time reversal-property [Eq. (6.41)] on this, we have

$$x^*(-n) \longleftrightarrow X^*(e^{j\omega})$$

Therefore,

$$\mathcal{F}[x_2(n)] = \mathcal{F}\left[\frac{x^*(n) + x(n)}{2}\right]$$

$$X_2(e^{j\omega}) = \frac{X^*(e^{j\omega}) + X(e^{j\omega})}{2}$$

$$= \Re\{X(e^{j\omega})\} = X_R(e^{j\omega})$$

$$X_2(e^{j\omega}) = X_R(e^{j\omega})$$

(c) Consider the given signal $x_3(n) = (n-1)^2 x(n) = n^2 x(n) - 2nx(n) + x(n)$. Using the differentiation in frequency-domain property [Eq. (6.48)], we have

$$nx(n) \longleftrightarrow j\frac{dX(e^{j\omega})}{d\omega}$$

Using the differentiation in frequency-domain property again, we have

$$n[nx(n)] \longleftrightarrow j\frac{d}{d\omega}\left(j\frac{dX(e^{j\omega})}{d\omega}\right)$$

$$n^2 x(n) \longleftrightarrow -\frac{d^2 X(e^{j\omega})}{d\omega}$$

Therefore,

$$\mathcal{F}[x_3(n)] = \mathcal{F}[(n-1)^2 x(n)]$$

$$= \mathcal{F}[n^2 x(n) - 2nx(n) + x(n)]$$

$$X_3(e^{j\omega}) = -\frac{d^2 X(e^{j\omega})}{d\omega} - 2j\frac{dX(e^{j\omega})}{d\omega} + X(e^{j\omega})$$

$$X_3(e^{j\omega}) = -\frac{d^2 X(e^{j\omega})}{d\omega} - 2j\frac{dX(e^{j\omega})}{d\omega} + X(e^{j\omega})$$

(d) Using the time-shifting property [Eq. (6.38)], we have

$$x(n+1) \longleftrightarrow X(e^{j\omega})e^{j\omega}$$

Using the time-expansion property [Eq. (6.45)] (choose $m = 1/2$)] on this, we have

$$\mathcal{F}[x_4(n)] = \mathcal{F}[x(2n+1)]$$

$$X_4(e^{j\omega}) = X(e^{j\omega/2})\,e^{j\omega/2}$$

$$X_4(e^{j\omega}) = X(e^{j\omega/2})\,e^{j\omega/2}$$

(e) Using the time-shifting property [Eq. (6.38)], we have

$$x(n+2) \longleftrightarrow X(e^{j\omega})\,e^{j2\omega}$$

Using the frequency-shifting property [Eq. (6.39)] on this, we have

$$x_5(n) = e^{j\pi/2n}x(n+2) \longleftrightarrow X(e^{j(\omega-\pi/2)})\,e^{j2(\omega-\pi/2)}$$

$$X_5(e^{j\omega}) = X(e^{j(\omega-\pi/2)})\,e^{j2(\omega-\pi/2)}$$

Example 6.33 Determine the signal $x(n)$ if its Fourier transform is as given in Fig. 6.24.

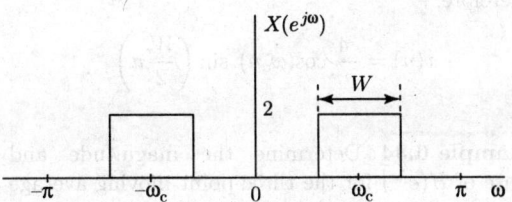

Fig. 6.24

Solution

The given DTFT can be written as

$$X(e^{j\omega}) = \begin{cases} 2, & \omega_c - \dfrac{W}{2} \le |\omega| \le \omega_c + \dfrac{W}{2} \\[2mm] 0, & \omega_c + \dfrac{W}{2} \le |\omega| \le \\[2mm] 0, & 0 \le |\omega| \le \omega_c - \dfrac{W}{2} \end{cases}$$

The inverse DTFT of $X(e^{j\omega})$ is given by

$$x(n) = \frac{1}{2\pi}\int_{-\pi}^{\pi} X(e^{j\omega})\,e^{j\omega n}\,d\omega$$

$$= \frac{1}{2\pi}\left[\int_{-\omega_c-(W/2)}^{-\omega_c+(W/2)} 2\,e^{j\omega n}\,d\omega + \int_{\omega_c-(W/2)}^{\omega_c+(W/2)} 2\,e^{j\omega n}\,d\omega\right]$$

$$= \frac{1}{\pi}\left[\frac{e^{j\omega n}}{jn}\Bigg|_{-\omega_c-(W/2)}^{-\omega_c+(W/2)} + \frac{e^{j\omega n}}{jn}\Bigg|_{\omega_c-(W/2)}^{\omega_c+(W/2)}\right]$$

$$= \frac{1}{jn\pi}\left[e^{j(-\omega_c+(W/2))n} - e^{j(-\omega_c-(W/2))n}\right.$$

$$\left. + e^{j(\omega_c+(W/2))n} - e^{j(\omega_c-(W/2))n}\right]$$

$$= \frac{2}{n\pi}\left[\frac{e^{j(\omega_c+(W/2))n} - e^{-j(\omega_c+(W/2))n}}{2j}\right.$$

$$\left. - \frac{e^{j(\omega_c-(W/2))n} - e^{-j(\omega_c-(W/2))n}}{2j}\right]$$

$$= \frac{2}{n\pi}\left[\sin\left(\omega_c + \frac{W}{2}\right)n - \sin\left(\omega_c - \frac{W}{2}\right)n\right]$$

$$= \frac{4}{n\pi}\cos(\omega_c n)\sin\left(\frac{W}{2}n\right)$$

Therefore,

$$x(n) = \frac{4}{n\pi}\cos(\omega_c n).\sin\left(\frac{W}{2}n\right)$$

Example 6.34 Determine the magnitude and phase of $H(e^{j\omega})$ for the three-point moving average system

$$y(n) = \frac{1}{3}[x(n+1) + x(n) + x(n-1)]$$

an plot these two functions for $0 \leq \omega \leq \pi$.

Solution

Consider the given difference equation

$$y(n) = \frac{1}{3}[x(n+1) + x(n) + x(n-1)]$$

Taking the DTFT of the above equation, we obtain

$$Y(e^{j\omega}) = \frac{1}{3}\left[X(e^{j\omega})e^{j\omega} + X(e^{j\omega}) + X(e^{j\omega})e^{-j\omega}\right]$$

$$\frac{Y(e^{j\omega})}{X(e^{j\omega})} = H(e^{j\omega}) = \frac{1}{3}[e^{j\omega} + 1 + e^{-j\omega}]$$

$$H(e^{j\omega}) = \frac{1}{3}[1 + 2\cos(\omega)]$$

The magnitude response is

$$|H(e^{j\omega})| = \frac{1}{3}|[1 + 2\cos(\omega)]|$$

and the phase response is

$$\angle H(e^{j\omega}) = \begin{cases} 0, & 0 \leq \omega \leq \dfrac{2\pi}{3} \\ \pi, & \dfrac{2\pi}{3} < \omega < \pi \end{cases}$$

Figure 6.25 illustrates the magnitude and phase phase of $H(e^{j\omega})$.

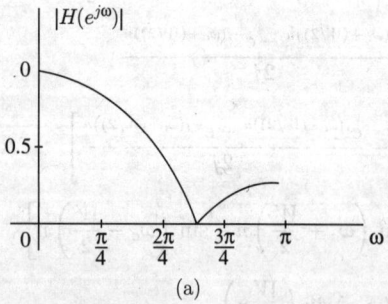

(a)

Fig. 6.25 (a)

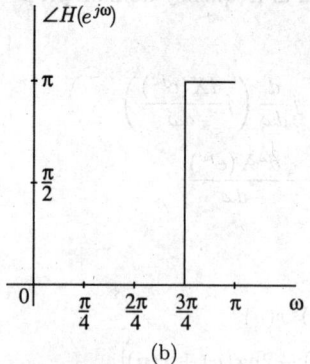

(b)

Fig. 6.25 (b)

Example 6.35 Let the sequence $x(n)$ be a real sequence and let

$$x(n) \longleftrightarrow X(e^{j\omega})$$

Prove that
(a)

$$|X(e^{j\omega})| = |X(e^{-j\omega})|$$

and
(b)

$$\angle X(e^{j\omega}) = -\angle X(e^{-j\omega})$$

Solution
Given that

$$x(n) \longleftrightarrow X(e^{j\omega})$$

In general the Fourier transform $X(e^{j\omega})$ is a complex function of the real variable ω and can be written in rectangular form as

$$X(e^{j\omega}) = X_R(e^{j\omega}) + jX_I(e^{j\omega})$$

where $X_R(e^{j\omega})$ and $X_I(e^{j\omega})$ are, respectively, the real and imaginary parts of $X(e^{j\omega})$.

The Fourier transform $X(e^{j\omega})$ can alternatively be expressed in the polar form as

$$X(e^{j\omega}) = |X(e^{j\omega})|e^{j\theta(\omega)}$$

where

$$\theta(\omega) = \angle X(e^{j\omega})$$

The quantity $|X(e^{j\omega})|$ is called the magnitude spectrum and the quantity $\theta(\omega) = \angle X(e^{j\omega})$ is called the phase spectrum.

The relations between the rectangular and polar forms of $X(e^{j\omega})$ is given by

$$X(e^{j\omega}) = X_R(e^{j\omega}) + jX_I(e^{j\omega}) = |X(e^{j\omega})|e^{j\theta(\omega)}$$

$$X_R(e^{j\omega}) + jX_I(e^{j\omega}) = |X(e^{j\omega})|\cos(\theta(\omega))$$
$$+ j|X(e^{j\omega})|\sin(\theta(\omega))$$

Therefore,

$$X_R(e^{j\omega}) = |X(e^{j\omega})|\cos(\theta(\omega)) = X_R(e^{-j\omega})$$
$$X_I(e^{j\omega}) = |X(e^{j\omega})|\sin(\theta(\omega)) = -X_I(e^{-j\omega})$$

(a) The magnitude spectrum $|X(e^{j\omega})|$ is given by

$$|X(e^{j\omega})| = \sqrt{X_R^2(e^{j\omega}) + X_I^2(e^{j\omega})}$$

Thus, for a real signal,

$$|X(e^{-j\omega})| = \sqrt{X_R^2(e^{-j\omega}) + X_I^2(e^{-j\omega})}$$

$$|X(e^{-j\omega})| = \sqrt{X_R^2(e^{j\omega}) + X_I^2(e^{j\omega})}$$

$$|X(e^{-j\omega})| = |X(e^{j\omega})|$$

The magnitude spectrum $|X(e^{j\omega})|$ is an even function of ω.

(b) We know that the Fourier transform $X(e^{j\omega})$ is a complex function of the real variable ω and can be written in the rectangular form as

$$X(e^{j\omega}) = X_R(e^{j\omega}) + jX_I(e^{j\omega})$$

$$\theta(\omega) = \angle X(e^{j\omega}) = \tan^{-1}\left[\frac{X_I(e^{j\omega})}{X_R(e^{j\omega})}\right]$$

For a real signal,

$$\angle X(e^{-j\omega}) = \tan^{-1}\left[\frac{X_I(e^{-j\omega})}{X_R(e^{-j\omega})}\right]$$

$$= \tan^{-1}\left[-\frac{X_I(e^{j\omega})}{X_R(e^{j\omega})}\right]$$

$$\angle X(e^{-j\omega}) = -\tan^{-1}\left[\frac{X_I(e^{j\omega})}{X_R(e^{j\omega})}\right]$$

$$\angle X(e^{-j\omega}) = -\angle X(e^{j\omega})$$

The above equation implies that the phase spectrum $\angle X(e^{j\omega})$ is an odd function of ω.

Example 6.36 Find the discrete-time Fourier transform of

(a) $x_1(n) = \cos(\omega_0 n)u(n)$

(b) $x_2(n) = \sin(\omega_0 n)u(n)$

Solution

(a) Consider the given signal

$$x_1(n) = \cos(\omega_0 n)u(n)$$

Using the multiplication property [Eq. (6.53)], we have

$$X_1(e^{j\omega}) = \mathcal{F}[\cos(\omega_0 n)] * \mathcal{F}[u(n)], \quad -\pi \le \omega < \pi$$

$$= \pi[\delta(\omega - \omega_0) + \delta(\omega + \omega_0)]$$

$$* \left[\frac{1}{1 - e^{-j\omega}} + \pi\delta(\omega)\right]$$

$$= \pi\left[\frac{1}{1 - e^{-j\omega_0}} + \frac{1}{1 - e^{j\omega_0}}\right]$$

$$+ \pi^2[\delta(\omega - \omega_0) + \delta(\omega + \omega_0)]$$

$$X_1(e^{j\omega}) = \pi + \pi^2[\delta(\omega - \omega_0) + \delta(\omega + \omega_0)]$$

(b) Consider the given signal

$$x_2(n) = \sin(\omega_0 n)u(n)$$

Using the multiplication property [Eq. (6.53)], we have

$$X_2(e^{j\omega}) = \mathcal{F}[\sin(\omega_0 n)] * \mathcal{F}[u(n)], \quad -\pi \le \omega < \pi$$

$$= \frac{\pi}{j}[\delta(\omega - \omega_0) - \delta(\omega + \omega_0)]$$

$$* \left[\frac{1}{1 - e^{-j\omega}} + \pi\delta(\omega)\right]$$

$$= \frac{\pi}{j}\left[\frac{1}{1 - e^{-j\omega_0}} - \frac{1}{1 - e^{j\omega_0}}\right]$$

$$+ \frac{\pi^2}{j}[\delta(\omega - \omega_0) - \delta(\omega + \omega_0)]$$

$$= \pi\frac{-\sin(\omega_0)}{1 - \cos(\omega_0)}$$

$$+ \pi^2[\delta(\omega - \omega_0) - \delta(\omega + \omega_0)]$$

$$= -\pi\frac{2\sin\left(\frac{\omega_0}{2}\right)\cos\left(\frac{\omega_0}{2}\right)}{2\sin^2\left(\frac{\omega_0}{2}\right)}$$

$$+ \pi^2[\delta(\omega - \omega_0) - \delta(\omega + \omega_0)]$$

$$X_2(e^{j\omega}) = -\pi\cot\left(\frac{\omega_0}{2}\right)$$
$$+ \pi^2[\delta(\omega - \omega_0) - \delta(\omega + \omega_0)]$$

Example 6.37 Determine the signal $x(n)$ for the following given discrete-time Fourier transform:

(a) $X(e^{j\omega}) = e^{-j\alpha\omega}$, for $-\omega_c \leq \omega \leq \omega_c$

(b) $X(e^{j\omega}) = e^{-j\omega}[1 + \cos(\omega)]$

Solution
(a) The inverse DTFT of $X(e^{j\omega})$ is given by

$$X(e^{j\omega}) = \frac{1}{2\pi} \int_{-\pi}^{\pi} X(e^{j\omega}) e^{j\omega n} d\omega$$

$$= \frac{1}{2\pi} \int_{-\omega_c}^{\omega_c} e^{-j\alpha\omega} e^{j\omega n} d\omega$$

$$= \frac{1}{2\pi} \int_{-\omega_c}^{\omega_c} e^{j\omega(n-\alpha)} d\omega$$

$$= \frac{1}{2\pi j(n-\alpha)} e^{j\omega(n-\alpha)} \Big|_{-\omega_c}^{\omega_c}$$

$$= \frac{1}{\pi(n-\alpha)} \frac{e^{j\omega_c(n-\alpha)} - e^{-j\omega_c(n-\alpha)}}{2j}$$

$$x(n) = \frac{\sin[\omega_c(n-\alpha)]}{\pi(n-\alpha)}$$

(b) Consider the given DTFT

$$X(e^{j\omega}) = e^{-j\omega}[1 + \cos(\omega)]$$

$$= e^{-j\omega}\left[1 + \frac{e^{j\omega} + e^{-j\omega}}{2}\right]$$

$$= e^{-j\omega} + 0.5 + 0.5\,e^{-j2\omega}$$

$$X(e^{j\omega}) = 0.5 + e^{-j\omega} + 0.5\,e^{-j2\omega}$$

Taking the inverse DTFT of the above equation, we obtain

$$x(n) = 0.5\delta(n) + \delta(n-1) + 0.5\delta(n-2)$$
$$= \{0.5,\ 1,\ 0.5\}$$
$$\quad\uparrow$$

Example 6.38 The following four facts are given about a particular signal $x(n)$ with Fourier transform $X(e^{j\omega})$:

1. $x(n) = 0$, for $n > 0$.
2. $x(0) > 0$.
3. $\Im\{X(e^{j\omega})\} = X_I(e^{j\omega}) = \sin(\omega) - \sin(2\omega)$
4. $\frac{1}{2\pi} \int_{-\pi}^{\pi} |X(e^{j\omega})|^2\, d\omega = 3$.

Determine $x(n)$.

Solution
We know that the DTFT of a real and an odd signal is purely imaginary and odd [Eq. (6.59)]. Therefore, we have

$$x_o(n) = \mathcal{O}\{x(n)\} \longleftrightarrow j\text{Im}\{X(e^{j\omega})\} = jX_I(e^{j\omega})$$

$$x_o(n) = \mathcal{F}^{-1}[jX_I(e^{j\omega})]$$
$$= \mathcal{F}^{-1}[j\sin(\omega) - j\sin(2\omega)]$$
$$= \mathcal{F}^{-1}\left[\frac{1}{2}\left(e^{j\omega} - e^{-j\omega} - e^{j2\omega} + e^{-j2\omega}\right)\right]$$
$$x_o(n) = \frac{1}{2}\Big[\delta(n+1) - \delta(n-1)$$
$$- \delta(n+2) + \delta(n-2)\Big]$$
$$= \left\{-\frac{1}{2}, \frac{1}{2}, 0, -\frac{1}{2}, \frac{1}{2}\right\}$$
$$\qquad\qquad\quad\uparrow$$

We also know that

$$x_o(n) = \frac{x(n) - x(-n)}{2}$$

and that $x(n) = 0$, for $n > 0$. Therefore,

$$x(n) = 2x_o(n), \quad \text{for } n < 0$$
$$x(n) = \delta(n+1) - \delta(n+2)$$

Now, we have to find $x(0)$. Using Parseval's relation [Eq. (6.60)], we have

$$\sum_{n=-\infty}^{\infty} |x(n)|^2 = \frac{1}{2\pi} \int_{-\pi}^{\pi} |X(e^{j\omega})|^2 d\omega$$

From fact number 4, we can write

$$\sum_{n=-\infty}^{0} |x(n)|^2 = 3$$

$$x^2(0) + \sum_{n=-\infty}^{-1} |x(n)|^2 = 3$$

$$x^2(0) + 2 = 3$$

$$x^2(0) = 1$$

$$x(0) = \pm 1$$

Since $x(0) > 0$, we conclude that $x(0) = 1$. Therefore,

$$x(n) = \delta(n) + \delta(n+1) - \delta(n+2)$$

$$= \{-1, 1, \underset{\uparrow}{1}\}$$

Example 6.39 Consider a signal $g(n)$ with Fourier transform $G(e^{j\omega})$. Suppose

$$g(n) = x_{(2)}(n)$$

where the signal $x(n)$ has a Fourier transform $X(e^{j\omega})$. Determine a real number α such that $0 < \alpha < 2\pi$ and $G(e^{j\omega}) = G(e^{j(\omega - \alpha)})$.

Solution
Using the time-expansion property [Eq. (6.45)], we have

$$g(n) = x_{(2)}(n) \longleftrightarrow G(e^{j\omega}) = X(e^{j2\omega})$$

Therefore, $G(e^{j\omega})$ is obtained by compressing $X(e^{j\omega})$ by a factor of 2. Since, we know that $X(e^{j\omega})$ is periodic with period 2π, we may conclude that $G(e^{j\omega})$ is periodic with period $\dfrac{1}{2\pi} = \pi$. Hence,

$$G(e^{j\omega}) = G(e^{j(\omega - \pi)}) = G(e^{j(\omega - \alpha)})$$

Therefore, $\alpha = \pi$.

Example 6.40 A linear time-invariant system is described by the following difference equation:

$$y(n) = ay(n-1) + bx(n), \quad 0 < a < 1$$

(a) Determine the magnitude response $|H(e^{j\omega})|$ and phase response $\angle H(e^{j\omega})$ of the system.

(b) Choose the parameter b such that the maximum value of $|H(e^{j\omega})|$ is unity.

(c) Determine the output of the system to the input signal

$$x(n) = 5 + 12\sin\left(\frac{\pi}{2}n\right) + 20\cos\left(\pi n + \frac{\pi}{4}\right)$$

for $a = 0.9$.

Solution
(a) Consider the given difference equation

$$y(n) = ay(n-1) + bx(n), \quad 0 < a < 1$$

Taking the Fourier transform of the above equation, we obtain

$$Y(e^{j\omega}) = aY(e^{j\omega})\,e^{-j\omega} + bX(e^{j\omega})$$

$$\frac{Y(e^{j\omega})}{X(e^{j\omega})} = H(e^{j\omega}) = \frac{b}{1 - a\,e^{-j\omega}}$$

$$H(e^{j\omega}) = \frac{b}{1 - a\cos(\omega) + ja\sin(\omega)}$$

Therefore,

$$|H(e^{j\omega})| = \frac{|b|}{\sqrt{(1 - a\cos(\omega))^2 + (a\sin(\omega))^2}}$$

$$|H(e^{j\omega})| = \frac{|b|}{\sqrt{1 + a^2 - 2a\cos(\omega)}}$$

and $\quad \angle H(e^{j\omega}) = \angle b - \tan^{-1}\dfrac{a\sin(\omega)}{1 - a\cos(\omega)}$

(b) Since the parameter a is positive, the denominator of $|H(e^{j\omega})|$ attains a minimum at $\omega = 0$. Therefore, $|H(e^{j\omega})|$ attains its maximum value at $\omega = 0$. At this frequency, we have

$$|H(e^{j0})| = \frac{|b|}{1 - a} = 1$$

$$|b| = 1 - a$$

$$b = \pm(1 - a)$$

(c) For $a = 0.9$ and $b = (1 - a) = 0.1$, the frequency response of the system is

$$H(e^{j\omega}) = \frac{0.1}{1 - 0.9\,e^{-j\omega}}$$

The first term in the input signal $x(n)$ is a constant signal corresponding to $\omega = 0$, thus

$$H(e^{j0}) = \frac{0.1}{1 - 0.9} = 1$$

The second term in $x(n)$ has a frequency $\pi/2$. At this frequency the frequency response of the system is

$$H(e^{j\pi/2}) = \frac{0.1}{1 - 0.9\,e^{-j\pi/2}}$$

$$= \frac{0.1}{1 + j0.9} = 0.074\,e^{-j42°}$$

Finally, the third term in $x(n)$ has a frequency $\omega = \pi$. At this frequency the frequency response of the system is

$$H(e^{j\pi}) = \frac{0.1}{1 - 0.9\,e^{-j\pi}}$$

$$= \frac{0.1}{1 + 0.9} = \frac{0.1}{1.9} = 0.053$$

Hence, the response of the system to $x(n)$ is

$$y(n) = 5H(e^{j0}) + H(e^{j\pi/2})12\sin\left(\frac{\pi}{2}n\right)$$

$$- H(e^{j\pi})20\cos\left(\pi n + \frac{\pi}{4}\right)$$

$$y(n) = 5 + 0.888\sin\left(\frac{\pi}{2}n - 42°\right)$$

$$- 1.06\cos\left(\pi n + \frac{\pi}{4}\right)$$

Example 6.41 Suppose we are given the following facts about an LTI system with impulse response $h(n)$ and frequency response $H(e^{j\omega})$:

1. $(1/4)^n\,u(n) \longrightarrow g(n)$,   where $g(n) = 0$ for $n \geq 2$ and $n < 0$
2. $H(e^{j\pi/2}) = 1$
3. $H(e^{j\omega}) = H(e^{j(\omega - \pi)})$

Determine $h(n)$.

Solution

Given that $g(n) = 0$ for $n \geq 2$ and $n < 0$. Therefore, the Fourier transform $G(e^{j\omega})$ of $g(n)$ is

$$G(e^{j\omega}) = \sum_{n=-\infty}^{\infty} g(n)\,e^{-j\omega n}$$

$$= \sum_{n=0}^{1} g(n)\,e^{-j\omega n}$$

$$G(e^{j\omega}) = g(0) + g(1)\,e^{-j\omega}$$

Given that the input to the system is $x(n) = (1/4)^n\,u(n)$, its Fourier transform is $X(e^{j\omega}) = \dfrac{1}{1 - \frac{1}{4}\,e^{-j\omega}}$. Also, when the input to the system is $x(n)$, the output is $g(n)$. Therefore,

$$H(e^{j\omega}) = \frac{G(e^{j\omega})}{X(e^{j\omega})}$$

$$= \left(g(0) + g(1)e^{-j\omega}\right)\left(1 - \frac{1}{4}e^{-j\omega}\right)$$

$$H(e^{j\omega}) = g(0) + \left[g(1) - \frac{1}{4}g(0)\right]e^{-j\omega} + \frac{1}{4}g(1)e^{-j2\omega}$$

$$H(e^{j\omega}) = h(0) + h(1)\,e^{-j\omega} + h(2)\,e^{-j2\omega}$$

Clearly, $h(n)$ is a three-point sequence. From the above equation, we have

$$H(e^{j(\omega - \pi)}) = h(0) + h(1)\,e^{-j(\omega - \pi)} + h(2)\,e^{-j2(\omega - \pi)}$$

$$= h(0) + h(1)e^{-j\omega}\,e^{j\pi} + h(2)\,e^{-j2\omega}\,e^{j2\pi}$$

$$H(e^{j(\omega - \pi)}) = h(0) - h(1)\,e^{-j\omega} + h(2)\,e^{-j2\omega}$$

Given that

$$H(e^{j\omega}) = H(e^{j(\omega - \pi)})$$

$$h(0) + h(1)\,e^{-j\omega} + h(2)\,e^{-j2\omega} = h(0) - h(1)\,e^{-j\omega}$$

$$+ h(2)\,e^{-j2\omega}$$

$$2h(1)\,e^{-j\omega} = 0$$

$$h(1) = 0$$

We also have

$$H(e^{j\pi/2}) = 1$$

$$h(0) + h(1)\,e^{-j\pi/2} + h(2)\,e^{-j2\pi/2} = 1$$

$$h(0) + 0 + h(2)\,e^{-j\pi} = 1$$

$$h(0) - h(2) = 1$$

Note that

$$g(n) = h(n) * x(n)$$

$$= \sum_{k=-\infty}^{\infty} h(k)x(n - k)$$

$$g(n) = \sum_{k=0}^{2} h(k)\left(\frac{1}{4}\right)^{n-k} u(n - k)$$

Evaluating this equation at $n = 2$, we have

$$g(2) = \sum_{k=0}^{2} h(k) \left(\frac{1}{4}\right)^{2-k} u(2-k) = 0$$

$$\frac{1}{16} h(0) + \frac{1}{4} h(1) = 0$$

Now solving the two equations

$$h(0) - h(2) = 1$$

$$\frac{1}{16} h(0) + \frac{1}{4} h(1) = 0$$

we obtain

$$h(0) = \frac{16}{17} \quad \text{and} \quad h(2) = -\frac{1}{17}$$

Therefore,

$$h(n) = \frac{16}{17}\delta(n) - \frac{1}{17}\delta(n-2)$$

$$= \left\{\frac{16}{17}, 0, -\frac{1}{17}\right\}$$

Example 6.42 An LTI system with impulse response $h_1(n) = (1/3)^n u(n)$ is connected in parallel with another causal LTI system with impulse response $h_2(n)$. The resulting parallel interconnection has the frequency response

$$H(e^{j\omega}) = \frac{-12 + 5 e^{-j\omega}}{12 - 7 e^{-j\omega} + e^{-j2\omega}}$$

Determine $h_2(n)$.

Solution
When the two systems are connected in parallel, the impulse response of the overall system is the sum of the impulse responses of the individual systems. Therefore,

$$h(n) = h_1(n) + h_2(n)$$

Taking the Fourier transform of the above equation, we obtain

$$H(e^{j\omega}) = H_1(e^{j\omega}) + H_2(e^{j\omega})$$

Given that $h_1(n) = (1/3)^n u(n)$, we obtain

$$H_1(e^{j\omega}) = \frac{1}{1 - \frac{1}{3} e^{-j\omega}}$$

Therefore,

$$H_2(e^{j\omega}) = H(e^{j\omega}) - H_1(e^{j\omega})$$

$$= \frac{-12 + 5 e^{-j\omega}}{12 - 7 e^{-j\omega} + e^{-j2\omega}} - \frac{1}{1 - 1/3 e^{-j\omega}}$$

$$= \frac{-12 + 5 e^{-j\omega}}{(4 - e^{-j\omega})(3 - e^{-j\omega})} - \frac{3}{3 - e^{-j\omega}}$$

$$= -\frac{8}{4 - e^{-j\omega}}$$

$$H_2(e^{j\omega}) = -\frac{2}{1 - \frac{1}{4} e^{-j\omega}}$$

Taking the inverse Fourier transform, we obtain

$$h_2(n) = -2 \left(\frac{1}{4}\right)^n u(n)$$

Example 6.43 Consider a discrete-time LTI system with frequency response

$$H(e^{j\omega}) = \left[\frac{1 + e^{-j2\omega} + 4 e^{-j4\omega}}{1 + \frac{1}{2} e^{-j2\omega}}\right] e^{-j(\omega - \pi/4)}$$

$$-\pi < \omega \leq \pi$$

Determine the Fourier transform of the output if the input is

$$x(n) = \cos\left(\frac{\pi n}{2}\right)$$

Solution
Consider the given input

$$x(n) = \cos\left(\frac{\pi n}{2}\right)$$

Taking the Fourier transform of the above equation, we obtain

$$X(e^{j\omega}) = \pi \left[\delta\left(\omega - \frac{\pi}{2}\right) + \delta\left(\omega + \frac{\pi}{2}\right)\right]$$

$$-\pi < \omega \leq \pi$$

The Fourier transform of the output is given by

$$Y(e^{j\omega}) = H(e^{j\omega})X(e^{j\omega})$$

$$= H(e^{j\omega})\pi\left[\delta\left(\omega - \frac{\pi}{2}\right) + \delta\left(\omega + \frac{\pi}{2}\right)\right]$$

$$Y(e^{j\omega}) = \pi H(e^{j\pi/2})\delta\left(\omega - \frac{\pi}{2}\right)$$

$$+ \pi H(e^{-j\pi/2})\delta\left(\omega + \frac{\pi}{2}\right)$$

Given that

$$H(e^{j\omega}) = \left[\frac{1 + e^{-j2\omega} + 4e^{-j4\omega}}{1 + \frac{1}{2}e^{-j2\omega}}\right]e^{-j(\omega - \pi/4)}$$

Therefore,

$$H(e^{j\pi/2}) = \left[\frac{1 + e^{-j2\pi/2} + 4e^{-j4\pi/2}}{1 + 1/2\,e^{-j2\pi/2}}\right]e^{-j(\pi/2 - \pi/4)}$$

$$= \left[\frac{1 + e^{-j\pi} + 4e^{-j2\pi}}{1 + 1/2\,e^{-j\pi}}\right]e^{-j\pi/4}$$

$$H(e^{j\pi/2}) = \left[\frac{1 - 1 + 4}{1 - 1/2}\right]e^{-j\pi/4}$$

$$H(e^{j\pi/2}) = 8\,e^{-j\pi/4}$$

Similarly, we can obtain

$$H(e^{-j\pi/2}) = 8\,e^{j3\pi/4}$$

Substituting the values of $H(e^{j\pi/2})$ and $H(e^{-j\pi/2})$ in the expression of $Y(e^{j\omega})$, we obtain

$$Y(e^{j\omega}) = 8\pi\left[e^{-j\pi/4}\delta\left(\omega - \frac{\pi}{2}\right) + e^{j3\pi/4}\delta\left(\omega + \frac{\pi}{2}\right)\right]$$

Example 6.44 A causal system is described by the difference equation

$$y(n) - ay(n - 1) = bx(n) + x(n - 1)$$

where a is real and less than 1 in magnitude. Find a value of b such that the frequency response of the

system satisfies

$$|H(e^{j\omega})| = 1, \quad \text{for all } \omega$$

This kind of system is called an all-pass system.

Solution

Consider the given difference equation,

$$y(n) - ay(n - 1) = bx(n) + x(n - 1)$$

Taking the Fourier transform of the above equation, we obtain

$$Y(e^{j\omega}) - aY(e^{j\omega})e^{-j\omega} = bX(e^{j\omega}) + X(e^{j\omega})e^{-j\omega}$$

$$H(e^{j\omega}) = \frac{Y(e^{j\omega})}{X(e^{j\omega})} = \frac{b + e^{-j\omega}}{1 - a\,e^{-j\omega}}$$

Taking the magnitude, we obtain

$$|H(e^{j\omega})| = \frac{|b + e^{-j\omega}|}{|1 - a\,e^{-j\omega}|}$$

$$= \frac{|b + \cos(\omega) - j\sin(\omega)|}{|1 - a\cos(\omega) + ja\sin(\omega)|}$$

$$|H(e^{j\omega})| = \frac{\sqrt{[b + \cos(\omega)]^2 + \sin^2(\omega)}}{\sqrt{[1 - a\cos(\omega)]^2 + a^2\sin^2(\omega)}}$$

$$1 = \frac{\sqrt{1 + b^2 + 2b\cos(\omega)}}{\sqrt{1 + a^2 - 2a\cos(\omega)}}$$

$$1 + b^2 + 2b\cos(\omega) = 1 + a^2 - 2a\cos(\omega)$$

From the above equation it is evident that $b = -a$.

Example 6.45 The centre of gravity of a signal $x(n)$ is defined as

$$C = \frac{\sum_{n=-\infty}^{\infty} nx(n)}{\sum_{n=-\infty}^{\infty} x(n)}$$

and provides a measure of 'time delay' of the signal.

(a) Express C in terms of $X(e^{j\omega})$.
(b) Compute C for the signal $x(n)$ whose DTFT is shown in Fig. 6.26.

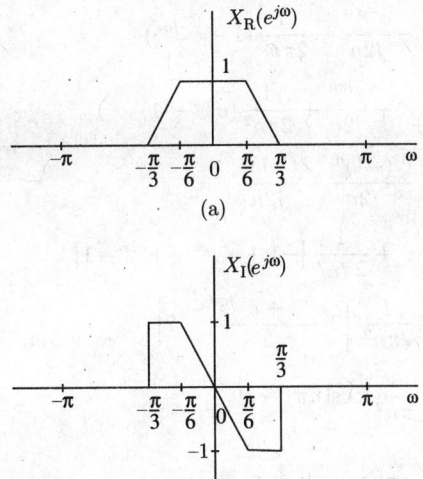

(a)

(b)

Fig. 6.26

Solution

(a) The DTFT $X(e^{j\omega})$ of a signal $x(n)$ is given by

$$X(e^{j\omega}) = \sum_{n=-\infty}^{\infty} x(n)\, e^{-j\omega n}$$

Differentiating both sides w.r.t. ω, we obtain

$$\frac{dX(e^{j\omega})}{d\omega} = \sum_{n=-\infty}^{\infty} [-jnx(n)]\, e^{-j\omega n}$$

$$j\frac{dX(e^{j\omega})}{d\omega} = \sum_{n=-\infty}^{\infty} [nx(n)]\, e^{-j\omega n}$$

$$j\frac{dX(e^{j\omega})}{d\omega}\bigg|_{\omega=0} = \sum_{n=-\infty}^{\infty} nx(n)$$

We also know that [Eq. (6.62)]

$$X(e^{j\omega}) = \sum_{n=-\infty}^{\infty} x(n)\, e^{-j\omega n}$$

$$X(e^{j\omega})\bigg|_{\omega=0} = \sum_{n=-\infty}^{\infty} x(n)$$

Therefore,

$$C = \frac{\sum_{n=-\infty}^{\infty} nx(n)}{\sum_{n=-\infty}^{\infty} x(n)} = \frac{j\,dX(e^{j\omega})/(d\omega)\big|_{\omega=0}}{X(e^{j\omega})\big|_{\omega=0}}$$

(b) We know that

$$X(e^{j\omega}) = X_R(e^{j\omega}) + jX_I(e^{j\omega})$$

$$\frac{dX(e^{j\omega})}{d\omega} = \frac{dX_R(e^{j\omega})}{d\omega} + j\frac{dX_I(e^{j\omega})}{d\omega}$$

$$\frac{dX(e^{j\omega})}{d\omega}\bigg|_{\omega=0} = \frac{dX_R(e^{j\omega})}{d\omega}\bigg|_{\omega=0} + j\frac{dX_I(e^{j\omega})}{d\omega}\bigg|_{\omega=0}$$

From Figs 6.26(a) and (b), we have

$$\frac{dX_R(e^{j\omega})}{d\omega}\bigg|_{\omega=0} = 0 \quad \text{and} \quad \frac{dX_I(e^{j\omega})}{d\omega}\bigg|_{\omega=0} = -\frac{6}{\pi}$$

Therefore,

$$\frac{dX(e^{j\omega})}{d\omega}\bigg|_{\omega=0} = 0 - j\frac{6}{\pi}$$

$$j\frac{dX(e^{j\omega})}{d\omega}\bigg|_{\omega=0} = \frac{6}{\pi}$$

and

$$X(e^{j\omega})|_{\omega=0} = X_R(e^{j\omega})_{\omega=0} + jX_I(e^{j\omega})_{\omega=0}$$

$$= 1 + j0$$

$$X(e^{j\omega})|_{\omega=0} = 1$$

Substituting $j\,dX(e^{j\omega})/(d\omega)\big|_{\omega=0} = 6/\pi$ and $X(e^{j\omega})|_{\omega=0} = 1$ in the expression of C, yields

$$C = \frac{j\,dX(e^{j\omega})/(d\omega)\big|_{\omega=0}}{X(e^{j\omega})\big|_{\omega=0}} = \frac{6}{\pi}$$

Example 6.46 An LTI system with unit sample response $h(n)$ and frequency response $H(e^{j\omega})$ is known to have the property that, when $-\pi \le \omega \le \pi$,

$$\cos(\omega_0 n) \longrightarrow \omega_0 \cos(\omega_0 n)$$

(a) Determine $H(e^{j\omega})$.
(b) Determine $h(n)$.

Solution

(a) Given that

$$\cos(\omega_0 n) \longrightarrow \omega_0 \cos(\omega_0 n)$$

$$\left(\frac{e^{j\omega_0 n}}{2} + \frac{e^{-j\omega_0 n}}{2}\right) \longrightarrow \omega_0 \left(\frac{e^{j\omega_0 n}}{2} + \frac{e^{-j\omega_0 n}}{2}\right)$$

$$\left(\frac{e^{j\omega_0 n}}{2} + \frac{e^{-j\omega_0 n}}{2}\right) \longrightarrow \left(\omega_0 \frac{e^{j\omega_0 n}}{2} + \omega_0 \frac{e^{-j\omega_0 n}}{2}\right)$$

Since the system is linear, we may conclude that

$$\frac{e^{j\omega_0 n}}{2} \longrightarrow \omega_0 \frac{e^{j\omega_0 n}}{2}$$

and

$$\frac{e^{-j\omega_0 n}}{2} \longrightarrow \omega_0 \frac{e^{-j\omega_0 n}}{2}$$

It is clear that when the input to the system is a complex exponential of frequency ω_0, the output is a complex exponential of the same frequency but scaled by $|\omega_0|$. Therefore, the frequency response of the system is

$$\frac{Y(e^{j\omega})}{X(e^{j\omega})} = H(e^{j\omega}) = |\omega| \quad 0 \le |\omega| \le \pi$$

$$H(e^{j\omega}) = \begin{cases} \omega, & 0 \le \omega \le \pi \\ -\omega, & -\pi \le \omega \le 0 \end{cases}$$

(b) Taking the inverse Fourier transform of the frequency response, we obtain

$$h(n) = \frac{1}{2\pi} \int_{-\pi}^{\pi} H(e^{j\omega})\, e^{j\omega n} d\omega$$

$$= \frac{1}{2\pi} \int_{-\pi}^{0} -\omega\, e^{j\omega n}\, d\omega + \frac{1}{2\pi} \int_{0}^{\pi} \omega\, e^{j\omega n}\, d\omega$$

$$= -\frac{1}{2\pi} \left[\frac{\omega\, e^{j\omega n}}{jn} \Big|_{-\pi}^{0} - \frac{e^{j\omega n}}{-n^2} \Big|_{-\pi}^{0} \right]$$

$$+ \frac{1}{2\pi} \left[\frac{\omega\, e^{j\omega n}}{jn} \Big|_{0}^{\pi} - \frac{e^{j\omega n}}{-n^2} \Big|_{0}^{\pi} \right]$$

$$= -\frac{1}{2\pi} \left[0 + \frac{\pi e^{-j\pi n}}{jn} + \frac{1}{n^2}(1 - e^{-jn\pi}) \right]$$

$$+ \frac{1}{2\pi} \left[\frac{\pi e^{j\pi n}}{jn} - 0 + \frac{1}{n^2}(e^{jn\pi} - 1) \right]$$

$$= -\frac{e^{-j\pi n}}{j2n} - \frac{1}{2\pi n^2}(1 - e^{-jn\pi})$$

$$+ \frac{e^{j\pi n}}{j2n} + \frac{1}{2\pi n^2}(e^{jn\pi} - 1)$$

$$h(n) = -\frac{(-1)^n}{j2n} + \frac{(-1)^n}{j2n}$$

$$+ \frac{1}{2\pi n^2}\left[-1 + e^{-jn\pi} + e^{jn\pi} - 1\right]$$

$$= \frac{1}{2\pi n^2}\left[2\frac{e^{jn\pi} + e^{-jn\pi}}{2} - 2\right]$$

$$h(n) = \frac{1}{\pi n^2}[\cos(n\pi) - 1]$$

Example 6.47 Let $x(n)$ be a discrete-time signal with DTFT $X(e^{j\omega})$. Determine and sketch the Fourier transform of the following signals:

(a) $y_1(n) = x(2n)$

(b) $y_2(n) = \begin{cases} x(n/2), & n \text{ even} \\ 0, & n \text{ odd} \end{cases}$

Solution

(a) Consider $y_1(n) = x(Mn) = x(2n)$, where $M = 2$. The DTFT of $y_1(n)$ is given by

$$Y_1(e^{j\omega}) = \sum_{n=-\infty}^{\infty} y_1(n)\, e^{-j\omega n}$$

$$Y_1(e^{j\omega}) = \sum_{n=-\infty}^{\infty} x(Mn)\, e^{-j\omega n}$$

Define an intermediate sequence

$$x_1(n) = \begin{cases} x(n), & n = 0, \pm M, \pm 2M, \ldots \\ 0, & \text{otherwise} \end{cases}$$

so that $y_1(n) = x(Mn) = x_1(Mn)$. Now

$$Y_1(e^{j\omega}) = \sum_{n=-\infty}^{\infty} x_1(Mn)\, e^{-j\omega n}$$

$$= \sum_{k=-\infty}^{\infty} x_1(k)\, e^{-j\omega k/M}$$

This step is valid because $x_1(k)$ is zero unless k is a multiple of M. So

$$Y_1(e^{jw}) = X_1(e^{jw/M})$$

It only remains to express $X_1(e^{jw})$ in terms of $X(e^{jw})$. For this note that

$$x_1(n) = c(n)x(n)$$

where $c(n)$ is a periodic sequence (period $N = M$) defined as

$$c(n) = \begin{cases} 1, & n = 0, \pm M, \pm 2M, \ldots \\ 0, & \text{otherwise} \end{cases}$$

Using DTFS, $c(n)$ can be expressed as

$$c(n) = \sum_{k=0}^{M-1} C_k\, e^{jk\omega_0 n}$$

where $\omega_0 = 2\pi/M$, and

$$C_k = \frac{1}{M} \sum_{m=0}^{M-1} c(n)\, e^{-jk\omega_0 n}$$

$$= \frac{1}{M} \sum_{m=0}^{M-1} \delta(n)\, e^{-jk\omega_0 n}$$

$$C_k = \frac{1}{M}$$

Thus, the periodic sequence $c(n)$ can be expressed as

$$c(n) = \frac{1}{M} \sum_{k=0}^{M-1} e^{j2\pi nk/M}$$

We can now obtain

$$X_1(e^{j\omega}) = \sum_{n=-\infty}^{\infty} x_1(n)\, e^{-j\omega n}$$

$$= \sum_{n=-\infty}^{\infty} c(n)x(n)\, e^{-j\omega n}$$

$$= \sum_{n=-\infty}^{\infty} \left(\frac{1}{M} \sum_{k=0}^{M-1} e^{j2\pi nk/M} \right) x(n)\, e^{-j\omega n}$$

$$= \frac{1}{M} \sum_{k=0}^{M-1} \left(\sum_{n=-\infty}^{\infty} x(n) e^{-j(\omega - 2\pi k/M)n} \right)$$

$$X_1(e^{j\omega}) = \frac{1}{M} \sum_{k=0}^{M-1} X(e^{j(\omega - 2\pi k/M)})$$

For $M = 2$, the above equation yields

$$X_1(e^{j\omega}) = \frac{1}{2} \sum_{k=0}^{1} X(e^{j(\omega - \pi k)})$$

$$X_1(e^{j\omega}) = \frac{1}{2}[X(e^{j\omega}) + X(e^{j\omega - \pi})]$$

$$= \frac{1}{2}[X(e^{j\omega}) + X(e^{j\omega} e^{-j\pi})]$$

$$X_1(e^{j\omega}) = \frac{1}{2}[X(e^{j\omega}) + X(-e^{j\omega})]$$

Thus, for $M = 2$, we get

$$Y_1(e^{j\omega}) = X_1(e^{j\omega/2})$$

$$= \frac{1}{2}[X(e^{j\omega/2}) + X(-e^{j\omega/2})]$$

(b) Consider the given signal

$$y_2(n) = \begin{cases} x(n/2), & n \text{ even} \\ 0, & n \text{ odd} \end{cases}$$

$$= \begin{cases} x(n/2), & \text{if } n \text{ is a multiple of 2} \\ 0, & \text{if } n \text{ is not a multiple of 2.} \end{cases}$$

The Fourier transform of $y_2(n)$ is given by

$$Y_2(e^{j\omega}) = \sum_{n=-\infty}^{\infty} y_2(n)\, e^{-j\omega n}$$

$$= \sum_{n=-\infty}^{\infty} x\left(\frac{n}{2}\right) e^{-j\omega n}$$

A change of variables is performed by letting $r = n/2$, which also yields $r = -\infty$ as $n = -\infty$, and $r = \infty$ as $n = \infty$. Therefore,

$$Y_2(e^{j\omega}) = \sum_{r=-\infty}^{\infty} x(r)\, e^{-j\omega 2r}$$

$$Y_2(e^{j\omega}) = X(e^{j2\omega})$$

Example 6.48 Consider a discrete-time LTI system with unit sample response

$$h(n) = \left(\frac{1}{2}\right)^n u(n) + \frac{1}{2}\left(\frac{1}{4}\right)^n u(n)$$

Determine a linear constant-coefficient difference equation relating the input and output of the system.

Solution
Consider the given unit sample response

$$h(n) = \left(\frac{1}{2}\right)^n u(n) + \frac{1}{2}\left(\frac{1}{4}\right)^n u(n)$$

Taking the Fourier transform of the unit sample response, we obtain

$$H(e^{j\omega}) = \frac{1}{1 - 1/2\, e^{-j\omega}} + \frac{1/2}{1 - 1/4\, e^{-j\omega}}$$

$$\frac{Y(e^{j\omega})}{X(e^{j\omega})} = \frac{(3/2) - (1/2)e^{-j\omega}}{1 - (3/4)\, e^{-j\omega} + (1/8)\, e^{-j2\omega}}$$

$$Y(e^{j\omega})\left[1 - \frac{3}{4}\, e^{-j\omega} + \frac{1}{8}\, e^{-j2\omega}\right]$$

$$= X(e^{j\omega})\left[\frac{3}{2} - \frac{1}{2}\, e^{-j\omega}\right]$$

$$Y(e^{j\omega}) - \frac{3}{4}\, e^{-j\omega}Y(e^{j\omega}) + \frac{1}{8}\, e^{-j2\omega}Y(e^{j\omega})$$

$$= \frac{3}{2}X(e^{j\omega}) - \frac{1}{2}\, e^{-j\omega}X(e^{j\omega})$$

Taking the inverse Fourier transform, we obtain

$$y(n) - \frac{3}{4}y(n-1) + \frac{1}{8}y(n-2)$$

$$= \frac{3}{2}x(n) - \frac{1}{2}x(n-1)$$

Example 6.49 Given input $x(n)$ and impulse response $h(n)$, as shown in Fig. 6.27, evaluate $y(n) = x(n) * h(n)$, using the DTFT.

Solution
The DTFT of the given input signal $x(n) = \{1, \underset{\uparrow}{1}, 1\}$ is given by

$$X(e^{j\omega}) = \sum_{n=-\infty}^{\infty} x(n)\, e^{-j\omega n}$$

$$= \sum_{n=-1}^{1} x(n)\, e^{-j\omega n}$$

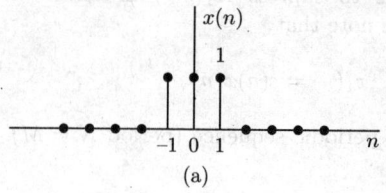

(a)

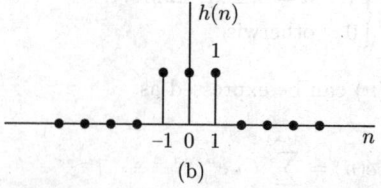

(b)

Fig. 6.27

$$= x(-1)\, e^{j\omega} + x(0) + x(1)\, e^{-j\omega}$$

$$X(e^{j\omega}) = 1 + e^{j\omega} + e^{-j\omega}$$

Similarly, the DTFT of the impulse response $h(n) = \{1, \underset{\uparrow}{1}, 1\}$ is given by

$$H(e^{j\omega}) = 1 + e^{j\omega} + e^{-j\omega}$$

Now using the convolution property of the DTFT, we have

$$y(n) = x(n) * h(n)$$

$$Y(e^{j\omega}) = X(e^{j\omega})H(e^{j\omega})$$

$$Y(e^{j\omega}) = [1 + e^{j\omega} + e^{-j\omega}][1 + e^{j\omega} + e^{-j\omega}]$$

$$Y(e^{j\omega}) = e^{j2\omega} + 2\, e^{j\omega} + 3 + 2\, e^{-j\omega} + e^{-j2\omega}$$

Taking the inverse Fourier transform, we obtain

$$y(n) = \delta(n+2) + 2\delta(n+1) + 3\delta(n)$$

$$+ 2\delta(n-1) + \delta(n-2)$$

SUMMARY

1. The Fourier transform of $x(n)$ is defined by

$$X(e^{j\omega}) = \sum_{n=-\infty}^{\infty} x(n)\,e^{-j\omega n}$$

2. The inverse Fourier transform of $X(e^{j\omega})$ is defined by

$$x(n) = \frac{1}{2\pi} \int_{-\pi}^{\pi} X(e^{j\omega})\,e^{j\omega n} d\omega$$

3. The magnitude of $X(e^{j\omega})$ plotted against ω is called the magnitude spectrum of $x(n)$ and $|X(e^{j\omega})|^2$ is called the energy spectrum.

4. The angle of $X(e^{j\omega})$ plotted versus ω is called the phase spectrum.

5. Parseval's theorem states that

$$E_x = \sum_{n=-\infty}^{\infty} |x(n)|^2 = \frac{1}{2\pi} \int_{-\pi}^{\pi} |X(e^{j\omega})|^2 d\omega$$

6. The power spectral density of $x(n)$ is defined by

$$G_x(e^{j\omega}) = \lim_{N\to\infty} \frac{|X_N(e^{j\omega})|^2}{2N+1}$$

where

$$x_N(n) \longleftrightarrow X_N(e^{j\omega})$$

and

$$x_N(n) = \begin{cases} x(n), & -N \le n \le N \\ 0, & \text{otherwise} \end{cases}$$

7. The DTFT is always periodic with periodic one in the f domain or period 2π in the ω domain.

8. If a signal is discrete in one domain, it is periodic in the other.

9. Convolution and multiplication of functions are dual in the time and frequency domains.

10. The DTFT of a periodic signal consists only of impulses.

11. The correlation function indicates how correlated two signals are as a function of how much one of them is shifted in time.

12. Correlation and convolution are closely related mathematical processes.

13. The correlation of a signal with a shifted version of itself is called autocorrelation.

14. Autocorrelation is closely related to signal energy or power and contains important information about how rapidly a signal varies in time.

15. ESD and PSD are the frequency-domain counterparts of autocorrelation and are related through the Fourier transform.

16. ESD and PSD indicated how the energy or power of a signal varies with frequency.

MULTIPLE-CHOICE QUESTIONS

1. The Fourier transform of a signal $x(n) = a^n u(n)$ is given by

   (a) $\dfrac{1}{1 - a\,e^{j\omega}}$

   (b) $\dfrac{1}{1 + a\,e^{j\omega}}$

   (c) $\dfrac{1}{1 - a\,e^{-j\omega}}$

   (d) $\dfrac{1}{a + e^{-j\omega}}$

2. The Fourier transform of a unit step function

   (a) does not exist

   (b) is another unit step

   (c) contains impulse functions

   (d) is $1/1 - e^{-j\omega}$

3. If the notation $\circledast$ is used to denote the convolution, and $x(n) \longleftrightarrow X(e^{\omega})$, $y(n) \longleftrightarrow Y(e^{\omega})$ makes the Fourier transform pair, then $x(n)y(n) \longleftrightarrow F(e^{\omega})$ is given by

   (a) $X(e^{\omega}) \circledast Y(e^{\omega})$

   (b) $X(e^{\omega})Y(e^{\omega})$

   (c) $1/2\pi X(e^{\omega})Y(e^{\omega})$

   (d) $1/2\pi[X(e^{\omega}) \circledast Y(e^{\omega})]$

4. The DTFT is periodic with period

   (a) π

(b) 2π

(c) $\pi/2$

(d) $\pi/4$.

5. The property of Fourier transform that states that the expansion in time domain is equivalent to compression in the frequency domain is

   (a) duality

   (b) time shifting

   (c) time expansion

   (d) frequency shifting

6. The DTFT of a function $x(n)$ is $X(e^{\omega})$. The FT of $nx(n)$ will be

   (a) $j\dfrac{dX(e^{\omega})}{d\omega}$

   (b) $\dfrac{dX(e^{\omega})}{d\omega}$

   (c) $j\omega\dfrac{dX(e^{\omega})}{d\omega}$

   (d) $\dfrac{X(e^{j\omega})}{j\omega}$

7. The Fourier transform of the exponential signal $e^{j\omega_0 n}$, for $0 < \omega \leq \pi$, is

   (a) a constant

   (b) a rectangular pulse

   (c) an impulse

   (d) an another exponential

8. The Fourier transform of a real and even signal $x(n)$ is

   (a) real and even function of frequency

   (b) real and odd function of frequency

   (c) complex, in general

   (d) purely imaginary

9. The Fourier transform of a real and odd signal $x(n)$ is

   (a) imaginary and odd function of frequency

   (b) real and odd function of frequency

   (c) complex, in general

   (d) purely imaginary

10. The frequency response of a system with $h(n) = \delta(n) - \delta(n-1)$ is given by

    (a) $\delta(\omega) - \delta(\omega - 1)$

    (b) $1 - e^{j\omega}$

    (c) $u(\omega) - u(\omega - 1)$

    (d) $1 - e^{-j\omega}$

11. A transmission is said to be .... if the response of the system is exact replica of the input signal.

    (a) LTI

    (b) distorted

    (c) distortionless

    (d) undefined

12. The group-delay function $\tau_g(\omega)$ is related to phase function $\phi(\omega)$ as

    (a) $\tau_g(\omega) = -\dfrac{d\phi(\omega)}{d\omega}$

    (b) $\tau_g(\omega) = \dfrac{d\phi(\omega)}{d\omega}$

    (c) $\tau_g(\omega) = -\dfrac{d^2\phi(\omega)}{d\omega^2}$

    (d) $\tau_g(\omega) = -\dfrac{\phi(\omega)}{\omega}$

13. For a nonperiodic discrete-time signal, the frequency-shift property states that if the DTFT of $x(n)$ is $X(e^{j\omega})$, then the DTFT of $x'(n)$ is $X(e^{j(\omega - \alpha)})$, where $x'(n)$ is given by

    (a) $e^{j\alpha n}x(n)$

    (b) $e^{-j\alpha n}x(n)$

    (c) $e^{jn}x(n - \alpha)$

    (d) $e^{-jn}x(n - \alpha)$

PROBLEMS

6.1 A signal $x(n)$ can be expressed as the sum of even and odd components:

$$x(n) = x_e(n) + x_o(n)$$

(a) If $x(n) \longleftrightarrow X(e^{j\omega})$, show that for real $x(n)$,

$$x_e(n) \longleftrightarrow \Re\{X(e^{j\omega})\}$$

and

$$x_o(n) \longleftrightarrow j\Im\{X(e^{j\omega})\}$$

(b) Verify these results by finding the DTFT of the even and odd components of the signal $(0.8)^n u(n)$.

6.2 An accumulator system has the property that an input $x(n)$ results in the output

$$y(n) = \sum_{k=-\infty}^{n} x(k)$$

(a) Find the impulse response $h(n)$ and the frequency response $H(e^{j\omega})$ for the accumulator.

(b) Use the result in part (a) to find the DTFT of $u(n)$.

6.3 $X(e^{j\omega})$ denotes the DTFT of a length-9 sequence $x(n)$ given by

$$x(n) = \{2, 3, -\underset{\uparrow}{1}, 0, -4, 3, 1, 2, 4\}, \quad -2 \le n \le 6$$

Evaluate the following functions of $X(e^{j\omega})$ without computing the transform itself:

(a) $X(e^{j0})$

(b) $X(e^{j\pi})$

(c) $\int_{-\pi}^{\pi} X(e^{j\omega}) \, d\omega$

(d) $\int_{-\pi}^{\pi} |X(e^{j\omega})|^2 \, d\omega$

(e) $\int_{-\pi}^{\pi} \left| \frac{dX(e^{j\omega})}{d\omega} \right|^2 \, d\omega$

6.4 Let $x(n)$ be a causal and absolutely summable real sequence with a DTFT $X(e^{j\omega})$. If $X_R(e^{j\omega})$ and $X_I(e^{j\omega})$ denote the real and imaginary parts of $X(e^{j\omega})$, show that they are related as

$$X_R(e^{j\omega}) = \frac{1}{2\pi} \int_{-\pi}^{\pi} X_I(e^{j\theta}) \cot\left(\frac{\omega - \theta}{2}\right) d\theta + x(0)$$

$$X_I(e^{j\omega}) = -\frac{1}{2\pi} \int_{-\pi}^{\pi} X_R(e^{j\theta}) \cot\left(\frac{\omega - \theta}{2}\right) d\theta$$

6.5 Consider the filter

$$y(n) = 0.9y(n-1) + bx(n)$$

(a) Determine b so that $|H(e^{j0})| = 1$.

(b) Determine the frequency at which $|H(e^{j\omega})| = 1/\sqrt{2}$.

(c) Is this filter low-pass, high-pass, or bandpass?

(d) Repeat parts (b) and (c) for the filter

$$y(n) = -0.9y(n-1) + 0.1x(n)$$

6.6 The input to a discrete-time system is given by

$$x(n) = \cos\left(\frac{\pi}{4}n\right) + \cos\left(\frac{3\pi}{4}n\right)$$

The impulse response of the system is specified as follows:

$$h(n) \longleftrightarrow H(e^{j\omega}) = \begin{cases} 1, & |\omega| < \pi/4 \\ 0, & \frac{\pi}{4} \ge |\omega| < \pi \end{cases}$$

Obtain the output $y(n)$ using the convolution property of the DTFT.

6.7 Compute the Fourier transform of each of the following signals:

(a) $x(n) = u(n-2) - u(n-6)$

(b) $x(n) = (1/2)^{-n} u(-n-1)$

(c) $x(n) = (1/2)^{|n|} u(-n-1)$

(d) $x(n) = (n-1)(1/3)^{|n|}$

6.8 Determine the signals having the following Fourier transforms:

(a) $X(e^{j\omega}) = \begin{cases} 0, & 0 \le |\omega| \le \omega_0 \\ 1, & \omega_0 < |\omega| \le \pi \end{cases}$

(b) $X(e^{j\omega}) = \cos^2(\omega)$

(c) $X(e^{j\omega}) = e^{-j\omega/2}$, for $-\pi \le \omega \le \pi$

(d) $X(e^{j\omega}) = \dfrac{e^{-j\omega} - 1/5}{1 - 1/5 \, e^{-j\omega}}$

6.9 A signal $x(n)$ has the following Fourier transform

$$X(e^{j\omega}) = \frac{1}{1 - a \, e^{-j\omega}}$$

Determine the Fourier transforms of the following signals:

(a) $x(2n+1)$

(b) $e^{j(\pi/2)n} x(n+2)$

(c) $x(-2n)$

(d) $x(n) \cos(0.3\pi n)$

(e) $x(n) * x(n-1)$

(f) $x(n) * x(-n)$

6.10 The following input–output pairs have been observed during the operation of various systems:

(a) $x(n) = (1/2)^n \longrightarrow y(n) = (1/8)^n$

(b) $x(n) = (1/2)^n u(n) \longrightarrow y(n) = (1/8)^n u(n)$

(c) $x(n) = e^{j\pi/5} \longrightarrow y(n) = 3\,e^{j\pi/5}$

Determine their frequency response if each of the above system is LTI.

6.11 Determine and sketch the magnitude and phase response of the following systems:

(a) $y(n) = \frac{1}{2}[x(n) + x(n-1)]$

(b) $y(n) = \frac{1}{2}[x(n) - x(n-1)]$

(c) $y(n) = \frac{1}{2}[x(n+1) - x(n-1)]$

(d) $y(n) = \frac{1}{2}[x(n+1) + x(n-1)]$

ANSWERS TO MULTIPLE-CHOICE QUESTIONS

1. (a) 2. (c) 3. (d) 4. (b) 5. (c) 6. (a) 7. (c) 8. (a) 9. (a)
10. (d) 11. (c) 12. (a) 13. (a)

Hilbert Transform

7.1 INTRODUCTION

The Fourier transform, which has occupied so much of our attention thus far, is particularly useful for evaluating the frequency content of an energy signal or, in a limiting sense, that of a power signal. As such, it provides the mathematical basis for analysing and designing frequency-selective filters for the separation of signals on the basis of their frequency contents. Another method of separating signals is based on *phase selectivity*, which uses phase shifts between the pertinent signals to achieve the desired separation. The simplest phase shift is that of 180°, which is merely a polarity reversal in the case of a sinusoidal signal. Shifting the phase angles of all components of a given signal by 180° requires the use of an ideal transformer. Another phase shift of interest is that of ±90°. In particular, when the phase angles of all components of a given signal are shifted by ±90°, the resulting function of time is known as the Hilbert transform of the signal.

The Hilbert transform has several important applications, which include the following:

1. It can be used to realize phase selectivity in the generation of a special kind of modulation known as single-side band modulation.
2. It provides the mathematical basis for the representation of bandpass signals.

The Hilbert transform applies to any signal that is Fourier transformable. Accordingly, it may be applied to energy signals as well as power signals.

7.2 CONTINUOUS-TIME HILBERT TRANSFORM

In Section 5.7.1, we saw that the constraint of *causality* of a signal implies unique relationships between the real and imaginary parts of the Fourier transform. Relationships of this type between the real and imaginary parts of complex functions are known as Hilbert transform relationships. Once again, to illustrate the relationships between the real and imaginary parts of the Fourier transform, we decompose $x(t)$ into its even and odd parts, i.e.,

$$x(t) = x_\mathrm{e}(t) + x_\mathrm{o}(t)$$

where

$$x_e(t) = \frac{1}{2}[x(t) + x(-t)]$$

and

$$x_o(t) = \frac{1}{2}[x(t) - x(-t)]$$

Now, if $x(t)$ is causal, it is possible to recover $x(t)$ from its even part $x_e(t)$ or from its odd part $x_o(t)$. Indeed, it can be easily seen that

$$x(t) = 2x_e(t)u(t) = 2x_o(t)u(t) \tag{7.1}$$

$$x_e(t) = x_o(t)\,\text{sgn}(t) \tag{7.2}$$

and

$$x_o(t) = x_e(t)\,\text{sgn}(t) \tag{7.3}$$

Since $x_o(t) = x_e(t)$ for $t > 0$, there is a strong relationship between $x_o(t)$ and $x_e(t)$. If $x(t)$ is absolutely integrable, the Fourier transform $X(\omega)$ exists, and

$$X(\omega) = X_R(\omega) + jX_I(\omega)$$

In addition, if $x(t)$ is real valued and causal, the conjugate property of the Fourier transform implies that

$$x_e(t) \longleftrightarrow X_R(\omega) \quad \text{and} \quad x_o(t) \longleftrightarrow jX_I(\omega)$$

Since $x(t)$ is completely specified by $x_e(t)$, it follows that $X(\omega)$ is completely determined if we know $X_R(\omega)$. Alternatively, $X(\omega)$ is completely determined from $X_I(\omega)$. In short, $X_R(\omega)$ and $X_I(\omega)$ are interdependent and cannot be specified independently if the signal is causal. Equivalently, the magnitude and phase responses of a causal signal are interdependent and hence cannot be specified independently.

Taking the Fourier transform of Eq. (7.2), we obtain

$$\mathcal{F}[x_e(t)] = \frac{1}{2\pi}\Big[\mathcal{F}[x_o(t)] * \mathcal{F}[\text{sgn}(t)]\Big]$$

$$X_R(\omega) = \frac{1}{2\pi}\Big[jX_I(\omega) * \frac{2}{j\omega}\Big]$$

$$X_R(\omega) = \frac{1}{2\pi}\int_{-\infty}^{\infty} jX_I(\theta)\frac{2}{j(\omega - \theta)}\,d\theta$$

$$X_R(\omega) = \frac{1}{\pi}\int_{-\infty}^{\infty}\frac{X_I(\theta)}{\omega - \theta}\,d\theta \tag{7.4}$$

Similarly, taking the Fourier transform of Eq. (7.3), we obtain

$$\mathcal{F}[x_o(t)] = \frac{1}{2\pi}\Big[\mathcal{F}[x_e(t)] * \mathcal{F}[\text{sgn}(t)]\Big]$$

$$jX_I(\omega) = \frac{1}{2\pi}\Big[X_R(\omega) * \frac{2}{j\omega}\Big]$$

$$jX_I(\omega) = \frac{1}{2\pi}\int\limits_{-\infty}^{\infty} X_R(\theta)\frac{2}{j(\omega-\theta)}d\theta$$

$$X_I(\omega) = -\frac{1}{\pi}\int\limits_{-\infty}^{\infty} \frac{X_R(\theta)}{\omega-\theta}d\theta \tag{7.5}$$

Equations (7.4) and (7.5), which are known as the Hilbert transform pair, hold for real and imaginary parts of a causal, stable, and real signal.

7.2.1 Hilbert Transform Relations for Complex Signals

Thus far, we have considered Hilbert transform relations for the Fourier transforms of causal signals. In this section, we consider *complex signals* for which the real and imaginary components can be related through a continuous convolution similar to the Hilbert transform relations derived in the previous section.

As mentioned previously, it is possible to base the derivation of the Hilbert transform relations on a notion of causality. Since we are interested in relating the real and imaginary parts of a complex signal, causality will be applied to the Fourier transform of the signal. Thus, with $x(t)$ denoting the signal and $X(\omega)$ its Fourier transform, we require that

$$X(\omega) = 0, \quad -\infty \leq \omega < 0 \tag{7.6}$$

(We could just as well assume that $X(\omega) = 0$ for $0 < \omega \leq \infty$.) The signal $x(t)$ corresponding to $X(\omega)$ must be complex, since, if $x(t)$ were real, $X(\omega)$ would be conjugate symmetric, i.e., $X(\omega) = X^*(-\omega)$. Therefore, we express $x(t)$ as

$$x(t) = x_r(t) + jx_i(t) \tag{7.7}$$

where $x_r(t)$ and $x_i(t)$ are real signals. The signal $x(t)$ is called an *analytical signal*.

If $X_r(\omega)$ and $X_i(\omega)$ denote the Fourier transforms of the real signals $x_r(t)$ and $x_i(t)$, respectively, then

$$X(\omega) = X_r(\omega) + jX_i(\omega) \tag{7.8}$$

and it follows that

$$X_r(\omega) = \frac{1}{2}[X(\omega) + X^*(-\omega)] \tag{7.9}$$

and

$$jX_{\mathrm{i}}(\omega) = \frac{1}{2}[X(\omega) - X^*(-\omega)] \tag{7.10}$$

Note that $X_{\mathrm{r}}(\omega)$ and $X_{\mathrm{i}}(\omega)$ are both complex-valued functions in general, and the complex transforms $X_{\mathrm{r}}(\omega)$ and $jX_{\mathrm{i}}(\omega)$ play a role similar to that played in the previous section by the even and odd parts, respectively, of a causal signal. However, $X_{\mathrm{r}}(\omega)$ is conjugate symmetric and $jX_{\mathrm{i}}(\omega)$ is conjugate antisymmetric, i.e.,

$$X_{\mathrm{r}}(\omega) = X_{\mathrm{r}}^*(-\omega) \quad \text{and} \quad jX_{\mathrm{i}}(\omega) = -jX_{\mathrm{i}}^*(-\omega) \tag{7.11}$$

If $X(\omega) = 0$ for $-\infty \le \omega < 0$, then there is no overlap between the portions of $X(\omega)$ and $X^*(-\omega)$. Thus, $X(\omega)$ can be recovered from either $X_{\mathrm{r}}(\omega)$ or $X_{\mathrm{i}}(\omega)$.

In particular,

$$X(\omega) = \begin{cases} 2X_{\mathrm{r}}(\omega), & 0 \le \omega \le \infty \\ 0, & -\infty \le \omega < 0 \end{cases}, \qquad X^*(-\omega) = \begin{cases} 0, & 0 \le \omega \le \infty \\ 2X_{\mathrm{r}}(\omega), & -\infty \le \omega < 0 \end{cases} \tag{7.12}$$

and

$$X(\omega) = \begin{cases} 2jX_{\mathrm{i}}(\omega), & 0 \le \omega \le \infty \\ 0, & -\infty \le \omega < 0 \end{cases}, \qquad X^*(-\omega) = \begin{cases} 0, & 0 \le \omega \le \infty \\ -2jX_{\mathrm{i}}(\omega), & -\infty \le \omega < 0 \end{cases} \tag{7.13}$$

Alternatively, from Eqs (7.12) and (7.13), we can relate $X_{\mathrm{r}}(\omega)$ and $X_{\mathrm{i}}(\omega)$ directly by

$$X_{\mathrm{i}}(\omega) = \begin{cases} -jX_{\mathrm{r}}(\omega), & 0 \le \omega \le \infty \\ jX_{\mathrm{r}}(\omega), & -\infty \le \omega < 0 \end{cases} \tag{7.14}$$

or

$$X_{\mathrm{i}}(\omega) = H(\omega)X_{\mathrm{r}}(\omega) \tag{7.15}$$

where

$$H(\omega) = \begin{cases} -j = 1e^{-j\pi/2}, & 0 < \omega \le \infty \\ j = 1e^{j\pi/2}, & -\infty \le \omega < 0 \end{cases} \tag{7.16}$$

or, equivalently,

$$H(\omega) = -j\,\mathrm{sgn}(\omega) \tag{7.17}$$

Equation (7.17) represents the frequency response of the Hilbert transformer and $x_{\mathrm{i}}(t)$ is the Hilbert transform of $x_{\mathrm{r}}(t)$.

Thus, according to the Eq. (7.15), $x_{\mathrm{i}}(t)$ can be obtained by processing $x_{\mathrm{r}}(t)$ with a continuous-time LTI system with frequency response $H(\omega)$ as given by Eq. (7.16). This frequency response has unity magnitude $|H(\omega)| = 1$, a phase angle of $-\pi/2$ for $0 \le \omega \le \infty$, and a phase angle of $+\pi/2$ for $-\infty \le \omega < 0$, as shown in Fig 7.1. Such a system is called an ideal 90° phase shifter or a *Hilbert transformer*. From Eq. (7.15), it follows that

$$X_{\mathrm{r}}(\omega) = \frac{1}{H(\omega)}X_{\mathrm{i}}(\omega) = -H(\omega)X_{\mathrm{i}}(\omega) \tag{7.18}$$

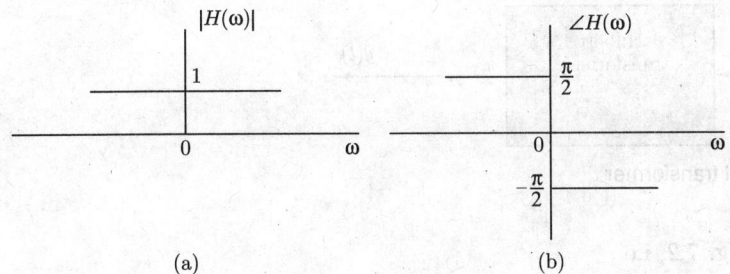

Fig. 7.1 (a) Magnitude spectrum and (b) phase response of the Hilbert transformer

Thus, $-x_{\mathrm{r}}(t)$ can also be obtained from $x_{\mathrm{i}}(t)$ with a 90° phase shifter. This is known as *inverse Hilbert transformer*.

The impulse response $h(t)$ of the Hilbert transformer corresponding to the frequency response $H(\omega)$ given in Eq. (7.17) can be obtained by using the duality property of the Fourier transform. We know that

$$\operatorname{sgn}(t) \longleftrightarrow \frac{2}{j\omega}$$

$$\frac{j}{2\pi}\operatorname{sgn}(t) \longleftrightarrow \frac{1}{\pi\omega}$$

$$g(t) \longleftrightarrow G(\omega)$$

$$g(t) = \frac{j}{2\pi}\operatorname{sgn}(t) \quad \text{and} \quad G(\omega) = \frac{1}{\pi\omega}$$

$$g(-\omega) = \frac{j}{2\pi}\operatorname{sgn}(-\omega) = -\frac{j}{2\pi}\operatorname{sgn}(\omega) \quad \text{and} \quad G(t) = \frac{1}{\pi t}$$

The duality property of the Fourier transform states that if

$$g(t) \longleftrightarrow G(\omega)$$

then

$$G(t) \longleftrightarrow 2\pi g(-\omega)$$

$$\frac{1}{\pi t} \longleftrightarrow -2\pi \frac{j}{2\pi}\operatorname{sgn}(\omega)$$

$$\frac{1}{\pi t} \longleftrightarrow -j\operatorname{sgn}(\omega)$$

Therefore, the impulse response of the Hilbert transformer is given by

$$h(t) = \frac{1}{\pi t} \tag{7.19}$$

Summary: Consider a real signal $g(t)$ with Fourier transform $G(\omega)$. The Hilbert transform of $g(t)$, which we shall denote by $\hat{g}(t) = HT[g(t)]$, can be obtained by passing the signal $g(t)$ through a continuous-time LTI system with impulse response $h(t) = 1/\pi t$, as

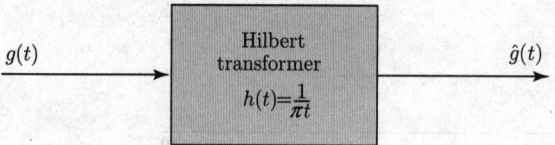

Fig. 7.2   Hilbert transformer

depicted in Fig. 7.2, i.e.,

$$\hat{g}(t) = g(t) * h(t)$$

$$\hat{g}(t) = g(t) * \frac{1}{\pi t}$$

$$\hat{g}(t) = \frac{1}{\pi} \int_{-\infty}^{\infty} \frac{g(\tau)}{t - \tau} d\tau \tag{7.20}$$

Using the convolution property of the Fourier transform, we obtain

$$\mathcal{F}[\hat{g}(t)] = \mathcal{F}[g(t)]\mathcal{F}[h(t)]$$

$$\hat{G}(\omega) = G(\omega)H(\omega)$$

Using Eq. (7.17), we obtain

$$\hat{G}(\omega) = -j \operatorname{sgn}(\omega)G(\omega) \tag{7.21}$$

Equation (7.21) states that given a signal $g(t)$, we may obtain its Hilbert transform $\hat{g}(t)$ by passing through a continuous-time LTI system with frequency response $H(\omega) = -j \operatorname{sgn}(\omega)$. This system may be considered as one that produces a phase shift of $-90°$ for all positive frequencies of the input signal and $+90°$ for all negative frequencies. The amplitudes of all frequency components in the signal, however, are unaffected by transmission through the system.

The *inverse Hilbert transform*, by means of which the original signal $g(t)$ or $G(\omega)$ is recovered from $\hat{g}(t)$ or $\hat{G}(\omega)$, is defined by [Eq. (7.18)],

$$G(\omega) = j \operatorname{sgn}(\omega)\hat{G}(\omega) \tag{7.22}$$

or

$$G(\omega) = -[-j \operatorname{sgn}(\omega)]\hat{G}(\omega)$$

Taking the inverse Fourier transform, we obtain

$$g(t) = -\frac{1}{\pi t} * \hat{g}(t) = \hat{g}(t) * -\frac{1}{\pi t}$$

$$g(t) = -\frac{1}{\pi} \int_{-\infty}^{\infty} \frac{\hat{g}(\tau)}{t - \tau} d\tau \tag{7.23}$$

7.3 PROPERTIES OF CONTINUOUS-TIME HILBERT TRANSFORM

The Hilbert transform differs from the Fourier transform in that it operates exclusively in the time domain. It has a number of properties, some of which are listed below. The signal $g(t)$ is assumed to be real valued, which is the usual domain of application of the Hilbert transform.

1. A signal $g(t)$ and its Hilbert transform $\hat{g}(t)$ have the same magnitude spectrum.

$$|\hat{G}(\omega)| = |G(\omega)| \tag{7.24}$$

Proof Given that

$$\hat{g}(t) = HT[g(t)]$$

From Eq. (7.21), we have

$$\hat{G}(\omega) = -j\,\mathrm{sgn}(\omega)G(\omega)$$

$$|\hat{G}(\omega)| = |-j\,\mathrm{sgn}(\omega)G(\omega)|$$

$$|\hat{G}(\omega)| = |-j\,\mathrm{sgn}(\omega)||G(\omega)|$$

Since $|-j\,\mathrm{sgn}(\omega)| = 1$ for all ω, we have

$$|\hat{G}(\omega)| = |G(\omega)|$$

As corollaries to this property, we may state that

(a) if a signal $g(t)$ is band limited, then its Hilbert transform $\hat{g}(t)$ is also band limited.
(b) the signal $g(t)$ and its Hilbert transform $\hat{g}(t)$ have the same energy spectral density if $g(t)$ is an energy signal, and the same power spectral density if it is a power signal.
(c) the signal $g(t)$ and its Hilbert transform $\hat{g}(t)$ have the same autocorrelation function. ■

2. If $\hat{g}(t)$ is the Hilbert transform of $g(t)$, then the Hilbert transform of $\hat{g}(t)$ is $-g(t)$, i.e., if

$$\hat{g}(t) = HT[g(t)]$$

then

$$-g(t) = HT[\hat{g}(t)] \tag{7.25}$$

Proof We know that

$$\hat{G}(\omega) = G(\omega)H(\omega)$$

From Eqs (7.16) and (7.17), we have

$$H(\omega) = -j \operatorname{sgn}(\omega) = \begin{cases} -j, & 0 < \omega \le \infty \\ j, & -\infty \le \omega < 0 \end{cases}$$

Therefore,

$$\frac{1}{H(\omega)} = -H(\omega) = j \operatorname{sgn}(\omega) = \begin{cases} j, & 0 < \omega \le \infty \\ -j, & -\infty \le \omega < 0 \end{cases}$$

We know that

$$\hat{G}(\omega) = G(\omega)H(\omega)$$

$$G(\omega) = \frac{1}{H(\omega)}\hat{G}(\omega)$$

$$G(\omega) = -H(\omega)\hat{G}(\omega)$$

$$-G(\omega) = H(\omega)\hat{G}(\omega)$$

Taking the inverse Fourier transform, we obtain

$$-g(t) = \hat{g}(t) * h(t) = \hat{g}(t) * \frac{1}{\pi t}$$

$$-g(t) = HT[\hat{g}(t)]$$

■

3. A signal $g(t)$ and its Hilbert transform $\hat{g}(t)$ are orthogonal, i.e., if $g(t)$ is an energy signal, then

$$\int_{-\infty}^{\infty} g(t)\hat{g}(t)\,dt = 0 \qquad (7.26)$$

Proof For a real signal $g(t)$ multiplied by its Hilbert transform $\hat{g}(t)$ we may write

$$\int_{-\infty}^{\infty} g(t)\hat{g}(t)\,dt = \frac{1}{2\pi}\int_{-\infty}^{\infty} G(\omega)\hat{G}^*(\omega)\,d\omega$$

$$= \frac{1}{2\pi}\int_{-\infty}^{\infty} G(\omega)\hat{G}(-\omega)\,d\omega$$

$$\int_{-\infty}^{\infty} g(t)\hat{g}(t)\,dt = \frac{1}{2\pi}\int_{-\infty}^{\infty} G(\omega)[-j\operatorname{sgn}(-\omega)G(-\omega)]\,d\omega$$

Since $\text{sgn}(-\omega) = -\text{sgn}(\omega)$ and $G(-\omega) = G^*(\omega)$, we have

$$\int\limits_{-\infty}^{\infty} g(t)\hat{g}(t)\, dt = \frac{1}{2\pi} \int\limits_{-\infty}^{\infty} G(\omega)[j\,\text{sgn}(\omega)G^*(\omega)]\, d\omega$$

$$= \frac{j}{2\pi} \int\limits_{-\infty}^{\infty} \text{sgn}(\omega)G(\omega)G^*(\omega)\, d\omega$$

$$\int\limits_{-\infty}^{\infty} g(t)\hat{g}(t)\, dt = \frac{j}{2\pi} \int\limits_{-\infty}^{\infty} \text{sgn}(\omega)|G(\omega)|^2\, d\omega$$

The integrand in the RHS of the above equation is an odd function of ω, being the product of the odd function $\text{sgn}(\omega)$ and the even function $|G(\omega)|^2$. Hence, the integral is zero [see Eq. (1.64)], yielding the final result

$$\int\limits_{-\infty}^{\infty} g(t)\hat{g}(t)\, dt = 0$$

This shows that an energy signal $g(t)$ and its Hilbert transform $\hat{g}(t)$ are orthogonal over the entire interval $(-\infty, \infty)$. Similarly, a power signal $g(t)$ and its Hilbert transform $\hat{g}(t)$ are orthogonal over one period, as shown by

$$\lim_{T\to\infty} \frac{1}{T} \int\limits_{-T/2}^{T/2} g(t)\hat{g}(t)\, dt = 0 \qquad (7.27)$$

∎

4. An energy signal $g(t)$ and its Hilbert transform $\hat{g}(t)$ have the same energy:

$$E_g = \int\limits_{-\infty}^{\infty} |g(t)|^2\, dt = \int\limits_{-\infty}^{\infty} |\hat{g}(t)|^2\, dt \qquad (7.28)$$

Proof Using the Parseval's theorem of energy signals, we have

$$\int\limits_{-\infty}^{\infty} |g(t)|^2\, dt = \frac{1}{2\pi} \int\limits_{-\infty}^{\infty} |G(\omega)|^2\, d\omega$$

Similarly, we have

$$\int\limits_{-\infty}^{\infty} |\hat{g}(t)|^2\, dt = \frac{1}{2\pi} \int\limits_{-\infty}^{\infty} |\hat{G}(\omega)|^2\, d\omega$$

$$= \frac{1}{2\pi} \int\limits_{-\infty}^{\infty} |-j\,\text{sgn}(\omega)G(\omega)|^2 \, d\omega$$

$$\int\limits_{-\infty}^{\infty} |\hat{g}(t)|^2 \, dt = \frac{1}{2\pi} \int\limits_{-\infty}^{\infty} |-j\,\text{sgn}(\omega)|^2 |G(\omega)|^2 \, d\omega$$

Since $|-j\,\text{sgn}(\omega)|^2 = 1$ for all ω, we obtain

$$\int\limits_{-\infty}^{\infty} |\hat{g}(t)|^2 \, dt = \frac{1}{2\pi} \int\limits_{-\infty}^{\infty} |G(\omega)|^2 \, d\omega$$

$$\int\limits_{-\infty}^{\infty} |\hat{g}(t)|^2 \, dt = \int\limits_{-\infty}^{\infty} |g(t)|^2 \, dt$$

Similarly, a power signal $g(t)$ and its Hilbert transform $\hat{g}(t)$ have the same power.

$$\lim_{T\to\infty} \frac{1}{T} \int\limits_{-T/2}^{T/2} |g(t)|^2 \, dt = \lim_{T\to\infty} \frac{1}{T} \int\limits_{-T/2}^{T/2} |\hat{g}(t)|^2 \, dt \qquad (7.29)$$

■

Example 7.1 Determine the Hilbert transform of $g(t) = \cos(\omega_c t)$.

Solution
Consider the given signal $g(t) = \cos(\omega_c t)$. The Fourier transform of $g(t)$ is given by

$$G(\omega) = \pi[\delta(\omega - \omega_c) + \delta(\omega + \omega_c)]$$

We know that [Eq. (7.21)],

$$\hat{G}(\omega) = -j\,\text{sgn}(\omega)G(\omega)$$

$$= -j\,\text{sgn}(\omega)\pi[\delta(\omega - \omega_c) + \delta(\omega + \omega_c)]$$

$$\hat{G}(\omega) = \frac{\pi}{j}[\delta(\omega - \omega_c) + \delta(\omega + \omega_c)]\text{sgn}(\omega)$$

It is clear from Fig. 7.3 that the function $[\delta(\omega - \omega_c) + \delta(\omega + \omega_c)]\,\text{sgn}(\omega) = [\delta(\omega - \omega_c) - \delta(\omega + \omega_c)]$. Therefore,

$$\hat{G}(\omega) = \frac{\pi}{j}[\delta(\omega - \omega_c) - \delta(\omega + \omega_c)]$$

Taking the inverse Fourier transform of the above equation, we obtain

$$\hat{g}(t) = \sin(\omega_c t)$$

Therefore,

$$\text{if} \quad g(t) = \cos(\omega_c t), \quad \text{then} \quad \hat{g}(t) = \sin(\omega_c t) \qquad (7.30)$$

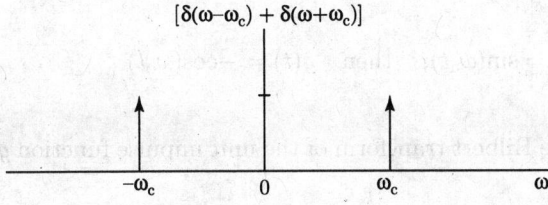

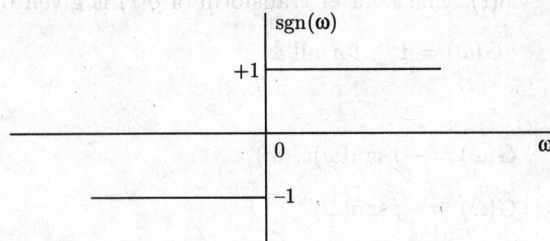

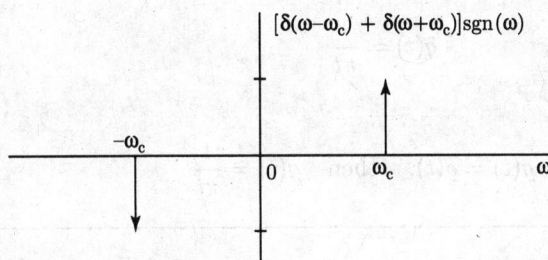

Fig. 7.3

Example 7.2 Determine the Hilbert transform of $g(t) = \sin(\omega_c t)$.

Solution

Consider the given signal $g(t) = \sin(\omega_c t)$. The Fourier transform of $g(t)$ is given by

$$G(\omega) = \frac{\pi}{j}[\delta(\omega - \omega_c) - \delta(\omega + \omega_c)]$$

We know that [Eq. (7.21)]

$$\hat{G}(\omega) = -j\,\text{sgn}(\omega)G(\omega)$$

$$= -j\,\text{sgn}(\omega)\frac{\pi}{j}[\delta(\omega - \omega_c) - \delta(\omega + \omega_c)]$$

$$\hat{G}(\omega) = -\pi[\delta(\omega - \omega_c) - \delta(\omega + \omega_c)]\,\text{sgn}(\omega)$$

$$\hat{G}(\omega) = -\pi[\delta(\omega - \omega_c) + \delta(\omega + \omega_c)]$$

Taking the inverse Fourier transform of the above equation, we obtain

$$\hat{g}(t) = -\cos(\omega_c t)$$

Therefore,

$$\text{if} \quad g(t) = \sin(\omega_c t), \quad \text{then} \quad \hat{g}(t) = -\cos(\omega_c t) \tag{7.31}$$

Example 7.3 Determine the Hilbert transform of the unit impulse function $g(t) = \delta(t)$.

Solution

Consider the given signal $g(t) = \delta(t)$. The Fourier transform of $g(t)$ is given by

$$G(\omega) = 1, \quad \text{for all } \omega$$

We know that [Eq. (7.21)]

$$\hat{G}(\omega) = -j \, \text{sgn}(\omega) G(\omega)$$

$$\hat{G}(\omega) = -j \, \text{sgn}(\omega)$$

Taking the inverse Fourier transform of the above equation, we obtain [Eq. (5.48)]

$$\hat{g}(t) = \frac{1}{\pi t}$$

Therefore,

$$\text{if} \quad g(t) = \delta(t), \quad \text{then} \quad \hat{g}(t) = \frac{1}{\pi t} \tag{7.32}$$

7.4 PRE-ENVELOPE OF CONTINUOUS-TIME SIGNALS

Let $g(t)$ be a real-valued signal. The *pre-envelope* of the signal $g(t)$ is a complex-valued function defined by

$$g_+(t) = g(t) + j\hat{g}(t) \tag{7.33}$$

where $\hat{g}(t)$ is the Hilbert transform of the signal $g(t)$. The pre-envelope is useful in handling bandpass signals and systems.

One of the important features of the pre-envelope $g_+(t)$ is the behaviour of its Fourier transform. Let $G_+(\omega)$ denote the Fourier transform of $g_+(t)$. Then, by taking the Fourier transform of the Eq. (7.33), we may write

$$G_+(\omega) = G(\omega) + j[-j \, \text{sgn}(\omega)]G(\omega)$$

$$= G(\omega) + \text{sgn}(\omega)G(\omega)$$

$$G_+(\omega) = G(\omega)[1 + \text{sgn}(\omega)]$$

$$G_+(\omega) = \begin{cases} 2G(\omega), & \omega > 0 \\ G(0), & \omega = 0 \\ 0, & \omega < 0 \end{cases} \tag{7.34}$$

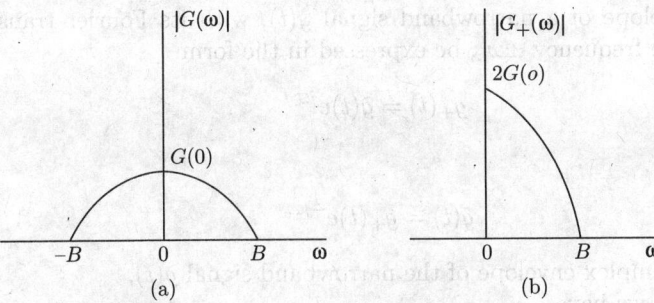

Fig. 7.4 (a) Magnitude spectrum of low-pass signal $g(t)$ and (b) magnitude spectrum of pre-envelope $g_+(t)$

where $G(0)$ is the value of $G(\omega)$ at $\omega = 0$. This means that the Fourier transform of pre-envelope vanishes for $\omega < 0$ as illustrated in Fig. 7.4 for the case of a low-pass signal. For the purpose of illustration in Fig. 7.4, we have used a low-pass signal with its spectrum limited to the band $-B \leq \omega \leq B$ and centred at the origin. Nevertheless, it should be emphasized that the pre-envelope can be defined for any signal, be it low-pass or bandpass, so long as it possesses a spectrum.

Equation (7.33) defines the pre-envelope $g_+(t)$ for positive frequencies; in a similar way, we may define the pre-envelope for negative frequencies as

$$g_-(t) = g(t) - j\hat{g}(t) \qquad (7.35)$$

The two pre-envelopes $g_+(t)$ and $g_-(t)$ are simply the complex conjugate of each other, as shown by

$$g_-(t) = g_+^*(t) \qquad (7.36)$$

The spectrum of the pre-envelope $g_+(t)$ is nonzero only for positive frequencies, hence the use of a plus sign as the subscript. In contrast, the spectrum of the other pre-envelope $g_-(t)$ is nonzero only for negative frequencies, as shown by the Fourier transform

$$G_-(\omega) = \begin{cases} 0, & \omega > 0 \\ G(0), & \omega = 0 \\ 2G(\omega), & \omega < 0 \end{cases} \qquad (7.37)$$

7.5 COMPLEX ENVELOPE AND BANDPASS SIGNALS

We say that a signal $g(t)$ is a bandpass signal if its Fourier transform $G(\omega)$ is non-negligible only in a band of frequencies of total extent $2B$, say, centred about some frequency $\pm\omega_c$. This is illustrated in Fig. 7.5(a). We refer to ω_c as a carrier frequency. In the majority of communication signals (modulated signals), we find that the bandwidth $2B$ is small compared with ω_c, and so we refer to such a signal as a *narrowband signal*.

Let the pre-envelope of a narrowband signal $g(t)$, with its Fourier transform $G(\omega)$ centred about some frequency $\pm\omega_c$, be expressed in the form

$$g_+(t) = \tilde{g}(t)e^{j\omega_c t} \tag{7.38}$$

or, equivalently,

$$\tilde{g}(t) = g_+(t)e^{-j\omega_c t} \tag{7.39}$$

where $\tilde{g}(t)$ is the complex envelope of the narrowband signal $g(t)$.

From Eq. (7.34), we have

$$G_+(\omega) = \begin{cases} 2G(\omega), & \omega > 0 \\ G(0), & \omega = 0 \\ 0, & \omega < 0 \end{cases}$$

The spectrum $G_+(\omega)$ is limited to the frequency band $(\omega_c - B) \le \omega \le (\omega_c + B)$, as illustrated in Fig. 7.5(b). Taking the Fourier transform of Eq. (7.39) and applying the frequency-shifting property, we obtain

$$\widetilde{G}(\omega) = G_+(\omega + \omega_c) \tag{7.40}$$

The spectrum $\widetilde{G}(\omega)$ of the complex envelope $\tilde{g}(t)$ is limited to the band $-B \le \omega \le B$ and centred at the origin as illustrated in Fig. 7.5(c). The complex envelope $\tilde{g}(t)$ of a

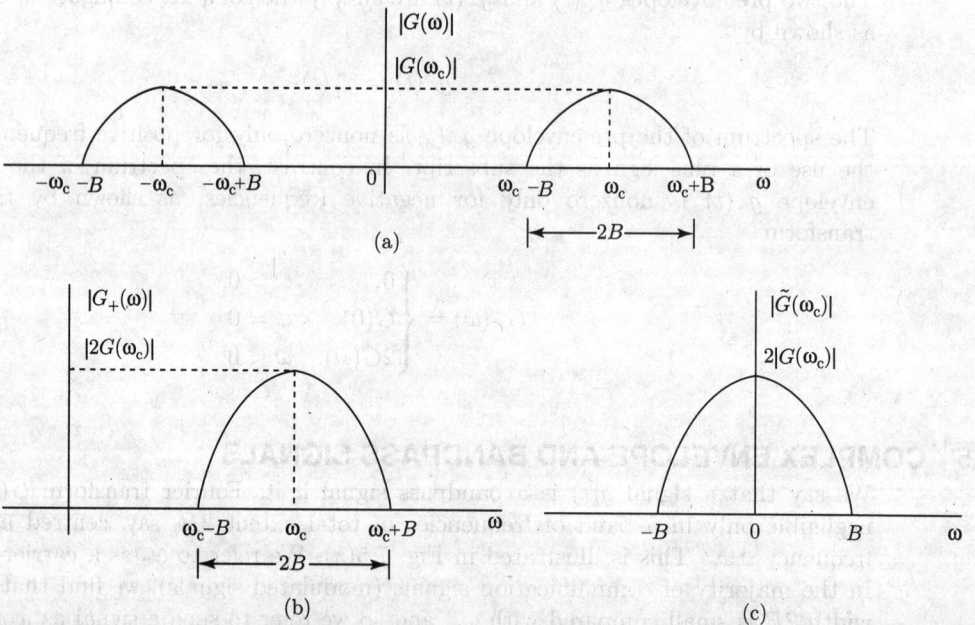

Fig. 7.5 Magnitude spectrum of (a) bandpass signal $g(t)$, (b) pre-envelope $q_+(t)$, and (c) complex envelope $\tilde{g}(t)$

bandpass signal $g(t)$ is a *low-pass signal*. From Eqs (7.33) and (7.38), we obtain

$$g(t) = \text{Re}[g_+(t)] \tag{7.41}$$

$$g(t) = \text{Re}[\tilde{g}(t)e^{j\omega_c t}] \tag{7.42}$$

In general, $\tilde{g}(t)$ is a complex-valued quantity; thus, we may express it in the form

$$\tilde{g}(t) = g_I(t) + jg_Q(t) \tag{7.43}$$

where $g_I(t)$ and $g_Q(t)$ are both real-valued low-pass functions. Substituting $\tilde{g}(t) = g_I(t) + jg_Q(t)$ into Eq. (7.42), we obtain

$$g(t) = Re\Big\{[g_I(t) + jg_Q(t)]\,[\cos(\omega_c t) + j\sin(\omega_c t)]\Big\}$$

$$g(t) = g_I(t)\cos(\omega_c t) - g_Q(t)\sin(\omega_c t) \tag{7.44}$$

Equation (7.44) represents the bandpass signal $g(t)$ in the *canonical form*. We refer to $g_I(t)$ as the *in-phase component* of the bandpass signal $g(t)$ and to $g_Q(t)$ as the *quadrature component* of the signal. Both $g_I(t)$ and $g_Q(t)$ are low-pass signals limited to the band $-B \le \omega \le B$. Hence, they may be derived from the bandpass signal $g(t)$.

Using Eqs (7.39) and (7.43), we obtain

$$g_I(t) = \text{Re}\{\tilde{g}(t)\}$$

$$g_I(t) = \text{Re}\Big\{g_+(t)\,e^{-j\omega_c t}\Big\}$$

Substituting $g_+(t) = g(t) + j\hat{g}(t)$ into the above equation, we obtain

$$g_I(t) = \text{Re}\Big\{[g(t) + j\hat{g}(t)]\,[\cos(\omega_c t) - j\sin(\omega_c t)]\Big\}$$

$$g_I(t) = g(t)\cos(\omega_c t) + \hat{g}(t)\sin(\omega_c t) \tag{7.45}$$

Equation (7.45) represents the in-phase component $g_I(t)$ in terms of the bandpass signal $g(t)$ and its Hilbert transform $\hat{g}(t)$.

Similarly, we may obtain the quadrature component $g_Q(t)$ in terms of the bandpass signal $g(t)$ and its Hilbert transform $\hat{g}(t)$. Using Eqs (7.39) and (7.43), we obtain

$$g_Q(t) = \text{Im}\{\tilde{g}(t)\}$$

$$g_Q(t) = \text{Im}\Big\{g_+(t)\,e^{-j\omega_c t}\Big\}$$

Substituting $g_+(t) = g(t) + j\hat{g}(t)$ into the above equation, we obtain

$$g_Q(t) = \text{Re}\Big\{[g(t) + j\hat{g}(t)][\cos(\omega_c t) - j\sin(\omega_c t)]\Big\}$$

$$g_Q(t) = \hat{g}(t)\cos(\omega_c t) - g(t)\sin(\omega_c t) \tag{7.46}$$

Equation (7.43) is the Cartesian form of expressing the complex envelope $\hat{g}(t)$. Alternatively, we may express it in the polar form

$$\tilde{g}(t) = a(t)\,e^{j\phi(t)} \tag{7.47}$$

where $a(t) = \sqrt{g_I^2(t) + g_Q^2(t)}$ and $\phi(t) = \tan^{-1}(g_Q(t)/g_I(t))$ are both real-valued low-pass functions. Based on this polar representation, the original bandpass signal $g(t)$ is defined by

$$g(t) = \text{Re}\{g_+(t)\}$$
$$= \text{Re}\{\tilde{g}(t)\, e^{j\omega_c t}\}$$
$$= \text{Re}\{a(t)e^{j\phi(t)}\, e^{j\omega_c t}\}$$
$$g(t) = \text{Re}\{a(t)\, e^{[j\omega_c t + \phi(t)]}\}$$
$$g(t) = a(t)\cos(\omega_c t + \phi(t))$$

where $a(t)$ is the *natural envelope* or simply the *envelope* of the bandpass signal $g(t)$ and $\phi(t)$ is the *phase* of the signal.

The envelope $a(t)$ equals the magnitude of the complex envelope $\tilde{g}(t)$ and also that of the pre-envelope $g_+(t)$, as shown by

$$a(t) = |\tilde{g}(t)| = |g_+(t)| \tag{7.48}$$

7.6 DISCRETE-TIME HILBERT TRANSFORM

In Section 6.9.1, we saw that the constraint of *causality* of a sequence implies unique relationships between the real and imaginary parts of the DTFT. Relationships of this type between the real and imaginary parts of complex functions are known as Hilbert transform relationships. Once again, to illustrate the relationships between the real and imaginary part of the Fourier transform, we decompose $x(n)$ into its even and odd parts, i.e.,

$$x(n) = x_e(n) + x_o(n)$$

where

$$x_e(n) = \frac{1}{2}[x(n) + x(-n)]$$

and

$$x_o(n) = \frac{1}{2}[x(n) - x(-n)]$$

Now, if $x(n)$ is causal, it is possible to recover $x(n)$ from its even part $x_e(n)$ for $0 \leq n \leq \infty$ or from its odd part $x_o(n)$ for $1 \leq n \leq \infty$. Indeed, it can be easily seen that

$$x(n) = 2x_e(n)u(n) - x_e(0)\delta(n), \quad n \geq 0 \tag{7.49}$$

and

$$x(n) = 2x_o(n)u(n) + x(0)\delta(n), \quad n \geq 1 \tag{7.50}$$

Since $x_o(n) = 0$ for $n = 0$, we cannot recover $x(0)$ from $x_o(n)$ and hence we must also know $x(0)$. In any case, it is apparent that $x_o(n) = x_e(n)$ for $n \geq 1$, so there is a strong relationship between $x_o(n)$ and $x_e(n)$.

If $x(n)$ is absolutely summable, then its DTFT $X(e^{j\omega})$ exists, and

$$X(e^{j\omega}) = X_{\mathrm{R}}(e^{j\omega}) + jX_{\mathrm{I}}(e^{j\omega})$$

In addition, if $x(n)$ is real valued and causal, the conjugate property of the Fourier transform implies that

$$x_{\mathrm{e}}(n) \longleftrightarrow X_{\mathrm{R}}(e^{j\omega}) \quad \text{and} \quad x_{\mathrm{o}}(n) \longleftrightarrow jX_{\mathrm{I}}(e^{j\omega})$$

Since $x(n)$ is completely specified by $x_{\mathrm{e}}(n)$, it follows that $X(e^{j\omega})$ is completely determined if we know $X_{\mathrm{R}}(e^{j\omega})$. Alternatively, $X(e^{j\omega})$ is completely determined from $X_{\mathrm{I}}(e^{j\omega})$. In short, $X_{\mathrm{R}}(e^{j\omega})$ and $X_{\mathrm{I}}(e^{j\omega})$ are interdependent and cannot be specified independently if the system is causal. Equivalently, the magnitude and phase responses of a causal filter are interdependent and hence cannot be specified independently.

Taking the Fourier transform of Eq. (7.49), we have

$$\mathcal{F}[x(n)] = 2\mathcal{F}[x_{\mathrm{e}}(n)u(n)] - x_{\mathrm{e}}(0)\mathcal{F}[\delta(n)]$$

$$X(e^{j\omega}) = X_{\mathrm{R}}(e^{j\omega}) + jX_{\mathrm{I}}(e^{j\omega}) = 2[X_{\mathrm{R}}(e^{j\omega}) * U(e^{j\omega})] - x_{\mathrm{e}}(0)$$

$$X_{\mathrm{R}}(e^{j\omega}) + jX_{\mathrm{I}}(e^{j\omega}) = 2\frac{1}{2\pi}\int\limits_{-\pi}^{\pi} X_{\mathrm{R}}(e^{j\theta})U(e^{j(\omega-\theta)})\, d\theta - x_{\mathrm{e}}(0)$$

$$X_{\mathrm{R}}(e^{j\omega}) + jX_{\mathrm{I}}(e^{j\omega}) = \frac{1}{\pi}\int\limits_{-\pi}^{\pi} X_{\mathrm{R}}(e^{j\theta})U(e^{j(\omega-\theta)})\, d\theta - x_{\mathrm{e}}(0) \qquad (7.51)$$

where $U(e^{j\omega})$ is the DTFT of the unit step sequence $u(n)$. The DTFT of $u(n)$ is given by

$$U(e^{j\omega}) = \pi\delta(\omega) + \frac{1}{1 - e^{-j\omega}}, \qquad -\pi < \omega < \pi$$

$$U(e^{j\omega}) = \pi\delta(\omega) + \frac{1}{2} - j\frac{1}{2}\cot\left(\frac{\omega}{2}\right) \qquad (7.52)$$

Substituting Eq. (7.52) into Eq. (7.51), we obtain

$$X_{\mathrm{R}}(e^{j\omega}) + jX_{\mathrm{I}}(e^{j\omega}) = \frac{1}{\pi}\int\limits_{-\pi}^{\pi} X_{\mathrm{R}}(e^{j\theta})\left[\pi\delta(\omega-\theta) + \frac{1}{2} - j\frac{1}{2}\cot\left(\frac{\omega-\theta}{2}\right)\right] d\theta - x_{\mathrm{e}}(0)$$

$$= \frac{1}{\pi}\int\limits_{-\pi}^{\pi} X_{\mathrm{R}}(e^{j\theta})\pi\delta(\omega-\theta)\, d\theta + \frac{1}{\pi}\int\limits_{-\pi}^{\pi} X_{\mathrm{R}}(e^{j\theta})\frac{1}{2}\, d\theta$$

$$- \frac{1}{\pi}\int\limits_{-\pi}^{\pi} X_{\mathrm{R}}(e^{j\theta})j\frac{1}{2}\cot\left(\frac{\omega-\theta}{2}\right) d\theta - x_{\mathrm{e}}(0)$$

$$= \int\limits_{-\pi}^{\pi} X_R(e^{j\theta}) \delta(\omega - \theta) d\theta + \frac{1}{2\pi} \int\limits_{-\pi}^{\pi} X_R(e^{j\theta}) \, d\theta$$

$$- j\frac{1}{2\pi} \int\limits_{-\pi}^{\pi} X_R(e^{j\theta}) \cot\left(\frac{\omega - \theta}{2}\right) d\theta - x_e(0)$$

$$X_R(e^{j\omega}) + jX_I(e^{j\omega}) = X_R(e^{j\theta})\Big|_{\theta=\omega} + x_e(0) - j\frac{1}{2\pi} \int\limits_{-\pi}^{\pi} X_R(e^{j\theta}) \cot\left(\frac{\omega - \theta}{2}\right) d\theta - x_e(0)$$

$$jX_I(e^{j\omega}) = -j\frac{1}{2\pi} \int\limits_{-\pi}^{\pi} X_R(e^{j\theta}) \cot\left(\frac{\omega - \theta}{2}\right) d\theta$$

$$X_I(e^{j\omega}) = -\frac{1}{2\pi} \int\limits_{-\pi}^{\pi} X_R(e^{j\theta}) \cot\left(\frac{\omega - \theta}{2}\right) d\theta \tag{7.53}$$

Similarly, taking the Fourier transform of Eq. (7.50), we obtain

$$\mathcal{F}[x(n)] = 2\mathcal{F}[x_o(n)u(n)] + x(0)\mathcal{F}[\delta(n)]$$

$$X(e^{j\omega}) = X_R(e^{j\omega}) + jX_I(e^{j\omega}) = 2[jX_I(e^{j\omega}) * U(e^{j\omega})] + x(0)$$

$$X_R(e^{j\omega}) + jX_I(e^{j\omega}) = 2j\frac{1}{2\pi} \int\limits_{-\pi}^{\pi} X_I(e^{j\theta})U(e^{j(\omega-\theta)}) \, d\theta + x(0)$$

$$X_R(e^{j\omega}) + jX_I(e^{j\omega}) = j\frac{1}{\pi} \int\limits_{-\pi}^{\pi} X_I(e^{j\theta})U(e^{j(\omega-\theta)}) \, d\theta + x(0) \tag{7.54}$$

where $U(e^{j\omega})$ is the DTFT of the unit step sequence $u(n)$. Substituting $U(e^{j\omega})$ from Eq. (7.52) into Eq. (7.54), we obtain

$$X_R(e^{j\omega}) + jX_I(e^{j\omega}) = j\frac{1}{\pi} \int\limits_{-\pi}^{\pi} X_I(e^{j\theta}) \left[\pi\delta(\omega - \theta) + \frac{1}{2} - j\frac{1}{2}\cot\left(\frac{\omega - \theta}{2}\right)\right] d\theta + x(0)$$

$$= j\frac{1}{\pi} \int\limits_{-\pi}^{\pi} X_I(e^{j\theta})\pi\delta(\omega - \theta) \, d\theta + j\frac{1}{\pi} \int\limits_{-\pi}^{\pi} X_I(e^{j\theta})\frac{1}{2} d\theta$$

$$- j\frac{1}{\pi} \int\limits_{-\pi}^{\pi} X_I(e^{j\theta})j\frac{1}{2}\cot\left(\frac{\omega - \theta}{2}\right) d\theta + x(0)$$

$$= j \int\limits_{-\pi}^{\pi} X_I(e^{j\theta}) \delta(\omega - \theta)\, d\theta + j\frac{1}{2\pi} \int\limits_{-\pi}^{\pi} X_I(e^{j\theta}) d\theta$$

$$+ \frac{1}{2\pi} \int\limits_{-\pi}^{\pi} X_I(e^{j\theta}) \cot\left(\frac{\omega - \theta}{2}\right) d\theta + x(0)$$

Since $X_I(e^{j\omega})$ is an odd function, we have $\int\limits_{-\pi}^{\pi} X_I(e^{j\theta}) d\theta = 0$ therefore,

$$X_R(e^{j\omega}) + jX_I(e^{j\omega}) = jX_I(e^{j\theta})\Big|_{\theta=\omega} + 0 + \frac{1}{2\pi} \int\limits_{-\pi}^{\pi} X_I(e^{j\theta}) \cot\left(\frac{\omega - \theta}{2}\right) d\theta + x(0)$$

$$X_R(e^{j\omega}) = \frac{1}{2\pi} \int\limits_{-\pi}^{\pi} X_I(e^{j\theta}) \cot\left(\frac{\omega - \theta}{2}\right) d\theta + x(0)$$

$$X_R(e^{j\omega}) = \frac{1}{2\pi} \int\limits_{-\pi}^{\pi} X_I(e^{j\theta}) \cot\left(\frac{\omega - \theta}{2}\right) d\theta + x(0) \tag{7.55}$$

Equations (7.53) and (7.55), which are known as the Hilbert transform pair, hold for real and imaginary parts of a causal, stable, and real sequence.

7.6.1 Hilbert Transform Relations for Complex Sequences

Thus far, we have considered Hilbert transform relations for the Fourier transforms of causal sequences. In this section, we consider *complex sequences* for which the real and imaginary components can be related through a discrete convolution similar to the Hilbert transform relations derived in the previous section.

As mentioned previously, it is possible to base the derivation of the Hilbert transform relations on a notion of causality. Since we are interested in relating the real and imaginary parts of a complex sequence, causality will be applied to the Fourier transform of the sequence. Thus, with $x(n)$ denoting the sequence and $X(e^{j\omega})$ its Fourier transform, we require that

$$X(e^{j\omega}) = 0, \quad -\pi \le \omega < 0 \tag{7.56}$$

(We could just as well assume that $X(e^{j\omega}) = 0$ for $0 < \omega \le \pi$.) The sequence $x(n)$ corresponding to $X(e^{j\omega})$ must be complex, since, if $x(n)$ were real, $X(e^{j\omega})$ would be conjugate symmetric, i.e., $X(e^{j\omega}) = X^*(e^{-j\omega})$. Therefore, we express $x(n)$ as

$$x(n) = x_r(n) + jx_i(n) \tag{7.57}$$

where $x_r(n)$ and $x_i(n)$ are real signals.

If $X_r(e^{j\omega})$ and $X_i(e^{j\omega})$ denote the Fourier transforms of the real signals $x_r(n)$ and $x_i(n)$, respectively, then

$$X(e^{j\omega}) = X_r(e^{j\omega}) + jX_i(e^{j\omega}) \tag{7.58}$$

and it follows that

$$X_r(e^{j\omega}) = \frac{1}{2}[X(e^{j\omega}) + X^*(e^{-j\omega})] \tag{7.59}$$

and

$$jX_i(e^{j\omega}) = \frac{1}{2}[X(e^{j\omega}) - X^*(e^{-j\omega})] \tag{7.60}$$

Note that $X_r(e^{j\omega})$ and $X_i(e^{j\omega})$ are both complex-valued functions in general, and the complex transforms $X_r(e^{j\omega})$ and $jX_i(e^{j\omega})$ play a role similar to that played in the previous section by the even and odd parts, respectively, of a causal sequence. However, $X_r(e^{j\omega})$ is conjugate symmetric and $jX_i(e^{j\omega})$ is conjugate antisymmetric, i.e.,

$$X_r(e^{j\omega}) = X_r^*(e^{-j\omega}) \quad \text{and} \quad jX_i(e^{j\omega}) = -jX_i^*(e^{-j\omega}) \tag{7.61}$$

If $X(e^{j\omega}) = 0$ for $-\pi \leq \omega < 0$, then there is no overlap between the portions of $X(e^{j\omega})$ and $X^*(e^{-j\omega})$. Thus, $X(e^{j\omega})$ can be recovered from either $X_r(e^{j\omega})$ or $X_i(e^{j\omega})$.

In particular,

$$X(e^{j\omega}) = \begin{cases} 2X_r(e^{j\omega}), & 0 \leq \omega < \pi \\ 0, & -\pi \leq \omega < 0 \end{cases}, \quad X^*(e^{-j\omega}) = \begin{cases} 0, & 0 \leq \omega < \pi \\ 2X_r(e^{j\omega}), & -\pi \leq \omega < 0 \end{cases} \tag{7.62}$$

and

$$X(e^{j\omega}) = \begin{cases} 2jX_i(e^{j\omega}), & 0 \leq \omega < \pi \\ 0, & -\pi \leq \omega < 0 \end{cases}, \quad X^*(e^{-j\omega}) = \begin{cases} 0, & 0 \leq \omega < \pi \\ -2jX_i(e^{j\omega}), & -\pi \leq \omega < 0 \end{cases} \tag{7.63}$$

Alternatively, from Eqs (7.62) and (7.63), we can relate $X_r(e^{j\omega})$ and $X_i(e^{j\omega})$ directly by

$$X_i(e^{j\omega}) = \begin{cases} -jX_r(e^{j\omega}), & 0 < \omega < \pi \\ jX_r(e^{j\omega}), & -\pi \leq \omega < 0 \end{cases} \tag{7.64}$$

or

$$X_i(e^{j\omega}) = H(e^{j\omega})X_r(e^{j\omega}) \tag{7.65}$$

where

$$H(e^{j\omega}) = -j\,\text{sgn}(\omega) = \begin{cases} -j = 1e^{-j(\pi/2)}, & 0 < \omega < \pi \\ j = 1e^{j(\pi/2)}, & -\pi < \omega < 0 \end{cases} \tag{7.66}$$

Equation (7.66) represents the frequency response of the Hilbert transformer and $x_i(n)$ is the Hilbert transform of $x_r(n)$.

Thus, according to Eq. (7.65), $x_i(n)$ can be obtained by processing $x_r(n)$ with a discrete-time LTI system with frequency response $H(e^{j\omega})$ as given by Eq. (7.66). This frequency response has unity magnitude $|H(e^{j\omega})| = 1$, a phase angle of $-\pi/2$ for $0 \leq \omega \leq \pi$, and a phase angle of $+\pi/2$ for $-\pi \leq \omega < 0$, as shown in Fig. 7.6. Such a system is called an ideal 90° phase shifter or a *Hilbert transformer*. From Eq. (7.65) it follows that

$$X_r(e^{j\omega}) = \frac{1}{H(e^{j\omega})}X_i(e^{j\omega}) = -H(e^{j\omega})X_i(e^{j\omega}) \tag{7.67}$$

Thus, $-x_r(n)$ can also be obtained from $x_i(n)$ with a 90° phase shifter. This is known as *inverse Hilbert transformer*.

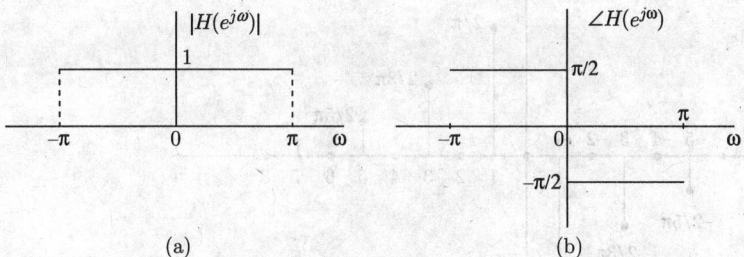

Fig. 7.6 (a) Magnitude response and (b) phase response of the Hilbert transformer

The impulse response $h(n)$ of the Hilbert transformer corresponding to the frequency response $H(e^{j\omega})$ given in Eq. (7.66) is

$$h(n) = \frac{1}{2\pi} \int_{-\pi}^{\pi} H(e^{j\omega}) e^{j\omega n} d\omega$$

$$= \frac{1}{2\pi} \int_{-\pi}^{0} j\, e^{j\omega n} d\omega + \frac{1}{2\pi} \int_{0}^{\pi} -j\, e^{j\omega n} d\omega$$

$$= \frac{j}{2\pi} \int_{-\pi}^{0} e^{j\omega n} d\omega - \frac{j}{2\pi} \int_{0}^{\pi} e^{j\omega n} d\omega$$

$$= \frac{j}{2\pi} \left[\left. \frac{e^{j\omega n}}{jn} \right|_{-\pi}^{0} - \left. \frac{e^{j\omega n}}{jn} \right|_{0}^{\pi} \right]$$

$$= \frac{1}{2n\pi} [1 - e^{-jn\pi} - e^{jn\pi} + 1]$$

$$h(n) = \frac{1}{2n\pi} \left[2 - 2\left(\frac{e^{jn\pi} + e^{-jn\pi}}{2} \right) \right]$$

$$h(n) = \frac{1}{2n\pi} [2 - 2\cos(n\pi)], \qquad n \neq 0$$

$$h(n) = \frac{1}{n\pi} [1 - \cos(n\pi)], \qquad n \neq 0$$

$$h(n) = \frac{2}{n\pi} \sin^2\left(\frac{n\pi}{2} \right), \qquad n \neq 0$$

Therefore, the impulse response of the Hilbert transformer is given by

$$h(n) = \begin{cases} \frac{2}{n\pi}\sin^2\left(\frac{n\pi}{2}\right), & n \neq 0 \\ 0, & n = 0 \end{cases} \qquad (7.68)$$

The impulse response is antisymmetric in nature as shown in Fig. 7.7.

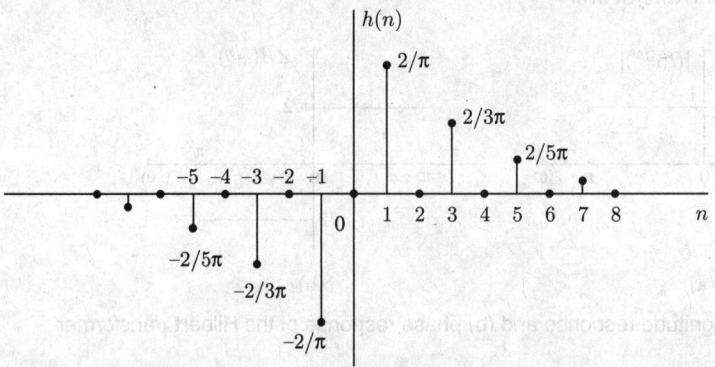

Fig. 7.7   Impulse response of the Hilbert transformer

Summary: Consider a real sequence $g(n)$ with Fourier transform $G(e^{j\omega})$. The Hilbert transform of $g(n)$, which we shall denote by $\hat{g}(n) = HT[g(n)]$, can be obtained by passing the sequence $g(n)$ through a discrete-time LTI system with impulse response $h(n)$ given in Eq. (7.68), i.e.,

$$\hat{g}(n) = g(n) * h(n)$$

$$\hat{g}(n) = \sum_{m=-\infty}^{\infty} g(m)h(n-m)$$

Using the convolution property of the Fourier transform, we obtain

$$\mathcal{F}[\hat{g}(n)] = \mathcal{F}[g(n)]\mathcal{F}[h(n)]$$

$$\hat{G}(e^{j\omega}) = G(e^{j\omega})H(e^{j\omega})$$

$$\hat{G}(e^{j\omega}) = -j\,\mathrm{sgn}(\omega)G(e^{j\omega}) \tag{7.69}$$

The *inverse Hilbert transform*, by means of which the original signal $g(n)$ or $G(e^{j\omega})$ is recovered from $\hat{g}(n)$ or $\hat{G}(e^{j\omega})$, is defined by [Eq. (7.67)]

$$G(e^{j\omega}) = -H(e^{j\omega})\hat{G}(e^{j\omega}) \tag{7.70}$$

$$G(e^{j\omega}) = j\,\mathrm{sgn}(\omega)\hat{G}(e^{j\omega}) \tag{7.71}$$

7.7 PROPERTIES OF DISCRETE-TIME HILBERT TRANSFORM

The Hilbert transform differs from the Fourier transform in that it operates exclusively in the time domain. It has a number of properties, some of which are listed below. The signal $g(n)$ is assumed to be real valued, which is the usual domain of application of the Hilbert transform.

1. A signal $g(n)$ and its Hilbert transform $\hat{g}(n)$ have the same magnitude spectrum.

$$|\hat{G}(e^{j\omega})| = |G(e^{j\omega})| \tag{7.72}$$

Proof Given that

$$\hat{g}(n) = HT[g(n)]$$

From Eq. (7.69), we have

$$\hat{G}(e^{j\omega}) = -j\,\text{sgn}(\omega)G(e^{j\omega})$$

$$|\hat{G}(e^{j\omega})| = |-j\,\text{sgn}(\omega)G(e^{j\omega})|$$

$$|\hat{G}(e^{j\omega})| = |-j\,\text{sgn}(\omega)||G(e^{j\omega})|$$

Since $|-j\,\text{sgn}(\omega)| = 1$ for all ω, we have

$$|\hat{G}(e^{j\omega})| = |G(e^{j\omega})|$$

As corollaries to this property, we may state that

(a) if a signal $g(n)$ is band limited, then its Hilbert transform $\hat{g}(n)$ is also band limited.
(b) the signal $g(n)$ and its Hilbert transform $\hat{g}(n)$ have the same energy spectral density if $g(n)$ is an energy signal, and the same power spectral density if it is a power signal.
(c) the signal $g(n)$ and its Hilbert transform $\hat{g}(n)$ have the same autocorrelation function.

2. If $\hat{g}(n)$ is the Hilbert transform of $g(n)$, then the Hilbert transform of $\hat{g}(n)$ is $-g(n)$, i.e., if

$$\hat{g}(n) = HT[g(n)]$$

then

$$-g(n) = HT[\hat{g}(n)] \qquad (7.73)$$

Proof We know that

$$\hat{G}(e^{j\omega}) = G(e^{j\omega})H(e^{j\omega})$$

From Eq. (7.66), we have

$$H(e^{j\omega}) = -j\,\text{sgn}(\omega) = \begin{cases} -j, & 0 < \omega < \pi \\ j, & -\pi < \omega < 0 \end{cases}$$

Therefore,

$$\frac{1}{H(e^{j\omega})} = -H(e^{j\omega}) = j\,\text{sgn}(\omega) = \begin{cases} j, & 0 < \omega < \pi \\ -j, & -\pi < \omega < 0 \end{cases}$$

We know that

$$\hat{G}(e^{j\omega}) = G(e^{j\omega})H(e^{j\omega})$$

$$G(e^{j\omega}) = \frac{1}{H(e^{j\omega})}\hat{G}(e^{j\omega})$$

$$G(e^{j\omega}) = -H(e^{j\omega})\hat{G}(e^{j\omega})$$

$$-G(e^{j\omega}) = H(e^{j\omega})\hat{G}(e^{j\omega})$$

Taking the inverse DTFT, we obtain

$$-g(n) = HT[\hat{g}(n)]$$

3. If $g(n)$ is an energy signal, then

$$\sum_{n=-\infty}^{\infty} g(n)\hat{g}(n) = 0 \qquad (7.74)$$

Proof For a real signal $g(n)$ multiplied by its Hilbert transform $\hat{g}(n)$ we may write

$$\sum_{n=-\infty}^{\infty} g(n)\hat{g}(n) = \frac{1}{2\pi} \int_{-\pi}^{\pi} G(e^{j\omega})\hat{G}^*(e^{j\omega})\, d\omega$$

$$= \frac{1}{2\pi} \int_{-\pi}^{\pi} G(e^{j\omega})\hat{G}(e^{-j\omega})\, d\omega$$

$$\sum_{n=-\infty}^{\infty} g(n)\hat{g}(n) = \frac{1}{2\pi} \int_{-\pi}^{\pi} G(e^{j\omega})[-j\,\text{sgn}(-\omega)G(e^{-j\omega})]\, d\omega$$

Since $\text{sgn}(-\omega) = -\text{sgn}(\omega)$ and $G(e^{-j\omega}) = G^*(e^{j\omega})$, we have

$$\sum_{n=-\infty}^{\infty} g(n)\hat{g}(n) = \frac{1}{2\pi} \int_{-\pi}^{\pi} G(e^{j\omega}))[j\,\text{sgn}(\omega)G^*(e^{j\omega})]\, d\omega$$

$$= \frac{j}{2\pi} \int_{-\pi}^{\pi} \text{sgn}(\omega)G(e^{j\omega})G^*(e^{j\omega})\, d\omega$$

$$\sum_{n=-\infty}^{\infty} g(n)\hat{g}(n) = \frac{j}{2\pi} \int_{-\pi}^{\pi} \text{sgn}(\omega)|G(e^{j\omega})|^2\, d\omega$$

The integrand in the RHS of the above equation is an odd function of ω, being the product of the odd function $\text{sgn}(\omega)$ and the even function $|G(e^{j\omega})|^2$.

Hence, the integral is zero [see Eq. (1.67)], yielding the final result

$$\sum_{n=-\infty}^{\infty} g(n)\hat{g}(n) = 0$$

Similarly, for a power signal $g(n)$,

$$\lim_{N \to \infty} \frac{1}{2N+1} \sum_{n=-N}^{N} g(n)\hat{g}(n) = 0 \tag{7.75}$$

∎

4. An energy signal $g(n)$ and its Hilbert transform $\hat{g}(n)$ have the same energy.

$$E_g = \sum_{n=-\infty}^{\infty} |g(n)|^2 = \sum_{n=-\infty}^{\infty} |\hat{g}(n)|^2 \tag{7.76}$$

Proof Using the Parseval's theorem of energy signals, we have

$$\sum_{n=-\infty}^{\infty} |g(n)|^2 = \frac{1}{2\pi} \int_{-\pi}^{\pi} |G(e^{j\omega})|^2 \, d\omega$$

Similarly, we have

$$\sum_{n=-\infty}^{\infty} |\hat{g}(n)|^2 = \frac{1}{2\pi} \int_{-\pi}^{\pi} |\hat{G}(e^{j\omega})|^2 \, d\omega$$

$$= \frac{1}{2\pi} \int_{-\pi}^{\pi} |-j \, \text{sgn}(\omega) G(e^{j\omega})|^2 \, d\omega$$

$$\sum_{n=-\infty}^{\infty} |\hat{g}(n)|^2 = \frac{1}{2\pi} \int_{-\pi}^{\pi} |-j \, \text{sgn}(\omega)|^2 |G(e^{j\omega})|^2 \, d\omega$$

Since $|-j \, \text{sgn}(\omega)|^2 = 1$ for all ω, we obtain

$$\sum_{n=-\infty}^{\infty} |\hat{g}(n)|^2 = \frac{1}{2\pi} \int_{-\pi}^{\pi} |G(e^{j\omega})|^2 d\omega$$

$$\sum_{n=-\infty}^{\infty} |\hat{g}(n)|^2 = \int_{-\pi}^{\pi} |g(n)|^2$$

Similarly, a power signal $g(n)$ and its Hilbert transform $\hat{g}(n)$ have the same power.

$$\lim_{N \to \infty} \frac{1}{2N+1} \sum_{n=-N}^{N} |g(n)|^2 = \lim_{N \to \infty} \frac{1}{2N+1} \sum_{n=-N}^{N} |\hat{g}(n)|^2 \qquad (7.77)$$

7.8   PRE-ENVELOPE OF DISCRETE-TIME SIGNALS

Let $g(n)$ be a real-valued signal. The *pre-envelope* of the signal $g(n)$ is a complex-valued function defined by

$$g_+(n) = g(n) + j\hat{g}(n) \qquad (7.78)$$

where $\hat{g}(n)$ is the Hilbert transform of the signal $g(n)$. The pre-envelope is useful in handling bandpass signals and systems.

One of the important features of the pre-envelope $g_+(n)$ is the behaviour of its Fourier transform. Let $G_+(e^{j\omega})$ denote the Fourier transform of $g_+(n)$. Then, by taking the Fourier transform of Eq. (7.78), we may write

$$G_+(e^{j\omega}) = G(e^{j\omega}) + j[-j\,\text{sgn}(\omega)]G(e^{j\omega})$$

$$= G(e^{j\omega}) + \text{sgn}(\omega)G(e^{j\omega})$$

$$G_+(e^{j\omega}) = G(e^{j\omega})[1 + \text{sgn}(\omega)]$$

$$G_+(e^{j\omega}) = \begin{cases} 2G(e^{j\omega}), & 0 < \omega < \pi \\ G(e^{j0}), & \omega = 0 \\ 0, & -\pi < \omega < 0 \end{cases} \qquad (7.79)$$

where $G(e^{j0})$ is the value of $G(e^{j\omega})$ at $\omega = 0$. This means that the Fourier transform of pre-envelope vanishes for $-\pi < \omega < 0$ as illustrated in Fig. 7.8 for the case of a low-pass sequence. For the purpose of illustration in Fig. 7.8, we have used a low-pass sequence with its spectrum limited to the band $-B \le \omega \le B$ and centred at the origin. Nevertheless, it should be emphasized that the pre-envelope can be defined for any sequence, be it low-pass or bandpass, so long as it possesses a spectrum.

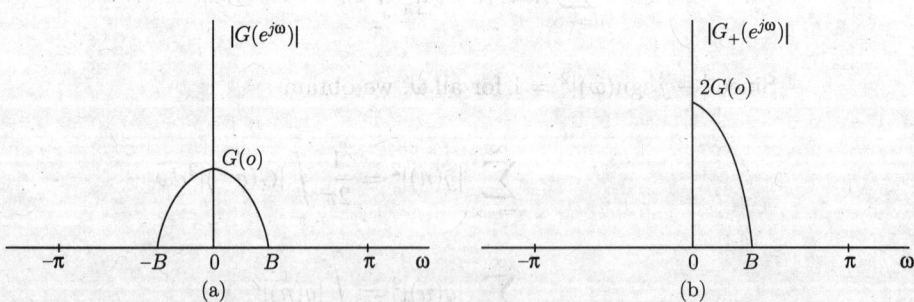

Fig. 7.8   Magnitude spectrum of (a) low-pass signal $g(n)$ and (b) pre-envelope $g_+(n)$

Equation (7.78) defines the pre-envelope $g_+(n)$ for positive frequencies; in a similar way, we may define the pre-envelope for negative frequencies as

$$g_-(n) = g(n) - j\hat{g}(n) \tag{7.80}$$

The two pre-envelopes $g_+(n)$ and $g_-(n)$ are simply the complex conjugate of each other, as shown by

$$g_-(n) = g_+^*(n) \tag{7.81}$$

The spectrum of the pre-envelope $g_+(n)$ is nonzero only for positive frequencies, hence the use of a plus sign as the subscript. In contrast, the spectrum of the other pre-envelope $g_-(n)$ is nonzero only for negative frequencies, as shown by the Fourier transform

$$G_-(e^{j\omega}) = \begin{cases} 0, & 0 < \omega < \pi \\ G(e^{j0}), & \omega = 0 \\ 2G(e^{j\omega}), & -\pi < \omega < 0 \end{cases} \tag{7.82}$$

7.9 COMPLEX ENVELOPE AND BANDPASS SIGNALS

Many of the applications of pre-envelope (analytic signals) concern narrowband communication. In such applications, it is sometimes convenient to represent a bandpass signal in terms of a low-pass signal.

Let the pre-envelope of a narrowband signal $g(n)$, with its Fourier transform $G(e^{j\omega})$ centred about some frequency $\pm\omega_c$, be expressed in the form

$$g_+(n) = \tilde{g}(n)\, e^{j\omega_c n} \tag{7.83}$$

or, equivalently,

$$\tilde{g}(n) = g_+(n)\, e^{-j\omega_c n} \tag{7.84}$$

where $\tilde{g}(n)$ is the complex envelope of the narrowband signal $g(n)$.

From Eq. (7.79), we have

$$G_+(e^{j\omega}) = \begin{cases} 2G(e^{j\omega}), & 0 < \omega < \pi \\ G(e^{j0}), & \omega = 0 \\ 0, & -\pi < \omega < 0 \end{cases}$$

The spectrum $G_+(e^{j\omega})$ is limited to the frequency band $(\omega_c - B) \le \omega \le (\omega_c + B)$, as illustrated in Fig. 7.9(b). Taking the Fourier transform of Eq. (7.84) and applying the frequency-shifting property, we obtain

$$\tilde{G}(e^{j\omega}) = G_+(e^{j(\omega+\omega_c)}) \tag{7.85}$$

The spectrum $\tilde{G}(e^{j\omega})$ of the complex envelope $\tilde{g}(n)$ is limited to the band $-B \le \omega \le B$ and centred at the origin as illustrated in Fig. 7.9(c). The complex envelope $\tilde{g}(n)$ of a bandpass signal $g(n)$ is a *low-pass signal*. From Eqs (7.78) and (7.83), we obtain

$$g(n) = \text{Re}[g_+(n)] \tag{7.86}$$

$$g(n) = \text{Re}[\tilde{g}(n)e^{j\omega_c n}] \tag{7.87}$$

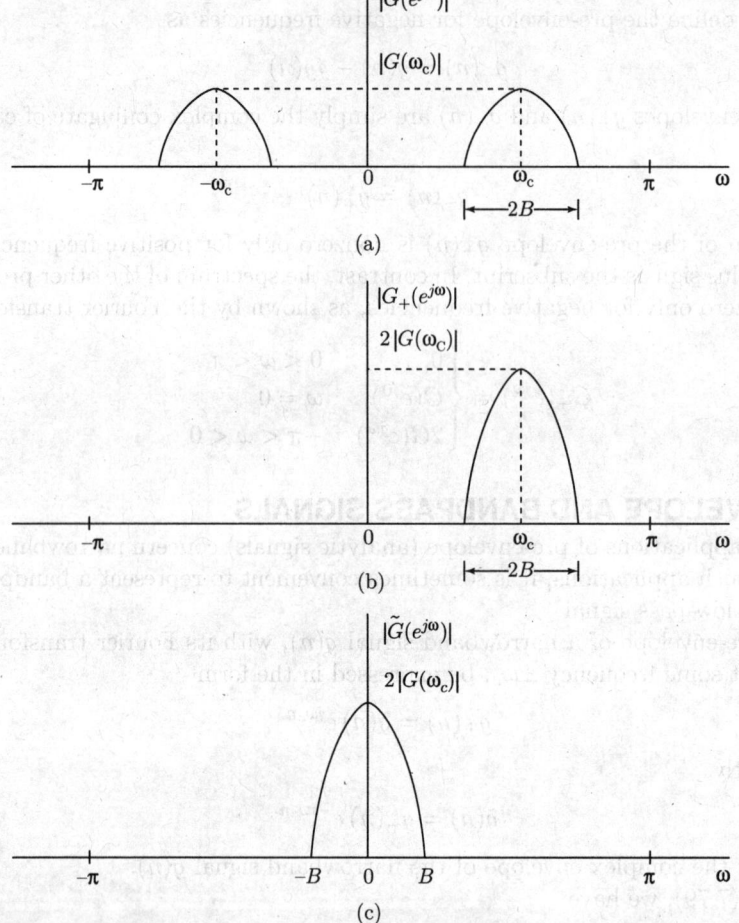

Fig. 7.9 Magnitude spectrum of (a) bandpass signal $g(n)$, (b) pre-envelope $g_+(n)$, and, (c) complex envelope $\tilde{g}(n)$ for $-\pi < \omega < \pi$

In general, $\tilde{g}(n)$ is a complex-valued quantity; thus, we may express it in the form

$$\tilde{g}(n) = g_I(n) + jg_Q(n) \tag{7.88}$$

where $g_I(n)$ and $g_Q(n)$ are both real-valued low-pass functions. Substituting $\tilde{g}(n) = g_I(n) + jg_Q(n)$ into Eq. (7.87), we obtain

$$g(n) = \text{Re}\Big\{[g_I(n) + jg_Q(n)]\,[\cos(\omega_c n) + j\sin(\omega_c n)]\Big\}$$

$$g(n) = g_I(n)\cos(\omega_c n) - g_Q(n)\sin(\omega_c n) \tag{7.89}$$

Equation (7.89) represents the bandpass signal $g(n)$ in the *canonical form*. We refer to $g_I(n)$ as the *in-phase component* of the bandpass signal $g(n)$ and to $g_Q(n)$ as the *quadrature component* of the signal. Both $g_I(n)$ and $g_Q(n)$ are low-pass signals limited to the band $-B \leq \omega \leq B$. Hence, they may be derived from the bandpass signal $g(n)$.

Using Eqs (7.84) and (7.88), we obtain

$$g_I(n) = \text{Re}\{\tilde{g}(n)\}$$

$$g_I(n) = \text{Re}\Big\{g_+(n)e^{-j\omega_c n}\Big\}$$

Substituting $g_+(n) = g(n) + j\hat{g}(n)$ into the above equation, we obtain

$$g_I(n) = \text{Re}\Big\{[g(n) + j\hat{g}(n)][\cos(\omega_c n) - j\sin(\omega_c n)]\Big\}$$

$$g_I(n) = g(n)\cos(\omega_c n) + \hat{g}(n)\sin(\omega_c n) \tag{7.90}$$

Equation (7.90) represents the in-phase component $g_I(n)$ in terms of the bandpass signal $g(n)$ and its Hilbert transform $\hat{g}(n)$.

Similarly, we may obtain the quadrature component $g_Q(n)$ in terms of the bandpass signal $g(n)$ and its Hilbert transform $\hat{g}(n)$. Using Eqs (7.84) and (7.88), we obtain

$$g_Q(n) = \text{Im}\{\tilde{g}(n)\}$$

$$g_Q(n) = \text{Im}\Big\{g_+(n)\,e^{-j\omega_c n}\Big\}$$

Substituting $g_+(n) = g(n) + j\hat{g}(n)$ into the above equation, we obtain

$$g_Q(n) = \text{Re}\Big\{[g(n) + j\hat{g}(n)]\,[\cos(\omega_c n) - j\sin(\omega_c n)]\Big\}$$

$$g_Q(n) = \hat{g}(n)\cos(\omega_c n) - g(n)\sin(\omega_c n) \tag{7.91}$$

Equation (7.88) is the Cartesian form of expressing the complex envelope $\hat{g}(n)$. Alternatively, we may express it in the polar form

$$\tilde{g}(n) = a(n)\,e^{j\phi(n)} \tag{7.92}$$

where $a(n) = \sqrt{g_I^2(n) + g_Q^2(n)}$ and $\phi(n) = \tan^{-1}(g_Q(n)/g_I(n))$ are both real-valued low-pass functions. Based on this polar representation, the original bandpass signal $g(n)$ is defined by

$$g(n) = \text{Re}\{g_+(n)\}$$

$$= \text{Re}\Big\{\tilde{g}(n)\,e^{j\omega_c n}\Big\}$$

$$= \text{Re}\Big\{a(n)\,e^{j\phi(n)}e^{j\omega_c n}\Big\}$$

$$g(n) = \text{Re}\Big\{a(n)e^{[j\omega_c n + \phi(n)]}\Big\}$$

$$g(n) = a(n)\cos(\omega_c n + \phi(n))$$

where $a(n)$ is the *natural envelope* or simply the *envelope* of the bandpass signal $g(n)$ and $\phi(n)$ is the *phase* of the signal.

The envelope $a(n)$ equals the magnitude of the complex envelope $\tilde{g}(n)$ and also that of the pre-envelope $g_+(n)$, as shown by

$$a(n) = |\tilde{g}(n)| = |g_+(n)| \tag{7.93}$$

SOLVED EXAMPLES

Example 7.4 Compute the Hilbert transform and pre-envelope of $x(t) = \text{sinc}(2t)$.

Solution
We know that

$$\text{sinc}(t) \longleftrightarrow \text{rect}\left(\frac{\omega}{2\pi}\right)$$

Using the time-scaling property of the Fourier transform, we obtain

$$x(t) = \text{sinc}(2t) \longleftrightarrow X(\omega) = \frac{1}{2}\text{rect}\left(\frac{\omega}{4\pi}\right)$$

The Fourier transform $X(\omega)$ is depicted in Fig. 7.10(a). The Fourier transform $\hat{X}(\omega)$ of the Hilbert transform $\hat{x}(t)$ is given by

$$\hat{X}(\omega) = -j\,\text{sgn}(\omega)X(\omega)$$

$$\hat{X}(\omega) = -j\,\text{sgn}(\omega)\frac{1}{2}\text{rect}\left(\frac{\omega}{4\pi}\right)$$

$\hat{X}(\omega)$ is shown in Fig. 7.10(c). Now, we shall take the inverse Fourier transform of $\hat{X}(\omega)$ to find the Hilbert transform $\hat{x}(t)$.

$$\hat{x}(t) = \frac{1}{2\pi}\int_{-\infty}^{\infty}\hat{X}(\omega)e^{j\omega t}d\omega$$

$$= \frac{1}{2\pi}\int_{-2\pi}^{0}\frac{j}{2}e^{j\omega t}d\omega + \frac{1}{2\pi}\int_{0}^{2\pi}\frac{-j}{2}e^{j\omega t}d\omega$$

$$= \frac{j}{4\pi}\left[\frac{e^{j\omega t}}{jt}\Big|_{-2\pi}^{0} - \frac{e^{j\omega t}}{jt}\Big|_{0}^{2\pi}\right]$$

$$= \frac{1}{4\pi t}[1 - e^{-j2\pi t} - e^{j2\pi t} + 1]$$

$$\hat{x}(t) = \frac{1}{2\pi t}[1 - \cos(2\pi t)]$$

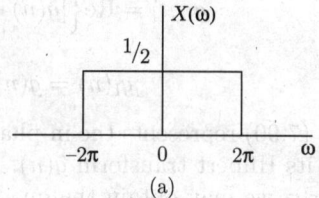
(a)

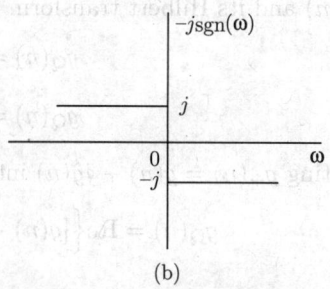
(b)

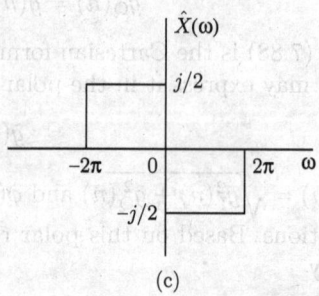
(c)

Fig. 7.10 Signal sketches for Example 7.4

The pre-envelope $x_+(t)$ of $x(t)$ is defined as

$$x_+(t) = x(t) + j\,\hat{x}(t)$$

$$= \text{sinc}(2t) + \frac{1}{2\pi t}[1 - \cos(2\pi t)]$$

$$= \frac{\sin(2\pi t)}{2\pi t} + \frac{1}{2\pi t}[1 - \cos(2\pi t)]$$

$$x_+(t) = \frac{1}{2\pi t}[\sin(2\pi t) + 1 - \cos(2\pi t)]$$

Example 7.5 Determine the Hilbert transform of the following signals:

(a) $x(t) = \text{rect}(t)$

(b) $x(t) = \dfrac{1}{1+t^2}$

(c) $x(t) = \dfrac{1}{t}$

Solution

(a) The Hilbert transform $\hat{x}(t)$ of $x(t)$ is given by

$$\hat{x}(t) = x(t) * \frac{1}{\pi t} = \text{rect}(t) * \frac{1}{\pi t} = \frac{1}{\pi} \int\limits_{-\infty}^{\infty} \frac{\text{rect}(\tau)}{t-\tau} d\tau$$

Since $\text{rect}(\tau) = \begin{cases} 1, & -1/2 < \tau < 1/2 \\ 0, & \text{elsewhere} \end{cases}$

we obtain

$$\hat{x}(t) = \frac{1}{\pi} \int\limits_{-1/2}^{1/2} \frac{1}{t-\tau} d\tau = -\frac{1}{\pi} \ln|t-\tau| \Big|_{-1/2}^{1/2}$$

$$= -\frac{1}{\pi} \left[\ln\left|t-\frac{1}{2}\right| - \ln\left|t+\frac{1}{2}\right| \right]$$

$$\hat{x}(t) = -\frac{1}{\pi} \ln\left| \frac{t-1/2}{t+1/2} \right|$$

(b) We know that [Eq. (5.47)]

$$X(\omega) = \mathcal{F}\left[\frac{1}{1+t^2} \right] = \pi e^{-|\omega|}$$

The Fourier transform $\hat{X}(\omega)$ of the Hilbert transform $\hat{x}(t)$ is given by

$$\hat{X}(\omega) = -j\,\text{sgn}(\omega)X(\omega) = -j\,\text{sgn}(\omega)\pi\,e^{-|\omega|}$$

Now, we shall take the inverse Fourier transform of $\hat{X}(\omega)$ to find the Hilbert transform $\hat{x}(t)$.

$$\hat{x}(t) = \frac{1}{2\pi} \int\limits_{-\infty}^{\infty} \hat{X}(\omega) e^{j\omega t} d\omega$$

$$= \frac{1}{2\pi} \int\limits_{-\infty}^{\infty} -j\,\text{sgn}(\omega)\pi\,e^{-|\omega|} e^{j\omega t} d\omega$$

$$\hat{x}(t) = \frac{1}{2\pi} \int\limits_{-\infty}^{0} j\pi\,e^{\omega}\,e^{j\omega t} d\omega + \frac{1}{2\pi} \int\limits_{0}^{\infty} -j\pi\,e^{-\omega}\,e^{j\omega t} d\omega$$

$$\hat{x}(t) = \frac{j}{2} \left[\underbrace{\int\limits_{-\infty}^{0} e^{\omega}\,e^{j\omega t} d\omega}_{I_1} - \underbrace{\int\limits_{0}^{\infty} e^{-\omega}\,e^{j\omega t} d\omega}_{I_2} \right]$$

$$\hat{x}(t) = \frac{j}{2}[I_1 - I_2]$$

where

$$I_1 = \int\limits_{-\infty}^{0} e^{\omega}\,e^{j\omega t} d\omega$$

$$= \frac{e^{\omega}\,e^{j\omega t}}{jt} \Big|_{-\infty}^{0} - \frac{1}{jt} \underbrace{\int\limits_{-\infty}^{0} e^{\omega}\,e^{j\omega t} d\omega}_{I_1}$$

$$I_1 = \frac{1}{jt}[1-0] - \frac{1}{jt}I_1$$

$$I_1 = \frac{1}{jt}[1 - I_1]$$

$$jtI_1 = 1 - I_1$$

$$I_1 = \frac{1}{1+jt}$$

Similarly, we can obtain

$$I_2 = \frac{1}{1-jt}$$

Substituting $I_1 = 1/(1+jt)$ and $I_2 = 1/(1-jt)$ into the expression of $\hat{x}(t)$, we obtain

$$\hat{x}(t) = \frac{j}{2}[I_1 - I_2] = \frac{j}{2}\left[\frac{1}{1+jt} - \frac{1}{1-jt} \right]$$

$$\hat{x}(t) = \frac{t}{1+t^2}$$

(c) We know that [Eq. (5.48)]

$$\mathcal{F}\left[\frac{1}{\pi t} \right] = -j\,\text{sgn}(\omega)$$

$$X(\omega) = \mathcal{F}\left[\frac{1}{t} \right] = -j\pi\,\text{sgn}(\omega)$$

The Fourier transform $\hat{X}(\omega)$ of the Hilbert transform $\hat{x}(t)$ is given by

$$\hat{X}(\omega) = -j\,\text{sgn}(\omega)X(\omega)$$

$$\hat{X}(\omega) = -j\,\text{sgn}(\omega)[-j\pi\,\text{sgn}(\omega)]$$

$$\hat{X}(\omega) = -\pi\,\text{sgn}^2(\omega)$$

Since $\text{sgn}^2(\omega) = 1$ for all ω, we obtain

$$\hat{X}(\omega) = -\pi$$

Taking the inverse Fourier transform of the above equation, we obtain

$$\hat{x}(t) = -\pi\delta(t)$$

Example 7.6 Consider the signal

$$g(t) = m(t)c(t)$$

where $m(t)$ is a low-pass signal while Fourier transform $M(\omega)$ vanishes for $|\omega| > B$, and $c(t)$ is a high-pass signal whose Fourier transform $C(\omega)$ vanishes for $|\omega| < B$. Show that the Hilbert transform of $g(t)$ is $\hat{g}(t) = m(t)\hat{c}(t)$, where $\hat{c}(t)$ is the Hilbert transform of $c(t)$.

Solution
Given that

$$g(t) = m(t)c(t)$$

Using the multiplication property of the Fourier transform, we obtain

$$G(\omega) = \frac{1}{2\pi}[M(\omega) * C(\omega)] = \frac{1}{2\pi}\int_{-\infty}^{\infty} M(\alpha)C(\omega - \alpha)\,d\alpha$$

Let $\hat{G}(\omega)$ be the Fourier transform of $\hat{g}(t)$. Hence,

$$\hat{G}(\omega) = -j\,\text{sgn}(\omega)G(\omega)$$

Using the formula of the inverse Fourier transform, we may then write

$$\hat{g}(t) = \frac{1}{2\pi}\int_{-\infty}^{\infty} \hat{G}(\omega)\,e^{j\omega t}\,d\omega$$

$$= \frac{1}{2\pi}\int_{-\infty}^{\infty} -j\,\text{sgn}(\omega)G(\omega)e^{j\omega t}\,d\omega$$

$$= \frac{1}{2\pi}\int_{-\infty}^{\infty} -j\,\text{sgn}(\omega)\left[\frac{1}{2\pi}\int_{-\infty}^{\infty} M(\alpha)C(\omega-\alpha)\,d\alpha\right]e^{j\omega t}\,d\omega$$

$$\hat{g}(t) = \frac{-j}{(2\pi)^2}\int_{-\infty}^{\infty}\int_{-\infty}^{\infty} \text{sgn}(\omega)M(\alpha)C(\omega - \alpha)e^{j\omega t}\,d\omega\,d\alpha$$

Let $\beta = \omega - \alpha$. Then we may write this relation as

$$\hat{g}(t) = \frac{-j}{(2\pi)^2}\int_{-\infty}^{\infty}\int_{-\infty}^{\infty} \text{sgn}(\alpha + \beta)M(\alpha)C(\beta)\,e^{j(\alpha+\beta)t}\,d\beta\,d\alpha$$

However, by definition, the product $M(\alpha)C(\beta)$ is nonzero only for $|\alpha| < B$ and $|\beta| > B$. With the values of α and β restricted in this manner, we find that $\text{sgn}(\alpha + \beta) = \text{sgn}(\beta)$. We may thus rewrite the above equation in the form

$$\hat{g}(t) = \frac{-j}{(2\pi)^2}\int_{-\infty}^{\infty}\int_{-\infty}^{\infty} \text{sgn}(\beta)M(\alpha)C(\beta)\,e^{j(\alpha+\beta)t}\,d\beta\,d\alpha$$

$$\hat{g}(t) = \frac{1}{2\pi}\int_{-\infty}^{\infty} M(\alpha)\,e^{j\alpha}\,d\alpha\,\frac{1}{2\pi}\int_{-\infty}^{\infty}[-j\,\text{sgn}(\beta)C(\beta)]e^{j\beta t}\,d\beta$$

$$\hat{g}(t) = \frac{1}{2\pi}\int_{-\infty}^{\infty} M(\alpha)e^{j\alpha}\,d\alpha\,\frac{1}{2\pi}\int_{-\infty}^{\infty} \hat{C}(\beta)e^{j\beta t}\,d\beta$$

$$\hat{g}(t) = m(t)\hat{c}(t)$$

Example 7.7 Determine the Hilbert transform of

(a) $x(t) = m(t)\cos(\omega_c t)$

(b) $x(t) = m(t)\sin(\omega_c t)$

Assume that $m(t)$ is band limited to the interval $-B \le \omega \le B$, where $B < \omega_c$.

Solution

(a) Consider the given signal $x(t) = m(t)\cos(\omega_c t)$. Taking its Fourier transform, we obtain

$$X(\omega) = M(\omega) * \pi[\delta(\omega - \omega_c) + \delta(\omega + \omega_c)]$$

$$X(\omega) = \pi[M(\omega - \omega_c) + M(\omega + \omega_c)]$$

Let $\hat{x}(t)$ be the Hilbert transform of $x(t)$ and $\hat{X}(\omega)$ be the Fourier transform of $\hat{x}(t)$. We may then write

$$\hat{X}(\omega) = -j\operatorname{sgn}(\omega)X(\omega)$$

$$= -j\operatorname{sgn}(\omega)\pi[M(\omega - \omega_c) + M(\omega + \omega_c)]$$

$$= \frac{\pi}{j}[M(\omega - \omega_c) + M(\omega + \omega_c)]\operatorname{sgn}(\omega)$$

$$\hat{X}(\omega) = \frac{\pi}{j}[M(\omega - \omega_c) - M(\omega + \omega_c)]$$

$$\hat{X}(\omega) = M(\omega) * \frac{\pi}{j}[\delta(\omega - \omega_c) - \delta(\omega + \omega_c)]$$

Taking the inverse Fourier transform of the above equation, we obtain

$$\hat{x}(t) = m(t)\sin(\omega_c t)$$

(b) Consider the given signal $x(t) = m(t)\sin(\omega_c t)$. Taking its Fourier transform, we obtain

$$X(\omega) = M(\omega) * \frac{\pi}{j}[\delta(\omega - \omega_c) - \delta(\omega + \omega_c)]$$

$$X(\omega) = \frac{\pi}{j}[M(\omega - \omega_c) - M(\omega + \omega_c)]$$

Let $\hat{x}(t)$ be the Hilbert transform of $x(t)$ and $\hat{X}(\omega)$ be the Fourier transform of $\hat{x}(t)$. We may then write

$$\hat{X}(\omega) = -j\operatorname{sgn}(\omega)X(\omega)$$

$$= -j\operatorname{sgn}(\omega)\frac{\pi}{j}[M(\omega - \omega_c) - M(\omega + \omega_c)]$$

$$= -\pi[M(\omega - \omega_c) - M(\omega + \omega_c)]\operatorname{sgn}(\omega)$$

$$\hat{X}(\omega) = -\pi[M(\omega - \omega_c) + M(\omega + \omega_c)]$$

$$\hat{X}(\omega) = -M(\omega) * \pi[\delta(\omega - \omega_c) + \delta(\omega + \omega_c)]$$

Taking the inverse Fourier transform of the above equation, we obtain

$$\hat{x}(t) = -m(t)\cos(\omega_c t)$$

Example 7.8 Determine the pre-envelope, complex envelope, and envelope of the radiofrequency (RF) pulse defined by

$$g(t) = A\operatorname{rect}\left(\frac{t}{T}\right)\cos(\omega_c t)$$

Solution

Consider the given RF pulse

$$g(t) = A\operatorname{rect}\left(\frac{t}{T}\right)\cos(\omega_c t)$$

Using the multiplication property of the Fourier transform, we obtain

$$\mathcal{F}[g(t)] = \frac{1}{2\pi}\left(\mathcal{F}\left[A\operatorname{rect}\left(\frac{t}{T}\right)\right] * \mathcal{F}[\cos(\omega_c t)]\right)$$

The Fourier transform of the rectangle signal $A\operatorname{rect}(t/T)$ is given by

$$\mathcal{F}\left[A\operatorname{rect}\left(\frac{t}{T}\right)\right] = AT\frac{\sin(\omega T/2)}{\omega T/2} = AT\operatorname{sinc}\left(\frac{\omega T}{2\pi}\right)$$

Therefore,

$$\mathcal{F}[g(t)] = \frac{1}{2\pi}\left(\mathcal{F}\left[A\operatorname{rect}\left(\frac{t}{T}\right)\right] * \mathcal{F}[\cos(\omega_c t)]\right)$$

$$G(\omega) = \frac{1}{2\pi}\left(AT\operatorname{sinc}\left(\frac{\omega T}{2\pi}\right)\right.$$

$$\left. * \pi[\delta(\omega - \omega_c) + \delta(\omega + \omega_c)]\right)$$

$$G(\omega) = \frac{1}{2}\left[AT\operatorname{sinc}\left(\frac{(\omega - \omega_c)T}{2\pi}\right)\right.$$

$$\left. + AT\operatorname{sinc}\left(\frac{(\omega + \omega_c)T}{2\pi}\right)\right]$$

$$G(\omega) = \frac{AT}{2}\left[\operatorname{sinc}\left(\frac{(\omega - \omega_c)T}{2\pi}\right)\right.$$

$$\left. + \operatorname{sinc}\left(\frac{(\omega + \omega_c)T}{2\pi}\right)\right]$$

$$G(\omega) = \begin{cases} \frac{AT}{2}\operatorname{sinc}\left(\frac{(\omega - \omega_c)T}{2\pi}\right), & \omega > 0 \\ 0, & \omega = 0 \\ \frac{AT}{2}\operatorname{sinc}\left(\frac{(\omega + \omega_c)T}{2\pi}\right), & \omega < 0 \end{cases}$$

From Eq. (7.34), we have

$$G_+(\omega) = \begin{cases} 2G(\omega), & \omega > 0 \\ G(0), & \omega = 0 \\ 0, & \omega < 0 \end{cases}$$

$$G_+(\omega) = \begin{cases} AT \operatorname{sinc}\left(\dfrac{(\omega - \omega_c)T}{2\pi}\right), & \omega > 0 \\ 0, & \omega \leq 0 \end{cases}$$

Taking the inverse Fourier transform of $G_+(\omega)$, we obtain the pre-envelope

$$g_+(t) = A \operatorname{rect}\left(\frac{t}{T}\right) e^{j\omega_c t}$$

We know that the complex envelope is given by [Eq. (7.38)]

$$g_+(t) = \tilde{g}(t) e^{j\omega_c t}$$

Therefore,

$$\tilde{g}(t) = A \operatorname{rect}\left(\frac{t}{T}\right)$$

and the envelope equals

$$a(t) = |g_+(t)| = |\tilde{g}(t)| = A \operatorname{rect}\left(\frac{t}{T}\right)$$

Example 7.9 Determine the Hilbert transform of the unit sample function $g(n) = \delta(n)$.

Solution

The DTFT of the given signal $g(n) = \delta(n)$ is given by

$$G(e^{j\omega}) = 1$$

We know that [Eq. (7.69)]

$$\hat{G}(e^{j\omega}) = -j \operatorname{sgn}(\omega) G(e^{j\omega})$$

$$= -j \operatorname{sgn}(\omega)$$

$$\hat{G}(e^{j\omega}) = \begin{cases} -j, & 0 < \omega < \pi \\ j, & -\pi < \omega < 0 \end{cases}$$

Taking the inverse DTFT of $\hat{G}(e^{j\omega})$, we obtain the Hilbert transform

$$\hat{g}(n) = \frac{1}{2\pi} \int_{-\infty}^{\infty} \hat{G}(e^{j\omega}) e^{j\omega n} d\omega$$

$$= \frac{1}{2\pi} \int_{-\pi}^{0} j\, e^{j\omega n} d\omega + \frac{1}{2\pi} \int_{0}^{\pi} -j\, e^{j\omega n} d\omega$$

$$= \frac{j}{2\pi} \int_{-\pi}^{0} e^{j\omega n} d\omega - \frac{j}{2\pi} \int_{0}^{\pi} e^{j\omega n} d\omega$$

$$= \frac{j}{2\pi} \left[\frac{e^{j\omega n}}{jn} \Big|_{-\pi}^{0} - \frac{e^{j\omega n}}{jn} \Big|_{0}^{\pi} \right]$$

$$= \frac{1}{2n\pi} [1 - e^{-jn\pi} - e^{jn\pi} + 1]$$

$$\hat{g}(n) = \frac{1}{2n\pi} \left[2 - 2\left(\frac{e^{jn\pi} + e^{-jn\pi}}{2} \right) \right]$$

$$\hat{g}(n) = \frac{1}{2n\pi} [2 - 2\cos(n\pi)], \qquad n \neq 0$$

$$\hat{g}(n) = \frac{1}{n\pi} [1 - \cos(n\pi)], \qquad n \neq 0$$

$$\hat{g}(n) = \frac{2}{n\pi} \sin^2\left(\frac{n\pi}{2}\right), \qquad n \neq 0$$

$$\hat{g}(n) = \begin{cases} \frac{2}{n\pi} \sin^2\left(\frac{n\pi}{2}\right), & n \neq 0 \\ 0, & n = 0 \end{cases}$$

Example 7.10 Determine the Hilbert transform of $g(n) = \cos(\omega_c n)$.

Solution

Consider the given signal $g(n) = \cos(\omega_c n)$. The Fourier transform of $g(n)$ is given by

$$G(e^{j\omega}) = \pi[\delta(\omega - \omega_c) + \delta(\omega + \omega_c)], \qquad -\pi < \omega < \pi$$

We know that [Eq. (7.69)]

$$\hat{G}(e^{j\omega}) = -j \operatorname{sgn}(\omega) G(e^{j\omega})$$

$$= -j \operatorname{sgn}(\omega) \pi[\delta(\omega - \omega_c) + \delta(\omega + \omega_c)]$$

$$\hat{G}(\omega) = \frac{\pi}{j} [\delta(\omega - \omega_c) + \delta(\omega + \omega_c)] \operatorname{sgn}(\omega)$$

$$\hat{G}(\omega) = \frac{\pi}{j} [\delta(\omega - \omega_c) - \delta(\omega + \omega_c)]$$

Taking the inverse DTFT of the above equation, we obtain

$$\hat{g}(n) = \sin(\omega_c n)$$

Example 7.11 Determine the Hilbert transform of $g(n) = \sin(\omega_c n)$.

Solution
Consider the given signal $g(n) = \sin(\omega_c n)$. The Fourier transform of $g(n)$ is given by

$$G(e^{j\omega}) = \frac{\pi}{j}[\delta(\omega - \omega_c) - \delta(\omega + \omega_c)], \qquad -\pi < \omega < \pi$$

We know that [Eq. (7.69)],

$$\hat{G}(e^{j\omega}) = -j\,\mathrm{sgn}(\omega)G(e^{j\omega})$$

$$= -j\,\mathrm{sgn}(\omega)\frac{\pi}{j}[\delta(\omega - \omega_c) - \delta(\omega + \omega_c)]$$

$$\hat{G}(\omega) = -\pi[\delta(\omega - \omega_c) - \delta(\omega + \omega_c)]\,\mathrm{sgn}(\omega)$$

$$\hat{G}(\omega) = -\pi[\delta(\omega - \omega_c) + \delta(\omega + \omega_c)]$$

Taking the inverse DTFT of the above equation, we obtain

$$\hat{g}(n) = -\cos(\omega_c n)$$

Example 7.12 Let a narrowband signal $g(t)$ be expressed in the form

$$g(t) = g_I(t)\cos(\omega_c t) - g_Q(t)\sin(\omega_c t)$$

Using $G_+(\omega)$ to denote the Fourier transform of the pre-envelope of $g_+(t)$, show that the Fourier transform of the in-phase component $g_I(t)$ and quadrature component $g_Q(t)$ are given by, respectively,

$$G_I(\omega) = \frac{1}{2}[G_+(\omega + \omega_c) + G_+^*(-\omega + \omega_c)]$$

$$G_Q(\omega) = \frac{1}{2j}[G_+(\omega + \omega_c) - G_+^*(-\omega + \omega_c)]$$

Solution
We know that

$$\tilde{g}(t) = g_+(t)\,e^{-j\omega_c t} = g_I(t) + jg_Q(t)$$

Taking the Fourier transform of the above equation, we obtain

$$\widetilde{G}(\omega) = G_+(\omega + \omega_c) = G_I(\omega) + j\,G_Q(\omega)$$

We also know that

$$g_I(t) = \frac{1}{2}[\tilde{g}(t) + \tilde{g}^*(t)]$$

$$G_I(\omega) = \frac{1}{2}[\widetilde{G}(\omega) + \widetilde{G}^*(-\omega)]$$

$$G_I(\omega) = \frac{1}{2}[G_+(\omega + \omega_c) + G_+^*(-\omega + \omega_c)]$$

Similarly, we obtain

$$g_Q(t) = \frac{1}{2j}[\tilde{g}(t) - \tilde{g}^*(t)]$$

$$G_Q(\omega) = \frac{1}{2j}[\widetilde{G}(\omega) - \widetilde{G}^*(-\omega)]$$

$$G_Q(\omega) = \frac{1}{2j}[G_+(\omega + \omega_c) - G_+^*(-\omega + \omega_c)]$$

Example 7.13 The in-phase component $g_I(t)$ and quadrature component $g_Q(t)$ of a narrowband signal $g(t)$ are given by, respectively,

$$g_I(t) = g(t)\cos(\omega_c t) + \hat{g}(t)\sin(\omega_c t)$$

$$g_Q(t) = \hat{g}(t)\cos(\omega_c t) - g(t)\sin(\omega_c t)$$

Show that

$$G_I(\omega) = \begin{cases} [G(\omega - \omega_c) + G(\omega + \omega_c)], & -B \le \omega \le B \\ 0, & \text{elsewhere} \end{cases}$$

and

$$G_Q(\omega) = \begin{cases} j[G(\omega - \omega_c) - G(\omega + \omega_c)], & -B \le \omega \le B \\ 0, & \text{elsewhere} \end{cases}$$

Assume that the spectrum $G(\omega)$ of $g(t)$ is limited to the interval $(\omega_c - B) \le |\omega|(\le \omega_c + B)$.

Solution
Consider the in-phase component

$$g_I(t) = g(t)\cos(\omega_c t) + \hat{g}(t)\sin(\omega_c t)$$

Taking the Fourier transform of the above equation, we obtain

$$G_{\mathrm{I}}(\omega) = \frac{1}{2\pi}\left(G(\omega) * \pi[\delta(\omega - \omega_c) + \delta(\omega + \omega_c)]\right)$$

$$+ \frac{1}{2\pi}\left([-j\,\mathrm{sgn}(\omega)G(\omega)]\right.$$

$$\left. * \frac{\pi}{j}[\delta(\omega - \omega_c) - \delta(\omega + \omega_c)]\right)$$

$$= \frac{1}{2}[G(\omega - \omega_c) + G(\omega + \omega_c)]$$

$$- \frac{1}{2}[\mathrm{sgn}(\omega - \omega_c)G(\omega - \omega_c)$$

$$- \mathrm{sgn}(\omega + \omega_c)G(\omega + \omega_c)]$$

$$G_{\mathrm{I}}(\omega) = \frac{1}{2}G(\omega - \omega_c)[1 - \mathrm{sgn}(\omega - \omega_c)]$$

$$+ \frac{1}{2}G(\omega + \omega_c)[1 + \mathrm{sgn}(\omega + \omega_c)]$$

Now, with the spectrum $G(\omega)$ of the original signal occupying the frequency band $(\omega_c - B) \le |\omega| \le (\omega_c + B)$, as illustrated in Fig. 7.11(a), we find that the corresponding shapes of $G(\omega - \omega_c)$ and $G(\omega + \omega_c)$ are shown in Figs 7.11(b) and (c), respectively. Figures 7.11(d), (e), and (f) shows the shapes of $\mathrm{sgn}(\omega)$, $\mathrm{sgn}(\omega - \omega_c)$, and $\mathrm{sgn}(\omega + \omega_c)$, respectively. Accordingly, we make the following observations:

1. For frequencies defined by $-B \le \omega \le B$, we have

$$\mathrm{sgn}(\omega - \omega_c) = -1$$

and

$$\mathrm{sgn}(\omega + \omega_c) = 1$$

Hence, substituting these results in the expression of $G_{\mathrm{I}}(\omega)$, we obtain

$$G_{\mathrm{I}}(\omega) = \frac{1}{2}G(\omega - \omega_c)[1 - (-1)] + \frac{1}{2}G(\omega + \omega_c)[1 + 1]$$

$$= \frac{1}{2}G(\omega - \omega_c)2 + \frac{1}{2}G(\omega + \omega_c)2$$

$$G_{\mathrm{I}}(\omega) = G(\omega - \omega_c) + G(\omega + \omega_c), \qquad -B \le \omega \le B$$

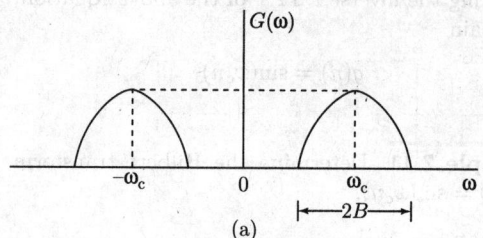

(a)

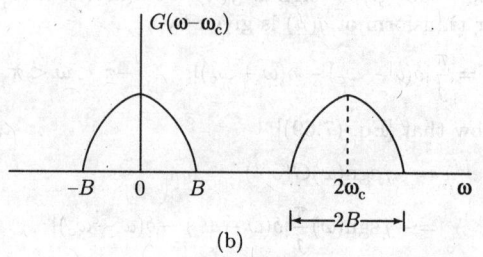

(b)

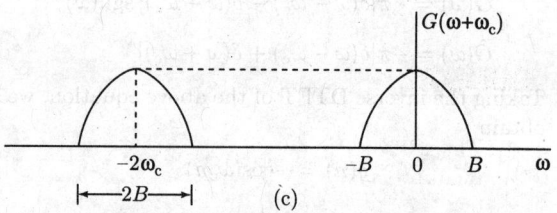

(c)

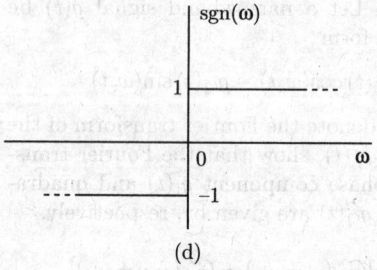

(d)

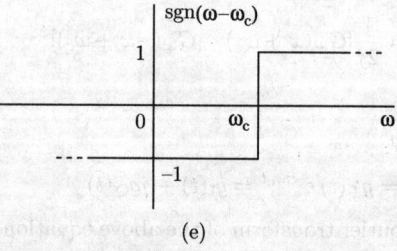

(e)

Fig. 7.11 (*Continued*)

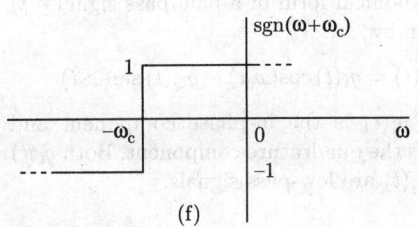

Fig. 7.11

2. For $(2\omega_c - B) \le \omega \le (2\omega_c + B)$, we have

$$\text{sgn}(\omega - \omega_c) = 1$$

and $$G(\omega + \omega_c) = 0$$

Hence, substituting these results in the expression of $G_I(\omega)$, we obtain

$$G_I(\omega) = 0$$

3. For $(-2\omega_c - B) \le \omega \le (-2\omega_c + B)$, we have

$$\text{sgn}(\omega + \omega_c) = -1$$

and $$G(\omega - \omega_c) = 0$$

Hence, substituting these results in the expression of $G_I(\omega)$, we obtain

$$G_I(\omega) = 0$$

Combining these results, we obtain

$$G_I(\omega) = \begin{cases} [G(\omega - \omega_c) + G(\omega + \omega_c)], & -B \le \omega \le B \\ 0, & \text{elsewhere} \end{cases}$$

Similarly, we can obtain

$$G_Q(\omega) = \begin{cases} j[G(\omega - \omega_c) - G(\omega + \omega_c)], & -B \le \omega \le B \\ 0, & \text{elsewhere} \end{cases}$$

SUMMARY

1. The Hilbert transformer is also known as quadrature filter or 90° phase shifter.

2. The Hilbert transformer produces a phase shift of $-90°$ for all positive frequencies of the input signal and $+90°$ for all negative frequencies.

3. The impulse response $h(t)$ of the continuous-time Hilbert transformer is given by

$$h(t) = \frac{1}{\pi t}$$

4. The impulse response $h(n)$ of the discrete-time Hilbert transformer is given by

$$h(n) = \begin{cases} \dfrac{2}{n\pi}\sin^2\left(\dfrac{n\pi}{2}\right), & n \ne 0 \\ 0, & n = 0 \end{cases}$$

which is antisymmetric in nature.

5. The frequency response of the continuous-time Hilbert transformer is given by

$$H(\omega) = -j\,\text{sgn}(\omega) = \begin{cases} -j = 1e^{-j\pi/2}, & 0 < \omega \le \infty \\ j = 1e^{j\pi/2}, & -\infty \le \omega < 0 \end{cases}$$

6. The frequency response of the discrete-time Hilbert transformer is given by

$$H(e^{j\omega}) = -j\,\text{sgn}(\omega) = \begin{cases} -j = 1e^{-j\pi/2}, & 0 < \omega < \pi \\ j = 1e^{j\pi/2}, & -\pi < \omega < 0 \end{cases}$$

7. The pre-envelope of a real-valued signal $g(t)$ is given by

$$g_+(t) = g(t) + j\hat{g}(t)$$

where $\hat{g}(t)$ is the Hilbert transform of the signal $g(t)$. Similarly, for discrete-time signal $g(n)$, the pre-envelope is given by

$$g_+(n) = g(n) + j\hat{g}(n)$$

8. The complex envelope and pre-envelope are related by

$$g_+(t) = \tilde{g}(t)e^{j\omega_c t}$$

Similarly, for discrete-time signals,

$$g_+(n) = \tilde{g}(n)e^{j\omega_c n}$$

9. The envelope $a(t)$ of a real-valued signal $g(t)$ equals the magnitude of the complex envelope $\tilde{g}(t)$ and also that of the pre-envelope $g_+(t)$, as shown by

$$a(t) = |\tilde{g}(t)| = |g_+(t)|$$

10. The canonical form of a bandpass signal $g(t)$ is given by

$$g(t) = g_I(t)\cos(\omega_c t) - g_Q(t)\sin(\omega_c t)$$

where $g_I(t)$ is the in-phase component and $g_Q(t)$ is the quadrature component. Both $g_I(t)$ and $g_Q(t)$ are low-pass signals.

MULTIPLE-CHOICE QUESTIONS

1. The Hilbert transform of the unit impulse function $\delta(t)$ is
   (a) $1/t$
   (b) $1/\pi t$
   (c) πt
   (d) t

2. The Hilbert transform of $1/t$ is
   (a) $1/t$
   (b) $1/\pi t$
   (c) $-\pi\delta(t)$
   (d) $\pi\delta(t)$

3. The impulse response $h(t)$ of the Hilbert transformer is
   (a) $1/t$
   (b) $1/\pi t$
   (c) πt
   (d) t

4. The frequency response $H(\omega)$ of the Hilbert transformer is
   (a) $-j\,\mathrm{sgn}(\omega)$
   (b) $j\,\mathrm{sgn}(\omega)$
   (c) $\mathrm{sgn}(\omega)$
   (d) $-j/\mathrm{sgn}(\omega)$

5. Let $\hat{g}(t)$ be the Hilbert transform of $g(t)$. The Hilbert transform of $\hat{g}(t)$ is
   (a) $\hat{g}(t)$
   (b) $g(t)$
   (c) $-g(t)$
   (d) $-1/g(t)$

6. The pre-envelope $g_+(t)$ of a real-valued signal $g(t)$ is
   (a) $g(t) + j\hat{g}(t)$
   (b) $g(t) - \hat{g}(t)$
   (c) $\hat{g}(t) + j\,g(t)$
   (d) $\hat{g}(t) - jg(t)$

7. The envelope of $g(t) = m(t)\cos(\omega_c t)$ is
   (a) $m(t)$
   (b) $-m(t)$
   (c) $|m(t)|$
   (d) $1/|m(t)|$

8. The Hilbert transform of $x(n) = \cos(\omega_c n)$ is
   (a) $-\cos(\omega_c n)$
   (b) $\sin(\omega_c n)$
   (c) $-\sin(\omega_c n)$
   (d) $1/\cos(\omega_c n)$

9. The complex envelope of the signal $x(t) = \mathrm{Re}[A(t)e^{j\omega_c t + \phi(t)}]$ equals
   (a) $A(t)$
   (b) $A(t)\cos[\phi(t)]$
   (c) $A(t)\sin[\phi(t)]$
   (d) $A(t)e^{j\phi(t)}$

PROBLEMS

7.1 If the unit impulse response of a causal LTI system contains no impulses at the origin, then show that with

$$H(\omega) = H_R(\omega) + jH_I(\omega)$$

$H_R(\omega)$ and $H_I(\omega)$ satisfy the following conditions:

$$H_R(\omega) = \frac{1}{\pi}\int_{-\infty}^{\infty}\frac{H_I(\theta)}{\omega - \theta}d\theta$$

$$H_I(\omega) = -\frac{1}{\pi}\int_{-\infty}^{\infty}\frac{H_R(\theta)}{\omega - \theta}d\theta$$

7.2 Let $x(n)$ be a causal and absolutely summable real sequence with a DTFT $X(e^{j\omega})$. If $X_{\mathrm{R}}(e^{j\omega})$ and $X_{\mathrm{I}}(e^{j\omega})$ denote the real and imaginary parts of $X(e^{j\omega})$, show that they are related as

$$X_{\mathrm{R}}(e^{j\omega}) = \frac{1}{2\pi} \int\limits_{-\pi}^{\pi} X_{\mathrm{I}}(e^{j\theta}) \cot\left(\frac{\omega - \theta}{2}\right) d\theta + x(0)$$

$$X_{\mathrm{I}}(e^{j\omega}) = -\frac{1}{2\pi} \int\limits_{-\pi}^{\pi} X_{\mathrm{R}}(e^{j\theta}) \cot\left(\frac{\omega - \theta}{2}\right) d\theta$$

7.3 Determine the Hilbert transform $\hat{g}(t)$ and the pre-envelope $g_+(t)$ of the following two signals:
(a) $g(t) = \mathrm{sinc}(t)$
(b) $g(t) = [1 + k\cos(\omega_m t)]\cos(\omega_c t)$

7.4 Let $\hat{g}(t)$ be the Hilbert transform of an energy signal $g(t)$. Show that
(a) $HT[\hat{g}(t)] = -g(t)$.

(b) $\displaystyle\int\limits_{-\infty}^{\infty} g(t)\hat{g}(t)\, dt = 0$

(c) $\displaystyle\int\limits_{-\infty}^{\infty} |g(t)|^2 \, dt = \int\limits_{-\infty}^{\infty} |\hat{g}(t)|^2 \, dt$

7.5 Determine the pre-envelope, complex envelope and envelope of the following two signals:
(a) $x(t) = [A_c + m(t)]\cos(\omega_c t)$

(b) $x(t) = m(t)\cos(\omega_c t)$

7.6 Consider a sequence $x(n)$ with DTFT $X(e^{j\omega})$. The sequence $x(n)$ is real valued and causal, and

$$X_{\mathrm{R}}(e^{j\omega}) = 2 - 2a\cos(\omega)$$

Determine $X_{\mathrm{I}}(e^{j\omega})$.

7.7 Consider a sequence $x(n)$ and its DTFT $X(e^{j\omega})$. The following is known:

$x(n)$ is real valued and causal,

$$X_{\mathrm{R}}(e^{j\omega}) = \frac{5}{4} - \cos(\omega)$$

Determine a sequence $x(n)$ consistent with the given information.

7.8 Find the Hilbert transform $\hat{x}(n)$ of the sequence $x(n) = \sin(\omega_c n)/n\pi$.

7.9 Let $x(n)$ denote a causal, complex-valued sequence with Fourier transform

$$X(e^{j\omega}) = X_{\mathrm{R}}(e^{j\omega}) + j\,X_{\mathrm{I}}(e^{j\omega})$$

If $X_{\mathrm{R}}(e^{j\omega}) = 1 + \cos(\omega) + \sin(\omega) - \sin(2\omega)$, determine $X_{\mathrm{I}}(e^{j\omega})$.

ANSWERS TO MULTIPLE-CHOICE QUESTIONS

1. (b)	2. (c)	3. (b)	4. (a)	5. (c)	6. (a)	7. (c)	8. (b)	9. (d)

Sampling

8.1 INTRODUCTION

Under certain conditions, a continuous-time signal can be completely represented by and recoverable from knowledge of its values, or samples, at points equally spaced in time. The importance of sampling lies in its role as a bridge between continuous-time signals and discrete-time signals. As we will see in this chapter, the fact that under certain conditions a continuous-time signal can be completely recovered from a sequence of its samples provides a mechanism for representing a continuous-time signal by a discrete-time signal. Sampling is also frequently performed on discrete-time signals to change the effective data rate, an operation termed subsampling. In this case, the sampling process discards certain values of the signal.

8.2 SAMPLING

Sampling is the process of converting a continuous-time signal $x(t)$ into a discrete-time signal $x(n)$ by measuring the amplitudes of the continuous-time signal $x(t)$ at integer multiples of a sampling interval T_s as shown in Fig. 8.1.

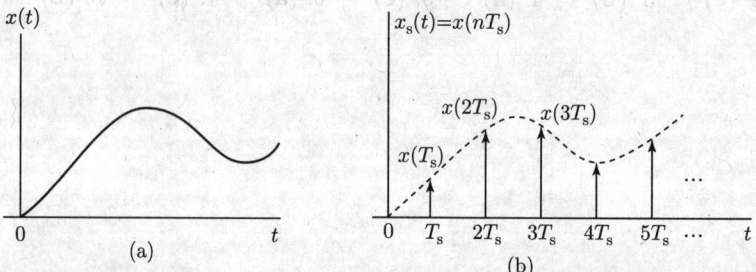

Fig. 8.1   (a) Continuous-time signal and (b) its sampled version

8.3 SAMPLING THEOREM FOR LOW-PASS SIGNALS

Sampling theorem states that a band-limited signal of finite energy, which has no frequency components higher than ω_h rad/s (or f_h Hz), may be completely recovered from a knowledge of its samples if the sampling frequency $\omega_s \geq 2\omega_h$ samples/rad/s

(or $f_s \geq 2f_h$ samples/s.). Thus, the sampling frequency

$$\omega_s \geq 2\omega_h \quad \text{or} \quad f_s \geq 2f_h \tag{8.1}$$

The minimum sampling rate, or minimum sampling frequency $\omega_s = 2\omega_h$ (or $f_s = 2f_h$), is referred to as the *Nyquist rate*; its reciprocal $T_s = 2\pi/\omega_s = 1/f_s = 1/2f_h$ (measured in seconds) is called the *Nyquist interval*.

We can also express the sampling theorem in terms of maximum sample spacing rather than a minimum sampling rate because $T_s = 1/f_s \leq 1/2f_h$; hence, the maximum sampling spacing for complete specification of the continuous-time signal is $T_s = 1/2f_h$. Sampling a signal at a rate that is less than or greater than the Nyquist rate is referred to as *undersampling* or *oversampling*, respectively.

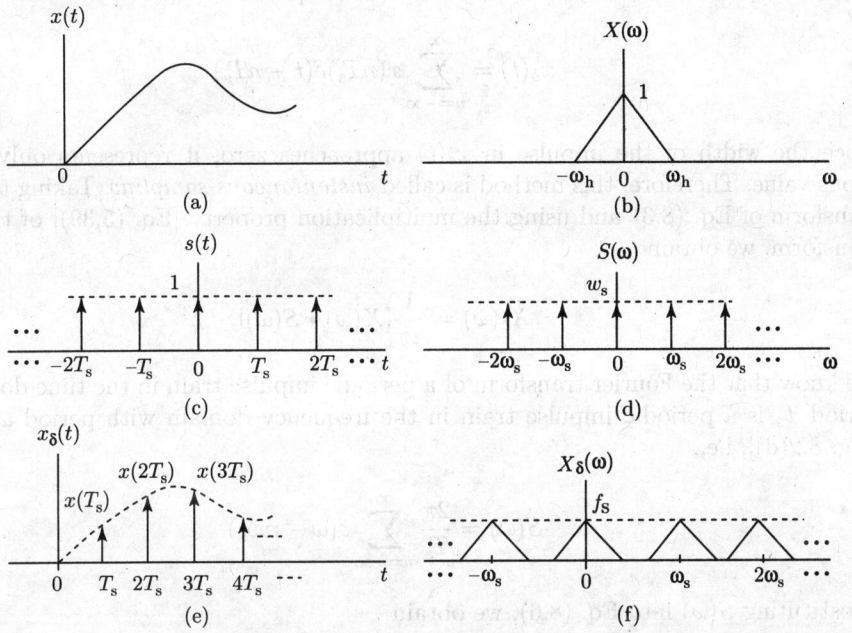

Fig. 8.2 (a) Signal $x(t)$, (b) spectrum of $x(t)$, (c) sampling signal, (d) spectrum of sampling signal, (e) sampled signal and (f) spectrum of the sampled signal.

Let $x(t)$ be a continuous-time signal as shown in Fig. 8.2(a). Let this signal be of finite energy and band limited, i.e., $x(t)$ does not contain any frequency components higher than ω_h rad/s. A sampling signal samples this signal regularly at the rate of ω_s samples/rad/s. The time space between any two successive samples is T_s seconds. Figure 8.2(c) shows the sampling signal (an impulse train) of frequency equal to sampling frequency ω_s, and Fig. 8.2(e) shows an instantaneous sampled version of $x(t)$.

The impulse train in Fig. 8.2(c) can be expressed as

$$s(t) = \sum_{n=-\infty}^{\infty} \delta(t - nT_s) \tag{8.2}$$

The sampled signal $x_\delta(t)$ in Fig. 8.2(e) is obtained by multiplying the signal $x(t)$ with impulse train $s(t)$. Therefore, $x_\delta(t)$ can be expressed mathematically as

$$x_\delta(t) = x(t)s(t) \tag{8.3}$$

$$x_\delta(t) = x(t) \sum_{n=-\infty}^{\infty} \delta(t - nT_s) \tag{8.4}$$

Because of the sampling property of the unit impulse [Eq. (1.14)], we know that multiplying $x(t)$ by a unit impulse samples the value of the signal at the point at which the impulse is located, i.e., $x(t)\delta(t - t_0) = x(t_0)\delta(t - t_0)$. Applying this to Eq. (8.4), we see, as illustrated in Fig. 8.2(c), that $x_\delta(t)$ is an impulse train with the amplitudes of the impulses equal to the samples of $x(t)$ at intervals spaced by T_s, i.e.,

$$x_\delta(t) = \sum_{n=-\infty}^{\infty} x(nT_s)\delta(t - nT_s) \tag{8.5}$$

Since the width of the impulse in $x_s(t)$ approaches zero, it represents only instantaneous value. Therefore, this method is called *instantaneous sampling*. Taking the Fourier transform of Eq. (8.3) and using the multiplication property [Eq. (5.39)] of the Fourier transform, we obtain

$$X_\delta(\omega) = \frac{1}{2\pi}[X(\omega) * S(\omega)] \tag{8.6}$$

We know that the Fourier transform of a periodic impulse train in the time domain with period T_s is a periodic impulse train in the frequency domain with period $\omega_s = 2\pi/T_s$ [Fig. 8.2(d)], i.e.,

$$S(\omega) = \frac{2\pi}{T_s} \sum_{n=-\infty}^{\infty} \delta(\omega - n\omega_s) \tag{8.7}$$

Substituting $S(\omega)$ into Eq. (8.6), we obtain

$$X_\delta(\omega) = \frac{1}{2\pi} \left[X(\omega) * \frac{2\pi}{T_s} \sum_{n=-\infty}^{\infty} \delta(\omega - n\omega_s) \right]$$

$$= \frac{1}{T_s} \left[X(\omega) * \sum_{n=-\infty}^{\infty} \delta(\omega - n\omega_s) \right]$$

Since convolution with an impulse simply shifts a signal [i.e., $X(\omega) * \delta(\omega - \omega_s) = X(\omega - \omega_s)$], it follows that

$$X_\delta(\omega) = \frac{1}{T_s} \sum_{n=-\infty}^{\infty} X(\omega - n\omega_s) \quad \text{or equivalently, in } f \text{ domain, } X_\delta(f) = \frac{1}{T_s} \sum_{n=-\infty}^{\infty} X(f - nf_s)$$

$$\tag{8.8}$$

or

$$X_\delta(\omega) = \frac{1}{T_s}[\cdots + X(\omega + 2\omega_s) + X(\omega + \omega_s) + X(\omega) + X(\omega - \omega_s) + X(\omega - 2\omega_s) + \cdots]$$

(8.9)

Equation (8.8) represents the spectrum of the sampled signal $x_\delta(t)$. Thus, the Fourier transform of the sampled signal is given by an infinite sum of shifted versions of the original signal's Fourier transform $X(\omega)$ scaled by $1/T_s$, as illustrated in Fig. 8.2(f).

Case I $\omega_s > 2\omega_h$. When $\omega_s > 2\omega_h$, there is no overlap between the shifted replicas of $X(\omega)$. Consequently $x(t)$ can be recovered exactly from $x_\delta(t)$ by means of a practical low-pass filter with gain T_s and a cutoff frequency ω_c greater than ω_h and less than $\omega_s - \omega_h$, i.e., $\omega_h < \omega_c < (\omega_s - \omega_h)$, as depicted in Fig. 8.3.

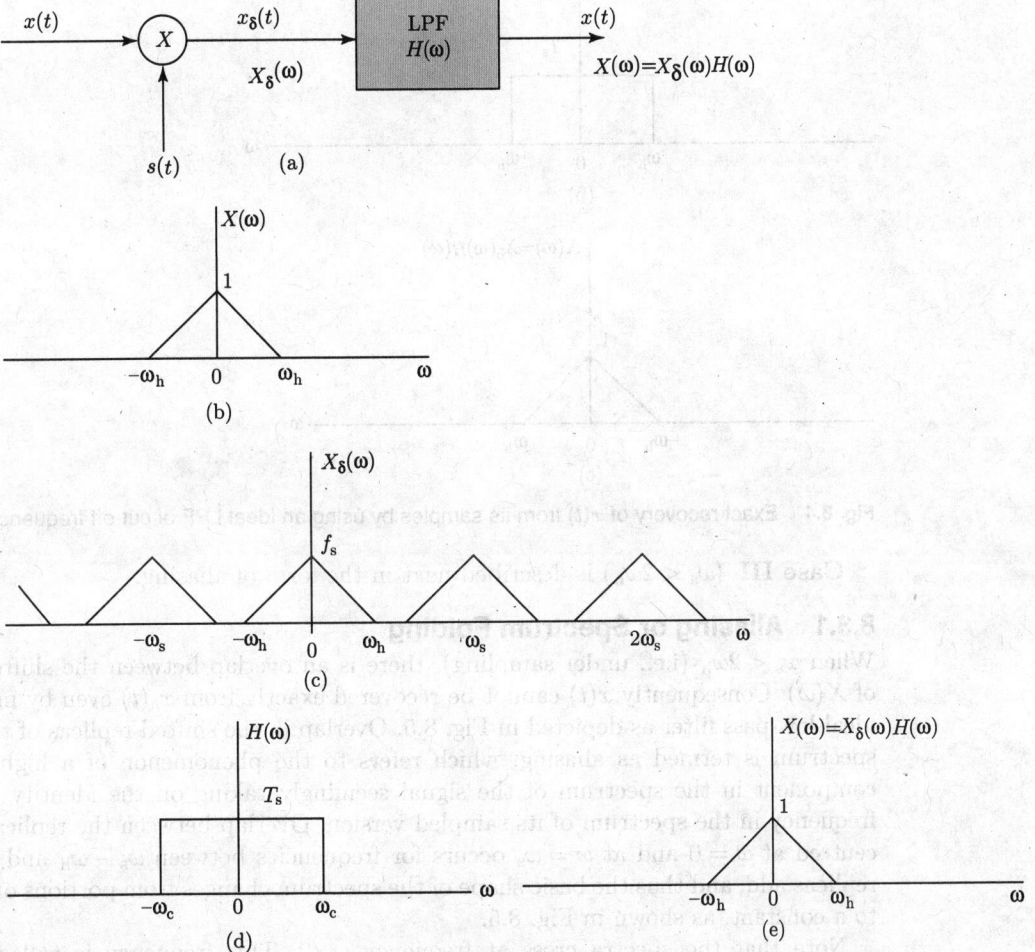

Fig. 8.3 Exact recovery of a continuous-time signal from its samples by using an ideal LPF

Case II $\omega_s = 2\omega_h$. When $\omega_s = 2\omega_h$, there is no overlap between the shifted replicas of $X(\omega)$. Consequently $x(t)$ can be recovered exactly from $x_\delta(t)$ by means of an ideal low-pass filter with gain T_s and a cutoff frequency $\omega_c = \omega_h$, as depicted in Fig. 8.4.

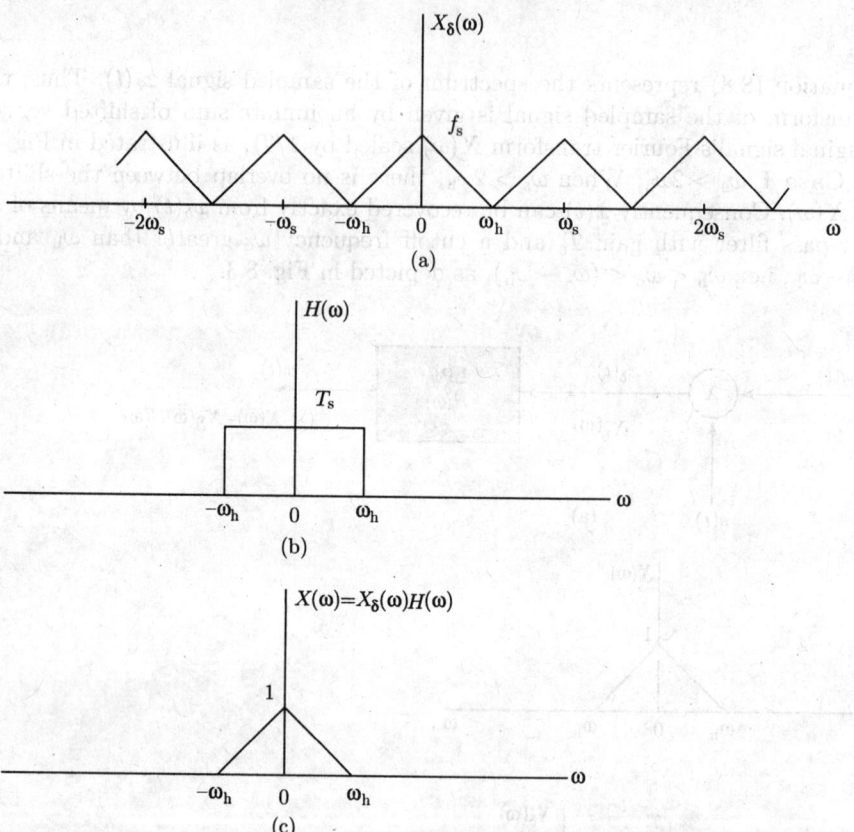

Fig. 8.4 Exact recovery of $x(t)$ from its samples by using an ideal LPF of cut-off frequency $\omega_c = \omega_h$

Case III $(\omega_s < 2\omega_h)$ is described next in the form of aliasing.

8.3.1 Aliasing or Spectrum Folding

When $\omega_s < 2\omega_h$ (i.e., under sampling), there is an overlap between the shifted replicas of $X(\omega)$. Consequently $x(t)$ cannot be recovered exactly from $x_\delta(t)$ even by means of an ideal low-pass filter as depicted in Fig. 8.5. Overlap in the shifted replicas of the original spectrum is termed as aliasing, which refers to the phenomenon of a high-frequency component in the spectrum of the signal seemingly taking on the identity of a lower frequency in the spectrum of its sampled version. Overlap between the replicas of $X(\omega)$ centred at $\omega = 0$ and at $\omega = \omega_s$ occurs for frequencies between $\omega_s - \omega_h$ and ω_h. These replicas add, and thus the basic shape of the spectrum changes from portions of a triangle to a constant, as shown in Fig. 8.5.

Note that the spectra cross at frequency $\omega_s/2$. This frequency is called the *folding frequency*. The spectrum, therefore, folds onto itself at the folding frequency. The

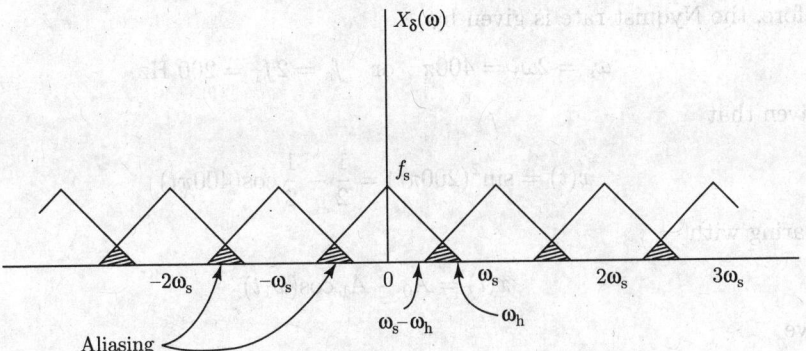

Fig. 8.5 Aliasing or spectrum folding

components of frequencies above $\omega_s/2$ reappear as components of frequencies below $\omega_s/2$. This tail inversion is known as spectral folding or aliasing.

To combat the effect of aliasing in practice, we may use two corrective measure, as described here:

1. Prior to sampling, a *low-pass pre-alias filter* is used to attenuate those high-frequency components of the signal that are not essential to the information conveyed by the signal.
2. The filtered signal is sampled at a rate slightly higher than the Nyquist rate.

The use of a sampling rate higher than the Nyquist rate also has the beneficial effect of easing the design of the reconstruction filter used to recover the original signal from its sampled version.

Example 8.1 Determine the Nyquist rate corresponding to each of the following signals:
(a) $x(t) = \sin(200\pi t)$
(b) $x(t) = \sin^2(200\pi t)$
(c) $x(t) = 1 + \cos(200\pi t) + \sin(400\pi t)$
(d) $x(t) = \cos(150\pi t)\sin(100\pi t)$
(e) $x(t) = \cos^3(200\pi t)$

Solution
(a) Given that

$$x(t) = \sin(200\pi t)$$

Comparing with

$$x(t) = \sin(\omega_1 t)$$

we have

$$\omega_1 = 200\pi \quad \text{or} \quad f_1 = 100$$

Therefore, the Nyquist rate is given by

$$\omega_s = 2\omega_1 = 400\pi \quad \text{or} \quad f_s = 2f_1 = 200 \text{ Hz}$$

(b) Given that

$$x(t) = \sin^2(200\pi t) = \frac{1}{2} - \frac{1}{2}\cos(400\pi t)$$

Comparing with

$$x(t) = A_0 - A_1\cos(\omega_1 t)$$

we have

$$\omega_1 = 400\pi \quad \text{or} \quad f_1 = 200$$

Therefore, the Nyquist rate is given by

$$\omega_s = 2\omega_1 = 800\pi \quad \text{or} \quad f_s = 2f_1 = 400 \text{ Hz}$$

(c) Given that

$$x(t) = 1 + \cos(200\pi t) + \sin(400\pi t)$$

Comparing with

$$x(t) = A_0 + A_1\cos(\omega_1 t) + A_2\sin(\omega_2 t)$$

we have

$$\omega_1 = 200\pi \quad \text{and} \quad \omega_2 = 400\pi$$

Maximum frequency present in the signal $x(t)$ is

$$\omega_{max} = \omega_2 = 400\pi \quad \text{or} \quad f_{max} = 200$$

Therefore, the Nyquist rate is given by

$$\omega_s = 2\omega_{max} = 800\pi \quad \text{or} \quad f_s = 2f_{max} = 400 \text{ Hz}$$

(d) Given that

$$x(t) = \cos(150\pi t)\sin(100\pi t)$$

$$= \frac{1}{2}\sin(250\pi t) - \frac{1}{2}\sin(50\pi t)$$

Comparing with

$$x(t) = A_1\sin(\omega_1 t) - A_2\sin(\omega_2 t)$$

we have

$$\omega_1 = 250\pi \quad \text{and} \quad \omega_2 = 50\pi$$

Maximum frequency present in the signal $x(t)$ is

$$\omega_{max} = \omega_1 = 250\pi \quad \text{or} \quad f_{max} = 125$$

Therefore, the Nyquist rate is given by

$$\omega_s = 2\omega_{max} = 500\pi \quad \text{or} \quad f_s = 2f_{max} = 250 \text{ Hz}$$

(e) Given that

$$x(t) = \cos^3(200\pi t) \qquad \left[\because \cos^3(\theta) = \frac{1}{4}[3\cos(\theta) + \cos(3\theta)] \right]$$

$$= \frac{3}{4}\cos(200\pi t) + \frac{1}{4}\cos(600\pi t)$$

Comparing with

$$x(t) = A_1\cos(\omega_1 t) + A_2\cos(\omega_2 t)$$

we have

$$\omega_1 = 200\pi \quad \text{and} \quad \omega_2 = 600\pi$$

Maximum frequency present in the signal $x(t)$ is

$$\omega_{max} = \omega_2 = 600\pi \quad \text{or} \quad f_{max} = 300$$

Therefore, the Nyquist rate is given by

$$\omega_s = 2\omega_{max} = 1200\pi \quad \text{or} \quad f_s = 2f_{max} = 600 \text{ Hz}$$

Example 8.2 Specify the Nyquist rate and Nyquist interval for each of the following signals:

(a) $x(t) = \text{sinc}(200t)$
(b) $x(t) = \text{sinc}^2(200t)$
(c) $x(t) = \text{sinc}(200t) + \text{sinc}^2(200t)$

Solution
(a) Given that

$$x(t) = \text{sinc}(200t) = \frac{\sin(200\pi t)}{200\pi t}$$

Comparing with

$$x(t) = \frac{\sin(\omega_1 t)}{\omega_1 t}$$

we have

$$\omega_1 = 200\pi \quad \text{or} \quad f_1 = 100$$

Therefore, the Nyquist rate is given by

$$\omega_s = 2\omega_1 = 400\pi \quad \text{or} \quad f_s = 2f_1 = 200 \text{ Hz}$$

(b) Given that

$$x(t) = \text{sinc}^2(200t) = \frac{\sin^2(200\pi t)}{(200\pi t)^2}$$

$$x(t) = \frac{1}{2(200\pi t)^2} - \frac{1}{2(200\pi t)^2}\cos(400\pi t)$$

Comparing with

$$x(t) = A_0 - A_1\cos(\omega_1 t)$$

we have

$$\omega_1 = 400\pi \quad \text{or} \quad f_1 = 200$$

Therefore, the Nyquist rate is given by

$$\omega_s = 2\omega_1 = 800\pi \quad \text{or} \quad f_s = 2f_1 = 400 \text{ Hz}$$

(c) Given that

$$x(t) = \text{sinc}(200t) + \text{sinc}^2(200t)$$

$$= \frac{\sin(200\pi t)}{200\pi t} + \frac{\sin^2(200\pi t)}{(200\pi t)^2}$$

$$x(t) = \frac{1}{2(200\pi t)^2} - \frac{1}{2(200\pi t)^2}\cos(400\pi t) + \frac{\sin(200\pi t)}{200\pi t}$$

Comparing with

$$x(t) = A_0 - A_1\cos(\omega_1 t) + A_2\sin(\omega_2 t)$$

we have

$$\omega_1 = 400\pi \quad \text{and} \quad \omega_2 = 200\pi$$

Maximum frequency present in the signal $x(t)$ is

$$\omega_{\max} = \omega_1 = 400\pi \quad \text{or} \quad f_{\max} = 200$$

Therefore, the Nyquist rate is given by

$$\omega_s = 2\omega_{\max} = 800\pi \quad \text{or} \quad f_s = 2f_{\max} = 400 \text{ Hz}$$

Example 8.3 A continuous-time signal $x(t)$ is obtained at the output of an ideal low-pass filter with cutoff frequency $\omega_c = 1000\pi$. If instantaneous sampling is performed on $x(t)$, which of the following sampling periods would guarantee that $x(t)$ can be recovered from its sampled version using an appropriate low-pass filter?
(a) $T_s = 0.5 \times 10^{-3}$
(b) $T_s = 2 \times 10^{-3}$
(c) $T_s = 10^{-4}$

Solution

Since $x(t)$ is obtained at the output of an ideal low-pass filter with cutoff frequency $\omega_c = 1000\pi$, the maximum frequency present in the signal $x(t)$ is $\omega_{max} = 1000\pi$ or $f_{max} = 500$ Hz. Therefore, the sampling frequency is given by

$$\omega_s \geq 2\omega_{max} = 2000\pi \quad \text{or} \quad f_s \geq 2f_{max} = 1000 \text{ Hz}$$

The Nyquist interval is given by

$$T_s \leq \frac{1}{2f_{max}} = \frac{1}{1000} = 10^{-3}$$

$$T_s \leq 10^{-3}$$

Only (a) and (c) satisfy this condition.

Example 8.4 Let $x(t)$ be a signal with Nyquist rate ω_0. Determine the Nyquist rate for each of the following signals:

(a) $x(t+1)$

(b) $x(t) + x(t-1)$

(c) $\dfrac{dx(t)}{dt}$

(d) $x^2(t)$

(e) $x(t)\cos(\omega_0 t)$

(f) $\displaystyle\int_{-\infty}^{t} x(\tau)\,d\tau$

Solution

If the signal $x(t)$ has a Nyquist rate $\omega_s = 2\omega_{max} = \omega_0$, then the maximum frequency present in the signal $x(t)$ is $\omega_{max} = \omega_0/2$ and its Fourier transform $X(\omega) = 0$ for $|\omega| > \omega_0/2$. Note that, when the frequency of the signal changes, the Nyquist rate changes accordingly.

(a) If

$$x(t) \longleftrightarrow X(\omega)$$

then

$$y(t) = x(t+1) \longleftrightarrow Y(\omega) = X(\omega)\,e^{j\omega}$$

Clearly, $Y(\omega) = 0$ for $|\omega| > \omega_0/2$ and the maximum frequency present in the signal $y(t) = x(t+1)$ is $\omega_{max} = \omega_0/2$. The time-shifting operation does not change the frequency of the signal. Therefore, the Nyquist rate for $x(t+1)$ is given by

$$\omega_s = 2\omega_{max} = \omega_0$$

(b) If

$$x(t) \longleftrightarrow X(\omega)$$

then

$$y(t) = x(t) + x(t-1) \longleftrightarrow Y(\omega) = X(\omega) + X(\omega)\,e^{j\omega} = X(\omega)[1 + e^{j\omega}]$$

Clearly, $Y(\omega) = 0$ for $|\omega| > \omega_0/2$ and the maximum frequency present in the signal $y(t) = x(t) + x(t-1)$ is $\omega_{max} = \omega_0/2$. Therefore, the Nyquist rate for $x(t) + x(t-1)$ is given by

$$\omega_s = 2\omega_{max} = \omega_0$$

(c) If

$$x(t) \longleftrightarrow X(\omega)$$

then

$$y(t) = \frac{dx(t)}{dt} \longleftrightarrow Y(\omega) = j\omega\, X(\omega)$$

Clearly, $Y(\omega) = 0$ for $|\omega| > \omega_0/2$ and the maximum frequency present in the signal $y(t) = dx(t)/dt$ is $\omega_{max} = \omega_0/2$. Therefore, the Nyquist rate for $dx(t)/dt$ is given by

$$\omega_s = 2\omega_{max} = \omega_0$$

(d) If

$$x(t) \longleftrightarrow X(\omega)$$

then

$$y(t) = x^2(t) = x(t)x(t) \longleftrightarrow Y(\omega) = \frac{1}{2\pi}[X(\omega) * X(\omega)]$$

Clearly, $Y(\omega) = 0$ for $|\omega| > (\omega_0/2 + \omega_0/2)$, i.e., $|\omega| > \omega_0$, and the maximum frequency present in the signal $y(t) = x^2(t)$ is $\omega_{max} = \omega_0$. (If we multiply two signals with the frequencies ω_1 and ω_2, the frequency of the resultant signal is $\omega_1 + \omega_2$.) Therefore, the Nyquist rate for $x^2(t)$ is given by

$$\omega_s = 2\omega_{max} = 2\omega_0$$

(e) If

$$x(t) \longleftrightarrow X(\omega)$$

then

$$y(t) = x(t)\cos(\omega_0 t) \longleftrightarrow Y(\omega) = \pi[X(\omega - \omega_0) + X(\omega + \omega_0)]$$

Clearly, $Y(\omega) = 0$ for $|\omega| > (\omega_0 + \omega_0/2)$, i.e., $|\omega| > 3\omega_0/2$, and the maximum frequency present in the signal $y(t) = x(t)\cos(\omega_0 t)$ is $\omega_{max} = 3\omega_0/2$. (If we multiply two signals with the frequencies ω_1 and ω_2, the frequency of the resultant signal is $\omega_1 + \omega_2$.) Therefore, the Nyquist rate for $x(t)\cos(\omega_0 t)$ is given by

$$\omega_s = 2\omega_{max} = \frac{3\omega_0}{2}$$

(f) If

$$x(t) \longleftrightarrow X(\omega)$$

then

$$y(t) = \int\limits_{-\infty}^{t} x(\tau)\, d\tau = x(t) * u(t) \longleftrightarrow Y(\omega) = \pi X(0)\delta(\omega) + \frac{X(\omega)}{j\omega}$$

Clearly, $Y(\omega) = 0$ for $|\omega| > \omega_0/2$ and the maximum frequency present in the signal $y(t) = \int\limits_{-\infty}^{t} x(\tau)\, d\tau$ is $\omega_{\max} = \omega_0/2$. Therefore, the Nyquist rate for $\int\limits_{-\infty}^{t} x(\tau)\, d\tau$ is given by

$$\omega_s = 2\omega_{\max} = \omega_0$$

Example 8.5 If the Nyquist rate for $x(t)$ is ω_s. Find the Nyquist rate for each of the following signals:

(a) $x(2t)$

(b) $x(t/3)$

(c) $x(t) * x(t)$

Solution

If the signal $x(t)$ has a Nyquist rate $\omega_s = 2\omega_{\max}$, then the maximum frequency present in the signal $x(t)$ is $\omega_{\max} = \omega_s/2$ and its Fourier transform $X(\omega) = 0$, for $|\omega| > \omega_s/2$. Note that when the frequency of the signal changes, the Nyquist rate changes accordingly.

(a) If

$$x(t) \longleftrightarrow X(\omega)$$

then

$$y(t) = x(2t) \longleftrightarrow Y(\omega) = \frac{1}{2}X\left(\frac{\omega}{2}\right)$$

Clearly, $Y(\omega) = 0$ for $|\omega| > \omega_0$ and the maximum frequency present in the signal $y(t) = x(2t)$ is $\omega_{\max} = \omega_0$. The time-scaling operation changes the frequency of the signal. Therefore, the Nyquist rate for $x(2t)$ is given by

$$\omega_s = 2\omega_{\max} = 2\omega_0$$

(b) If

$$x(t) \longleftrightarrow X(\omega)$$

then

$$y(t) = x\left(\frac{t}{3}\right) \longleftrightarrow Y(\omega) = 3X(3\omega)$$

Clearly, $Y(\omega) = 0$ for $|\omega| > \omega_0/6$ and the maximum frequency present in the signal $y(t) = x(t/3)$ is $\omega_{\max} = \omega_0/6$. Therefore, the Nyquist rate for $x(t/3)$ is given by

$$\omega_s = 2\omega_{\max} = \frac{\omega_0}{3}$$

(c) If

$$x(t) \longleftrightarrow X(\omega)$$

then

$$y(t) = x(t) * x(t) \longleftrightarrow Y(\omega) = X(\omega)X(\omega) = X^2(\omega)$$

Clearly, $Y(\omega) = 0$ for $|\omega| > \omega_0$ and the maximum frequency present in the signal $y(t) = x(t) * x(t)$ is $\omega_{max} = \omega_0$. Therefore, the Nyquist rate for $x(t) * x(t)$ is given by

$$\omega_s = 2\omega_{max} = 2\omega_0$$

8.4 SAMPLING TECHNIQUES

In the last section we studied what is called instantaneous sampling or ideal sampling. The samples were of height equal to the signal $x(t)$ at sampling instants $t = nT_s$, $n = 0, \pm 1, \pm 2, \ldots$. But this type of sampling is not possible in practice. Practical sampling differs from ideal sampling in the following aspects:

1. The sampling signal is a pulse train rather than an impulse train, i.e., the sampled pulses have finite duration rather than impulses. The amplitude of these pulses is also finite.
2. The practical reconstruction filters are not ideal filters.

Natural sampling and flat top sampling are practical sampling methods. Flat top sampling is preferred because of its noise immunity. There are three types of sampling techniques, namely,

1. ideal or instantaneous or impulse sampling,
2. natural or chopper sampling, and
3. flat top sampling.

8.5 IMPULSE SAMPLING OR IDEAL SAMPLING OR INSTANTANEOUS SAMPLING

Ideal sampling is same as instantaneous sampling. Figure 8.6 shows the switching sampler. If the closing time t of the switch approaches zero, the output $x_s(t)$ gives only instantaneous value. The instantaneous sampling gives train of impulses. The area of each impulse in the sampled version is equal to instantaneous value of the input signal $x(t)$. In ideal sampling, the sampling signal is a periodic impulse train that can be expressed as

$$s(t) = \sum_{n=-\infty}^{\infty} \delta(t - nT_s)$$

The sampled signal $x_\delta(t)$ is obtained by multiplying the signal $x(t)$ with impulse train $s(t)$. Therefore, $x_\delta(t)$ can be expressed mathematically as

$$x_\delta(t) = x(t)s(t)$$

Since the width of the impulse in $x_\delta(t)$ approaches zero, it represents only instantaneous value. Therefore, this method is called *instantaneous sampling*. Taking the Fourier transform of the above equation and using the multiplication property [Eq. (5.39)] of the Fourier transform, we obtain

$$X_\delta(\omega) = \frac{1}{2\pi}[X(\omega) * S(\omega)] \tag{8.10}$$

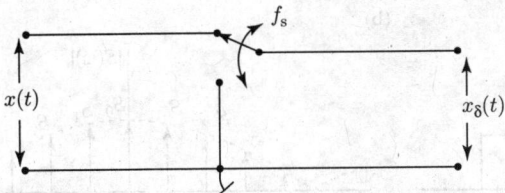

Fig. 8.6 Switching sampler

We know that the Fourier transform of a periodic impulse train in the time domain with period T_s is a periodic impulse train in the frequency domain with period $\omega_s = 2\pi/T_s$, i.e.,

$$S(\omega) = \frac{2\pi}{T_s} \sum_{n=-\infty}^{\infty} \delta(\omega - n\omega_s) \tag{8.11}$$

Substituting $S(\omega)$ into Eq. (8.10), we obtain

$$X_\delta(\omega) = \frac{1}{2\pi} \left[X(\omega) * \frac{2\pi}{T_s} \sum_{n=-\infty}^{\infty} \delta(\omega - n\omega_s) \right]$$

$$= \frac{1}{T_s} \left[X(\omega) * \sum_{n=-\infty}^{\infty} \delta(\omega - n\omega_s) \right]$$

Since convolution with an impulse simply shifts a signal [i.e., $X(\omega) * \delta(\omega - \omega_s) = X(\omega - \omega_s)$], it follows that

$$X_\delta(\omega) = \frac{1}{T_s} \sum_{n=-\infty}^{\infty} X(\omega - n\omega_s) \tag{8.12}$$

The above equation gives the spectrum of an ideally sampled signal. Instantaneous sampling is possible only in theory because it is not possible to generate impulse train. The spectrum of the sampled signal is shown in Fig. 8.2(f).

8.6 NATURAL SAMPLING OR CHOPPER SAMPLING

Although instantaneous sampling is a convenient model, a more practical way of accomplishing the sampling of a band-limited continuous-time signal $x(t)$ is to multiply $x(t)$, shown in Fig. 8.7(a), by the sampling signal (pulse train or switching waveform) $s(t)$, shown in Fig. 8.7(c). Each pulse in the sampling signal $s(t)$ has width T and amplitude $1/T$. Multiplication by $s(t)$ can be viewed as the opening and closing of a switch as shown

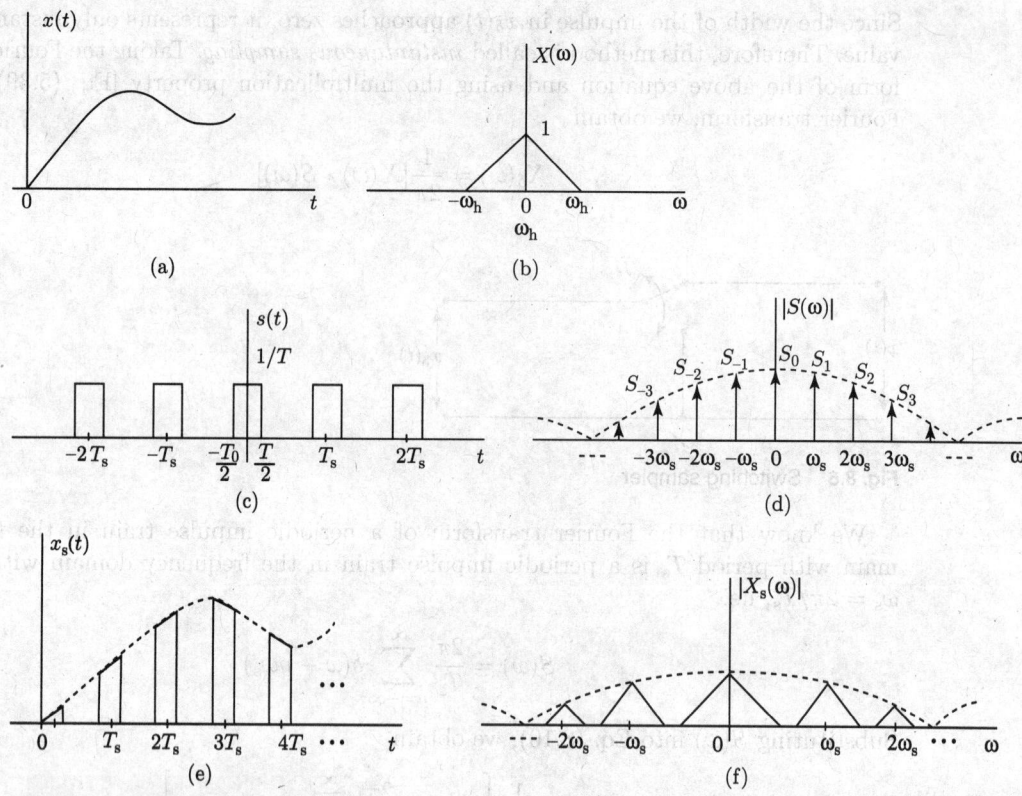

Fig. 8.7 Natural sampling

in Fig. 8.8. When $s(t)$ goes high, a switch 's' is closed and the sampled signal $x_s(t) = x(t)$, and when $s(t)$ goes low, the switch 's' is open and the sampled signal $x_s(t) = 0$. As before, the sampling frequency is designated ω_s (or f_s), and T_s is the sampling interval. The resulting sampled signal $x_s(t)$ is illustrated in Fig. 8.7(e) and is expressed as

$$x_s(t) = x(t)s(t) \tag{8.13}$$

The sampling here is termed *natural sampling*, since the top of each pulse in the sampled signal $x_s(t)$ retains the shape of the signal $x(t)$ during the pulse interval. Natural sampling is sometimes called chopper sampling because the waveform of the sampled signal $x_s(t)$ appears to be chopped off from the original signal $x(t)$.

Here, the sampling signal $s(t)$ is periodic with period T_s and frequency $\omega_s = 2\pi/T_s$. We can express the periodic pulse train as a Fourier series [Eq. (3.26)] in the form

$$s(t) = \sum_{n=-\infty}^{\infty} S_n\, e^{jn\omega_s t} \tag{8.14}$$

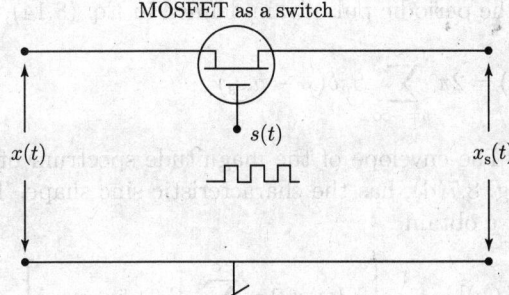

Fig. 8.8 Natural sampler

where

$$S_{\mathrm{n}} = \frac{1}{T_{\mathrm{s}}} \int\limits_{-T_{\mathrm{s}}/2}^{T_{\mathrm{s}}/2} s(t)\, e^{-jn\omega_{\mathrm{s}}t}\, dt$$

We can defined the sampling signal $s(t)$ for one period as

$$s(t) = \begin{cases} \dfrac{1}{T}, & -T/2 \le t \le T/2 \\[2mm] 0, & T/2 < |t| < T \end{cases}$$

Therefore,

$$S_{\mathrm{n}} = \frac{1}{T_{\mathrm{s}}} \int\limits_{-T/2}^{T/2} \frac{1}{T}\, e^{-jn\omega_{\mathrm{s}}t}\, dt = \frac{1}{T_{\mathrm{s}}T} \frac{e^{-jn\omega_{\mathrm{s}}t}}{-jn\omega_{\mathrm{s}}}\bigg|_{-T/2}^{T/2}$$

$$= \frac{1}{T_{\mathrm{s}}T} \frac{e^{-jn\omega_{\mathrm{s}}T/2} - e^{jn\omega_{\mathrm{s}}T/2}}{-jn\omega_{\mathrm{s}}}$$

$$= \frac{1}{n\pi(T/T_{\mathrm{s}})T_{\mathrm{s}}} \frac{e^{jn\omega_{\mathrm{s}}T/2} - e^{-jn\omega_{\mathrm{s}}T/2}}{2j}$$

$$= \frac{1}{n\pi(T/T_{\mathrm{s}})T_{\mathrm{s}}} \sin\left(n\omega_{\mathrm{s}}\frac{T}{2}\right)$$

$$S_{\mathrm{n}} = \frac{1}{T_{\mathrm{s}}} \frac{\sin(n\pi T/T_{\mathrm{s}})}{n\pi T/T_{\mathrm{s}}}$$

$$S_{\mathrm{n}} = \frac{1}{T_{\mathrm{s}}} \operatorname{sinc}\left(\frac{nT}{T_{\mathrm{s}}}\right) \tag{8.15}$$

Taking the Fourier transform of the periodic pulse train defined in Eq. (8.14), we obtain

$$S(\omega) = 2\pi \sum_{n=-\infty}^{\infty} S_n \delta(\omega - n\omega_s) \tag{8.16}$$

where S_n is given by Eq. (8.15). The envelope of the magnitude spectrum of the pulse train, seen as a dashed line in Fig. 8.7(d), has the characteristic sinc shape. Taking the Fourier transform of Eq. (8.13), we obtain

$$X_s(\omega) = \frac{1}{2\pi}[X(\omega) * S(\omega)] = \frac{1}{2\pi}\left[X(\omega) * 2\pi \sum_{n=-\infty}^{\infty} S_n \delta(\omega - n\omega_s)\right]$$

$$X_s(\omega) = \sum_{n=-\infty}^{\infty} S_n X(\omega - n\omega_s) \tag{8.17}$$

$$X_s(\omega) = \frac{1}{T_s} \sum_{n=-\infty}^{\infty} \text{sinc}\left(\frac{nT}{T_s}\right) X(\omega - n\omega_s) \tag{8.18}$$

Similar to the unit impulse sampling case, Eq. (8.18) and Fig. 8.7(f) illustrate that $X_s(\omega)$ is a replication of $X(\omega)$, periodically repeated in frequency every ω_s rad/s. In this natural-sampled case, however, the spectrum $X_s(\omega)$ is weighted by sinc function compared with a constant value in the impulse sampled case.

Note that *in the limit*, as the pulse width $T \longrightarrow 0$, the *pulse train* $\longrightarrow$ *impulse train*, $\text{sinc}(nT/T_s) \longrightarrow 1$, $S_n \longrightarrow 1/T_s$ for all n, and Eq. (8.18) converges to Eq. (8.12). Hence, as the pulse width $T \longrightarrow 0$, the natural sampling approaches to impulse sampling.

8.7 FLAT-TOP SAMPLING

Flat-top sampling is somewhat similar to natural sampling, where the continuous-time signal $x(t)$ is multiplied by a periodic rectangular pulse train. However, in natural sampling the top of each rectangular pulse in the sampled signal $x_s(t)$ varies with the signal $x(t)$, whereas in flat-top sampled signal the top of each rectangular pulse remains constant and equal to instantaneous value of the signal $x(t)$ at the start of sampling. There are two operations involved in the generation of a flat-top sampled signal:

1. *Instantaneous sampling* of the continuous-time signal $x(t)$ every T_s seconds, where the sampling rate $f_s = 1/T_s$ is chosen in accordance with the sampling theorem.

2. *Lengthening* the duration of each sample so obtained to some constant value T.

These two operations are jointly referred to as 'sample and hold'. One important reason for intentionally lengthening the duration of each pulse is to reduce the bandwidth. Figure 8.9(a) shows the functional diagram of the sample and hold circuit generating flat-top samples and Fig. 8.9(c) shows the flat-top sampled signal $x_s(t)$.

We can see from Fig. 8.9(c) that only the starting edge of the pulse represents instantaneous value of the signal $x(t)$. The flat-top sampled signal $x_s(t)$ is mathematically equivalent to the convolution of instantaneous sampled signal $x_\delta(t)$ and a rectangular

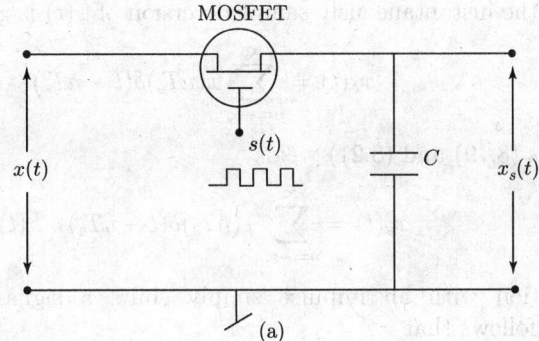

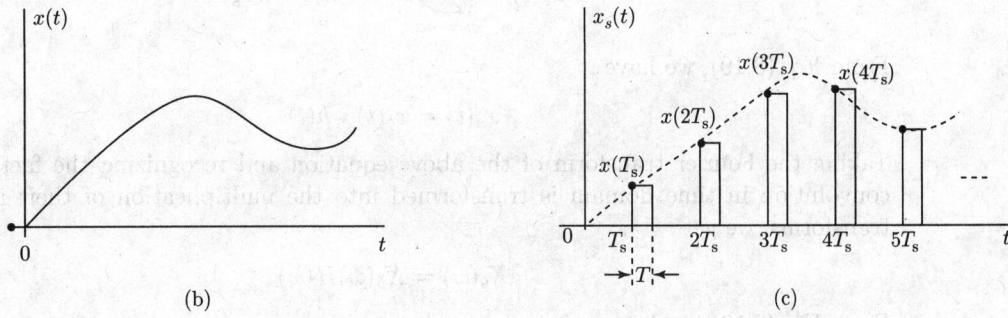

Fig. 8.9 (a) Sample and hold circuit, (b) continuous-time signal, and (c) flat-top sampled signal $x_s(t)$

pulse $h(t)$. Therefore,

$$x_s(t) = x_\delta(t) * h(t) \tag{8.19}$$

Let $h(t)$ be a standard rectangular pulse of unit amplitude and duration T (shown in Fig. 8.10), defined as follows:

$$h(t) = \begin{cases} 1, & 0 < t < T \\ 0, & \text{otherwise} \end{cases} \tag{8.20}$$

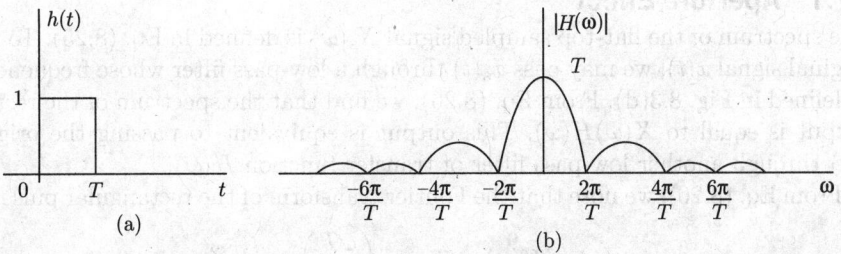

Fig. 8.10 (a) Rectangular pulse $h(t)$ and (b) its magnitude spectrum

By definition, the instantaneously sampled version of $x(t)$ is given by [Eq. (8.5)]

$$x_\delta(t) = \sum_{n=-\infty}^{\infty} x(nT_s)\delta(t - nT_s) \tag{8.21}$$

Combining Eqs (8.19) and (8.21) yields

$$x_s(t) = \sum_{n=-\infty}^{\infty} x(nT_s)\delta(t - nT_s) * h(t)$$

Since convolution with an impulse simply shifts a signal [i.e., $h(t) * \delta(t - nT_s) = h(t - nT_s)$], it follows that

$$x_s(t) = \sum_{n=-\infty}^{\infty} x(nT_s)h(t - nT_s) \tag{8.22}$$

From Eq. (8.19), we have

$$x_s(t) = x_\delta(t) * h(t)$$

Taking the Fourier transform of the above equation and recognizing the fact that the convolution in time domain is transformed into the multiplication of their respective transforms, we get

$$X_s(\omega) = X_\delta(\omega)H(\omega) \tag{8.23}$$

From Eq. (8.12), we have

$$X_\delta(\omega) = \frac{1}{T_s} \sum_{n=-\infty}^{\infty} X(\omega - n\omega_s) \tag{8.24}$$

Substituting Eq. (8.24) in Eq. (8.23) yields

$$X_s(\omega) = \frac{1}{T_s} \sum_{n=-\infty}^{\infty} X(\omega - n\omega_s)H(\omega) \tag{8.25}$$

$$X_s(\omega) = \frac{1}{T_s}[\cdots + X(\omega + \omega_s)H(\omega) + X(\omega)H(\omega) + X(\omega - \omega_s)H(\omega) + \cdots] \tag{8.26}$$

Equation (8.25) represents the spectrum of the flat-top sampled signal.

8.7.1 Aperture Effect

The spectrum of the flat-top sampled signal $X_s(\omega)$ is defined in Eq. (8.25). To recover the original signal $x(t)$, we may pass $x_s(t)$ through a low-pass filter whose frequency response is defined in Fig. 8.3(d). From Eq. (8.26), we find that the spectrum of the resulting filter output is equal to $X(\omega)H(\omega)$. This output is equivalent to passing the original signal $x(t)$ through another low-pass filter of transfer function $H(\omega)$.

From Eq. (8.20), we note that the Fourier transform of the rectangular pulse is given by

$$H(\omega) = T\,\text{sinc}\left(\frac{\omega T}{2\pi}\right) e^{-j\omega T/2} \tag{8.27}$$

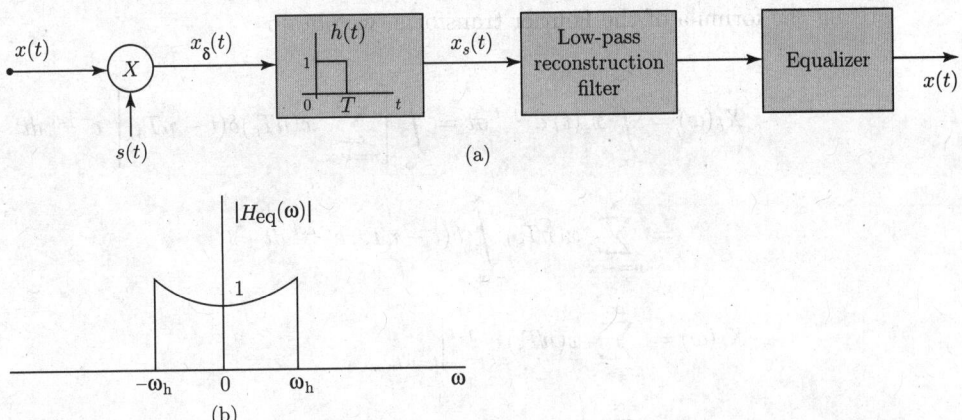

Fig. 8.11 (a) Overall system for flat top sampling and reconstruction and (b) magnitude spectrum of equalizer.

which is shown in Fig. 8.10(b). Thus, we can see from Fig. 8.10(b) that by using flat-top samples an amplitude distortion is introduced in the reconstructed signal $x(t)$. The high frequency roll-off of $H(\omega)$ acts like a low-pass filter and attenuates upper portion (high frequencies) of the spectrum $X(\omega)$. The distortion caused by the nonuniform spectral gain $H(\omega)$ on high frequencies of $x(t)$ is called the *aperture effect*.

This distortion may be corrected by connecting an equalizer in cascade with the low-pass reconstruction filter, as shown in Fig. 8.11(a). The equalizer has the effect of decreasing the in-band loss of the reconstruction filter as the frequency increases in such a manner so as to compensate for the aperture effect. Ideally, the frequency response of the equalizer is given by

$$H_{\text{eq}}(\omega) = \frac{1}{H(\omega)} = \frac{e^{j\omega T/2}}{T \operatorname{sinc}(\omega T/2\pi)} \tag{8.28}$$

The magnitude spectrum $|H_{\text{eq}}(\omega)|$ is shown in Fig. 8.11(b).

8.8 RECONSTRUCTION OF A SIGNAL FROM ITS SAMPLES USING INTERPOLATION

In ideal sampling, the sampling signal is a periodic impulse train that can be expressed as

$$s(t) = \sum_{n=-\infty}^{\infty} \delta(t - nT_{\text{s}})$$

The sampled signal $x_\delta(t)$ is obtained by multiplying the signal $x(t)$ with impulse train $s(t)$. Therefore, $x_\delta(t)$ can be expressed mathematically as

$$x_\delta(t) = x(t)s(t) = x(t) \sum_{n=-\infty}^{\infty} \delta(t - nT_{\text{s}}) = \sum_{n=-\infty}^{\infty} x(nT_{\text{s}})\delta(t - nT_{\text{s}})$$

Using the formula of the Fourier transform, we obtain

$$X_\delta(\omega) = \int_{-\infty}^{\infty} x_s(t)\, e^{-j\omega t}\, dt = \int_{-\infty}^{\infty} \left[\sum_{n=-\infty}^{\infty} x(nT_s)\delta(t - nT_s) \right] e^{-j\omega t}\, dt$$

$$= \sum_{n=-\infty}^{\infty} x(nT_s) \int_{-\infty}^{\infty} \delta(t - nT_s)\, e^{-j\omega t}\, dt$$

$$X_\delta(\omega) = \sum_{n=-\infty}^{\infty} x(nT_s)\, e^{-j\omega t}\bigg|_{t=nT_s}$$

$$X_\delta(\omega) = \sum_{n=-\infty}^{\infty} x(nT_s)\, e^{-j\omega n T_s} \tag{8.29}$$

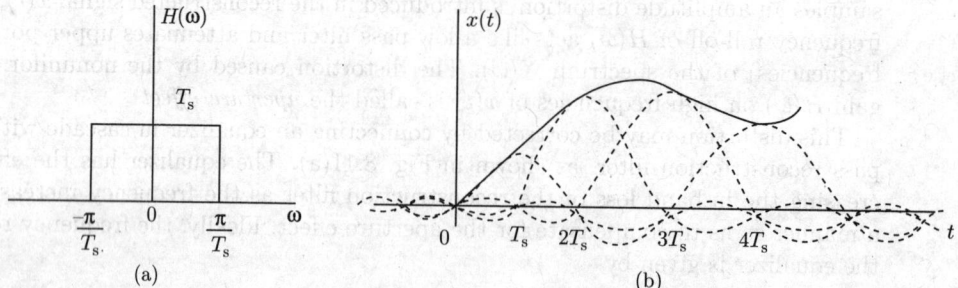

Fig. 8.12 (a) Frequency response of an ideal low-pass filter and (b) the reconstructed signal.

The process of reconstructing a continuous-time signal $x(t)$ from its sampled version is also known as *interpolation*. In Section 8.3, we saw that a signal $x(t)$ band limited to f_h Hz can be reconstructed (interpolated) exactly from its samples if the sampling frequency $f_s \geq 2f_h$ or the sampling interval $T_s \leq 1/2f_h$. This reconstruction is accomplished by passing the sampled signal $x_s(t)$ through an ideal low-pass filter of gain T_s and having a bandwidth of any value between f_h and $(f_s - f_h)$ Hz [or between ω_h and $\omega_s - \omega_h$ as depicted in Fig. 8.3(d)]. From a practical view point, a good choice is the middle value $f_s/2 = 1/2T_s$ Hz or π/T_s rad/s. This value allows for small deviations in the ideal filter characteristics on either side of the cutoff frequency. With this choice of cutoff frequency and gain T_s, the ideal low-pass filter [as shown in Fig. 8.12(a)] required for signal reconstruction (or interpolation) is

$$H(\omega) = T_s \operatorname{rect}\left(\frac{\omega T_s}{2\pi}\right) = \begin{cases} T_s, & -\pi/T_s < \omega < \pi/T_s \\ 0, & \text{elsewhere} \end{cases} \tag{8.30}$$

In frequency domain, the output of the low-pass filter is given by

$$X(\omega) = X_\delta(\omega)H(\omega)$$

Using inverse Fourier transform formula, we get

$$x(t) = \frac{1}{2\pi}\int\limits_{-\infty}^{\infty} X(\omega)\,e^{j\omega t}d\omega = \frac{1}{2\pi}\int\limits_{-\infty}^{\infty} X_\delta(\omega)H(\omega)\,e^{j\omega t}d\omega = \frac{1}{2\pi}\int\limits_{-\pi/T_s}^{\pi/T_s} T_s X_\delta(\omega)\,e^{j\omega t}d\omega$$

Substituting $X_\delta(\omega)$ from Eq. (8.12) into the above equation gives

$$x(t) = \frac{T_s}{2\pi}\int\limits_{-\pi/T_s}^{\pi/T_s}\left[\sum_{n=-\infty}^{\infty} x(nT_s)\,e^{-j\omega nT_s}\right]e^{j\omega t}d\omega$$

$$x(t) = \frac{T_s}{2\pi}\sum_{n=-\infty}^{\infty} x(nT_s)\int\limits_{-\pi/T_s}^{\pi/T_s} e^{j\omega(t-nT_s)}d\omega$$

$$= \frac{T_s}{2\pi}\sum_{n=-\infty}^{\infty} x(nT_s)\frac{e^{j\omega(t-nT_s)}}{j(t-nT_s)}\bigg|_{-\pi/T_s}^{\pi/T_s}$$

$$= \frac{T_s}{\pi(t-nT_s)}\sum_{n=-\infty}^{\infty} x(nT_s)\frac{e^{j\pi/T_s(t-nT_s)}-e^{-j\pi/T_s(t-nT_s)}}{2j}$$

$$x(t) = \sum_{n=-\infty}^{\infty} x(nT_s)\frac{\sin\left(\pi/T_s(t-nT_s)\right)}{\pi/T_s(t-nT_s)}$$

$$x(t) = \sum_{n=-\infty}^{\infty} x(nT_s)\,\mathrm{sinc}\left(\frac{1}{T_s}(t-nT_s)\right) \tag{8.31}$$

For the case of Nyquist sampling rate, $T_s = 1/2f_h$, the above equation simplifies to

$$x(t) = \sum_{n=-\infty}^{\infty} x(nT_s)\,\mathrm{sinc}\left(2f_h(t-nT_s)\right) \tag{8.32}$$

Equation (8.32) is the *interpolation formula*, which yields values of $x(t)$ between samples as a weighted sum of all the sample values as shown in Fig. 8.12(b).

The interpolation process here is expressed in the frequency domain as a filtering operation. Now we shall examine this process from the time-domain viewpoint.

Time domain view (Interpolation): The ideal interpolation filter frequency response obtained in Eq. (8.30) is illustrated in Fig. 8.12(a). The impulse response of this interpolating filter is given by

$$h(t) = \mathcal{F}^{-1}[H(\omega)] = \mathcal{F}^{-1}\left[T_s \operatorname{rect}\left(\frac{\omega T_s}{2\pi}\right)\right]$$

$$h(t) = \operatorname{sinc}\left(\frac{t}{T_s}\right) \tag{8.33}$$

This $h(t)$ is depicted in Fig. 8.13. We can observe that $h(t) = 0$ at all Nyquist sampling instants $t = \pm nT_s$ except at $t = 0$. When the sampled signal $x_\delta(t)$ is applied to this filter, the output is $x(t)$. Each sample in $x_\delta(t)$, being an impulse, generates a sinc pulse of height equal to the strength of the sample, as illustrated in Fig. 8.12(b). The output of the filter $x(t)$ is obtained by the convolution of the input $x_\delta(t)$ with the impulse response $h(t)$:

$$x(t) = x_\delta(t) * h(t) = \sum_{n=-\infty}^{\infty} x(nT_s)\delta(t - nT_s) * h(t)$$

$$x(t) = \sum_{n=-\infty}^{\infty} x(nT_s)h(t - nT_s) \tag{8.34}$$

$$x(t) = \sum_{n=-\infty}^{\infty} x(nT_s)\operatorname{sinc}\left(\frac{1}{T_s}(t - nT_s)\right) \tag{8.35}$$

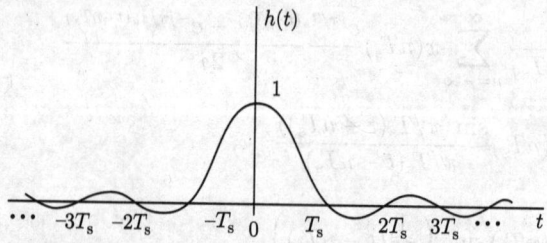

Fig. 8.13   Impulse response of the interpolating (low-pass) filter.

8.8.1   Zero-order Hold Interpolation

The zero-order hold can be viewed as a form of interpolation between sample values in which the impulse response $h(t)$ of the interpolating filter is a rectangular pulse as depicted in Fig. 8.14(a).

$$h(t) = u(t) - u(t - T_s) = \operatorname{rect}\left(\frac{t - T_s/2}{T_s}\right) = \begin{cases} 1, & 0 < t < T_s \\ 0, & \text{otherwise} \end{cases} \tag{8.36}$$

When the sampled signal $x_\delta(t)$ is applied to this filter, the output is $x_r(t)$. Each sample in $x_\delta(t)$, being an impulse, generates a rectangular pulse of height equal to the strength of the sample, as illustrated in Fig. 8.14(b). The output of the filter $x_r(t)$ is obtained by the convolution of the input $x_\delta(t)$ with the impulse response $h(t)$:

$$x_r(t) = x_\delta(t) * h(t) = \sum_{n=-\infty}^{\infty} x(nT_s)\delta(t - nT_s) * h(t)$$

$$x_r(t) = \sum_{n=-\infty}^{\infty} x(nT_s)h(t - nT_s) \tag{8.37}$$

Substitution of $h(t)$ from Eq. (8.36) in Eq. (8.37) yields

$$x_r(t) = \sum_{n=-\infty}^{\infty} x(nT_s)\, \text{rect}\left(\frac{t - nT_s - (T_s/2)}{T_s}\right) \tag{8.38}$$

The filter output is a staircase approximation of $x(t)$. Figure 8.14(c) shows the magnitude of the transfer function of the zero-order hold interpolating filter superimposed on the desired transfer function of the exact interpolating filter.

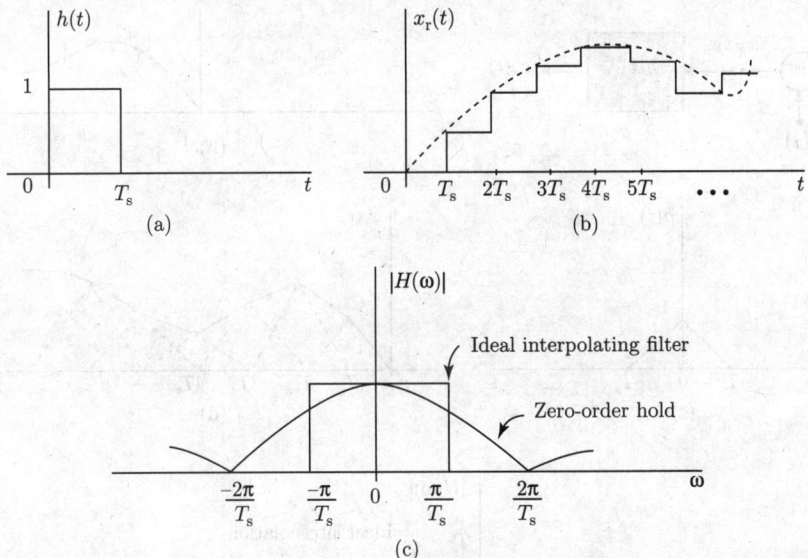

Fig. 8.14 (a) Impulse response of a zero-order hold interpolating filter, (b) output of the interpolating filter and (c) frequency response for the zero-order hold and for the ideal interpolating filter.

8.8.2 First-order Hold Interpolation (or Linear Interpolation)

The first-order hold can be viewed as a form of interpolation between sample values in which the impulse response $h(t)$ of the interpolating filter is a triangular pulse, as depicted in Fig. 8.15(c).

$$h(t) = r(t + T_s) - 2\,r(t) + r(t - T_s) = \Delta\left(\frac{t}{T_s}\right) = \begin{cases} \left(1 + \dfrac{t}{T_s}\right), & -T_s \le t \le 0 \\[2mm] \left(1 - \dfrac{t}{T_s}\right), & 0 \le t \le T_s \\[2mm] 0, & \text{otherwise} \end{cases} \quad (8.39)$$

When the sampled signal $x_\delta(t)$ is applied to this filter, the output is $x_r(t)$. Each sample in $x_\delta(t)$, being an impulse, generates a triangular pulse of height equal to the strength of the sample, as illustrated in Fig. 8.15(d). The output of the filter $x_r(t)$ is obtained by the convolution of the input $x_\delta(t)$ with the impulse response $h(t)$:

$$x_r(t) = x_\delta(t) * h(t) = \sum_{n=-\infty}^{\infty} x(nT_s)\delta(t - nT_s) * h(t)$$

$$x_r(t) = \sum_{n=-\infty}^{\infty} x(nT_s)h(t - nT_s) \quad (8.40)$$

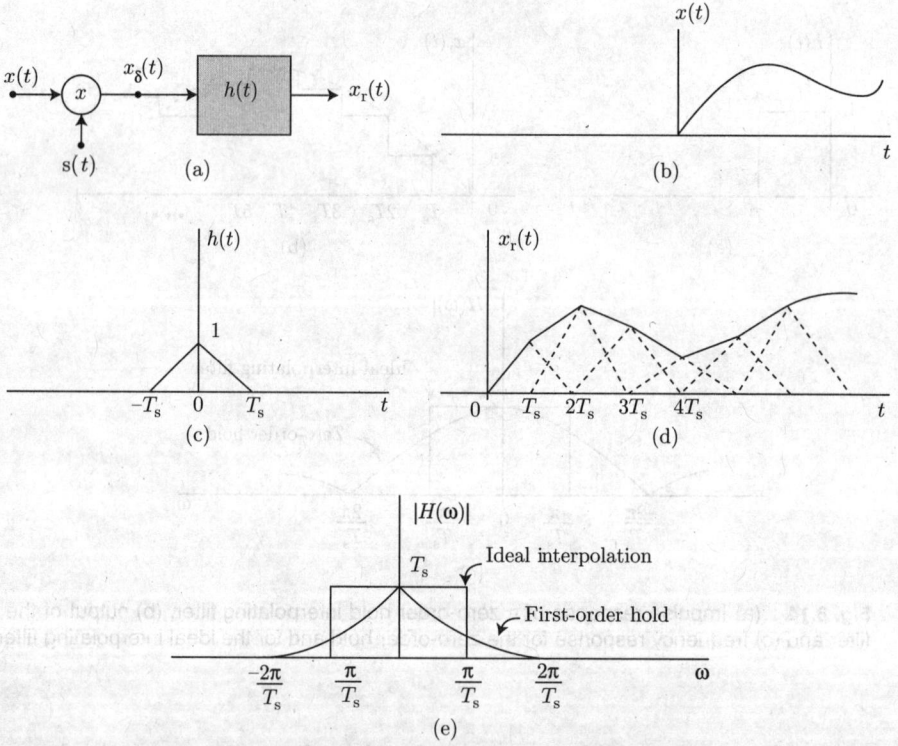

Fig. 8.15   (a) System for sampling and interpolation, (b) input signal $x(t)$, (c) impulse response of first-order interpolating filter, (d) output of the interpolating filter and (e) frequency response for the first-order hold and for the ideal interpolating filter.

Substitution of $h(t)$ from Eq. (8.39) in Eq. (8.40) yields

$$x_{\mathrm{r}}(t) = \sum_{n=-\infty}^{\infty} x(nT_{\mathrm{s}})\Delta\left(\frac{t - nT_{\mathrm{s}}}{T_{\mathrm{s}}}\right) \qquad (8.41)$$

The filter output is a linear approximation of $x(t)$. Figure 8.15(e) shows the magnitude of the transfer function of the first-order hold interpolating filter, superimposed on the desired transfer function of the exact interpolating filter.

8.9 SAMPLING OF SINUSOIDAL SIGNALS

The sampling establishes a relationship between the time variables t and n of continuous-time and discrete-time signals, respectively. Indeed, these variables are linearly related through the sampling period T_{s} or, equivalently, through the sampling rate $f_{\mathrm{s}} = 1/T_{\mathrm{s}}$, as

$$t = nT_{\mathrm{s}} = \frac{n}{f_{\mathrm{s}}} \qquad (8.42)$$

As a consequence of Eq. (8.42), there exists a relationship between the frequency variable f (or ω) for continuous-time signals and the frequency f_{d} (or ω_{d}) for discrete-time signals. To establish this relationship, consider a continuous-time sinusoidal signal of the form

$$x(t) = A\cos(2\pi f t) \qquad (8.43)$$

which, when sampled periodically at a rate $f_{\mathrm{s}} = 1/T_{\mathrm{s}}$ samples/s, yields

$$x(t)|_{t=nT_{\mathrm{s}}} = x(nT_{\mathrm{s}}) = x(n) = A\cos(2\pi f n T_{\mathrm{s}})$$

$$x(n) = A\cos\left(2\pi n \frac{f}{f_{\mathrm{s}}}\right) \qquad (8.44)$$

$$x(n) = A\cos\left(2\pi n f_{\mathrm{d}}\right) \qquad (8.45)$$

where

$$f_{\mathrm{d}} = \frac{f}{f_{\mathrm{s}}} \qquad (8.46)$$

or, equivalently, as

$$2\pi f_{\mathrm{d}} = \frac{2\pi f}{f_{\mathrm{s}}}$$

$$\omega_{\mathrm{d}} = \frac{\omega}{f_{\mathrm{s}}} = \omega T_{\mathrm{s}} \qquad (8.47)$$

The range of the frequency variable f or ω for continuous-time sinusoids are

$$-\infty < f < \infty \quad \text{or} \quad -\infty < \omega < \infty \qquad (8.48)$$

However, the situation is different for discrete-time sinusoids. According to the sampling theorem, the sampling frequency $f_{\mathrm{s}} \geq 2f$. Substituting this value of f_{s} in Eq. (8.46), we obtain

$$f_d \leq \frac{f}{2f}$$

$$f_d \leq \frac{1}{2}$$

or, equivalently

$$\omega_d \leq \pi$$

Therefore, the range of the frequency variable f_d or ω_d for discrete-time sinusoids are

$$-\frac{1}{2} \leq f_d \leq \frac{1}{2} \quad \text{or} \quad -\pi \leq \omega_d \leq \pi \tag{8.49}$$

By substituting $f_d = f/f_s$ from Eq. (8.46) and $\omega_d = \omega/f_s$ from Eq. (8.47) into Eq. (8.49), we find that the frequency of the continuous-time sinusoids when sampled at a rate $f_s = 1/T_s$ must fall in the range

$$-\frac{1}{2T_s} = -\frac{f_s}{2} \leq f \leq \frac{f_s}{2} = \frac{1}{2T_s} \tag{8.50}$$

or, equivalently,

$$-\frac{\pi}{T_s} = -\pi f_s \leq \omega \leq \pi f_s = \frac{\pi}{T_s} \tag{8.51}$$

From these relations we observe that the fundamental difference between continuous-time and discrete-time signals is in their range of values of the frequency variables f and f_d, or ω and ω_d. Periodic sampling of a continuous-time signal implies a mapping of the infinite frequency range for the variable f (or ω) into a finite frequency range for the variable f_d (or ω_d). Since the highest frequency in a discrete-time signal is $\omega_d = \pi$ or $f_d = 1/2$, it follows that, with a sampling rate f_s, the corresponding highest values of f and ω are

$$f_{max} = \frac{f_s}{2} = \frac{1}{2T_s} \tag{8.52}$$

$$\omega_{max} = \pi f_s = \frac{\pi}{T_s} \tag{8.53}$$

Therefore, sampling introduces an ambiguity since the highest frequency in a continuous-time signal that can be uniquely distinguished when such a signal is sampled at a rate $f_s = 1/T_s$ is $f_{max} = f_s/2$, or $\omega_{max} = \pi f_s$. To see what happens to frequencies above $f_s/2$, let us consider the following example.

Example 8.6 Consider the two continuous-time sinusoidal signals

$$x_1(t) = \cos(20\pi t)$$

$$x_2(t) = \cos(100\pi t)$$

which are sampled at a rate $f_s = 40$ Hz. Find the corresponding discrete-time signals.

Solution

To convert a continuous-time sinusoidal signal into a discrete-time signal, substitute $t = nT_s$. Therefore,

$$x_1(t)|_{t=nT_s} = x_1(nT_s) = x_1(n) = \cos(20\pi nT_s) = \cos\left(2\pi \frac{10}{40}n\right) = \cos\left(\frac{\pi}{2}n\right)$$

$$x_2(t)|_{t=nT_s} = x_2(nT_s) = x_2(n) = \cos(100\pi nT_s) = \cos\left(2\pi \frac{50}{40}n\right) = \cos\left(\frac{5\pi}{2}n\right)$$

However,

$$x_2(n) = \cos\left(\frac{5\pi}{2}n\right) = \cos\left(2\pi n + \frac{\pi n}{2}\right) = \cos\left(\frac{\pi n}{2}\right) = x_1(n)$$

Hence, $x_2(n) = x_1(n)$. Thus, the sinusoidal signals are identical and, consequently, indistinguishable. If we are given the sampled values generated by $\cos(\pi n/2)$, there is some ambiguity as to whether these sampled values correspond to $x_1(t)$ or $x_2(t)$. Since $x_2(t)$ yields exactly the same values as $x_1(t)$ when the two are sampled at $f_s = 40$ samples/s, we say that the frequency $f_2 = 50$ Hz is an *alias* of the frequency $f_1 = 10$ Hz at the sampling frequency $f_s = 40$ Hz.

In general, the sampling of a continuous-time sinusoidal signal

$$x(t) = A\cos(2\pi f_0 t) \tag{8.54}$$

with a sampling rate $f_s = 1/T_s$ results in a discrete-time signal

$$x(n) = x(nT_s) = A\cos(2\pi f_0 nT_s) = A\cos\left(2\pi \frac{f_0}{f_s}n\right) = A\cos(2\pi f_d n) \tag{8.55}$$

where $f_d = f_0/f_s$ is the frequency of the discrete-time sinusoid. If we assume that $-f_s/2 \le f_0 \le f_s/2$, the frequency f_d of $x(n)$ is in the range $-1/2 \le f_d \le 1/2$, which is the frequency range for discrete-time signals. In this case, the relationship between f_0 and f_d is one-to-one, and hence it is possible to identify (or reconstruct) the continuous-time signal $x(t)$ from the samples $x(n)$.

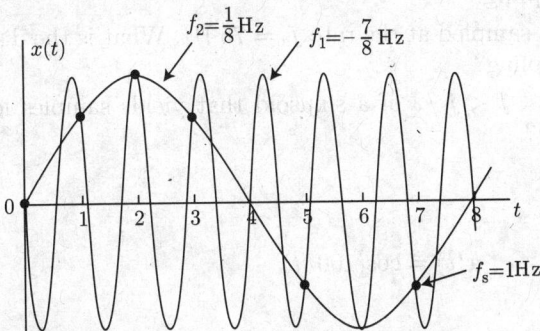

Fig. 8.16 Illustration of aliasing

On the other hand, if the sinusoids

$$x(t) = A\cos(2\pi f_k t) \tag{8.56}$$

where

$$f_k = f_0 + kf_s, \quad k = \pm 1, \pm 2, \ldots \tag{8.57}$$

are sampled at a rate f_s, it is clear that the frequency f_k is outside the fundamental frequency range $-f_s/2 \le f \le f_s/2$. Consequently, the sampled signal is

$$x(n) = x(nT_s) = A\cos(2\pi f_k nT_s) = A\cos\left(2\pi \frac{f_0 + kf_s}{f_s} n\right)$$

$$= A\cos\left(2\pi \frac{f_0}{f_s} n + 2\pi kn\right)$$

$$= A\cos\left(2\pi \frac{f_0}{f_s} n\right)$$

$$= A\cos(2\pi f_d n)$$

which is identical to the discrete-time signal in Eq. (8.55) obtained by sampling the signal $x(t)$ given in Eq. (8.54). Thus an infinite number of continuous-time sinusoids is represented by the same set of samples. Consequently, if we are given the sequence $x(n)$, an ambiguity exists as to which continuous-time signal $x(t)$ these values represent. Equivalently, we can say that the frequencies $f_k = f_0 + kf_s$, $k = \pm 1, \pm 2, \ldots$ are indistinguishable from the frequency f_0 after sampling and hence they are aliases of f_0.

An example of aliasing is illustrated in Fig. 8.16, where two sinusoids with frequencies $f_1 = -7/8$ Hz and $f_2 = 1/8$ Hz yield identical samples when a sampling rate of $f_s = 1$ Hz is used.

Example 8.7 Consider the continuous-time signal

$$x(t) = \cos(100\pi t)$$

(a) Determine the minimum sampling rate required to avoid aliasing.

(b) Suppose that the signal is sampled at the rate $f_s = 200$ Hz. What is the discrete-time signal obtained after sampling?

(c) Suppose that the signal is sampled at the rate $f_s = 75$ Hz. What is the discrete-time signal obtained after sampling?

(d) What is the frequency $0 < f < f_s/2$ of a sinusoid that yields samples identical to those obtained in part (c)?

Solution

(a) Given that

$$x(t) = \cos(100\pi t)$$

Comparing with

$$x(t) = \cos(\omega_1 t)$$

we have

$$\omega_1 = 100\pi \quad \text{or} \quad f_1 = 50 \text{ Hz}$$

Therefore, the Nyquist rate is given by

$$\omega_s = 2\omega_1 = 200\pi \quad \text{or} \quad f_s = 2f_1 = 100 \text{ Hz}$$

(b) If the signal is sampled at $f_s = 200$ Hz, the discrete-time signal is

$$x(t)|_{t=nT_s} = x(nT_s) = x(n) = \cos(100\pi n T_s)$$

$$= \cos\left(\frac{100\pi}{200}n\right)$$

$$x(n) = \cos\left(\frac{\pi}{2}n\right)$$

(c) If the signal is sampled at $f_s = 75$ Hz, the discrete-time signal is

$$x(t)|_{t=nT_s} = x(nT_s) = x(n) = \cos(100\pi n T_s)$$

$$= \cos\left(\frac{100\pi}{75}n\right)$$

$$x(n) = \cos\left(\frac{4\pi}{3}n\right)$$

$$= \cos\left(2n\pi - \frac{2\pi}{3}n\right)$$

$$x(n) = \cos\left(\frac{2\pi}{3}n\right)$$

The frequency of the discrete-time signal $x(n)$ is $f_d = 1/3$ cycles/sample.

(d) For the sampling rate $f_s = 75$ Hz, we have

$$f_d = \frac{f}{f_s}$$

or, equivalently, we have

$$f = f_d f_s$$

$$f = \frac{1}{3}75 = 25 \text{ Hz}$$

Clearly, the sinusoidal signal

$$y(t) = \cos(2\pi f t)$$

$$y(t) = \cos(50\pi t)$$

sampled at $f_s = 75$ Hz yields identical samples. Hence, $f = 50$ Hz is an alias of $f = 25$ Hz for the sampling rate $f_s = 75$ Hz.

Example 8.8 The signals

$$x_1(t) = 10\cos(100\pi t) \quad \text{and} \quad x_2(t) = 10\cos(50\pi t)$$

are both sampled with $f_s = 75$ Hz. Show that the two sequences of samples so obtained are identical.

Solution
If the signal is sampled at $f_s = 75$ Hz, the discrete-time signal is

$$x_1(t)|_{t=nT_s} = x_1(nT_s) = x_1(n) = 10\cos(100\pi n T_s)$$

$$= 10\cos\left(\frac{100\pi}{75}n\right)$$

$$x_1(n) = 10\cos\left(\frac{4\pi}{3}n\right)$$

$$= 10\cos\left(2n\pi - \frac{2\pi}{3}n\right)$$

$$x_1(n) = 10\cos\left(\frac{2\pi}{3}n\right)$$

Similarly, we have

$$x_2(t)|_{t=nT_s} = x_2(nT_s) = x_2(n) = 10\cos(50\pi n T_s)$$

$$= 10\cos\left(\frac{50\pi}{75}n\right)$$

$$x_2(n) = 10\cos\left(\frac{2\pi}{3}n\right) = x_1(n)$$

8.10 SAMPLING THEOREM FOR REAL-VALUED BANDPASS SIGNALS

In previous sections, we have discussed the sampling theorem for low-pass signals. However, when the signal is a bandpass signal, then a different criterion must be used to sample the signal. The spectrum of baseband (low-pass) signal includes the origin, whereas the spectrum of bandpass signals occupies a range of frequencies between f_L and f_H, where $f_L > 0$. The quantity $B = f_H - f_L$ is a measure of the bandwidth of the bandpass signal.

The bandpass sampling theorem states that if a real-valued bandpass signal $x(t)$ has a spectrum of bandwidth $\omega_B = 2\pi B = 2\pi(f_H - f_L)$ and an upper frequency limit $\omega_H = 2\pi f_H$, then the original signal $x(t)$ can be recovered from the sampled signal $x_s(t)$ by bandpass filtering if $f_s = 2f_H/k$, where $k = \lfloor f_H/B \rfloor$ is the integer part of f_H/B.

Let $x(t)$ be a real-valued bandpass signal $x(t)$ with $X(f) = 0$ for $|f| < f_L$ and $|f| > f_H$. A sampling signal samples this signal regularly at the rate of f_s samples/s. The time space between any two successive samples is T_s seconds. The sampling signal $s(t)$ is an impulse train of frequency equal to sampling frequency f_s, which can be expressed mathematically as

$$s(t) = \sum_{n=-\infty}^{\infty} \delta(t - nT_s) \qquad (8.58)$$

The sampled signal $x_s(t)$ is obtained by multiplying the signal $x(t)$ with impulse train $s(t)$. Therefore, $x_s(t)$ can be expressed mathematically as

$$x_s(t) = x(t)s(t) \qquad (8.59)$$

$$x_s(t) = x(t) \sum_{n=-\infty}^{\infty} \delta(t - nT_s) \qquad (8.60)$$

Because of the sampling property of the unit impulse [Eq. (1.14)], we know that multiplying $x(t)$ by a unit impulse samples the value of the signal at the point at which the impulse is located, i.e., $x(t)\delta(t - t_0) = x(t_0)\delta(t - t_0)$. Applying this to Eq. (8.60), we get

$$x_s(t) = \sum_{n=-\infty}^{\infty} x(nT_s)\delta(t - nT_s) \qquad (8.61)$$

Taking the Fourier transform of Eq. (8.59) and using the multiplication property of the Fourier transform, we obtain

$$X_s(f) = [X(f) * S(f)] \qquad (8.62)$$

We know that the Fourier transform of a periodic impulse train in the time domain with period T_s is a periodic impulse train in the frequency domain with period $f_s = 1/T_s$, i.e.,

$$S(f) = \frac{1}{T_s} \sum_{n=-\infty}^{\infty} \delta(f - nf_s) \qquad (8.63)$$

Substituting $S(f)$ into Eq. (8.62), we obtain

$$X_s(f) = \left[X(f) * \frac{1}{T_s} \sum_{n=-\infty}^{\infty} \delta(f - nf_s) \right]$$

$$= \frac{1}{T_s} \left[X(f) * \sum_{n=-\infty}^{\infty} \delta(f - nf_s) \right]$$

Since convolution with an impulse simply shifts a signal [i.e., $X(f) * \delta(f - f_s) = X(f - f_s)$], it follows that

$$X_s(f) = \frac{1}{T_s} \sum_{n=-\infty}^{\infty} X(f - nf_s) \qquad (8.64)$$

or

$$X_s(f) = \frac{1}{T_s}[\cdots + X(f+2f_s) + X(f+f_s) + X(f) + X(f-f_s) + X(f-2f_s) + \cdots] \tag{8.65}$$

Equation (8.65) represents the spectrum of the bandpass sampled signal $x_s(t)$. Thus, the Fourier transform of the bandpass sampled signal is given by an infinite sum of shifted versions of the original signal's Fourier transform $X(f)$ scaled by $1/T_s$.

Case I Let $x(t)$ be a real-valued bandpass signal $x(t)$ with the spectrum $X(f) = 0$ for $|f| < f_L$ and $|f| > f_H$, the bandwidth $B = f_H - f_L$, and the centre frequency $f_c = (f_H + f_L)/2$. Now, *If $f_c > B/2$ and f_H is an integer multiple of the bandwidth B, no aliasing will occur if $x(t)$ is sampled at a sampling frequency*

$$f_s = 2B \tag{8.66}$$

i.e., the sampling frequency is equal to twice the bandwidth of the signal.

If f_H is an integer multiple of B, we may express f_L and f_H as follows:

$$f_L = (m-1)B \quad \text{and} \quad f_H = mB$$

With a sampling frequency of $f_s = 2B$, the sampled signal has a spectrum [Eq. (8.64)]:

$$X_s(f) = \frac{1}{T_s} \sum_{n=-\infty}^{\infty} X(f - n2B)$$

Suppose that $f_L = 8$ kHz and $f_H = 10$ KHz; therefore, $B = f_H - f_L = 2$ kHz and $f_c = (f_H + f_L)/2 = 9$ kHz. Let $x(t)$ have a spectrum as shown in Fig. 8.17(a). The spectrum of the bandpass sampled signal with $f_s = 2B = 4$ kHz is given by

$$X_s(f) = \frac{1}{T_s} \sum_{n=-\infty}^{\infty} X(f - 4n)$$

which is shown in Fig. 8.17(b).

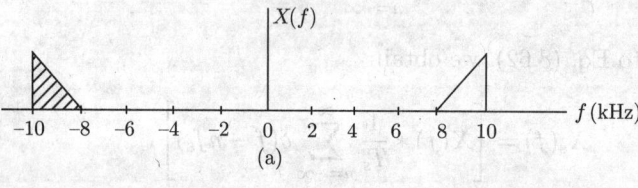

(a)

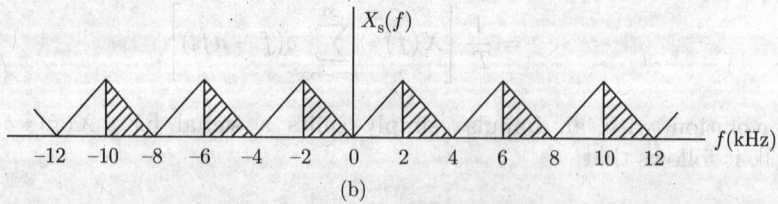

(b)

Fig. 8.17 (a) Spectrum of a real-valued bandpass signal and (b) spectrum of the bandpass sampled signal.

Case II f_H *is not an integer multiple of* B. If f_H is not an integer multiple of B, we may always increase B until this is the case. Specifically, let

$$k = \left\lfloor \frac{f_H}{B} \right\rfloor$$

where $\lfloor \cdot \rfloor$ is defined to be the integer part. Now, if we simply increase B to B' where

$$k = \frac{f_H}{B'}$$

we have the case described in Case I where f_H is an integer multiple of the bandwidth. Thus, $x(t)$ may be sampled without aliasing a sampling frequency of

$$f_s = 2B' = 2\frac{f_H}{k} = \frac{2f_H}{\lfloor f_H/B \rfloor} \tag{8.67}$$

Suppose that $f_L = 15$ kHz and $f_H = 25$ kHz, therefore, $B = f_H - f_L = 10$ kHz, and $f_c = (f_H + f_L)/2 = 20$ kHz. Let $x(t)$ have a spectrum as shown in Fig. 8.18(a). $x(t)$ can be sampled without aliasing if the sampling frequency is

$$f_s = 2\frac{f_H}{k} = \frac{2f_H}{\lfloor f_H/B \rfloor} = \frac{2 \times 25}{\lfloor 25/10 \rfloor} = \frac{50}{\lfloor 2.5 \rfloor} = \frac{50}{2} = 25 \quad \text{kHz}$$

The spectrum of the bandpass sampled signal with $f_s = 25$ kHz is given by

$$X_s(f) = \frac{1}{T_s} \sum_{n=-\infty}^{\infty} X(f - 25n)$$

which is shown in Fig. 8.18(b).

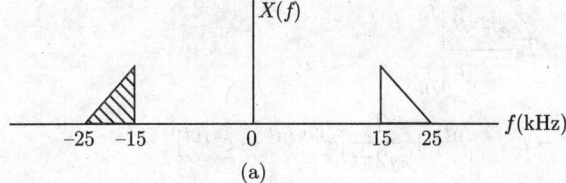

(a)

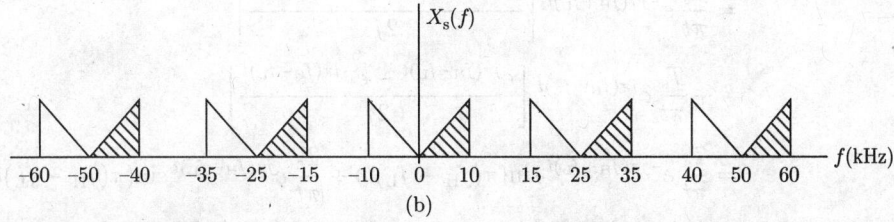

(b)

Fig. 8.18   (a) Spectrum of a real-valued bandpass signal and (b) spectrum of the bandpass sampled signal.

8.10.1 Reconstruction of Real Bandpass Signal

If the condition $f_s = 2B = 2(f_H - f_L)$ is satisfied, $x(t)$ may be uniquely reconstructed from $x_s(t)$ by using a bandpass filter with a frequency response $H(f)$ as shown in Fig. 8.19.

$$H(f) = \begin{cases} T_s, & f_L \leq |f| \leq f_H \\ 0, & \text{otherwise} \end{cases} \tag{8.68}$$

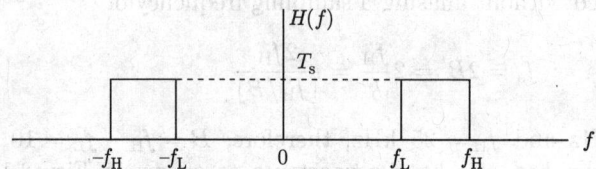

Fig. 8.19 Frequency response of bandpass reconstruction filter.

Taking the inverse Fourier transform of the above equation to get the impulse response $h(t)$ of the bandpass filter yields

$$h(t) = \int\limits_{-\infty}^{\infty} H(f)\, e^{j2\pi ft} df$$

$$= \int\limits_{-f_H}^{-f_L} T_s\, e^{j2\pi ft} df + \int\limits_{f_L}^{f_H} T_s\, e^{j2\pi ft} df$$

$$= T_s \frac{e^{j2\pi ft}}{j2\pi t}\Bigg|_{-f_H}^{-f_L} + T_s \frac{e^{j2\pi ft}}{j2\pi t}\Bigg|_{f_L}^{f_H}$$

$$= \frac{T_s}{j2\pi t}[e^{-j2\pi f_L t} - e^{-j2\pi f_H t}] + \frac{T_s}{j2\pi t}[e^{j2\pi f_H t} - e^{\pi f_L t}]$$

$$= \frac{T_s}{\pi t} e^{-j\pi(f_H+f_L)t} \left[\frac{e^{j\pi(f_H-f_L)t} - e^{-j\pi(f_H-f_L)t}}{2j}\right]$$

$$+ \frac{T_s}{\pi t} e^{j\pi(f_H+f_L)t} \left[\frac{e^{j\pi(f_H-f_L)t} - e^{-j\pi(f_H-f_L)t}}{2j}\right]$$

$$= \frac{T_s}{\pi t} e^{-j\pi(f_H+f_L)t} \sin(\pi(f_H - f_L)t) + \frac{T_s}{\pi t} e^{j\pi(f_H+f_L)t} \sin(\pi(f_H - f_L)t)$$

$$= \frac{1}{2} \frac{\sin(\pi f_s/2t)}{\pi f_s/2t} e^{-j2\pi((f_H+f_L))/2t} + \frac{1}{2} \frac{\sin(\pi f_s/2t)}{\pi \frac{f_s}{2}t} e^{j2\pi((f_H+f_L))2t}$$

$$h(t) = \frac{1}{2}\text{sinc}\left(\frac{f_s}{2}t\right)e^{-j2\pi f_c t} + \frac{1}{2}\text{sinc}\left(\frac{f_s}{2}t\right)e^{j2\pi f_c t}$$

$$h(t) = \text{sinc}\left(\frac{f_s}{2}t\right)\left[\frac{e^{-j2\pi f_c t} + e^{j2\pi f_c t}}{2}\right]$$

$$h(t) = \text{sinc}\left(\frac{f_s}{2}t\right)\cos(2\pi f_c t) \tag{8.69}$$

Therefore, the output of the reconstruction bandpass filter, which produces the real bandpass signal $x(t)$ is

$$x(t) = x_s(t) * h(t) = \sum_{n=-\infty}^{\infty} x(nT_s)\delta(t - nT_s) * h(t)$$

$$= \sum_{n=-\infty}^{\infty} x(nT_s)h(t - nT_s)$$

$$x(t) = \sum_{n=-\infty}^{\infty} x(nT_s)\text{sinc}\left(\frac{f_s}{2}(t - nT_s)\right)\cos(2\pi f_c(t - nT_s))$$

$$x(t) = \sum_{n=-\infty}^{\infty} x(nT_s)\text{sinc}\left(\frac{1}{2T_s}(t - nT_s)\right)\cos(2\pi f_c(t - nT_s)) \tag{8.70}$$

8.11 SAMPLING THEOREM FOR COMPLEX BANDPASS SIGNALS

A complex bandpass continuous-time signal $x(t)$ has a Fourier transform $X(f)$ that is nonzero over the frequency range $[f_L, f_H]$, as shown in Fig. 8.20(a). *The bandpass sampling theorem for complex signals states that if a complex bandpass signal $x(t)$ has a spectrum of bandwidth $\omega_B = 2\pi B = 2\pi(f_H - f_L)$ and an upper frequency limit $\omega_H = 2\pi f_H$, then the original signal $x(t)$ can be recovered from the sampled signal $x_s(t)$ by bandpass filtering if the sampling frequency*

$$f_s \geq f_H - f_L \tag{8.71}$$

Let $x(t)$ be a complex bandpass continuous-time signal $x(t)$ with a Fourier transform $X(f)$ that is nonzero over the frequency range $[f_L, f_H]$, as shown in Fig. 8.20(a). Because the highest frequency in $x(t)$ is f_H, the Nyquist rate is $2f_H$. However, note that if $x(t)$ is multiplied by a complex exponential of frequency $f_c = (f_H + f_L)/2$,

$$y(t) = x(t)e^{-j\frac{f_H+f_L}{2}t}$$

then $Y(f) = X(f + f_c)$, and $y(t)$ is a complex low-pass signal with a spectrum shown in Fig. 8.20(b), where $f_0 = (f_H - f_L)/2$. Thus, the Nyquist rate for $y(t)$ is $f_s = 2f_0 = f_H - f_L$.

A sampling signal samples the complex bandpass signal $x(t)$ regularly at the rate of f_s samples/s. The time space between any two successive samples is T_s seconds. The sampling signal $s(t)$ is an impulse train of frequency equal to sampling frequency f_s, which can be expressed mathematically as

$$s(t) = \sum_{n=-\infty}^{\infty} \delta(t - nT_s) \tag{8.72}$$

The sampled signal $x_s(t)$ is obtained by multiplying the signal $x(t)$ with impulse train $s(t)$. Therefore, $x_s(t)$ can be expressed mathematically as

$$x_s(t) = x(t)s(t) \tag{8.73}$$

$$x_s(t) = x(t) \sum_{n=-\infty}^{\infty} \delta(t - nT_s)$$

$$x_s(t) = \sum_{n=-\infty}^{\infty} x(nT_s)\delta(t - nT_s) \tag{8.74}$$

Taking the Fourier transform of Eq. (8.73) and using the multiplication property of the Fourier transform, we obtain

$$X_s(f) = [X(f) * S(f)] \tag{8.75}$$

We know that the Fourier transform of a periodic impulse train in the time domain with period T_s is a periodic impulse train in the frequency domain with period $f_s = 1/T_s$, i.e.,

$$S(f) = \frac{1}{T_s} \sum_{n=-\infty}^{\infty} \delta(f - nf_s) \tag{8.76}$$

Substituting $S(f)$ into Eq. (8.75), we obtain

$$X_s(f) = \left[X(f) * \frac{1}{T_s} \sum_{n=-\infty}^{\infty} \delta(f - nf_s) \right]$$

$$= \frac{1}{T_s} \left[X(f) * \sum_{n=-\infty}^{\infty} \delta(f - nf_s) \right]$$

Since convolution with an impulse simply shifts a signal [i.e., $X(f) * \delta(f - f_s) = X(f - f_s)$], it follows that

$$X_s(f) = \frac{1}{T_s} \sum_{n=-\infty}^{\infty} X(f - nf_s)$$

$$= \frac{1}{T_s} [\cdots + X(f + 2f_s) + X(f + f_s) + X(f) + X(f - f_s) + X(f - 2f_s) + \cdots]$$

as illustrated in Fig. 8.20(c).

In order for there to be no interference between the shifted spectra, it is necessary that

$$(f_H - f_s) \le f_L$$

or

$$f_s \ge (f_H - f_L)$$

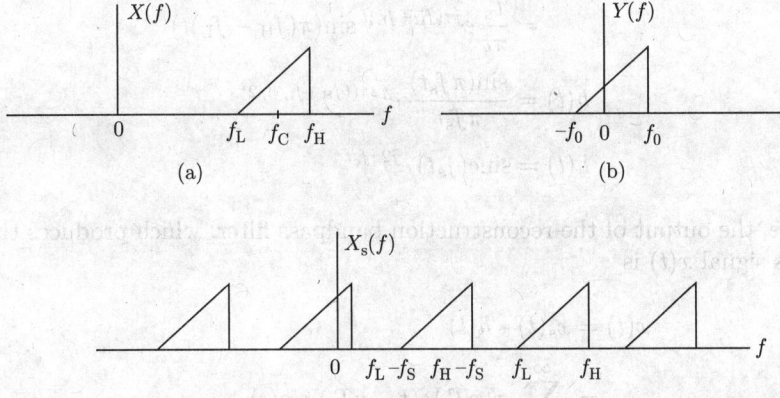

Fig. 8.20 (a) Spectrum of a complex bandpass signal, (b) spectrum of a complex low-pass signal, and (c) spectrum of a complex bandpass sampled signal.

8.11.1 Reconstruction of Complex Bandpass Signal

If the condition $f_\mathrm{s} \geq (f_\mathrm{H} - f_\mathrm{L})$ is satisfied, $x(t)$ may be uniquely reconstructed from $x_\mathrm{s}(t)$ by using a bandpass filter with a frequency response $H(f)$, as shown in Fig. 8.21:

$$H(f) = \begin{cases} T_\mathrm{s}, & f_\mathrm{L} \leq f \leq f_\mathrm{H} \\ 0, & \text{otherwise} \end{cases} \tag{8.77}$$

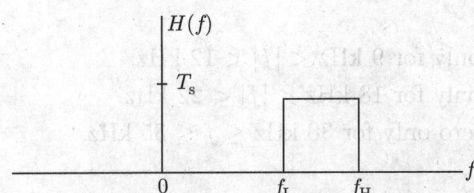

Fig. 8.21 Frequency response of a complex bandpass reconstruction filter.

Taking the inverse Fourier transform of the above equation to get the impulse response $h(t)$, we obtain

$$h(t) = \int_{-\infty}^{\infty} H(f)\, e^{j2\pi ft}\, df = \int_{f_\mathrm{L}}^{f_\mathrm{H}} T_\mathrm{s}\, e^{j2\pi ft} df = T_\mathrm{s} \frac{e^{j2\pi ft}}{j2\pi t} \bigg|_{f_\mathrm{L}}^{f_\mathrm{H}}$$

$$= \frac{T_\mathrm{s}}{j2\pi t} [e^{j2\pi f_\mathrm{H} t} - e^{j2\pi f_\mathrm{L} t}]$$

$$= \frac{T_\mathrm{s}}{\pi t}\, e^{j\pi(f_\mathrm{H}+f_\mathrm{L})t} \left[\frac{e^{j\pi(f_\mathrm{H}-f_\mathrm{L})t} - e^{-j\pi(f_\mathrm{H}-f_\mathrm{L})t}}{2j} \right]$$

$$= \frac{T_s}{\pi t} e^{j\pi(f_H+f_L)t} \sin(\pi(f_H - f_L)t)$$

$$h(t) = \frac{\sin(\pi f_s t)}{\pi f_s t} e^{j2\pi((f_H+f_L))/2t}$$

$$h(t) = \operatorname{sinc}(f_s t) e^{j2\pi f_c t} \tag{8.78}$$

Therefore, the output of the reconstruction bandpass filter, which produces the complex bandpass signal $x(t)$ is

$$x(t) = x_s(t) * h(t)$$

$$= \sum_{n=-\infty}^{\infty} x(nT_s)\delta(t - nT_s) * h(t)$$

$$x(t) = \sum_{n=-\infty}^{\infty} x(nT_s)h(t - nT_s)$$

$$x(t) = \sum_{n=-\infty}^{\infty} x(nT_s)\operatorname{sinc}(f_s(t-nT_s)) e^{j2\pi f_c(t-nT_s)} \tag{8.79}$$

Example 8.9 Determine the minimum sampling frequency f_s for each of the following bandpass signals:

(a) $x(t)$ is real with $X(f)$ nonzero only for 9 kHz $< |f| <$ 12 kHz.
(b) $x(t)$ is real with $X(f)$ nonzero only for 18 kHz $< |f| <$ 22 kHz.
(c) $x(t)$ is complex with $X(f)$ nonzero only for 30 kHz $\leq f \leq$ 35 kHz.

Solution
(a) $x(t)$ is real with $X(f)$ nonzero only for 9 kHz $< |f| <$ 12 kHz. For this signal, the bandwidth $B = f_H - f_L = 3$ kHz and $f_H = 12$ kHz$= 4B$ is an integer multiple of B. Therefore, the minimum sampling frequency is $f_s = 2B = 6$ kHz.

(b) $x(t)$ is real with $X(f)$ nonzero only for 18 kHz $< |f| <$ 22 kHz. For this signal, the bandwidth $B = f_H - f_L = 4$ kHz and $f_H = 22$ kHz is not an integer multiple of B. Therefore, the sampling frequency is

$$f_s = \frac{2f_H}{\lfloor f_H/B \rfloor} = \frac{2 \times 22k}{\lfloor 22/4 \rfloor} = \frac{2 \times 22k}{\lfloor 5.5 \rfloor} = \frac{2 \times 22k}{5} = 8.8 \,\text{kHz}$$

(c) For a complex bandpass signal with a spectrum that is nonzero for $f_L \leq f \leq f_H$, the minimum sampling frequency is $f_s = (f_H - f_L)$. Thus, for the given signal $x(t)$ the sampling frequency $f_s = 35 - 30 = 5$ kHz.

Example 8.10 Given the signal

$$x(t) = 10\cos(2000\pi t)\cos(8000\pi t)$$

(a) What is the minimum sampling rate based on the low-pass uniform sampling theorem?

(b) Repeat (a) based on the bandpass sampling theorem.

Solution

(a) Consider the given signal

$$x(t) = 10\cos(2000\pi t)\cos(8000\pi t)$$

$$= 5\cos(6000\pi t) + 5\cos(10000\pi t)$$

Comparing with

$$x(t) = A_1\cos(2\pi f_1 t) + A_2\cos(2\pi f_2 t)$$

we have

$$f_1 = 3000 = 3\,\text{kHz} \quad\text{and}\quad f_2 = 5000 = 5\,\text{kHz}$$

The maximum frequency present in the signal $x(t)$ is

$$f_{\max} = f_2 = 5000 = 5\ \text{kHz}$$

Therefore, the Nyquist rate is given by

$$f_s = 2f_{\max} = 10\,\text{kHz}$$

(b) For the given signal $x(t)$, we have $f_L = 3$ kHz, $f_H = 5$ kHz, and $B = f_H - f_L = 2$ kHz. Therefore, the sampling frequency is

$$f_s = \frac{2f_H}{\lfloor f_H/B\rfloor} = \frac{2\times 5\text{k}}{\lfloor 5/2\rfloor} = \frac{2\times 5\text{k}}{\lfloor 2.5\rfloor} = \frac{2\times 5\text{k}}{2} = 5\,\text{kHz}$$

Example 8.11 Consider the bandpass signal $x(t)$ with a spectrum shown in Fig. 8.22(a). Check sampling theorem by sketching the spectrum of the ideally sampled signal $x_s(t)$ when (i) $f_s = 25$ kHz, (ii) $f_s = 45$ kHz, and (iii) $f_s = 50$ kHz. Indicate if and how the signal can be recovered.

Solution

From Fig. 8.22(a), $f_L = 15$ kHz, $f_H = 25$ kHz, and $B = (f_H - f_L) = 25 - 15 = 10$ kHz. Therefore, the sampling frequency is

$$f_s = \frac{2f_H}{\lfloor f_H/B\rfloor} = \frac{2\times 25\text{k}}{\lfloor 25/10\rfloor} = \frac{2\times 25\text{k}}{\lfloor 2.5\rfloor} = \frac{2\times 25\text{k}}{2} = 25\,\text{kHz}$$

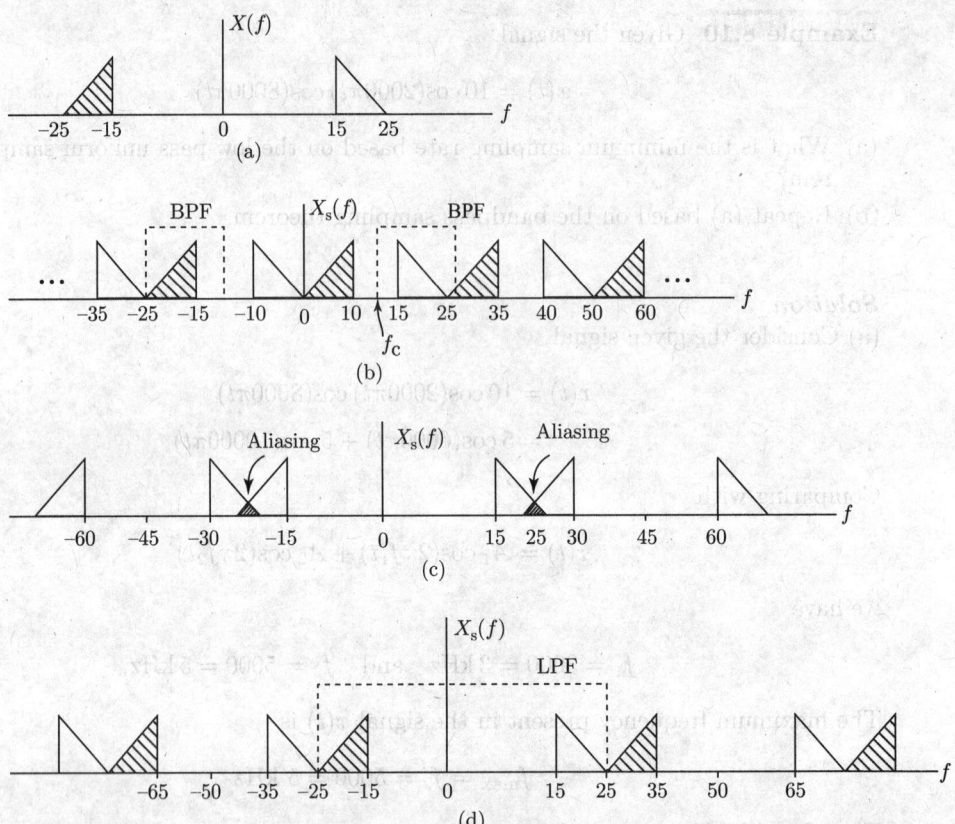

Fig. 8.22 (a) Spectrum of bandpass signal, (b) bandpass sampled signal spectrum for $f_s = 25$ kHz, (c) bandpass sampled signal spectrum for $f_s = 45$ kHz and (d) bandpass sampled signal spectrum for $f_s = 50$ kHz.

(i) From Fig. 8.22(b), we see that $x(t)$ can be recovered from the sampled signal by using a bandpass filter of the frequency response

$$H(f) = \begin{cases} T_s, & f_c \leq |f| \leq 25\,\text{kHz} \quad \text{with} \quad 10\,\text{kHz} \leq f_c < 15\,\text{kHz} \\ 0, & \text{otherwise} \end{cases}$$

(ii) From Fig. 8.22(c), we see that $x(t)$ cannot be recovered from the sampled signal $x_s(t)$.

(iii) From Fig. 8.22(d), we see that $x(t)$ can be recovered from the sampled signal by using an ideal low-pass filter of the frequency response

$$H(f) = \begin{cases} T_s, & |f| \leq 25\,\text{kHz} \\ 0, & \text{otherwise} \end{cases}$$

8.12 SAMPLING OF DISCRETE-TIME SIGNALS

In analogy with continuous-time sampling, the sampling of a discrete-time signal can be represented as shown in Fig. 8.23. Here, the sampled signal $x_s(n)$ is equal to the original signal $x(n)$ at integer multiples of the sampling period N_s and is zero at the intermediate samples, i.e.,

$$x_s(n) = \begin{cases} x(n), & \text{if } n \text{ is an integer multiple of } N_s \\ 0, & \text{otherwise} \end{cases} \tag{8.80}$$

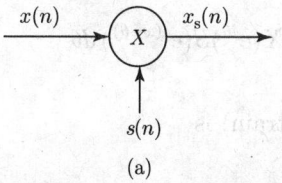

(a)

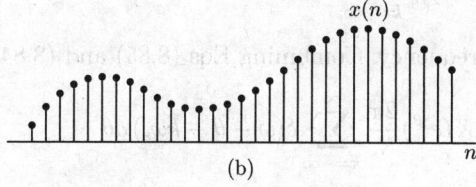

(b)

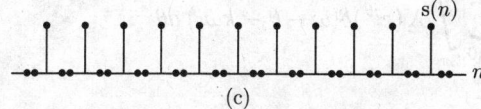

(c)

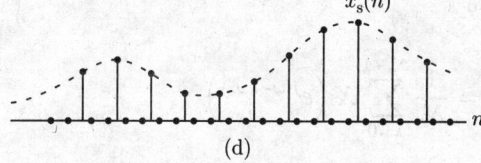

(d)

Fig. 8.23 Discrete-time sampling

In other words, the sampled signal $x_s(n)$ is equal to the multiplication of the original signal $x(n)$ with a sampling signal (an impulse train) $s(n)$, i.e.,

$$x_s(n) = x(n)s(n) \tag{8.81}$$

where

$$s(n) = \sum_{k=-\infty}^{\infty} \delta(n - kN_s) \tag{8.82}$$

Therefore, we have

$$x_s(n) = x(n)s(n) = x(n) \sum_{k=-\infty}^{\infty} \delta(n - kN_s)$$

$$x_s(n) = \sum_{k=-\infty}^{\infty} x(kN_s)\delta(n - kN_s) \tag{8.83}$$

Taking the DTFT of Eq. (8.81), and using the multiplication property [Eq. (6.53)], we get

$$X_s(e^{j\omega}) = X(e^{j\omega}) \circledast S(e^{j\omega}) = \frac{1}{2\pi} \int_{2\pi} X(e^{j\theta}) S(e^{j(\omega-\theta)}) \, d\theta \tag{8.84}$$

The DTFT of the sampling signal $s(n)$ (an impulse train) is

$$S(e^{j\omega}) = \frac{2\pi}{N_s} \sum_{k=-\infty}^{\infty} \delta(\omega - k\omega_s) \tag{8.85}$$

where $\omega_s = 2\pi/N_s$ is the sampling frequency. Combining Eqs (8.85) and (8.84), we have

$$X_s(e^{j\omega}) = \frac{1}{2\pi} \int_{2\pi} X(e^{j\theta}) \frac{2\pi}{N_s} \sum_{k=-\infty}^{\infty} \delta(\omega - \theta - k\omega_s) \, d\theta$$

$$= \frac{1}{N_s} \sum_{k=0}^{N_s-1} \int_{2\pi} X(e^{j\theta})\delta(\omega - \theta - k\omega_s) \, d\theta$$

$$X_s(e^{j\omega}) = \frac{1}{N_s} \sum_{k=0}^{N_s-1} X(e^{j\theta})\Big|_{\theta=\omega-k\omega_s}$$

$$X_s(e^{j\omega}) = \frac{1}{N_s} \sum_{k=0}^{N_s-1} X\left(e^{j(\omega-k\omega_s)}\right) \tag{8.86}$$

Equation (8.86) represents the spectrum of the sampled signal $x_s(n)$ and is illustrated in Fig. 8.24. If $(\omega_s - \omega_h) > \omega_h$ or, equivalently $\omega_s > 2\omega_h$, there is no aliasing (i.e., the nonzero portions of $X(e^{j\omega})$ do not overlap) as shown in Fig. 8.24(c), whereas if $\omega_s < 2\omega_h$, as in Fig. 8.24(d), aliasing error results.

Thus, in the absence of aliasing (i.e., with $\omega_s > 2\omega_h$), the original signal $x(n)$ can be recovered from $x_s(n)$ by passing the sampled signal $x_s(n)$ through a low-pass filter with gain equal to N_s and cutoff frequency $\omega_c = \omega_s/2$, as illustrated in Fig. 8.25. Therefore, the frequency response of the reconstruction filter is given by

$$H(e^{j\omega}) = \begin{cases} N_s, & |\omega| \leq \omega_c \\ 0, & \omega_c < |\omega| \leq \pi \end{cases} \tag{8.87}$$

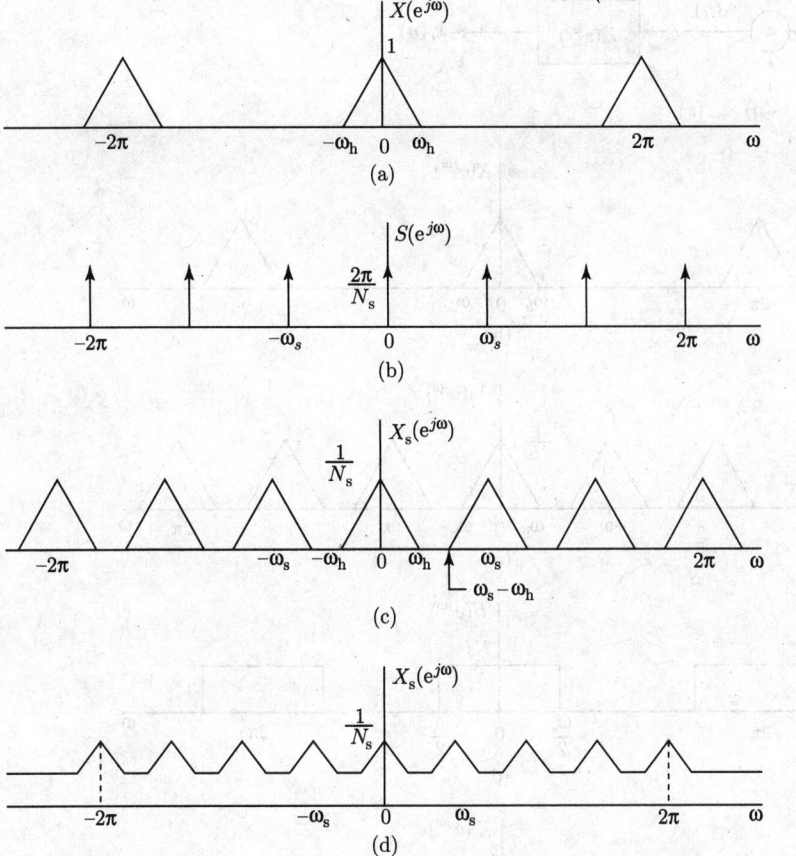

Fig. 8.24 Effect in the frequency domain of impulse train sampling of a discrete-time signal

The reconstruction of $x(n)$ by using a low-pass filter applied to $x_s(n)$ can be interpreted in the time domain as an interpolation formula. By taking the IDTFT of $H(e^{j\omega})$, we have the impulse response $h(n)$ of the low-pass filter as

$$h(n) = N_s \frac{\omega_c}{\pi} \frac{\sin(n\omega_c)}{n\omega_c} \tag{8.88}$$

The reconstructed signal is then

$$x_r(n) = x_s(n) * h(n)$$

$$= \sum_{k=-\infty}^{\infty} x(kN_s)\delta(n - kN_s) * h(n)$$

$$x_r(n) = \sum_{k=-\infty}^{\infty} x(kN_s)h(n - kN_s) \tag{8.89}$$

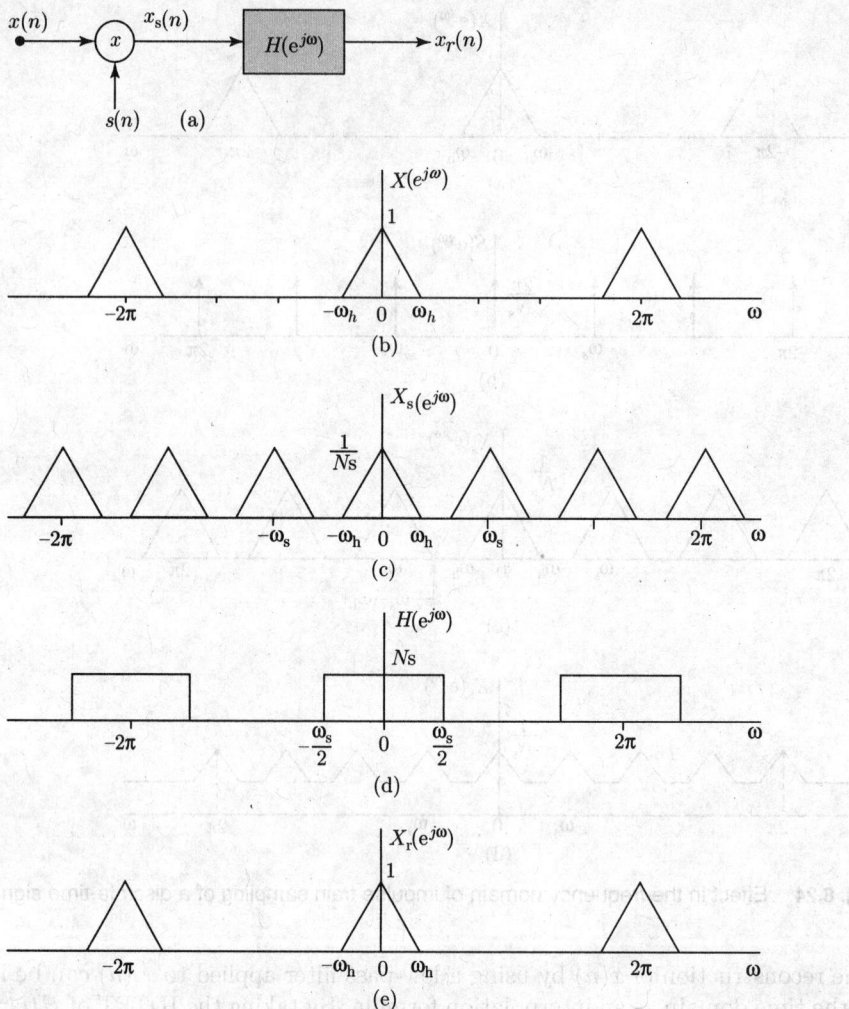

Fig. 8.25 Exact recovery of a discrete-time signal from its samples using an ideal LPF

$$x_r(n) = N_s \frac{\omega_c}{\pi} \sum_{k=-\infty}^{\infty} x(kN_s) \frac{\sin\left((n - kN_s)\omega_c\right)}{(n - kN_s)\omega_c} \tag{8.90}$$

with $\omega_c = \omega_s/2 = \pi/N_s$, we have

$$x_r(n) = \sum_{k=-\infty}^{\infty} x(kN_s) \frac{\sin\left(\pi/N_s(n - kN_s)\right)}{\pi/N_s(n - kN_s)} \tag{8.91}$$

Equation (8.91) represents ideal band-limited interpolation and requires the implementation of an ideal low-pass filter.

8.12.1 Decimation or Down-sampling

The sampled signal in Eq. (8.80) or in Eq. (8.83) has $N_s - 1$ samples out of every N_s samples as zeros, i.e., in between the sampling instants, $x_s(n)$ is known to be zero. We define a new sequence $x_d(n)$ that retains only the nonzero values of $x_s(n)$, i.e.,

$$x_d(n) = x_s(nN_s) = x(nN_s) \tag{8.92}$$

The operation of extracting every N_sth sample is commonly referred to as *decimation*. Thus, the decimation operation by an integer factor $N_s > 1$ on a signal $x(n)$ consists of keeping every N_sth sample of $x(n)$ and removing $N_s - 1$ between samples, generating an output sequence $x_d(n)$ according to the relation given in Eq. (8.92). This results in a signal $x_d(n)$ whose sampling rate is $1/N_s$th that of $x(n)$. Basically, all input samples with indices equal to an integer multiple of N_s are retained at the output, and all others are discarded. Figure 8.26 illustrates the decimation operation for a factor $N_s = 3$.

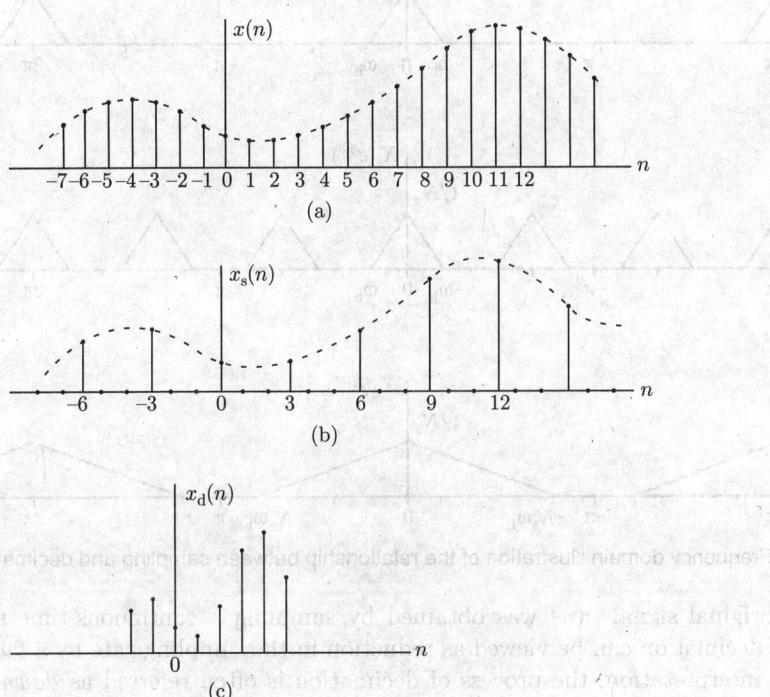

Fig. 8.26 Relationship between $x_s(n)$ and $x_d(n)$

The DTFT of the decimated signal $x_d(n)$ is given by

$$X_d(e^{j\omega}) = \sum_{n=-\infty}^{\infty} x_d(n)\, e^{-j\omega n}$$

$$= \sum_{n=-\infty}^{\infty} x_s(nN_s)\, e^{-j\omega n}$$

Let $m = nN_s$, or equivalently $n = m/N_s$, we can write

$$X_d(e^{j\omega}) = \sum_{m=-\infty}^{\infty} x_s(m)\, e^{-j\omega\frac{m}{N_s}}$$

$$x_d(e^{j\omega}) = X_s(e^{j\frac{\omega}{N_s}}) \tag{8.93}$$

This relationship is illustrated in Fig. 8.27, and from this figure we observe that the spectrum for the sampled signal and the decimated signal differ only in a frequency scaling. The spectrum of the decimated signal is spreaded over a larger portion of the frequency band.

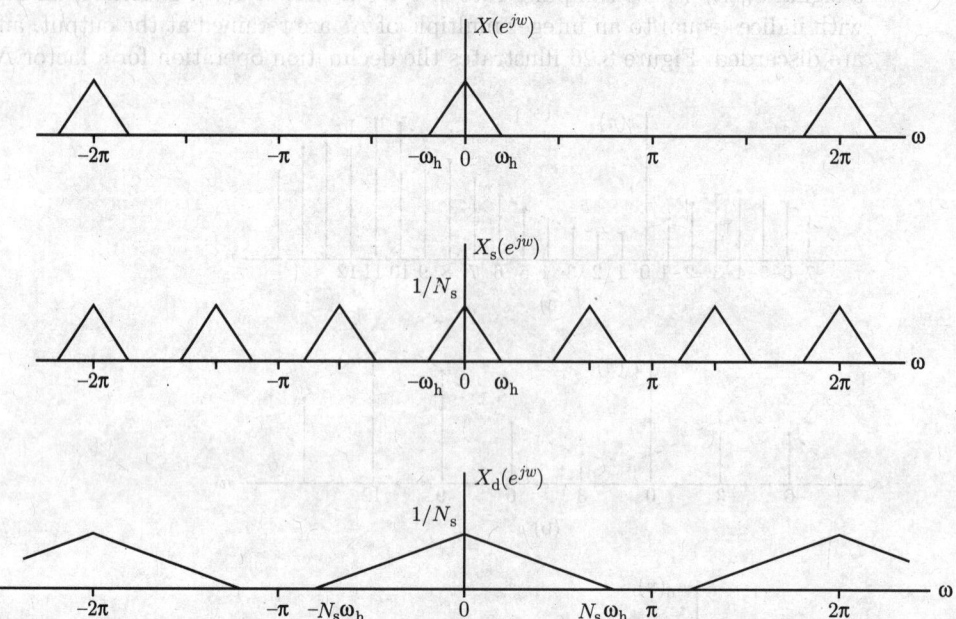

Fig. 8.27 Frequency domain illustration of the relationship between sampling and decimation

If the original signal $x(n)$ was obtained by sampling a continuous-time signal, the process of decimation can be viewed as reduction in the sampling rate by a factor of N_s. With this interpretation, the process of decimation is often referred as *down-sampling*. Let $N_s = M$ be the integer sampling rate reduction factor for the signal $x(n)$,

$$\frac{T_s'}{T_s} = M \tag{8.94}$$

The new sampling frequency f_s' becomes

$$f_s' = \frac{1}{T_s'} = \frac{1}{MT_s}$$

$$f_s' = \frac{f_s}{M} \tag{8.95}$$

8.12.2 Interpolation or Up-sampling

There are situations in which it is useful to convert a sequence to a higher equivalent sampling rate. This process is referred to as *up-sampling* or *interpolation*. This process is the reverse of down-sampling. In up-sampling by an integer factor $L > 1$, $L - 1$ equidistant zero-valued samples are inserted between each two consecutive samples of the input signal $x(n)$ to generate an output signal $x_u(n)$ according to the relation

$$X_u(n) = \begin{cases} X\left(\dfrac{n}{L}\right), & n = 0, \pm L, \pm 2L, \cdots \\ 0, & \text{otherwise} \end{cases} \tag{8.96}$$

The sampling rate of $x_u(n)$ is L times larger than that of the original signal $x(n)$. The DTFT of the interpolated signal $x_u(n)$ is given by

$$x_u(e^{j\omega}) = \sum_{n=-\infty}^{\infty} x_u(n)\, e^{-j\omega n} = \sum_{n=-\infty}^{\infty} x\left(\frac{n}{L}\right) e^{-j\omega n}$$

Let $m = n/L$, or equivalently $n = mL$, we can write

$$x_u(e^{j\omega}) = \sum_{m=-\infty}^{\infty} x(m)\, e^{-j\omega L m}$$

$$x_u(e^{j\omega}) = X(e^{j\omega L}) \tag{8.97}$$

Equation (8.97) represents that the spectrum $X(e^{j\omega})$ is compressed by a factor of L. The spectrum of the interpolated signal $x_u(e^{j\omega})$ contains the images of the original signal spectrum $X(e^{j\omega})$ placed at the harmonics of the sampling frequency $\pm 2\pi/L$, $\pm 4\pi/L$. In other words, there will be $L - 1$ additional images of the input spectrum in the baseband. Thus, a spectrum $X(e^{j\omega})$ band-limited to the low-frequency region does not look like a low-frequency spectrum after up-sampling because of the insertion of zero-valued samples between the nonzero samples of $x_u(n)$. Low-pass filtering of $x_u(n)$ removes the $L - 1$ images and, in effect, "fills in" on the zero-valued samples in $x_u(n)$ with interpolated sample values.

After low-pass filtering

$$X_i(e^{jw}) = H(e^{jw}) x_u(e^{jwL}) \tag{8.98}$$

For ideal low-pass filter with cutoff frequency π/L and gain L, we get

$$X_i(e^{jw}) = \begin{cases} L x_u(e^{jwL}), & |w| < \dfrac{\pi}{L} \\ 0, & \dfrac{\pi}{L} \le |\omega| \le \pi \end{cases} \tag{8.99}$$

The overall process is summarized in Fig. 8.28.

If the original signal $x(n)$ was obtained by sampling a continuous-time signal, the process of increasing the sampling rate of a signal is known as interpolation. With this interpretation, the process of interpolation is often referred as *up-sampling*. Let $L > 1$ be an integer interpolating factor of the signal $x(n)$,

$$\frac{T_s'}{T_s} = \frac{1}{L} \tag{8.100}$$

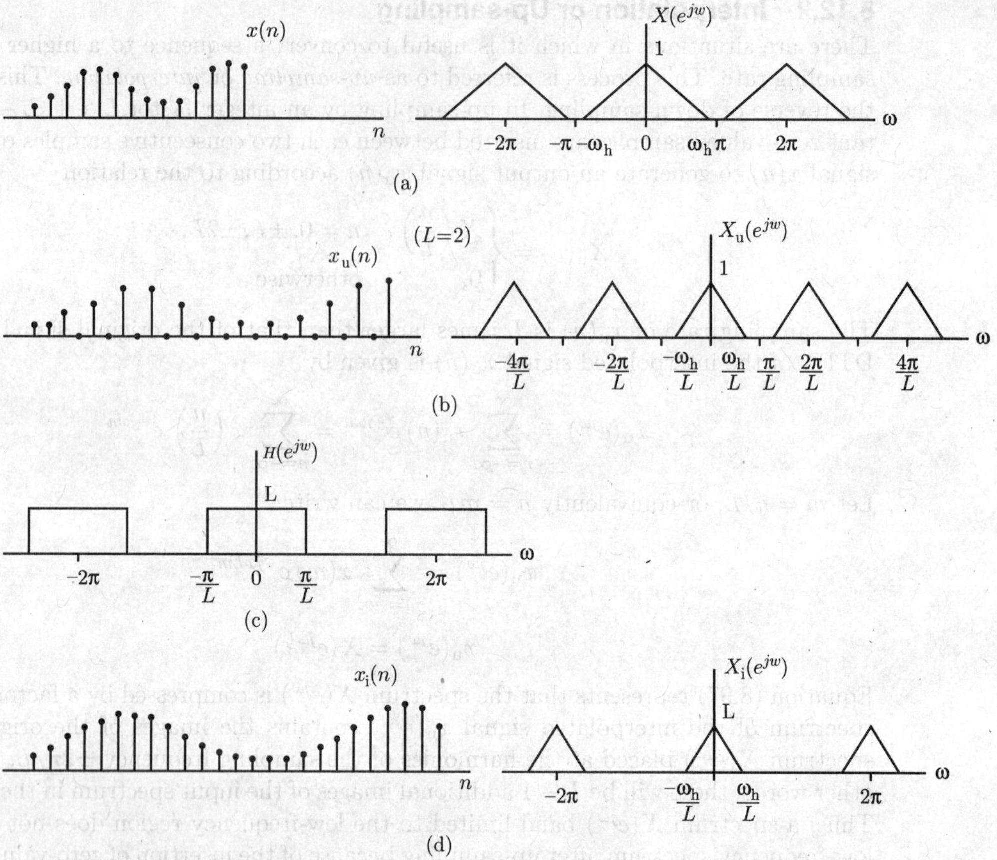

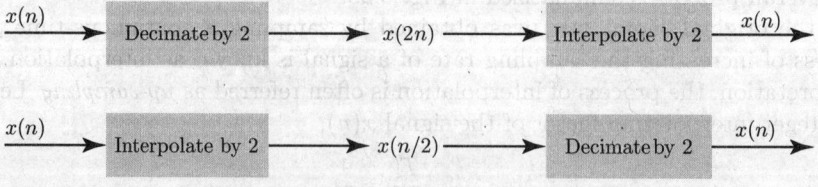

x(n)

X(e^{jw})

(a)

x_u(n)

(L=2)

X_u(e^{jw})

(b)

H(e^{jw})

(c)

x_i(n)

X_i(e^{jw})

(d)

Fig. 8.28 Up-sampling: (a) original signal and its DTFT, (b) up-sampled signal (L = 2) and its DTFT, (c) frequency response of the LPF, and (d) interpolated signal and its spectra

The new sampling frequency f'_s becomes

$$f'_s = \frac{1}{T'_s} = \frac{L}{T_s}$$

$$f'_s = L f_s \tag{8.101}$$

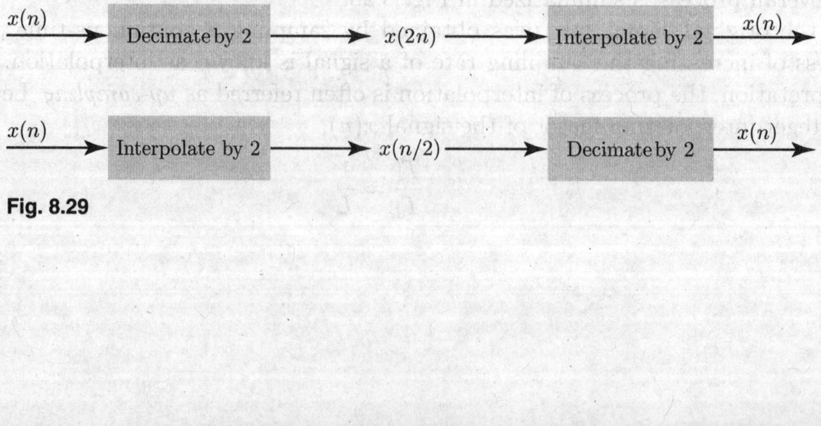

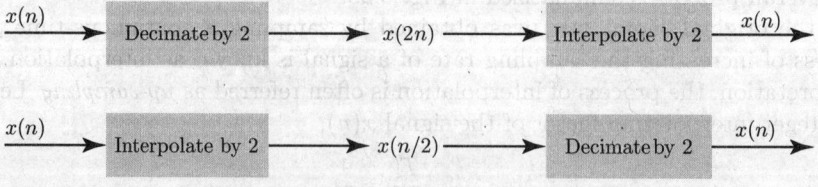

Fig. 8.29

Important Note: It may appear that decimation (discarding signal samples) and interpolation (inserting signal samples) are inverse operations, but this is not always the case. Consider the two sets of operations shown in Fig. 8.29. Both sets of operations start with $x(n)$ and appear to recover $x(n)$, suggesting that interpolation and decimation are inverse operations. In fact, only the second sequence of operations (interpolation followed by decimation) recovers $x(n)$ exactly. To see why, let $x(n) = \{\underset{\uparrow}{1}, 2, 8, 4, 6\}$. Using zero interpolation, for example, the two sequences of operations result in

$$\{\underset{\uparrow}{1}, 2, 8, 4, 6\} \xrightarrow[n \to 2n]{\text{Decimate by 2}} \{\underset{\uparrow}{1}, 8, 6\} \xrightarrow[n \to n/2]{\text{Zero interpolation}} \{\underset{\uparrow}{1}, 0, 8, 0, 6\}$$

$$\{\underset{\uparrow}{1}, 2, 8, 4, 6\} \xrightarrow[n \to n/2]{\text{Zero interpolation}} \{\underset{\uparrow}{1}, 0, 2, 0, 8, 0, 4, 0, 6\} \xrightarrow[n \to 2n]{\text{Decimate by 2}}$$
$$\longrightarrow \{\underset{\uparrow}{1}, 2, 8, 4, 6\}$$

We see that decimation is indeed the inverse of interpolation, but the converse is not necessarily true.

8.12.3 Fractional Delays

Fractional (typically half-sample) delays are sometimes required in practice and can be implemented using interpolation and decimation. If we require that interpolation be followed by decimation and integer shifts, the correct result is obtained by using interpolation followed by an integer shift and decimation. To generate the signal $y(n) = x(n - M/N) = x(Nn - M/N)$ from $x(n)$, we use the sequences of operations shown in Fig. 8.30. The idea is to ensure that each operation (interpolation, shift, and decimation) involves integers.

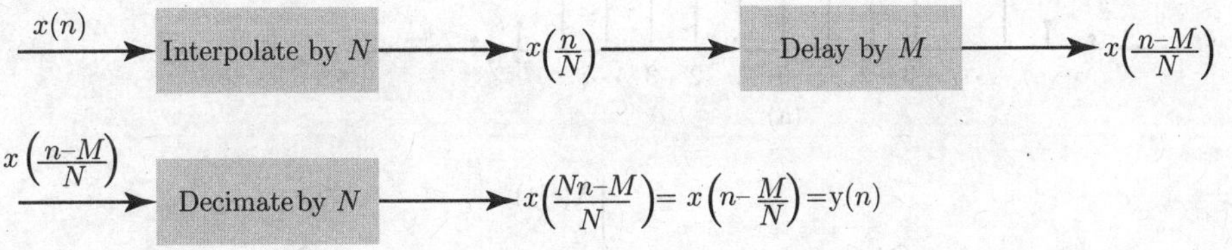

Fig. 8.30   Sequence of operations for fractional delay.

Example 8.12   Impulse-train sampling of $x(n)$ is used to obtain

$$g(n) = \sum_{k=-\infty}^{\infty} x(n)\delta(n - kN_{\text{s}})$$

If $X(e^{j\omega}) = 0$ for $3\pi/7 \leq |\omega| \leq \pi$, determine the largest value for the sampling interval N_{s} which ensures that no aliasing takes place while sampling $x(n)$.

Solution

The maximum frequency of the signal $x(n)$ is $\omega_h = 3\pi/7$. The sampling frequency is

$$\omega_s = 2\omega_h$$

$$\omega_s = \frac{6\pi}{7}$$

$$\frac{7}{3} = \frac{2\pi}{\omega_s}$$

Comparing with $N/m = 2\pi/\omega_s$, we get the sampling period $N_s = 7$.

Example 8.13 Consider a discrete-time signal $x(n)$ from which we form new signals $x_s(n)$ and $x_d(n)$, where $x_s(n)$ corresponds to sampling $x(n)$ with a sampling period of 2 and $x_d(n)$ corresponds to decimating $x(n)$ by a factor of 2, so that

$$x_s(n) = \begin{cases} x(n), & n = 0, \pm 2, \pm 4, \cdots \\ 0, & n = \pm 1, \pm 3, \cdots \end{cases}$$

and

$$x_d(n) = x(2n)$$

(a) If $x(n)$ is illustrated in Fig. 8.31(a), sketch the sequence $x_s(n)$ and $x_d(n)$.

(b) If $X(e^{j\omega})$ is as shown in Fig. 8.31(b), sketch $X_s(e^{j\omega})$ and $X_d(e^{j\omega})$.

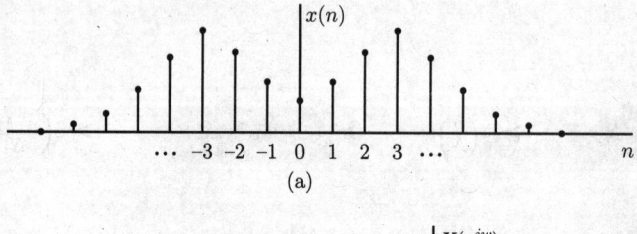

(a)

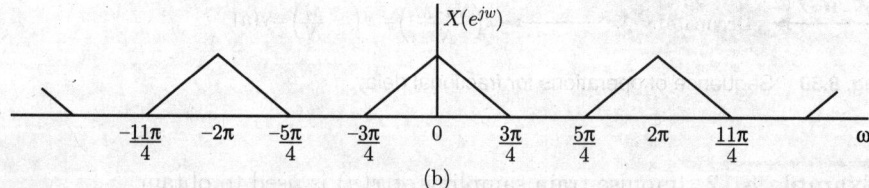

(b)

Fig. 8.31

Solution

(a) The signals $x_s(n)$ and $x_d(n)$ are sketched in Figs 8.32(a) and (b).

(b) Since $x_s(n)$ corresponds to sampling $x(n)$ with a sampling period of $N_s = 2$, therefore, $\omega_s = 2\pi/2 = \pi$. With sampling frequency $\omega_s = \pi$, the spectrum of the sampled signal is

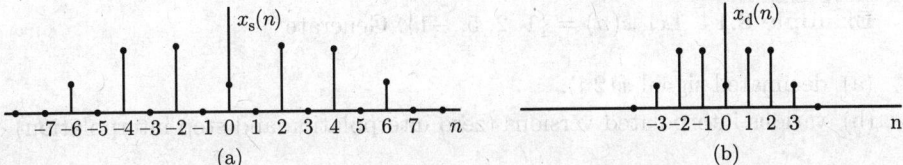

Fig. 8.32

given by [Eq. (8.86)]

$$X_s(e^{j\omega}) = \frac{1}{N_s} \sum_{k=0}^{N_s-1} X\left(e^{j(\omega-k\omega_s)}\right)$$

$$X_s(e^{j\omega}) = \frac{1}{2} \sum_{k=0}^{1} X\left(e^{j(\omega-k\pi)}\right)$$

$X_s(e^{j\omega})$ is sketched in Fig. 8.33(a).

With $N_s = 2$, the spectrum of the decimated signal is given by [Eq. (8.93)]

$$X_d(e^{j\omega}) = X_s(e^{j\omega/N_s}) = \frac{1}{N_s} \sum_{k=0}^{N_s-1} X\left(e^{j(\omega/N_s-k\omega_s)}\right)$$

$$X_d(e^{j\omega}) = X_s(e^{j\omega/2}) = \frac{1}{2} \sum_{k=0}^{1} X\left(e^{j(\omega/2-k\pi)}\right)$$

$X_d(e^{j\omega})$ is sketched in Fig. 8.33(b).

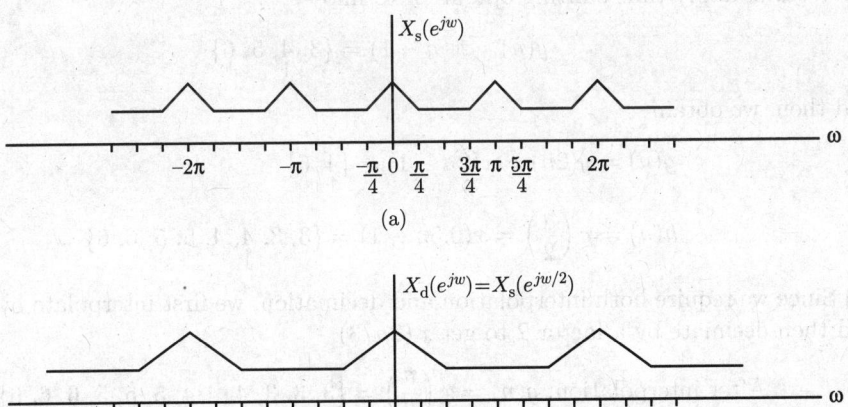

Fig. 8.33

Example 8.14 Let $x(n) = \{1, \underset{\uparrow}{2}, 5, -1\}$. Generate

(a) decimated signal $x(2n)$.

(b) various interpolated versions (zero interpolation and step interpolation) of $x(n/3)$.

Solution

(a) To generate $y(n) = x(2n)$, we remove samples at the odd indices to obtain

$$x(n) = \{\underset{\uparrow}{2}, -1\}$$

(b) *Zero-interpolation:* In zero-interpolation by a factor $L > 1$, $L - 1$ zero-valued samples are inserted between each two consecutive samples of the input signal $x(n)$. Here, for $L = 3$, the zero-interpolated signal is

$$\{1, 0, 0, \underset{\uparrow}{2}, 0, 0, 5, 0, 0, -1, 0, 0\}$$

Step-interpolation: In step-interpolation by a factor $L > 1$, every sample value is repeated (or held) $L - 1$ successive times. Here, for $L = 3$, the step-interpolated signal is

$$\{1, 1, 1, \underset{\uparrow}{2}, 2, 2, 5, 5, 5, -1, -1, -1\}$$

Example 8.15 Let $x(n) = \{3, 4, \underset{\uparrow}{5}, 6\}$.

(a) Find $g(n) = x(2n - 1)$ and the step-interpolated signal $h(n) = x(0.5n - 1)$.

(b) Find $y(n) = x(2n/3)$ assuming step-interpolation where needed.

Solution

(a) We first apply time-shifting operation to find

$$y(n) = x(n - 1) = \{3, \underset{\uparrow}{4}, 5, 6\}$$

and then, we obtain

$$g(n) = y(2n) = x(2n - 1) = \{\underset{\uparrow}{4}, 6\}$$

$$h(n) = y\left(\frac{n}{2}\right) = x(0.5n - 1) = \{3, 3, \underset{\uparrow}{4}, 4, 5, 5, 6, 6\}$$

(b) Since we require both interpolation and decimation, we first interpolate by a factor 3 and then decimate by a factor 2 to get $x(2n/3)$.

After interpolation: $g(n) = x\left(\dfrac{n}{3}\right) = \{3, 3, 3, 4, 4, 4, \underset{\uparrow}{5}, 5, 5, 6, 6, 6\}$

After decimation: $y(n) = g(2n) = x\left(\dfrac{2n}{3}\right) = \{3, 3, 4, \underset{\uparrow}{5}, 5, 6\}$

8.13 RELATIONSHIP BETWEEN DTFT AND CTFT

Consider a continuous-time signal $x(t)$, its instantaneous sampled version $x_\delta(t)$, and its equivalent discrete-time signal $x(n)$. Since we will be dealing with Fourier transform in both continuous time and discrete time, in this section we only distinguish the continuous-time and discrete-time frequency variables by using ω in continuous time and Ω in discrete time. Assume that the continuous-time Fourier transforms of $x(t)$ and $x_\delta(t)$ are $X(\omega)$ and $X_\delta(\omega)$, respectively, while the discrete-time Fourier transform of $x(n)$ is $X(e^{j\Omega})$.

Consider the sampled signal $x_\delta(t)$ [Eq. (8.5)], i.e.,

$$x_\delta(t) = \sum_{n=-\infty}^{\infty} x(nT_s)\delta(t - nT_s)$$

$$\mathcal{F}[x_\delta(t)] = \mathcal{F}\left[\sum_{n=-\infty}^{\infty} x(nT_s)\delta(t - nT_s)\right]$$

$$X_\delta(\omega) = \sum_{n=-\infty}^{\infty} x(nT_s)\mathcal{F}[\delta(t - nT_s)]$$

$$X_\delta(\omega) = \sum_{n=-\infty}^{\infty} x(nT_s)\,e^{-j\omega nT_s} \tag{8.102}$$

Now consider the DTFT of $x(n)$, i.e.,

$$X(e^{j\Omega}) = \sum_{n=-\infty}^{\infty} x(n)\,e^{-j\Omega n}$$

Since $x(n) = x(nT_s)$, we have

$$X(e^{j\Omega}) = \sum_{n=-\infty}^{\infty} x(nT_s)\,e^{-j\Omega n} \tag{8.103}$$

Comparing Eqs (8.102) and (8.103), we see that $X(e^{j\Omega})$ and $X_\delta(\omega)$ are related through

$$X(e^{j\Omega}) = X_\delta(\omega)\Big|_{\omega=\frac{\Omega}{T_s}} = X_\delta\left(\frac{\Omega}{T_s}\right) \tag{8.104}$$

From Eq. (8.8), $X(\omega)$ and $X_\delta(\omega)$ are related through

$$X_\delta(\omega) = \frac{1}{T_s}\sum_{n=-\infty}^{\infty} X(\omega - n\omega_s) \tag{8.105}$$

Consequently,

$$X(e^{j\Omega}) = \frac{1}{T_s}\sum_{n=-\infty}^{\infty} X\left(\frac{1}{T_s}(\Omega - 2n\pi)\right) \tag{8.106}$$

This relationship among $X(\omega)$, $X_\delta(\omega)$, and $X(e^{j\Omega})$ is illustrated in Fig. 8.34.

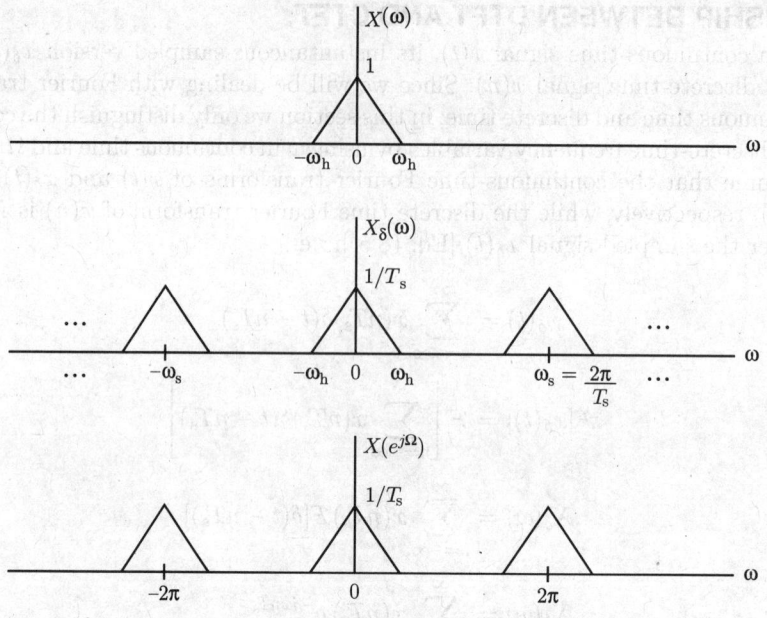

Fig. 8.34 Relationship between $X(\omega)$, $X_\delta(\omega)$, and $X(e^{j\Omega})$

SOLVED EXAMPLES

Example 8.16 Determine the conditions on the sampling interval T_s so that each $x(t)$ is uniquely represented by the discrete-time sequence $x(n) = x(nT_s)$.

(a) $x(t) = \cos(\pi t) + 3\sin(2\pi t) + \sin(4\pi t)$

(b) $x(t) = \cos(2\pi t)\,\text{sinc}(t) + 3\sin(6\pi t)\,\text{sinc}(2t)$

(c) The signal $x(t)$ with FT given in Fig. 8.35

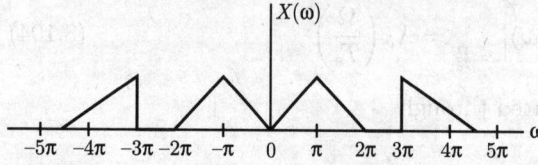

Fig. 8.35

Solution

(a) Given that

$$x(t) = \cos(\pi t) + 3\sin(2\pi t) + \sin(4\pi t)$$

Comparing with

$$x(t) = A_1 \cos(\omega_1 t) + A_2 \sin(\omega_2 t) + A_3 \sin(\omega_3 t)$$

we have

$$\omega_1 = \pi, \quad \omega_2 = 2\pi, \quad \text{and} \quad \omega_3 = 4\pi$$

The maximum frequency present in the given signal $x(t)$ is

$$\omega_{\max} = \omega_3 = 4\pi \quad \text{or} \quad f_{\max} = \frac{\omega_{\max}}{2\pi} = 2$$

Therefore, the sampling frequency is given by

$$\omega_s \geq 2\omega_{\max} = 8\pi \,\text{rad/s}$$

or

$$f_s \geq 2f_{\max} = 4 \,\text{Hz}$$

and hence, the sampling interval

$$T_s \leq \frac{1}{4}$$

(b) Given that

$$x(t) = \cos(2\pi t)\,\text{sinc}(t) + 3\sin(6\pi t)\,\text{sinc}(2t)$$

$$= \cos(2\pi t)\frac{\sin(\pi t)}{\pi t} + 3\sin(6\pi t)\frac{\sin(2\pi t)}{2\pi t}$$

$$x(t) = \frac{1}{2\pi t}[\sin(3\pi t) - \sin(\pi t)]$$

$$+ \frac{3}{4\pi t}[\cos(4\pi t) - \cos(8\pi t)]$$

The maximum frequency present in the given signal $x(t)$ is

$$\omega_{\max} = 8\pi \quad \text{or} \quad f_{\max} = \frac{\omega_{\max}}{2\pi} = 4$$

Therefore, the sampling frequency is given by

$$f_s \geq 2f_{\max} = 8 \text{ Hz}$$

and hence, the sampling interval

$$T_s \leq \frac{1}{8}$$

(c) As shown in Fig. 8.35, the maximum frequency component present in the signal $x(t)$ is

$$\omega_{\max} = 4.5\pi \quad \text{or} \quad f_{\max} = \frac{\omega_{\max}}{2\pi} = \frac{4.5}{2}$$

Therefore, the sampling frequency is given by

$$f_s \geq 2f_{\max} = 4.5 \text{ Hz}$$

and hence, the sampling interval

$$T_s \leq \frac{1}{4.5}$$

Example 8.17 A signal

$$x(t) = \cos(200\pi t) + 2\cos(320\pi t)$$

is ideally sampled at $f_s = 300$ Hz. If the sampled signal is passed through an ideal low-pass filter with a gain $1/f_s$ and cutoff frequency of 250 Hz, what frequency components will appear in the output?

Solution
The Fourier transform $X(f)$ of the given signal $x(t)$ is

$$X(f) = \frac{1}{2}[\delta(f - 100) + \delta(f + 100)]$$

$$+ [\delta(f - 160) + \delta(f + 160)]$$

With sampling frequency $f_s = 300$ Hz, the Fourier transform of the sampled signal is given as

[Eq. (8.8)]

$$X_\delta(f) = f_s \sum_{n=-\infty}^{\infty} X(f - nf_s)$$

$$= 300 \sum_{n=-\infty}^{\infty} X(f - 300n)$$

$$X_\delta(f) = \cdots + 300X(f + 300) + 300X(f)$$

$$+ 300X(f - 300) + \cdots$$

The sampled signal spectrum $X_\delta(f)$ is shown in Fig. 8.36.

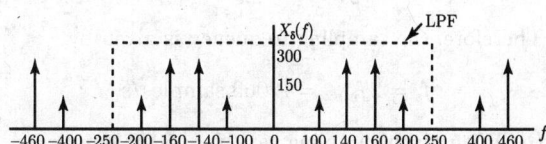

Fig. 8.36

If the sampled signal is passed through an ideal low-pass filter with a gain $1/f_s$ and cutoff frequency of 250 Hz, the frequency components that will appear at the output of the low-pass filter are 100, 140, 160, and 200 Hz. The output $Y(f)$ of the low-pass filter can be written as

$$Y(f) = \frac{1}{2}[\delta(f - 100) + \delta(f + 100)]$$

$$+ [\delta(f - 140) + \delta(f + 140)]$$

$$+ [\delta(f - 160) + \delta(f + 160)]$$

$$+ \frac{1}{2}[\delta(f - 200) + \delta(f + 200)]$$

Taking the inverse Fourier transform of the above equation, we obtain

$$y(t) = \cos(200\pi t) + 2\cos(280\pi t)$$

$$+ 2\cos(320\pi t) + \cos(400\pi t)$$

Example 8.18 Consider the continuous-time signal

$$x(t) = 3\cos(2000\pi t) + 5\sin(6000\pi t)$$

$$+ 10\cos(12000\pi t)$$

(a) What is the Nyquist rate for this signal?

(b) Assume now that we sample this signal using a sampling rate $f_s = 5000$ samples/s. What is the discrete-time signal obtained after sampling?

(c) What is the continuous-time signal $y(t)$ we can reconstruct from the samples ($f_s = 5000$ samples/s) if we use ideal interpolation?

Solution

(a) The maximum frequency component present in the signal $x(t)$ is

$$f_{max} = \frac{\omega_{max}}{2\pi} = \frac{12000\pi}{2\pi} = 6000\,\text{Hz}$$

Therefore, the sampling frequency is given by

$$f_s = 2f_{max} = 12000\,\text{samples/s}$$

(b) With given sampling rate $f_s = 5000$ samples/s, the sampling interval $T_s = 1/f_s = 1/5000$. Therefore, the discrete-time signal obtained after sampling is

$$x(n) = x(nT_s) = x(t)|_{t=nT_s}$$

$$= 3\cos(2000\pi nT_s) + 5\sin(6000\pi nT_s)$$
$$+ 10\cos(12000\pi nT_s)$$

$$= 3\cos\left(2000\pi n\frac{1}{5000}\right)$$

$$+ 5\sin\left(6000\pi n\frac{1}{5000}\right)$$

$$+ 10\cos\left(12000\pi n\frac{1}{5000}\right)$$

$$x(n) = 3\cos\left(\frac{2}{5}\pi n\right) + 5\sin\left(\frac{6}{5}\pi n\right)$$

$$+ 10\cos\left(\frac{12}{5}\pi n\right)$$

(c) The Fourier transform $X(f)$ of the given signal $x(t)$ is (with f in kilohertz)

$$X(f) = \frac{3}{2}[\delta(f-1) + \delta(f+1)]$$

$$+ \frac{5}{2}[\delta(f-3) + \delta(f+3)]$$

$$+ 5[\delta(f-6) + \delta(f+6)]$$

With sampling frequency $f_s = 5$ kHz, the Fourier transform of the sampled signal is given as [Eq. (8.8)]

$$X_\delta(f) = f_s \sum_{n=-\infty}^{\infty} X(f - nf_s)$$

$$= 5 \sum_{n=-\infty}^{\infty} X(f - 5n)$$

$$X_\delta(f) = \cdots + 5X(f+5) + 5X(f)$$
$$+ 5X(f-5) + \cdots$$

The sampled signal spectrum $X_\delta(f)$ is shown in Fig. 8.37.

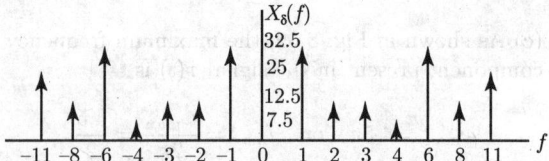

Fig. 8.37

For an ideal interpolation, the sampled signal is passed through an ideal low-pass filter with a gain $1/12$ and cutoff frequency of 6 kHz. The frequency response of the ideal low-pass filter is

$$H(f) = \begin{cases} \dfrac{1}{12}, & -6 \le f \le 6 \\ 0, & \text{elsewhere} \end{cases}$$

The frequency components that will appear at the output of the low-pass filter are 1, 2, 3, 4, and 6 kHz. The output $Y(f)$ of the low-pass filter can be written as

$$Y(f) = \frac{32.5}{12}[\delta(f-1) + \delta(f+1)]$$

$$+ \frac{12.5}{12}[\delta(f-2) + \delta(f+2)]$$

$$+ \frac{12.5}{12}[\delta(f-3) + \delta(f+3)]$$

$$+ \frac{7.5}{12}[\delta(f-4) + \delta(f+4)]$$

$$+ \frac{32.5}{12}[\delta(f-6) + \delta(f+6)]$$

Taking the inverse Fourier transform of the above equation, we obtain

$$y(t) = \frac{32.5}{6} \cos(2000\pi t) + \frac{12.5}{6} \cos(4000\pi t)$$

$$+ \frac{12.5}{6} \cos(6000\pi t) + \frac{7.5}{6} \cos(8000\pi t)$$

$$+ \frac{32.5}{6} \cos(12000\pi t)$$

Example 8.19 Let $x(t)$ and $y(t)$ be two low-pass signals with bandwidth of 150 Hz and 350 Hz, respectively, and let $z(t) = x(t)y(t)$. The signal $z(t)$ is sampled using an ideal sampler at intervals of T_s seconds. What is the maximum value that T_s can take without introducing aliasing?

Solution
If

$$x(t) \longleftrightarrow X(f) \quad \text{and} \quad y(t) \longleftrightarrow Y(f)$$

then

$$z(t) = x(t)y(t) \longleftrightarrow Z(f) = [X(f) * Y(f)]$$

Clearly, $Z(f) = 0$ for $|f| > (150 + 350)$, i.e., $|f| > 500$ and the maximum frequency present in the signal $z(t) = x(t)y(t)$ is $f_{max} = 500$. [If $x(t)$ and $y(t)$ have bandwidths B_x and B_y Hz, respectively, the bandwidth of $z(t) = x(t)y(t)$ is $B_z = B_x + B_y$ Hz. This results from the application of the width property of convolution (see Section 2.3.6).] Therefore, the Nyquist rate (minimum sampling frequency) for $z(t)$ is given by

$$f_s = 2f_{max} = 1000 \text{ samples/s}$$

Therefore, the maximum value that T_s can take without introducing aliasing is

$$T_s = \frac{1}{f_s} = \frac{1}{1000} = 1 \text{ ms}$$

Example 8.20 A waveform $x(t) = 10 + 10 \sin(500t)$ is to be sampled periodically and reproduced from these samples. Find the maximum allowable time interval between sample values. How many sample values are required to be stored in order to produce 2 s of this waveform?

Solution
Given that

$$x(t) = 10 + 10 \sin(500t)$$

The maximum frequency present in the given signal $x(t)$ is

$$f_{max} = \frac{500}{2\pi} = 79.58 \text{ Hz}$$

Therefore, the Nyquist rate (minimum sampling frequency) is given by

$$f_s = 2f_{max} = 159.16 \text{ Hz}$$

Thus, the maximum allowable time interval between the sample values is given by

$$T_s = \frac{1}{f_s} = \frac{1}{159.16} = 6.28 \text{ ms}$$

The number of sample values required to be stored in order to produce 1 s of this waveform is given by

$$\text{Number of samples} = \frac{2\,\text{s}}{6.28\,\text{ms}} = 318.5$$

Example 8.21 A signal $x(t) = \sin(\pi t)/\pi t$ is sampled by $s(t) = \sum_{n=-\infty}^{\infty} \delta(t - n/2)$. Determine and sketch the sampled signal and its Fourier transform.

Solution
Let $x_s(t)$ be the sampled signal. Therefore,

$$x_s(t) = x(t)s(t)$$

$$= \frac{\sin(\pi t)}{\pi t} \sum_{n=-\infty}^{\infty} \delta\left(t - \frac{n}{2}\right)$$

$$x_s(t) = \sum_{n=-\infty}^{\infty} \frac{\sin(\pi n/2)}{\pi n/2} \delta\left(t - \frac{n}{2}\right)$$

The Fourier transform $X(f)$ of the given signal $x(t)$ is

$$X(f) = \text{rect}(f) = \begin{cases} 1, & -\frac{1}{2} < f < \frac{1}{2} \\ 0, & \text{otherwise} \end{cases}$$

Here, the sampling interval $T_s = 1/2$ and the sampling frequency $f_s = 2$ samples/s. The Fourier transform of the sampled signal is given by

$$X_s(f) = f_s \sum_{n=-\infty}^{\infty} X(f - nf_s)$$

$$= 2 \sum_{n=-\infty}^{\infty} X(f - 2n)$$

$$X_s(f) = \cdots + 2X(f + 2) + 2X(f)$$
$$+ 2X(f - 2) + \cdots$$

$X(f)$ and $X_s(f)$ are illustrated in Figs 8.38(a) and (b), respectively.

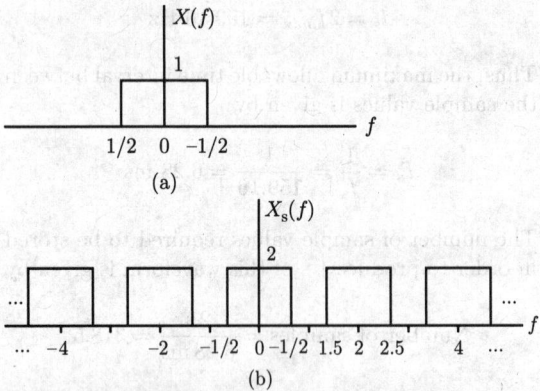

(a)

(b)

Fig. 8.38

Example 8.22 The signal

$$x(t) = \cos(5\pi t) + 0.5\cos(10\pi t)$$

is instantaneously sampled. The interval between samples is T_s.

(a) Find the maximum allowable value for T_s.

(b) If the sampling signal is

$$s(t) = 5 \sum_{n=-\infty}^{\infty} \delta(t - 0.1n)$$

and the sampled signal $x_s(t) = x(t)s(t)$ consists of train of impulses, each with a different strength

$$x_s(t) = \sum_{n=-\infty}^{\infty} I_n \delta(t - 0.1n)$$

Find I_0, I_1, and I_2.

(c) To reconstruct the signal, $x_s(t)$ is passed through a rectangular low-pass filter. Find the minimum filter bandwidth to reconstruct the signal without distortion.

Solution

(a) The maximum frequency component present in the signal $x(t)$ is

$$f_{max} = \frac{\omega_{max}}{2\pi} = \frac{10\pi}{2\pi} = 5\,\text{Hz}$$

Therefore, the Nyquist rate (minimum sampling frequency) is given by

$$f_s = 2f_{max} = 10\,\text{samples/s}$$

and hence, the maximum allowable value for T_s is given by

$$T_s = \frac{1}{f_s} = \frac{1}{10} = 0.1\,\text{s}$$

(b) Given that

$$s(t) = 5 \sum_{n=-\infty}^{\infty} \delta(t - 0.1n)$$

and the sampled signal

$$x_s(t) = x(t)s(t)$$

$$= 5x(t) \sum_{n=-\infty}^{\infty} \delta(t - 0.1n)$$

Using the time-shifting property of the impulse function, we obtain

$$x_s(t) = \sum_{n=-\infty}^{\infty} 5x(0.1n)\delta(t - 0.1n)$$

Comparing with

$$x_s(t) = \sum_{n=-\infty}^{\infty} I_n\delta(t - 0.1n)$$

we get

$$I_n = 5x(0.1n)$$

Therefore,

$$I_0 = 5x(0) = 5[\cos(0) + 0.5\cos(0)]$$

$$= 5[1 + 0.5] = 5 \times 1.5 = 7.5$$

$$I_1 = 5x(0.1) = 5[\cos(0.5\pi) + 0.5\cos(\pi)]$$

$$= 5[0 - 0.5] = 5 \times -0.5 = -2.5$$

$$I_2 = 5x(0.2) = 5[\cos(\pi) + 0.5\cos(2\pi)]$$

$$= 5[-1 + 0.5] = 5 \times -0.5 = -2.5$$

(c) The Fourier transform of the given signal $x(t)$ is

$$X(f) = 0.5[\delta(f - 2.5) + \delta(f + 2.5)]$$

$$+ 0.25[\delta(f - 5) + \delta(f + 5)]$$

The Fourier transform of the sampling signal $s(t)$ is

$$S(f) = 50 \sum_{n=-\infty}^{\infty} \delta(f - 10n)$$

Taking the Fourier transform of the sampled signal $x_s(t)$, we obtain

$$X_s(f) = X(f) * S(f)$$

$$= X(f) * 50 \sum_{n=-\infty}^{\infty} \delta(f - 10n)$$

$$X_s(f) = 50 \sum_{n=-\infty}^{\infty} X(f - 10n)$$

$$X_s(f) = \cdots + 50X(f + 10) + 50X(f)$$

$$+ 50X(f - 10) + \cdots$$

$X_s(f)$ is illustrated in Fig. 8.39.

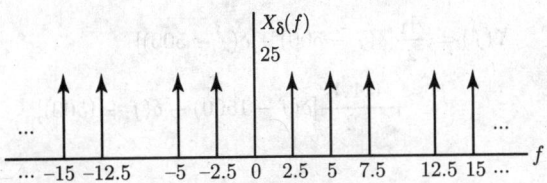

Fig. 8.39

To reconstruct the original signal without distortion, the minimum filter bandwidth must be 5 Hz so that the filter will pass the two original frequency components (2.5 and 5 Hz).

Example 8.23 It is given that the Fourier transform of the pulse shown in Fig. 8.40(a) is

$$X(f) = \tau \, \text{sinc}^2(\tau f) \quad \text{where} \quad \text{sinc}(x) \triangleq \frac{\sin(\pi x)}{\pi x}$$

The acoustic pulse received by the sound receiver on a submarine is of the form

$$y(t) = 30 \, \text{sinc}^2(100t)$$

(a) Sketch the spectrum of the pulse.

(b) If this pulse is to be processed in discrete-time, what is the mean sampling frequency, f_s, that can be used.

(c) Also sketch the spectrum of the sampled signal $y_s(t)$ if sampling is done at the following rates: (i) $f_s' = f_s$, (ii) $f_s' = 2f_s$, and (iii) $f_s' = f_s/2$.

Solution

(a) The given signal $x(t)$ can be expressed as

$$x(t) = \triangle\left(\frac{t}{\tau}\right) = \begin{cases} \left(1 + \dfrac{t}{\tau}\right), & -\tau \le t < 0 \\ \left(1 - \dfrac{t}{\tau}\right), & 0 \le t \le \tau \end{cases}$$

Given that

$$x(t) = \triangle\left(\frac{t}{\tau}\right) \longleftrightarrow X(f) = \tau \, \text{sinc}^2(\tau f)$$

Using the duality property of the Fourier transform, we have if

$$x(t) \longleftrightarrow X(f)$$

then

$$X(t) \longleftrightarrow x(-f)$$

$$\tau \, \text{sinc}^2(\tau t) \longleftrightarrow \triangle\left(-\frac{f}{\tau}\right)$$

$$\tau \, \text{sinc}^2(\tau t) \longleftrightarrow \triangle\left(\frac{f}{\tau}\right)$$

$$100 \, \text{sinc}^2(100t) \longleftrightarrow \triangle\left(\frac{f}{100}\right)$$

$$\frac{30}{100} 100 \, \text{sinc}^2(100t) \longleftrightarrow \frac{30}{100} \triangle\left(\frac{f}{100}\right)$$

$$y(t) = 30 \, \text{sinc}^2(100t) \longleftrightarrow Y(f) = 0.3 \triangle\left(\frac{f}{100}\right)$$

$Y(f)$ is illustrated in Fig. 8.40(b).

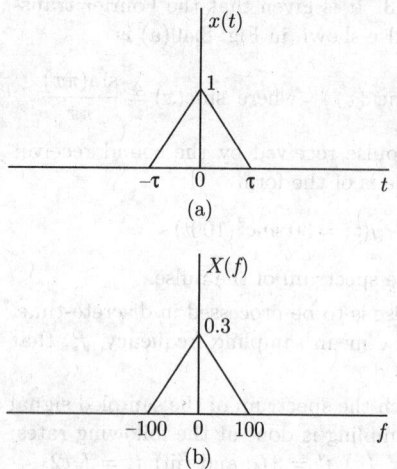

Fig. 8.40

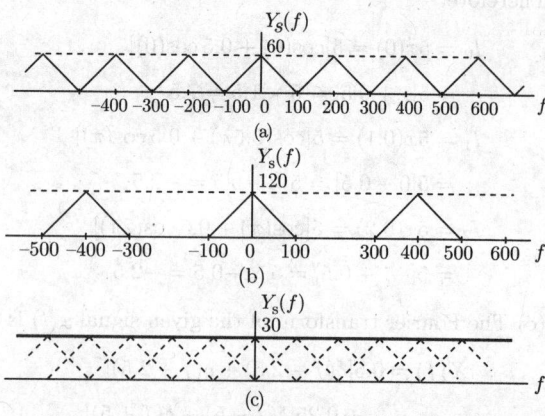

Fig. 8.41

(b) Given that

$$y(t) = 30\,\mathrm{sinc}^2(100t)$$

$$= 30\,\frac{\sin^2(100\pi t)}{(100\pi t)^2}$$

$$x(t) = \frac{1}{2(100\pi t)^2} - \frac{1}{2(100\pi t)^2}\cos(200\pi t)$$

Comparing with

$$x(t) = A_0 - A_1\cos(\omega_1 t)$$

we have

$$\omega_1 = 200\pi \quad \text{or} \quad f_1 = 100$$

Therefore, the Nyquist rate (minimum sampling frequency) is given by

$$\omega_s = 2\omega_1 = 400\pi \quad \text{or} \quad f_s = 2f_1 = 200 \text{ Hz}$$

(c) The spectrum of the sampled signal $y_s(t)$ is given by

$$Y_s(f) = f_s' \sum_{n=-\infty}^{\infty} Y(f - nf_s')$$

Figures 8.41(a), (b), and (c) show the spectrum of the sampled signal for the sampling rate $f_s' = f_s = 200$, $f_s' = 2f_s = 400$, and $f_s' = f_s/2 = 100$ Hz, respectively.

Example 8.24 A waveform consisting of a sinusoid of a frequency 500 Hz and its third, harmonic of amplitude 0.1 times that of the fundamental is ideally sampled at the rate of 10000 samples/s. Sketch the spectrum of the sampled waveform showing the first four components.

Solution
A signal that consists of a sinusoid of a frequency 500 Hz and its third harmonic of amplitude 0.1 times that of the fundamental is mathematically expressed as

$$x(t) = A_1\cos(1000\pi t) + 0.1A_1\cos(3 \times 1000\pi t)$$

$$= A_1\cos(1000\pi t) + 0.1A_1\cos(3000\pi t)$$

Taking the Fourier transform of the above equation, we obtain

$$X(f) = \frac{A_1}{2}[\delta(f - 500) + \delta(f + 500)]$$

$$+ \frac{0.1A_1}{2}[\delta(f - 1500) + \delta(f + 1500)]$$

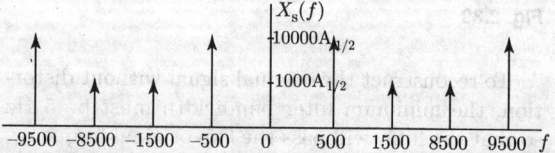

Fig. 8.42

With $f_s = 10000$, the spectrum of the sampled signal is given by

$$X_s(f) = f_s \sum_{n=-\infty}^{\infty} X(f - nf_s)$$

$$X_s(f) = 10000 \sum_{n=-\infty}^{\infty} X(f - 10000n)$$

Fig. 8.42 illustrates the first four components of $X_s(f)$.

Example 8.25 The spectrum of a signal $x(t)$ is shown in Fig. 8.43. This signal is naturally sampled with a periodic train of rectangular pulses of duration of duration $T = 50/3$ ms. Plot the spectrum of the sampled signal for frequencies up to 50 Hz for the following two conditions:

(a) The sampling rate is equal to the Nyquist rate.
(b) The sampling rate is equal to 10 samples per second.

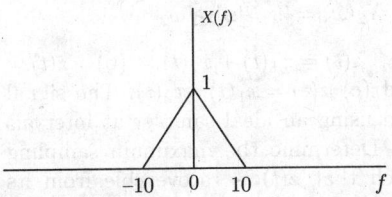

Fig. 8.43

Solution
Given the width of the sampling pulse is $T = 50/3$. The Nyquist rate is $f_s = 20$ Hz. The spectrum of the naturally sampled signal is given by [Eq. (8.18)]

$$X_s(f) = \frac{1}{T_s} \sum_{n=-\infty}^{\infty} \text{sinc}\left(\frac{nT}{T_s}\right) X(f - nf_s)$$

Consider the term $\text{sinc}(nT/T_s)$.

$$\text{sinc}\left(\frac{nT}{T_s}\right) = \frac{\sin(n\pi T/T_s)}{n\pi T/T_s}$$

Therefore, the $\text{sinc}(nT/T_s)$ function goes to zero when

$$\sin\left(\frac{n\pi T}{T_s}\right) = 0 = \sin(m\pi) \quad m = \pm 1, \pm 2, ..., etc.$$

$$\frac{n\pi T}{T_s} = m\pi$$

$$nf_s = m\frac{1}{T} = m\frac{3 \times 10^3}{50} = \pm 60 \text{ Hz}, \pm 120 \text{ Hz}, \cdots$$

(a) The spectrum of the sampled signal at the Nyquist rate $f_s = 20$ Hz is as shown in Fig. 8.44(a).
(b) The spectrum of the sampled signal at the sampling rate $f_s = 10$ Hz is as shown in Fig. 8.44(b).

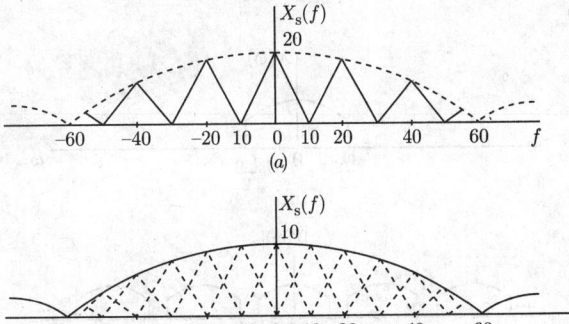

Fig. 8.44

Example 8.26 Let $x(t)$ be a signal with Nyquist rate ω_0. Also, let

$$y(t) = x(t)s(t-1)$$

where

$$s(t) = \sum_{n=-\infty}^{\infty} \delta(t - nT_s) \quad \text{and} \quad T_s < \frac{2\pi}{\omega_0}$$

Specify the constraints on the magnitude and phase of the frequency response of a filter that gives $x(t)$ as its output when $y(t)$ is the input.

Solution
Given that $x(t)$ is a signal with Nyquist rate ω_0; therefore, maximum frequency present in the $x(t)$ is

$\omega_0/2$, and also given that

$$s(t) = \sum_{n=-\infty}^{\infty} \delta\left(t - nT_s\right)$$

Taking the Fourier transform of the above equation, we obtain

$$S(\omega) = \frac{2\pi}{T_s} \sum_{n=-\infty}^{\infty} \delta\left(\omega - n\frac{2\pi}{T_s}\right)$$

Using the time-shifting property of the Fourier transform, we obtain

$$\mathcal{F}[s(t-1)] = e^{-j\omega} S(\omega)$$

$$= e^{-j\omega} \frac{2\pi}{T_s} \sum_{n=-\infty}^{\infty} \delta\left(\omega - n\frac{2\pi}{T_s}\right)$$

$$\mathcal{F}[s(t-1)] = \frac{2\pi}{T_s} \sum_{n=-\infty}^{\infty} \delta\left(\omega - n\frac{2\pi}{T_s}\right) e^{-jn2\pi/T_s}$$

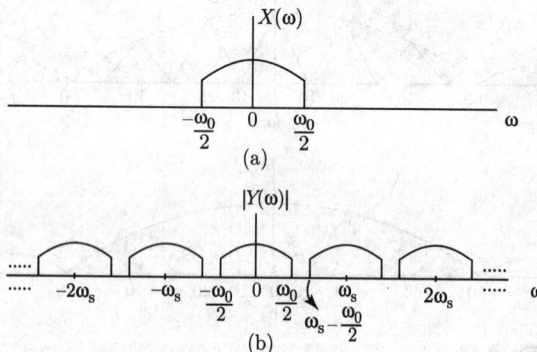

(a)

(b)

Fig. 8.45

Since $\qquad y(t) = x(t)s(t-1)$

Using the multiplication property of the Fourier transform, we have

$$Y(\omega) = \frac{1}{2\pi}\left[X(\omega) * \mathcal{F}[s(t-1)]\right]$$

$$= \frac{1}{2\pi}\left[X(\omega)\right.$$

$$\left. * \frac{2\pi}{T_s} \sum_{n=-\infty}^{\infty} \delta\left(\omega - n\frac{2\pi}{T_s}\right) e^{-jn2\pi/T_s}\right]$$

$$Y(\omega) = \frac{1}{T_s} \sum_{n=-\infty}^{\infty} X\left(\omega - n\frac{2\pi}{T_s}\right) e^{-jn2\pi/T_s}$$

$$Y(\omega) = \frac{1}{T_s}\left[\cdots + X\left(\omega + \frac{2\pi}{T_s}\right) e^{j2\pi/T_s}\right.$$

$$\left. + X(\omega) + X\left(\omega - \frac{2\pi}{T_s}\right) e^{-j2\pi/T_s} + \cdots\right]$$

$|Y(\omega)|$ is depicted in Fig. 8.45(b). In order to recover $x(t)$ from $y(t)$, we need to be able to isolate one replica of $X(\omega)$ from $Y(\omega)$. From Fig. 8.45(b), it is clear that it is possible if we pass $Y(\omega)$ through a low-pass filter of the frequency response

$$H(\omega) = \begin{cases} T_s, & |\omega| < \omega_c \\ 0, & \text{otherwise} \end{cases}$$

where $\omega_0/2 < \omega_c < (2\pi/T_s) - (\omega_0/2)$ is the cutoff frequency of the low-pass filter.

Example 8.27 Let $x_1(t)$ and $x_2(t)$ be two low-pass signals band limited to ω_1 and ω_2, respectively, i.e.,

$$X_1(\omega) = 0, \quad |\omega| \geq \omega_1$$

$$X_2(\omega) = 0, \quad |\omega| \geq \omega_2$$

and let (a) $z(t) = x_1(t) + x_2(t)$, (b) $z(t) = x_1(t)x_2(t)$, and (c) $z(t) = x_1(t) * x_2(t)$. The signal $z(t)$ is sampled using an ideal sampler at intervals of T_s seconds. Determine the maximum sampling interval T_s such that $z(t)$ is recoverable from its sampled version $z_s(t)$ through the use of an ideal low-pass filter.

Solution
(a) Given that $z(t) = x_1(t) + x_2(t)$. The maximum frequency present in the signal $z(t)$ is ω_1 (if $\omega_1 > \omega_2$) or ω_2 (if $\omega_2 > \omega_1$), i.e.,

$$\omega_{\max} = \begin{cases} \omega_1, & \text{if } \omega_1 > \omega_2 \\ \omega_2, & \text{if } \omega_2 > \omega_1 \end{cases}$$

The Nyquist rate (minimum sampling frequency) is $\omega_s = 2\omega_{\max}$. Therefore,

$$T_s = \frac{2\pi}{\omega_s} = \frac{2\pi}{2\omega_{\max}} = \frac{\pi}{\omega_{\max}}$$

$$T_\mathrm{s} = \begin{cases} \dfrac{\pi}{\omega_1}, & \text{if } \omega_1 > \omega_2 \\[2mm] \dfrac{\pi}{\omega_2}, & \text{if } \omega_2 > \omega_1 \end{cases}$$

(b) If

$$x_1(t) \longleftrightarrow X_1(\omega)$$
$$x_2(t) \longleftrightarrow X_2(\omega)$$

then

$$z(t) = x_1(t)x_2(t) \longleftrightarrow Z(\omega) = \frac{1}{2\pi}[X_1(\omega) * X_2(\omega)]$$

Clearly, $Z(\omega) = 0$ for $|\omega| > (\omega_1 + \omega_2)$ and the maximum frequency present in the signal $z(t) = x_1(t)x_2(t)$ is $\omega_\mathrm{max} = (\omega_1 + \omega_2)$. (If we multiply two signals with the frequencies ω_1 and ω_2, the frequency of the resultant signal is $\omega_1 + \omega_2$.) Therefore, the Nyquist rate for $z(t)$ is given by

$$\omega_\mathrm{s} = 2\omega_\mathrm{max} = 2(\omega_1 + \omega_2)$$

and hence,

$$T_\mathrm{s} = \frac{2\pi}{\omega_\mathrm{s}} = \frac{2\pi}{2(\omega_1 + \omega_2)} = \frac{\pi}{(\omega_1 + \omega_2)}$$

(c) If

$$x_1(t) \longleftrightarrow X_1(\omega)$$
$$x_2(t) \longleftrightarrow X_2(\omega)$$

then

$$z(t) = x_1(t) * x_2(t) \longleftrightarrow Z(\omega) = X_1(\omega)X_2(\omega)$$

or

$$Z(\omega) = \begin{cases} 0, & |\omega| \geq \omega_1 \quad \text{if } \omega_1 < \omega_2 \\ 0, & |\omega| \geq \omega_2 \quad \text{if } \omega_2 < \omega_1 \end{cases}$$

Therefore, the maximum frequency present in the signal $z(t) = x_1(t) * x_2(t)$ is

$$\omega_\mathrm{max} = \begin{cases} \omega_1, & \text{if } \omega_1 < \omega_2 \\ \omega_2, & \text{if } \omega_2 < \omega_1 \end{cases}$$

Thus, the Nyquist rate (minimum sampling frequency) is $\omega_\mathrm{s} = 2\omega_\mathrm{max}$, and hence,

$$T_\mathrm{s} = \frac{2\pi}{\omega_\mathrm{s}} = \frac{2\pi}{\omega_{2\omega_\mathrm{max}}} = \frac{\pi}{\omega_\mathrm{max}}$$

Therefore,

$$T_\mathrm{s} = \begin{cases} \dfrac{\pi}{\omega_1}, & \text{if } \omega_1 < \omega_2 \\[2mm] \dfrac{\pi}{\omega_2}, & \text{if } \omega_2 < \omega_1 \end{cases}$$

Example 8.28 Consider the signal

$$x(t) = \left(\frac{\sin(50\pi t)}{\pi t} \right)^2$$

which we wish to sample with a sampling frequency of $\omega_\mathrm{s} = 150\pi$ to obtain a signal $x_\mathrm{s}(t)$ with Fourier transform $X_\mathrm{s}(\omega)$. Determine the maximum value of ω_0 for which it is guaranteed that

$$X_\mathrm{s}(\omega) = 75\, X(\omega) \quad \text{for} \quad |\omega| \leq \omega_0$$

where $X(\omega)$ is the Fourier transform of $x(t)$.

Solution
We know that [Eq. (5.22)]

$$A \triangle \left(\frac{t}{T} \right) \longleftrightarrow AT \operatorname{sinc}^2 \left(\frac{\omega T}{2\pi} \right)$$

$$\frac{1}{T} \triangle \left(\frac{t}{T} \right) \longleftrightarrow \left(\frac{\sin(\omega T/2)}{\omega T/2} \right)^2$$

Substituting $T/2 = 50\pi$, we obtain

$$\frac{1}{100\pi} \triangle \left(\frac{t}{100\pi} \right) \longleftrightarrow \left(\frac{\sin(50\pi\omega)}{50\pi\omega} \right)^2$$

$$\frac{50}{2\pi} \triangle \left(\frac{t}{100\pi} \right) \longleftrightarrow \left(\frac{\sin(50\pi\omega)}{\pi\omega} \right)^2$$

$$y(t) \longleftrightarrow Y(\omega)$$

$$y(t) = \frac{50}{2\pi} \triangle \left(\frac{t}{100\pi} \right)$$

and

$$Y(\omega) = \left(\frac{\sin(50\pi\omega)}{\pi\omega} \right)^2$$

Therefore,

$$y(-\omega) = \frac{50}{2\pi} \triangle \left(-\frac{\omega}{100\pi} \right) = \frac{50}{2\pi} \triangle \left(\frac{\omega}{100\pi} \right)$$

and

$$Y(t) = \left(\frac{\sin(50\pi t)}{\pi t}\right)^2$$

The duality property of the Fourier transform states that if

$$y(t) \longleftrightarrow Y(\omega)$$

then

$$Y(t) \longleftrightarrow 2\pi y(-\omega)$$

$$\left(\frac{\sin(50\pi t)}{\pi t}\right)^2 \longleftrightarrow 2\pi \frac{50}{2\pi} \triangle\left(\frac{\omega}{100\pi}\right)$$

$$x(t) = \left(\frac{\sin(50\pi t)}{\pi t}\right)^2 \longleftrightarrow X(\omega) = 50 \triangle\left(\frac{\omega}{100\pi}\right)$$

Therefore,

$$X(\omega) = 50 \triangle\left(\frac{\omega}{100\pi}\right)$$

The Fourier transform $X(\omega)$ of $x(t)$ is as shown in Fig. 8.46(a). We know from Eq. (8.8) that the spectrum of the instantaneous sampled signal is

$$G(\omega) = \frac{1}{T_s} \sum_{n=-\infty}^{\infty} X(\omega - n\omega_s)$$

where $T_s = 2\pi/\omega_s = 1/75$. $G(\omega)$ is as shown in Fig. 8.46(b). Clearly,

$$G(\omega) = \frac{1}{T_s} X(\omega) = 75X(\omega) \quad \text{for } |\omega| \le 50\pi$$

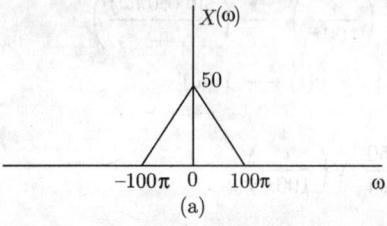

(a)

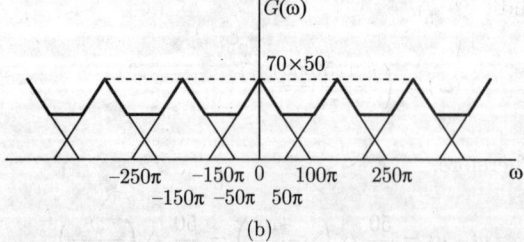

(b)

Fig. 8.46

Example 8.29 Let $x(t)$ be a continuous-time signal whose Fourier transform has the property that $X(\omega) = 0$ for $|\omega| \ge 2000\pi$. A discrete-time signal

$$x(n) = x(n \times 0.5 \times 10^{-3})$$

is obtained. For each of the following constraints on the Fourier transform $X(e^{j\omega})$ of $x(n)$, determine the corresponding constraints on $X(\omega)$:

(a) $X(e^{j\omega})$ is real.

(b) The maximum value of $X(e^{j\omega})$ over all ω is 1.

(c) $X(e^{j\omega}) = 0$ for $3\pi/4 \le |\omega| \le \pi$.

(d) $X(e^{j\omega}) = X(e^{j(\omega-\pi)})$.

Solution

We know from Eq. (8.106) that the relationship between DTFT and CTFT is

$$X(e^{j\omega}) = \frac{1}{T_s} \sum_{n=-\infty}^{\infty} X\left(\frac{1}{T_s}(\omega - 2n\pi)\right)$$

(a) Since $X(e^{j\omega})$ is just formed by shifting and summing replicas of $X(\omega)$, we may argue that if $X(e^{j\omega})$ is real, then $X(\omega)$ must also be real.

(b) $X(e^{j\omega})$ consists of replicas of $X(\omega)$ which are scaled by $1/T_s$. Therefore, if the maximum value of $X(e^{j\omega})$ over all ω is 1, then the maximum value of $X(\omega)$ over all ω is $T_s = 0.5 \times 10^{-3}$.

(c) The region $3\pi/4 \le |\omega| \le \pi$ in the discrete-time domain corresponds to the region $3\pi/4T_s \le |\omega| \le \dfrac{\pi}{T_s}$ in the continuous-time domain. Therefore, if

$$X(e^{j\omega}) = 0 \quad \text{for } \frac{3\pi}{4} \le |\omega| \le \pi$$

then

$$X(\omega) = 0 \quad \text{for } 1500\pi \le |\omega| \le 2000\pi$$

But since we already have $X(\omega) = 0$ for $|\omega| \ge 2000\pi$, we have $X(\omega) = 0$ for $|\omega| \ge 1500\pi$.

(d) In this case, since π in discrete-time frequency domain corresponds to $\pi/T_s = 2000\pi$ in continuous-time frequency domain, therefore $X(\omega) = X(\omega - 2000\pi)$.

SUMMARY

1. Sampling is the process of converting a continuous-time signal $x(t)$ into a discrete-time signal $x(n)$ by measuring the amplitudes of the continuous-time signal $x(t)$ at integer multiples of a sampling interval T_s.

2. Sampling theorem states that a band-limited signal of finite energy, which has no frequency components higher than ω_h rad/s (or f_h Hz), may be completely recovered from a knowledge of its samples if the sampling frequency $\omega_s \geq 2\omega_h$ samples/rad/s (or $f_s \geq 2f_h$ samples/s). According to the sampling theorem, the sampling frequency

$$\omega_s \geq 2\omega_h \quad \text{or} \quad f_s \geq 2f_h$$

3. The minimum sampling frequency $f_s = 2f_h$ is called the Nyquist rate of sampling.

4. The maximum sampling interval $T_s = 1/2f_h$ is called the Nyquist interval.

5. The spectrum of the sampled signal is a periodically repeated version of the spectrum of the signal sampled.

6. If the signal is sampled at a rate more than twice its highest frequency, there will be no aliasing, and the original signal can be recovered without any distortion.

7. The process of reconstruction of original signal $x(t)$ from its sampled version $x_s(t)$ is known as interpolation.

8. The ideal interpolating function is the sinc function, but since it is noncausal, other methods like zero-order hold and first-order hold must be used in practice.

9. There are three types of sampling techniques, namely,

(a) ideal or instantaneous or impulse sampling,

(b) natural or chopper sampling, and

(c) flat-top sampling.

10. The bandpass sampling theorem states that if a *real-valued* bandpass signal $x(t)$ has a spectrum of bandwidth $\omega_B = 2\pi B = 2\pi(f_H - f_L)$ and an upper frequency limit $\omega_H = 2\pi f_H$, then the original signal $x(t)$ can be recovered from the sampled signal $x_s(t)$ by bandpass filtering if $f_s = 2f_H/k$, where $k = \lfloor f_h/B \rfloor$ is the integer part of f_h/B.

11. Let $x(t)$ be a real-valued bandpass signal $x(t)$ with the spectrum $X(f) = 0$ for $|f| < f_L$ and $|f| > f_H$, the bandwidth $B = f_H - f_L$, and the center frequency $f_c = f_H + f_L/2$. Now, if $f_c > B/2$ and f_H is an integer multiple of the bandwidth B, no aliasing will occur if $x(t)$ is sampled at a sampling frequency

$$f_s = 2B$$

i.e., the sampling frequency is equal to twice the bandwidth of the signal.

12. The bandpass sampling theorem for *complex signals* states that if a complex bandpass signal $x(t)$ has a spectrum of bandwidth $\omega_B = 2\pi B = 2\pi(f_H - f_L)$ and an upper-frequency limit $\omega_H = 2\pi f_H$, then the original signal $x(t)$ can be recovered from the sampled signal $x_s(t)$ by bandpass filtering if the sampling frequency

$$f_s \geq f_H - f_L$$

MULTIPLE-CHOICE QUESTIONS

1. What is the Nyquist rate of the following signal? $x(t) = 3\cos(50\pi t) + 10\sin(300\pi t) - \cos(100\pi t)$.

   (a) 50 Hz
   (b) 100 Hz
   (c) 200 Hz
   (d) 300 Hz

2. The signals $x_1(t)$ and $x_2(t)$ are both band limited to $(-\omega_1, +\omega_1)$ and $(-\omega_2, +\omega_2)$, respectively. The Nyquist sampling rate for the signal $x_1(t)x_2(t)$ will be

   (a) $2\omega_1$, if $\omega_1 > \omega_2$
   (b) $2\omega_2$, if $\omega_1 < \omega_2$
   (c) $2(\omega_1 + \omega_2)$

(d) $\dfrac{\omega_1 + \omega_2}{2}$

3. A bandpass signal extends from 1 o 2 KHz. The minimum sampling frequency needed to retain all information in the sampled signal is

   (a) 1 kHz
   (b) 2 kHz
   (c) 3 kHz
   (d) 4 kHz

4. Zero-order hold used in practical reconstruction of continuous-time signals is mathematically represented as a weighted-sum of rectangular pulses shifted by

   (a) any multiples of the sampling interval.
   (b) integer multiples of the sampling interval.
   (c) one sampling interval
   (d) 1-s intervals.

5. A continuous-time signal $x(t)$ is sampled using an impulse train. In terms of $X(\omega)$, the Fourier transform of $x(t)$, the spectrum of the sampled signal can be expressed as

   (a) $\sum_{n=-\infty}^{\infty}[X(\omega + n\omega_s)\delta(\omega)]$

(b) $\sum_{n=-\infty}^{\infty}[X(n\omega) * \delta(\omega + n\omega_s)]$
(c) $\sum_{n=-\infty}^{\infty}[X(\omega) * \delta(\omega + n\omega_s)]$
(d) $\sum_{n=-\infty}^{\infty}[X(\omega)\delta(\omega + n\omega_s)]$

6. Aperture effect occurs in communication due to

   (a) sampling at less than Nyquist rate.
   (b) flat-top sampling.
   (c) finite bandwidth of transmission channel.
   (d) short duration of samples.

7. A baseband signal has a spectral range that extends from 20 Hz to 82 KHz. Find the acceptable range of the sampling frequency.

   (a) (a) $> 40Hz$, $< 82KHz$
   (b) $40Hz$ to 82KHz
   (c) $\geq$ 164 KHz
   (d) $\leq$ 164 KHz

8. Flat-top sampling leads to

   (a) aperture effect.
   (b) aliasing.
   (c) loss of signal.
   (d) loss of higher frequency components.

PROBLEMS

8.1 Find the Nyquist rate and Nyquist interval for each of the following signals:

1. $x_1(t) = 3\cos(50\pi t) + 5\sin(300\pi t)$
2. $x_2(t) = 1 + 2\cos(20\pi t) + 3\sin(30\pi t)$
3. $x_3(t) = \text{sinc}^2(50t)$
4. $x_4(t) = \text{sinc}(200t)\sin(150\pi t)$

8.2 Consider sampling the signal $x(t) = (\sin(2\pi t))/\pi t$. Sketch the FT of the sampled signal for the following sampling intervals:

1. $T_s = 1/8$
2. $T_s = 1/3$
3. $T_s = 1/2$
4. $T_s = 2/3$

8.3 The signal $z(t)$ is generated by convolving two band-limited signals $x_1(t)$ and $x_2(t)$, i.e.,

$$z(t) = x_1(t) * x_2(t)$$

where

$$X_1(\omega) = 0 \quad \text{for } |\omega| > 1000\pi$$
$$X_2(\omega) = 0 \quad \text{for } |\omega| > 2000\pi$$

Impulse-train sampling is performed on $z(t)$ to obtain

$$z_s(t) = \sum_{n=-\infty}^{\infty} z(t)\delta(t - nT_s)$$

Specify the range of values for the sampling period T_s which ensures that $z(t)$ is recoverable from $z_s(t)$.

8.4 Consider a real, odd, and periodic signal $x(t)$ whose Fourier series representation may be expressed as

$$x(t) = \sum_{k=0}^{5} \left(\frac{1}{2}\right)^k \sin(k\pi t)$$

Let $x_s(t)$ represent the instantaneous sampled signal with a sampling period of $T_s = 0.2$.

(a) Does aliasing occur when the instantaneous sampling is performed on $x(t)$?

(b) If $x_s(t)$ is passed through an ideal low-pass filter with cutoff frequency $\omega_c = \pi/T_s$ and passband gain T_s, determine the Fourier series representation of the output signal $y(t)$.

8.5 Let $x(n) = \{4, 0, \underset{\uparrow}{2}, -1, -3\}$. Find and sketch the following signals.

1. The decimated signal $x_d(n) = x(2n)$.
2. The zero-interpolated signal $y(n) = x(n/2)$.
3. The step-interpolated signal $y(n) = x(n/2)$.

8.6

(a) Describe the sequence of operations required to generate $x(n - (2/3))$ from $x(n)$.

(b) Let $x(n) = \{\underset{\uparrow}{1}, 4, 7, 10, 13\}$. Sketch $x(n)$ and $x(n - (2/3))$. Use step interpolation where required.

(c) Generalize the results of part (a) to generate $x\left(n - \frac{M}{L}\right)$ from $x(n)$. Any restriction on M and L.

8.7 Each of the following sinusoids is sampled at $f_s = 100$ Hz. Determine if aliasing has occurred and set up an expression for each sampled signal using a digital frequency in the principal range ($|f| < 1/2$).

1. $x_1(t) = \cos(320\pi t)$
2. $x_2(t) = \cos(140\pi t)$
3. $x_3(t) = \sin(60\pi t)$

8.8 Suppose that a discrete-time sequence $x(n)$ is band limited so that

$$X(e^{j\omega}) = 0 \quad 0.3\pi < |\omega| < \pi$$

This sequence is then sampled to form the sequence

$$y(n) = x(nN)$$

where N is an integer. Find the largest value for N for which $x(n)$ may be uniquely recovered from $y(n)$.

8.9 Find two different continuous-time signals that will produce the sequence

$$x(n) = \cos(0.15\pi n)$$

when sampled with a sampling frequency of 8 KHz.

ANSWERS TO MULTIPLE-CHOICE QUESTIONS

    1. **(d)**   2. **(c)**   3. **(b)**   4. **(c)**   5. **(c)**   6. **(b)**   7. **(c)**   8. **(a)**

Laplace Transform

9.1 INTRODUCTION

In Chapters 3 to 6, we saw how frequency-domain methods are extremely useful in the study of signals and LTI systems. In these chapters, we demonstrated that Fourier analysis reduces the convolution operation required to compute the output of LTI systems to just the product of the Fourier transform of the input signal and the frequency response of the system. One of the problems we can run into is that many of the input signals we would like to use do not have Fourier transforms. Examples are $e^{at}u(t)$, $a > 0$; e^{-at}, $-\infty < t < \infty$; $tu(t)$; and other signals that are not absolutely integrable. The Fourier transform does not exist for signals that are not absolutely integrable, so the Fourier transform based methods cannot be employed in this class of problems. The difficulty could be resolved by generalizing the Fourier transform so that the signal $x(t)$ is expressed as a sum of complex exponentials e^{st}, where $s = \sigma + j\omega$ and thus is not restricted to the imaginary axis only. This is equivalent to multiplying the signal by an exponential convergent factor. For example, $e^{-\sigma t} e^{at}u(t)$ satisfies Dirichlet's conditions for $\sigma > \alpha$. This observation leads to generalization of the continuous-time Fourier transform, known as Laplace transform, named after the French mathematician Pierre Simon de Laplace.

The Laplace transform possesses a distinct set of properties for analysing signals and LTI systems. Many of these properties parallel those of the Fourier transform. The Laplace transform can be applied to the analysis of many unstable systems and consequently plays an important role in the investigation of the stability or instability of systems.

The Laplace transform comes in two varieties: *Bilateral*, or two-sided Laplace transform, and *unilateral*, or one-sided Laplace transform. The bilateral Laplace transform offers insight into the nature of system characteristics such as stability, causality, and frequency response. The unilateral Laplace transform is a convenient tool for solving differential equations with initial conditions.

9.2   THE BILATERAL (TWO-SIDED) LAPLACE TRANSFORM

Consider applying a complex exponential input $x(t) = e^{st}$ to an LTI system with impulse response $h(t)$. The system output is given by

$$y(t) = h(t) * x(t) = \int_{-\infty}^{\infty} h(\tau) x(t-\tau)\, d\tau$$

$$= \int_{-\infty}^{\infty} h(\tau)\, e^{s(t-\tau)}\, d\tau$$

$$= e^{st} \int_{-\infty}^{\infty} h(\tau)\, e^{-s\tau}\, d\tau$$

$$y(t) = H(s)\, e^{st}$$

where

$$H(s) = \int_{-\infty}^{\infty} h(\tau)\, e^{-s\tau}\, d\tau$$

or equivalently,

$$H(s) = \int_{-\infty}^{\infty} h(t)\, e^{-st}\, dt \qquad (9.1)$$

$H(s)$ is known as the transfer function of the LTI system. We know that a signal for which the system output is a constant times the input is referred to as an *eigenfunction* of the system, and the amplitude factor is referred to as the system's *eigenvalue*. Hence, we identify e^{st} as an eigenfunction of the LTI system and $H(s)$ as the corresponding eigenvalue. For general values of the complex variable s, $H(s)$ is referred to as the *bilateral Laplace transform* or simply the *Laplace transform* of the impulse response $h(t)$. The bilateral Laplace transform in Eq. (9.1) involves an integration from $-\infty$ to $+\infty$, while the unilateral Laplace transform has a form similar to that in Eq. (9.1), but with limits of integration from 0^- to $+\infty$.

9.2.1   Inverse Laplace Transform

Substituting $s = \sigma + j\omega$ into Eq. (9.1) and using t as the variable of integration, we obtain

$$H(\sigma + j\omega) = \int_{-\infty}^{\infty} h(t)\, e^{-(\sigma+j\omega)t}\, dt = \int_{-\infty}^{\infty} [h(t) e^{-\sigma t}]\, e^{-j\omega t}\, dt$$

The above equation indicates that $H(\sigma + j\omega)$ is the Fourier transform of $[h(t)e^{-\sigma t}]$. Hence, the inverse Fourier transform of $H(\sigma + j\omega)$ must be $[h(t)\,e^{-\sigma t}]$, i.e.,

$$h(t)\,e^{-\sigma t} = \frac{1}{2\pi} \int\limits_{-\infty}^{\infty} H(\sigma + j\omega)\,e^{j\omega t}\,d\omega$$

$$h(t) = e^{\sigma t} \frac{1}{2\pi} \int\limits_{-\infty}^{\infty} H(\sigma + j\omega)\,e^{j\omega t}\,d\omega$$

$$h(t) = \frac{1}{2\pi} \int\limits_{-\infty}^{\infty} H(\sigma + j\omega)\,e^{(\sigma + j\omega)t}\,d\omega$$

A change of variables is performed by letting $s = \sigma + j\omega$, which also yields $d\omega = ds/j$, $s \rightarrow (\sigma - j\infty)$ as $\omega \rightarrow -\infty$, and $s \rightarrow (\sigma + j\infty)$ as $\omega \rightarrow \infty$. Therefore,

$$h(t) = \frac{1}{2\pi j} \int\limits_{\sigma - j\infty}^{\sigma + j\infty} H(s)\,e^{st}\,ds \tag{9.2}$$

Equation (9.2) expresses $h(t)$ as a function of $H(s)$. We say that $h(t)$ is the *inverse Laplace transform* of $H(s)$.

We have obtained the Laplace transform of the impulse response of a system. This relationship holds for an arbitrary signal. The Laplace transform of a general signal $x(t)$ is defined as

$$\mathcal{L}[x(t)] = X(s) = \int\limits_{-\infty}^{\infty} x(t)\,e^{-st}\,dt \tag{9.3}$$

and the inverse Laplace transform of $X(s)$ is

$$x(t) = \frac{1}{2\pi j} \int\limits_{\sigma - j\infty}^{\sigma + j\infty} X(s)\,e^{st}\,ds \tag{9.4}$$

This equation states that $x(t)$ can be represented as a weighted integral of complex exponentials e^{st}. In practice, we usually do not evaluate this integral directly, since it requires techniques of contour integration. Instead, we determine inverse Laplace transform by exploiting the one-to-one relationship between $x(t)$ and $X(s)$.

We denote the transform relationship between $x(t)$ and $X(s)$ as

$$x(t) \longleftrightarrow X(s) \tag{9.5}$$

Finding the inverse Laplace transform by using Eq. (9.4) requires integration in the complex plane, a subject beyond the scope of this book. However, for the case of rational transforms, the inverse Laplace transform can be determined without directly evaluating Eq. (9.4) by using the technique of partial fraction expansion.

9.3 RELATIONSHIP BETWEEN LAPLACE TRANSFORM AND FOURIER TRANSFORM

Consider a continuous-time signal $x(t)$. Its Laplace transform is defined as

$$\mathcal{L}[x(t)] = X(s) = \int\limits_{-\infty}^{\infty} x(t)\, e^{-st}\, dt$$

Substituting $s = \sigma + j\omega$ into the above equation, we obtain

$$\mathcal{L}[x(t)] = \int\limits_{-\infty}^{\infty} x(t)\, e^{-(\sigma+j\omega)t}\, dt = \int\limits_{-\infty}^{\infty} [x(t)\, e^{-\sigma t}]\, e^{-j\omega t}\, dt = \mathcal{F}[x(t)\, e^{-\sigma t}]$$

$$\mathcal{L}[x(t)] = \mathcal{F}[x(t)\, e^{-\sigma t}] \tag{9.6}$$

Thus, the Laplace transform of $x(t)$ is the Fourier transform of $x(t)\, e^{-\sigma t}$. Now, if $\sigma = 0$, we obtain

$$\mathcal{L}[x(t)] = \mathcal{F}[x(t)], \quad \text{for } \sigma = 0 \tag{9.7}$$

9.4 REGION OF CONVERGENCE (ROC) FOR LAPLACE TRANSFORMS

Equation (9.6) indicates that the Laplace transform of $x(t)$ is the Fourier transform of $x(t)\, e^{-\sigma t}$. Hence, the Laplace transform is guaranteed to converge if $x(t)\, e^{-\sigma t}$ is absolutely integrable. That is, we must have

$$\int\limits_{-\infty}^{\infty} |x(t)\, e^{-\sigma t}|\, dt < \infty \tag{9.8}$$

This guarantees that $X(s)$ will be finite, since

$$\mathcal{L}[x(t)] = X(s) = \int\limits_{-\infty}^{\infty} x(t)\, e^{-st} dt$$

Substituting $s = \sigma + j\omega$ into the above equation, we obtain

$$X(s) = \int_{-\infty}^{\infty} x(t) e^{-(\sigma+j\omega)t} \, dt$$

$$X(s) = \int_{-\infty}^{\infty} [x(t) e^{-\sigma t}] e^{-j\omega t} \, dt$$

$$|X(s)| = \left| \int_{-\infty}^{\infty} [x(t) e^{-\sigma t}] e^{-j\omega t} \, dt \right|$$

$$|X(s)| \le \int_{-\infty}^{\infty} \left| x(t) e^{-\sigma t} e^{-j\omega t} \right| dt$$

$$|X(s)| \le \int_{-\infty}^{\infty} \left| x(t) e^{-\sigma t} \right| dt$$

So if

$$\int_{-\infty}^{\infty} \left| x(t) e^{-\sigma t} \right| dt < \infty$$

then

$$|X(s)| < \infty$$

The range of $\Re\{s\} = \sigma$ for which the Laplace transform converges is termed the *region of convergence* (ROC). That is, the ROC consists of those values of σ for which the Fourier transform of $x(t) e^{-\sigma t}$ converges. Note that the Laplace transform exists for some signals that do not have a Fourier transform. By limiting ourselves to a certain range of σ, we may ensure that $x(t) e^{-\sigma t}$ is absolutely integrable, even though $x(t)$ is not absolutely integrable by itself.

The ROC can also provide us with information about whether $x(t)$ is Fourier transformable or not. Since the Fourier transform is obtained from the bilateral Laplace transform by setting $\sigma = 0$, the ROC in this case is a single line (the $j\omega$-axis). Therefore, if the ROC for $X(s)$ includes the $j\omega$ axis, $x(t)$ is Fourier transformable, and $X(\omega)$ can be obtained by replacing s in $X(s)$ by $j\omega$.

9.5 *s*-PLANE

It is convenient to represent the complex frequency s graphically in terms of a complex plane called the s-plane, as depicted in Fig. 9.1. The horizontal axis represents the real part of s (i.e., σ), and the vertical axis represents the imaginary past of s (i.e., $j\omega$).

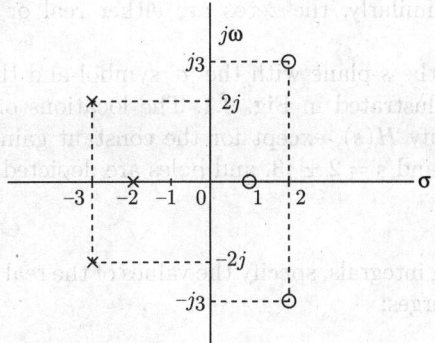

Fig. 9.1 *s*-plane

The $j\omega$-axis divides the s-plane into two equal halves. The region of the s-plane to the left of the $j\omega$-axis is termed the *left half of the s-plane*, while the region of the s-plane to the right of the $j\omega$-axis is termed the *right half of the s-plane*. The real part of s (i.e., σ) is negative in the left half of the s-plane and positive in the right half of the s-plane.

9.5.1 Poles and Zeros

The most commonly encountered form of the Laplace transform in engineering is a ratio of two polynomials in s, i.e.,

$$H(s) = \frac{b_m s^m + b_{m-1} s^{m-1} + \cdots + b_1 s + b_0}{a_n s^n + a_{n-1} s^{n-1} + \cdots + a_1 s + a_0}$$

It is often convenient to factor the polynomials in the numerator and denominator, and to write the transfer function in terms of those factors:

$$H(s) = \frac{N(s)}{D(s)} = K \frac{(s - z_1)(s - z_2) \cdots (s - z_{m-1})(s - z_m)}{(s - p_1)(s - p_2) \cdots (s - p_{n-1})(s - p_n)} \qquad (9.9)$$

where the numerator and denominator polynomials, $N(s)$ and $D(s)$, have real coefficients and $K = b_m/a_n$. As written in the above equation, the z_i's are the roots of the equation $N(s) = 0$, and are defined as zeros, and the p_i's are the roots of the equation $D(s) = 0$, and are defined as poles. In Eq. (9.9) the factors in the numerator and denominator are written so that when $s = z_i$ the numerator $N(s) = 0$ and the transfer function vanishes, i.e.,

$$\lim_{s \to z_i} H(s) = 0$$

and similarly when $s = p_i$ the denominator polynomial $D(s) = 0$ and the value of the transfer function becomes unbounded, that is,

$$\lim_{s \to p_i} H(s) = \infty$$

All the coefficients of polynomials $N(s)$ and $D(s)$ are real, and therefore the poles and zeros must either be purely real or appear in complex conjugate pairs. The existence of a single complex pole without a corresponding conjugate pole would generate complex coefficients in the polynomial $D(s)$. Similarly, the zeros are either real or appear in complex conjugate pairs.

We denote the locations of zeros in the s-plane with the 'o' symbol and the location of the poles with the '×' symbol, as illustrated in Fig. 9.1. The locations of the poles and zeros in the s-plane uniquely specify $H(s)$, except for the constant gain factor K. In Fig. 9.1, zeros are depicted at $s = 1$ and $s = 2 \pm j3$, and poles are depicted at $s = -2$ and $s = -3 \pm 2j$.

Example 9.1 For each of the following integrals, specify the values of the real parameter σ which ensures that the integral converges:

1. $\displaystyle\int_0^\infty e^{-5t}\, e^{-(\sigma+j\omega)t}\, dt$

2. $\displaystyle\int_{-\infty}^0 e^{-5t}\, e^{-(\sigma+j\omega)t}\, dt$

3. $\displaystyle\int_{-5}^5 e^{-5t}\, e^{-(\sigma+j\omega)t}\, dt$

4. $\displaystyle\int_{-\infty}^\infty e^{-5t}\, e^{-(\sigma+j\omega)t}\, dt$

5. $\displaystyle\int_{-\infty}^\infty e^{-5|t|}\, e^{-(\sigma+j\omega)t}\, dt$

Solution

1. Consider the given integral

$$\int_0^\infty e^{-5t}\, e^{-(\sigma+j\omega)t}\, dt = \int_0^\infty e^{-(\sigma+5)t}\, e^{-j\omega t}\, dt$$

$$= -\frac{1}{(5+\sigma+j\omega)}\, e^{-(\sigma+5)t}\, e^{-j\omega t}\Big|_0^\infty$$

Note that $|e^{-j\omega t}| = 1$ regardless of the value of ωt. Therefore, as $t \to \infty$, $e^{-(\sigma+5)t}\, e^{-j\omega t} \to 0$ only if $(\sigma+5) > 0$ or, $\sigma > -5$, and $e^{-(\sigma+5)t}\, e^{-j\omega t} \to \infty$ only if $(\sigma+5) < 0$ or, $\sigma < -5$. Thus,

$$\lim_{t\to\infty} e^{-(\sigma+5)t}\, e^{-j\omega t} = \begin{cases} 0, & (\sigma+5) > 0 \longrightarrow \sigma > -5 \\ \infty, & (\sigma+5) < 0 \longrightarrow \sigma < -5 \end{cases}$$

Therefore,

$$\int_0^\infty e^{-(\sigma+5)t}\, e^{-j\omega t}\, dt = -\frac{1}{(5+\sigma+j\omega)}[0-1]$$

$$= \frac{1}{(5+\sigma+j\omega)} \quad \text{for } \sigma > -5$$

2. Consider the given integral

$$\int_{-\infty}^{0} e^{-5t} e^{-(\sigma+j\omega)t} \, dt = \int_{-\infty}^{0} e^{-(\sigma+5)t} e^{-j\omega t} dt$$

$$= -\frac{1}{(5+\sigma+j\omega)} e^{-(\sigma+5)t} e^{-j\omega t} \Big|_{-\infty}^{0}$$

Note that $|e^{-j\omega t}| = 1$ regardless of the value of ωt. Therefore, as $t \to -\infty$, $e^{-(\sigma+5)t} e^{-j\omega t} \to 0$ only if $(\sigma+5) < 0$ or, $\sigma < -5$, and $e^{-(\sigma+5)t} e^{-j\omega t} \to \infty$ only if $(\sigma+5) > 0$ or, $\sigma > -5$. Thus,

$$\lim_{t \to -\infty} e^{-(\sigma+5)t} e^{-j\omega t} = \begin{cases} 0, & (\sigma+5) < 0 \longrightarrow \sigma < -5 \\ \infty, & (\sigma+5) > 0 \longrightarrow \sigma > -5 \end{cases}$$

Therefore,

$$\int_{-\infty}^{0} e^{-(\sigma+5)t} e^{-j\omega t} \, dt = -\frac{1}{(5+\sigma+j\omega)}[1-0]$$

$$= -\frac{1}{(5+\sigma+j\omega)} \quad \text{for } \sigma < -5$$

3. The given integral may be written as

$$\int_{-5}^{5} e^{-(5+\sigma)t} e^{-j\omega t} \, dt$$

Since this integral has a finite range of limit, it has a finite value for all finite values of σ.

4. The given integral may be written as

$$\int_{-\infty}^{\infty} e^{-(5+\sigma)t} e^{-j\omega t} \, dt = \int_{-\infty}^{0} e^{-(5+\sigma)t} e^{-j\omega t} \, dt + \int_{0}^{\infty} e^{-(5+\sigma)t} e^{-j\omega t} \, dt$$

The first integral diverges for $\sigma < -5$. The second integral diverges for $\sigma > -5$. If $\sigma = -5$, then both the integrals still diverge. Therefore, the integral does not converge for any value of σ.

5. The given integral may be written as

$$\int_{-\infty}^{0} e^{-(-5+\sigma)t} e^{-j\omega t} \, dt + \int_{0}^{\infty} e^{-(5+\sigma)t} e^{-j\omega t} \, dt$$

The first integral converges for $\sigma < 5$. The second integral converges for $\sigma > -5$. Therefore, the integral converges for $|\sigma| < 5$.

Example 9.2 Determine the Laplace transform of the causal signal [shown in Fig. 9.2(a)]

$$x(t) = e^{-at}u(t)$$

and depict the ROC and the locations of poles and zeros in the s-plane.

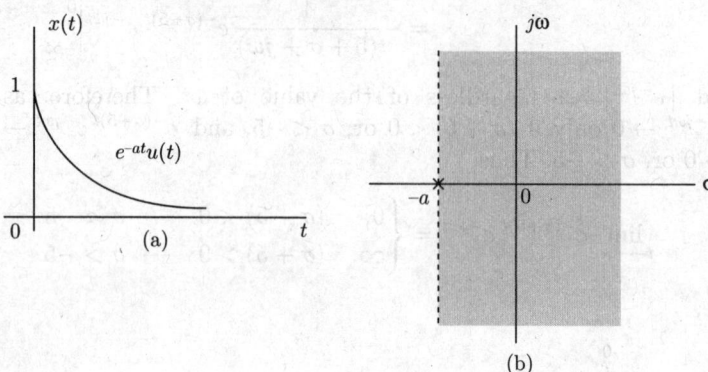

Fig. 9.2 (a) Signal $x(t) = e^{-at}u(t)$ and (b) its ROC

Solution
By definition

$$X(s) = \int_{-\infty}^{\infty} x(t)\, e^{-st}\, dt = \int_{-\infty}^{\infty} e^{-at}u(t)\, e^{-st}\, dt$$

Since, $u(t) = \begin{cases} 1, & t > 0 \\ 0, & t < 0 \end{cases}$, we obtain

$$X(s) = \int_{0}^{\infty} e^{-at}\, e^{-st}\, dt = \int_{0}^{\infty} e^{-(s+a)t}\, dt$$

$$X(s) = -\frac{1}{s+a}\, e^{-(s+a)t}\Big|_{0}^{\infty}$$

Substitute $s = \sigma + j\omega$ into the above equation to obtain

$$X(s) = -\frac{1}{\sigma + j\omega + a}\, e^{-(\sigma+a)t}\, e^{-j\omega t}\Big|_{0}^{\infty}$$

Now, if $(\sigma + a) > 0$ or $\sigma > -a$, then $e^{-(\sigma+a)t} \to 0$ as $t \to \infty$, and

$$X(s) = -\frac{1}{\sigma + j\omega + a}[0 - 1] \quad \sigma > -a$$

$$X(s) = \frac{1}{\sigma + j\omega + a} \quad \sigma > -a$$

$$X(s) = \frac{1}{s+a} \quad \Re\{s\} > -a$$

Therefore,

$$e^{-at}u(t) \longleftrightarrow \frac{1}{s+a} \quad \Re\{s\} > -a \tag{9.10}$$

The ROC for $X(s)$ is $\Re\{s\} > -a$, as shown in the shaded area in Fig. 9.2(b).

Example 9.3 Determine the Laplace transform of an anticausal signal [shown in Fig. 9.3(a)]

$$x(t) = -e^{-at}u(-t)$$

and depict the ROC and the locations of poles and zeros in the s-plane.

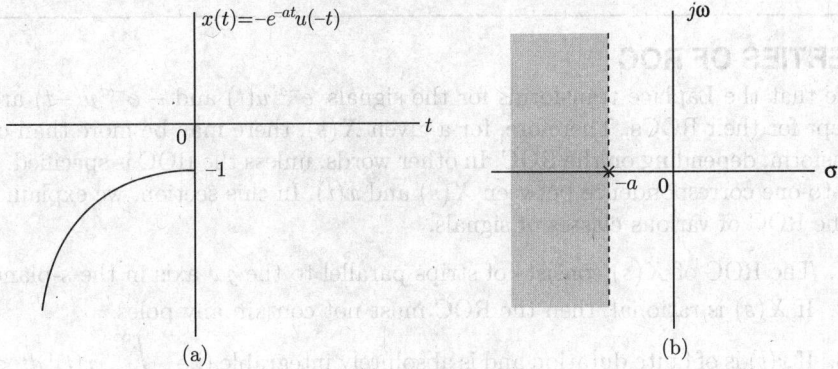

(a)   (b)

Fig. 9.3 (a) Signal $x(t) = -e^{-at}u(-t)$ and (b) its ROC

Solution
By definition

$$X(s) = \int_{-\infty}^{\infty} x(t)\, e^{-st}\, dt = \int_{-\infty}^{\infty} -e^{-at}u(-t)\, e^{-st}\, dt$$

Since, $u(-t) = \begin{cases} 0, & t > 0 \\ 1, & t < 0 \end{cases}$, we obtain

$$X(s) = -\int_{-\infty}^{0} e^{-at}\, e^{-st}\, dt = -\int_{-\infty}^{0} e^{-(s+a)t}\, dt$$

$$X(s) = \frac{1}{s+a} e^{-(s+a)t} \Big|_{-\infty}^{0}$$

Substitute $s = \sigma + j\omega$ into the above equation to obtain

$$X(s) = \frac{1}{\sigma + j\omega + a} e^{-(\sigma+a)t}\, e^{-j\omega t} \Big|_{-\infty}^{0}$$

Now, if $(\sigma + a) < 0$ or $\sigma < -a$, then $e^{-(\sigma+a)t} \to 0$ as $t \to -\infty$, and

$$X(s) = \frac{1}{\sigma + j\omega + a}[1 - 0] \qquad \sigma < -a$$

$$X(s) = \frac{1}{\sigma + j\omega + a} \qquad \sigma < -a$$

$$X(s) = \frac{1}{s + a} \qquad \Re\{s\} < -a$$

Therefore,

$$-e^{-at}u(-t) \longleftrightarrow \frac{1}{s+a} \quad \Re\{s\} < -a \tag{9.11}$$

The ROC for $X(s)$ is $\Re\{s\} < -a$, as shown in the shaded area in Fig. 9.3(b).

9.6   PROPERTIES OF ROC

Note that the Laplace transforms for the signals $e^{-at}u(t)$ and $-e^{-at}u(-t)$ are identical except for their ROCs. Therefore, for a given $X(s)$, there may be more than one inverse transform, depending on the ROC. In other words, unless the ROC is specified, there is no one-to-one correspondence between $X(s)$ and $x(t)$. In this section, we explain properties of the ROC of various classes of signals.

1. The ROC of $X(s)$ consists of strips parallel to the $j\omega$-axis in the s-plane.
2. If $X(s)$ is rational, then the ROC must not contain any poles.
3. If $x(t)$ is of finite duration and is absolutely integrable (i.e., $\int_{-\infty}^{\infty} |x(t)|\, dt < \infty$), then the ROC is the entire s-plane.
4. If $x(t)$ is right-sided and of infinite duration (i.e., $x(t) = 0$ for all $t < T_1$ for some finite T_1), then the ROC is the region in the s-plane to the right of the rightmost pole.
5. If $x(t)$ is left-sided and of infinite duration (i.e., $x(t) = 0$ for all $t > T_2$ for some finite T_2), then the ROC is the region in the s-plane to the left of the leftmost pole.
6. If $x(t)$ is two-sided and of infinite duration (i.e., the signal is of infinite extent for both $t < 0$ and $t > 0$), then the ROC will consist of a strip in the s-plane.

Example 9.4 Determine the Laplace transform of

$$x(t) = e^{-2t}u(t) - e^{-3t}u(t)$$

and depict the ROC and the locations of poles and zeros in the s-plane.

Solution
Consider the given signal

$$x(t) = e^{-2t}u(t) - e^{-3t}u(t)$$

$$\mathcal{L}[x(t)] = \mathcal{L}[e^{-2t}u(t) - e^{-3t}u(t)]$$

$$X(s) = \mathcal{L}[e^{-2t}u(t)] + \mathcal{L}[e^{-3t}u(t)]$$

From Eq. (9.10), we have

$$e^{-at}u(t) \longleftrightarrow \frac{1}{s+a} \qquad \Re\{s\} > -a$$

Therefore,

$$e^{-2t}u(t) \longleftrightarrow \frac{1}{s+2} \qquad \Re\{s\} > -2$$

$$e^{-3t}u(t) \longleftrightarrow \frac{1}{s+3} \qquad \Re\{s\} > -3$$

The set of values of $\Re\{s\}$ for which the Laplace transforms of both terms converge is $\Re\{s\} > -2$, and thus we obtain

$$X(s) = \frac{1}{s+2} - \frac{1}{s+3} \qquad \Re\{s\} > -2$$

$$X(s) = \frac{1}{s^2 + 5s + 6} \qquad \Re\{s\} > -2$$

The ROC for $X(s)$ is $\Re\{s\} > -2$, as shown in the shaded area in Fig. 9.4. For a right-sided infinite duration signal, the ROC is the region in the s-plane to the right of the rightmost pole.

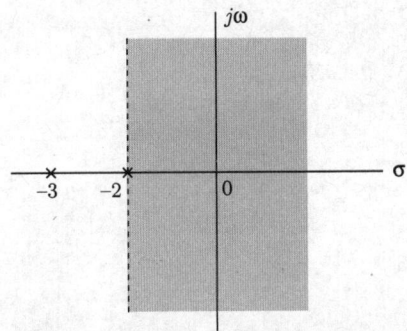

Fig. 9.4 Pole-zero plot and ROC for Example 9.4

Example 9.5 Determine the Laplace transform of

$$x(t) = -e^{-2t}u(-t) + e^{-3t}u(-t)$$

and depict the ROC and the locations of poles and zeros in the s-plane.

Solution
Consider the given signal

$$x(t) = -e^{-2t}u(-t) + e^{-3t}u(-t)$$

$$\mathcal{L}[x(t)] = \mathcal{L}[-e^{-2t}u(-t) + e^{-3t}u(-t)]$$

$$X(s) = \mathcal{L}[-e^{-2t}u(-t)] - \mathcal{L}[-e^{-3t}u(-t)]$$

From Eq. (9.11), we have

$$-e^{-at}u(-t) \longleftrightarrow \frac{1}{s+a} \qquad \Re\{s\} < -a$$

Therefore,

$$-e^{-2t}u(-t) \longleftrightarrow \frac{1}{s+2} \qquad \Re\{s\} < -2$$

$$-e^{-3t}u(-t) \longleftrightarrow \frac{1}{s+3} \qquad \Re\{s\} < -3$$

The set of values of $\Re\{s\}$ for which the Laplace transforms of both terms converge is $\Re\{s\} < -3$, and thus we obtain

$$X(s) = \frac{1}{s+2} - \frac{1}{s+3} \qquad \Re\{s\} < -3$$

$$X(s) = \frac{1}{s^2 + 5s + 6} \qquad \Re\{s\} < -3$$

The ROC for $X(s)$ is $\Re\{s\} < -3$, as shown in the shaded area in Fig. 9.5. For a left-sided infinite duration signal, the ROC is the region in the s-plane to the left of the leftmost pole.

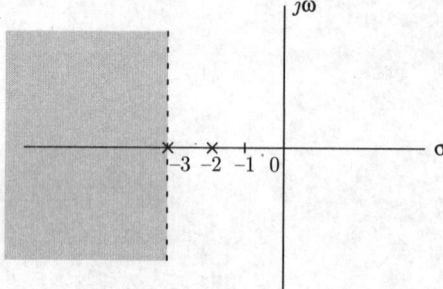

Fig. 9.5 Pole-zero plot and ROC for Example 9.5

Example 9.6 Determine the Laplace transform of the finite duration signal

$$x(t) = \begin{cases} e^{-at}, & 0 < t < T \\ 0, & \text{otherwise} \end{cases}$$

Solution
By definition

$$X(s) = \int_{-\infty}^{\infty} x(t)\, e^{-st}\, dt = \int_{0}^{T} e^{-at}\, e^{-st}\, dt$$

$$= \int_{0}^{T} e^{-(s+a)t}\, dt$$

$$X(s) = \frac{1}{s+a}[1 - e^{-(s+a)T}]$$

The ROC is the entire s-plane. For a finite duration signal, the ROC is the entire s-plane. At $s = -a$, both numerator and denominator are zero in the above expression. To determine $X(s)$ at $s = -a$, we can use the L'Hôpital's rule to obtain

$$\lim_{s \to -a} X(s) = \lim_{s \to -a} \left[\frac{d/dt(1 - e^{-(s+a)T})}{d/dt(s + a)} \right]$$

$$X(-a) = \lim_{s \to -a} T \, e^{-(s+a)T}$$

$$X(-a) = T$$

Example 9.7 Determine the Laplace transform of

(a) a unit impulse function $x(t) = \delta(t)$

(b) a unit step function $x(t) = u(t)$

(c) a unit ramp signal $x(t) = r(t)$

Solution

(a) Given that $x(t) = \delta(t)$. By definition

$$X(s) = \int_{-\infty}^{\infty} x(t) \, e^{-st} \, dt = \int_{-\infty}^{\infty} \delta(t) \, e^{-st} \, dt = e^{-st} \Big|_{t=0} = 1$$

Therefore,

$$\delta(t) \longleftrightarrow 1 \quad \text{ROC is the entire } s\text{-plane} \tag{9.12}$$

(b) Given that $x(t) = u(t)$. By definition

$$X(s) = \int_{-\infty}^{\infty} x(t) \, e^{-st} \, dt = \int_{-\infty}^{\infty} u(t) \, e^{-st} \, dt = \int_{0}^{\infty} e^{-st} \, dt$$

$$= -\frac{1}{s} e^{-st} \Big|_{0}^{\infty}$$

$$= -\frac{1}{s}[0 - 1] \quad \Re\{s\} > 0$$

$$X(s) = \frac{1}{s} \quad \Re\{s\} > 0$$

Therefore,

$$u(t) \longleftrightarrow \frac{1}{s} \quad \Re\{s\} > 0 \tag{9.13}$$

(c) Given that $x(t) = r(t) = tu(t)$. By definition

$$X(s) = \int_{-\infty}^{\infty} x(t) e^{-st} \, dt = \int_{-\infty}^{\infty} tu(t) e^{-st} \, dt = \int_{0}^{\infty} t e^{-st} \, dt$$

$$= -\frac{1}{s} t e^{-st} \Big|_{0}^{\infty} + \frac{1}{s} \int_{0}^{\infty} e^{-st} \, dt$$

$$= 0 - \frac{1}{s^2} e^{-st} \Big|_{0}^{\infty}$$

$$= -\frac{1}{s^2} [0 - 1] \quad \Re\{s\} > 0$$

$$X(s) = \frac{1}{s^2} \quad \Re\{s\} > 0$$

Therefore,

$$r(t) \longleftrightarrow \frac{1}{s^2} \quad \Re\{s\} > 0 \tag{9.14}$$

Example 9.8 Determine the Laplace transform of the following signals:
(a) $x(t) = \sin(\omega_0 t) u(t)$
(b) $x(t) = \cos(\omega_0 t) u(t)$

Solution
(a) Given that

$$x(t) = \sin(\omega_0 t) u(t)$$

$$x(t) = \frac{1}{2j} [e^{j\omega_0 t} u(t) - e^{-j\omega_0 t} u(t)]$$

Taking the Laplace transform of the above equation, we get

$$\mathcal{L}[x(t)] = \frac{1}{2j} \left(\mathcal{L}[e^{j\omega_0 t} u(t)] - \mathcal{L}[e^{-j\omega_0 t} u(t)] \right)$$

From Eq. (9.10), we have

$$e^{-at} u(t) \longleftrightarrow \frac{1}{s+a} \quad \Re\{s\} > -a$$

Therefore,

$$e^{j\omega_0 t} u(t) \longleftrightarrow \frac{1}{s - j\omega_0} \quad \Re\{s\} > 0$$

$$e^{-j\omega_0 t} u(t) \longleftrightarrow \frac{1}{s + j\omega_0} \quad \Re\{s\} > 0$$

The set of values of $\Re\{s\}$ for which the Laplace transforms of both terms converge is $\Re\{s\} > 0$, and thus we obtain

$$X(s) = \frac{1}{2j}\left(\frac{1}{s - j\omega_0} - \frac{1}{s + j\omega_0}\right) = \frac{1}{2j}\left(\frac{s + j\omega_0 - s + j\omega_0}{s^2 + \omega_0^2}\right) = \frac{\omega_0}{s^2 + \omega_0^2}$$

Therefore,

$$\sin(\omega_0 t)u(t) \longleftrightarrow \frac{\omega_0}{s^2 + \omega_0^2} \qquad \Re\{s\} > 0 \qquad\qquad (9.15)$$

The ROC for $X(s)$ is $\Re\{s\} > 0$, as shown in the shaded area in Fig. 9.6(a).

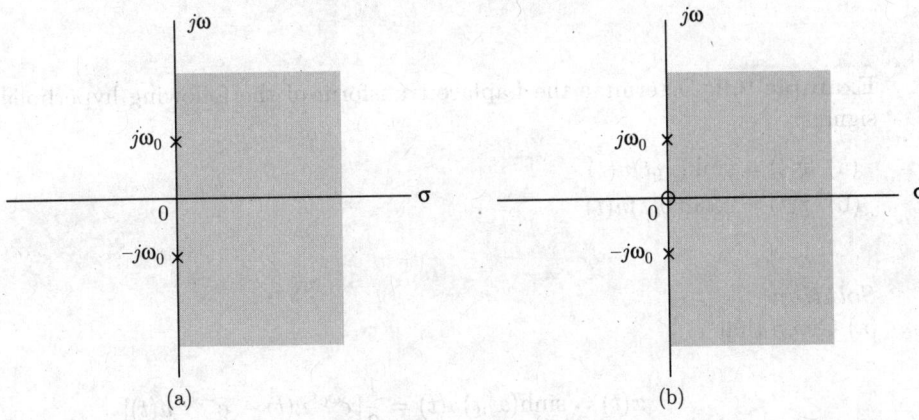

Fig. 9.6 Pole-zero plot and ROC for (a) $x(t) = \sin(\omega_0 t)u(t)$ and (b) $x(t) = \cos(\omega_0 t)u(t)$

(b) Given that

$$x(t) = \cos(\omega_0 t)u(t)$$

$$x(t) = \frac{1}{2}[e^{j\omega_0 t}u(t) + e^{-j\omega_0 t}u(t)]$$

Taking the Laplace transform of the above equation, we get

$$\mathcal{L}[x(t)] = \frac{1}{2}\left(\mathcal{L}[e^{j\omega_0 t}u(t)] + \mathcal{L}[e^{-j\omega_0 t}u(t)]\right)$$

From Eq. (9.10), we have

$$e^{-at}u(t) \longleftrightarrow \frac{1}{s + a} \qquad \Re\{s\} > -a$$

Therefore,

$$e^{j\omega_0 t}u(t) \longleftrightarrow \frac{1}{s - j\omega_0} \qquad \Re\{s\} > 0$$

$$e^{-j\omega_0 t}u(t) \longleftrightarrow \frac{1}{s + j\omega_0} \qquad \Re\{s\} > 0$$

The set of values of $\Re\{s\}$ for which the Laplace transforms of both terms converge is $\Re\{s\} > 0$, and thus we obtain

$$X(s) = \frac{1}{2}\left(\frac{1}{s - j\omega_0} + \frac{1}{s + j\omega_0}\right) = \frac{1}{2}\left(\frac{s + j\omega_0 + s - j\omega_0}{s^2 + \omega_0^2}\right) = \frac{s}{s^2 + \omega_0^2}$$

Therefore,

$$\cos(\omega_0 t)u(t) \longleftrightarrow \frac{s}{s^2 + \omega_0^2} \qquad \Re\{s\} > 0 \tag{9.16}$$

The ROC for $X(s)$ is $\Re\{s\} > 0$, as shown in the shaded area in Fig. 9.6(b).

Example 9.9 Determine the Laplace transform of the following hyperbolic sinusoidal signals:

(a) $x(t) = \sinh(\omega_0 t)u(t)$
(b) $x(t) = \cosh(\omega_0 t)u(t)$

Solution
(a) Given that

$$x(t) = \sinh(\omega_0 t)u(t) = \frac{1}{2}[e^{\omega_0 t}u(t) - e^{-\omega_0 t}u(t)]$$

Taking the Laplace transform of the above equation, we get

$$\mathcal{L}[x(t)] = \frac{1}{2}\left(\mathcal{L}[e^{\omega_0 t}u(t)] - \mathcal{L}[e^{-\omega_0 t}u(t)]\right)$$

From Eq. (9.10), we have

$$e^{-at}u(t) \longleftrightarrow \frac{1}{s + a} \qquad \Re\{s\} > -a$$

Therefore,

$$e^{\omega_0 t}u(t) \longleftrightarrow \frac{1}{s - \omega_0} \qquad \Re\{s\} > \omega_0$$

$$e^{-\omega_0 t}u(t) \longleftrightarrow \frac{1}{s + \omega_0} \qquad \Re\{s\} > -\omega_0$$

The set of values of $\Re\{s\}$ for which the Laplace transforms of both terms converge is $\Re\{s\} > \omega_0$, and thus we obtain

$$X(s) = \frac{1}{2}\left(\frac{1}{s - \omega_0} - \frac{1}{s + \omega_0}\right) = \frac{1}{2}\left(\frac{s + \omega_0 - s + \omega_0}{s^2 - \omega_0^2}\right) = \frac{\omega_0}{s^2 - \omega_0^2}$$

Therefore,

$$\sinh(\omega_0 t)u(t) \longleftrightarrow \frac{\omega_0}{s^2 - \omega_0^2} \qquad \Re\{s\} > \omega_0 \tag{9.17}$$

The ROC for $X(s)$ is $\Re\{s\} > \omega_0$, as shown in the shaded area in Fig. 9.7(a).

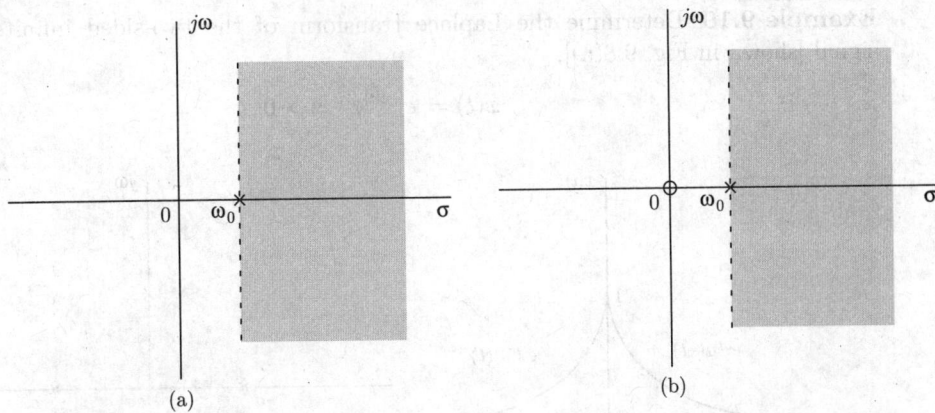

Fig. 9.7 Pole-zero plot and ROC for (a) $x(t) = \sinh(\omega_0 t)u(t)$ and (b) $x(t) = \cosh(\omega_0 t)u(t)$

(b) Given that

$$x(t) = \cosh(\omega_0 t)u(t) = \frac{1}{2}[e^{\omega_0 t}u(t) + e^{-\omega_0 t}u(t)]$$

Taking the Laplace transform of the above equation, we get

$$\mathcal{L}[x(t)] = \frac{1}{2}\left(\mathcal{L}[e^{\omega_0 t}u(t)] + \mathcal{L}[e^{-\omega_0 t}u(t)]\right)$$

From Eq. (9.10), we have

$$e^{-at}u(t) \longleftrightarrow \frac{1}{s+a} \qquad \Re\{s\} > -a$$

Therefore,

$$e^{\omega_0 t}u(t) \longleftrightarrow \frac{1}{s - \omega_0} \qquad \Re\{s\} > \omega_0$$

$$e^{-\omega_0 t}u(t) \longleftrightarrow \frac{1}{s + \omega_0} \qquad \Re\{s\} > -\omega_0$$

The set of values of $\Re\{s\}$ for which the Laplace transforms of both terms converge is $\Re\{s\} > \omega_0$, and thus we obtain

$$X(s) = \frac{1}{2}\left(\frac{1}{s - \omega_0} + \frac{1}{s + \omega_0}\right) = \frac{1}{2}\left(\frac{s + \omega_0 + s - \omega_0}{s^2 - \omega_0^2}\right) = \frac{s}{s^2 - \omega_0^2}$$

Therefore,

$$\cosh(\omega_0 t)u(t) \longleftrightarrow \frac{s}{s^2 - \omega_0^2} \qquad \Re\{s\} > \omega_0 \tag{9.18}$$

The ROC for $X(s)$ is $\Re\{s\} > \omega_0$, as shown in the shaded area in Fig. 9.7(b).

Example 9.10 Determine the Laplace transform of the two-sided infinite duration signal [shown in Fig. 9.8(a)]:

$$x(t) = e^{-a|t|}, \quad a > 0$$

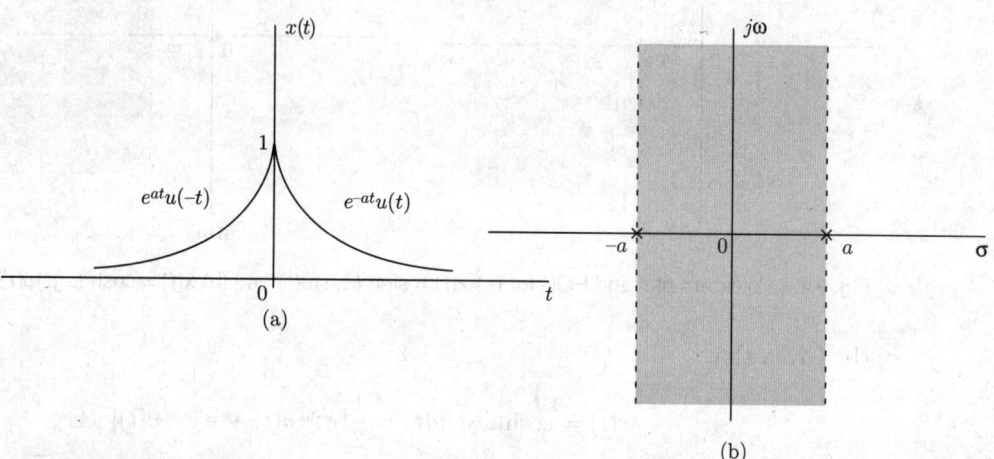

Fig. 9.8 (a) Signal $x(t) = e^{-a|t|}$, $a > 0$ and (b) its ROC

Solution
The signal $x(t)$ is defined as

$$x(t) = e^{-a|t|} = \begin{cases} e^{at}, & t < 0 \\ e^{-at}, & t \geq 0 \end{cases}$$

$$= e^{at}u(-t) + e^{-at}u(t)$$

From Eq. (9.10), we have

$$e^{-at}u(t) \longleftrightarrow \frac{1}{s+a} \qquad \Re\{s\} > -a$$

From Eq. (9.11), we have

$$-e^{-at}u(-t) \longleftrightarrow \frac{1}{s+a} \qquad \Re\{s\} < -a$$

Therefore,

$$e^{at}u(-t) \longleftrightarrow -\frac{1}{s-a} \qquad \Re\{s\} < a$$

The set of values of $\Re\{s\}$ for which the Laplace transforms of both terms converge is $-a < \Re\{s\} < a$, and thus we obtain

$$X(s) = \frac{1}{s+a} - \frac{1}{s-a} = \frac{-2a}{s^2 - a^2}$$

Therefore,

$$e^{-a|t|} \longleftrightarrow \frac{-2a}{s^2 - a^2} \qquad -a < \Re\{s\} < a \qquad (9.19)$$

The ROC for $X(s)$ is $\Re\{s\} > 0$, as shown in the shaded area in Fig. 9.8(b). For a two-sided infinite duration signal, the ROC is a strip in the s-plane.

Example 9.11 Find the Laplace transform of the signal $x(t) = t^n u(t)$.

Solution
By definition

$$X(s) = \mathcal{L}[t^n u(t)] = \int_{-\infty}^{\infty} x(t)\, e^{-st}\, dt = \int_{-\infty}^{\infty} t^n u(t)\, e^{-st}\, dt$$

$$= \int_{0}^{\infty} t^n\, e^{-st}\, dt$$

$$= -\frac{1}{s} t^n\, e^{-st}\Big|_{0}^{\infty} + \frac{n}{s} \int_{0}^{\infty} t^{n-1}\, e^{-st}\, dt \quad \text{for } n \geq 1$$

$$= 0 + \frac{n}{s} \mathcal{L}[t^{n-1} u(t)]$$

$$\mathcal{L}[t^n u(t)] = \frac{n}{s} \mathcal{L}[t^{n-1} u(t)] \quad \text{for } n \geq 1$$

Similarly, we can obtain

$$\mathcal{L}[t^{n-1} u(t)] = \frac{n-1}{s} \mathcal{L}[t^{n-2} u(t)] \quad \text{for } n \geq 2$$

$$\mathcal{L}[t^{n-2} u(t)] = \frac{n-2}{s} \mathcal{L}[t^{n-3} u(t)] \quad \text{for } n \geq 3$$

$$\vdots$$

Therefore,

$$\mathcal{L}[t^n u(t)] = \frac{n}{s} \frac{n-1}{s} \frac{n-2}{s} \frac{n-3}{s} \cdots \frac{2}{s} \frac{1}{s} \mathcal{L}[t^{n-n} u(t)] = \frac{n!}{s^n} \frac{1}{s}$$

$$\mathcal{L}[t^n u(t)] = \frac{n!}{s^{n+1}} \quad \Re\{s\} > 0$$

Hence,

$$t^n u(t) \longleftrightarrow \frac{n!}{s^{n+1}} \quad \Re\{s\} > 0 \tag{9.20}$$

or, equivalently,

$$\frac{t^n}{n!} u(t) \longleftrightarrow \frac{1}{s^{n+1}} \quad \Re\{s\} > 0 \tag{9.21}$$

and

$$\frac{t^{n-1}}{(n-1)!} u(t) \longleftrightarrow \frac{1}{s^n} \quad \Re\{s\} > 0 \tag{9.22}$$

Example 9.12 Consider the signal

$$x(t) = e^{-5t} u(t-1)$$

and denote its Laplace transform by $X(s)$.

(a) Evaluate $X(s)$ and find its ROC.

(b) Determine the values of the finite numbers A and t_0 such that the Laplace transform $G(s)$ of

$$g(t) = A e^{-5t} u(-t - t_0)$$

has the same algebraic form as $X(s)$. What is the ROC corresponding to $G(s)$?

Solution
(a) By definition

$$X(s) = \int_{-\infty}^{\infty} x(t) e^{-st} \, dt = \int_{-\infty}^{\infty} e^{-5t} u(t-1) e^{-st} \, dt$$

Since

$$u(t-1) = \begin{cases} 1, & t > 1 \\ 0, & t < 1 \end{cases}$$

we have

$$X(s) = \int_{1}^{\infty} e^{-5t} e^{-st} \, dt = \int_{1}^{\infty} e^{-(s+5)t} \, dt$$

$$= \frac{e^{-(s+5)t}}{-(s+5)} \Big|_{1}^{\infty}$$

$$= -1(s+5)[0 - e^{-(s+5)}] \quad \Re\{s\} > -5$$

$$X(s) = \frac{e^{-(s+5)}}{s+5} \quad \Re\{s\} > -5$$

(b) By definition

$$G(s) = \int_{-\infty}^{\infty} g(t)\, e^{-st}\, dt = \int_{-\infty}^{\infty} A\, e^{-5t} u(-t - t_0)\, e^{-st}\, dt$$

Since

$$u(-t - t_0) = \begin{cases} 1, & (-t - t_0) > 0 \longrightarrow t < -t_0 \\ 0, & (-t - t_0) < 0 \longrightarrow t > -t_0 \end{cases}$$

we have

$$G(s) = \int_{-\infty}^{-t_0} A\, e^{-5t}\, e^{-st}\, dt = \int_{-\infty}^{-t_0} A\, e^{-(s+5)t}\, dt$$

$$= A \frac{e^{-(s+5)t}}{-(s+5)} \bigg|_{-\infty}^{-t_0}$$

$$= \frac{-A}{s+5}[0 - e^{(s+5)t_0}] \quad \Re\{s\} < -5$$

$$G(s) = \frac{A\, e^{(s+5)t_0}}{s+5} \quad \Re\{s\} < -5$$

Clearly, $G(s)$ has the same algebraic form as $X(s)$ for $A = 1$ and $t_0 = -1$.

Example 9.13 Find the inverse Laplace transform of

$$X(s) = \frac{-3}{(s+2)(s-1)}$$

if the ROC is

(a) $\Re\{s\} > 1$
(b) $\Re\{s\} < -2$
(c) $-2 < \Re\{s\} < 1$

Solution
The partial fraction expansion of $X(s)$ yields

$$X(s) = \frac{1}{s+2} - \frac{1}{s-1}$$

$X(s)$ has poles at -2 and 1.
(a) The ROC and the locations of the poles are depicted in Fig. 9.9(a). The ROC, $\Re\{s\} > 1$, is to the right of the rightmost pole, so both poles correspond to causal

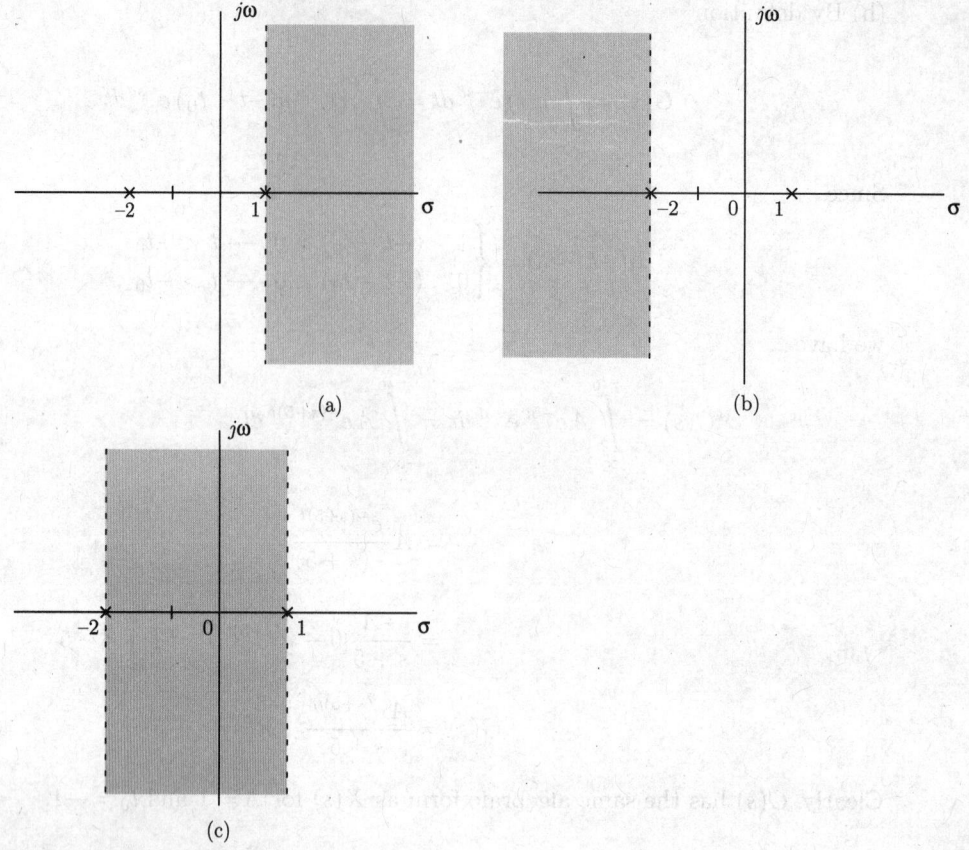

Fig. 9.9 ROC for Example 9.13

(right-sided) signals. Therefore,

$$e^{-2t}u(t) \longleftrightarrow \frac{1}{s+2}$$

$$e^{t}u(t) \longleftrightarrow \frac{1}{s-1}$$

and

$$x(t) = e^{-2t}u(t) - e^{t}u(t)$$

(b) The ROC and the locations of the poles are depicted in Fig. 9.9(b). The ROC, $\Re\{s\} < -2$, is to the left of the leftmost pole, so both poles correspond to anticausal (left-sided) signals. Therefore,

$$-e^{-2t}u(-t) \longleftrightarrow \frac{1}{s+2}$$

$$-e^{t}u(-t) \longleftrightarrow \frac{1}{s-1}$$

and

$$x(t) = -e^{-2t}u(-t) + e^{t}u(-t)$$

(c) The ROC and the locations of the poles are depicted in Fig. 9.9(c). The ROC, $-2 < \Re\{s\} < 1$, is a strip. The pole of the first term is at -2. The ROC lies to the right of this pole, so this pole corresponds to a causal (right-sided) signal. Therefore,

$$e^{-2t}u(t) \longleftrightarrow \frac{1}{s+2}$$

The second term has a pole at $s = 1$. Here the ROC is to the left of this pole, so this pole corresponds to the anticausal (left-sided) signal. Therefore,

$$-e^{t}u(-t) \longleftrightarrow \frac{1}{s-1}$$

and hence we obtain

$$x(t) = e^{-2t}u(t) + e^{t}u(-t)$$

Example 9.14 Find the inverse Laplace transform of

$$X(s) = \frac{-5s - 7}{(s+1)(s-1)(s+2)}$$

if the ROC is

(a) $\Re\{s\} > 1$
(b) $\Re\{s\} < -2$
(c) $-1 < \Re\{s\} < 1$
(d) $-2 < \Re\{s\} < -1$

Solution
Using the partial fraction expansion, we obtain

$$X(s) = \frac{1}{s+1} - \frac{2}{s-1} + \frac{1}{s+2}$$

$X(s)$ has poles at -1, 1, and -2.
(a) The ROC and the locations of the poles are depicted in Fig. 9.10(a). The ROC, $\Re\{s\} > 1$, is to the right of the rightmost pole, so all the three poles correspond to causal (right-sided) signals. Therefore,

$$e^{-t}u(t) \longleftrightarrow \frac{1}{s+1}$$

$$2e^{t}u(t) \longleftrightarrow \frac{2}{s-1}$$

$$e^{-2t}u(t) \longleftrightarrow \frac{1}{s+2}$$

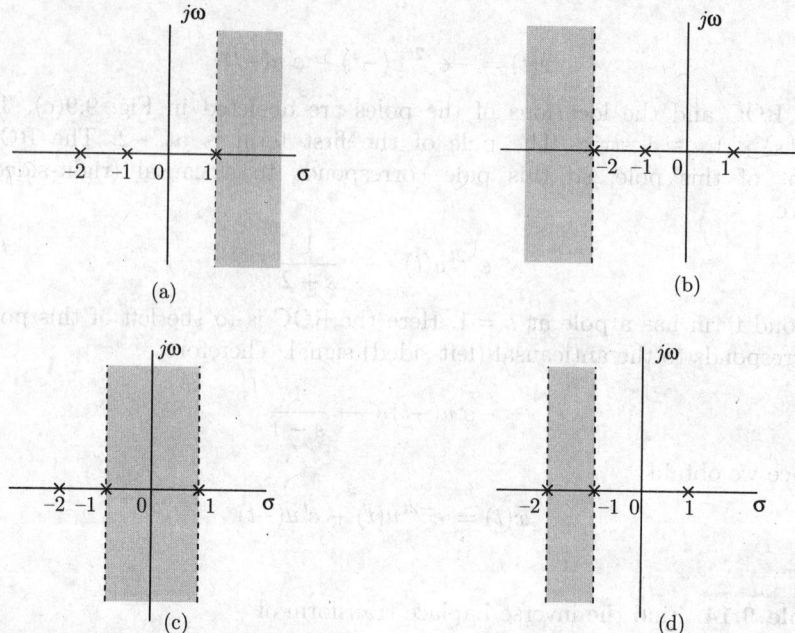

Fig. 9.10 ROC for Example 9.14

and hence,

$$x(t) = e^{-t}u(t) - 2\,e^{t}u(t) + e^{-2t}u(t)$$

(b) The ROC and the locations of the poles are depicted in Fig. 9.10(b). The ROC, $\Re\{s\} < -2$, is to the left of the leftmost pole, so all the three poles correspond to anti-causal (left-sided) signals. Therefore,

$$-e^{-t}u(-t) \longleftrightarrow \frac{1}{s+1}$$

$$-2\,e^{t}u(-t) \longleftrightarrow \frac{2}{s-1}$$

$$-e^{-2t}u(-t) \longleftrightarrow \frac{1}{s+2}$$

and hence,

$$x(t) = -e^{-t}u(-t) + 2\,e^{t}u(-t) - e^{-2t}u(-t)$$

(c) The ROC and the locations of the poles are depicted in Fig. 9.10(c). The ROC, $-1 < \Re\{s\} < 1$, is a strip. The pole of the first term is at -1. The ROC lies to the right of this pole, so this pole corresponds to a causal (right-sided) signal. Therefore,

$$e^{-t}u(t) \longleftrightarrow \frac{1}{s+1}$$

The second term has a pole at $s = 1$. Here the ROC is to the left of this pole, so this pole corresponds to an anticausal (left-sided) signal. Therefore,

$$-2\,e^t u(-t) \longleftrightarrow \frac{2}{s-1}$$

The third term has a pole at $s = -2$. Here the ROC is to the right of this pole, so this pole corresponds to a causal (right-sided) signal. Therefore,

$$e^{-2t}u(t) \longleftrightarrow \frac{1}{s+2}$$

and hence we obtain

$$x(t) = e^{-t}u(t) + 2\,e^t u(-t) + e^{-2t}u(t)$$

(d) The ROC and the locations of the poles are depicted in Fig. 9.10(d). The ROC, $-2 < \Re\{s\} < -1$, is a strip. The pole of the first term is at -1. The ROC lies to the left of this pole, so this pole corresponds to an anticausal (left-sided) signal. Therefore,

$$-e^{-t}u(-t) \longleftrightarrow \frac{1}{s+1}$$

The second term has a pole at $s = 1$. Here the ROC is to the left of this pole, so this pole corresponds to an anti-causal (left-sided) signal. Therefore,

$$-2\,e^t u(-t) \longleftrightarrow \frac{2}{s-1}$$

The third term has a pole at $s = -2$. Here the ROC is to the right of this pole, so this pole corresponds to causal (right-sided) signal. Therefore,

$$e^{-2t}u(t) \longleftrightarrow \frac{1}{s+2}$$

and hence we obtain

$$x(t) = -\,e^{-t}u(-t) + 2\,e^t u(-t) + e^{-2t}u(t)$$

9.7 PROPERTIES OF THE LAPLACE TRANSFORM

The properties of the bilateral Laplace transform (or Laplace transform) are similar to those of the Fourier transform. The derivations of many of these properties are analogous to those of the corresponding properties for the Fourier transform. The properties described in this section are specifically for the bilateral Laplace transform. The unilateral and bilateral transforms have many properties in common, although there are important differences.

9.7.1 Linearity

If

$$x_1(t) \longleftrightarrow X_1(s) \quad \text{with ROC} = R_1$$

and

$$x_2(t) \longleftrightarrow X_2(s) \quad \text{with ROC} = R_2$$

then

$$ax_1(t) + bx_2(t) \longleftrightarrow aX_1(s) + bX_2(s) \quad \text{with ROC containing } R_1 \cap R_2 \qquad (9.23)$$

As indicated, the ROC of $X(s)$ is at least the intersection of R_1 and R_2, which could be empty, in which case $X(s)$ has no ROC, i.e., $x(t)$ has no Laplace transform.

Proof The Laplace transform of $ax_1(t) + bx_2(t)$ is given by

$$\mathcal{L}[ax_1(t) + bx_2(t)] = \int_{-\infty}^{\infty} [ax_1(t) + bx_2(t)] \, e^{-st} \, dt$$

$$= a \int_{-\infty}^{\infty} x_1(t) \, e^{-st} \, dt + b \int_{-\infty}^{\infty} x_2(t) \, e^{-st} \, dt$$

$$= aX_1(s) + bX_2(s) \qquad \blacksquare$$

9.7.2 Time Shifting

If

$$x(t) \longleftrightarrow X(s) \quad \text{with ROC} = R$$

then

$$x(t - t_0) \longleftrightarrow X(s) \, e^{-st_0} \quad \text{with ROC} = R \qquad (9.24)$$

Proof The Laplace transform of $x(t - t_0)$ is given by

$$\mathcal{L}[x(t - t_0)] = \int_{-\infty}^{\infty} x(t - t_0) \, e^{-st} \, dt$$

A change of variables is performed by letting $\tau = t - t_0$, which also yields $d\tau = dt$, $\tau \to -\infty$ as $t \to -\infty$, and $\tau \to \infty$ as $t \to \infty$. Therefore,

$$\mathcal{L}[x(t - t_0)] = \int_{-\infty}^{\infty} x(\tau) \, e^{-s(\tau + t_0)} \, d\tau$$

$$= e^{-st_0} \int_{-\infty}^{\infty} x(\tau) \, e^{-s\tau} \, d\tau$$

$$\mathcal{L}[x(t - t_0)] = X(s) \, e^{-st_0} \qquad \blacksquare$$

Example 9.15 Determine the Laplace transform of the following signals:

(a) $x_1(t) = \delta(t+T) + \delta(t-T)$
(b) $x_2(t) = u(t+T) - u(t-T)$
(c) $x_3(t) = r(t+T) - r(t-T)$

The signals $x_1(t)$, $x_2(t)$, and $x_3(t)$ are shown in Figs 9.11(a), (b), and (c), respectively.

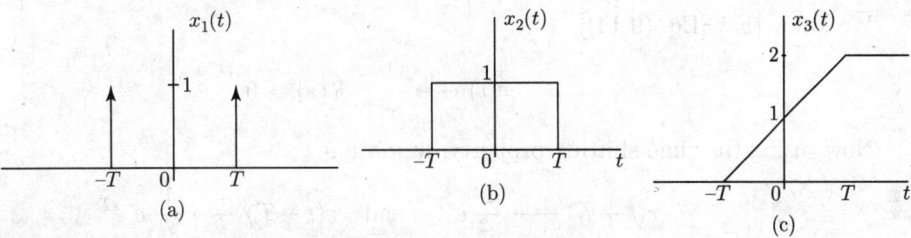

Fig. 9.11 Signal (a) $x_1(t)$, (b) $x_2(t)$, and (c) $x_3(t)$ for Example 9.15

Solution:
(a) Given that

$$x_1(t) = \delta(t+T) + \delta(t-T)$$
$$\mathcal{L}[x_1(t)] = X_1(s) = \mathcal{L}[\delta(t+T)] + \mathcal{L}[\delta(t-T)]$$

We know that [Eq. (9.12)]

$$\delta(t) \longleftrightarrow 1 \quad \text{ROC is the entire } s\text{-plane}$$

Now, using the time-shifting property, we obtain

$$\delta(t+T) \longleftrightarrow e^{sT} \quad \text{and} \quad \delta(t-T) \longleftrightarrow e^{-sT}$$

Therefore,

$$X_1(s) = e^{sT} + e^{-sT} \quad \text{ROC is the entire } s\text{-plane}$$

(b) Given that

$$x_2(t) = u(t+T) - u(t-T)$$
$$\mathcal{L}[x_2(t)] = X_2(s) = \mathcal{L}[u(t+T)] - \mathcal{L}[u(t-T)]$$

We know that [Eq. (9.13)]

$$u(t) \longleftrightarrow \frac{1}{s} \quad \Re\{s\} > 0$$

Now, using the time-shifting property, we obtain

$$u(t+T) \longleftrightarrow \frac{1}{s} e^{sT} \quad \text{and} \quad u(t-T) \longleftrightarrow \frac{1}{s} e^{-sT}$$

Therefore,

$$X_2(s) = \frac{1}{s}(e^{sT} - e^{-sT}) \quad \Re\{s\} > 0$$

(c) Given that

$$x_3(t) = r(t+T) - r(t-T)$$
$$\mathcal{L}[x_3(t)] = X_3(s) = \mathcal{L}[r(t+T)] - \mathcal{L}[r(t-T)]$$

We know that [Eq. (9.14)]

$$r(t) \longleftrightarrow \frac{1}{s^2} \quad \Re\{s\} > 0$$

Now, using the time-shifting property, we obtain

$$r(t+T) \longleftrightarrow \frac{1}{s^2} e^{sT} \quad \text{and} \quad r(t-T) \longleftrightarrow \frac{1}{s^2} e^{-sT}$$

Therefore,

$$X_3(s) = \frac{1}{s^2}(e^{sT} - e^{-sT}) \quad \Re\{s\} > 0$$

Example 9.16 A zero-order hold system has the impulse response $h(t)$ as shown in Fig. 9.12. Find the Laplace transform of this impulse response.

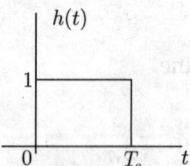

Fig. 9.12 Impulse response of a zero-order hold system

Solution:
From Fig. 9.12, the impulse response $h(t)$ can be written as

$$h(t) = u(t) - u(t - T_s)$$

where $u(t)$ is the unit step function. Taking the Laplace transform of the above equation, we obtain

$$H(s) = \mathcal{L}[h(t)] = \mathcal{L}[u(t)] - \mathcal{L}[u(t - T_s)]$$

$$H(s) = \frac{1}{s} - \frac{1}{s} e^{-sT}$$

$$H(s) = \frac{1}{s}(1 - e^{-sT})$$

9.7.3 Shifting in the *s*-Domain
If

$$x(t) \longleftrightarrow X(s) \quad \text{with ROC} = R$$

then

$$x(t)\,e^{s_0 t} \longleftrightarrow X(s - s_0) \quad \text{with ROC} = R + \Re\{s_0\} \tag{9.25}$$

Proof The Laplace transform of $x(t)\,e^{s_0 t}$ is given by

$$\mathcal{L}[x(t)\,e^{s_0 t}] = \int_{-\infty}^{\infty} [x(t)\,e^{s_0 t}]\,e^{-st}\,dt$$

$$= \int_{-\infty}^{\infty} x(t)\,e^{-(s-s_0)t}\,dt$$

$$\mathcal{L}[x(t)\,e^{s_0 t}] = X(s - s_0)$$

That is, the ROC associated with $X(s - s_0)$ is that of $X(s)$, shifted by $\Re\{s_0\}$. Note that if $X(s)$ has a pole or zero at $s = a$, then $X(s - s_0)$ has a pole or zero at $(s - s_0) = a$, i.e., $s = (a + s_0)$. ∎

Example 9.17 Find the Laplace transform and ROC of the following damped sinusoidal signals:

(a) $g_1(t) = e^{-at}\cos(\omega_0 t)u(t)$
(b) $g_2(t) = e^{-at}\sin(\omega_0 t)u(t)$

Solution:
(a) Consider the given signal

$$g_1(t) = e^{-at}\cos(\omega_0 t)u(t) = e^{-at}x(t)$$

where $x(t) = \cos(\omega_0 t)u(t)$ and its Laplace transform is given by [Eq. (9.16)]

$$X(s) = \frac{s}{s^2 + \omega_0^2} \quad \Re\{s\} > 0$$

Now, taking the Laplace transform of the given signal $g_1(t)$ and using shifting in the *s*-domain property, we obtain

$$\mathcal{L}[g_1(t)] = \mathcal{L}[e^{-at}x(t)]$$

$$G_1(s) = X(s + a) = \frac{s + a}{(s + a)^2 + \omega_0^2} \quad \Re\{s\} > -a$$

Therefore,

$$e^{-at}\cos(\omega_0 t)u(t) \longleftrightarrow \frac{s+a}{(s+a)^2 + \omega_0^2} \quad \Re\{s\} > -a \tag{9.26}$$

The ROC of $g_1(t) = e^{-at}\cos(\omega_0 t)u(t)$ is shown in Fig. 9.13(a).

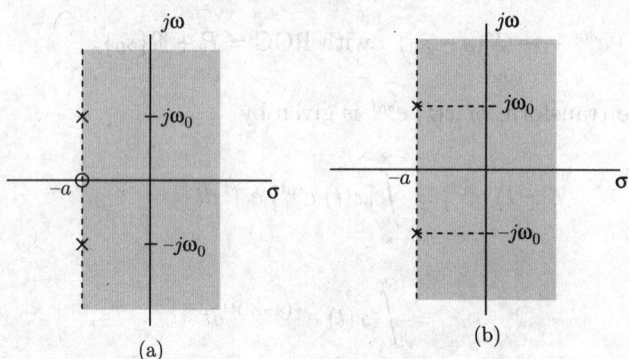

(a)   (b)

Fig. 9.13 Pole-zero plot and ROC for (a) $g_1(t) = e^{-at}\cos(\omega_0 t)u(t)$ and (b) $g_2(t) = e^{-at}\sin(\omega_0 t)u(t)$

(b) Consider the given signal

$$g_2(t) = e^{-at}\sin(\omega_0 t)u(t) = e^{-at}x(t)$$

where $x(t) = \sin(\omega_0 t)u(t)$ and its Laplace transform is given by [Eq. (9.15)]

$$X(s) = \frac{\omega_0}{s^2 + \omega_0^2} \quad \Re\{s\} > 0$$

Now, taking the Laplace transform of the given signal $g_2(t)$ and using shifting in the s-domain property, we obtain

$$\mathcal{L}[g_2(t)] = \mathcal{L}[e^{-at}x(t)]$$

$$G_2(s) = X(s+a) = \frac{\omega_0}{(s+a)^2 + \omega_0^2} \quad \Re\{s\} > -a$$

Therefore,

$$e^{-at}\sin(\omega_0 t)u(t) \longleftrightarrow \frac{\omega_0}{(s+a)^2 + \omega_0^2} \quad \Re\{s\} > -a \tag{9.27}$$

The ROC of $g_1(t) = e^{-at}\sin(\omega_0 t)u(t)$ is shown in Fig. 9.13(b).

Example 9.18 Find the Laplace transform and ROC of the signal $g(t) = t^n e^{-at}u(t)$.

Solution:
Consider the given signal

$$g(t) = t^n e^{-at}u(t) = e^{-at}x(t)$$

where $x(t) = t^n u(t)$ and its Laplace transform is given by [Eq. (9.20)]

$$X(s) = \frac{n!}{s^{n+1}} \quad \Re\{s\} > 0$$

Now, taking the Laplace transform of the given signal $g(t)$ and using shifting in the s-domain property, we obtain

$$\mathcal{L}[g(t)] = \mathcal{L}[e^{-at}x(t)]$$

$$G(s) = X(s+a) = \frac{n!}{(s+a)^{n+1}} \quad \Re\{s\} > -a$$

Therefore,

$$t^n\, e^{-at} u(t) \longleftrightarrow \frac{n!}{(s+a)^{n+1}} \quad \Re\{s\} > -a \qquad (9.28)$$

or, equivalently,

$$\frac{t^n}{n!}\, e^{-at} u(t) \longleftrightarrow \frac{1}{(s+a)^{n+1}} \quad \Re\{s\} > -a \qquad (9.29)$$

and

$$\frac{t^{n-1}}{(n-1)!}\, e^{-at} u(t) \longleftrightarrow \frac{1}{(s+a)^n} \quad \Re\{s\} > -a \qquad (9.30)$$

9.7.4 Time Scaling

If

$$x(t) \longleftrightarrow X(s) \quad \text{with ROC} = R$$

then

$$x(at) \longleftrightarrow \frac{1}{|a|} X\left(\frac{s}{a}\right) \quad \text{with ROC} = aR \qquad (9.31)$$

Aside from the amplitude factor of $1/a$, scaling in time domain by a factor of a corresponds to scaling in the s-domain by a factor of $1/a$. Also, for any value of s in the ROC of $x(t)$, the value s/a will be in the ROC of $x(at)$; that is, the ROC associated with $x(at)$ is a compressed ($a > 1$) or expanded ($a < 1$) version of the ROC of $x(t)$.

Proof The Laplace transform of $x(at)$ is

$$\mathcal{L}[x(at)] = \int_{-\infty}^{\infty} x(at)\, e^{-st}\, dt$$

Case I: For a positive real constant 'a'

$$\mathcal{L}[x(at)] = \int_{-\infty}^{\infty} x(at)\, e^{-st}\, dt$$

A change of variables is performed by letting $\tau = at$, which also yields $d\tau = a\,dt$, $\tau \to -\infty$ as $t \to -\infty$, and $\tau \to \infty$ as $t \to \infty$. Therefore,

$$\mathcal{L}[x(at)] = \frac{1}{a} \int_{-\infty}^{\infty} x(\tau)\, e^{-s/a\tau}\, d\tau$$

$$= \frac{1}{a} X \left(\frac{s}{a}\right)$$

Case II: For a negative real constant '$-a$'

$$\mathcal{F}[x(-at)] = \int_{-\infty}^{\infty} x(-at)\, e^{-st}\, dt$$

A change of variables is performed by letting $\tau = -at$, which also yields $d\tau = -a\,dt$, $\tau \to \infty$ as $t \to -\infty$, and $\tau \to -\infty$ as $t \to \infty$. Therefore,

$$\mathcal{L}[x(-at)] = -\frac{1}{a} \int_{\infty}^{-\infty} x(\tau)\, e^{(s/a)\tau}\, d\tau$$

$$= \frac{1}{a} \int_{-\infty}^{\infty} x(\tau)\, e^{-(-s/a)\tau}\, d\tau$$

$$= \frac{1}{a} X \left(-\frac{s}{a}\right)$$

Combining the two cases, we obtain

$$\mathcal{L}[x(at)] = \frac{1}{|a|} X \left(\frac{s}{a}\right)$$

Example 9.19 Determine the Laplace transform and ROC of the following signals:

(a) $g_1(t) = \delta(2t)$
(b) $g_2(t) = r(2t)$

Solution:
(a) We know that [Eq. (1.11)]

$$g_1(t) = \delta(2t) = \frac{1}{2}\delta(t)$$

Taking the Laplace transform, we obtain

$$G_1(s) = \mathcal{L}[\delta(2t)] = \mathcal{L}\left[\frac{1}{2}\delta(t)\right] = \frac{1}{2}\mathcal{L}[\delta(t)] = \frac{1}{2} \quad \text{ROC is the entire } s\text{-plane}$$

(b) **Method 1**   Let $x(t) = r(t)$. Therefore,

$$g_2(t) = x(2t)$$

We know that [Eq. (9.14)]

$$X(s) = \mathcal{L}[r(t)] = \frac{1}{s^2} \quad \Re\{s\} > 0$$

Now, using the time-scaling property, we obtain

$$\mathcal{L}[g_2(t)] = \mathcal{L}[x(2t)] = \frac{1}{2} X \left(\frac{s}{2}\right)$$

$$= \frac{1}{2} \frac{1}{(s/2)^2}$$

$$G_2(s) = \frac{2}{s^2} \quad \Re\{s\} > 0$$

Method 2   Given that

$$g_2(t) = r(2t) = 2tu(2t) = 2tu(t)$$

$$\mathcal{L}[g_2(t)] = 2\mathcal{L}[tu(t)]$$

$$G_2(s) = \frac{2}{s^2} \quad \Re\{s\} > 0$$

Example 9.20 Determine the Laplace transform of the following signals:

(a) $g_1(t) = \delta(2t - 3)$

(b) $g_2(t) = u(2t - 1)$

(c) $g_3(t) = r\left(\frac{1}{3}t - 2\right)$

Solution:

(a) We know that [Eq. (9.12)]

$$\delta(t) \longleftrightarrow 1$$

Using the time-shifting property, we obtain

$$\delta(t - 3) \longleftrightarrow e^{-3s}$$

Now, using the time-scaling property, we obtain

$$\delta(2t - 3) \longleftrightarrow \frac{1}{2} e^{-3/2s}$$

(b) We know that [Eq. (9.13)]

$$u(t) \longleftrightarrow \frac{1}{s}$$

Using the time-shifting property, we obtain

$$u(t - 1) \longleftrightarrow \frac{1}{s} e^{-s}$$

Now, using the time-scaling property, we obtain

$$u(2t - 1) \longleftrightarrow \frac{1}{2} \frac{1}{s/2} e^{-s/2}$$

$$u(2t - 1) \longleftrightarrow \frac{1}{s} e^{-s/2}$$

(c) We know that [Eq. (9.14)]

$$r(t) \longleftrightarrow \frac{1}{s^2}$$

Using the time-shifting property, we obtain

$$r(t - 2) \longleftrightarrow \frac{1}{s^2} e^{-2s}$$

Now, using the time-scaling property, we obtain

$$r\left(\frac{1}{3}t - 2\right) \longleftrightarrow 3\frac{1}{(3s)^2} e^{-6s}$$

$$r\left(\frac{1}{3}t - 2\right) \longleftrightarrow \frac{1}{3s^2} e^{-6s}$$

9.7.5 Scaling in the *s*-Domain

If

$$x(t) \longleftrightarrow X(s) \quad \text{with ROC} = R$$

then

$$\frac{1}{|a|} x\left(\frac{t}{a}\right) \longleftrightarrow X(as) \quad \text{with ROC} = \frac{R}{a} \tag{9.32}$$

Aside from the amplitude factor of $1/a$, scaling in s-domain by a factor of a corresponds to scaling in the time-domain by a factor of $1/a$. Also, for any value of s in the ROC of $x(t)$, the value as will be in the ROC of $x(t/a)$; that is, the ROC associated with $x(t/a)$ is a compressed ($a < 1$) or expanded ($a > 1$) version of the ROC of $x(t)$.

Proof The Laplace transform of $1/|a| x(t/a)$ is

$$\mathcal{L}\left[\frac{1}{|a|} x\left(\frac{t}{a}\right)\right] = \int\limits_{-\infty}^{\infty} \frac{1}{|a|} x\left(\frac{t}{a}\right) e^{-st} dt$$

Case I For a positive real constant 'a',

$$\mathcal{L}\left[\frac{1}{a}x\left(\frac{t}{a}\right)\right] = \int_{-\infty}^{\infty} \frac{1}{a}x\left(\frac{t}{a}\right)e^{-st}\,dt$$

A change of variables is performed by letting $\tau = t/a$, which also yields $t = a\tau$, $dt = a\,d\tau$, $\tau \to -\infty$ as $t \to -\infty$, and $\tau \to \infty$ as $t \to \infty$. Therefore,

$$\mathcal{L}\left[\frac{1}{a}x\left(\frac{t}{a}\right)\right] = \int_{-\infty}^{\infty} x(\tau)\,e^{-as\tau}\,d\tau = X(as)$$

Case II For a negative real constant '$-a$',

$$\mathcal{F}\left[\frac{1}{-a}x\left(-\frac{t}{a}\right)\right] = \int_{-\infty}^{\infty} \frac{1}{-a}x\left(-\frac{t}{a}\right)e^{-st}\,dt$$

A change of variables is performed by letting $\tau = -t/a$, which also yields $t = -a\tau$, $dt = -a\,d\tau$, $\tau \to \infty$ as $t \to -\infty$, and $\tau \to -\infty$ as $t \to \infty$. Therefore,

$$\mathcal{L}\left[\frac{1}{-a}x\left(-\frac{t}{a}\right)\right] = \int_{\infty}^{-\infty} x(\tau)\,e^{(as)\tau}\,d\tau$$

$$= -\int_{-\infty}^{\infty} x(\tau)\,e^{-(-as)\tau}\,d\tau$$

$$\mathcal{L}\left[\frac{1}{-a}x\left(-\frac{t}{a}\right)\right] = -X(-as)$$

$$\mathcal{L}\left[\frac{1}{a}x\left(-\frac{t}{a}\right)\right] = X(-as)$$

Combining the two cases, we obtain

$$\mathcal{L}\left[\frac{1}{|a|}x\left(\frac{t}{a}\right)\right] = X(as)$$

■

9.7.6 Time Reversal

If

$$x(t) \longleftrightarrow X(s) \quad \text{with ROC} = R$$

then

$$x(-t) \longleftrightarrow X(-s) \quad \text{with ROC} = -R \tag{9.33}$$

Proof Substituting $a = -1$ in Eq. (9.31), we have

$$\mathcal{L}[x(-t)] = \frac{1}{|-1|} X\left(\frac{s}{-1}\right)$$

$$\mathcal{L}[x(-t)] = X(-s)$$

Thus, time reversal of $x(t)$ results in a reversal of the ROC. An interesting consequence of the time-reversal property is that if $x(t)$ is even, then its Laplace transform is also even, i.e.,

$$\text{if} \quad x(-t) = x(t) \quad \text{then} \quad X(-s) = X(s) \tag{9.34}$$

Similarly, if $x(t)$ is odd, then so is its Laplace transform, that is

$$\text{if} \quad x(-t) = -x(t), \quad \text{then} \quad X(-s) = -X(s) \tag{9.35}$$

∎

Example 9.21 Determine the Laplace transform of the following signals:

(a) $g_2(t) = u(-2t - 1)$

(b) $g_3(t) = r(-3t - 2)$

Solution:

(a) We know that [Eq. (9.13)]

$$u(t) \longleftrightarrow \frac{1}{s}$$

Using the time-shifting property, we obtain

$$u(t - 1) \longleftrightarrow \frac{1}{s} e^{-s}$$

Using the time-scaling property, we obtain

$$u(2t - 1) \longleftrightarrow \frac{1}{2} \frac{1}{s/2} e^{-s/2}$$

Now, using the time-reversal property, we obtain

$$u(-2t - 1) \longleftrightarrow \frac{1}{-s} e^{-s/2}$$

$$u(-2t - 1) \longleftrightarrow -\frac{1}{s} e^{s/2}$$

(b) We know that [Eq. (9.14)]

$$r(t) \longleftrightarrow \frac{1}{s^2}$$

Using the time-shifting property, we obtain

$$r(t-2) \longleftrightarrow \frac{1}{s^2} e^{-2s}$$

Using the time-scaling property, we obtain

$$r(3t-2) \longleftrightarrow \frac{1}{3} \frac{1}{(s/3)^2} e^{-2s/3}$$

$$r(3t-2) \longleftrightarrow \frac{3}{s^2} e^{-2s/3}$$

Now, using the time-reversal property, we obtain

$$r(-3t-2) \longleftrightarrow \frac{3}{(-s)^2} e^{-(-2s/3)}$$

$$r(-3t-2) \longleftrightarrow \frac{3}{s^2} e^{2s/3}$$

9.7.7 Differentiation in the Time Domain

If

$$x(t) \longleftrightarrow X(s) \quad \text{with ROC} = R$$

then

$$\frac{dx(t)}{dt} \longleftrightarrow sX(s) \quad \text{with ROC containing } R \tag{9.36}$$

Proof From Eq. (9.4), we have

$$x(t) = \mathcal{L}^{-1}[X(s)] = \frac{1}{2\pi j} \int_{\sigma-j\infty}^{\sigma+j\infty} X(s) e^{st} \, ds$$

Differentiating both sides, we get

$$\frac{dx(t)}{dt} = \frac{1}{2\pi j} \int_{\sigma-j\infty}^{\sigma+j\infty} [sX(s)] e^{st} \, ds$$

$$\frac{dx(t)}{dt} = \mathcal{L}^{-1}[sX(s)]$$

Therefore,

$$\frac{dx(t)}{dt} \longleftrightarrow sX(s)$$

Similarly, the differentiation property can be extended to yield

$$\frac{d^n x(t)}{dt^n} \longleftrightarrow s^n X(s) \qquad (9.37)$$

∎

Example 9.22 Using the bilateral time-shift and differentiation properties find the Laplace transform of

$$x(t) = \frac{d^2}{dt^2}\left(e^{-3(t-2)}u(t-2)\right)$$

Solution:
We know that

$$e^{-3t}u(t) \longleftrightarrow \frac{1}{s+3} \quad \Re\{s\} > -3$$

Using the time-shifting property [Eq. (9.24)], we obtain

$$e^{-3(t-2)}u(t-2) \longleftrightarrow \frac{1}{s+3}\, e^{-2s} \quad \Re\{s\} > -3$$

Now, using differentiation in the time-domain property [Eq. (9.37)], we obtain

$$\frac{d^2}{dt^2}\, e^{-3(t-2)}u(t-2) \longleftrightarrow \frac{s^2}{s+3}\, e^{-2s} \quad \Re\{s\} > -3$$

9.7.8 Differentiation in the *s*-Domain
If

$$x(t) \longleftrightarrow X(s) \quad \text{with ROC} = R$$

then

$$tx(t) \longleftrightarrow -\frac{dX(s)}{ds} \quad \text{with ROC} = R \qquad (9.38)$$

Proof By definition

$$X(s) = \int\limits_{-\infty}^{\infty} x(t)\, e^{-st}\, dt$$

Differentiating both sides w.r.t. *s*, we get

$$\frac{dX(s)}{ds} = \int\limits_{-\infty}^{\infty} [-tx(t)]\, e^{-st}\, dt$$

$$\frac{dX(s)}{ds} = \mathcal{L}[-tx(t)]$$

Therefore,

$$tx(t) \longleftrightarrow -\frac{dX(s)}{ds}$$

Similarly, differentiation in the s-domain property can be extended to yield

$$t^n x(t) \longleftrightarrow (-1)^n \frac{d^n X(s)}{ds^n} \tag{9.39}$$

∎

Example 9.23 Determine the Laplace transform and ROC for the following signals:

(a) $g_1(t) = t e^{-at} u(t)$

(b) $g_2(t) = t \sin(\omega_0 t) u(t)$

(c) $g_3(t) = t \cos(\omega_0 t) u(t)$

Solution

(a) Consider the given signal

$$g_1(t) = te^{-at}u(t) = tx(t)$$

where $x(t) = e^{-at}u(t)$ and its Laplace transform is given by [Eq. (9.10)]

$$X(s) = \frac{1}{s+a} \qquad \Re\{s\} > -a$$

Now, taking the Laplace transform of the given signal $g_1(t)$ and using differentiation in the s-domain property, we obtain

$$\mathcal{L}[g_1(t)] = \mathcal{L}[tx(t)]$$

$$G_1(s) = -\frac{dX(s)}{ds} = -\frac{d(1/s+a)}{ds} = \frac{1}{(s+a)^2}$$

Therefore,

$$t e^{-at} u(t) \longleftrightarrow \frac{1}{(s+a)^2} \qquad \Re\{s\} > -a \tag{9.40}$$

(b) Consider the given signal

$$g_2(t) = t\sin(\omega_0 t)u(t) = tx(t)$$

where $x(t) = \sin(\omega_0 t)u(t)$ and its Laplace transform is given by [Eq. (9.15)]

$$X(s) = \frac{\omega_0}{s^2 + \omega_0^2} \qquad \Re\{s\} > 0$$

Now, taking the Laplace transform of the given signal $g_2(t)$ and using differentiation in the s-domain property, we obtain

$$\mathcal{L}[g_2(t)] = \mathcal{L}[tx(t)]$$

$$G_2(s) = -\frac{dX(s)}{ds} = -\frac{d\left(\omega_0/s^2 + \omega_0^2\right)}{ds} = \frac{2\omega_0 s}{(s^2 + \omega_0^2)^2}$$

Therefore,

$$t\sin(\omega_0 t)u(t) \longleftrightarrow \frac{2\omega_0 s}{(s^2 + \omega_0^2)^2} \qquad \Re\{s\} > 0 \tag{9.41}$$

(c) Consider the given signal

$$g_3(t) = t\cos(\omega_0 t)u(t) = tx(t)$$

where $x(t) = \cos(\omega_0 t)u(t)$ and its Laplace transform is given by [Eq. (9.16)]

$$X(s) = \frac{s}{s^2 + \omega_0^2} \qquad \Re\{s\} > 0$$

Now, taking the Laplace transform of the given signal $g_3(t)$ and using differentiation in the s-domain property, we obtain

$$\mathcal{L}[g_3(t)] = \mathcal{L}[tx(t)]$$

$$G_3(s) = -\frac{dX(s)}{ds} = -\frac{d\left(s/s^2 + \omega_0^2\right)}{ds} = \frac{s^2 - \omega_0^2}{(s^2 + \omega_0^2)^2}$$

Therefore,

$$t\cos(\omega_0 t)u(t) \longleftrightarrow \frac{s^2 - \omega_0^2}{(s^2 + \omega_0^2)^2} \qquad \Re\{s\} > 0 \tag{9.42}$$

9.7.9 Convolution Property
If

$$x_1(t) \longleftrightarrow X_1(s) \quad \text{with ROC} = R_1$$

and

$$x_2(t) \longleftrightarrow X_2(s) \quad \text{with ROC} = R_2$$

then

$$x_1(t) * x_2(t) \longleftrightarrow X_1(s)X_2(s) \quad \text{with ROC containing } R_1 \cap R_2 \tag{9.43}$$

The Laplace transform maps the convolution of two signals into the product of their Laplace transforms. The ROC of $X_1(s)X_2(s)$ includes the intersection of the ROCs of $X_1(s)$ and $X_2(s)$ and may be larger if pole-zero cancellation occurs in the product.

Proof The Laplace transform of $x_1(t) * x_2(t)$ is given by

$$\mathcal{L}[x_1(t) * x_2(t)] = \int\limits_{-\infty}^{\infty} [x_1(t) * x_2(t)]\, e^{-st}\, dt$$

$$= \int\limits_{-\infty}^{\infty} \left(\int\limits_{-\infty}^{\infty} x_1(\tau) x_2(t - \tau) d\tau \right) e^{-st}\, dt$$

Interchanging the order of integration and noting that $x_1(\tau)$ does not depend on t, we obtain

$$\mathcal{L}[x_1(t) * x_2(t)] = \int\limits_{-\infty}^{\infty} x_1(\tau) \left(\int\limits_{-\infty}^{\infty} x_2(t - \tau)\, e^{-st}\, dt \right) d\tau$$

Using the time-shifting property [Eq. (9.24)], the bracketed term is $X_2(s)\, e^{-s\tau}$. Substituting this into the above equation yields

$$\mathcal{L}[x_1(t) * x_2(t)] = \int\limits_{-\infty}^{\infty} x_1(\tau) \left(X_2(s)\, e^{-s\tau} \right) d\tau$$

$$= X_2(s) \int\limits_{-\infty}^{\infty} x_1(\tau)\, e^{-s\tau}\, d\tau$$

$$\mathcal{L}[x_1(t) * x_2(t)] = X_2(s) X_1(s) = X_1(s) X_2(s)$$

Therefore,

$$x_1(t) * x_2(t) \longleftrightarrow X_1(s) X_2(s)$$

The convolution property states that convolution in the time domain corresponds to multiplication in the s-domain. ∎

Example 9.24 Consider a signal $y(t)$ which is related to two signals $x_1(t)$ and $x_2(t)$ by

$$y(t) = x_1(t - 2) * x_2(-t + 3)$$

where

$$x_1(t) = e^{-2t} u(t) \quad \text{and} \quad x_2(t) = e^{-3t} u(t)$$

Use properties of the Laplace transform to determine the Laplace transform $Y(s)$ of $y(t)$.

Solution:
We know that [Eq. (9.10)]

$$x_1(t) = e^{-2t} u(t) \longleftrightarrow \frac{1}{s + 2} \qquad \Re\{s\} > -2$$

Using the time-shifting property [Eq. (9.24)], we obtain

$$x_1(t-2) = e^{-2(t-2)}u(t-2) \longleftrightarrow \frac{e^{-2s}}{s+2} \qquad \Re\{s\} > -2$$

Similarly, we obtain

$$x_2(t) = e^{-3t}u(t) \longleftrightarrow \frac{1}{s+3} \qquad \Re\{s\} > -3$$

Using the time-shifting property [Eq. (9.24)], we obtain

$$x_1(t+3) = e^{-3(t+3)}u(t+3) \longleftrightarrow \frac{e^{3s}}{s+3} \qquad \Re\{s\} > -3$$

Now, using the time-reversal property [Eq. (9.33)], we obtain

$$x_1(-t+3) = e^{-3(-t+3)}u(-t+3) \longleftrightarrow \frac{e^{-3s}}{-s+3} \qquad \Re\{s\} < 3$$

Now, using the convolution property [Eq. (9.43)], we obtain

$$\mathcal{L}[y(t)] = \mathcal{L}[x_1(t-2)]\mathcal{L}[x_2(-t+3)]$$

$$Y(s) = \left(\frac{e^{-2s}}{s+2}\right)\left(\frac{e^{-3s}}{-s+3}\right) \qquad -2 < \Re\{s\} < 3$$

$$Y(s) = \frac{e^{-5s}}{(s+2)(3-s)} \qquad -2 < \Re\{s\} < 3$$

9.7.10 Multiplication Property

If

$$x_1(t) \longleftrightarrow X_1(s) \quad \text{with ROC} = R_1$$

and

$$x_2(t) \longleftrightarrow X_2(s) \quad \text{with ROC} = R_2$$

then

$$x_1(t)x_2(t) \longleftrightarrow \frac{1}{2\pi j}[X_1(s) * X_2(s)] = \frac{1}{2\pi j} \int_{\sigma_1-j\infty}^{\sigma_1+j\infty} X_1(s_1)X_2(s-s_1)\,ds_1$$

$$\text{with ROC containing } R_1 \cap R_2 \qquad (9.44)$$

The Laplace transform maps the multiplication of two signals into the convolution of their Laplace transforms.

Proof The Laplace transform of $x_1(t)x_2(t)$ is given by

$$\mathcal{L}[x_1(t)x_2(t)] = \int\limits_{-\infty}^{\infty} [x_1(t)x_2(t)] \, e^{-st} \, dt$$

$$= \int\limits_{-\infty}^{\infty} \left(\frac{1}{2\pi j} \int\limits_{\sigma_1-j\infty}^{\sigma_1+j\infty} X_1(s_1) \, e^{s_1 t} ds_1 \right) x_2(t) \, e^{-st} \, dt$$

Interchanging the order of integration and noting that $X_1(s_1)$ does not depend on t, we have

$$\mathcal{L}[x_1(t)x_2(t)] = \frac{1}{2\pi j} \int\limits_{\sigma_1-j\infty}^{\sigma_1+j\infty} X_1(s_1) \left(\int\limits_{-\infty}^{\infty} [x_2(t) \, e^{s_1 t}] \, e^{-st} \, dt \right) ds_1$$

Using shifting in the s-domain property [Eq. (9.23)], the bracketed term is $X_2(s - s_1)$. Substituting this into the above equation yields

$$\mathcal{L}[x_1(t)x_2(t)] = \frac{1}{2\pi j} \int\limits_{\sigma_1-j\infty}^{\sigma_1+j\infty} X_1(s_1)X_2(s - s_1) \, ds_1 = \frac{1}{2\pi j} \, [X_1(s) * X_2(s)]$$

Therefore,

$$x_1(t)x_2(t) \longleftrightarrow \frac{1}{2\pi j} \, [X_1(s) * X_2(s)]$$

∎

9.7.11   Integration in the Time Domain

If

$$x(t) \longleftrightarrow X(s) \quad \text{with ROC} = R$$

then

$$\int\limits_{-\infty}^{t} x(\tau) \, d\tau \longleftrightarrow \frac{1}{s} X(s) \quad \text{with ROC containing } R \cap \{\Re\{s\} > 0\} \tag{9.45}$$

Proof If a signal $x(t)$ is convolved with a unit step function $u(t)$, we get

$$x(t) * u(t) = \int\limits_{-\infty}^{\infty} x(\tau)u(t - \tau) \, d\tau$$

Since

$$u(t - \tau) = \begin{cases} 1, & \tau < t \\ 0, & \tau > t \end{cases}$$

we have

$$x(t) * u(t) = \int_{-\infty}^{t} x(\tau) \, d\tau$$

The convolution of a signal with a unit step function is the same as the cumulative integral of the signal. Now we can prove the integration property of the Laplace transform,

$$\int_{-\infty}^{t} x(\tau) \, d\tau = x(t) * u(t)$$

$$\mathcal{L}\left[\int_{-\infty}^{t} x(\tau) \, d\tau\right] = \mathcal{L}[x(t) * u(t)]$$

Using the convolution property [Eq. (9.43)] and Eq. (9.13), we obtain

$$\mathcal{L}\left[\int_{-\infty}^{t} x(\tau) \, d\tau\right] = X(s)U(s) = \frac{1}{s}X(s)$$

Therefore,

$$\int_{-\infty}^{t} x(\tau) \, d\tau \longleftrightarrow \frac{1}{s}X(s)$$

Example 9.25 Determine the bilateral Laplace transform and the corresponding ROC for each of the following signals:

(a) $x(t) = e^{-t}\dfrac{d}{dt}\left(e^{-(t+1)}u(t+1)\right)$

(b) $x(t) = \displaystyle\int_{-\infty}^{t} e^{2\tau}\sin(\tau)u(-\tau) \, d\tau$

Solution:
(a) We know that

$$e^{-t}u(t) \longleftrightarrow \frac{1}{s+1} \qquad \Re\{s\} > -1$$

Using the time-shifting property [Eq. (9.24)], we obtain

$$e^{-(t+1)}u(t+1) \longleftrightarrow \frac{1}{s+1} e^{s} \qquad \Re\{s\} > -1$$

Now, using differentiation in the time-domain property [Eq. (9.36)], we obtain

$$\frac{d}{dt}\left(e^{-(t+1)}u(t+1)\right) \longleftrightarrow \frac{s}{s+1}\, e^s \qquad \Re\{s\} > -1$$

Now, using shifting in the s-domain property [Eq. (9.25)], we obtain

$$e^{-t}\frac{d}{dt}\left(e^{-(t+1)}u(t+1)\right) \longleftrightarrow \frac{s+1}{s+2}\, e^{s+1} \qquad \Re\{s\} > -2$$

(b) We know that

$$\sin(t)u(-t) \longleftrightarrow \frac{-1}{s^2+1} \qquad \Re\{s\} < 0$$

Using shifting in the s-domain property [Eq. (9.25)], we obtain

$$e^{2t}\sin(t)u(-t) \longleftrightarrow \frac{-1}{(s-2)^2+1} \qquad \Re\{s\} < 2$$

Now, using integration in the time-domain property [Eq. (9.45)], we obtain

$$x(t) * u(t) = \int_{-\infty}^{t} e^{2\tau}\sin(\tau)u(-\tau)\, d\tau \longleftrightarrow \frac{-1}{s[(s-2)^2+1]} \qquad 0 < \Re\{s\} < 2$$

9.7.12 Conjugation Property

If

$$x(t) \longleftrightarrow X(s)$$

then

$$x^*(t) \longleftrightarrow X^*(s^*) \qquad (9.46)$$

Proof The Laplace transform of $x^*(t)$ is

$$\mathcal{L}[x^*(t)] = \int_{-\infty}^{\infty} x^*(t)\, e^{-st}\, dt = \left[\int_{-\infty}^{\infty} x(t)\, e^{-s^*t}\, dt\right]^* = [X(s^*)]^* = X^*(s^*)$$

Therefore, if $x(t)$ is real, that is, $x(t) = x^*(t)$, then

$$X(s) = X^*(s^*) \qquad (9.47)$$

$$X^*(s) = X(s^*) \qquad (9.48)$$

Consequently, if $x(t)$ is real and if $X(s)$ has a pole or zero at $s = s_0$, then $X(s)$ also has a pole or zero at the complex conjugate point $s = s_0^*$. ∎

9.8 LAPLACE TRANSFORM OF CAUSAL PERIODIC SIGNALS

The Laplace transform of a causal periodic signal can be determined from the knowledge of the Laplace transform of its first cycle (period). Consider a causal periodic signal $x(t)$ with period T_0 as shown in Fig. 9.14.

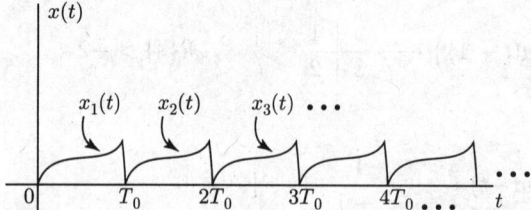

Fig. 9.14 Causal periodic signal

Let $x_1(t)$, $x_2(t)$, $x_3(t), \ldots$, be the signals representing the 1st, 2nd, 3rd, $\ldots$, cycles of a causal periodic signal $x(t)$. Therefore, the causal periodic signal $x(t)$ can be written as

$$x(t) = x_1(t) + x_2(t) + x_3(t) + x_4(t) + \cdots$$
$$= x_1(t) + x_1(t - T_0) + x_1(t - 2T_0) + x_1(t - 3T_0) + \cdots$$

Assume $x_1(t) \longleftrightarrow X_1(s)$. Using the time-shifting property [Eq. (9.24)], the Laplace transform of the above equation becomes

$$X(s) = X_1(s) + X_1(s)\, e^{-sT_0} + X_1(s)\, e^{-s2T_0} + X_1(s)\, e^{-s3T_0} + \cdots$$

$$= X_1(s)[1 + e^{-sT_0} + e^{-s2T_0} + e^{-s3T_0} + \cdots]$$

$$= X_1(s) \sum_{n=0}^{\infty} e^{-snT_0}$$

$$= X_1(s) \sum_{n=0}^{\infty} \left(e^{-sT_0}\right)^n$$

$$X(s) = X_1(s) \frac{1}{1 - e^{-sT_0}}$$

Therefore, the Laplace transform of a periodic signal $x(t)$ is given by

$$X(s) = \frac{X_1(s)}{1 - e^{-sT_0}} \qquad (9.49)$$

Example 9.26 Find the laplace transform of the square wave shown in Fig. 9.15(a).

Solution:

We know that the Laplace transform of a causal periodic signal is given by [Eq. (9.49)]

$$X(s) = \frac{X_1(s)}{1 - e^{-sT_0}}$$

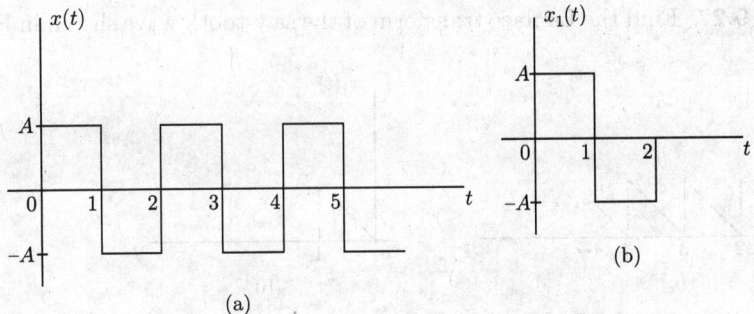

Fig. 9.15 (a) Causal periodic sequare wave $x(t)$ and (b) first cycle of $x(t)$

where $X_1(s)$ is the Laplace transform of the first cycle $x_1(t)$ of the causal periodic signal $x(t)$. The given square wave is a causal periodic signal with period $T_0 = 2$. The first cycle $x_1(t)$ of $x(t)$ is as shown in Fig. 9.15(b). Using the step function $u(t)$, $x_1(t)$ can be written as

$$x_1(t) = Au(t) - 2Au(t-1) + Au(t-2)$$

Taking the Laplace transform of the above equation, we obtain

$$\mathcal{L}[x_1(t)] = A\mathcal{L}[u(t)] - 2A\mathcal{L}[u(t-1)] + A\mathcal{L}[u(t-2)]$$

$$X_1(s) = \frac{A}{s} - \frac{2A}{s}\,e^{-s} + \frac{A}{s}\,e^{-2s} = \frac{A}{s}[1 - 2\,e^{-s} + e^{-2s}] = \frac{A}{s}(1 - e^{-s})^2$$

Substituting $X_1(s)$ and $T_0 = 2$ in the expression of $X(s)$, we get

$$X(s) = \frac{X_1(s)}{1 - e^{-sT_0}} = \frac{A}{s}\frac{(1 - e^{-s})^2}{1 - e^{-2s}}$$

$$= \frac{A}{s}\frac{(1 - e^{-s})^2}{(1 + e^{-s})(1 - e^{-s})}$$

$$= \frac{A}{s}\frac{1 - e^{-s}}{1 + e^{-s}}$$

$$= \frac{A}{s}\frac{e^{-s/2}(e^{s/2} - e^{-s/2})}{e^{-s/2}(e^{s/2} + e^{-s/2})}$$

$$X(s) = \frac{A}{s}\left(\frac{e^{s/2} - e^{-s/2}}{e^{s/2} + e^{-s/2}}\right)$$

$$X(s) = \frac{A}{s}\tanh\left(\frac{s}{2}\right)$$

Example 9.27 Find the Laplace transform of the saw-tooth wave shown in Fig. 9.16(a).

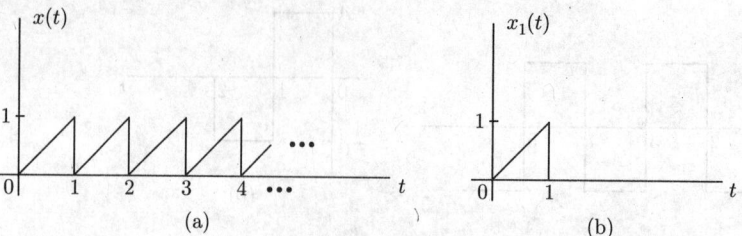

(a) (b)

Fig. 9.16 (a) Causal periodic saw-tooth wave $x(t)$ and (b) first cycle of $x(t)$

Solution

We know that the Laplace transform of a causal periodic signal is given by [Eq. (9.49)]

$$X(s) = \frac{X_1(s)}{1 - e^{-sT_0}}$$

where $X_1(s)$ is the Laplace transform of the first cycle $x_1(t)$ of the causal periodic signal $x(t)$. The given saw-tooth wave is a causal periodic signal with period $T_0 = 1$. The first cycle $x_1(t)$ of $x(t)$ is as shown in Fig. 9.16(b). Using the step function $u(t)$ and ramp function $r(t)$, $x_1(t)$ can be written as

$$x_1(t) = r(t) - r(t-1) - u(t-1)$$

Taking the Laplace transform of the above equation, we obtain

$$\mathcal{L}[x_1(t)] = \mathcal{L}[r(t)] - \mathcal{L}[r(t-1)] - \mathcal{L}[u(t-1)]$$

$$X_1(s) = \frac{1}{s^2} - \frac{1}{s^2}\,e^{-s} - \frac{1}{s}\,e^{-s}$$

$$X_1(s) = \frac{1}{s^2}[1 - e^{-s} - s\,e^{-s}]$$

Substituting $X_1(s)$ and $T_0 = 1$ in the expression of $X(s)$, we get

$$X(s) = \frac{X_1(s)}{1 - e^{-sT_0}} = \frac{1}{s^2}\frac{1 - e^{-s} - s\,e^{-s}}{1 - e^{-s}}$$

Example 9.28 Find the Laplace transform of the triangular wave shown in Fig. 9.17(a).

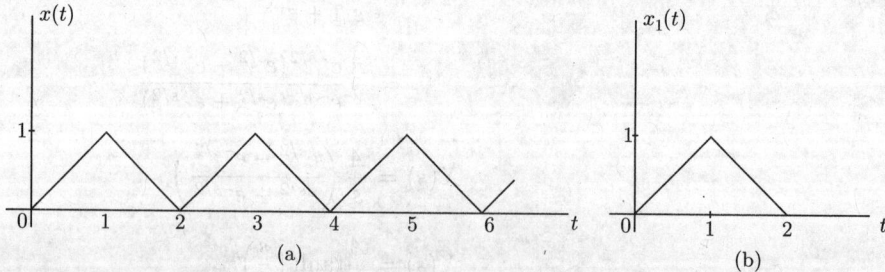

(a) (b)

Fig. 9.17 (a) Causal periodic triangular wave $x(t)$ and (b) first cycle of $x(t)$

Solution

We know that the Laplace transform of a causal periodic signal is given by [Eq. (9.49)]

$$X(s) = \frac{X_1(s)}{1 - e^{-sT_0}}$$

where $X_1(s)$ is the Laplace transform of the first cycle $x_1(t)$ of the causal periodic signal $x(t)$. The given saw-tooth wave is a causal periodic signal with period $T_0 = 2$. The first cycle $x_1(t)$ of $x(t)$ is as shown in Fig. 9.17(b). Using the ramp function $r(t)$, $x_1(t)$ can be written as

$$x_1(t) = r(t) - 2r(t-1) + r(t-2)$$

Taking the Laplace transform of the above equation, we obtain

$$\mathcal{L}[x_1(t)] = \mathcal{L}[r(t)] - 2\mathcal{L}[r(t-1)] + \mathcal{L}[r(t-2)]$$

$$X_1(s) = \frac{1}{s^2} - \frac{2}{s^2}\,e^{-s} + \frac{1}{s^2}\,e^{-2s} = \frac{1}{s^2}[1 - 2\,e^{-s} + e^{-2s}] = \frac{1}{s^2}(1 - e^{-s})^2$$

Substituting $X_1(s)$ and $T_0 = 2$ in the expression of $X(s)$, we get

$$X(s) = \frac{X_1(s)}{1 - e^{-sT_0}} = \frac{1}{s^2}\frac{(1 - e^{-s})^2}{1 - e^{-2s}}$$

$$= \frac{1}{s^2}\frac{(1 - e^{-s})^2}{(1 + e^{-s})(1 - e^{-s})}$$

$$= \frac{1}{s^2}\frac{1 - e^{-s}}{1 + e^{-s}}$$

$$= \frac{1}{s^2}\frac{e^{-s/2}(e^{s/2} - e^{-s/2})}{e^{-s/2}(e^{s/2} + e^{-s/2})}$$

$$X(s) = \frac{1}{s^2}\left(\frac{e^{s/2} - e^{-s/2}}{e^{s/2} + e^{-s/2}}\right)$$

$$X(s) = \frac{1}{s^2}\tanh\left(\frac{s}{2}\right)$$

9.9 ANALYSIS AND CHARACTERIZATION OF LTI SYSTEMS USING THE LAPLACE TRANSFORM

If $x(t)$ and $y(t)$ are the input and output of an LTI system with impulse response $h(t)$ [Fig. 9.18], then

$$y(t) = x(t) * h(t)$$

Application of the time convolution property to the above equation yields

$$\mathcal{L}[y(t)] = \mathcal{L}[x(t)] * \mathcal{L}[h(t)]$$

$$Y(s) = X(s)H(s) \qquad (9.50)$$

or

$$H(s) = \frac{Y(s)}{X(s)} \tag{9.51}$$

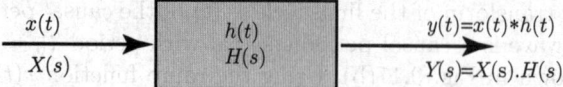

Fig. 9.18 LTI system

$H(s)$ is commonly referred to as the *system function* or *transfer function*. The transfer function $H(s)$ of an LTI system is equal to the ratio of the Laplace transform of the output signal to the Laplace transform of the input signal when all initial conditions are zero. For $s = j\omega$, $H(s)$ is the frequency response of the LTI system.

9.9.1 The Transfer Function and Differential-equation System Description

The transfer function may be related directly to the differential-equation description of an LTI system by using the bilateral Laplace transform. The relationship between the input and output of an Nth-order LTI system is described by the differential equation

$$\sum_{k=0}^{N} a_k \frac{d^k y(t)}{dt^k} = \sum_{k=0}^{M} b_k \frac{d^k x(t)}{dt^k}$$

Taking the Laplace transform of the above equation, we obtain

$$\sum_{k=0}^{N} a_k s^k Y(s) = \sum_{k=0}^{M} b_k s^k X(s)$$

$$H(s) = \frac{Y(s)}{X(s)} = \frac{\sum_{k=0}^{M} b_k s^k}{\sum_{k=0}^{N} a_k s^k}$$

$H(s)$ is a ratio of polynomials in s and is thus termed a rational transfer function.

9.9.2 Impulse Response and Step Response

The impulse response $h(t)$ is defined as the output of an LTI system due to a unit impulse signal input $\delta(t)$ applied at time $t = 0$. Also,

$$h(t) \longleftrightarrow H(s)$$

where $H(s)$ is the system function or transfer function.

The step response $s(t)$ is defined as the output of an LTI system due to a unit step input signal, i.e., $x(t) = u(t)$. The step response of an LTI system with impulse response $h(t)$ is given by

$$s(t) = h(t) * x(t)$$
$$= h(t) * u(t)$$

$$= \int_{-\infty}^{\infty} h(\tau)u(t-\tau)\,d\tau$$

$$s(t) = \int_{-\infty}^{t} h(\tau)\,d\tau$$

Therefore,

$$\mathcal{L}[s(t)] = \mathcal{L}\left[\int_{-\infty}^{t} h(\tau)\,d\tau\right]$$

$$S(s) = \frac{H(s)}{s} \tag{9.52}$$

Conversely,

$$h(t) = \frac{ds(t)}{dt}$$

$$\mathcal{L}[h(t)] = \mathcal{L}\left[\frac{ds(t)}{dt}\right]$$

$$H(s) = s\,S(s) \tag{9.53}$$

Example 9.29 Find the transfer function and the impulse response of a causal LTI system described by the differential equation

$$\frac{d^2y(t)}{dt^2} + 2\frac{dy(t)}{dt} + y(t) = \frac{dx(t)}{dt} - 2x(t)$$

Solution
Given that

$$\frac{d^2y(t)}{dt^2} + 2\frac{dy(t)}{dt} + y(t) = \frac{dx(t)}{dt} - 2x(t)$$

Taking the Laplace transform of the above equation, we obtain

$$s^2Y(s) + 2sY(s) + Y(s) = s^2X(s) - 2X(s)$$

$$Y(s)(s^2 + 2s + 1) = X(s)(s - 2)$$

$$H(s) = \frac{Y(s)}{X(s)} = \frac{s - 2}{s^2 + 2s + 1}$$

$$H(s) = \frac{s - 2}{(s+1)^2} = \frac{s - 3 + 1}{(s+1)^2} = \frac{s + 1}{(s+1)^2} - \frac{3}{(s+1)^2}$$

$$H(s) = \frac{1}{s + 1} - \frac{3}{(s+1)^2}$$

We know that

$$e^{-t}u(t) \longleftrightarrow \frac{1}{s+1}$$

$$3t\, e^{-t}u(t) \longleftrightarrow \frac{3}{(s+1)^2}$$

Therefore, the impulse response

$$h(t) = e^{-t}u(t) - 3t\, e^{-t}u(t)$$

Example 9.30 The unit step response of an LTI system is $s(t) = 2\, e^{-t}u(t)$. Determine its system function and the impulse response.

Solution
Given that the unit step response

$$s(t) = 2\, e^{-t}u(t)$$

$$\mathcal{L}[s(t)] = \mathcal{L}[2\, e^{-t}u(t)]$$

$$S(s) = \frac{2}{s+1}$$

The relationship between $\mathcal{L}[h(t)] = H(s)$ and $\mathcal{L}[s(t)] = S(s)$ is given by

$$H(s) = s\, S(s) = \frac{2s}{s+1} = \frac{2s+2-2}{s+1} = \frac{2(s+1)}{s+1} - \frac{2}{s+1}$$

$$H(s) = 2 - \frac{2}{s+1}$$

The inverse transform of this equation yields

$$h(t) = 2\delta(t) - 2\, e^{-t}u(t)$$

Example 9.31 An LTI system has a unit step response given by $s(t) = (1 - e^{-t} - t\, e^{-t})u(t)$. For a certain input $x(t)$, the output is observed to be equal to $y(t) = (2 - 3\, e^t + e^{-3t})u(t)$. What is $x(t)$?

Solution
Given that

$$y(t) = (2 - 3\, e^{-t} + e^{-3t})u(t)$$

$$Y(s) = \frac{2}{s} - \frac{3}{s+1} + \frac{1}{s+3}$$

$$Y(s) = \frac{2(s^2 + 4s + 3) - 3s(s+3) + s(s+1)}{s(s+1)(s+3)} = \frac{6}{s(s+1)(s+3)}$$

Consider the given unit step response

$$s(t) = (1 - e^{-t} - t\, e^{-t})u(t)$$

$$s(t) = u(t) - e^{-t}u(t) - t\, e^{-t}u(t)$$

The Laplace transform of this equation yields

$$S(s) = \frac{1}{s} - \frac{1}{s+1} - \frac{1}{(s+1)^2}$$

We know that [Eq. (9.53)]

$$H(s) = s\, S(s) = s\left(\frac{1}{s} - \frac{1}{s+1} - \frac{1}{(s+1)^2}\right)$$

$$H(s) = 1 - \frac{s}{s+1} - \frac{s}{(s+1)^2}$$

$$= \frac{(s+1)^2 - s(s+1) - s}{(s+1)^2}$$

$$= \frac{s^2 + 2s + 1 - s^2 - s - s}{(s+1)^2}$$

$$H(s) = \frac{1}{(s+1)^2} = \frac{Y(s)}{X(s)}$$

Therefore,

$$X(s) = \frac{Y(s)}{H(s)} = \frac{6/(s(s+1)(s+3))}{1/((s+1)^2)}$$

$$= \frac{6(s+1)}{s(s+3)}$$

$$X(s) = \frac{2}{s} + \frac{4}{s+3}$$

The inverse transform of this equation yields

$$x(t) = 2u(t) + 4\, e^{-3t}u(t)$$

9.9.3 Causality

For a causal LTI system, the impulse response $h(t) = 0$ for $t < 0$ and thus is right-sided. We know that if a signal is right-sided and of infinite duration, then the ROC is the region in the s-plane to the right of the rightmost pole. Consequently, *the ROC associated with* the system function for a causal system is a right-half plane. However, the converse of this statement is not necessarily true. If $H(s)$ is rational, then we can determine whether the system is causal simply by checking to see if its ROC is a right-half plane, that is, *for a system with a rational system function, causality of the system is equivalent to the ROC being the right-half plane to the right of the rightmost pole.*

In an exactly analogous manner, we can deal with the concept of anticausality. For an anticausal LTI system, the impulse response $h(t) = 0$ for $t > 0$ and is thus left-sided. We know that if a signal is left-sided and of infinite duration, then the ROC is the region in the s-plane to the left of the leftmost pole. Consequently, *the ROC associated with the system function for an anticausal system is a left-half plane.* However, the converse of this statement is not necessarily true.

Example 9.32 For the following system functions, check whether the corresponding LTI system is causal, anticausal, or noncausal.

(a) $H_1(s) = \dfrac{1}{s^2 + 5s + 6}$ $\Re\{s\} > -2$

(b) $H_2(s) = \dfrac{1}{s^2 + 5s + 6}$ $\Re\{s\} < -3$

(c) $H_3(s) = \dfrac{1}{s^2 + 5s + 6}$ $-2 < \Re\{s\} < -3$

(d) $H_4(s) = \dfrac{e^{2s}}{s + 1}$ $\Re\{s\} > -1$

Solution
(a) Consider the given system function

$$H_1(s) = \frac{1}{s^2 + 5s + 6} = \frac{1}{(s + 2)(s + 3)} \qquad \Re\{s\} > -2$$

Since the given system function $H_1(s)$ is rational and its ROC [shown in Fig. 9.19(a)] is to the right of the rightmost pole, the corresponding LTI system is causal. Also it can be verified by its impulse response. Given that

$$H_1(s) = \frac{1}{(s + 2)(s + 3)}$$

$$= \frac{1}{s + 2} - \frac{1}{s + 3}$$

Taking the inverse Laplace transform, we obtain

$$h_1(t) = e^{-2t}u(t) - e^{-3t}u(t)$$

Since $h_1(t) = 0$ for $t < 0$, this system is causal.

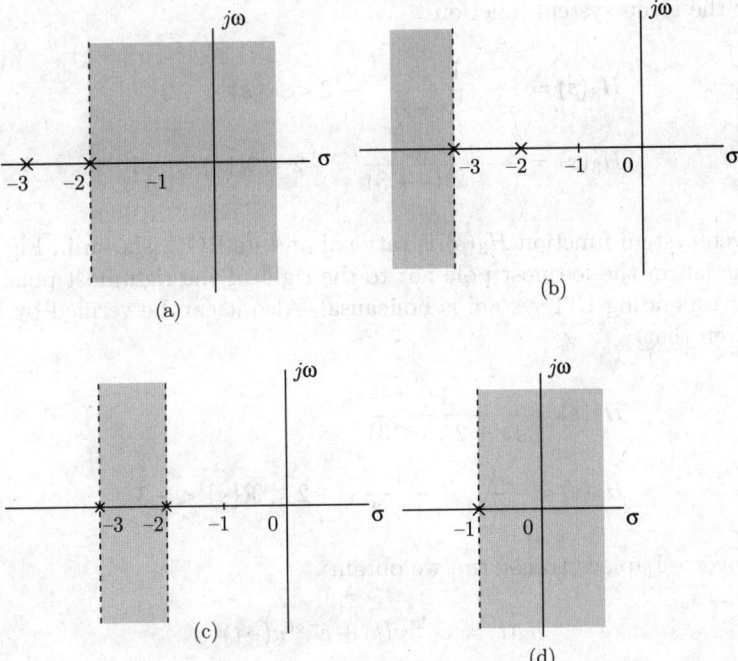

Fig. 9.19 (a) ROC for $H_1(s)$, (b) ROC for $H_2(s)$, (c) ROC for $H_3(s)$, and (d) ROC for $H_4(s)$

(b) Consider the given system function

$$H_2(s) = \frac{1}{s^2 + 5s + 6} \quad \Re\{s\} < -3$$

$$H_2(s) = \frac{1}{(s+2)(s+3)} \quad \Re\{s\} < -3$$

Since the given system function $H_2(s)$ is rational and its ROC [shown in Fig. 9.19(b)] is to the left of the leftmost pole, the corresponding LTI system is anticausal. Also it can be verified by its impulse response. Given that

$$H_2(s) = \frac{1}{(s+2)(s+3)}$$

$$= \frac{1}{s+2} - \frac{1}{s+3}$$

Taking the inverse Laplace transform, we obtain

$$h_2(t) = -e^{-2t}u(-t) + e^{-3t}u(-t)$$

Since $h_2(t) = 0$ for $t > 0$, this system is anticausal.

(c) Consider the given system function

$$H_3(s) = \frac{1}{s^2 + 5s + 6} \quad -2 < \Re\{s\} < -3$$

$$H_3(s) = \frac{1}{(s+2)(s+3)} \quad -2 < \Re\{s\} < -3$$

Since the given system function $H_3(s)$ is rational and its ROC [shown in Fig. 9.19(c)] is neither to the left of the leftmost pole nor to the right of the rightmost pole, but it is a strip, the corresponding LTI system is noncausal. Also it can be verified by its impulse response. Given that

$$H_3(s) = \frac{1}{(s+2)(s+3)}$$

$$H_3(s) = \frac{1}{s+2} - \frac{1}{s+3} \quad -2 < \Re\{s\} < -3$$

Taking the inverse Laplace transform, we obtain

$$h_3(t) = e^{-2t}u(t) + e^{-3t}u(-t)$$

Since $h_3(t)$ is two-sided, this system is noncausal.

(d) Consider the given system function

$$H_4(s) = \frac{e^{2s}}{s+1} \quad \Re\{s\} > -1$$

Since the given system function $H_4(s)$ is not rational, we cannot determine whether the system is causal by checking its ROC. For this system, the ROC [shown in Fig. 9.19(a)] is to the right of the rightmost pole; therefore, the impulse response must be right-sided. We know that

$$e^{-at}u(t) \longleftrightarrow \frac{1}{s+a} \quad \Re\{s\} > -a$$

$$e^{-t}u(t) \longleftrightarrow \frac{1}{s+1} \quad \Re\{s\} > -1$$

$$e^{-(t+2)}u(t+2) \longleftrightarrow \frac{e^{2s}}{s+1} \quad \Re\{s\} > -1$$

Therefore,

$$h_4(t) = e^{-(t+2)}u(t+2)$$

Since $h_4(t)$ is two-sided, this system is noncausal.

9.9.4 Stability

An LTI system is stable if the impulse response $h(t)$ is *absolutely integrable*, that is,

$$\int\limits_{-\infty}^{\infty} |h(t)|\, dt < \infty$$

This implies that the Fourier transform exists, and thus the ROC includes the $j\omega$-axis in the s-plane. Thus, *an LTI system is stable if and only if the ROC of its system function $H(s)$ includes the $j\omega$-axis.*

9.9.5 Stability of a Causal LTI System

Consider a causal LTI system with a rational system function $H(s)$. Since the system is causal, the ROC is to the right of the rightmost pole. Consequently, for this system to be stable (i.e., for the ROC to include the $j\omega$-axis), the rightmost pole of $H(s)$ must be to the left of the $j\omega$-axis. That is, *a causal system with rational system function $H(s)$ is stable if and only if all the poles of $H(s)$ lie in the left-half of the s-plane.*

Example 9.33 For the following system functions, check whether the corresponding LTI system is causal and stable.

(a) $H_1(s) = \dfrac{1}{s^2 - s - 6} \quad \Re\{s\} > 3$

(b) $H_2(s) = \dfrac{1}{s^2 - s - 6} \quad \Re\{s\} < -2$

(c) $H_3(s) = \dfrac{1}{s^2 - s - 6} \quad -2 < \Re\{s\} < 3$

Solution

(a) Consider the given system function

$$H_1(s) = \frac{1}{s^2 - s - 6} = \frac{1}{(s-3)(s+2)} \quad \Re\{s\} > 3$$

Since the given system function $H_1(s)$ is rational and its ROC [shown in Fig. 9.20(a)] is to the right of the rightmost pole, the corresponding LTI system is causal. Also, note that the ROC does not include the $j\omega$-axis, and, consequently, the corresponding system is unstable. This can be verified by its impulse response. Given that

$$H_1(s) = \frac{1}{(s-3)(s+2)}$$

$$= \frac{1/5}{s-3} - \frac{1/5}{s+2}$$

Taking the inverse Laplace transform, we obtain

$$h_1(t) = \frac{1}{5}[e^{3t}u(t) - e^{-2t}u(t)]$$

Note that $h_1(t) = 0$ for $t < 0$, and $\displaystyle\int\limits_{-\infty}^{\infty} h_1(t)\, dt = \infty$, i.e., the impulse response is not absolutely integrable, so this system is causal and unstable.

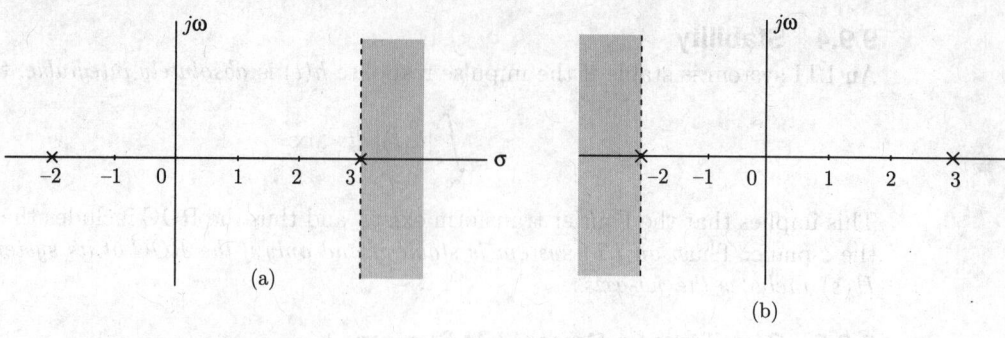

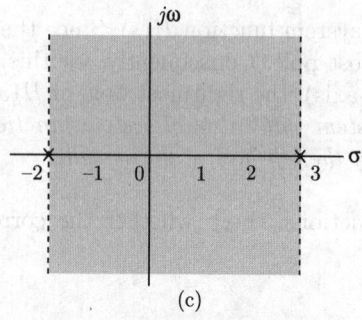

Fig. 9.20 ROC for (a) causal and unstable system $H_1(s)$, (b) anticausal and unstable system $H_2(s)$ and (c) noncausal and stable system $H_3(s)$.

(b) Consider the given system function

$$H_2(s) = \frac{1}{s^2 - s - 6} \quad \Re\{s\} < -2$$

$$H_2(s) = \frac{1}{(s - 3)(s + 2)} \quad \Re\{s\} < -2$$

Since the given system function $H_2(s)$ is rational and its ROC [shown in Fig. 9.20(b)] is to the left of the leftmost pole, the corresponding LTI system is anticausal. Also, note that the ROC does not include the $j\omega$-axis, and, consequently, the corresponding system is unstable. This can be verified by its impulse response. Given that

$$H_2(s) = \frac{1}{(s - 3)(s + 2)} \quad \Re\{s\} < -2$$

$$= \frac{1/5}{s - 3} - \frac{1/5}{s + 2}$$

Taking the inverse Laplace transform, we obtain

$$h_2(t) = \frac{1}{5}[-e^{3t}u(-t) + e^{-2t}u(-t)]$$

Note that $h_2(t) = 0$ for $t > 0$, and $\displaystyle\int_{-\infty}^{\infty} h_2(t)\, dt = \infty$, i.e., the impulse response is not absolutely integrable, so this system is anticausal and unstable.

(c) Consider the given system function

$$H_3(s) = \frac{1}{s^2 - s - 6} \qquad -2 < \Re\{s\} < 3$$

$$H_3(s) = \frac{1}{(s-3)(s+2)} \qquad -2 < \Re\{s\} < 3$$

Since the given system function $H_3(s)$ is rational and its ROC [shown in Fig. 9.20(c)] is a strip, the corresponding LTI system is noncausal. Also, note that the ROC includes the $j\omega$-axis, and, consequently, the corresponding system is stable. This can be verified by its impulse response. Given that

$$H_3(s) = \frac{1}{(s-3)(s+2)} \qquad -2 < \Re\{s\} < 3$$

$$= \frac{1/5}{s-3} - \frac{1/5}{s+2}$$

Taking the inverse Laplace transform, we obtain

$$h_3(t) = \frac{1}{5}[-e^{3t}u(-t) - e^{-2t}u(t)]$$

Note that $h_3(t)$ is two-sided, and $\displaystyle\int_{-\infty}^{\infty} h_3(t)\,dt < \infty$, i.e., the impulse response is absolutely integrable, so this system is noncausal and stable.

Example 9.34 For the following system functions, check whether the corresponding LTI system is causal and stable.

$$H(s) = \frac{1}{s^2 + 5s + 6} \qquad \Re\{s\} > -2$$

Solution
Consider the given system function

$$H(s) = \frac{1}{s^2 + 5s + 6} = \frac{1}{(s+2)(s+3)} \qquad \Re\{s\} > -2$$

Since the given system function $H(s)$ is rational and its ROC [shown in Fig. 9.21] is to the right of the rightmost pole, the corresponding LTI system is causal. Also, note that the ROC includes the $j\omega$-axis, and, consequently, the corresponding system is stable. This can be verified by its impulse response. Given that

$$H(s) = \frac{1}{(s+2)(s+3)} \qquad \Re\{s\} > -2$$

$$= \frac{1}{s+2} - \frac{1}{s+3}$$

Taking the inverse Laplace transform, we obtain

$$h(t) = e^{-2t}u(t) - e^{-3t}u(t)$$

Note that $h(t) = 0$ for $t < 0$, and $\displaystyle\int_{-\infty}^{\infty} h(t)\,dt < \infty$, i.e., the impulse response is absolutely integrable, so this system is causal and stable.

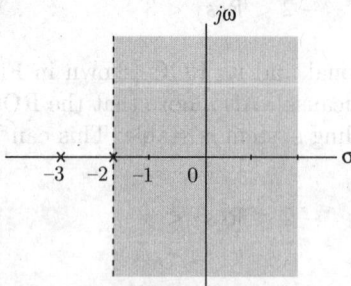

Fig. 9.21   ROC for Example 9.34

Example 9.35   The transfer function of the system is given by

$$H(s) = \frac{2}{s+3} + \frac{1}{s-2}$$

Determine the impulse response if the system is (a) stable and (b) causal. State whether the system will be stable and causal simultaneously.

Solution
Given that

$$H(s) = \frac{2}{s+3} + \frac{1}{s-2}$$

The system has poles at -3 and 2.

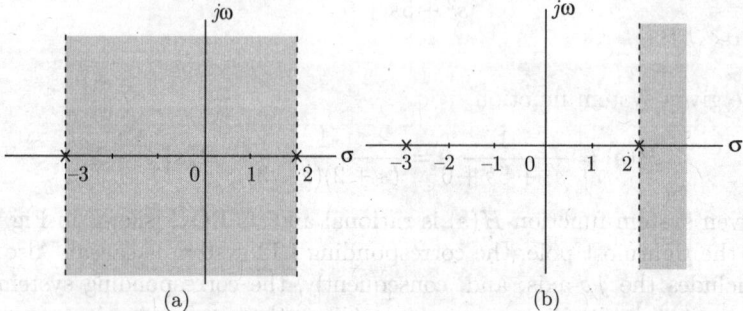

Fig. 9.22   ROC for the system function $H(s)$ of Example 9.35; (a) stable system, (b) causal system

(a) For this system to be stable, the ROC must include the $j\omega$-axis. To include the $j\omega$-axis, its ROC must lie between -3 and 2, i.e., $-3 < \Re\{s\} < 2$ [shown in Fig. 9.22(a)]. The pole of the first term is at -3. The ROC lies to the right of this pole, so this pole corresponds to causal (right-sided) signal. Therefore,

$$2\,e^{-3t}u(t) \longleftrightarrow \frac{2}{s+3}$$

The second term has a pole at $s = 2$. Here the ROC is to the left of this pole, so this pole corresponds to the anticausal (left-sided) signal. Therefore,

$$-e^{2t}u(-t) \longleftrightarrow \frac{1}{s-2}$$

and hence we obtain

$$h(t) = 2\,e^{-3t}u(t) - e^{2t}u(-t)$$

(b) For this system to be causal, the ROC lies to the right of the rightmost pole, i.e., $\Re\{s\} > 2$ [shown in Fig. 9.22(b)]. The ROC, $\Re\{s\} > 2$, is to the right of the rightmost pole, so all the poles correspond to causal (right-sided) signals. Therefore,

$$2\,e^{-3t}u(t) \longleftrightarrow \frac{2}{s+3}$$

$$e^{2t}u(t) \longleftrightarrow \frac{1}{s-2}$$

and hence we obtain

$$h(t) = 2\,e^{-3t}u(t) + e^{2t}u(t)$$

For a system to be stable and causal simultaneously, all the poles must lie in the left half of the s-plane. Since this system has one pole ($s = 2$) in the right half of the s-plane, it cannot be stable and causal simultaneously.

Example 9.36 Find the inverse Laplace transform of

$$H(s) = \frac{4s^2 + 15s + 8}{(s+2)^2(s-1)}$$

assuming that (a) $h(t)$ is causal and (b) the Fourier transform of $h(t)$ exists, i.e., $h(t)$ is absolutely integrable.

Solution
Given that

$$H(s) = \frac{4s^2 + 15s + 8}{(s+2)^2(s-1)}$$

Using partial fraction expansion, we obtain

$$H(s) = \frac{1}{s+2} + \frac{2}{(s+2)^2} + \frac{3}{s-1}$$

The system has a double pole at -2 and a single pole at 1.
(a) For this system to be causal, the ROC lies to the right of the rightmost pole, i.e., $\Re\{s\} > 1$ [shown in Fig. 9.23(a)]. Since the ROC, $\Re\{s\} > 1$, is to the right of the rightmost

pole, all the poles correspond to causal (right-sided) signals. Therefore,

$$e^{-2t}u(t) \longleftrightarrow \frac{1}{s+2}$$

$$2t\,e^{-2t}u(t) \longleftrightarrow \frac{1}{(s+2)^2}$$

$$3e^{t}u(t) \longleftrightarrow \frac{3}{s-1}$$

and hence we obtain

$$h(t) = e^{-2t}u(t) + 2t\,e^{-2t}u(t) + 3\,e^{t}u(t)$$

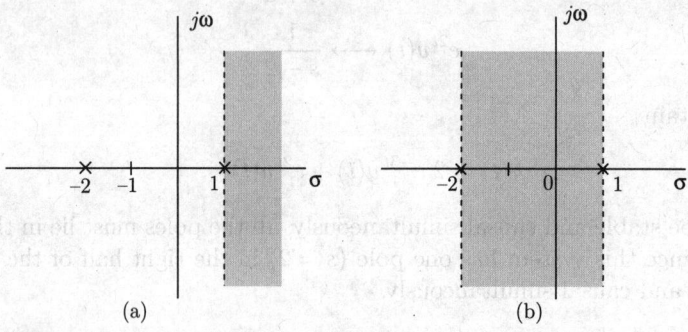

(a)   (b)

Fig. 9.23

(b) For $h(t)$ to be Fourier transformable, i.e., for $h(t)$ to be absolutely summable (stable system), the ROC must include the $j\omega$-axis. To include the $j\omega$-axis, its ROC is in the region $-2 < \Re\{s\} < 1$ [shown in Fig. 9.23(b)]. The pole of the first and second term is at -2. The ROC lies to the right of this pole, so this pole corresponds to a causal (right-sided) signal. Therefore,

$$e^{-2t}u(t) \longleftrightarrow \frac{1}{s+2}$$

$$2t\,e^{-2t}u(t) \longleftrightarrow \frac{1}{(s+2)^2}$$

The third term has a pole at $s = 1$. Here the ROC is to the left of this pole, so this pole corresponds to an anticausal (left-sided) signal. Therefore,

$$-3\,e^{t}u(-t) \longleftrightarrow \frac{3}{s-1}$$

and hence we obtain

$$h(t) = e^{-2t}u(t) + 2t\,e^{-2t}u(t) - 3\,e^{t}u(-t)$$

Example 9.37 Consider a continuous-time LTI system for which the input $x(t)$ and output $y(t)$ are related by the differential equation

$$\frac{d^2 y(t)}{dt^2} - \frac{dy(t)}{dt} - 2y(t) = x(t)$$

Let $X(s)$ and $Y(s)$ denote the Laplace transforms of $x(t)$ and $y(t)$, respectively, and let $H(s)$ denote the Laplace transform of $h(t)$, the system impulse response.

(a) Determine $H(s)$ as a ratio of two polynomials in s. Sketch the pole-zero pattern of $H(s)$.

(b) Determine $h(t)$ for each of the following cases:

1. The system is stable.
2. The system is causal.
3. The system is neither stable nor causal.

Solution

(a) Consider the given differential equation

$$\frac{d^2 y(t)}{dt^2} - \frac{dy(t)}{dt} - 2y(t) = x(t)$$

Taking the Laplace transform of the above equation, we obtain

$$s^2 Y(s) - sY(s) - 2Y(s) = X(s)$$

$$Y(s)[s^2 - s - 2] = X(s)$$

$$H(s) = \frac{Y(s)}{X(s)} = \frac{1}{s^2 - s - 2}$$

$$H(s) = \frac{1}{(s - 2)(s + 1)}$$

The system has poles at -1 and 2. The pole-zero pattern is shown in Fig. 9.24(a).

(b) Using partial fraction expansion

$$H(s) = \frac{1/3}{s - 2} - \frac{1/3}{s + 1}$$

1. For this system to be stable, the ROC must include the $j\omega$-axis. To include the $j\omega$-axis, its ROC must be in the region $-1 < \Re\{s\} < 2$ [shown in Fig. 9.24(b)].

The first term has a pole at $s = 2$. Here the ROC is to the left of this pole, so this pole corresponds to the anti-causal (left-sided) signal. Therefore,

$$-\frac{1}{3} e^{2t} u(-t) \longleftrightarrow \frac{1/3}{s - 2}$$

The pole of the second term is at -1. The ROC lies to the right of this pole, so this pole corresponds to a causal (right-sided) signal. Therefore,

$$\frac{1}{3} e^{-t} u(t) \longleftrightarrow \frac{1/3}{s + 1}$$

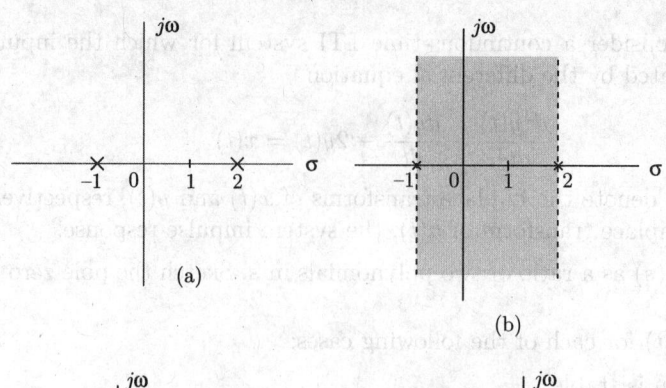

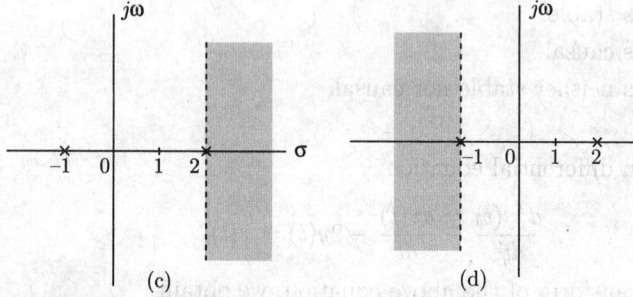

Fig. 9.24 (a) Pole-zero pattern of *H*(*s*) and ROC for (b) stable system, (c) causal system, and (d) neither stable nor causal system of Example 9.37

and hence we obtain

$$h(t) = -\frac{1}{3}\, e^{2t}u(-t) - \frac{1}{3}\, e^{-t}u(t)$$

2. For this system to be causal, the ROC lies to the right of the rightmost pole, i.e., $\Re\{s\} > 2$ [shown in Fig. 9.24(c)].

Since the ROC, $\Re\{s\} > 2$, is to the right of the rightmost pole, all the poles correspond to causal (right-sided) signals. Therefore,

$$\frac{1}{3}\, e^{2t}u(t) \longleftrightarrow \frac{1/3}{s - 2}$$

$$\frac{1}{3}\, e^{-t}u(t) \longleftrightarrow \frac{1/3}{s + 1}$$

and hence we obtain

$$h(t) = \frac{1}{3}\, e^{2t}u(t) - \frac{1}{3}\, e^{-t}u(t)$$

3. For this system to be neither stable nor causal, the ROC must not include the $j\omega$-axis and also not be to the left of the leftmost pole. So for this case, the ROC will be $\Re\{s\} < -1$ [shown in Fig. 9.24(d)]. Since the ROC, $\Re\{s\} < -1$, is to the left of the

leftmost pole, all the poles correspond to anti-causal (left-sided) signals. Therefore,

$$-\frac{1}{3}\,e^{2t}u(-t) \longleftrightarrow \frac{1/3}{s-2}$$

$$-\frac{1}{3}\,e^{-t}u(-t) \longleftrightarrow \frac{1/3}{s+1}$$

and hence we obtain

$$h(t) = -\frac{1}{3}\,e^{2t}u(-t) + \frac{1}{3}\,e^{-t}u(-t)$$

Example 9.38 Find the response $y(t)$ of a noncausal system with the transfer function

$$H(s) = \frac{-1}{s-1} \quad \Re\{s\} < 1$$

to the input $x(t) = e^{-2t}u(t)$.

Solution
Given that

$$x(t) = e^{-2t}u(t)$$

$$\mathcal{L}[x(t)] = X(s) = \frac{1}{s+2} \quad \Re\{s\} > -2$$

We know that [Eq. (9.50)]

$$Y(s) = X(s)H(s) = \frac{-1}{(s-1)(s+2)}$$

The ROC of $Y(s)$ is in the region $-2 < \Re\{s\} < 1$. Using partial fraction expansion, we obtain

$$Y(s) = \frac{-1/3}{s-1} + \frac{1/3}{s+2} \quad -2 < \Re\{s\} < 1$$

Taking the inverse Laplace transform, we obtain

$$y(t) = \frac{1}{3}[e^{t}u(-t) + e^{-2t}u(t)]$$

Example 9.39 Find the response $y(t)$ of an LTI system with the transfer function

$$H(s) = \frac{1}{s+5} \quad \Re\{s\} > -5$$

and the input

$$x(t) = e^{-t}u(t) + e^{-2t}u(-t)$$

Solution

Consider the given input

$$x(t) = e^{-t}u(t) + e^{-2t}u(-t) = x_1(t) + x_2(t)$$

We know that

$$x_1(t) = e^{-t}u(t) \longleftrightarrow X_1(s) = \frac{1}{s+1} \quad \Re\{s\} > -1$$

$$x_2(t) = e^{-2t}u(-t) \longleftrightarrow X_2(s) = -\frac{1}{s+2} \quad \Re\{s\} < -2$$

Although the Laplace transform of each of the individual terms have an ROC, there is no common ROC, and thus $x(t)$ has no Laplace transform.

In this case, we must determine separately the system response to each of the two input components, $x_1(t) = e^{-t}u(t)$ and $x_2(t) = e^{-2t}u(-t)$. If $y_1(t)$ and $y_2(t)$ are the system responses to $x_1(t)$ and $x_2(t)$, respectively, then

$$Y_1(s) = X_1(s)H(s) = \frac{1}{(s+1)(s+5)} \quad \Re\{s\} > -1$$

$$Y_1(s) = \frac{1/4}{s+1} - \frac{1/4}{s+5} \quad \Re\{s\} > -1$$

$$y_1(t) = \frac{1}{4}[e^{-t}u(t) + e^{-5t}u(t)]$$

and

$$Y_2(s) = X_2(s)H(s) = \frac{-1}{(s+2)(s+5)} \quad -5 < \Re\{s\} < -2$$

$$Y_2(s) = \frac{-1/3}{s+2} + \frac{1/3}{s+5} \quad -5 < \Re\{s\} < -2$$

$$y_2(t) = \frac{1}{3}[e^{-2t}u(-t) + e^{-5t}u(t)]$$

Therefore, the response $y(t)$ is

$$y(t) = y_1(t) + y_2(t)$$

$$= \frac{1}{4}[e^{-t}u(t) + e^{-5t}u(t)] + \frac{1}{3}[e^{-2t}u(-t) + e^{-5t}u(t)]$$

$$y(t) = \frac{1}{4} e^{-t}u(t) + \frac{1}{12} e^{-5t}u(t) + \frac{1}{3} e^{-2t}u(-t)$$

9.10 UNILATERAL (ONE-SIDED) LAPLACE TRANSFORM

The unilateral Laplace transform of a signal $x(t)$ is defined as

$$\mathcal{L}_{u}[x(t)] = X(s) = \int_{0^-}^{\infty} x(t)\, e^{-st}\, dt \tag{9.54}$$

where the lower limit of integration, 0^-, implies that we do include discontinuities and impulses that occur at $t = 0$ in the interval of integration. Hence $X(s)$ depends on $x(t)$ for $t \geq 0$. The evaluation of the inverse unilateral Laplace transform is also the same as for bilateral Laplace transforms. We shall denote the relationship between $x(t)$ and $X(s)$ as

$$x(t) \overset{\mathcal{L}_{u}}{\longleftrightarrow} X(s) \tag{9.55}$$

The difference in the definitions of the unilateral and bilateral Laplace transforms lies in the lower limit of integral. The bilateral Laplace transform in Eq. (9.3) involves an integration from $-\infty$ to $+\infty$, while the unilateral Laplace transform in Eq. (9.54) involves an integration from 0^- to $+\infty$. Consequently, two signals that differ for $t < 0$, but that are identical for $t \geq 0$, will have different bilateral Laplace transforms but identical unilateral Laplace transforms. Similarly, any signal that is identically zero for $t < 0$ will have identical bilateral and unilateral transforms.

The unilateral Laplace transform is used in analysing causal systems and, particularly, systems specified by linear constant-coefficient differential equations with nonzero initial conditions (i.e., systems that are not initially at rest).

For a unilateral Laplace transform, there is a unique inverse transform of $X(s)$; consequently, there is no need to specify the ROC explicitly (ROC for a unilateral Laplace transform must always be a right-half plane). For this reason, we shall generally ignore any mention of the ROC for a unilateral Laplace transform.

Example 9.40 Find (a) bilateral Laplace transform and (b) unilateral Laplace transform of the signal

$$x(t) = e^{-a(t+1)}u(t+1)$$

Solution
(a) We know that

$$e^{-at}u(t) \longleftrightarrow \frac{1}{s+a} \quad \Re\{s\} > -a$$

Using the time-shifting property, we get

$$e^{-a(t+1)}u(t+1) \longleftrightarrow \frac{e^{s}}{s+a} \quad \Re\{s\} > -a$$

Therefore,

$$\mathcal{L}[x(t)] = X(s) = \frac{e^{s}}{s+a} \quad \Re\{s\} > -a$$

(b) From Eq. (9.54), we have

$$\mathcal{L}_u[x(t)] = X(s) = \int_{0^-}^{\infty} x(t)\, e^{-st}\, dt = \int_{0^-}^{\infty} e^{-a(t+1)} u(t+1)\, e^{-st}\, dt$$

$$= \int_{0^-}^{\infty} e^{-a(t+1)}\, e^{-st}\, dt$$

$$= e^{-a} \int_{0^-}^{\infty} e^{-(s+a)t}\, dt$$

$$\mathcal{L}_u[x(t)] = X(s) = \frac{e^{-a}}{s+a} \qquad \mathcal{R}\{s\} > -a$$

Example 9.41   Find the inverse Laplace transform of

$$X(s) = \frac{s + 3 + 5\, e^{-2s}}{(s+1)(s+2)}$$

Solution
Given that

$$X(s) = \frac{s + 3 + 5\, e^{-2s}}{(s+1)(s+2)} = \frac{s+3}{(s+1)(s+2)} + \frac{5\, e^{-2s}}{(s+1)(s+2)}$$

$$= X_1(s) + X_2(s)\, e^{-2s}$$

where

$$X_1(s) = \frac{s+3}{(s+1)(s+2)} = \frac{2}{s+1} - \frac{1}{s+2}$$

and

$$X_2(s) = \frac{5}{(s+1)(s+2)} = \frac{5}{s+1} - \frac{5}{s+2}$$

Taking the inverse Laplace transform of $X_1(s)$ and $X_2(s)$, we obtain

$$x_1(t) = [2\, e^{-t} - e^{-2t}] u(t)$$

$$x_2(t) = [5\, e^{-t} - 5\, e^{-2t}] u(t)$$

Also, because

$$X(s) = X_1(s) + X_2(s)\, e^{-2s}$$

We can write

$$x(t) = x_1(t) + x_2(t-2)$$

$$= [2\,e^{-t} - e^{-2t}]u(t) + [5\,e^{-(t-2)} - 5\,e^{-2(t-2)}]u(t-2)$$

Example 9.42 Determine the unilateral Laplace transform of each of the following signals:

(a) $x(t) = e^{-2t}u(t+1)$
(b) $x(t) = \delta(t+1) + \delta(t) + e^{-2(t+3)}u(t+1)$

Solution
(a) From Eq. (9.54), we have

$$\mathcal{L}_u[x(t)] = X(s) = \int\limits_{0^-}^{\infty} x(t)\,e^{-st}\,dt = \int\limits_{0^-}^{\infty} e^{-2t}u(t+1)\,e^{-st}\,dt$$

$$= \int\limits_{0^-}^{\infty} e^{-2t}\,e^{-st}\,dt = \int\limits_{0^-}^{\infty} e^{-(s+2)t}\,dt$$

$$= -\frac{e^{-(s+2)t}}{s+2}\bigg|_{0^-}^{\infty} = -\frac{1}{s+2}[0-1] = \frac{1}{s+2}$$

$$\mathcal{L}_u[x(t)] = X(s) = \frac{1}{s+2}$$

(b) From Eq. (9.54), we have

$$\mathcal{L}_u[x(t)] = X(s) = \int\limits_{0^-}^{\infty} x(t)\,e^{-st}\,dt$$

$$= \int\limits_{0^-}^{\infty} [\delta(t+1) + \delta(t) + e^{-2(t+3)}u(t+1)]\,e^{-st}\,dt$$

$$= \int\limits_{0^-}^{\infty} \delta(t+1)\,e^{-st}\,dt + \int\limits_{0^-}^{\infty} \delta(t)\,e^{-st}\,dt + \int\limits_{0^-}^{\infty} e^{-2(t+3)}u(t+1)\,e^{-st}\,dt$$

$$= 0 + e^{-st}\bigg|_{t=0} + \int\limits_{0^-}^{\infty} e^{-2(t+3)}\,e^{-st}\,dt$$

$$= 1 + e^{-6} \int_{0-}^{\infty} e^{-(s+2)t} \, dt$$

$$= 1 - \frac{e^{-(s+2)t}}{s+2} \Big|_{0-}^{\infty}$$

$$= 1 - \frac{1}{s+2}[0 - 1]$$

$$= 1 + \frac{1}{s+2}$$

$$\mathcal{L}_u[x(t)] = X(s) = \frac{s+3}{s+2}$$

9.11 RELATIONSHIP BETWEEN BILATERAL AND UNILATERAL LAPLACE TRANSFORMS

The bilateral Laplace transform can be evaluated using the unilateral Laplace transform in two steps:

1. Split $x(t)$ into its causal and anticausal components $x_1(t)$ and $x_2(t)$, respectively.

$$x_1(t) = x(t)u(t)$$

$$x_2(t) = x(t)u(-t)$$

2. The signals $x_1(t)$ and $x_2(-t)$ are both causal. Determine $\mathcal{L}_u[x_1(t)]$ and add it with $\mathcal{L}_u[x_2(-t)]$, with s replaced by $-s$, i.e.,

$$X(s) = \mathcal{L}_u[x_1(t)] + \mathcal{L}_u[x_2(-t)]\Big|_{s \to -s}$$

or, equivalently,

$$\mathcal{L}[x(t)] = \mathcal{L}_u[x(t)u(t)] + \mathcal{L}_u[x(-t)u(t)]\Big|_{s \to -s} \tag{9.56}$$

Proof We separate $x(t)$ into two components $x_1(t)$ and $x_2(t)$, representing the causal component and the anticausal component of $x(t)$, respectively:

$$x_1(t) = x(t)u(t)$$

$$x_2(t) = x(t)u(-t)$$

The bilateral Laplace transform of $x(t)$ is given by

$$X(s) = \int\limits_{-\infty}^{\infty} x(t)\, e^{-st}\, dt$$

$$= \int\limits_{-\infty}^{0^-} x_2(t)\, e^{-st}\, dt + \int\limits_{0^-}^{\infty} x_1(t)\, e^{-st}\, dt$$

$$X(s) = X_2(s) + X_1(s)$$

Now, consider $X_2(s)$, given by

$$X_2(s) = \int\limits_{-\infty}^{0^-} x_2(t)\, e^{-st}\, dt$$

Using the substitution $t = -\tau$ yields

$$X_2(s) = \int\limits_{0^+}^{\infty} x_2(-\tau)\, e^{s\tau}\, d\tau$$

If $x(t)$ has any singularities at the origin, they are included in $x_1(t)$. Consequently, $x_2(t) = 0$ at $t = 0$. Hence,

$$X_2(s) = \int\limits_{0^-}^{\infty} x_2(-\tau)\, e^{s\tau}\, d\tau$$

$$X_2(-s) = \int\limits_{0^-}^{\infty} x_2(-\tau)\, e^{-s\tau}\, d\tau$$

$$X_2(-s) = \int\limits_{0^-}^{\infty} x_2(-t)\, e^{-st}\, dt$$

$$= \mathcal{L}_u[x_2(-t)]$$

Therefore,

$$X_2(s) = X_2(-s)\Big|_{s \to -s} = \mathcal{L}_u[x_2(-t)]\Big|_{s \to -s}$$

Also, because

$$X(s) = X_1(s) + X_2(s)$$

$$X(s) = \mathcal{L}_u[x_1(t)] + \mathcal{L}_u[x_2(-t)]\Big|_{s \to -s}$$

■

9.12 PROPERTIES OF UNILATERAL LAPLACE TRANSFORM

The unilateral and bilateral Laplace transforms have many properties in common, although there are important differences. The properties of linearity, scaling, s-domain shifting, conjugation, convolution, and differentiation in the s-domain properties are identical for the bilateral and unilateral Laplace transforms. The derivation of each of these properties is identical to that of its bilateral counterpart.

9.12.1 Linearity
If

$$x_1(t) \xleftrightarrow{\mathcal{L}_u} X_1(s) \quad \text{and} \quad x_2(t) \xleftrightarrow{\mathcal{L}_u} X_2(s)$$

then

$$ax_1(t) + bx_2(t) \xleftrightarrow{\mathcal{L}_u} aX_1(s) + bX_2(s) \tag{9.57}$$

9.12.2 Time Scaling
If

$$x(t) \xleftrightarrow{\mathcal{L}_u} X(s)$$

then

$$x(at) \xleftrightarrow{\mathcal{L}_u} \frac{1}{a}X\left(\frac{s}{a}\right) \quad a > 0 \tag{9.58}$$

9.12.3 Shifting in the s-Domain
If

$$x(t) \xleftrightarrow{\mathcal{L}_u} X(s)$$

then

$$e^{s_0 t}x(t) \xleftrightarrow{\mathcal{L}_u} X(s - s_0) \tag{9.59}$$

9.12.4 Conjugation
If

$$x(t) \xleftrightarrow{\mathcal{L}_u} X(s)$$

then

$$x^*(t) \xleftrightarrow{\mathcal{L}_u} X^*(s^*) \tag{9.60}$$

9.12.5 Differentiation in s-Domain
If

$$x(t) \xleftrightarrow{\mathcal{L}_u} X(s)$$

then

$$tx(t) \xleftrightarrow{\mathcal{L}_u} -\frac{dX(s)}{ds} \qquad (9.61)$$

Example 9.43 Find the inverse Laplace transform of

$$X(s) = \ln\left(1 + \frac{\omega^2}{s^2}\right)$$

Solution

Given that

$$X(s) = \ln\left(1 + \frac{\omega^2}{s^2}\right)$$

Differentiating w.r.t. s, we get

$$\frac{dX(s)}{ds} = \frac{1}{1 + (\omega^2/s^2)}(-2)\frac{\omega^2}{s^3}$$

$$\frac{dX(s)}{ds} = -\frac{2\omega^2}{s(s^2 + \omega^2)}$$

$$-\frac{dX(s)}{ds} = \frac{2(\omega^2 + s^2 - s^2)}{s(s^2 + \omega^2)}$$

$$= \frac{2(\omega^2 + s^2)}{s(s^2 + \omega^2)} - \frac{2s^2}{s(s^2 + \omega^2)}$$

$$-\frac{dX(s)}{ds} = \frac{2}{s} - 2\frac{s}{s^2 + \omega^2}$$

Using the differentiation property [Eq. (9.61)], we obtain

$$\mathcal{L}_u^{-1}\left[-\frac{dX(s)}{ds}\right] = tx(t) = \mathcal{L}_u^{-1}\left[\frac{2}{s} - 2\frac{s}{s^2 + \omega^2}\right]$$

$$tx(t) = 2u(t) - 2\cos(\omega t)u(t)$$

$$x(t) = \frac{2}{t}[1 - \cos(\omega t)]u(t)$$

9.12.6 Convolution

If $x_1(t) = x_2(t) = 0$, for all $t < 0$, and if

$$x_1(t) \xleftrightarrow{\mathcal{L}_u} X_1(s) \quad \text{and} \quad x_2(t) \xleftrightarrow{\mathcal{L}_u} X_2(s)$$

then

$$x_1(t) * x_2(t) \xleftrightarrow{\mathcal{L}_u} X_1(s)X_2(s) \qquad (9.62)$$

It is important to note that the convolution property for unilateral transform applies only if the signals $x_1(t)$ and $x_2(t)$ are both zero for $t < 0$. That is, while we have seen that the bilateral Laplace transform of $x_1(t) * x_2(t)$ always equals to the product of the bilateral transforms of $x_1(t)$ and $x_2(t)$, the unilateral transform of $x_1(t) * x_2(t)$ in general does not equal to the product of the unilateral transforms if either $x_1(t)$ or $x_2(t)$ is nonzero for $t < 0$. .

Example 9.44 Use the convolution property of the Laplace transform to determine

$$y(t) = e^{at} u(t) * e^{bt} u(t)$$

Solution

Given that

$$y(t) = e^{at} u(t) * e^{bt} u(t)$$

Using the convolution property of the Laplace transform [Eq. (9.60)], we obtain

$$Y(s) = \mathcal{L}_u[e^{at} u(t)] \mathcal{L}_u[e^{bt} u(t)]$$

$$Y(s) = \frac{1}{s-a} \frac{1}{s-b}$$

Using partial fraction expansion, we obtain

$$Y(s) = \frac{1}{a-b} \left[\frac{1}{s-a} - \frac{1}{s-b} \right]$$

Taking the inverse Laplace transform of this equation, we obtain

$$y(t) = \frac{1}{a-b} [e^{at} u(t) - e^{bt} u(t)]$$

9.12.7 Multiplication

If $x_1(t) = x_2(t) = 0$, for all $t < 0$, and if

$$x_1(t) \xleftrightarrow{\mathcal{L}_u} X_1(s) \quad \text{and} \quad x_2(t) \xleftrightarrow{\mathcal{L}_u} X_2(s)$$

then

$$x_1(t) x_2(t) \xleftrightarrow{\mathcal{L}_u} \frac{1}{2\pi j} [X_1(s) * X_2(s)] = \frac{1}{2\pi j} \int_{\sigma_1 - j\infty}^{\sigma_1 + j\infty} X_1(s_1) X_2(s - s_1) \, ds_1 \tag{9.63}$$

9.12.8 Integration in the s-Domain

If $x(t)$ satisfies the condition of existence $\int\limits_{0^-}^{\infty} |x(t)\, e^{-\sigma t}|\, dt < \infty$, and the limit of $x(t)/t$, as t approaches 0 from the right, exists, and

$$x(t) \xleftrightarrow{\mathcal{L}_{\mathrm{u}}} X(s)$$

then

$$\frac{x(t)}{t} \xleftrightarrow{\mathcal{L}_{\mathrm{u}}} \int\limits_{s}^{\infty} X(s)\, ds \tag{9.64}$$

Proof By definition

$$\mathcal{L}_{\mathrm{u}}[x(t)] = X(s) = \int\limits_{0^-}^{\infty} x(t)\, e^{-st}\, dt$$

Integrating both sides w.r.t. s, and taking the limit from s to ∞, we get

$$\int\limits_{s}^{\infty} X(s)\, ds = \int\limits_{s}^{\infty} \left(\int\limits_{0^-}^{\infty} x(t)\, e^{-st}\, dt \right) ds$$

$$= \int\limits_{0^-}^{\infty} x(t) \left(\int\limits_{s}^{\infty} e^{-st} ds \right) dt$$

$$= \int\limits_{0^-}^{\infty} x(t) \left(\frac{e^{-st}}{t} \right) dt$$

$$\int\limits_{s}^{\infty} X(s)\, ds = \int\limits_{0^-}^{\infty} \left[\frac{x(t)}{t} \right] e^{-st}\, dt$$

$$\int\limits_{s}^{\infty} X(s)\, ds = \mathcal{L}_{\mathrm{u}} \left[\frac{x(t)}{t} \right]$$

Therefore,

$$\frac{x(t)}{t} \xleftrightarrow{\mathcal{L}_{\mathrm{u}}} \int\limits_{s}^{\infty} X(s)\, ds$$

∎

Example 9.45 Find the Laplace transform of the signal

$$g(t) = \frac{1}{t}\sin(\omega t)u(t)$$

Solution

Given that

$$g(t) = \frac{1}{t}\sin(\omega t)u(t) = \frac{x(t)}{t}$$

where $x(t) = \sin(\omega t)u(t)$ and its Laplace transform is given by

$$X(s) = \frac{\omega}{s^2 + \omega^2}$$

Also,

$$g(t) = \frac{x(t)}{t}$$

Now, using integration in the s-domain property [Eq. (9.64)], we obtain

$$\mathcal{L}_u[g(t)] = G(s) = \mathcal{L}_u\left[\frac{x(t)}{t}\right] = \int_s^\infty X(s)\, ds = \int_s^\infty \frac{\omega}{s^2 + \omega^2}\, ds$$

$$G(s) = \tan^{-1}\left(\frac{s}{\omega}\right)\Big|_s^\infty$$

$$G(s) = \tan^{-1}(\infty) - \tan^{-1}\left(\frac{s}{\omega}\right)$$

$$G(s) = \frac{\pi}{2} - \tan^{-1}\left(\frac{s}{\omega}\right) \tag{9.65}$$

$$\tan^{-1}\left(\frac{s}{\omega}\right) = \frac{\pi}{2} - G(s)$$

$$\frac{s}{\omega} = \tan\left(\frac{\pi}{2} - G(s)\right)$$

$$\frac{s}{\omega} = \cot\left(G(s)\right)$$

$$\frac{1}{\cot\left(G(s)\right)} = \frac{1}{\frac{s}{\omega}}$$

$$\tan\left(G(s)\right) = \frac{\omega}{s}$$

$$G(s) = \tan^{-1}\left(\frac{\omega}{s}\right)$$

Therefore,

$$\frac{1}{t}\sin(\omega t)u(t) \longleftrightarrow \tan^{-1}\left(\frac{\omega}{s}\right) \tag{9.66}$$

Example 9.46 Show that

$$\frac{e^{-at} - e^{-bt}}{t} u(t) \overset{\mathcal{L}_u}{\longleftrightarrow} \ln\left(\frac{s+b}{s+a}\right)$$

Solution

Let

$$g(t) = \frac{e^{-at} - e^{-bt}}{t} u(t) = \frac{x(t)}{t}$$

where $x(t) = [e^{-at} - e^{-bt}]u(t)$ and its Laplace transform is given by

$$X(s) = \frac{1}{s+a} - \frac{1}{s+b}$$

Also,

$$g(t) = \frac{x(t)}{t}$$

Now, using integration in the s-domain property [Eq. (9.64)], we obtain

$$\mathcal{L}_u[g(t)] = G(s) = \mathcal{L}_u\left[\frac{x(t)}{t}\right] = \int_s^\infty X(s)\, ds = \lim_{N\to\infty} \int_s^N X(s)\, ds$$

$$G(s) = \lim_{N\to\infty} \int_s^N \left(\frac{1}{s+a} - \frac{1}{s+b}\right) ds$$

$$= \lim_{N\to\infty} \left[\ln(s+a)\Big|_s^N - \ln(s+b)\Big|_s^N\right]$$

$$= \lim_{N\to\infty} \left[\ln(N+a) - \ln(s+a) - \ln(N+b) + \ln(s+b)\right]$$

$$= \lim_{N\to\infty} \left[\ln\left(\frac{N+a}{N+b}\right) + \ln\left(\frac{s+b}{s+a}\right)\right]$$

$$= \lim_{N\to\infty} \left[\ln\left(\frac{1+\frac{a}{N}}{1+\frac{b}{N}}\right)\right] + \ln\left(\frac{s+b}{s+a}\right)$$

$$G(s) = \ln(1) + \ln\left(\frac{s+b}{s+a}\right)$$

$$G(s) = \ln\left(\frac{s+b}{s+a}\right)$$

Therefore,

$$\frac{e^{-at} - e^{-bt}}{t} u(t) \longleftrightarrow \ln\left(\frac{s+b}{s+a}\right) \tag{9.67}$$

Example 9.47 Show that

$$\frac{e^{bt} - e^{at}}{t} u(t) \xleftrightarrow{\mathcal{L}_u} \ln\left(\frac{s-a}{s-b}\right)$$

Solution

Let

$$g(t) = \frac{e^{bt} - e^{at}}{t} u(t) = \frac{x(t)}{t}$$

where $x(t) = [e^{bt} - e^{at}]u(t)$ and its Laplace transform is given by

$$X(s) = \frac{1}{s-b} - \frac{1}{s-a}$$

Also,

$$g(t) = \frac{x(t)}{t}$$

Now, using integration in the s-domain property [Eq. (9.64)], we obtain

$$\mathcal{L}_u[g(t)] = G(s) = \mathcal{L}_u\left[\frac{x(t)}{t}\right] = \int_s^\infty X(s)\, ds = \lim_{N\to\infty} \int_s^N X(s)\, ds$$

$$G(s) = \lim_{N\to\infty} \int_s^N \left(\frac{1}{s-b} - \frac{1}{s-a}\right) ds$$

$$= \lim_{N\to\infty} \left[\ln(s-b)\Big|_s^N - \ln(s-a)\Big|_s^N\right]$$

$$G(s) = \lim_{N\to\infty} \left[\ln(N-b) - \ln(s-b) - \ln(N-a) + \ln(s-a)\right]$$

$$= \lim_{N\to\infty} \left[\ln\left(\frac{N-b}{N-a}\right) + \ln\left(\frac{s-a}{s-b}\right)\right]$$

$$= \lim_{N\to\infty} \left[\ln\left(\frac{1-\frac{b}{N}}{1-\frac{a}{N}}\right)\right] + \ln\left(\frac{s-a}{s-b}\right)$$

$$G(s) = \ln(1) + \ln\left(\frac{s-a}{s-b}\right)$$

$$G(s) = \ln\left(\frac{s-a}{s-b}\right)$$

Therefore,

$$\frac{e^{bt} - e^{at}}{t} u(t) \xleftrightarrow{\mathcal{L}_u} \ln\left(\frac{s-a}{s-b}\right) \tag{9.68}$$

9.12.9 Time Shifting

If

$$x(t) \xleftrightarrow{\mathcal{L}_{u}} X(s)$$

then

$$x(t - t_0) \xleftrightarrow{\mathcal{L}_{u}} X(s)\, e^{-st_0} \qquad\qquad (9.69)$$

Observe that $x(t)$ starts at $t = 0$, and, therefore, $x(t - t_0)$ starts at $t = t_0$. So, this property may be restated as follows. If

$$x(t)u(t) \xleftrightarrow{\mathcal{L}_{u}} X(s)$$

then

$$x(t - t_0)u(t - t_0) \xleftrightarrow{\mathcal{L}_{u}} X(s)\, e^{-st_0} \qquad\qquad (9.70)$$

Proof By definition

$$\mathcal{L}_{u}\left[x(t - t_0)u(t - t_0)\right] = \int_{0^-}^{\infty} x(t - t_0)u(t - t_0)\, e^{-st}\, dt$$

A change of variables is performed by letting $\tau = t - t_0$, which also yields $d\tau = dt$, $\tau \to -t_0$ as $t \to 0^-$, and $\tau \to \infty$ as $t \to \infty$. Therefore,

$$\mathcal{L}[x(t - t_0)u(t - t_0)] = \int_{-t_0}^{\infty} x(\tau)u(\tau)\, e^{-s(\tau + t_0)}\, d\tau$$

Since

$$u(\tau) = \begin{cases} 1, & \tau > 0 \\ 0, & \tau < 0 \end{cases}$$

we obtain

$$\mathcal{L}[x(t - t_0)u(t - t_0)] = \int_{0}^{\infty} x(\tau)\, e^{-s(\tau + t_0)}\, d\tau$$

$$= e^{-st_0} \int_{0}^{\infty} x(\tau)\, e^{-s\tau}\, d\tau$$

$$\mathcal{L}[x(t - t_0)u(t - t_0)] = X(s)\, e^{-st_0} \qquad\qquad \blacksquare$$

9.12.10 Differentiation in the Time Domain

If

$$x(t) \xleftrightarrow{\mathcal{L}_{u}} X(s)$$

then

$$\frac{dx(t)}{dt} \xleftrightarrow{\mathcal{L}_u} sX(s) - x(0^-) \tag{9.71}$$

Repeated application of this property yields

$$\frac{d^2x(t)}{dt^2} \xleftrightarrow{\mathcal{L}_u} s^2X(s) - sx(0^-) - \dot{x}(0^-) \tag{9.72}$$

where $\dot{x}(0^-) = dx(t)/dt\big|_{t=0^-}$. In general,

$$\frac{d^nx(t)}{dt^n} \xleftrightarrow{\mathcal{L}_u} s^nX(s) - \sum_{k=1}^{n} s^{n-k}x^{(k-1)}(0^-) \tag{9.73}$$

where $x^{(r)}(0^-) = \dfrac{dx^r(t)}{dt^r}\bigg|_{t=0^-}$.

Proof From Eq. (9.54), we have

$$\mathcal{L}_u\left[\frac{dx(t)}{dt}\right] = \int_{0^-}^{\infty} \frac{dx(t)}{dt}\, e^{-st}\, dt$$

Integrating by parts, we obtain

$$\mathcal{L}_u\left[\frac{dx(t)}{dt}\right] = x(t)\, e^{-st}\bigg|_{0^-}^{\infty} + s\int_{0^-}^{\infty} x(t)\, e^{-st}\, dt$$

For the Laplace transform $X(s)$ to exist, it is necessary that $x(t)e^{-st} \to 0$ as $t \to \infty$ for $\Re\{s\} > 0$. Thus,

$$\mathcal{L}_u\left[\frac{dx(t)}{dt}\right] = [0 - x(0^-)] + sX(s)$$

$$\mathcal{L}_u\left[\frac{dx(t)}{dt}\right] = sX(s) - x(0^-)$$

9.12.11 Integration in the Time Domain
If

$$x(t) \xleftrightarrow{\mathcal{L}_u} X(s)$$

then

$$\int_{0^-}^{t} x(\tau)\, d\tau \xleftrightarrow{\mathcal{L}_u} \frac{X(s)}{s} \qquad \text{for } x(t) = 0, \quad t < 0 \tag{9.74}$$

and

$$\int_{-\infty}^{t} x(\tau)\, d\tau \xleftrightarrow{\mathcal{L}_u} \frac{X(s)}{s} + \frac{\int_{-\infty}^{0^-} x(\tau)\, d\tau}{s} \qquad (9.75)$$

Proof We know that if $x(t) = 0$ for $t < 0$, then

$$\int_{0^-}^{\infty} x(\tau)\, d\tau = x(t) * u(t)$$

Taking the unilateral Laplace transform on both sides of the above equation, we obtain

$$\mathcal{L}_u \left[\int_{0^-}^{\infty} x(\tau)\, d\tau \right] = \mathcal{L}_u[x(t) * u(t)]$$

$$= \frac{X(s)}{s}$$

Therefore,

$$\int_{0^-}^{\infty} x(\tau)\, d\tau \xleftrightarrow{\mathcal{L}_u} \frac{X(s)}{s}$$

To prove Eq. (9.75), observe that

$$\int_{-\infty}^{t} x(\tau)\, d\tau = \int_{-\infty}^{0^-} x(\tau)\, d\tau + \int_{0^-}^{t} x(\tau)\, d\tau$$

Note that the first term on the RHS is a constant for $t \geq 0$. Taking the Laplace transform of the above equation, we obtain

$$\mathcal{L}_u \left[\int_{-\infty}^{t} x(\tau)\, d\tau \right] = \mathcal{L}_u \left[\int_{-\infty}^{0^-} x(\tau)\, d\tau \right] + \mathcal{L}_u \left[\int_{0^-}^{t} x(\tau)\, d\tau \right]$$

$$= \frac{\int_{-\infty}^{0^-} x(\tau)\, d\tau}{s} + \frac{X(s)}{s}$$

Therefore,

$$\int_{-\infty}^{t} x(\tau)\, d\tau \xleftrightarrow{\mathcal{L}_u} \frac{X(s)}{s} + \frac{\int_{-\infty}^{0^-} x(\tau)\, d\tau}{s}$$

9.12.12 Initial-value Theorem

The initial-value theorem allows us to determine the initial value $x(0^+)$ of $x(t)$ directly from $X(s)$. The theorem states that if $x(t) = 0$, for $t < 0$, and if it contains no impulses or higher-order singularities at the origin, then

$$x(0^+) = \lim_{s \to \infty} sX(s) \qquad (9.76)$$

The initial-value theorem does not apply to rational functions $X(s)$ in which the order of the numerator polynomial is greater than or equal to that of the denominator polynomial.

Proof Method 1   Since $x(t) = 0$ for $t < 0$, $x(t) = x(t)u(t)$. Now, expanding $x(t)$ as a Taylor series at $t = 0^+$, we obtain

$$x(t) = \left[x(0^+) + x^{(1)}(0^+)t + \cdots + x^{(n)}(0^+)\frac{t^n}{n!} + \cdots \right] u(t)$$

where $x^{(n)}(0^+) = d^n x(t)/dt^n \big|_{t=0^+}$. Taking the Laplace transform of both sides yields

$$X(s) = \frac{x(0^+)}{s} + \frac{x^{(1)}(0^+)}{s^2} + \cdots + \frac{x^{(n)}(0^+)}{s^{n+1}} + \cdots \qquad (9.77)$$

Multiplying both sides of Eq. (9.77) by s and taking the limit as $s \to \infty$, we obtain

$$\lim_{s \to \infty} sX(s) = x(0^+) + \lim_{s \to \infty} \frac{x^{(1)}(0^+)}{s} + \cdots + \lim_{s \to \infty} \frac{x^{(n)}(0^+)}{s^n} + \cdots$$

$$\lim_{s \to \infty} sX(s) = x(0^+)$$

As a generalization, multiplying both sides by s^{n+1} and taking the limit as $s \to \infty$ yields

$$x^{(n)}(0^+) = \lim_{s \to \infty} [s^{n+1}X(s) - s^n x(0^+) - s^{n-1}x^{(1)}(0^+) - \cdots - sx^{(n-1)}(0^+)] \qquad (9.78)$$

This more general form of the initial-value theorem is simplified if $x^{(n)}(0^+) = 0$ for $n < N$. In that case,

$$x^{(N)}(0^+) = \frac{dx^N(t)}{dt^N}\bigg|_{t=0^+} = \lim_{s \to \infty} s^{N+1}X(s) \qquad (9.79)$$

This property is useful, since it allows us to compute the initial value of the signal $x(t)$ and its derivative directly from the Laplace transform $X(s)$ without having to find the inverse $x(t)$.

Method 2   Taking the unilateral Laplace transform of $dx(t)/dt$, we obtain

$$\mathcal{L}_{\mathrm{u}}\left[\frac{dx(t)}{dt}\right] = \int_{0^-}^{\infty} \frac{dx(t)}{dt} e^{-st} \, dt$$

Using differentiation in the time-domain property [Eq. (9.71)], we obtain

$$sX(s) - x(0^-) = \int_{0^-}^{\infty} \frac{dx(t)}{dt} e^{-st} \, dt$$

$$= \int_{0^-}^{0^+} \frac{dx(t)}{dt} e^{-st} \, dt + \int_{0^+}^{\infty} \frac{dx(t)}{dt} e^{-st} \, dt$$

$$= x(t) \Big|_{0^-}^{0^+} + \int_{0^+}^{\infty} \frac{dx(t)}{dt} e^{-st} \, dt$$

$$sX(s) - x(0^-) = x(0^+) - x(0^-) + \int_{0^+}^{\infty} \frac{dx(t)}{dt} e^{-st} \, dt$$

$$sX(s) = x(0^+) + \int_{0^+}^{\infty} \frac{dx(t)}{dt} e^{-st} \, dt$$

$$\lim_{s \to \infty} sX(s) = x(0^+) + \lim_{s \to \infty} \int_{0^+}^{\infty} \frac{dx(t)}{dt} e^{-st} \, dt$$

$$\lim_{s \to \infty} sX(s) = x(0^+)$$

■

Example 9.48 Use the initial-value theorem to find the initial value of the signal corresponding to the Laplace transform

$$X(s) = \frac{s+1}{s(s+2)}$$

Verify that the answer obtained is correct.

Solution
From the initial-value theorem [Eq. (9.76)], we have

$$x(0^+) = \lim_{s \to \infty} sX(s) = \lim_{s \to \infty} s \left(\frac{s+1}{s(s+2)} \right)$$

$$= \lim_{s \to \infty} \left(\frac{s+1}{s+2} \right)$$

$$x(0^+) = \lim_{s \to \infty} \left(\frac{1 + (1/s)}{1 + (2/s)} \right)$$

$$x(0^+) = 1$$

Using partial fraction expansion, the Laplace transform can be written as

$$X(s) = \frac{1/2}{s} + \frac{1/2}{s+2}$$

The inverse transform of this equation yields

$$x(t) = \frac{1}{2}u(t) + \frac{1}{2}e^{-2t}u(t)$$

Therefore,

$$x(t)|_{t=0} = x(0^+) = \frac{1}{2} + \frac{1}{2} = 1$$

9.12.13 Final-value Theorem

The final-value theorem allows us to determine the final value $x(\infty)$ of $x(t)$ directly from $X(s)$. The theorem states that if $x(t) = 0$, for $t < 0$, and if it contains no impulses or higher-order singularities at the origin, then

$$\lim_{t\to\infty} x(t) = x(\infty) = \lim_{s\to 0} sX(s) \qquad (9.80)$$

The final-value theorem is applicable only if the poles of $sX(s)$ are in the left half of the s-plane, with at most a single pole at $s = 0$. If $X(s)$ has a pole in the right half of the s-plane, $x(t)$ contains an exponentially growing term and $x(\infty)$ does not exist. If there is a pole on the imaginary axis, then $x(t)$ contains an oscillating term and $x(\infty)$ does not exist. However, if there is a pole at the origin, then $x(t)$ contains a constant term, and hence $x(\infty)$ exists and is a constant.

Proof Taking the unilateral Laplace transform of $dx(t)/dt$, we obtain

$$\mathcal{L}_u\left[\frac{dx(t)}{dt}\right] = \int_{0^-}^{\infty} \frac{dx(t)}{dt}\, e^{-st}\, dt$$

Using differentiation in the time-domain property [Eq. (9.71)], we obtain

$$sX(s) - x(0^-) = \int_{0^-}^{\infty} \frac{dx(t)}{dt}\, e^{-st}\, dt$$

$$\lim_{s\to 0}[sX(s) - x(0^-)] = \lim_{s\to 0} \int_{0^-}^{\infty} \frac{dx(t)}{dt}\, e^{-st}\, dt$$

$$= x(t)\Big|_{0^-}^{\infty}$$

$$\lim_{s\to 0} sX(s) - x(0^-) = x(\infty) - x(0^-)$$

$$\lim_{s\to 0} sX(s) = x(\infty)$$

Example 9.49 The final value of $x(t) = [2 + e^{-3t}]u(t)$ is obviously $x(\infty) = 2$. Show that this final value can be found with the final-value theorem.

Solution
Given that

$$x(t) = [2 + e^{-3t}]u(t)$$

The Laplace transform of the above equation yields

$$X(s) = \frac{2}{s} + \frac{1}{s+3}$$

Using the final-value theorem [Eq. (9.80)], we have

$$\lim_{t \to \infty} x(t) = x(\infty) = \lim_{s \to 0} sX(s) = \lim_{s \to 0} s\left[\frac{2}{s} + \frac{1}{s+3}\right]$$

$$x(\infty) = \lim_{s \to 0}\left[2 + \frac{s}{s+3}\right]$$

$$x(\infty) = 2$$

Example 9.50 Determine the initial and final values of $x(t)$ if its Laplace transform is given by

$$X(s) = \frac{10(2s+3)}{s(s^2 + 2s + 5)}$$

Solution
From the initial-value theorem [Eq. (9.76)], we have

$$x(0^+) = \lim_{s \to \infty} sX(s) = \lim_{s \to \infty} s\frac{10(2s+3)}{s(s^2 + 2s + 5)} = \lim_{s \to \infty} \frac{10(2s+3)}{s^2 + 2s + 5} = 0$$

From the final-value theorem [Eq. (9.80)], we have

$$\lim_{t \to \infty} x(t) = x(\infty) = \lim_{s \to 0} sX(s) = \lim_{s \to 0} s\left[\frac{10(2s+3)}{s(s^2 + 2s + 5)}\right] = \lim_{s \to 0}\left[\frac{10(2s+3)}{s^2 + 2s + 5}\right] = 6$$

Example 9.51 Find the final value of the response $y(t)$ of a system whose transfer function is

$$H(s) = \frac{s+3}{s^2 + 4s + 5}$$

when the system is excited by
  (a) a unit step, i.e., $x(t) = u(t)$
  (b) a unit impulse, i.e., $x(t) = \delta(t)$

Solution

(a) Given that $x(t) = u(t)$, so $X(s) = 1/s$. We know that [Eq. (9.50)]

$$Y(s) = X(s)H(s) = \frac{1}{s}\frac{s+3}{s^2+4s+5}$$

Since all the poles of $sY(s)$ lie in the left half of the s-plane, the final-value theorem is applicable. From the final-value theorem [Eq. (9.80)], we have

$$\lim_{t\to\infty} y(t) = y(\infty) = \lim_{s\to 0} sY(s) = \lim_{s\to 0} s\left[\frac{1}{s}\frac{s+3}{s^2+4s+5}\right]$$

$$y(\infty) = \lim_{s\to 0}\left[\frac{s+3}{s^2+4s+5}\right]$$

$$y(\infty) = \frac{3}{5}$$

(b) Given that $x(t) = \delta(t)$, so $X(s) = 1$. We know that [Eq. (9.50)]

$$Y(s) = X(s)H(s) = \frac{s+3}{s^2+4s+5}$$

Since all the poles of $sY(s)$ lie in the left half of the s-plane, the final-value theorem is applicable. From the final-value theorem [Eq. (9.80)], we have

$$\lim_{t\to\infty} y(t) = y(\infty) = \lim_{s\to 0} sY(s)$$

$$= \lim_{s\to 0} s\left[\frac{s+3}{s^2+4s+5}\right]$$

$$y(\infty) = 0$$

9.13 SOLUTION OF DIFFERENTIAL AND INTEGRO-DIFFERENTIAL EQUATIONS

One of the most common applications of the Laplace transform is to solve linear constant-coefficient differential equations. Such equations are used to model continuous-time LTI systems. Solving of these equations depends on the differentiation property of the Laplace transform. Because $dy^n(t)/dt^n \longleftrightarrow s^n Y(s)$, the Laplace transform of a differential equation is an algebraic equation that can be readily solved for $Y(s)$. The procedure is straightforward and systematic, and we summarize it in the following three steps:

1. For a given set of initial conditions, take the Laplace transform of both sides of the differential equation to obtain an algebraic equation in $Y(s)$.
2. Solve the algebraic equation for $Y(s)$.
3. Take the inverse Laplace transform to obtain $y(t)$.

Example 9.52 Solve the second-order linear differential equation

$$\frac{d^2y(t)}{dt^2} + 5\frac{dy(t)}{dt} + 6y(t) = \frac{dx(t)}{dt} + x(t)$$

for the initial conditions $y(0^-) = 2$ and $dy(t)/dt\big|_{t=0^-} = \dot{y}(0^-) = 1$ and the input $x(t) = e^{-4t}u(t)$.

Solution
If $x(t) = e^{-4t}u(t)$, then $x(0^-) = 0$ and

$$X(s) = \frac{1}{s+4}$$

Now, consider the given differential equation

$$\frac{d^2y(t)}{dt^2} + 5\frac{dy(t)}{dt} + 6y(t) = \frac{dx(t)}{dt} + x(t)$$

Taking the Laplace transform of the above equation, we obtain

$$[s^2Y(s) - sy(0^-) - \dot{y}(0^-)] + 5[sY(s) - y(0^-)] + 6Y(s) = [sX(s) - x(0^-)] + X(s)$$

$$[s^2Y(s) - 2s - 1] + 5[sY(s) - 2] + 6Y(s) = [sX(s) - 0] + X(s)$$

$$(s^2 + 5s + 6)Y(s) - (2s + 11) = X(s)(s + 1)$$

$$(s^2 + 5s + 6)Y(s) = (2s + 11) + \frac{s+1}{s+4}$$

$$(s + 2)(s + 3)Y(s) = \frac{2s^2 + 20s + 45}{s+4}$$

$$Y(s) = \frac{2s^2 + 20s + 45}{(s+2)(s+3)(s+4)}$$

Expanding the RHS into partial fractions yields

$$Y(s) = \frac{13/2}{s+2} - \frac{3}{s+3} - \frac{3/2}{s+4}$$

Taking the inverse Laplace transform of the above equation, we obtain

$$y(t) = \left[\frac{13}{2}e^{-2t} - 3e^{-3t} - \frac{3}{2}e^{-4t}\right]u(t)$$

9.13.1 Zero-input Response and Zero-state Response

The Laplace transform method for solving differential equations offers a clear separation between the *zero-input response* of the system to initial conditions and the *zero-state response* of the system associated with the input. *The zero-input response represents the system output when the input is zero, i.e., $x(t) = 0$, and thus it is the result of internal system conditions (such as energy storage, initial conditions) alone. In contrast, the* *zero-state response* represents the system response to the external input $x(t)$ when the initial conditions are zero. The total response is defined as

$$\text{Total response} = \text{Zero-input response} + \text{Zero-state response} \tag{9.81}$$

Natural Response and Forced Reponse: In total response, the response due to the poles of the system is called *natural response* of the system. The response due to the poles of the input is called the *forced response* of the system.

Example 9.53 Use the unilateral Laplace transform to determine the output of a system represented by the differential equation

$$\frac{d^2y(t)}{dt^2} + 5\frac{dy(t)}{dt} + 6y(t) = \frac{dx(t)}{dt} + 6x(t)$$

in response to the input $x(t) = u(t)$. Assume that the initial conditions on the system are $y(0^-) = 1$ and $\dot{y}(0^-) = 2$. Identify the zero-state response $y_{zs}(t)$, of the system, and the zero-input response $y_{zi}(t)$.

Solution
If $x(t) = u(t)$, then $x(0^-) = 0$ and $X(s) = 1/s$. Now, consider the given differential equation

$$\frac{d^2y(t)}{dt^2} + 5\frac{dy(t)}{dt} + 6y(t) = \frac{dx(t)}{dt} + 6x(t)$$

Taking the Laplace transform of the above equation, we obtain

$$[s^2Y(s) - sy(0^-) - \dot{y}(0^-)] + 5[sY(s) - y(0^-)] + 6Y(s) = [sX(s) - x(0^-)] + 6X(s)$$

$$[s^2Y(s) - s - 2] + 5[sY(s) - 1] + 6Y(s) = [sX(s) - 0] + 6X(s)$$

$$(s^2 + 5s + 6)Y(s) - (s + 7) = X(s)(s + 6)$$

$$(s^2 + 5s + 6)Y(s) = \underbrace{(s + 7)}_{\text{initial condition terms}} + \underbrace{X(s)(s + 6)}_{\text{input terms}}$$

$$Y(s) = \underbrace{\frac{s + 7}{s^2 + 5s + 6}}_{\text{zero-input component}} + \underbrace{\frac{s + 6}{s(s^2 + 5s + 6)}}_{\text{zero-state component}}$$

$$Y(s) = \frac{s + 7}{(s + 2)(s + 3)} + \frac{s + 6}{s(s + 2)(s + 3)}$$

Using partial fraction expansion, we obtain

$$Y(s) = \left[\frac{5}{s + 2} - \frac{4}{s + 3}\right] + \left[\frac{1}{s} - \frac{2}{s + 2} + \frac{1}{s + 3}\right]$$

The inverse transform of this equation yields

$$y(t) = [5 e^{-2t} u(t) - 4 e^{-3t} u(t)] + [u(t) - 2 e^{-2t} u(t) + e^{-3t} u(t)]$$

$$y(t) = \underbrace{[5 e^{-2t} - 4 e^{-3t}] u(t)}_{\text{Zero-input response or natural response}} + \underbrace{[1 - 2 e^{-2t} + e^{-3t}] u(t)}_{\text{Zero-state response or forced response}}$$

$$y(t) = y_{zi}(t) + y_{zs}(t)$$

where the zero-input response is

$$y_{zi}(t) = [5 e^{-2t} - 4 e^{-3t}] u(t)$$

and the zero-state response is

$$y_{zs}(t) = [1 - 2 e^{-2t} + e^{-3t}] u(t)$$

Example 9.54 Determine the forced response and natural response of an LTI system described by the following differential equation with specified input and initial conditions:

$$\frac{d^2 y(t)}{dt^2} + 5 \frac{dy(t)}{dt} + 6y(t) = -4x(t) - 3 \frac{dx(t)}{dt} \quad x(t) = e^{-t} u(t)$$

$$y(0^-) = -1 \quad \dot{y}(0^-) = 5$$

Solution
Consider the given differential equation

$$\frac{d^2 y(t)}{dt^2} + 5 \frac{dy(t)}{dt} + 6y(t) = -4x(t) - 3 \frac{dx(t)}{dt},$$

Taking the bilateral Laplace transform of the above equation to get the system function $H(s) = \dfrac{Y(s)}{X(s)}$.

$$s^2 Y(s) + 5sY(s) + 6Y(s) = -4X(s) - 3sX(s)$$

$$(s^2 + 5s + 6)Y(s) + s = -X(s)(4 + 3s)$$

$$H(s) = \frac{Y(s)}{X(s)} = -\frac{4 + 3s}{s^2 + 5s + 6} = -\frac{4 + 3s}{(s + 2)(s + 3)}.$$

The system function has poles at $p_1 = -2$, and $p_2 = -3$. The response due to these poles is known as natural response.

If $x(t) = e^{-t} u(t)$, then $x(0^-) = 0$ and

$$X(s) = \frac{1}{s + 1}$$

Now, consider the given differential equation

$$\frac{d^2 y(t)}{dt^2} + 5 \frac{dy(t)}{dt} + 6y(t) = -4x(t) - 3 \frac{dx(t)}{dt}$$

Taking the Laplace transform of the above equation, we obtain

$$[s^2Y(s) - sy(0^-) - \dot{y}(0^-)] + 5[sY(s) - y(0^-)] + 6Y(s) = -4X(s) - 3[sX(s) - x(0^-)]$$

$$[s^2Y(s) + s - 5] + 5[sY(s) + 1] + 6Y(s) = -4X(s) - 3[sX(s) - 0]$$

$$(s^2 + 5s + 6)Y(s) + s = -X(s)(4 + 3s)$$

$$(s^2 + 5s + 6)Y(s) = \underbrace{-s}_{\text{Initial condition terms}} \quad \underbrace{-X(s)(4 + 3s)}_{\text{Input terms}}$$

$$Y(s) = \underbrace{-\frac{s}{s^2 + 5s + 6}}_{\text{Zero-input response}} \quad \underbrace{-\frac{4 + 3s}{(s+1)(s^2 + 5s + 6)}}_{\text{Zero-state response}}$$

$$Y(s) = -\frac{s}{(s+2)(s+3)} - \frac{4 + 3s}{(s+1)(s+2)(s+3)}$$

Using partial fraction expansion, we obtain

$$Y(s) = \left[\frac{2}{s+2} - \frac{3}{s+3}\right] + \left[-\frac{1/2}{s+1} - \frac{2}{s+2} + \frac{5/2}{s+3}\right]$$

The inverse transform of this equation yields

$$y(t) = [2\,e^{-2t}u(t) - 3\,e^{-3t}u(t)] + \left[-\frac{1}{2}\,e^{-t}u(t) - 2\,e^{-2t}u(t) + \frac{5}{2}\,e^{-3t}u(t)\right]$$

$$y(t) = \underbrace{[2\,e^{-2t} - 3\,e^{-3t}]u(t)}_{\text{Natural response}} + \underbrace{\left[-\frac{1}{2}\,e^{-t} - 2\,e^{-2t} + \frac{5}{2}\,e^{-3t}\right]u(t)}_{\text{Forced response}}$$

$$y(t) = y_n(t) + y_f(t)$$

where the natural response is

$$y_n(t) = [2\,e^{-2t} - 3\,e^{-3t}]u(t)$$

and the forced response is

$$y_f(t) = \left[-\frac{1}{2}\,e^{-t} - 2\,e^{-2t} + \frac{5}{2}\,e^{-3t}\right]u(t)$$

Example 9.55 Consider the system S characterized by the differential equation

$$\frac{d^3y(t)}{dt^3} + 6\frac{d^2y(t)}{dt^2} + 11\frac{dy(t)}{dt} + 6y(t) = x(t)$$

(a) Determine the zero-state response of this system for the input $x(t) = e^{-4t}u(t)$.
(b) Determine the zero-input response of the system for $t > 0^-$, given that

$$y(0^-) = 1, \quad \dot{y}(0^-) = -1, \quad \ddot{y}(0^-) = 1$$

(c) Determine the total response of the system when the input is $x(t) = e^{-4t}u(t)$ and the initial conditions are the same as those specified in part (b).

Solution

(a) The zero-state response represents the system response to the external input $x(t)$ when the initial conditions are zero.

Given that $x(t) = e^{-4t}u(t)$, its Laplace transform $X(s) = 1/(s+4)$. Now, consider the given differential equation

$$\frac{d^3y(t)}{dt^3} + 6\frac{d^2y(t)}{dt^2} + 11\frac{dy(t)}{dt} + 6y(t) = x(t)$$

Taking the Laplace transform of the given differential equation, we obtain

$$s^3Y(s) + 6s^2Y(s) + 11sY(s) + 6Y(s) = X(s)$$

$$Y(s)[s^3 + 6s^2 + 11s + 6] = \frac{1}{s+4}$$

$$Y(s) = \frac{1}{(s+4)(s^3 + 6s^2 + 11s + 6)}$$

$$Y(s) = \frac{1}{(s+1)(s+2)(s+3)(s+4)}$$

Using partial fraction expansion, we obtain

$$Y(s) = \frac{1/6}{s+1} + \frac{1/2}{s+2} - \frac{1/2}{s+3} - \frac{1/6}{s+4}$$

Taking the inverse Laplace transform, we get the zero-state response

$$y_{zs}(t) = \frac{1}{6}e^{-t}u(t) + \frac{1}{2}e^{-2t}u(t) - \frac{1}{2}e^{-3t}u(t) - \frac{1}{6}e^{-4t}u(t)$$

(b) The zero-input response represents the system output when the input is zero, i.e., $x(t) = 0$ or, equivalently, $X(s) = 0$. Now, consider the given differential equation

$$\frac{d^3y(t)}{dt^3} + 6\frac{d^2y(t)}{dt^2} + 11\frac{dy(t)}{dt} + 6y(t) = x(t)$$

Taking the Laplace transform of the given differential equation, we obtain

$$[s^3Y(s) - s^2y(0^-) - s\dot{y}(0^-) - \ddot{y}(0^-)] + 6[s^2Y(s) - sy(0^-) - \dot{y}(0^-)]$$
$$+ 11[sY(s) - y(0^-)] + 6Y(s) = X(s)$$

$$[s^3Y(s) - s^2 + s - 1] + 6[s^2Y(s) - s + 1] + 11[sY(s) - 1] + 6Y(s) = 0$$

$$Y(s)[s^3 + 6s^2 + 11s + 6] - s^2 - 5s - 6 = 0$$

$$Y(s)[s^3 + 6s^2 + 11s + 6] = s^2 + 5s + 6$$

$$Y(s) = \frac{s^2 + 5s + 6}{s^3 + 6s^2 + 11s + 6}$$

$$Y(s) = \frac{(s+2)(s+3)}{(s+1)(s+2)(s+3)}$$

$$Y(s) = \frac{1}{s+1}$$

Taking the inverse Laplace transform, we get the zero-input response

$$y_n(t) = e^{-t}u(t)$$

(c) The total response is the sum of the zero-state and zero-input responses.

$$y(t) = y_f(t) + y_n(t) = \frac{1}{6}\,e^{-t}u(t) + \frac{1}{2}\,e^{-2t}u(t) - \frac{1}{2}\,e^{-3t}u(t) - \frac{1}{6}\,e^{-4t}u(t) + e^{-t}u(t)$$

$$y(t) = \frac{7}{6}\,e^{-t}u(t) + \frac{1}{2}\,e^{-2t}u(t) - \frac{1}{2}\,e^{-3t}u(t) - \frac{1}{6}\,e^{-4t}u(t)$$

9.14 BLOCK DIAGRAM REPRESENTATION

As we saw earlier, the Laplace transform is a useful tool for computing the system transfer function if the system is described by its differential equation or if the output is expressed explicitly in terms of the input. The situation changes considerably in cases where a large number of components or elements are interconnected to form the complete system. In such cases, it is convenient to represent the system by suitably interconnected subsystems, each of which can be separately and easily analysed. A linear system can be characterized by its transfer function $H(s)$. Figure 9.25(a) shows a block diagram of a system with a transfer function $H(s)$ and its input and output $X(s)$ and $Y(s)$, respectively. Subsystems may be interconnected by using cascade (series), parallel, and feedback interconnections, the three elementary types.

9.14.1 Cascade Interconnection

When the systems are connected in series, as depicted in Fig. 9.25(b), then the transfer function of the overall system is the product of the individual transfer functions. This result can also be proved by observing that in Fig. 9.25(b)

$$H(s) = \frac{Y(s)}{X(s)} = \frac{W(s)}{X(s)}\frac{Y(s)}{W(s)} = H_1(s)H_2(s)$$

We can extend this result to any number of systems connected in cascade. It follows from this discussion that the systems connected in cascade can be interchanged without affecting the overall transfer function. This commutation property of LTI systems follows from the commutative and associative properties of convolution (see Section 2.3).

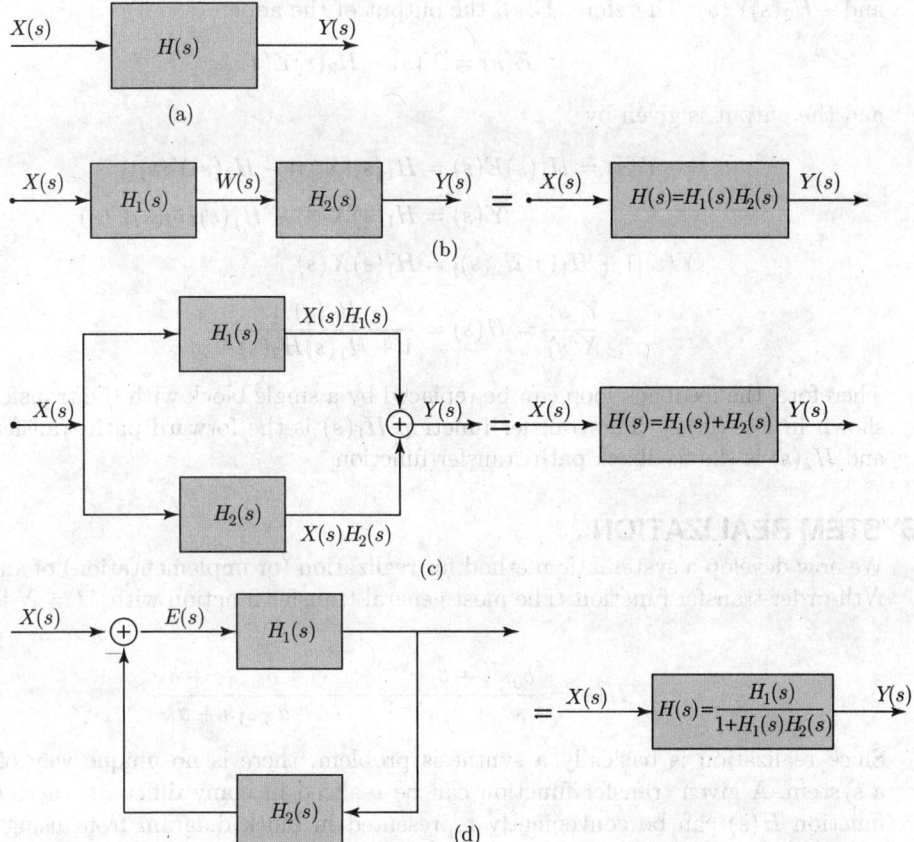

Fig. 9.25 Elementary connection of block diagrams

9.14.2 Parallel Interconnection

When the systems are connected in parallel, as depicted in Fig. 9.25(c), then the transfer function of the overall system is the sum of the individual transfer functions. This result can also be proved by observing that in Fig. 9.25(c)

$$Y(s) = X(s)H_1(s) + X(s)H_2(s)$$
$$Y(s) = X(s)[H_1(s) + H_2(s)]$$
$$\frac{Y(s)}{X(s)} = H(s) = H_1(s) + H_2(s)$$

We can extend this result to any number of systems connected in parallel.

9.14.3 Feedback Interconnection

When the output is fed back to the input, as shown in Fig. 9.25(d), the overall transfer function $H(s) = Y(s)/X(s)$ can be computed as follows. The input to the adder are $X(s)$

and $-H_2(s)Y(s)$. Therefore, $E(s)$, the output of the adder, is

$$E(s) = X(s) - H_2(s)Y(s)$$

and the output is given by

$$Y(s) = H_1(s)E(s) = H_1(s)[X(s) - H_2(s)Y(s)]$$
$$Y(s) = H_1(s)X(s) - H_1(s)H_2(s)Y(s)$$
$$Y(s)[1 + H_1(s)H_2(s)] = H_1(s)X(s)$$
$$\frac{Y(s)}{X(s)} = H(s) = \frac{H_1(s)}{1 + H_1(s)H_2(s)} \tag{9.82}$$

Therefore, the feedback loop can be replaced by a single block with the transfer function shown in Eq. (9.82). The transfer function $H_1(s)$ is the forward path transfer function and $H_2(s)$ is the feedback path transfer function.

9.15 SYSTEM REALIZATION

We now develop a systematic method for realization (or implementation) of an arbitrary Nth-order transfer function. The most general transfer function with $M = N$ is given by

$$H(s) = \frac{b_0 s^N + b_1 s^{N-1} + \cdots + b_{N-1}s + b_N}{s^N + a_1 s^{N-1} + \cdots + a_{N-1}s + a_N}$$

Since realization is basically a synthesis problem, there is no unique way of realizing a system. A given transfer function can be realized in many different ways. A transfer function $H(s)$ can be conveniently represented in block diagram from using the basic building blocks representing the integrators or differentiators, the multiplier, the adder, and the pick-off node as shown in Fig. 9.26.

We avoid use of differentiators because

1. a differentiator represents an unstable system because a bounded input like the step input $u(t)$ results in an unbounded output $\delta(t)$.

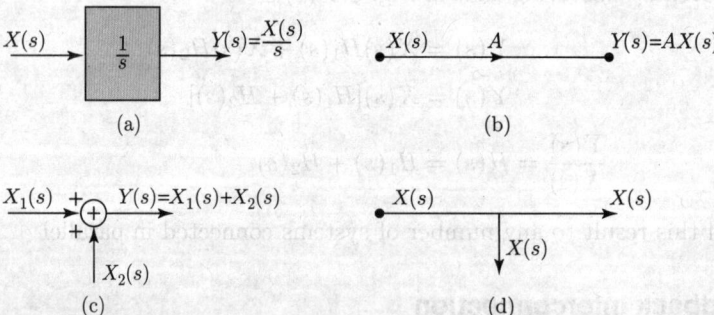

Fig. 9.26 (a) Basic building block representing (a) integrator, (b) multiplier, (c) adder, and (d) pick-off node

2. the noise is enhanced by a differentiator. Noise is a wideband signal containing components of all frequencies from 0 to a very high frequency approaching ∞. Hence, noise contains a significant amount of rapidly varying components and we know that the derivative of any rapidly varying signal is high.

There are several advantages in representing the system in block diagram form:

1. It is easy to write down the computational algorithm by inspection.
2. It is easy to analyse the block diagram to determine the explicit relation between the output and the input.
3. It is easy to manipulate a block diagram to derive other equivalent block diagrams yielding different computational algorithms.
4. It is easy to determine the hardware requirements.
5. It is easier to develop block diagram representations from the transfer function directly leading to a variety of equivalent representations.

9.15.1 Direct Form I Realization

An Nth-order transfer function is characterized by $2N + 1$ unique coefficients and, in general, requires $2N + 1$ multipliers and $2N$ two-input adders for implementation. Structures in which the multiplier coefficients are precisely the coefficients of the transfer function are called *direct form* structures.

Consider for simplicity a third-order system characterized by a transfer function

$$H(s) = \frac{b_0 s^3 + b_1 s^2 + b_2 s + b_3}{s^3 + a_1 s^2 + a_2 s + a_3}$$

$$H(s) = \frac{b_0 + (b_1/s) + (b_2/s^2) + (b_3/s^3)}{1 + (a_1/s) + (a_2/s^2) + (a_3/s^3)}$$

We can express $H(s)$ as

$$H(s) = \underbrace{\left(b_0 + \frac{b_1}{s} + \frac{b_2}{s^2} + \frac{b_3}{s^3}\right)}_{H_1(s)} \underbrace{\left(\frac{1}{1 + a_1/s + (a_2/s^2) + (a_3/s^3)}\right)}_{H_2(s)}$$

We can realize $H(s)$ as a cascade of transfer function $H_1(s)$ followed by $H_2(s)$, as depicted in Fig. 9.27(a), where

$$H_1(s) = \frac{W(s)}{X(s)} = b_0 + \frac{b_1}{s} + \frac{b_2}{s^2} + \frac{b_3}{s^3} \tag{9.83}$$

$$W(s) = \left(b_0 + \frac{b_1}{s} + \frac{b_2}{s^2} + \frac{b_3}{s^3}\right) X(s) \tag{9.84}$$

and
$$H_2(s) = \frac{Y(s)}{W(s)} = \frac{1}{1 + (a_1/s) + (a_2/s^2) + (a_3/s^3)} \tag{9.85}$$

$$Y(s) = W(s) - \left(\frac{a_1}{s} + \frac{a_2}{s^2} + \frac{a_3}{s^3}\right) Y(s) \tag{9.86}$$

We shall first realize $H_1(s)$ given by Eq. (9.83). Equation (9.84) shows that the output $W(s)$ can be synthesized by adding the input $b_0X(s)$ to $b_1X(s)/s$, $b_2X(s)/s^2$, and $b_3X(s)/s^3$. Because the transfer function of an integrator is $1/s$, the signals $X(s)/s$, $X(s)/s^2$, and $X(s)/s^3$ can be obtained by successive integration of the input $x(t)$. The left-half section of Fig. 9.27(b) shows the realization of $H_1(s)$.

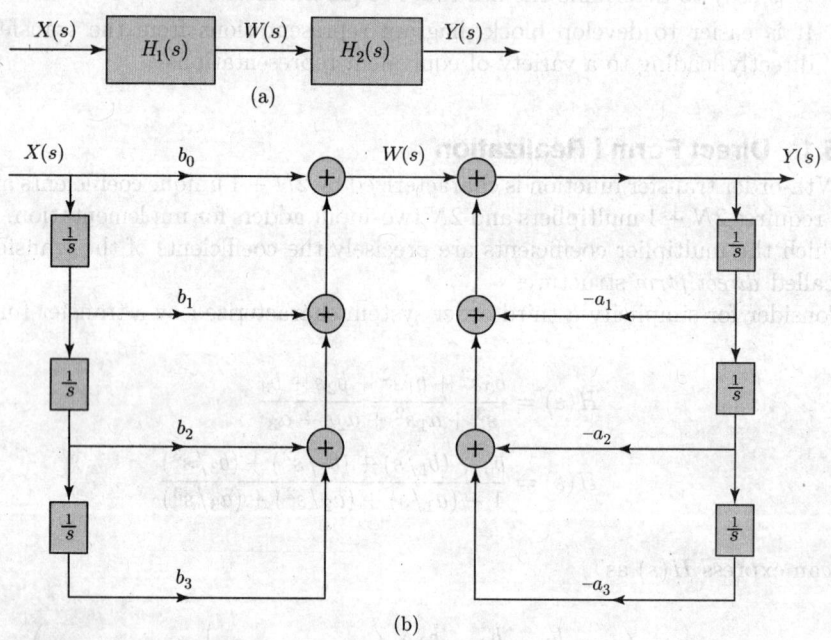

Fig. 9.27 (a) Realization of a function in two steps and (b) direct form I realization of a third-order continuous-time LTI system

We next consider the realization of $H_2(s)$ given by Eq. (9.85). Equation (9.86) shows that the output $Y(s)$ can be synthesized by subtracting $a_1Y(s)/s$, $a_2Y(s)/s^2$, and $a_3Y(s)/s^3$ from $W(s)$. To obtain signals $Y(s)/s$, $Y(s)/s^2$, and $Y(s)/s^3$, we assume that we already have the desired output $Y(s)$. Successive integration of $Y(s)$ yields the needed signals $Y(s)/s$, $Y(s)/s^2$, and $Y(s)/s^3$. The right-half section of Fig. 9.27(b) shows the realization of $H_2(s)$. We can generalize this procedure, known as the *direct form I* realization, for any value of N. This procedure requires $2N$ integrators to realize an Nth-order transfer function. This realization is *noncanonic* since it employs six integrators to implement a third-order transfer function. *A realization is canonic if the number of integrators used in the realization is equal to the order of the transfer function realized. Thus, canonic realization has no redundant integrators.*

9.15.2 Direct Form II (or Canonic) Realization

In the direct form I, we realize $H(s)$ by implementing $H_1(s)$ followed by $H_2(s)$, as shown in Fig. 9.27(a). We can also realize $H(s)$, as shown in Fig. 9.28(a), where $H_2(s)$ is followed by $H_1(s)$. This procedure is known as the *direct form II* realization.

(a)

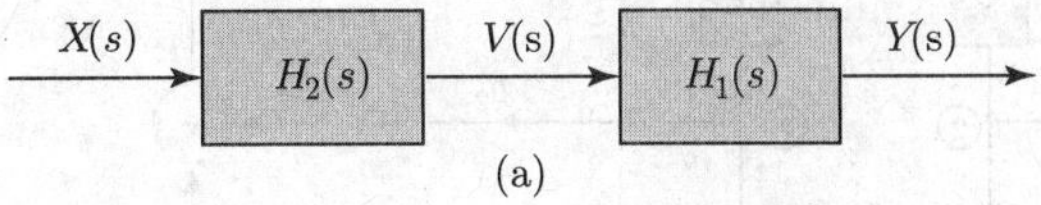

Fig. 9.28(a)

The direct form I realization [Fig. 9.27(b)] implements zeros first [the left-half section represented by $H_1(s)$] followed by realization of poles [the right-half section represented by $H_2(s)$] of $H(s)$. In contrast, the direct form II realization implements poles first followed by zeros.

Again consider for simplicity a third-order system characterized by a transfer function

$$H(s) = \frac{b_0 s^3 + b_1 s^2 + b_2 s + b_3}{s^3 + a_1 s^2 + a_2 s + a_3}$$

$$H(s) = \frac{b_0 + (b_1/s) + (b_2/s^2) + (b_3/s^3)}{1 + (a_1/s) + (a_2/s^2) + (a_3/s^3)}$$

We can express $H(s)$ as

$$H(s) = \underbrace{\left(\frac{1}{1 + (a_1/s) + (a_2/s^2) + (a_3/s^3)}\right)}_{H_2(s)} \underbrace{\left(b_0 + \frac{b_1}{s} + \frac{b_2}{s^2} + \frac{b_3}{s^3}\right)}_{H_1(s)}$$

We can realize $H(s)$ as a cascade of transfer function $H_2(s)$ followed by $H_1(s)$, as depicted in Fig. 9.28(a), where

$$H_2(s) = \frac{V(s)}{X(s)} = \frac{1}{1 + (a_1/s) + (a_2/s^2) + (a_3/s^3)} \tag{9.87}$$

$$V(s) = X(s) - \left(\frac{a_1}{s} + \frac{a_2}{s^2} + \frac{a_3}{s^3}\right) V(s) \tag{9.88}$$

and

$$H_1(s) = \frac{Y(s)}{V(s)} = b_0 + \frac{b_1}{s} + \frac{b_2}{s^2} + \frac{b_3}{s^3} \tag{9.89}$$

$$Y(s) = \left(b_0 + \frac{b_1}{s} + \frac{b_2}{s^2} + \frac{b_3}{s^3}\right) V(s) \tag{9.90}$$

The left-half section of Fig. 9.28(b) shows the realization of $H_2(s)$ and the right-half section shows the realization of $H_1(s)$.

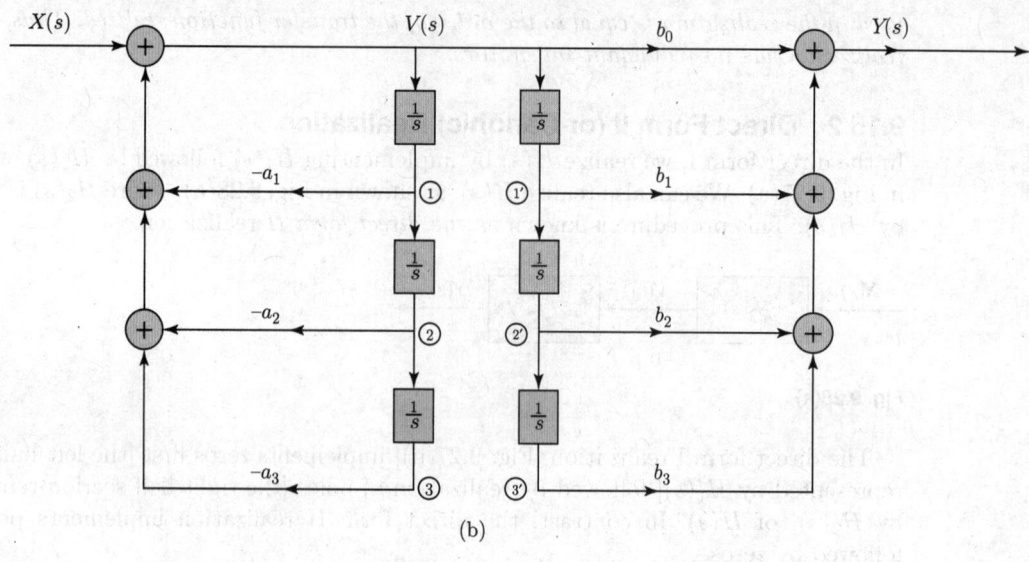

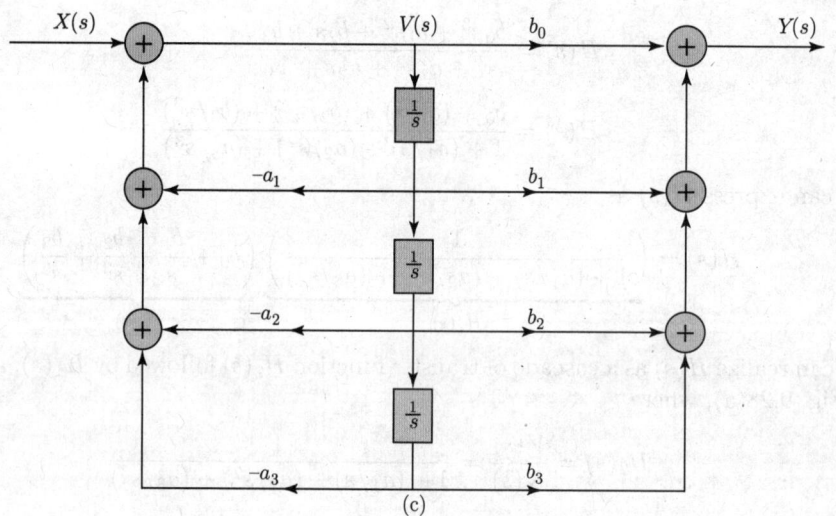

Fig. 9.28 Direct form II realization of a third-order LTI system

We observe that in Fig. 9.28(b), the signal variables at nodes 1 and 1' are the same, and hence the two top integrators can be shared. Likewise, the signal variables at nodes 2 and 2' are the same, which permits the sharing of the two middle integrators. Following the same argument, we can share the integrators, leading to the final structure shown in Fig. 9.28(c).

This implementation halves the number of integrators to N, and is thus more efficient in hardware utilization. This is the direct form II realization. This realization is *canonic* since it employs N integrators to implement an Nth-order transfer function.

Example 9.56 Find the canonic realization of the following transfer functions:

(a) $H(s) = \dfrac{5}{s+7}$

(b) $H(s) = \dfrac{s}{s+7}$

(c) $H(s) = \dfrac{s+5}{s+7}$

(d) $H(s) = \dfrac{4s+28}{s^2+6s+5}$

Solution
(a) Given that

$$H(s) = \frac{Y(s)}{X(s)} = \frac{V(s)}{X(s)}\frac{Y(s)}{V(s)} = \frac{5}{s+7} = \frac{\frac{5}{s}}{1+7/s} = \underbrace{\left(\frac{1}{1+7/s}\right)}_{H_2(s)}\underbrace{\left(\frac{5}{s}\right)}_{H_1(s)}$$

where

$$H_2(s) = \frac{V(s)}{X(s)} = \frac{1}{1+7/s}$$

$$V(s) = X(s) - 7\frac{V(s)}{s}$$

and

$$H_1(s) = \frac{Y(s)}{V(s)} = \frac{5}{s}$$

$$Y(s) = 5\frac{V(s)}{s}$$

The given transfer function $H(s) = 5/(s+7)$ is of the first order; therefore, we need only one integrator for its realization, as shown in Fig. 9.29(a).

(b) Given that

$$H(s) = \frac{s}{s+7}$$

$$H(s) = \frac{Y(s)}{X(s)} = \frac{1}{1+7/s}$$

$$Y(s) = X(s) - 7\frac{Y(s)}{s}$$

The given transfer function $H(s) = s/(s+7)$ is of the first order; therefore, we need only one integrator for its realization, as shown in Fig. 9.29(b).

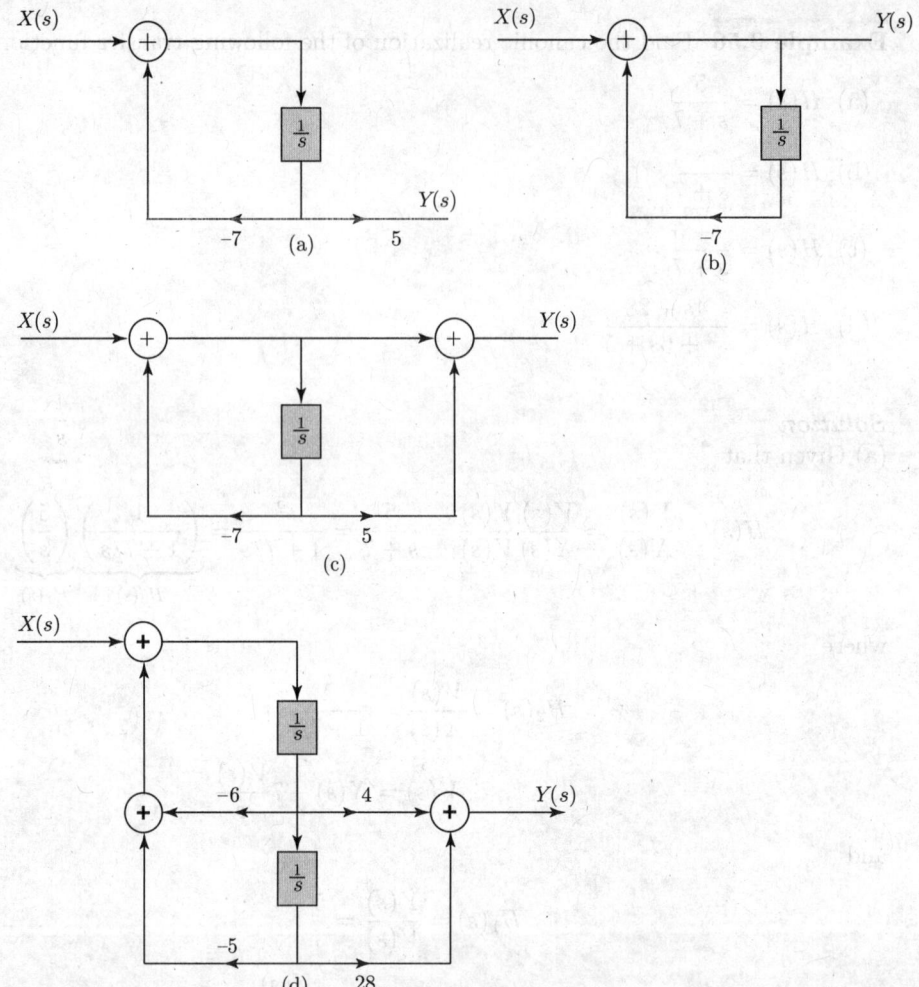

Fig. 9.29

(c) Given that

$$H(s) = \frac{Y(s)}{X(s)} = \frac{V(s)}{X(s)}\frac{Y(s)}{V(s)} = \frac{s+5}{s+7} = \frac{1+\frac{5}{s}}{1+\frac{7}{s}} = \underbrace{\left(\frac{1}{1+\frac{7}{s}}\right)}_{H_2(s)}\underbrace{\left(1+\frac{5}{s}\right)}_{H_1(s)}$$

where

$$H_2(s) = \frac{V(s)}{X(s)} = \frac{1}{1+\frac{7}{s}}$$

$$V(s) = X(s) - 7\frac{V(s)}{s}$$

and

$$H_1(s) = \frac{Y(s)}{V(s)} = 1 + \frac{5}{s}$$

$$Y(s) = V(s) + 5\frac{V(s)}{s}$$

The given transfer function $H(s) = (s+5)/(s+7)$ is of the first order; therefore, we need only one integrator for its realization, as shown in Fig. 9.29(c).

(d) Given that

$$H(s) = \frac{Y(s)}{X(s)} = \frac{V(s)}{X(s)}\frac{Y(s)}{V(s)} = \frac{4s + 28}{s^2 + 6s + 5} = \frac{\frac{4}{s} + \left(\frac{28}{s^2}\right)}{1 + \left(\frac{6}{s}\right) + \frac{5}{s^2}}$$

$$= \underbrace{\left(\frac{1}{1 + \left(\frac{6}{s}\right) + \left(\frac{5}{s^2}\right)}\right)}_{H_2(s)} \underbrace{\left(\frac{4}{s} + \frac{28}{s^2}\right)}_{H_1(s)}$$

where

$$H_2(s) = \frac{V(s)}{X(s)} = \frac{1}{1 + \left(\frac{6}{s}\right) + \left(\frac{5}{s^2}\right)}$$

$$V(s) = X(s) - 6\frac{V(s)}{s} - 5\frac{V(s)}{s^2}$$

and

$$H_1(s) = \frac{Y(s)}{V(s)} = \frac{4}{s} + \frac{28}{s^2}$$

$$Y(s) = 4\frac{V(s)}{s} + 28\frac{V(s)}{s^2}$$

The given transfer function $H(s) = \dfrac{4s + 28}{s^2 + 6s + 5}$ is of the second order; therefore, we need only two integrators for its realization, as shown in Fig. 9.29(d).

9.15.3 Cascade and Parallel Realization

By expressing the numerator polynomial $N(s)$ and the denominator polynomial $D(s)$ of the transfer function $H(s)$ as a product of polynomials of lower degree, a digital filter is often realized as a cascade of low-order systems. Consider, for example, $H(s) = N(s)/D(s)$ expressed as

$$H(s) = \frac{N_1(s)N_2(s)N_3(s)}{D_1(s)D_2(s)D_3(s)}$$

Various different cascade realizations of $H(s)$ can be obtained by different pole-zero polynomial pairings. Some examples of such realizations are shown in Fig. 9.30(a). Additional cascade realizations are obtained by simply changing the ordering of the systems. Figure 9.30(b) illustrates examples of different structures obtained by changing the ordering of the systems.

A system can be realized in a parallel form by making use of partial fraction expansion of the transfer function. From a practical viewpoint, parallel and cascade forms are preferable because parallel and certain cascade forms are numerically less sensitive than a canonic direct form to small parameter variations in the system.

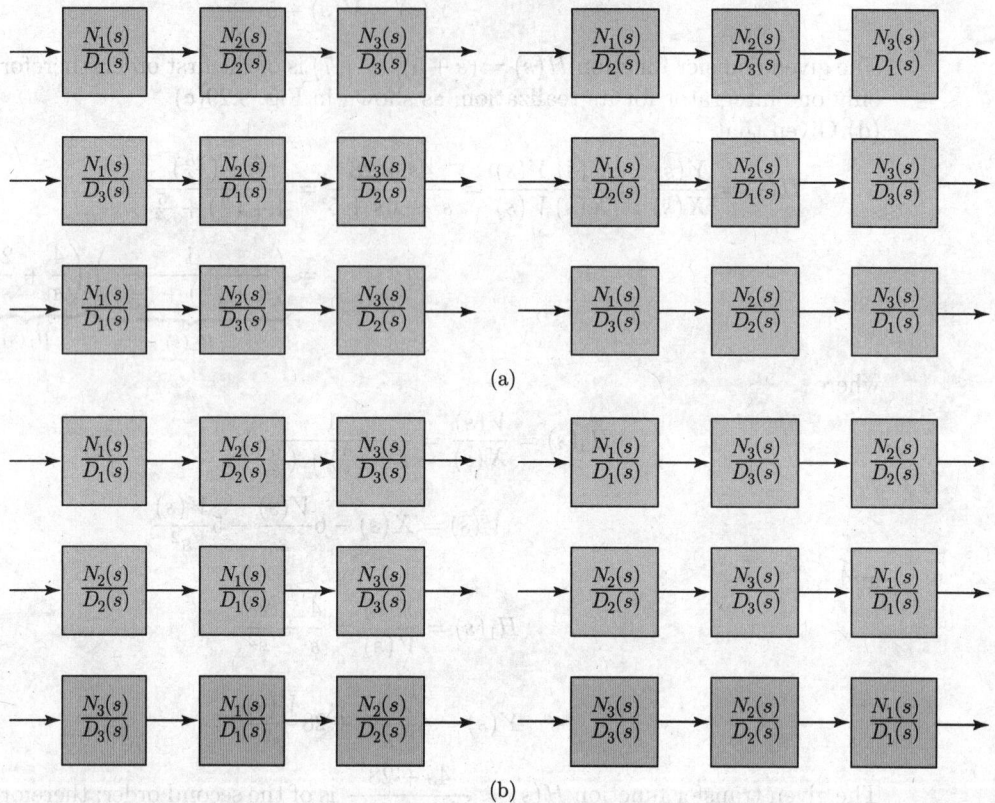

(a)

(b)

Fig. 9.30 Different cascade realizations obtained by (a) different pole-zero pairings and (b) changing the ordering of the systems

---

Example 9.57 Find the canonic, cascade, and parallel realizations of

$$H(s) = \frac{s+3}{s^2 + 7s + 10}$$

Solution

Canonic realization Given that

$$H(s) = \frac{Y(s)}{X(s)} = \frac{V(s)}{X(s)} \frac{Y(s)}{V(s)} = \frac{s+3}{s^2 + 7s + 10} = \frac{(1/s) + (3/s^2)}{1 + (7/s) + (10/s^2)}$$

$$= \underbrace{\left(\frac{1}{1 + (7/s) + (10/s^2)} \right)}_{H_2(s)} \underbrace{\left(\frac{1}{s} + \frac{3}{s^2} \right)}_{H_1(s)}$$

where

$$H_2(s) = \frac{V(s)}{X(s)} = \frac{1}{1 + (7/s) + (10/s^2)}$$

$$V(s) = X(s) - 7\frac{V(s)}{s} - 10\frac{V(s)}{s^2}$$

and

$$H_1(s) = \frac{Y(s)}{V(s)} = \frac{1}{s} + \frac{3}{s^2}$$

$$Y(s) = \frac{V(s)}{s} + 3\frac{V(s)}{s^2}$$

The given transfer function $H(s) = (s+3)/(s^2 + 7s + 10)$ is of the second order; therefore, we need only two integrators for its realization, as shown in Fig. 9.31(a).

Cascade realization Given that

$$H(s) = \frac{s+3}{s^2 + 7s + 10} = \frac{s+3}{(s+2)(s+5)} = \left(\frac{s+3}{s+2}\right)\left(\frac{1}{s+5}\right) = H_1(s)H_2(s)$$

where

$$H_1(s) = \frac{s+3}{s+2} = \frac{1 + 3/s}{1 + 2/s}$$

and

$$H_2(s) = \frac{1}{s+5} = \frac{1/s}{1 + 5/s}$$

We can realize $H(s)$ as a cascade of $H_1(s)$ and $H_2(s)$. Both $H_1(s)$ and $H_2(s)$ are first-order transfer functions and can be implemented by using canonic realization, as shown in Fig. 9.31(b).

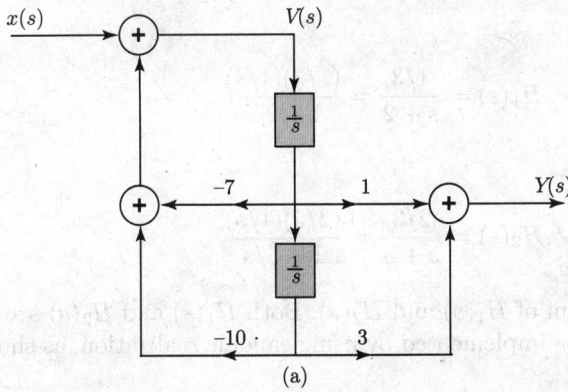

(a)

Fig. 9.31(a)

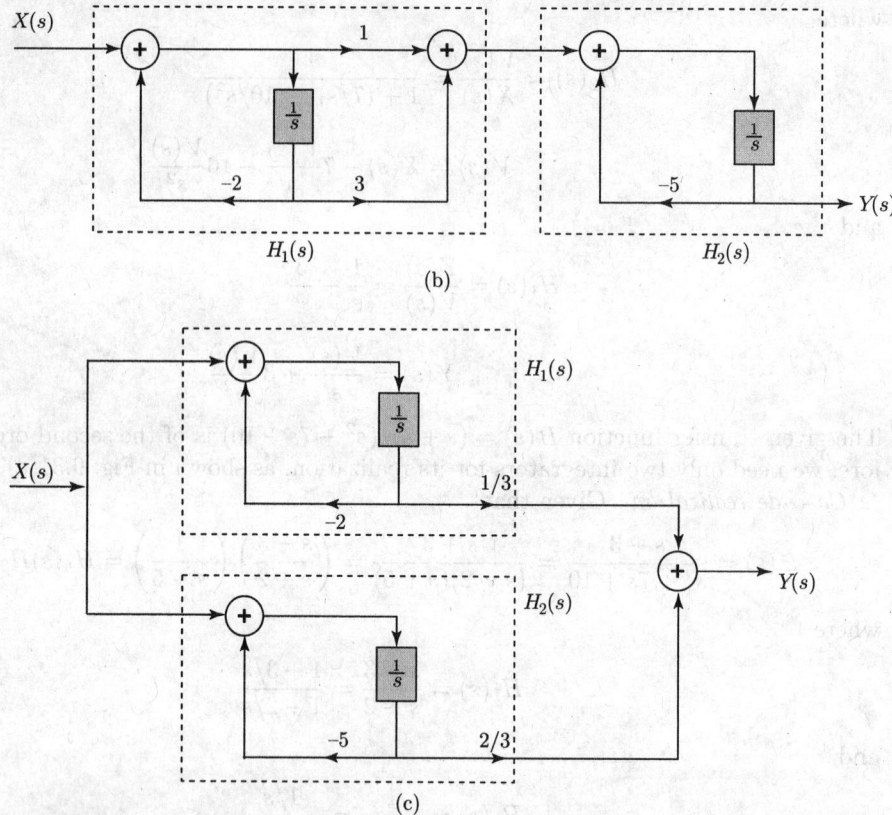

Fig. 9.31(b), (c)

Parallel realization We can express $H(s)$ as a sum of partial fractions as

$$H(s) = \frac{s+3}{(s+2)(s+5)} = \frac{1/3}{s+2} + \frac{2/3}{s+5} = H_1(s) + H_2(s)$$

where

$$H_1(s) = \frac{1/3}{s+2} = \frac{(1/3)(1/s)}{1+2/s}$$

and

$$H_2(s) = \frac{2/3}{s+5} = \frac{(2/3)(1/s)}{1+5/s}$$

We can realize $H(s)$ as a sum of $H_1(s)$ and $H_2(s)$. Both $H_1(s)$ and $H_2(s)$ are first-order transfer functions and can be implemented by using canonic realization, as shown in Fig. 9.31(c).

9.15.4 Transposed Realization

Two realizations are said to be *equivalent* if they have the same transfer function. A simple way to generate an equivalent structure from a given realization is via the *transpose operation*, which is as follows:

1. Reverse all paths.
2. Replace pick-off nodes by adders, and vice versa.
3. Interchange the input and output nodes.

Figure 9.32 shows the transposed version of the canonic direct form realization shown in Fig. 9.27(c) found according to the rules just listed.

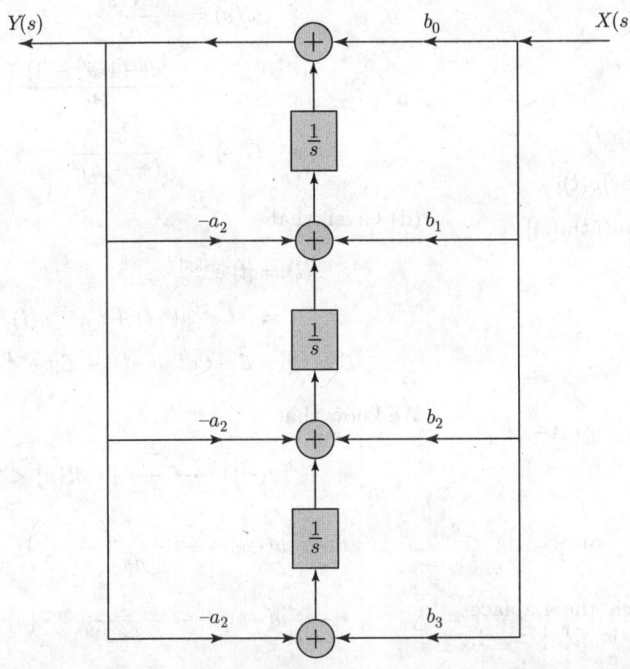

Fig. 9.32 Transposed version of the canonic direct from-II realization.

SOLVED EXAMPLES

Example 9.58 Determine the Laplace transform and the associated ROC for each of the following functions of time:

(a) $x(t) = e^{-2t}u(t) + e^{-3t}u(t)$

(b) $x(t) = e^{-4t}u(t) + e^{-5t}\sin(5t)u(t)$

(c) $g(t) = t\, e^{-2|t|}$

(d) $x(t) = |t|\, e^{-2|t|}$

(e) $x(t) = |t|\, e^{2t}u(-t)$

(f) $x(t) = \delta(t) + u(t)$

(g) $x(t) = \delta(3t) + u(3t)$

Solution

(a) Consider the given signal

$$x(t) = e^{-2t}u(t) + e^{-3t}u(t)$$

$$\mathcal{L}[x(t)] = \mathcal{L}[e^{-2t}u(t) - e^{-3t}u(t)]$$

$$X(s) = \mathcal{L}[e^{-2t}u(t)] + \mathcal{L}[e^{-3t}u(t)]$$

From Eq. (9.10), we have

$$e^{-at}u(t) \longleftrightarrow \frac{1}{s+a} \quad \Re\{s\} > -a$$

Therefore,

$$e^{-2t}u(t) \longleftrightarrow \frac{1}{s+2} \quad \Re\{s\} > -2$$

$$e^{-3t}u(t) \longleftrightarrow \frac{1}{s+3} \quad \Re\{s\} > -3$$

The set of values of $\Re\{s\}$ for which the Laplace transforms of both terms converge is $\Re\{s\} > -2$, and thus, we obtain

$$X(s) = \frac{1}{s+2} + \frac{1}{s+3} \quad \Re\{s\} > -2$$

$$X(s) = \frac{2s+5}{s^2+5s+6} \quad \Re\{s\} > -2$$

(b) Consider the given signal

$$x(t) = e^{-4t}u(t) + e^{-5t}\sin(5t)u(t)$$

$$\mathcal{L}[x(t)] = \mathcal{L}[e^{-4t}u(t) + e^{-5t}\sin(5t)u(t)]$$

$$X(s) = \mathcal{L}[e^{-4t}u(t)] + \mathcal{L}[e^{-5t}\sin(5t)u(t)]$$

From Eq. (9.10), we obtain

$$e^{-4t}u(t) \longleftrightarrow \frac{1}{s+4} \quad \Re\{s\} > -4$$

and from Eq. (9.27), we have

$$e^{-at}\sin(\omega_0 t)u(t) \longleftrightarrow \frac{\omega_0}{(s+a)^2 + \omega_0^2} \quad \Re\{s\} > -a$$

Therefore,

$$e^{-5t}\sin(5t)u(t) \longleftrightarrow \frac{5}{(s+5)^2 + 25} \quad \Re\{s\} > -5$$

The set of values of $\Re\{s\}$ for which the Laplace transforms of both terms converge is $\Re\{s\} > -4$, and thus, we obtain

$$X(s) = \frac{1}{s+4} + \frac{5}{(s+5)^2 + 25} \quad \Re\{s\} > -4$$

$$X(s) = \frac{s^2 + 15s + 70}{(s+4)(s^2 + 10s + 50)} \quad \Re\{s\} > -4$$

(c) Consider the given signal

$$g(t) = t\,e^{-2|t|} = tx(t)$$

where

$$x(t) = e^{-2|t|}$$

From Eq. (9.19), we have

$$e^{-a|t|} \longleftrightarrow \frac{-2a}{s^2 - a^2} \quad -a < \Re\{s\} < a$$

Therefore,

$$\underbrace{e^{-2|t|}}_{x(t)} \longleftrightarrow \underbrace{\frac{-4}{s^2 - 4}}_{X(s)} \quad -2 < \Re\{s\} < 2$$

Now, taking the Laplace transform of the given signal $g(t)$ and using differentiation in the s-domain property, we obtain

$$\mathcal{L}[g(t)] = \mathcal{L}[tx(t)]$$

$$G(s) = -\frac{dX(s)}{ds}$$

$$= -\frac{d(-4/s^2 - 4)}{ds}$$

$$G(s) = -\frac{8s}{(s^2 - 4)^2}$$

(d) Given that

$$x(t) = |t|\,e^{-2|t|}$$

$$= -t\,e^{2t}u(-t) + t\,e^{-2t}u(t)$$

$$\mathcal{L}[x(t)] = \mathcal{L}[-t\,e^{2t}u(-t)] + \mathcal{L}[t\,e^{-2t}u(t)]$$

We know that

$$-e^{2t}u(-t) \longleftrightarrow \frac{1}{s-2} \quad \Re\{s\} < 2$$

$$t[-e^{2t}u(-t)] \longleftrightarrow -\frac{d}{ds}\left(\frac{1}{s-2}\right)$$

$$t[-e^{2t}u(-t)] \longleftrightarrow \frac{1}{(s-2)^2} \quad \Re\{s\} < 2$$

and

$$e^{-2t}u(t) \longleftrightarrow \frac{1}{s+2} \quad \Re\{s\} > -2$$

$$t[e^{-2t}u(t)] \longleftrightarrow -\frac{d}{ds}\left(\frac{1}{s+2}\right)$$

$$t[e^{-2t}u(t)] \longleftrightarrow \frac{1}{(s+2)^2} \quad \Re\{s\} > -2$$

Therefore,

$$X(s) = \frac{1}{(s-2)^2} + \frac{1}{(s+2)^2} \quad -2 < \Re\{s\} < 2$$

$$X(s) = \frac{2s^2 + 8}{(s^2 - 4)^2} \quad -2 < \Re\{s\} < 2$$

(e) Given that

$$x(t) = |t| \, e^{2t} u(-t)$$
$$= -t \, e^{2t} u(-t)$$

We know that

$$-e^{2t}u(-t) \longleftrightarrow \frac{1}{s-2} \quad \Re\{s\} < 2$$

$$t[-e^{2t}u(-t)] \longleftrightarrow -\frac{d}{ds}\left(\frac{1}{s-2}\right)$$

$$t[-e^{2t}u(-t)] \longleftrightarrow \frac{1}{(s-2)^2} \quad \Re\{s\} < 2$$

Therefore,

$$X(s) = \frac{1}{(s-2)^2} \quad \Re\{s\} < 2$$

(f) Given that

$$x(t) = \delta(t) + u(t)$$
$$\mathcal{L}[x(t)] = \mathcal{L}[\delta(t)] + \mathcal{L}[u(t)]$$
$$X(s) = 1 + \frac{1}{s} \quad \Re\{s\} > 0$$
$$X(s) = \frac{s+1}{s} \quad \Re\{s\} > 0$$

(g) Given that

$$x(t) = \delta(3t) + u(3t)$$
$$= \frac{1}{3}\delta(t) + u(t)$$
$$\mathcal{L}[x(t)] = \frac{1}{3}\mathcal{L}[\delta(t)] + \mathcal{L}[u(t)]$$
$$X(s) = \frac{1}{3} + \frac{1}{s} \quad \Re\{s\} > 0$$
$$X(s) = \frac{s+3}{3s} \quad \Re\{s\} > 0$$

Example 9.59 Determine the Laplace transform of the following signals:

(a) $x(t) = \cos^3(3t)u(t)$

(b) $x(t) = (1 + 0.5\sin(t))\sin(1000t)u(t)$

Solution

(a) Given that

$$x(t) = \cos^3(3t)u(t)$$

$$x(t) = \frac{1}{4}\cos(9t)u(t) + \frac{3}{4}\cos(3t)u(t)$$

Since $\cos^3(x) = 1/4 \cos(3x) + 3/4 \cos(x)$, we have

$$\mathcal{L}[x(t)] = \mathcal{L}\left[\frac{1}{4}\cos(9t)u(t)\right] + \mathcal{L}\left[\frac{3}{4}\cos(3t)u(t)\right]$$

$$X(s) = \frac{1}{4}\frac{s}{s^2+81} + \frac{3}{4}\frac{s}{s^2+9}$$

$$X(s) = \frac{1}{4}\left(\frac{s}{s^2+81} + \frac{3s}{s^2+9}\right)$$

(b) Given that

$$x(t) = [1 + 0.5\sin(t)]\sin(1000t)u(t)$$

$$= \sin(1000t)u(t)$$

$$\quad + 0.5\sin(t)\sin(1000t)u(t)$$

$$x(t) = \sin(1000t)u(t) + 0.25\cos(999t)u(t)$$

$$\quad - 0.25\cos(1001t)u(t)$$

$$\mathcal{L}[x(t)] = \mathcal{L}[\sin(1000t)u(t)]$$

$$\quad + 0.25\mathcal{L}[\cos(999t)u(t)]$$

$$\quad - 0.25\mathcal{L}[\cos(1001t)u(t)]$$

$$X(s) = \frac{1000}{s^2+10^6} + 0.25\frac{s}{s^2+(999)^2}$$

$$\quad - 0.25\frac{s}{s^2+(1001)^2}$$

Example 9.60 Find the unilateral Laplace transform of

$$x(t) = \frac{1}{\sqrt{t}}u(t) = \begin{cases} \frac{1}{\sqrt{t}}, & t > 0 \\ 0, & t < 0 \end{cases}$$

Solution

Given that

$$x(t) = \frac{1}{\sqrt{t}}u(t) = t^{-1/2}u(t)$$

We know that [Eq. (9.54)]

$$X(s) = \int_{0^-}^{\infty} x(t) \, e^{-st} \, dt$$

$$= \int_{0^-}^{\infty} t^{-1/2} u(t) \, e^{-st} \, dt$$

$$X(s) = \int_{0}^{\infty} t^{-1/2} \, e^{-st} \, dt$$

A change of variables is performed by letting $\tau = st$, which also yields $d\tau = dt$, $\tau \to 0$ as $t \to 0$, and $\tau \to \infty$ as $t \to \infty$. Therefore,

$$X(s) = \frac{1}{\sqrt{s}} \int_{0}^{\infty} \tau^{-1/2} \, e^{-\tau} \, d\tau$$

$$= \frac{1}{\sqrt{s}} \Gamma\left(\frac{1}{2}\right)$$

$$X(s) = \sqrt{\frac{\pi}{s}}$$

Example 9.61 An absolutely integrable signal $x(t)$ is known to have a pole at $s = 2$. Answer the following questions:
(a) Could $x(t)$ be of finite duration?
(b) Could $x(t)$ be left sided?
(c) Could $x(t)$ be right sided?
(d) Could $x(t)$ be two sided?

Solution
(a) No. We know that for a finite duration signal, the ROC is the entire s-plane. Therefore, there can be no poles in finite s-plane for a finite duration signal. Since, the given signal $x(t)$ has a pole at $s = 2$, clearly, in this problem this is not the case.

(b) Yes. Since the signal is absolutely integrable, the ROC must include the $j\omega$-axis. Furthermore, $X(s)$ has a pole at $s = 2$. Therefore, one valid ROC for the signal would be $\Re\{s\} < 2$. Since the ROC lies left to a pole, it would correspond to a left-sided signal.

(c) No. Since the signal is absolutely integrable, the ROC must include the $j\omega$-axis. Furthermore, $X(s)$

has a pole at $s = 2$. For a right-sided signal ROC is of the form $\Re\{s\} > a$ for some constant a and ROC does not contain any pole. Therefore, $x(t)$ cannot be right-sided signal.

(d) Yes. Since the signal is absolutely integrable, the ROC must include the $j\omega$-axis. Furthermore, $X(s)$ has a pole at $s = 2$. Therefore, a valid ROC for the signal could be a strip in the region $a < \Re\{s\} < 2$ such that $a < 0$ and we know that this would correspond to a two-sided signal.

Example 9.62 How many signals have a Laplace transform that may be expressed as

$$X(s) = \frac{s - 1}{(s + 2)(s + 3)(s^2 + s + 1)}$$

in its region of convergence?

Solution
We may find different signals with the given Laplace transform by choosing different region of convergence. Consider the given Laplace transform

$$X(s) = \frac{s - 1}{(s + 2)(s + 3)(s^2 + s + 1)}$$

$$= \frac{s - 1}{(s + 2)(s + 3)\left(s + \frac{1}{2} - j\sqrt{3}/2\right)\left(s + \frac{1}{2} + j\sqrt{3}/2\right)}$$

The poles of the given Laplace transform are

$$s_1 = -2 \quad s_2 = -3$$

$$s_3 = -\frac{1}{2} + j\frac{\sqrt{3}}{2} \quad s_4 = -\frac{1}{2} - j\frac{\sqrt{3}}{2}$$

The pole-zero diagram is shown in Fig. 9.33. Based on the location of these poles, we may choose from the following ROCs:

(i) $\Re\{s\} > -1/2$
(ii) $-2 < \Re\{s\} < -1/2$
(iii) $-3 < \Re\{s\} < -2$
(iv) $\Re\{s\} < -3$

Therefore, we may find four different signals with the given Laplace transform.

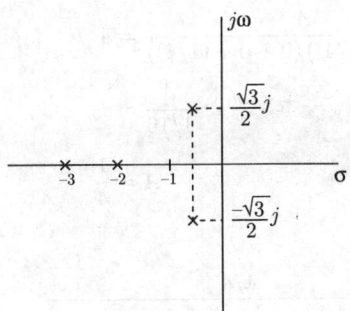

Fig. 9.33 Pole-zero plot for Example 9.62

Example 9.63 Let $x(t)$ be a signal that has a rational Laplace transform with exactly two poles located at $s = -1$ and $s = -3$. If $g(t) = e^{2t}x(t)$ and $G(\omega) = \mathcal{F}[g(t)]$ converges, determine whether $x(t)$ is left sided, right sided, or two sided.

Solution
Using shifting in the s-domain property, we know that if

$$x(t) \longleftrightarrow X(s) \quad \text{with ROC} = \text{R}$$

then

$$\underbrace{e^{2t}x(t)}_{g(t)} \longleftrightarrow \underbrace{X(s-2)}_{G(s)} \quad \text{with ROC} = \text{R+2}$$

That is, the ROC of $G(s)$ is the ROC of $X(s)$ shifted to the right by 2.

We are also given that $X(s)$ has exactly two poles, located at

$$s_1 = -1 \quad \text{and} \quad s_2 = -3$$

Since, $G(s) = x(s - 2)$, $G(s)$ also has exactly two poles, located at

$$s_1 = -1 + 2 = 1 \quad \text{and} \quad s_2 = -3 + 2 = -1$$

Since, we are given that $G(\omega) = \mathcal{F}[g(t)]$ converges, we may infer that the $j\omega$-axis lies in the ROC of $G(s)$. That is, the ROC of $G(s)$ is the region $-1 < \Re\{s\} < 1$ in the s-plane. Therefore, we may conclude that $g(t)$ is a two-sided signal and hence $x(t) = e^{-2t}g(t)$ will also be a two-sided signal.

Example 9.64 Suppose we are given the following three facts about the signal $x(t)$:

1. $x(t) = 0$, for $t < 0$.

2. $x\left(\dfrac{k}{80}\right) = 0$, for $k = 1, 2, 3, \ldots$.

3. $x\left(\dfrac{1}{160}\right) = e^{-120}$.

Let $X(s)$ denote the Laplace transform of $x(t)$, and determine which of the following statements is consistent with the given information about $x(t)$:
(a) $X(s)$ has only one pole in the finite s-plane.
(b) $X(s)$ has only two poles in the finite s-plane.
(c) $X(s)$ has more than two poles in the finite s-plane.

Solution
(a) If $X(s)$ has only one pole in the finite s-plane, then $x(t)$ would be of the form $A\,e^{-at}u(t)$. Clearly, such a signal violates condition 2. Therefore, this statement is inconsistent with the given information.
(b) If $X(s)$ has only two poles in the finite s-plane, then $x(t)$ would be of the form $A\,e^{-at}\sin(\omega_0 t)u(t)$. Clearly, such a signal could be made to satisfy all three conditions:
(1) $x(t) = 0$, for $t < 0$ because of $u(t)$.
(2) $x\left(\dfrac{k}{80}\right) = 0$, for $k = 1, 2, 3, \ldots$, if we choose $\omega_0 = 80\pi$.
(3) $\qquad x(t) = A\,e^{-at}\sin(\omega_0 t)u(t)$

$$x\left(\frac{1}{160}\right) = A\,e^{-a1/160}\sin\left(80\pi\frac{1}{160}\right)u\left(\frac{1}{160}\right)$$

$$e^{-120} = A\,e^{-a1/160}\sin\left(\frac{\pi}{2}\right)$$

$$e^{-120} = A\,e^{-a1/160}$$

Thus,

$$a\frac{1}{160} = 120$$

$$a = 19200$$

Therefore, this statement is consistent with the given information.
(c) If $X(s)$ has more than two poles in the finite s-plane, then $x(t)$ would be of the form $Ae^{-at}\sin(\omega_0 t)u(t) + B\,e^{-bt}\sin(\omega_0 t)u(t)$. Clearly, such a signal could be made to satisfy all the three conditions. Therefore, this statement is consistent with the given information.

Example 9.65 Suppose the following facts are given about the signal $x(t)$ with Laplace transform $X(s)$:

1. $x(t)$ is real and even

2. $X(s)$ has four poles and no zeros in the finite
   s-plane

3. $X(s)$ has a pole at $s = \dfrac{1}{2} e^{j\pi/4}$

4. $\displaystyle\int_{-\infty}^{\infty} x(t)dt = 4$

Determine $X(s)$ and its ROC.

Solution

Since $X(s)$ has four poles and no zeros in the finite
s-plane, we may assume that $X(s)$ is of the form

$$X(s) = \frac{A}{(s-a)(s-b)(s-c)(s-d)}$$

Since $x(t)$ is real, the poles of $X(s)$ must occur
in conjugate reciprocal pairs. Therefore, we assume
that $b = a^*$ and $d = c^*$. This results in

$$X(s) = \frac{A}{(s-a)(s-a^*)(s-c)(s-c^*)}$$

Since the signal $x(t)$ is also even, the Laplace transform $X(s)$ must also be even [Eq. (9.34)]. This implies that the poles have to be symmetric about the
$j\omega$-axis. Therefore, we may assume that $c = -a^*$.
This results in

$$X(s) = \frac{A}{(s-a)(s-a^*)(s+a^*)(s+a)}$$

The location of one of the poles is $a = 1/2\, e^{j\pi/4}$.
Therefore, we have

$$X(s) = \frac{A}{\left(s - (1/2)\, e^{j\pi/4}\right)\left(s - (1/2)\, e^{-j\pi/4}\right)}$$

$$\times \frac{1}{\left(s + (1/2)\, e^{-j\pi/4}\right)\left(s + (1/2)\, e^{j\pi/4}\right)}$$

$$X(s) = \frac{A}{\left(s^2 - (s/\sqrt{2}) + \frac{1}{4}\right)\left(s^2 + (s/\sqrt{2}) + \frac{1}{4}\right)}$$

We know that

$$X(s) = \int_{-\infty}^{\infty} x(t)\, e^{-st}\, dt$$

$$X(s)\Big|_{s=0} = X(0) = \int_{-\infty}^{\infty} x(t)\, dt$$

$$\frac{A}{(0 - 0 + (1/4))\, (0 + 0 + (1/4))} = 4$$

$$\frac{A}{1/16} = 4$$

$$A = \frac{1}{4}$$

Therefore,

$$X(s) = \frac{1/4}{\left(s^2 - (s/\sqrt{20} + \frac{1}{4}\right)\left(s^2 + (s/\sqrt{2}) + \frac{1}{4}\right)}$$

Example 9.66 Consider two right-sided signals
$x(t)$ and $y(t)$ related through differential equations

$$\frac{dx(t)}{dt} = -2y(t) + \delta(t)$$

and

$$\frac{dy(t)}{dt} = 2x(t)$$

Determine $Y(s)$ and $X(s)$, along with their ROC.

Solution

Consider the given differential equations

$$\frac{dx(t)}{dt} = -2y(t) + \delta(t) \quad \text{and} \quad \frac{dy(t)}{dt} = 2x(t)$$

Taking the Laplace transform of both sides of the
two differential equations, we have

$$sX(s) = -2Y(s) + 1 \quad \text{and} \quad sY(s) = 2X(s)$$

solving for $X(s)$ and $Y(s)$, we obtain

$$Y(s) = \frac{2}{s^2 + 4} \quad \text{and} \quad X(s) = \frac{s}{s^2 + 4}$$

The ROC for both $X(s)$ and $Y(s)$ is $\Re\{s\} > 0$ because both are right-sided signals.

Example 9.67 A causal LTI system S with impulse response $h(t)$ has its input $x(t)$ and output
$y(t)$ related through a linear constant-coefficient differential equation of the form

$$\frac{d^3y(t)}{dt^3} + (1+\alpha)\frac{d^2y(t)}{dt^2}$$

$$+ \alpha(1+\alpha)\frac{dy(t)}{dt} + \alpha^2 y(t) = x(t)$$

(a) If

$$g(t) = \frac{dh(t)}{dt} + h(t)$$

how many poles does $G(s)$ have?
(b) For what real values of the parameter α is S guaranteed to be stable.

Solution
Consider the given differential equation

$$\frac{d^3y(t)}{dt^3} + (1+\alpha)\frac{d^2y(t)}{dt^2}$$

$$+ \alpha(1+\alpha)\frac{dy(t)}{dt} + \alpha^2 y(t) = x(t)$$

Taking the Laplace transform of both sides of the given differential equation, we obtain

$$s^3 Y(s) + (1+\alpha)s^2 Y(s)$$

$$+ \alpha(1+\alpha)sY(s) + \alpha^2 Y(s) = X(s)$$

$$Y(s)[s^3 + (1+\alpha)s^2 + \alpha(1+\alpha)s + \alpha^2] = X(s)$$

$$H(s) = \frac{Y(s)}{X(s)} = \frac{1}{s^3 + (1+\alpha)s^2 + \alpha(1+\alpha)s + \alpha^2}$$

$$H(s) = \frac{1}{(s+1)(s^2 + \alpha s + \alpha^2)}$$

(a) Consider the given differential equation

$$g(t) = \frac{dh(t)}{dt} + h(t)$$

Taking the Laplace transform of both sides of the given differential equation, we obtain

$$G(s) = sH(s) + H(s)$$

$$= H(s)(s+1)$$

$$G(s) = \frac{1}{s^2 + \alpha s + \alpha^2}$$

Clearly, $G(s)$ has two poles.
(b) We know that

$$H(s) = \frac{1}{(s+1)(s^2 + \alpha s + \alpha^2)}$$

$$= \frac{1}{(s+1)\left(s + (\alpha/2) - j(\sqrt{3}\alpha/2)\right)}$$

$$\times \frac{1}{\left(s + (\alpha/2) + j(\sqrt{3}\alpha/2)\right)}$$

Therefore, $H(s)$ has poles at -1, $-(\alpha/2) + j(\sqrt{3}\alpha/2)$, and $-(\alpha/2) - j(\sqrt{3}\alpha/2)$. We know that for a causal system to be stable, all the poles of the system must lie in the left half of the s-plane. For this to be true, $-(\alpha/2) < 0$, i.e., $\alpha > 0$.

Example 9.68 We are given the following five facts about a real signal $x(t)$ with Laplace transform $X(s)$:

1. $X(s)$ has exactly two poles
2. $X(s)$ has no zeros in the finite s-plane
3. $X(s)$ has a pole at $s = -1 + j$
4. $e^{2t}x(t)$ is not absolutely integrable
5. $X(0) = 8$

Determine $X(s)$ and specify its ROC.

Solution
Since $X(s)$ has exactly two poles in the finite s-plane, we may assume that $X(s)$ is of the form

$$X(s) = \frac{A}{(s-a)(s-b)}$$

Since $x(t)$ is real, the poles of $X(s)$ must occur in conjugate reciprocal pairs. Therefore, we assume that $b = a^*$. This results in

$$X(s) = \frac{A}{(s-a)(s-a^*)}$$

The location of one of the poles is $a = -1 + j$. Therefore, we have

$$X(s) = \frac{A}{(s+1-j)(s+1+j)}$$

We know that

$$X(s) = \int_{-\infty}^{\infty} x(t) e^{-st} \, dt$$

$$X(s)\Big|_{s=0} = X(0) = \int_{-\infty}^{\infty} x(t) \, dt$$

From fact 5, we have

$$X(0) = \frac{A}{(0+1-j)(0+1+j)} = \frac{A}{2} = 8$$

$$A = 16$$

Therefore,

$$X(s) = \frac{16}{(s+1-j)(s+1+j)}$$

$$= \frac{16}{s^2 + 2s + 2}$$

From the pole locations we know that there are two possible choices of ROC, either $\Re\{s\} < -1$ or $\Re\{s\} > -1$. From fact 4, we have

$$y(t) = e^{2t}x(t) \longleftrightarrow Y(s) = X(s-2)$$

The ROC of $Y(s)$ is the ROC of $X(s)$ shifted by 2 to the right. The poles of $Y(s)$ are located at $-1+j+2 = 1+j$ and $-1-j+2 = 1-j$. Since it is given that $y(t)$ is not absolutely integrable, the ROC of $Y(s)$ should not include the $j\omega$-axis. This is possible only if ROC of $X(s)$ is $\Re\{s\} > -1$.

Example 9.69 Show that for an LTI system, when the input is $x(t) = e^{s_0 t}$, the output is of the form $y(t) = H(s_0) e^{s_0 t}$. How is $H(s_0)$ related to the impulse response of the system?

Solution

We know that the input and output of an LTI system are related by

$$y(t) = h(t) * x(t) = \int_{-\infty}^{\infty} h(\tau)x(t-\tau)\, d\tau$$

$$= \int_{-\infty}^{\infty} h(\tau)\, e^{s_0(t-\tau)}\, d\tau$$

$$= e^{s_0 t} \int_{-\infty}^{\infty} h(\tau)\, e^{-s_0 \tau}\, d\tau$$

$$y(t) = H(s_0)\, e^{s_0 t}$$

where

$$H(s_0) = \int_{-\infty}^{\infty} h(\tau)\, e^{-s_0 \tau}\, d\tau = \mathcal{L}[h(t)]\Big|_{s=s_0}$$

Example 9.70 Determine the impulse response $h(t)$ of a system having a double-order pole at $s = -a$ and a zero at $s = -b$, where $a, b > 0$ and $b - a = B$. It is also given that $h(0) = 2$.

Solution

Since $h(t)$ of the system has a double-order pole at $s = -a$ and a zero at $s = -b$, we may assume that $H(s)$ is of the form

$$X(s) = \frac{K(s+b)}{(s+a)^2}$$

$$X(s) = \frac{Ks}{(s+a)^2} + \frac{Kb}{(s+a)^2}$$

$$X(s) = KsG(s) + KbG(s)$$

where

$$G(s) = \frac{1}{(s+a)^2}$$

Taking its inverse Laplace transform, we get

$$g(t) = t\, e^{-at} u(t)$$

Also, because

$$X(s) = KsG(s) + KbG(s)$$

Taking its inverse Laplace transform, we get

$$x(t) = K\frac{dg(t)}{dt} + Kbg(t)$$

$$= K\frac{d}{dt}t\, e^{-at}u(t) + Kbt\, e^{-at}u(t)$$

$$= K\left[t\left(e^{-at}\delta(t) - a\, e^{-at}u(t)\right) + e^{-at}u(t)\right]$$

$$\quad + Kbt\, e^{-at}u(t)$$

$$= K\left[0 - Kat\, e^{-at}u(t) + K\, e^{-at}u(t)\right]$$

$$\quad + Kbt\, e^{-at}u(t)$$

$$= K(b-a)t\, e^{-at}u(t) + K\, e^{-at}u(t)$$

$$h(t) = KBt\, e^{-at}u(t) + K\, e^{-at}u(t)$$

$$h(t) = K(Bt+1)\, e^{-at}u(t)$$

Also, given that

$$h(0) = 2 \quad K = 2$$

Therefore,

$$h(t) = 2(Bt+1)\, e^{-at}u(t)$$

Example 9.71 Determine the impulse response $h(t)$ of the system $H(s)$ of an LTI system from the following facts:

1. When the input to the system is $x(t) = e^{2t}$ and the output is $y(t) = 1/6\, e^{2t}$
2. When $h(t)$ satisfies the differential equation

$$\frac{dh(t)}{dt} + 2h(t) = e^{-4t}u(t) - bu(t)$$

where b is an unknown constant. Your answer must not contain any unknown constant.

Solution
We know that if we apply a complex exponential input $x(t) = e^{s_0 t}$ to an LTI system with impulse response $h(t)$, the system output will be $y(t) = H(s_0)\, e^{s_0 t}$, i.e.,

$$x(t) \longrightarrow y(t)$$
$$e^{s_0 t} \longrightarrow H(s_0)\, e^{s_0 t}$$

Given that

$$e^{2t} \longrightarrow \frac{1}{6}\, e^{2t} = H(2)\, e^{2t}$$

Therefore, $H(2) = 1/6$. Now, consider the given differential equation

$$\frac{dh(t)}{dt} + 2h(t) = e^{-4t}u(t) - bu(t)$$

Taking the Laplace transform of the above equation, we obtain

$$sH(s) + 2H(s) = \frac{1}{s+4} - \frac{b}{s}$$

$$(s+2)H(s) = \frac{s - b(s+4)}{s(s+4)}$$

$$H(s) = \frac{s - b(s+4)}{s(s+2)(s+4)}$$

Also, given that

$$H(s)\Big|_{s=2} = H(2) = \frac{2 - 6b}{2 \times 4 \times 6} = \frac{1}{6}$$

$$b = -1$$

Therefore,

$$H(s) = \frac{s + (s+4)}{s(s+2)(s+4)}$$

$$= \frac{2(s+2)}{s(s+2)(s+4)}$$

$$H(s) = \frac{2}{s(s+4)}$$

Using partial fraction expansion, we obtain

$$H(s) = \frac{1/2}{s} - \frac{1/2}{s+4}$$

The inverse Laplace transform of the above equation yields

$$h(t) = \frac{1}{2}[1 - e^{-4t}]u(t)$$

Example 9.72 Solve, by using Laplace transform, the following set of simultaneous differential equations for $x(t)$:

$$2\frac{dx(t)}{dt} + 4x(t) + \frac{dy(t)}{dt} + 7y(t) = 5u(t)$$

$$\frac{dx(t)}{dt} + x(t) + \frac{dy(t)}{dt} + 3y(t) = 5\delta(t)$$

The initial conditions are $x(0^-) = y(0^-) = 0$.

Solution
Consider the given differential equation

$$2\frac{dx(t)}{dt} + 4x(t) + \frac{dy(t)}{dt} + 7y(t) = 5u(t)$$

Taking the Laplace transform of the above equation, we obtain

$$2sX(s) + 4X(s) + sY(s) + 7Y(s) = \frac{5}{s}$$

$$(2s+4)X(s) + (s+7)Y(s) = \frac{5}{s}$$

$$X(s) = \frac{5}{s(2s+4)} - \frac{s+7}{2s+4}Y(s)$$

Now, consider the second differential equation

$$\frac{dx(t)}{dt} + x(t) + \frac{dy(t)}{dt} + 3y(t) = 5\delta(t)$$

Taking the Laplace transform of the above equation, we obtain

$$sX(s) + X(s) + sY(s) + 3Y(s) = 5$$

$$(s+1)X(s) + (s+3)Y(s) = 5$$

$$Y(s) = \frac{5}{s+3} - \frac{s+1}{s+3}X(s)$$

Substituting this expression of $Y(s)$ into the expression of $X(s)$, we obtain

$$X(s) = \frac{5}{s(2s+4)} - \frac{s+7}{2s+4}Y(s)$$

$$= \frac{5}{s(2s+4)} - \frac{s+7}{2s+4}\left[\frac{5}{s+3} - \frac{s+1}{s+3}X(s)\right]$$

$$X(s) = \frac{5}{s(2s+4)} - \frac{5(s+7)}{(s+3)(2s+4)}$$

$$+ \frac{(s+1)(s+7)}{(s+3)(2s+4)}X(s)$$

$$X(s)\left[1 - \frac{(s+1)(s+7)}{(s+3)(2s+4)}\right] = \frac{5}{s(2s+4)}$$

$$- \frac{5(s+7)}{(s+3)(2s+4)}$$

$$X(s)\left[(s+3)(2s+4) - (s+1)(s+7)\right]$$

$$= \frac{5(s+3)}{s} - 5(s+7)$$

$$X(s)[s^2 + 2s + 5] = -30 - 5s + \frac{15}{s}$$

$$X(s)[(s+1)^2 + 4] = -25 - 5(s+1) + \frac{15}{s}$$

$$X(s) = -\frac{25}{(s+1)^2+4} - \frac{5(s+1)}{(s+1)^2+4}$$

$$+ \frac{15}{s[(s+1)^2+4]}$$

Applying partial fraction expansion on the last term of the RHS of the above equation, we get

$$X(s) = -\frac{25}{(s+1)^2+4} - \frac{5(s+1)}{(s+1)^2+4}$$

$$+ \frac{3}{s} + \frac{-3s-6}{(s+1)^2+4}$$

$$= -\frac{25}{(s+1)^2+4} - \frac{5(s+1)}{(s+1)^2+4}$$

$$+ \frac{3}{s} - \frac{3(s+1)}{(s+1)^2+4} - \frac{3}{(s+1)^2+4}$$

$$X(s) = \frac{3}{s} - \frac{28}{(s+1)^2+4} - \frac{8(s+1)}{(s+1)^2+4}$$

$$X(s) = \frac{3}{s} - 14\frac{2}{(s+1)^2+4} - 8\frac{(s+1)}{(s+1)^2+4}$$

The inverse Laplace transform of the above equation yields

$$x(t) = 3u(t) - 14\,e^{-t}\sin(2t)u(t) - 8\,e^{-t}\cos(2t)u(t)$$

Example 9.73 An LTI system has an impulse response $h(t) = e^{-at}u(t)$, and when it is excited by an input signal $x(t)$, its output is $y(t) = (e^{-bt} - e^{-ct})u(t)$. Determine its input $x(t)$.

Solution

Given that the impulse response

$$h(t) = e^{-at}u(t)$$

Its Laplace transform is given by

$$H(s) = \frac{1}{s+a}$$

Also, given that the system output

$$y(t) = (e^{-bt} - e^{-ct})u(t)$$

The Laplace transform of the above equation yields

$$Y(s) = \frac{1}{s+b} - \frac{1}{s+c} = \frac{c-b}{(s+b)(s+c)}$$

We know that

$$X(s) = \frac{Y(s)}{H(s)}$$

$$X(s) = \frac{c-b(s+a)}{(s+b)(s+c)}$$

Using partial fraction expansion, we obtain

$$X(s) = \frac{a-b}{s+b} + \frac{c-a}{s+c}$$

The inverse Laplace transform of the above equation yields

$$x(t) = (a-b)\,e^{-bt}u(t) + (c-a)\,e^{-ct}u(t)$$

Example 9.74 Consider the LTI system shown in Fig. 9.34(a) for which we are given the following information:

$$X(s) = \frac{s+2}{s-2}$$

$$x(t) = 0, \quad t > 0$$

and

$$y(t) = -\frac{2}{3}\,e^{2t}u(t) + \frac{1}{3}\,e^{-t}u(t)$$

$y(t)$ is shown in Fig. 9.34(b).

(a) Determine $H(s)$ and its region of convergence.

(b) Determine $h(t)$.

(c) Using the system function $H(s)$ found in part (a), determine the output $y(t)$ if the input is $x(t) = e^{3t}$, $-\infty < t < \infty$.

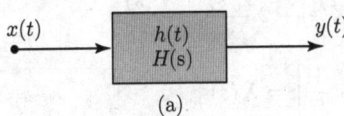

(a)

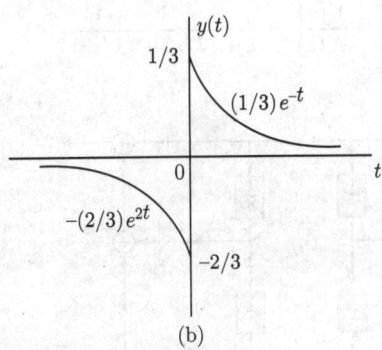

(b)

Fig. 9.34

Solution

(a) Given that

$$y(t) = -\frac{2}{3}\,e^{2t}u(t) + \frac{1}{3}\,e^{-t}u(t)$$

Taking the Laplace transform of the above equation, we get

$$Y(s) = \frac{2/3}{s-2} + \frac{1/3}{s+1}, \quad -1 < \Re\{s\} < 2$$

$$Y(s) = \frac{s}{(s-2)(s+1)}, \quad -1 < \Re\{s\} < 2$$

Given that $x(t) = 0$, for $t > 0$, i.e., $x(t)$ is left sided. Therefore, the ROC for $X(s)$ is $\Re\{s\} < 2$.

We know that

$$H(s) = \frac{Y(s)}{X(s)} = \frac{s}{(s+1)(s+2)}$$

We know that the ROC of $Y(s)$ has to be the intersection of the ROCs of $X(s)$ and $H(s)$. This leads to conclude that the ROC of $H(s)$ is $\Re\{s\} > -1$.

(b) The partial fraction expansion of $H(s)$ is

$$H(s) = \frac{-1}{s+1} + \frac{2}{s+2}$$

Therefore,

$$h(t) = -e^{-t}u(t) + 2\,e^{-2t}u(t)$$

(c) We know that

$$e^{s_0 t} \longrightarrow H(s_0)\,e^{s_0 t}$$

Therefore,

$$x(t) = e^{3t} \longrightarrow H(3)\,e^{3t} = y(t)$$

Hence,

$$y(t) = H(3)\,e^{3t} = \frac{3}{(3+1)(3+2)}\,e^{3t} = \frac{3}{20}\,e^{3t}$$

Example 9.75 A causal LTI system S has the block diagram representation shown in Fig. 9.35. Determine a differential equation relating the input $x(t)$ to the output $y(t)$ of this system.

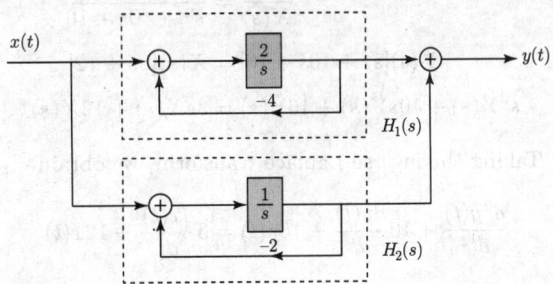

Fig. 9.35

Solution

The overall system shown in Fig. 9.35 may be treated as two feedback systems $H_1(s)$ and $H_2(s)$ connected in parallel as shown in Fig. 9.36. Using

Eq. (9.82) [and Fig. 9.25(d)], the system function of the upper feedback system is

$$H_1(s) = \frac{2/s}{1 + 4(2/s)} = \frac{2}{s+8}$$

Similarly, the system function of lower feedback system is

$$H_2(s) = \frac{1/s}{1 + 2(1/s)} = \frac{1}{s+2}$$

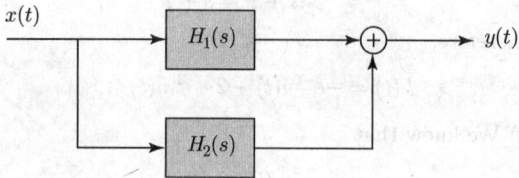

Fig. 9.36

Since $H_1(s)$ and $H_2(s)$ are connected in parallel, the system function of the overall system is

$$H(s) = H_1(s) + H_2(s)$$

$$= \frac{2}{s+8} + \frac{1}{s+2}$$

$$H(s) = \frac{3s + 12}{s^2 + 10s + 16}$$

Also because $H(s) = Y(s)/X(s)$, we may write,

$$\frac{Y(s)}{X(s)} = \frac{3s + 12}{s^2 + 10s + 16}$$

$$Y(s)[s^2 + 10s + 16] = X(s)[3s + 12]$$

$$s^2 Y(s) + 10sY(s) + 16Y(s) = 3sX(s) + 12X(s)$$

Taking the inverse Laplace transform, we obtain

$$\frac{d^2 y(t)}{dt^2} + 10\frac{dy(t)}{dt} + 16y(t) = 3\frac{dx(t)}{dt} + 12x(t)$$

Example 9.76 The input $x(t)$ and output $y(t)$ of a causal LTI system are related through the block diagram representation shown in Fig. 9.37.
(a) Determine a differential equation relating $y(t)$ and $x(t)$.
(b) Is this system stable?

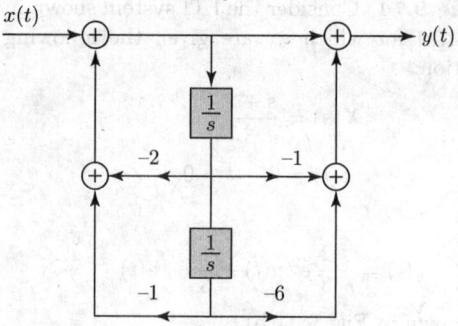

Fig. 9.37

Solution
(a) We may redraw the block diagram as shown in Fig. 9.38. From the left-half section of the Fig. 9.38, it is clear that

$$V(s) = X(s) - 2\frac{V(s)}{s} - \frac{V(s)}{s^2}$$

$$V(s)\left[1 + \frac{2}{s} + \frac{1}{s^2}\right] = X(s)$$

$$\frac{V(s)}{X(s)} = \frac{1}{1 + (2/s) + (1/s^2)}$$

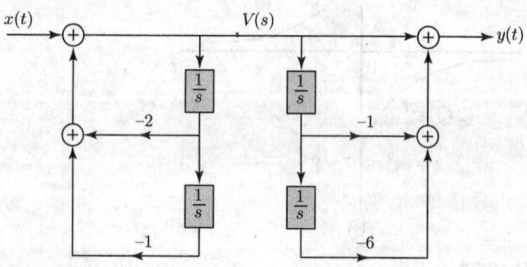

Fig. 9.38

Now, from the right-half section of the Fig. 9.38, it is clear that

$$Y(s) = V(s) - \frac{V(s)}{s} - 6\frac{V(s)}{s^2}$$

$$Y(s) = V(s)\left[1 - \frac{1}{s} - \frac{6}{s^2}\right]$$

$$\frac{Y(s)}{V(s)} = 1 - \frac{1}{s} - \frac{6}{s^2}$$

Therefore,

$$H(s) = \frac{Y(s)}{X(s)} = \left(\frac{V(s)}{X(s)}\right)\left(\frac{Y(s)}{V(s)}\right)$$

$$= \left(\frac{1}{1 + (2/s) + (1/s^2)}\right)$$

$$\left(1 - \frac{1}{s} - \frac{6}{s^2}\right)$$

$$= \frac{1 - (1/s) - (6/s^2)}{1 + (2/s) + 1/s^2}$$

$$H(s) = \frac{Y(s)}{X(s)} = \frac{s^2 - s - 6}{s^2 + 2s + 1}$$

$$Y(s)[s^2 + 2s + 1] = X(s)[s^2 - s - 6]$$

$$s^2 Y(s) + 2s Y(s) + Y(s)$$
$$= s^2 X(s) - s X(s) - 6X(s)$$

The inverse Laplace transform of the above equation yields

$$\frac{d^2 y(t)}{dt^2} + 2\frac{dy(t)}{dt} + y(t) = \frac{d^2 x(t)}{dt^2} - \frac{dx(t)}{dt} - 6x(t)$$

(b) Consider the overall system function

$$H(s) = \frac{s^2 - s - 6}{s^2 + 2s + 1}$$

$$= \frac{(s - 3)(s + 2)}{(s + 1)^2}$$

We know that for a causal system to be stable, all the poles must lie in the left half of the s-plane. The two poles of the system are at -1. We may conclude that this system is stable.

Example 9.77 Find the transfer function and frequency response (magnitude and phase response) for the following:
(a) An ideal delay of T seconds.
(b) An ideal differentiator.
(c) An ideal integrator.
Also plot the magnitude response and phase response.

Solution

(a) *Ideal delay* For an ideal delay of T seconds, the input $x(t)$ and output $y(t)$ are related by

$$y(t) = x(t - T)$$

The Laplace transform of this equation yields

$$Y(s) = X(s)\, e^{-sT}$$

$$H(s) = \frac{Y(s)}{X(s)} = e^{-sT}$$

The frequency response is given by

$$H(s)\Big|_{s=j\omega} = H(j\omega) = e^{-j\omega T} = |H(j\omega)|\, e^{j\angle H(j\omega)}$$

Consequently,

$$|H(j\omega)| = 1 \quad \text{and} \quad \angle H(j\omega) = -\omega T$$

These magnitude and phase responses are shown in Fig. 9.39(a). The magnitude response is constant (unity) for all frequencies. The phase response increases linearly with frequency with a slope of $-T$.

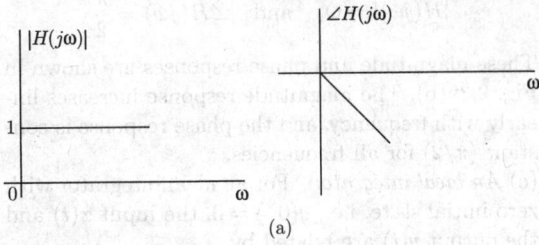

(a)

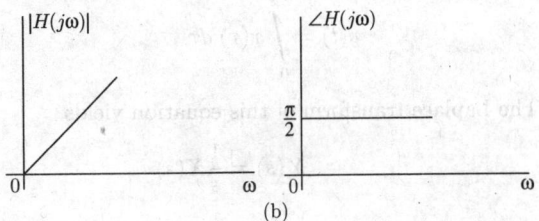

(b)

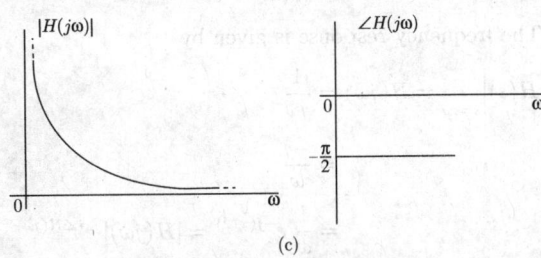

(c)

Fig. 9.39 Magnitude and phase response of (a) an ideal delay, (b) an ideal differentiator and (c) an ideal integrator

(b) *An ideal differentiator*:   For an ideal differentiator, the input $x(t)$ and the output $y(t)$ are related by

$$y(t) = \frac{dx(t)}{dt}$$

The Laplace transform of this equation yields

$$Y(s) = sX(s)$$

$$H(s) = \frac{Y(s)}{X(s)} = s$$

The frequency response is given by

$$H(s)\Big|_{s=j\omega} = H(j\omega) = j\omega$$

$$= \omega\, e^{j(\pi/2)} = |H(j\omega)|\, e^{j\angle H(j\omega)}$$

Consequently,

$$|H(j\omega)| = \omega, \quad \text{and} \quad \angle H(j\omega) = \frac{\pi}{2}$$

These magnitude and phase responses are shown in Fig. 9.39(b). The magnitude response increases linearly with frequency, and the phase response is constant $(\pi/2)$ for all frequencies.

(c) *An ideal integrator*   For an ideal integrator with zero initial state, i.e., $y(0^-) = 0$, the input $x(t)$ and the output $y(t)$ are related by

$$y(t) = \int_0^t x(\tau)\, d\tau$$

The Laplace transform of this equation yields

$$Y(s) = \frac{1}{s}X(s)$$

$$H(s) = \frac{Y(s)}{X(s)} = \frac{1}{s}$$

The frequency response is given by

$$H(s)\Big|_{s=j\omega} = H(j\omega) = \frac{1}{j\omega}$$

$$= \frac{-j}{\omega}$$

$$= \frac{1}{\omega}\, e^{-j(\pi/2)} = |H(j\omega)|\, e^{j\angle H(j\omega)}$$

Consequently,

$$|H(j\omega)| = \frac{1}{\omega} \quad \text{and} \quad \angle H(j\omega) = -\frac{\pi}{2}$$

These magnitude and phase responses are shown in Fig. 9.39(c). The magnitude response is inversely proportional to frequency, and the phase response is constant $(-\pi/2)$ for all frequencies.

Example 9.78 Find   the   inverse   unilateral Laplace transform of

(a) $X(s) = \dfrac{2s^3 - 9s^2 + 4s + 10}{s^2 - 3s - 4}$

(b) $X(s) = \dfrac{s^2 + s - 3}{s^2 + 3s + 2}$

Solution

(a) Since the order of the numerator polynomial is greater than the order of the denominator polynomial, the given rational Laplace transform is improper. We use long division to express $X(s)$ as the sum of a proper rational function and a polynomial in s:

$$
\begin{array}{r}
2s - 3 \\
s^2 - 3s - 4 \overline{\big)\, 2s^3 - 9s^2 + 4s + 10} \\
2s^3 - 6s^2 - 8s \\
\hline
-3s^2 + 12s + 10 \\
-3s^2 + 9s + 12 \\
\hline
3s - 2
\end{array}
$$

Thus, we may write

$$X(s) = 2s - 3 + \frac{3s - 2}{s^2 - 3s - 4}$$

$$X(s) = 2s - 3 + \frac{3s - 2}{(s + 1)(s - 4)}$$

Using a partial fraction expansion to expand the rational function, we obtain

$$X(s) = 2s - 3 + \frac{1}{s + 1} + \frac{2}{s - 4}$$

The inverse Laplace transform of this equation yields

$$x(t) = 2\frac{d\delta(t)}{dt} - 3\delta(t) + e^{-t}u(t) + 2\, e^{4t}u(t)$$

(b) Since the order of the numerator denominator polynomial is equal, the given rational Laplace transform is improper. We use long division to express $X(s)$ as the sum of a proper rational function and a polynomial in s:

$$\begin{array}{r} 1 \\ s^2+3s+2\overline{\smash{\big)}\,s^2+s-3} \\ \underline{s^2+3s+2} \\ -2s-5 \end{array}$$

Thus, we may write

$$X(s) = 1 + \frac{-2s-5}{s^2+3s+2}$$

$$X(s) = 1 - \frac{2s+5}{(s+1)(s+2)}$$

Using a partial fraction expansion to expand the rational function, we obtain

$$X(s) = 1 - \frac{3}{s+1} + \frac{1}{s+2}$$

The inverse Laplace transform of this equation yields

$$x(t) = \delta(t) - 3\,e^{-t}u(t) + e^{-2t}u(t)$$

Example 9.79 Find the inverse unilateral Laplace transform of

$$X(s) = \frac{4s^2+6}{s^3+s^2-2}$$

Solution
Consider the given Laplace transform

$$X(s) = \frac{4s^2+6}{s^3+s^2-2}$$

$$X(s) = \frac{4s^2+6}{(s-1)(s^2+2s+2)}$$

The partial fraction expansion for $X(s)$ takes the form

$$X(s) = \frac{A}{s-1} + \frac{Bs+C}{s^2+2s+2} \qquad (9.91)$$

Multiply both sides of the above equation by $s-1$ and evaluate at $s=1$ to obtain

$$A = (s-1)X(s)\Big|_{s=1}$$

$$A = (s-1)\frac{4s^2+6}{(s-1)(s^2+2s+2)}\Big|_{s=1}$$

$$A = 2$$

Now consider Eq. (9.91),

$$X(s) = \frac{A}{s-1} + \frac{Bs+C}{s^2+2s+2}$$

$$\frac{4s^2+6}{(s-1)(s^2+2s+2)} = \frac{2}{s-1} + \frac{Bs+C}{s^2+2s+2}$$

$$4s^2+6 = 2(s^2+2s+2) + (Bs+C)(s-1)$$

$$= (2+B)s^2 + (4-B+C)s + (4-C)$$

Equating coefficients of s^2 gives $B=2$, and equating coefficients of s^0 gives $C=-2$. Hence,

$$X(s) = \frac{2}{s-1} + \frac{2s-2}{s^2+2s+2}$$

$$= \frac{2}{s-1} + \frac{2s-2+2-2}{(s+1)^2+1}$$

$$= \frac{2}{s-1} + 2\frac{s+1}{(s+1)^2+1} - 4\frac{1}{(s+1)^2+1}$$

The inverse Laplace transform of this equation yields

$$x(t) = 2\,e^t u(t) + 2\,e^{-t}\cos(t)u(t) - 4\,e^{-t}\sin(t)u(t)$$

Example 9.80 Find the inverse unilateral Laplace transform of

$$X(s) = \frac{1}{e^{s+3}}\frac{s^2}{(s+1)(s+2)}$$

Solution
Given that

$$X(s) = \frac{1}{e^{s+3}}\frac{s^2}{(s+1)(s+2)}$$

$$= \frac{1}{e^{s+3}}G(s)$$

where

$$G(s) = \frac{s^2}{(s+1)(s+2)}$$

$$= \frac{s^2}{s^2+3s+2}$$

Since the order of the numerator denominator polynomial is equal, the rational Laplace transform $G(s)$ is improper. We use long division to express $G(s)$ as the sum of a proper rational function and a polynomial in s:

$$s^2 + 3s + 2 \overline{)\,s^2}$$
$$\underline{s^2 + 3s + 2}$$
$$-3s - 2$$

Thus, we may write

$$G(s) = 1 + \frac{-3s - 2}{s^2 + 3s + 2}$$

$$G(s) = 1 - \frac{3s + 2}{(s + 1)(s + 2)}$$

Using a partial fraction expansion to expand the rational function, we obtain

$$G(s) = 1 + \frac{1}{s + 1} - \frac{4}{s + 2}$$

The inverse Laplace transform of this equation yields

$$g(t) = \delta(t) + e^{-t}u(t) - 4\,e^{-2t}u(t)$$

Also, because

$$X(s) = \frac{1}{e^{s+3}}G(s)$$

$$= \frac{1}{e^3}\,e^{-s}G(s)$$

Therefore,

$$x(t) = \frac{1}{e^3}g(t - 1)$$

$$= \frac{1}{e^3}[\delta(t - 1) + e^{-(t-1)}u(t - 1)$$

$$- 4\,e^{-2(t-1)}u(t - 1)]$$

Example 9.81 Find the Laplace transform of the half-wave rectification of $\sin(\omega t)$ shown in Fig. 9.40(a).

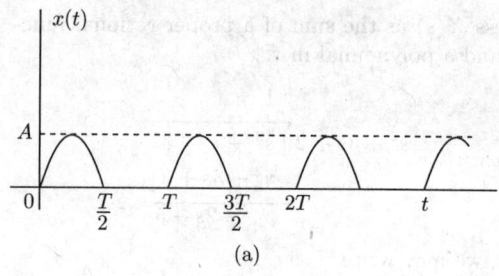

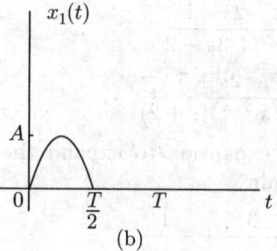

Fig. 9.40 (a) Half-wave rectified sine wave and (b) its first cycle

Solution
We know that the Laplace transform of a causal periodic signals is given by [Eq. (9.49)]

$$X(s) = \frac{X_1(s)}{1 - e^{-sT_0}}$$

where $X_1(s)$ is the Laplace transform of the first cycle $x_1(t)$ of the causal periodic signal $x(t)$. The given half-wave rectified sinusoidal wave is a causal periodic signal with period $T_0 = T$. The first cycle $x_1(t)$ of $x(t)$ is as shown in Fig. 9.40(b). The first cycle $x_1(t)$ can be written as

$$x_1(t) = \begin{cases} A\sin(\omega t), & 0 \le t \le \dfrac{T}{2} \\ 0, & \dfrac{T}{2} \le t \le T \end{cases}$$

$$x_1(t) = A\sin(\omega t)u(t)$$

$$+ A\sin\left(\omega\left(t - \frac{T}{2}\right)\right)u\left(t - \frac{T}{2}\right)$$

Taking the Laplace transform of the above equation, we obtain

$$\mathcal{L}[x_1(t)] = A\mathcal{L}[\sin(\omega t)u(t)]$$

$$+ A\mathcal{L}\left[\sin\left(\omega\left(t - \frac{T}{2}\right)\right)u\left(t - \frac{T}{2}\right)\right]$$

$$X_1(s) = A\frac{\omega}{s^2 + \omega^2} + A\frac{\omega}{s^2 + \omega^2} e^{-sT/2}$$

$$= \frac{A\omega}{s^2 + \omega^2}\left[1 + e^{-sT/2}\right]$$

Substituting $X_1(s)$ into the expression of $X(s)$, we get

$$X(s) = \frac{X_1(s)}{1 - e^{-sT_0}}$$

$$= \frac{A\omega}{s^2 + \omega^2}\frac{1 + e^{-sT/2}}{1 - e^{-sT}}$$

$$= \frac{A\omega}{s^2 + \omega^2}\frac{1 + e^{-sT/2}}{(1 + e^{-sT/2})(1 - e^{-sT/2})}$$

$$X(s) = \frac{A\omega}{s^2 + \omega^2}\left(\frac{1}{1 - e^{-sT/2}}\right)$$

Example 9.82 Find the Laplace transform of the full-wave rectification of $\sin(\omega_0 t)$ shown in Fig. 9.41(a).

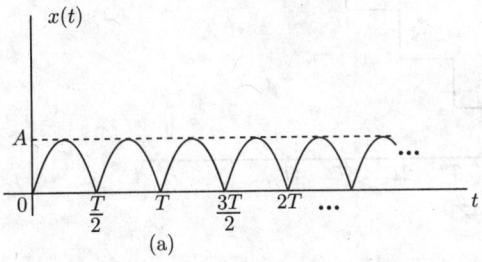

(a)

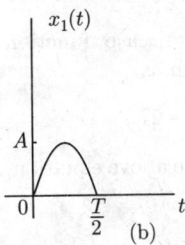

(b)

Fig. 9.41 (a) Full-wave rectified sine wave and (b) its first cycle.

Solution

We know that the Laplace transform of a causal periodic signals is given by [Eq. (9.49)]

$$X(s) = \frac{X_1(s)}{1 - e^{-sT_0}}$$

where $X_1(s)$ is the Laplace transform of the first cycle $x_1(t)$ of the causal periodic signal $x(t)$. The given half-wave rectified sinusoidal wave is a causal periodic signal with period $T_0 = T/2$. The first cycle $x_1(t)$ of $x(t)$ is as shown in Fig. 9.41(b). The first cycle $x_1(t)$ can be written as

$$x_1(t) = A\sin(\omega t) \quad 0 \le t \le \frac{T}{2} \quad \omega = \frac{2\pi}{T}$$

$$x_1(t) = A\sin(\omega t)u(t)$$

$$+ A\sin\left(\omega\left(t - \frac{T}{2}\right)\right)u\left(t - \frac{T}{2}\right)$$

Taking the Laplace transform of the above equation, we obtain

$$\mathcal{L}[x_1(t)] = A\mathcal{L}[\sin(\omega t)u(t)]$$

$$+ A\mathcal{L}\left[\sin\left(\omega\left(t - \frac{T}{2}\right)\right)u\left(t - \frac{T}{2}\right)\right]$$

$$X_1(s) = A\frac{\omega}{s^2 + \omega^2} + A\frac{\omega}{s^2 + \omega^2} e^{-sT/2}$$

$$= \frac{A\omega}{s^2 + \omega^2}\left[1 + e^{-sT/2}\right]$$

Substituting $X_1(s)$ into the expression of $X(s)$, we get

$$X(s) = \frac{X_1(s)}{1 - e^{-sT_0}}$$

$$= \frac{A\omega}{s^2 + \omega^2}\frac{1 + e^{-sT/2}}{1 - e^{-sT/2}}$$

$$= \frac{A\omega}{s^2 + \omega^2}\frac{e^{-sT/4}\left(e^{sT/4} + e^{-sT/4}\right)}{e^{-sT/4}\left(e^{sT/4} - e^{-sT/4}\right)}$$

$$X(s) = \frac{A\omega}{s^2 + \omega^2}\frac{\left(e^{sT/4} + e^{-sT/4}\right)}{\left(e^{sT/4} - e^{-sT/4}\right)}$$

$$X(s) = \frac{A\omega}{s^2 + \omega^2}\coth\left(\frac{sT}{4}\right)$$

Example 9.83 Using convolution, find the inverse Laplace transform $h(t)$ of

(a) $H(s) = \dfrac{1}{(s^2 + 1)^2}$

(b) $H(s) = \dfrac{1}{s^2(s - a)}$

Solution
(a) Given that

$$H(s) = \frac{1}{(s^2 + 1)^2}$$

$$= \left(\frac{1}{s^2 + 1}\right)\left(\frac{1}{s^2 + 1}\right)$$

Taking the inverse Laplace transform of this equation yields

$$\mathcal{L}^{-1}[H(s)] = \mathcal{L}^{-1}\left[\frac{1}{s^2 + 1}\right] * \mathcal{L}^{-1}\left[\frac{1}{s^2 + 1}\right]$$

$$h(t) = \sin(t)u(t) * \sin(t)u(t)$$

$$= \int_{-\infty}^{\infty} \sin(\tau)u(\tau)\sin(t - \tau)u(t - \tau)\, d\tau$$

$$= \int_{0}^{t} \sin(\tau)\sin(t - \tau)\, d\tau$$

$$= \int_{0}^{t} \frac{1}{2}[\cos(2\tau - t) - \cos(t)]d\tau$$

$$= \frac{1}{2}\int_{0}^{t} \cos(2\tau - t)\, d\tau - \frac{1}{2}\int_{0}^{t} \cos(\tau)\, d\tau$$

$$h(t) = \frac{1}{2}\sin(t) - \frac{1}{2}t\cos(t) \quad t \geq 0$$

(b) Given that

$$H(s) = \frac{1}{s^2(s - a)}$$

$$= \left(\frac{1}{s^2}\right)\left(\frac{1}{s - a}\right)$$

Taking the inverse Laplace transform of this equation yields

$$\mathcal{L}^{-1}[H(s)] = \mathcal{L}^{-1}\left[\frac{1}{s^2}\right] * \mathcal{L}^{-1}\left[\frac{1}{s - a}\right]$$

$$h(t) = tu(t) * e^{at}u(t)$$

$$h(t) = \int_{-\infty}^{\infty} \tau u(\tau)\, e^{a(t - \tau)}u(t - \tau)\, d\tau$$

$$= \int_{0}^{t} \tau\, e^{a(t - \tau)}\, d\tau$$

$$= e^{at}\int_{0}^{t} \tau\, e^{-a\tau}\, d\tau$$

$$h(t) = \frac{1}{a^2}[e^{at} - at - 1] \quad t \geq 0$$

Example 9.84 Find the Laplace transform of the staircase function shown in Fig. 9.42.

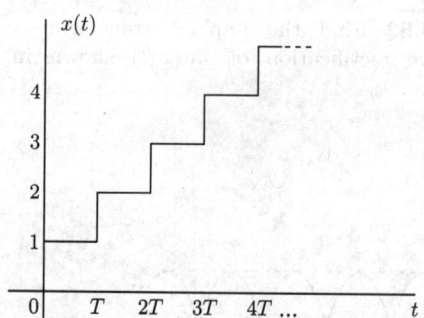

Fig. 9.42

Solution
Using step function $u(t)$, the staircase function (shown in Fig. 9.42) can be written as

$$x(t) = u(t) + u(t - T) + u(t - 2T) + \cdots$$

Taking the Laplace transform of the above equation, we obtain

$$X(s) = \frac{1}{s} + \frac{1}{s}e^{-sT} + \frac{1}{s}e^{-2sT} + \cdots$$

$$= \frac{1}{s}[1 + e^{-sT} + e^{-2sT} + \cdots]$$

$$X(s) = \frac{1}{s(1 - e^{-sT})}$$

Example 9.85 Find the Laplace transform of the signals shown in Fig. 9.43.

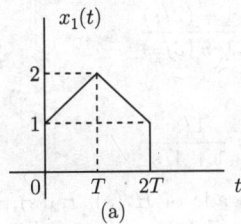

(a)

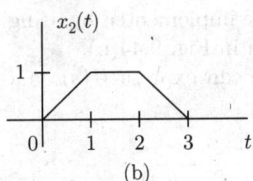

(b)

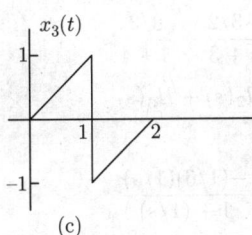

(c)

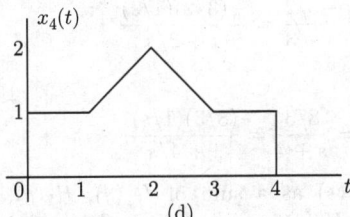

(d)

Fig. 9.43

Solution

(a) Using step function $u(t)$ and ramp function $r(t)$, the given signal $x_1(t)$ can be written as

$$x_1(t) = u(t) + r(t) - r(t-1) - r(t-1)$$
$$+ r(t-2) - u(t-2)$$
$$= u(t) + r(t) - 2r(t-1) + r(t-2)$$
$$- u(t-2)$$

Taking the Laplace transform of the above equation, we obtain

$$X_1(s) = \frac{1}{s} + \frac{1}{s^2} - \frac{2}{s^2} e^{-s} + \frac{1}{s^2} e^{-2s} - \frac{1}{s} e^{-2s}$$
$$= \frac{1}{s}[1 - e^{-2s}] + \frac{1}{s^2}[1 - 2e^{-s} + e^{-2s}]$$

(b) Using ramp function $r(t)$, the given signal $x_2(t)$ can be written as

$$x_2(t) = r(t) - r(t-1) - r(t-2) + r(t-3)$$

Taking the Laplace transform of the above equation, we obtain

$$X_2(s) = \frac{1}{s^2} - \frac{1}{s^2} e^{-s} - \frac{1}{s^2} e^{-2s} + \frac{1}{s^2} e^{-3s}$$
$$= \frac{1}{s^2}[1 - e^{-s} - e^{-2s} + e^{-3s}]$$

(c) Using step function $u(t)$ and ramp function $r(t)$, the given signal $x_3(t)$ can be written as

$$x_3(t) = r(t) - 2u(t-1) - r(t-2)$$

Taking the Laplace transform of the above equation, we obtain

$$X_3(s) = \frac{1}{s^2} - \frac{2}{s} e^{-s} - \frac{1}{s^2} e^{-2s}$$
$$= \frac{1}{s^2}[1 - 2s\,e^{-s} - e^{-2s}]$$

(d) Using step function $u(t)$ and ramp function $r(t)$, the given signal $x_4(t)$ can be written as

$$x_4(t) = u(t) + r(t-1) - r(t-2) - r(t-2)$$
$$+ r(t-3) - u(t-4)$$
$$= u(t) + r(t-1) - 2r(t-2) + r(t-3)$$
$$- u(t-4)$$

Taking the Laplace transform of the above equation, we obtain

$$X_4(s) = \frac{1}{s} + \frac{1}{s^2} e^{-s} - \frac{2}{s^2} e^{-2s} + \frac{1}{s^2} e^{-3s} - \frac{1}{s} e^{-4s}$$
$$= \frac{1}{s}[1 - e^{-4s}] + \frac{1}{s^2}[e^{-s} - 2e^{-2s} + e^{-3s}]$$

Example 9.86 Realize

$$H(s) = \frac{s(s+2)}{(s+1)(s+3)(s+4)}$$

by canonic direct (direct form II), series, parallel, and the transposed form of canonic direct.

Solution

(a) *Canonic realization:* Given that

$$H(s) = \frac{s(s+2)}{(s+1)(s+3)(s+4)}$$

$$= \frac{s^2 + 2s}{s^3 + 8s^2 + 19s + 12}$$

$$= \frac{(1/s) + (2/s^2)}{1 + (8/s) + (19/s^2) + (12/s^3)}$$

$$= \underbrace{\left(\frac{1}{1 + (8/s) + (19/s^2) + (12/s^3)}\right)}_{H_2(s)} \underbrace{(1/s + 2/s^2)}_{H_1(s)}$$

where

$$H_2(s) = \frac{V(s)}{X(s)}$$

$$= \frac{1}{1 + (8/s) + (19/s^2) + (12/s^3)}$$

$$V(s) = X(s) - 8\frac{V(s)}{s} - 19\frac{V(s)}{s^2} - 12\frac{V(s)}{s^3}$$

and

$$H_1(s) = \frac{Y(s)}{V(s)} = \frac{1}{s} + \frac{2}{s^2}$$

$$Y(s) = \frac{V(s)}{s} + 2\frac{V(s)}{s^2}$$

The given transfer function $H(s) = (s(s+2))/(s+1)(s+3)(s+4)$ is of the third order; therefore, we need only three integrator for its realization as shown in Fig. 9.44(a).

(b) *Cascade realization* Given that

$$H(s) = \frac{s(s+2)}{(s+1)(s+3)(s+4)}$$

$$H(s) = \left(\frac{s}{s+1}\right)\left(\frac{s+2}{s+3}\right)\left(\frac{1}{s+4}\right) = H_1(s)H_2(s)$$

where

$$H_1(s) = \frac{s}{s+1} = \frac{1}{1 + (1/s)}$$

$$H_2(s) = \frac{s+2}{s+3} = \frac{1 + (2/s)}{1 + (3/s)}$$

and

$$H_3(s) = \frac{1}{s+4} = \frac{1/s}{1 + 4/s}$$

We can realize $H(s)$ as a cascade of $H_1(s)$, $H_2(s)$, and $H_3(s)$. $H_1(s)$, $H_2(s)$, and $H_3(s)$ are first-order transfer functions and can be implemented by using canonic realization as shown in Fig. 9.44(b).

(c) *Parallel realization* We can express $H(s)$ as a sum of partial fractions as

$$H(s) = \frac{s(s+2)}{(s+1)(s+3)(s+4)}$$

$$= \frac{-1/6}{s+1} + \frac{-3/2}{s+3} + \frac{8/3}{s+4}$$

$$H(s) = H_1(s) + H_2(s) + H_3(s)$$

where

$$H_1(s) = \frac{-1/6}{s+1} = \frac{-(1/6)(1/s)}{1 + (1/s)}$$

$$H_2(s) = \frac{-3/2}{s+3} = \frac{-(3/2)(1/s)}{1 + 3/s}$$

and

$$H_3(s) = \frac{8/3}{s+4} = \frac{(8/3)(1/s)}{1 + 4/s}$$

We can realize $H(s)$ as a sum of $H_1(s)$, $H_2(s)$, and $H_2(s)$. $H_1(s)$, $H_2(s)$, and $H_3(s)$ are first-order transfer functions and can be implemented by using canonic realization as shown in Fig. 9.44(c).

(d) *Transposed realization of canonic direct form* Fig. 9.44(d) shows the transposed version of the canonic direct form realization in Fig. 9.44(a) found by

(i) reversing all paths.

(ii) replacing pick-off nodes by adders and vice versa.

(iii) interchanging the input and output nodes.

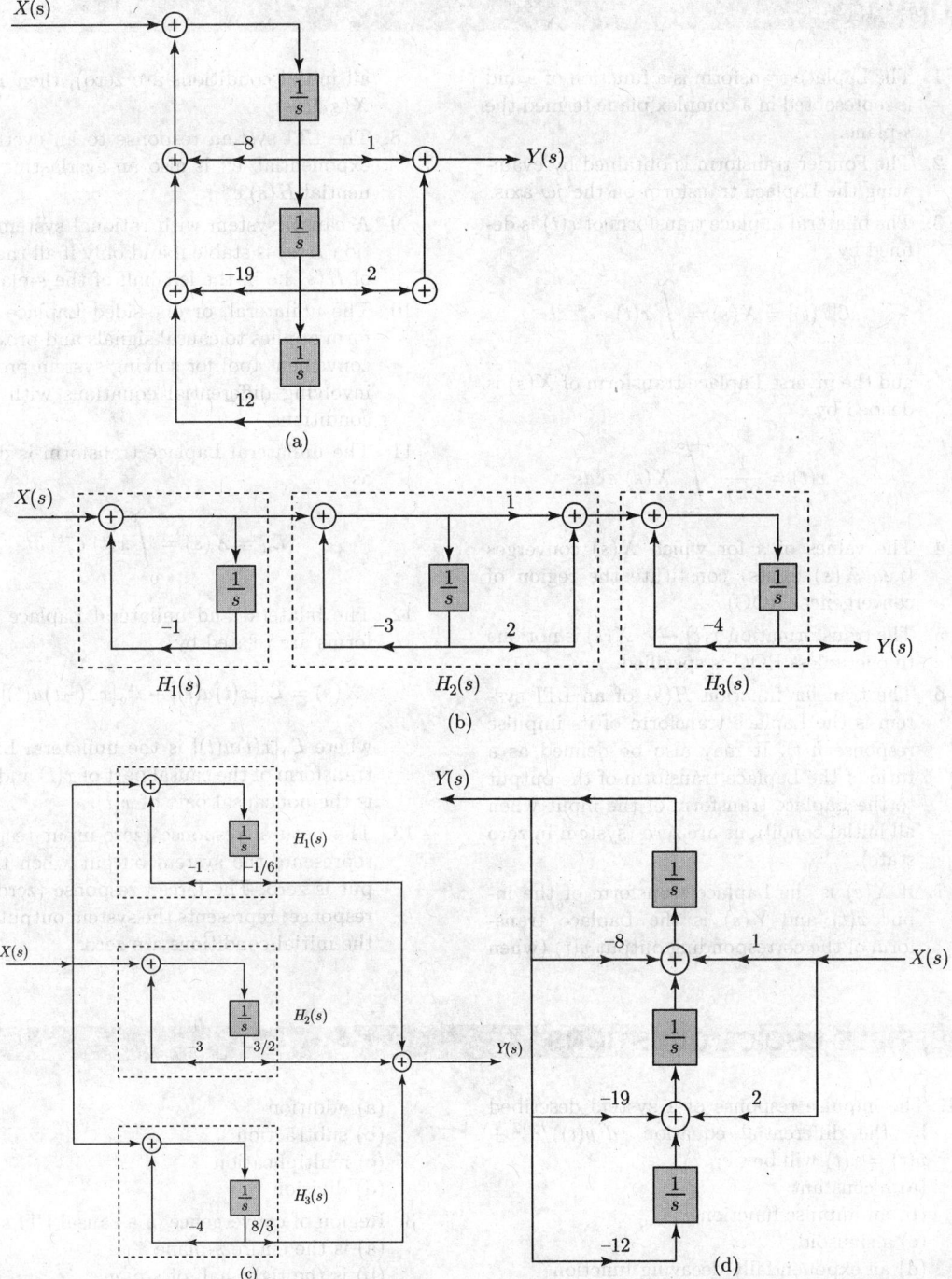

Fig. 9.44 Realization of $H(s)$ for Example 9.86 using (a) canonic direct form-II, (b) series, (c) parallel and (d) transposed form of canonic direct form-II

SUMMARY

1. The Laplace transform is a function of s and is represented in a complex plane termed the s-plane.

2. The Fourier transform is obtained by evaluating the Laplace transform on the $j\omega$-axis.

3. The bilateral Laplace transform of $x(t)$ is defined by

$$\mathcal{L}[x(t)] = X(s) = \int\limits_{-\infty}^{\infty} x(t) \, e^{-st} \, dt$$

and the inverse Laplace transform of $X(s)$ is defined by

$$x(t) = \frac{1}{2\pi j} \int\limits_{\sigma-j\infty}^{\sigma+j\infty} X(s) \, e^{st} ds$$

4. The values of s for which $X(s)$ converges (i.e., $X(s)$ exists) constitute the region of convergence (ROC).

5. The transformation $x(t) \longleftrightarrow X(s)$ is not one to one unless ROC is specified.

6. The transfer function $H(s)$ of an LTI system is the Laplace transform of its impulse response $h(t)$. It may also be defined as a ratio of the Laplace transform of the output to the Laplace transform of the input when all initial conditions are zero (system in zero state).

7. If $X(s)$ is the Laplace transform of the input $x(t)$ and $Y(s)$ is the Laplace transform of the corresponding output $y(t)$ (when all initial conditions are zero), then $Y(s) = X(s)H(s)$.

8. The LTI system response to an everlasting exponential e^{st} is also an everlasting exponential $H(s) \, e^{st}$.

9. A causal system with rational system function $H(s)$ is stable if and only if all the poles of $H(s)$ lie in the left-half of the s-plane.

10. The unilateral, or one-sided Laplace transform applies to causal signals and provides a convenient tool for solving system problems involving differential equations with initial conditions.

11. The unilateral Laplace transform is defined as

$$\mathcal{L}_{u} = X(s) = \int\limits_{0^-}^{\infty} x(t) \, e^{-st} \, dt$$

12. The bilateral and unilateral Laplace transforms are related by

$$X(s) = \mathcal{L}_{u}[x(t)u(t)] + \mathcal{L}_{u}[x_-(-t)u(t)]\Big|_{s \to -s}$$

where $\mathcal{L}_{u}[x(t)u(t)]$ is the unilateral Laplace transform of the causal part of $x(t)$ and $x_-(t)$ is the noncausal part of $x(t)$.

13. The natural response (zero-input response) represents the system output when the input is zero. The forced response (zero-state response) represents the system output when the initial conditions are zero.

MULTIPLE-CHOICE QUESTIONS

1. The impulse response of a system described by the differential equation $(d^2 y(t))/dt^2 + y(t) = x(t)$ will be
   (a) a constant
   (b) an impulse function
   (c) a sinusoid
   (d) an exponentially decaying function

2. Laplace transform converts convolution of time-signals to

   (a) addition
   (b) subtraction
   (c) multiplication
   (d) division

3. Region of convergence of a causal LTI system
   (a) is the entire s-plane
   (b) is the right-half of s-plane
   (c) left-half of s-plane
   (d) does not exist

4. Laplace transform of t^n is always equal to
   (a) n/s^n
   (b) $n!/s^n$
   (c) $n!/s^{n+1}$
   (d) all

5. The final value of $x(t) \xleftrightarrow{\mathcal{L}_u} X(s) =$ $(5s + 6)/(s^3 + 6s^2 + 3s)$ is
   (a) 2
   (b) 1
   (c) 5/6
   (d) 5

6. The Laplace transform of $u(t)$ is
   (a) $1/s$
   (b) s
   (c) $1/s^2$
   (d) 1

7. The Laplace transform of $u(t)$ is $A(s)$ and the Fourier transform of $u(t)$ is $B(j\omega)$. Then
   (a) $B(j\omega) = A(s)\Big|_{s=j\omega}$
   (b) $A(s) = 1/s$, but $B(j\omega) \neq \dfrac{1}{j\omega}$
   (c) $A(s) \neq 1/s$, but $B(j\omega) = \dfrac{1}{j\omega}$
   (d) $A(s) \neq 1/s$, but $B(j\omega) \neq \dfrac{1}{j\omega}$

8. The Laplace transform of $x(t - t_0)$ is
   (a) $e^{st_0} X(s)$
   (b) $e^{-st_0} X(s)$
   (c) $X(s - s_0)$
   (d) none of the above

9. If $x(t) \xleftrightarrow{\mathcal{L}_u} X(s)$, then $\mathcal{L}_u \left[\dfrac{dx(t)}{dt} \right]$ is given by
   (a) $\dfrac{dX(s)}{ds}$
   (b) $\dfrac{X(s)}{s} - \dfrac{x(0^-)}{s}$
   (c) $sX(s) - x(0^-)$
   (d) $sX(s) - sx(0^-)$

10. If the Laplace transform of $x(t)$ is $\omega/(s^2 + \omega^2)$, then the value of $\lim_{t\to\infty} x(t)$

(a) cannot be determined
(b) is zero
(c) is unity
(d) is infinity

11. The unit impulse response of an LTI system is the unit step function $u(t)$. For $t > 0$, the response of the system to an excitation $e^{-at}u(t)$, $a > 0$, will be
    (a) ae^{-at}
    (b) $(1 - e^{-at})/a$
    (c) $a(1 - e^{-at})$
    (d) $1 - e^{-at}$

12. If R_x is the ROC of $x(t)$ and R_y is the ROC of $y(t)$, then the ROC of $x(t) * y(t)$ is
    (a) $R_x + R_y$
    (b) $R_x - R_y$
    (c) $R_x \cap R_y$
    (d) $R_x \cup R_y$

13. If the notation $*$ is used to denote the convolution and $x(t) \longleftrightarrow X(s)$ and $y(t) \longleftrightarrow Y(s)$ make the Laplace transform pair, then $x(t)y(t) \longleftrightarrow F(s)$ given by
    (a) $X(s) * Y(s)$
    (b) $X(s)Y(s)$
    (c) $\dfrac{1}{2\pi}[X(s) * Y(s)]$
    (d) $\dfrac{1}{2\pi j}[X(s) * Y(s)]$

14. Region of convergence of $X(s)$ contain
    (a) zeros
    (b) poles
    (c) no zero
    (d) no pole

15. If the system is causal and stable, the system poles must lie
    (a) on the $j\omega$-axis
    (b) on the left-half of the s-plane
    (c) on the right-half of the s-plane
    (d) both (a) and (c)

PROBLEMS

9.1 Find the Laplace transform of the following signals:
(a) $x_1(t) = 3u(t) - 3u(t - 3)$
(b) $x_2(t) = r(t) - r(t - 1) - u(t - 2)$
(c) $x_3(t) = 10u(t) - 10\,e^{-t}u(t)$

9.2 Find the Laplace transform of the following signals:
(a) $x_1(t) = (t - 1)u(t)$
(b) $x_2(t) = e^{-(t+4)}u(t)$
(c) $x_3(t) = e^{(t-3)}u(t - 4)$

9.3 Find the Laplace transform of the following signals:

(a) $x_1(t) = e^{-5t}[u(t) - u(t-5)]$

(b) $x_2(t) = [e^{-3t} - e^{-10t} + 2\cos(3t)]u(t)$

9.4 Use the Laplace transform to determine the output $y(t)$ of the system represented by the differential equation:

$$\frac{d^2y(t)}{dt^2} + 4\frac{dy(t)}{dt} + 3y(t) = e^{-t}$$

$$\dot{y}(0^-) = y(0^-) = 1$$

9.5 State the initial-value and final-value theorems of Laplace transforms. Compute the initial and final values for

$$X(s) = \frac{3s+4}{s(s+1)(s+2)^2}$$

9.6 Consider the signal $x(t) = e^{-t}u(t) + e^{-2t}u(t)$. Express its Laplace transform in the form $X(s) = K(N(s)/D(s))$, where K = system constant. Identify the ROC. Indicate the poles and zeros in the s-plane.

9.7 Determine the forced and natural responses of the system described by the following differential equations with the specified input and initial conditions:

$$\frac{dy(t)}{dt} + 3y(t) = 4x(t) \quad x(t) = \cos(2t)u(t)$$

$$y(0^-) = -2$$

$$\frac{d^2y(t)}{dt^2} + 4y(t) = 8x(t) \quad x(t) = u(t),$$

$$y(0^-) = 1 \quad \dot{y}(0^-) = 2$$

9.8 The differential equation characterizing a continuous LTI system is given by

$$2\frac{d^2y(t)}{dt^2} + 3\frac{dy(t)}{dt} + y(t) = u(t)$$

With initial conditions $y(0^-) = -1$ and $\dot{y}(0^-) = 1$. Determine $Y(s)$ as the sum of zero-input response and zero-state response. Also determine $y(t)$.

9.9 Using the Laplace transform, solve the following differential equations and in each case determine the zero-input and zero-state components of the solution:

(a)

$$\frac{d^2y(t)}{dt^2} + 3\frac{dy(t)}{dt} + 2y(t) = \frac{dx(t)}{dt}$$

if $y(0^-) = \dot{y}(0^-) = 0$ and $x(t) = u(t)$.

(b)

$$\frac{d^2y(t)}{dt^2} + 4\frac{dy(t)}{dt} + 4y(t) = \frac{dx(t)}{dt} + x(t)$$

if $y(0^-) = 2$, $\dot{y}(0^-) = 1$ and $x(t) = e^{-t}u(t)$.

9.10 For a system with transfer function

$$H(s) = \frac{2s+3}{s^2 + 2s + 5}$$

(a) Find the zero-state response for input $x(t)$ of (i) $10u(t)$ and (ii) $u(t-5)$.

(b) For this system, write the differential equation relating the output $y(t)$ to the input $x(t)$.

9.11 An LTI system has a step response given by $s(t) = e^{-t}u(t) - e^{-2t}u(t)$. Determine the output of this system $y(t)$ given an input $x(t) = \delta(t - \pi) - \cos(\sqrt{3}t)u(t)$.

9.12 Find the inverse Laplace transform of the following functions:

(a) $X(s) = \dfrac{2s+5}{(s+2)(s+3)} \quad -3 < \Re\{s\} < -2$

(b) $X(s) = \dfrac{2s-5}{(s-2)(s-3)} \quad 2 < \Re\{s\} < 3$

(c) $X(s) = \dfrac{2s+3}{(s+1)(s+2)} \quad \Re\{s\} > -1$

(d) $X(s) = \dfrac{2s+3}{(s+1)(s+2)} \quad \Re\{s\} < -2$

9.13 An absolutely integrable signal $x(t)$ has a pole at $s = \pi$. It is possible that other poles may be present.

(a) Can $x(t)$ be left-sided? Explain.

(b) Can $x(t)$ be right-sided? Explain.

(c) Can $x(t)$ be two-sided? Explain.

(d) Can $x(t)$ be of finite duration? Explain.

9.14 The Laplace transform of a signal $x(t)$ that is zero for $t < 0$ is

$$X(s) = \frac{s^3 + 2s^2 + 3s + 2}{s^4 + 2s^3 + 2s^2 + 2s + 2}$$

Determine the Laplace transform of the following signals:

(a) $y(t) = 3x\left(\dfrac{t}{3}\right)$

(b) $y(t) = tx(t)$

(c) $y(t) = tx(t-1)$

(d) $y(t) = \dfrac{dx(t)}{dt}$

(e) $y(t) = (t-1)x(t-1) + \dfrac{dx(t)}{dt}$

(f) $y(t) = \int\limits_{0}^{t} x(\tau)\, d\tau$

9.15 Determine the initial and final values of each of the signals whose Laplace transform are as follows without computing the inverse Laplace transform. If there is no final value, state why not.

(a) $X(s) = \dfrac{1}{s+a}$

(b) $X(s) = \dfrac{1}{(s+a)^n}$

(c) $X(s) = \dfrac{6}{s(s^2+25)}$

(d) $X(s) = \dfrac{s+2}{s+3}$

(e) $X(s) = \dfrac{s^2+s+3}{s^3+4s^2+2s+2}$

9.16 Find the following convolutions using Laplace transform:

(a) $e^{at}u(t) * e^{bt}u(t)$, $a \neq b$.

(b) $e^{at}u(t) * e^{at}u(t)$

(c) $tu(t) * e^{at}u(t)$

(d) $\text{rect}\left(\dfrac{t}{T}\right) * \text{rect}\left(\dfrac{t}{T}\right)$

9.17 Realize

(a)

$$H(s) = \dfrac{s(s+2)}{(s+1)(s+3)(s+4)}$$

(b)

$$H(s) = \dfrac{s^3}{(s+1)^2(s+2)(s+3)}$$

by canonic, cascade, and parallel forms.

9.18 Realize the transfer function in Problem 9.17

by using the transposed form of the realizations found in Problem 9.17.

9.19 Let

$$g(t) = x(t) + \alpha x(-t)$$

where

$$x(t) = \beta\, e^{-t}u(t)$$

and the Laplace transform of $g(t)$ is

$$G(s) = \dfrac{s}{s^2-1} \qquad -1 < \Re\{s\} < 1$$

Determine the values of the constant α and β.

9.20 Obtain the time function $f(t)$ whose Laplace transform is

$$F(s) = \dfrac{s^2+3s+1}{(s+1)^3(s+2)^2}$$

9.21 Find the inverse Laplace transform of the following:

(a) $X(s) = \dfrac{s^2}{(s+a)^2+b^2}$

(b) $X(s) = ln\left(\dfrac{s+1}{s+2}\right)$

9.22 The autocorrelation function $r_{xx}(t)$ of a signal is given by

$$r_{xx}(t) = \int\limits_{-\infty}^{\infty} x(\tau)x(\tau+t)\, d\tau$$

Derive an expression for $R_{ss}(s) = \mathcal{L}[r_{xx}(\tau)]$ in terms of $X(s)$, where $X(s) = \mathcal{L}[x(t)]$.

ANSWERS TO MULTIPLE-CHOICE QUESTIONS

1. **(c)**	2. **(c)**	3. **(b)**	4. **(b)**	5. **(a)**	6. **(a)**	7. **(b)**	8. **(b)**	9. **(c)**
10. **(a)**	11. **(b)**	12. **(c)**	13. **(d)**	14. **(d)**	15. **(b)**			

z-Transform

10.1 INTRODUCTION

In Chapter 9, we developed the Laplace transform as an extension of the continuous-time Fourier transform. This extension was motivated in part by the fact that it can be applied to a broader class of signals compared to Fourier transform because there are many signals for which the Fourier transform does not converge but the Laplace transform does.

In this chapter, we study the z-transform, which is the discrete-time counterpart of the Laplace transform. Just as the Laplace transform provides us a frequency-domain technique for analysing signals for which the CTFT does not exist, the z-transform enables us to analyse certain discrete-time signals that do not have a discrete-time Fourier transform (DTFT). The z-transform plays the same role in the analysis of discrete-time signals and LTI systems as the Laplace transform does in the analysis of continuous-time signals and LTI systems.

The z-transform comes into two varieties: *bilateral*, or two-sided z-transform, and *unilateral*, or one-sided z-transform. The bilateral z-transform offers insight into the nature of system characteristics such as stability, causality, and frequency response. The unilateral z-transform is a convenient tool for solving difference equations with initial conditions.

10.2 BILATERAL (TWO-SIDED) z-TRANSFORM

Consider applying a complex exponential input $x(n) = z^n$ to an LTI system with impulse response $h(n)$. The system output is given by

$$y(n) = h(n) * x(n) = \sum_{k=-\infty}^{\infty} h(k)x(n-k)$$

$$y(n) = \sum_{k=-\infty}^{\infty} h(k)z^{n-k}$$

$$= z^n \sum_{k=-\infty}^{\infty} h(k)z^{-k}$$

$$y(n) = H(z)z^n$$

where

$$H(z) = \sum_{k=-\infty}^{\infty} h(k) z^{-k}$$

or equivalently,

$$H(z) = \sum_{n=-\infty}^{\infty} h(n) z^{-n} \tag{10.1}$$

$H(z)$ is known as the transfer function or system function of the LTI system. We know that a signal for which the system output is a constant times, the input is referred to as an *eigenfunction* of the system and the amplitude factor is referred to as the system's *eigenvalue*. Hence, we identify z^n as an eigenfunction of the LTI system and $H(z)$ as the corresponding eigenvalue. For general values of the complex variable z, $H(z)$ is referred to as the *bilateral z-transform* or simply the *z-transform* of the impulse response $h(n)$. The bilateral z-transform in Eq. (10.1) involves a summation from $-\infty$ to $+\infty$, while the unilateral z-transform has a form similar to that in Eq. (10.1), but with limits of summation from 0 to $+\infty$.

10.2.1 Inverse z-Transform

Substituting $z = re^{j\omega}$ into Eq. (10.1) and using n as the variable of summation, we obtain

$$H(re^{j\omega}) = \sum_{n=-\infty}^{\infty} h(n) \left(re^{j\omega} \right)^n$$

$$H(re^{j\omega}) = \sum_{n=-\infty}^{\infty} [h(n) r^{-n}] e^{-j\omega n}$$

The above equation indicates that $H(re^{j\omega})$ is the DTFT of $[h(n)r^{-n}]$. Hence, the inverse Fourier transform of $H(re^{j\omega})$ must be $[h(n)r^{-n}]$, i.e.,

$$h(n) r^{-n} = \frac{1}{2\pi} \int_{2\pi} H(re^{j\omega}) e^{j\omega n} \, d\omega$$

$$h(n) = r^n \frac{1}{2\pi} \int_{2\pi} H(re^{j\omega}) e^{j\omega n} d\omega$$

$$h(n) = \frac{1}{2\pi} \int_{2\pi} H(re^{j\omega}) \left(re^{j\omega} \right)^n d\omega \tag{10.2}$$

A change of variables is performed by letting $z = re^{j\omega}$, which also yields $dz = jre^{j\omega} d\omega = rz \, d\omega$, or $d\omega = 1/jz^{-1} \, dz$. The integration in Eq. (10.2) is over a 2π interval in ω, which, in terms of z, corresponds to one traversal around the circle $|z| = r$. Consequently, in terms of an integration in the z-plane, Eq. (10.2) can be rewritten as

$$h(n) = \frac{1}{2\pi j} \oint_C H(z) z^{n-1} \, dz \tag{10.3}$$

where C denotes the closed contour in the region of convergence of $H(z)$, taken in a counterclockwise direction. Equation (10.3) expresses $h(n)$ as a function of $H(z)$. We say that $h(n)$ is the *inverse z-transform* of $H(z)$.

We have obtained the z-transform of the impulse response of a system. This relationship holds for an arbitrary signal. The z-transform of a general signal $x(n)$ is defined as

$$\mathcal{Z}[x(n)] = X(z) = \sum_{n=-\infty}^{\infty} x(n)z^{-n} \tag{10.4}$$

and the inverse z-transform of $X(z)$ is

$$x(n) = \frac{1}{2\pi j} \oint_C X(z)z^{n-1}\,dz \tag{10.5}$$

In practice, we usually do not evaluate this integral directly, since it requires techniques of contour integration in the complex plane. There are, however, a number of alternative procedures for obtaining the inverse z-transform. As with Laplace transforms, one particularly useful procedure for rational z-transforms consists of expanding the algebraic expression into a partial fraction expansion and recognizing the signal associated with the individual terms.

We denote the transform relationship between $x(n)$ and $X(z)$ as

$$x(n) \longleftrightarrow X(z) \tag{10.6}$$

10.3 RELATIONSHIP BETWEEN z-TRANSFORM AND DISCRETE-TIME FOURIER TRANSFORM

Consider a discrete-time signal $x(n)$. Its z-transform is defined as

$$\mathcal{Z}[x(n)] = X(z) = \sum_{n=-\infty}^{\infty} x(n)z^{-n}$$

Substituting $z = re^{j\omega}$ into the above equation, we obtain

$$\mathcal{Z}[x(n)] = \sum_{n=-\infty}^{\infty} x(n) \left(re^{j\omega}\right)^{-n}$$

$$\mathcal{Z}[x(n)] = \sum_{n=-\infty}^{\infty} [x(n)r^{-n}]\,e^{-j\omega n}$$

$$\mathcal{Z}[x(n)] = \mathcal{F}[x(n)r^{-n}] \tag{10.7}$$

Thus, the z-transform of $x(n)$ is the Fourier transform of $x(n)r^{-n}$.

Now, if $r = 1$, we obtain

$$\mathcal{Z}[x(n)] = \mathcal{F}[x(n)], \quad \text{for } r = 1 \tag{10.8}$$

The relationship between the z-transform and the Fourier transform for discrete-time signals parallels closely the corresponding discussion in Sections 9.3 and 9.4 for continuous-time signals, but with some important differences. In the continuous-time case, the Laplace transform reduces to the CTFT on the imaginary axis (i.e., $s = j\omega$). In contrast, the z-transform reduces to the DTFT when $|z| = 1$ (i.e., $z = e^{j\omega}$).

10.4 z-PLANE

It is convenient to represent the complex number z as a location in a complex plane termed the z-plane (Fig. 10.1). The point $z = re^{j\omega}$ is located at a distance r from the origin and at an angle ω from the positive real axis. Note that if $r = 1$, the DTFT is obtained from the z-transform by substituting $z = e^{j\omega}$ into Eq. (10.4), i.e.,

$$X(e^{j\omega}) = X(z)\Big|_{z=e^{j\omega}} \tag{10.9}$$

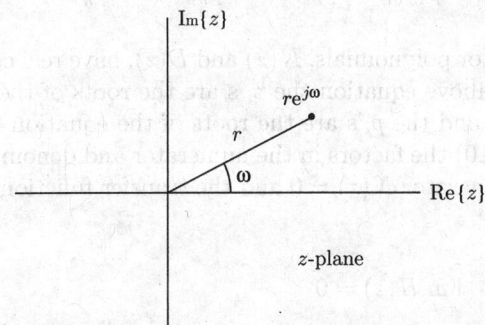

Fig. 10.1 The z-plane; a point is located at a distance r from the origin and an angle ω relative to the real axis

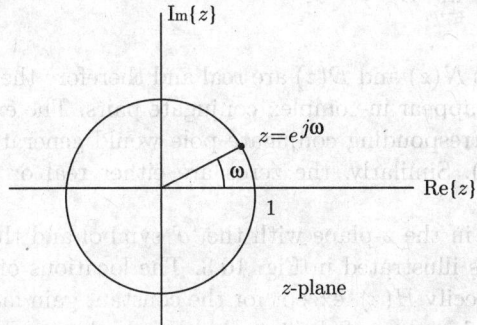

Fig. 10.2 The unit circle, $z = e^{j\omega}$, in the z-plane

Thus, the z-transform reduces to the DTFT on the contour in the complex z-plane corresponding to a circle with a radius of unity, as indicated in Fig. 10.2, and plays a role in the discussion of the z-transform similar to the role of the imaginary axis in the s-plane for the Laplace transform. The frequency ω in the DTFT corresponds to the point on the unit circle at an angle of ω with respect to the positive real axis. We say that the DTFT corresponds to the z-transform evaluated on the unit circle.

10.4.1 Poles and Zeros

The most commonly encountered form of the z-transform in engineering is a ratio of two polynomials in z, i.e.,

$$H(z) = \frac{b_m z^m + b_{m-1} z^{m-1} + \cdots + b_1 z + b_0}{a_n z^n + a_{n-1} z^{n-1} + \cdots + a_1 z + a_0}$$

It is often convenient to factor the polynomials in the numerator and denominator, and to write the transfer function in terms of those factors:

$$H(z) = \frac{N(z)}{D(z)} = K \frac{(z - z_1)(z - z_2) \cdots (z - z_{m-1})(z - z_m)}{(z - p_1)(z - p_2) \cdots (z - p_{n-1})(z - p_n)} \qquad (10.10)$$

where the numerator and denominator polynomials, $N(z)$ and $D(z)$, have real coefficients and $K = b_m/a_n$. As written in the above equation the z_i's are the roots of the equation $N(z) = 0$, and are defined as zeros, and the p_i's are the roots of the equation $D(z) = 0$, and are defined as poles. In Eq. (10.10) the factors in the numerator and denominator are written so that when $z = z_i$, the numerator $N(z) = 0$ and the transfer function vanishes, i.e.,

$$\lim_{z \to z_i} H(z) = 0$$

and similarly when $z = p_i$ the denominator polynomial $D(z) = 0$ and the value of the transfer function becomes unbounded

$$\lim_{z \to p_i} H(z) = \infty$$

All of the coefficients of polynomials $N(z)$ and $D(z)$ are real and therefore the poles and zeros must be either purely real, or appear in complex conjugate pairs. The existence of a single complex pole without a corresponding conjugate pole would generate complex coefficients in the polynomial $D(z)$. Similarly, the zeros are either real or appear in complex conjugate pairs.

We denote the locations of zeros in the z-plane with the 'o' symbol and the location of the poles with the '$\times$' symbol, as illustrated in Fig. 10.3. The locations of the poles and zeros in the z-plane uniquely specify $H(z)$, except for the constant gain factor K. In Fig. 10.3, zeros are depicted at $z = 1$ and $z = 2 \pm j3$, and poles are depicted at $z = -2$ and $z = -3 \pm 2j$.

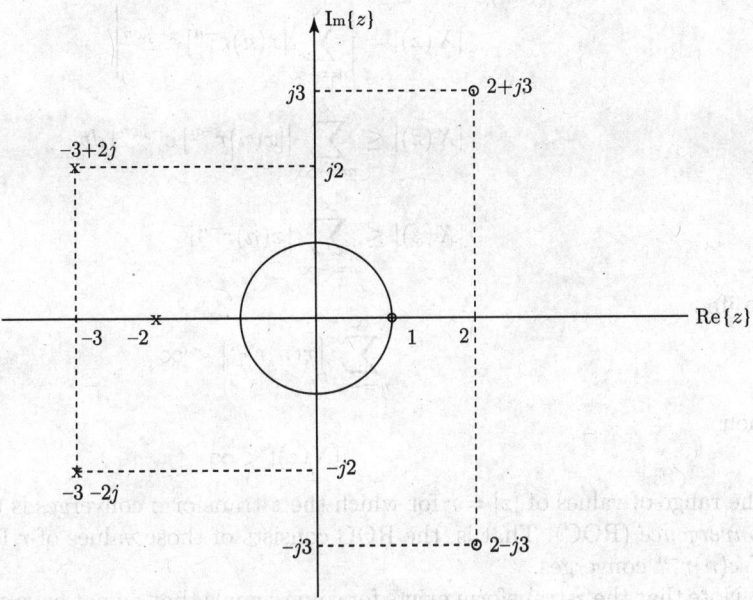

Fig. 10.3 Poles and zeros in the *z*-plane

10.5 REGION-OF-CONVERGENCE FOR *z*-TRANSFORMS

Equation (10.7) indicates that the *z*-transform of $x(n)$ is the Fourier transform of $x(n)r^{-n}$, i.e.,

$$\mathcal{Z}[x(n)] = \sum_{n=-\infty}^{\infty} [x(n)r^{-n}] e^{-j\omega n}$$

Hence, *z*-transform is guaranteed to converge if $x(n)r^{-n}$ is absolutely summable. That is, we must have

$$\sum_{n=-\infty}^{\infty} |x(n)r^{-n}| < \infty \tag{10.11}$$

This guarantees that $X(z)$ will be finite since

$$\mathcal{Z}[x(n)] = X(z) = \sum_{n=-\infty}^{\infty} x(n)z^{-n}$$

Substituting $z = re^{j\omega}$ into the above equation, we obtain

$$X(z) = \sum_{n=-\infty}^{\infty} x(n) \left(re^{j\omega}\right)^{-n}$$

$$X(z) = \sum_{n=-\infty}^{\infty} [x(n)r^{-n}] e^{-j\omega n}$$

$$|X(z)| = \left| \sum_{n=-\infty}^{\infty} [x(n)r^{-n}] e^{-j\omega n} \right|$$

$$|X(z)| \le \sum_{n=-\infty}^{\infty} \left| [x(n)r^{-n}] e^{-j\omega n} \right| dt$$

$$|X(z)| \le \sum_{n=-\infty}^{\infty} \left| x(n)r^{-n} \right|$$

So if

$$\sum_{n=-\infty}^{\infty} \left| x(n)r^{-n} \right| < \infty$$

then

$$|X(z)| < \infty$$

The range of values of $|z| = r$ for which the z-transform converges is termed the *region of convergence* (ROC). That is, the ROC consists of those values of r for which the DTFT of $x(n)r^{-n}$ converges.

Note that the z-transform exists for some signals that do not have a DTFT. By limiting ourselves to a certain range of values of r, we may ensure that $x(n)r^{-n}$ is absolutely summable, even though $x(n)$ is not absolutely summable by itself.

The ROC can also provide us with information about whether $x(n)$ is Fourier transformable or not. Since the DTFT is obtained from the bilateral z-transform by setting $r = 1$, the ROC in this case is a circle of radius 1, i.e., unit circle. Therefore, if the ROC for $X(z)$ includes the unit circle, $x(n)$ is Fourier transformable, and $X(e^{j\omega})$ can be obtained by replacing z in $X(z)$ by $e^{j\omega}$.

Example 10.1 Determine the constraint on $r = |z|$ for each of the following sums to converge:

(a) $\displaystyle\sum_{n=-1}^{\infty} \left(\frac{1}{2}\right)^{n+1} z^{-n}$

(b) $\displaystyle\sum_{n=1}^{\infty} \left(\frac{1}{2}\right)^{-n+1} z^{n}$

(c) $\displaystyle\sum_{n=0}^{\infty} \left(\frac{1 + (-1)^n}{2}\right) z^{-n}$

Solution

(a) Consider the given summation

$$\sum_{n=-1}^{\infty} \left(\frac{1}{2}\right)^{n+1} z^{-n} = \frac{1}{2}z + \frac{1}{2}\sum_{n=0}^{\infty} \left(\frac{1}{2}\right)^{n} z^{-n}$$

$$= \frac{1}{2}z + \frac{1}{2}\sum_{n=0}^{\infty} \left(\frac{1}{2}z^{-1}\right)^{n}$$

The summation of the second term on the RHS will converge if $\left|1/2z^{-1}\right| < 1$, or $|z| > 1/2$. Therefore, the given summation will converge if $1/2 < |z| < \infty$.

(b) Consider the given summation

$$\sum_{n=1}^{\infty} \left(\frac{1}{2}\right)^{-n+1} z^n = \frac{1}{2} \sum_{n=1}^{\infty} \left(\frac{1}{2}\right)^{-n} z^n$$

$$= \frac{1}{2} \sum_{n=1}^{\infty} (2z)^n$$

Therefore, the given summation will converge if $|2z| < 1$, or $|z| < 1/2$.

(c) Consider the given summation

$$\sum_{n=0}^{\infty} \left(\frac{1+(-1)^n}{2}\right) z^{-n} = \frac{1}{2} \sum_{n=0}^{\infty} z^{-n} + \frac{1}{2} \sum_{n=0}^{\infty} (-1)^n z^{-n}$$

$$= \frac{1}{2} \sum_{n=0}^{\infty} \left(z^{-1}\right)^n + \frac{1}{2} \sum_{n=0}^{\infty} \left(-z^{-1}\right)^n$$

The first summation will converge if $\left|z^{-1}\right| < 1$, or $|z| > 1$. The summation of the second term on the RHS will converge if $\left|-z^{-1}\right| < 1$, or $|z| > 1$. Therefore, the given summation will converge if $|z| > 1$.

Example 10.2 Determine the *z*-transform of the causal signal [shown in Fig. 10.4(a)]

$$x(n) = a^n u(n)$$

and depict the ROC and the locations of poles and zeros in the *z*-plane.

Solution
By definition

$$X(z) = \sum_{n=-\infty}^{\infty} x(n)z^{-n} = \sum_{n=-\infty}^{\infty} a^n u(n)z^{-n}$$

Since

$$u(n) = \begin{cases} 1, & n \geq 0 \\ 0, & n < 0 \end{cases}$$

we obtain

$$X(z) = \sum_{n=0}^{\infty} a^n z^{-n} = \sum_{n=0}^{\infty} \left(az^{-1}\right)^n$$

$$X(z) = \frac{1}{1 - az^{-1}}, \quad \text{for } |az^{-1}| < 1 \text{ or equivalently } |z| > |a|$$

$$X(z) = \frac{z}{z - a}, \quad |z| > |a|$$

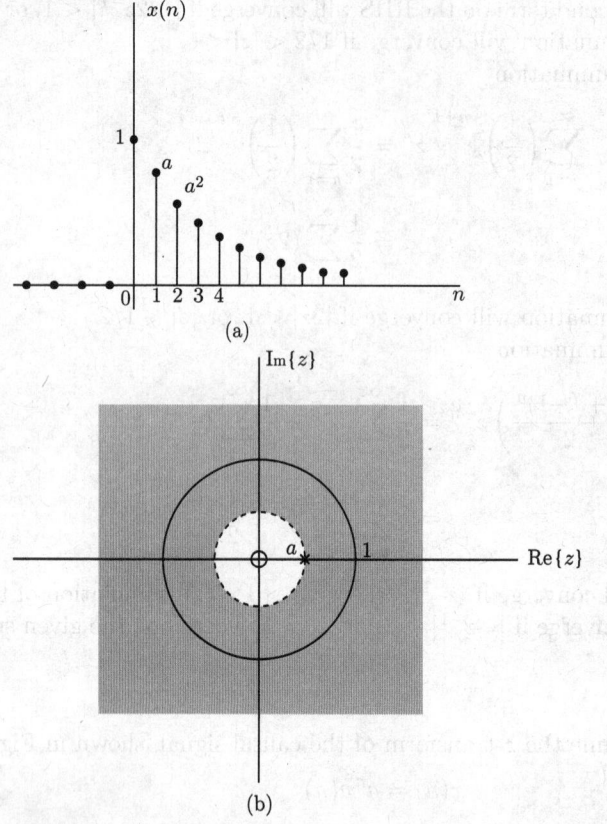

Fig. 10.4 (a) The signal $x(n) = a^n u(n)$ for $0 < a < 1$ and (b) the pole-zero plot and ROC

Therefore,

$$a^n u(n) \longleftrightarrow \frac{1}{1 - az^{-1}}, \quad |z| > |a| \qquad (10.12)$$

The ROC for $X(z)$ is $|z| > |a|$, as shown in the shaded area in Fig. 10.4(b).

Example 10.3 Determine the z-transform of the anticausal signal [shown in Fig. 10.5(a)]

$$x(n) = -a^n u(-n - 1)$$

and depict the ROC and the locations of poles and zeros in the z-plane.

Solution
By definition

$$X(z) = \sum_{n=-\infty}^{\infty} x(n)z^{-n} = \sum_{n=-\infty}^{\infty} -a^n u(-n - 1)z^{-n}$$

Since

$$u(-n-1) = \begin{cases} 1, & (-n-1) \geq 0 \longrightarrow n \leq -1 \\ 0, & (-n-1) < 0 \longrightarrow n > -1 \end{cases}$$

we obtain

$$X(z) = -\sum_{n=-\infty}^{-1} a^n z^{-n} = -\sum_{n=1}^{\infty} a^{-n} z^n = -\sum_{n=1}^{\infty} \left(a^{-1}z\right)^n = -\sum_{n-1=0}^{\infty} \left(a^{-1}z\right)^n$$

A change of variables is performed by letting $m = n - 1$, which also yields $n = m + 1$, $m = 0$ as $n = 1$, and $m = \infty$ as $n = \infty$. Therefore,

$$X(z) = -\sum_{m=0}^{\infty} \left(a^{-1}z\right)^{m+1}$$

$$= -(a^{-1}z) \sum_{m=0}^{\infty} \left(a^{-1}z\right)^m$$

$$= -(a^{-1}z)\frac{1}{1 - a^{-1}z}, \quad \text{for } |a^{-1}z| < 1 \text{ or equivalently } |z| < |a|$$

$$= \frac{1}{1 - az^{-1}} \quad |z| < |a|$$

$$X(z) = \frac{z}{z - a} \quad |z| < |a|$$

Therefore,

$$-a^n u(-n-1) \longleftrightarrow \frac{1}{1 - az^{-1}}, \quad |z| < |a| \tag{10.13}$$

The ROC for $X(z)$ is $|z| < |a|$, as shown in the shaded area in Fig. 10.5(b).

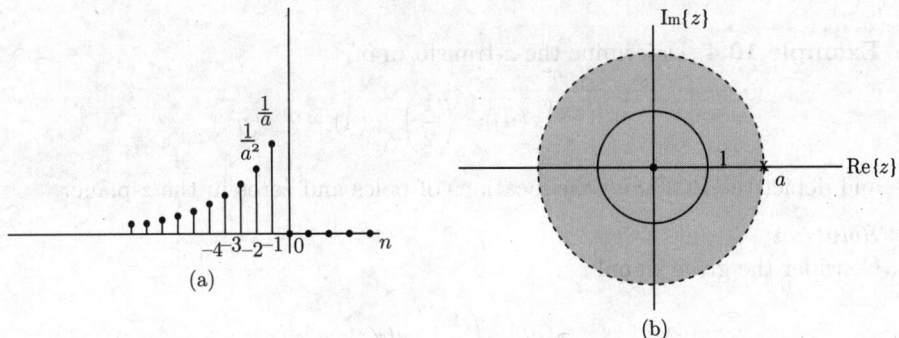

Fig. 10.5

10.6 PROPERTIES OF ROC

Note that the z-transforms for signals $a^n u()$ and $-a^n u(-n-1)$ are identical except for their regions of convergence. Therefore, for a given $X(z)$, there may be more than one inverse transform, depending on the ROC. In other words, unless the ROC is specified, there is no one-to-one correspondence between $X(z)$ and $x(n)$. In this section, we explain properties of the ROC of various classes of signals.

1. The ROC of $X(z)$ consists of a ring in the z-plane 'centred' about the origin.

2. If $X(z)$ is rational, then the ROC must not contain any poles.

3. If $x(n)$ is of finite duration and is absolutely summable (i.e., $\sum_{n=-\infty}^{\infty} |x(n)| < \infty$), then the ROC is the entire z-plane, except possibly $z = 0$ and/or $z = \infty$.

4. If $x(n)$ is right-sided and of infinite duration (i.e., $x(n) = 0$ for all $n < N_1$ for some finite N_1), then the ROC is the region in the z-plane outside the outermost pole, i.e., outside the circle of radius equal to the largest magnitude of the pole of $X(z)$. Furthermore, if $x(n)$ is causal (i.e., if it is right-sided and equal to 0 for $n < 0$), then the ROC also includes $z = \infty$.

5. If $x(n)$ is left-sided and of infinite duration (i.e., $x(n) = 0$ for all $n > N_2$ for some finite N_2), then the ROC is the region in the z-plane inside the innermost pole , i.e., inside the circle of radius equal to the smallest magnitude of the poles of $X(z)$ other than any at $z = 0$ and extending inward to and possibly including $z = 0$. In particular, if $x(n)$ is anticausal (i.e., if it is left-sided and equal to 0 for $n > 0$), then the ROC also includes $z = 0$.

6. If $x(n)$ is two-sided and of infinite duration (i.e., the signal is of infinite extent for both $n < 0$ and $n > 0$), then the ROC will consist of a ring in the z-plane.

7. For a finite-duration right-sided signal (i.e., $x(n) = 0$ for $n < 0$ and $n > N_1$ for some finite N_1), the ROC will be the entire z-plane except $z = 0$.

8. For a finite-duration two-sided signal (i.e., the signal is of finite extent for both $n < 0$ and $n > 0$), the ROC will be the entire z-plane except $z = 0$ and $z = \infty$.

9. For a finite-duration left-sided signal (i.e., $x(n) = 0$ for $n > 0$ and $n < -N_1$ for some finite positive N_1), the ROC will be the entire z-plane except $z = \infty$.

Example 10.4 Determine the z-transform of

$$x(n) = \left(\frac{1}{2}\right)^n u(n) + 2^n u(n)$$

and depict the ROC and the locations of poles and zeros in the z-plane.

Solution
Consider the given signal

$$x(n) = \left(\frac{1}{2}\right)^n u(n) + 2^n u(n)$$

$$\mathcal{Z}[x(n)] = \mathcal{Z}\left[\left(\frac{1}{2}\right)^n u(n) + 2^n u(n)\right]$$

$$X(z) = \mathcal{Z}\left[\left(\frac{1}{2}\right)^n u(n)\right] + \mathcal{Z}[2^n u(n)]$$

From Eq. (10.12), we have

$$a^n u(n) \longleftrightarrow \frac{1}{1 - az^{-1}}, \quad |z| > |a|$$

therefore,

$$\left(\frac{1}{2}\right)^n u(n) \longleftrightarrow \frac{1}{1 - \frac{1}{2}z^{-1}}, \quad |z| > \frac{1}{2}$$

$$2^n u(n) \longleftrightarrow \frac{1}{1 - 2z^{-1}}, \quad |z| > 2$$

The set of values of $|z|$ for which the z-transforms of both terms converge is $|z| > 2$, and thus, we obtain

$$X(z) = \frac{1}{1 - \frac{1}{2}z^{-1}} + \frac{1}{1 - 2z^{-1}}, \quad |z| > 2$$

$$= \frac{2 - \frac{5}{2}z^{-1}}{1 - \frac{5}{2}z^{-1} + z^{-2}}, \quad |z| > 2$$

The ROC for $X(z)$ is $|z| > 2$. The pole-zero pot and the ROC for the z-transform of each of the individual terms and for the combined signal are shown in Fig. 10.6. For a right-sided and of infinite duration signal, the ROC is the region in the z-plane outside the outermost pole.

Example 10.5 Determine the z-transform of

$$x(n) = -\left(\frac{1}{2}\right)^n u(-n - 1) + 2^n u(-n - 1)$$

and depict the ROC and the locations of poles and zeros in the z-plane.

Solution
Consider the given signal

$$x(n) = -\left(\frac{1}{2}\right)^n u(-n - 1) + 2^n u(-n - 1)$$

$$\mathcal{Z}[x(n)] = \mathcal{Z}\left[-\left(\frac{1}{2}\right)^n u(-n - 1) + 2^n u(-n - 1)\right]$$

$$X(z) = \mathcal{Z}\left[-\left(\frac{1}{2}\right)^n u(-n - 1)\right] - \mathcal{Z}[-2^n u(-n - 1)]$$

From Eq. (10.13), we have

$$-a^n u(-n - 1) \longleftrightarrow \frac{1}{1 - az^{-1}}, \quad |z| < |a|$$

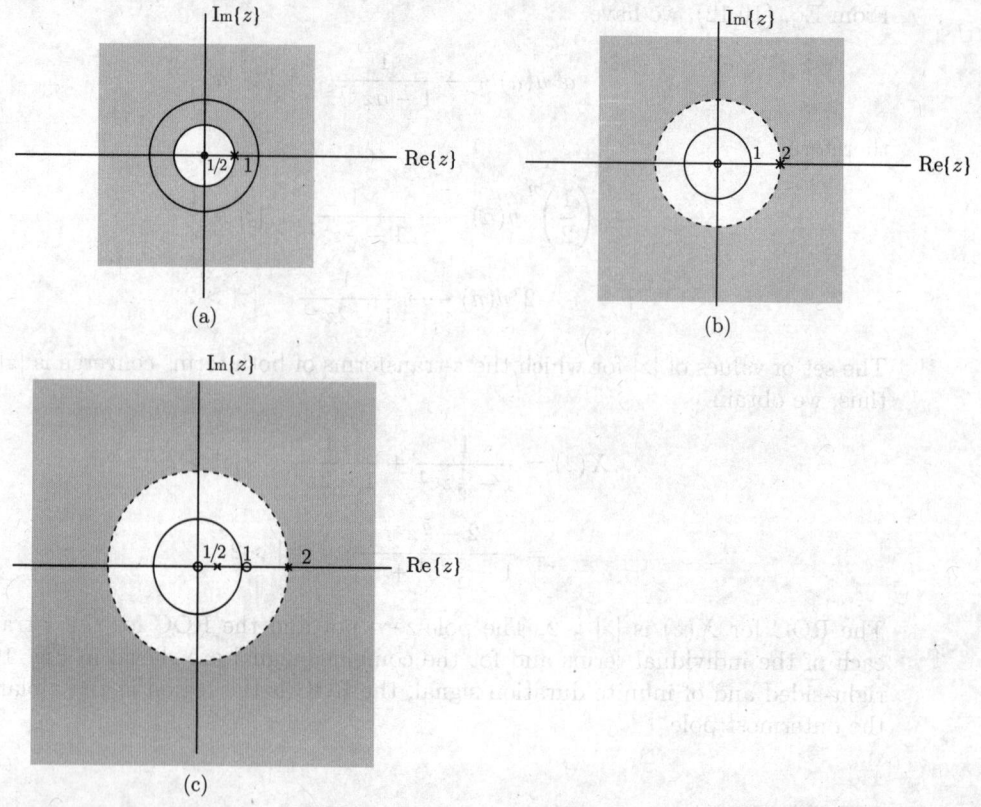

Fig. 10.6 Pole-zero plot and ROC for the individual terms and the sum in Example 10.4: (a) $1/\left(1 - \frac{1}{2}z^{-1}\right), |z| > \frac{1}{2}$; (b) $1/(1 - 2z^{-1}), |z| > 2$; (c) $(2 - \frac{5}{2}z^{-1})/(1 - \frac{5}{2}z^{-1} + z^{-2}), |z| > 2$.

therefore,

$$-\left(\frac{1}{2}\right)^n u(-n-1) \longleftrightarrow \frac{1}{1 - \frac{1}{2}z^{-1}}, \quad |z| < \frac{1}{2}$$

$$-2^n u(-n-1) \longleftrightarrow \frac{1}{1 - 2z^{-1}}, \quad |z| < 2$$

The set of values of $|z|$ for which the z-transforms of both terms converge is $|z| < 1/2$, and thus, we obtain

$$X(z) = \frac{1}{1 - \frac{1}{2}z^{-1}} - \frac{1}{1 - 2z^{-1}}, \quad |z| < \frac{1}{2}$$

$$= \frac{-\frac{3}{2}z^{-1}}{1 - \frac{5}{2}z^{-1} + z^{-2}}, \quad |z| < \frac{1}{2}$$

The ROC for $X(z)$ is $|z| < 1/2$. The pole-zero pot and the ROC for the z-transform of each of the individual terms and for the combined signal are shown in Fig. 10.7. For a left-sided and of infinite duration signal, the ROC is the region in the z-plane inside the innermost pole.

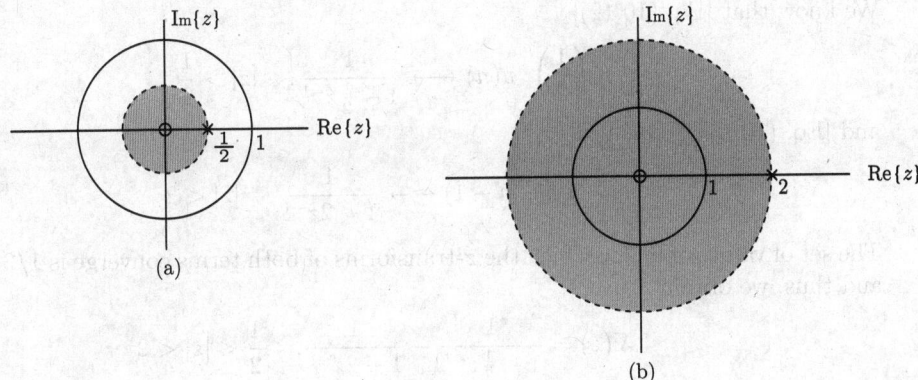

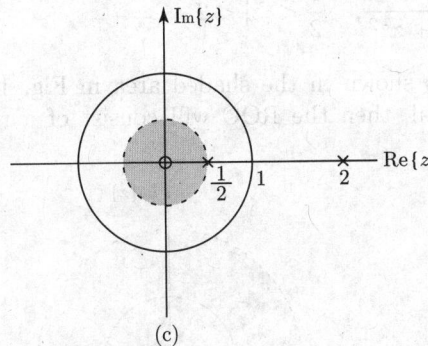

Fig. 10.7 Pole-zero plot and ROC for the individual terms and the sum in Example 10.5:
(a) $1/(1 - \frac{1}{2}z^{-1})$, $|z| < \frac{1}{2}$; (b) $1/(1 - 2z^{-1})$, $|z| < 2$; (c) $(-\frac{3}{2}z^{-1})/(1 - \frac{5}{2}z^{-1} + z^{-2})$, $|z| < \frac{1}{2}$

Example 10.6 Determine the z-transform of

$$x(n) = \left(\frac{1}{2}\right)^n u(n) + 2^n u(-n - 1)$$

and depict the ROC and the locations of poles and zeros in the z-plane.

Solution
Consider the given signal

$$x(n) = \left(\frac{1}{2}\right)^n u(n) + 2^n u(-n - 1)$$

$$\mathcal{Z}[x(n)] = \mathcal{Z}\left[\left(\frac{1}{2}\right)^n u(n) + 2^n u(-n - 1)\right]$$

$$X(z) = \mathcal{Z}\left[\left(\frac{1}{2}\right)^n u(n)\right] - \mathcal{Z}[-2^n u(-n - 1)]$$

We know that [Eq. (10.12)]

$$\left(\frac{1}{2}\right)^n u(n) \longleftrightarrow \frac{1}{1 - \frac{1}{2}z^{-1}}, \quad |z| > \frac{1}{2}$$

and [Eq. (10.13)]

$$-2^n u(-n-1) \longleftrightarrow \frac{1}{1 - 2z^{-1}}, \quad |z| < 2$$

The set of values of $|z|$ for which the z-transforms of both terms converge is $1/2 < |z| < 2$, and thus, we obtain

$$X(z) = \frac{1}{1 - \frac{1}{2}z^{-1}} - \frac{1}{1 - 2z^{-1}}, \quad \frac{1}{2} < |z| < 2$$

$$= \frac{-\frac{3}{2}z^{-1}}{1 - \frac{5}{2}z^{-1} + z^{-2}}, \quad \frac{1}{2} < |z| < 2$$

The ROC for $X(z)$ is $1/2 < |z| < 2$, as shown in the shaded area in Fig. 10.8. For a two-sided and of infinite duration signal, then the ROC will consist of a ring in the z-plane.

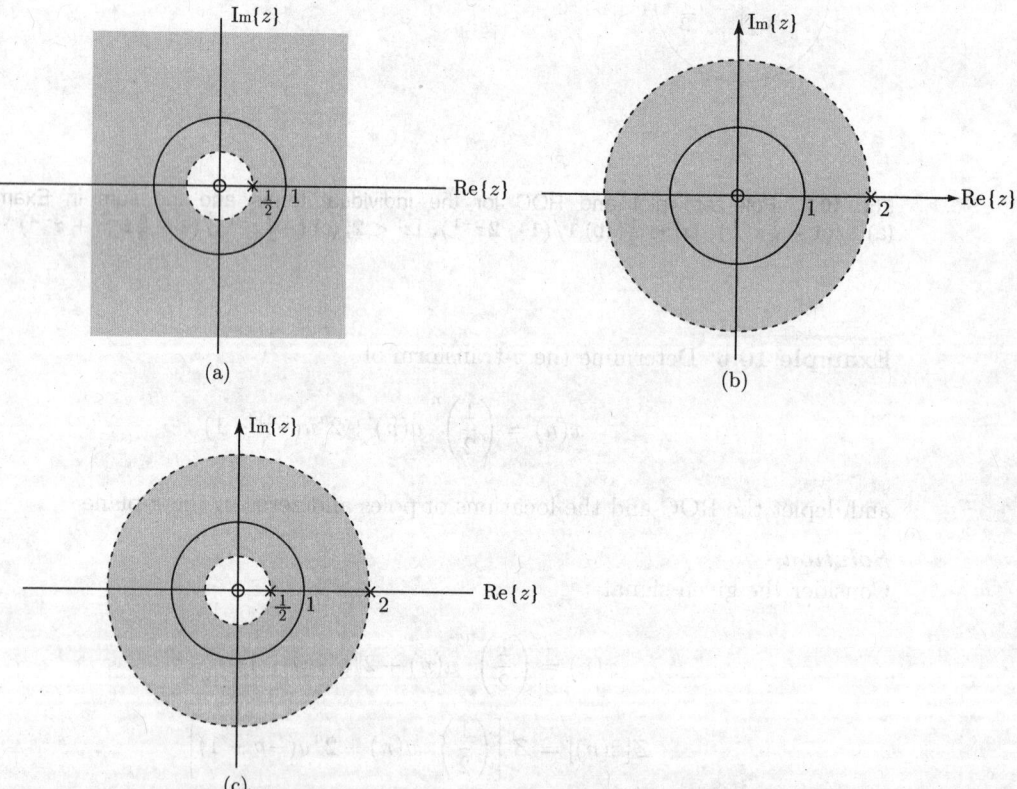

Fig. 10.8 Pole-zero plot and ROC for the individual terms and the sum in Example 10.6: (a) $1/(1 - \frac{1}{2}z^{-1})$, $|z| > \frac{1}{2}$; (b) $1/(1 - 2z^{-1})$, $|z| < 2$; (c) $(-\frac{3}{2}z^{-1})/(1 - \frac{5}{2}z^{-1} + z^{-2})$, $\frac{1}{2} < |z| < 2$

Example 10.7 Determine the z-transform of

(a) a unit impulse function $x(n) = \delta(n)$

(b) a unit step function $x(n) = u(n)$

(c) $x(n) = -u(-n-1)$

(d) $x(n) = u(-n)$

Solution

(a) By definition

$$X(z) = \sum_{n=-\infty}^{\infty} x(n)z^{-n} = \sum_{n=-\infty}^{\infty} \delta(n)z^{-n}$$

$$= z^{-n}\Big|_{n=0}$$

$X(z) = 1$ ROC is the entire z-plane including $z = 0$ and $z = \infty$

Therefore,

$\delta(n) \longleftrightarrow 1$ ROC is the entire z-plane including $z = 0$ and $z = \infty$.   (10.14)

(b) By definition

$$X(z) = \sum_{n=-\infty}^{\infty} x(n)z^{-n} = \sum_{n=-\infty}^{\infty} u(n)z^{-n} = \sum_{n=0}^{\infty} z^{-n}$$

$$= \frac{1}{1 - z^{-1}} \quad |z^{-1}| < 1$$

$$X(z) = \frac{z}{z - 1} \quad |z| > 1$$

Therefore,

$$u(n) \longleftrightarrow \frac{1}{1 - z^{-1}} \quad |z| > 1 \tag{10.15}$$

(c) By definition

$$X(z) = \sum_{n=-\infty}^{\infty} x(n)z^{-n} = \sum_{n=-\infty}^{\infty} -u(-n-1)z^{-n}$$

Since

$$u(-n-1) = \begin{cases} 1, & (-n-1) \geq 0 \longrightarrow n \leq -1 \\ 0, & n > -1 \end{cases}$$

we obtain

$$X(z) = -\sum_{n=-\infty}^{-1} z^{-n} = -\sum_{n=1}^{\infty} z^{n} = -\sum_{n=1}^{\infty} (z)^{n} = -\sum_{n-1=0}^{\infty} (z)^{n}$$

A change of variables is performed by letting $m = n - 1$, which also yields $n = m + 1$, $m = 0$ as $n = 1$, and $m = \infty$ as $n = \infty$. Therefore,

$$X(z) = -\sum_{m=0}^{\infty} (z)^{m+1} = -z \sum_{m=0}^{\infty} (z)^m$$

$$= -z \frac{1}{1-z}, \quad \text{for } |z| < 1$$

$$= \frac{1}{1-z^{-1}}, \quad |z| < 1$$

$$X(z) = \frac{z}{z-1}, \quad |z| < 1$$

Therefore,

$$-u(-n-1) \longleftrightarrow \frac{1}{1-z^{-1}} \quad |z| < 1 \tag{10.16}$$

(d) By definition

$$X(z) = \sum_{n=-\infty}^{\infty} x(n)z^{-n} = \sum_{n=-\infty}^{\infty} u(-n)z^{-n} = \sum_{n=-\infty}^{0} z^{-n}$$

$$= \sum_{n=0}^{\infty} z^{n}$$

$$X(z) = \frac{1}{1-z} \quad |z| < 1$$

$$X(z) = \frac{-z^{-1}}{1-z^{-1}} \quad |z| < 1$$

Therefore,

$$u(-n) \longleftrightarrow \frac{-z^{-1}}{1-z^{-1}} \quad |z| < 1 \tag{10.17}$$

Example 10.8 Determine the z-transform of the two-sided infinite duration signal

$$x(n) = a^{|n|}$$

for the following two cases:

(a) $0 < a < 1$

(b) $1 < a < \infty$

Solution

Case I $(0 < a < 1)$. This two-sided signal is illustrated in Fig. 10.9(a) for $0 < a < 1$. The signal $x(n)$ can be expressed as the sum of a right-sided and a left-sided signal, i.e.,

$$x(n) = a^{|n|}$$

$$= \begin{cases} a^n, & n \geq 0 \\ a^{-n}, & n < 0 \end{cases}$$

$$x(n) = a^n u(n) + a^{-n} u(-n-1)$$

$$\mathcal{Z}[x(n)] = \mathcal{Z}[a^n u(n) + a^{-n} u(-n-1)]$$

$$X(z) = \mathcal{Z}[a^n u(n)] - \mathcal{Z}[-a^{-n} u(-n-1)]$$

From Eq. (10.12), we have

$$a^n u(n) \longleftrightarrow \frac{1}{1 - az^{-1}}, \quad |z| > a$$

From Eq. (10.13), we have

$$- \left(a^{-1}\right)^n u(-n-1) \longleftrightarrow \frac{1}{1 - a^{-1}z^{-1}}, \quad |z| < \frac{1}{a}$$

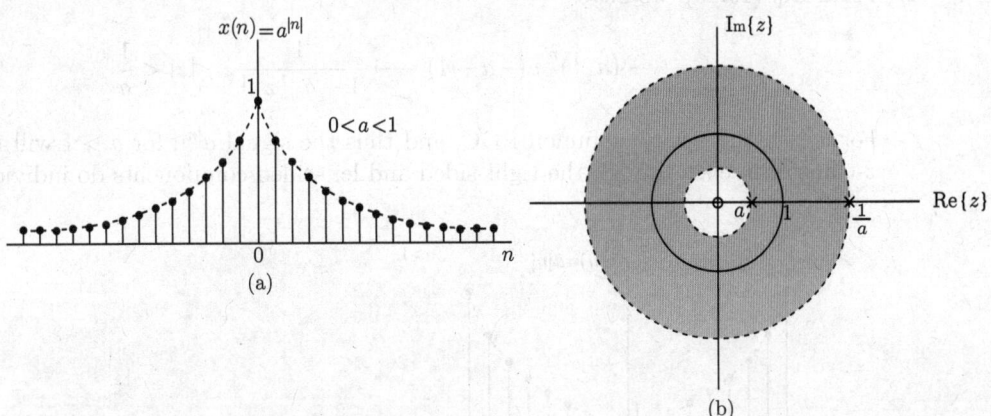

Fig. 10.9   (a) The signal $x(n) = a^{|n|}$ for $0 < a < 1$ and (b) its pole-zero plot and ROC

The set of values of $|z|$ for which the z-transforms of both terms converge is $a < |z| < 1/a$, and thus, we obtain

$$X(z) = \frac{1}{1 - az^{-1}} - \frac{1}{1 - a^{-1}z^{-1}}, \quad a < |z| < \frac{1}{a}$$

$$= \frac{a^2 - 1}{a} \frac{z}{(z - a)\left(z - \frac{1}{a}\right)}, \quad a < |z| < \frac{1}{a}$$

Therefore,

$$a^{|n|} \longleftrightarrow \frac{a^2 - 1}{a} \frac{z}{(z - a)\left(z - \frac{1}{a}\right)}, \quad 0 < a < 1 \quad \text{and} \quad a < |z| < \frac{1}{a} \tag{10.18}$$

The ROC for $X(z)$ is $a < |z| < 1/a$, as shown in the shaded area in Fig. 10.9(b). For a two-sided and of infinite duration signal, the ROC is a ring in the z-plane.

Case II $(1 < a < \infty)$. This two-sided signal is illustrated in Fig. 10.10 for $1 < a < \infty$. The signal $x(n)$ can be expressed as the sum of a right-sided and a left-sided signal, i.e.,

$$x(n) = a^{|n|}$$

$$= \begin{cases} a^n, & n \geq 0 \\ a^{-n}, & n < 0 \end{cases}$$

$$x(n) = a^n u(n) + a^{-n} u(-n-1)$$

$$\mathcal{Z}[x(n)] = \mathcal{Z}[a^n u(n) + a^{-n} u(-n-1)]$$

$$X(z) = \mathcal{Z}[a^n u(n)] - \mathcal{Z}[-a^{-n} u(-n-1)]$$

From Eq. (10.12), we have

$$a^n u(n) \longleftrightarrow \frac{1}{1 - az^{-1}}, \quad |z| > a$$

From Eq. (10.13), we have

$$-\left(a^{-1}\right)^n u(-n-1) \longleftrightarrow \frac{1}{1 - a^{-1}z^{-1}}, \quad |z| < \frac{1}{a}$$

For $a > 1$, there is no common ROC, and thus the signal $a^{|n|}$, for $a > 1$ will not have a z-transform, even though the right-sided and left-sided components do individually.

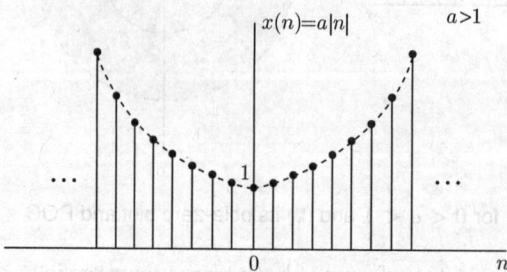

Fig. 10.10 The signal $x(n) = a^{|n|}$ for $a > 1$

Example 10.9 Determine the z-transform and the pole-zero plot for the signal

$$x(n) = \begin{cases} a^n, & 0 \leq n \leq N-1 \\ 0, & \text{elsewhere} \end{cases}$$

where $a > 0$.

Solution
By definition

$$X(z) = \sum_{n=-\infty}^{\infty} x(n)z^{-n} = \sum_{n=0}^{N-1} a^n z^{-n} = \sum_{n=0}^{N-1} \left(az^{-1}\right)^n = \frac{\left(1 - az^{-1}\right)^N}{1 - az^{-1}}$$

$$X(z) = \frac{\left(1 - az^{-1}\right)^N}{1 - az^{-1}}$$

$$X(z) = \frac{1}{z^{N-1}} \frac{z^N - a^N}{z - a}$$

since $a > 0$, the equation $z^N - a^N = 0$, or $z^N = a^N$, has N roots at

$$z_k = ae^{j2\pi k/N} \quad k = 0, 1, 2, \ldots, N - 1$$

The zero $z_0 = a$ (i.e., root for $k = 0$) cancels the pole at $z = a$. Thus

$$X(z) = \frac{(z - z_1)(z - z_2)\cdots(z - z_{N-1})}{z^{N-1}}$$

which has $N - 1$ zeros and $N - 1$ poles, located as shown in Fig. 10.11. Since $x(n)$ is of finite duration, the ROC is the entire z-plane except at $z = 0$ because of the $N - 1$ poles are located at the origin.

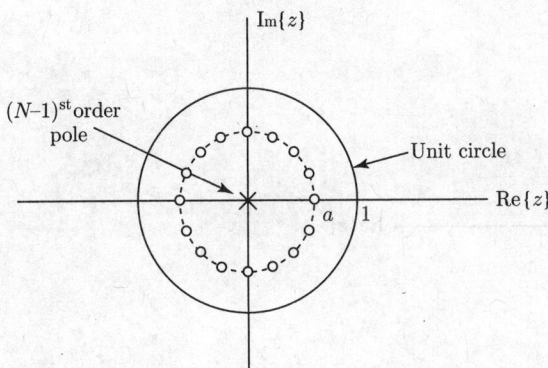

Fig. 10.11 Pole-zero pattern for $N = 16$ and $0 < a < 1$

Example 10.10 Determine the z-transform and the pole-zero plot for the signal

$$x(n) = \begin{cases} 1, & 0 \le n \le N - 1 \\ 0, & \text{elsewhere} \end{cases}$$

Solution

By definition

$$X(z) = \sum_{n=-\infty}^{\infty} x(n)z^{-n} = \sum_{n=0}^{N-1} z^{-n}$$

$$= \begin{cases} N, & \text{if } z = 1 \\ \dfrac{1 - z^{-N}}{1 - z^{-1}}, & \text{if } z \neq 1 \end{cases}$$

$$X(z) = \begin{cases} N, & \text{if } z = 1 \\ \dfrac{1}{z^{N-1}} \dfrac{z^N - 1}{z - 1}, & \text{if } z \neq 1 \end{cases}$$

The equation $z^N - 1 = 0$, or $z^N = 1$, has N roots at

$$z_k = e^{j2\pi k/N} \quad k = 0, 1, 2, \ldots, N - 1$$

The zero $z_0 = 1$ (i.e., root for $k = 0$) cancels the pole at $z = 1$. Thus,

$$X(z) = \frac{(z - z_1)(z - z_2) \cdots (z - z_{N-1})}{z^{N-1}}$$

which has $N - 1$ zeros and $N - 1$ poles, located as shown in Fig. 10.12. Since $x(n)$ is of finite duration, the ROC is the entire z-plane except at $z = 0$ because of the $N - 1$ poles are located at the origin.

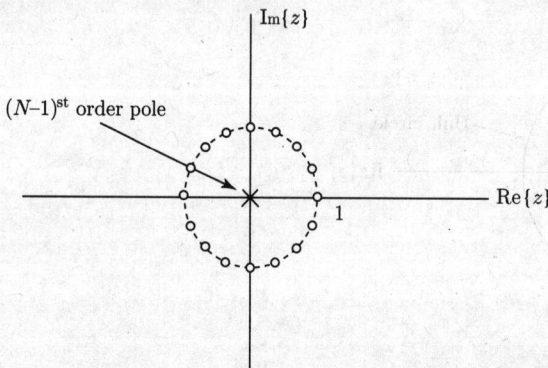

Fig. 10.12 Pole-zero pattern for $N = 16$

Example 10.11 Determine the z-transform of the following signals:

(a) $x(n) = \sin(\omega_0 n)u(n)$

(b) $x(n) = \cos(\omega_0 n)u(n)$

Solution

(a) Consider the given signal

$$x(n) = \sin(\omega_0 n)u(n)$$

$$x(n) = \frac{1}{2j}[e^{j\omega_0 n} - e^{-j\omega_0 n}]u(n)$$

$$\mathcal{Z}[x(n)] = \frac{1}{2j}\mathcal{Z}[e^{j\omega_0 n}u(n) - e^{-j\omega_0 n}u(n)]$$

$$X(z) = \frac{1}{2j}\mathcal{Z}[e^{j\omega_0 n}u(n)] - \frac{1}{2j}\mathcal{Z}[e^{-j\omega_0 n}u(n)]$$

$$= \frac{1}{2j}\mathcal{Z}\left[\left(e^{j\omega_0}\right)^n u(n)\right] - \frac{1}{2j}\mathcal{Z}\left[\left(e^{-j\omega_0}\right)^n u(n)\right]$$

From Eq. (10.12), we have

$$a^n u(n) \longleftrightarrow \frac{1}{1 - az^{-1}}, \quad |z| > |a|$$

Therefore,

$$\left(e^{j\omega_0}\right)^n u(n) \longleftrightarrow \frac{1}{1 - e^{j\omega_0}z^{-1}}, \quad |z| > |e^{j\omega_0}| \longrightarrow |z| > 1$$

$$\left(e^{-j\omega_0}\right)^n u(n) \longleftrightarrow \frac{1}{1 - e^{-j\omega_0}z^{-1}}, \quad |z| > |e^{-j\omega_0}| \longrightarrow |z| > 1$$

The set of values of $|z|$ for which the z-transforms of both terms converge is $|z| > 1$, and thus, we obtain

$$X(z) = \frac{1}{2j}\left[\frac{1}{1 - e^{j\omega_0}z^{-1}} - \frac{1}{1 - e^{-j\omega_0}z^{-1}}\right], \quad |z| > 1$$

$$= \frac{1}{2j}\left[\frac{1 - e^{-j\omega_0}z^{-1} - 1 + e^{j\omega_0}z^{-1}}{(1 - e^{j\omega_0}z^{-1})(1 - e^{-j\omega_0}z^{-1})}\right], \quad |z| > 1$$

$$= \frac{z^{-1}(e^{j\omega_0} - e^{-j\omega_0})/2j}{1 - 2z^{-1}((e^{j\omega_0} + e^{-j\omega_0})/2) + z^{-2}}, \quad |z| > 1$$

$$X(z) = \frac{z^{-1}\sin(\omega_0)}{1 - 2z^{-1}\cos(\omega_0) + z^{-2}}, \quad |z| > 1$$

Therefore,

$$\sin(\omega_0 n)u(n) \longleftrightarrow \frac{z^{-1}\sin(\omega_0)}{1 - 2z^{-1}\cos(\omega_0) + z^{-2}}, \quad |z| > 1 \qquad (10.19)$$

The ROC for $X(z)$ is $|z| > 1$, as shown in the shaded area in Fig. 10.13.

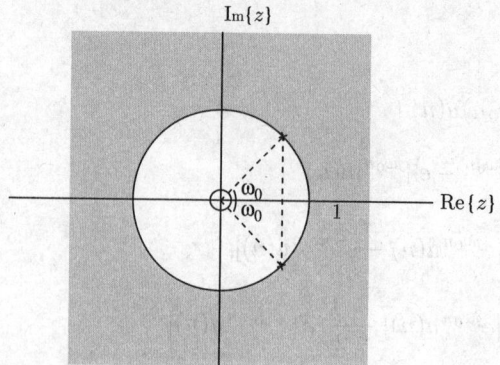

Fig. 10.13 Pole-zero pattern and ROC for Example 10.11

(b) Consider the given signal

$$x(n) = \cos(\omega_0 n)u(n)$$

$$x(n) = \frac{1}{2}[e^{j\omega_0 n} + e^{-j\omega_0 n}]u(n)$$

$$\mathcal{Z}[x(n)] = \frac{1}{2}\mathcal{Z}[e^{j\omega_0 n}u(n) + e^{-j\omega_0 n}u(n)]$$

$$X(z) = \frac{1}{2}\mathcal{Z}[e^{j\omega_0 n}u(n)] + \frac{1}{2}\mathcal{Z}[e^{-j\omega_0 n}u(n)]$$

$$= \frac{1}{2}\mathcal{Z}\left[\left(e^{j\omega_0}\right)^n u(n)\right] + \frac{1}{2}\mathcal{Z}\left[\left(e^{-j\omega_0}\right)^n u(n)\right]$$

From Eq. (10.12), we have

$$a^n u(n) \longleftrightarrow \frac{1}{1 - az^{-1}}, \quad |z| > |a|$$

Therefore,

$$\left(e^{j\omega_0}\right)^n u(n) \longleftrightarrow \frac{1}{1 - e^{j\omega_0}z^{-1}}, \quad |z| > |e^{j\omega_0}| \longrightarrow |z| > 1$$

$$\left(e^{-j\omega_0}\right)^n u(n) \longleftrightarrow \frac{1}{1 - e^{-j\omega_0}z^{-1}}, \quad |z| > |e^{-j\omega_0}| \longrightarrow |z| > 1$$

The set of values of $|z|$ for which the z-transforms of both terms converge is $|z| > 1$, and thus, we obtain

$$X(z) = \frac{1}{2}\left[\frac{1}{1 - e^{j\omega_0}z^{-1}} + \frac{1}{1 - e^{-j\omega_0}z^{-1}}\right], \quad |z| > 1$$

$$= \frac{1}{2}\left[\frac{1 - e^{-j\omega_0}z^{-1} + 1 - e^{j\omega_0}z^{-1}}{(1 - e^{j\omega_0}z^{-1})(1 - e^{-j\omega_0}z^{-1})}\right], \quad |z| > 1$$

$$= \frac{1}{2}\left[\frac{2 - 2z^{-1}(e^{j\omega_0} + e^{-j\omega_0})/2}{1 - 2z^{-1}((e^{j\omega_0} + e^{-j\omega_0})/2) + z^{-2}}\right], \quad |z| > 1$$

$$= \frac{1}{2}\left[\frac{2 - 2z^{-1}\cos(\omega_0)}{1 - 2z^{-1}\cos(\omega_0) + z^{-2}}\right], \quad |z| > 1$$

$$X(z) = \frac{1 - z^{-1}\cos(\omega_0)}{1 - 2z^{-1}\cos(\omega_0) + z^{-2}}, \quad |z| > 1$$

Therefore,

$$\cos(\omega_0 n)u(n) \longleftrightarrow \frac{1 - z^{-1}\cos(\omega_0)}{1 - 2z^{-1}\cos(\omega_0) + z^{-2}}, \quad |z| > 1 \qquad (10.20)$$

The ROC for $X(z)$ is $|z| > 1$, as shown in the shaded area in Fig. 10.13.

Example 10.12 Determine the z-transform and ROC of the following finite-duration signals:

(a) $x_1(n) = \{\underset{\uparrow}{1}, 2, 6, -2, 0, 3\}$

(b) $x_2(n) = \{1, 2, \underset{\uparrow}{6}, -2, 0, 3\}$

(c) $x_3(n) = \{\underset{\uparrow}{0}, 0, 1, 2, 6, -2, 3\}$

(d) $x_4(n) = \{1, 2, 6, -2, 0, \underset{\uparrow}{3}\}$

Solution
(a) By definition

$$X_1(z) = \sum_{n=-\infty}^{\infty} x_1(n)z^{-n} = \sum_{n=0}^{5} x(n)z^{-n}$$

$$= x(0) + x(1)z^{-1} + x(2)z^{-2} + x(3)z^{-3} + x(4)z^{-4} + x(5)z^{-5}$$

$$X_1(z) = 1 + 2z^{-1} + 6z^{-2} - 2z^{-3} + 3z^{-5}$$

ROC is the entire z-plane except $z = 0$ [because $X(z)$ becomes unbounded for $z = 0$]. For a finite-duration right-sided signal [i.e., $x(n) = 0$ for $n < 0$ and $n > N_1$ for some finite N_1], the ROC will be the entire z-plane except $z = 0$.
(b) By definition

$$X_2(z) = \sum_{n=-\infty}^{\infty} x_2(n)z^{-n} = \sum_{n=-2}^{3} x(n)z^{-n}$$

$$= x(-2)z^2 + x(-1)z + x(0) + x(1)z^{-1} + x(2)z^{-2} + x(3)z^{-3}$$

$$X_2(z) = z^2 + 2z + 6 - 2z^{-1} + 3z^{-3}$$

ROC is the entire z-plane except $z = 0$ and $z = \infty$ [because $X(z)$ becomes unbounded for both $z = 0$ and $z = \infty$]. For a finite-duration two-sided signal (i.e., the signal is of finite extent for both $n < 0$ and $n > 0$), the ROC will be the entire z-plane except $z = 0$ and $z = \infty$.

(c) By definition

$$X_3(z) = \sum_{n=-\infty}^{\infty} x_3(n)z^{-n} = \sum_{n=0}^{6} x(n)z^{-n}$$

$$= x(0) + x(1)z^{-1} + x(2)z^{-2} + x(3)z^{-3} + x(4)z^{-4} + x(5)z^{-5} + x(6)z^{-6}$$

$$X_3(z) = z^{-2} + 2z^{-3} + 6z^{-4} - 2z^{-5} + 3z^{-6}$$

ROC is the entire z-plane except $z = 0$ (because $X(z) = \infty$ for $z = 0$).

(d) By definition

$$X_4(z) = \sum_{n=-\infty}^{\infty} x_4(n)z^{-n} = \sum_{n=-5}^{0} x(n)z^{-n}$$

$$= x(-5)z^5 + x(-4)z^4 + x(-3)z^3 + x(-2)z^2 + x(-1)z + x(0)$$

$$X_4(z) = z^5 + 2z^4 + 6z^3 - 2z^2 + 3$$

ROC is the entire z-plane except $z = \infty$ [because $X(z)$ becomes unbounded for $z = \infty$]. For a finite-duration left-sided signal [i.e., $x(n) = 0$ for $n > 0$ and $n < -N_1$ for some finite positive N_1], the ROC will be the entire z-plane except $z = \infty$.

10.7 s- TO z-PLANE MAPPING

The relationship between the Laplace transform (s-plane) and the z-transform (z-plane) can be established by considering the sampled signal $x_s(t)$ obtained by sampling a continuous-time signal $x(t)$. Let T be the sampling interval. The sampled signal is given by [Eq. (8.5)]

$$x_s(t) = \sum_{n=-\infty}^{\infty} x(nT)\delta(t - nT)$$

Taking the Laplace transform yields

$$X_s(s) = \int_{-\infty}^{\infty} x_s(t)e^{-st}\, dt = \int_{-\infty}^{\infty} \left[\sum_{n=-\infty}^{\infty} x(nT)\delta(t - nT) \right] e^{-st}\, dt$$

which, upon interchanging integration and summation, is

$$X_s(s) = \sum_{n=-\infty}^{\infty} x(nT) \int_{-\infty}^{\infty} \delta(t - nT)\, e^{-st}\, dt = \sum_{n=-\infty}^{\infty} x(nT)\, e^{-st}\Big|_{t=nT}$$

$$X_s(s) = \sum_{n=-\infty}^{\infty} x(n)\, e^{-snT}$$

$$X_s(s) = \sum_{n=-\infty}^{\infty} x(n) \left(e^{sT}\right)^{-n}$$

By introducing a new variable $z = e^{sT}$, this equation can be expressed as

$$X_s(s)\Big|_{z=e^{sT}} = \sum_{n=-\infty}^{\infty} x(n) z^{-n}$$

$$= X(z)$$

It is cleared from the above discussion that the *z*-transform can be considered to be the Laplace transform with a change of variable

$$z = e^{sT} \quad \text{or equivalently} \quad s = \frac{1}{T}\ln(z) \tag{10.21}$$

Recall from Chapter 9 that the Laplace variable *s* was given by

$$s = \sigma + j\omega$$

where σ was a constant used to ensure convergence of the integral defining the Laplace transform and thus the exitance of the Laplace transform itself. From Eq. (10.21),

$$z = e^{(\sigma + j\omega)t}$$

$$= e^{\sigma t}\, e^{j\omega t}$$

so that the magnitude of *z* is given by

$$|z| = e^{\sigma t} \tag{10.22}$$

Now, consider the following three cases:

Case I $\sigma = 0$. The imaginary axis ($s = j\omega$) in the *s*-plane, $\sigma = 0$, corresponds to $|z| = 1$ in the *z*-plane. That is, the transformation $z = e^{sT}$ transforms the imaginary axis ($s = j\omega$) in the *s*-plane into a unit circle ($|z| = 1$) in the *z*-plane.

Case II $\sigma < 0$. The left-half of the *s*-plane, $\sigma < 0$, corresponds to $|z| < 1$ in the *z*-plane. That is, the transformation $z = e^{sT}$ maps the left-half of the *s*-plane into the inside of the unit circle in the *z*-plane.

Case III $\sigma < 0$. The right-half of the *s*-plane, $\sigma > 0$, corresponds to $|z| > 1$ in the *z*-plane. That is, the transformation $z = e^{sT}$ maps the right-half of the *s*-plane into the outside of the unit circle in the *z*-plane.

This mapping of the Laplace variable *s* into the *z*-plane through $z = e^{sT}$ is illustrated in Fig. 10.14.

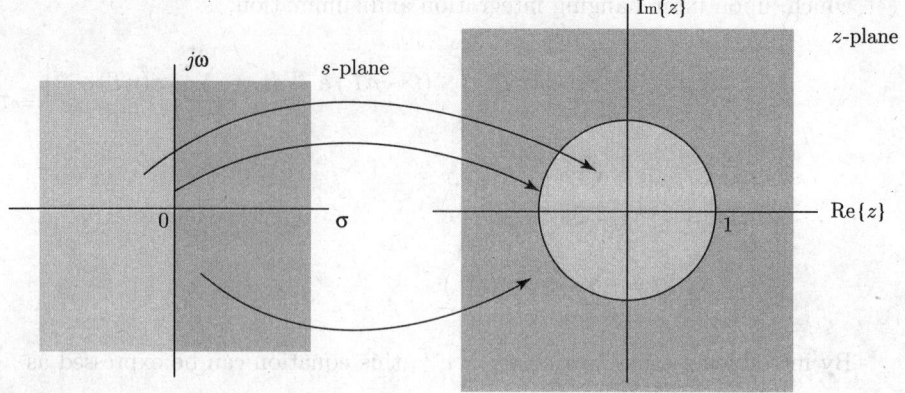

Fig. 10.14 Mapping from s-plane to z-plane

Example 10.13 Determine the z-transform of the signal, $x(n)$, obtained by sampling the continuous-time signal

$$x(t) = e^{-at}u(t)$$

every T seconds.

Solution

The signal, $x(n)$, is obtained by sampling the continuous-time signal

$$x(t) = e^{-at}u(t)$$

every T seconds. Therefore,

$$x(t)\Big|_{t=nT} = x(nT) = x(n) = e^{-anT}u(nT)$$

$$x(n) = e^{-anT}u(n)$$

By definition

$$X(z) = \sum_{n=-\infty}^{\infty} x(n)z^{-n} = \sum_{n=-\infty}^{\infty} e^{-anT}u(n)z^{-n}$$

$$= \sum_{n=0}^{\infty} e^{-anT}z^{-n} = \sum_{n=0}^{\infty} \left(e^{-aT}z^{-1}\right)^n$$

$$X(z) = \frac{1}{1 - e^{-aT}z^{-1}} \quad \text{for} \quad |e^{-aT}z^{-1}| < 1 \longrightarrow |z| > |e^{-aT}|$$

10.8 PROPERTIES OF THE z-TRANSFORM

Properties of the z-transform are as following:

10.8.1 Linearity

If

$$x_1(n) \longleftrightarrow X_1(z) \quad \text{with ROC} = R_1$$

and

$$x_2(n) \longleftrightarrow X_2(z) \quad \text{with ROC} = R_2$$

then

$$ax_1(n) + bx_2(n) \longleftrightarrow aX_1(z) + bX_2(z) \quad \text{with ROC containing } R_1 \cap R_2 \qquad (10.23)$$

Proof The z-transform of $ax_1(n) + bx_2(n)$ is given by

$$\mathcal{Z}[ax_1(n) + bx_2(n)] = \sum_{n=-\infty}^{\infty} [ax_1(n) + bx_2(n)]z^{-n}$$

$$= a\sum_{n=-\infty}^{\infty} x_1(n)z^{-n} + b\sum_{n=-\infty}^{\infty} x_2(n)z^{-n}$$

$$= aX_1(z) + bX_2(z)$$

As indicated, the ROC of the linear combination is at least the intersection of R_1 and R_2. If the linear combination is such that there is no pole-zero cancellation, then the ROC will be exactly equal to the overlap of the individual regions of convergence. But, if the linear combination is such that some zeros are introduced that cancels poles, then the ROC may be larger. A simple example of this occurs when $x_1(n)$ and $x_2(n)$ are both of infinite duration, but the linear combination is of finite duration. In this case the ROC of the linear combination is the entire z-plane, with the possible exception of zero and/or infinity. This phenomenon is illustrated in the next example. ∎

Example 10.14 Determine the z-transform and the ROC of the signal

$$x(n) = a^n u(n) - a^n u(n-1)$$

Solution
By definition

$$X(z) = \sum_{n=-\infty}^{\infty} x(n)z^{-n} = \sum_{n=-\infty}^{\infty} [a^n u(n) - a^n u(n-1)]z^{-n}$$

$$= \sum_{n=-\infty}^{\infty} a^n u(n)z^{-n} - \sum_{n=-\infty}^{\infty} a^n u(n-1)z^{-n}$$

$$= \sum_{n=0}^{\infty} \left(az^{-1} \right)^n - \sum_{n=1}^{\infty} \left(az^{-1} \right)^n$$

$$= \frac{1}{1 - az^{-1}} - \frac{az^{-1}}{1 - az^{-1}} \qquad \text{for} \quad |az^{-1}| < 1 \longrightarrow |z| > |a|$$

$$X(z) = \frac{1 - az^{-1}}{1 - az^{-1}} = 1 \qquad \text{ROC is the entire } z\text{-plane.}$$

The sequences $a^n u(n)$ and $a^n u(n-1)$ both have a region of convergence defined by $|z| > |a|$. The signal $x(n) = a^n u(n) - a^n u(n-1) = \delta(n)$ is a finite duration signal, and its z-transform is $X(z) = 1$, which has a ROC that is the entire z-plane.

10.8.2 Time Shifting

If

$$x(n) \longleftrightarrow X(z) \quad \text{with ROC} = R$$

then

$$x(n - n_0) \longleftrightarrow z^{-n_0} X(z) \quad \begin{array}{l} \text{with ROC} = R, \text{ except for the possible} \\ \text{addition and deletion of } z = 0 \text{ or } z = \infty. \end{array} \qquad (10.24)$$

Proof The z-transform of $x(n - n_0)$ is given by

$$\mathcal{Z}[x(n - n_0)] = \sum_{n=-\infty}^{\infty} x(n - n_0) z^{-n}$$

A change of variables is performed by letting $m = n - n_0$, which also yields $n = m + n_0$, $m = -\infty$ as $n = -\infty$, and $m = \infty$ as $n = \infty$. Therefore,

$$\mathcal{Z}[x(n - n_0)] = \sum_{m=-\infty}^{\infty} x(m) z^{-(m + n_0)}$$

$$= z^{-n_0} \sum_{m=-\infty}^{\infty} x(m) z^{-m}$$

$$\mathcal{Z}[x(n - n_0)] = X(z) z^{-n_0}$$

■

Example 10.15 Determine the z-transform and ROC of the following signals:

(a) $x_1(n) = \delta(n - k)$

(b) $x_2(n) = \delta(n + k)$

(c) $x_3(n) = 2\delta(n + 2) + 3\delta(n) - 5\delta(n - 1) + 3\delta(n - 2)$

Solution

(a) We know that

$$\delta(n) \longleftrightarrow 1 \quad \text{ROC is the entire } z\text{-plane.}$$

Therefore, using the time-shifting property, we obtain

$$\delta(n-k) \longleftrightarrow z^{-k} \quad \text{ROC is the entire } z\text{-plane except } z = 0.$$

(b) We know that

$$\delta(n) \longleftrightarrow 1 \quad \text{ROC is the entire } z\text{-plane.}$$

Therefore, using the time-shifting property, we obtain

$$\delta(n+k) \longleftrightarrow z^{k} \quad \text{ROC is the entire } z\text{-plane except } z = \infty.$$

(c) Consider the given signal

$$x_3(n) = 2\delta(n+2) + 3\delta(n) - 5\delta(n-1) + 3\delta(n-2)$$

$$\mathcal{Z}[x_3(n)] = 2\mathcal{Z}[\delta(n+2)] + 3\mathcal{Z}[\delta(n)] - 5\mathcal{Z}[\delta(n-1)] + 3\mathcal{Z}[\delta(n-2)]$$

$$X_3(z) = 2z^2 + 3 - 5z^{-1} + 3z^{-2} \quad \text{ROC is the entire } z\text{-plane except } z = 0 \text{ and } z = \infty.$$

Example 10.16 Determine the *z*-transform and ROC of the signal

$$x(n) = u(n) - u(n-10)$$

Solution

Consider the given signal

$$x(n) = u(n) - u(n-10)$$

$$\mathcal{Z}[x(n)] = \mathcal{Z}[u(n)] - \mathcal{Z}[u(n-10)]$$

$$X(z) = \frac{1}{1-z^{-1}} - \frac{z^{-10}}{1-z^{-1}} = \frac{1-z^{-10}}{1-z^{-1}} = \frac{1}{z^{10}}\left(\frac{z^{10}-1}{z-1}\right)$$

$$X(z) = \frac{1}{z^{10}}\left(\frac{z^{10}-1}{z-1}\right) \quad \text{ROC is the entire } z\text{-plane except } z = 0.$$

10.8.3 Scaling in the *z*-Domain

If

$$x(n) \longleftrightarrow X(z) \quad \text{with ROC} = R$$

then

$$z_0^n x(n) \longleftrightarrow X\left(\frac{z}{z_0}\right) \quad \text{with ROC} = |z_0|R \qquad (10.25)$$

The notation $|z_0|R$ implies that the ROC R is scaled by $|z_0|$. If R is $a < |z| < b$, then the new ROC is $|z_0|a < |z| < b|z_0|$. Also, if $X(z)$ has a pole (or zero) at $z = a$, then $X(z/z_0)$ has a pole (or zero) at $z = z_0a$. This indicates that the poles and zeros of $X(z)$ have their radii changed by $|z_0|$, and their angles are changed by $\angle\{z_0\}$. If z_0 has unit magnitude, then the radius is unchanged; if z_0 is a positive real number, then the angle is unchanged.

Proof The z-transform of $z_0^n x(n)$ is given by

$$\mathcal{Z}[z_0^n x(n)] = \sum_{n=-\infty}^{\infty} [z_0^n x(n)]z^{-n}$$

$$= \sum_{n=-\infty}^{\infty} x(n)\left(\frac{z}{z_0}\right)^{-n}$$

$$\mathcal{Z}[z_0^n x(n)] = X\left(\frac{z}{z_0}\right)$$

∎

Example 10.17 Determine the z-transform of the following signals:

(a) $g_1(n) = a^n \sin(\omega_0 n)u(n)$
(b) $g_2(n) = a^n \cos(\omega_0 n)u(n)$

Solution
(a) Consider the given signal

$$g_1(n) = a^n \sin(\omega_0 n)u(n) = a^n x(n)$$

where $x(n) = \sin(\omega_0 n)u(n)$ and its z-transform is given by [Eq. (10.19)],

$$X(z) = \frac{z^{-1}\sin(\omega_0)}{1 - 2z^{-1}\cos(\omega_0) + z^{-2}}, \quad |z| > 1$$

Now, taking the z-transform of the given signal $g_1(n)$ and using the scaling in the z-domain property, we obtain

$$\mathcal{Z}[g_1(n)] = \mathcal{Z}[a^n x(n)]$$

$$G_1(z) = X\left(\frac{z}{a}\right)$$

$$G_1(z) = \frac{(z/a)^{-1}\sin(\omega_0)}{1 - 2(z/a)^{-1}\cos(\omega_0) + (z/a)^{-2}}, \quad |z| > |a|$$

$$= \frac{az^{-1}\sin(\omega_0)}{1 - 2az^{-1}\cos(\omega_0) + a^2z^{-2}}, \quad |z| > |a|$$

Therefore,

$$a^n \sin(\omega_0 n)u(n) \longleftrightarrow \frac{az^{-1}\sin(\omega_0)}{1 - 2az^{-1}\cos(\omega_0) + a^2 z^{-2}}, \quad |z| > |a| \qquad (10.26)$$

(b) Consider the given signal

$$g_2(n) = a^n \cos(\omega_0 n)u(n) = a^n x(n)$$

where $x(n) = \cos(\omega_0 n)u(n)$ and its z-transform is given by [Eq. (10.20)],

$$X(z) = \frac{1 - z^{-1}\cos(\omega_0)}{1 - 2z^{-1}\cos(\omega_0) + z^{-2}}, \quad |z| > 1$$

Now, taking the z-transform of the given signal $g_2(n)$ and using the scaling in the z-domain property, we obtain

$$\mathcal{Z}[g_2(n)] = \mathcal{Z}[a^n x(n)]$$

$$G_2(z) = X\left(\frac{z}{a}\right)$$

$$G_2(z) = \frac{1 - (z/a)^{-1}\cos(\omega_0)}{1 - 2(z/a)^{-1}\cos(\omega_0) + (z/a)^{-2}}, \quad |z| > |a|$$

$$= \frac{1 - az^{-1}\cos(\omega_0)}{1 - 2az^{-1}\cos(\omega_0) + a^2 z^{-2}}, \quad |z| > |a|$$

Therefore,

$$a^n \cos(\omega_0 n)u(n) \longleftrightarrow \frac{1 - az^{-1}\cos(\omega_0)}{1 - 2az^{-1}\cos(\omega_0) + a^2 z^{-2}}, \quad |z| > |a| \qquad (10.27)$$

10.8.4 Time Reversal
If

$$x(n) \longleftrightarrow X(z) \quad \text{with ROC} = R$$

then

$$x(-n) \longleftrightarrow X\left(\frac{1}{z}\right) = X(z^{-1}) \quad \text{with ROC} = \frac{1}{R} \qquad (10.28)$$

The time-reversal, or reflection, corresponds to replacing z by z^{-1}. Hence if R is of the form $a < |z| < b$, the ROC of the reflected signal is $a < 1/|z| < b$, or $1/b < |z| < 1/a$.

Proof The z-transform of $x(-n)$ is given by

$$\mathcal{Z}[x(-n)] = \sum_{n=-\infty}^{\infty} x(-n)z^{-n}$$

A change of variables is performed by letting $m = -n$, which also yields $m = \infty$ as $n = -\infty$, and $m = -\infty$ as $n = \infty$. Therefore,

$$\mathcal{Z}[x(-n)] = \sum_{m=\infty}^{-\infty} x(m)z^m$$

$$= \sum_{m=-\infty}^{\infty} x(m) \left(\frac{1}{z}\right)^{-m}$$

$$\mathcal{Z}[x(-n)] = X\left(\frac{1}{z}\right) = X(z^{-1})$$

An interesting consequence of the time-reversal property is that, if $x(n)$ is real and even, i.e.,

$$\text{if} \quad x(n) = x(-n), \quad \text{then} \quad X(z) = X(z^{-1}) = X\left(\frac{1}{z}\right) \tag{10.29}$$

If $X(z)$ has a zero at $z = z_0$, i.e.,

$$X(z_0) = 0$$

then

$$X\left(\frac{1}{z_0}\right) = 0$$

which implies that $X(z)$ will also have a zero at $z = 1/z_0$. The same holds true for poles. That is, if there is a pole at z_0, there must be a pole at $z = 1/z_0$. Similarly, if $x(n)$ is odd, i.e.,

$$\text{if} \quad x(n) = -x(-n) \quad \text{then} \quad X(z) = -X(z^{-1}) = -X\left(\frac{1}{z}\right) \tag{10.30}$$

■

10.8.5 Differentiation in the z-Domain

If

$$x(n) \longleftrightarrow X(z) \quad \text{with ROC} = R$$

then

$$nx(n) \longleftrightarrow -z\frac{dX(z)}{dz} \quad \text{with ROC} = R. \tag{10.31}$$

Multiplication by n in the time domain corresponds to differentiation with respect to z and multiplication of the result by $-z$ in the z-domain. This operation does not change the location of the poles and hence, the ROC remains same.

Proof The z-transform of $x(n)$ is given by

$$X(z) = \sum_{n=-\infty}^{\infty} x(n)z^{-n}$$

Differentiating both sides of the above equation, we obtain

$$\frac{dX(z)}{dz} = \sum_{n=-\infty}^{\infty} x(n)\frac{dz^{-n}}{dz}$$

$$= \sum_{n=-\infty}^{\infty} [-nx(n)]z^{-n-1}$$

$$\frac{dX(z)}{dz} = -\frac{1}{z}\sum_{n=-\infty}^{\infty} [nx(n)]z^{-n}$$

$$-z\frac{dX(z)}{dz} = \sum_{n=-\infty}^{\infty} [nx(n)]z^{-n}$$

$$-z\frac{dX(z)}{dz} = \mathcal{Z}[nx(n)]$$

■

Example 10.18 Determine the z-transform of the sequence

$$g(n) = na^n u(n)$$

Solution
Consider the given signal

$$g(n) = na^n u(n) = nx(n)$$

where $x(n) = a^n u(n)$ and its z-transform is given by [Eq. (10.12)],

$$X(z) = \frac{1}{1 - az^{-1}}, \quad |z| > |a|$$

Now, taking the z-transform of the given signal $g(n)$ and using the differentiation in the z-domain property, we obtain

$$\mathcal{Z}[g(n)] = \mathcal{Z}[nx(n)]$$

$$G(z) = -z\frac{dX(z)}{dz}$$

$$= -z\frac{d}{dz}\frac{1}{1 - az^{-1}}$$

$$= -z\frac{-az^{-2}}{(1 - az^{-1})^2}$$

$$G(z) = \frac{az^{-1}}{(1 - az^{-1})^2}, \quad |z| > |a|$$

Therefore,

$$na^n u(n) \longleftrightarrow \frac{az^{-1}}{(1 - az^{-1})^2}, \quad |z| > |a| \tag{10.32}$$

or, equivalently

$$\frac{1}{a}na^n u(n) \longleftrightarrow \frac{z^{-1}}{(1 - az^{-1})^2}, \quad |z| > |a|$$

and, using the time-shifting property, we obtain

$$\frac{1}{a}(n + 1)a^{n+1} u(n + 1) \longleftrightarrow z\frac{z^{-1}}{(1 - az^{-1})^2}, \quad |z| > |a|$$

$$(n + 1)a^n u(n + 1) \longleftrightarrow \frac{1}{(1 - az^{-1})^2}, \quad |z| > |a|$$

It is worth noting that, although the LHS is multiplied by a step function that begins at $n = -1$, the sequence $(n + 1)a^n u(n + 1)$ is still zero prior to $n = 0$, since the factor $n + 1$ is zero at $n = -1$. Thus,

$$(n + 1)a^n u(n) \longleftrightarrow \frac{1}{(1 - az^{-1})^2}, \quad |z| > |a| \tag{10.33}$$

Example 10.19 Determine the z-transform of the sequence

$$g(n) = nu(n)$$

Solution
Consider the given signal

$$g(n) = nu(n) = nx(n)$$

where $x(n) = u(n)$ and its z-transform is given by [Eq. (10.15)],

$$X(z) = \frac{1}{1 - z^{-1}}, \quad |z| > 1$$

Now, taking the z-transform of the given signal $g(n)$ and using the differentiation in the z-domain property, we obtain

$$\mathcal{Z}[g(n)] = \mathcal{Z}[nx(n)]$$

$$G(z) = -z\frac{dX(z)}{dz}$$

$$G(z) = -z \frac{d}{dz} \frac{1}{1 - z^{-1}}$$

$$= -z \frac{-z^{-2}}{(1 - z^{-1})^2}$$

$$= \frac{z^{-1}}{(1 - z^{-1})^2}, \quad |z| > 1$$

Therefore,

$$nu(n) \longleftrightarrow \frac{z^{-1}}{(1 - z^{-1})^2}, \quad |z| > 1 \tag{10.34}$$

Now, using the time-shifting property, we obtain

$$(n + 1)u(n + 1) \longleftrightarrow z \frac{z^{-1}}{(1 - z^{-1})^2}, \quad |z| > 1$$

$$(n + 1)u(n + 1) \longleftrightarrow \frac{1}{(1 - z^{-1})^2}, \quad |z| > 1$$

It is worth noting that, although the LHS is multiplied by a step function that begins at $n = -1$, the sequence $(n + 1)u(n + 1)$ is still zero prior to $n = 0$, since the factor $n + 1$ is zero at $n = -1$. Thus,

$$(n + 1)u(n) \longleftrightarrow \frac{1}{(1 - z^{-1})^2}, \quad |z| > 1 \tag{10.35}$$

Example 10.20 Determine the signal $x(n)$ whose z-transform is given by

$$X(z) = \log(1 + az^{-1}), \quad |z| > |a|$$

Solution
By taking the first derivative of $X(z)$, we can convert the given z-transform to a rational expression. Therefore, by taking the first derivative of $X(z)$, we obtain

$$\frac{dX(z)}{dz} = \frac{-az^{-2}}{1 + az^{-1}}$$

Thus,

$$-z \frac{dX(z)}{dz} = \frac{-az^{-1}}{1 + az^{-1}}$$

Taking the inverse z-transform yields

$$\mathcal{Z}^{-1}\left[-z \frac{dX(z)}{dz} \right] = \mathcal{Z}^{-1}\left[\frac{-az^{-1}}{1 + az^{-1}} \right]$$

$$nx(n) = \mathcal{Z}^{-1}\left[\frac{-az^{-1}}{1 + az^{-1}} \right]$$

We know that

$$a^n u(n) \longleftrightarrow \frac{1}{1 - az^{-1}}, \quad |z| > |a|$$

Therefore,

$$(-a)^n u(n) \longleftrightarrow \frac{1}{1 + az^{-1}}, \quad |z| > |a|$$

$$a(-a)^n u(n) \longleftrightarrow \frac{a}{1 + az^{-1}}, \quad |z| > |a|$$

Using the time-shifting property, we obtain

$$a(-a)^{n-1} u(n-1) \longleftrightarrow \frac{az^{-1}}{1 + az^{-1}}, \quad |z| > |a|$$

$$-(-a)^n u(n-1) \longleftrightarrow \frac{az^{-1}}{1 + az^{-1}}, \quad |z| > |a|$$

Consequently,

$$nx(n) = -(-a)^n u(n-1)$$

$$x(n) = \frac{-(-a)^n}{n} u(n-1) = \frac{1}{n}(-1)^{n+1} a^n u(n-1)$$

Example 10.21 Find the z-transform of the signal $g(n) = |n|a^{|n|}$

Solution
Consider the given signal

$$g(n) = |n|a^{|n|} = \begin{cases} na^n, & n \ge 0 \\ -na^{-n}, & n \le 0 \end{cases}$$

$$g(n) = na^n u(n) - na^{-n} u(-n) = x(n) + x(-n)$$

where $x(n) = na^n u(n)$ and its z-transform is given by [Eq. (10.32)]

$$X(z) = \frac{az^{-1}}{(1 - az^{-1})^2}, \quad |z| > |a|$$

Using the time-reversal property, the z-transform of $x(-n) = -na^{-n}u(-n)$ is given by

$$\mathcal{Z}[x(-n)] = X\left(\frac{1}{z}\right) = \frac{a(1/z)^{-1}}{\left(1 - a(1/z)^{-1}\right)^2}, \quad \frac{1}{|z|} > |a|$$

$$= \frac{az}{(1 - az)^2}, \quad |z| < \frac{1}{|a|}$$

Now, consider

$$g(n) = x(n) + x(-n)$$

Taking *z*-transform yields

$$G(z) = X(z) + X\left(\frac{1}{z}\right)$$

$$= \frac{az^{-1}}{(1 - az^{-1})^2} + \frac{az}{(1 - az)^2}, \quad |a| < |z| < \frac{1}{|a|}$$

$$= \frac{(1 - az)^2 az^{-1} + (1 - az^{-1})^2 az}{(1 - az^{-1})^2 (1 - az)^2}, \quad |a| < |z| < \frac{1}{|a|}$$

$$G(z) = \frac{a(1 + a^2)(z + z^{-1}) - 4a^2}{(1 - az^{-1})^2 (1 - az)^2}, \quad |a| < |z| < \frac{1}{|a|}$$

10.8.6 Time Expansion

The continuous-time concept of time scaling does not directly extend to discrete time, since the discrete-time index is defined only for integer values. However, the discrete-time concept of time expansion can be defined and does play an important role in discrete-time signal and system analysis. Let m be a positive integer, and define the signal

$$x_{(m)}(n) = \begin{cases} x\left(\dfrac{n}{m}\right), & \text{if } n \text{ is a multiple of } m \\ 0, & \text{if } n \text{ is not a multiple of } m. \end{cases} \tag{10.36}$$

$x_m(n)$ can be obtained from $x(n)$ by placing $m - 1$ zeros between successive values of the original signal. Intuitively, we can think of $x_{(m)}(n)$ as a slowed-down version of $x(n)$. Now, if

$$x(n) \longleftrightarrow X(z) \quad \text{with ROC} = R$$

then,

$$x_{(m)}(n) \longleftrightarrow X(z^m) \quad \text{with ROC} = R^{1/m} \tag{10.37}$$

That is, if R is $a < |z| < b$, then the new ROC is $a < |z|^m < b$, or $a^{1/m} < |z| < b^{1/m}$. Also, if $X(z)$ has a pole (or zero) at $z = a$, then $X(z^m)$ has a pole (or zero) at $z = a^{1/m}$.

Proof The *z*-transform of $x_{(m)}(n)$ is given by

$$\mathcal{Z}[x_{(m)}(n)] = \sum_{n=-\infty}^{\infty} x_{(m)}(n)z^{-n} = \sum_{n=-\infty}^{\infty} x\left(\frac{n}{m}\right)z^{-n}$$

A change of variables is performed by letting $r = n/m$, which also yields $r = -\infty$ as $n = -\infty$, and $r = \infty$ as $n = \infty$. Therefore,

$$\mathcal{Z}[x_{(m)}(n)] = \sum_{r=-\infty}^{\infty} x(r)z^{-mr} = \sum_{r=-\infty}^{\infty} x(r)\left(z^m\right)^{-r} = X(z^m)$$

■

Example 10.22 Find the z-transform and ROC of the signal

$$g(n) = a^{\frac{n}{3}} u\left(\frac{n}{3}\right) = \begin{cases} a^{\frac{n}{3}}, & n = 0, 3, 6, \cdots \\ 0, & \text{elsewhere} \end{cases}$$

where $|a| < 1$.

Solution

Consider the given signal

$$g(n) = a^{\frac{n}{3}} u\left(\frac{n}{3}\right) = x\left(\frac{n}{3}\right) = x_{(3)}(n)$$

where $x(n) = a^n u(n)$, and its z-transform is given by

$$a^n u(n) \longleftrightarrow \frac{1}{1 - az^{-1}}, \quad |z| > |a|$$

Now, using the time-expansion property, we obtain

$$x_{(3)}(n) \longleftrightarrow X(z^3)$$

$$a^{\frac{n}{3}} u\left(\frac{n}{3}\right) \longleftrightarrow \frac{1}{1 - az^{-3}}, \quad |z| > |a|^{\frac{1}{3}}$$

Example 10.23 Find the z-transform and ROC of the signal

$$g(n) = a^n u\left(\frac{n}{10}\right) = \begin{cases} a^n, & n = 0, 10, 20, \cdots \\ 0, & \text{elsewhere} \end{cases}$$

where $|a| < 1$.

Solution

Consider the given signal

$$g(n) = a^n u\left(\frac{n}{10}\right) = a^{10n/10} u\left(\frac{n}{10}\right) = x\left(\frac{n}{10}\right) = x_{(10)}(n)$$

where $x(n) = \left(a^{10}\right)^n u(n)$, and its z-transform is given by

$$\left(a^{10}\right)^n u(n) \longleftrightarrow \frac{1}{1 - a^{10}z^{-1}}, \quad |z| > |a|^{10}$$

Now, using the time-expansion property, we obtain

$$x_{(10)}(n) \longleftrightarrow X(z^{10})$$

$$a^{10n/10} u\left(\frac{n}{10}\right) \longleftrightarrow \frac{1}{1 - a^{10}z^{-10}}, \quad |z| > |a|^{101/10}$$

$$a^n u\left(\frac{n}{10}\right) \longleftrightarrow \frac{1}{1 - a^{10}z^{-10}}, \quad |z| > |a|$$

Example 10.24 Find the z-transform of a causal signal

$$g(n) = u\left(\frac{n}{2}\right) = \begin{cases} 1, & n = 0, 2, 4, \cdots \\ 0, & \text{elsewhere} \end{cases} = \begin{cases} 1, & n \text{ even} \\ 0, & n \text{ odd} \end{cases}$$

Solution
Method 1 Consider the given signal

$$g(n) = u\left(\frac{n}{2}\right) = x\left(\frac{n}{2}\right) = x_{(2)}(n)$$

where $x(n) = u(n)$, and its z-transform is given by

$$u(n) \longleftrightarrow \frac{1}{1 - z^{-1}}, \quad |z| > 1$$

Now, using the time-expansion property, we obtain

$$x_{(2)}(n) \longleftrightarrow X(z^2)$$

$$u\left(\frac{n}{2}\right) \longleftrightarrow \frac{1}{1 - z^{-2}}, \quad |z| > 1$$

Method 2 By definition

$$G(z) = \sum_{n=-\infty}^{\infty} g(n)z^{-n} = \sum_{\substack{n=0 \\ n \text{ even}}}^{\infty} (1)z^{-n}$$

Substituting $n = 2r$, where r is varying from 0 to ∞, we obtain

$$G(z) = \sum_{r=0}^{\infty} z^{-2r} = \sum_{r=0}^{\infty} \left(z^{-2}\right)^r$$

$$G(z) = \frac{1}{1 - z^{-2}}, \quad |z^{-2}| < 1 \longrightarrow |z| > 1$$

Example 10.25 Find the z-transform of a signal

$$g(n) = u\left(\frac{n-1}{2}\right) = u\left(\frac{1}{2}(n-1)\right) = \begin{cases} 1, & n = 1, 3, 5, \cdots \\ 0, & \text{elsewhere} \end{cases} = \begin{cases} 1, & n \text{ odd} \\ 0, & n \text{ even} \end{cases}$$

Solution
Method 1 Consider the given signal

$$g(n) = u\left(\frac{n-1}{2}\right) = x\left(\frac{n-1}{2}\right)$$

where $x(n) = u(n)$, and its z-transform is given by

$$u(n) \longleftrightarrow \frac{1}{1 - z^{-1}}, \quad |z| > 1$$

Using the time-expansion property, we obtain

$$x_{(2)}(n) \longleftrightarrow X(z^2)$$

$$u\left(\frac{n}{2}\right) \longleftrightarrow \frac{1}{1 - z^{-2}}, \quad |z| > 1$$

Now, using the time-shifting property, we obtain

$$u\left(\frac{n-1}{2}\right) \longleftrightarrow \frac{z^{-1}}{1 - z^{-2}}, \quad |z| > 1$$

Method 2 By definition

$$G(z) = \sum_{n=-\infty}^{\infty} g(n)z^{-n} = \sum_{\substack{n=0 \\ n \text{ odd}}}^{\infty} (1)z^{-n}$$

Substituting $n = 2r + 1$, where r is varying from 0 to ∞, we obtain

$$G(z) = \sum_{r=0}^{\infty} z^{-(2r+1)} = z^{-1} \sum_{r=0}^{\infty} \left(z^{-2}\right)^r$$

$$G(z) = \frac{z^{-1}}{1 - z^{-2}}, \quad |z^{-2}| < 1 \longrightarrow |z| > 1$$

10.8.7 Convolution Property

If

$$x_1(n) \longleftrightarrow X_1(z) \quad \text{with ROC} = R_1$$

and

$$x_2(n) \longleftrightarrow X_2(z) \quad \text{with ROC} = R_2$$

then

$$x_1(n) * x_2(n) \longleftrightarrow X_1(z)X_2(z) \quad \text{with ROC containing } R_1 \cap R_2 \qquad (10.38)$$

The z-transform maps the convolution of two signals into the product of their z-transforms.

Proof The z-transform of $x_1(n) * x_2(n)$ is given by

$$\mathcal{Z}[x_1(n) * x_2(n)] = \sum_{n=-\infty}^{\infty} [x_1(n) * x_2(n)]z^{-n}$$

$$= \sum_{n=-\infty}^{\infty} \left(\sum_{m=-\infty}^{\infty} x_1(m)x_2(n - m) \right) z^{-n}$$

By interchanging the order of summation, we have

$$\mathcal{Z}[x_1(n) * x_2(n)] = \sum_{m=-\infty}^{\infty} x_1(m) \left(\sum_{n=-\infty}^{\infty} x_2(n-m)z^{-n} \right)$$

Applying the time-shifting property [Eq. (10.24)], the bracketed term is $X_2(z)z^{-m}$. Substituting this into the above equation yields

$$\mathcal{Z}[x_1(n) * x_2(n)] = \sum_{m=-\infty}^{\infty} x_1(m) \left(X_2(z)z^{-m} \right)$$

$$= X_2(z) \sum_{m=-\infty}^{\infty} x_1(m)z^{-m}$$

$$\mathcal{Z}[x_1(n) * x_2(n)] = X_2(z)X_1(z) = X_1(z)X_2(z)$$

The convolution property states that convolution in the time domain corresponds to multiplication in the z-domain. Convolution of the two signals, using z-transform, requires the following steps:

1. Compute the z-transforms of the signal to be convolved, i.e.,

$$X_1(z) = \mathcal{Z}[x_1(n)]$$
$$X_2(z) = \mathcal{Z}[x_2(n)]$$

2. Multiply the two z-transforms, i.e.,

$$X(z) = X_1(z)X_2(z)$$

3. Find the inverse z-transform of $X(z)$, i.e.,

$$x(n) = \mathcal{Z}^{-1}[X(z)]$$

■

Example 10.26 Evaluate the convolution of a unit step function $u(n)$ with itself using z-transform.

Solution
The convolution of a unit step function $u(n)$ with itself can be expressed as

$$x(n) = u(n) * u(n)$$

Using the convolution property, we obtain

$$\mathcal{Z}[x(n)] = \mathcal{Z}[u(n)]\mathcal{Z}[u(n)]$$

$$X(z) = \left(\frac{1}{1-z^{-1}} \right) \left(\frac{1}{1-z^{-1}} \right)$$

$$X(z) = \left(\frac{1}{1-z^{-1}} \right)^2$$

The inverse z-transform yields [Eq. (10.35)]

$$x(n) = (n+1)u(n)$$

Example 10.27 Compute the convolution $x(n)$ of the signals

$$x_1(n) = \{\underset{\uparrow}{1},\ -2,\ 1\}$$

$$x_2(n) = \begin{cases} 1, & 0 \le n \le 5 \\ 0, & \text{elsewhere} \end{cases}$$

Solution

By definition

$$X_1(z) = \sum_{n=-\infty}^{\infty} x_1(n)z^{-n} = \sum_{n=0}^{2} x(n)z^{-n}$$

$$= x(0) + x(1)z^{-1} + x(2)z^{-2}$$

$$= 1 - 2z^{-1} + z^{-2}$$

$$X_1(z) = (1 - z^{-1})^2$$

ROC is the entire z-plane except $z = 0$ [because $X(z)$ becomes unbounded for $z = 0$]. Similarly, we have

$$X_2(z) = \sum_{n=-\infty}^{\infty} x_2(n)z^{-n} = \sum_{n=0}^{5} z^{-n}$$

$$X_2(z) = \frac{1 - z^{-6}}{1 - z^{-1}}$$

ROC is the entire z-plane except $z = 0$ [because $X(z)$ becomes unbounded for $z = 0$]. Now, consider

$$x(n) = x_1(n) * x_2(n)$$

Using the convolution property, we obtain

$$X(z) = X_1(z)X_2(z)$$

$$= (1 - z^{-1})^2 \frac{1 - z^{-6}}{1 - z^{-1}}$$

$$= (1 - z^{-1})(1 - z^{-6})$$

$$X(z) = 1 - z^{-1} - z^{-6} + z^{-7}$$

The inverse z-transform yields

$$x(n) = \delta(n) - \delta(n-1) - \delta(n-6) + \delta(n-7)$$
$$= \{\underset{\uparrow}{1}, -1, 0, 0, 0, 0, -1, 1\}$$

10.8.8 Correlation Property
If

$$x_1(n) \longleftrightarrow X_1(z) \quad \text{with ROC} = R_1$$

and

$$x_2(n) \longleftrightarrow X_2(z) \quad \text{with ROC} = R_2$$

then

$$r_{x_1 x_2}(m) \sum_{n=-\infty}^{\infty} x_1(n) x_2(n-m) \longleftrightarrow R_{x_1 x_2}(z) = X_1(z) X_2\left(\frac{1}{z}\right)$$

$$\text{with ROC containing } R_1 \cap \frac{1}{R_2} \tag{10.39}$$

Proof We know that

$$r_{x_1 x_2}(m) = \sum_{n=-\infty}^{\infty} x_1(n) x_2(n-m)$$
$$= x_1(m) * x_2(-m)$$

Taking the z-transform of the above equation and using the convolution property, we obtain

$$\mathcal{Z}[r_{x_1 x_2}(m)] = \mathcal{Z}[x_1(m) * x_2(-m)]$$
$$R_{x_1 x_2}(z) = X_1(z) X_2\left(\frac{1}{z}\right)$$

Example 10.28 Determine the autocorrelation sequence of the signal

$$x(n) = a^n u(n), \quad -1 < a < 1$$

Solution

Since the autocorrelation sequence of a signal is its correlation with itself, Eq. (10.39) gives

$$R_{xx}(z) = \mathcal{Z}[r_{xx}(m)] = X(z)X\left(\frac{1}{z}\right)$$

We know that

$$\mathcal{Z}[a^n u(n)] = X(z) = \frac{1}{1 - az^{-1}}, \quad |z| > |a|$$

and

$$X\left(\frac{1}{z}\right) = \frac{1}{1 - az}, \quad |z| < \frac{1}{|a|}$$

Thus

$$R_{xx}(z) = \left(\frac{1}{1 - az^{-1}}\right)\left(\frac{1}{1 - az}\right) = \frac{1}{1 - a(z + z^{-1}) + a^2}, \quad |a| < |z| < \frac{1}{|a|}$$

Since the ROC of $R_{xx}(z)$ is a ring in the z-plane, $r_{xx}(n)$ is a two-sided signal, even if $x(n)$ is causal. Now, consider

$$R_{xx}(z) = \left(\frac{1}{1 - az^{-1}}\right)\left(\frac{1}{1 - az}\right), \quad |a| < |z| < \frac{1}{|a|}$$

$$= \left(\frac{z}{z - a}\right)\left(\frac{1}{1 - az}\right)$$

$$\frac{R_{xx}(z)}{z} = \left(\frac{1}{z - a}\right)\left(\frac{1}{1 - az}\right)$$

Using the partial fraction expansion, we obtain

$$\frac{R_{xx}(z)}{z} = \frac{1/(1 - a^2)}{z - a} + \frac{a/(1 - a^2)}{1 - az}$$

$$R_{xx}(z) = \frac{1}{1 - a^2}\left[\frac{z}{z - a} + \frac{az}{1 - az}\right]$$

$$= \frac{1}{1 - a^2}\left[\frac{1}{1 - az^{-1}} - \frac{1}{1 - (1/a)z^{-1}}\right]$$

The inverse z-transform yields

$$r_{xx}(m) = \frac{1}{1 - a^2}\left[a^m u(m) + \left(\frac{1}{a}\right)^m u(-m - 1)\right]$$

$$= \frac{1}{1 - a^2} a^{|m|} \quad (-\infty < m < \infty)$$

10.8.9 Accumulation Property

If

$$x(n) \longleftrightarrow X(z) \quad \text{with ROC} = R$$

then

$$\sum_{k=-\infty}^{n} x(k) \longleftrightarrow \frac{1}{1 - z^{-1}} X(z) \quad \text{with ROC at least} = R \cap \{|z| > 1\} \qquad (10.40)$$

Proof Convolving a signal $x(n)$ with a unit step function $u(n)$, we obtain

$$x(n) * u(n) = \sum_{k=-\infty}^{\infty} x(k) u(n - k)$$

we know that

$$u(n - k) = \begin{cases} 1, & (n - k) \geq 0 \longrightarrow k \leq n \\ 0, & (n - k) < 0 \longrightarrow k > n \end{cases}$$

we have

$$x(n) * u(n) = \sum_{k=-\infty}^{n} x(k)$$

Now we can prove the accumulation property of the z-transform:

$$\sum_{k=-\infty}^{n} x(k) = x(n) * u(n)$$

$$\mathcal{Z}\left[\sum_{k=-\infty}^{n} x(k)\right] = \mathcal{Z}[x(n) * u(n)]$$

Using the convolution property [Eq. (10.38)], we obtain

$$\mathcal{Z}\left[\sum_{k=-\infty}^{n} x(k)\right] = X(z)\mathcal{Z}[u(n)]$$

$$= X(z)\left(\frac{1}{1 - z^{-1}}\right)$$

Therefore,

$$\sum_{k=-\infty}^{n} x(k) \longleftrightarrow X(z)\left(\frac{1}{1 - z^{-1}}\right)$$

Example 10.29 Find the z-transform of the signal $g(n)$ where

$$g(n) = \sum_{k=-n}^{n} a^{|k|}, \quad n \geq 0$$

and $g(n) = 0$ for $n < 0$. Assume that $|a| < 1$.

Solution

Consider the given signal

$$g(n) = \sum_{k=-n}^{n} a^{|k|}, \quad n \geq 0$$

$$= \sum_{k=-n}^{-1} a^{-k} + \sum_{k=0}^{n} a^{k}$$

$$= \left(\sum_{k=1}^{n} a^{k} + a^{0}\delta(n) \right) - a^{0}\delta(n) + \sum_{k=0}^{n} a^{k}$$

$$= \sum_{k=0}^{n} a^{k} + \sum_{k=0}^{n} a^{k} - a^{0}\delta(n)$$

$$= 2\sum_{k=0}^{n} a^{k} - 1$$

$$= 2 \sum_{k=-\infty}^{n} a^{k}u(k) - \delta(n)$$

$$g(n) = 2 \sum_{k=-\infty}^{n} x(k) - \delta(n)$$

where $x(k) = a^{k}u(k)$ and its z-transform is given by

$$X(z) = \frac{1}{1 - az^{-1}}, \quad |z| > |a|$$

Again, consider

$$g(n) = 2 \sum_{k=-\infty}^{n} x(k) - \delta(n)$$

Using the accumulation property, we obtain

$$G(z) = 2\frac{X(z)}{1 - z^{-1}} - 1, \quad |z| > 1$$

$$= 2\frac{1}{(1 - az^{-1})(1 - z^{-1})} - 1, \quad |z| > 1$$

$$G(z) = \frac{1 + (1 + a)z^{-1} - az^{-2}}{(1 - az^{-1})(1 - z^{-1})}, \quad |z| > 1$$

10.8.10 First Difference

If

$$x(n) \longleftrightarrow X(z) \quad \text{with ROC} = R$$

then

$$x(n) - x(n-1) \longleftrightarrow (1 - z^{-1})X(z) \quad \text{with ROC at least} = R \cap \{|z| > 0\} \qquad (10.41)$$

A common use of this property is in situations where evaluation of the *z*-transform is easier for the first difference than for the original sequence.

Proof Given that

$$x(n) \longleftrightarrow X(z)$$

Using the time-shifting property, we get

$$x(n - 1) \longleftrightarrow X(z)z^{-1}$$

Now, using the linearity property, we get

$$x(n) - x(n-1) \longleftrightarrow X(z) - X(z)z^{-1}$$
$$x(n) - x(n-1) \longleftrightarrow (1 - z^{-1})X(z)$$

■

10.8.11 Conjugation and Conjugate Symmetry

If

$$x(n) \longleftrightarrow X(z) \quad \text{with ROC} = R$$

then

$$x^*(n) \longleftrightarrow X^*(z^*) \quad \text{with ROC} = R \qquad (10.42)$$

Proof By definition, the z-transform of $x^*(n)$ is

$$\mathcal{Z}[x^*(n)] = \sum_{n=-\infty}^{\infty} x^*(n) z^{-n} = \left[\sum_{n=-\infty}^{\infty} x(n) (z^*)^{-n} \right]^*$$

$$= [X(z^*)]^*$$

$$\mathcal{Z}[x^*(n)] = X^*(z^*)$$

Case-I If $x(n)$ is real, i.e., if

$$x^*(n) = x(n)$$

Then

$$\mathcal{Z}[x^*(n)] = \mathcal{Z}[x(n)]$$

$$X^*(z^*) = X(z)$$

$$X(z) = X^*(z^*) \tag{10.43}$$

Thus, if $X(z)$ has a pole (or zero) at $z = z_0$, it must have a pole (or zero) at the complex conjugate point $z = z_0^*$.

If $x(n)$ *is real and even*, i.e., if

$$x(n) = x^*(n) = x(-n)$$

Then from Eqs (10.29) and (10.43), we have

$$X(z) = X^*(z^*) = X\left(\frac{1}{z}\right) \tag{10.44}$$

Thus, if $X(z)$ has a pole (or zero) at $z = z_0 = re^{j\theta}$, then there must be a pole (or zero) at $z = 1/z_0 = 1/r\, e^{-j\theta}$, $z = z_0^* = re^{-j\theta}$, and $z = 1/z_0^* = 1/r\, e^{j\theta}$. Thus, a general pole constellation (or zero constellation) is a quadruplet

$$z_0 = re^{j\theta}, \qquad \frac{1}{z_0} = \frac{1}{r} e^{-j\theta}, \qquad z_0^* = re^{-j\theta}, \qquad \frac{1}{z_0^*} = \frac{1}{r} e^{j\theta}$$

as shown in Fig. 10.15.

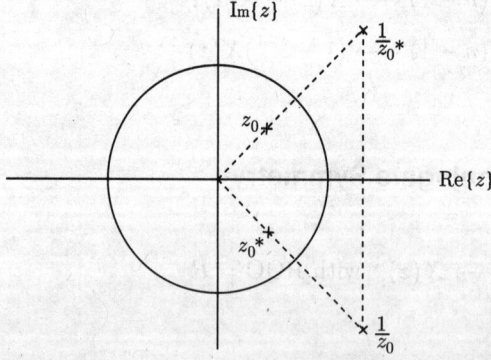

Fig. 10.15 A general pole constellation

Case-II *If $x(n)$ is real and odd, i.e., if*

$$x(n) = x^*(n) = -x(-n)$$

Then from Eqs (10.30) and (10.43), we have

$$X(z) = X^*(z^*) = -X\left(\frac{1}{z}\right) \tag{10.45}$$

Thus, if $X(z)$ has a pole (or zero) at $z = z_0 = re^{j\theta}$, then there must be a pole (or zero) at $z = 1/z_0 = 1/r\,e^{-j\theta}$, $z = z_0^* = re^{-j\theta}$, and $z = 1/z_0^* = 1/r\,e^{j\theta}$. Thus, a general pole constellation (or zero constellation) is a quadruplet

$$z_0 = re^{j\theta}, \qquad \frac{1}{z_0} = \frac{1}{r}e^{-j\theta}, \qquad z_0^* = re^{-j\theta}, \qquad \frac{1}{z_0^*} = \frac{1}{r}e^{j\theta}$$

as shown in Fig. 10.15. ∎

10.9 *z*-TRANSFORM OF CAUSAL PERIODIC SIGNALS

The z-transform of a causal periodic signal can be determined from the knowledge of the z-transform of its first cycle (period). Consider a causal periodic signal $x(n)$ with period N_0 as shown in Fig. 10.16.

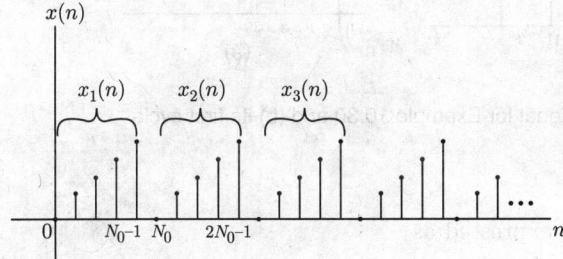

Fig. 10.16 Causal periodic signal

Let $x_1(n)$, $x_2(n)$, $x_3(n), \ldots,$ be the signals representing the 1st, 2nd, 3rd, $\ldots$, cycles of a causal periodic signal $x(n)$. Therefore, the causal periodic signal $x(n)$ can be written as

$$x(n) = x_1(n) + x_2(n) + x_3(n) + x_4(n) + \cdots$$

$$= x_1(n) + x_1(n - N_0) + x_1(n - 2N_0) + x_1(n - 3N_0) + \cdots = \sum_{k=0}^{\infty} x_1(n - kN_0)$$

Assume $x_1(n) \longleftrightarrow X_1(z)$. Using the time-shifting property [Eq. (10.24)], the z-transform of the above equation becomes

$$X(z) = X_1(z) + X_1(z)z^{-N_0} + X_1(z)z^{-2N_0} + X_1(z)z^{-3N_0} + \cdots$$

$$= X_1(z)[1 + z^{-N_0} + z^{-2N_0} + z^{-3N_0} + \cdots]$$

$$= X_1(z) \sum_{m=0}^{\infty} z^{-mN_0}$$

$$= X_1(z) \sum_{m=0}^{\infty} \left(z^{-N_0}\right)^m$$

$$X(z) = X_1(z)\frac{1}{1 - z^{-N_0}}, \quad |z^{-N_0}| < 1 \longrightarrow |z| > 1$$

$$X(z) = \frac{X_1(z)}{1 - z^{-N_0}}, \quad |z| > 1 \tag{10.46}$$

Example 10.30 Consider the signal shown in Fig. 10.17(a). The signal repeats periodically wit a period $N_0 = 4$ for $n \geq 0$ and is zero for $n < 0$. Find the z-transform of this signal along with the ROC.

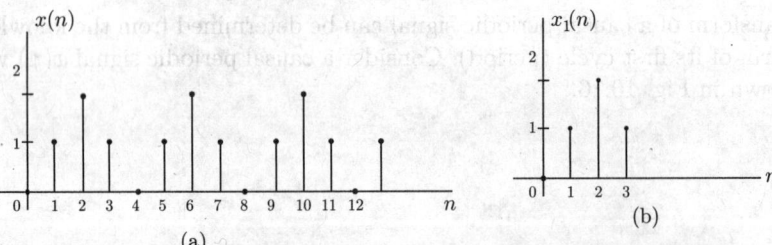

Fig. 10.17 (a) Causal periodic signal for Example 10.30 and (b) its first cycle

Solution
The given signal $x(n)$ can be expressed as

$$x(n) = \sum_{k=0}^{\infty} x_1(n - kN_0)$$

where $N_0 = 4$, and $x_1(n)$ is shown in Fig. 10.17(b), which can be expressed as

$$x_1(n) = \delta(n - 1) + 2\delta(n - 2) + \delta(n - 3)$$

Taking the z-transform of the above equation, we obtain

$$X_1(z) = z^{-1} + 2z^{-2} + z^{-3}$$

$$= z^{-1}[1 + 2z^{-1} + z^{-2}]$$

We know that the z-transform of a causal periodic signal $x(n)$ is defined as [Eq. (10.46)]

$$X(z) = \frac{X_1(z)}{1 - z^{-N_0}}$$

$$= \frac{z^{-1}[1 + 2z^{-1} + z^{-2}]}{1 - z^{-4}}, \quad |z| > 1$$

10.10 INVERSION OF THE z-TRANSFORM

There are three methods that are used for the evaluation of inverse z-transform:

1. Contour integration method (or residue method).
2. Long division method (or power series expansion method).
3. Partial fraction expansion method.

10.10.1 Contour Integration Method (or Residue Method)

As we saw in Section 10.2.1, the inverse z-transform is formally given by

$$x(n) = \frac{1}{2\pi j} \oint_C X(z) z^{n-1} \, dz \tag{10.47}$$

where the integral is a contour integral over a closed path C that encloses the origin and lies within the region of convergence of $X(z)$. The contour is any circle of radius R in the region of convergence. Going around this contour (counterclockwise) once is the same as increasing the frequency ω (the angle indicated in the Fig. 10.18) from 0 to 2π.

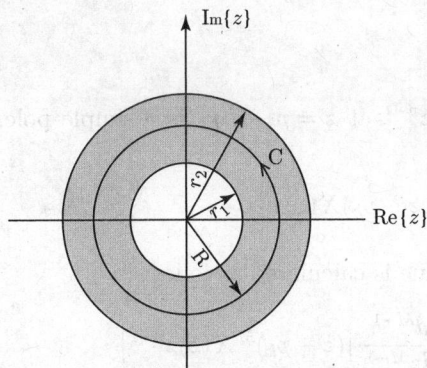

Fig. 10.18 Contour C for integral in Eq. (10.47)

Residues The contour integral can be evaluated using standard techniques from complex variable theory. There is a theorem due to Cauchy which says that if $f(z)$ is single-valued and analytic on and inside a closed contour C except for a finite number of poles $p_1, p_2, \ldots, p_N$ inside C, then

$$\frac{1}{2\pi j} \oint_C f(z)\, dz = \sum_{k=1}^{N} R_k \tag{10.48}$$

where R_k is the so-called residue of $f(z)$ at the pole p_k. When p_k is a simple pole (as against a multiple pole such as a double pole), then the residue is calculated as follows:

$$R_k = \lim_{z \to p_k} (z - p_k) f(z) \tag{10.49}$$

If p_k is a pole of order M, then the residue is calculated using a more complicated formula:

$$R_k = \lim_{z \to p_k} \frac{1}{(M-1)!} \frac{d^{M-1}}{dz^{M-1}} \left[(z - p_k)^M f(z) \right] \tag{10.50}$$

So, given a rational function $X(z)$ and a specified region of convergence $R_1 < |z| < R_2$, we can compute the inverse z-transform $x(n)$ as follows:

1. Make a list of the poles of $f(z) = z^{n-1} X(z)$ in the region $0 \le |z| < R1$.
2. Compute the residues of these poles and add them up. The result is precisely $x(n)$.

Therefore, in the case of inverse z-transform, we have

$$x(n) = \frac{1}{2\pi j} \oint_C X(z) z^{n-1}\, dz$$

$$= \sum_{\text{all poles } \{p_k\} \text{ inside } C} [\text{residue of } X(z) z^{n-1} \text{ at } z = p_k]$$

$$= \sum_{k=1}^{N} R_k$$

where R_k is the residue of $f(z) = X(z) z^{n-1}$ at $z = p_k$. If p_k is a simple pole, then the residue is calculated as follows:

$$R_k = \lim_{z \to p_k} (z - p_k) X(z) z^{n-1} \tag{10.51}$$

If p_k is a pole of order M, then the residue is calculated as follows:

$$R_k = \lim_{z \to p_k} \frac{1}{(M-1)!} \frac{d^{M-1}}{dz^{M-1}} \left[(z - p_k)^M X(z) z^{n-1} \right] \tag{10.52}$$

This method is called the contour integration method for computing the inverse z-transform. It can be somewhat cumbersome because of the poles created at $z = 0$ by the factor z^{n-1} in Eq. (10.47) (when $n < -1$). These are multiple poles and the multiplicity $n - 1$ depends on the time index n. Calculating the residues for these can be laborious.

Example 10.31 Evaluate the inverse z-transform of

$$X(z) = \frac{1}{1 - 2z^{-1}}, \quad |z| > 2$$

using the contour integration method.

Solution
By definition [Eq. (10.47)]

$$x(n) = \frac{1}{2\pi j} \oint_C X(z) z^{n-1} dz = \frac{1}{2\pi j} \oint_C \frac{z^{n-1}}{1 - 2z^{-1}} dz$$

$$= \frac{1}{2\pi j} \oint_C f(z)\, dz$$

where C is a circle at radius greater than 2 and

$$f(z) = \frac{z^{n-1}}{1 - 2z^{-1}} = \frac{z^n}{z - 2}$$

Now, consider two cases.

Case I If $n \geq 0$, $f(z)$ has zeros and only pole inside C is $z = 2$. Hence

$$x(n) = \frac{1}{2\pi j} \oint_C f(z)\, dz = \text{Residue of } f(z) \text{ at } z = 2$$

$$= \lim_{z \to 2} (z - 2) f(z)$$

$$= \lim_{z \to 2} (z - 2) \frac{z^n}{z - 2}$$

$$= \lim_{z \to 2} z^n$$

$$x(n) = 2^n \quad n \geq 0$$

Case II If $n < 0$, $f(z)$ has an nth-order pole at $z = 0$, which is also inside C. Thus there are contributions from both poles. For $n = -1$, we have $f(z) = (1/z(z - 2))$, and

$$x(-1) = \frac{1}{2\pi j} \oint_C f(z)\, dz = [\text{residue of } f(z) \text{ at } z = 0] + [\text{residue of } f(z) \text{ at } z = 2]$$

$$= \lim_{z \to 0} z f(z) + \lim_{z \to 2} (z - 2) f(z)$$

$$= \lim_{z \to 0} z \frac{1}{z(z - 2)} + \lim_{z \to 2} (z - 2) \frac{1}{z(z - 2)}$$

$$= \lim_{z \to 0} \frac{1}{z - 2} + \lim_{z \to 2} \frac{1}{z} = 0$$

Similarly, for $n = -2$, we have $f(z) = (1/z^2(z-2))$, and

$$x(-2) = \frac{1}{2\pi j} \oint_C f(z)\,dz = [\text{residue of } f(z) \text{ at } z = 0] + [\text{residue of } f(z) \text{ at } z = 2]$$

$$= \lim_{z \to 0} \frac{d}{dz} z^2 f(z) + \lim_{z \to 2} (z-2)f(z)$$

$$= \lim_{z \to 0} \frac{d}{dz} \left(z^2 \frac{1}{z^2(z-2)} \right) + \lim_{z \to 2} (z-2) \frac{1}{z^2(z-2)}$$

$$= \lim_{z \to 0} \frac{d}{dz} \left(\frac{1}{z-2} \right) + \lim_{z \to 2} \frac{1}{z^2}$$

$$x(-2) = 0$$

By continuing in the same way, we can show that $x(n) = 0$, for $n < 0$. Hence, we may conclude that

$$x(n) = \begin{cases} 2^n, & n \geq 0 \\ 0, & n < 0 \end{cases}$$

$$= 2^n u(n)$$

Example 10.32 Evaluate the inverse z-transform of

$$X(z) = \frac{1}{1 - 1.5z^{-1} + 0.5z^{-2}}, \quad |z| > 1$$

using the contour integration method.

Solution
Given that

$$X(z) = \frac{1}{1 - 1.5z^{-1} + 0.5z^{-2}}, \quad |z| > 1$$

$$= \frac{z^2}{z^2 - 1.5z + 0.5}$$

$$= \frac{z^2}{(z - 0.5)(z - 1)}$$

By definition [Eq. (10.47)]

$$x(n) = \frac{1}{2\pi j} \oint_C X(z) z^{n-1}\,dz = \frac{1}{2\pi j} \oint_C \frac{z^2 z^{n-1}}{(z - 0.5)(z - 1)}\,dz$$

$$= \frac{1}{2\pi j} \oint_C f(z)\,dz$$

where C is a circle at radius greater than 1 and

$$f(z) = \frac{z^2 z^{n-1}}{(z-0.5)(z-1)} = \frac{z^{n+1}}{(z-0.5)(z-1)}$$

Since the ROC, $|z| > 1$, is the region in the z-plane outside the outermost pole located at $z = 1$, both poles correspond to causal (right-sided) signals. For $n \geq 0$, $f(z)$ has zeros and the poles inside C are $z = 0.5$ and $z = 1$. Hence

$$x(n) = \frac{1}{2\pi j} \oint_C f(z)\, dz = [\text{residue of } f(z) \text{ at } z = 0.5] + [\text{residue of } f(z) \text{ at } z = 1]$$

$$= \lim_{z \to 0.5}(z - 0.5)f(z) + \lim_{z \to 1}(z - 1)f(z)$$

$$= \lim_{z \to 0.5}(z - 0.5)\frac{z^{n+1}}{(z-0.5)(z-1)} + \lim_{z \to 1}(z - 1)\frac{z^{n+1}}{(z-0.5)(z-1)}$$

$$= \lim_{z \to 0.5}\frac{z^{n+1}}{(z-1)} + \lim_{z \to 1}\frac{z^{n+1}}{(z-0.5)}$$

$$= \frac{(0.5)^{n+1}}{-0.5} + \frac{(1)^{n+1}}{0.5}$$

$$x(n) = -\left(\frac{1}{2}\right)^n + 2, \quad n \geq 0$$

$$x(n) = \left[2 - \left(\frac{1}{2}\right)^n\right]u(n)$$

10.10.2 Power Series Expansion Method (or Long Division Method)

The z-transform is a power series expansion,

$$X(z) = \sum_{n=-\infty}^{\infty} x(n)z^{-n} = \cdots + x(-2)z^2 + x(-1)z + x(0) + x(1)z^{-1} + x(2)z^{-2} + \cdots$$

where the signal values $x(n)$ are the coefficients of z^{-n} in the expansion. Therefore, the basic idea in this method is the following: Given a z-transform $X(z)$ with its corresponding ROC, we can express $X(z)$ as a power series in z^{-1} or z of the form defined in Eq. (10.4). The values of the signal $x(n)$ are then given by the coefficient associated with z^{-n}. This inversion method is suitable to signals that are one-sided, i.e., discrete-time signals with ROCs of the form $|z| < a$ or $|z| > a$. If the ROC is $|z| > a$, then we express $X(z)$ as a power series in z^{-1}, so that we obtain a right-sided signal. If the ROC is $|z| < a$, then we express $X(z)$ as a power series in z, so that we obtain a left-sided signal.

Long division may be used to obtain the power series whenever $X(z)$ is a ratio of polynomials and long division is simple to perform. However, long division may not lead to a closed form expression for $x(n)$.

Example 10.33 Find the inverse z-transform of

$$X(z) = \frac{1 + z^{-1}}{1 - (1/3)z^{-1}}$$

when

(a) ROC: $|z| > 1/3$.

(b) ROC: $|z| < 1/3$, using a power series expansion.

Solution

Given that

$$X(z) = \frac{1 + z^{-1}}{1 - (1/3)z^{-1}}$$

(a) Since the ROC is $|z| > 1/3$, we express $X(z)$ as a power series in z^{-1}, so that we obtain a right-sided signal. We divide the numerator by the denominator to obtain

$$
\begin{array}{r}
1 + \frac{4}{3}z^{-1} + \frac{4}{9}z^{-2} + \frac{4}{27}z^{-3} + \cdots \\[4pt]
\hline
1 - \frac{1}{3}z^{-1} \overline{)\, 1 + z^{-1} } \\[2pt]
1 - \frac{1}{3}z^{-1} \\[2pt]
\hline
\frac{4}{3}z^{-1} \\[2pt]
\frac{4}{3}z^{-1} - \frac{4}{9}z^{-2} \\[2pt]
\hline
\frac{4}{9}z^{-2} \\[2pt]
\frac{4}{9}z^{-2} - \frac{4}{27}z^{-3} \\[2pt]
\hline
\frac{4}{27}z^{-3}
\end{array}
$$

We can write, therefore,

$$X(z) = 1 + \frac{4}{3}z^{-1} + \frac{4}{9}z^{-2} + \frac{4}{27}z^{-3} + \cdots$$

Comparing with

$$X(z) = x(0) + x(1)z^{-1} + x(2)z^{-2} + x(3)z^{-3} + \cdots$$

we obtain

$$x(0) = 1, \qquad x(1) = \frac{4}{3}, \qquad x(2) = \frac{4}{9}, \qquad x(3) = \frac{4}{27}, \cdots$$

or equivalently, we may express $x(n)$ as

$$x(n) = \left\{ \underset{\uparrow}{0}, \frac{4}{3}, \frac{4}{9}, \frac{4}{27}, \cdots \right\}$$

(b) Since the ROC is $|z| < 1/3$, we express $X(z)$ as a power series in z, so that we obtain a left-sided signal. We divide the numerator by the denominator to obtain

$$
\begin{array}{r}
-3 - 12z - 36z^2 + \cdots \\
-\tfrac{1}{3}z^{-1} + 1 \overline{)\, z^{-1} + 1} \\
z^{-1} - 3 \\
\hline
4 \\
4 - 12z \\
\hline
12z \\
12z - 36z^2 \\
\hline
36z^2
\end{array}
$$

We can write, therefore,

$$X(z) = -3 - 12z - 36z^2 + \cdots$$
$$= \cdots + (-36)z^2 + (-12)z + (-3)$$

Comparing the above equation with

$$X(z) = \cdots + x(-2)z^2 + x(-1)z + x(0)$$

we obtain

$$x(-2) = -36, \qquad x(-1) = -12, \qquad x(0) = -3$$

or equivalently, we may express $x(n)$ as

$$x(n) = \left\{ \cdots, \ -36, \ -12, \ \underset{\uparrow}{-3} \right\}$$

From this example, we note that, in general, the method of long division will not provide answers for $x(n)$ when n is large because the division becomes tedious. Although, the method provides a direct evaluation of $x(n)$, a closed-form solution is not possible, except if the resulting pattern is simple enough to infer the general term $x(n)$. Hence, this method is used only if one wished to determine the values of the first few samples of the signal.

Example 10.34 Determine the inverse z-transform of the function

$$X(z) = \frac{z}{z - 0.5}, \qquad |z| > 0.5$$

Solution

Given that

$$X(z) = \frac{z}{z - 0.5}$$

$$= \frac{1}{1 - 0.5z^{-1}}, \quad |z| > 0.5$$

Since the ROC is $|z| > 0.5$, we express $X(z)$ as a power series in z^{-1} so that we obtain a right-sided signal. We divide the numerator by the denominator to obtain

$$
\begin{array}{r}
1 + 0.5z^{-1} + (0.5)^2 z^{-2} + (0.5)^3 z^{-3} + \cdots \\
\hline
1 - 0.5z^{-1} \,\big)\; 1 \\
1 - 0.5z^{-1} \\
\hline
0.5z^{-1} \\
0.5z^{-1} - (0.5)^2 z^{-2} \\
\hline
(0.5)^2 z^{-2} \\
(0.5)^2 z^{-2} - (0.5)^3 z^{-3} \\
\hline
(0.5)^3 z^{-3}
\end{array}
$$

We can write, therefore,

$$X(z) = 1 + 0.5z^{-1} + (0.5)^2 z^{-2} + (0.5)^3 z^{-3} + \cdots$$

$$= x(0) + x(1)z^{-1} + x(2)z^{-2} + x(3)z^{-3} + \cdots$$

so that

$$x(0) = 1, \qquad x(1) = 0.5, \qquad x(2) = (0.5)^2, \qquad x(3) = (0.5)^3, \cdots$$

It can be easily seen that this corresponds to the signal

$$x(n) = (0.5)^n u(n)$$

An advantage of the power series approach is the ability to find the inverse z-transforms for signals that are not a ratio of polynomials in z.

Example 10.35 Determine the inverse z-transform of

$$X(z) = \log(1 + az^{-1}), \quad |z| > |a|$$

using the power series expansion.

Solution

According to the logarithmic series expansion

$$\log(1 + x) = \sum_{n=1}^{\infty} \frac{-1}{n}(-x)^n = x - \frac{x^2}{2} + \frac{x^3}{3} - \cdots$$

Therefore,

$$X(z) = \log(1 + az^{-1}) = az^{-1} - \frac{(az^{-1})^2}{2} + \frac{(az^{-1})^3}{3} - \cdots$$

$$= \sum_{n=1}^{\infty} \frac{-1}{n}(-az^{-1})^n, \quad |z| > |a|$$

$$= \sum_{n=1}^{\infty} \frac{-1}{n}(-a)^n z^{-n}, \quad |z| > |a|$$

$$= \sum_{n=1}^{\infty} \frac{-1}{n}(-1)^n a^n z^{-n}, \quad |z| > |a|$$

$$= \sum_{n=1}^{\infty} \frac{1}{n}(-1)^{n+1} a^n z^{-n}, \quad |z| > |a|$$

$$X(z) = \sum_{n=-\infty}^{\infty} \left[\frac{1}{n}(-1)^{n+1} a^n u(n-1) \right] z^{-n}, \quad |z| > |a|$$

Comparing the above equation with

$$X(z) = \sum_{n=-\infty}^{\infty} x(n) z^{-n}$$

we get

$$x(n) = \frac{1}{n}(-1)^{n+1} a^n u(n-1)$$

Example 10.36 Determine the inverse z-transform of

$$X(z) = e^z \quad \text{with ROC all } z \text{ except } |z| = \infty.$$

Solution

According to the power series expansion of e^x, we have

$$e^x = \sum_{n=0}^{\infty} \frac{x^n}{n!}$$

Therefore,

$$X(z) = e^z = \sum_{n=0}^{\infty} \frac{z^n}{n!} = \sum_{n=-\infty}^{0} \frac{z^{-n}}{(-n)!}$$

$$= \sum_{n=-\infty}^{\infty} \frac{1}{(-n)!} u(-n) z^{-n}$$

Comparing the above equation with

$$X(z) = \sum_{n=-\infty}^{\infty} x(n)z^{-n}$$

we get

$$x(n) = \frac{1}{(-n)!} u(-n)$$

Example 10.37 Find the inverse z-transform of

$$X(z) = \sin(z)$$

Solution
According to the power series expansion of $\sin(x)$, we have

$$\sin(x) = x - \frac{x^3}{3!} + \frac{x^5}{5!} - \frac{x^7}{7!} + \cdots$$

Therefore,

$$X(z) = \sin(z) = z - \frac{z^3}{3!} + \frac{z^5}{5!} - \frac{z^7}{7!} + \cdots$$

$$= \cdots + \left(-\frac{1}{7!}\right) z^7 + \left(\frac{1}{5!}\right) z^5 + \left(-\frac{1}{3!}\right) z^3 + z$$

Comparing the above equation with

$$X(z) = \cdots + x(-7)z^7 + x(-5)z^5 + x(-3)z^3 + z$$

we obtain

$$\cdots, \qquad x(-7) = -\frac{1}{7!}, \qquad x(-5) = \frac{1}{5!}, \qquad x(-3) = -\frac{1}{3!}, \qquad x(-1) = 1$$

or equivalently, we may express $x(n)$ as

$$x(n) = \left\{ \cdots, -\frac{1}{7!}, 0, \frac{1}{5!}, 0, -\frac{1}{3!}, 0, 1, \underset{\uparrow}{0} \right\}$$

10.10.3 Partial Fraction Expansion Method

In analysing discrete-time systems, we are usually more interested finding a general expression for the discrete-time signal corresponding to a rational z-transform rather than in finding just the first few samples by power series expansion. To find the general expression, we can use partial fraction expansion and the known inverse z-transform pairs of low-degree rational functions.

If $X(z)$ ia a rational function, then

$$X(z) = \frac{N(z)}{D(z)} = \frac{b_0 + b_1 z^{-1} + \cdots + b_M z^{-M}}{a_0 + a_1 z^{-1} + \cdots + a_N z^{-N}} \qquad (10.53)$$

Without loss of generality, we assume that $a_0 = 1$, so that Eq. (10.53) can be expressed as

$$X(z) = \frac{N(z)}{D(z)} = \frac{b_0 + b_1 z^{-1} + \cdots + b_M z^{-M}}{1 + a_1 z^{-1} + \cdots + a_N z^{-N}} \qquad (10.54)$$

Note that if $a_0 \neq 1$, we can obtain Eq. (10.54) from Eq. (10.53) by dividing both numerator and denominator by a_0.

A rational function of the form given in Eq. (10.54) is called proper if $a_N \neq 0$ and $M < N$. An improper rational function ($M \geq N$) can always be written as the sum of a polynomial and a proper rational function. Here, we assume that $X(z)$ given in Eq. (10.53) is a proper rational function.

To simplify our discussion, we eliminate negative powers of z by multiplying both the numerator and denominator of Eq. (10.54) by z^N. This results in

$$X(z) = \frac{b_0 z^N + b_1 z^{N-1} + \cdots + b_M z^{N-M}}{z^N + a_1 z^{N-1} + \cdots + a_N} \qquad (10.55)$$

which contains only positive powers of z. Since $N > M$, the function

$$\frac{X(z)}{z} = \frac{b_0 z^{N-1} + b_1 z^{N-2} + \cdots + b_M z^{N-M-1}}{z^N + a_1 z^{N-1} + \cdots + a_N} \qquad (10.56)$$

is also always proper.

Our task in performing a partial fraction expansion is to express Eq. (10.56) or, equivalently, Eq. (10.54) as a sum of simple fractions. For this purpose, we first factor the denominator polynomial in Eq. (10.56) into factors that contain the poles $p_1, p_2, \cdots, p_N$ of $X(z)$.

$$\frac{X(z)}{z} = \frac{b_0 z^{N-1} + b_1 z^{N-2} + \cdots + b_M z^{N-M-1}}{(z - p_1)(z - p_2) \cdots (z - p_N)} \qquad (10.57)$$

Now consider two cases:

Case-I (Distinct poles) Suppose that the poles $p_1, p_2, \cdots, p_N$ are all different (distinct), then using the partial fraction expansion, we may write

$$\frac{X(z)}{z} = \frac{A_1}{z - p_1} + \frac{A_2}{z - p_2} + \cdots + \frac{A_N}{z - p_N} \qquad (10.58)$$

The problem is to determine the coefficients A_1, A_2, A_N. We can determine the coefficients A_1, A_2, A_N by multiplying both sides of Eq. (10.58) by each of the terms $(z - p_k)$, $k = 1, 2, \cdots, N$, and evaluating the resulting expressions at the corresponding pole positions, $p_1, p_2, \cdots, p_N$. Thus, we have, in general,

$$\frac{(z - p_k)X(z)}{z} = \frac{(z - p_k)A_1}{z - p_1} + \cdots + A_k + \cdots + \frac{(z - p_k)A_N}{z - p_N} \qquad (10.59)$$

Consequently, with $z = p_k$, Eq. (10.59) yields the kth coefficient as

$$A_k = \left. \frac{(z - p_k)X(z)}{z} \right|_{z=p_k}, \quad k = 1, 2, \cdots, N \tag{10.60}$$

The expansion given in Eq. (10.58), and the formula given in Eq. (10.60), hold for both real and complex poles. The only constraint is that all poles be distinct.

Now that we have performed the partial fraction expansion, we are ready to take the final step in the inversion of $X(z)$. From the partial fraction expansion in Eq. (10.58), it follows that

$$X(z) = A_1 \frac{1}{1 - p_1 z^{-1}} + A_2 \frac{1}{1 - p_2 z^{-1}} + \cdots + A_N \frac{1}{1 - p_N z^{-1}} \tag{10.61}$$

The inverse z-transform, $x(n) = \mathcal{Z}^{-1}[X(z)]$, can be obtained by inverting each term in Eq. (10.61) by using the formula

$$\mathcal{Z}^{-1}\left[\frac{1}{1 - p_k z^{-1}}\right] = \begin{cases} (p_k)^n u(n), & \text{if ROC } |z| > |p_k| \text{ (causal signals)} \\ -(p_k)^n u(-n - 1), & \text{if ROC } |z| < |p_k| \text{ (anticausal signals)} \end{cases} \tag{10.62}$$

Case-II Multiple order poles If $X(z)$ has a pole of multiplicity r, i.e., it contains in its denominator the factor $(z - p_k)^r$, then the expansion given in Eq. (10.58) is no longer true. If a pole p_k is repeated r times, then there are r terms in the partial fraction expansion associated with that pole. The partial fraction expansion must contain the terms

$$\frac{A_{1k}}{z - p_k} + \frac{A_{2k}}{(z - p_k)^2} + \cdots + \frac{A_{rk}}{(z - p_k)^r} = \sum_{i=1}^{r} \frac{A_{ik}}{(z - p_k)^i}$$

The coefficients A_{ik} are computed from the equation

$$A_{ik} = \frac{1}{(r - i)!} \left. \left[\frac{d^{r-i}}{dz^{r-i}} \left((z - p_k)^r \frac{X(z)}{z} \right) \right] \right|_{z=p_k} \tag{10.63}$$

Now that we have performed the partial fraction expansion, we are ready to take the final step in the inversion of $X(z)$. The inverse z-transform of terms of the form $(A/(z - p_k)^r)$ is required. In the case of a double pole, the following z-transform pair is useful:

$$\mathcal{Z}^{-1}\left[\frac{p_k z^{-1}}{(1 - p_k z^{-1})^2}\right] = n p_k^n u(n) \tag{10.64}$$

provided that the ROC is $|z| > |p_k|$.

Example 10.38 Find the inverse z-transform of

$$X(z) = \frac{1 - z^{-1} + z^{-2}}{(1 - (1/2)z^{-1})(1 - 2z^{-1})(1 - z^{-1})} \quad \text{with ROC } 1 < |z| < 2.$$

Solution

Given that

$$X(z) = \frac{1 - z^{-1} + z^{-2}}{\left(1 - \frac{1}{2}z^{-1}\right)(1 - 2z^{-1})(1 - z^{-1})}$$

For simplification, we eliminate negative powers of z by multiplying both the numerator and denominator of the above equation by z^3. This results in

$$X(z) = \frac{z^3 - z^2 + z}{\left(z - \frac{1}{2}\right)(z - 2)(z - 1)}$$

$$\frac{X(z)}{z} = \frac{z^2 - z + 1}{\left(z - \frac{1}{2}\right)(z - 2)(z - 1)}$$

We use partial fraction expansion to write

$$\frac{X(z)}{z} = \frac{A_1}{\left(z - \frac{1}{2}\right)} + \frac{A_2}{z - 2} + \frac{A_3}{z - 1}$$

where

$$A_1 = \left(z - \frac{1}{2}\right)\frac{X(z)}{z}\Bigg|_{z=1/2} = 1$$

$$A_2 = (z - 2)\frac{X(z)}{z}\Bigg|_{z=2} = 2$$

$$A_3 = (z - 1)\frac{X(z)}{z}\Bigg|_{z=1} = -2$$

Therefore,

$$\frac{X(z)}{z} = \frac{1}{z - \frac{1}{2}} + \frac{2}{z - 2} - \frac{2}{z - 1}$$

$$X(z) = \frac{z}{z - \frac{1}{2}} + \frac{2z}{z - 2} - \frac{2z}{z - 1}$$

$$= \frac{1}{1 - \frac{1}{2}z^{-1}} + \frac{2}{1 - 2z^{-1}} - \frac{2}{1 - z^{-1}}$$

The ROC and the locations of the poles are depicted in Fig. 10.19.

The ROC, $1 < |z| < 2$, is a ring in the z-plane. The pole of the first term is at $z = 1/2$. The ROC has a radius greater than this pole, so this pole corresponds to causal (right-sided) signal. Therefore,

$$\left(\frac{1}{2}\right)^n u(n) \longleftrightarrow \frac{1}{1 - \frac{1}{2}z^{-1}}$$

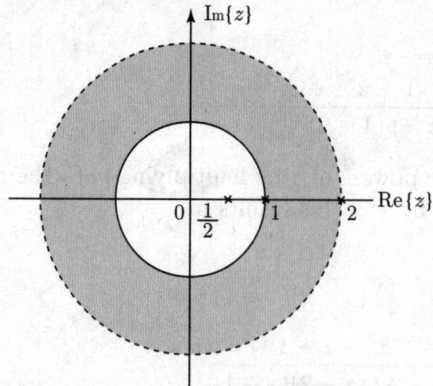

Fig. 10.19 ROC and locations of poles for Example 10.38

The second term has a pole at $z = 2$. Here the ROC has a radius less than this pole, so this pole corresponds to the anticausal (left-sided) signal. Therefore,

$$-2(2)^n u(-n-1) \longleftrightarrow \frac{2}{1 - 2z^{-1}}$$

The third term has a pole at $z = 1$. Here the ROC has a radius greater than this pole, so this pole corresponds to causal (right-sided) signal. Therefore,

$$2u(n) \longleftrightarrow \frac{2}{1 - z^{-1}}$$

and hence, we obtain

$$x(n) = \left(\frac{1}{2}\right)^n u(n) - 2(2)^n u(-n-1) - 2u(n)$$

Example 10.39 Determine the inverse z-transform of

$$X(z) = \frac{1}{1 - 1.5z^{-1} + 0.5z^{-2}}$$

if

(a) ROC $|z| > 1$.
(b) ROC $|z| < 0.5$.
(c) ROC $0.5 < |z| < 1$.

Solution
Given that

$$X(z) = \frac{1}{1 - 1.5z^{-1} + 0.5z^{-2}}$$

For simplification, we eliminate negative powers of z by multiplying both the numerator and denominator of the above equation by z^2. This results in

$$X(z) = \frac{z^2}{z^2 - 1.5z + 0.5}$$

$$\frac{X(z)}{z} = \frac{z}{z^2 - 1.5z + 0.5}$$

$$= \frac{z}{(z-1)(z-0.5)}$$

We use partial fraction expansion to write

$$\frac{X(z)}{z} = \frac{A_1}{z-1} + \frac{A_2}{z-0.5}$$

where

$$A_1 = (z-1)\frac{X(z)}{z}\bigg|_{z=1} = 2$$

$$A_2 = (z-0.5)\frac{X(z)}{z}\bigg|_{z=0.5} = -1$$

Therefore,

$$\frac{X(z)}{z} = \frac{2}{z-1} - \frac{1}{z-0.5}$$

$$X(z) = \frac{2z}{z-1} - \frac{z}{z-0.5}$$

$$= \frac{2}{1-z^{-1}} - \frac{1}{1-0.5z^{-1}}$$

$X(z)$ has poles at 1 and 0.5.

(a) The ROC and the locations of the poles are depicted in Fig. 10.20(a). The ROC, $|z| > 1$, is the region in the z-plane outside the outermost pole located at $z = 1$, so both poles correspond to causal (right-sided) signals. Therefore,

$$2u(n) \longleftrightarrow \frac{2}{1-z^{-1}}$$

$$(0.5)^n u(n) \longleftrightarrow \frac{1}{1-0.5z^{-1}}$$

and hence,

$$x(n) = 2u(n) - (0.5)^n u(n)$$

$$= [2 - 0.5^n]u(n)$$

(b) The ROC and the locations of the poles are depicted in Fig. 10.20(b). The ROC, $|z| < 0.5$, is the region in the z-plane inside the innermost pole located at $z = 0.5$, so

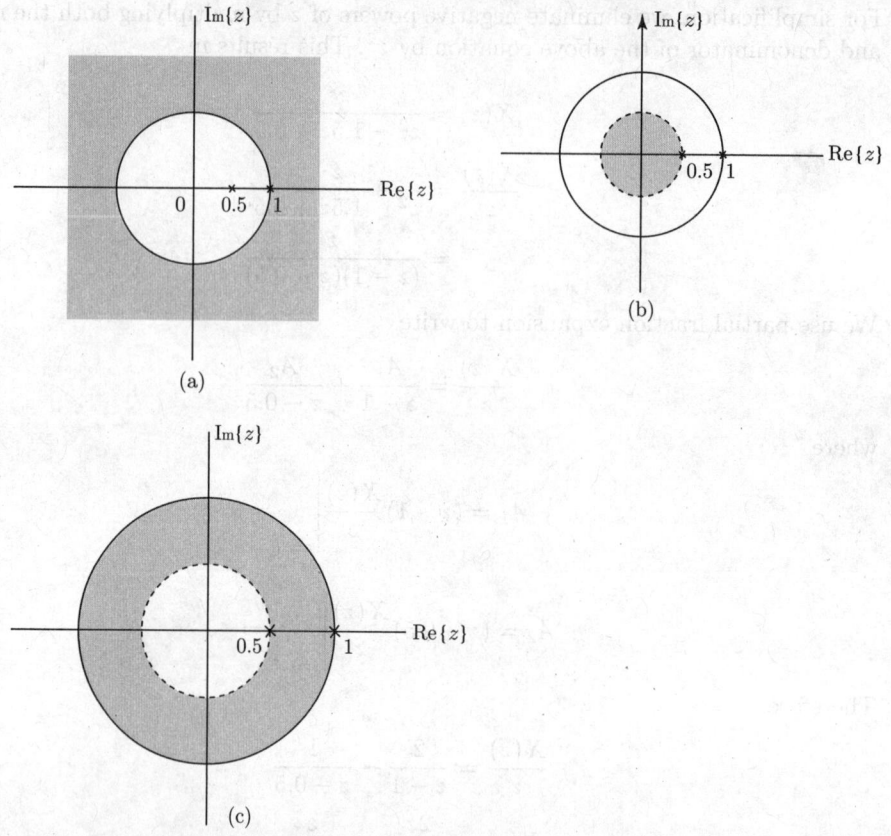

Fig. 10.20 ROC and locations of pole for Example 10.39

both poles correspond to anticausal (left-sided) signals. Therefore,

$$-2u(-n-1) \longleftrightarrow \frac{2}{1-z^{-1}}$$

$$-(0.5)^n u(-n-1) \longleftrightarrow \frac{1}{1-0.5z^{-1}}$$

and hence,

$$x(n) = -2u(-n-1) + (0.5)^n u(-n-1)$$

$$= [-2 + 0.5^n]u(-n-1)$$

(c) The ROC and the locations of the poles are depicted in Fig. 10.20(c). The ROC, $0.5 < |z| < 1$, is a ring in the z-plane. The pole of the first term is at $z = 1$. The ROC lies inside of this pole, so this pole corresponds to anti-causal (left-sided) signal. Therefore,

$$-2u(-n-1) \longleftrightarrow \frac{2}{1-z^{-1}}$$

The second term has a pole at $z = 0.5$. Here the ROC lies outside of this pole, so this pole corresponds to the causal (right-sided) signal. Therefore,

$$(0.5)^n u(n) \longleftrightarrow \frac{1}{1-0.5z^{-1}}$$

and hence, we obtain

$$x(n) = -2u(-n-1) - (0.5)^n u(n)$$

Example 10.40 Determine the causal signal $x(n)$ having the z-transform

$$X(z) = \frac{1}{(1+z^{-1})(1-z^{-1})^2}$$

Solution
Given that

$$X(z) = \frac{1}{(1+z^{-1})(1-z^{-1})^2}$$

For simplification, we eliminate negative powers of z by multiplying both the numerator and denominator of the above equation by z^3. This results in

$$X(z) = \frac{z^3}{(z+1)(z-1)^2}$$

$$\frac{X(z)}{z} = \frac{z^2}{(z+1)(z-1)^2}$$

We use partial fraction expansion to write

$$\frac{X(z)}{z} = \frac{A_1}{z+1} + \frac{A_2}{z-1} + \frac{A_3}{(z-1)^2}$$

where

$$A_1 = (z+1)\frac{X(z)}{z}\bigg|_{z=-1} = \frac{1}{4}$$

$$A_2 = \left[\frac{d}{dz}\left((z-1)^2\frac{X(z)}{z}\right)\right]\bigg|_{z=1} = \frac{3}{4}$$

$$A_3 = (z-1)^2\frac{X(z)}{z}\bigg|_{z=1} = \frac{1}{2}$$

Therefore,

$$\frac{X(z)}{z} = \frac{\frac{1}{4}}{z+1} + \frac{\frac{3}{4}}{z-1} + \frac{\frac{1}{2}}{(z-1)^2}$$

$$X(z) = \frac{1}{4}\frac{z}{z+1} + \frac{3}{4}\frac{z}{z-1} + \frac{1}{2}\frac{z}{(z-1)^2}$$

$$= \frac{1}{4}\frac{1}{1+z^{-1}} + \frac{3}{4}\frac{1}{1-z^{-1}} + \frac{1}{2}\frac{z^{-1}}{(1-z^{-1})^2}$$

Taking the inverse z-transform of the above equation yields

$$x(n) = \frac{1}{4}(-1)^n u(n) + \frac{3}{4}u(n) + \frac{1}{2}nu(n)$$

$$= \left[\frac{1}{4}(-1)^n + \frac{3}{4} + \frac{1}{2}n\right]u(n)$$

Example 10.41 Find the inverse z-transform of

$$X(z) = \frac{z^3 - 10z^2 - 4z + 4}{2z^2 - 2z - 4} \quad \text{with ROC } |z| < 1$$

Solution
Given that

$$X(z) = \frac{z^3 - 10z^2 - 4z + 4}{2z^2 - 2z - 4}$$

Since the order of the numerator polynomial is greater than the order of the denominator polynomial, the given rational z-transform is improper. We use long division to express $X(z)$ as the sum of a proper rational function and a polynomial in z:

$$
\begin{array}{r}
0.5z - 4.5 \\
2z^2 - 2z - 4 \overline{) \; z^3 - 10z^2 - 4z + 4} \\
\underline{z^3 - z^2 \quad - 2z} \\
-9z^2 - 2z + 4 \\
\underline{-9z^2 + 9z + 18} \\
-11z - 14
\end{array}
$$

Thus, we may write

$$X(z) = 0.5z - 4.5 - \frac{11z + 14}{2z^2 - 2z - 4}$$

$$= 0.5z - 4.5 - \frac{5.5z + 7}{z^2 - z - 2}$$

$$X(z) = 0.5z - 4.5 - \frac{5.5z + 7}{(z+1)(z-2)}$$

Using a partial fraction expansion to expand the rational function, we obtain

$$X(z) = 0.5z - 4.5 + \frac{0.5}{z+1} - \frac{6}{z-2}$$

$$= 0.5z - 4.5 + \frac{0.5z^{-1}}{1 + z^{-1}} - \frac{6z^{-1}}{1 - 2z^{-1}}$$

We know that

$$-(-1)^n u(-n-1) \longleftrightarrow \frac{1}{1+z^{-1}}, \quad |z| < 1$$

$$-0.5(-1)^{n-1}u(-n) \longleftrightarrow \frac{0.5z^{-1}}{1+z^{-1}} \quad \text{(using the time-shifting property)}$$

similarly, we obtain

$$-6(2)^{n-1}u(-n) \longleftrightarrow \frac{6z^{-1}}{1-2z^{-1}}$$

Also, because

$$X(z) = 0.5z - 4.5 + \frac{0.5z^{-1}}{1+z^{-1}} - \frac{6z^{-1}}{1-2z^{-1}}$$

The inverse z-transform yields

$$x(n) = 0.5\delta(n+1) - 4.5\delta(n) - 0.5(-1)^{n-1}u(-n) + 6(2)^{n-1}u(-n)$$

10.11 ANALYSIS AND CHARACTERIZATION OF LTI SYSTEMS USING THE z-TRANSFORM

If $x(n)$ and $y(n)$ are the input and output of an LTI system [shown in Fig. 10.21] with impulse response $h(n)$, then

$$y(n) = x(n) * h(n)$$

Application of the convolution property to the above equation yields

$$\mathcal{Z}[y(n)] = \mathcal{Z}[x(n)] * \mathcal{Z}[h(n)]$$

$$Y(z) = X(z)H(z) \tag{10.65}$$

or

$$H(z) = \frac{Y(z)}{X(z)} \tag{10.66}$$

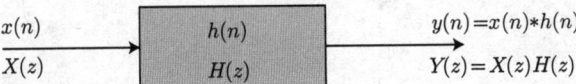

Fig. 10.21 LTI system

$H(z)$ is commonly referred to as the *system function* or *transfer function*. The transfer function $H(z)$ of a discrete-time LTI system is equal to the ratio of the z-transform of the output to the z-transform of the input signal when all initial conditions are zero. For $z = e^{j\omega}$, $H(z)$ is the frequency response of the discrete-time LTI system.

10.11.1 Transfer Function and Difference-equation System Description

The transfer function may be related directly to the difference-equation description of a discrete-time LTI system by using the bilateral z-transform. The relationship between the input and output of an Nth-order LTI system is described by the following difference equation:

$$\sum_{k=0}^{N} a_k\, y(n-k) = \sum_{k=0}^{M} b_k\, x(n-k)$$

Taking the z-transform of the above equation, we obtain

$$\sum_{k=0}^{N} a_k\, z^{-k} Y(z) = \sum_{k=0}^{M} b_k\, z^{-k} X(z)$$

$$H(z) = \frac{Y(z)}{X(z)} = \frac{\sum_{k=0}^{M} b_k z^k}{\sum_{k=0}^{N} a_k z^{-k}}$$

$H(z)$ is a ratio of polynomials in z and is thus termed a rational transfer function.

10.11.2 Impulse Response and Step Response

The impulse response $h(n)$ is defined as the output of an LTI system due to a unit impulse signal input $\delta(n)$ applied at time $n = 0$. Also,

$$h(n) \longleftrightarrow H(z)$$

where $H(z)$ is the system function or transfer function.

The step response $s(n)$ is defined as the output of an LTI system due to a unit step input signal, i.e., $x(n) = u(n)$. The step response of an LTI system with impulse response $h(n)$ is given by

$$s(n) = h(n) * x(n) = h(n) * u(n)$$

$$= \sum_{m=-\infty}^{\infty} h(m) u(n-m)$$

$$s(n) = \sum_{m=-\infty}^{n} h(m)$$

Therefore,

$$\mathcal{Z}[s(n)] = \mathcal{Z}\left[\sum_{m=-\infty}^{n} h(m)\right]$$

Using the accumulation property, we obtain

$$S(z) = \frac{H(z)}{1 - z^{-1}} \tag{10.67}$$

Conversely,

$$s(n) = \sum_{m=-\infty}^{n} h(m)$$

$$= h(n) + \sum_{m=-\infty}^{n-1} h(m)$$

$$s(n) = h(n) + s(n-1)$$

$$h(n) = s(n) - s(n-1)$$

$$\mathcal{Z}[h(n)] = \mathcal{Z}\left[s(n) - s(n-1)\right]$$

$$H(z) = [1 - z^{-1}]S(z) \tag{10.68}$$

Example 10.42 Determine the transfer function and impulse response for the causal LTI system described by the difference equation

$$y(n) - \frac{1}{2}y(n-1) = x(n) + 2x(n-1)$$

Solution
Given that

$$y(n) - \frac{1}{2}y(n-1) = x(n) + 2x(n-1)$$

Taking the z-transform of the difference equation, we obtain

$$Y(z)\left[1 - \frac{1}{2}z^{-1}\right] = X(z)[1 - 2z^{-1}]$$

Hence the system function is

$$\frac{Y(z)}{X(z)} = H(z) = \frac{1 - 2z^{-1}}{1 - \frac{1}{2}z^{-1}}$$

$$= \frac{1}{1 - \frac{1}{2}z^{-1}} - \frac{2z^{-1}}{1 - \frac{1}{2}z^{-1}}$$

The inverse z-transform of the above equation yields

$$h(n) = \left(\frac{1}{2}\right)^{n} u(n) - 2\left(\frac{1}{2}\right)^{n-1} u(n-1)$$

This is the unit impulse response (or unit sample response) of the system.

Example 10.43 Determine the step response of an LTI system whose impulse response is given by

$$h(n) = (-1)^n[u(n+2) - u(n-3)]$$

Solution

Given that

$$
\begin{aligned}
h(n) &= (-1)^n[u(n+2) - u(n-3)] \\
&= (-1)^n u(n+2) - (-1)^n u(n-3) \\
&= (-1)^{-2}(-1)^{n+2}u(n+2) - (-1)^3(-1)^{n-3}u(n-3) \\
h(n) &= (-1)^{n+2}u(n+2) + (-1)^{n-3}u(n-3)
\end{aligned}
$$

Using the time-shifting property of the z-transform, we obtain

$$
\begin{aligned}
H(z) &= \frac{z^2}{1+z^{-1}} + \frac{z^{-3}}{1+z^{-1}} \\
&= \frac{z^2 + z^{-3}}{1+z^{-1}}
\end{aligned}
$$

The relationship between the z-transform, $S(z)$, of the step response $s(n)$ and the system function $H(z)$ is given by [Eq. (10.67)]

$$
\begin{aligned}
S(z) &= \frac{H(z)}{1-z^{-1}} = \frac{z^2 + z^{-3}}{(1-z^{-1})(1+z^{-1})} \\
S(z) &= \frac{z^2 + z^{-3}}{1 - z^{-2}} \\
&= \frac{z^4 + z^{-1}}{z^2 - 1}
\end{aligned}
$$

Since the order of the numerator polynomial is greater than the order of the denominator polynomial, the rational z-transform $S(z)$ is improper. We use long division to express $S(z)$ as the sum of a proper rational function and a polynomial in z:

$$
\begin{array}{r}
z^2 + 1 \\
z^2 - 1 \overline{\smash{)}\ z^4 + z^{-1} } \\
\underline{z^4 - z^2 } \\
z^2 + z^{-1} \\
\underline{z^2 - 1 } \\
z^{-1} + 1
\end{array}
$$

Thus, we may write

$$S(z) = z^2 + 1 + \frac{z^{-1} + 1}{z^2 - 1}$$

$$= 1 + z^2 + \frac{1}{z(z-1)}$$

$$S(z) = 1 + z^2 + \frac{z^{-2}}{1 - z^{-1}}$$

The inverse z-transform of the above yields the step response.

$$s(n) = \delta(n) + \delta(n+2) + u(n-2)$$

Example 10.44 A causal discrete-time LTI system is described by

$$y(n) - \frac{3}{4}y(n-1) + \frac{1}{8}y(n-2) = x(n)$$

where $x(n)$ and $y(n)$ are the input and output of the system, respectively.

(a) Determine the system function $H(z)$ for causal system function.
(b) Find the impulse response $h(n)$ of the system.
(c) Find the step response of the system.

Solution
(a) Given that

$$y(n) - \frac{3}{4}y(n-1) + \frac{1}{8}y(n-2) = x(n)$$

Taking the z-transform of the difference equation, we obtain

$$Y(z)\left[1 - \frac{3}{4}z^{-1} + \frac{1}{8}z^{-2}\right] = X(z)$$

Hence the system function is

$$\frac{Y(z)}{X(z)} = H(z) = \frac{1}{1 - \frac{3}{4}z^{-1} + \frac{1}{8}z^{-2}}$$

(b) Consider the system function

$$H(z) = \frac{1}{1 - \frac{3}{4}z^{-1} + \frac{1}{8}z^{-2}} = \frac{z^2}{z^2 - \frac{3}{4}z + \frac{1}{8}}$$

$$\frac{H(z)}{z} = \frac{z}{\left(z - \frac{1}{4}\right)\left(z - \frac{1}{2}\right)}$$

We use the partial fraction expansion to write

$$\frac{H(z)}{z} = \frac{A_1}{z - \frac{1}{4}} + \frac{A_2}{z - \frac{1}{2}}$$

where

$$A_1 = \left(z - \frac{1}{4}\right) \frac{H(z)}{z}\bigg|_{z=1/4} = -1$$

$$A_2 = \left(z - \frac{1}{2}\right) \frac{H(z)}{z}\bigg|_{z=1/2} = 2$$

Therefore,

$$\frac{H(z)}{z} = \frac{-1}{z - \frac{1}{4}} + \frac{2}{z - \frac{1}{2}}$$

$$H(z) = \frac{-z}{z - \frac{1}{4}} + \frac{2z}{z - \frac{1}{2}}$$

$$= \frac{-1}{1 - \frac{1}{4}z^{-1}} + \frac{2}{1 - \frac{1}{2}z^{-1}}$$

The inverse z-transform yields

$$h(n) = -\left(\frac{1}{4}\right)^n u(n) + 2\left(\frac{1}{2}\right)^n u(n)$$

this is the unit impulse response of the system.
(c) The relationship between $S(z) = \mathcal{Z}[s(n)]$ and the system function $H(z)$ is given by [Eq. (10.67)]

$$S(z) = \frac{H(z)}{1 - z^{-1}} = \frac{1}{(1 - z^{-1})\left(1 - \frac{3}{4}z^{-1} + \frac{1}{8}z^{-2}\right)}$$

$$S(z) = \frac{z^3}{(z - 1)\left(z - \frac{1}{2}\right)\left(z - \frac{1}{4}\right)}$$

$$\frac{S(z)}{z} = \frac{z^2}{(z - 1)\left(z - \frac{1}{2}\right)\left(z - \frac{1}{4}\right)}$$

We use the partial fraction expansion to write

$$\frac{S(z)}{z} = \frac{A_1}{z - 1} + \frac{A_2}{z - \frac{1}{2}} + \frac{A_3}{z - \frac{1}{4}}$$

where

$$A_1 = (z-1)\frac{S(z)}{z}\bigg|_{z=1} = \frac{8}{3}$$

$$A_2 = \left(z-\frac{1}{2}\right)\frac{S(z)}{z}\bigg|_{z=1/2} = -2$$

$$A_3 = \left(z-\frac{1}{4}\right)\frac{S(z)}{z}\bigg|_{z=1/4} = \frac{1}{3}$$

Therefore,

$$\frac{S(z)}{z} = \frac{8/3}{z-1} - \frac{2}{z-\frac{1}{2}} + \frac{1/3}{z-\frac{1}{4}}$$

$$S(z) = \frac{8}{3}\frac{z}{z-1} - 2\frac{z}{z-\frac{1}{2}} + \frac{1}{3}\frac{z}{z-\frac{1}{4}}$$

$$= \frac{8}{3}\frac{1}{1-z^{-1}} - 2\frac{1}{1-\frac{1}{2}z^{-1}} + \frac{1}{3}\frac{1}{1-\frac{1}{4}z^{-1}}$$

The inverse z-transform yields

$$s(n) = \frac{8}{3}u(n) - 2\left(\frac{1}{2}\right)^n u(n) + \frac{1}{3}\left(\frac{1}{4}\right)^n u(n)$$

Example 10.45 Consider the system

$$H(z) = \frac{z^{-1} + \frac{1}{2}z^{-2}}{1 - \frac{3}{5}z^{-1} + \frac{2}{25}z^{-2}}$$

Determine
  (a) the impulse response.
  (b) the zero-state step response.
  (c) the step-response if $y(-1) = 1$ and $y(-2) = 2$.

Solution
(a) Given that

$$H(z) = \frac{z^{-1} + \frac{1}{2}z^{-2}}{1 - \frac{3}{5}z^{-1} + \frac{2}{25}z^{-2}} = z^{-1}\left[\frac{1 + \frac{1}{2}z^{-1}}{1 - \frac{3}{5}z^{-1} + \frac{2}{25}z^{-2}}\right]$$

$$= z^{-1}\left[\frac{1 + \frac{1}{2}z^{-1}}{\left(1 - \frac{1}{5}z^{-1}\right)\left(1 - \frac{2}{5}z^{-1}\right)}\right]$$

Using the partial fraction expansion, we obtain

$$H(z) = z^{-1} \left[\frac{-\frac{7}{2}}{1 - \frac{1}{5}z^{-1}} + \frac{\frac{9}{2}}{1 - \frac{2}{5}z^{-1}} \right]$$

$$= -\frac{7}{2} \frac{z^{-1}}{1 - \frac{1}{5}z^{-1}} + \frac{9}{2} \frac{z^{-1}}{1 - \frac{2}{5}z^{-1}}$$

The inverse z-transform of the above equation yields

$$h(n) = -\frac{7}{2} \left(\frac{1}{5} \right)^{n-1} u(n-1) + \frac{9}{2} \left(\frac{2}{5} \right)^{n-1} u(n-1)$$

$$= \left[-\frac{7}{2} \left(\frac{1}{5} \right)^{n-1} + \frac{9}{2} \left(\frac{2}{5} \right)^{n-1} \right] u(n-1)$$

(b) Zero-state step response is the response of the system when all the initial conditions are zero and the input is the unit step function, i.e., $x(n) = u(n)$. The z-transform of the input is given by

$$X(z) = \frac{1}{1 - z^{-1}}$$

We know that

$$Y(z) = H(z)X(z) = \left[\frac{z^{-1} + \frac{1}{2}z^{-2}}{\left(1 - \frac{1}{5}z^{-1} \right) \left(1 - \frac{2}{5}z^{-1} \right)} \right] \frac{1}{1 - z^{-1}}$$

$$= \frac{z^{-1} + \frac{1}{2}z^{-2}}{\left(1 - z^{-1} \right) \left(1 - \frac{1}{5}z^{-1} \right) \left(1 - \frac{2}{5}z^{-1} \right)}$$

Using the partial fraction expansion, we obtain

$$Y(z) = \frac{\frac{25}{8}}{1 - z^{-1}} + \frac{\frac{7}{8}}{1 - \frac{1}{5}z^{-1}} + \frac{-3}{1 - \frac{2}{5}z^{-1}}$$

The inverse z-transform of the above equation yields

$$y(n) = \frac{25}{8}u(n) + \frac{7}{8} \left(\frac{1}{5} \right)^n u(n) - 3 \left(\frac{2}{5} \right)^n u(n)$$

$$= \left[\frac{25}{8} + \frac{7}{8} \left(\frac{1}{5} \right)^n - 3 \left(\frac{2}{5} \right)^n \right] u(n)$$

(c) With $x(n) = u(n)$, we have $x(-1) = x(-2) = 0$ and $X(z) = \frac{1}{1-z^{-1}}$. Now, consider the given system function

$$H(z) = \frac{Y(z)}{X(z)} = \frac{z^{-1} + \frac{1}{2}z^{-2}}{1 - \frac{3}{5}z^{-1} + \frac{2}{25}z^{-2}}$$

$$Y(z) \left[1 - \frac{3}{5}z^{-1} + \frac{2}{25}z^{-2} \right] = X(z) \left[z^{-1} + \frac{1}{2}z^{-2} \right]$$

$$Y(z) - \frac{3}{5}z^{-1}Y(z) + \frac{2}{25}z^{-2}Y(z) = z^{-1}X(z) + \frac{1}{2}z^{-2}X(z)$$

The inverse z-transform of the above equation yields the following difference equation:

$$y(n) - \frac{3}{5}y(n-1) + \frac{2}{25}y(n-2) = x(n-1) + \frac{1}{2}x(n-2)$$

Now, taking the unilateral z-transform of the above equation, we obtain

$$Y(z) - \frac{3}{5}\left[z^{-1}Y(z) + y(-1)\right] + \frac{2}{25}\left[z^{-2}Y(z) + z^{-1}y(-1) + y(-2)\right]$$

$$= \left[z^{-1}X(z) + x(-1)\right] + \frac{1}{2}\left[z^{-2}X(z) + z^{-1}x(-1) + x(-2)\right]$$

$$Y(z) - \frac{3}{5}\left[z^{-1}Y(z) + 1\right] + \frac{2}{25}\left[z^{-2}Y(z) + z^{-1} + 2\right]$$

$$= \left[z^{-1}X(z) + 0\right] + \frac{1}{2}\left[z^{-2}X(z) + 0 + 0\right]$$

$$Y(z)\left[1 - \frac{3}{5}z^{-1} + \frac{2}{25}z^{-2}\right] - \frac{3}{5} + \frac{2}{25}z^{-1} + \frac{4}{25} = z^{-1}X(z) + \frac{1}{2}z^{-2}X(z)$$

$$Y(z)\left[1 - \frac{3}{5}z^{-1} + \frac{2}{25}z^{-2}\right] = \left(\frac{3}{5} - \frac{2}{25}z^{-1} - \frac{4}{25}\right) + X(z)\left[z^{-1} + \frac{1}{2}z^{-2}\right]$$

$$= \left(\frac{11}{25} - \frac{2}{25}z^{-1}\right) + X(z)\left[z^{-1} + \frac{1}{2}z^{-2}\right]$$

$$Y(z) = \frac{\frac{1}{25}(11 - 2z^{-1})}{1 - \frac{3}{5}z^{-1} + \frac{2}{25}z^{-2}} + \frac{z^{-1}\left(1 + \frac{1}{2}z^{-1}\right)}{\left(1 - z^{-1}\right)\left(1 - \frac{3}{5}z^{-1} + \frac{2}{25}z^{-2}\right)}$$

$$Y(z) = \left(\frac{\frac{1}{25}(11 - 2z^{-1})}{\left(1 - \frac{1}{5}z^{-1}\right)\left(1 - \frac{2}{5}z^{-1}\right)}\right)$$

$$+ \left(\frac{z^{-1}\left(1 + \frac{1}{2}z^{-1}\right)}{\left(1 - z^{-1}\right)\left(1 - \frac{1}{5}z^{-1}\right)\left(1 - \frac{2}{5}z^{-1}\right)}\right)$$

Using the partial fraction expansion, we obtain

$$Y(z) = \left(\frac{-\frac{1}{25}}{1 - \frac{1}{5}z^{-1}} + \frac{\frac{12}{25}}{1 - \frac{2}{5}z^{-1}}\right) + \left(\frac{\frac{25}{8}}{1 - z^{-1}} + \frac{\frac{7}{8}}{1 - \frac{1}{5}z^{-1}} + \frac{-3}{1 - \frac{2}{5}z^{-1}}\right)$$

The inverse z-transform of this equation yields

$$y(n) = \left[-\frac{1}{25}\left(\frac{1}{5}\right)^n u(n) + \frac{12}{25}\left(\frac{2}{5}\right)^n u(n)\right] + \left[\frac{25}{8}u(n) + \frac{7}{8}\left(\frac{1}{5}\right)^n u(n) - 3\left(\frac{2}{5}\right)^n u(n)\right]$$

$$y(n) = \left[\frac{25}{8} + \frac{167}{200}\left(\frac{1}{5}\right)^n - \frac{63}{25}\left(\frac{2}{5}\right)^n\right]u(n)$$

10.11.3 Causality

For a causal LTI system, the impulse response $h(n) = 0$ for $n < 0$ and thus is right-sided. We know that if a signal is right-sided and of infinite duration, then the ROC is the region in the z-plane exterior of a circle outside the outermost pole. Consequently, *the ROC associated with the system function for a causal system is outside a circle of radius equal to the largest magnitude of the poles of $H(z)$ including infinity.* However, the converse of this statement is not necessarily true.

If $H(z)$ is rational, then we can determine whether the system is causal simply by checking to see if its ROC is outside of a circle, i.e., *a discrete-time LTI system with rational system function $H(z)$ is causal if and only if (a) the ROC is the exterior of a circle outside the outermost pole and (b) with $H(z)$ expressed as a ratio of polynomials in z, the order of the numerator cannot be greater than the order of the denominator.*

In an exactly analogous manner, we can deal with the concept of anticausality. For an anticausal LTI system, the impulse response $h(n) = 0$ for $n > 0$ and thus is left-sided. We know that if a signal is left-sided and of infinite duration, then the ROC is the region in the z-plane inside the innermost pole. Consequently, *the ROC associated with the system function for an anticausal system is inside a circle of radius equal to the smallest magnitude of the poles of $H(z)$.* However, the converse of this statement is not necessarily true.

Example 10.46 For the following system functions, check whether the corresponding LTI system is causal, anticausal, or noncausal.

(a) $H_1(z) = \dfrac{3 - 4z^{-1}}{1 - 3.5z^{-1} + 1.5z^{-2}}, \quad |z| > 3$

(b) $H_2(z) = \dfrac{3 - 4z^{-1}}{1 - 3.5z^{-1} + 1.5z^{-2}}, \quad |z| < 0.5$

(c) $H_3(z) = \dfrac{3 - 4z^{-1}}{1 - 3.5z^{-1} + 1.5z^{-2}}, \quad 0.5 < |z| < 3$

Solution
(a) Consider the given system function

$$H_1(z) = \frac{3 - 4z^{-1}}{1 - 3.5z^{-1} + 1.5z^{-2}}, \quad |z| > 3$$

$$= \frac{3 - 4z^{-1}}{(1 - 0.5z^{-1})(1 - 3z^{-1})}$$

$$= \frac{1}{1 - 0.5z^{-1}} + \frac{2}{1 - 3z^{-1}}$$

Since the given system function $H_1(z)$ is rational and its ROC [shown in Fig. 10.22(a)] is the region in the z-plane outside the outermost pole, the corresponding LTI system is causal. Also it can be verified by its impulse response. Given that

$$H_1(z) = \frac{1}{1 - 0.5z^{-1}} + \frac{2}{1 - 3z^{-1}}$$

Taking the inverse z-transform, we obtain

$$h_1(n) = (0.5)^n u(n) + 2(3)^n u(n)$$

Since $h_1(n) = 0$ for $n < 0$, this system is causal.

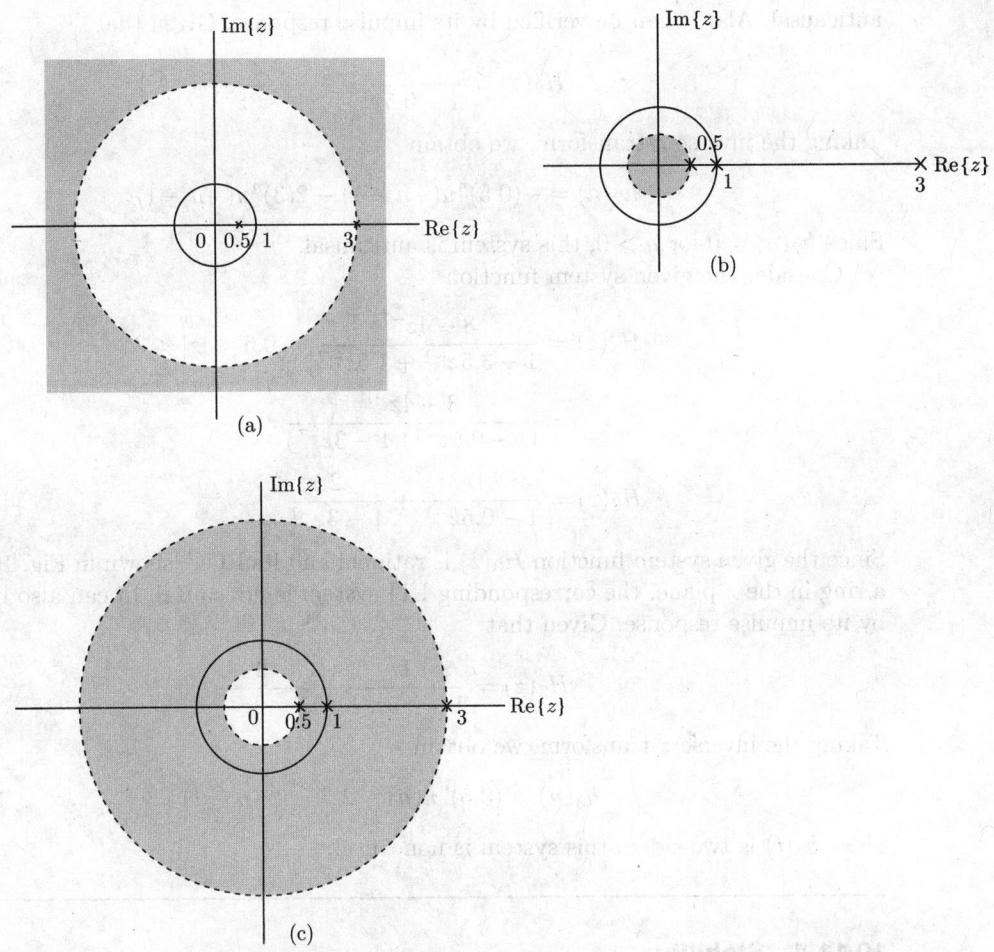

Fig. 10.22 Pole-zero plot and ROC for Example 10.46

(b) Consider the given system function

$$H_2(z) = \frac{3 - 4z^{-1}}{1 - 3.5z^{-1} + 1.5z^{-2}}, \quad |z| < 0.5$$

$$= \frac{3 - 4z^{-1}}{(1 - 0.5z^{-1})(1 - 3z^{-1})}$$

$$= \frac{1}{1 - 0.5z^{-1}} + \frac{2}{1 - 3z^{-1}}$$

Since the given system function $H_2(z)$ is rational and its ROC [shown in Fig. 10.22(b)] is the region in the z-plane inside the innermost pole, the corresponding LTI system is anticausal. Also it can be verified by its impulse response. Given that

$$H_2(z) = \frac{1}{1 - 0.5z^{-1}} + \frac{2}{1 - 3z^{-1}}$$

Taking the inverse z-transform, we obtain

$$h_2(n) = -(0.5)^n u(-n-1) - 2(3)^n u(-n-1)$$

Since $h_2(n) = 0$ for $n > 0$, this system is anticausal.

(c) Consider the given system function

$$H_3(z) = \frac{3 - 4z^{-1}}{1 - 3.5z^{-1} + 1.5z^{-2}}, \quad 0.5 < |z| < 3$$

$$= \frac{3 - 4z^{-1}}{(1 - 0.5z^{-1})(1 - 3z^{-1})}$$

$$H_3(z) = \frac{1}{1 - 0.5z^{-1}} + \frac{2}{1 - 3z^{-1}}$$

Since the given system function $H_3(z)$ is rational and its ROC [shown in Fig. 10.22(c)] is a ring in the z-plane, the corresponding LTI system is noncausal. In can also be verified by its impulse response. Given that

$$H_3(z) = \frac{1}{1 - 0.5z^{-1}} + \frac{2}{1 - 3z^{-1}}$$

Taking the inverse z-transform, we obtain

$$h_3(n) = (0.5)^n u(n) - 2(3)^n u(-n-1)$$

Since $h_3(t)$ is two-sided, this system is noncausal.

10.11.4 Stability

An LTI system is stable if the impulse response $h(n)$ is *absolutely summable*, i.e.,

$$\sum_{n=-\infty}^{\infty} |h(n)| < \infty$$

This implies that the discrete-time Fourier transform exists, and thus the ROC includes the unit circle in the z-plane. Thus, *an LTI system is stable if and only if the ROC of its system function $H(z)$ includes the unit circle.*

10.11.5 Stability of a Causal LTI System

Consider a causal LTI system with a rational system function $H(z)$. Since the system is causal, the ROC is outside the outermost pole. Consequently, for this system to be stable (i.e., for the ROC to include the unit circle), the outermost pole of $H(z)$ must be inside the unit circle, i.e., the largest magnitude of the poles of $H(z)$ must be less than one. That is, *a causal system with rational system function $H(z)$ is stable if and only if all the poles of $H(z)$ lie inside the unit circle.*

Example 10.47 Check whether the corresponding LTI system with system function

$$H(z) = \frac{-1 - 0.4z^{-1}}{1 - 2.8z^{-1} + 1.6z^{-2}}$$

is stable and causal, if the ROC is

(a) $|z| > 2$
(b) $|z| < 0.8$
(c) $0.8 < |z| < 2$

Solution
Consider the given system function

$$H(z) = \frac{-1 - 0.4z^{-1}}{1 - 2.8z^{-1} + 1.6z^{-2}}$$

$$= \frac{-z^2 - 0.4z}{z^2 - 2.8z + 1.6}$$

$$H(z) = \frac{-z(z + 0.4)}{(z - 0.8)(z - 2)}$$

$$\frac{H(z)}{z} = \frac{-(z + 0.4)}{(z - 0.8)(z - 2)}$$

Using the partial fraction expansion, we obtain

$$\frac{H(z)}{z} = \frac{1}{z - 0.8} - \frac{2}{z - 2}$$

$$H(z) = \frac{z}{z - 0.8} - \frac{2z}{z - 2}$$

This system has poles at $z = 0.8$, and $z = 2$.
(a) Since the given system function $H(z)$ is rational and its ROC, $|z| > 2$ [shown in Fig. 10.23(a)] is the region in the z-plane outside the outermost pole, the corresponding LTI system is causal. Also, note that the ROC does not include the unit circle, and consequently, the corresponding system is unstable. This can be verified by its impulse response. Given that

$$H(z) = \frac{z}{z - 0.8} - \frac{2z}{z - 2}$$

$$= \frac{1}{1 - 0.8z^{-1}} - \frac{2}{1 - 2z^{-1}}$$

Taking the inverse z-transform, we obtain

$$h(n) = (0.8)^n u(n) - (2)^n u(n)$$

Note that $h(n) = 0$, for $n < 0$, and $\sum_{n=-\infty}^{\infty} h(n) = \infty$, i.e., the impulse response is not absolutely summable, so this system is causal and unstable.

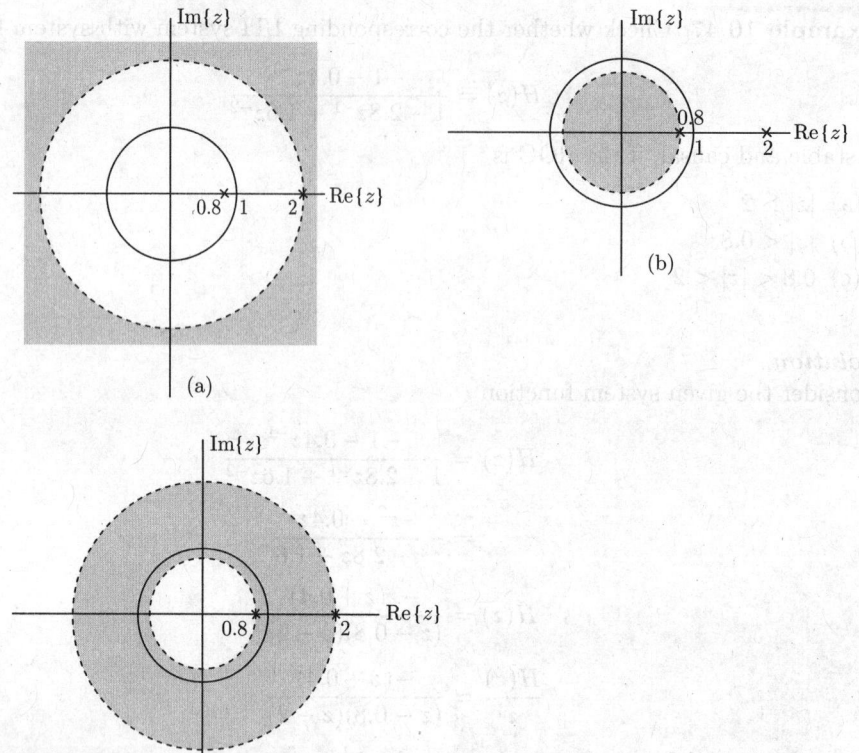

Fig. 10.23 Pole-zero plot and ROC for Example 10.47

(b) Since the given system function $H(z)$ is rational and its ROC, $|z| < 0.8$ [shown in Fig. 10.23(b)], is the region in the z-plane inside the innermost pole, the corresponding LTI system is anticausal. Also, note that the ROC does not include the unit circle, and consequently, the corresponding system is unstable. This can be verified by its impulse response. Given that

$$H(z) = \frac{z}{z - 0.8} - \frac{2z}{z - 2}$$

$$= \frac{1}{1 - 0.8z^{-1}} - \frac{2}{1 - 2z^{-1}}$$

Taking the inverse z-transform, we obtain

$$h(n) = -(0.8)^n u(-n - 1) + (2)^n u(-n - 1)$$

Note that $h(n) = 0$, for $n > 0$, and $\sum_{n=-\infty}^{\infty} h(n) = \infty$, i.e., the impulse response is not absolutely summable, so this system is anticausal and unstable.

(c) Since the given system function $H(z)$ is rational and its ROC, $0.8 < |z| < 2$ [shown in Fig. 10.23(c)], is a ring in the z-plane, the corresponding LTI system is noncausal. Also, note that the ROC includes the unit circle, and consequently, the corresponding system is stable. This can be verified by its impulse response. Given that

$$H(z) = \frac{z}{z - 0.8} - \frac{2z}{z - 2}$$

$$= \frac{1}{1 - 0.8z^{-1}} - \frac{2}{1 - 2z^{-1}}$$

Taking the inverse z-transform, we obtain

$$h(n) = (0.8)^n u(n) + (2)^n u(-n - 1)$$

Note that $h(n)$ is two-sided, and $\sum_{n=-\infty}^{\infty} h(n) < \infty$, i.e., the impulse response is absolutely summable, so this system is noncausal and stable.

Example 10.48 An LTI system is characterized by the system function

$$H(z) = \frac{3 - 4z^{-1}}{1 - 3.5z^{-1} + 1.5z^{-2}}$$

Specify the ROC of $H(z)$ and determine $h(n)$ for the following conditions:

(a) The system is causal and unstable.
(b) The system is noncausal and stable.
(c) The system is anticausal and unstable.

Solution
Given that

$$H(z) = \frac{3 - 4z^{-1}}{1 - 3.5z^{-1} + 1.5z^{-2}}$$

$$= \frac{3z^2 - 4z}{z^2 - 3.5z + 1.5}$$

$$\frac{H(z)}{z} = \frac{3z - 4}{z^2 - 3.5z + 1.5}$$

$$= \frac{3z - 4}{(z - 0.5)(z - 3)}$$

Using the partial fraction expansion, we obtain

$$\frac{H(z)}{z} = \frac{1}{z - 0.5} + \frac{2}{z - 3}$$

$$H(z) = \frac{z}{z - 0.5} + \frac{2z}{z - 3}$$

$$= \frac{1}{1 - 0.5z^{-1}} + \frac{2}{1 - 3z^{-1}}$$

This system has poles at $z = 0.5$ and $z = 3$.

(a) For this system to be causal and unstable, the ROC of $H(z)$ is the region in the z-plane outside the outermost pole and it must not include the unit circle. Therefore, the ROC is the region, $|z| > 3$ [shown in Fig. 10.24(a)].

Since the ROC, $|z| > 3$, is the region in the z-plane outside the outermost pole, all the poles correspond to causal (right-sided) signals. Now, consider

$$H(z) = \frac{1}{1 - 0.5z^{-1}} + \frac{2}{1 - 3z^{-1}}, \quad |z| > 3$$

The inverse z-transform yields

$$h(n) = (0.5)^n u(n) + 2(3)^n u(n)$$

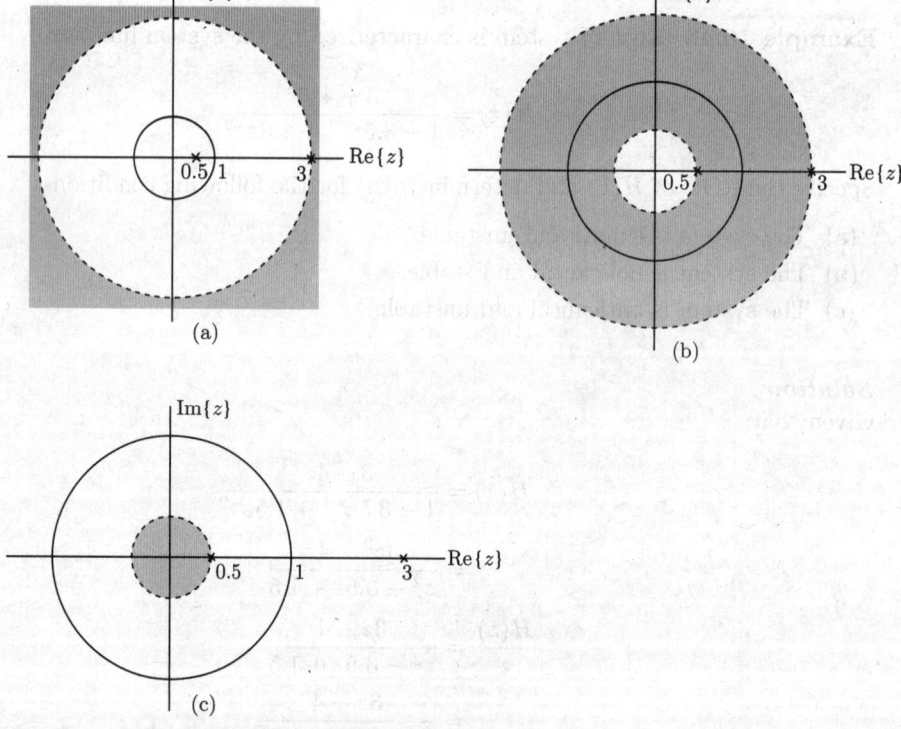

Fig. 10.24 (a) Causal and unstable system, (b) noncausal and stable system, and (c) anticausal and unstable system

(b) For this system to be noncausal and stable, the ROC of $H(z)$ is a ring in the z-plane and it must include the unit circle. Therefore, the ROC is the region, $0.5 < |z| < 3$ [shown in Fig. 10.24(b)].

The pole of the first term is at 0.5. The ROC has a radius greater than the pole at $z = 0.5$, so this pole corresponds to causal (right-sided) signal. Therefore,

$$(0.5)^n u(n) \longleftrightarrow \frac{1}{1 - 0.5z^{-1}}$$

The second term has a pole at $z = 3$. The ROC has a radius less than the pole at $z = 3$, so this pole corresponds to the anticausal (left-sided) signal. Therefore,

$$-2(3)^n u(-n - 1) \longleftrightarrow \frac{2}{1 - 3z^{-1}}$$

and hence, we obtain

$$h(n) = (0.5)^n u(n) - 2(3)^n u(-n - 1)$$

(c) For this system to be anticausal and unstable, the ROC of $H(z)$ is the region in the z-plane inside the innermost pole and it must not include the unit circle. Therefore, the ROC is the region, $|z| < 0.5$ [shown in Fig. 10.24(c)].

Since the ROC, $|z| < 0.5$, is the region in the z-plane inside the innermost pole, all the poles correspond to anticausal (right-sided) signals. Now, consider

$$H(z) = \frac{1}{1 - 0.5z^{-1}} + \frac{2}{1 - 3z^{-1}}, \quad |z| < 0.5$$

The inverse z-transform yields

$$h(n) = -(0.5)^n u(-n - 1) - 2(3)^n u(-n - 1)$$

Example 10.49 Find the inverse z-transform of

$$X(z) = \frac{3(1 - z^{-1})}{1 - 2.5z^{-1} + z^{-2}}$$

assuming that (a) the signal is causal and (b) the signal has a DTFT, i.e., $x(n)$ is absolutely summable.

Solution
Given that

$$X(z) = \frac{3(1 - z^{-1})}{1 - 2.5z^{-1} + z^{-2}}$$

$$= \frac{3(z^2 - z)}{z^2 - 2.5z + 1}$$

$$\frac{X(z)}{z} = \frac{3(z - 1)}{z^2 - 2.5z + 1}$$

$$= \frac{3(z - 1)}{(z - 0.5)(z - 2)}$$

Using the partial fraction expansion, we obtain

$$\frac{X(z)}{z} = \frac{1}{z - 0.5} + \frac{2}{z - 2}$$

$$X(z) = \frac{z}{z - 0.5} + \frac{2z}{z - 2}$$

$$= \frac{1}{1 - 0.5z^{-1}} + \frac{2}{1 - 2z^{-1}}$$

This system has poles at $z = 0.5$ and $z = 2$.

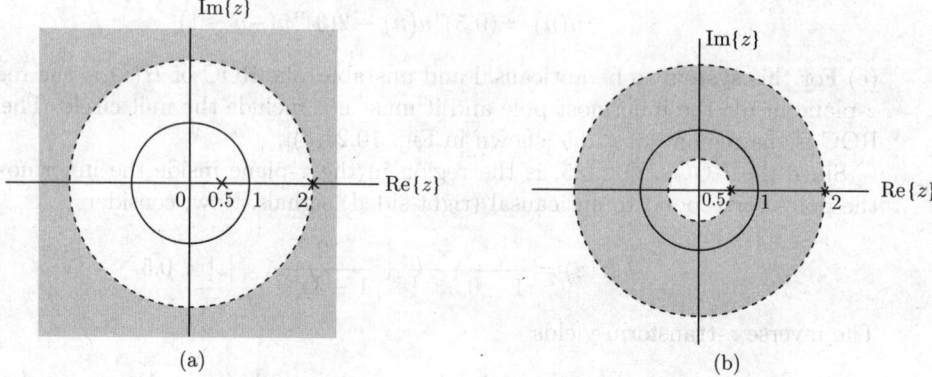

Fig. 10.25 (a) Causal signal, (b) absolutely summable signal

(a) For $x(n)$ to be causal, the ROC of $X(z)$ is the region in the z-plane outside the outermost pole, i.e., $|z| > 2$ [shown in Fig. 10.25(a)].

Since the ROC, $|z| > 2$, is the region in the z-plane outside the outermost pole, all the poles correspond to causal (right-sided) signals. Now, consider

$$X(z) = \frac{1}{1 - 0.5z^{-1}} + \frac{2}{1 - 2z^{-1}}, \quad |z| > 2$$

The inverse z-transform yields

$$x(n) = (0.5)^n u(n) + 2(2)^n u(n)$$

(b) For $x(n)$ to be Fourier transformable, i.e., for $x(n)$ to be absolutely summable, the ROC of $X(z)$ must include the unit circle. Therefore, its ROC is the region $0.5 < |z| < 2$ [shown in Fig. 10.25(b)].

The pole of the first term is at 0.5. The ROC has a radius greater than the pole at $z = 0.5$, so this pole corresponds to causal (right-sided) signal. Therefore,

$$(0.5)^n u(n) \longleftrightarrow \frac{1}{1 - 0.5z^{-1}}$$

The second term has a pole at $z = 2$. The ROC has a radius less than the pole at $z = 2$, so this pole corresponds to the anti-causal (left-sided) signal. Therefore,

$$-2(2)^n u(-n-1) \longleftrightarrow \frac{2}{1 - 2z^{-1}}$$

and hence, we obtain

$$x(n) = (0.5)^n u(n) - 2(2)^n u(-n-1)$$

10.12 UNILATERAL (ONE-SIDED) z-TRANSFORM

The unilateral z-transform of a signal $x(n)$ is defined as

$$\mathcal{Z}_u[x(n)] = X(z) = \sum_{n=0}^{\infty} x(n)z^{-n} \qquad (10.69)$$

We shall denote the relationship between $x(n)$ and $X(z)$ as

$$x(n) \overset{\mathcal{Z}_u}{\longleftrightarrow} X(z) \qquad (10.70)$$

The difference in the definitions of the unilateral and bilateral z-transform lies in the lower limit of the summation. The bilateral z-transform in Eq. (10.4) involves a summation from $-\infty$ to $+\infty$, while the unilateral z-transform in Eq. (10.69) involves a summation from 0 to $+\infty$. Due to this choice of lower limit, the unilateral z-transform has the following characteristics:

1. Two signals that differ for $n < 0$, but that are identical for $n \geq 0$, will have different bilateral z-transforms, but identical unilateral z-transforms.

2. Any signal $x(n)$ that is identically zero for $n < 0$ [i.e., the signal $x(n)u(n)$] has identical bilateral and unilateral z-transforms.

3. The unilateral z-transform is used in analysing causal systems and, particularly, systems specified by linear constant-coefficient difference equations with nonzero initial conditions (i.e., systems that are not initially at rest).

4. For a unilateral z-transform, there is a unique inverse transform of $X(z)$; consequently, there is no need to specify the ROC explicitly (ROC for a unilateral z-transform must always the exterior of a circle). For this reason, we shall generally ignore any mention of the ROC for unilateral z-transform.

Example 10.50 Find (a) bilateral z-transform and (b) unilateral z-transform of the signal

$$x(t) = a^{n+1}u(n+1)$$

Solution

(a) We know that

$$a^n u(n) \longleftrightarrow \frac{1}{1 - az^{-1}}, \quad |z| > |a|$$

Using the time-shifting property, we obtain

$$a^{n+1}u(n+1) \longleftrightarrow \frac{z}{1 - az^{-1}}, \quad |z| > |a|$$

Therefore,

$$\mathcal{Z}[x(n)] = X(z) = \frac{z}{1 - az^{-1}}, \quad |z| > |a|$$

(b) By definition [Eq. (10.69)]

$$\mathcal{Z}_u[x(n)] = X(z) = \sum_{n=0}^{\infty} x(n)z^{-n} = \sum_{n=0}^{\infty} a^{n+1}u(n+1)z^{-n}$$

$$= \sum_{n=0}^{\infty} a^{n+1}z^{-n}$$

$$= a\sum_{n=0}^{\infty} a^n z^{-n}$$

$$= a\sum_{n=0}^{\infty} (az^{-1})^n$$

$$\mathcal{Z}_u[x(n)] = X(z) = \frac{a}{1 - az^{-1}}$$

Example 10.51 Determine the unilateral z-transform of each of the following signals:

(a) $x(n) = a^n u(n+1)$

(b) $x(n) = \delta(n+1) + \delta(n) + a^{n+3}u(n+1)$

Solution

(a) From Eq. (10.69), we have

$$\mathcal{Z}_u[x(n)] = X(z) = \sum_{n=0}^{\infty} x(n)z^{-n} = \sum_{n=0}^{\infty} a^n u(n+1)z^{-n}$$

$$= \sum_{n=0}^{\infty} a^n z^{-n}$$

$$= \sum_{n=0}^{\infty} \left(az^{-1}\right)^n$$

$$= \frac{1}{1-az^{-1}}$$

$$\mathcal{Z}_u[x(n)] = X(z) = \frac{1}{1-az^{-1}}$$

(b) From Eq. (10.69), we have

$$\mathcal{Z}_u[x(n)] = X(z) = \sum_{n=0}^{\infty} x(n)z^{-n} = \sum_{n=0}^{\infty} [\delta(n+1) + \delta(n) + a^{n+3}u(n+1)]z^{-n}$$

$$= \sum_{n=0}^{\infty} \delta(n+1)z^{-n} + \sum_{n=0}^{\infty} \delta(n)z^{-n} + \sum_{n=0}^{\infty} a^{n+3}u(n+1)z^{-n}$$

$$= 0 + z^{-n}\Big|_{n=0} + \sum_{n=0}^{\infty} a^{n+3}z^{-n}$$

$$= 1 + a^3 \sum_{n=0}^{\infty} \left(az^{-1}\right)^n$$

$$X(z) = 1 + \frac{a^3}{1-az^{-1}}$$

Example 10.52 Determine the unilateral z-transform of each of the following sequences:

(a) $x(n) = \left(\dfrac{1}{4}\right)^n u(3-n)$

(b) $x(n) = 2^n u(-n) + \left(\dfrac{1}{4}\right)^n u(n-1)$

Solution

(a) By definition [Eq. (10.69)], we have

$$\mathcal{Z}_u[x(n)] = X(z) = \sum_{n=0}^{\infty} x(n)z^{-n} = \sum_{n=0}^{\infty} \left(\frac{1}{4}\right)^n u(3-n)z^{-n}$$

Since

$$u(3-n) = \begin{cases} 1, & (3-n) \geq 0 \longrightarrow n \leq 3 \\ 0, & (3-n) < 0 \longrightarrow n > 3 \end{cases}$$

Therefore,

$$X(z) = \sum_{n=0}^{3} \left(\frac{1}{4}\right)^n z^{-n} = 1 + \frac{1}{4}z^{-1} + \frac{1}{16}z^{-2} + \frac{1}{64}z^{-3}$$

(b) By definition [Eq. (10.69)], we have

$$\mathcal{Z}_u[x(n)] = X(z) = \sum_{n=0}^{\infty} x(n)z^{-n} = \sum_{n=0}^{\infty} \left[2^n u(-n) + \left(\frac{1}{4}\right)^n u(n-1) \right] z^{-n}$$

$$= \sum_{n=0}^{\infty} 2^n u(-n)z^{-n} + \sum_{n=0}^{\infty} \left(\frac{1}{4}\right)^n u(n-1)z^{-n}$$

$$= \left(2^n z^{-n} \right)\Big|_{n=0} + \sum_{n=1}^{\infty} \left(\frac{1}{4}z^{-1}\right)^n$$

$$= 1 + \frac{\frac{1}{4}z^{-1}}{1 - \frac{1}{4}z^{-1}}$$

$$X(z) = \frac{1}{1 - \frac{1}{4}z^{-1}}$$

Example 10.53 Determine the unilateral z-transform of the following signals:

(a) $x(n) = \{\underset{\uparrow}{1}, 2, 5, 4, 0, 3\}$

(b) $x(n) = \{1, 2, \underset{\uparrow}{5}, 4, 0, 3\}$

(c) $x(n) = \{\underset{\uparrow}{0}, 0, 1, 2, 5, 4, 0, 3\}$

Solution
(a) By definition [Eq. (10.69)], we have

$$\mathcal{Z}_u[x(n)] = X(z) = \sum_{n=0}^{\infty} x(n)z^{-n} = \sum_{n=0}^{5} x(n)z^{-n}$$

$$= x(0) + x(1)z^{-1} + x(2)z^{-2} + x(3)z^{-3} + x(4)z^{-4} + x(5)z^{-5}$$

$$X(z) = 1 + 2z^{-1} + 5z^{-2} + 4z^{-3} + 3z^{-5}$$

(b) By definition [Eq. (10.69)], we have

$$\mathcal{Z}_u[x(n)] = X(z) = \sum_{n=0}^{\infty} x(n)z^{-n} = \sum_{n=0}^{3} x(n)z^{-n}$$

$$= x(0) + x(1)z^{-1} + x(2)z^{-2} + x(3)z^{-3}$$

$$X(z) = 5 + 4z^{-1} + 3z^{-3}$$

(c) By definition [Eq. (10.69)], we have

$$\mathcal{Z}_u[x(n)] = X(z) = \sum_{n=0}^{\infty} x(n)z^{-n}$$

$$= \sum_{n=0}^{7} x(n)z^{-n}$$

$$= x(0) + x(1)z^{-1} + x(2)z^{-2} + x(3)z^{-3} + x(4)z^{-4} + x(5)z^{-5} + x(6)z^{-6} + x(7)z^{-7}$$

$$X(z) = z^{-2} + 2z^{-3} + 5z^{-4} + 4z^{-5} + 3z^{-7}$$

10.13 PROPERTIES OF UNILATERAL z-TRANSFORM

The unilateral z-transform has many important properties, some of which are identical to their bilateral counterparts and several of which differ in significant ways. In particular, linearity, scaling in the z-domain, time expansion, conjugation, and differentiation in the z-domain properties are identical for the bilateral and unilateral z-transforms. The derivation of each of these properties is identical to that of its bilateral counterpart. One bilateral property, namely, the time-reversal property has no meaningful counterpart for the unilateral transform.

10.13.1 Linearity
If

$$x_1(n) \overset{\mathcal{Z}_u}{\longleftrightarrow} X_1(z) \quad \text{and} \quad x_2(n) \overset{\mathcal{Z}_u}{\longleftrightarrow} X_2(z)$$

then

$$ax_1(n) + bx_2(n) \overset{\mathcal{Z}_u}{\longleftrightarrow} aX_1(z) + bX_2(z) \tag{10.71}$$

10.13.2 Scaling in the z-Domain
If

$$x(n) \overset{\mathcal{Z}_u}{\longleftrightarrow} X(z)$$

then

$$z_0^n x(n) \overset{\mathcal{Z}_u}{\longleftrightarrow} X\left(\frac{z}{z_0}\right) \tag{10.72}$$

10.13.3 Differentiation in the z-Domain

If

$$x(n) \xleftrightarrow{\;Z_u\;} X(z)$$

then

$$nx(n) \xleftrightarrow{\;Z_u\;} -z\frac{dX(z)}{dz} \tag{10.73}$$

10.13.4 Time Expansion

Let m be a positive integer, and define the signal

$$x_{(m)}(n) = \begin{cases} x\left(\frac{n}{m}\right), & \text{if } n \text{ is a multiple of } m \\ 0, & \text{if } n \text{ is not a multiple of } m \end{cases} \tag{10.74}$$

$x_m(n)$ can be obtained from $x(n)$ by placing $m-1$ zeros between successive values of the original signal. Intuitively, we can think of $x_{(m)}(n)$ as a slowed-down version of $x(n)$. Now, if

$$x(n) \xleftrightarrow{\;Z_u\;} X(z)$$

then

$$x_{(m)}(n) \xleftrightarrow{\;Z_u\;} X(z^m) \tag{10.75}$$

10.13.5 Conjugation Property

If

$$x(n) \xleftrightarrow{\;Z_u\;} X(z)$$

then

$$x^*(n) \xleftrightarrow{\;Z_u\;} X^*(z^*) \tag{10.76}$$

10.13.6 Convolution Property

If $x_1(n) = x_2(n) = 0$ for all $n < 0$, and if

$$x_1(n) \xleftrightarrow{\;Z_u\;} X_1(z)$$

and

$$x_2(n) \xleftrightarrow{\;Z_u\;} X_2(z)$$

then

$$x_1(n) * x_2(n) \xleftrightarrow{\;Z_u\;} X_1(z)X_2(z) \tag{10.77}$$

It is important to note that the convolution property for unilateral z-transform applies only if the signals $x_1(n)$ and $x_2(n)$ are both zero for $n < 0$. That is, while we have seen that the bilateral z-transform of $x_1(n) * x_2(n)$ always equals to the product of the bilateral z-transforms of $x_1(n)$ and $x_2(n)$, the unilateral z-transform of $x_1(n) * x_2(n)$ in general does not equal to the product of the unilateral transforms if either $x_1(n)$ or $x_2(n)$ is nonzero for $n < 0$.

Example 10.54 Use the convolution property to show that $u(n) * u(n-1) = nu(n)$.

Solution

Let $x(n) = u(n) * u(n-1)$. Taking the z-transform of $x(n)$ and using the convolution property, we get

$$\mathcal{Z}[X(z)] = \mathcal{Z}[u(n) * u(n-1)]$$

$$X(z) = \mathcal{Z}[u(n)]\mathcal{Z}[u(n-1)]$$

$$= \left(\frac{1}{1-z^{-1}}\right)\left(\frac{z^{-1}}{1-z^{-1}}\right)$$

$$X(z) = \frac{z^{-1}}{(1-z^{-1})^2}$$

The inverse z-transform of the above equation yields [Eq. (10.34)]

$$x(n) = nu(n)$$

10.13.7 Accumulation Property

If $x(n) = 0$ for $n < 0$ and if

$$x(n) \xleftrightarrow{\mathcal{Z}_u} X(z)$$

then

$$\sum_{k=0}^{n} x(k) \xleftrightarrow{\mathcal{Z}_u} \frac{X(z)}{1-z^{-1}} \tag{10.78}$$

Proof We know that if $x(n) = 0$ for $n < 0$, then

$$\sum_{k=0}^{n} x(k) = x(n) * u(n)$$

Taking the unilateral z-transform on both sides of the above equation, we obtain

$$\mathcal{Z}_u\left[\sum_{k=0}^{n} x(k)\right] = \mathcal{Z}_u[x(n) * u(n)]$$

$$= \frac{X(z)}{1-z^{-1}}$$

Therefore,

$$\sum_{k=0}^{n} x(k) \xleftrightarrow{\mathcal{Z}_u} \frac{X(z)}{1-z^{-1}}$$

10.13.8 Time-delay (Right-shift) Property

If

$$x(n) \overset{Z_u}{\longleftrightarrow} X(z)$$

then, for integer values of n_0

$$x(n - n_0) \overset{Z_u}{\longleftrightarrow} z^{-n_0} X(z) + z^{-n_0} \sum_{n=1}^{n_0} x(-n) z^n, \quad n_0 > 0 \qquad (10.79)$$

In case $x(n)$ is causal [i.e., $x(n) = x(n)u(n) = 0$ for $n < 0$], and if

$$x(n)u(n) \overset{Z_u}{\longleftrightarrow} X(z)$$

then

$$x(n - n_0)u(n - n_0) \overset{Z_u}{\longleftrightarrow} z^{-n_0} X(z) \qquad (10.80)$$

Proof By definition [Eq. (10.69)], we have

$$\mathcal{Z}_u[x(n - n_0)] = \sum_{n=0}^{\infty} x(n - n_0) z^{-n}$$

A change of variables is performed by letting $m = n - n_0$, which also yields $n = m + n_0$, $m = -n_0$ as $n = 0$, and $m = \infty$ as $n = \infty$. Therefore,

$$\mathcal{Z}_u[x(n - n_0)] = \sum_{m=-n_0}^{\infty} x(m) z^{-(m+n_0)} = z^{-n_0} \sum_{m=-n_0}^{\infty} x(m) z^{-m}$$

$$= z^{-n_0} \left[\sum_{m=-n_0}^{-1} x(m) z^{-m} + \sum_{m=0}^{\infty} x(m) z^{-m} \right]$$

$$= z^{-n_0} \left[\sum_{m=1}^{n_0} x(-m) z^m + X(z) \right]$$

$$\mathcal{Z}_u[x(n - n_0)] = z^{-n_0} \sum_{n=1}^{n_0} x(-n) z^n + z^{-n_0} X(z)$$

To prove Eq. (10.80), we have

$$\mathcal{Z}_u[x(n - n_0)u(n - n_0)] = \sum_{n=0}^{\infty} x(n - n_0)u(n - n_0) z^{-n}$$

$$= \sum_{n=n_0}^{\infty} x(n - n_0) z^{-n}$$

A change of variables is performed by letting $m = n - n_0$, which also yields $n = m + n_0$, $m = 0$ as $n = n_0$, and $m = \infty$ as $n = \infty$. Therefore,

$$\mathcal{Z}_u[x(n - n_0)u(n - n_0)] = \sum_{m=0}^{\infty} x(m)z^{-(m+n_0)}$$

$$= z^{-n_0} \sum_{m=0}^{\infty} x(m)z^{-m}$$

$$\mathcal{Z}_u[x(n - n_0)u(n - n_0)] = z^{-n_0} X(z)$$

∎

Example 10.55 Find the unilateral *z*-transform of the following signals:

(a) $x(n - 1)$

(b) $x(n - 2)$

(c) $x(n - 1)u(n - 1)$

Solution

(a) We know that [Eq. (10.79)]

$$\mathcal{Z}_u[x(n - n_0)] = z^{-n_0} \sum_{n=1}^{n_0} x(-n)z^n + z^{-n_0} X(z)$$

Substituting $n_0 = 1$ into the above equation, we obtain

$$\mathcal{Z}_u[x(n - 1)] = z^{-1} \sum_{n=1}^{1} x(-n)z^n + z^{-1} X(z)$$

$$= z^{-1} x(-1)z + z^{-1} X(z)$$

$$= z^{-1} X(z) + x(-1)$$

Therefore,

$$\mathcal{Z}_u[x(n - 1)] = z^{-1} X(z) + x(-1) \tag{10.81}$$

(b) We know that [Eq. (10.79)]

$$\mathcal{Z}_u[x(n - n_0)] = z^{-n_0} \sum_{n=1}^{n_0} x(-n)z^n + z^{-n_0} X(z)$$

Substituting $n_0 = 2$ into the above equation, we obtain

$$\mathcal{Z}_u[x(n - 2)] = z^{-2} \sum_{n=1}^{2} x(-n)z^n + z^{-2} X(z)$$

$$= z^{-2} x(-1)z + z^{-2} x(-2)z^2 + z^{-2} X(z)$$

$$= z^{-2} X(z) + z^{-1} x(-1) + x(-2)$$

Therefore,

$$\mathcal{Z}_u[x(n-2)] = z^{-2}X(z) + z^{-1}x(-1) + x(-2) \tag{10.82}$$

(c) We know that [Eq. (10.80)]

$$\mathcal{Z}_u[x(n-n_0)u(n-n_0)] = z^{-n_0}X(z)$$

Substituting $n_0 = 1$ into the above equation, we obtain

$$\mathcal{Z}_u[x(n-1)u(n-1)] = z^{-1}X(z)$$

Example 10.56 If $X(z)$ denotes the unilateral z-transform of $x(n)$, determine, in terms of $X(z)$, the unilateral z-transform of

(a) $x(n-3)$

(b) $\sum_{k=-\infty}^{n} x(k)$

Solution

(a) We know that [Eq. (10.79)]

$$\mathcal{Z}_u[x(n-n_0)] = z^{-n_0} \sum_{n=1}^{n_0} x(-n)z^n + z^{-n_0}X(z)$$

Substituting $n_0 = 3$ into the above equation, we obtain

$$\mathcal{Z}_u[x(n-3)] = z^{-3} \sum_{n=1}^{3} x(-n)z^n + z^{-3}X(z)$$

$$= z^{-3}x(-1)z + z^{-3}x(-2)z^2 + z^{-3}x(-3)z^3 + z^{-3}X(z)$$

$$\mathcal{Z}_u[x(n-3)] = z^{-3}X(z) + z^{-2}x(-1) + z^{-1}x(-2) + x(-3)$$

(b) Consider the given signal

$$\sum_{k=-\infty}^{n} x(k) = \sum_{m=0}^{\infty} x(n-m)$$

Taking the unilateral z-transform of the above equation, we obtain

$$\mathcal{Z}_u\left[\sum_{k=-\infty}^{n} x(k)\right] = \mathcal{Z}_u\left[\sum_{m=0}^{\infty} x(n-m)\right] = \sum_{m=0}^{\infty} \mathcal{Z}_u[x(n-m)]$$

Using the time-delay property [Eq. (10.79)], we obtain

$$\mathcal{Z}_\mathrm{u}\left[\sum_{k=-\infty}^{n} x(k)\right] = \sum_{m=0}^{\infty}\left[z^{-m}\sum_{n=1}^{m} x(-n)z^n + z^{-m}X(z)\right]$$

$$= \sum_{m=0}^{\infty} z^{-m}X(z) + \sum_{m=0}^{\infty}\left[z^{-m}\sum_{n=1}^{m} x(-n)z^n\right]$$

$$\mathcal{Z}_\mathrm{u}\left[\sum_{k=-\infty}^{n} x(k)\right] = \frac{X(z)}{1-z^{-1}} + \sum_{m=0}^{\infty}\left[z^{-m}\sum_{n=1}^{m} x(-n)z^n\right]$$

10.13.9 Time-Advance (Left-shift) Property

If

$$x(n) \xleftrightarrow{\mathcal{Z}_\mathrm{u}} X(z)$$

then for integer values of n_0

$$x(n+n_0) \xleftrightarrow{\mathcal{Z}_\mathrm{u}} z^{n_0}X(z) - z^{n_0}\sum_{n=0}^{n_0-1} x(n)z^{-n}, \quad n_0 > 0 \qquad (10.83)$$

Proof By definition [Eq. (10.69)], we have

$$\mathcal{Z}_\mathrm{u}[x(n+n_0)] = \sum_{n=0}^{\infty} x(n+n_0)z^{-n}$$

A change of variables is performed by letting $m = n + n_0$, which also yields $n = m - n_0$, $m = n_0$ as $n = 0$, and $m = \infty$ as $n = \infty$. Therefore,

$$\mathcal{Z}_\mathrm{u}[x(n+n_0)] = \sum_{m=n_0}^{\infty} x(m)z^{-(m-n_0)} = z^{n_0}\sum_{m=n_0}^{\infty} x(m)z^{-m}$$

$$= z^{n_0}\left[\sum_{m=n_0}^{\infty} x(m)z^{-m} + \sum_{m=0}^{n_0-1} x(m)z^{-m} - \sum_{m=0}^{n_0-1} x(m)z^{-m}\right]$$

$$= z^{n_0}\left[\sum_{m=0}^{\infty} x(m)z^{-m} - \sum_{m=0}^{n_0-1} x(m)z^{-m}\right]$$

$$= z^{n_0}\left[X(z) - \sum_{n=0}^{n_0-1} x(n)z^{-n}\right]$$

$$\mathcal{Z}_\mathrm{u}[x(n+n_0)] = z^{n_0}X(z) - z^{n_0}\sum_{n=0}^{n_0-1} x(n)z^{-n}$$

Example 10.57 Find the unilateral z-transform of the following signals:

(a) $x(n+1)$

(b) $x(n+2)$

Solution

(a) We know that [Eq. (10.83)]

$$\mathcal{Z}_u[x(n+n_0)] = z^{n_0} X(z) - z^{n_0} \sum_{n=0}^{n_0-1} x(n) z^{-n}$$

Substituting $n_0 = 1$ into the above equation, we obtain

$$\mathcal{Z}_u[x(n+1)] = zX(z) - z \sum_{n=0}^{1-1} x(n) z^{-n}$$

$$= zX(z) - z \sum_{n=0}^{0} x(n) z^{-n}$$

$$= zX(z) - zx(0)$$

Therefore,

$$\mathcal{Z}_u[x(n+1)] = zX(z) - zx(0) \tag{10.84}$$

(b) We know that [Eq. (10.83)]

$$\mathcal{Z}_u[x(n+n_0)] = z^{n_0} X(z) - z^{n_0} \sum_{n=0}^{n_0-1} x(n) z^{-n}$$

Substituting $n_0 = 2$ into the above equation, we obtain

$$\mathcal{Z}_u[x(n+2)] = z^2 X(z) - z^2 \sum_{n=0}^{2-1} x(n) z^{-n}$$

$$= z^2 X(z) - z^2 \sum_{n=0}^{1} x(n) z^{-n}$$

$$= z^2 X(z) - z^2 x(0) - zx(1)$$

Therefore,

$$\mathcal{Z}_u[x(n+2)] = z^2 X(z) - z^2 x(0) - zx(1) \tag{10.85}$$

10.13.10 First Difference

If $x(n)$ is causal [i.e., $x(n) = 0$ for $n < 0$], and if

$$x(n) \xleftrightarrow{Z_u} X(z)$$

then

$$x(n) - x(n-1) \xleftrightarrow{Z_u} (1 - z^{-1})X(z) \qquad (10.86)$$

Proof Given that

$$x(n) \xleftrightarrow{Z_u} X(z)$$

Using the time-shifting property [Eq. (10.80)], we get

$$x(n-1) \xleftrightarrow{Z_u} z^{-1}X(z)$$

Now, using the linearity property, we get

$$x(n) - x(n-1) \xleftrightarrow{Z_u} X(z) - z^{-1}X(z)$$

$$\xleftrightarrow{Z_u} (1 - z^{-1})X(z)$$

10.13.11 Initial-value Theorem

If $x(n)$ is causal [i.e., $x(n) = 0$ for $n < 0$], then

$$x(0) = \lim_{z \to \infty} X(z) \qquad (10.87)$$

As one consequence of the initial-value theorem, for a causal sequence, if $x(0)$ is finite, then $\lim_{z \to \infty} X(z)$ is finite. Consequently, with $X(z)$ expressed as a ratio of polynomials in z, the order of the numerator polynomial cannot be greater than the order of the denominator polynomial; or, equivalently, the number of finite zeros of $X(z)$ cannot be greater than the number of finite poles.

Proof By definition [Eq. (10.4)], we have

$$X(z) = \sum_{n=-\infty}^{\infty} x(n)z^{-n}$$

Since $x(n) = 0$ for $n < 0$, we obtain

$$X(z) = \sum_{n=0}^{\infty} x(n)z^{-n}$$

$$X(z) = x(0) + x(1)z^{-1} + x(2)z^{-2} + \cdots$$

$$\lim_{z \to \infty} X(z) = \lim_{z \to \infty} [x(0) + x(1)z^{-1} + x(2)z^{-2} + \cdots]$$

$$\lim_{z \to \infty} X(z) = x(0)$$

Example 10.58 Use the initial-value theorem to find the initial value of the signals corresponding to the following z-transforms:

(a) $X_1(z) = \dfrac{2 + z^{-1}}{(1 - z^{-1})(1 + 0.5z^{-1})}$

(b) $X_2(z) = \dfrac{1 - 3z^{-1}}{(1 - 0.1z^{-1})(1 + 0.6z^{-1})}$

(c) $X_3(z) = \dfrac{0.5 + 0.25z^{-1}}{1 - 1.3z^{-1} + 0.2z^{-2} + 0.1z^{-3}}$

Solution
(a) Given that

$$X_1(z) = \frac{2 + z^{-1}}{(1 - z^{-1})(1 + 0.5z^{-1})}$$

$$= \frac{2 + 1/z}{(1 - 1/z)(1 + 0.5/z)}$$

Using the initial-value theorem [Eq. (10.87)], we have

$$x_1(0) = \lim_{z \to \infty} X_1(z) = \lim_{z \to \infty} \frac{2 + 1/z}{(1 - 1/z)(1 + 0.5/z)}$$

$$= \frac{2 + 0}{(1 - 0)(1 + 0)}$$

$$x_1(0) = 2$$

(b) Given that

$$X_2(z) = \frac{1 - 3z^{-1}}{(1 - 0.1z^{-1})(1 + 0.6z^{-1})}$$

$$= \frac{1 - 3/z}{(1 - 0.1/z)(1 + 0.6/z)}$$

Using the initial-value theorem [Eq. (10.87)], we have

$$x_2(0) = \lim_{z \to \infty} X_2(z) = \lim_{z \to \infty} \frac{1 - 3/z}{(1 - 0.1/z)(1 + 0.6/z)}$$

$$= \frac{1 - 0}{(1 - 0)(1 + 0)}$$

$$x_2(0) = 1$$

(c) Given that

$$X_3(z) = \frac{0.5 + 0.25z^{-1}}{1 - 1.3z^{-1} + 0.2z^{-2} + 0.1z^{-3}}$$

$$= \frac{0.5 + (0.25/z)}{1 - (1.3/z) + (0.2/z^2) + (0.1/z^3)}$$

Using the initial-value theorem [Eq. (10.87)], we have

$$x_3(0) = \lim_{z \to \infty} X_3(z) = \lim_{z \to \infty} \frac{0.5 + (0.25/z)}{1 - (1.3/z) + (0.2/z^2) + (0.1/z^3)}$$

$$= \frac{0.5 + 0}{1 - 0 + 0 + 0}$$

$$x_3(0) = 0.5$$

10.13.12 Final-value Theorem

The final-value theorem states that if $\lim_{n \to \infty} x(n)$ exists and if

$$x(n) \xleftrightarrow{\mathcal{Z}_u} X(z)$$

then

$$\lim_{n \to \infty} x(n) = x(\infty) = \lim_{z \to 1}(1 - z^{-1})X(z) \tag{10.88}$$

The limit in Eq. (10.88) exists if all the poles of $(1 - z^{-1})X(z)$ lie inside the unit circle, i.e., all the poles of $(1 - z^{-1})X(z)$ have magnitude less than one.

Proof For no apparent reason, let's take the unilateral z-transform of the signal $x(n) - x(n-1)$,

$$\mathcal{Z}_u[x(n) - x(n-1)] = \sum_{n=0}^{\infty}[x(n) - x(n-1)]z^{-n}$$

$$\mathcal{Z}_u[x(n)] - \mathcal{Z}_u[x(n-1)] = \lim_{N \to \infty}\left[\sum_{n=0}^{N}[x(n) - x(n-1)]z^{-n}\right]$$

$$X(z) - [z^{-1}X(z) + x(-1)] = \lim_{N \to \infty}\left[\sum_{n=0}^{N}[x(n) - x(n-1)]z^{-n}\right]$$

$$\lim_{z \to 1}\left[(1 - z^{-1})X(z) - x(-1)\right] = \lim_{z \to 1}\lim_{N \to \infty}\left[\sum_{n=0}^{N}[x(n) - x(n-1)]z^{-n}\right]$$

$$\lim_{z \to 1}(1 - z^{-1})X(z) - x(-1) = \lim_{N \to \infty}\left[\sum_{n=0}^{N}[x(n) - x(n-1)]\right]$$

$$= \lim_{N \to \infty}\left[[x(0) - x(-1)] + [x(1) - x(0)] + [x(2) - x(1)] + \cdots \right.$$

$$+ [x(N-2) - x(N-3)] + [x(N-1) - x(N-2)]$$

$$\left. + [x(N) - x(N-1)] \right]$$

$$\lim_{z \to 1}(1 - z^{-1})X(z) - x(-1) = \lim_{N \to \infty}\left[-x(-1) + x(N)\right]$$

$$\lim_{z \to 1}(1 - z^{-1})X(z) - x(-1) = -x(-1) + \lim_{N \to \infty} x(N)$$

$$\lim_{z \to 1}(1 - z^{-1})X(z) = \lim_{N \to \infty} x(N)$$

$$= x(\infty)$$

Example 10.59 Find the final value of the signals corresponding to the following z-transforms:

(a) $X_1(z) = \dfrac{1 + z^{-1}}{1 - 0.25z^{-2}}$

(b) $X_2(z) = \dfrac{2z^{-1}}{1 - 1.8z^{-1} + 0.8z^{-2}}$

(c) $X_3(z) = \dfrac{1}{1 + 2z^{-1} - 3z^{-2}}$

Solution

(a) Given that

$$X_1(z) = \frac{1 + z^{-1}}{1 - 0.25z^{-2}}$$

$$(1 - z^{-1})X_1(z) = \frac{(1 - z^{-1})(1 + z^{-1})}{(1 - 0.5z^{-1})(1 + 0.5z^{-1})}$$

It has two poles at $z = 0.5$ and $z = -0.5$. Note that both the poles of $(1 - z^{-1})X_1(z)$ lie inside the unit circle; therefore,

$$x_1(\infty) = \lim_{z \to 1}(1 - z^{-1})X_1(z)$$

$$= \lim_{z \to 1}\frac{(1 - z^{-1})(1 + z^{-1})}{(1 - 0.5z^{-1})(1 + 0.5z^{-1})}$$

$$x_1(\infty) = \frac{(0)(2)}{(0.5)(1.5)} = 0$$

(b) Given that

$$X_2(z) = \frac{2z^{-1}}{1 - 1.8z^{-1} + 0.8z^{-2}}$$

$$(1 - z^{-1})X_2(z) = \frac{2z^{-1}(1 - z^{-1})}{(1 - z^{-1})(1 - 0.8z^{-1})}$$

$$= \frac{2z^{-1}}{1 - 0.8z^{-1}}$$

It has one pole at $z = 0.8$, which lies inside the unit circle; therefore,

$$x_2(\infty) = \lim_{z \to 1}(1 - z^{-1})X_2(z)$$

$$= \lim_{z \to 1} \frac{2z^{-1}}{1 - 0.8z^{-1}}$$

$$x_2(\infty) = \frac{2}{0.2} = 10$$

(c) Given that

$$X_3(z) = \frac{1}{1 + 2z^{-1} - 3z^{-2}}$$

$$(1 - z^{-1})X_3(z) = \frac{(1 - z^{-1})}{(1 - z^{-1})(1 + 3z^{-1})}$$

$$= \frac{1}{1 + 3z^{-1}}$$

It has one pole at $z = -3$, which lies outside the unit circle; therefore, the final-value theorem cannot be used to find $x_3(\infty)$.

Example 10.60 Using the final-value theorem, find the steady-state value of

$$x(n) = [(0.5)^n - 0.5]u(n)$$

and verify.

Solution
Given that

$$x(n) = [(0.5)^n - 0.5]u(n)$$

$$= (0.5)^n u(n) - 0.5u(n)$$

Taking the z-transform of the above equation, we obtain

$$X(z) = \frac{1}{1 - 0.5z^{-1}} - \frac{0.5}{1 - z^{-1}}$$

$$= \frac{0.5 - 0.75z^{-1}}{(1 - 0.5z^{-1})(1 - z^{-1})}$$

$$(1 - z^{-1})X(z) = \frac{0.5 - 0.75z^{-1}}{1 - 0.5z^{-1}}$$

It has one pole at $z = 0.5$, which lies inside the unit circle; therefore,

$$x(\infty) = \lim_{z \to 1}(1 - z^{-1})X(z)$$

$$= \lim_{z \to 1} \frac{0.5 - 0.75z^{-1}}{1 - 0.5z^{-1}}$$

$$x(\infty) = \frac{0.5 - 0.75}{1 - 0.5} = -0.5$$

To verify it, consider the time-domain signal

$$x(n) = [(0.5)^n - 0.5]u(n)$$

$$\lim_{n \to \infty} x(n) = \lim_{n \to \infty} [(0.5)^n - 0.5]u(n)$$

$$x(\infty) = -0.5$$

10.13.13 Solving Difference Equations Using the Unilateral z-Transform

One of the most common applications of the unilateral z-transform is to solve linear constant-coefficient difference equations subject to nonzero initial conditions. Such equations are used to model discrete-time LTI systems. Solving of these equations depends on the time-delay and time-advance properties of the unilateral z-transform. As in the case of the Laplace transform with differential equations, the z-transform converts difference equations into algebraic equations that are readily solved to find the solution in the z-domain. The procedure is straightforward and systematic, and we summarizes it in the following three steps:

1. For a given set of initial conditions, take the z-transform of both sides of the difference equation to obtain an algebraic equation in $Y(z)$.
2. Solve the algebraic equation for $Y(z)$.
3. Take the inverse z-transform to obtain $y(n)$.

Example 10.61 A discrete-time system is characterized by the difference equation:

$$y(n) - \frac{1}{2}y(n-1) = x(n) - \frac{1}{2}x(n-1)$$

Find $y(n)$ for $n \geq 0$ when $x(n) = u(n)$ and the initial conditions is given as $y(-1) = 1$. Or, determine the step response of the system

$$y(n) - \frac{1}{2}y(n-1) = x(n) - \frac{1}{2}x(n-1)$$

when the initial condition is $y(-1) = 1$.

Solution

For $x(n) = u(n)$, we have

$$X(z) = \frac{1}{1 - z^{-1}}$$

Now, consider the given difference equation

$$y(n) - \frac{1}{2}y(n-1) = x(n) - \frac{1}{2}x(n-1)$$

$$= u(n) - \frac{1}{2}u(n-1)$$

Taking the unilateral z-transform of the above equation, we obtain

$$\mathcal{Z}_u\left[y(n) - \frac{1}{2}y(n-1)\right] = \mathcal{Z}_u\left[u(n) - \frac{1}{2}u(n-1)\right]$$

$$Y(z) - \frac{1}{2}[z^{-1}Y(z) + y(-1)] = \frac{1}{1 - z^{-1}} - \frac{\frac{1}{2}z^{-1}}{1 - z^{-1}}$$

$$Y(z)\left[1 - \frac{1}{2}z^{-1}\right] - \frac{1}{2}y(-1) = \frac{1 - \frac{1}{2}z^{-1}}{1 - z^{-1}}$$

$$Y(z) = \frac{1}{1 - z^{-1}} + \frac{\frac{1}{2}}{1 - \frac{1}{2}z^{-1}}$$

The inverse z-transform of the above equation yields

$$y(n) = u(n) + \frac{1}{2}\left(\frac{1}{2}\right)^n u(n)$$

$$= \left[1 + \left(\frac{1}{2}\right)^{n+1}\right]u(n)$$

Example 10.62 A second-order discrete-time system is characterized by the difference equation

$$y(n) - 3y(n-1) + 2y(n-2) = x(n) - 2x(n-1)$$

Find $y(n)$ for $n \geq 0$ when $x(n) = u(n)$ and the initial conditions are given as $y(-1) = y(-2) = 1$.

Solution

For $x(n) = u(n)$, we have

$$X(z) = \frac{1}{1 - z^{-1}}$$

Now, consider the given difference equation

$$y(n) - 3y(n-1) + 2y(n-2) = x(n) - 2x(n-1) = 2u(n) - u(n-1)$$

Taking the unilateral z-transform of the above equation, we obtain

$$\mathcal{Z}_u[y(n) - 3y(n-1) - 2y(n-2)] = \mathcal{Z}_u[u(n) - 2u(n-1)]$$

$$Y(z) - 3[z^{-1}Y(z) + y(-1)] + 2[z^{-2}Y(z) + z^{-1}y(-1) + y(-2)] = \frac{1}{1 - z^{-1}} - \frac{2z^{-1}}{1 - z^{-1}}$$

$$Y(z)[1 - 3z^{-1} + 2z^{-2}] - 3y(-1) + 2z^{-1}y(-1) + 2y(-2) = \frac{1 - 2z^{-1}}{1 - z^{-1}}$$

$$Y(z)[(1 - z^{-1})(1 - 2z^{-1})] - 3 + 2z^{-1} + 2 = \frac{1 - 2z^{-1}}{1 - z^{-1}}$$

$$Y(z) = \frac{1 - 2z^{-1}}{(1 - z^{-1})^2(1 - 2z^{-1})}$$

$$+ \frac{1 - 2z^{-1}}{(1 - z^{-1})(1 - 2z^{-1})}$$

$$Y(z) = \frac{1}{(1 - z^{-1})^2} + \frac{1}{1 - z^{-1}}$$

The inverse z-transform of the above equation yields

$$y(n) = (n+1)u(n) + u(n) = (n+2)u(n)$$

10.13.14 Zero-input Response and Zero-state Response

The z-transform method for solving difference equations offers a clear separation between the *zero-input response* of the system to initial conditions and the *zero-state response* of the system associated with the input. *The zero-input response represents the system output when the input is zero, i.e., $x(n) = 0$, and thus it is the result of internal system conditions (such as energy storage, initial conditions) alone. In contrast, the zero-state response represents the system response to the external input $x(n)$ when the initial conditions are zero.* The total response is defined as

$$\text{Total response} = \text{Zero-input response} + \text{Zero-state response}$$
$$\text{(natural response)} \qquad \text{(forced response)} \qquad (10.89)$$

Natural Response and Forced Response: In total response, the response due to the poles of the system is called *natural response* of the system. The response due to the poles of the input is called the *forced response* of the system.

Example 10.63 Use the unilateral *z*-transform to determine the output of a system represented by the difference equation

$$y(n+2) - \frac{5}{6}y(n+1) + \frac{1}{6}y(n) = 5x(n+1) - x(n)$$

in response to the input $x(n) = u(n)$. Assume that the initial conditions on the system are $y(-1) = 2$ and $y(-2) = 0$. Identify the forced response (zero-state response) $y_{\mathrm{f}}(n)$ of the system and the natural response (zero-input response) $y_{\mathrm{n}}(n)$.

Solution
For $x(n) = u(n)$, $x(-1) = x(-2) = 0$, and $X(z) = (1/1 - z^{-1})$. Now, consider the given difference equation

$$y(n+2) - \frac{5}{6}y(n+1) + \frac{1}{6}y(n) = 5x(n+1) - x(n)$$

Because the given difference equation is in time advance operator form, the use of the left-shift property [Eq. (10.83)] may seem appropriate for its solution. Unfortunately, as seen from Eq. (10.83), this property require a knowledge of auxiliary conditions $y(0)$, $y(1), \ldots, y(N-1)$ rather than of the initial conditions $y(-1), y(-2), \ldots, y(-n)$, which are generally given. This difficulty can be overcome by expressing the given difference equation in delay operator form (obtained by replacing n with $n-2$) and then using the right-shift property. The given difference equation in delay operator form is

$$y(n) - \frac{5}{6}y(n-1) + \frac{1}{6}y(n-2) = 5x(n-1) - x(n-2)$$

Taking the unilateral *z*-transform of the above equation, we obtain

$$Y(z) - \frac{5}{6}\left[z^{-1}Y(z) + y(-1)\right] + \frac{1}{6}\left[z^{-2}Y(z) + z^{-1}y(-1) + y(-2)\right]$$
$$= 5[z^{-1}X(z) + x(-1)] - [z^{-2}X(z) + z^{-1}x(-1) + x(-2)]$$

$$Y(z) - \frac{5}{6}\left[z^{-1}Y(z) + 2\right] + \frac{1}{6}\left[z^{-2}Y(z) + 2z^{-1} + 0\right] = 5[z^{-1}X(z) + 0]$$
$$- [z^{-2}X(z) + 0 + 0]$$

$$Y(z)\left[1 - \frac{5}{6}z^{-1} + \frac{1}{6}z^{-2}\right] - \frac{5}{3} + \frac{1}{3}z^{-1} = 5z^{-1}X(z) - z^{-2}X(z)$$

$$Y(z)\left[1 - \frac{5}{6}z^{-1} + \frac{1}{6}z^{-2}\right] = \underbrace{\left(\frac{5}{3} - \frac{1}{3}z^{-1}\right)}_{\text{initial condition terms}} + \underbrace{X(z)[5z^{-1} - z^{-2}]}_{\text{input terms}}$$

$$Y(z) = \underbrace{\frac{\frac{1}{3}(5 - z^{-1})}{1 - \frac{5}{6}z^{-1} + \frac{1}{6}z^{-2}}}_{\text{zero-input component}} + \underbrace{\frac{z^{-1}(5 - z^{-1})}{(1 - z^{-1})\left(1 - \frac{5}{6}z^{-1} + \frac{1}{6}z^{-2}\right)}}_{\text{zero-state response}}$$

$$Y(z) = \left(\frac{\frac{1}{3}(5 - z^{-1})}{\left(1 - \frac{1}{3}z^{-1}\right)\left(1 - \frac{1}{2}z^{-1}\right)}\right) + \left(\frac{z^{-1}(5 - z^{-1})}{(1 - z^{-1})\left(1 - \frac{1}{3}z^{-1}\right)\left(1 - \frac{1}{2}z^{-1}\right)}\right)$$

Using partial fraction expansion, we obtain

$$Y(z) = \left(\frac{3}{1-\frac{1}{2}z^{-1}} - \frac{\frac{4}{3}}{1-\frac{1}{3}z^{-1}}\right) + \left(\frac{12}{1-z^{-1}} - \frac{18}{1-\frac{1}{2}z^{-1}} + \frac{6}{1-\frac{1}{3}z^{-1}}\right)$$

The inverse z-transform of this equation yields

$$y(n) = \left[3\left(\frac{1}{2}\right)^n u(n) - \frac{4}{3}\left(\frac{1}{3}\right)^n u(n)\right] + \left[12u(n) - 18\left(\frac{1}{2}\right)^n u(n) + 6\left(\frac{1}{3}\right)^n u(n)\right]$$

$$y(n) = \underbrace{\left[3\left(\frac{1}{2}\right)^n - \frac{4}{3}\left(\frac{1}{3}\right)^n\right]u(n)}_{\text{zero-input response or natural response}} + \underbrace{\left[12 - 18\left(\frac{1}{2}\right)^n + 6\left(\frac{1}{3}\right)^n\right]u(n)}_{\text{zero-state response or forced response}}$$

$$y(n) = y_{zi}(n) + y_{zs}(n)$$

where the zero-input response is

$$y_{zi}(n) = \left[3\left(\frac{1}{2}\right)^n - \frac{4}{3}\left(\frac{1}{3}\right)^n\right]u(n)$$

and the zero-state response is

$$y_{zs}(n) = \left[12 - 18\left(\frac{1}{2}\right)^n + 6\left(\frac{1}{3}\right)^n\right]u(n)$$

The poles of the given system are located at $p_1 = \frac{1}{2}$ and $p_2 = \frac{1}{3}$. The response $y(n)$ can be separated into two parts: the natural response $y_n(n)$, and the forced response $y_f(n)$.

$$y(n) = \left[3\left(\frac{1}{2}\right)^n u(n) - \frac{4}{3}\left(\frac{1}{3}\right)^n u(n)\right] + \left[12u(n) - 18\left(\frac{1}{2}\right)^n u(n) + 6\left(\frac{1}{3}\right)^n u(n)\right]$$

$$= \left[3\left(\frac{1}{2}\right)^n u(n) - \frac{4}{3}\left(\frac{1}{3}\right)^n u(n) - 18\left(\frac{1}{2}\right)^n u(n) + 6\left(\frac{1}{3}\right)^n u(n)\right] + 12u(n)$$

$$y(n) = \underbrace{\left[-15\left(\frac{1}{2}\right)^n + \frac{14}{3}\left(\frac{1}{3}\right)^n\right]u(n)}_{\text{natural response} = y_n(n)} + \underbrace{12u(n)}_{\text{forced response} = y_f(n)}$$

Example 10.64 Determine the transient response and steady-state response of the system characterized by the difference equation

$$y(n) = 0.5y(n-1) + x(n)$$

when the input signal is $x(n) = 10\cos(n\pi/4)\,u(n)$. The system is initially at rest (i.e., it is relaxed).

Solution
Consider the given difference equation

$$y(n) = 0.5y(n-1) + x(n)$$

Taking its z-transform, we obtain

$$\frac{Y(z)}{X(z)} = H(z) = \frac{1}{1 - 0.5z^{-1}}$$

This system has a pole at $z = 0.5$. Taking the z-transform of the input, we obtain [Eq. (10.20)]

$$X(z) = \frac{10\left(1 - z^{-1}\cos(\pi/4)\right)}{1 - 2z^{-1}\cos(\pi/4) + z^{-2}} = \frac{10\left[1 - \frac{1}{\sqrt{2}}z^{-1}\right]}{1 - \sqrt{2}z^{-1} + z^{-2}}$$

The output of the system is given by [Eq. (10.65)]

$$Y(z) = H(z)X(z) = \frac{10\left[1 - (1/\sqrt{2})z^{-1}\right]}{(1 - 0.5z^{-1})(1 - \sqrt{2}z^{-1} + z^{-2})}$$

$$= \frac{10\left[1 - (1/\sqrt{2})z^{-1}\right]}{(1 - 0.5z^{-1})(1 - e^{j\pi/4}z^{-1})(1 - e^{-j\pi/4}z^{-1})}$$

Using partial fraction expansion, we obtain

$$Y(z) = \underbrace{\frac{6.3}{1 - 0.5z^{-1}}}_{\text{pole of the system}} + \underbrace{\frac{6.78\,e^{-j28.7}}{1 - e^{j\pi/4}z^{-1}} + \frac{6.78\,e^{j28.7}}{1 - e^{-j\pi/4}z^{-1}}}_{\text{poles of the forcing function, i.e., input}}$$

The inverse z-transform of this equation yields

$$y(n) = \underbrace{6.3(0.5)^n u(n)}_{\text{transient response}} + \underbrace{\left[6.78\,e^{-j28.7}\,e^{jn\pi/4} + 6.78\,e^{j28.7}\,e^{-jn\pi/4}\right]u(n)}_{\text{steady-state response}}$$

In this example, the system is initially relaxed, i.e., all the initial conditions are zero. Therefore, the natural or transient response is

$$y_n(n) = 6.3(0.5)^n u(n)$$

and the forced or steady-state response is

$$y_f(n) = \left[6.78e^{-j28.7}e^{j\frac{n\pi}{4}} + 6.78e^{-j28.7}e^{-j\frac{n\pi}{4}}\right]u(n) = 13.56\cos\left(\frac{\pi}{4}n - 28.7°\right)u(n)$$

10.14 BLOCK DIAGRAM REPRESENTATION

As we saw earlier, the z-transform is a useful tool for computing the system transfer function $H(z)$ if the system is described by its difference equation or if the output is expressed explicitly in terms of the input. The situation changes considerably in cases where a large number of components or elements are interconnected to form the complete system. In such cases, it is convenient to represent the system by suitably interconnected subsystems, each of which can be separately and easily analysed. A linear system can be characterized by its transfer function $H(z)$. Figure 10.26(a) shows a block diagram of a system with a transfer function $H(z)$ and its input and output $X(z)$ and $Y(z)$, respectively. Subsystems may be interconnected by using cascade (series), parallel, and feedback interconnections, the three elementary types.

10.14.1 Cascade Interconnection

When the systems are connected in series, as depicted in Fig. 10.26(b), then the transfer function of the overall system is the product of the individual transfer functions. This result can also be proved by observing that in Fig. 10.26(b):

$$H(z) = \frac{Y(z)}{X(z)} = \frac{W(z)}{X(z)} \frac{Y(z)}{W(z)} = H_1(z)H_2(z)$$

We can extend this result to any number of systems connected in cascade. It follows from this discussion that the systems connected in cascade can be interchanged without affecting the overall transfer function. This commutation property of LTI systems follows from the commutative and associative property of convolution (see Section 2.3).

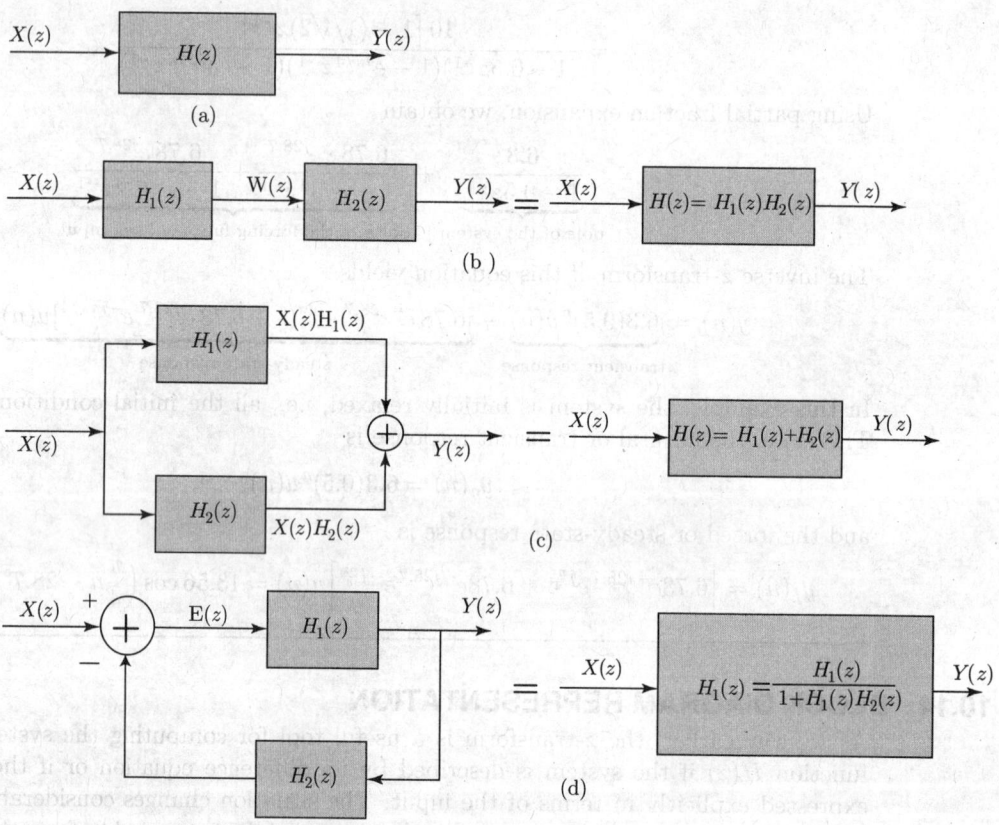

Fig. 10.26 Elementary connection of block diagrams

10.14.2 Parallel Interconnection

When the systems are connected in parallel, as depicted in Fig. 10.26(c), then the transfer function of the overall system is the sum of the individual transfer functions. This result can also be proved by observing that in Fig. 10.26(c):

$$Y(z) = X(z)H_1(z) + X(z)H_2(z)$$

$$Y(z) = X(z)[H_1(z) + H_2(z)]$$

$$\frac{Y(z)}{X(z)} = H(z) = H_1(z) + H_2(z)$$

We can extend this result to any number of systems connected in parallel.

10.14.3 Feedback Interconnection

When the output is fed back to the input, as shown in Fig. 10.26(d), the overall transfer function $H(z) = Y(z)/X(z)$ can be computed as follows. The input to the adder are $X(z)$ and $-H_2(z)Y(z)$. Therefore, $E(z)$, the output of the adder, is

$$E(z) = X(z) - H_2(z)Y(z)$$

But

$$Y(z) = H_1(z)E(z)$$

$$= H_1(z)[X(z) - H_2(z)Y(z)]$$

$$Y(z) = H_1(z)X(z) - H_1(z)H_2(z)Y(z)$$

$$Y(z)[1 + H_1(z)H_2(z)] = H_1(z)X(z)$$

$$\frac{Y(z)}{X(z)} = H(z) = \frac{H_1(z)}{1 + H_1(z)H_2(z)} \tag{10.90}$$

Therefore the feedback loop can be replaced by a single block with the transfer function shown in Eq. (10.90). The transfer function $H_1(z)$ is the forward path transfer function, and $H_2(z)$ is the feedback path transfer function.

10.15 SYSTEM REALIZATION

We now develop a systematic method for realization (or implementation) of an arbitrary Nth-order transfer function. The most general Nth-order transfer function is given by

$$H(z) = \frac{b_0 z^N + b_1 z^{N-1} + \cdots + b_{N-1} z + b_N}{z^N + a_1 z^{N-1} + \cdots + a_{N-1} z + a_N}$$

Since realization is basically a synthesis problem, there is no unique way of realizing a system. A given transfer function can be realized in many different ways. A transfer function $H(z)$ can be conveniently represented in block diagram from using the basic building blocks representing the unit delay, the multiplier, the adder, and the pick-off node as shown in Fig. 10.27.

There are several advantages in representing the system in block diagram form:

1. It is easy to write down the computational algorithm by inspection.
2. It is easy to analyse the block diagram to determine the explicit relation between the output and input.

3. It is easy to manipulate a block diagram to derive other equivalent block diagrams yielding different computational algorithms.

4. It is easy to determine the hardware requirements.

5. It is easier to develop block diagram representations from the transfer function directly leading to a variety of equivalent representations.

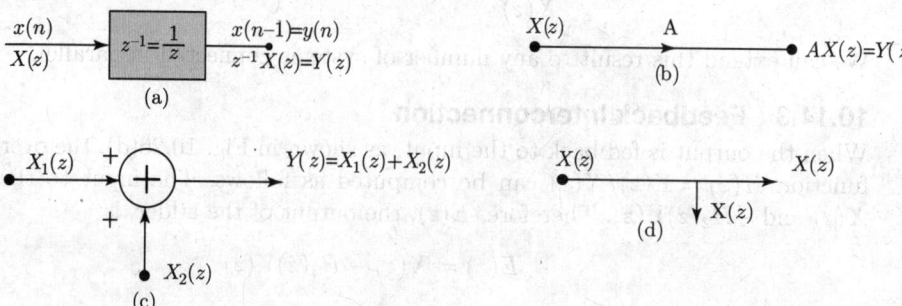

(a)

(b)

(c)

(d)

Fig. 10.27 Basic building blocks representing (a) unit-delay, (b) multiplier, (c) adder, and (d) pick-off node

10.15.1 Direct Form I Realization

An Nth-order transfer function is characterized by $2N + 1$ unique coefficients and, in general, requires $2N + 1$ multipliers and $2N$ two-input adders for implementation. Structures in which the multipliers coefficients are precisely the coefficients of the transfer function are called *direct form* structures.

Consider for simplicity a third-order system characterized by a transfer function:

$$H(z) = \frac{b_0 z^3 + b_1 z^2 + b_2 z + b_3}{z^3 + a_1 z^2 + a_2 z + a_3}$$

$$H(z) = \frac{b_0 + (b_1/z) + (b_2/z^2) + (b_3/z^3)}{1 + (a_1/z) + (a_2/z^2) + (a_3/z^3)}$$

We can express $H(z)$ as

$$H(z) = \underbrace{\left(b_0 + \frac{b_1}{z} + \frac{b_2}{z^2} + \frac{b_3}{z^3}\right)}_{H_1(z)} \underbrace{\left(\frac{1}{1 + (a_1/z) + (a_2/z^2) + (a_3/z^3)}\right)}_{H_2(z)}$$

We can realize $H(z)$ as a cascade of transfer function $H_1(z)$ followed by $H_2(z)$, as depicted in Fig. 10.28(a), where

$$H_1(z) = \frac{W(z)}{X(z)} = b_0 + \frac{b_1}{z} + \frac{b_2}{z^2} + \frac{b_3}{z^3} \qquad (10.91)$$

$$W(z) = \left(b_0 + \frac{b_1}{z} + \frac{b_2}{z^2} + \frac{b_3}{z^3}\right) X(z) \qquad (10.92)$$

and

$$H_2(z) = \frac{Y(z)}{W(z)} = \frac{1}{1 + (a_1/z) + (a_2/z^2) + (a_3/z^3)} \qquad (10.93)$$

$$Y(z) = W(z) - \left(\frac{a_1}{z} + \frac{a_2}{z^2} + \frac{a_3}{z^3}\right)Y(z) \qquad (10.94)$$

We shall first realize $H_1(z)$ given by Eq. (10.91). Equation (10.92) shows that the output $W(z)$ can be synthesized by adding the input $b_0 X(z)$ to $b_1(X(z)/z)$, $b_2(X(z)/z^2)$, and $b_3(X(z)/z^3)$. Because the transfer function of a unit delay is $z^{-1} = 1/z$, the signals $X(z)/z$, $X(z)/z^2$, and $X(z)/z^3$ can be obtained by a successive delay of the input $x(n)$. The left-half section of Fig. 10.28(b) shows the realization of $H_1(z)$.

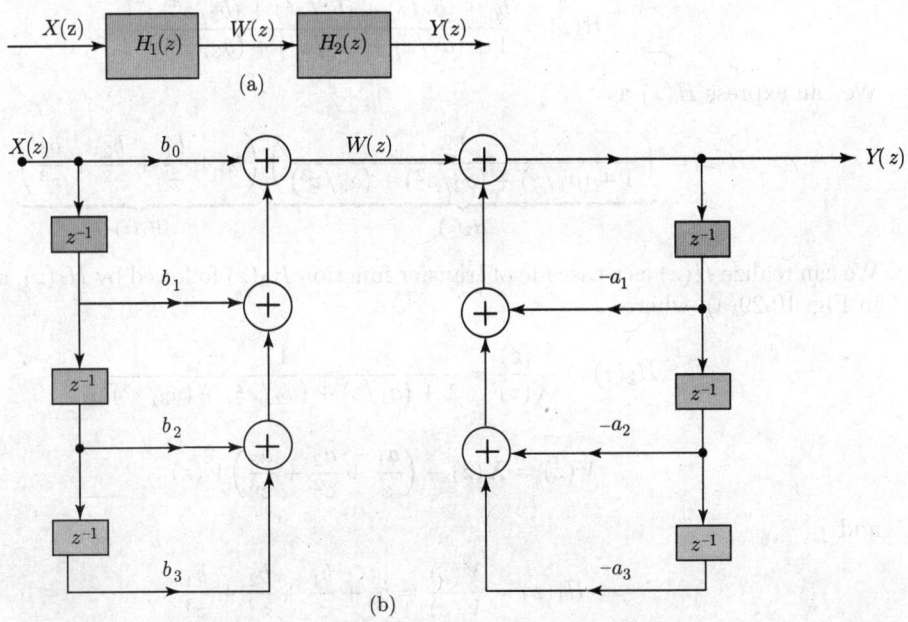

Fig. 10.28   (a) Realization of a transfer function in two steps and (b) direct form-I realization of a third-order discrete-time LTI system

We next consider the realization of $H_2(z)$ given by Eq. (10.93). Equation (10.94) shows that the output $Y(z)$ can be synthesized by subtracting $a_1(Y(z)/z)$, $a_2(Y(z)/z^2)$, and $a_3(Y(z)/z^3)$ from $W(z)$. To obtain signals $(Y(z)/z)$, $(Y(z)/z^2)$, and $(Y(z)/z^3)$, we assume that we already have the desired output $Y(z)$. Successive delay of $Y(z)$ yields the needed signal $Y(z)/z$, $Y(z)/z^2$, and $Y(z)/z^3$. The right-half section of Fig. 10.28(b) shows the realization of $H_2(z)$. We can generalize this procedure, known as the *direct form I* realization, for any value of N. This procedure requires $2N$ delay units to realize an Nth-order transfer function. This realization is *noncanonic* since it employs six delay units to implement a third-order transfer function. *A realization is canonic if the number of delay units used in the realization is equal to the order of the transfer function realized. Thus canonic realization has no redundant delay units.*

10.15.2 Direct Form II (or Canonic) Realization

In the direct form I, we realize $H(z)$ by implementing $H_1(z)$ followed by $H_2(z)$, as shown in Fig. 10.28(a). We can also realize $H(z)$, as shown in Fig. 10.29(a), where $H_2(z)$ is followed by $H_1(z)$. This procedure is known as the *direct form II* realization.

The direct form I realization [Fig. 10.28(b)] implements zeros first [the left-half section represented by $H_1(z)$] followed by realization of poles [the right-half section represented by $H_2(z)$] of $H(z)$. In contrast, the direct form II realization implements poles first followed by zeros.

Again consider for simplicity a third-order system characterized by a transfer function:

$$H(z) = \frac{b_0 z^3 + b_1 z^2 + b_2 z + b_3}{z^3 + a_1 z^2 + a_2 z + a_3}$$

$$H(z) = \frac{b_0 + (b_1/z) + (b_2/z^2) + (b_3/z^3)}{1 + (a_1/z) + (a_2/z^2) + (a_3/z^3)}$$

We can express $H(z)$ as

$$H(z) = \underbrace{\left(\frac{1}{1 + (a_1/z) + (a_2/z^2) + (a_3/z^3)} \right)}_{H_2(z)} \underbrace{\left(b_0 + \frac{b_1}{z} + \frac{b_2}{z^2} + \frac{b_3}{z^3} \right)}_{H_1(z)}$$

We can realize $H(z)$ as a cascade of transfer function $H_2(z)$ followed by $H_1(z)$, as depicted in Fig. 10.29(a), where

$$H_2(z) = \frac{V(z)}{X(z)} = \frac{1}{1 + (a_1/z) + (a_2/z^2) + (a_3/z^3)} \tag{10.95}$$

$$V(z) = X(z) - \left(\frac{a_1}{z} + \frac{a_2}{z^2} + \frac{a_3}{z^3} \right) V(z) \tag{10.96}$$

and

$$H_1(z) = \frac{Y(z)}{V(z)} = b_0 + \frac{b_1}{z} + \frac{b_2}{z^2} + \frac{b_3}{z^3} \tag{10.97}$$

$$Y(z) = \left(b_0 + \frac{b_1}{z} + \frac{b_2}{z^2} + \frac{b_3}{z^3} \right) V(z) \tag{10.98}$$

The left-half section of Fig. 10.29(b) shows the realization of $H_2(z)$ and the right-half section shows the realization of $H_1(z)$.

We observe that in Fig. 10.29(b) the signal variables at nodes 1 and 1' are the same, and hence, the two top delay units can be shared. Likewise, the signal variables at nodes 2 and 2' are the same, which permits the sharing of the two middle delay units. Following the same argument, we can share the delay units leading to the final structure shown in Fig. 10.29(c).

This implementation halves the number of delay units to N, and thus more efficient in hardware utilization. This is the direct form II realization. This realization is *canonic* since it employs N delay units to implement a Nth-order transfer function.

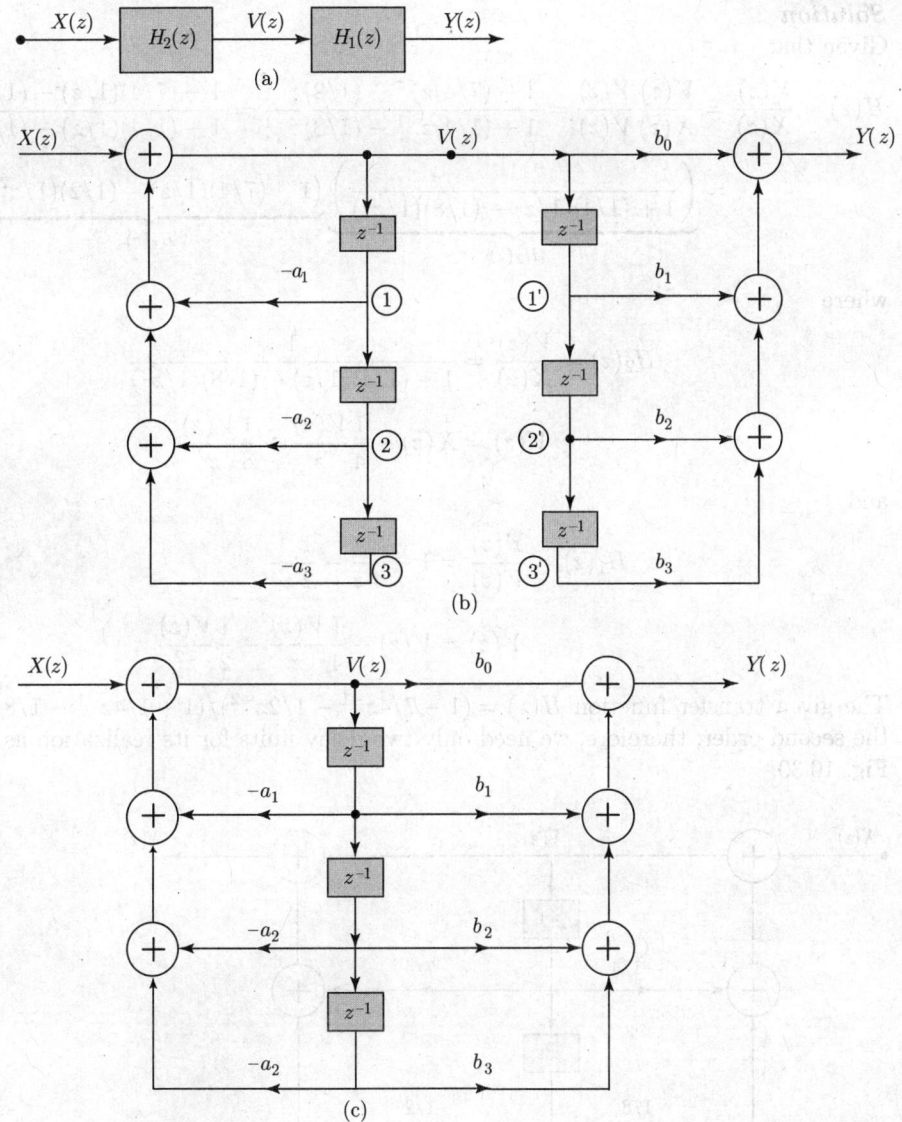

Fig. 10.29 Direct form II realization of a third-order LTI system

Example 10.65 Find the canonic (direct form II) realization of

$$H(z) = \frac{1 - (\frac{7}{4})z^{-1} - (\frac{1}{2})z^{-2}}{1 + (\frac{1}{4})z^{-1} - (\frac{1}{8})z^{-2}}$$

Solution

Given that

$$H(z) = \frac{Y(z)}{X(z)} = \frac{V(z)}{X(z)}\frac{Y(z)}{V(z)} = \frac{1 - (7/4)z^{-1} - (1/2)z^{-2}}{1 + (1/4)z^{-1} - (1/8)z^{-2}} = \frac{1 - (7/4)(1/z) - (1/2)(1/z^2)}{1 + (1/4)(1/z) - (1/8)(1/z^2)}$$

$$= \underbrace{\left(\frac{1}{1 + (1/4)(1/z) - (1/8)(1/z^2)}\right)}_{H_2(z)} \underbrace{\left(1 - (7/4)(1/z) - (1/2)(1/z^2)\right)}_{H_1(z)}$$

where

$$H_2(z) = \frac{V(z)}{X(z)} = \frac{1}{1 + (1/4)(1/z) - (1/8)(1/z^2)}$$

$$V(z) = X(z) - \frac{1}{4}\frac{V(z)}{z} + \frac{1}{8}\frac{V(z)}{z^2}$$

and

$$H_1(z) = \frac{Y(z)}{V(z)} = 1 - \frac{7}{4}\frac{1}{z} - \frac{1}{2}\frac{1}{z^2}$$

$$Y(z) = V(z) - \frac{7}{4}\frac{V(z)}{z} - \frac{1}{2}\frac{V(z)}{z^2}$$

The given transfer function $H(z) = (1 - 7/4z^{-1} - 1/2z^{-2})/(1 + 1/4z^{-1} - 1/8z^{-2})$ is of the second order; therefore, we need only two delay units for its realization as shown in Fig. 10.30.

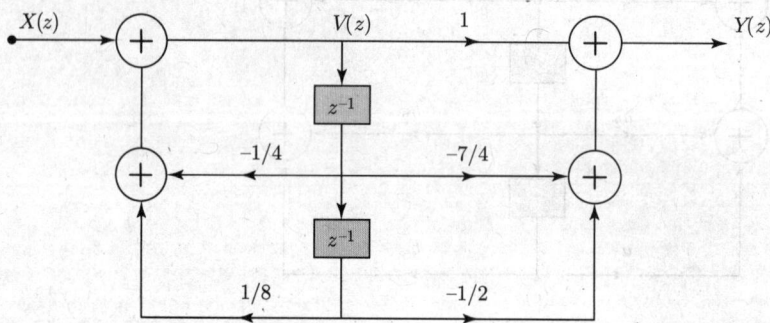

Fig. 10.30 Direct form II realization of $H(z)$ of Example 10.65

10.15.3 Cascade and Parallel Realization

By expressing the numerator polynomial $N(z)$ and the denominator polynomial $D(z)$ of the transfer function, $H(z)$ as a product of polynomials of lower degree is often realized as a cascade of low-order systems. Consider, for example, $H(z) = N(z)/D(z)$ expressed as

$$H(z) = \frac{N_1(z)N_2(z)N_3(z)}{D_1(z)D_2(z)D_3(z)}$$

Various different cascade realizations of $H(z)$ can be obtained by different pole-zero polynomial pairings. Some examples of such realizations are shown in Fig. 10.31(a). Additional cascade realizations are obtained by simply changing the ordering of the systems. Figure 10.31(b) illustrates examples of different structures obtained by changing the ordering of the systems.

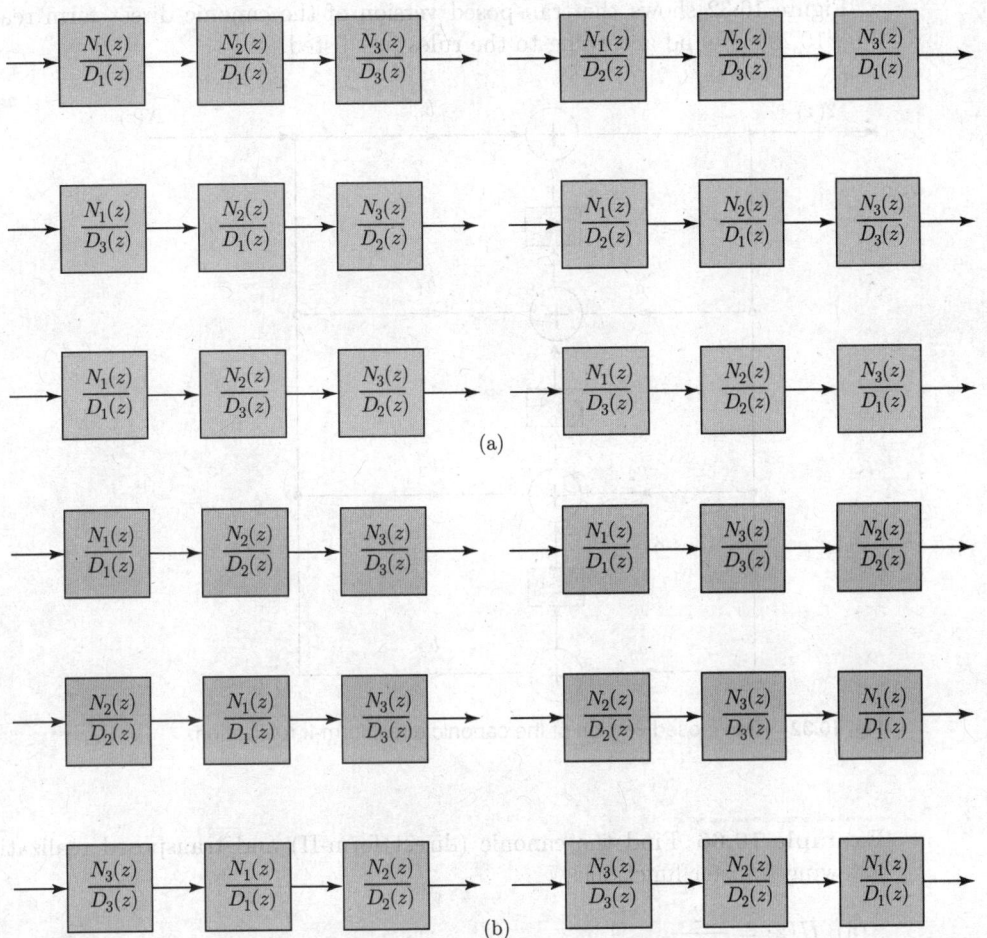

Fig. 10.31 Different cascade realizations obtained by (a) different pole-zero pairings and (b) changing the ordering of the systems

A system can be realized in a parallel form by making use of partial fraction expansion of the transfer function. From a practical viewpoint, parallel and cascade forms are preferable because parallel and certain cascade forms are numerically less sensitive than canonic direct form to small parameter variations in the system.

10.15.4 Transposed Realization

Two realizations are said to be *equivalent* if they have the same transfer function. A simple way to generate an equivalent structure from a given realization is via the *transpose operation*, which is as follows:

1. Reverse all paths.
2. Replace pick-off nodes by adders, and vice versa.
3. Interchange the input and output nodes.

Figure 10.32 shows the transposed version of the canonic direct form realization in Fig. 10.29(c) found according to the rules just listed.

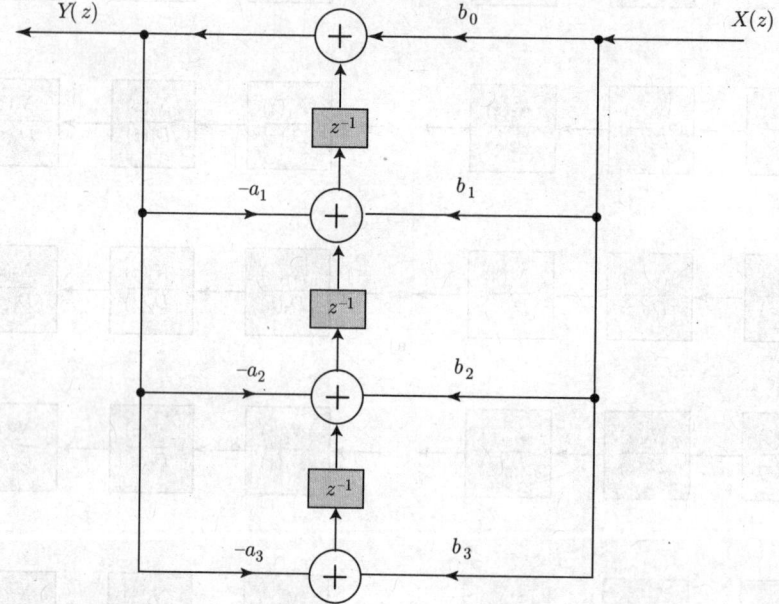

Fig. 10.32   Transposed version of the canonic direct form-II realization.

Example 10.66 Find the canonic (direct form-II) and transposed realization of the following transfer functions:

(a) $H(z) = \dfrac{2}{z+3}$

(b) $H(z) = \dfrac{z}{z+7}$

(c) $H(z) = \dfrac{4z+28}{z+1}$

(d) $H(z) = \dfrac{4z+28}{z^2+6z+5}$

Solution
(a) Given that

$$H(z) = \frac{Y(z)}{X(z)} = \frac{V(z)}{X(z)}\frac{Y(z)}{V(z)} = \frac{2}{z+3} = \frac{\frac{2}{z}}{1+3/z} = \underbrace{\frac{1}{1+3/z}}_{H_2(z)} \underbrace{\frac{2}{z}}_{H_1(z)}$$

where

$$H_2(z) = \frac{V(z)}{X(z)} = \frac{1}{1+\dfrac{3}{z}}$$

$$V(z) = X(z) - 3\frac{V(z)}{z}$$

and

$$H_1(z) = \frac{Y(z)}{V(z)} = \frac{2}{z}$$

$$Y(z) = 2\frac{V(z)}{z}$$

The given transfer function $H(z) = (2/z + 3)$ is of the first order; therefore, we need only one delay unit for its realization. Figure 10.33(a) shows the canonic (direct form II), and Fig. 10.33(b) its transpose.

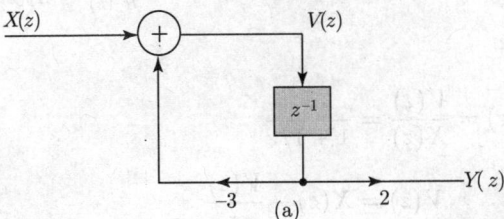

(a)

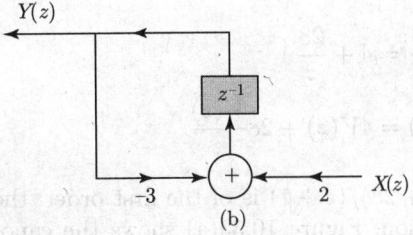

(b)

Fig. 10.33

(b) Given that

$$H(z) = \frac{z}{z+7}$$

$$H(z) = \frac{Y(z)}{X(z)} = \frac{1}{1+7/z}$$

$$Y(z) = X(z) - 7\frac{Y(z)}{z}$$

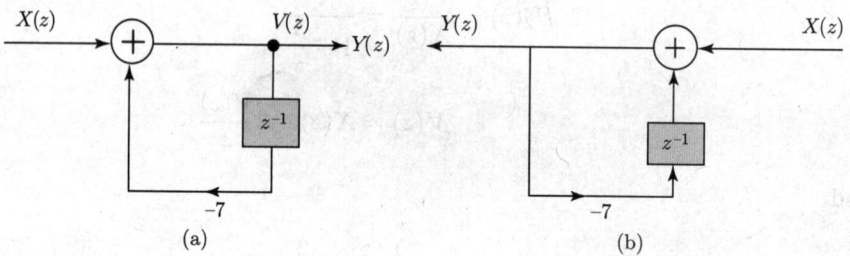

Fig. 10.34

The given transfer function $H(z) = (z/z + 7)$ is of the first order; therefore, we need only one delay unit for its realization. Figure 10.34(a) shows the canonic (direct form II), and Fig. 10.34(b) its transpose.

(c) Given that

$$H(z) = \frac{Y(z)}{X(z)} = \frac{V(z)}{X(z)}\frac{Y(z)}{V(z)} = \frac{4z+28}{z+1} = \frac{4+28/z}{1+1/z} = \underbrace{\frac{1}{1+1/z}}_{H_2(z)}\underbrace{4+\frac{28}{z}}_{H_1(z)}$$

where

$$H_2(z) = \frac{V(z)}{X(z)} = \frac{1}{1+1/z}$$

$$V(z) = X(z) - \frac{V(z)}{z}$$

and

$$H_1(z) = \frac{Y(z)}{V(z)} = 4 + \frac{28}{z}$$

$$Y(z) = 4V(z) + 28\frac{V(z)}{z}$$

The given transfer function $H(z) = (4z + 28)/(z + 1)$ is of the first order; therefore, we need only one delay unit for its realization. Figure 10.35(a) shows the canonic (direct form II), and Fig. 10.35(b) its transpose.

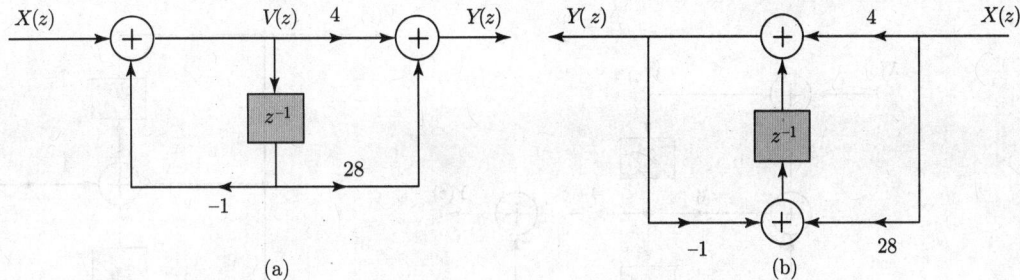

Fig. 10.35

(d) Given that

$$H(z) = \frac{Y(z)}{X(z)} = \frac{V(z)}{X(z)}\frac{Y(z)}{V(z)} = \frac{4z + 28}{z^2 + 6z + 5} = \frac{(4/z) + (28/z^2)}{1 + (6/z) + (5/z^2)}$$

$$= \underbrace{\left(\frac{1}{1 + (6/z) + (5/z^2)}\right)}_{H_2(z)}\underbrace{\left(\frac{4}{z} + \frac{28}{z^2}\right)}_{H_1(z)}$$

where

$$H_2(z) = \frac{V(z)}{X(z)} = \frac{1}{1 + (6/z) + (5/z^2)}$$

$$V(z) = X(z) - 6\frac{V(z)}{z} - 5\frac{V(z)}{z^2}$$

and

$$H_1(z) = \frac{Y(z)}{V(z)} = \frac{4}{z} + \frac{28}{z^2}$$

$$Y(z) = 4\frac{V(z)}{z} + 28\frac{V(z)}{z^2}$$

The given transfer function $H(z) = (4z + 28)/(z^2 + 6z + 5)$ is of the second order; therefore, we need only two delay units for its realization. Figure 10.36(a) shows the canonic (direct form II), and Fig. 10.36(b) its transpose.

SOLVED EXAMPLES

Example 10.67 Determine the z-transform for each of the following signals. Sketch the pole-zero plot and indicate the ROC. Indicate whether or not the DTFT of the signals exist.

(a) $x(n) = a^n u(-n - n_0), \quad a > 1$

(b) $x(n) = \left(\frac{1}{2}\right)^{n+1} u(n + 3)$

(c) $x(n) = n\left(\frac{1}{2}\right)^{|n|}$

(d) $x(n) = |n|\left(\frac{1}{2}\right)^{|n|}$

Solution

(a) By definition, we have

$$X(z) = \sum_{n=-\infty}^{\infty} x(n)z^{-n}$$

$$= \sum_{n=-\infty}^{\infty} a^n u(-n - n_0)z^{-n}$$

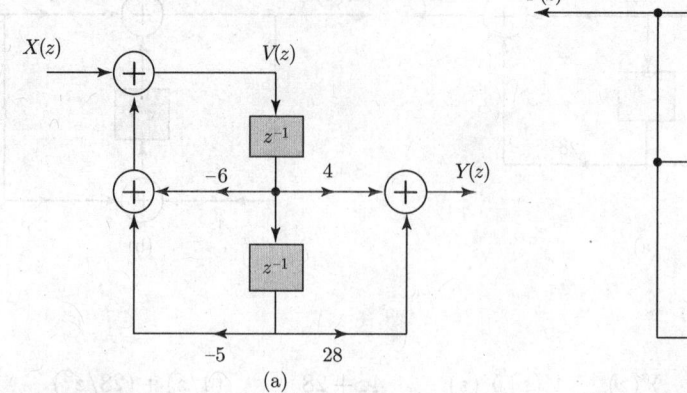

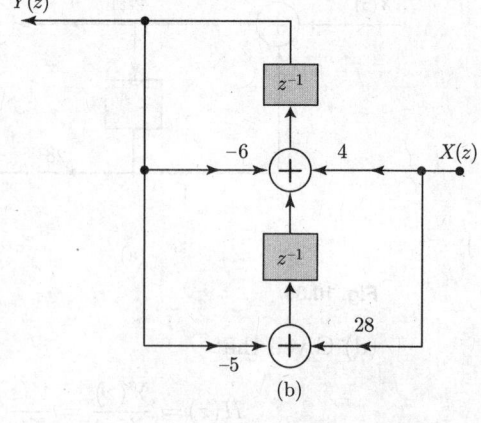

Fig. 10.36

Since

$$u(-n - n_0) = \begin{cases} 1, & (-n - n_0) \geq 0 \longrightarrow n \leq n_0 \\ 0, & (-n - n_0) < 0 \longrightarrow n > -n_0 \end{cases}$$

we obtain

$$X(z) = \sum_{n=-\infty}^{-n_0} a^n z^{-n}$$

$$= \sum_{n=\infty}^{n_0} a^{-n} z^n$$

$$X(z) = \sum_{n=n_0}^{\infty} \left(a^{-1} z\right)^n$$

A change of variables is performed by letting $m = n - n_0$, which also yields $n = m + n_0$, $m = 0$ as $n = n_0$, and $m = -\infty$ as $n = -\infty$. Therefore,

$$X(z) = \sum_{m=0}^{\infty} \left(a^{-1} z\right)^{m+n_0}$$

$$= \left(a^{-1} z\right)^{n_0} \sum_{m=0}^{\infty} \left(a^{-1} z\right)^m$$

$$= \left(a^{-1} z\right)^{n_0} \frac{1}{1 - a^{-1} z}, \quad |a^{-1} z| < 1$$

$$X(z) = -\frac{\left(a^{-1} z\right)^{n_0 - 1}}{1 - a z^{-1}}, \quad |z| < |a|$$

The pole-zero plot and the ROC is shown in Fig. 10.37(a). Since the ROC includes the unit circle, the DTFT of the signal exist.

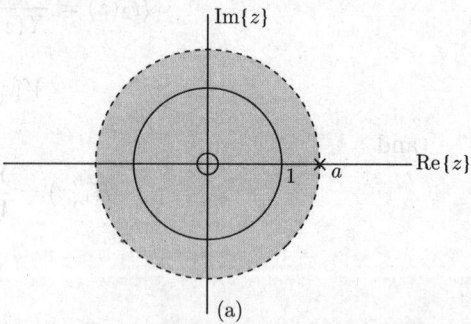

Fig. 10.37(a) Pole-zero plot and ROC for Example 10.67(a)

(b) Given that

$$x(n) = \left(\frac{1}{2}\right)^{n+1} u(n+3)$$

$$= 4\left(\frac{1}{2}\right)^{n+3} u(n+3)$$

We know that [Eq. (10.12)]

$$\left(\frac{1}{2}\right)^n u(n) \longleftrightarrow \frac{1}{1 - \frac{1}{2} z^{-1}}, \quad |z| > \frac{1}{2}$$

Using time shifting property, we obtain

$$4\left(\frac{1}{2}\right)^{n+3} u(n+3) \longleftrightarrow \frac{4z^3}{1-\frac{1}{2}z^{-1}}, \quad |z| > \frac{1}{2}$$

The pole-zero plot and the ROC is shown in Fig. 10.37(b). Since the ROC includes the unit circle, the DTFT of the signal exist.

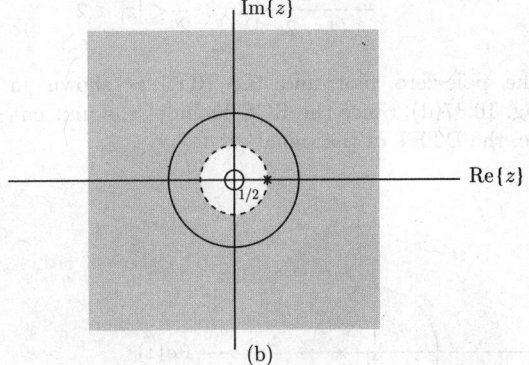

Fig. 10.37(b)

(c) Given that

$$x(n) = n\left(\frac{1}{2}\right)^{|n|}$$

$$= \begin{cases} n\left(\frac{1}{2}\right)^n, & n \geq 0 \\ n\left(\frac{1}{2}\right)^{-n}, & n < 0 \end{cases}$$

$$= n\left(\frac{1}{2}\right)^n u(n) + n\left(\frac{1}{2}\right)^{-n} u(-n-1)$$

$$x(n) = n\left(\frac{1}{2}\right)^n u(n) + n\left(2\right)^n u(-n-1)$$

We know that

$$\left(\frac{1}{2}\right)^n u(n) \longleftrightarrow \frac{1}{1-\frac{1}{2}z^{-1}}, \quad |z| > \frac{1}{2}$$

and

$$(2)^n u(-n-1) \longleftrightarrow -\frac{1}{1-2z^{-1}}, \quad |z| < 2$$

Using differentiation in the z-domain property [Eq. (10.31)], we obtain

$$n\left(\frac{1}{2}\right)^n u(n) \longleftrightarrow -z\frac{d}{dz}\left(\frac{1}{1-\frac{1}{2}z^{-1}}\right)$$

$$\longleftrightarrow \frac{\frac{1}{2}z^{-1}}{\left(1-\frac{1}{2}z^{-1}\right)^2}$$

and

$$n(2)^n u(-n-1) \longleftrightarrow -z\frac{d}{dz}\left(-\frac{1}{1-2z^{-1}}\right)$$

$$\longleftrightarrow -\frac{2z^{-1}}{\left(1-2z^{-1}\right)^2}$$

Therefore,

$$X(z) = \mathcal{Z}\left[n\left(\frac{1}{2}\right)^n u(n)\right] + \mathcal{Z}\left[n\left(2\right)^n u(-n-1)\right]$$

$$= \frac{\frac{1}{2}z^{-1}}{\left(1-\frac{1}{2}z^{-1}\right)^2} - \frac{2z^{-1}}{\left(1-2z^{-1}\right)^2}, \quad \frac{1}{2} < |z| < 2$$

The pole-zero plot and the ROC is shown in Fig. 10.37(c). Since the ROC includes the unit circle, the DTFT of the signal exist.

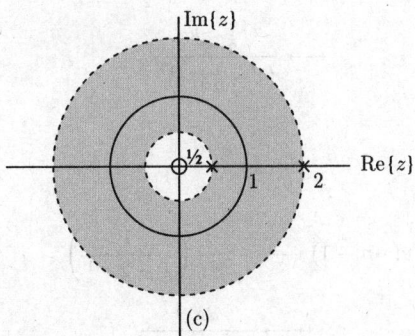

Fig. 10.37(c) Pole-zero plot and ROC for Example 10.67(c)

(d) Given that

$$x(n) = |n| \left(\frac{1}{2}\right)^{|n|}$$

$$= \begin{cases} n \left(\frac{1}{2}\right)^n, & n \geq 0 \\ -n \left(\frac{1}{2}\right)^{-n}, & n < 0 \end{cases}$$

$$= n \left(\frac{1}{2}\right)^n u(n) - n \left(\frac{1}{2}\right)^{-n} u(-n-1)$$

$$x(n) = n \left(\frac{1}{2}\right)^n u(n) - n \left(2\right)^n u(-n-1)$$

We know that

$$\left(\frac{1}{2}\right)^n u(n) \longleftrightarrow \frac{1}{1 - \frac{1}{2}z^{-1}}, \quad |z| > \frac{1}{2}$$

and

$$-(2)^n u(-n-1) \longleftrightarrow \frac{1}{1 - 2z^{-1}}, \quad |z| < 2$$

Using differentiation in the z-domain property [Eq. (10.31)], we obtain

$$n \left(\frac{1}{2}\right)^n u(n) \longleftrightarrow -z \frac{d}{dz} \left(\frac{1}{1 - \frac{1}{2}z^{-1}}\right)$$

$$\longleftrightarrow \frac{\frac{1}{2}z^{-1}}{\left(1 - \frac{1}{2}z^{-1}\right)^2}$$

and

$$-n(2)^n u(-n-1) \longleftrightarrow -z \frac{d}{dz} \left(\frac{1}{1 - 2z^{-1}}\right)$$

$$\longleftrightarrow \frac{2z^{-1}}{\left(1 - 2z^{-1}\right)^2}$$

Therefore,

$$X(z) = \mathcal{Z}\left[n \left(\frac{1}{2}\right)^n u(n)\right]$$

$$+ \mathcal{Z}\left[-n(2)^n u(-n-1)\right]$$

$$= \frac{\frac{1}{2}z^{-1}}{\left(1 - \frac{1}{2}z^{-1}\right)^2}$$

$$+ \frac{2z^{-1}}{\left(1 - 2z^{-1}\right)^2}, \quad \frac{1}{2} < |z| < 2$$

The pole-zero plot and the ROC is shown in Fig. 10.37(d). Since the ROC includes the unit circle, the DTFT of the signal exist.

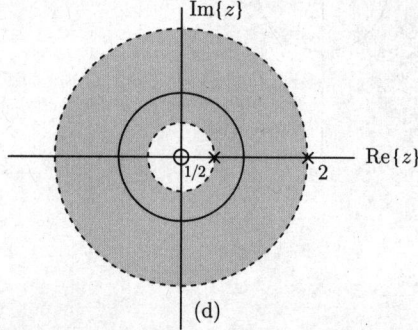

Fig. 10.37(d) Pole-zero plot and ROC for Example 10.67(d)

Example 10.68 Determine the z-transform of the signal

$$x(n) = \begin{cases} n, & 0 \leq n \leq N-1 \\ N, & N \leq n \end{cases}$$

Solution
Given that

$$x(n) = \begin{cases} n, & 0 \leq n \leq (N-1) \\ N, & N \leq n \end{cases}$$

$$= n[u(n) - u(n-N)] + Nu(n-N)$$

$$= nu(n) - nu(n-N) + Nu(n-N)$$

$$x(n) = nu(n) - (n-N)u(n-N)$$

We know that [Eq. (10.34)]

$$nu(n) \longleftrightarrow \frac{z^{-1}}{(1 - z^{-1})^2}, \quad |z| > 1$$

Using the time-shifting property, we obtain

$$(n - N)u(n - N) \longleftrightarrow z^{-N}\frac{z^{-1}}{(1 - z^{-1})^2}, \quad |z| > 1$$

Therefore,

$$X(z) = \frac{z^{-1}}{(1 - z^{-1})^2} - z^{-N}\frac{z^{-1}}{(1 - z^{-1})^2}$$

$$X(z) = \frac{z^{-1}(1 - z^{-N})}{(1 - z^{-1})^2}, \quad |z| > 1$$

Example 10.69 Let

$$x(n) = (-1)^n u(n) + a^n u(-n - n_0)$$

Determine the constraints on the complex number a and the integer n_0, given that the ROC of $X(z)$ is the region $1 < |z| < 2$.

Solution

Given that

$$x(n) = (-1)^n u(n) + a^n u(-n - n_0)$$

We know that

$$(-1)^n u(n) \longleftrightarrow \frac{1}{1 + z^{-1}}, \quad |z| > 1$$

and

$$a^n u(-n - n_0) \longleftrightarrow -\frac{(a^{-1}z)^{n_0 - 1}}{1 - az^{-1}}, \quad |z| < |a|$$

Therefore,

$$X(z) = \mathcal{Z}\left[(-1)^n u(n)\right] + \mathcal{Z}\left[a^n u(-n - n_0)\right]$$

$$X(z) = \frac{1}{1 + z^{-1}} - \frac{(a^{-1}z)^{n_0 - 1}}{1 - az^{-1}}, \quad 1 < |z| < |a|$$

Given that the ROC of $X(z)$ is the region $1 < |z| < 2$, therefore, $a = 2$ and n_0 can take on any value.

Example 10.70 Let $x(n)$ be an absolutely summable signal with rational z-transform $X(z)$. If $X(z)$ is known to have a pole at $z = 1/2$, could $x(n)$ be

(a) a finite-duration signal?

(b) a left-sided signal?

(c) a right-sided signal?

(d) a two-sided signal?

Solution

(a) No. We know that for a finite-length signal, the ROC is the entire z-plane. Therefore, there can be no poles in the finite z-plane for a finite-length signal.

(b) No. Since the signal is absolutely summable, the ROC must include the unit circle. Since it is given that the signal has a pole at $z = 1/2$, the ROC can never be the region $|z| < 1/2$. Therefore, the signal cannot be left-sided.

(c) Yes. Since the signal is absolutely summable, the ROC must include the unit circle. Since it is given that the signal has a pole at $z = 1/2$, a valid ROC would be the region $|z| > 1/2$. Therefore, the signal could be right-sided.

(d) Yes. Since the signal is absolutely summable, the ROC must include the unit circle. Clearly, we may define the ROC which is a ring in the z-plane and includes the unit circle. Therefore, the signal could be two-sided.

Example 10.71 Suppose we are given the following five facts about a particular LTI system with impulse response $h(n)$ and z-transform $H(z)$:

1. $h(n)$ is real.

2. $h(n)$ is right-sided.

3. $\lim_{z \to \infty} H(z) = 1$.

4. $H(z)$ has two zeros.

5. $H(z)$ has one of its poles at a non-real location on the circle defined by $|z| = 3/4$.

Answer the following two questions:

(a) Is this system causal?

(b) Is this system stable.

Solution

(a) Since $\lim_{z \to \infty} H(z) = 1$, $H(z)$ has no poles at infinity. Furthermore, since $h(n)$ is given to be right-sided, $h(n)$ has to be causal.

(b) Since $\lim_{z \to \infty} H(z) = 1$ and $h(n)$ is causal, the numerator and denominator polynomials of $H(z)$ have the same order. Since $H(z)$ has two zeros, we may conclude that it also has two poles. Since $h(n)$ is real, the poles must occur in conjugate pairs. Also, it is given that one of the poles lies on the circle defined by $|z| = 3/4$. Therefore, the other pole also lies on the same circle. Clearly, the ROC for $H(z)$ is $|z| > 3/4$, and will include the unit circle. Therefore, we may conclude that the system is stable.

Example 10.72 Determine the inverse z-transform of

$$X(z) = \frac{1}{1024}\left[\frac{1024 - z^{-10}}{1 - \frac{1}{2}z^{-1}}\right], \quad |z| > 0$$

Solution

Given that

$$X(z) = \frac{1}{1024}\left[\frac{1024 - z^{-10}}{1 - \frac{1}{2}z^{-1}}\right], \quad |z| > 0$$

$$X(z) = \frac{1}{1 - \frac{1}{2}z^{-1}} - \frac{1}{1024}\frac{z^{-10}}{1 - \frac{1}{2}z^{-1}}$$

The inverse z-transform of the above equation yields

$$x(n) = \left(\frac{1}{2}\right)^n u(n) - \frac{1}{1024}\left(\frac{1}{2}\right)^{n-10} u(n-10)$$

$$= \left(\frac{1}{2}\right)^n u(n)$$

$$- \frac{1}{1024}\left(\frac{1}{2}\right)^{-10}\left(\frac{1}{2}\right)^n u(n-10)$$

$$x(n) = \left(\frac{1}{2}\right)^n [u(n) - u(n-10)]$$

Example 10.73 If $H(z) = 1/(1 - \frac{1}{4}z^{-2})$ and $h(n) = A_1\alpha_1^n u(n) + A_2\alpha_2^n u(n)$, determine the values of A_1, α_1, A_2, and α_2.

Solution

Given that

$$H(z) = \frac{1}{1 - \frac{1}{4}z^{-2}}$$

$$= \frac{1}{\left(1 - \frac{1}{2}z^{-1}\right)\left(1 + \frac{1}{2}z^{-1}\right)}$$

Using the partial fraction expansion, we obtain

$$H(z) = \frac{\frac{1}{2}}{1 - \frac{1}{2}z^{-1}} + \frac{\frac{1}{2}}{1 + \frac{1}{2}z^{-1}}$$

The inverse z-transform of the above equation yields

$$h(n) = \frac{1}{2}\left(\frac{1}{2}\right)^n u(n) + \frac{1}{2}\left(-\frac{1}{2}\right)^n u(n)$$

$$= A_1\alpha_1^n u(n) + A_2\alpha_2^n u(n)$$

So,

$$A_1 = \frac{1}{2} \quad \alpha_1 = \frac{1}{2} \quad A_2 = \frac{1}{2} \quad \alpha_2 = \frac{1}{2}$$

Example 10.74

A causal signal $x(n)$ has the z-transform

$$X(z) = \sin(z^{-1})(1 + 3z^{-2} + 2z^{-4})$$

Find $x(11)$.

Solution

Given that

$$X(z) = \sin(z^{-1})(1 + 3z^{-2} + 2z^{-4})$$

$$= \left(z^{-1} - \frac{z^{-3}}{3!} + \frac{z^{-5}}{5!} - \frac{z^{-7}}{7!}\right.$$

$$\left. + \frac{z^{-9}}{9!} - \frac{z^{-11}}{11!} + \cdots\right)(1 + 3z^{-2} + 2z^{-4})$$

$$= \sum_{n=0}^{\infty} x(n)z^{-n}$$

Since $x(11)$ is simply the coefficient in front of z^{-11} in this power series expansion of $X(z)$, therefore

$$x(11) = -\frac{1}{11!} + \frac{3}{9!} - \frac{2}{7!}$$

Example 10.75 Determine the signal $x(n)$ with z-transform

$$X(z) = e^z + e^{1/z}$$

Solution

Given that

$$X(z) = e^z + e^{1/z}$$

$$= e^z + e^{z^{-1}}$$

$$= \left(1 + z + \frac{z^2}{2!} + \frac{z^3}{3!} + \cdots\right)$$

$$+ \left(1 + z^{-1} + \frac{z^{-2}}{2!} + \frac{z^{-3}}{3!} + \cdots\right)$$

$$= \left(1 + \sum_{n=-\infty}^{-1} \frac{1}{n!} z^{-n}\right) + \left(\sum_{n=0}^{\infty} \frac{1}{n!} z^{-n}\right)$$

$$= 1 + \left(\sum_{n=0}^{\infty} \frac{1}{n!} z^{-n} + \sum_{n=-\infty}^{-1} \frac{1}{n!} z^{-n}\right)$$

$$X(z) = 1 + \sum_{n=-\infty}^{\infty} \frac{1}{n!} z^{-n}$$

The inverse z-transform of the above equation yields

$$x(n) = \delta(n) + \frac{1}{n!}$$

Example 10.76 Let

$$y(n) = \left(\frac{1}{9}\right)^n u(n)$$

Determine two distinct signals such that each has a z-transform $X(z)$, which satisfies both of the following conditions:

1. $\dfrac{X(z) + X(-z)}{2} = Y(z^2)$

2. $X(z)$ has only one pole and only one zero in the z-plane.

Solution

Given that

$$y(n) = \left(\frac{1}{9}\right)^n u(n)$$

Taking the z-transform of the above equation, we obtain

$$Y(z) = \frac{1}{1 - \frac{1}{9}z^{-1}}, \quad |z| > \frac{1}{9}$$

$$Y(z^2) = \frac{1}{1 - \frac{1}{9}z^{-2}}$$

$$= \frac{1}{1 - \left(\frac{1}{3}z^{-1}\right)^2}$$

$$= \frac{1}{\left(1 - \frac{1}{3}z^{-1}\right)\left(1 + \frac{1}{3}z^{-1}\right)}$$

$$Y(z^2) = \frac{\frac{1}{2}}{1 - \frac{1}{3}z^{-1}} + \frac{\frac{1}{2}}{1 + \frac{1}{3}z^{-1}}$$

$$Y(z^2) = \frac{1}{2}\left(\frac{1}{1 - \frac{1}{3}z^{-1}} + \frac{1}{1 + \frac{1}{3}z^{-1}}\right), \quad |z| > \frac{1}{3}$$

Case-I Comparing the above equation with

$$Y(z^2) = \frac{1}{2}[X(z) + X(-z)]$$

we get

$$X(z) = \frac{1}{1 - \frac{1}{3}z^{-1}}, \quad |z| > \frac{1}{3}$$

The inverse z-transform of the above equation yields

$$x(n) = \left(\frac{1}{3}\right)^n u(n)$$

Case-II Consider the expression

$$Y(z^2) = \frac{1}{2}\left(\frac{1}{1 - \frac{1}{3}z^{-1}} + \frac{1}{1 + \frac{1}{3}z^{-1}}\right), \quad |z| > \frac{1}{3}$$

Comparing the above equation with

$$Y(z^2) = \frac{1}{2}[X(-z) + X(z)]$$

we get

$$X(z) = \frac{1}{1 + \frac{1}{3}z^{-1}}, \quad |z| > \frac{1}{3}$$

The inverse z-transform of the above equation yields

$$x(n) = \left(-\frac{1}{3}\right)^n u(n)$$

Example 10.77 Determine all possible signals that can have the following z-transform.

$$X(z) = \frac{1}{1 - 1.5z^{-1} + 0.5z^{-2}}$$

Solution
Given that

$$X(z) = \frac{1}{1 - 1.5z^{-1} + 0.5z^{-2}}$$

$$= \frac{z^2}{z^2 - 1.5z + 0.5}$$

$$\frac{X(z)}{z} = \frac{z}{(z - 0.5)(z - 1)}$$

Using the partial fraction expansion, we obtain

$$\frac{X(z)}{z} = \frac{-1}{z - 0.5} + \frac{2}{z - 1}$$

$$X(z) = \frac{-z}{z - 0.5} + \frac{2z}{z - 1}$$

$$X(z) = -\frac{1}{1 - 0.5z^{-1}} + \frac{2}{1 - z^{-1}}$$

$X(z)$ has two poles at $z = 0.5$ and $z = 1$. There are three possible choices of the ROCs: (i) $|z| > 1$, (ii) $0.5 < |z| < 1$, and (iii) $|z| < 0.5$.

Case-I ROC: $|z| > 1$ [shown in Fig. 10.38(a)]. Since the ROC, $|z| > 1$, is the region in the z-plane outside the outermost pole, all the poles correspond to causal (right-sided) signals. Now, consider

$$X(z) = -\frac{1}{1 - 0.5z^{-1}} + \frac{2}{1 - z^{-1}}, \quad |z| > 1$$

The inverse z-transform yields

$$x(n) = -(0.5)^n u(n) + 2u(n)$$

$$= [2 - (0.5)^n]u(n)$$

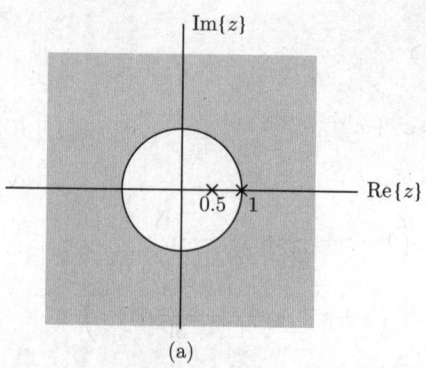

(a)

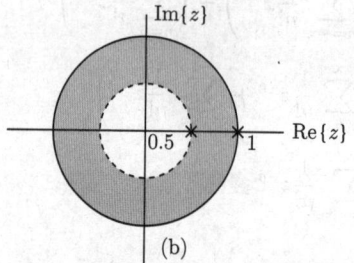

(b)

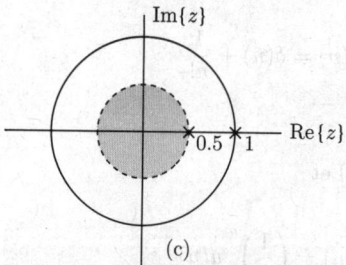

(c)

Fig. 10.38 Possible ROCs for Example 10.77

Case-II ROC: $0.5 < |z| < 1$ [shown in Fig. 10.38(b)]. The pole of the first term is at 0.5. The ROC has a radius greater than the pole at $z = 0.5$, so this pole corresponds to causal (right-sided) signal. Therefore,

$$-(0.5)^n u(n) \longleftrightarrow -\frac{1}{1 - 0.5z^{-1}}$$

The second term has a pole at $z = 1$. The ROC has a radius less than the pole at $z = 1$, so this pole corresponds to the anticausal (left-sided) signal. Therefore,

$$-2u(-n - 1) \longleftrightarrow \frac{2}{1 - z^{-1}}$$

and hence, we obtain

$$x(n) = -(0.5)^n u(n) - 2u(-n-1)$$

Case-III ROC: $|z| < 0.5$ [shown in Fig. 10.38(c)]. Since the ROC, $|z| < 0.5$, is the region in the z-plane inside the innermost pole, all the poles correspond to anticausal (right-sided) signals. Now, consider

$$X(z) = -\frac{1}{1-0.5z^{-1}} + \frac{2}{1-z^{-1}}, \quad |z| > 1$$

The inverse z-transform yields

$$x(n) = (0.5)^n u(-n-1) - 2u(-n-1)$$
$$= -[2 - (0.5)^n]u(-n-1)$$

Example 10.78

(a) State and prove the initial-value theorem for anticausal signals (i.e., $x(n) = 0$, $n > 0$).

(b) Show that if $x(n)$ is causal [i.e., $x(n) = 0$ for $n < 0$], then

$$x(1) = \lim_{z \to \infty} z[X(z) - x(0)]$$

Solution

(a) If $x(n)$ is anticausal [i.e., $x(n) = 0$ for $n > 0$], then

$$x(0) = \lim_{z \to 0} X(z)$$

Proof If $x(n)$ is anticausal [i.e., $x(n) = 0$ for $n > 0$], then its z-transform is given by

$$X(z) = \sum_{n=-\infty}^{0} x(n)z^{-n}$$

$$= x(0) + x(-1)z + x(-2)z + \cdots$$

$$\lim_{z \to 0} X(z) = \lim_{z \to 0}[x(0) + x(-1)z + x(-2)z + \cdots]$$

$$= x(0)$$

(b) If $x(n)$ is causal [i.e., $x(n) = 0$ for $n < 0$], then its z-transform is given by

$$X(z) = \sum_{n=0}^{\infty} x(n)z^{-n}$$

$$= x(0) + x(1)z^{-1} + x(2)z^{-2}$$
$$+ x(3)z^{-3} + \cdots$$

$$X(z) - x(0) = x(1)z^{-1} + x(2)z^{-2}$$
$$+ x(3)z^{-3} + \cdots$$

$$z[X(z) - x(0)] = x(1) + x(2)z^{-1}$$
$$+ x(3)z^{-2} + \cdots$$

$$\lim_{z \to \infty} z[X(z) - x(0)] = \lim_{z \to \infty}[x(1) + x(2)z^{-1}$$
$$+ x(3)z^{-2} + \cdots]$$

$$\lim_{z \to \infty} z[X(z) - x(0)] = x(1)$$

Example 10.79 Let $y(n)$ be a signal that is generated from a signal $x(n)$ as follows:

$$y(n) = \sum_{k=-\infty}^{n} k x(k)$$

(a) Show that $y(n)$ satisfies the time-varying difference equation

$$y(n) = y(n-1) + nx(n)$$

and show that

$$Y(z) = \frac{-z^2}{z-1} \frac{dX(z)}{dz}$$

where $X(z)$ and $Y(z)$ are the z-transforms of $x(n)$ and $y(n)$, respectively.

(b) Use this property to find the z-transform of

$$y(n) = \sum_{k=0}^{n} k \left(\frac{1}{3}\right)^k, \quad n \geq 0$$

Solution

(a) From the definition of $y(n)$, we see that

$$y(n-1) = \sum_{k=-\infty}^{n-1} k x(k)$$

Given that

$$y(n) = \sum_{k=-\infty}^{n} kx(k)$$

$$= nx(n) + \sum_{k=-\infty}^{n-1} kx(k)$$

$$y(n) = nx(n) + y(n-1)$$

$$y(n) - y(n-1) = x(n)$$

Taking the z-transform on both sides, we get

$$Y(z)[1 - z^{-1}] = -z\frac{dX(z)}{dz}$$

$$Y(z) = \frac{-z}{1 - z^{-1}}\frac{dX(z)}{dz}$$

$$= \frac{-z^2}{z - 1}\frac{dX(z)}{dz}$$

(b) To find the z-transform of the given signal, note that

$$y(n) = \sum_{k=-\infty}^{n} kx(k)$$

where

$$x(n) = \left(\frac{1}{3}\right)^n u(n)$$

Because the z-transform of $x(n)$ is

$$X(z) = \frac{1}{1 - \frac{1}{3}z^{-1}}, \quad |z| > \frac{1}{3}$$

then

$$Y(z) = \frac{-z^2}{z - 1}\frac{dX(z)}{dz}$$

$$= \frac{-z^2}{z - 1}\frac{-\frac{1}{3}z^{-2}}{\left(1 - \frac{1}{3}z^{-1}\right)^2}$$

$$Y(z) = \frac{\frac{1}{3}z^{-1}}{\left(1 - z^{-1}\right)\left(1 - \frac{1}{3}z^{-1}\right)^2}, \quad |z| > 1$$

Example 10.80 When the input to an LTI system is

$$x(n) = \left(\frac{1}{3}\right)^n u(n) + (2)^n u(-n - 1)$$

and the corresponding output is

$$y(n) = 5\left(\frac{1}{3}\right)^n u(n) - 5\left(\frac{2}{3}\right)^n u(n)$$

(a) Find the system function $H(z)$ of the system. Plot the poles and zeros of $H(z)$ and indicate the ROC.

(b) Find the impulse response $h(n)$ of the system.

(c) Write a difference equation that is satisfied by the given input and output.

(d) Is the system stable? Is it causal?

Solution

(a) Given that

$$x(n) = \left(\frac{1}{3}\right)^n u(n) + (2)^n u(-n - 1)$$

Taking the z-transform of the above equation, we obtain

$$X(z) = \frac{1}{1 - \frac{1}{3}z^{-1}} - \frac{1}{1 - 2z^{-1}}, \quad \frac{1}{3} < |z| < 2$$

$$= \frac{-\frac{5}{3}z^{-1}}{\left(1 - \frac{1}{3}z^{-1}\right)\left(1 - 2z^{-1}\right)}$$

Now, consider the given output

$$y(n) = 5\left(\frac{1}{3}\right)^n u(n) - 5\left(\frac{2}{3}\right)^n u(n)$$

Taking the z-transform of the above equation, we obtain

$$Y(z) = \frac{5}{1 - \frac{1}{3}z^{-1}} - \frac{5}{1 - \frac{2}{3}z^{-1}}, \quad |z| > \frac{2}{3}$$

$$= \frac{-\frac{5}{3}z^{-1}}{\left(1 - \frac{1}{3}z^{-1}\right)\left(1 - \frac{2}{3}z^{-1}\right)}$$

Now

$$H(z) = \frac{Y(z)}{X(z)} = \frac{1 - 2z^{-1}}{1 - \frac{2}{3}z^{-1}}, \quad |z| > \frac{2}{3}$$

The pole-zero plot of $H(z)$ is shown in Fig. 10.39

(b) Consider

$$H(z) = \frac{1 - 2z^{-1}}{1 - \frac{2}{3}z^{-1}}, \quad |z| > \frac{2}{3}$$

$$= \frac{1}{1 - \frac{2}{3}z^{-1}} - \frac{2z^{-1}}{1 - \frac{2}{3}z^{-1}}$$

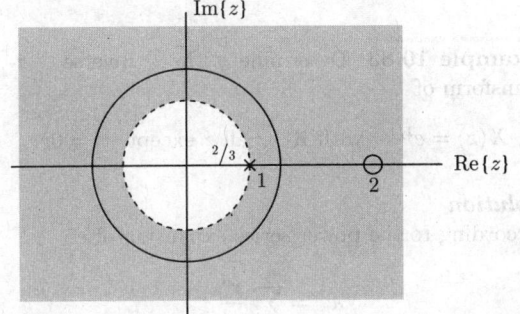

Fig. 10.39

The inverse z-transform of the above equation yields

$$h(n) = \left(\frac{2}{3}\right)^n u(n) - 2\left(\frac{2}{3}\right)^{n-1} u(n-1)$$

$$= \left(\frac{2}{3}\right)^n [u(n) - 3u(n-1)]$$

(c) Since

$$H(z) = \frac{Y(z)}{X(z)} = \frac{1 - 2z^{-1}}{1 - \frac{2}{3}z^{-1}}$$

$$Y(z)\left[1 - \frac{2}{3}z^{-1}\right] = X(z)\left[1 - 2z^{-1}\right]$$

$$Y(z) - \frac{2}{3}z^{-1}Y(z) = X(z) - 2z^{-1}X(z)$$

The inverse z-transform of the above equation leads to

$$y(n) - \frac{2}{3}y(n-1) = x(n) - 2x(n-1)$$

(d) The system is stable because the ROC includes the unit circle. It is also causal since the impulse response $h(n) = 0$ for $n < 0$.

Example 10.81 We are given the following five facts about a discrete-time signal $x(n)$ with z-transform $X(z)$:

1. $x(n)$ is real and right-sided.
2. $X(z)$ has exactly two poles.
3. $X(z)$ has two zeros at the origin.
4. $X(z)$ has a pole at $z = (1/2)\,e^{j\pi/3}$.
5. $X(1) = 8/3$.

Determine $X(z)$ and specify its ROC.

Solution

From fact 2 and 4, we know that $x(n)$ has exactly two poles and one of the pole is located at $z = (1/2)\,e^{j\pi/3}$. Now from fact 1, we also know that $x(n)$ is real. Therefore the poles and zeros of $X(z)$ must occur in complex conjugate pairs. From fact 4, we know that one of the pole is located at $z = (1/2)\,e^{j\pi/3}$, we may conclude that $X(z)$ must have another pole at $z = (1/2)\,e^{-j\pi/3}$.

Now, since $X(z)$ has exactly two poles, we have to assume that $X(z)$ has two or less zeros. If $X(z)$ has more than two zeros, than $X(z)$ must have poles at infinity. From fact 3, we know that $X(z)$ has two zeros at the origin, therefore $X(z)$ must be of the form

$$X(z) = \frac{Kz^2}{\left(z - \frac{1}{2}e^{j\pi/3}\right)\left(z - \frac{1}{2}e^{-j\pi/3}\right)}$$

From fact 5 we know that $X(1) = 8/3$, therefore,

$$X(z)\Big|_{z=1} = X(1)$$

$$\frac{K}{\left(1 - \frac{1}{2}e^{j\pi/3}\right)\left(1 - \frac{1}{2}e^{-j\pi/3}\right)} = \frac{8}{3}$$

$$\frac{K}{1 - \cos(\pi/3) + \frac{1}{4}} = \frac{8}{3}$$

$$\frac{K}{\frac{3}{4}} = \frac{8}{3}$$

$$K = \frac{8}{3} \times \frac{3}{4} = 2$$

Hence,

$$X(z) = \frac{2z^2}{\left(z - \frac{1}{2}e^{j\pi/3}\right)\left(z - \frac{1}{2}e^{-j\pi/3}\right)}$$

$$= \frac{2z^2}{z^2 - z\cos(\pi/3) + \frac{1}{4}}$$

$$X(z) = \frac{2z^2}{z^2 - \frac{1}{2}z + \frac{1}{4}}$$

Since $x(n)$ is right-sided, the ROC must be $|z| > 1/2$.

Example 10.82 Apply the final value theorem to determine $x(\infty)$ for the signal

$$x(n) = \begin{cases} 1, & \text{if } n \text{ is even} \\ 0, & \text{otherwise} \end{cases}$$

Solution
Given that

$$x(n) = \begin{cases} 1, & \text{if } n \text{ is even} \\ 0, & \text{otherwise} \end{cases}$$

From the definition of the unilateral z-transform, we have

$$X(z) = \sum_{n=0}^{\infty} x(n)z^{-n}$$

$$= \sum_{\substack{n=0 \\ n \text{ even}}}^{\infty} (1)z^{-n}$$

Substituting $n = 2r$, where r is varying from 0 to ∞, we obtain

$$X(z) = \sum_{r=0}^{\infty} z^{-2r}$$

$$= \sum_{r=0}^{\infty} (z^{-2})^r$$

$$X(z) = \frac{1}{1 - z^{-2}}, \quad |z^{-2}| < 1 \longrightarrow |z| > 1$$

From the final-value theorem, we have

$$x(\infty) = \lim_{z \to 1}(1 - z^{-1})X(z)$$

$$= \lim_{z \to 1}\frac{1 - z^{-1}}{1 - z^{-2}}$$

$$= \lim_{z \to 1}\frac{1 - z^{-1}}{(1 - z^{-1})(1 + z^{-1})}$$

$$= \lim_{z \to 1}\frac{1}{1 + z^{-1}}$$

$$x(\infty) = \frac{1}{2}$$

Example 10.83 Determine the inverse z-transform of

$$X(z) = e^{1/z} \quad \text{with ROC all } z \text{ except } |z| = 0.$$

Solution
According to the power series expansion of e^x

$$e^x = \sum_{n=0}^{\infty}\frac{x^n}{n!}$$

Therefore,

$$X(z) = e^{1/z} = \sum_{n=0}^{\infty}\frac{1/z^n}{n!}$$

$$= \sum_{n=0}^{\infty}\frac{z^{-n}}{n!}$$

$$= \sum_{n=-\infty}^{\infty}\frac{1}{n!}u(n)z^{-n}$$

comparing the above equation with

$$X(z) = \sum_{n=-\infty}^{\infty} x(n)z^{-n}$$

we get

$$x(n) = \frac{1}{n!}u(n)$$

Example 10.84 Determine the inverse z-transform of

$$X(z) = e^{z^2} \quad \text{with ROC all } z \text{ except } |z| = \infty.$$

Solution
According to the power series expansion of e^x,

$$e^x = \sum_{n=0}^{\infty} \frac{x^n}{n!}$$

Therefore,

$$X(z) = e^{z^2} = \sum_{n=0}^{\infty} \frac{(z^2)^n}{n!}$$

$$= \sum_{n=-\infty}^{0} \frac{(z^2)^{-n}}{(-n)!}$$

$$X(z) = \sum_{n=-\infty}^{\infty} \frac{1}{(-n)!} u(-n) z^{-2n}$$

A change of variables is performed by letting $m = 2n$, which also yields $n = m/2$, $m = -\infty$ as $n = -\infty$, and $m = \infty$ as $n = \infty$. Therefore,

$$X(z) = \sum_{m=-\infty}^{\infty} \frac{1}{(-m/2)!} u\left(-\frac{m}{2}\right) z^{-m}$$

$$= \sum_{n=-\infty}^{\infty} \frac{1}{(-n/2)!} u\left(-\frac{n}{2}\right) z^{-n}$$

comparing the above equation with

$$X(z) = \sum_{n=-\infty}^{\infty} x(n) z^{-n}$$

we get

$$x(n) = \frac{1}{(-n/2)!} u\left(-\frac{n}{2}\right)$$

Since,

$$u\left(-\frac{n}{2}\right) = \begin{cases} 0, & n > 0 \text{ or } n \text{ odd} \\ 1, & \text{otherwise} \end{cases}$$

we get

$$x(n) = \begin{cases} 0, & n > 0 \text{ or } n \text{ odd} \\ \frac{1}{(-n/2)!}, & \text{otherwise} \end{cases}$$

Example 10.85 Consider

$$x(n) \longleftrightarrow X(z) = \frac{10}{1 + \frac{1}{2} z^{-1}}$$

For the following two cases (i) $|z| > \frac{1}{2}$ and (ii) $|z| < \frac{1}{2}$, without explicitly computing $x(n)$, determine whether the DTFT of the corresponding time-signal exists. Identify the DTFT, if it exists.

Solution
Case-I The ROC, $|z| > 1/2$, is shown in Fig. 10.40(a). Since the ROC includes the unit circle, the DTFT of the corresponding time-signal exists. The DTFT $X(e^{j\omega})$ can be obtained from $X(z)$ by substituting $z = e^{j\omega}$. Given that

$$X(z) = \frac{10}{1 + \frac{1}{2} z^{-1}}$$

$$X(z)\Big|_{z=e^{j\omega}} = X(e^{j\omega}) = \frac{10}{1 + \frac{1}{2} e^{-j\omega}}$$

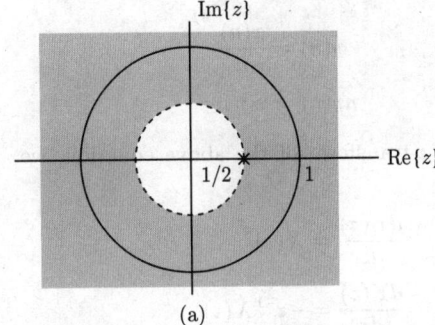

(a)

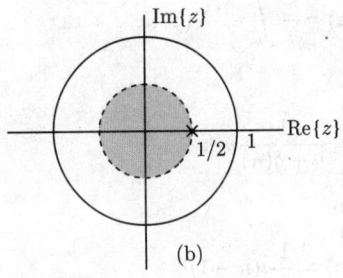

(b)

Fig. 10.40

Case-II The ROC, $|z| < 1/2$, is shown in Fig. 10.40(b). Since the ROC does not include the unit circle, the DTFT of the corresponding time-signal does not exist.

Example 10.86 Let $x(n)$ be a discrete-time signal with $x(n) = 0$ for $n \le 0$ and z-transform $X(z)$. Furthermore, given $x(n)$, let the discrete-time signal

$y(n)$ be defined by

$$y(n) = \begin{cases} \frac{1}{n}x(n), & n > 0 \\ 0, & \text{otherwise} \end{cases}$$

(a) Compute $Y(z)$ in terms of $X(z)$.

(b) Using the result of part (a), find the z-transform of

$$w(n) = \frac{1}{n + \delta(n)}u(n-1)$$

Solution

(a) Given that

$$y(n) = \begin{cases} \frac{1}{n}x(n), & n > 0 \\ 0, & \text{otherwise} \end{cases}$$

Therefore for $n > 0$, we have

$$y(n) = \frac{x(n)}{n}$$

$$ny(n) = x(n)$$

Taking the z-transform of the above equation, we obtain

$$-z\frac{dY(z)}{dz} = X(z)$$

$$\frac{dY(z)}{dz} = -z^{-1}X(z)$$

$$Y(z) = -\int z^{-1}X(z)\,dz$$

(b) Given that

$$w(n) = \frac{1}{n + \delta(n)}u(n-1)$$

For $n > 0$, we obtain

$$w(n) = \frac{1}{n}u(n-1)$$

To apply the result of part (a), we let $x(n) = u(n-1)$ and $w(n) = y(n)$. Therefore,

$$W(z) = -\int z^{-1}X(z)\,dz.$$

$$= -\int z^{-1}\frac{z^{-1}}{1 - z^{-1}}\,dz$$

$$= -\int \frac{1}{z(z-1)}\,dz$$

$$W(z) = -\int \left[\frac{-1}{z} + \frac{1}{z-1}\right]dz$$

$$= -\int \frac{-1}{z}\,dz - \int \frac{1}{z-1}\,dz$$

$$= \ln(z) - \ln(z-1)$$

$$W(z) = \ln\left(\frac{z}{z-1}\right)$$

Example 10.87   Determine the signal $h(n)$ whose z-transform is

$$H(z) = \frac{1}{1 - 2r\cos(\theta)z^{-1} + r^2z^{-2}}, \quad |z| > r > 0$$

Solution

Given that

$$H(z) = \frac{1}{1 - 2r\cos(\theta)z^{-1} + r^2z^{-2}}, \quad |z| < 1$$

$$= \frac{1}{1 - 2r\left(\frac{e^{j\theta} + e^{-j\theta}}{2}\right)z^{-1} + r^2z^{-2}}$$

$$= \frac{1}{1 - re^{j\theta}z^{-1} - re^{-j\theta}z^{-1} + r^2z^{-2}}$$

$$H(z) = \frac{1}{(1 - re^{j\theta}z^{-1})(1 - re^{-j\theta}z^{-1})}$$

$$H(z) = \frac{z^2}{(z - re^{j\theta})(z - re^{-j\theta})}$$

$$\frac{H(z)}{z} = \frac{z}{(z - re^{j\theta})(z - re^{-j\theta})}$$

Using the partial fraction expansion, we obtain

$$\frac{H(z)}{z} = \frac{\frac{e^{j\theta}}{e^{j\theta} - e^{-j\theta}}}{z - re^{j\theta}} + \frac{-\frac{e^{-j\theta}}{e^{j\theta} - e^{-j\theta}}}{z - re^{-j\theta}},$$

$$= \frac{1}{e^{j\theta} - e^{-j\theta}}\left[\frac{e^{j\theta}}{z - re^{j\theta}} - \frac{e^{-j\theta}}{z - re^{-j\theta}}\right]$$

$$H(z) = \frac{1}{2j\sin(\theta)}\left[\frac{ze^{j\theta}}{z - re^{j\theta}} - \frac{ze^{j\theta}}{z - re^{-j\theta}}\right]$$

$$= \frac{1}{2j\sin(\theta)}\left[\frac{e^{j\theta}}{1 - re^{j\theta}z^{-1}} - \frac{e^{-j\theta}}{1 - re^{-j\theta}z^{-1}}\right]$$

Taking the inverse z-transform of the above equation, we obtain

$$h(n) = \frac{1}{2j\sin(\theta)}\left[e^{j\theta}\left(re^{j\theta}\right)^n\right.$$

$$\left. - e^{-j\theta}\left(re^{-j\theta}\right)^n\right]u(n)$$

$$= \frac{r^n}{2j\sin(\theta)}\left[e^{j(n+1)\theta} - e^{-j(n+1)\theta}\right]u(n)$$

$$= \frac{r^n}{\sin(\theta)}\left[\frac{e^{j(n+1)\theta} - e^{-j(n+1)\theta}}{2j}\right]u(n)$$

$$h(n) = \frac{r^n\sin[(n+1)\theta]}{\sin(\theta)}u(n)$$

Example 10.88 An LTI system is given by the difference equation

$$y(n) + 2y(n-1) + y(n-2) = x(n)$$

Determine the unit impulse response.

Solution

Consider the given difference equation

$$y(n) + 2y(n-1) + y(n-2) = x(n)$$

Taking the bilateral z-transform of the above equation, we obtain

$$Y(z) + 2z^{-1}Y(z) + z^{-2}Y(z) = X(z)$$

$$Y(z)[1 + 2z^{-1} + z^{-2}] = X(z)$$

$$\frac{Y(z)}{X(z)} = H(z) = \frac{1}{1 + 2z^{-1} + z^{-2}}$$

$$H(z) = \frac{1}{(1 + z^{-1})^2}$$

We know that

$$(-1)^n u(n) \longleftrightarrow \frac{1}{1 + z^{-1}}$$

Now, using the differentiation property, we obtain

$$n(-1)^n u(n) \longleftrightarrow -z\frac{d}{dz}\left(\frac{1}{1 + z^{-1}}\right)$$

$$\longleftrightarrow -\frac{z^{-1}}{(1 + z^{-1})^2}$$

$$-(n+1)(-1)^{n+1}u(n+1) \longleftrightarrow \frac{1}{(1 + z^{-1})^2}$$

$$-(n+1)(-1)^{n+1}u(n) \longleftrightarrow \frac{1}{(1 + z^{-1})^2}$$

Therefore, the impulse response is given by

$$h(n) = -(n+1)(-1)^{n+1}u(n)$$

Example 10.89 Find the z-transform of the digital signal obtained by sampling the analog signal $x(t) = e^{-4t}\sin(4t)u(t)$ at intervals of $T = 0.1$ seconds.

Solution
The signal $x(n)$ is obtained by sampling the continuous-time signal

$$x(t) = e^{-4t}\sin(4t)u(t)$$

every $T = 0.1$ seconds. Therefore,

$$x(t)\big|_{t=nT} = x(nT)$$

$$x(n) = e^{-4nT}\sin(4nT)u(nT)$$

$$= e^{-4nT}\sin(4nT)u(n)$$

$$x(n) = e^{-0.4n}\sin(0.4n)u(n)$$

We know that [Eq. (10.19)]

$$\mathcal{Z}[\sin(0.4n)u(n)]$$

$$= \frac{z^{-1}\sin(0.4)}{1 - 2z^{-1}\cos(0.4) + z^{-2}}, \quad |z| > 1$$

Now using scaling in z-domain property [Eq. (10.26)], we obtain

$$\mathcal{Z}[(e^{-0.4})^n\sin(0.4n)u(n)]$$

$$= \frac{e^{-0.4}z^{-1}\sin(0.4)}{1 - 2e^{-0.4}z^{-1}\cos(0.4) + e^{-0.8}z^{-2}}, \quad |z| > 1$$

Example 10.90

(a) $x(n)$ is a real right-sided signal having a z-transform $X(z)$. $X(z)$ has two poles, one of which is at $a\,e^{j\phi}$ and two zeros, one of which is at $re^{-j\theta}$. It is also known that $\sum_{n=-\infty}^{\infty} x(n) = 1$. Determine $X(z)$ as a ratio of polynomials in z^{-1}.

(b) If $a = 1/2$, $r = 2$, $\theta = \phi = \pi/4$ in part (a), determine the magnitude of $X(z)$ on the unit circle.

Solution

(a) Since $x(n)$ is real, it must have complex conjugate poles and zeros. $X(z)$ has two poles, one of which is at $a\,e^{j\phi}$, the other pole must be at $a\,e^{-j\phi}$. Also $x(t)$ has two zeros, one of which is at $re^{-j\theta}$, the other zero must be at $re^{j\theta}$. Therefore, $X(z)$ is of the form

$$X(z) = K\frac{\left(z - re^{j\theta}\right)\left(z - re^{-j\theta}\right)}{\left(z - a\,e^{j\phi}\right)\left(z - a\,e^{-j\phi}\right)}$$

$$= K\frac{z^2 - 2zr\cos(\theta) + r^2}{z^2 - 2za\cos(\phi) + a^2}$$

$$X(z) = K\frac{1 - 2r\cos(\theta)z^{-1} + r^2 z^{-2}}{1 - 2a\cos(\phi)z^{-1} + a^2 z^{-2}}$$

where K is determined by the given the fact

$$\sum_{n=-\infty}^{\infty} x(n) = X(z)\big|_{z=1} = 1$$

$$K\frac{1 - 2r\cos(\theta) + r^2}{1 - 2a\cos(\phi) + a^2} = 1$$

$$K = \frac{1 - 2a\cos(\phi) + a^2}{1 - 2r\cos(\theta) + r^2}$$

Therefore,

$$X(z) = \left(\frac{1 - 2a\cos(\phi) + a^2}{1 - 2r\cos(\theta) + r^2}\right)$$

$$\frac{1 - 2r\cos(\theta)z^{-1} + r^2 z^{-2}}{1 - 2a\cos(\phi)z^{-1} + a^2 z^{-2}}$$

(b) Substituting $a = 1/2$, $r = 2$, and $\theta = \phi = \pi/4$ in the expression of $X(z)$, we obtain

$$X(z) = \left(\frac{1 - \cos(\pi/4) + 1/4}{1 - 4\cos(\pi/4) + 4}\right)$$

$$\frac{1 - 4\cos(\pi/4)z^{-1} + 4z^{-2}}{1 - \cos(\pi/4)z^{-1} + 1/4z^{-2}}$$

Substituting $z = 1$ in the expression of $X(z)$ to determine the magnitude of $X(z)$ on the unit circle.

$$X(1) = \left(\frac{1 - \cos(\pi/4) + 1/4}{1 - 4\cos(\pi/4) + 4}\right)$$

$$\left(\frac{1 - 4\cos(\pi/4) + 4}{1 - \cos(\pi/4) + 1/4}\right)$$

$$X(1) = 1$$

Example 10.91   Find the z-transform and its ROC for the signal

$$x(n) = a^n[u(n) - u(n - N)], \quad a \neq 0$$

Also determine and sketch the poles and zeros of the z-transform for $N = 4$.

Solution

Consider the given signal

$$x(n) = a^n[u(n) - u(n - N)], \quad a \neq 0$$

$$= a^n u(n) - a^n u(n - N)$$

$$x(n) = a^n u(n) - a^N a^{n-N} u(n - N)$$

Taking the z-transform of the above equation and using the time-shifting property, we obtain

$$X(z) = \frac{1}{1 - az^{-1}} - a^N \frac{z^{-N}}{1 - az^{-1}}$$

$$= \frac{1}{1 - az^{-1}} - \frac{(az^{-1})^N}{1 - az^{-1}}$$

$$X(z) = \frac{1 - (az^{-1})^N}{1 - az^{-1}}$$

For $N = 4$, we have

$$X(z) = \frac{1 - (az^{-1})^4}{1 - az^{-1}}$$

$$= \frac{\left[1 + (az^{-1})^2\right]\left[1 - (az^{-1})^2\right]}{1 - az^{-1}}$$

$$= \frac{\left[1 + (az^{-1})^2\right](1 - az^{-1})(1 + az^{-1})}{1 - az^{-1}}$$

$$X(z) = \left[1 + (az^{-1})^2\right](1 + az^{-1})$$

$$= \frac{(z^2 + a^2)(z + a)}{z^3}$$

It has three zeros at $z = -a$, and $z = \pm ja$, and three poles at $z = 0$. The pole zero-diagram for $N = 4$ is shown in Fig. 10.41.

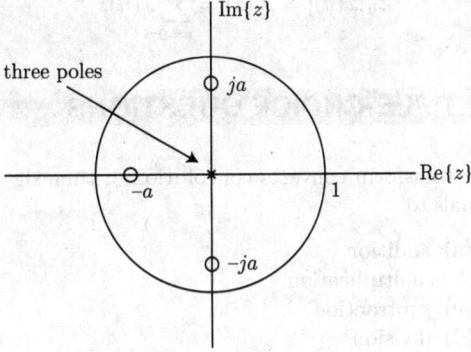

Fig. 10.41 Pole-zero diagram for $N = 4$ for Example 10.91

SUMMARY

1. The z-transform is the discrete-time counterpart of the Laplace transform.

2. The bilateral z-transform of the discrete-time signal $x(n)$ is

$$X(z) = \sum_{n=-\infty}^{\infty} x(n)z^{-n}$$

3. The unilateral z-transform of the discrete-time signal $x(n)$ is

$$X(z) = \sum_{n=0}^{\infty} x(n)z^{-n}$$

4. The region of convergence (ROC) of the z-transform consists of those values of z for which the sum converges.

5. For causal signals, the ROC in the z-plane lies outside a circle containing all the poles of $X(z)$. For anticausal signals, the ROC is inside the circle such that all poles of $X(z)$ are external to this circle. If $x(n)$ consists of both a causal and an anticausal part, then the ROC is an annular region, such that the poles outside this region correspond to the anticausal part of $x(n)$, and the poles inside the annulus correspond to the causal part.

6. The transformation $x(n) \longleftrightarrow X(z)$ is not one-to-one unless ROC is specified.

7. The transfer function $H(z)$ of an LTI system is the z-transform of its impulse response $h(n)$. It may also be defined as a ratio of the bilateral z-transform of the output to the bilateral z-transform of the input when all initial conditions are zero (system is in zero state).

8. If $X(z)$ is the z-transform of the input $x(n)$ and $Y(z)$ is the z-transform of the corresponding output $y(n)$ (when all initial conditions are zero), then $Y(z) = X(z)H(z)$.

9. The LTI system response to an everlasting exponential z_0^n is also an everlasting exponential $H(z_0)z_0^n$.

10. A causal system with rational system function $H(z)$ is stable if and only if all the poles of $H(z)$ lie inside the unit circle in the z-plane.

11. The unilateral, or one-sided, z-transform applies to causal signals and provides a convenient tool for solving system problems involving difference equations with initial conditions.

12. The unilateral z-transform is defined as

$$\mathcal{Z}_{\mathrm{u}}[x(n)] = X(z) = \sum_{n=0}^{\infty} x(n) z^{-n}$$

13. The natural response (zero-input response) represents the system output when the input is zero. The forced response (zero-state response) represents the system output when the initial conditions are zero.

MULTIPLE-CHOICE QUESTIONS

1. z-transform converts convolution of time signals to

   (a) addition
   (b) multiplication
   (c) subtraction
   (d) division

2. Given the z-transform, the corresponding DTFT, if it exists, is obtained by replacing z by

   (a) $j\omega$
   (b) $-j\omega$
   (c) $e^{j\omega}$
   (d) $e^{-j\omega}$

3. For a system with input $x(n) = \delta(n-1)$ and impulse response $h(n) = \delta(n+1)$, the z-transform of the output, the z-transform of the output is

   (a) 0
   (b) 1
   (c) z
   (d) z^{-1}

4. The region of convergence of a causal finite duration discrete-time signal is

   (a) the entire z-plane except $z = 0$
   (b) the entire z-plane except $z = \infty$
   (c) the entire z-plane
   (d) a ring in z-plane

5. The step response of a LTI system when the impulse response $h(n)$ is unit step $u(n)$ is

   (a) $n+1$
   (b) n
   (c) $n-1$
   (d) n^2

6. The z-transform of $\delta(n-m)$ is

   (a) z^{-n}
   (b) z^{-m}

   (c) $\dfrac{1}{z-n}$
   (d) $\dfrac{1}{z-m}$

7. The ROC of the z-transform of the signal $2^n u(n) - 3^n u(-n-1)$

   (a) is $|z| > 1$
   (b) is $|z| < 1$
   (c) is $2 < |z| < 3$
   (d) does not exist

8. The ROC of the z-transform of the signal $2^n u(n) - 2^n u(n-1)$

   (a) is $|z| > 2$
   (b) is $|z| < 2$
   (c) is entire z-plane
   (d) does not exist

9. The number of possible regions of convergence of the function

$$\frac{(e^{-2} - 2) z}{(z - e^{-2})(z - 2)}$$

   is

   (a) 1
   (b) 2
   (c) 3
   (d) 4

10. If R_1 is the ROC of $x_1(n)$ and R_2 is the ROC of $x_2(n)$, then the ROC of $x_1(n)$ convoluted $x_2(n)$ is

   (a) $R_1 + R_2$
   (b) $R_1 - R_2$
   (c) $R_1 \cap R_2$
   (d) $R_1 \cup R_2$

11. The z-transform of the signal

$$x(n) = \sum_{k=-\infty}^{0} \delta(n-k)$$

   has the following region of convergence:

   (a) $|z| > 1$

(b) $|z| = 1$
(c) $|z| < 1$
(d) $0 < |z| < 1$

12. Two signals $x_1(n)$ and $x_2(n)$ are related by $x_2(n) = x_1(-n)$. In the z-domain, their ROCs are

   (a) the same
   (b) reciprocal of each other
   (c) negative of each other
   (d) complements of each other

13. A LTI discrete-time system is given by the difference equation

$$y(n) - ay(n - 1) = x(n)$$

where $|a| < 1$, if all the initial conditions are zero then the system is

   (a) unstable
   (b) stable
   (c) critically stable
   (d) undamped and oscillatory

14. A finite length signal has $X(z) = 0.5 + 0.2z^{-1} + 0.7z^{-2} + 0.5z^{-3}$, its ROC is

   (a) the entire z-plane except $z = 0$
   (b) outside the unit circle
   (c) inside the unit circle
   (d) on the unit circle

15. The convolution property of the z-transforms states that the inverse z-transform of $H(z)X(z)$ is given by

   (a) $\sum_{k=0}^{n-1} h(k)x(n - k)$
   (b) $\sum_{k=0}^{\infty} h(k)x(k - n)$
   (c) $\sum_{k=-\infty}^{\infty} h(k)x(n - k)$
   (d) $\sum_{k=-\infty}^{0} h(n - k)x(n)$

16. If the z-transform of $x(n)$ is $X(z)$ with ROC $|z| > R$, then, the z-transform of $a^n x(n)$, $a > 0$ and its ROC are

   (a) $X\left(\dfrac{z}{a}\right)$, $|z| > aR$
   (b) $X\left(\dfrac{z}{a}\right)$, $|z| > R/a$
   (c) $X(az)$, $|z| < aR$
   (d) (a)$X(az)$, $|z| < R/a$

17. The region of convergence of the signal $x(n) = \{1, 2, \underset{\uparrow}{8}, 4, 6\}$ is

   (a) all z, except $z = 0$ and $z = \infty$
   (b) all z, except $z = 0$
   (c) all z, except $z = \infty$
   (d) all z.

18. The transfer function of a stable system is

$$H(z) = \frac{1}{1 - 0.5z^{-1}} + \frac{1}{1 - 2z^{-1}}$$

Its impulse response will be

   (a) $(0.5)^n u(n) + 2^n u(n)$
   (b) $-(0.5)^n u(-n - 1) - 2^n u(n)$
   (c) $(0.5)^n u(n) - 2^n u(-n - 1)$
   (d) $-(0.5)^n u(-n - 1) - 2^n u(-n - 1)$

PROBLEMS

10.1 An LTI system is characterized by the difference equation

$$x(n - 2) - 9x(n - 1) + 18x(n) = 0$$

with initial conditions $x(-1) = 1$ and $x(-2) = 9$. Find $x(n)$ by using z-transform and state the properties of z-transform used in your calculation.

10.2 Use power-series expansion to determine the time-domain signal $x(n)$, given

$$x(n) \longleftrightarrow X(z) = \frac{1}{1 - z^{-2}}$$

for the two cases: (i) $|z| > 1$ (causal) and (ii) $|z| < 1$ (noncausal).

10.3 Determine

   (i) the z-transform of $x(n - n_0)$, starting from the definition.
   (ii) the input to the system, using z-transform, given output $y(n) = \delta(n - 2)$ and impulse response $h(n) = (1/2)^n u(n)$.

10.4 Solve the following equation if the auxiliary conditions are $y(0) = 1$, $y(1) = 2$, and the input $x(n) = u(n)$:

$$y(n) + 3y(n - 1) + 2y(n - 2) = x(n - 1) + 3x(n - 2)$$

10.5 Consider a real-valued sequence $x(n)$ with rational z-transform $X(z)$.

1. From the definition of the z-transform, show that

$$X(z) = X^*(z^*)$$

2. From your result in part (a), show that if a pole (or zero) of $X(z)$ occurs at $z = z_0$, then a pole (or zero) must also occur at $z = z_0^*$.

3. Show that for a real, even sequence, if there is a pole (or zero) of $H(z)$ at $z = re^{j\theta}$, then there is also a pole (or zero) of $H(z)$ at $z = (1/r)\,e^{j\theta}$ and at $z = (1/r)\,e^{-j\theta}$.

10.6 Consider a sequence $x_1(n)$ with z-transform $X_1(z)$ and a sequence $x_2(n)$ with z-transform $X_2(z)$, where

$$x_2(n) = x_1(-n)$$

Show that $X_2(z) = X_1(1/z)$, and from this, show that if $X_1(z)$ has a pole (or zero) at $z = z_0$, then $X_2(z)$ has a pole (or zero) at $z = 1/z_0$.

10.7 Compute the unit step response $s(n)$ of the system with impulse response

$$h(n) = \begin{cases} 3^n, & n < 0 \\ \left(\dfrac{2}{5}\right)^n, & n \geq 0 \end{cases}$$

10.8 Use the unilateral (one-sided) z-transform to determine $y(n)$, $n \geq 0$ in the following cases: (a) $y(n) + \frac{1}{2}y(n-1) - \frac{1}{4}y(n-2) = 0$; $y(-1) = y(-2) = 1$ (b) $y(n) - 1.5y(n-1) + 0.5y(n-2) = 0$; $y(-1) = 1, y(-2) = 0$ (c) $y(n) = \frac{1}{2}y(n-1) + x(n)$; $x(n) = \left(\frac{1}{3}\right)^n u(n)$, $y(-1) = 1$

10.9 Consider the system

$$H(z) = \frac{1 - 2z^{-1} + 2z^{-2} - z^{-3}}{(1 - z^{-1})(1 - 0.5z^{-1})(1 - 0.2z^{-1})}$$

1. Sketch the pole-zero pattern. Is the system stable?

2. Determine the impulse response of the system.

10.10 Use the convolution property to

1. express the z-transform of

$$y(n) = \sum_{k=-\infty}^{n} x(k)$$

in terms of $X(z)$.

2. determine the z-transform of $x(n) = (n+1)u(n)$.

10.11 Using long division, determine the inverse z-transform of

$$X(z) = \frac{1 + 2z^{-1}}{1 - 2z^{-1} + z^{-2}}$$

if (a) $x(n)$ is causal and (b) $x(n)$ is anticausal.

10.12 Determine the z-transform and sketch the ROC of the following signals.

(a) $x(n) = (-3)^n u(-n-1)$

(b) $x(n) = \begin{cases} 1, & -5 \leq n \leq 5 \\ 0, & \text{otherwise} \end{cases}$

(c) $x(n) = \begin{cases} \left(\dfrac{1}{3}\right)^n, & n \geq 0 \\ 3^n, & n < 0 \end{cases}$

10.13 Let $x(n)$ be a finite length sequence that is nonzero only for $0 \leq n \leq N-1$, and consider the one-sided periodic signal, $y(n)$, that is formed by periodically extending $x(n)$ as follows:

$$y(n) = \sum_{k=0}^{\infty} x(n - kN)$$

Express the z-transform of $y(n)$ in terms of $X(z)$ and find the region of convergence of $Y(z)$.

10.14 (a) When input $x(n) = u(n) + \left(-\frac{1}{2}\right)^n u(n)$ is applied to a linear causal time-invariant system, the output is

$$y(n) = 6\left(-\frac{1}{4}\right)^n u(n) - 6\left(-\frac{1}{3}\right)^n u(n)$$

Find the transfer function of the system. (b) What is the difference equation representation of the system?

10.15 Determine the ROC of $aX(z) + bY(z)$, given that

$$X(z) = \frac{z}{(z - 0.5)(z - 1.5)}, \quad 0.5 < |z| < 1.5$$

$$Y(z) = \frac{0.25z}{(z - 0.25)(z - 0.5)}, \quad |z| > 0.5$$

For what relationship between a and b the ROC will be the largest?

10.16 A causal system is described by

$$H(z) = \frac{1 + z^{-1}}{(1 - az^{-1})(1 - bz^{-1})}$$

For what values of a and b will the system be (i) causal and (ii) noncausal.

10.17 Determine the inverse z-transform of the following $X(z)$ by the partial fraction expansion method.

$$X(z) = \frac{z+2}{2z^2 - 7z + 2}$$

if the ROCs are (a) $|z| > 3$, (b) $|z| < 1/2$, and (c) $1/2 < |z| < 3$.

ANSWERS TO MULTIPLE-CHOICE QUESTIONS

1. **(b)**   2. **(c)**   3. **(b)**   4. **(a)**   5. **(a)**   6. **(b)**   7. **(c)**   8. **(c)**   9. **(c)**
10. **(c)**   11. **(c)**   12. **(b)**   13. **(b)**   14. **(a)**   15. **(c)**   16. **(a)**   17. **(a)**   18. **(c)**

State Space Analysis

11.1 INTRODUCTION

So far we have studied LTI systems based on their input–output relationships, which are known as the external descriptions of the systems, e.g., transfer function. The transfer function is only defined under zero initial conditions and it is only applicable to linear time-invariant systems and there too it is generally restricted to single-input single-output systems. In this chapter we discuss the method of state space representations of systems, which are known as the internal descriptions of the systems. The state space approach is a direct time-domain approach that can be used for analysis and design of linear and nonlinear, time-invariant and time-variant, and multiinput multioutput systems. The state space representation of the system has many advantages.

11.2 ADVANTAGES OF STATE SPACE REPRESENTATIONS

Some advantages of state space representations are:

1. It provides an insight into the behaviour of the system.
2. It can be applied to linear or nonlinear, time-invariant, or time-variant systems.
3. It allows us to handle systems with multiple inputs and outputs in a unified way.
4. It is easier to apply where the Laplace transform or z-transform cannot be applied.
5. It is suitable for digital computer computation because it is a time-domain approach.

11.3 THE CONCEPT OF STATE

11.3.1 State and State Variables

The state of a dynamic system (or system with memory) at time t_0 (or n_0) is defined as the minimal information that is sufficient to determine the state and the output of the system for all times $t \geq t_0$ (or $n \geq n_0$) when the input to the system is also known for all times $t \geq t_0$ (or $n \geq n_0$). The variables that contain this information are called the *state variables*. The state of the system is related to the memory of the system. Given the state of the system at t_0 (or n_0) and the input from t_0 to t_1 (or from n_0 to n_1), we can find both the output and the state at t_1 (or n_1). Note that this definition of the state of the system applies only to causal systems (systems in which the output does not anticipate future values of the input).

11.3.2 State Vector

If we need N variables to completely describe the behaviour of a given system, then these N state variables may be considered as N components of a vector $\mathbf{q}$. Such a vector is called a *state vector*. For example, a state vector for a continuous-time system is given by

$$\mathbf{q}(t) = \begin{bmatrix} q_1(t) \\ q_2(t) \\ \vdots \\ q_N(t) \end{bmatrix} \tag{11.1}$$

where $q_1(t)$, $q_2(t)$, ..., $q_N(t)$ are the state variables of the continuous-time system. The state variables $q_1(t)$, $q_2(t)$, ..., $q_N(t)$ are the minimum number of variables of a system such that their initial values at any instant t_0 are sufficient to determine the behaviour of the system for all time $t \geq t_0$ when the input to the system is known for $t \geq t_0$.

Similarly for discrete-time system, the state vector is given by

$$\mathbf{q}(n) = \begin{bmatrix} q_1(n) \\ q_2(n) \\ \vdots \\ q_N(n) \end{bmatrix} \tag{11.2}$$

11.3.3 State Space

The N-dimensional space whose coordinate axes consists of the q_1-axis, q_2-axis, ..., q_N-axis is called the *state space*. Any state can be represented by a point in the state space.

11.3.4 Selection of State Variables

Since the state variables of a system can be interpreted as the 'memory elements' of the system, for continuous-time systems which are formed by integrators, multipliers, and adders, we choose the outputs of the integrators as the state variables of the system. For discrete-time systems that are formed by unit-delay elements, multipliers, and adders, we choose the outputs of the unit-delay elements as the state variables of the system. For a continuous-time system containing physical energy-storing elements, the outputs of these memory elements can be chosen to be the state variables of the systems. Note that the choice of state variables of a system is not unique. There are infinitely many choices for any given system. If the system is described by the differential or difference equation, the state variables can be chosen as shown in the following sections.

11.4 STATE SPACE REPRESENTATION OF CONTINUOUS-TIME LTI SYSTEMS

11.4.1 Systems Described by Differential Equations

Suppose that a single-input single-output continuous-time LTI system is described by an Nth order differential equation

$$\frac{d^N y(t)}{dt^N} + a_1 \frac{d^{N-1} y(t)}{dt^{N-1}} + \cdots + a_N y(t) = x(t) \tag{11.3}$$

One possible set of initial conditions is $y(0), y^{(1)}(0), \cdots, y^{(N-1)}(0)$, where $y^{(k)}(t) = (d^k y(t))/dt^k$. Thus, let us define N state variables $q_1(t), q_2(t), \cdots, q_N(t)$ as

$$q_1(t) = y(t)$$

$$q_2(t) = y^{(1)}(t)$$

$$\vdots$$

$$q_N(t) = y^{(N-1)}(t) \tag{11.4}$$

Then from Eqs (11.3) and (11.4) we have

$$\dot{q}_1(t) = q_2(t)$$

$$\dot{q}_2(t) = q_3(t)$$

$$\vdots$$

$$\dot{q}_N(t) = -a_N q_1(t) - a_{N-1} q_2(t) - \cdots - a_1 q_N(t) + x(t) \tag{11.5}$$

and

$$y(t) = q_1(t) \tag{11.6}$$

where $\dot{q}_k(t) = (dq_k(t))/dt$. Equations (11.5) and (11.6) can be expressed in matrix form as

$$\begin{bmatrix} \dot{q}_1(t) \\ \dot{q}_2(t) \\ \vdots \\ \dot{q}_N(t) \end{bmatrix} = \begin{bmatrix} 0 & 1 & 0 & \cdots & 0 \\ 0 & 0 & 1 & \cdots & 0 \\ \vdots & \vdots & \vdots & \ddots & \vdots \\ -a_N & -a_{N-1} & -a_{N-2} & \cdots & -a_1 \end{bmatrix} \begin{bmatrix} q_1(t) \\ q_2(t) \\ \vdots \\ q_N(t) \end{bmatrix} + \begin{bmatrix} 0 \\ 0 \\ \vdots \\ 1 \end{bmatrix} x(t) \tag{11.7}$$

$$y(t) = \begin{bmatrix} 1 & 0 & \cdots & 0 \end{bmatrix} \begin{bmatrix} q_1(t) \\ q_2(t) \\ \vdots \\ q_N(t) \end{bmatrix} \tag{11.8}$$

We now define an $N \times 1$ matrix $\mathbf{q}(t)$, which we call the state vector:

$$\mathbf{q}(t) = \begin{bmatrix} q_1(t) \\ q_2(t) \\ \vdots \\ q_N(t) \end{bmatrix} \tag{11.9}$$

The derivative of a matrix is obtained by taking the derivative of each element of the matrix. Thus,

$$\frac{d\mathbf{q}(t)}{dt} = \dot{\mathbf{q}}(t) = \begin{bmatrix} \dot{q}_1(t) \\ \dot{q}_2(t) \\ \vdots \\ \dot{q}_N(t) \end{bmatrix} \tag{11.10}$$

Then Eqs (11.7) and (11.8) can be rewritten compactly as

$$\dot{\mathbf{q}}(t) = \mathbf{A}\mathbf{q}(t) + \mathbf{b}x(t) \tag{11.11}$$

$$y(t) = \mathbf{c}\mathbf{q}(t) \tag{11.12}$$

where

$$\mathbf{A} = \begin{bmatrix} 0 & 1 & 0 & \cdots & 0 \\ 0 & 0 & 1 & \cdots & 0 \\ \vdots & \vdots & \vdots & \ddots & \vdots \\ -a_N & -a_{N-1} & -a_{N-2} & \cdots & -a_1 \end{bmatrix} \qquad \mathbf{b} = \begin{bmatrix} 0 \\ 0 \\ \vdots \\ 1 \end{bmatrix} \qquad \mathbf{c} = \begin{bmatrix} 1 & 0 & \cdots & 0 \end{bmatrix}$$

Equations (11.11) and (11.12) are called an N-dimensional state space representation (or state equations) of the system, and the $N \times N$ matrix $\mathbf{A}$ is termed the system matrix. In general, state equations of a single-input single-output continuous-time LTI system are given by

$$\dot{\mathbf{q}}(t) = \mathbf{A}\mathbf{q}(t) + \mathbf{b}x(t) \tag{11.13}$$

$$y(t) = \mathbf{c}\mathbf{q}(t) + dx(t) \tag{11.14}$$

Example 11.1 A system is described by the differential equation

$$\frac{d^3y(t)}{dt^3} + 6\frac{d^2y(t)}{dt^2} + 11\frac{dy(t)}{dt} + 10y(t) = 8x(t)$$

where $y(t)$ is the output and $x(t)$ is the input to the system. Obtain state space representation of the system.

Solution
Select the state variables as

$$q_1(t) = y(t) \qquad q_2(t) = \dot{y}(t) \qquad q_3(t) = \ddot{y}(t)$$

then

$$\dot{q}_1(t) = q_2(t) \qquad \dot{q}_2(t) = q_3(t) \qquad \dot{q}_3(t) = -10q_1(t) - 11q_2(t) - 6q_3(t) + 8x(t)$$

The above equations can be rewritten compactly as

$$\underbrace{\begin{bmatrix} \dot{q}_1(t) \\ \dot{q}_2(t) \\ \dot{q}_3(t) \end{bmatrix}}_{\dot{\mathbf{q}}(t)} = \underbrace{\begin{bmatrix} 0 & 1 & 0 \\ 0 & 0 & 1 \\ -10 & -11 & -6 \end{bmatrix}}_{\mathbf{A}} \underbrace{\begin{bmatrix} q_1(t) \\ q_2(t) \\ q_3(t) \end{bmatrix}}_{\mathbf{q}(t)} + \underbrace{\begin{bmatrix} 0 \\ 0 \\ 8 \end{bmatrix}}_{\mathbf{b}} x(t)$$

The output is

$$y(t) = q_1(t) = \underbrace{\begin{bmatrix} 1 & 0 & 0 \end{bmatrix}}_{\mathbf{c}} \underbrace{\begin{bmatrix} q_1(t) \\ q_2(t) \\ q_3(t) \end{bmatrix}}_{\mathbf{q}(t)}$$

Therefore,

$$\dot{\mathbf{q}}(t) = \mathbf{A}\mathbf{q}(t) + \mathbf{b}x(t)$$

$$y(t) = \mathbf{c}\mathbf{q}(t)$$

11.4.2 Multiple-Input Multiple-Output System

If a continuous-time LTI system has m inputs, p outputs, and N state variables, then a state space representation of the system can be expressed as

$$\dot{\mathbf{q}}(t) = \mathbf{A}\mathbf{q}(t) + \mathbf{B}\mathbf{x}(t) \tag{11.15}$$

$$\mathbf{y}(t) = \mathbf{C}\mathbf{q}(t) + \mathbf{D}\mathbf{x}(t) \tag{11.16}$$

where

$$\mathbf{q}(t) = \begin{bmatrix} q_1(t) \\ q_2(t) \\ \vdots \\ q_N(t) \end{bmatrix} \qquad \mathbf{x}(t) = \begin{bmatrix} x_1(t) \\ x_2(t) \\ \vdots \\ x_m(t) \end{bmatrix} \qquad \mathbf{y}(t) = \begin{bmatrix} y_1(t) \\ y_2(t) \\ \vdots \\ y_p(t) \end{bmatrix}$$

and

$$\mathbf{A} = \begin{bmatrix} a_{11} & a_{12} & \cdots & a_{1N} \\ a_{21} & a_{22} & \cdots & a_{2N} \\ \vdots & \vdots & \ddots & \vdots \\ a_{N1} & a_{N2} & \cdots & a_{NN} \end{bmatrix}_{N \times N} \qquad \mathbf{B} = \begin{bmatrix} b_{11} & b_{12} & \cdots & b_{1m} \\ b_{21} & b_{22} & \cdots & b_{2m} \\ \vdots & \vdots & \ddots & \vdots \\ b_{N1} & b_{N2} & \cdots & b_{Nm} \end{bmatrix}_{N \times m}$$

$$\mathbf{C} = \begin{bmatrix} c_{11} & c_{12} & \cdots & c_{1N} \\ c_{21} & c_{22} & \cdots & c_{2N} \\ \vdots & \vdots & \ddots & \vdots \\ c_{p1} & c_{p2} & \cdots & c_{pN} \end{bmatrix}_{p \times N} \qquad \mathbf{D} = \begin{bmatrix} d_{11} & d_{12} & \cdots & d_{1m} \\ d_{21} & d_{22} & \cdots & d_{2m} \\ \vdots & \vdots & \ddots & \vdots \\ d_{p1} & d_{p2} & \cdots & d_{pm} \end{bmatrix}_{p \times m}$$

Example 11.2 A system is described by the following differential equation. Represent the system in state space.

$$\frac{d^3y(t)}{dt^3} + 3\frac{d^2y(t)}{dt^2} + 4\frac{dy(t)}{dt} + 4y(t) = x_1(t) + 3x_2(t) + 4x_3(t)$$

and outputs are

$$y_1(t) = 4\frac{dy(t)}{dt} + 3x_1(t) \qquad y_2(t) = \frac{d^2y(t)}{dt^2} + 4x_2(t) + x_3(t)$$

Solution
Select the state variables as

$$q_1(t) = y(t) \qquad q_2(t) = \dot{y}(t) \qquad q_3(t) = \ddot{y}(t)$$

then

$$\dot{q}_1(t) = q_2(t) \qquad \dot{q}_2(t) = q_3(t) \qquad \dot{q}_3(t) = -4q_1(t) - 4q_2(t) - 3q_3(t) + x_1(t) + 3x_2(t)$$
$$+ 4x_3(t)$$

The above equations can be rewritten compactly as

$$\underbrace{\begin{bmatrix} \dot{q}_1(t) \\ \dot{q}_2(t) \\ \dot{q}_3(t) \end{bmatrix}}_{\dot{q}(t)} = \underbrace{\begin{bmatrix} 0 & 1 & 0 \\ 0 & 0 & 1 \\ -4 & -4 & -3 \end{bmatrix}}_{A} \underbrace{\begin{bmatrix} q_1(t) \\ q_2(t) \\ q_3(t) \end{bmatrix}}_{q(t)} + \underbrace{\begin{bmatrix} 0 & 0 & 0 \\ 0 & 0 & 0 \\ 1 & 3 & 4 \end{bmatrix}}_{B} \underbrace{\begin{bmatrix} x_1(t) \\ x_2(t) \\ x_3(t) \end{bmatrix}}_{x(t)}$$

The outputs are

$$y_1(t) = 4q_2(t) + 3x_1(t) \qquad y_2(t) = q_3(t) + 4x_2(t) + x_3(t)$$

The above equations can be rewritten compactly as

$$\underbrace{\begin{bmatrix} y_1(t) \\ y_2(t) \end{bmatrix}}_{y(t)} = \underbrace{\begin{bmatrix} 0 & 4 & 0 \\ 0 & 0 & 1 \end{bmatrix}}_{C} \underbrace{\begin{bmatrix} q_1(t) \\ q_2(t) \\ q_3(t) \end{bmatrix}}_{q(t)} + \underbrace{\begin{bmatrix} 3 & 0 & 0 \\ 0 & 4 & 1 \end{bmatrix}}_{D} \underbrace{\begin{bmatrix} x_1(t) \\ x_2(t) \\ x_3(t) \end{bmatrix}}_{bfx(t)}$$

Therefore,

$$\dot{q}(t) = \mathbf{A}q(t) + \mathbf{B}x(t)$$
$$y(t) = \mathbf{C}q(t) + \mathbf{D}x(t)$$

11.4.3 Electrical Circuits

We can determine the state equations of electrical circuits by using the following steps:

1. Choose all independent capacitor voltages and inductor currents to be the state variables.

2. Choose a set of loop currents, and express the state variables and their first derivatives in terms of these loop currents.

3. Write loop equations, and eliminate all variables other than state variables (and their first derivatives) from the equations derived in Steps 2 and 3.

Example 11.3 Find the state-variable description of the circuit depicted in Fig. 11.1 if the input is the applied voltage $x(t)$ and the output is the current $y(t)$ through the resistor R_2.

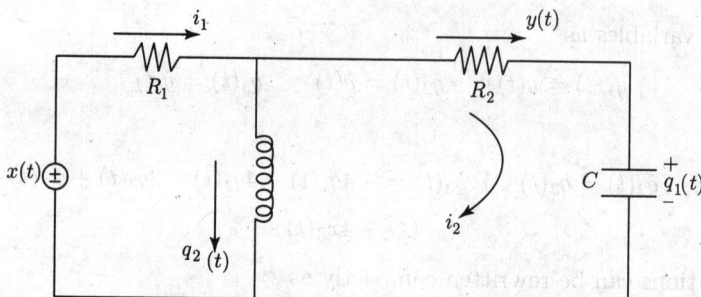

Fig. 11.1

Solution

Step 1 There is one inductor and one capacitor in the network. Therefore, we shall choose the capacitor voltage $q_1(t)$ and the inductor current $q_2(t)$ as the state variables.

Step 2 The relationship between the loop currents and the state variables can be written by inspection:

$$C\dot{q}_1(t) = i_2$$

$$q_2(t) = i_1 - i_2$$

The output equation is

$$y(t) = i_2$$

Step 3 The loop equations are

$$i_1 R_1 + L\dot{q}_2(t) - x(t) = 0$$

$$i_2 R_2 + q_1(t) - L\dot{q}_2(t) = 0$$

After some simple algebraic manipulation, we obtain

$$\dot{q}_1(t) = -\frac{1}{(R_1 + R_2)C}q_1(t) - \frac{R_1}{(R_1 + R_2)C}q_2(t) + \frac{1}{(R_1 + R_2)C}x(t)$$

$$\dot{q}_2(t) = \frac{R_1}{(R_1 + R_2)L}q_1(t) - \frac{R_1 R_2}{(R_1 + R_2)L}q_2(t) + \frac{R_2}{(R_1 + R_2)L}x(t)$$

The above equations can be rewritten compactly as

$$\underbrace{\begin{bmatrix} \dot{q}_1(t) \\ \dot{q}_2(t) \end{bmatrix}}_{\dot{\mathbf{q}}(t)} = \underbrace{\begin{bmatrix} -\dfrac{1}{(R_1+R_2)C} & \dfrac{R_1}{(R_1+R_2)C} \\[2ex] \dfrac{R_1}{(R_1+R_2)L} & \dfrac{R_1 R_2}{(R_1+R_2)L} \end{bmatrix}}_{\mathbf{A}} \underbrace{\begin{bmatrix} q_1(t) \\ q_2(t) \end{bmatrix}}_{\mathbf{q}(t)} + \underbrace{\begin{bmatrix} \dfrac{1}{(R_1+R_2)C} \\[2ex] \dfrac{R_2}{(R_1+R_2)L} \end{bmatrix}}_{\mathbf{b}} x(t)$$

$$y(t) = -\frac{1}{R_1+R_2}q_1(t) - \frac{R_1}{R_1+R_2}q_2(t) + \frac{1}{R_1+R_2}x(t)$$

$$y(t) = \underbrace{\begin{bmatrix} -\dfrac{1}{R_1+R_2} & -\dfrac{R_1}{R_1+R_2} \end{bmatrix}}_{\mathbf{c}} \underbrace{\begin{bmatrix} q_1(t) \\ q_2(t) \end{bmatrix}}_{\mathbf{q}(t)} + \underbrace{\frac{1}{R_1+R_2}}_{d} x(t)$$

11.4.4   System Described by Transfer Function

It is relatively easy to determine the state equations of a system specified by its transfer function. Consider, for example, a first-order system with the transfer function

$$H(s) = \frac{1}{s+a}$$

The system realization appears in Fig. 11.2. The integrator output $q(t)$ serves as a natural state variable space since, in practical realization, initial conditions are placed on the integrator output. The integrator input is naturally $\dot{q}(t)$. From Fig. 11.2, we have

$$\dot{q}(t) = -aq(t) + x(t)$$

$$y(t) = q(t)$$

We know that a given transfer function can be realized in several ways. Consequently, we will be able to obtain different state-space description of the same system by using different realizations.

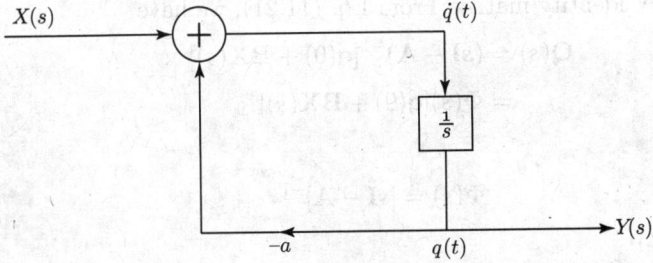

Fig. 11.2

11.5   SOLUTION OF STATE EQUATIONS FOR CONTINUOUS-TIME LTI SYSTEMS

The state equations of a linear system are N simultaneous linear differential equations of the first order. These equations can be solved in both time and s-domains (Laplace transform). The latter is relatively easier to deal with than the time-domain solution.

11.5.1   Laplace Transform Method

Consider an N-dimensional state space representation

$$\dot{\mathbf{q}}(t) = \mathbf{A}\mathbf{q}(t) + \mathbf{B}\mathbf{x}(t) \tag{11.17}$$

$$\mathbf{y}(t) = \mathbf{C}\mathbf{q}(t) + \mathbf{D}\mathbf{x}(t) \tag{11.18}$$

where $\mathbf{A}$, $\mathbf{B}$, $\mathbf{C}$, and $\mathbf{D}$ are $N \times N$, $N \times m$, $p \times N$, $p \times m$ matrices, respectively. Let $\mathbf{q}(0)$ be the initial state. Taking the unilateral Laplace transform of Eqs (11.17) and (11.18) and using Eq. (9.71), we get

$$s\mathbf{Q}(s) - \mathbf{q}(0) = \mathbf{A}\mathbf{Q}(s) + \mathbf{B}\mathbf{X}(s) \tag{11.19}$$

$$Y(s) = \mathbf{C}\mathbf{Q}(s) + \mathbf{D}\mathbf{X}(s) \tag{11.20}$$

where

$$\mathbf{Q}(s) = \mathcal{L}_u[\mathbf{q}(t)] = \begin{bmatrix} Q_1(s) \\ Q_2(s) \\ \vdots \\ Q_N(s) \end{bmatrix} \qquad \mathbf{X}(s) = \mathcal{L}_u[\mathbf{x}(t)] = \begin{bmatrix} X_1(s) \\ X_2(s) \\ \vdots \\ X_m(s) \end{bmatrix}$$

$$\mathbf{Y}(s) = \mathcal{L}_u[\mathbf{y}(t)] = \begin{bmatrix} Y_1(s) \\ Y_2(s) \\ \vdots \\ Y_p(s) \end{bmatrix}$$

where $Q_k(s) = \mathcal{L}_u[q_k(t)]$, $X_k(s) = \mathcal{L}_u[x_k(t)]$, and $Y_k(s) = \mathcal{L}_u[y_k(t)]$. Consider Eq. (11.19)

$$s\mathbf{Q}(s) - \mathbf{q}(0) = \mathbf{A}\mathbf{Q}(s) + \mathbf{B}\mathbf{X}(s)$$

$$s\mathbf{Q}(s) - \mathbf{A}\mathbf{Q}(s) = \mathbf{q}(0) + \mathbf{B}\mathbf{X}(s)$$

$$(s\mathbf{I} - \mathbf{A})\mathbf{Q}(s) = \mathbf{q}(0) + \mathbf{B}\mathbf{X}(s) \tag{11.21}$$

where $\mathbf{I}$ is the $N \times N$ identity matrix. From Eq. (11.21), we have

$$\mathbf{Q}(s) = (s\mathbf{I} - \mathbf{A})^{-1}[\mathbf{q}(0) + \mathbf{B}\mathbf{X}(s)] \tag{11.22}$$

$$= \mathbf{\Phi}(s)[\mathbf{q}(0) + \mathbf{B}\mathbf{X}(s)] \tag{11.23}$$

where

$$\mathbf{\Phi}(s) = [s\mathbf{I} - \mathbf{A}]^{-1} \tag{11.24}$$

Thus from Eq. (11.23)

$$\mathbf{Q}(s) = \mathbf{\Phi}(s)\mathbf{q}(0) + \mathbf{\Phi}(s)\mathbf{B}\mathbf{X}(s) \tag{11.25}$$

Taking the inverse Laplace transform, we get

$$\mathbf{q}(t) = \underbrace{\mathcal{L}^{-1}[\mathbf{\Phi}(s)]\mathbf{q}(0)}_{\text{zero-input component}} + \underbrace{\mathcal{L}^{-1}[\mathbf{\Phi}(s)\mathbf{B}\mathbf{X}(s)]}_{\text{zero-state component}} \qquad (11.26)$$

Now consider the output Eq. (11.20),

$$\mathbf{Y}(s) = \mathbf{C}\mathbf{Q}(s) + \mathbf{D}\mathbf{X}(s)$$

Upon substituting Eq. (11.25) into this equation, we have

$$\mathbf{Y}(s) = \mathbf{C}[\mathbf{\Phi}(s)\mathbf{q}(0) + \mathbf{\Phi}(s)\mathbf{B}\mathbf{X}(s)] + \mathbf{D}\mathbf{X}(s)$$

$$\mathbf{Y}(s) = \underbrace{\mathbf{C}\mathbf{\Phi}(s)\mathbf{q}(0)}_{\text{zero-input response}} + \underbrace{[\mathbf{C}\mathbf{\Phi}(s)\mathbf{B} + \mathbf{D}]\mathbf{X}(s)}_{\text{zero-state response}} \qquad (11.27)$$

The zero-state response [i.e., the response $\mathbf{Y}(s)$ when $\mathbf{q}(0) = \mathbf{0}$] is given by

$$\mathbf{Y}(s) = [\mathbf{C}\mathbf{\Phi}(s)\mathbf{B} + \mathbf{D}]\mathbf{X}(s)$$

$$\frac{\mathbf{Y}(s)}{\mathbf{X}(s)} = \mathbf{H}(s) = \mathbf{C}\mathbf{\Phi}(s)\mathbf{B} + \mathbf{D} \qquad (11.28)$$

The matrix $\mathbf{H}(s)$ is a $p \times m$ matrix [p is the number of outputs and m is the number of inputs]. The ijth element $H_{ij}(s)$ of $\mathbf{H}(s)$ is the transfer function that relates the output $y_i(t)$ to the input $x_j(t)$.

Characteristic Roots (or Eigenvalues) of a Matrix

Consider Eq. (11.24)

$$\mathbf{\Phi}(s) = [s\mathbf{I} - \mathbf{A}]^{-1}$$

$$= \frac{1}{\det[s\mathbf{I} - \mathbf{A}]} \text{adj}[s\mathbf{I} - \mathbf{A}]$$

$$\mathbf{\Phi}(s) = \frac{1}{|s\mathbf{I} - \mathbf{A}|} \text{adj}[s\mathbf{I} - \mathbf{A}]$$

Note that the denominator of every element of $\mathbf{\Phi}(s)$ is $|s\mathbf{I} - \mathbf{A}|$ because $\mathbf{\Phi}(s) = [s\mathbf{I} - \mathbf{A}]^{-1}$ and the inverse of a matrix has its determinant in the denominator. Since $\mathbf{C}$, $\mathbf{B}$, and $\mathbf{D}$ are matrices with constant elements, we see from Eq. (11.28) that the denominator of $\mathbf{\Phi}(s)$ will also be the denominator of $\mathbf{H}(s)$. Hence, the denominator of every element of $\mathbf{H}(s)$ is $|s\mathbf{I} - \mathbf{A}|$. In other words, the zeros of the polynomial $|s\mathbf{I} - \mathbf{A}|$ are also the poles of all transfer functions of the system. Therefore, the zeros of the polynomial $|s\mathbf{I} - \mathbf{A}|$ are the characteristic roots of the system. Hence, the characteristic roots of the system are the roots of the equation

$$|s\mathbf{I} - \mathbf{A}| = 0 \qquad (11.29)$$

Since $|s\mathbf{I} - \mathbf{A}|$ is an Nth-order polynomial in s with N zeros $\lambda_1, \lambda_2, \cdots, \lambda_N$, we can write Eq. (11.29) as

$$|s\mathbf{I} - \mathbf{A}| = s^N + a_1 s^{N-1} + \cdots + a_{N-1}s + a_N$$

$$= (s - \lambda_1)(s - \lambda_2)\cdots(s - \lambda_N) = 0$$

Equation (11.28) is known as the characteristic equation of the matrix $\mathbf{A}$, and $\lambda_1, \lambda_2, \cdots, \lambda_N$ are the characteristic roots (or eigenvalues) of $\mathbf{A}$. If all the eigenvalues $\lambda_k, 1 \le k \le N$ of $\mathbf{A}$ are distinct and $\Re\{\lambda_k\} < 0$, then the system is BIBO stable.

Example 11.4 Find the state vector $\mathbf{q}(t)$ for the system whose state equation is given by

$$\dot{\mathbf{q}}(t) = \begin{bmatrix} -1 & 1 \\ 0 & -2 \end{bmatrix} \mathbf{q}(t) + \begin{bmatrix} 0 \\ 1 \end{bmatrix} \mathbf{x}(t)$$

where $\mathbf{x}(t) = u(t)$ and the initial conditions are $q_1(0) = -1$ and $q_2(0) = 0$.

Solution

Comparing the given state equation with $\dot{\mathbf{q}}(t) = \mathbf{A}\mathbf{q}(t) + \mathbf{B}\mathbf{x}(t)$, we get

$$\mathbf{A} = \begin{bmatrix} -1 & 1 \\ 0 & -2 \end{bmatrix} \qquad \mathbf{B} = \begin{bmatrix} 0 \\ 1 \end{bmatrix}$$

From Eq. (11.23), we have

$$\mathbf{Q}(s) = \mathbf{\Phi}(s)[\mathbf{q}(0) + \mathbf{B}\mathbf{X}(s)]$$

where $\mathbf{\Phi}(s) = [s\mathbf{I} - \mathbf{A}]^{-1}$. Let us first find $\mathbf{\Phi}(s)$. We have

$$[s\mathbf{I} - \mathbf{A}] = s\begin{bmatrix} 1 & 0 \\ 0 & 1 \end{bmatrix} - \begin{bmatrix} -1 & 1 \\ 0 & -2 \end{bmatrix} = \begin{bmatrix} s+1 & -1 \\ 0 & s+2 \end{bmatrix}$$

and

$$\mathbf{\Phi}(s) = [s\mathbf{I} - \mathbf{A}]^{-1} = \frac{1}{(s+1)(s+2)}\begin{bmatrix} s+2 & 1 \\ 0 & s+1 \end{bmatrix} = \begin{bmatrix} \dfrac{1}{s+1} & \dfrac{1}{(s+1)(s+2)} \\ 0 & \dfrac{1}{s+2} \end{bmatrix}$$

Given that $\mathbf{x}(t) = u(t)$, thus $\mathbf{X}(s) = 1/s$. Therefore, we have

$$\mathbf{B}\mathbf{X}(s) = \begin{bmatrix} 0 \\ 1 \end{bmatrix}\frac{1}{s} = \begin{bmatrix} 0 \\ \frac{1}{s} \end{bmatrix} \quad \text{and} \quad \mathbf{q}(0) = \begin{bmatrix} -1 \\ 0 \end{bmatrix}$$

Therefore

$$\mathbf{q}(0) + \mathbf{B}\mathbf{X}(s) = \begin{bmatrix} -1 \\ \frac{1}{s} \end{bmatrix}$$

and

$$\mathbf{Q}(s) = \mathbf{\Phi}(s)[\mathbf{q}(0) + \mathbf{B}\mathbf{X}(s)]$$

$$= \begin{bmatrix} \dfrac{1}{s+1} & \dfrac{1}{(s+1)(s+2)} \\[3mm] 0 & \dfrac{1}{s+2} \end{bmatrix} \begin{bmatrix} -1 \\[2mm] \dfrac{1}{s} \end{bmatrix}$$

$$= \begin{bmatrix} \dfrac{-s^2 - 2s + 1}{s(s+1)(s+2)} \\[4mm] \dfrac{1}{s(s+2)} \end{bmatrix} = \begin{bmatrix} \dfrac{\frac{1}{2}}{s} - \dfrac{2}{s+1} + \dfrac{\frac{1}{2}}{s+2} \\[4mm] \dfrac{\frac{1}{2}}{s} - \dfrac{\frac{1}{2}}{s+2} \end{bmatrix}$$

The inverse Laplace transform of this equation yields

$$\begin{bmatrix} q_1(t) \\[2mm] q_2(t) \end{bmatrix} = \begin{bmatrix} \left(\dfrac{1}{2} - 2e^{-t} + \dfrac{1}{2}e^{-2t} \right) u(t) \\[4mm] \left(\dfrac{1}{2} - \dfrac{1}{2}e^{-2t} \right) u(t) \end{bmatrix}$$

11.5.2   Solution in the Time Domain

If the state equation is

$$\dot{\mathbf{q}}(t) = \mathbf{A}\mathbf{q}(t) + \mathbf{B}\mathbf{x}(t) \tag{11.30}$$

The solution of the vector differential Eq. (11.30) is

$$\mathbf{q}(t) = e^{\mathbf{A}t}\mathbf{q}(0) + \int_0^t e^{\mathbf{A}(t-\tau)}\mathbf{B}\mathbf{x}(\tau)\,d\tau \tag{11.31}$$

and if the output equation is given by

$$\mathbf{y}(t) = \mathbf{C}\mathbf{q}(t) + \mathbf{D}\mathbf{x}(t) \tag{11.32}$$

The solution of the vector differential Eq. (11.32) is

$$\mathbf{y}(t) = \mathbf{C}\,e^{\mathbf{A}t}\mathbf{q}(0) + \int_0^t \mathbf{C}\,e^{\mathbf{A}(t-\tau)}\mathbf{B}\mathbf{x}(\tau)\,d\tau + \mathbf{D}\mathbf{x}(t) \tag{11.33}$$

Proof Following

$$e^{at} = 1 + at + \frac{a^2}{2!}t^2 + \frac{a^3}{3!}t^3 + \cdots + \frac{a^n}{n!}t^n + \cdots$$

$$= \sum_{k=0}^{\infty} \frac{a^k}{k!}t^k$$

we define

$$e^{\mathbf{A}t} = 1 + \mathbf{A}t + \frac{\mathbf{A}^2}{2!}t^2 + \frac{\mathbf{A}^3}{3!}t^3 + \cdots + \frac{\mathbf{A}^n}{n!}t^n + \cdots \tag{11.34}$$

$$= \sum_{k=0}^{\infty} \frac{\mathbf{A}^k}{k!}t^k \tag{11.35}$$

where $k! = k(k-1)\dots 2.1$. If $t = 0$, then Eq. (11.34) reduces to

$$e^{\mathbf{0}} = \mathbf{I} \tag{11.36}$$

where $\mathbf{0}$ is an $N \times N$ zero matrix whose enterers are all zeros. As in if we premultiply or postmultiply the infinite series for $e^{\mathbf{A}t}$ [Eq. (11.35)], by an infinite series for $e^{-\mathbf{A}t}$, we find that

$$(e^{-\mathbf{A}t})(e^{\mathbf{A}t}) = (e^{\mathbf{A}t})(e^{-\mathbf{A}t}) = e^{\mathbf{0}} = \mathbf{I} \tag{11.37}$$

Thus,

$$e^{-\mathbf{A}t} = (e^{\mathbf{A}t})^{-1} \tag{11.38}$$

which indicates that $e^{-\mathbf{A}t}$ is the inverse of $e^{\mathbf{A}t}$.

The differentiation of Eq. (11.34) with respect to t yields

$$\frac{d}{dt}e^{\mathbf{A}t} = 0 + \mathbf{A} + \frac{\mathbf{A}^2}{2!}2t + \frac{\mathbf{A}^3}{3!}3t^2 + \cdots$$

$$= \mathbf{A} + \mathbf{A}^2 t + \frac{\mathbf{A}^3}{2!}t^2 + \cdots$$

$$= \mathbf{A}\left[\mathbf{I} + \mathbf{A}t + \frac{\mathbf{A}^2}{2!}t^2 + \cdots\right]$$

$$= \left[\mathbf{I} + \mathbf{A}t + \frac{\mathbf{A}^2}{2!}t^2 + \cdots\right]\mathbf{A}$$

$$\frac{d}{dt}e^{\mathbf{A}t} = \mathbf{A}\,e^{\mathbf{A}t} = e^{\mathbf{A}t}\mathbf{A} \tag{11.39}$$

We know that

$$\frac{d}{dt}[\mathbf{U}\mathbf{V}] = \frac{d\mathbf{U}}{dt}\mathbf{V} + \mathbf{U}\frac{d\mathbf{V}}{dt}$$

Using this relationship, we observe that

$$\frac{d}{dt}[e^{-\mathbf{A}t}\mathbf{q}(t)] = \left[\frac{d}{dt}e^{-\mathbf{A}t}\right]\mathbf{q}(t) + e^{-\mathbf{A}t}\dot{\mathbf{q}}(t)$$

$$= -e^{-\mathbf{A}t}\mathbf{A}\mathbf{q}(t) + e^{-\mathbf{A}t}\dot{\mathbf{q}}(t) \tag{11.40}$$

Now premultiplying both sides of Eq. (11.30) by $e^{-\mathbf{A}t}$, we obtain

$$e^{-\mathbf{A}t}\dot{\mathbf{q}}(t) = e^{-\mathbf{A}t}\mathbf{A}\mathbf{q}(t) + e^{-\mathbf{A}t}\mathbf{B}\mathbf{x}(t)$$

$$e^{-\mathbf{A}t}\dot{\mathbf{q}}(t) - e^{-\mathbf{A}t}\mathbf{A}\mathbf{q}(t) = e^{-\mathbf{A}t}\mathbf{B}\mathbf{x}(t) \tag{11.41}$$

Using Eq. (11.40), Eq. (11.41) can be rewritten as

$$\frac{d}{dt}[e^{-\mathbf{A}t}\mathbf{q}(t)] = e^{-\mathbf{A}t}\mathbf{B}\mathbf{x}(t) \tag{11.42}$$

Integrating both sides of Eq. (11.42) from 0 to t, we get

$$e^{-\mathbf{A}t}\mathbf{q}(t)\Big|_0^t = \int_0^t e^{-\mathbf{A}\tau}\mathbf{B}\mathbf{x}(\tau)\,d\tau$$

$$e^{-\mathbf{A}t}\mathbf{q}(t) - \mathbf{q}(0) = \int_0^t e^{-\mathbf{A}\tau}\mathbf{B}\mathbf{x}(\tau)\,d\tau$$

$$e^{-\mathbf{A}t}\mathbf{q}(t) = \mathbf{q}(0) + \int_0^t e^{-\mathbf{A}\tau}\mathbf{B}\mathbf{x}(\tau)\,d\tau \tag{11.43}$$

Premultiplying both sides of Eq. (11.43) by $e^{\mathbf{A}t}$ and using Eq. (11.37), we obtain

$$\mathbf{q}(t) = \underbrace{e^{\mathbf{A}t}\mathbf{q}(0)}_{\text{zero-input component}} + \underbrace{\int_0^t e^{\mathbf{A}(t-\tau)}\mathbf{B}\mathbf{x}(\tau)\,d\tau}_{\text{zero-state component}} \tag{11.44}$$

If the initial state is $\mathbf{q}(t_0)$ and we have $\mathbf{x}(t)$ for $t \geq t_0$, then

$$\mathbf{q}(t) = \underbrace{e^{\mathbf{A}(t-t_0)}\mathbf{q}(t_0)}_{\text{zero-input component}} + \underbrace{\int_{t_0}^t e^{\mathbf{A}(t-\tau)}\mathbf{B}\mathbf{x}(\tau)\,d\tau}_{\text{zero-state component}} \tag{11.45}$$

The result of Eq. (11.44) can be expressed more conveniently in terms of the matrix convolution. We can define the convolution of two matrices in a manner similar to the multiplication of two matrices, except that the multiplication of two elements is replaced by their convolution. For example,

$$\begin{bmatrix} g_1 & g_2 \\ g_3 & g_4 \end{bmatrix} * \begin{bmatrix} h_1 & h_2 \\ h_3 & h_4 \end{bmatrix} = \begin{bmatrix} (g_1 * h_1 + g_2 * h_3) & (g_1 * h_2 + g_2 * h_4) \\ (g_3 * h_1 + g_4 * h_3) & (g_3 * h_2 + g_4 * h_4) \end{bmatrix}$$

By using this definition of matrix convolution, we can express Eq. (11.44) as

$$\mathbf{q}(t) = e^{\mathbf{A}t}\mathbf{q}(0) + e^{\mathbf{A}t} * \mathbf{B}\mathbf{x}(t) \tag{11.46}$$

Note that the limits of the convolution integral [Eq. (11.44)] are from 0 to t. Hence, all the elements of $e^{\mathbf{A}t}$ in the convolution term of Eq. (11.46) are implicitly assumed to be multiplied by $u(t)$. The matrix function $e^{\mathbf{A}t}$ is known as the *state transition matrix* (STM).

Now consider the output Eq. (11.32)

$$\mathbf{y}(t) = \mathbf{C}\mathbf{q}(t) + \mathbf{D}\mathbf{x}(t)$$

Substituting Eq. (11.44) into the above equation, we obtain

$$\mathbf{y}(t) = \mathbf{C}\left[e^{\mathbf{A}t}\mathbf{q}(0) + \int_0^t e^{\mathbf{A}(t-\tau)}\mathbf{B}\mathbf{x}(\tau)\,d\tau\right] + \mathbf{D}\mathbf{x}(t)$$

$$= \mathbf{C}\,e^{\mathbf{A}t}\mathbf{q}(0) + \int_0^t \mathbf{C}\,e^{\mathbf{A}(t-\tau)}\mathbf{B}\mathbf{x}(\tau)\,d\tau + \mathbf{D}\mathbf{x}(t) \tag{11.47}$$

By using the definition of the matrix convolution, we can express Eq. (11.47) as

$$\mathbf{y}(t) = \mathbf{C}\left[e^{\mathbf{A}t}\mathbf{q}(0) + e^{\mathbf{A}t} * \mathbf{B}\mathbf{x}(t)\right] + \mathbf{D}\mathbf{x}(t) \tag{11.48}$$

Since the elements of **B** are constants,

$$e^{\mathbf{A}t} * \mathbf{B}\mathbf{x}(t) = e^{\mathbf{A}t}\mathbf{B} * \mathbf{x}(t)$$

With this result Eq. (11.48) becomes

$$\mathbf{y}(t) = \mathbf{C}\left[e^{\mathbf{A}t}\mathbf{q}(0) + e^{\mathbf{A}t}\mathbf{B} * \mathbf{x}(t)\right] + \mathbf{D}\mathbf{x}(t) \tag{11.49}$$

We know that $\delta(t) * x(t) = x(t)$. Let us define an $m \times m$ diagonal matrix $\boldsymbol{\delta}(t)$ such that all its diagonal elements are unit impulse functions. It is then obvious that $\boldsymbol{\delta}(t) * \mathbf{x}(t) = \mathbf{x}(t)$ and Eq. (11.49) can be expressed as

$$\mathbf{y}(t) = \mathbf{C}\left[e^{\mathbf{A}t}\mathbf{q}(0) + e^{\mathbf{A}t}\mathbf{B} * \mathbf{x}(t)\right] + \mathbf{D}\boldsymbol{\delta}(t) * \mathbf{x}(t)$$

$$= \mathbf{C}\,e^{\mathbf{A}t}\mathbf{q}(0) + \left[\mathbf{C}\,e^{\mathbf{A}t}\mathbf{B} + \mathbf{D}\boldsymbol{\delta}(t)\right] * \mathbf{x}(t) \tag{11.50}$$

With the notation $\mathcal{L}_u^{-1}[\boldsymbol{\Phi}(s)] = \boldsymbol{\phi}(t) = e^{\mathbf{A}t}$, Eq. (11.50) may be expressed as

$$\mathbf{y}(t) = \underbrace{\mathbf{C}\boldsymbol{\phi}(t)\mathbf{q}(0)}_{\text{zero-input response}} + \underbrace{\left[\mathbf{C}\boldsymbol{\phi}(t)\mathbf{B} + \mathbf{D}\boldsymbol{\delta}(t)\right] * \mathbf{x}(t)}_{\text{zero-state response}} \tag{11.51}$$

The zero-state response, i.e., the response when $\mathbf{q}(0) = \mathbf{0}$, is

$$\mathbf{y}(t) = \left[\mathbf{C}\boldsymbol{\phi}(t)\mathbf{B} + \mathbf{D}\boldsymbol{\delta}(t)\right] * \mathbf{x}(t)$$

$$= \mathbf{h}(t) * \mathbf{x}(t) \tag{11.52}$$

where

$$\mathbf{h}(t) = \mathbf{C}\boldsymbol{\phi}(t)\mathbf{B} + \mathbf{D}\boldsymbol{\delta}(t) \tag{11.53}$$

The matrix $\mathbf{h}(t)$ is a $p \times m$ matrix known as the *impulse response matrix*. The ijth element $h_{ij}(t)$ of $\mathbf{h}(t)$ is the impulse response (zero-state response) that relates the output $y_i(t)$ to the input $x_j(t) = \delta(t)$ and when all other inputs are zero. It can also be seen from Eqs (11.28) and (11.52) that

$$\mathcal{L}[\mathbf{h}(t)] = \mathbf{H}(s)$$

∎

11.5.3 Evaluation of State Transition Matrix $\phi(t) = e^{\mathbf{A}t}$

It is clear from Eqs (11.46) and (11.51) that in order to determine $\mathbf{q}(t)$ and $\mathbf{y}(t)$, we have to first obtain $e^{\mathbf{A}t}$. There are several methods to obtain the STM $\phi(t) = e^{\mathbf{A}t}$.

Method 1 (Infinite Power Series Summation Method) We can find the STM $\phi(t) = e^{\mathbf{A}t}$ by using power series method:

$$e^{\mathbf{A}t} = 1 + \mathbf{A}t + \frac{\mathbf{A}^2}{2!}t^2 + \frac{\mathbf{A}^3}{3!}t^3 + \cdots + \frac{\mathbf{A}^n}{n!}t^n + \cdots$$

Although this method is straightforward, the major problem is that it is usually not possible to recognize a closed form corresponding to this solution.

Method 2 (Finite Power Series Summation Method) The characteristic equation of the matrix $\mathbf{A}$ is

$$|\lambda\mathbf{I} - \mathbf{A}| = 0 \tag{11.54}$$

Since $|\lambda\mathbf{I} - \mathbf{A}|$ is an Nth-order polynomial in λ with N eigenvalues $\lambda_1, \lambda_2, \ldots, \lambda_N$, we can write Eq. (11.54) as

$$Q(\lambda) = |\lambda\mathbf{I} - \mathbf{A}| = \lambda^N + a_{N-1}\lambda^{N-1} + \cdots + a_2\lambda^2 + a_1\lambda + a_0 = 0 \tag{11.55}$$

$Q(\lambda)$ is called the characteristic polynomial of the matrix $\mathbf{A}$. The N zeros of the characteristic polynomial are the eigenvalues of $\mathbf{A}$.

Consider a function $f(\lambda)$ in the form of an infinite power series:

$$f(\lambda) = \alpha_0 + \alpha_1\lambda + \alpha_2\lambda^2 + \cdots = \sum_{k=0}^{\infty} \alpha_k\lambda^k \tag{11.56}$$

Since λ, being an eigenvalue of $\mathbf{A}$, satisfies the characteristic equation [Eq. (11.55)], we can write

$$\lambda^N = -a_0 - a_1\lambda - a_2\lambda^2 - \cdots - a_{N-1}\lambda^{N-1} \tag{11.57}$$

Multiplying both sides by λ and then substituting the expression for λ^N [Eq. (11.56)] on the right and rearranging, we get

$$\lambda^{N+1} = b_0 + b_1\lambda + b_2\lambda^2 + \cdots + b_{N-1}\lambda^{N-1} \tag{11.58}$$

By continuing this process, we can express any positive integral power of λ as a linear combination of $\lambda, \lambda^2, \ldots, \lambda^{N-1}$. Hence, the infinite series on the RHS of Eq. (11.56) can always be expressed in terms of $\lambda, \lambda^2, \ldots, \lambda^{N-1}$ and a constant as

$$f(\lambda) = \beta_0 + \beta_1\lambda + \beta_2\lambda^2 + \cdots + \beta_{N-1}\lambda^{N-1} = \sum_{k=0}^{N-1} \beta_k\lambda^k \tag{11.59}$$

If we assume that there are N distinct eigenvalues $\lambda_1, \lambda_2, \ldots, \lambda_N$, then Eq. (11.59) hold for these N values of λ. The substitution of these values in Eq. (11.59) yields N simultaneous equations:

$$
\begin{bmatrix} f(\lambda_1) \\ f(\lambda_2) \\ \vdots \\ f(\lambda_N) \end{bmatrix} = \begin{bmatrix} 1 & \lambda_1 & \lambda_1^2 & \cdots & \lambda_1^{N-1} \\ 1 & \lambda_2 & \lambda_1^2 & \cdots & \lambda_1^{N-1} \\ \vdots & \vdots & \vdots & \ddots & \vdots \\ 1 & \lambda_N & \lambda_N^2 & \cdots & \lambda_N^{N-1} \end{bmatrix} \begin{bmatrix} \beta_0 \\ \beta_1 \\ \vdots \\ \beta_{N-1} \end{bmatrix}
\tag{11.60}
$$

and

$$
\begin{bmatrix} \beta_0 \\ \beta_1 \\ \vdots \\ \beta_{N-1} \end{bmatrix} = \begin{bmatrix} 1 & \lambda_1 & \lambda_1^2 & \cdots & \lambda_1^{N-1} \\ 1 & \lambda_2 & \lambda_1^2 & \cdots & \lambda_1^{N-1} \\ \vdots & \vdots & \vdots & \ddots & \vdots \\ 1 & \lambda_N & \lambda_N^2 & \cdots & \lambda_N^{N-1} \end{bmatrix}^{-1} \begin{bmatrix} f(\lambda_1) \\ f(\lambda_2) \\ \vdots \\ f(\lambda_N) \end{bmatrix}
\tag{11.61}
$$

The Cayley–Hamilton theorem states that every $N \times N$ matrix $\mathbf{A}$ satisfies its own characteristic equation. In other words, Eq. (11.55) is valid if λ is replaced by $\mathbf{A}$:

$$
Q(\mathbf{A}) = \mathbf{A}^N + a_{N-1}\mathbf{A}^{N-1} + \cdots + a_1\mathbf{A} + a_0\mathbf{I} = 0
\tag{11.62}
$$

Rewriting Eq. (11.62), we have

$$
\mathbf{A}^N = -a_0\mathbf{I} - a_1\mathbf{A} - \cdots - a_{N-1}\mathbf{A}^{N-1}
\tag{11.63}
$$

Multiplying both sides by $\mathbf{A}$ and then substituting the expression for $\mathbf{A}^n$ [Eq. (11.63)] on the right and rearranging, we get

$$
\mathbf{A}^{N+1} = b_0\mathbf{I} + b_1\mathbf{A} + \cdots + b_{N-1}\mathbf{A}^{N-1}
\tag{11.64}
$$

By continuing this process, we can express any positive integral power of $\mathbf{A}$ as a linear combination of $\mathbf{I}, \mathbf{A}, \ldots, \mathbf{A}^{N-1}$. Hence, the infinite series given by

$$
f(\mathbf{A}) = \alpha_0\mathbf{I} + \alpha_1\mathbf{A} + \alpha_2\mathbf{A}^2 + \cdots = \sum_{k=0}^{\infty} \alpha_k\mathbf{A}^k
\tag{11.65}
$$

can always be expressed in terms of $\mathbf{I}, \mathbf{A}, \mathbf{A}^2, \ldots, \mathbf{A}^{N-1}$ as

$$
f(\mathbf{A}) = \beta_0\mathbf{I} + \beta_1\mathbf{A} + \beta_2\mathbf{A}^2 + \cdots + \beta_{N-1}\mathbf{A}^{N-1} = \sum_{k=0}^{N-1} \beta_k\mathbf{A}^k
\tag{11.66}
$$

in which the coefficients β_i's are found from Eq. (11.61). If all eigenvalues of $\mathbf{A}$ are not distinct, then Eq. (11.59) will not yield N equations. Assume that an eigenvalue λ_i has multiplicity r and all other eigenvalues are distinct. In this case differentiating both sides of Eq. (11.59) r times with respect to λ and setting $\lambda = \lambda_i$, we obtain r equations corresponding to λ_i:

$$
\left. \frac{d^{n-1}}{d\lambda^{n-1}} f(\lambda) \right|_{\lambda=\lambda_i} = \left. \frac{d^{n-1}}{d\lambda^{n_1}} \left(\sum_{k=0}^{N-1} \beta_k \lambda^k \right) \right|_{\lambda=\lambda_i} \qquad n = 1, 2, \ldots, r
\tag{11.67}
$$

Combining Eqs (11.61) and (11.67), we can determine all coefficients β_k in Eq. (11.66).

Similarly, the STM $f(\mathbf{A}) = \phi(t) = e^{\mathbf{A}t}$ [infinite power series described in Eq. (11.34)] can be expressed as

$$e^{\mathbf{A}t} = \beta_0 \mathbf{I} + \beta_1 \mathbf{A} + \beta_2 \mathbf{A}^2 + \cdots + \beta_{N-1} \mathbf{A}^{N-1} \tag{11.68}$$

where the coefficients β_is are found from Eq. (11.61).

Method 3 (Diagonalization Method) Let $\mathbf{A}$ be an $N \times N$ matrix. If

$$\mathbf{A}\mathbf{x} = \lambda\mathbf{x} \tag{11.69}$$

for some scalar λ and nonzero column vector $\mathbf{x}$, then λ is called an *eigenvalue* (or *characteristic value*) of $\mathbf{A}$ and $\mathbf{x}$ is called an *eigenvector* associated with λ. If all eigenvalues of $\mathbf{A}$ are distinct and if $\mathbf{x}_1, \mathbf{x}_2, \ldots, \mathbf{x}_N$ are eigenvectors associated with the eigenvalues $\lambda_1, \lambda_2, \cdots, \lambda_N$, then

$$\mathbf{A}\mathbf{x}_1 = \lambda_1\mathbf{x}_1, \qquad \mathbf{A}\mathbf{x}_2 = \lambda_2\mathbf{x}_2, \qquad \mathbf{A}\mathbf{x}_N = \lambda_N\mathbf{x}_N$$

Let

$$\mathbf{P} = \begin{bmatrix} \mathbf{x}_1 & \mathbf{x}_2 & \cdots & \mathbf{x}_N \end{bmatrix} \tag{11.70}$$

Then

$$\mathbf{A}\mathbf{P} = \mathbf{A} \begin{bmatrix} \mathbf{x}_1 & \mathbf{x}_2 & \cdots & \mathbf{x}_N \end{bmatrix}$$

$$= \begin{bmatrix} \mathbf{A}\mathbf{x}_1 & \mathbf{A}\mathbf{x}_2 & \cdots & \mathbf{A}\mathbf{x}_N \end{bmatrix}$$

$$= \begin{bmatrix} \lambda_1\mathbf{x}_1 & \lambda_2\mathbf{x}_2 & \cdots & \lambda_N\mathbf{x}_N \end{bmatrix}$$

$$\mathbf{P}\mathbf{\Lambda} = \begin{bmatrix} \mathbf{x}_1 & \mathbf{x}_2 & \cdots & \mathbf{x}_N \end{bmatrix} \begin{bmatrix} \lambda_1 & 0 & \cdots & 0 \\ 0 & \lambda_2 & \cdots & 0 \\ \vdots & \vdots & \ddots & \vdots \\ 0 & 0 & \cdots & \lambda_N \end{bmatrix} = \mathbf{P}\mathbf{\Lambda} \tag{11.71}$$

where

$$\mathbf{\Lambda} = \begin{bmatrix} \lambda_1 & 0 & \cdots & 0 \\ 0 & \lambda_2 & \cdots & 0 \\ \vdots & \vdots & \ddots & \vdots \\ 0 & 0 & \cdots & \lambda_N \end{bmatrix} \tag{11.72}$$

By definition $\mathbf{P}$ has N linearly independent column vectors. Thus, $\mathbf{P}$ is nonsingular and $\mathbf{P}^{-1}$ exists, and hence

$$\mathbf{P}^{-1}\mathbf{A}\mathbf{P} = \mathbf{\Lambda} = \begin{bmatrix} \lambda_1 & 0 & \cdots & 0 \\ 0 & \lambda_2 & \cdots & 0 \\ \vdots & \vdots & \ddots & \vdots \\ 0 & 0 & \cdots & \lambda_N \end{bmatrix} \tag{11.73}$$

We call $\mathbf{P}$ the *diagonalization* matrix, or *eigenvector* matrix, and $\boldsymbol{\Lambda}$ the *eigenvalue* matrix. Some important points regarding the eigenvalues and eigenvectors of $\mathbf{A}$ are as follows:

1. If the eigenvalues of $\mathbf{A}$ are $\lambda_1, \lambda_2, \ldots, \lambda_N$, then the eigenvalues of $\mathbf{A}^n$ are $\lambda_1^n, \lambda_2^n, \ldots, \lambda_N^n$.

2. Each eigenvector of $\mathbf{A}$ is still an eigenvector of $\mathbf{A}^n$.

3. If $\mathbf{P}$ diagonalizes $\mathbf{A}$, i.e., $\mathbf{P}^{-1}\mathbf{A}\mathbf{P} = \boldsymbol{\Lambda}$, then it also diagonalizes $\mathbf{A}^n$, i.e.,

$$\mathbf{P}^{-1}\mathbf{A}^n\mathbf{P} = \boldsymbol{\Lambda}^n = \begin{bmatrix} \lambda_1^n & 0 & \cdots & 0 \\ 0 & \lambda_2^n & \cdots & 0 \\ \vdots & \vdots & \ddots & \vdots \\ 0 & 0 & \cdots & \lambda_N^n \end{bmatrix} \qquad (11.74)$$

Since $\mathbf{P}$ diagonalizes $\mathbf{A}$, i.e., $\mathbf{P}^{-1}\mathbf{A}\mathbf{P} = \boldsymbol{\Lambda}$, we have

$$\mathbf{A} = \mathbf{P}\boldsymbol{\Lambda}\mathbf{P}^{-1}$$

and

$$\mathbf{A}^2 = [\mathbf{P}\boldsymbol{\Lambda}\mathbf{P}^{-1}][\mathbf{P}\boldsymbol{\Lambda}\mathbf{P}^{-1}] = \mathbf{P}\boldsymbol{\Lambda}^2\mathbf{P}^{-1} \qquad (11.75)$$

$$\mathbf{A}^3 = [\mathbf{P}\boldsymbol{\Lambda}^2\mathbf{P}^{-1}][\mathbf{P}\boldsymbol{\Lambda}\mathbf{P}^{-1}] = \mathbf{P}\boldsymbol{\Lambda}^3\mathbf{P}^{-1} \qquad (11.76)$$

Thus, we obtain

$$f(\mathbf{A}) = \mathbf{P}f(\boldsymbol{\Lambda})\mathbf{P}^{-1} = \mathbf{P}\begin{bmatrix} f(\lambda_1) & 0 & \cdots & 0 \\ 0 & f(\lambda_2) & \cdots & 0 \\ \vdots & \vdots & \ddots & \vdots \\ 0 & 0 & \cdots & f(\lambda_N) \end{bmatrix}\mathbf{P}^{-1} \qquad (11.77)$$

Hence, the STM $f(\mathbf{A}) = \boldsymbol{\phi}(t) = e^{\mathbf{A}t}$ [infinite power series described in Eq. (11.34)] can be expressed as

$$\boldsymbol{\phi}(t) = e^{\mathbf{A}t} = \mathbf{P}\begin{bmatrix} e^{\lambda_1 t} & 0 & \cdots & 0 \\ 0 & e^{\lambda_2 t} & \cdots & 0 \\ \vdots & \vdots & \ddots & \vdots \\ 0 & 0 & \cdots & e^{\lambda_N t} \end{bmatrix}\mathbf{P}^{-1} \qquad (11.78)$$

Method 4 (Spectral Decomposition Method) The *minimal* (or minimum) polynomial $m(\lambda)$ of an $N \times N$ matrix $\mathbf{A}$ is the polynomial of lowest degree having 1 as its leading coefficient such that $m(\mathbf{A}) = \mathbf{0}$. Since $\mathbf{A}$ satisfies its characteristic equation, the degree of $m(\lambda)$ is not greater than N. It can be shown that if the minimal polynomial $m(\lambda)$ of an $N \times N$ matrix $\mathbf{A}$ has the form

$$m(\lambda) = (\lambda - \lambda_1)(\lambda - \lambda_2) \cdots (\lambda - \lambda_N) \qquad (11.79)$$

then $\mathbf{A}$ can be represented by

$$\mathbf{A} = \lambda_1\mathbf{E}_1 + \lambda_2\mathbf{E}_2 + \cdots + \lambda_N\mathbf{E}_N \qquad (11.80)$$

where E_j, $j = 1, 2, \ldots, N$, are called *constituent matrices* and have the following properties:

1. $\mathbf{I} = \mathbf{E}_1 + \mathbf{E}_2 + \cdots + \mathbf{E}_N$
2. $\mathbf{E}_i \mathbf{E}_j = \mathbf{0}, \quad i \neq j$
3. $\mathbf{E}_i^2 = \mathbf{E}_i$
4. $\mathbf{A}\mathbf{E}_i = \mathbf{E}_i \mathbf{A} = \lambda_i \mathbf{E}_i$

Any matrix $\mathbf{B}$ for which $\mathbf{B}^2 = \mathbf{B}$ is called *idempotent*. Thus, the constituent matrices E_j are called idempotent matrices. The set of eigenvalues of $\mathbf{A}$ is called the spectrum of $\mathbf{A}$ and Eq. (11.80) is called the *spectral decomposition* of $\mathbf{A}$. Using the properties of the constituent matrices, we have

$$\mathbf{A}^2 = \lambda_1^2 \mathbf{E}_1 + \lambda_2^2 \mathbf{E}_2 + \cdots + \lambda_N^2 \mathbf{E}_N$$

$$\vdots$$

$$\mathbf{A}^n = \lambda_1^n \mathbf{E}_1 + \lambda_2^n \mathbf{E}_2 + \cdots + \lambda_N^n \mathbf{E}_N \qquad (11.81)$$

and

$$f(\mathbf{A}) = f(\lambda_1)\mathbf{E}_1 + f(\lambda_2)\mathbf{E}_2 + \cdots + f(\lambda_N)\mathbf{E}_N \qquad (11.82)$$

Thus, the STM $\phi(t) = e^{\mathbf{A}t}$ can be expressed as

$$\phi(t) = e^{\mathbf{A}t} = e^{\lambda_1 t}\mathbf{E}_1 + e^{\lambda_2 t}\mathbf{E}_2 + \cdots + e^{\lambda_N t}\mathbf{E}_N \qquad (11.83)$$

The constituent matrices E_j can be evaluated as follows. The partial fraction expansion of

$$\frac{1}{m(\lambda)} = \frac{1}{(\lambda - \lambda_1)(\lambda - \lambda_2) \cdots (\lambda - \lambda_N)}$$

$$= \frac{k_1}{\lambda - \lambda_1} + \frac{k_2}{\lambda - \lambda_2} + \cdots + \frac{k_N}{\lambda - \lambda_N}$$

leads to

$$k_j = \frac{1}{\displaystyle\prod_{\substack{m=1 \\ m \neq j}}^{N} (\lambda_j - \lambda_m)}$$

Then

$$\frac{1}{m(\lambda)} = \frac{k_1 g_1(\lambda) + k_2 g_2(\lambda) + \cdots + k_N g_N(\lambda)}{(\lambda - \lambda_1)(\lambda - \lambda_2) \cdots (\lambda - \lambda_N)}$$

where

$$g_j(\lambda) = \prod_{\substack{m=1 \\ m \neq j}}^{N} (\lambda - \lambda_m)$$

Let $e_j(\lambda) = k_j g_j(\lambda)$. Then the constituent matrices $\mathbf{E}_j$ can be evaluated as

$$\mathbf{E}_j = e_j(\mathbf{A}) = \frac{\displaystyle\prod_{\substack{m=1 \\ m \neq j}}^{N} (\mathbf{A} - \lambda_m \mathbf{I})}{\displaystyle\prod_{\substack{m=1 \\ m \neq j}}^{N} (\lambda_j - \lambda_m)} \tag{11.84}$$

Method 5 (Laplace Transform Method) Using the Laplace transform, we can evaluate $\phi(t) = e^{\mathbf{A}t}$. Comparing Eqs (11.26) and (11.44), we see that

$$\phi(t) = e^{\mathbf{A}t} = \mathcal{L}_u^{-1}[\boldsymbol{\Phi}(s)] = \mathcal{L}_u^{-1}\left[[s\mathbf{I} - \mathbf{A}]^{-1}\right] \tag{11.85}$$

11.5.4 Properties of STM $\phi(t) = e^{\mathbf{A}t}$

The STM possesses several properties, some of which are as follows:

1. $\phi(0) = \mathbf{I}$
2. $\phi(t) = e^{\mathbf{A}t} = (e^{-\mathbf{A}t})^{-1} = [\phi(-t)]^{-1}$
3. $\phi(t_2 - t_0) = \phi(t_2 - t_1)\phi(t_1 - t_0)$
4. $\phi(t - t_0) = \phi(t)\phi^{-1}(t_0)$
5. $\phi(t + t_0) = \phi(t)\phi(t_0)$

11.6 STATE SPACE REPRESENTATION OF DISCRETE-TIME LTI SYSTEMS

11.6.1 Systems Described by Difference Equations

Suppose that a single-input single-output discrete-time LTI system is described by an Nth-order difference equation

$$y(n) + a_1 y(n-1) + \cdots + a_N y(n-N) = x(n) \tag{11.86}$$

We know from previous discussion that if $x(n)$ is given for $n \geq 0$, then Eq. (11.86) requires N initial conditions $y(-1)$, $y(-2)$, ..., $y(-N)$ to uniquely determine the complete solution for $n > 0$. That is, N values are required to specify the state of the system at any time.

Let us define N state variables $q_1(n)$, $q_2(n)$, ..., $q_N(n)$ as

$$q_1(n) = y(n-N)$$

$$q_2(n) = y\big(n - (N-1)\big) = y(n - N + 1)$$

$$\vdots$$

$$q_N(n) = Y(N-1) \tag{11.87}$$

Then from Eqs (11.86) and (11.87), we have

$$q_1(n+1) = q_2(n)$$
$$q_2(n+1) = q_3(n)$$
$$\vdots$$
$$q_N(n+1) = -a_N q_1(n) - a_{N-1} q_2(n) - \cdots - a_1 q_N(n) + x(n) \qquad (11.88)$$

and

$$y(n) = -a_N q_1(n) - a_{N-1} q_2(n) - \cdots - a_1 q_N(n) + x(n) \qquad (11.89)$$

In matrix form, Eqs (11.88) and (11.89) can be expressed as

$$\underbrace{\begin{bmatrix} q_1(n+1) \\ q_2(n+1) \\ \vdots \\ q_N(n+1) \end{bmatrix}}_{\mathbf{q}(n+1)} = \underbrace{\begin{bmatrix} 0 & 1 & 0 & \cdots & 0 \\ 0 & 0 & 1 & \cdots & 0 \\ \vdots & \vdots & \vdots & \ddots & \vdots \\ -a_N & -a_{N-1} & -a_{N-2} & \cdots & -a_1 \end{bmatrix}}_{\mathbf{A}} \underbrace{\begin{bmatrix} q_1(n) \\ q_2(n) \\ \vdots \\ q_N(n) \end{bmatrix}}_{\mathbf{q}(n)} + \underbrace{\begin{bmatrix} 0 \\ 0 \\ \vdots \\ 1 \end{bmatrix}}_{\mathbf{b}} x(n) \qquad (11.90)$$

$$y(n) = \underbrace{\begin{bmatrix} -a_N & -a_{N-1} & \cdots & -a_1 \end{bmatrix}}_{\mathbf{c}} \underbrace{\begin{bmatrix} q_1(n) \\ q_2(n) \\ \vdots \\ q_N(n) \end{bmatrix}}_{\mathbf{q}(n)} + \underbrace{[1]}_{d} x(n) \qquad (11.91)$$

Then Eqs (11.90) and (11.91) can be written compactly as

$$\mathbf{q}(n+1) = \mathbf{A}\mathbf{q}(n) + \mathbf{b}x(n) \qquad (11.92)$$
$$y(n) = \mathbf{c}\mathbf{q}(n) + d\,x(n) \qquad (11.93)$$

Equations (11.92) and (11.93) are called an N-dimensional state space representation (or state equations) of the system, and the $N \times N$ matrix $\mathbf{A}$ is termed the system matrix.

Example 11.5 Find state equations of a discrete-time system described by

$$y(n) - \frac{3}{4}y(n-1) + \frac{1}{8}y(n-2) = x(n)$$

Solution

Choose the state variables $q_1(n)$ and $q_2(n)$ as

$$q_1(n) = y(n-2)$$
$$q_2(n) = y(n-1)$$

Then,

$$q_1(n + 1) = q_2(n)$$

$$q_2(n + 1) = -\frac{1}{8}q_1(n) + \frac{3}{4}q_2(n) + x(n)$$

$$y(n) = -\frac{1}{8}q_1(n) + \frac{3}{4}q_2(n) + x(n)$$

The above equations can be rewritten compactly as

$$\underbrace{\begin{bmatrix} q_1(n+1) \\ q_2(n+1) \end{bmatrix}}_{\mathbf{q}(n+1)} = \underbrace{\begin{bmatrix} 0 & 1 \\ -\frac{1}{8} & \frac{3}{4} \end{bmatrix}}_{\mathbf{A}} \underbrace{\begin{bmatrix} q_1(n) \\ q_2(n) \end{bmatrix}}_{\mathbf{q}(n)} + \underbrace{\begin{bmatrix} 0 \\ 1 \end{bmatrix}}_{\mathbf{b}} x(n)$$

and

$$y(n) = \underbrace{\begin{bmatrix} -\frac{1}{8} & \frac{3}{4} \end{bmatrix}}_{\mathbf{C}} \underbrace{\begin{bmatrix} q_1(n) \\ q_2(n) \end{bmatrix}}_{\mathbf{q}(n)} + x(n)$$

11.6.2 Multiple-Input Multiple-Output System

If a discrete-time LTI system has m inputs and p outputs and N state variables, then a state space representation of the system can be expressed as

$$\mathbf{q}(n + 1) = \mathbf{A}\mathbf{q}(n) + \mathbf{B}\mathbf{x}(n) \tag{11.94}$$

$$\mathbf{y}(n) = \mathbf{C}\mathbf{q}(n) + \mathbf{D}\mathbf{x}(n) \tag{11.95}$$

where

$$\mathbf{q}(n) = \begin{bmatrix} q_1(n) \\ q_2(n) \\ \vdots \\ q_N(n) \end{bmatrix} \quad \mathbf{x}(n) = \begin{bmatrix} x_1(n) \\ x_2(n) \\ \vdots \\ x_m(n) \end{bmatrix} \quad \mathbf{y}(n) = \begin{bmatrix} y_1(n) \\ y_2(n) \\ \vdots \\ y_p(n) \end{bmatrix}$$

and

$$\mathbf{A} = \begin{bmatrix} a_{11} & a_{12} & \cdots & a_{1N} \\ a_{21} & a_{22} & \cdots & a_{2N} \\ \vdots & \vdots & \ddots & \vdots \\ a_{N1} & a_{N2} & \cdots & a_{NN} \end{bmatrix}_{N \times N} \quad \mathbf{B} = \begin{bmatrix} b_{11} & b_{12} & \cdots & b_{1m} \\ b_{21} & b_{22} & \cdots & b_{2m} \\ \vdots & \vdots & \ddots & \vdots \\ b_{N1} & b_{N2} & \cdots & b_{Nm} \end{bmatrix}_{N \times m}$$

$$\mathbf{C} = \begin{bmatrix} c_{11} & c_{12} & \cdots & c_{1N} \\ c_{21} & c_{22} & \cdots & c_{2N} \\ \vdots & \vdots & \ddots & \vdots \\ c_{p1} & c_{p2} & \cdots & c_{pN} \end{bmatrix}_{p \times N} \quad \mathbf{D} = \begin{bmatrix} d_{11} & d_{12} & \cdots & d_{1m} \\ d_{21} & d_{22} & \cdots & d_{2m} \\ \vdots & \vdots & \ddots & \vdots \\ d_{p1} & d_{p2} & \cdots & d_{pm} \end{bmatrix}_{p \times m}$$

11.7 SOLUTION OF STATE EQUATIONS FOR DISCRETE-TIME LTI SYSTEMS

11.7.1 Solution in the Time Domain

Consider an N-dimensional state space representation

$$\mathbf{q}(n+1) = \mathbf{A}\mathbf{q}(n) + \mathbf{B}\mathbf{x}(n) \tag{11.96}$$

$$\mathbf{y}(n) = \mathbf{C}\mathbf{q}(n) + \mathbf{D}\mathbf{x}(n) \tag{11.97}$$

where $\mathbf{A}$, $\mathbf{B}$, $\mathbf{C}$, and $\mathbf{D}$ are $N \times N$, $N \times m$, $p \times N$, $p \times m$ matrices, respectively. One method of finding $\mathbf{q}(n)$, given the initial state $\mathbf{q}(0)$, is to solve Eq. (11.96) iteratively. Thus,

$$\mathbf{q}(1) = \mathbf{A}\mathbf{q}(0) + \mathbf{B}\mathbf{x}(0)$$

$$\mathbf{q}(2) = \mathbf{A}\mathbf{q}(1) + \mathbf{B}\mathbf{x}(1) = \mathbf{A}[\mathbf{A}\mathbf{q}(0) + \mathbf{B}\mathbf{x}(0)] + \mathbf{B}\mathbf{x}(1)$$

$$= \mathbf{A}^2\mathbf{q}(0) + \mathbf{A}\mathbf{B}\mathbf{x}(0) + \mathbf{B}\mathbf{x}(1)$$

By continuing this process, we obtain

$$\mathbf{q}(n) = \mathbf{A}^n\mathbf{q}(0) + \mathbf{A}^{n-1}\mathbf{B}\mathbf{x}(0) + \mathbf{A}^{n-2}\mathbf{B}\mathbf{x}(1) + \cdots + \mathbf{B}\mathbf{x}(n-1) \quad n > 0$$

$$\mathbf{q}(n) = \mathbf{A}^n\mathbf{q}(0) + \sum_{k=0}^{n-1} \mathbf{A}^{n-1-k}\mathbf{B}\mathbf{x}(k) \quad n > 0 \tag{11.98}$$

The upper limit on the summation in Eq. (11.98) is nonnegative. Hence, $n \geq 1$, and the summation is recognized as the convolution sum

$$\sum_{k=0}^{n-1} \mathbf{A}^{n-1-k}\mathbf{B}\mathbf{x}(k) = \mathbf{A}^{n-1}u(n-1) * \mathbf{B}\mathbf{x}(n)$$

Consequently

$$\mathbf{q}(n) = \underbrace{\mathbf{A}^n\mathbf{q}(0)}_{\text{zero-input component}} + \underbrace{\mathbf{A}^{n-1}u(n-1) * \mathbf{B}\mathbf{x}(n)}_{\text{zero-state component}} \tag{11.99}$$

Now, consider the output Eq. (11.97)

$$\mathbf{y}(n) = \mathbf{C}\mathbf{q}(n) + \mathbf{D}\mathbf{x}(n)$$

Substituting Eq. (11.98) into the above equation, we obtain

$$\mathbf{y}(n) = \mathbf{C}\left[\mathbf{A}^n\mathbf{q}(0) + \sum_{k=0}^{n-1} \mathbf{A}^{n-1-k}\mathbf{B}\mathbf{x}(k)\right] + \mathbf{D}\mathbf{x}(n)$$

$$= \mathbf{C}\mathbf{A}^n\mathbf{q}(0) + \sum_{k=0}^{n-1} \mathbf{C}\mathbf{A}^{n-1-k}\mathbf{B}\mathbf{x}(k) + \mathbf{D}\mathbf{x}(n)$$

$$\mathbf{y}(n) = \mathbf{C}\mathbf{A}^n\mathbf{q}(0) + \mathbf{C}\mathbf{A}^{n-1}u(n-1) * \mathbf{B}\mathbf{x}(n) + \mathbf{D}\mathbf{x}(n) \tag{11.100}$$

11.7.2   *z*-Transform Method

Consider an N-dimensional state space representation

$$\mathbf{q}(n+1) = \mathbf{A}\mathbf{q}(n) + \mathbf{B}\mathbf{x}(n) \tag{11.101}$$

$$\mathbf{y}(n) = \mathbf{C}\mathbf{q}(n) + \mathbf{D}\mathbf{x}(n) \tag{11.102}$$

where $\mathbf{A}$, $\mathbf{B}$, $\mathbf{C}$, and $\mathbf{D}$ are $N \times N$, $N \times m$, $p \times N$, $p \times m$ matrices, respectively. Let $\mathbf{q}(0)$ be the initial state. Taking the unilateral z-transform of Eqs (11.101) and (11.102) and using Eq. (10.79), we get

$$z\mathbf{Q}(z) - z\mathbf{q}(0) = \mathbf{A}\mathbf{Q}(z) + \mathbf{B}\mathbf{X}(z) \tag{11.103}$$

$$Y(z) = \mathbf{C}\mathbf{Q}(z) + \mathbf{D}\mathbf{X}(z) \tag{11.104}$$

where

$$\mathbf{Q}(z) = \mathcal{Z}_u[\mathbf{q}(n)] = \begin{bmatrix} Q_1(z) \\ Q_2(z) \\ \vdots \\ Q_N(z) \end{bmatrix} \qquad \mathbf{X}(z) = \mathcal{Z}_u[\mathbf{x}(n)] = \begin{bmatrix} X_1(z) \\ X_2(z) \\ \vdots \\ X_m(z) \end{bmatrix}$$

$$\mathbf{Y}(z) = \mathcal{Z}_u[\mathbf{y}(n)] = \begin{bmatrix} Y_1(z) \\ Y_2(z) \\ \vdots \\ Y_p(z) \end{bmatrix}$$

where $Q_k(z) = \mathcal{Z}_u[q_k(n)]$, $X_k(z) = \mathcal{Z}_u[x_k(n)]$, and $Y_k(z) = \mathcal{Z}_u[y_k(n)]$. Consider Eq. (11.103),

$$z\mathbf{Q}(z) - z\mathbf{q}(0) = \mathbf{A}\mathbf{Q}(z) + \mathbf{B}\mathbf{X}(z)$$

$$z\mathbf{Q}(z) - \mathbf{A}\mathbf{Q}(z) = z\mathbf{q}(0) + \mathbf{B}\mathbf{X}(z)$$

$$[z\mathbf{I} - \mathbf{A}]\mathbf{Q}(z) = z\mathbf{q}(0) + \mathbf{B}\mathbf{X}(z) \tag{11.105}$$

where $\mathbf{I}$ is the $N \times N$ identity matrix. From Eq. (11.105), we have

$$\mathbf{Q}(z) = [z\mathbf{I} - \mathbf{A}]^{-1}[z\mathbf{q}(0) + \mathbf{B}\mathbf{X}(z)]$$

$$= [z\mathbf{I} - \mathbf{A}]^{-1}z\mathbf{q}(0) + [z\mathbf{I} - \mathbf{A}]^{-1}\mathbf{B}\mathbf{X}(z)$$

$$= [\mathbf{I} - z^{-1}\mathbf{A}]^{-1}\mathbf{q}(0) + [z\mathbf{I} - \mathbf{A}]^{-1}\mathbf{B}\mathbf{X}(z) \tag{11.106}$$

Taking the inverse z-transform, we get

$$\mathbf{q}(n) = \underbrace{\mathcal{Z}_u^{-1}\Big[[\mathbf{I} - z^{-1}\mathbf{A}]^{-1}\Big]\mathbf{q}(0)}_{\text{zero-input component}} + \underbrace{\mathcal{Z}_u^{-1}\Big[[z\mathbf{I} - \mathbf{A}]^{-1}\mathbf{B}\mathbf{X}(z)\Big]}_{\text{zero-state component}} \tag{11.107}$$

A comparison of Eq. (11.98) with Eq. (11.107) shows that

$$\mathbf{A}^n = \mathcal{Z}_u^{-1}\Big[[\mathbf{I} - z^{-1}\mathbf{A}]^{-1}\Big] \tag{11.108}$$

Now consider the output Eq. (11.104),

$$Y(z) = \mathbf{C}\mathbf{Q}(z) + \mathbf{D}\mathbf{X}(z)$$

Upon substituting Eq. (11.106) into this equation, we have

$$\mathbf{Y}(z) = \mathbf{C}\Big[[\mathbf{I} - z^{-1}\mathbf{A}]^{-1}\mathbf{q}(0) + [z\mathbf{I} - \mathbf{A}]^{-1}\mathbf{B}\mathbf{X}(z)\Big] + \mathbf{D}\mathbf{X}(z)$$

$$\mathbf{Y}(z) = \underbrace{\mathbf{C}[\mathbf{I} - z^{-1}\mathbf{A}]^{-1}\mathbf{q}(0)}_{\text{zero-input response}} + \underbrace{\Big[\mathbf{C}[z\mathbf{I} - \mathbf{A}]^{-1}\mathbf{B} + \mathbf{D}\Big]\mathbf{X}(z)}_{\text{zero-state response}} \qquad (11.109)$$

$$\mathbf{Y}(z) = \mathbf{C}[\mathbf{I} - z^{-1}\mathbf{A}]^{-1}\mathbf{q}(0) + \mathbf{H}(z)\mathbf{X}(z) \qquad (11.110)$$

where

$$\mathbf{H}(z) = \mathbf{C}[z\mathbf{I} - \mathbf{A}]^{-1}\mathbf{B} + \mathbf{D} \qquad (11.111)$$

The matrix $\mathbf{H}(z)$ is a $p \times m$ transfer function matrix [p is the number of outputs and m is the number of inputs]. The ijth element $H_{ij}(z)$ of $\mathbf{H}(z)$ is the transfer function that relates the output $y_i(n)$ to the input $x_j(n)$. The unit impulse response matrix $\mathbf{h}(n)$ is given by

$$\mathbf{h}(n) = \mathcal{Z}_u^{-1}[\mathbf{H}(z)]$$

The matrix $\mathbf{h}(n)$ is a $p \times m$ matrix known as the *impulse response matrix*. The ijth element $h_{ij}(n)$ of $\mathbf{h}(n)$ is the impulse response (zero-state response) that relates the output $y_i(n)$ to the input $x_j(n) = \delta(n)$ and when all other inputs are zero.

11.7.3 Determination of $\mathbf{A}^n$

There are several methods to determine $\mathbf{A}^n$

Method 1 (Series Summation Method) From Eqs (11.63) and (11.64) and the argument that we can express any positive integral power of $\mathbf{A}$ as a linear combination of $\mathbf{I}$, $\mathbf{A}$, ..., $\mathbf{A}^{N-1}$. We get

$$\mathbf{A}^n = \beta_0\mathbf{I} + \beta_1\mathbf{A} + \beta_2\mathbf{A}^2 + \cdots + \beta_{N-1}\mathbf{A}^{N-1} \qquad (11.112)$$

Method 2 (Diagonalization Method) This method of finding $\mathbf{A}^n$ is based on the *diagonalization* of a matrix $\mathbf{A}$. If eigenvalues $\lambda_1, \lambda_2, \ldots, \lambda_N$ of $\mathbf{A}$ are all distinct, then $\mathbf{A}^n$ can be expressed as [Eq. (11.74)]

$$\mathbf{A}^n = \mathbf{P}\boldsymbol{\Lambda}^n\mathbf{P}^{-1} = \mathbf{P}\begin{bmatrix} \lambda_1^n & 0 & \cdots & 0 \\ 0 & \lambda_2^n & \cdots & 0 \\ \vdots & \vdots & \ddots & \vdots \\ 0 & 0 & \cdots & \lambda_N^n \end{bmatrix}\mathbf{P}^{-1} \qquad (11.113)$$

where $\mathbf{P}$ is given by Eq. (11.70).

Method 3 (Spectral Decomposition Method) This method of finding $\mathbf{A}^n$ is based on the *spectral decomposition* of a matrix $\mathbf{A}$. If eigenvalues $\lambda_1, \lambda_2, \ldots, \lambda_N$ of $\mathbf{A}$ are all distinct, then $\mathbf{A}^n$ can be expressed as [Eq. (11.81)]

$$\mathbf{A}^n = \lambda_1^n \mathbf{E}_1 + \lambda_2^n \mathbf{E}_2 + \cdots + \lambda_N^n \mathbf{E}_N = \sum_{k=1}^{N} \lambda_k^n E_k \qquad (11.114)$$

where E_j, $j = 1, 2, \ldots, N$ are called *constituent matrices*, which can be evaluated using Eq. (11.84).

Method 4 (z-Transform Method) Using the z-transform, we can evaluate $\mathbf{A}^n$. Comparing Eqs (11.98) and (11.107), we see that

$$\mathbf{A}^n = \mathcal{Z}_u^{-1}\left[[\mathbf{I} - z^{-1}\mathbf{A}]^{-1}\right] \qquad (11.115)$$

SOLVED EXAMPLES

Example 11.6 Consider the RLC circuit shown in Fig. 11.3. Let the output $y(t)$ be the loop current. Find a state space representation of the circuit.

Example 11.7 Write the state equation for the circuit shown in Fig. 11.4.

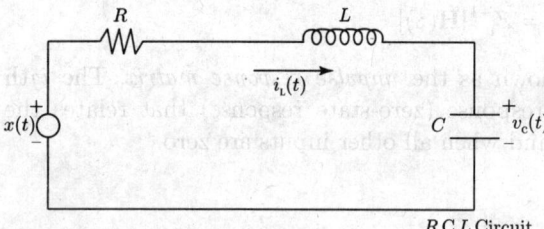

Fig. 11.3

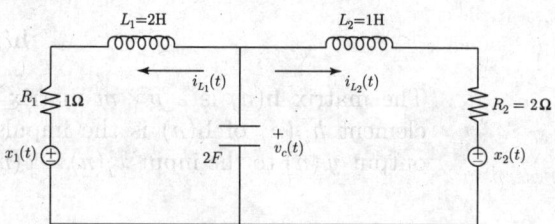

Fig. 11.4

Solution

We choose the state variables $q_1(t) = i_L(t)$ and $q_2(t) = v_c(t)$. Then by KVL we get

$$L\dot{q}_1(t) + Rq_1(t) + q_2(t) = x(t)$$

$$C\dot{q}_2(t) = q_1(t)$$

$$y(t) = q_1(t)$$

Rearranging and writing in matrix form, we get

$$\dot{\mathbf{q}}(t) = \begin{bmatrix} -\dfrac{R}{L} & -\dfrac{1}{L} \\ \dfrac{1}{C} & 0 \end{bmatrix} \mathbf{q}(t) + \begin{bmatrix} \dfrac{1}{L} \\ 0 \end{bmatrix} x(t)$$

$$y(t) = \begin{bmatrix} 1 & 0 \end{bmatrix} \mathbf{q}(t)$$

Solution

Step 1 There are two inductors and one capacitor in the network. Therefore, we shall choose the inductor currents $q_1(t) = i_{L_1}(t)$, $q_2(t) = i_{L_2}(t)$ and the capacitor voltage $q_3(t) = v_c(t)$ as the state variables.

Step 2 The relationship between the loop currents and the state variables can be written by inspection:

$$q_1(t) = i_{L_1}(t), \quad q_2(t) = i_{L_2}(t), \quad C\dot{q}_3(t) = -i_{L_1} - i_{L_2}$$

Step 3 The loop equations are

$$2\dot{q}_1(t) + q_1(t) - q_3(t) + x_1(t) = 0$$

$$\dot{q}_2(t) + 2q_2(t) - q_3(t) + x_2(t) = 0$$

After some algebraic manipulation, we obtain

$$\dot{q}_1(t) = -\frac{1}{2}q_1(t) + \frac{1}{2}q_3(t) - \frac{1}{2}x_1(t)$$

$$\dot{q}_2(t) = -2q_2(t) + q_3(t) - x_2(t)$$

$$\dot{q}_3(t) = -\frac{1}{2}q_1(t) - \frac{1}{2}q_2(t)$$

The above equations can be written compactly as

$$\underbrace{\begin{bmatrix} \dot{q}_1(t) \\ \dot{q}_2(t) \\ \dot{q}_3(t) \end{bmatrix}}_{\dot{\mathbf{q}}(t)} = \underbrace{\begin{bmatrix} -\frac{1}{2} & 0 & \frac{1}{2} \\ 0 & -2 & 1 \\ -\frac{1}{2} & -\frac{1}{2} & 0 \end{bmatrix}}_{\mathbf{A}} \underbrace{\begin{bmatrix} q_1(t) \\ q_2(t) \\ q_3(t) \end{bmatrix}}_{\mathbf{q}(t)}$$

$$+ \underbrace{\begin{bmatrix} -\frac{1}{2} & 0 \\ 0 & -1 \\ 0 & 0 \end{bmatrix}}_{\mathbf{B}} \underbrace{\begin{bmatrix} x_1(t) \\ x_2(t) \end{bmatrix}}_{\mathbf{x}(t)}$$

Example 11.8 Find state equations of a continuous-time LTI system with system function

$$H(s) = \frac{b_0 s^3 + b_1 s^2 + b_2 s + b_3}{s^3 + a_1 s^2 + a_2 s + a_3}$$

Solution
From the definition of the system function

$$H(s) = \frac{Y(s)}{X(s)} = \frac{b_0 s^3 + b_1 s^2 + b_2 s + b_3}{s^3 + a_1 s^2 + a_2 s + a_3}$$

Divide the transfer function into two parts

$$H(s) = \frac{Q(s)}{X(s)} \frac{Y(s)}{Q(s)} = \frac{b_0 s^3 + b_1 s^2 + b_2 s + b_3}{s^3 + a_1 s^2 + a_2 s + a_3}$$

where

$$\frac{Q(s)}{X(s)} = \frac{1}{s^3 + a_1 s^2 + a_2 s + a_3}$$

$$Q(s)[s^3 + a_1 s^2 + a_2 s + a_3] = X(s)$$

The inverse Laplace transform of the above equation yields

$$\dddot{q}(t) + a_1\ddot{q}(t) + a_2\dot{q}(t) + a_3q(t) = x(t)$$

Select the state variables as

$$q_1(t) = q(t) \qquad q_2(t) = \dot{q}(t) \qquad q_3(t) = \ddot{q}(t)$$

then

$$\dot{q}_1(t) = q_2(t) \qquad \dot{q}_2(t) = q_3(t)$$

$$\dot{q}_3(t) = -a_3q_1(t) - a_2q_2(t) - a_1q_3(t) + x(t)$$

The above equations can be rewritten compactly as

$$\underbrace{\begin{bmatrix} \dot{q}_1(t) \\ \dot{q}_2(t) \\ \dot{q}_3(t) \end{bmatrix}}_{\dot{\mathbf{q}}(t)} = \underbrace{\begin{bmatrix} 0 & 1 & 0 \\ 0 & 0 & 1 \\ -a_3 & -a_2 & -a_1 \end{bmatrix}}_{\mathbf{A}} \underbrace{\begin{bmatrix} q_1(t) \\ q_2(t) \\ q_3(t) \end{bmatrix}}_{\mathbf{q}(t)}$$

$$+ \underbrace{\begin{bmatrix} 0 \\ 0 \\ 1 \end{bmatrix}}_{\mathbf{b}} x(t)$$

Now, consider

$$\frac{Y(s)}{Q(s)} = b_0 s^3 + b_1 s^2 + b_2 s + b_3$$

$$Y(s) = Q(s)[b_0 s^3 + b_1 s^2 + b_2 s + b_3]$$

The inverse Laplace transform of the above equation yields

$$y(t) = b_0 \dddot{q}(t) + b_1 \ddot{q}(t) + b_2 \dot{q}(t) + b_3 q(t)$$

Using the state variables defined above, we obtain

$$y(t) = b_0\dot{q}_3(t) + b_1 q_3(t) + b_2 q_2(t) + b_3 q_1(t)$$

$$= b_0[-a_3q_1(t) - a_2q_2(t) - a_1q_3(t) + x(t)]$$

$$\qquad + b_1 q_3(t) + b_2 q_2(t) + b_3 q_1(t)$$

$$= [b_3 - b_0 a_3]q_1(t) + [b_2 - b_0 a_2]q_2(t)$$

$$\qquad + [b_1 - b_0 a_1]q_3(t) + b_0 x(t)$$

The above equations can be rewritten compactly as

$$y(t)$$

$$= \underbrace{\begin{bmatrix} b_3 - b_0 a_3 & b_2 - b_0 a_2 & b_1 - b_0 a_1 \end{bmatrix}}_{\mathbf{C}} \underbrace{\begin{bmatrix} q_1(t) \\ q_2(t) \\ q_3(t) \end{bmatrix}}_{\mathbf{q}(t)}$$

$$+ b_0 x(t)$$

Example 11.9 Consider a continuous-time LTI system with system function

$$H(s) = \frac{3s + 7}{(s+1)(s+2)(s+5)}$$

Find a state space representation of the system such that its system matrix **A** is diagonal.

Solution
Given that

$$H(s) = \frac{3s + 7}{(s+1)(s+2)(s+5)}$$

Using the partial fraction expansion, we obtain

$$H(s) = \frac{1}{s+1} - \frac{1/3}{s+2} - \frac{2/3}{s+5}$$
$$= H_1(s) + H_2(s) + H_3(s)$$

where

$$H_1(s) = \frac{1}{s+1}$$

$$H_2(s) = -\frac{\frac{1}{3}}{s+2}$$

$$H_3(s) = -\frac{\frac{2}{3}}{s+5}$$

Let

$$H_k(s) = \frac{Y_k(s)}{X(s)} = \frac{z_k}{s - p_k}$$

$$Y_k(s)[s - p_k] = z_k X(s)$$

$$Y_k(s) = \frac{1}{s}[p_k Y_k(s) + z_k X(s)]$$

The simulation diagram of $Y_k(s)$ is shown in Fig. 11.5. Thus $H(s) = H_1(s) + H_2(s) + H_3(s)$ can be simulated by the diagram in Fig. 11.6 obtained by parallel combination of three systems.

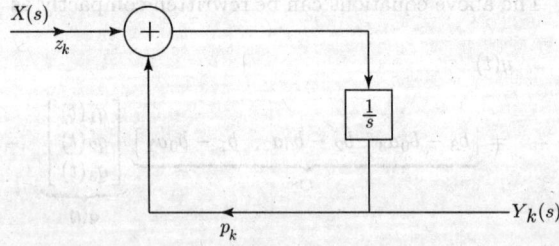

Fig. 11.5

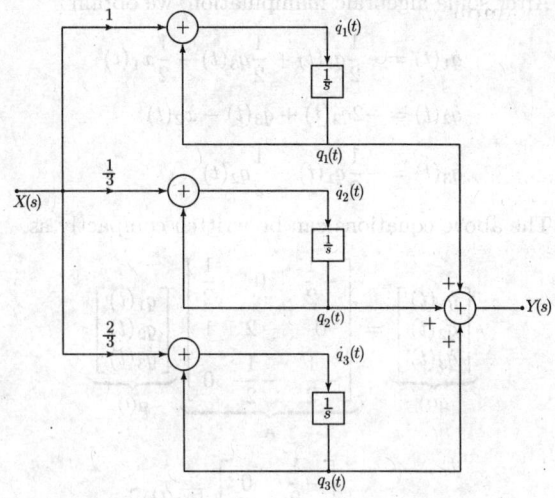

Fig. 11.6

Choosing the outputs of integrators as state variables as shown in Fig. 11.6, we get

$$\dot{q}_1(t) = -q_1(t) + x(t)$$

$$\dot{q}_2(t) = -2q_2(t) - \frac{1}{3}x(t)$$

$$\dot{q}_3(t) = -5q_3(t) - \frac{2}{3}x(t)$$

$$y(t) = q_1(t) + q_2(t) + q_3(t)$$

In matrix form

$$\underbrace{\begin{bmatrix} \dot{q}_1(t) \\ \dot{q}_2(t) \\ \dot{q}_3(t) \end{bmatrix}}_{\dot{\mathbf{q}}(t)} = \underbrace{\begin{bmatrix} -1 & 0 & 0 \\ 0 & -2 & 0 \\ 0 & 0 & -5 \end{bmatrix}}_{\mathbf{A}} \underbrace{\begin{bmatrix} q_1(t) \\ q_2(t) \\ q_3(t) \end{bmatrix}}_{\mathbf{q}(t)}$$

$$+ \underbrace{\begin{bmatrix} 1 \\ -1/3 \\ -2/3 \end{bmatrix}}_{\mathbf{b}} x(t)$$

$$y(t) = \underbrace{\begin{bmatrix} 1 & 1 & 1 \end{bmatrix}}_{\mathbf{C}} \underbrace{\begin{bmatrix} q_1(t) \\ q_2(t) \\ q_3(t) \end{bmatrix}}_{\mathbf{q}(t)}$$

Example 11.10 Find the STM $\phi(t) = e^{\mathbf{A}(t)}$ for

$$\mathbf{A} = \begin{bmatrix} 0 & 1 \\ -3 & -4 \end{bmatrix}$$

using the Cayley-Hamilton theorem (finite series summation) method.

Solution
First we find

$$[\lambda \mathbf{I} - \mathbf{A}] = \lambda \begin{bmatrix} 1 & 0 \\ 0 & 1 \end{bmatrix} - \begin{bmatrix} 0 & 1 \\ -3 & -4 \end{bmatrix}$$

$$= \begin{bmatrix} \lambda & -1 \\ 3 & \lambda + 4 \end{bmatrix}$$

The characteristic polynomial $Q(\lambda)$ of $\mathbf{A}$ is given by

$$Q(\lambda) = |\lambda \mathbf{I} - \mathbf{A}| = 0$$

$$\begin{vmatrix} \lambda & -1 \\ 3 & \lambda + 4 \end{vmatrix} = 0$$

$$\lambda^2 + 4\lambda + 3 = 0$$

$$(\lambda + 1)(\lambda + 3) = 0$$

Thus, the eigenvalues of $\mathbf{A}$ are $\lambda_1 = -1$ and $\lambda_2 = -3$. Hence, by Eq. (11.68) we have

$$e^{\mathbf{A}t} = \beta_0 \mathbf{I} + \beta_1 \mathbf{A} = \begin{bmatrix} \beta_0 & \beta_1 \\ -3\beta_1 & \beta_0 - 4\beta_1 \end{bmatrix}$$

From Eq. (11.61), we have

$$\begin{bmatrix} \beta_0 \\ \beta_1 \end{bmatrix} = \begin{bmatrix} 1 & \lambda_1 \\ 1 & \lambda_2 \end{bmatrix}^{-1} \begin{bmatrix} e^{\lambda_1 t} \\ e^{\lambda_2 t} \end{bmatrix}$$

$$= \begin{bmatrix} 1 & -1 \\ 1 & -3 \end{bmatrix}^{-1} \begin{bmatrix} e^{-t} \\ e^{-3t} \end{bmatrix}$$

$$= \frac{1}{2} \begin{bmatrix} 3 & -1 \\ 1 & -1 \end{bmatrix} \begin{bmatrix} e^{-t} \\ e^{-3t} \end{bmatrix}$$

$$\begin{bmatrix} \beta_0 \\ \beta_1 \end{bmatrix} = \frac{1}{2} \begin{bmatrix} 3e^{-t} - e^{-3t} \\ e^{-t} - e^{-3t} \end{bmatrix}$$

Hence,

$$e^{\mathbf{A}t} = \frac{1}{2} \begin{bmatrix} 3e^{-t} - e^{-3t} & e^{-t} - e^{-3t} \\ -3e^{-t} + 3e^{-3t} & -e^{-t} + 3e^{-3t} \end{bmatrix}$$

$$= e^{-t} \frac{1}{2} \begin{bmatrix} 3 & 1 \\ -3 & -1 \end{bmatrix} + e^{-3t} \frac{1}{2} \begin{bmatrix} -1 & -1 \\ 3 & 3 \end{bmatrix}$$

Example 11.11 Repeat Example 11.10 using the diagonalization method.

Solution
Let $\mathbf{x}$ be an eigenvector of $\mathbf{A}$ associated with λ. Then,

$$\mathbf{A}\mathbf{x} = \lambda \mathbf{x}$$

$$[\lambda \mathbf{I} - \mathbf{A}]\mathbf{x} = \mathbf{0}$$

$$\begin{bmatrix} \lambda & -1 \\ 3 & \lambda + 4 \end{bmatrix} \begin{bmatrix} x_1 \\ x_2 \end{bmatrix} = 0$$

Rewriting the above equation in component form, we get

$$\lambda x_1 - x_2 = 0 \tag{11.116}$$

$$3x_1 + (\lambda + 4)x_2 = 0 \tag{11.117}$$

Eigenvector corresponding to λ_1 This vector is obtained from Eqs (11.116) and (11.117) with $\lambda = \lambda_1 = -1$, i.e.,

$$-x_1 - x_2 = 0$$

$$3x_1 + 3x_2 = 0$$

A solution is $x_2 = -x_1$. If we choose $x_1 = 1$, we obtain the eigenvector

$$\mathbf{x_1} = \begin{bmatrix} 1 \\ -1 \end{bmatrix}$$

Eigenvector corresponding to λ_2 This vector is obtained from Eqs (11.116) and (11.117) with $\lambda = \lambda_2 = -3$, i.e.,

$$-3x_1 - x_2 = 0$$

$$3x_1 + x_2 = 0$$

A solution is $x_2 = -3x_1$. If we choose $x_1 = 1$, we obtain the eigenvector

$$\mathbf{x_2} = \begin{bmatrix} 1 \\ -3 \end{bmatrix}$$

Therefore,

$$\mathbf{P} = \begin{bmatrix} \mathbf{x_1} & \mathbf{x_2} \end{bmatrix} = \begin{bmatrix} 1 & 1 \\ -1 & -3 \end{bmatrix}$$

Then

$$\mathbf{P}^{-1} = \begin{bmatrix} \dfrac{3}{2} & \dfrac{1}{2} \\ -\dfrac{1}{2} & -\dfrac{1}{2} \end{bmatrix}$$

and by Eq. (11.78) we obtain

$$e^{\mathbf{A}t} = \begin{bmatrix} 1 & 1 \\ -1 & -3 \end{bmatrix} \begin{bmatrix} e^{\lambda_1 t} & 0 \\ 0 & e^{\lambda_2 t} \end{bmatrix} \begin{bmatrix} \dfrac{3}{2} & \dfrac{1}{2} \\ -\dfrac{1}{2} & -\dfrac{1}{2} \end{bmatrix}$$

$$= \frac{1}{2} \begin{bmatrix} 3e^{-t} - e^{-3t} & e^{-t} - e^{-3t} \\ -3e^{-t} + 3e^{-3t} & -e^{-t} + 3e^{-3t} \end{bmatrix}$$

$$e^{\mathbf{A}t} = e^{-t}\frac{1}{2}\begin{bmatrix} 3 & 1 \\ -3 & -1 \end{bmatrix} + e^{-3t}\frac{1}{2}\begin{bmatrix} -1 & -1 \\ 3 & 3 \end{bmatrix}$$

Example 11.12 Repeat Example 11.10 using the spectral decomposition method.

Solution
Since all eigenvalues of **A** are distinct, by Eq. (11.84) we have

$$\mathbf{E}_1 = \frac{1}{\lambda_1 - \lambda_2}[\mathbf{A} - \lambda_2\mathbf{I}] = \frac{1}{2}[\mathbf{A} + 3\mathbf{I}]$$

$$= \frac{1}{2}\begin{bmatrix} 3 & 1 \\ -3 & -1 \end{bmatrix}$$

$$\mathbf{E}_2 = \frac{1}{\lambda_2 - \lambda_1}[\mathbf{A} - \lambda_1\mathbf{I}] = -\frac{1}{2}[\mathbf{A} + \mathbf{I}]$$

$$= \frac{1}{2}\begin{bmatrix} -1 & -1 \\ 3 & 3 \end{bmatrix}$$

Then by Eq. (11.83) we obtain

$$e^{\mathbf{A}t} = e^{-t}\mathbf{E}_1 + e^{-3t}\mathbf{E}_2$$

$$= e^{-t}\frac{1}{2}\begin{bmatrix} 3 & 1 \\ -3 & -1 \end{bmatrix} + e^{-3t}\frac{1}{2}\begin{bmatrix} -1 & -1 \\ 3 & 3 \end{bmatrix}$$

Example 11.13 Repeat Example 11.10 using the Laplace transform method.

Solution
First, we must find $[s\mathbf{I} - \mathbf{A}]^{-1}$:

$$[s\mathbf{I} - \mathbf{A}]^{-1} = \begin{bmatrix} s & -1 \\ 3 & s+4 \end{bmatrix}^{-1}$$

$$= \frac{1}{(s+1)(s+3)}\begin{bmatrix} s+4 & 1 \\ -3 & s \end{bmatrix}$$

$$= \begin{bmatrix} \dfrac{s+4}{(s+1)(s+3)} & \dfrac{1}{(s+1)(s+3)} \\ -\dfrac{3}{(s+1)(s+3)} & \dfrac{s}{(s+1)(s+3)} \end{bmatrix}$$

$$= \frac{1}{2}\begin{bmatrix} \dfrac{3}{s+1} - \dfrac{1}{s+3} & \dfrac{1}{s+1} - \dfrac{1}{s+3} \\ -\dfrac{3}{s+1} + \dfrac{3}{s+3} & -\dfrac{1}{s+1} + \dfrac{3}{s+3} \end{bmatrix}$$

Then, by Eq. (11.85) we obtain

$$e^{\mathbf{A}t} = \mathcal{L}_u^{-1}\left[[s\mathbf{I} - \mathbf{A}]^{-1}\right]$$

$$= \frac{1}{2}\begin{bmatrix} 3e^{-t} - e^{-3t} & e^{-t} - e^{-3t} \\ -3e^{-t} + 3e^{-3t} & -e^{-t} + 3e^{-3t} \end{bmatrix}$$

Example 11.14 Find state equations of a discrete-time system described by

$$y(n) - \frac{3}{4}y(n-1) + \frac{1}{8}y(n-2)$$

$$= x(n) + \frac{1}{2}x(n-1)$$

Solution
Given that

$$y(n) - \frac{3}{4}y(n-1) + \frac{1}{8}y(n-2)$$

$$= x(n) + \frac{1}{2}x(n-1)$$

Because of the existence of the term $1/2x(n-1)$ on the RHS of the above equation, the selection of $y(n-2)$ and $y(n-1)$ as state variables will not yield the desired state equations of the system. Taking the z-transform of the above equation and rearranging, we obtain

$$\frac{Y(z)}{X(z)} = \frac{W(z)}{X(z)}\frac{Y(z)}{W(z)} = \frac{1 + (1/2)z^{-1}}{1 - (3/4)z^{-1} + (1/8)z^{-2}}$$

where

$$\frac{W(z)}{X(z)} = \frac{1}{1 - (3/4)z^{-1} + (1/8)z^{-2}}$$

$$W(z)\left[1 - \frac{3}{4}z^{-1} + \frac{1}{8}z^{-2}\right] = X(z)$$

The inverse z-transform of the above equation yields

$$w(n) - \frac{3}{4}w(n-1) + \frac{1}{8}w(n-2) = x(n)$$

Select the state variables as

$$q_1(n) = w(n-2)$$
$$q_2(n) = w(n-1)$$

then

$$q_1(n+1) = q_2(n)$$
$$q_2(n+1) = -\frac{1}{8}q_1(n) + \frac{3}{4}q_2(n) + x(n)$$

The above equations can be rewritten compactly as

$$\underbrace{\begin{bmatrix} q_1(n+1) \\ q_2(n+1) \end{bmatrix}}_{q(n+1)} = \underbrace{\begin{bmatrix} 0 & 1 \\ -\dfrac{1}{8} & \dfrac{3}{4} \end{bmatrix}}_{A} \underbrace{\begin{bmatrix} q_1(n) \\ q_2(n) \end{bmatrix}}_{q(n)}$$

$$+ \underbrace{\begin{bmatrix} 0 \\ 1 \end{bmatrix}}_{b} x(n)$$

Now consider

$$\frac{Y(z)}{W(z)} = 1 + \frac{1}{2}z^{-1}$$

$$Y(z) = W(z)\left[1 + \frac{1}{2}z^{-1}\right]$$

The inverse z-transform of the above equation yields

$$y(n) = w(n) + \frac{1}{2}w(n-1)$$

Using the state variables defined above, we obtain

$$y(n) = q_2(n+1) + \frac{1}{2}q_2(n)$$

$$= -\frac{1}{8}q_1(n) + \frac{3}{4}q_2(n) + x(n) + \frac{1}{2}q_2(n)$$

$$= -\frac{1}{8}q_1(n) + \frac{5}{4}q_2(n) + x(n)$$

The above equations can be rewritten compactly as

$$y(n) = \underbrace{\begin{bmatrix} -\dfrac{1}{8} & \dfrac{5}{4} \end{bmatrix}}_{C} \underbrace{\begin{bmatrix} q_1(n) \\ q_2(n) \end{bmatrix}}_{q(n)} + x(n)$$

Example 11.15 The system equations are given by

$$\dot{\mathbf{q}}(t) = \mathbf{A}\mathbf{q}(t) + \mathbf{B}x(t)$$
$$y(t) = \mathbf{C}\mathbf{q}(t)$$

where

$$\mathbf{A} = \begin{bmatrix} 0 & 1 \\ -2 & -3 \end{bmatrix} \quad \mathbf{B} = \begin{bmatrix} 0 \\ 1 \end{bmatrix} \quad \mathbf{C} = \begin{bmatrix} 1 & 0 \end{bmatrix}$$

Find the transfer function of the system.

Solution
Here $\mathbf{D} = \mathbf{0}$. First we determine $\mathbf{\Phi}(s) = [s\mathbf{I} - \mathbf{A}]^{-1}$.

$$\mathbf{\Phi}(s) = [s\mathbf{I} - \mathbf{A}]^{-1} = \begin{bmatrix} s & -1 \\ 2 & s+3 \end{bmatrix}^{-1}$$

$$= \begin{bmatrix} \dfrac{s+3}{s^2+3s+2} & \dfrac{1}{s^2+3s+2} \\ -\dfrac{2}{s^2+3s+2} & \dfrac{s}{s^2+3s+2} \end{bmatrix}$$

Hence the transfer function matrix is given by

$$\mathbf{H}(s)$$
$$= \mathbf{C}\mathbf{\Phi}(s)\mathbf{B} + \mathbf{D}$$

$$= \begin{bmatrix} 1 & 0 \end{bmatrix} \begin{bmatrix} \dfrac{s+3}{s^2+3s+2} & \dfrac{1}{s^2+3s+2} \\ -\dfrac{2}{s^2+3s+2} & \dfrac{s}{s^2+3s+2} \end{bmatrix} \begin{bmatrix} 0 \\ 1 \end{bmatrix} + \mathbf{0}$$

$$= \frac{1}{s^2+3s+2}$$

Example 11.16 An LTI system is characterized by the homogeneous state equation

$$\begin{bmatrix} \dot{q}_1(t) \\ \dot{q}_2(t) \end{bmatrix} = \begin{bmatrix} 1 & 0 \\ 1 & 1 \end{bmatrix} \begin{bmatrix} q_1(t) \\ q_2(t) \end{bmatrix}$$

Compute the solution of homogeneous equation, assuming the initial state vector

$$\mathbf{q}(0) = \begin{bmatrix} 1 \\ 0 \end{bmatrix}$$

Solution

First we determine $\Phi(s) = [s\mathbf{I} - \mathbf{A}]^{-1}$.

$$\Phi(s) = [s\mathbf{I} - \mathbf{A}]^{-1} = \begin{bmatrix} s-1 & 0 \\ -1 & s-1 \end{bmatrix}^{-1}$$

$$\Phi(s) = \begin{bmatrix} \dfrac{1}{s-1} & 0 \\ \dfrac{1}{(s-1)^2} & \dfrac{1}{s-1} \end{bmatrix}$$

The inverse Laplace transform of the above equation yields

$$\phi(t) = e^{\mathbf{A}t} = \begin{bmatrix} e^t & 0 \\ te^t & e^t \end{bmatrix}$$

From Eq. (11.44), we have

$$\mathbf{q}(t) = e^{\mathbf{A}t}\mathbf{q}(0)$$

$$= \begin{bmatrix} e^t & 0 \\ te^t & e^t \end{bmatrix}\begin{bmatrix} 1 \\ 0 \end{bmatrix}$$

$$= \begin{bmatrix} e^t \\ te^t \end{bmatrix}$$

Example 11.17 A state model of the system is given by

$$\begin{bmatrix} \dot{q}_1(t) \\ \dot{q}_2(t) \end{bmatrix} = \underbrace{\begin{bmatrix} 0 & 1 \\ -2 & -3 \end{bmatrix}}_{\mathbf{A}}\begin{bmatrix} q_1(t) \\ q_2(t) \end{bmatrix} + \underbrace{\begin{bmatrix} 0 \\ 1 \end{bmatrix}}_{\mathbf{B}}x(t)$$

$$y(t) = \underbrace{\begin{bmatrix} 1 & 0 \end{bmatrix}}_{\mathbf{C}}\begin{bmatrix} q_1(t) \\ q_2(t) \end{bmatrix}$$

Find

(a) the characteristic equation and eigenvalues of the system.

(b) the transfer function $\mathbf{H}(s) = \mathbf{Y}(s)/\mathbf{X}(s)$.

(c) the state transition matrix $\phi(t) = e^{\mathbf{A}t}$.

(d) the state response for a unit step input under zero initial conditions.

(e) the time response $y(t)$.

Solution

(a) The characteristic equation is given by

$$|\lambda\mathbf{I} - \mathbf{A}| = 0$$

$$\begin{vmatrix} \lambda & -1 \\ 2 & \lambda+3 \end{vmatrix} = 0$$

$$\lambda^2 + 3\lambda + 2 = 0$$

$$(\lambda+1)(\lambda+2) = 0$$

and hence the eigenvalues are: $\lambda_1 = -1$, $\lambda_2 = -2$.

(b) Here $\mathbf{D} = \mathbf{0}$. First we determine $\Phi(s) = [s\mathbf{I} - \mathbf{A}]^{-1}$.

$$\Phi(s) = [s\mathbf{I} - \mathbf{A}]^{-1} = \begin{bmatrix} s & -1 \\ 2 & s+3 \end{bmatrix}^{-1}$$

$$= \begin{bmatrix} \dfrac{s+3}{s^2+3s+2} & \dfrac{1}{s^2+3s+2} \\ -\dfrac{2}{s^2+3s+2} & \dfrac{s}{s^2+3s+2} \end{bmatrix}$$

Hence the transfer function matrix is given by

$$\mathbf{H}(s)$$

$$= \mathbf{C}\Phi(s)\mathbf{B} + \mathbf{D}$$

$$= \begin{bmatrix} 1 & 0 \end{bmatrix}\begin{bmatrix} \dfrac{s+3}{s^2+3s+2} & \dfrac{1}{s^2+3s+2} \\ -\dfrac{2}{s^2+3s+2} & \dfrac{s}{s^2+3s+2} \end{bmatrix}\begin{bmatrix} 0 \\ 1 \end{bmatrix} + \mathbf{0}$$

$$= \frac{1}{s^2+3s+2}$$

(c) The STM $\phi(t) = e^{\mathbf{A}t}$ is given by

$$\phi(t) = e^{\mathbf{A}t} = \mathcal{L}_u^{-1}\left[[s\mathbf{I} - \mathbf{A}]^{-1}\right]$$

$$= \mathcal{L}_u^{-1}\begin{bmatrix} \dfrac{s+3}{s^2+3s+2} & \dfrac{1}{s^2+3s+2} \\ -\dfrac{2}{s^2+3s+2} & \dfrac{s}{s^2+3s+2} \end{bmatrix}$$

$$= \mathcal{L}_u^{-1}\begin{bmatrix} \dfrac{s+3}{(s+1)(s+2)} & \dfrac{1}{(s+1)(s+2)} \\ -\dfrac{2}{(s+1)(s+2)} & \dfrac{s}{(s+1)(s+2)} \end{bmatrix}$$

$$= \begin{bmatrix} 2e^{-t} - e^{-2t} & e^{-t} - e^{-2t} \\ -2e^{-t} + 2e^{-2t} & -e^{-t} + 2e^{-2t} \end{bmatrix}$$

(d) First determine $e^{\mathbf{A}(t-\tau)}\mathbf{B}$.

$$e^{\mathbf{A}(t-\tau)}\mathbf{B}$$

$$= \begin{bmatrix} 2e^{-(t-\tau)} - e^{-2(t-\tau)} & e^{-(t-\tau)} - e^{-2(t-\tau)} \\ -2e^{-(t-\tau)} + 2e^{-2(t-\tau)} & -e^{-(t-\tau)} + 2e^{-2(t-\tau)} \end{bmatrix}$$

$$\begin{bmatrix} 0 \\ 1 \end{bmatrix}$$

$$= \begin{bmatrix} e^{-(t-\tau)} - e^{-2(t-\tau)} \\ -e^{-(t-\tau)} + 2e^{-2(t-\tau)} \end{bmatrix}$$

The state response $\mathbf{q}(t)$ for a unit step input $x(t) = u(t)$ under zero initial conditions $\mathbf{q}(0) = \mathbf{0}$ is given by [Eq. (11.44)]

$$\mathbf{q}(t) = \underbrace{e^{\mathbf{A}t}\mathbf{q}(0)}_{\text{zero-input component}} + \underbrace{\int_0^t e^{\mathbf{A}(t-\tau)}\mathbf{B}\mathbf{x}(\tau)\,d\tau}_{\text{zero-state component}}$$

$$= 0 + \int_0^t \begin{bmatrix} e^{-(t-\tau)} - e^{-2(t-\tau)} \\ -e^{-(t-\tau)} + 2e^{-2(t-\tau)} \end{bmatrix} u(\tau)\,d\tau$$

$$\begin{bmatrix} q_1(t) \\ q_2(t) \end{bmatrix} = \int_0^t \begin{bmatrix} e^{-(t-\tau)} - e^{-2(t-\tau)} \\ -e^{-(t-\tau)} + 2e^{-2(t-\tau)} \end{bmatrix} d\tau$$

$$= \begin{bmatrix} \frac{1}{2} - e^{-t} + \frac{1}{2}e^{-2t} \\ e^{-t} - e^{-2t} \end{bmatrix}$$

(e) Given that

$$y(t) = \underbrace{\begin{bmatrix} 1 & 0 \end{bmatrix}}_{\mathbf{C}} \begin{bmatrix} q_1(t) \\ q_2(t) \end{bmatrix}$$

$$y(t) = q_1(t)$$

$$= \frac{1}{2} - e^{-t} + \frac{1}{2}e^{-2t}$$

Example 11.18 Find $\mathbf{A}^n$ for

$$\mathbf{A} = \begin{bmatrix} -1 & 1 \\ 0 & 2 \end{bmatrix}$$

by the finite series summation (Cayley-Hamilton theorem) method.

Solution

First we find

$$[\lambda \mathbf{I} - \mathbf{A}] = \lambda \begin{bmatrix} 1 & 0 \\ 0 & 1 \end{bmatrix} - \begin{bmatrix} -1 & 1 \\ 0 & 2 \end{bmatrix}$$

$$= \begin{bmatrix} \lambda + 1 & -1 \\ 0 & \lambda - 2 \end{bmatrix}$$

The characteristic polynomial $Q(\lambda)$ of $\mathbf{A}$ is given by

$$Q(\lambda) = |\lambda \mathbf{I} - \mathbf{A}| = 0$$

$$\begin{vmatrix} \lambda + 1 & -1 \\ 0 & \lambda - 2 \end{vmatrix} = 0$$

$$(\lambda + 1)(\lambda - 2) = 0$$

Thus, the eigenvalues of $\mathbf{A}$ are $\lambda_1 = -1$ and $\lambda_2 = 2$. Hence, by Eq. (11.112) we have

$$\mathbf{A}^n = \beta_0 \mathbf{I} + \beta_1 \mathbf{A} = \begin{bmatrix} \beta_0 - \beta_1 & \beta_1 \\ 0 & \beta_0 + 2\beta_1 \end{bmatrix}$$

From Eq. (11.61), we have

$$\begin{bmatrix} \beta_0 \\ \beta_1 \end{bmatrix} = \begin{bmatrix} 1 & \lambda_1 \\ 1 & \lambda_2 \end{bmatrix}^{-1} \begin{bmatrix} \lambda_1^n \\ \lambda_2^n \end{bmatrix}$$

$$= \begin{bmatrix} 1 & -1 \\ 1 & 2 \end{bmatrix}^{-1} \begin{bmatrix} (-1)^n \\ 2^n \end{bmatrix}$$

$$= \frac{1}{3} \begin{bmatrix} 2 & 1 \\ -1 & 1 \end{bmatrix} \begin{bmatrix} (-1)^n \\ 2^n \end{bmatrix}$$

$$\begin{bmatrix} \beta_0 \\ \beta_1 \end{bmatrix} = \frac{1}{3} \begin{bmatrix} 2(-1)^n + 2^n \\ -(-1)^n + 2^n \end{bmatrix}$$

Hence,

$$\mathbf{A}^n = \frac{1}{3} \begin{bmatrix} 3(-1)^n & -(-1)^n + 2^n \\ 0 & 3 \times 2^n \end{bmatrix}$$

SUMMARY

1. The state equation of an LTI system in a state-variable form is

$$\dot{\mathbf{q}}(t) = \mathbf{A}\mathbf{q}(t) + \mathbf{B}\mathbf{x}(t)$$

2. The output equation of an LTI system in a state-variable form is

$$\mathbf{y}(t) = \mathbf{C}\mathbf{q}(t) + \mathbf{D}\mathbf{x}(t)$$

3. The matrix $\phi(t) = e^{\mathbf{A}t}$ is called the state-transition matrix (STM).

4. The STM $\phi(t) = e^{\mathbf{A}t}$ can be evaluated using the Cayley-Hamilton theorem, which states that any matrix $\mathbf{A}$ satisfies its own characteristic equation.

5. A continuous-time system is stable if and only if all the eigenvalues of the matrix $\mathbf{A}$ have negative real parts.

6. The transfer function of a continuous-time system is given by

$$\mathbf{H}(s) = \mathbf{C}\mathbf{\Phi}(s)\mathbf{B} + \mathbf{D}$$

7. The state equations of discrete-time system are given by

$$\mathbf{q}(n+1) = \mathbf{A}\mathbf{q}(n) + \mathbf{B}\mathbf{x}(n)$$
$$\mathbf{y}(n) = \mathbf{C}\mathbf{q}(n) + \mathbf{D}\mathbf{x}(n)$$

8. A discrete-time LTI system is stable if all the eigenvalues of $\mathbf{A}$ are inside the unit circle.

9. The transfer function of a continuous-time system is given by

$$\mathbf{H}(z) = \mathbf{C}[z\mathbf{I} - \mathbf{A}]^{-1}\mathbf{B} + \mathbf{D}$$

MULTIPLE-CHOICE QUESTIONS

1. Static variable describes
   (a) static behaviour of the system
   (b) dynamic behaviour of the system
   (c) transient behaviour of the system
   (d) steady-state behaviour of the system

2. Characteristic equation of the system is
   (a) $|s\mathbf{I} - \mathbf{A}| = 0$
   (b) $[s\mathbf{I} - \mathbf{A}]^{-1} = 0$
   (c) $\mathrm{Adj}[s\mathbf{I} - \mathbf{A}] = 0$
   (d) $[s\mathbf{I} - \mathbf{A}]^{-1} = 1$

3. Tick the correct statement
   (a) state model of a system is unique
   (b) state model of a system is not unique
   (c) state model of a system is arbitrary
   (d) state model does not exist

4. Transfer function obtained from the state model is
   (a) unique
   (b) non-unique
   (c) does not depend on the state model
   (d) arbitrary

5. Which statement is wrong for the STM
   $\Phi(t) = e^{\mathbf{A}t}$
   (a) $\Phi(0) = \mathbf{I}$
   (b) $\Phi^{-1}(t) = -\Phi(t)$

   (c) $\Phi(t_1)\Phi(t_2) = \Phi(t_1 + t_2)$
   (d) $\Phi^{-1}(t) = \Phi(-t)$

6. Transfer function method can only be applied to the system with
   (a) zero initial conditions
   (b) zero input condition
   (c) both (a) and (b)
   (d) any condition

7. In the state space method, number of variables are
   (a) fixed
   (b) arbitrary number
   (c) depends on the order of the system
   (d) both (a) and (c)

8. Number of state variable of discrete-time described by

$$y(n) - \frac{3}{4}y(n-1) + \frac{1}{3}y(n-2) = x(n)$$

   is

   (a) 2
   (b) 3
   (c) 4
   (d) 1

9. Eigenvalues of

$$\mathbf{A} = \begin{bmatrix} 0 & 1 \\ -\dfrac{1}{3} & \dfrac{4}{3} \end{bmatrix}$$

are

(a) 1, 2

(b) 1, $\frac{1}{2}$

(c) 1, $\frac{1}{3}$

(d) 1, 1

10. For a discrete-time system to be BIBO stable

(a) all the eigenvalues should be one

(b) all the eigenvalues should be distinct

(c) all the eigenvalues should be distinct and less than one

(d) all the eigenvalues should be greater than one

11. Continuous-time system will be BIBO stable if

(a) all the eigenvalues are one

(b) all the eigenvalues are distinct and their real parts are negative

(c) all the eigenvalues are negative

(d) all the eigenvalues are zero

PROBLEMS

11.1 Write a set of state equations corresponding to the differential equation

$$3\frac{d^3y(t)}{dt^3} - 2\frac{d^2y(t)}{dt^2} - \frac{dy(t)}{dt} + 4y(t) = 2x(t)$$

Also express them as a state equation.

11.2 Repeat Problem 11.1 for the differential equation

$$\frac{d^4y(t)}{dt^4} - 2\frac{d^2y(t)}{dt^2} - 5y(t) = -3x(t)$$

11.3 Consider the network shown in Fig. 11.7. The initial voltages across the capacitor C_1 and C_2 are 1/2 V and 1 V, respectively. Using the state variable method, find the voltages across these capacitors for $t > 0$. Assume that $R_1 = R_2 = R_3 = 1\Omega$ and $C_1 = C_2 = 1$ F.

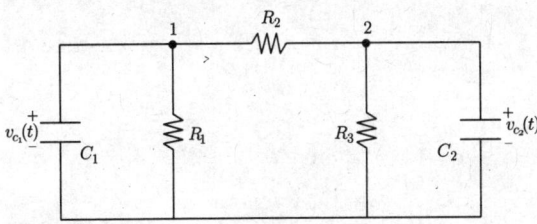

Fig. 11.7

11.4 Define state variables and find the corresponding state and output equations in matrix form for the electric circuit shown in Fig. 11.8.

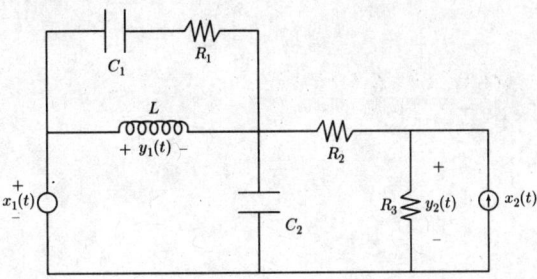

Fig. 11.8

11.5 Find the STM $\phi(t) = e^{\mathbf{A}(t)}$ for

$$\mathbf{A} = \begin{bmatrix} 0 & 1 \\ -6 & -5 \end{bmatrix}$$

using the Cayley-Hamilton theorem (finite series summation) method.

11.6 Repeat Problem 11.5, using the diagonalization method.

11.7 Repeat Problem 11.5 using the spectral decomposition method.

11.8 Repeat Problem 11.5 using the Laplace transform method.

11.9 Find $\mathbf{A}^n$ for

$$\mathbf{A} = \begin{bmatrix} 0 & 1 \\ -\dfrac{1}{3} & \dfrac{4}{3} \end{bmatrix}$$

11.10 Use the eigenvalue method to find the state transition matrix and impulse-response matrix for the system with the following parameter matrices:

$$A = \begin{bmatrix} -4 & 1 \\ -2 & -1 \end{bmatrix}$$

$$B = \begin{bmatrix} 1 & 2 & 0 \\ 0 & -1 & 1 \end{bmatrix}$$

$$C = \begin{bmatrix} -3 & 1 \end{bmatrix} \qquad D = \begin{bmatrix} -1 & 0 & 1 \end{bmatrix}$$

11.11 Repeat Problem 11.10 for the system with the following parameter matrices:

$$A = \begin{bmatrix} -3 & 2 \\ 1 & -4 \end{bmatrix}$$

$$B = \begin{bmatrix} 2 \\ 1 \end{bmatrix}$$

$$C = \begin{bmatrix} 1 & 0 \\ 0 & 1 \\ 1 & 0 \end{bmatrix} \qquad D = \begin{bmatrix} 0 \\ 0 \\ 1 \end{bmatrix}$$

11.12 Repeat Problem 11.10 using Laplace transform method.

ANSWERS TO MULTIPLE-CHOICE QUESTIONS

1. (b) 2. (a) 3. (b) 4. (a) 5. (b) 6. (a) 7. (d) 8. (a) 9. (c)
10. (c) 11. (b)

MATLAB Programs

12.1 INTRODUCTION

MATLAB is an interactive, matrix-based system for scientific and engineering numeric computation and visualization. It integrates computation, visualization, and programming in an easy-to-use environment where problems and solutions are expressed in familiar mathematical notations. MATLAB has extensive facilities for displaying vectors and matrices as graphs as well as annotating and printing these graphs. It includes high-level functions for two-dimensional and three-dimensional data visualization, image processing, animation, and presentation graphics. The name MATLAB stands for MATrix LABoratory since its basic data element is a matrix (array).

12.2 MATLAB VARIABLES—SCALARS, VECTORS, AND MATRICES

MATLAB stores variables in the form of matrices which are MN, where M is the number of rows and N the number of columns. A 1×1 matrix is a scalar, a $1 \times N$ matrix is a row vector, and an $M \times 1$ matrix is a column vector. All elements of a matrix can be real or complex numbers; $\sqrt{-1}$ can be written as either 'i' or 'j' provided they are not redefined by the user. A matrix is written with a square bracket '[ ]' with spaces separating adjacent columns and semicolons separating adjacent rows. For example, consider the following assignments of the variable x:

Real scalar $\gg$ x = 5

Complex scalar $\gg$ x = 5 +10j (or $\gg$ x = 5 +10i)

Row vector $\gg$ x = [1 2 3] (or x = [1, 2, 3])

Column vector $\gg$ x = [1; 2; 3]

3×3 matrix $\gg$ x = [1 2 3; 4 5 6; 7 8 9]

There are a few notes of caution. Complex elements of a matrix should not be typed with spaces, i.e., '-1+2j' is fine as a matrix element, '-1 +2j' is not. Also, '-1+2j' is interpreted correctly, whereas '-1+j2' is not (MATLAB interprets the 'j2' as the name of a variable. You can always write '-1+j*2'.

12.2.1 Complex Number Operations

Some of the important operations on complex numbers are given below:

$$\text{Complex scalar} \gg \texttt{x = 3+4j}$$

$$\text{Real part of x} \gg \texttt{real(x)} \Rightarrow 3$$

$$\text{Imaginary part of x} \gg \texttt{imag(x)} \Rightarrow 4$$

$$\text{Magnitude of x} \gg \texttt{abs(x)} \Rightarrow 5$$

$$\text{Angle of x} \gg \texttt{angle(x)} \Rightarrow 0.9273$$

$$\text{Complex conjugate of x} \gg \texttt{conj(x)} \Rightarrow \texttt{3-4i}$$

12.2.2 Generating Vectors

Vectors can be generated using the `':'` command. For example, to generate a vector `'x'` that takes on the values 0 to 10 in increments of 0.5, type the following which would generate a 1×21 matrix:

$$\gg \texttt{x = [0:0.5:10];}$$

Other commands used to generate vectors include `'linspace'`, which generates a vector by specifying the first and last number and the number of equally spaced entries between the first and the last number, and `'logspace'`, which is the same except that entries are spaced logarithmically between the first and the last entry.

12.2.3 Accessing Vector Elements

Elements of a matrix are accessed by specifying the row and column. For example, in the matrix specified by `A = [1 2 3; 4 5 6; 7 8 9]`, the element in the first row and third column can be accessed by

$$\gg \texttt{x = A(1,3)} \text{ which yields } 3$$

The entire second row can be accessed by

$$\gg \texttt{y = A(2,:)} \text{ which yields } [4 \ 5 \ 6]$$

where the `':'` means 'take all the entries in the column'. A submatrix of A consisting of rows 1 and 2 and all the three columns is specified by

$$\gg \texttt{z = A(1:2,1:3)} \text{ which yields } [1 \ 2 \ 3; \ 4 \ 5 \ 6]$$

12.3 MATRIX OPERATIONS

MATLAB contains a number of arithmetic, relational, and logical operations on matrices.

12.3.1 Arithmetic Matrix Operations

The basic arithmetic operations on matrices (and of course on scalars, which are special cases of matrices) are

+	addition	\	left division
−	subtraction	^	exponentiation (power)
*	multiplication	'	conjugate transpose
/	right division		

An error message occurs if the sizes of matrices are incompatible for the operation. Division is defined as follows: The solution to $A * x = b$ is x = A\b and the solution to $x * A = b$ is x = b/A, provided A is invertible and all the matrices are compatible. Addition and subtraction involve element-by-element arithmetic operations, whereas matrix multiplication and division do not. However, MATLAB provides for element-by-element operations as well by prepending a '.' before the operator as shown below:

.*	multiplication	.^	exponentiation (power)
./	right division	.'	transpose (unconjugated)
.\	left division		

The difference between matrix multiplication and element-by-element multiplication can be seen in the following example:

```
>> A = [1 2; 3 4]
A =
 1 2
 3 4
>> B = A*A
B =
  7 10
 15 22
>> C = A.*A
C =
 1  4
 9 16
```

12.3.2 Relational Operations

The basic relational operations are

<	less than	>=	greater than or equal to
<=	less than or equal to	==	equal to
>	greater than	~=	not equal to

These are element-by-element operations that return a matrix of ones (1 = true) and zeros (0 = false). Be careful of the distinction between '=' and '=='.

12.4 FLOW CONTROL OPERATIONS

MATLAB contains the usual set of flow control structures, e.g., for, while, and if, plus the logical operators, e.g., & (and), | (or), and ~ (not).

12.5 MATH FUNCTIONS

MATLAB comes with a large number of built-in functions that operate on matrices on an element-by-element basis. These include:

sin	sine
cos	cosine
tan	tangent
asin	inverse sine
acos	inverse cosine
atan	inverse tangent
exp	exponential
log	natural logarithm
log10	common logarithm
sqrt	square root
abs	absolute value
sign	signum

12.6 SIMPLE PLOTTING COMMANDS

The simple two-dimensional plotting commands include

plot	plot in linear coordinates as a continuous function
stem	plot in linear coordinates as discrete samples
loglog	logarithmic x and y axes
semilogx	linear y and logarithmic x axes
semilogy	linear x and logarithmic y axes
bar	bar graph
errorbar	error bar graph
hist	histogram
polar	polar coordinates

SOLVED EXAMPLES

Generation of Basic Signals

Example 12.1 Write a program to generate the following signals:

(a) Unit step function $u(t)$

(b) Unit impulse function $\delta(t)$

(c) A rectangular pulse of width 2

(d) Unit ramp function $r(t)$

Solution

```
%***** Generation of Basic Signals ***
clc;clear all;close all;
t=-5:0.005:5;
%**** Unit Step Function ********
u=0.5*(sign(t)+1);
%**** Unit Impulse Function ****
impls=0.5*(sign(t)+1)-0.5*(sign(t-
0.005)+1);
```

```
%****Rectangular Pulse*******
rectpulse=(sign(t)+1)-(sign(t-2)+1);
%***Ramp Signal *******
tt=0:50;
rmp=0.5*tt.*(sign(tt)+1);
figure;subplot(411);
plot(t,u);AXIS([-5 5 0 1.5]);
title('Unit Step Function');
subplot(412);
plot(t,impls);AXIS([-5 5 0 1.5]);
title('Impulse Function');
subplot(413);
plot(t,rectpulse);AXIS([-5 5 0 2.5]);
title('Rectangular Pulse');
subplot(414);plot(tt,rmp);
xlabel('Time-->');ylabel('Amplitude--
>');
title('Unit Ramp Function');
```

The result is shown in Fig. 12.1.

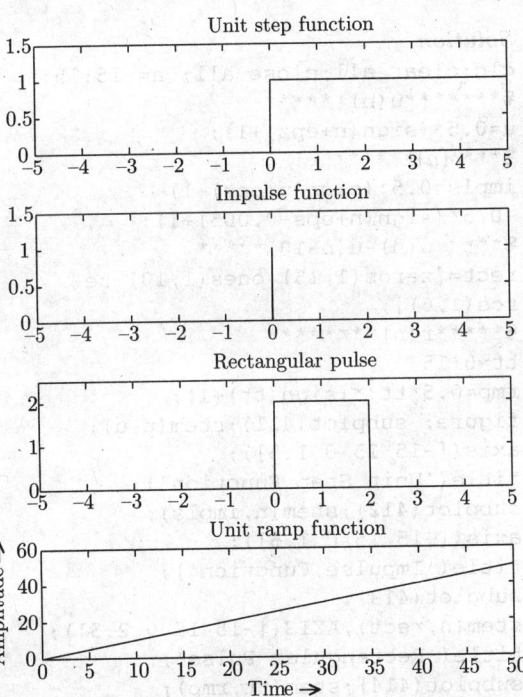

Fig. 12.1   Continuous-time basic signals

Example 12.2 Visualizing operations on independent variables. Write a program to generate the following signals:

(a) $u(t-1)$

(b) $u(2t+1)$

(c) $\delta(t-2.5)$

(d) $r(t-3)$

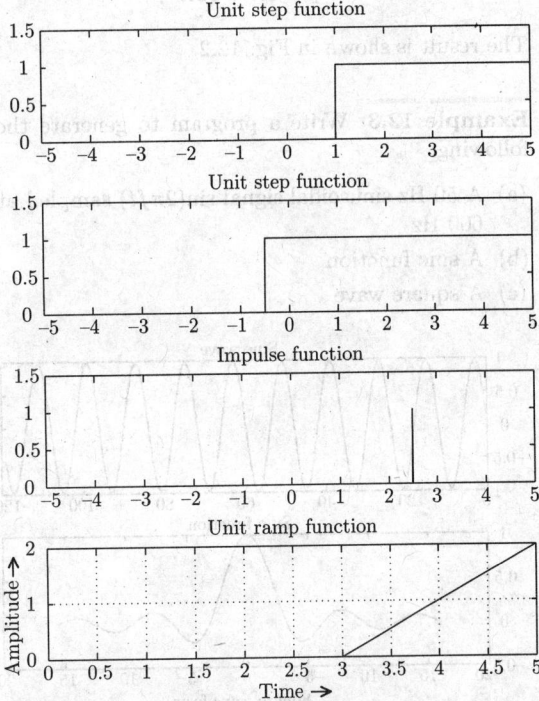

Fig. 12.2   Operations on independent variables

Solution
```
clc;clear all;close all;
t=-5:0.005:5;
u1=0.5*(sign(t-1)+1);%u(t-1)
u2=0.5*(sign(2*t+1)+1); %u(2t+1)
%**δ(t-2.5)****
ims=0.5*(sign(t-2.5)+1)···
-0.5*(sign(t-2.5-0.005)+1);
%*** Ramp Signal *********
tt=0:5; rmp=0.5*(tt-3).*(sign(tt-
3)+1);
figure; subplot(411);
plot(t,u1); axis([-5 5 0 1.5]);
title('Unit Step Function');
```

```
subplot(412);plot(t,u2);
axis([-5 5 0 1.5]);
title('Unit Step Function');
subplot(413);plot(t,ims);
axis([-5 5 0 1.5]);
title('Impulse Function');
subplot(414);plot(tt,rmp);grid on;
xlabel('Time-->');ylabel('Amplitude--
>');
title('Unit Ramp Function');
```

The result is shown in Fig. 12.2.

Example 12.3 Write a program to generate the following:

(a) A 50 Hz sinusoidal signal $\sin(2\pi ft)$ sampled at 600 Hz

(b) A sinc function

(c) A square wave

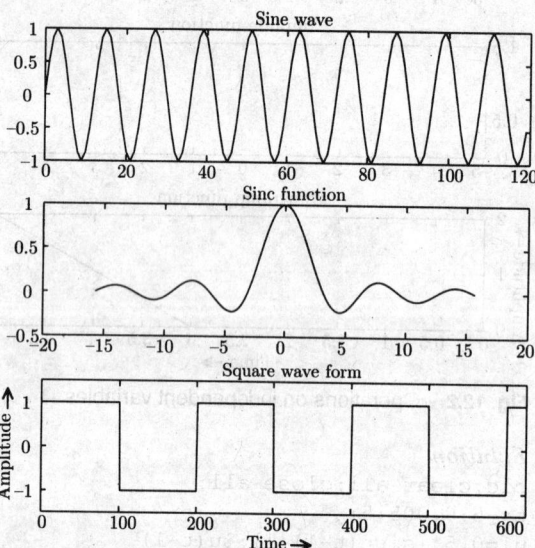

Fig. 12.3 Basic signals

Solution
```
%*** Sinusoidal Signal ****
clc;clear all;close all;
N=120;n=0:N-1;f=50;Ts=1/600;
x=sin(2*pi*f*n*Ts);
%***Sinc Function *****
t=-5*pi:pi/50:5*pi;
sincpulse=sin(t)./t;
```

```
%*** Square Wave ****
tt= 0:.0001:.0625;t1=1:626;
y = square(2*pi*50*tt);
figure; subplot(311);plot(n,x);
title('Sine Wave');
subplot(312);plot(t,sincpulse);
title('Sinc Function'); subplot(313);
plot(t1,y);axis([0 625 -1.5 1.5])
xlabel('Time-->');ylabel('Amplitude--
>');
title('Square Waveform');
```

The result is shown in Fig. 12.3

Example 12.4 Write a program to generate the following discrete-time signals:

(a) Unit step function $u(n)$

(b) Unit impulse function or unit sample function $\delta(n)$

(c) A rectangular pulse of width 10

(d) Unit ramp function $r(n)$

Solution
```
clc;clear all;close all; n=-15:15;
%*******u(n)******
u=0.5*(sign(n+eps)+1);
%***δ(n)******
impls=0.5*(sign(n+eps)+1)···
-0.5*(sign(n+eps-0.005)+1);
%**** u(n)-u(n-10)*****
rect=[zeros(1,15) ones(1,10) ze-
ros(1,6)];
%*****r(n)********
tt=0:15;
rmp=0.5*tt.*(sign(tt)+1);
figure; subplot(411);stem(n,u);
axis([-15 15 0 1.5]);
title('Unit Step Function');
subplot(412);stem(n,impls);
axis([-15 15 0 1.5]);
title('Impulse Function');
subplot(413);
stem(n,rect);AXIS([-15 15 0 2.5]);
title('Rectangular Pulse');
subplot(414);stem(tt,rmp);
xlabel('Time-->');ylabel('Amplitude--
>');
title('Unit Ramp Function');
```

The result is shown in Fig. 12.4.

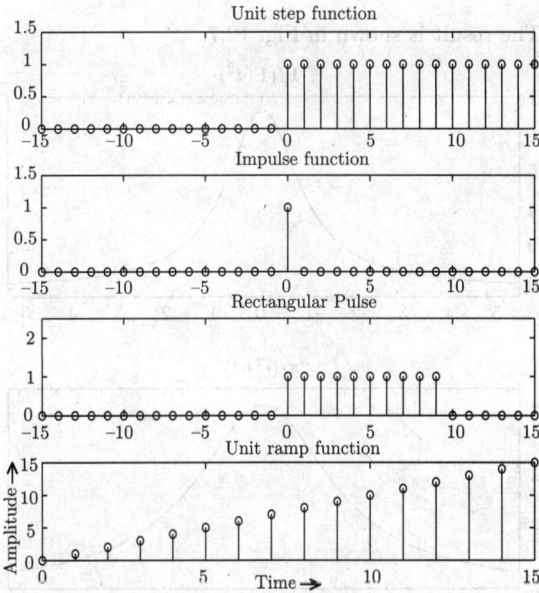

Fig. 12.4   Discrete-time basic signals

Example 12.5   Let $x(n) = u(n) - u(n - 10)$. Decompose $x(n)$ into even and odd components.

Solution
```
clc;clear all;close all;
n=-10:15; x=[zeros(1,10) ones(1,10)
zeros(1,6)];
m=-fliplr(n); m1=min([m,n]);m2=
max([m,n]);
m=m1:m2; nm=n(1)-m(1);
n1=1:length(n); x1=zeros(1,length(m));
x1(n1+nm)=x; x=x1;
xe=0.5*(x+fliplr(x));% Even part of
x(n)
x0=0.5*(x-fliplr(x));% Odd part of
x(n)
figure; t=-15:15;
subplot(311);stem(t,x);
axis([-15 15 0 1.5]);title('Original
Signal');
subplot(312);stem(t,xe);
axis([-15 15 0 1.5]);title('Even
Part')
```

```
subplot(313);stem(t,x0);
axis([-15 15 -1 1]);
xlabel('Time-->');ylabel('Amplitude--
>');
title('Odd part');
```

The result is shown in Fig. 12.5.

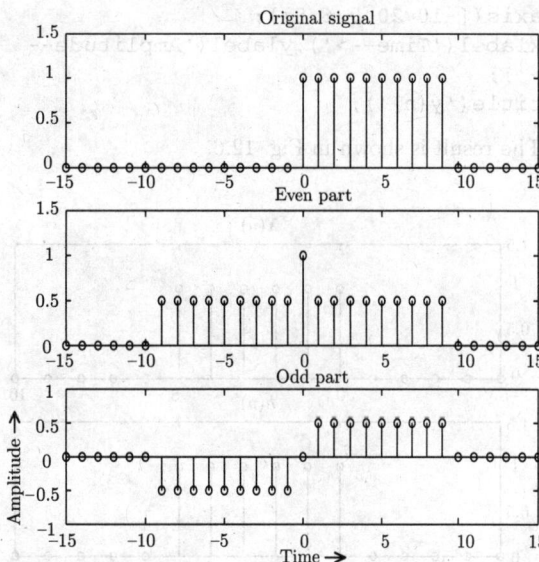

Fig. 12.5   Even and odd parts of $x(n)$

Convolution

Example 12.6   Let $x(n) = h(n) = u(n) - u(n - 6)$. Write a program to determine $y(n) = x(n) * h(n)$.

Solution
```
clc;clear all;close all;
nx=-5:10;nh=-5:10;
x=[zeros(1,5) ones(1,6) zeros(1,5)];
h=[zeros(1,5) ones(1,6) zeros(1,5)];
Nx=length(x); Nh=length(h);
Ny=Nx+Nh-1;
newx=[x zeros(1,Ny-Nx)];
newh=[h zeros(1,Ny-Nh)];
for n=0:Ny-1
sum=0;
for k=0:n
sum=sum+newx(k+1)*newh(n-k+1);
```

```
end
y(n+1)=sum;
end
n1=min(nx)+min(nh):max(nx)+max(nh);
figure; subplot(311);stem(nx,x);
axis([-5 10 0 1.5]);title('x(n)');
subplot(312);stem(nh,h);
axis([-5 10 0 1.5]);title('h(n)');
subplot(313);stem(n1,y);
axis([-10 20 0 6.5]);
xlabel('Time-->');ylabel('Amplitude--
>');
title('y(n)');
```

The result is shown in Fig. 12.6.

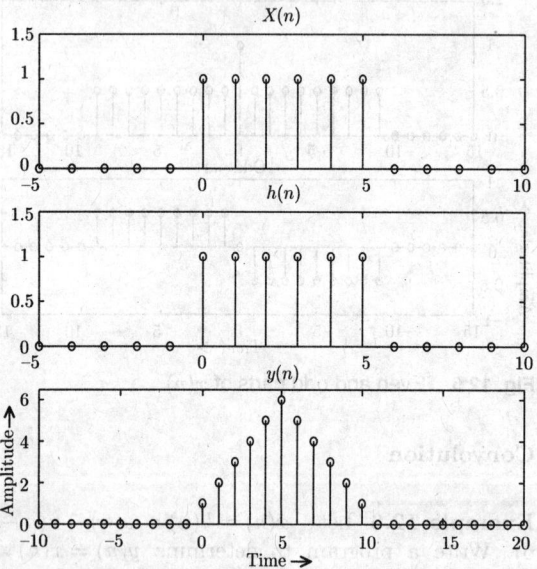

Fig. 12.6 Convolution of a rectangular signal with itself

Example 12.7 Determine the convolution of $x(t) = \dfrac{1}{1+t^2}$ with itself.

Solution
```
clc; clear all;close all;
syms t tau%for symbolic calculation
x=1/(1+t^2);
%***Convolution Integral**
z=int(subs(x,tau)*subs(x,t-tau),tau,-
inf,inf);
```

```
z=simplify(z);
figure; subplot(211); ezplot(x);
subplot(212);ezplot(z);
```

The result is shown in Fig. 12.7.

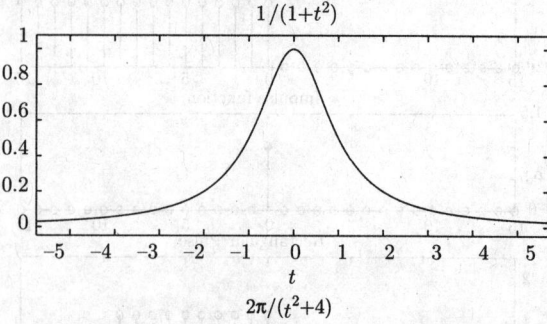

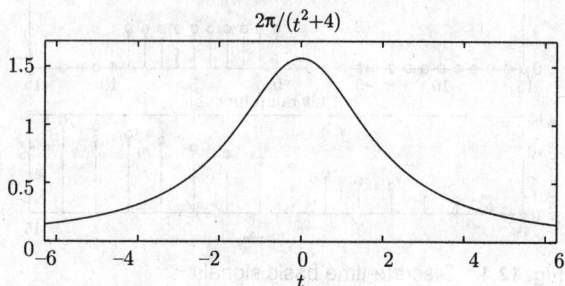

Fig. 12.7 Convolution of $x(t) = \dfrac{1}{1+t^2}$ with itself

Correlation

Example 12.8 Compute the crosscorrelation between the following two sequences:
$$x(n) = \{1, 2, \underset{\uparrow}{-1}, 3\}$$
$$h(n) = \{\underset{\uparrow}{3}, -2, 1, 4\}$$

Solution
```
clc;clear all;close all;
x=[1 2 -1 3];nx=-2:1;
h=[3 -2 1 4];nh=0:3;
hh=fliplr(h);
Nx=length(x);
Nh=length(h);
Ny=Nx+Nh-1;
newx=[x zeros(1,Ny-Nx)];
newh=[hh zeros(1,Ny-Nh)];
for n=0:Ny-1
sum=0;
for k=0:n
sum=sum+newx(k+1)*newh(n-k+1);
```

```
end
y(n+1)=sum;
end
n1=min(nx)+min(nh):max(nx)+max(nh);
figure;
subplot(311);stem(nx,x);
title('x(n)');
subplot(312); stem(nh,h);
title('h(n)');
subplot(313);stem(n1,y);
xlabel('Time-->');ylabel('Amplitude--
>');
title('y(n)');
```

The result is shown in Fig. 12.8.

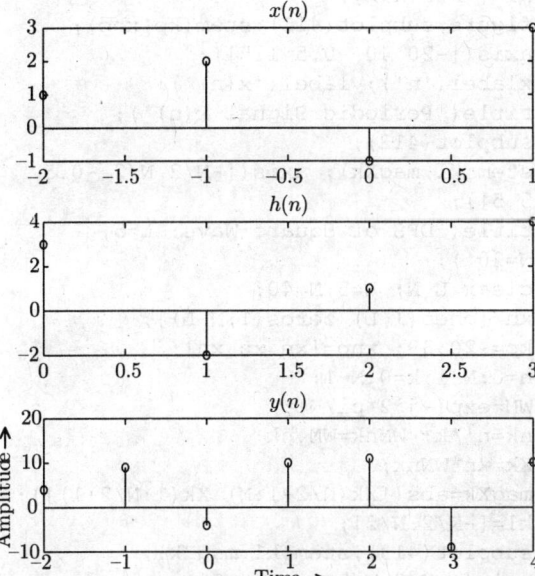

Fig. 12.8 Correlation between $x(n)$ and $x(n)$ sequences

Example 12.9 Let

$$x(n) = \{3, 11, 7, \underset{\uparrow}{0}, -1, 4, 2\}$$

be a prototype sequence, and let $y(n)$ be its noise-corrupted-and-shifted version

$$y(n) = x(n-2) + w(n)$$

where $w(n)$ is a Gaussian noise sequence with zero mean and variance 1. Compute the crosscorrelation between $y(n)$ and $x(n)$.

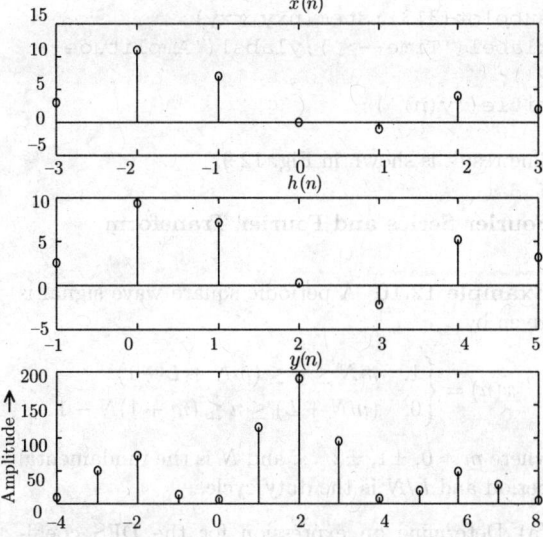

Fig. 12.9 Correlation between $x(n)$ and $y(n)$ sequences

Solution

```
clc;clear all;close all;
x=[3 11 7 0 -1 4 2];nx=-3:3;
xn2=[3 11 7 0 -1 4 2];ny=nx+2;%x(n-2)
%***Gaussian Noise***
w=randn(1,length(xn2));
%**y(n)=x(n-2)+w(n)**
y=xn2+w;
yy=fliplr(y);
Nx=length(x);
Ny=length(y);
Nxy=Nx+Ny-1;
newx=[x zeros(1,Nxy-Nx)];
newy=[yy zeros(1,Nxy-Ny)];
for n=0:Nxy-1
sum=0;
for k=0:n
sum=sum+newx(k+1)*newy(n-k+1);
end
rxy(n+1)=sum;
end
nxy=min(nx)+min(ny):max(nx)+max(ny);
figure;
```

```
subplot(311);stem(nx,x);
title('x(n)');
subplot(312);stem(ny,y);
title('h(n)');
subplot(313);stem(nxy,rxy);
xlabel('Time-->');ylabel('Amplitude--
>');
title('y(n)');
```

The result is shown in Fig. 12.9.

Fourier Series and Fourier Transform

Example 12.10 A periodic square wave signal is given by

$$x(n) = \begin{cases} 1, & mN \leq n \leq (mN + L - 1) \\ 0, & (mN + L) \leq n \leq (m+1)N - 1 \end{cases}$$

where $m = 0, \pm 1, \pm 2, \ldots$ and N is the fundamental period and L/N is the duty cycle.

(a) Determine an expression for the DFS coefficients $|X_k|$ in terms of L and N.

(b) Plot the magnitude $|X_k|$ for $L = 5$, $N = 20$; $L = 5$, $N = 40$; $L = 9$, $N = 60$.

(c) Comment on the results.

Solution

By definition $\omega_0 = 2\pi/N$

$$X_k = \frac{1}{N} \sum_{n=0}^{N-1} x(n) e^{-j\omega_0 kn}$$

$$= \frac{1}{N} \sum_{n=0}^{L-1} e^{-j\frac{2\pi}{N}kn}$$

$$= \begin{cases} \frac{1}{N}L, & k = 0, \pm N, \pm 2N, \ldots \\ \frac{1}{N}\frac{1-e^{-j2\pi kL/N}}{1-e^{-j2\pi k/N}}, & \text{otherwise} \end{cases}$$

The last expression can be simplified to

$$\frac{1}{N}\frac{1-e^{-j2\pi kL/N}}{1-e^{-j2\pi k/N}}$$

$$= \frac{1}{N}\frac{e^{-j\pi kL/N}}{e^{-j\pi k/N}}\frac{e^{j\pi kL/N}-e^{-j\pi kL/N}}{e^{j\pi k/N}-e^{-j\pi k/N}}$$

$$= \frac{1}{N}e^{-j\pi k(L-1)/N}\frac{\sin(\pi kL/N)}{\sin(\pi k/N)}$$

The magnitude of X_k is given by

$$|X_k|, = \begin{cases} \frac{1}{N}L, & k = 0, \pm N, \pm 2N, \ldots \\ \left|\frac{1}{N}\frac{\sin(\pi kL/N)}{\sin(\pi k/N)}\right|, & \text{otherwise} \end{cases}$$

```
clc;clear all;close all;
L=5;N=20;
xn=[ones(1,L) zeros(1,N-L)];
kp=-20:39;
xnp=[xn xn xn];
n=0:N-1;k=0:N-1;
WN=exp(-j*2*pi/N);
nk=n'*k; WNnk=WN.ñk; Xk=xn*WNnk;
magXk=abs([Xk(N/2+1:N) Xk(1:N/2+1)]);
k1=[-N/2:N/2];
figure;subplot(411);stem(kp,xnp);
axis([-20 40 -0.5 1.5]);
xlabel('n');ylabel('x(n)');
title('Periodic Signal x(n)');
subplot(412);
stem(k1,magXk); axis([-N/2 N/2 -0.5
5.5]);
title('DFS of Square Wave: L=5,
N=20');
clear L N; L=5;N=40;
xn=[ones(1,L) zeros(1,N-L)];
kp=-20:39; xnp=[xn xn xn];
n=0:N-1;k=0:N-1;
WN=exp(-j*2*pi/N);
nk=n'*k; WNnk=WN.ñk;
Xk=xn*WNnk;
magXk=abs([Xk(N/2+1:N) Xk(1:N/2+1)]);
k1=[-N/2:N/2];
subplot(413);stem(k1,magXk);
axis([-N/2 N/2 -0.5 5.5]);
title('DFS of Square Wave:
L=5, N=40');
clear L N; L=9;N=60;
xn=[ones(1,L) zeros(1,N-L)];
kp=-20:39; xnp=[xn xn xn];
n=0:N-1;k=0:N-1;
WN=exp(-j*2*pi/N);
nk=n'*k; WNnk=WN.ñk;
Xk=xn*WNnk;
magXk=abs([Xk(N/2+1:N)
Xk(1:N/2+1)]);
k1=[-N/2:N/2];
subplot(414);stem(k1,magXk);
axis([-N/2 N/2 -0.5 9.5]);
```

```
xlabel('k');ylabel('|Xk|);
title('DFS of square wave: L=9,
N=60');
```

The result is shown in Fig. 12.10.

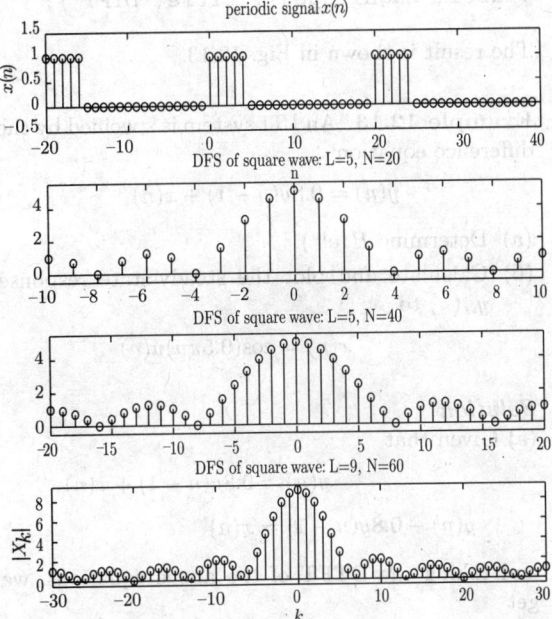

Fig. 12.10 Discrete-time Fourier series

Example 12.11 Let

$$x(t) = 2 \operatorname{rect}\left(\frac{t}{4}\right) = \begin{cases} 2, & -2 < t < 2 \\ 0, & \text{otherwise} \end{cases}$$

Determine and plot its continuous-time Fourier transform.

Solution
```
clc; clear all;close all;
syms t omega%for symbolic calculation
x=2; expw=exp(-j*omega *t);
%***Fourier Transform**
z=int(x*expw,omega,-2,2);
z=simplify(z);
figure(1);
subplot(211); ezplot('2', [-2 2]);
subplot(212) ezplot(z);
```

The result is shown in Fig. 12.11.

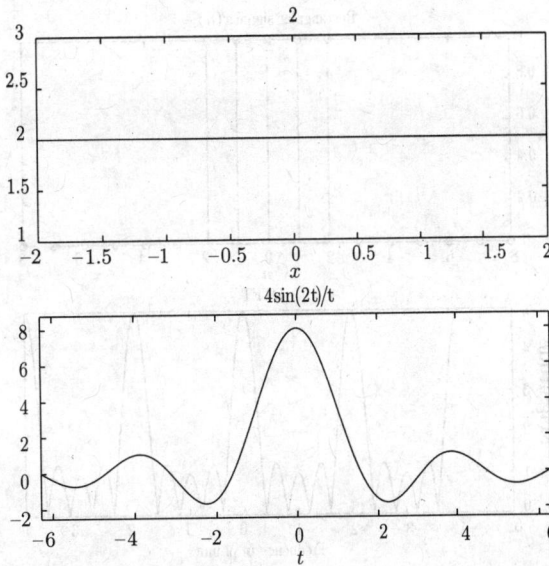

Fig. 12.11 Fourier transform of a rectangular pulse

Example 12.12 Determine and plot the discrete-time Fourier transform (DTFT) of the following:
(a) A rectangular pulse given by

$$x(n) = \begin{cases} 1, & -2 \le n \le 2 \\ 0, & \text{otherwise} \end{cases}$$

(b) A sinusoidal signal

$$x(n) = \cos\left(\frac{\pi n}{2}\right), \quad 0 \le n \le 100$$

Solution
(a)
```
clc; clear all;close all;
n=-2:2;x=ones(1,5); k=-400:400;
w=(pi/100)*k;
X=x*(exp(-j*pi/100)).^(n'*k);
magX=abs(X);
figure;subplot(211);
stem(-8:8, [zeros(1,6) x zeros(1,6)]);
xlabel('n');ylabel('x');
title('Rectangle Signal x(n)')
subplot(212); plot(w/pi,magX);
xlabel('Frequency in pi Units');
ylabel('Magnitude'); title('DTFT');
```

The result is shown in Fig. 12.12.

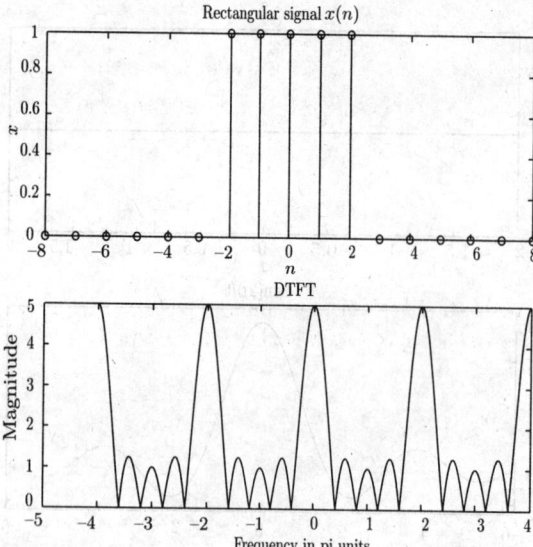

Fig. 12.12 DTFT of a rectangular pulse

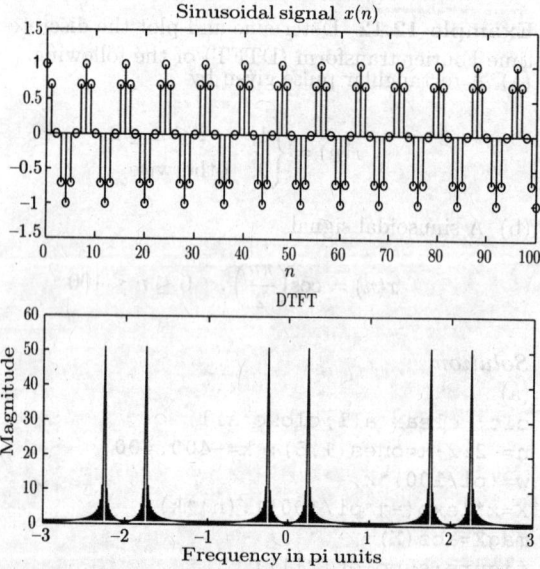

Fig. 12.13 DTFT of a sinusoidal signal

(b)
```
clc; clear all;close all;
n=0:100;x=cos(pi*n/4);
k=-300:300;w=(pi/100)*k;
X=x*(exp(-j*pi/100)).^(n'*k);
magX=abs(X);
```

```
figure;subplot(211);
stem(n,x);axis([0 100 -1.5 1.5]);
xlabel('n');ylabel('x');
title('Sinusoidal Signal x(n)');
subplot(212); plot(w/pi,magX);
xlabel('Frequency in pi Units');
ylabel('Magnitude'); title('DTFT');
```

The result is shown in Fig. 12.13.

Example 12.13 An LTI system is specified by the difference equation

$$y(n) = 0.8y(n-1) + x(n)$$

(a) Determine $H(e^{j\omega})$.

(b) Calculate and plot the steady-state response $y_{ss}(n)$ to

$$x(n) = \cos(0.5\pi n)u(n)$$

Solution
(a) Given that

$$y(n) = 0.8y(n-1) + x(n)$$
$$y(n) - 0.8y(n-1) = x(n)$$

and taking the DTFT of the above equations, we get

$$\frac{Y(e^{j\omega})}{X(e^{j\omega})} = H(e^{j\omega}) = \frac{1}{1 - 0.8e^{-j\omega}}$$

$$= \frac{e^{j\omega}}{e^{j\omega} - 0.8}$$

(b) Given that the input is $x(n) = \cos(0.5\pi n)$ with frequency $\omega_0 = 0.05\pi$,

$$H(e^{j\omega})\Big|_{\omega_0=0.05\pi} = \frac{1}{1 - 0.8e^{-j0.05\pi}}$$

$$= 4.092e^{-j0.5377}$$

Therefore, the response to a periodic signal is

$$y_{ss}(n) = 4.0928\cos(0.05\pi n - 0.5377)$$

$$= 4.0928\cos[0.05\pi(n - 3.42)]$$

This means that at the output the sinusoid is scaled by 4.0928 and shifted by 3.42 samples.

```
clc; clear all;close all;
b=1,a=[1 -0.8];
n=[0:100];x=cos(0.05*pi*n);
```

```
y=filter(b,a,x);
subplot(211);stem(n,x);
xlabel('n');ylabel('x(n)');
title('Input Signal');
subplot(212);stem(n,y);
xlabel('n');ylabel('y(n)');
title('Output Signal');
```

The result is shown in Fig. 12.14.

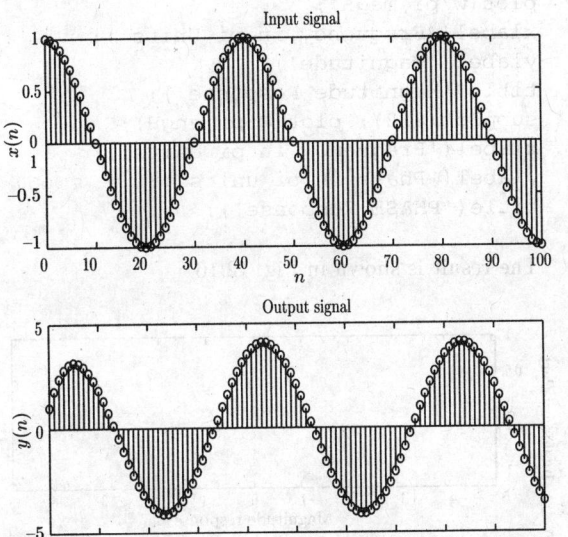

Fig. 12.14 Steady-state response

Sampling Theorem

Example 12.14 Let $x(t) = e^{-1000|t|}$.

(a) Sample $x(t)$ at $f_s = 5000$ samples/sec to obtain $x(n)$.

(b) From the samples $x(n)$, reconstruct $x(t)$.

Solution

(a) We know that

$$e^{-a|t|} \longleftrightarrow \frac{2a}{a^2 + \omega^2}$$

$$e^{-1000|t|} \longleftrightarrow \frac{2 \times 1000}{1000^2 + \omega^2}$$

$$\longleftrightarrow \frac{0.002}{1 + \left(\dfrac{\omega}{1000}\right)^2}$$

Using the approximation $e^{-5} \approx 0$, we know that $x(t)$ can be approximated by a finite-duration signal over $-0.005 \le t \le 0.005$. Similarly, $X(\omega) = \dfrac{0.002}{1 + \left(\dfrac{\omega}{1000}\right)^2} \approx 0$ for $\omega \ge 2\pi(2000)$. Hence, we choose

$$\Delta(t) = 5 \times 10^{-5} \ll \frac{1}{2 \times 2000} = 25 \times 10^{-5}$$

We will obtain $x(n)$ by sampling $x(t)$ at $T_s = 1/f_s = 0.0002$ sec over $-0.005 \le t \le 0.005$, which gives $x(n)$ over $-25 \le n \le 25$.

(b)
```
clc; clear all;close all;
Ts=0.0002; n=-25:1:25; nTs=n*Ts;
x=exp(-1000*abs(nTs));
%** Reconstruction**
Dt=0.0002;t=-0.005:Dt:0.005; fs=5000;
xt=x*sinc(fs*(ones(length(n),1)*t···
-nTs'*ones(1,length(t))));
figure;;stem(n,x);
hold on; plot(-25:25,xt);
xlabel('Time');ylabel('x(t)');
title('Reconstructed Signal');
```

The result is shown in Fig. 12.15.

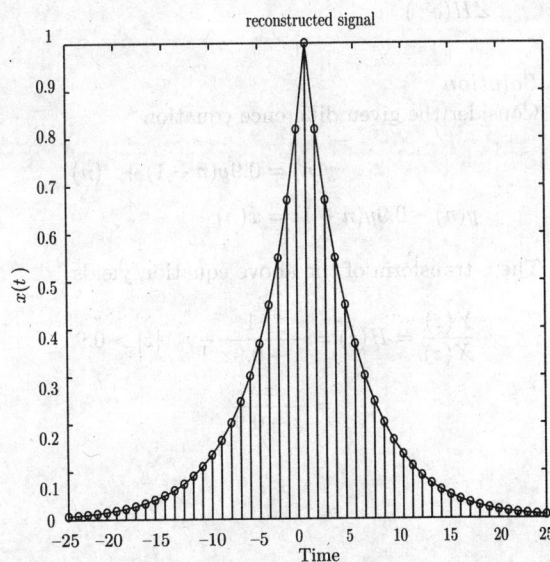

Fig. 12.15 Sampling and interpolation

Laplace Transform

Example 12.15 Using MATLAB's symbolic math toolbox, determine the Laplace transform of the following signals:

(a) $x_1(t) = te^{-at}u(t)$

(b) $x_2(t) = [\sin(at) + \cos(bt)](u(t)$

Solution

```
clc; clear all;close all;
syms a b t
x1t=t*exp(-a*t);
x2t=sin(a*t)+cos(b*t);
X1s=laplace(x1t);
X2s=laplace(x2t);
X2s=simplify(X2s);
disp(X1s);disp(X2s);
```

z-Transform

Example 12.16 Given a causal system

$$y(n) = 0.9y(n-1) + x(n)$$

(a) Find $H(z)$ and sketch its pole-zero plot.

(b) Plot the frequency response $|H(e^{j\omega})|$ and $\angle H(e^{j\omega})$.

Solution
Consider the given difference equation

$$y(n) = 0.9y(n-1) + x(n)$$

$$y(n) - 0.9y(n-1) = x(n)$$

The z-transform of the above equation yields

$$\frac{Y(z)}{X(z)} = H(z) = \frac{1}{1 - 0.9z^{-1}}, \quad |z| > 0.9$$

$$= \frac{z}{z - 0.9}$$

The system has one pole at $z = 0.9$ and one zero at the origin.

```
clc; clear all;close all;
b=[1];a=[1 -0.9];
zplane(b,a);
w=[-200:1:200]*pi/100;
H=freqz(b,a,w);
magH=abs(H);angH=angle(H);
figure; subplot(211);
plot(w/pi,magH);
xlabel('Frequency in pi Units');
ylabel('Magnitude');
title('Magnitude Response');
subplot(212); plot(w/pi,angH);
xlabel('Frequency in pi Units');
ylabel('Phase in pi units');
title('PHASE Response');
```

The result is shown in Fig. 12.16.

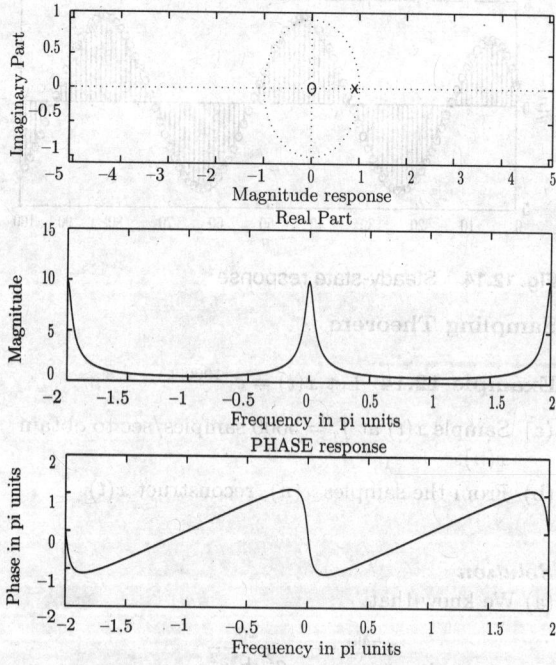

Fig. 12.16 Pole-zero plot and frequency response

Mathematical Relations

A.1 TRIGONOMETRIC IDENTITIES

$$e^{\pm j\theta} = \cos(\theta) \pm j\sin(\theta)$$

$$\cos(\theta) = \frac{e^{j\theta} + e^{-j\theta}}{2} = \sin\left(\theta + \frac{\pi}{2}\right)$$

$$\sin(\theta) = \frac{e^{j\theta} - e^{-j\theta}}{2j} = \cos\left(\theta - \frac{\pi}{2}\right)$$

$$\sin^2(\theta) + \cos^2(\theta) = 1$$

$$\cos^2(\theta) - \sin^2(\theta) = \cos(2\theta)$$

$$\cos^2(\theta) = \frac{1}{2}[1 + \cos(2\theta)]$$

$$\sin^2(\theta) = \frac{1}{2}[1 - \cos(2\theta)]$$

$$\cos^3(\theta) = \frac{1}{4}[3\cos(\theta) + \cos(3\theta)]$$

$$\sin^3(\theta) = \frac{1}{4}[3\sin(\theta) - \sin(3\theta)]$$

$$\sin(a \pm b) = \sin(a)\cos(b) \pm \cos(a)\sin(b)$$

$$\cos(a \pm b) = \cos(a)\cos(b) \mp \sin(a)\sin(b)$$

$$\tan(a \pm b) = \frac{\tan(a) \pm \tan(b)}{1 \mp \tan(a)\tan(b)}$$

$$\sin(a)\sin(b) = \frac{1}{2}[\cos(a - b) - \cos(a + b)]$$

$$\cos(a)\cos(b) = \frac{1}{2}[\cos(a - b) + \cos(a + b)]$$

$$\sin(a)\cos(b) = \frac{1}{2}[\sin(a - b) + \sin(a + b)]$$

$$\sin(a) + \sin(b) = 2\sin\left(\frac{a + b}{2}\right)\cos\left(\frac{a - b}{2}\right)$$

$$\cos(a) + \cos(b) = 2\cos\left(\frac{a+b}{2}\right)\cos\left(\frac{a-b}{2}\right)$$

$$\cos(a) - \cos(b) = 2\sin\left(\frac{a+b}{2}\right)\sin\left(\frac{b-a}{2}\right)$$

$$A\cos(\alpha) + B\sin(\beta) = \sqrt{(A^2 + B^2)}\cos\left(\alpha - \tan^{-1}\frac{B}{A}\right)$$

$$\sinh(a) = \frac{e^a - e^{-a}}{2}$$

$$\cosh(a) = \frac{e^a + e^{-a}}{2}$$

$$\tanh(a) = \frac{\sinh(a)}{\cosh(a)}$$

$$\cosh^2(a) - \sinh^2(a) = 1$$

$$\cosh(a) + \sinh(a) = e^a$$

$$\cosh(a) - \sinh(a) = e^{-a}$$

A.2 POWER SERIES EXPANSION

$$(1 + x)^n = 1 + nx + \frac{n(n-1)}{2!}x^2 + \cdots + \binom{n}{k}x^k + \cdots + x^n$$

$$e^x = 1 + x + \frac{x^2}{2!} + \cdots + \frac{x^n}{n!} + \cdots$$

$$\ln(1 + x) = x - \frac{x^2}{2} + \frac{x^3}{3} - \cdots (-1)^{n+1}\frac{x^n}{n} + \cdots, \qquad |x| < 1$$

$$\sin(x) = x - \frac{x^3}{3!} + \frac{x^5}{5!} + \cdots + (-1)^n\frac{x^{2n+1}}{(2n+1)!} + \cdots$$

$$\cos(x) = 1 - \frac{x^2}{2!} + \frac{x^4}{4!} - \frac{x^6}{6!} + \cdots + (-1)^n\frac{x^{2n}}{(2n)!} + \cdots$$

$$\tan(x) = x + \frac{x^3}{3} + \frac{2}{15}x^5 + \cdots$$

$$a^x = 1 + x\ln(a) + \frac{(x\ln(a))^2}{2!} + \cdots + \frac{(x\ln(a))^n}{n!} + \cdots$$

$$\sinh(x) = x + \frac{x^3}{3!} + \frac{x^5}{5!} + \cdots + \frac{x^{2n+1}}{(2n+1)!} + \cdots$$

$$\cos(x) = 1 + \frac{x^2}{2!} + \frac{x^4}{4!} + \frac{x^6}{6!} + \cdots + \frac{x^{2n}}{(2n)!} + \cdots$$

A.3 SUMS OF POWERS OF NATURAL NUMBERS

$$\sum_{k=1}^{N} k = \frac{N(N+1)}{2}$$

$$\sum_{k=1}^{N} k^2 = \frac{N(N+1)(2N+1)}{6}$$

$$\sum_{k=1}^{N} k^3 = \frac{N^2(N+1)^2}{4}$$

A.3.1 Series of Exponentials

$$\sum_{n=0}^{N} a^n = \begin{cases} \dfrac{1 - a^{N+1}}{1 - a}, & a \neq 1 \\ N + 1, & a = 1 \end{cases}$$

$$\sum_{n=0}^{N} e^{j2\pi kn/N} = \begin{cases} N, & k = 0, N \\ 0, & 1 \leq k \leq N - 1 \end{cases}$$

$$\sum_{n=0}^{\infty} a^n = \frac{1}{1 - a}, \qquad |a| < 1$$

$$\sum_{n=0}^{\infty} n a^n = \frac{1}{(1 - a)^2}, \qquad |a| < 1$$

$$\sum_{n=0}^{\infty} n^2 a^n = \frac{a^2 + a}{(1 - a)^3}, \qquad |a| < 1$$

A.4 DERIVATIVES

$$\frac{d}{dt}[t^n] = n t^{n-1}$$

$$\frac{d}{dt}[\ln(t)] = \frac{1}{t}$$

$$\frac{d}{dt}[e^{at}] = a\, e^{at}$$

$$\frac{d}{dt}[\sin(at)] = a \cos(at)$$

$$\frac{d}{dt}[\cos(at)] = -a \sin(at)$$

$$\frac{d}{dt}[\tan(at)] = a \sec^2(at)$$

$$\frac{d}{dt}[\sin^{-1}(at)] = \frac{a}{\sqrt{1 - (at)^2}}$$

$$\frac{d}{dt}[\cos^{-1}(at)] = -\frac{a}{\sqrt{1-(at)^2}}$$

$$\frac{d}{dt}[\tan^{-1}(at)] = \frac{a}{1+(at)^2}$$

$$\frac{d}{dt}[x(t)y(t)] = x(t)\frac{d}{dt}[y(t)] + y(t)\frac{d}{dt}[x(t)]$$

$$\frac{d}{dt}\left[\frac{x(t)}{y(t)}\right] = \frac{\left\{y(t)\dfrac{d[x(t)]}{dt} - x(t)\dfrac{d}{dt}[y(t)]\right\}}{y^2(t)}$$

A.5 DEFINITE INTEGRALS

$$\int_0^\infty \left[\frac{a}{a^2+t^2}\right]dt = \frac{\pi}{2}, \qquad a > 0$$

$$\int_0^\infty t^n e^{-at}\, dt = \frac{n!}{a^{n+1}}, \qquad a > 0, n \text{ is a positive integer}$$

$$\int_0^\infty e^{-a^2 t^2}\, dt = \frac{\sqrt{\pi}}{2a}, \qquad a > 0$$

$$\int_0^\pi \sin^2(nt)\, dt = \int_0^\pi \cos^2(nt)\, dt = \frac{\pi}{2}, \qquad n \text{ is an integer}$$

$$\int_0^\pi \sin(nt)\sin(mt)\, dt = \int_0^\pi \cos(nt)\cos(mt)\, dt = 0, \qquad n \text{ and } m \text{ are unequal integers}$$

$$\int_0^\pi \sin(nt)\cos(mt)\, dt = \begin{cases} \dfrac{2n}{n^2-m^2}, & n-m \text{ is an odd integer} \\ 0, & n-m \text{ is an even integer} \end{cases}$$

$$\int_0^\infty \mathrm{sinc}(at)\, dt = \frac{1}{2a}, \qquad a > 0$$

$$\int_0^\infty \mathrm{sinc}^2(at)\, dt = \frac{1}{2a}, \qquad a > 0$$

$$\int_0^\infty e^{-at}\cos(bt)\, dt = \frac{a}{a^2+b^2}, \qquad a > 0$$

$$\int_0^\infty e^{-at}\sin(bt)\, dt = \frac{b}{a^2+b^2}, \qquad a > 0$$

$$\int_a^b u\,dv = uv\Big|_a^b - \int_a^b v\,du$$

A.6 INDEFINITE INTEGRALS

$$\int t^n\,dt = \frac{t^{n+1}}{n+1}, \qquad n \neq 1$$

$$\int \frac{1}{t}\,dt = \ln(t)$$

$$\int e^{at}\,dt = \frac{e^{at}}{a}, \qquad a \neq 0$$

$$\int \sin(at)\,dt = -\frac{\cos(at)}{a}$$

$$\int \cos(at)\,dt = \frac{\sin(at)}{a}$$

$$\int \sin^2(at)\,dt = \frac{2at - \sin(2at)}{4a}$$

$$\int \cos^2(at)\,dt = \frac{2at + \sin(2at)}{4a}$$

$$\int t^n \sin(at)\,dt = \frac{1}{a}\left[-t^n \cos(at) + n\int t^{n-1}\cos(at)\,dt \right]$$

$$\int t^n \cos(at)\,dt = \frac{1}{a}\left[t^n \sin(at) - n\int t^{n-1}\sin(at)\,dt \right]$$

A.7 EXPONENTIAL AND LOGARITHMIC FUNCTIONS

$$e^a\,e^b = e^{a+b}$$

$$\frac{e^a}{e^b} = e^{a-b}$$

$$\ln(ab) = \ln(a) + \ln(b)$$

$$\ln\left(\frac{a}{b}\right) = \ln(a) - \ln(b)$$

$$\ln(a^b) = b\ln(a)$$

$$\log_b(N) = \log_a(N)\log_b(a) = \frac{\log_a(N)}{\log_a(b)}$$

A.8 TAYLOR SERIES

$$f(x) = f(a) + \frac{(x-a)}{1!}\dot{f}(a) + \frac{(x-a)^2}{2!}\ddot{f}(a) + \cdots$$

Complex Numbers

B.1 REPRESENTATION OF COMPLEX NUMBERS

The complex number z can be expressed in several ways.

Cartesian or rectangular form:

$$z = a + jb \tag{B.1}$$

where $j = \sqrt{-1}$ and a and b are real numbers referring to the real part and the imaginary part of z. a and b are expressed as

$$a = \mathrm{Re}\{z\} \quad \text{and} \quad b = \mathrm{Im}\{z\} \tag{B.2}$$

where Re denotes the 'real part of' and Im denotes the 'imaginary part of'.

Polar form:

$$z = r\,e^{j\theta} \tag{B.3}$$

where $r > 0$ is the *magnitude* of z and θ is the *angle* or *phase* of z. These quantities are often written as

$$r = |z| \quad \text{and} \quad \theta = \angle z \tag{B.4}$$

Figure B.1 is the graphical representation of z. Using Euler's formula,

$$z = r\,e^{j\theta}$$
$$a + jb = r[\cos(\theta) + j\sin(\theta)] \tag{B.5}$$

Thus, the relationships between the Cartesian and polar representations of z are

$$a = r\cos(\theta) \quad \text{and} \quad b = r\sin(\theta) \tag{B.6}$$
$$r = \sqrt{a^2 + b^2} \quad \text{and} \quad \theta = \tan^{-1}\left(\frac{b}{a}\right) \tag{B.7}$$

B.2 ADDITION, MULTIPLICATION, AND DIVISION

If $z_1 = a_1 + jb_1$ and $z_2 = a_2 + jb_2$, then

$$z_1 + z_2 = (a_1 + a_2) + j(b_1 + b_2) \tag{B.8}$$

$$z_1 z_2 = (a_1 a_2 - b_1 b_2) + j(a_1 b_2 + b_1 a_2) \tag{B.9}$$

$$\frac{z_1}{z_2} = \frac{a_1 + jb_1}{a_2 + jb_2} = \frac{(a_1 + jb_1)(a_2 - jb_2)}{(a_2 + jb_2)(a_2 - jb_2)}$$

$$= \frac{(a_1 a_2 + b_1 b_2) + j(-a_1 b_2 + b_1 a_2)}{a_2^2 + b_2^2} \tag{B.10}$$

If $z_1 = r_1 e^{j\theta_1}$ and $z_2 = r_2 e^{j\theta_2}$, then

$$z_1 z_2 = (r_1 r_2) \, e^{j(\theta_1 + \theta_2)} \tag{B.11}$$

$$\frac{z_1}{z_2} = \left(\frac{r_1}{r_2}\right) e^{j(\theta_1 - \theta_2)} \tag{B.12}$$

B.3 COMPLEX CONJUGATE

The complex conjugate of z is denoted by z^* and is given by

$$z^* = a - jb = r \, e^{-j\theta} \tag{B.13}$$

Some useful relationships are as follows:

1. $zz^* = r^2$

2. $\dfrac{z}{z^*} = e^{j2\theta}$

3. $z + z^* = 2\text{Re}\{z\}$

4. $z - z^* = j2\text{Im}\{z\}$

5. $(z_1 + z_2)^* = z_1^* + z_2^*$

6. $(z_1 z_2)^* = z_1^* z_2^*$

7. $\left(\dfrac{z_1}{z_2}\right)^* = \dfrac{z_1^*}{z_2^*}$

B.4 POWERS AND ROOTS OF COMPLEX NUMBERS

The nth power of complex number $z = r \, e^{j\theta}$ is

$$z^n = r^n \, e^{jn\theta} = r^n [\cos(n\theta) + j \sin(n\theta)] \tag{B.14}$$

from which we have DeMoivre's relation

$$[\cos(\theta) + j \sin(\theta)]^n = \cos(n\theta) + j \sin(n\theta) \tag{B.15}$$

The nth root of a complex number z is the number w such that

$$w^n = z = r \, e^{j\theta} \tag{B.16}$$

Thus, to find the nth root of a complex number z, we must solve

$$w^n - r \, e^{j\theta} = 0$$

which is an equation of degree n and has n roots. These roots are given by

$$w_k = r^{1/n} \, e^{j[\theta + 2(k-1)\pi]/n}, \qquad k = 1, 2, \ldots, n \tag{B.17}$$

Partial Fraction Expansion

A rational function $X(s)$ is ratio of two polynomials $N(s)$ and $D(s)$:

$$X(s) = \frac{N(s)}{D(s)} = k \frac{(s - z_1)(s - z_2) \cdots (s - z_m)}{(s - p_1)(s - p_2) \cdots (s - p_n)}$$

where m is the highest power of numerator polynomial $N(s)$ and n is the highest power of denominator polynomial $D(s)$. Depending upon the highest power, rational function can be of two types:

1. If $m > n$, rational function $X(s)$ is called improper rational function.
2. If $m < n$, $X(s)$ is proper rational function.

For improper rational function, $X(s)$ is first separated by long division such that

$$X(s) = Q(s) + \frac{R(s)}{D(s)}$$

where $R(s)/D(s)$ will be a proper rational function. Then partial fraction is continued on the line of proper rational function.

For proper rational function, $X(s)$ may have three different types of denominator roots (poles):

1. All the poles are simple.
2. Poles are complex conjugate and simple.
3. Multiple poles at same point.

Case I (All poles are simple) Suppose that the poles p_1, p_2, ..., p_n are all different (distinct), then, using the partial fraction expansion, we may write

$$X(s) = \frac{A_1}{s - p_1} + \frac{A_2}{s - p_2} + \cdots + \frac{A_n}{s - p_n} \tag{C.1}$$

The problem is to determine the coefficients A_1, A_2, A_n. We can determine the coefficients A_1, A_2, A_n by multiplying both sides of Eq. (C.1) by each of the terms $(s - p_k)$, $k = 1, 2, \cdots, n$, and by evaluating the resulting expressions at the corresponding pole positions, $p_1, p_2, \cdots, p_n$. Thus, we have, in general,

$$(s - p_k)X(s) = \frac{(s - p_k)A_1}{s - p_1} + \cdots + A_k + \cdots + \frac{(s - p_k)A_n}{s - p_n} \tag{C.2}$$

Consequently, with $s = p_k$, Eq. (C.2) yields the kth coefficient as

$$A_k = (s - p_k)X(s)\Big|_{s=p_k} \qquad k = 1, 2, \cdots, n \tag{C.3}$$

The expansion given in Eq. (C.1) and the formula given in Eq. (C.3) hold for both real and complex poles. The only constraint is that all poles be distinct.

Case II (Multiple order poles) If $X(s)$ has a pole of multiplicity r, i.e., it contains in its denominator the factor $(s - p_k)^r$, then the expansion given in Eq. (C.1) is no longer true. If a pole p_k is repeated r times, then there are r terms in the partial fraction expansion associated with that pole. The partial fraction expansion must contain the terms

$$\frac{A_{1k}}{s - p_k} + \frac{A_{2k}}{(s - p_k)^2} + \cdots + \frac{A_{rk}}{(s - p_k)^r} = \sum_{i=1}^{r} \frac{A_{ik}}{(s - p_k)^i}$$

The coefficients A_{ik} are computed from the equation

$$A_{ik} = \frac{1}{(r - i)!} \left[\frac{d^{r-i}}{ds^{r-i}} \left((s - p_k)^r X(s) \right) \right]\Bigg|_{s=p_k} \tag{C.4}$$

Example C.1 Obtain partial fraction expansion of

$$X(s) = \frac{s}{(s + 1)(s + 2)}$$

Solution

Partial fraction expansion of $X(s)$ is given by

$$X(s) = \frac{A_1}{s + 1} + \frac{A_2}{s + 2}$$

where, from Eq. (C.3)

$$A_1 = (s + 1)\frac{s}{(s + 1)(s + 2)}\Big|_{s=-1}$$

$$= \frac{-1}{-1 + 2} = -1$$

and

$$A_2 = (s + 2)\frac{s}{(s + 1)(s + 2)}\Big|_{s=-2}$$

$$= \frac{-2}{-2 + 1} = 2$$

Thus,

$$X(s) = \frac{-1}{s+1} + \frac{2}{s+2}$$

Example C.2 Obtain partial fraction expansion of

$$X(s) = \frac{s+1}{(s+3)(s+2)^2}$$

Solution

Partial fraction expansion of $X(s)$ is given by

$$X(s) = \frac{A_1}{s+3} + \frac{A_{12}}{s+2} + \frac{A_{22}}{(s+2)^2}$$

where from Eq. (C.3)

$$A_1 = (s+3)\frac{s+1}{(s+3)(s+2)^2}\bigg|_{s=-3}$$

$$= \frac{-2}{(-1)^2} = -2$$

$$A_{12} = \frac{1}{1!}\frac{d}{ds}\left[(s+2)^2\frac{s+1}{(s+3)(s+2)^2}\right]\bigg|_{s=-2}$$

$$= \frac{d}{ds}\left[\frac{s+1}{s+3}\right]\bigg|_{s=-2} = 2$$

and

$$A_{22} = \left[(s+2)^2\frac{s+1}{(s+3)(s+2)^2}\right]\bigg|_{s=-2} = -1$$

Thus,

$$X(s) = \frac{-2}{s+3} + \frac{2}{s+2} - \frac{1}{(s+2)^2}$$

Model Question Papers

Paper-I

Note: Attempt all questions. All questions carry equal marks.

Max. Marks: 100

1. Attempt any **four** parts of the following:

   (a) Sketch and check whether the given signal

   $$x(t) = \sum_{n=-\infty}^{\infty} e^{-(2t-n)} u(2t - n)$$

   is periodic? If yes, compute its average power.

   (b) Sketch the even and odd parts of (i) a unit impulse function, (ii) a unit step function, and (iii) a unit ramp function.

   (c) Sketch the function

   $$x(t) = u\left(\sin\frac{\pi}{T}t\right) - u\left(-\sin\frac{\pi}{T}t\right)$$

   (d) Under what conditions, will the system characterized by

   $$y(n) = \sum_{k=n_0}^{\infty} e^{-ak} x(n - k)$$

   be linear, time invariant, causal, stable, and memoryless.

   (e) Let $h(n)$ be the impulse response of the LTI causal system described by the difference equation

   $$y(n) = ay(n - 1) + x(n)$$

   and let $h(n) * h_I(n) = \delta(n)$. Find $h_I(n)$.

   (f) $x(n)$, $h(n)$, and $y(n)$ are, respectively, the input signal, unit impulse response, and output signal of a linear time-invariant causal system and it is given that

   $$y(n - 2) = x(n - n_1) * x(n - n_2)$$

   where $*$ denotes convolution. Find the possible sets of values of n_1 and n_2.

2. Attempt any **four** parts of the following:

(a) Determine the odd and even part of the signal

$$x(t) = \sqrt{2} \cos\left(\omega t + \frac{\pi}{4}\right)$$

(b) Determine the output $y(n) = x(3n - 2)$, given

$$x(n) = \begin{cases} 1, & |n| \le 2 \\ 0, & |n| > 2 \end{cases}$$

(c) Find the convolution of $x(n) * \delta(n - 2)$, given

$$x(n) = \delta(n + 2) + 2\delta(n) + 3\delta(n - 2)$$

(d) Find whether the system with impulse response

$$h(n) = u(n) - u(-n - 1)$$

is causal and stable.

(e) Determine the Fourier transform of the signal

$$x(t) = \sin(\pi t)\, e^{-2t} u(t)$$

(f) A signal

$$x(t) = \cos(5\pi t) + 0.5 \cos(10\pi t)$$

is ideally sampled at intervals T_s seconds to obtain $x_s(t)$. Find the maximum allowable valuer of T_s. To reconstruct the signal without distortion, find the minimum filter bandwidth required when $x_s(t)$ is passed through a rectangular lowpass filter.

3. Attempt any **two** parts of the following:

(a) Find the Fourier series representation for the signal

$$x(t) = 3 \cos(0.6\pi t) + 2 \sin(1.2\pi t) + \cos(2.1\pi t)$$

for all t. Sketch the magnitude and phase spectra.

(b) $X(e^{j\omega})$ denotes the DTFT of a length-9 sequence $x(n)$ given by

$$x(n) = \{2, 3, -1, 0, -4, 3, 1, 2, 4\}, \quad -2 \le n \le 6$$

Evaluate the following functions of $X(e^{j\omega})$ without computing the transform itself:

(i) $X(e^{j0})$,   (ii) $X(e^{j\pi})$,   (iii) $\int_{-\pi}^{\pi} X(e^{j\omega})\, d\omega$,   (iv) $\int_{-\pi}^{\pi} |X(e^{j\omega})|^2\, d\omega$,

(v) $\int_{-\pi}^{\pi} \left| \frac{dX(e^{j\omega})}{d\omega} \right|^2 d\omega$

(c) Consider a continuous-time signal $x(t)$ with $x(t) \longleftrightarrow X(\omega)$.
(i) Show that $X(t) \longleftrightarrow 2\pi x(-\omega)$.

(ii) Find $x(t)$ from

$$X(\omega) = \frac{1}{(1 + j\omega)^2}$$

using the convolution property of the Fourier transform.

4. Attempt any **two** parts of the following:

(a) Obtain the continuous-time signal $x(t)$ whose Laplace transform is

$$X(s) = \frac{s^2 + 3s + 1}{(s + 1)^3(s + 2)^2}$$

(b) The differential equation characterizing a continuous LTI system is given by

$$2\frac{d^2y(t)}{dt^2} + 3\frac{dy(t)}{dt} + y(t) = u(t)$$

With initial conditions $y(0^-) = -1$ and $\dot{y}(0^-) = 1$. Determine $Y(s)$ as the sum of zero-input response and zero-state response. Also determine $y(t)$.

(c) Show that

$$\frac{e^{-at} - e^{-bt}}{t} u(t) \overset{\mathcal{L}}{\longleftrightarrow} \ln\left(\frac{s + b}{s + a}\right)$$

5. Attempt any **two** parts of the following:

(a) Find the z-transform of $X(z)$ and sketch the pole zero with the ROC for each of the following sequences:

(i) $x(n) = \left(\dfrac{1}{2}\right)^n u(n) + \left(\dfrac{1}{3}\right)^n u(n)$

(ii) $x(n) = \left(\dfrac{1}{3}\right)^n u(n) + \left(\dfrac{1}{2}\right)^n u(-n - 1)$

(b) Determine the inverse transform of

$$X(z) = \frac{z}{3z^2 - 4z + 1}$$

if the region of convergence are (i) $|z| > 1$, (ii) $|z| < 1/3$, and (iii) $1/3 < |z| < 1$.

(c) (i) Consider a sequence $x_1(n)$ with z-transform $X_1(z)$ and a sequence $x_2(n)$ with z-transform $X_2(z)$, where

$$x_2(n) = x_1(-n)$$

Show that $X_2(z) = X_1(1/z)$, and from this, show that if $X_1(z)$ has a pole (or zero) at $z = z_0$, then $X_2(z)$ has a pole (or zero) at $z = 1/z_0$.

(ii) Compute the unit step response $s(n)$ of the system with impulse response

$$h(n) = \begin{cases} 3^n, & n < 0 \\ \left(\dfrac{2}{5}\right)^n, & n \geq 0 \end{cases}$$

Paper-II

Note: Attempt all questions. All questions carry equal marks.

Max. Marks: 100

1. Attempt any **four** parts of the following:

   (a) Determine and sketch the magnitude and phase response of the system characterized by the difference equation

   $$y(n) = \frac{1}{3}[x(n-1) + x(n) + x(n+1)]$$

   in the range $0 < \omega < 2\pi$.

   (b) Determine the following convolution:

   $$x(t) = \delta(t) * \delta(t-T) * \delta(t-2t) * \delta(t+3T)$$

   (c) Determine the Nyquist rate corresponding to each of the following signals:
   - (i) $x(t) = \sin(200\pi t)$
   - (ii) $x(t) = \text{sinc}(200t)$
   - (iii) $x(t) = 1 + \cos(200\pi t) + \sin(400\pi t)$
   - (iv) $x(t) = \text{sinc}(200t) + \text{sinc}^2(200t)$
   - (v) $x(t) = \cos^3(200\pi t)$

   (d) Sketch

   (i) $x_1(t) = \delta(\cos t)$ and (ii) $x_2(t) = \text{sgn}\left(\sin\frac{\pi}{T}t\right)$

   (e) Show that the Fourier series coefficients of the signal

   $$z(t) = x(t)y(t) = \sum_{n=-\infty}^{\infty} X_n\, e^{jn\omega_0 t}$$

   are given by the discrete convolution

   $$Z_n = \sum_{k=-\infty}^{\infty} X_k Y_{n-k}$$

   (f) An arbitrary real-valued continuous-time signal $x(t)$ occupies the entire interval $-\infty < x(t) < \infty$. Show that the energy of the signal $x(t)$ is equal to the sum of the energy of the even component $x_e(t)$ and the energy of the odd component $x_o(t)$.

2. Attempt any **four** parts of the following:

   (a) Find the DTFT

   $$x(n) = u(n)$$

(b) Determine signal energy and power of the following signals:
   (i) $x(n) = u(n)$
   (ii) $x(t) = e^{-3t}$
(c) Determine whether the following signals are periodic, and for those which are, find the fundamental period:
   (i) $x(n) = (-1)^n$
   (ii) $x(n) = (-1)^{n^2}$
   (iii) $x(t) = \sum_{k=-\infty}^{\infty}(-1)^k \delta(t - 2k)$
(d) Show that
   (i) the convolution of an odd and an even function is an odd function.
   (ii) the convolution of two odd functions is an even function.
   (iii) the convolution of two even functions is an even function.
(e) Determine and sketch the magnitude and phase response of the LTI causal system described by the differential equation

$$\frac{dy(t)}{dt} + y(t) = \frac{dx(t)}{dt} - x(t)$$

(f) Determine the response of the system with impulse response

$$h(n) = \left(\frac{1}{2}\right)^n u(n)$$

when the input is the complex exponential sequence

$$x(n) = A\, e^{jn\pi/2}, \quad -\infty < n < \infty$$

3. Attempt any **two** parts of the following:

(a) A signal $x(t) = (\sin(\pi t))/\pi t$ is sampled by $s(t) = \sum_{n=-\infty}^{\infty} \delta(t - n/2)$. Determine and sketch the sampled signal $x_s(t)$ and its Fourier transform $X_s(\omega)$.
(b) Determine the discrete-time Fourier transform (DTFT) of
   (i) $x_1(n) = \sin(\omega_0 n)$
   (ii) $x_2(n) = \sin(\omega_0 n)u(n)$.
(c) Consider the signal

$$x(t) = \begin{cases} 0, & |t| > 1 \\ (t+1)/2, & -1 \le t \le 1 \end{cases}$$

   (i)   Determine the closed form expression for $X(\omega)$.
   (ii) Take the real part of your answer to part (i), and verify that it is the Fourier transform of the even part of $x(t)$.
   (iii) What is the Fourier transform of the odd part of $x(t)$?

4. Attempt any **two** parts of the following:

(a) Show that Laplace transform converts time differentiation into multiplication by s and integration into division by s. Consider zero-initial conditions. Hence, find $\mathcal{L}[\cos(\omega_0 t)]$ given

$$\mathcal{L}[\sin(\omega_0 t)] = X(s) = \frac{\omega_0}{s^2 + \omega_0^2}$$

(b) (i) Find out the response of a continuous time system to unit step input given the impulse response

$$h(t) = \frac{1}{RC} e^{-t/RC} u(t)$$

(ii) Find the Laplace transform of the signal

$$g(t) = \frac{1}{t} \sin(\omega t) u(t)$$

(c) Consider the system S characterized by the differential equation

$$\frac{d^3 y(t)}{dt^3} + 6\frac{d^2 y(t)}{dt^2} + 11\frac{dy(t)}{dt} + 6y(t) = x(t)$$

(i) Determine the zero-state response (or forced response) of this system for the input $x(t) = e^{-4t} u(t)$.

(ii) Determine the zero-input response (or natural response) of the system for $t > 0^-$, given that

$$y(0^-) = 1, \quad \dot{y}(0^-) = -1, \quad \ddot{y}(0^-) = 1$$

(iii) Determine the total response of the system when the input is $x(t) = e^{-4t} u(t)$ and the initial conditions are the same as those specified in part (ii).

5. Attempt any **two** parts of the following:

(a) Let $x(n)$ be a finite length sequence that is nonzero only for $0 \leq n \leq (N-1)$, and consider the one-sided periodic signal, $y(n)$, that is formed by periodically extending $x(n)$ as follows:

$$y(n) = \sum_{k=0}^{\infty} x(n - kN)$$

Express the z-transform of $y(n)$ in terms of $X(z)$ and find the region of convergence of $Y(z)$.

(b) Determine the ROC of $aX(z) + bY(z)$, given that

$$X(z) = \frac{z}{(z-0.5)(z-1.5)}, \qquad 0.5 < |z| < 1.5$$

$$Y(z) = \frac{0.25z}{(z-0.25)(z-0.5)}, \qquad |z| > 0.5$$

For what relationship between a and b the ROC will be the largest?

(c) Determine the inverse z-transform of

(i) $X(z) = \log(1 + az^{-1})$, $\quad |z| > |a|$ using the power series expansion.

(ii) $X(z) = e^z$ with ROC all z except $|z| = \infty$.

(iii) $X(z) = \sin(z)$

Bibliography

1. Haykin, S. and B.V. Veen, *Signals & Systems*, 2nd ed., John Wiley & Sons, New Delhi, 2003.
2. Kreyszig, E., *Advanced Engineering Mathematics*, 8th ed., John Wiley & Sons, New York, 2004.
3. Lathi, B.P., *Modern Digital and Analog Communication Systems*, 3rd ed., Oxford University Press, New Delhi, 2001.
4. Lathi, B.P., *Linear Systems and Signals*, 2nd ed., Oxford University Press, New Delhi, 2004.
5. Mitra, S.K., *Digital Signal Processing: A Computer Based Approach*, 2nd ed., Tata McGraw-Hill, New York, 2001.
6. Oppenheim, A.V. and R.W. Schafer, *Discrete-Time Signal Processing*, 2nd ed., Pearson Education, New Delhi, 2001.
7. Oppenheim, A.V., A.S., Willsky, and S.H. Nawab, *Signals & Systems*, 2nd ed., Prentice-Hall, New Delhi, 2002.
8. Proakis, J.G. and D.G. Manolakis, *Digital Signal Processing: Priciples, Algorithms, and Applications*, 3rd ed., Prentice-Hall, New Delhi, 2002.
9. Roberts, M.J., *Fundamnetals of Signals & Systems*, Special Indian edition, India, 2007.

Index

NOTES

NOTES

NOTES

NOTES